DIGITAL
COMMUNICATIONS

DIGITAL COMMUNICATIONS
Fundamentals and Applications

BERNARD SKLAR

The Aerospace Corporation, El Segundo, California
and
University of California, Los Angeles

P T R Prentice Hall
Englewood Cliffs, New Jersey 07632

Library of Congress Cataloging-in-Publication Data

SKLAR, BERNARD (date)
  Digital communications.

  Bibliography: p.
  Includes index.
  1. Digital communications. I. Title.
TK5103.7.S55 1988   621.38′0413   87-1316
ISBN 0-13-211939-0

Editorial/production supervision and
  interior design: Reynold Rieger
Cover design: Wanda Lubelska Design
Manufacturing buyers: Gordon Osbourne and Paula Benevento

 © 1988 by P T R Prentice Hall
Prentice-Hall, Inc.
A Paramount Communications Company
Englewood Cliffs, New Jersey 07632

Printed in the United States of America

20 19 18 17 16 15 14 13

ISBN 0-13-211939-0

Prentice-Hall International (UK) Limited, *London*
Prentice-Hall of Australia Pty. Limited, *Sydney*
Prentice-Hall Canada Inc., *Toronto*
Prentice-Hall Hispanoamericana, S.A., *Mexico*
Prentice-Hall of India Private Limited, *New Delhi*
Prentice-Hall of Japan, Inc., *Tokyo*
Simon & Schuster Asia Pte. Ltd., *Singapore*
Editora Prentice-Hall do Brasil, Ltda., *Rio de Janeiro*

To my mother, Ruth Sklar,
the memory of my father, Julius Sklar,
my wife, Gwen, and our children,
Debra, Sharon, and Dean

Contents

2   FORMATTING AND BASEBAND TRANSMISSION   51

3 BANDPASS MODULATION AND DEMODULATION   117

5 CHANNEL CODING: PART 1   245

8   SYNCHRONIZATION

Maurice A. King, Jr.

429

9   MULTIPLEXING AND MULTIPLE ACCESS

475

Contents

11   SOURCE CODING 595

Fredric J. Harris

12   ENCRYPTION AND DECRYPTION   668

A   A REVIEW OF FOURIER TECHNIQUES   710

B   FUNDAMENTALS OF STATISTICAL DECISION THEORY   733

Preface

This book is intended to provide a comprehensive coverage of digital communication systems for senior-level undergraduates, first-year graduate students, and practicing engineers. Even though the emphasis of the book is on digital communications, necessary analog fundamentals are included, since analog waveforms are used for the radio transmission of digital signals.

The key feature of a digital communication system is that it deals with a finite set of discrete messages, in contrast to an analog communication system in which messages are defined on a continuum. The objective at the receiver of the digital system is *not* to reproduce a waveform with precision; it is, instead, to determine from a noise-perturbed signal which of the finite set of waveforms had been sent by the transmitter. In fulfillment of this objective, an impressive assortment of signal processing techniques has arisen over the past two decades.

The book develops these important techniques in the context of a unified structure. The structure, in block diagram form, appears at the beginning of each chapter; blocks in the diagram are emphasized, as appropriate, to correspond to the subject of that chapter. Major purposes of the book are (1) to add organization and structure to a field that has grown rapidly in the last two decades, and (2) to ensure awareness of the "big picture" even while delving into the details. The signals and key processing steps are traced from the information source through the transmitter, channel, receiver, and ultimately to the information sink. Signal transformations are organized according to functional classes: formatting and source coding, modulation, channel coding, multiplexing and multiple access, spreading, encryption, and synchronization. Throughout the book, emphasis is placed on system goals and the need to trade off basic system parameters such as signal-to-noise ratio, probability of error, and bandwidth (spectral) expenditure.

ORGANIZATION OF THE BOOK

It is assumed that the reader is familiar with Fourier methods and convolution. Appendix A reviews these techniques, emphasizing those properties that are particularly useful in the study of communication theory. It is also assumed that the reader has a knowledge of basic probability and has some familiarity with random variables. Appendix B builds on these disciplines for a short treatment on statistical decision theory with emphasis on hypothesis testing—so important in the understanding of detection theory. Chapter 1 introduces the overall digital communication system and the basic signal transformations that are highlighted in subsequent chapters. Some basic ideas of random variables and the additive white Gaussian noise (AWGN) model are reviewed. Also, the relationship between power spectral density and autocorrelation, and the basics of signal transmission through linear systems, are established. Chapter 2 covers the signal processing step, known as formatting, the step that renders an information signal compatible with a digital system. Chapter 2 also emphasizes the *transmission* of baseband signals. Chapter 3 deals with bandpass modulation and demodulation techniques. The detection of digital signals in Gaussian noise is stressed, and receiver optimization is examined. Chapter 4 deals with link analysis, an important subject for providing overall system insight; it considers some subtleties usually neglected at the college level. Chapters 5 and 6 deal with channel coding—a cost-effective way of providing improvement in system error performance. Chapter 5 emphasizes linear block coding, and Chapter 6 emphasizes convolutional coding.

Chapter 7 considers various modulation/coding system trade-offs dealing with probability of bit error performance, bandwidth efficiency, and signal-to-noise ratio. Chapter 8 deals with synchronization for digital systems. It covers phase-locked-loop implementation for achieving carrier synchronization; bit synchronization, frame synchronization, and network synchronization; and some fundamentals of synchronization as applied to satellite links.

Chapter 9 treats multiplexing and multiple access. It explores techniques that are available for utilizing the communication resource efficiently. Chapter 10 introduces spread-spectrum techniques and their application in such areas as multiple access, ranging, and interference rejection. This technology is particularly important for most military communication systems. The subject of source coding in Chapter 11 deals with data formatting, as is done in Chapter 2; the main difference between formatting and source coding is that source coding additionally involves data redundancy reduction. Rather than considering source coding immediately after formatting, source coding has purposely been treated in a later chapter. It is felt that the reader should be involved with the fundamental processing steps, such as modulation and channel coding, early in the book, before examining some of the special considerations of source coding. Chapter 12 covers some basic encryption/decryption ideas. It includes some classical encryption concepts, as well as some of the proposals for a class of encryption systems called public key cryptosystems.

If the book is used for a two-term course, a simple partitioning is suggested: the first six chapters to be taught in the first term, and the last six chapters in the second term. If the book is used for a one-term only course, it is suggested that the course material be selected from the following chapters: 1, 2, 3, 4, 5, 6, 8, and 10.

ACKNOWLEDGMENTS

This book is an outgrowth of my teaching activities at the University of California, Los Angeles, and my work in the Communications Division at The Aerospace Corporation. A number of people have contributed in many ways and it is a pleasure to acknowledge them. Dr. Maurice King, my colleague at Aerospace, carefully reviewed and made important contributions to each chapter. His continual assistance has been invaluable. He also contributed Chapter 8, Synchronization. Professor Fred Harris of San Diego State University suggested many improvements and contributed Chapter 11, Source Coding. I want to pay special thanks to Dr. Marvin Simon of the Jet Propulsion Laboratory for providing me with much encouragement and many valuable suggestions.

I also want to thank Professor Jim Omura of UCLA for sharing with me his considerable knowledge of encryption and thereby helping me improve Chapter 12. Professor Raymond Pickholtz of George Washington University gave me lots of beneficial advice throughout the writing process. Professors William Lindsey and Andreas Polydoros of the University of Southern California suggested important improvements. Professor James Modestino of Rensselaer Polytechnic Institute, Dr. Adam Lender of Lockheed Palo Alto Research Laboratory, and Professor Ron Iltis of the University of California, Santa Barbara, each provided valuable reviews. Dr. Todd Citron of Hughes Aircraft, Dr. Joe Odenwalder of MA/COM Linkabit, and Dr. Unjeng Cheng of Axiomatics were extremely helpful in the chapters on channel coding. Mr. Don Martin and Mr. Ned Feldman of The Aerospace Corporation made numerous suggestions and contributions. I also want to pay special thanks to Professor Wayne Stark of the University of Michigan, whose unique critical talents enhanced the manuscript's continuity.

The block diagrams in Figures 1.2 and 1.3, at each chapter opening, and on the cover of the book, first appeared in the two part paper: © 1983 IEEE; B. Sklar, "A Structured Overview of Digital Communications—A Tutorial Review," *IEEE Communications Magazine*, August and October, 1983. Permission from IEEE to reprint these figures throughout the book is gratefully acknowledged.

My students at UCLA and those at Aerospace used early versions of chapters of this book and made many helpful contributions. I am indebted to all those students who have taken my courses and thus helped me with this project. I also want to express my appreciation to my management at Aerospace, Mr. Hal

McDonnell and Mr. Fred Jones, for their indulgence and moral support. I want to acknowledge and thank Ms. Cynthia Dickson for her diligence and speed in typing the entire manuscript.

Finally, I want to thank my wife, Gwen, for her very unselfish support, her understanding, and her endurance of the many months I had time for only *one* devotion—the writing of this book.

BERNARD SKLAR

Tarzana, California

CHAPTER 1

Signals and Spectra

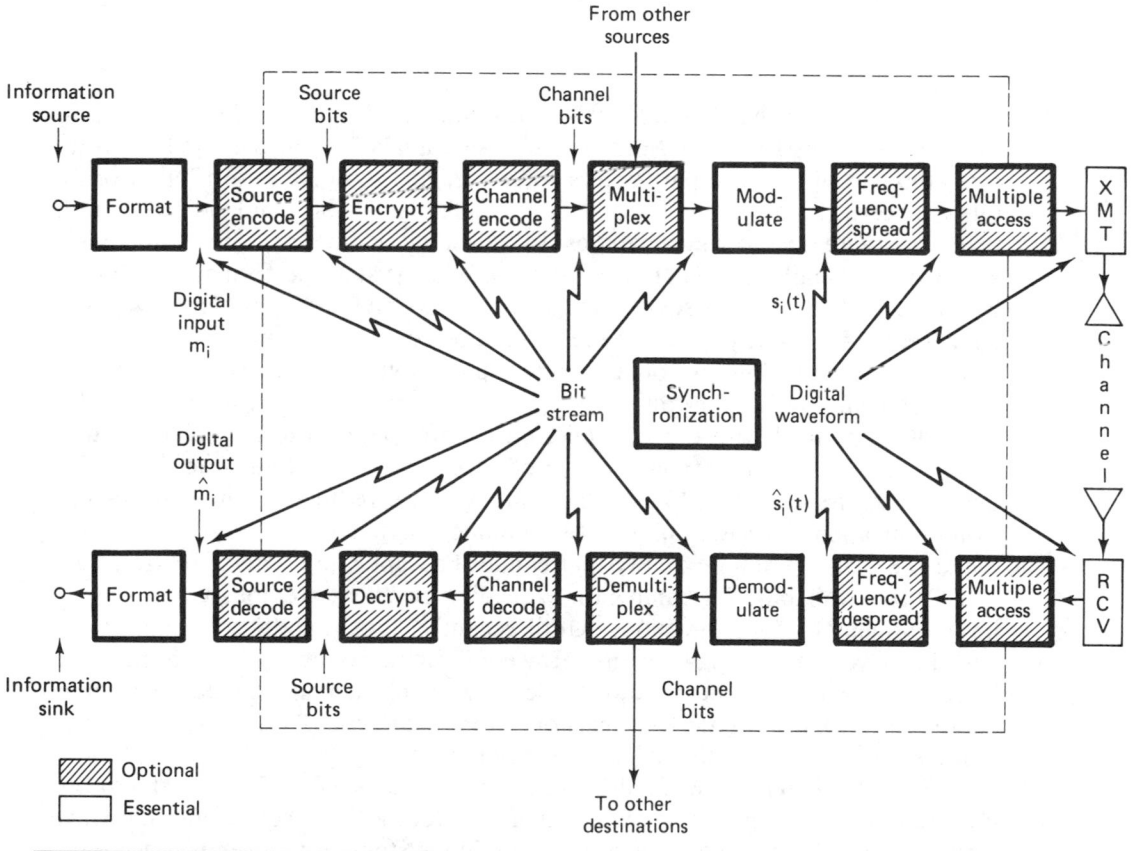

This book presents the ideas and techniques fundamental to digital communication systems. Emphasis is placed on system design goals and on the need for trade-offs among basic system parameters such as signal-to-noise ratio (SNR), probability of error, and bandwidth expenditure. Transmission bandwidth is a finite resource; there is a growing awareness that bandwidth must be conserved, shared, and used efficiently. In general, we shall see that system performance can often be improved through the use of increased transmission bandwidth. However, such an increase is not always possible, because of physical limitations or the constraint of government regulations concerning the allocation and conservation of the usable electromagnetic spectrum.

We shall deal with the transmission of information (voice, video, or data) over a path (channel) that may consist of wires, waveguides, or free space. Frequently, the treatment will be in the context of a satellite communications link. Communication via satellites has two unique characteristics: (1) the ability to cover the globe with a flexibility that cannot be duplicated with terrestrial links, and (2) the availability of bandwidth exceeding anything previously available for intercontinental communications. Until recently, most satellite communication systems have been analog in nature. However, digital communication is becoming increasingly attractive because of the ever-growing demand for data communication and because digital transmission offers data processing options and flexibilities not available with analog transmission.

The principal feature of a digital communication system (DCS) is that during a finite interval of time, it sends a waveform from a finite set of possible waveforms, in contrast to an analog communication system, which sends a waveform from an infinite variety of waveform shapes with theoretically infinite resolution. In a DCS, the objective at the receiver is *not* to reproduce a transmitted waveform with precision; it is, instead, to determine from a noise-perturbed signal which waveform from the finite set of waveforms had been sent by the transmitter. An important measure of system performance in a DCS is the probability of error (P_E).

1.1 DIGITAL COMMUNICATION SIGNAL PROCESSING

1.1.1 Why Digital?

Why are communication systems, military and commercial alike, "going digital"? There are many reasons. The primary advantage is the ease with which digital signals, compared to analog signals, are regenerated. Figure 1.1 illustrates an ideal binary digital pulse propagating along a transmission line. The shape of the waveform is affected by two basic mechanisms: (1) as all transmission lines and circuits have some nonideal transfer function, there is a distorting effect on the ideal pulse; and (2) unwanted electrical noise or other interference further distorts the pulse waveform. Both of these mechanisms cause the pulse shape to degrade as a function of line length, as shown in Figure 1.1. During the time that the transmitted pulse can still be reliably identified (before it is degraded to an ambiguous state by the transmission line), the pulse is amplified by a digital amplifier that recovers its original ideal shape. The pulse is thus "reborn" or regenerated. Circuits that perform this function at regular intervals along a transmission system are called *regenerative repeaters*.

Digital circuits are less subject to distortion and interference than are analog circuits. Since binary digital circuits operate in one of two states, fully on or fully off, to be meaningful a disturbance must be large enough to change the circuit operating point from one state to the other. Such two-state operation facilitates signal regeneration and thus prevents noise and other disturbances from accu-

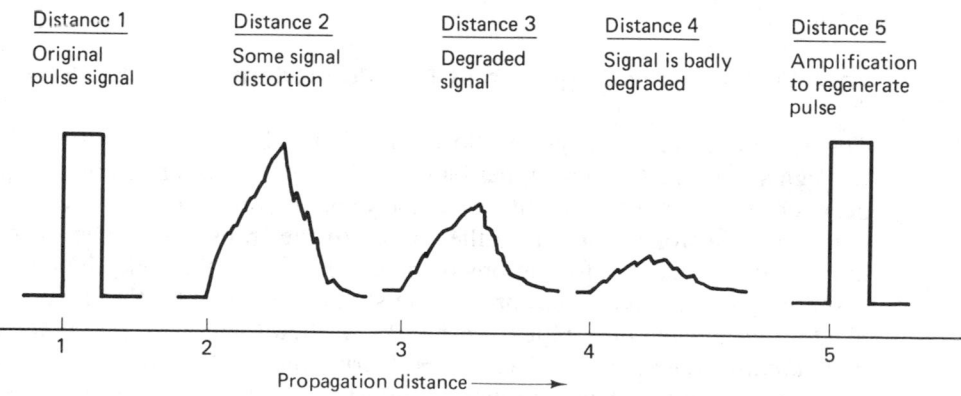

Figure 1.1 Pulse degradation and regeneration.

mulating in transmission. Analog signals, however, are *not* two-state signals; they can take an *infinite variety* of shapes. With analog circuits, even a small disturbance can render the reproduced waveform unacceptably distorted. Once the analog signal is distorted, the distortion cannot be removed by amplification. Since, with analog signals, accumulated noise is irrevocably bound to the signal, analog signals cannot be completely regenerated. Extremely low error rates producing high signal fidelity are possible through error detection and correction with digital techniques, but similar procedures are not available with analog.

There are other important advantages to digital communications. Digital circuits are *more reliable* and can be produced at lower cost than analog circuits. Also, digital hardware lends itself to *more flexible* implementation than analog hardware [e.g., microprocessors, digital switching, and large-scale integrated (LSI) circuits]. The combining of digital signals using time-division multiplexing (TDM) is *simpler* than the combining of analog signals using frequency-division multiplexing (FDM). Different types of digital signals (data, telegraph, telephone, television) can be treated as identical signals in transmission and switching—*a bit is a bit*. Also, for convenient switching, digital messages can be handled in autonomous groups called *packets*. Digital techniques lend themselves naturally to signal processing functions that protect against interference and jamming, or that provide encryption and privacy; such techniques are discussed in Chapters 10 and 12, respectively. Also, much data communication is computer to computer, or digital instrument or terminal to computer. Such digital terminations are naturally best served by digital communication links.

Most system choices entail trade-offs; system options are rarely all good or all bad. Thus far we have discussed only the *benefits* of digital transmission. What do you suppose are the *costs* or *liabilities*? A major disadvantage of digital transmission is that it typically requires a *greater system bandwidth* to communicate the same information in a digital format as compared to an analog format. Throughout this book we emphasize that bandwidth is a valuable resource, not always available. Bandwidth-efficient signaling techniques are discussed in Chapters 2 and 7. Another cost of digital transmission is that digital detection requires system synchronization (Chapter 8), whereas analog signals generally have no such requirement.

1.1.2 Typical Block Diagram and Transformations

The functional block diagram shown in Figure 1.2 illustrates the signal flow through a typical DCS. The upper blocks—format, source encode, encrypt, channel encode, multiplex, modulate, frequency spread, and multiple access—indicate the signal transformations from the source to the transmitter. The lower blocks indicate the signal transformations from the receiver to the sink; the lower blocks essentially reverse the signal processing steps performed by the upper blocks. It used to be that the only blocks within the dashed lines were the *modulator* and *demodulator,* together called a *modem*. During the past two decades, other signal processing functions were frequently incorporated within the same assembly as the modulator and demodulator. Consequently, at present, the term "modem"

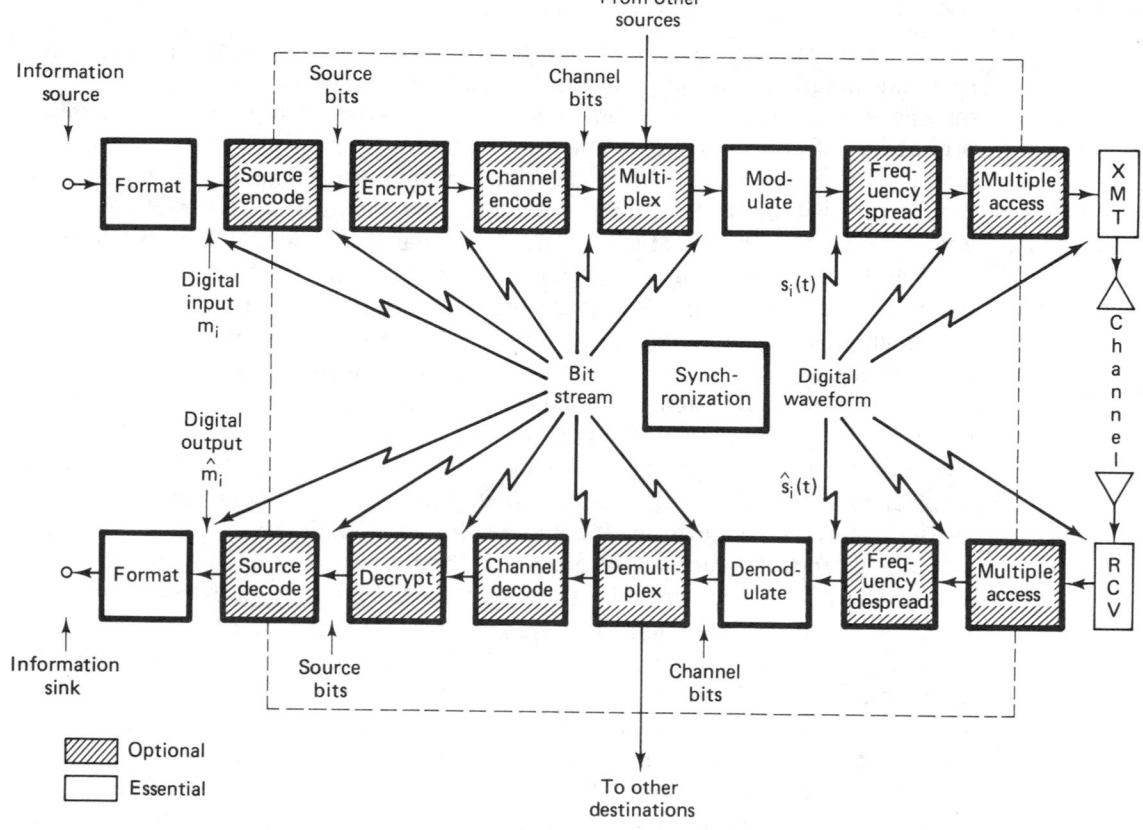

Figure 1.2 Block diagram of a typical digital communication system. (Reprinted with permission from B. Sklar, "A Structured Overview of Digital Communications," *IEEE Commun. Mag.*, August 1983, Fig. 1, p. 5. © 1983 IEEE.)

often encompasses all the processing steps shown within the dashed lines of Figure 1.2; when this is the case, the modem can be thought of as the "brains" of the system. Note that what constitutes a modem is not a precise concept; some of the blocks have purposely been shown *on* the dashed line rather than either inside or outside the modem. The transmitter and receiver can be thought of as the "muscles" of the system. The transmitter usually consists of a frequency up-conversion stage, a high-power amplifier, and an antenna. The receiver portion usually consists of an antenna, a low-noise amplifier (LNA), and a down-converter stage, typically to an intermediate frequency (IF).

Of all the signal processing steps, only formatting, modulation, and demodulation are essential for a DCS; the other processing steps within the modem are design options for specific system needs. *Formatting* transforms the source information into *digital symbols*; it makes the information compatible with the signal processing within a digital communication system. *Modulation* is the process by which the symbols are converted to *waveforms* that are compatible with the transmission channel.

The source encoding step produces analog-to-digital (A/D) conversion (for analog sources) *and* removes *redundant or unneeded* information. Encryption prevents unauthorized users from understanding messages and from injecting false messages into the system. Channel coding, for a given data rate, can reduce the probability of error (P_E), or reduce the signal-to-noise ratio (SNR) requirement, at the expense of bandwidth or decoder complexity. Channel coding can also reduce the system bandwidth requirement at the expense of SNR or P_E performance. Frequency spreading can produce a signal that is less vulnerable to interference (both natural and intentional) and can be used to enhance the privacy of the communicators. Multiplexing and multiple access procedures combine signals that might have different characteristics or might originate from different sources, so that they can share a portion of the communications resource.

The flow of the signal processing steps shown in Figure 1.2 represents a typical arrangement; however, the blocks are sometimes implemented in a different order. For example, multiplexing can take place prior to channel encoding, or prior to modulation, or—with a two-step modulation process (subcarrier and carrier)—it can be performed between the two modulation steps. Similarly, spreading can take place anywhere along the transmission chain; its precise location depends on the particular technique used. Figure 1.2 illustrates the reciprocal aspect of the procedure; any signal processing step that takes place in the transmitting chain must be reversed in the receiving chain. The figure also indicates that from the source to the modulator a message, also called a *baseband signal* or a *bit stream,* is characterized by a sequence of digital symbols. After modulation, the message takes the form of a digitally encoded waveform or *digital waveform.* Similarly, in the reverse direction, a received message appears as a digital waveform until it is demodulated. Thereafter it takes the form of a bit stream for all further signal processing steps. At various points along the signal route, noise corrupts the waveform $s(t)$ so that its reception must be termed an estimate $\hat{s}(t)$. Such noise and its deleterious effects on system performance are considered in Chapter 4.

Figure 1.3 shows the basic signal processing functions, which may be viewed as transformations from one signal space to another. The transformations are classified into seven basic groups:

1. Formatting and source coding
2. Modulation/demodulation
3. Channel coding
4. Multiplexing and multiple access
5. Spreading
6. Encryption
7. Synchronization

Although this organization has some inherent overlap, it provides a useful structure for the book. Beginning with Chapter 2, the seven basic transformations are considered individually. In Chapter 2 we discuss the basic formatting techniques for transforming the source information into digital symbols, as well as

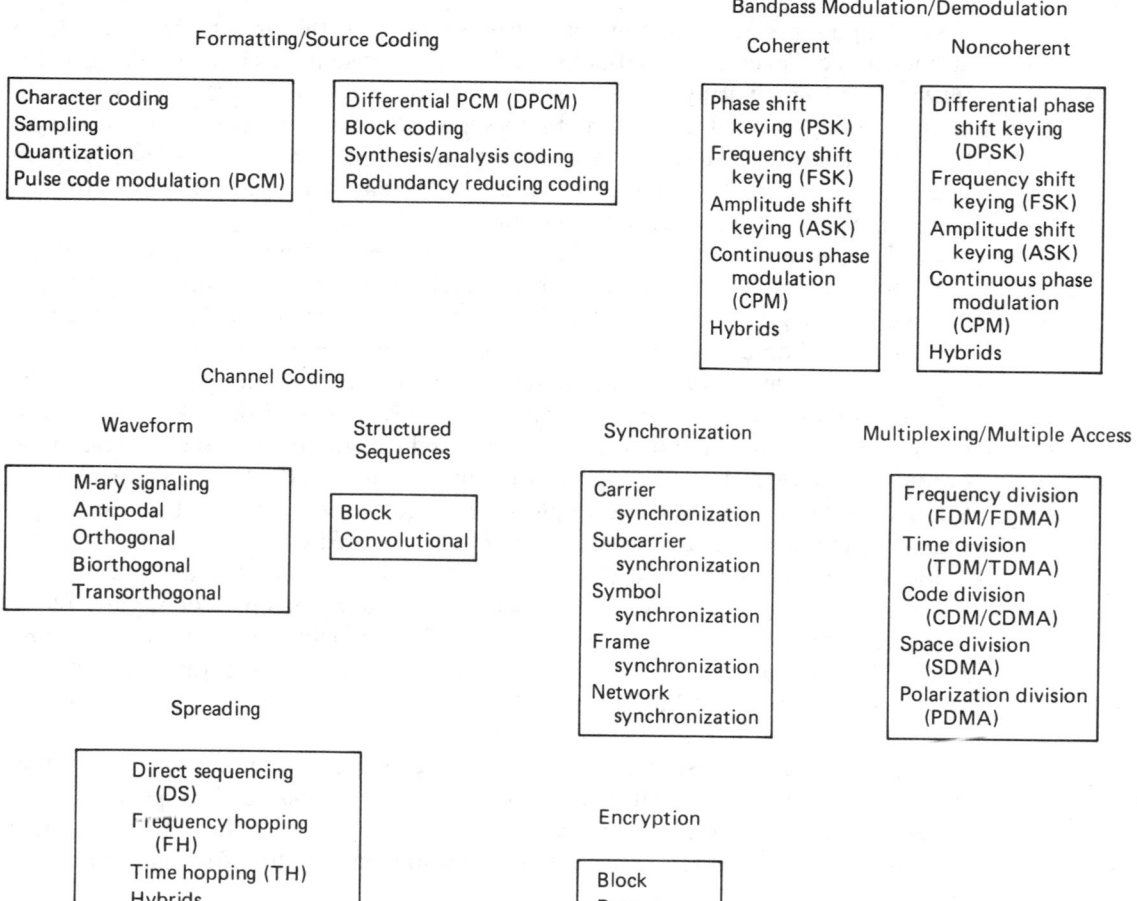

Figure 1.3 Basic digital communication transformations. (Reprinted with permission from B. Sklar, "A Structured Overview of Digital Communications," *IEEE Commun. Mag.*, August 1983, Fig. 2, p. 6. © 1983 IEEE.)

the selection of waveforms for making the symbols compatible with baseband transmission. As seen in Figure 1.3, formatting and source coding are grouped together; they are similar in that they involve data digitization. Since the term "source coding" has taken on the connotation of data redundancy reduction in addition to digitization, it is treated later, as a special formatting case, in Chapter 11.

In Figure 1.3, bandpass modulation/demodulation is partitioned into two basic categories, coherent and noncoherent. The process of *demodulation* involves the detection of the baseband information. Digital demodulation is typically accomplished with the aid of reference waveforms. When the references contain all the signal attributes, particularly phase information, the process is termed *coherent*; when phase information is not used, the process is termed *noncoherent*. Both techniques are detailed in Chapter 3.

Chapter 4 is devoted to link analysis. In the past, this area has received little attention in colleges or in textbooks, probably because it was considered straightforward and not worthy of detailed discussion. However, of the many specifications, analyses, and tabulations that support a developing communication system, link analysis stands out in its ability to provide overall system insight. In Chapter 4 we bring together all the link fundamentals that are essential for the analysis of most communication systems.

Channel coding deals with the techniques used to enhance digital signals so that they are less vulnerable to such channel impairments as noise, fading, and jamming. In Figure 1.3 channel coding is partitioned into two basic categories, waveform coding and structured sequences. *Waveform coding* involves the use of new waveforms, yielding improved detection performance over that of the original waveforms. *Structured sequences* involve the use of redundant bits to determine whether or not an error has occurred due to noise on the channel. One of these techniques, known as automatic repeat request (ARQ), simply recognizes the occurrence of an error and requests that the sender retransmit the message; other techniques, called forward error correction (FEC), are capable of automatically correcting the errors (within specified limitations). Under the heading of structured sequences, we shall discuss the two prevalent techniques, block coding and convolutional coding. In Chapter 5 we consider waveform coding and linear block coding. In Chapter 6 we consider convolutional coding, Viterbi decoding (and other decoding algorithms), hard versus soft decoding procedures, and interleaving and deinterleaving.

In Chapter 7 we summarize the design goals for a communication system and present various modulation and coding trade-offs that need to be considered in the design of a system. We discuss theoretical limitations such as the Nyquist criterion and the Shannon limit. We also examine bandwidth-efficient modulation schemes.

Chapter 8 deals with synchronization. In digital communications, synchronization involves the estimation of both time and frequency. The subject is partitioned as shown in Figure 1.3. Coherent systems need to synchronize their frequency reference with the carrier (and possibly subcarrier) in both frequency and phase. For noncoherent systems, phase synchronization is not needed. The fundamental time-synchronization process is symbol synchronization. The demodulator needs to know when to start and end the symbol detection procedure; a timing error will degrade detection performance. The next time-synchronization level, frame synchronization, allows the reconstruction of the message; and network synchronization allows coordination with other users in order to use the resource efficiently. In Chapter 8 we are concerned with the alignment of the timing of spatially separated periodic processes; the alignment is illustrated for the case of a satellite communications link.

Chapter 9 deals with multiplexing and multiple access. The two terms mean very similar things. Both involve the idea of resource sharing. The main difference between the two is that *multiplexing* takes place locally (e.g., on a printed circuit board, within an assembly, or even within a facility), and *multiple access* takes place remotely (e.g., multiple users share the use of a satellite transponder). Multiplexing involves an algorithm that is known a priori; usually, it is hard-wired into the system. Multiple access, on the other hand, is generally adaptive and may require overhead to enable the algorithm to operate. In Chapter 9 we discuss the classical ways of sharing the resource: frequency division, time division, and code division. We also consider some of the multiple access techniques that have emerged as a result of satellite communications.

Chapter 10 introduces a transformation of primary importance in military communications called spreading. The chapter deals with the spread-spectrum techniques that are emerging as important for achieving interference protection, privacy, or flexible access of the communications resource.

Chapter 11 treats source coding—techniques that deal with the task of forming efficient descriptions of source information. Source coding can be applied to digital data and to waveform signals; it can reduce data redundancy and thus reduce data rates. We will see that the advantage of source coding is a reduction of the system resources (i.e., bandwidth) required to describe the information.

The final chapter of the book, Chapter 12, deals with encryption and decryption, whose basic goals are privacy and authentication. *Privacy* refers to preventing unauthorized persons from extracting information (eavesdropping) from the channel. *Authentication* refers to preventing unauthorized persons from injecting spurious signals (spoofing) into the channel. In this chapter we highlight the data encryption standard (DES) and some current ideas for a class of encryption systems called public key cryptosystems.

1.1.3 Basic Digital Communication Nomenclature

Some of the basic digital signal nomenclature that frequently appears in digital communication literature is as follows:

Information source: the device producing information to be communicated by means of the DCS. Information sources can be analog or discrete. The output of an analog source can have any value in a continuous range of amplitudes, whereas the output of a discrete information source takes its value from a finite set. Analog information sources can be transformed into digital sources through the use of sampling and quantization. Sampling and quantization techniques called formatting and source coding (see Figure 1.3) are described in Chapters 2 and 11.

Textual message: a sequence of characters (see Figure 1.4a). For digital transmission, the message will be a sequence of digits or symbols from a finite symbol set or alphabet.

Character: a member of an alphabet or set of symbols (see Figure 1.4b). Characters may be mapped into a sequence of binary digits. There are several standardized codes used for character encoding, including the American Standard Code for Information Interchange (ASCII), Extended Binary Coded Decimal Interchange Code (EBCDIC), Hollerith, Baudot, Murray, and Morse.

(a)
HOW ARE YOU?
OK
$9, 567, 216.73

(b)
A
9
&

(c)
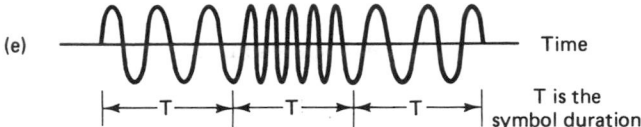

H | O | W

0 0 0 1 0 0 1 1 1 1 1 0 0 1 1 1 1 0 1 0 1

(d)
1   Binary symbol (k = 1, M = 2)
10   Quaternary symbol (k = 2, M = 4)
011 8-ary symbol (k = 3, M = 8)

(e)
Time

T is the symbol duration

Figure 1.4 Nomenclature examples. (a) Textual messages. (b) Characters. (c) Bit stream (7-bit ASCII). (d) Symbols m_i, $i = 1, \ldots, M$, $M = 2^k$. (e) Bandpass digital waveform $s_i(t)$, $i = 1, \ldots, M$.

Binary digit (bit): the fundamental information unit for all digital systems. The term *bit* is also used as a unit of information content; this second usage is described in Chapter 7.

Bit stream: a sequence of binary digits (ones and zeros). Sometimes, a sequence of two-level pulses is used as a convenient illustration of the bit stream. The bit stream in Figure 1.4c uses a 7-bit ASCII character code for representing the message "HOW." A bit stream is often termed a *baseband signal,* which implies that its spectral content extends from (or near) dc up to some finite value, usually less than a few megahertz.

Symbol (digital message): groups of k bits considered as a unit or character m_i, from a finite symbol set or alphabet (see Figure 1.4d). The size of the alphabet, M, is $M = 2^k$ (i.e., k is the number of bits in the symbol). For transmission, each m_i symbol ($i = 1, \ldots, M$) will be represented by a corresponding waveform $s_1(t), s_2(t), \ldots, s_M(t)$. The symbol, m_i, is sent by transmitting the digital waveform, $s_i(t)$, for T seconds, the symbol time duration. The next symbol is sent during the next time interval, T. The fact that the symbol set transmitted by the DCS is finite is a primary difference between a DCS and an analog system. The DCS receiver need only decide which of the M waveforms was transmitted; however, an analog receiver must be capable of accurately estimating a continuous range of waveforms.

Digital waveform: a voltage or current waveform (a pulse for baseband transmission, or a sinusoid for bandpass transmission) that represents a digital symbol. The waveform characteristics (amplitude, width, position for pulses, or amplitude, frequency, phase for sinusoids) allow its identification as one of the symbols in the finite symbol alphabet. Figure 1.4e shows an example of a bandpass digital waveform. Even though the waveform is sinusoidal, and consequently has an analog appearance, it is called a *digital waveform* because it is encoded with digital information. In the figure, during each time interval, T, a preassigned frequency indicates the value of a digit.

Data rate: data rate in bits per second (bits/s) is given by $R = k/T = (1/T)$ $\log_2 M$ bits/s, where k bits identify a symbol from an $M = 2^k$-symbol alphabet, and T is the k-bit symbol duration.

1.1.4 Digital versus Analog Performance Criteria

A principal difference between analog and digital communication systems has to do with the way in which we evaluate their performance. Analog systems draw their waveforms from a continuum, which therefore forms an infinite set; that is, a receiver must deal with an infinite number of possible waveshapes. The figure of merit for the performance of analog communication systems is a fidelity criterion, such as signal-to-noise ratio, percent distortion, or expected mean-square error between the transmitted and received waveforms.

By contrast, a digital communication system transmits signals that represent digits. These digits form a finite set or alphabet, and the set is known a priori to the receiver. A figure of merit for digital communication systems is the probability of incorrectly detecting a digit, or the probability of error (P_E).

1.2 CLASSIFICATION OF SIGNALS

1.2.1 Deterministic and Random Signals

A signal can be classified as *deterministic,* meaning that there is no uncertainty with respect to its value at any time, or as *random,* meaning that there is some degree of uncertainty before the signal actually occurs. Deterministic signals or waveforms are modeled by explicit mathematical expressions, such as $x(t) = 5 \cos 10t$. For a random waveform it is *not* possible to write such an explicit expression. However, when examined over a long period, a random waveform, also referred to as a *random process*, may exhibit certain regularities that can be described in terms of probabilities and statistical averages. Such a model, in the form of a probabilistic description of the random process, is particularly useful for characterizing signals and noise in communication systems.

1.2.2 Periodic and Nonperiodic Signals

A signal $x(t)$ is called *periodic in time* if there exists a constant $T_0 > 0$ such that

$$x(t) = x(t + T_0) \qquad \text{for } -\infty < t < \infty \qquad (1.1)$$

where t denotes time. The smallest value of T_0 that satisfies this condition is called the *period* of $x(t)$. The period T_0 defines the duration of one complete cycle of $x(t)$. A signal for which there is no value of T_0 that satisfies Equation (1.1) is called a *nonperiodic signal*.

1.2.3 Analog and Discrete Signals

An *analog signal, $x(t)$,* is a continuous function of time; that is, $x(t)$ is uniquely defined for all t. An electrical analog signal arises when a physical waveform (e.g., speech) is converted into an electrical signal by means of a transducer. By comarison, a *discrete signal, $x(kT)$,* is one that exists only at discrete times; it is characterized by a sequence of numbers defined for each time, kT, where k is an integer and T is a fixed time interval.

1.2.4 Energy and Power Signals

An electrical signal can be represented as a voltage, $v(t)$, or a current, $i(t)$, with instantaneous power $p(t)$ across a resistor $\mathcal{R}$ defined by

$$p(t) = \frac{v^2(t)}{\mathcal{R}} \qquad (1.2)$$

or

$$p(t) = i^2(t)\mathcal{R} \qquad (1.3)$$

In communication systems, power is often normalized by assuming $\mathcal{R}$ to be 1 Ω, although $\mathcal{R}$ may be another value in the actual circuit. If the actual value of the power is needed, it is obtained by "denormalization" of the normalized value. For the normalized case, Equations (1.2) and (1.3) have the same form. Therefore, regardless of whether the signal is a voltage or current waveform, the normalization convention allows us to express the instantaneous power as

$$p(t) = x^2(t) \qquad (1.4)$$

where $x(t)$ is either a voltage or a current signal. The energy dissipated during the time interval $(-T/2, T/2)$ by a real signal with instantaneous power expressed by Equation (1.4) can then be written as

$$E_x^T = \int_{-T/2}^{T/2} x^2(t)\, dt \qquad (1.5)$$

and the average power dissipated by the signal during the interval is

$$P_x^T = \frac{1}{T} \int_{-T/2}^{T/2} x^2(t) \, dt \tag{1.6}$$

The performance of a communication system depends on the detected signal *energy*; higher-energy signals are detected more reliably (with fewer errors) than are lower-energy signals—the transmitted *energy does the work*. On the other hand, power is the *rate* at which energy is delivered. It is important for different reasons. The power determines the voltages that must be applied to a transmitter and the intensities of the electromagnetic fields that one must contend with in radio systems (i.e., fields in waveguides that connect the transmitter to the antenna, and fields around the radiating elements of the antenna).

In analyzing communication signals it is often desirable to deal with the *waveform energy*. We classify $x(t)$ as an *energy signal* if, and only if, it has nonzero but finite energy ($0 < E_x < \infty$) for all time, where

$$\begin{aligned} E_x &= \lim_{T \to \infty} \int_{-T/2}^{T/2} x^2(t) \, dt \\ &= \int_{-\infty}^{\infty} x^2(t) \, dt \end{aligned} \tag{1.7}$$

In the real world we always transmit signals having finite energy ($0 < E_x < \infty$). However, in order to describe *periodic signals,* which by definition [Equation (1.1)] exist for all time and thus have infinite energy, and in order to deal with random signals that have infinite energy, it is convenient to define a class of signals called *power signals.* A signal is defined to be a power signal if, and only if, it has finite but nonzero power ($0 < P_x < \infty$) for all time, where

$$P_x = \lim_{T \to \infty} \frac{1}{T} \int_{-T/2}^{T/2} x^2(t) \, dt \tag{1.8}$$

The energy and power classifications are mutually exclusive. An energy signal has finite energy but *zero average power,* whereas a power signal has finite average power but *infinite energy.* A waveform in a system may be constrained in either its power or energy values. As a general rule, periodic signals and random signals are classified as power signals, while signals that are both deterministic and non-periodic are classified as energy signals [1, 2].

As mentioned earlier, signal energy and power are both important parameters in specifying a communication system. The classification of a signal as either an energy signal or a power signal is a convenient model to facilitate the mathematical treatment of various signals and noise.

1.2.5 The Unit Impulse Function

A useful function in communication theory is the unit impulse or *Dirac delta function,* $\delta(t)$. The impulse function is an abstraction—an infinitely large amplitude pulse, with zero pulse width, and unity weight (area under the pulse), concentrated at the point where its argument is zero. The unit impulse is characterized by the following relationships:

$$\int_{-\infty}^{\infty} \delta(t)\, dt = 1 \tag{1.9}$$

$$\delta(t) = 0 \qquad \text{for } t \neq 0 \tag{1.10}$$

$$\delta(t) \text{ is unbounded at } t = 0 \tag{1.11}$$

$$\int_{-\infty}^{\infty} x(t)\delta(t - t_0)\, dt = x(t_0) \tag{1.12}$$

The unit impulse function, $\delta(t)$, is not a function in the usual sense. When operations involve $\delta(t)$, the convention is to interpret $\delta(t)$ as a unit-area pulse of finite amplitude and nonzero duration, after which the limit is considered as the pulse duration approaches zero. $\delta(t - t_0)$ can be depicted graphically as a spike located at $t = t_0$ with height equal to its integral or area. Thus $A\delta(t - t_0)$ with A constant represents an impulse function whose area or weight is equal to A, that is zero everywhere except at $t = t_0$.

Equation (1.12) is known as the *sifting* or *sampling property* of the unit impulse function; the unit impulse multiplier selects a sample of the function $x(t)$ evaluated at $t = t_0$.

1.3 SPECTRAL DENSITY

The *spectral density* of a signal characterizes the distribution of the signal's energy or power in the frequency domain. This concept is particularly important when considering filtering in communication systems. We need to be able to evaluate the signal and noise at the filter output. The energy spectral density (ESD) or the power spectral density (PSD) is used in the evaluation.

1.3.1 Energy Spectral Density

The total energy of a real-valued energy signal $x(t)$, defined over the interval $(-\infty, \infty)$, is described by Equation (1.7). Using Parseval's theorem [1], we can relate the energy of such a signal expressed in the time domain to the energy expressed in the frequency domain, as follows:

$$E_x = \int_{-\infty}^{\infty} x^2(t)\, dt = \int_{-\infty}^{\infty} |X(f)|^2\, df \tag{1.13}$$

where $X(f)$ is the Fourier transform of the nonperiodic signal $x(t)$ (for a review of Fourier techniques, see Appendix A). Let $\Psi_x(f)$ denote the squared magnitude spectrum, defined as

$$\Psi_x(f) = |X(f)|^2 \tag{1.14}$$

The quantity $\Psi_x(f)$ is the waveform *energy spectral density* (ESD) of the signal $x(t)$. Therefore, from Equation (1.13), we can express the total energy of the signal $x(t)$ by integrating the spectral density with respect to frequency, as follows:

$$E_x = \int_{-\infty}^{\infty} \Psi_x(f) \, df \qquad (1.15)$$

This equation states that the energy of a signal is equal to the area under the $\Psi_x(f)$ versus frequency curve. Energy spectral density describes the signal energy per unit bandwidth measured in joules/hertz. There are equal energy contributions from both positive and negative frequency components, since for a real signal, $x(t)$, $|X(f)|$ is an even function of frequency. Therefore, the energy spectral density is symmetrical in frequency about the origin, and thus the total energy of the signal $x(t)$ can be expressed as

$$E_x = 2 \int_{0}^{\infty} \Psi_x(f) \, df \qquad (1.16)$$

1.3.2 Power Spectral Density

The average power, P_x, of a real-valued power signal, $x(t)$, is defined in Equation (1.8). If $x(t)$ is a *periodic signal* with period T_0, it is classified as a power signal. The expression for the average power of a periodic signal takes the form of Equation (1.6), where the time average is taken over the signal period T_0, as follows:

$$P_x = \frac{1}{T_0} \int_{-T_0/2}^{T_0/2} x^2(t) \, dt \qquad (1.17a)$$

Parseval's theorem for a real-valued periodic signal [1] takes the form

$$P_x = \frac{1}{T_0} \int_{-T_0/2}^{T_0/2} x^2(t) \, dt = \sum_{n=-\infty}^{\infty} |c_n|^2 \qquad (1.17b)$$

where the $|c_n|$ terms are the complex Fourier series coefficients of the periodic signal (see Appendix A).

To apply Equation (1.17b), we need only know the magnitude of the coefficients, $|c_n|$. The *power spectral density* (PSD) function, $G_x(f)$, of the periodic signal, $x(t)$, is a real, even, and nonnegative function of frequency that gives the distribution of the power of $x(t)$ in the frequency domain, defined as

$$G_x(f) = \sum_{n=-\infty}^{\infty} |c_n|^2 \, \delta(f - nf_0) \qquad (1.18)$$

Equation (1.18) defines the power spectral density of a periodic signal, $x(t)$, as a succession of the weighted delta functions. Therefore, the PSD of a periodic signal is a discrete function of frequency. Using the PSD defined in Equation (1.18), we can now write the average normalized power of a real-valued signal, as follows:

$$P_x = \int_{-\infty}^{\infty} G_x(f)\, df = 2 \int_{0}^{\infty} G_x(f)\, df \tag{1.19}$$

Equation (1.18) describes the PSD of periodic (power) signals only. If $x(t)$ is a nonperiodic signal it *cannot* be expressed by a Fourier series, and if it is a nonperiodic power signal (having infinite energy) it *may not* have a Fourier transform. However, we may still express the power spectral density of such signals in the *limiting sense*. If we form a *truncated version*, $x_T(t)$, of the nonperiodic power signal, $x(t)$, by observing it only in the interval $(-T/2, T/2)$, then $x_T(t)$ has finite energy, and has a proper Fourier transform, $X_T(f)$. It can be shown [2] that the power spectral density of the nonperiodic $x(t)$ can then be defined in the limit as

$$G_x(f) = \lim_{T \to \infty} \frac{1}{T} |X_T(f)|^2 \tag{1.20}$$

Example 1.1   Average Normalized Power

(a) Find the average normalized power in the waveform, $x(t) = A \cos 2\pi f_0 t$, using time averaging.
(b) Repeat part (a) using the summation of spectral coefficients.

Solution

(a) Using Equation (1.17a), we have

$$P_x = \frac{1}{T_0} \int_{-T_0/2}^{T_0/2} A^2 \cos^2 2\pi f_0 t\, dt$$

$$= \frac{A^2}{2T_0} \int_{-T_0/2}^{T_0/2} (1 + \cos 4\pi f_0 t)\, dt$$

$$= \frac{A^2}{2T_0} (T_0) = \frac{A^2}{2}$$

(b) Using Equations (1.18) and (1.19) gives us

$$G_x(f) = \sum_{n=-\infty}^{\infty} |c_n|^2 \delta(f - nf_0)$$

$$\left. \begin{array}{l} c_1 = c_{-1} = \dfrac{A}{2} \\[2em] c_n = 0 \quad \text{for } n = 0,\ \pm 2,\ \pm 3,\ \ldots \end{array} \right\} \quad \text{(see Appendix A)}$$

$$G_x(f) = \left(\frac{A}{2}\right)^2 \delta(f - f_0) + \left(\frac{A}{2}\right)^2 \delta(f + f_0)$$

$$P_x = \int_{-\infty}^{\infty} G_x(f)\, df = \frac{A^2}{2}$$

1.4 AUTOCORRELATION

1.4.1 Autocorrelation of an Energy Signal

Correlation is a matching process; *autocorrelation* refers to the matching of a signal with a delayed version of itself. The autocorrelation function, $R_x(\tau)$, of a real-valued energy signal, $x(t)$, is defined as

$$R_x(\tau) = \int_{-\infty}^{\infty} x(t)x(t + \tau)\, dt \qquad \text{for } -\infty < \tau < \infty \qquad (1.21)$$

The autocorrelation function, $R_x(\tau)$, provides a measure of how closely the signal matches a copy of itself as the copy is shifted τ units in time. The variable τ plays the role of a scanning or searching parameter. $R_x(\tau)$ is not a function of time; it is only a function of the time difference, τ, between the waveform and its shifted copy.

The autocorrelative function of a real-valued *energy* signal has the following properties:

1. $R_x(\tau) = R_x(-\tau)$ symmetrical in τ about zero

2. $R_x(\tau) \leq R_x(0)$ for all τ maximum value occurs at the origin

3. $R_x(\tau) \leftrightarrow \Psi_x(f)$ autocorrelation and ESD form a Fourier transform pair, as designated by the double-headed arrows

4. $R_x(0) = \int_{-\infty}^{\infty} x^2(t)\, dt$ value at the origin is equal to the energy of the signal

If items 1 through 3 are satisfied, $R_x(\tau)$ satisfies the properties of an autocorrelation function. Property 4 can be derived from property 3 and thus need not be included as a basic test.

1.4.2 Autocorrelation of a Periodic (Power) Signal

The autocorrelation function of a real-valued power signal $x(t)$ is defined as

$$R_x(\tau) = \lim_{T \to \infty} \frac{1}{T} \int_{-T/2}^{T/2} x(t)x(t + \tau)\, dt \qquad \text{for } -\infty < \tau < \infty \qquad (1.22)$$

When the power signal, $x(t)$, is periodic with period T_0, the time average in Equation (1.22) may be taken over a *single period*, T_0, and the autocorrelation function can be expressed as follows:

$$R_x(\tau) = \frac{1}{T_0} \int_{-T_0/2}^{T_0/2} x(t)x(t + \tau)\, dt \qquad \text{for } -\infty < \tau < \infty \qquad (1.23)$$

The autocorrelation function of a real-valued *periodic* signal has properties similar to those of an energy signal, as follows:

1. $R_x(\tau) = R_x(-\tau)$ symmetrical in τ about zero

2. $R_x(\tau) \leq R_x(0)$ for all τ maximum value occurs at the origin

3. $R_x(\tau) \leftrightarrow G_x(f)$ autocorrelation and PSD form a Fourier transform pair

4. $R_x(0) = \dfrac{1}{T_0} \displaystyle\int_{-T_0/2}^{T_0/2} x^2(t)\, dt$ value at the origin is equal to the average power of the signal

1.5 RANDOM SIGNALS

The main objective of a communication system is the transfer of information over a channel. All useful message signals appear random; that is, the receiver does not know, a priori, which of the possible message waveforms will be transmitted. Also, the noise that accompanies the message signals is due to random electrical signals. Therefore, we need to be able to form efficient descriptions of random signals.

1.5.1 Random Variables

Let a *random variable, $X(A)$*, represent the functional relationship between a random event, A, and a real number. For notational convenience we shall designate the random variable by X, and let the functional dependence upon A be implicit. The random variable may be discrete or continuous. The *distribution function, $F_X(x)$*, of the random variable, X, is given by

$$F_X(x) = P(X \leq x) \tag{1.24}$$

where $P(X \leq x)$ is the probability that the value taken by the random variable, X, is less than or equal to a real number, x. The distribution function, $F_X(x)$, has the following properties:

1. $0 \leq F_X(x) \leq 1$

2. $F_X(x_1) \leq F_X(x_2)$ if $x_1 \leq x_2$

3. $F_X(-\infty) = 0$

4. $F_X(+\infty) = 1$

Another useful function relating to the random variable, X, is the *probability density function* (pdf), denoted $p_X(x)$, where

$$p_X(x) = \frac{dF_X(x)}{dx} \tag{1.25}$$

As in the case of the distribution function, the pdf is a function of a real number, x. The name "density function" arises from the fact that the probability of the event $x_1 \leq X \leq x_2$ equals

$$P(x_1 \leq X \leq x_2) = P(X \leq x_2) - P(X \leq x_1)$$

$$= F_X(x_2) - F_X(x_1)$$

$$= \int_{x_1}^{x_2} p_X(x)\, dx$$

The probability density function has the following properties

1. $p_X(x) \geq 0$

2. $\int_{-\infty}^{\infty} p_X(x)\, dx = F_X(+\infty) - F_X(-\infty) = 1$

Thus, a probability density function is always a nonnegative function with a total area of one. Throughout the book we use the designation, $p_X(x)$, for the probability density function of a *continuous* random variable. For ease of notation, we will often omit the subscript, X, and write simply, $p(x)$. We will use the designation $P(X = x_i)$ for the probability of a random variable, X, where X can take on *discrete* values only.

1.5.1.1 Ensemble Averages

The *mean value, m_X,* or *expected value* of a random variable, X, is defined by

$$m_X = E\{X\} = \int_{-\infty}^{\infty} x p_X(x)\, dx \tag{1.26}$$

where $E\{\cdot\}$ is called the *expected value operator*. The *nth moment* of a probability distribution of a random variable, X, is defined by

$$E\{X^n\} = \int_{-\infty}^{\infty} x^n p_X(x)\, dx \tag{1.27}$$

For the purposes of communication system analysis, the most important moments of X are the first two moments. Thus, $n = 1$ in Equation (1.27) gives m_X as discussed above, whereas $n = 2$ gives the mean-square value of X, as follows:

$$E\{X^2\} = \int_{-\infty}^{\infty} x^2 p_X(x)\, dx \tag{1.28}$$

We can also define *central moments*, which are the moments of the difference between X and m_X. The second central moment, called the *variance* of X, is defined as

$$\text{var}\,(X) = E\{(X - m_X)^2\} = \int_{-\infty}^{\infty} (x - m_X)^2 p_X(x)\, dx \tag{1.29}$$

The variance of X is also denoted as σ_X^2, and its square root, σ_X, is called the *standard deviation* of X. Variance is a measure of the "randomness" of the random variable X. By specifying the variance of a random variable, we are constraining the width of its probability density function. The variance and the mean-square value are related by

$$\sigma_X^2 = \mathbf{E}\{X^2 - 2m_X X + m_X^2\}$$
$$= \mathbf{E}\{X^2\} - 2m_X\mathbf{E}\{X\} + m_X^2$$
$$= \mathbf{E}\{X^2\} - m_X^2$$

Thus, the variance is equal to the difference between the mean-square value and the square of the mean.

1.5.2 Random Processes

A random process, $X(A, t)$, can be viewed as a function of two variables, *an event A*, and *time*. Figure 1.5 illustrates a random process. In the figure there are *N sample functions* of time, $\{X_j(t)\}$. Each of the sample functions can be regarded as the output of a different noise generator. For a specific event A_j, we have a

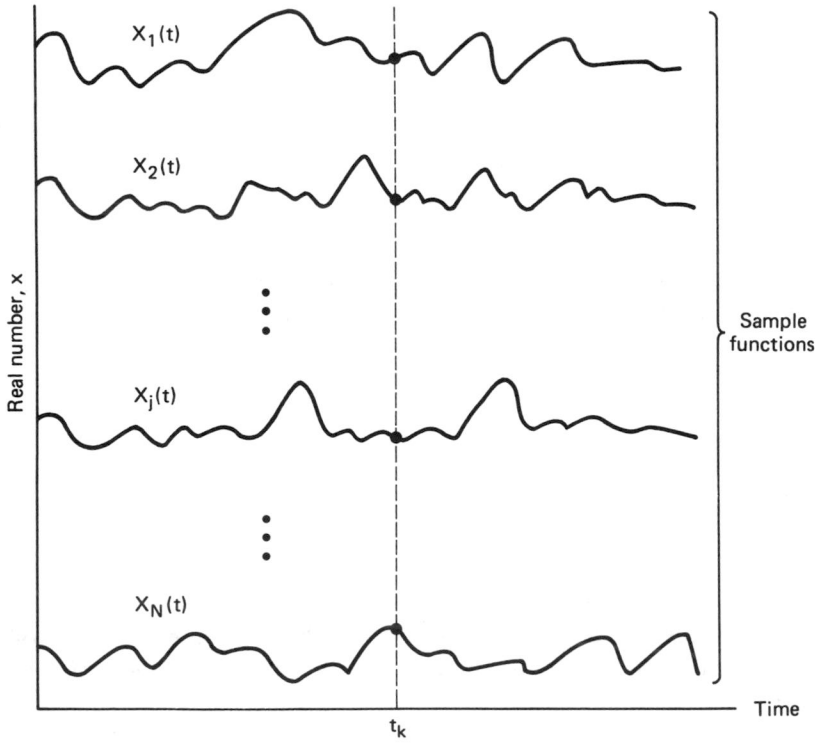

Figure 1.5 Random noise process.

single time function, $X(A_j, t) = X_j(t)$ (i.e., a sample function). The totality of all sample functions is called an *ensemble*. For a specific time t_k, $X(A, t_k)$ is a *random variable* $X(t_k)$, whose value depends on the event. Finally, for a specific event, $A = A_j$ and a specific time $t = t_k$, $X(A_j, t_k)$ is simply a *number*. For notational convenience we shall designate the random process by $X(t)$, and let the functional dependence upon A be implicit.

1.5.2.1 Statistical Averages of a Random Process

Because the value of a random process at any future time is unknown (since the identity of the event A is unknown), a random process whose distribution functions are continuous can be described statistically with a probability density function (pdf). In general the form of the pdf of a random process will be different for different times. In most situations it is not practical to determine empirically the probability distribution of a random process. However, a partial description consisting of the mean and autocorrelation function are often adequate for the needs of communication systems. We define the mean of the random process, $X(t)$, as

$$E\{X(t_k)\} = \int_{-\infty}^{\infty} x p_{X_k}(x) \, dx = m_X(t_k) \tag{1.30}$$

where $X(t_k)$ is the random variable obtained by observing the random process at time t_k, and the pdf of $X(t_k)$, the density over the ensemble of events at time t_k, is designated $p_{X_k}(x)$.

We define the autocorrelation function of the random process, $X(t)$, to be a function of two variables, t_1 and t_2, as shown by

$$R_X(t_1, t_2) = E\{X(t_1)X(t_2)\} \tag{1.31}$$

where $X(t_1)$ and $X(t_2)$ are random variables obtained by observing $X(t)$ at times t_1 and t_2. respectively. The autocorrelation function is a measure of the degree to which two time samples of the same random process are related.

1.5.2.2 Stationarity

A random process $X(t)$ is said to be *stationary* in the *strict sense* if none of its statistics are affected by a shift in the time origin. A random process is said to be *wide-sense stationary* (WSS) if two of its statistics, its mean and autocorrelation function, do not vary with a shift in the time origin. Thus, a process is WSS if

$$E\{X(t)\} = m_X = \text{a constant} \tag{1.32}$$

and

$$R_X(t_1, t_2) = R_X(t_1 - t_2) \tag{1.33}$$

Strict-sense stationary implies wide-sense stationary, but not vice versa. Most of the useful results in communication theory are predicated on random information signals and noise being wide-sense stationary. From a practical point of view it is not necessary for a random process to be stationary for all time, but only for some observation interval of interest.

For stationary processes, the autocorrelation function in Equation (1.33) does not depend on time but only on the difference between t_1 and t_2. That is, all pairs of values of $X(t)$ at points in time separated by $\tau = t_1 - t_2$ have the same correlation value. Thus, for stationary systems, we can denote $R_X(t_1, t_2)$ simply as $R_X(\tau)$.

1.5.2.3 Autocorrelation of a Wide-Sense Stationary Random Process

Just as the variance provides a measure of randomness for random variables, the autocorrelation function provides a similar measure for random processes. For a wide-sense stationary process, the autocorrelation function is only a function of the *time difference* $\tau = t_1 - t_2$, that is,

$$R_X(\tau) = E\{X(t)X(t + \tau)\} \qquad \text{for } -\infty < \tau < \infty \tag{1.34}$$

For a zero mean WSS processes, $R_X(\tau)$ indicates the extent to which the random values of the process separated by τ seconds in time are statistically correlated. In other words, $R_X(\tau)$ gives us an idea of the frequency response that is associated with a random process. If $R_X(\tau)$ changes slowly as τ increases from zero to some value, it indicates that, on the average, sample values of $X(t)$ taken at $t = t_1$ and $t = t_1 + \tau$ are nearly the same. Thus, we would expect a frequency domain representation of $X(t)$ to contain a preponderance of low frequencies. On the other hand if $R_X(\tau)$ decreases rapidly as τ is increased, we would expect $X(t)$ to change rapidly with time and thereby contain mostly high frequencies.

Properties of the autocorrelation function of a real-valued wide-sense stationary process are:

1. $R_X(\tau) = R_X(-\tau)$ symmetrical in τ about zero

2. $R_X(\tau) \leq R_X(0)$ for all τ maximum value occurs at the origin

3. $R_X(\tau) \leftrightarrow G_X(f)$ autocorrelation and power spectral density form a Fourier transform pair

4. $R_X(0) = E\{X^2(t)\}$ value at the origin is equal to the average power of the signal

1.5.3 Time Averaging and Ergodicity

To compute m_X and $R_X(\tau)$ by ensemble averaging, we would have to average across all the sample functions of the process and would need to have complete knowledge of the first- and second-order joint probability density functions. Such knowledge is generally not available.

When a random process belongs to a special class, known as an *ergodic process,* its time averages equal its ensemble averages, and the statistical properties of the process can be determined by *time averaging over a single sample function* of the process. For a random process to be ergodic it must be stationary in the strict sense. (The converse is not necessary.) However, for communication systems, where we are satisfied to meet the conditions of wide-sense stationarity, we are interested only in the mean and autocorrelation functions.

We can say that a random process is *ergodic in the mean* if

$$m_X = \lim_{T \to \infty} 1/T \int_{-T/2}^{T/2} X(t) \, dt \tag{1.35}$$

and it is *ergodic in the autocorrelation function* if

$$R_X(\tau) = \lim_{T \to \infty} 1/T \int_{-T/2}^{T/2} X(t)X(t + \tau) \, dt \tag{1.36}$$

Testing for the ergodicity of a random process is usually very difficult. In practice one makes an intuitive judgment as to whether it is reasonable to interchange the time and ensemble averages. A reasonable assumption in the analysis of most communication signals (in the absence of transient effects) is that the random waveforms are ergodic in the mean and the autocorrelation function. Since time averages equal ensemble averages for ergodic processes, fundamental electrical engineering parameters, such as dc value, rms value, and average power can be related to the moments of an ergodic random process. A summary of these relationships is:

1. The quantity $m_X = E\{X(t)\}$ is equal to the dc level of the signal.
2. The quantity m_X^2 is equal to the normalized power in the dc component.
3. The second moment of $X(t)$, $E\{X^2(t)\}$, is equal to the total average normalized power.
4. The quantity $\sqrt{E\{X^2(t)\}}$ is equal to the root-mean-square (rms) value of the voltage or current signal.
5. The variance, σ_X^2, is equal to the average normalized power in the time-varying or ac component of the signal.
6. If the process has zero mean (i.e., $m_X = m_X^2 = 0$), then $\sigma_X^2 = E\{X^2\}$, and the variance is the same as the mean-square value, or the variance represents the total power in the normalized load.
7. The standard deviation, σ_X, is the rms value of the ac component of the signal.
8. If $m_X = 0$, then σ_X is the rms value of the signal.

1.5.4 Power Spectral Density of a Random Process

A random process, $X(t)$, can generally be classified as a power signal having a power spectral density (PSD), $G_X(f)$, of the form shown in Equation (1.20). $G_X(f)$ is particularly useful in communications systems, because it describes the distribution of a signal's power in the frequency domain. The PSD enables us to evaluate the signal power that will pass through a network having known frequency characteristics. We summarize the principal features of PSD functions as follows:

1. $G_X(f) \geq 0$ and is always real valued

2. $G_X(f) = G_X(-f)$ for $X(t)$ real-valued

3. $G_X(f) \leftrightarrow R_X(\tau)$ PSD and autocorrelation form a Fourier transform pair

4. $P_X = \int_{-\infty}^{\infty} G_X(f)\, df$ relationship between average normalized power and PSD

Figure 1.6a illustrates a single sample waveform from a WSS random process, $X(t)$. The waveform is a binary random sequence with unit-amplitude positive and negative (bipolar) pulses. The positive and negative pulses occur with equal probability. The duration of each binary digit is T seconds, and the average or dc value of the random sequence is zero. Figure 1.6b shows the same sequence displaced τ_1 seconds in time; this sequence is therefore denoted $X(t - \tau_1)$. Let us assume that $X(t)$ is ergodic in the autocorrelation function so that we can use time averaging instead of ensemble averaging to find $R_X(\tau)$. The value of $R_X(\tau_1)$ is obtained by taking the product of the two sequences $X(t)$ and $X(t - \tau_1)$ and finding the average value using Equation (1.36). Equation (1.36) is accurate for ergodic processes *only in the limit*. However, integration over an integer number of periods can provide us with an estimate of $R_X(\tau)$. Notice that $R_X(\tau_1)$ can be obtained by a positive or negative shift of $X(t)$. Figure 1.6c illustrates such a calculation, using the single sample sequence (Figure 1.6a) and its shifted replica (Figure 1.6b). The cross-hatched areas under the product curve $X(t)X(t - \tau_1)$ contribute to positive values of the product, and the dotted areas contribute to negative values. The sequences can be further shifted by $\tau_2, \tau_3, \ldots$, each shift yielding a point on the overall autocorrelation function $R_X(\tau)$ shown in Figure 1.6d. Every random bit stream has an autocorrelation plot of the general shape shown in Figure 1.6d. The plot peaks at $R_X(0)$ [the best match occurs when τ equals zero, since $R(\tau) \leq R(0)$ for all τ], and it declines as τ increases. Figure 1.6d shows points corresponding to $R_X(0)$ and $R_X(\tau_1)$.

The analytical expression for the autocorrelation function $R_X(\tau)$ shown in Figure 1.6d, is [1]

$$R_X(\tau) = \begin{cases} 1 - \dfrac{|\tau|}{T} & \text{for } |\tau| \leq T \\ 0 & \text{for } |\tau| > T \end{cases} \tag{1.37}$$

The autocorrelation function allows us to express a random signal's power spectral density directly. Since the PSD and the autocorrelation function are Fourier transforms of each other, the PSD, $G_X(f)$, of the random binary sequence can be found,

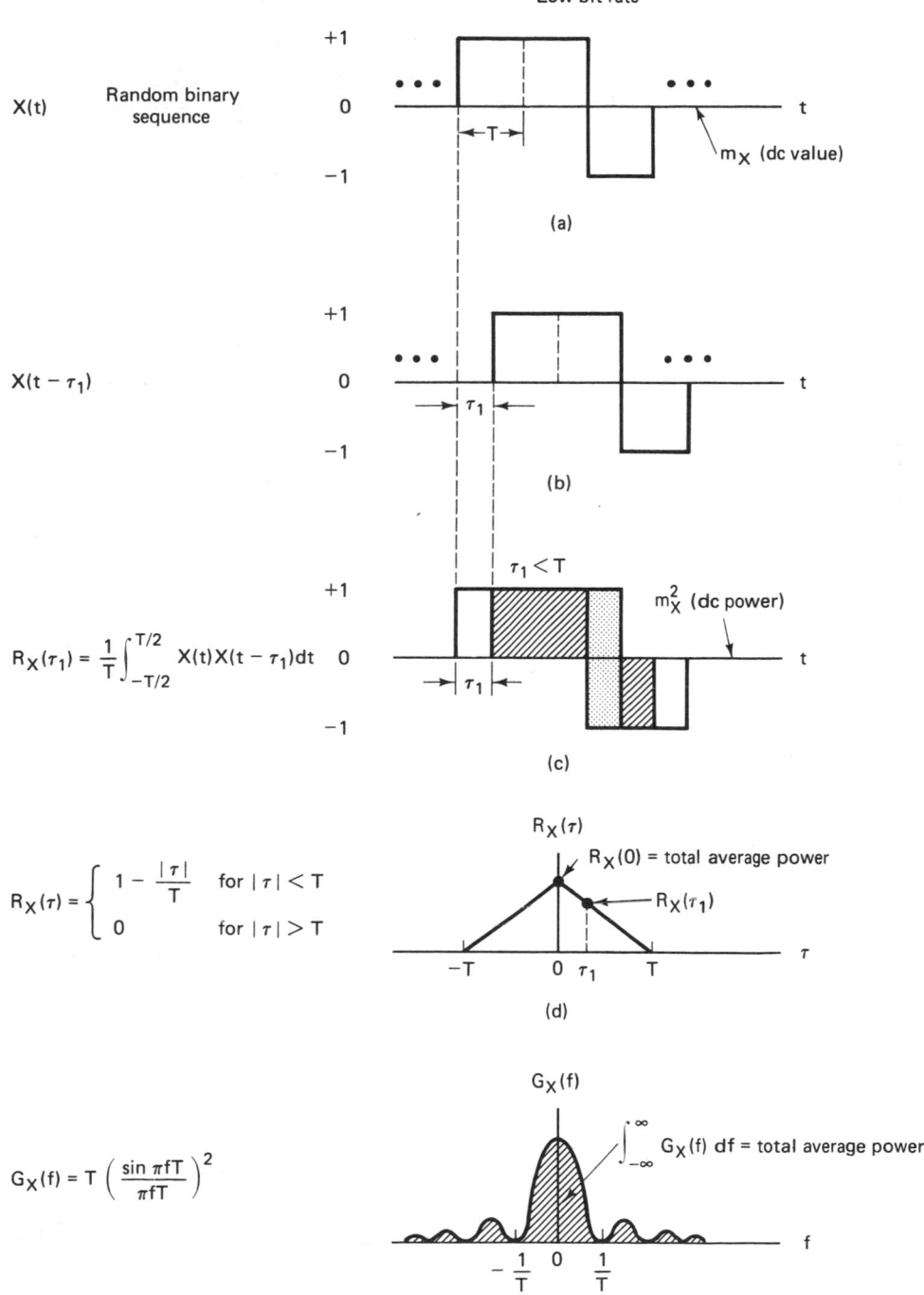

Figure 1.6 Autocorrelation and power spectral density.

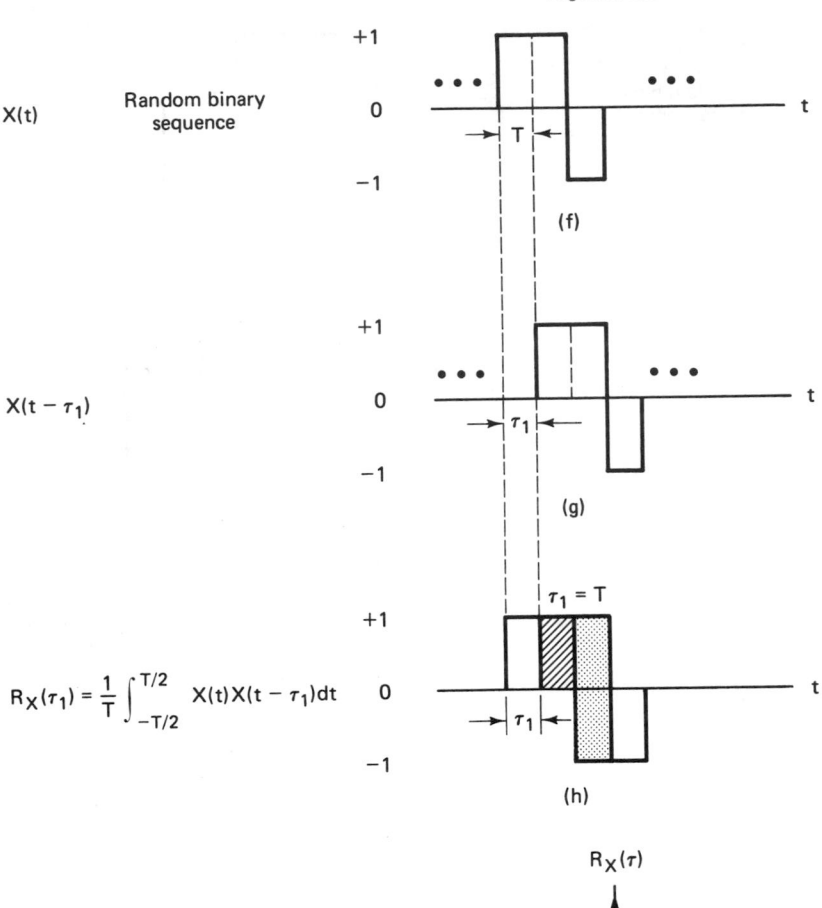

High bit rate

X(t) Random binary sequence

+1
0
−1

$\rightarrow$ T $\leftarrow$

t

(f)

$X(t - \tau_1)$

+1
0
−1

$\rightarrow \tau_1 \leftarrow$

t

(g)

$R_X(\tau_1) = \dfrac{1}{T} \displaystyle\int_{-T/2}^{T/2} X(t) X(t - \tau_1) dt$

$\tau_1 = T$

+1
0
−1

$\rightarrow |\tau_1| \leftarrow$

t

(h)

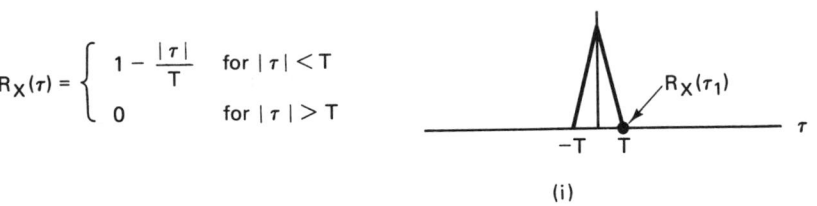

$R_X(\tau) = \begin{cases} 1 - \dfrac{|\tau|}{T} & \text{for } |\tau| < T \\ 0 & \text{for } |\tau| > T \end{cases}$

$R_X(\tau)$

$R_X(\tau_1)$

−T   T

τ

(i)

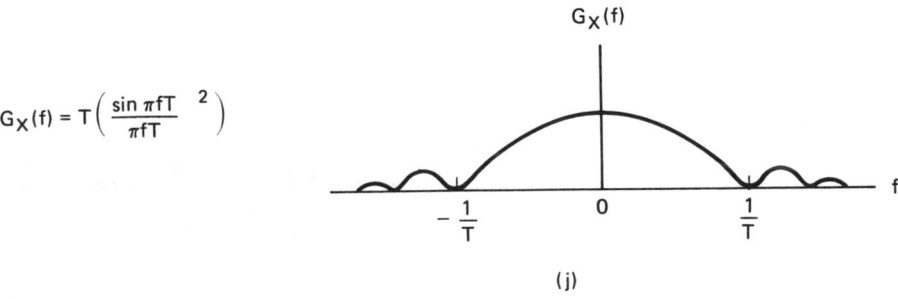

$G_X(f) = T \left(\dfrac{\sin \pi f T}{\pi f T} \right)^2$

$G_X(f)$

$-\dfrac{1}{T}$ 0 $\dfrac{1}{T}$

f

(j)

Figure 1.6 *(Continued)*

Signals and Spectra   Chap. 1

using Table A.1, as the transform of $R_X(\tau)$ in Equation (1.37). $G_X(f)$ is shown below, and its general shape is illustrated in Figure 1.6e.

$$G_X(f) = T \left(\frac{\sin \pi fT}{\pi fT} \right)^2 = T \operatorname{sinc}^2 fT \qquad (1.38)$$

where

$$\operatorname{sinc} y = \frac{\sin \pi y}{\pi y} \qquad (1.39)$$

Notice that the area under the PSD curve represents the average power in the signal. One convenient measure of *bandwidth* is the width of the main spectral lobe. Figure 1.6e illustrates that the bandwidth of a signal is inversely related to the symbol duration or pulse width. Figures 1.6f–j repeat the steps shown in Figures 1.6a–e, except that the bit duration is shorter. Notice that the shape of the shorter-bit-duration $R_X(\tau)$ is narrower, shown in Figure 1.6i, than it is for the longer-bit-duration $R_X(\tau)$, shown in Figure 1.6d. In Figure 1.6i, $R_X(\tau_1) = 0$; in other words, a shift of τ_1 in the case of the shorter-bit-duration example is enough to produce a zero match, or a complete decorrelation between the shifted sequences. Since the pulse duration, T, is shorter in Figure 1.6f, and the bit rate is higher than in Figure 1.6a, the bandwidth occupancy in Figure 1.6j is greater than the lower-bit-rate bandwidth occupancy shown in Figure 1.6e.

1.5.5 Noise in Communication Systems

The term *noise* refers to *unwanted* electrical signals that are always present in electrical systems. The presence of noise superimposed on a signal tends to obscure or mask the signal; it limits the receiver's ability to make correct symbol decisions, and thereby limits the rate of information transmission. Noise arises from a variety of sources, both man-made and natural. *Man-made noise* includes such sources as spark-plug ignition noise, switching transients, and other radiating electromagnetic signals. *Natural noise* includes electrical circuit and component noise, atmospheric disturbances, and galactic sources.

Good engineering design can eliminate much of the noise or its undesirable effect through filtering, shielding, the choice of modulation, and the selection of an optimum receiver site. For example, sensitive radio astronomy measurements are typically located at remote desert locations, far from man-made noise sources. However, there is one natural source of noise, called *thermal* or *Johnson noise,* that cannot be eliminated. Thermal noise [4, 5] is caused by the thermal motion of electrons in all dissipative components—resistors, wires, and so on. The same electrons that are responsible for electrical conduction are also responsible for thermal noise.

We can describe thermal noise as a zero-mean *Gaussian* random process. A Gaussian process, $n(t)$, is a random function whose value, n, at any arbitrary time, t, is statistically characterized by the Gaussian probability density function, $p(n)$:

$$p(n) = \frac{1}{\sigma\sqrt{2\pi}} \exp\left[-\frac{1}{2}\left(\frac{n}{\sigma}\right)^2\right] \qquad (1.40)$$

where σ^2 is the variance of n. The *normalized* or *standardized Gaussian density function* of a zero-mean process is obtained by assuming that $\sigma = 1$. This normalized pdf is shown sketched in Figure 1.7.

We will often represent a random signal as the sum of a Gaussian noise random variable and a dc signal:

$$z = a + n$$

where z is the random signal, a the dc component, and n the Gaussian noise random variable. The pdf $p(z)$ is then expressed as

$$p(z) = \frac{1}{\sigma\sqrt{2\pi}} \exp\left[-\frac{1}{2}\left(\frac{z-a}{\sigma}\right)^2\right] \qquad (1.41)$$

where, as before, σ^2 is the variance of n. The Gaussian distribution is often used as the system noise model because of a theorem, called the *central limit theorem* [3], which states that under very general conditions the probability distribution of the sum of j statistically independent random variables approaches the Gaussian distribution as $j \rightarrow \infty$, no matter what the individual distribution functions may be. Therefore, even though individual noise mechanisms might have other than

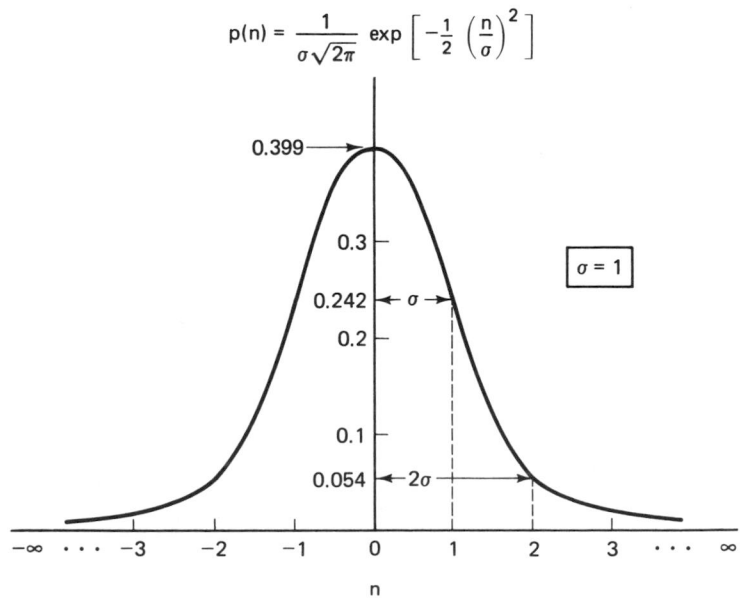

Figure 1.7 Normalized ($\sigma = 1$) Gaussian probability density function.

Signals and Spectra   Chap. 1

Gaussian distributions, the aggregate of many such mechanisms will tend toward the Gaussian distribution.

1.5.5.1 White Noise

The primary spectral characteristic of thermal noise is that its power spectral density is *the same* for all frequencies of interest in most communication systems; in other words, a thermal noise source emanates an equal amount of noise power per unit bandwidth at all frequencies—from dc to about 10^{12} Hz. Therefore, a simple model for thermal noise assumes that its power spectral density $G_n(f)$ is flat for all frequencies, as shown in Figure 1.8a, and is denoted as follows:

$$G_n(f) = \frac{N_0}{2} \quad \text{watts/hertz} \tag{1.42}$$

where the factor of 2 is included to indicate that $G_n(f)$ is a *two-sided* power spectral density. When the noise power has such a uniform spectral density, we refer to it as *white noise*. The adjective "white" is used in the sense that white light contains equal amounts of all frequencies within the visible band of electromagnetic radiation.

The autocorrelation function of white noise is given by the inverse Fourier transform of the noise power spectral density (see Table A.1) denoted as follows:

$$R_n(\tau) = \mathscr{F}^{-1}\{G_n(f)\} = \frac{N_0}{2} \delta(\tau) \tag{1.43}$$

Thus the autocorrelation of white noise is a delta function weighted by the factor $N_0/2$ and occurring at $\tau = 0$, as seen in Figure 1.8b. Note that $R_n(\tau)$ is zero for $\tau \neq 0$; that is, any two different samples of white noise, no matter how close together in time they are taken, are uncorrelated.

The average power, P_n, of white noise is *infinite* because its bandwidth is infinite. This can be seen by combining Equations (1.19) and (1.42) to yield.

$$P_n = \int_{-\infty}^{\infty} \frac{N_0}{2}\, df = \infty \tag{1.44}$$

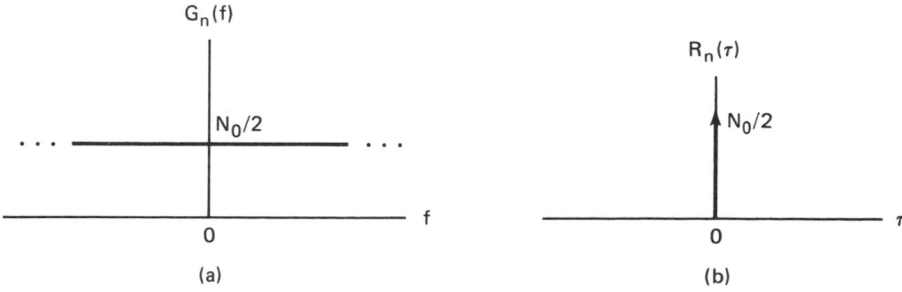

Figure 1.8 (a) Power spectral density of white noise. (b) Autocorrelation function of white noise.

Although white noise is a useful abstraction, no noise process can truly be white; however, the noise encountered in many real systems can be assumed to be approximately white. We can only observe such noise after it has passed through a real system which will have a finite bandwidth. Thus, as long as the bandwidth of the noise is appreciably larger than that of the system, the noise can be considered to have an infinite bandwidth.

The delta function in Equation (1.43) means that the noise signal, $n(t)$, is totally decorrelated from its time-shifted version, for any $\tau > 0$. Equation (1.43) indicates that *any* two different samples of a white noise process are uncorrelated. Since thermal noise is a Gaussian process and the samples are uncorrelated, the noise samples are also independent [3]. Therefore, the effect on the detection process of a channel with *additive white Gaussian noise* (AWGN) is that the noise affects each transmitted symbol *independently*. Such a channel is called a *memoryless channel*. The term "additive" means that the noise is simply superimposed or added to the signal—that there are no multiplicative mechanisms at work.

Since thermal noise is present in all communication systems and is the prominent noise source for most systems, the thermal noise characteristics—additive, white, and Gaussian—are most often used to model the noise in communication systems. Since zero-mean Gaussian noise is completely characterized by its *variance,* this model is particularly simple to use in the detection of signals and in the design of optimum receivers. In this book we shall assume, unless otherwise stated, that the system is corrupted by *additive zero-mean white Gaussian noise,* even though this is sometimes an oversimplification.

1.6 SIGNAL TRANSMISSION THROUGH LINEAR SYSTEMS

Having developed a set of models for signals and noise, we now consider the characterization of systems and their effects on such signals and noise. Since a system can be characterized equally well in the time domain or the frequency domain, techniques will be developed in both domains to analyze the response of a linear system to an arbitrary input signal. The signal, applied to the input of the system, as shown in Figure 1.9, can be described either as a time-domain signal, $x(t)$, or by its Fourier transform, $X(f)$. The use of time-domain analysis yields the time-domain output, $y(t)$, and in the process, $h(t)$, the characteristic or *impulse response* of the network, will be defined. When the input is considered in the frequency domain, we shall define a *frequency transfer function, $H(f)$,* for the system, which will determine the frequency-domain output, $Y(f)$. The system is assumed to be linear and time invariant. It is also assumed that there is no stored energy in the system at the time the input is applied.

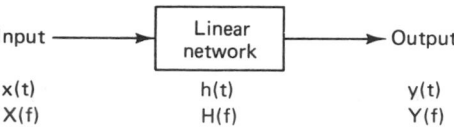

Figure 1.9 Linear system and its key parameters.

1.6.1 Impulse Response

The linear time-invariant system or network illustrated in Figure 1.9 is characterized in the time domain by an impulse response, $h(t)$, which is the response when the input is equal to a unit impulse $\delta(t)$; that is,

$$h(t) = y(t) \qquad \text{when } x(t) = \delta(t) \tag{1.45}$$

The response of the network to an arbitrary input $x(t)$ is then found by the convolution of $x(t)$ with $h(t)$, where $*$ denotes the convolution operation (see Section A.5):

$$y(t) = x(t) * h(t) = \int_{-\infty}^{\infty} x(\tau)h(t - \tau)\, d\tau \tag{1.46}$$

The system is assumed to be *causal*, which means that there can be *no* output prior to the time, $t = 0$, when the input is applied. Therefore, the lower limit of integration can be changed to zero, and we can express the output $y(t)$ as

$$y(t) = \int_{0}^{\infty} x(\tau)h(t - \tau)\, d\tau \tag{1.47}$$

Equations (1.46) and (1.47) are called the *superposition integral* or the *convolution integral*.

1.6.2 Frequency Transfer Function

The frequency-domain output signal, $Y(f)$, is obtained by taking the Fourier transform of both sides of Equation (1.46). Since convolution in the time-domain transforms to multiplication in the frequency domain (and vice versa), Equation (1.46) yields

$$Y(f) = X(f)H(f) \tag{1.48}$$

or

$$H(f) = \frac{Y(f)}{X(f)} \tag{1.49}$$

provided, of course, that $X(f) \neq 0$ for all f. Here $H(f) = \mathcal{F}\{h(t)\}$, the Fourier transform of the impulse response function, is called the *frequency transfer function* or the *frequency response* of the network. In general, $H(f)$ is complex and can be written as

$$H(f) = |H(f)|\, e^{j\theta(f)} \tag{1.50}$$

where $|H(f)|$ is the magnitude response. The phase response, $\theta(f)$, is defined as

$$\theta(f) = \tan^{-1} \frac{\text{Im } \{H(f)\}}{\text{Re } \{H(f)\}} \tag{1.51}$$

where the terms "Re" and "Im" denote "the real part of" and "the imaginary part of," respectively.

The frequency transfer function of a linear time-invariant network can easily be measured in the laboratory with a sinusoidal generator at the input of the network and an oscilloscope at the output. When the input waveform $x(t)$ is expressed as

$$x(t) = A \cos 2\pi f_0 t$$

the output of the network will be

$$y(t) = A |H(f_0)| \cos [2\pi f_0 t + \theta(f_0)] \qquad (1.52)$$

The input frequency, f_0, is stepped through the values of interest; at each step, the amplitude and phase at the output are measured.

1.6.2.1 Random Processes and Linear Systems

If a random process forms the input to a time-invariant linear system, the output will also be a random process. That is, each sample function of the input process yields a sample function of the output process. The input power spectral density, $G_X(f)$, and the output power spectral density, $G_Y(f)$, are related as follows:

$$G_Y(f) = G_X(f) |H(f)|^2 \qquad (1.53)$$

Equation (1.53) provides a simple way of finding the power spectral density out of a time-invariant linear system when the input is a random process.

In Chapters 2 and 3 we consider the detection of signals in Gaussian noise. We will utilize a fundamental property of a Gaussian process applied to a linear system, stated as follows: It can be shown that if a Gaussian process, $X(t)$, is applied to a time-invariant linear filter, the random process, $Y(t)$, developed at the output of the filter is also Gaussian [6].

1.6.3 Distortionless Transmission

What is required of a network for it to behave like an *ideal* transmission line? The output signal from an ideal transmission line may have some time delay compared to the input, and it may have a different amplitude than the input (just a scale change), but otherwise it must have no distortion—it must have the same shape as the input. Therefore, for ideal distortionless transmission, we can describe the output signal as

$$y(t) = Kx(t - t_0) \qquad (1.54)$$

where K and t_0 are constants. Taking the Fourier transform of both sides (see Section A.3.1), we write

$$Y(f) = KX(f)e^{-j2\pi f t_0} \qquad (1.55)$$

Substituting the expression (1.55) for $Y(f)$ into Equation (1.49), we see that the required system transfer function for distortionless transmission is

$$H(f) = Ke^{-j2\pi f t_0} \qquad (1.56)$$

Therefore, to achieve *ideal distortionless transmission,* the overall system response must have a constant magnitude response, and its phase shift must be linear with frequency. It is not enough that the system amplify or attenuate all frequency components equally. All of the signal's frequency components must also arrive with identical time delay in order to add up correctly. Since time delay, t_0, is related to phase shift, θ, and radian frequency, $\omega = 2\pi f$, as follows,

$$t_0 \text{ (seconds)} = \frac{\theta \text{ (radians)}}{2\pi f \text{ (radians/second)}} \tag{1.57}$$

it is clear that phase shift must be proportional to frequency in order for the time delay of all components to be identical. In practice, a signal will be distorted in passing through some parts of a system. Phase or amplitude correction (*equalization*) networks may be introduced elsewhere in the system to correct for this distortion. It is the overall input–output characteristic of the system that determines its performance.

1.6.3.1 Ideal Filter

One cannot build the ideal network described in Equation (1.56). The problem is that Equation (1.56) implies an infinite bandwidth capability, where the bandwidth of a system is defined as the interval of positive frequencies over which the magnitude $|H(f)|$ remains within a specified value. In Section 1.7 various measures of bandwidth are enumerated. As an approximation to the ideal infinite-bandwidth network, let us choose a truncated network that passes, without distortion, all frequency components between f_ℓ and f_u, where f_ℓ is the lower cutoff frequency and f_u is the upper cutoff frequency, as shown in Figure 1.10. Each of these networks is called an *ideal filter.* Outside the range $f_\ell < f < f_u$, which is called the *passband,* the ideal filter is assumed to have a response of zero magnitude. The effective width of the passband is specified by the filter bandwidth $W_f = (f_u - f_\ell)$ hertz.

When $f_\ell \neq 0$ and $f_u \neq \infty$, the filter is called a *bandpass filter* (BPF), shown in Figure 1.10a. When $f_\ell = 0$ and f_u has a finite value, the filter is called a *low-pass filter* (LPF), shown in Figure 1.10b. When f_ℓ has a nonzero value and when $f_u \to \infty$, the filter is called a *high-pass filter* (HPF), shown in Figure 1.10c.

Following Equation (1.56), for the ideal low-pass filter transfer function with bandwidth $W_f = f_u$ hertz, shown in Figure 1.10b, we can write the transfer function as follows (letting $K = 1$):

$$H(f) = |H(f)| \, e^{-j\theta(f)} \tag{1.58}$$

where

$$|H(f)| = \begin{cases} 1 & \text{for } |f| < f_u \\ 0 & \text{for } |f| \geq f_u \end{cases} \tag{1.59}$$

and

$$e^{-j\theta(f)} = e^{-j2\pi f t_0} \tag{1.60}$$

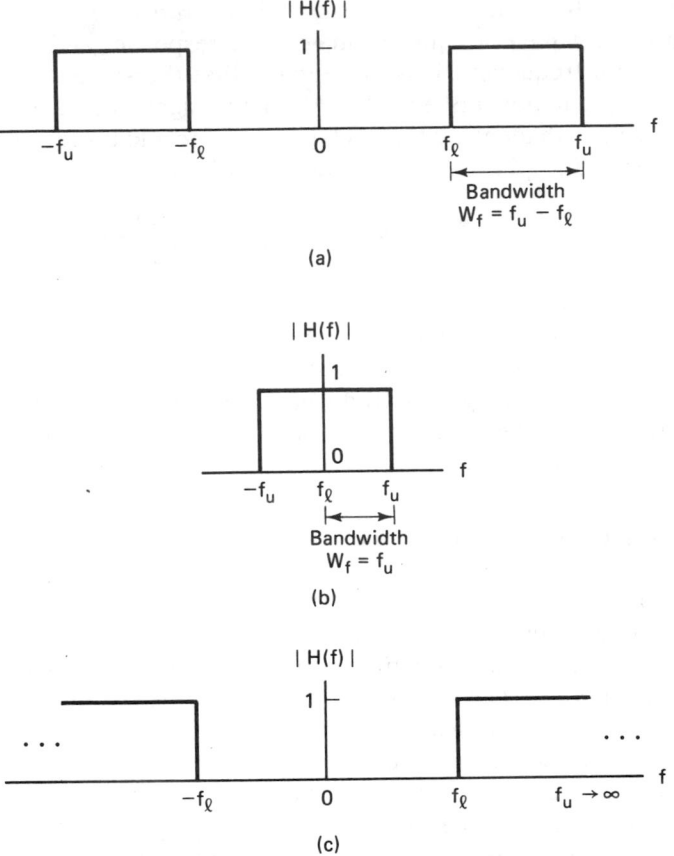

Figure 1.10 Ideal filter transfer function. (a) Ideal bandpass filter. (b) Ideal low-pass filter. (c) Ideal high-pass filter.

The impulse response $h(t)$ of the ideal low-pass filter, illustrated in Figure 1.11, is

$$h(t) = \mathcal{F}^{-1}\{H(f)\} = \int_{-\infty}^{\infty} H(f)e^{j2\pi ft}\, df \qquad (1.61)$$

$$= \int_{-f_u}^{f_u} e^{-j2\pi ft_0}e^{j2\pi ft}\, df$$

$$= \int_{-f_u}^{f_u} e^{j2\pi f(t-t_0)}\, df$$

$$= 2f_u \frac{\sin 2\pi f_u(t-t_0)}{2\pi f_u(t-t_0)}$$

$$= 2f_u \operatorname{sinc} 2f_u(t-t_0) \qquad (1.62)$$

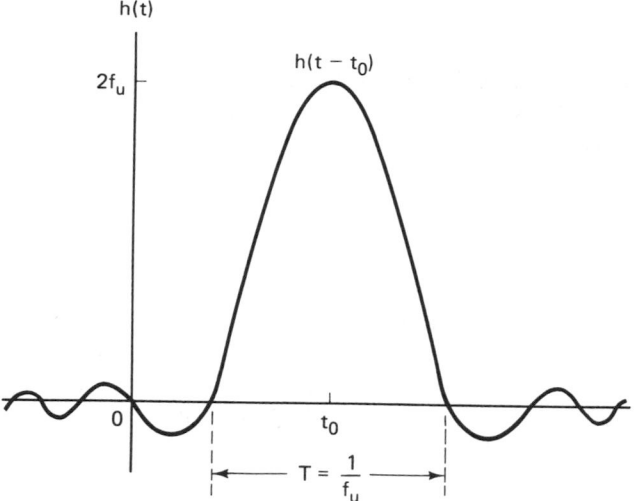

Figure 1.11 Impulse response of the ideal low-pass filter.

where sinc x is as defined in Equation (1.39). The impulse response shown in Figure 1.11 is noncausal, which means that it has a nonzero output prior to the application of an input at time $t = 0$. Therefore, it should be clear that the ideal filter described in Equation (1.58) is not realizable.

Example 1.2 Effect of an Ideal Filter on White Noise

White noise with power spectral density $G_n(f) = N_0/2$, shown in Figure 1.8a, forms the input to the ideal low-pass filter shown in Figure 1.10b. Find the power spectral density, $G_Y(f)$, and the autocorrelation function, $R_Y(\tau)$, of the output signal.

Solution

$$G_Y(f) = G_n(f) |H(f)|^2$$

$$= \begin{cases} \dfrac{N_0}{2} & \text{for } |f| < f_u \\ 0 & \text{otherwise} \end{cases}$$

The autocorrelation is the inverse Fourier transform of the power spectral density and is given by (see Table A.1)

$$R_Y(\tau) = N_0 f_u \frac{\sin 2\pi f_u \tau}{2\pi f_u \tau}$$

$$= N_0 f_u \text{ sinc } 2f_u \tau$$

Comparing this result with Equation (1.62), we see that $R_Y(\tau)$ has the same shape as the impulse response of the ideal low-pass filter shown in Figure 1.11. In this example the ideal low-pass filter transforms the autocorrelation function of white noise (defined by the delta function) into a sinc function. After filtering, we no longer have white noise. The output noise signal will have zero correlation with shifted copies of itself, only at shifts of $\tau = n/2f_u$, where n is any integer other than zero.

1.6.3.2 Realizable Filters

The very simplest example of a realizable low-pass filter is made up of resistance ($\mathcal{R}$) and capacitance (C), as shown in Figure 1.12a; it is called an $\mathcal{R}C$ *filter*, and its transfer function can be expressed as [7]

$$H(f) = \frac{1}{1 + j2\pi f \mathcal{R}C} = \frac{1}{\sqrt{1 + (2\pi f \mathcal{R}C)^2}} \, e^{-j\theta(f)} \tag{1.63}$$

where $\theta(f) = \tan^{-1} 2\pi f \mathcal{R}C$. The magnitude characteristic, $|H(f)|$, and the phase characteristic, $\theta(f)$ are plotted in Figures 1.12b and c, respectively. The low-pass filter bandwidth is defined to be its half-power point; this point is the frequency at which the output signal power has fallen to one-half of its peak value, or the frequency at which the magnitude of the output voltage has fallen to $1/\sqrt{2}$ of its peak value.

The half-power point is generally expressed in decibel (dB) units as the -3-dB point, or the point which is 3 dB down from the peak, where the decibel

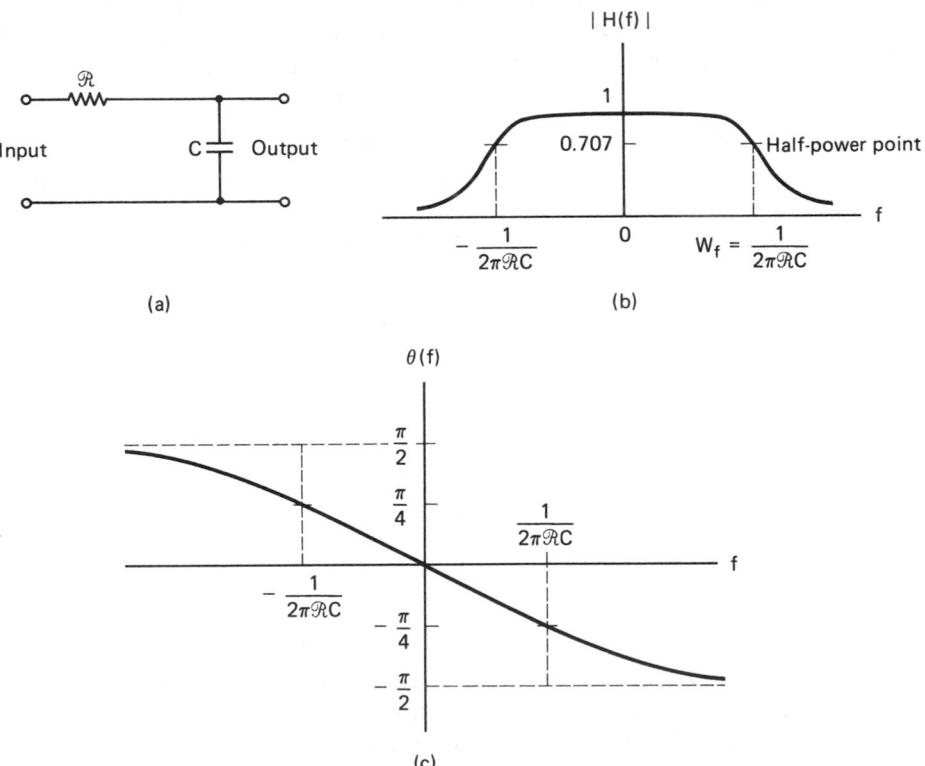

(a)

(b)

(c)

Figure 1.12 $\mathcal{R}C$ filter and its transfer function. (a) $\mathcal{R}C$ filter. (b) Magnitude characteristic of the $\mathcal{R}C$ filter. (c) Phase characteristic of the $\mathcal{R}C$ filter.

is defined as the ratio of two amounts of power, P_1 and P_2, existing at two points. By definition

$$\text{number of dB} = 10 \log_{10} \frac{P_2}{P_1} = 10 \log_{10} \frac{V_2^2/\mathcal{R}_2}{V_1^2/\mathcal{R}_1} \qquad (1.64a)$$

where V_1 and V_2 are voltages and $\mathcal{R}_1$ and $\mathcal{R}_2$ are resistances. For communication systems, *normalized power* is generally used for analysis; in this case, $\mathcal{R}_1$ and $\mathcal{R}_2$ are set equal to 1 Ω, so that

$$\text{number of dB} = 10 \log_{10} \frac{P_2}{P_1} = 10 \log_{10} \frac{V_2^2}{V_1^2} \qquad (1.64b)$$

The amplitude response, $|H(f)|$, can be expressed in decibels by

$$|H(f)|_{\text{dB}} = 20 \log_{10} \frac{V_2}{V_1} = 20 \log_{10} |H(f)| \qquad (1.64c)$$

where V_1 and V_2 are the input and output voltages, respectively, and where the input and output resistances have been assumed equal.

From Equation (1.63) it is easy to verify that the half-power point of the low-pass $\mathcal{R}C$ filter corresponds to $\omega = 1/\mathcal{R}C$ radians per second or $f = 1/(2\pi\mathcal{R}C)$ hertz. Thus the bandwidth W_f in hertz is $1/(2\pi\mathcal{R}C)$. The filter *shape factor* is a measure of how well a realizable filter approximates the ideal filter. It is typically defined as the ratio of the filter bandwidths at the -60-dB and -6-dB amplitude response points. A sharp-cutoff bandpass filter can be made with a shape factor as low as about 2. By comparison, the shape factor of the simple $\mathcal{R}C$ low-pass filter is almost 600.

There are several useful approximations to the ideal low-pass filter characteristic. One of these, the *Butterworth filter,* approximates the ideal low-pass filter with the following function:

$$|H_n(f)| = \frac{1}{\sqrt{1 + (f/f_u)^{2n}}} \qquad n \geq 1 \qquad (1.65)$$

where f_u is the upper -3-dB cutoff frequency. The magnitude function, $|H(f)|$, is sketched (single sided) for several values of n in Figure 1.13. Note that as n gets larger, the magnitude characteristics approach that of the ideal filter. Butterworth filters are popular because they are the best approximation to the ideal, in the sense of *maximal flatness* in the filter passband.

Example 1.3   Effect of an $\mathcal{R}C$ Filter on White Noise

White noise with spectral density, $G_n(f) = N_0/2$, shown in Figure 1.8a, forms the input to the $\mathcal{R}C$ filter shown in Figure 1.12a. Find the power spectral density, $G_Y(f)$, and the autocorrelation function, $R_Y(\tau)$, of the output signal.

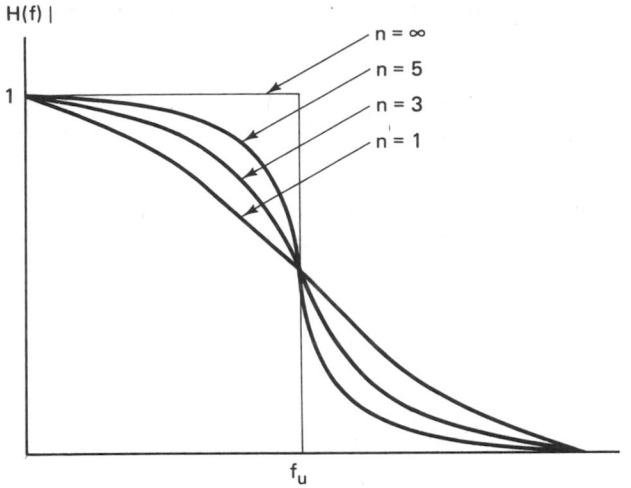

$|H(f)|$

$n = \infty$

$n = 5$

$n = 3$

$n = 1$

1

f_u f

Figure 1.13 Butterworth filter magnitude response.

Solution

$$G_Y(f) = G_n(f) \, |H(f)|^2$$

$$= \frac{N_0}{2} \, \frac{1}{1 + (2\pi f \mathcal{R}C)^2}$$

$$R_Y(\tau) = \mathcal{F}^{-1}\{G_Y(f)\}$$

Using Table A.1, the inverse Fourier transform of $G_Y(f)$ is

$$R_Y(\tau) = \frac{N_0}{4\mathcal{R}C} \exp\left(-\frac{|\tau|}{\mathcal{R}C}\right)$$

As might have been predicted, we no longer have white noise after filtering. The $\mathcal{R}C$ filter transforms the input autocorrelation function of white noise (defined by the delta function) into an exponential function. For a narrowband filter (a large $\mathcal{R}C$ product), the output noise will exhibit higher correlation between noise samples of a fixed time shift than will the output noise from a wideband filter.

1.6.4 Signals, Circuits, and Spectra

Signals have been described in terms of their spectra. Similarly, networks or circuits have been described in terms of their spectral characteristics or frequency transfer functions. How is a signal's bandwidth affected as a result of the signal passing through a filter circuit? Figure 1.14 illustrates two cases of interest. In Figure 1.14a (case 1), the input signal has a narrowband spectrum, and the filter transfer function is a wideband function. From Equation (1.48) we see that the output signal spectrum is simply the product of these two spectra. In Figure 1.14a we can verify that multiplication of the two spectral functions will result in a spectrum with a bandwidth approximately equal to the smaller of the two bandwidths (when one of the two spectral functions goes to zero, the multiplication yields zero). Therefore, for case 1, the output signal spectrum is constrained by

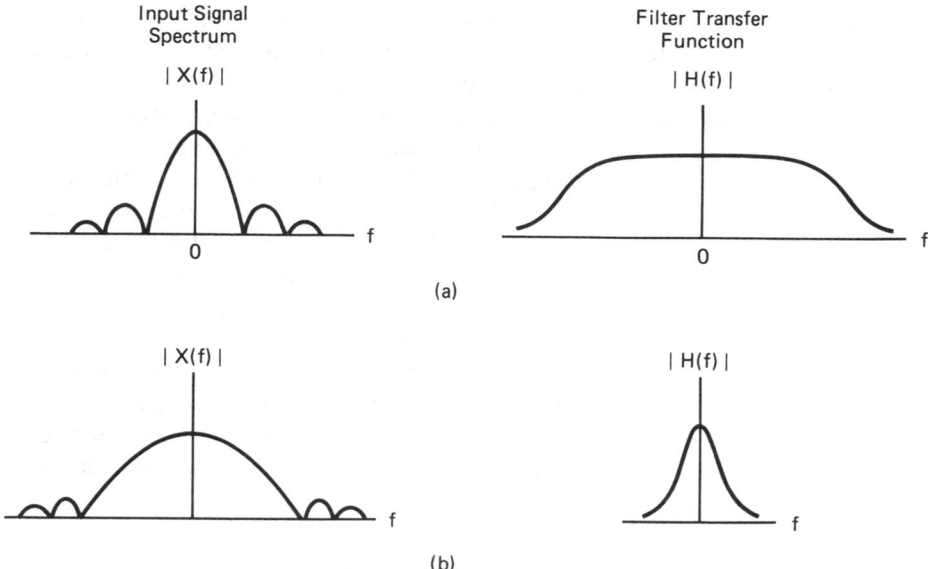

Input Signal Spectrum

| X(f) |

Filter Transfer Function

| H(f) |

(a)

| X(f) |

| H(f) |

(b)

Figure 1.14 Spectral characteristics of the input signal and the circuit contribute to the spectral characteristics of the output signal. (a) Case 1: Output bandwidth is constrained by input signal bandwidth. (b) Case 2: Output bandwidth is constrained by filter bandwidth.

the input signal spectrum alone. Similarly, we see that for case 2, in Figure 1.14b, where the input signal is a wideband signal but the filter has a narrowband transfer function, the bandwidth of the output signal is constrained by the filter bandwidth; the output signal will be a filtered (distorted) rendition of the input signal.

The effect of a filter on a waveform can also be viewed in the time domain. The output, $y(t)$, resulting from convolving an ideal input pulse, $x(t)$ (having amplitude V_m and pulse width T), with the impulse response of a low-pass $\mathcal{R}C$ filter can be written as [8]

$$y(t) = \begin{cases} V_m(1 - e^{-t/\mathcal{R}C}) & \text{for } 0 \leq t \leq T \\ V'_m e^{-(t - T)/\mathcal{R}C} & \text{for } t > T \end{cases} \tag{1.66}$$

where

$$V'_m = V_m(1 - e^{-T/\mathcal{R}C}) \tag{1.67}$$

Let us define the pulse bandwidth, W_p, and the $\mathcal{R}C$ filter bandwidth, W_f, as

$$W_p = \frac{1}{T} \tag{1.68}$$

and

$$W_f = \frac{1}{2\pi\mathcal{R}C} \tag{1.69}$$

Sec. 1.6   Signal Transmission Through Linear Systems

39

The ideal input pulse, $x(t)$, and its magnitude spectrum $|X(f)|$, are shown in Figure 1.15. The $\mathcal{R}C$ filter and its magnitude characteristic, $|H(f)|$, are shown in Figures 1.12a and b, respectively. Following Equations (1.66) to (1.69), three cases are illustrated in Figure 1.16. Example 1 illustrates the case where $W_p \ll W_f$. Notice that the output response, $y(t)$, is a reasonably good approximation of the input pulse, $x(t)$, shown in dashed lines. This represents an example of *good fidelity*. In example 2, where $W_p \simeq W_f$, we can still recognize that a pulse had been transmitted from the output, $y(t)$. Finally, example 3 illustrates the case where $W_p \gg W_f$. Here the presence of the pulse is barely perceptible from the output, $y(t)$. Can you think of an application where the large filter bandwidth or good fidelity of example 1 is called for? A *precise ranging application*, perhaps, where the pulse time of arrival translates into distance, necessitates a pulse with a steep rise time. Which example characterizes the binary digital communications application? *It is example 2.* As we pointed out earlier regarding Figure 1.1, one of the principal features of binary digital communications is that each received pulse

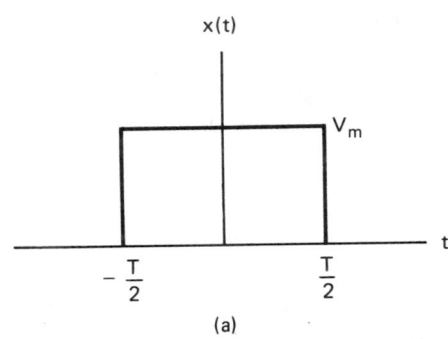

(a)

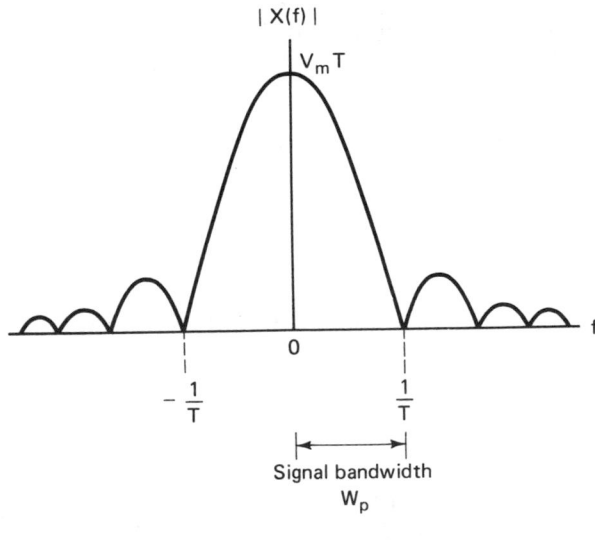

(b)

Figure 1.15 (a) Ideal pulse. (b) Magnitude spectrum of the ideal pulse.

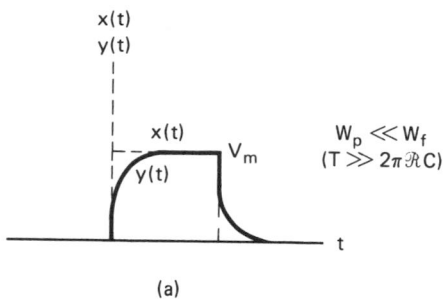

(a)

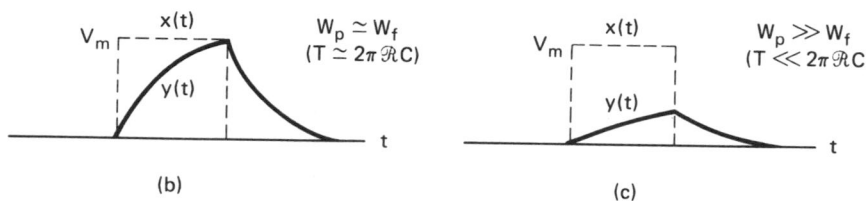

(b)   (c)

Figure 1.16 Three examples of filtering an ideal pulse. (a) Example 1: Good-fidelity output. (b) Example 2: Good-recognition output. (c) Example 3: Poor-recognition output.

need only be accurately *perceived* as being in one of its two states; a high-fidelity signal need not be maintained. Example 3 has been included for completeness; it would not be used as a design criterion for a practical system.

1.7 BANDWIDTH OF DIGITAL DATA

1.7.1 Baseband versus Bandpass

An easy way to translate the spectrum of a low-pass or baseband signal, $x(t)$, to a higher frequency is to multiply or *heterodyne* the baseband signal with a carrier wave, $\cos 2\pi f_c t$, as shown in Figure 1.17a. The resulting waveform, $x_c(t)$, is called a *double-sideband* (DSB) *modulated signal* and is expressed as

$$x_c(t) = x(t) \cos 2\pi f_c t \tag{1.70}$$

From the frequency shifting theorem (see Section A.3.2) the spectrum of the DSB signal, $x_c(t)$, is given by $X_c(f)$:

$$X_c(f) = \tfrac{1}{2}[X(f - f_c) + X(f + f_c)] \tag{1.71}$$

The magnitude spectrum $|X(f)|$ of the baseband signal, $x(t)$, having a bandwidth f_m, and the magnitude spectrum, $|X_c(f)|$, of the DSB signal, $x_c(t)$, having a bandwidth W_{DSB}, are shown in Figure 1.17b and c, respectively. In the plot of $|X_c(f)|$,

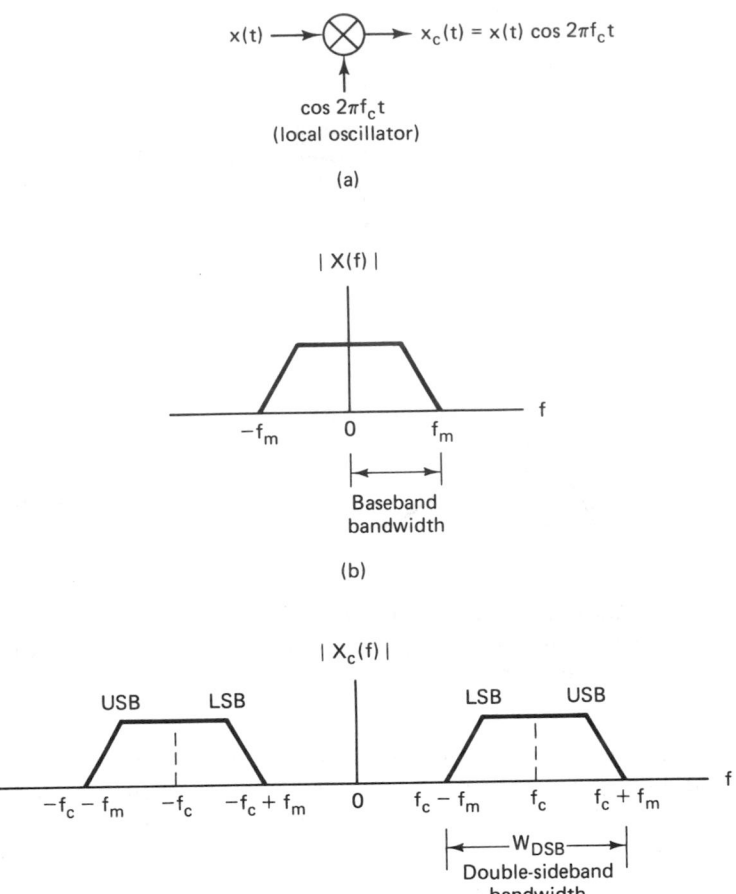

Figure 1.17 Comparison of baseband and double-sideband spectra. (a) Heterodyning. (b) Baseband spectrum. (c) Double-sideband spectrum.

spectral components corresponding to positive baseband frequencies, appear in the range f_c to $(f_c + f_m)$. This part of the DSB spectrum is called the *upper sideband* (USB). Spectral components corresponding to negative baseband frequencies appear in the range $(f_c - f_m)$ to f_c. This part of the DSB spectrum is called the *lower sideband* (LSB). Mirror images of the USB and LSB spectra appear in the negative-frequency half of the plot. The *carrier wave* is sometimes referred to as a *local oscillator* (LO) *signal,* a *mixing signal,* or a *heterodyne signal.* Generally, the carrier wave frequency is much higher than the bandwidth of the baseband signal; that is,

$$f_c \gg f_m$$

From Figure 1.17 we can readily compare the bandwidth f_m, required to transmit the baseband signal, with the bandwidth W_{DSB}, required to transmit the DSB signal; we see that

$$W_{\text{DSB}} = 2f_m \qquad (1.72)$$

That is, we need twice as much transmission bandwidth to transmit a DSB version of the signal than we do to transmit its baseband counterpart.

1.7.2 The Bandwidth Dilemma

Many important theorems of communication and information theory are based on the assumption of *strictly bandlimited* channels, which means that no signal power whatever is allowed outside the defined band. We are faced with the dilemma that strictly bandlimited signals are not realizable since they imply signals with infinite duration; nonbandlimited signals, having energy at arbitrarily high frequencies, appear just as unreasonable. It is no wonder that there is no single universal definition of bandwidth.

All bandwidth criteria have in common the attempt to specify a measure of the width, W, of a nonnegative real-valued power spectral density defined for all frequencies $|f| < \infty$. Figure 1.18 illustrates some of the most common definitions of bandwidth; in general, the various criteria are not interchangeable. The single-

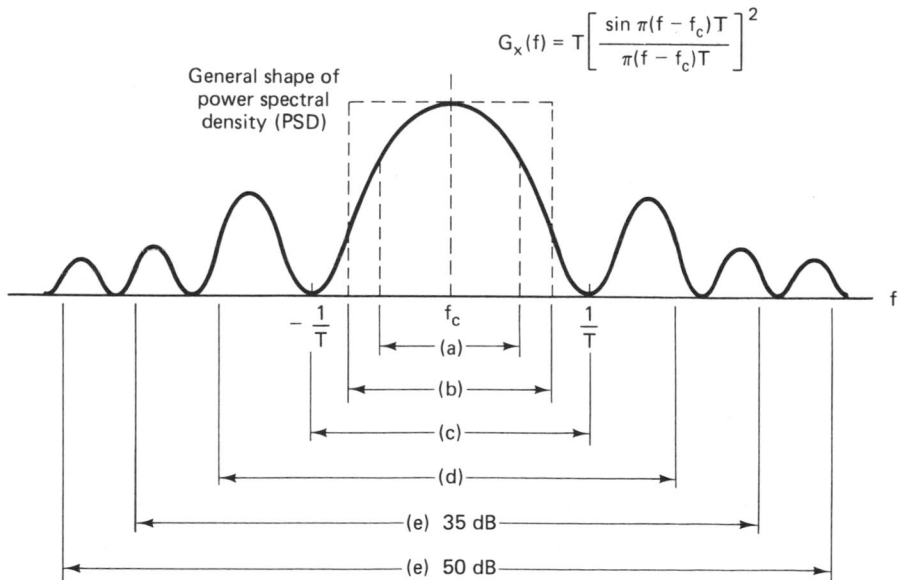

Figure 1.18 Bandwidth of digital data. (a) Half-power. (b) Noise equivalent. (c) Null to null. (d) 99% of power. (e) Bounded PSD (defines attenuation outside bandwidth) at 35 and 50 dB.

sided power spectral density, $G_x(f)$, for a single heterodyned pulse, $x_c(t)$, takes the analytical form

$$G_x(f) = T \left[\frac{\sin \pi(f - f_c)T}{\pi(f - f_c)T} \right]^2 \qquad (1.73)$$

where f_c is the carrier wave frequency and T is the pulse duration. This power spectral density, whose general appearance is sketched in Figure 1.18, also characterizes a *random pulse sequence,* assuming that the averaging time is long relative to the pulse duration. The plot consists of a main lobe and smaller symmetrical sidelobes. The general shape of the plot is valid for most digital modulation formats; some formats, however, do not have well-defined lobes. The bandwidth criteria depicted in Figure 1.18 are as follows:

(a) *Half-power bandwidth.* This is the interval between frequencies at which $G_x(f)$ has dropped to half-power, or 3 dB below the peak value.

(b) *Equivalent rectangular or noise equivalent bandwidth.* The noise equivalent bandwidth was originally conceived to permit rapid computation of output noise power from an amplifier with a wideband noise input; the concept can similarly be applied to a signal bandwidth. The noise equivalent bandwidth W_N of a signal is defined by the relationship $W_N = P_x/G_x(f_c)$, where P_x is the total signal power over all frequencies and $G_x(f_c)$ is the value of $G_x(f)$ at the band center (assumed to be the maximum value over all frequencies).

(c) *Null-to-null bandwidth.* The most popular measure of bandwidth for digital communications is the width of the main spectral lobe, where most of the signal power is contained. This criterion lacks complete generality since some modulation formats lack well-defined lobes.

(d) *Fractional power containment bandwidth.* This bandwidth criterion has been adopted by the Federal Communications Commission (FCC Rules and Regulations Section 2.202) and states that the occupied bandwidth is the band that leaves exactly 0.5% of the signal power above the upper band limit and exactly 0.5% of the signal power below the lower band limit. Thus 99% of the signal power is inside the occupied band.

(e) *Bounded power spectral density.* A popular method of specifying bandwidth is to state that everywhere outside the specified band, $G_x(f)$ must have fallen at least to a certain stated level below that found at the band center. Typical attenuation levels might be 35 or 50 dB.

(f) *Absolute bandwidth.* This is the interval between frequencies, outside of which the spectrum is zero. This is a useful abstraction. However, for all realizable waveforms, the absolute bandwidth is infinite.

Example 1.4   Strictly Bandlimited Signals

The concept of a signal that is strictly limited to a band of frequencies is not realizable. Prove this by showing that a *strictly bandlimited* signal must also be a signal of *infinite time duration.*

Solution

Let $x(t)$ be a signal, with Fourier transform $X(f)$, that is strictly limited to the band of frequencies centered at $\pm f_c$ and of width $2W$. We may express $X(f)$ in terms of an ideal filter transfer function, $H(f)$, illustrated in Figure 1.19a, as follows:

$$X(f) = X'(f)H(f) \tag{1.74}$$

where, $X'(f)$ is the Fourier transform of a signal $x'(t)$, not necessarily bandlimited, where

$$H(f) = \text{rect}\left(\frac{f - f_c}{2W}\right) + \text{rect}\left(\frac{f + f_c}{2W}\right) \tag{1.75}$$

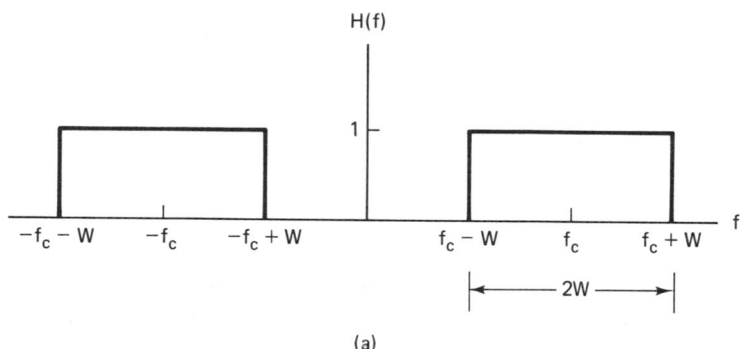

(a)

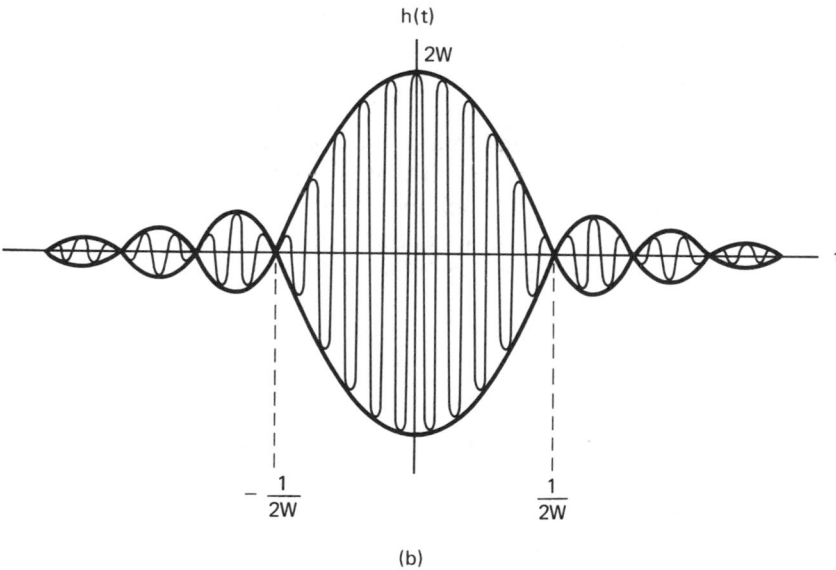

(b)

Figure 1.19 Transfer function and impulse response for a strictly bandlimited signal. (a) Ideal bandpass filter. (b) Ideal bandpass impulse response.

and where

$$\text{rect} \left(\frac{f}{2W} \right) = \begin{cases} 1 & \text{for } -W < f < W \\ 0 & \text{for } |f| > W \end{cases}$$

We can express $X(f)$ in terms of $X'(f)$ as

$$X(f) = \begin{cases} X'(f) & \text{for } (f_c - W) \leq |f_c| \leq (f_c + W) \\ 0 & \text{otherwise} \end{cases}$$

Multiplication in the frequency domain, as seen in Equation (1.74), transforms to convolution in the time domain as follows:

$$x(t) = x'(t) * h(t) \qquad (1.76)$$

where $h(t)$, the inverse Fourier transform of $H(f)$, can be written as (see Tables A.1 and A.2)

$$h(t) = 2W \, (\text{sinc } 2Wt) \cos 2\pi f_c t$$

and is illustrated in Figure 1.19b. We note that $h(t)$ is of *infinite duration*. It follows, therefore, that $x(t)$ obtained in Equation (1.76) by convolving $x'(t)$ with $h(t)$ is also of infinite duration and therefore is *not realizable*.

1.8 CONCLUSION

In this chapter, the goals of the book have been outlined and the basic nomenclature has been defined. The fundamental concepts of time-varying signals, such as classification, spectral density, and autocorrelation, have been reviewed. Also, random signals have been considered, and white Gaussian noise, the primary noise model in most communication systems, has been characterized, statistically and spectrally. Finally, we have treated the important area of signal transmission through linear systems and have examined some of the realizable approximations to the ideal case. We have also established that the concept of an absolute bandwidth is an abstraction, and that in the real world we are faced with the need to choose a definition of bandwidth that is useful for our particular application. In the remainder of the book, each of the signal processing steps introduced in this chapter will be explored in the context of the typical system block diagram appearing at the beginning of each chapter.

REFERENCES

1. Haykin, S., *Communication Systems,* John Wiley & Sons, Inc., New York, 1983.

2. Shanmugam, K. S., *Digital and Analog Communication Systems,* John Wiley & Sons, Inc., New York, 1979.

3. Papoulis, A., *Probability, Random Variables, and Stochastic Processes,* McGraw-Hill Book Company, New York, 1965.

4. Johnson, J. B., "Thermal Agitation of Electricity in Conductors," *Phys. Rev.,* vol. 32, July 1928, pp. 97–109.

5. Nyquist, H., "Thermal Agitation of Electric Charge in Conductors," *Phys. Rev.,* vol. 32, July 1928, pp. 110–113.

6. Van Trees, H. L., *Detection, Estimation, and Modulation Theory,* Part 1, John Wiley & Sons, New York, 1968.

7. Schwartz, M., *Information Transmission, Modulation, and Noise,* McGraw-Hill Book Company, New York, 1970.

8. Millman, J., and Taub, H., *Pulse, Digital, and Switching Waveforms,* McGraw-Hill Book Company, New York, 1965.

PROBLEMS

1.1. Classify the following signals as energy signals or power signals. Find the normalized energy or normalized power of each.

(a) $x(t) = A \cos 2\pi f_0 t$ for $-\infty < t < \infty$

(b) $x(t) = \begin{cases} A \cos 2\pi f_0 t & \text{for } -T_0/2 \leq t \leq T_0/2, \text{ where } T_0 = 1/f_0 \\ 0 & \text{elsewhere} \end{cases}$

(c) $x(t) = \begin{cases} A \exp(-at) & \text{for } t > 0, a > 0 \\ 0 & \text{elsewhere} \end{cases}$

(d) $x(t) = \cos t + 5 \cos 2t$ for $-\infty < t < \infty$

1.2. Determine the energy spectral density of a square pulse $x(t) = \text{rect}(t/T)$, where $\text{rect}(t/T)$ equals 1, for $-T/2 \leq t \leq T/2$, and equals 0, elsewhere. Calculate the normalized energy E_x in the pulse.

1.3. Find an expression for the average normalized power in a periodic signal in terms of its complex Fourier series coefficients.

1.4. Using time averaging, find the average normalized power in the waveform $x(t) = 10 \cos 10t + 20 \cos 20t$.

1.5. Repeat Problem 1.4 using the summation of spectral coefficients.

1.6. Determine which, if any, of the following functions have the properties of autocorrelation functions. Justify your determination. [*Note:* $\mathcal{F}\{R(\tau)\}$ must be a nonnegative function. Why?]

(a) $x(\tau) = \begin{cases} 1 & \text{for } -1 \leq \tau \leq 1 \\ 0 & \text{otherwise} \end{cases}$

(b) $x(\tau) = \delta(\tau) + \sin 2\pi f_0 \tau$

(c) $x(\tau) = \exp(|\tau|)$

(d) $x(\tau) = 1 - |\tau|$ for $-1 \leq \tau \leq 1$

1.7. Determine which, if any, of the following functions have the properties of power spectral density functions. Justify your determination.

(a) $X(f) = \delta(f) + \cos^2 2\pi f$

(b) $X(f) = 10 + \delta(f - 10)$

(c) $X(f) = \exp(-2\pi |f - 10|)$

(d) $X(f) = \exp[-2\pi(f^2 - 10)]$

1.8. Find the autocorrelation function of $x(t) = A \cos(2\pi f_0 t + \phi)$ in terms of its period, $T_0 = 1/f_0$. Find the average normalized power of $x(t)$, using $P_x = R(0)$.

1.9. **(a)** Use the results of Problem 1.8 to find the autocorrelation function, $R(\tau)$, of waveform $x(t) = 10 \cos 10t + 20 \cos 20t$.

(b) Use the relationship $P_x = R(0)$ to find the average normalized power in $x(t)$. Compare the answer with the answers to Problems 1.4 and 1.5.

1.10. For the function $x(t) = 1 + \cos 2\pi f_0 t$, calculate **(a)** the average value of $x(t)$; **(b)** the ac power of $x(t)$; **(c)** the rms value of $x(t)$.

1.11. Consider a random process given by $X(t) = A \cos(2\pi f_0 t + \phi)$, where A and f_0 are constants and ϕ is a random variable that is uniformly distributed over $(0, 2\pi)$. If $X(t)$ is an ergodic process, the time averages of $X(t)$ in the limit as $t \rightarrow \infty$ are equal to the corresponding ensemble averages of $X(t)$.

(a) Use time averaging over an integer number of periods to calculate the approximations to the first and second moments of $X(t)$.

(b) Use Equations (1.26) and (1.28) to calculate the ensemble-average approximations to the first and second moments of $X(t)$. Compare the results with your answers in part (a).

1.12. The Fourier transform of a signal, $x(t)$ is defined by $X(f) = \text{sinc } f$, where the sinc function is as defined in Equation (1.39). Find the autocorrelation function, $R_x(\tau)$, of the signal $x(t)$.

1.13. Use the sampling property of the unit impulse function to evaluate the following integrals.

(a) $\int_{-\infty}^{\infty} \cos 6t \delta(t - 3) \, dt$

(b) $\int_{-\infty}^{\infty} 10\delta(t)(1 + t)^{-1} \, dt$

(c) $\int_{-\infty}^{\infty} \delta(t + 4)(t^2 + 6t + 1) \, dt$

(d) $\int_{-\infty}^{\infty} \exp(-t^2)\delta(t - 2) \, dt$

1.14. Find $X_1(f) * X_2(f)$ for the spectra shown in Figure P1.1.

1.15. The two-sided power spectral density, $G_x(f) = 10^{-6}f^2$, of a waveform $x(t)$ is shown in Figure P1.2.

(a) Find the normalized average power in $x(t)$ over the frequency band from 0 to 10 kHz.

(b) Find the normalized average power contained in the frequency band from 5 to 6 kHz.

1.16. Decibels are logarithmic measures of *power ratios*, as described in Equation (1.64a). Sometimes, a similar formulation is used to express nonpower measurements in decibels (referenced to some designated unit). As an example, calculate how many decibels of hamburger meat you would buy to feed 2 hamburgers each to a group of 100 people. Assume that you and the butcher have agreed on the unit of "$\frac{1}{2}$ pound of meat" (the amount in one hamburger) as a reference unit.

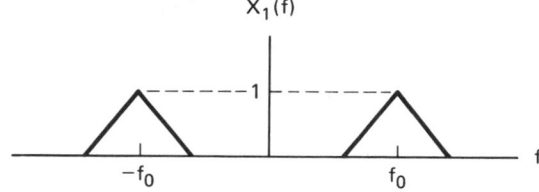

$X_1(f)$

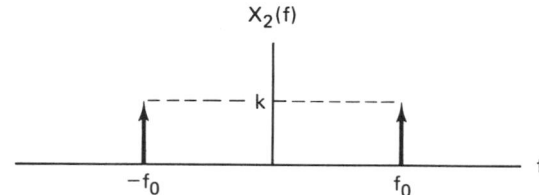

$X_2(f)$

Figure P1.1

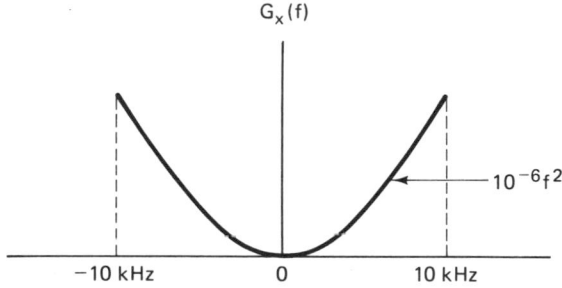

$G_x(f)$

$10^{-6}f^2$

-10 kHz   0   10 kHz

Figure P1.2

1.17. Consider the Butterworth low-pass amplitude response given in Equation (1.65).

(a) Find the value of n so that $|H(f)|^2$ is constant to within ± 1 dB over the range $|f| \leq 0.9f_u$.

(b) Show that as n approaches infinity, the amplitude response approaches that of an ideal low-pass filter.

1.18. Consider the network in Figure 1.9, whose frequency transfer function is $H(f)$. An impulse $\delta(t)$ is applied at the input. Show that the response $y(t)$ at the output is the inverse Fourier transform of $H(f)$.

1.19. An example of a *holding circuit,* commonly used in pulse systems, is shown in Figure P1.3. Determine the impulse response of this circuit.

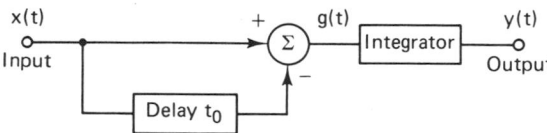

Figure P1.3

1.20. Given the spectrum

$$G_x(f) = 10^{-4}\left\{\frac{\sin\left[\pi(f - 10^6)10^{-4}\right]}{\pi(f - 10^6)10^{-4}}\right\}^2$$

Find the value of the signal bandwidth using the following bandwidth definitions:

(a) Half-power bandwidth.
(b) Noise equivalent bandwidth.
(c) Null-to-null bandwidth.
(d) 99% of power bandwidth.
(e) Bandwidth beyond which the attenuation is 35 dB.
(f) Absolute bandwidth.

CHAPTER 2

Formatting
and
Baseband Transmission

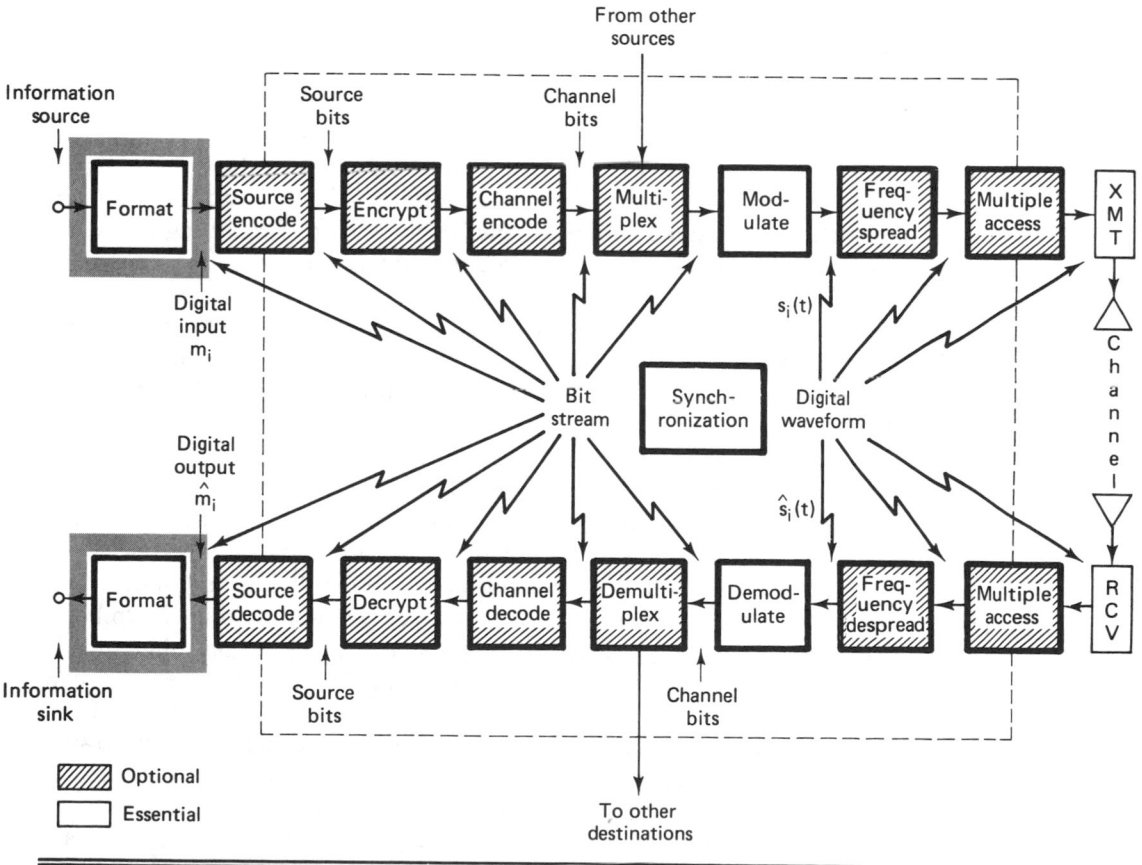

The first essential signal processing step, *formatting*, makes the source signal compatible with digital processing. *Transmit formatting* is a transformation from source information to digital symbols (in the receive chain, formatting is the reverse transformation). When there is data redundancy reduction or data compression, in addition to formatting, the process is termed *source coding*. Some authors consider formatting to be a special case of source coding. We treat formatting (and baseband transmission) in this chapter, and treat source coding as a special case of the *efficient description* of source information in Chapter 11. In Figure 2.1 the main formatting topics are highlighted—character coding, sampling, quantization, and pulse code modulation (PCM).

A signal whose spectrum extends from (or near) dc up to some finite value, usually less than a few megahertz, is called a *baseband* or *low-pass* signal. Such a signal is implied whenever we use the term "information," "message," or "data." For the transmission of baseband signals by a digital communication system, the information is *formatted* so that it is represented by digital symbols. Then, pulse waveforms are assigned that represent these symbols; this step is referred to as *pulse modulation* or *baseband modulation*. These waveforms can then be transmitted over a cable.

Baseband signals are not appropriate for propagation through many transmission media. Baseband signals whose spectrum has been shifted to a frequency band that is more appropriate for propagation through a transmission medium are called *bandpass modulation signals* or simply *bandpass signals*. Bandpass signals have their spectral content clustered in a band of frequencies near a value called

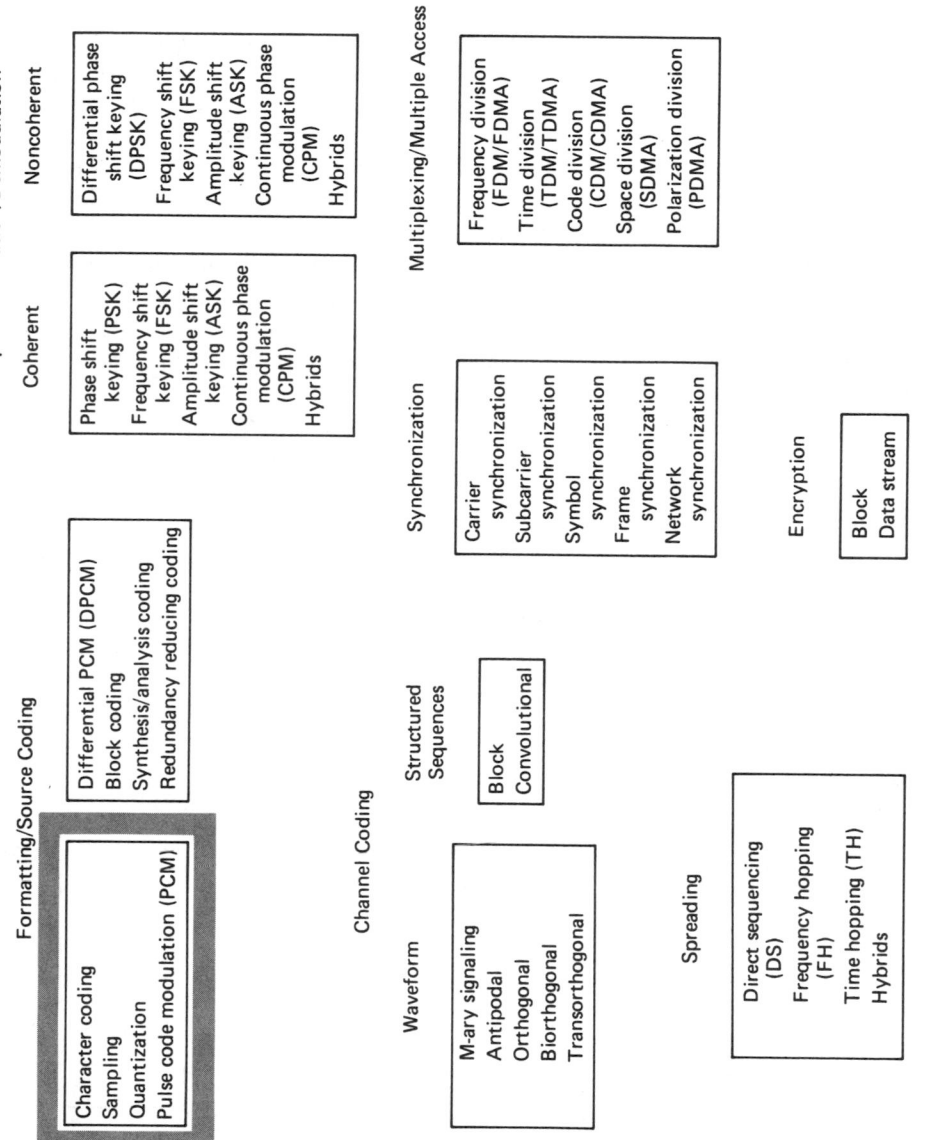

Figure 2.1 Basic digital communication transformations.

the *carrier frequency*. In Chapter 3 we deal with the modulation and demodulation of these bandpass signals.

2.1 BASEBAND SYSTEMS

In Figure 1.2 we presented a block diagram of a typical digital communication system. A version of this functional diagram, focusing primarily on the formatting and transmission of *baseband* signals, is shown in Figure 2.2. Data already in a digital format would bypass the formatting function. Textual information is transformed into binary digits by use of a coder. Analog information is formatted using three separate processes: sampling, quantization, and coding. In all cases, the formatting step results in a sequence of binary digits.

These digits are to be transmitted through a *baseband channel*, such as a pair of wires or a coaxial cable. However, no channel can be used for the transmission of binary digits without first transforming the digits to *waveforms* that are compatible with the channel. For baseband channels, compatible waveforms are pulses.

In Figure 2.2, the conversion from binary digits to pulse waveforms takes place in the block labeled *waveform encoder*, also called a *baseband modulator*. The output of the waveform encoder is typically a sequence of pulses with characteristics that correspond to the binary digits being sent. After transmission through the channel, the received waveforms are detected to produce an estimate of the transmitted digits, and then the final step, (reverse) formatting, recovers an estimate of the source information.

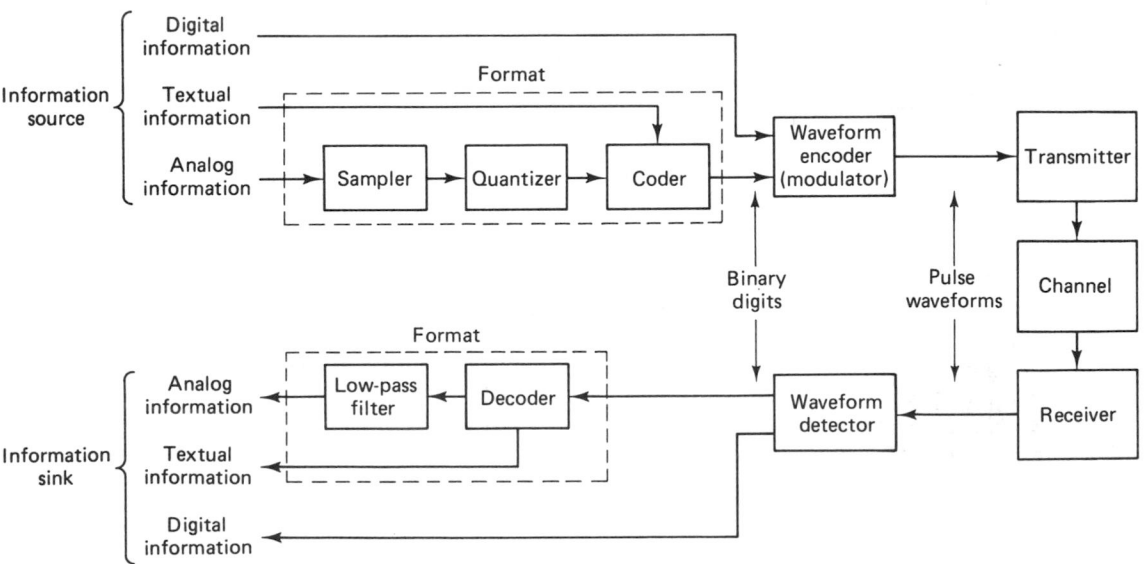

Figure 2.2   Formatting and transmission of baseband signals.

2.2 FORMATTING TEXTUAL DATA (CHARACTER CODING)

The original form of most communicated data (except for computer-to-computer transmissions) is either textual or analog. If the data consist of alphanumeric text, they will be character encoded with one of several standard formats, examples of which are, the American Standard Code for Information Interchange (ASCII), the Extended Binary Coded Decimal Interchange Code (EBCDIC), Baudot, and Hollerith. The textual material is thereby transformed into a digital format. The ASCII format is shown in Figure 2.3; the EBCDIC format is shown in Figure 2.4. The bit numbers signify the order of serial transmission, where bit number 1 is the first signaling element. Character coding, then, is the step that transforms text into binary digits (bits). Sometimes, existing character codes are modified to meet specialized needs. For example, the 7-bit ASCII code (Figure 2.3) can be modified to include an added bit for error detection purposes (see Chapter 5). On the other hand, sometimes the code is truncated to a 6-bit ASCII version, which provides capability for only 64 characters instead of the 128 characters allowed by 7-bit ASCII.

2.3 MESSAGES, CHARACTERS, AND SYMBOLS

Textual messages are comprised of a sequence of alphanumeric characters. When digitally transmitted the characters are first encoded into a sequence of bits, called a *bit stream* or *baseband signal*. Groups of k bits can then be combined to form new digits, or *symbols*, from a finite symbol set or alphabet of $M = 2^k$ such symbols. A system using a symbol set size of M is referred to as an *M-ary system*. The value of k or M represents an important initial choice in the design of any digital communication system. For $k = 1$, the system is termed *binary*, the size of the symbol set is $M = 2$, and the modulator uses one of the two different waveforms to represent the binary "one" and the other to represent the binary "zero." For this special case, the symbol and the bit are the same. For $k = 2$, the system is termed *quaternary* or 4-*ary* ($M = 4$). At each symbol time, the modulator uses one of the four different waveforms that represents the symbol. The partitioning of the sequence of message bits is determined by the specification of the symbol set size, M. The following example should help clarify the relationship between the terms "message," "character," "symbol," "bit," and "digital waveform."

2.3.1 Example of Messages, Characters, and Symbols

Figure 2.5 shows examples of bit stream partitioning, based on the system specification for the values of k and M. The textual message in the figure is the word "THINK." Using 6-bit ASCII character coding (bit numbers 1 to 6 from Figure 2.3) yields a bit stream comprised of 30 bits. In Figure 2.5a, the symbol set size, M, has been chosen to be 8 (each symbol represents an 8-ary digit). The bits are therefore partitioned into groups of three ($k = \log_2 8$); the resulting 10 numbers

Figure 2.3 Seven-bit American standard code for information interchange (ASCII).

Bits 1 2 3 4	5 6 7 →	0 0 0	1 0 0	0 1 0	1 1 0	0 0 1	1 0 1	0 1 1	1 1 1
0 0 0 0		NUL	DLE	SP	0	@	P	`	p
1 0 0 0		SOH	DC1	!	1	A	Q	a	q
0 1 0 0		STX	DC2	"	2	B	R	b	r
1 1 0 0		ETX	DC3	#	3	C	S	c	s
0 0 1 0		EOT	DC4	$	4	D	T	d	t
1 0 1 0		ENQ	NAK	%	5	E	U	e	u
0 1 1 0		ACK	SYN	&	6	F	V	f	v
1 1 1 0		BEL	ETB	'	7	G	W	g	w
0 0 0 1		BS	CAN	(	8	H	X	h	x
1 0 0 1		HT	EM	)	9	I	Y	i	y
0 1 0 1		LF	SUB	*	:	J	Z	j	z
1 1 0 1		VT	ESC	+	;	K	[	k	{
0 0 1 1		FF	FS	,	<	L	\	l	|
1 0 1 1		CR	GS	-	=	M	]	m	}
0 1 1 1		SO	RS	.	>	N	^	n	~
1 1 1 1		SI	US	/	?	O	_	o	DEL

NUL	Null, or all zeros		DC1	Device control 1
SOH	Start of heading		DC2	Device control 2
STX	Start of text		DC3	Device control 3
ETX	End of text		DC4	Device control 4
EOT	End of transmission		NAK	Negative acknowledge
ENQ	Enquiry		SYN	Synchronous idle
ACK	Acknowledge		ETB	End of transmission block
BEL	Bell, or alarm		CAN	Cancel
BS	Backspace		EM	End of medium
HT	Horizontal tabulation		SUB	Substitute
LF	Line feed		ESC	Escape
VT	Vertical tabulation		FS	File separator
FF	Form feed		GS	Group separator
CR	Carriage return		RS	Record separator
SO	Shift out		US	Unit separator
SI	Shift in		SP	Space
DLE	Data link escape		DEL	Delete

Figure 2.4 EBCDIC character code set.

Bits 1234 \ 5678	0000	0001	0010	0011	0100	0101	0110	0111	1000	1001	1010	1011	1100	1101	1110	1111
0000	NUL	SOH	STX	ETX	PF	HT	LC	DEL			SMM	VT	FF	CR	SO	SI
0001	DLE	DC1	DC2	DC3	RES	NL	BS	IL	CAN	EM	CC		IFS	IGS	IRS	IUS
0010	DS	SOS	FS		BYP	LF	EOB	PRE			SM			ENQ	ACK	BEL
0011			SYN		PN	RS	UC	EOT					DC4	NAK		SUB
0100	SP										¢	.	<	(	+	|
0101	&										!	$	*	)	;	¬
0110	-	/									¦	,	%	_	>	?
0111											:	#	@	'	=	"
1000		a	b	c	d	e	f	g	h	i						
1001		j	k	l	m	n	o	p	q	r						
1010			s	t	u	v	w	x	y	z						
1011																
1100		A	B	C	D	E	F	G	H	I						
1101		J	K	L	M	N	O	P	Q	R						
1110			S	T	U	V	W	X	Y	Z						
1111	0	1	2	3	4	5	6	7	8	9						

Abbreviations

- PF — Punch off
- HT — Horizontal tab
- LC — Lower case
- DEL — Delete
- SP — Space
- UC — Upper case
- RES — Restore
- NL — New line
- BS — Backspace
- IL — Idle
- PN — Punch on
- EOT — End of transmission
- BYP — Bypass
- LF — Line feed
- EOB — End of block
- PRE — Prefix (ESC)
- RS — Reader stop
- SM — Start message
- DS — Digit select
- SOS — Start of significance
- IFS — Interchange file separator
- IGS — Interchange group separator
- IRS — Interchange record separator
- IUS — Interchange unit separator
- Others — Same as ASCII

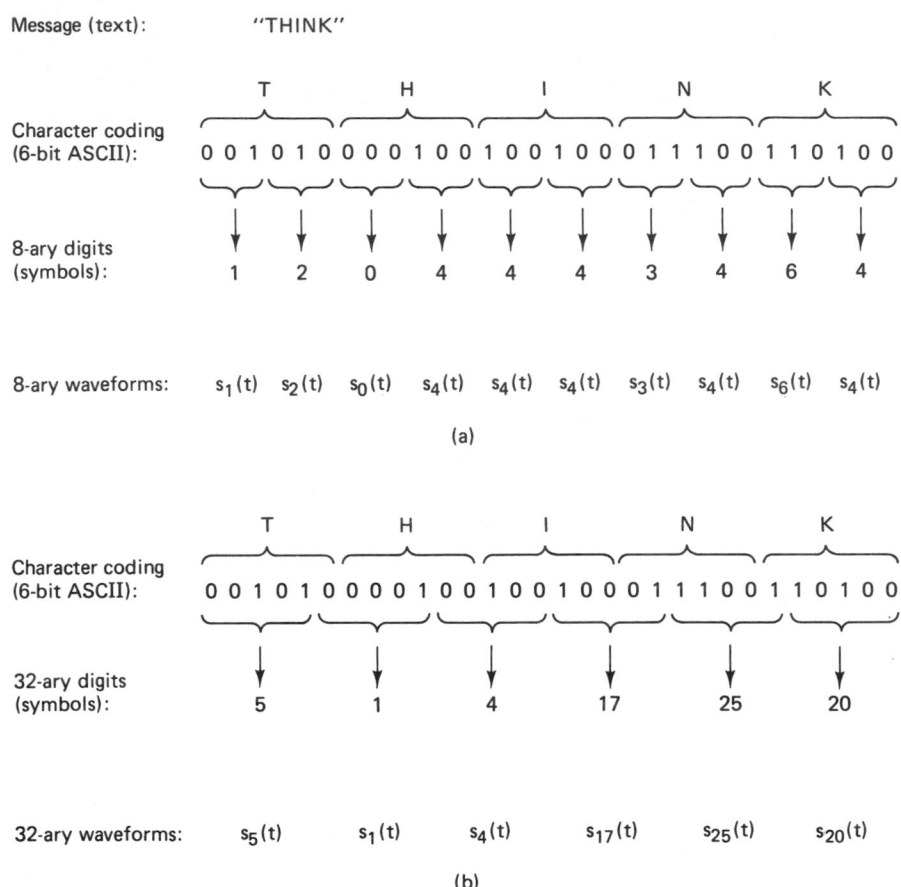

Message (text):   "THINK"

	T	H	I	N	K

Character coding (6-bit ASCII):
0 0 1 0 1 0 0 0 0 1 0 0 1 0 0 1 0 0 0 1 1 1 0 0 1 1 0 1 0 0

8-ary digits (symbols):   1   2   0   4   4   4   3   4   6   4

8-ary waveforms: $s_1(t)$ $s_2(t)$ $s_0(t)$ $s_4(t)$ $s_4(t)$ $s_4(t)$ $s_3(t)$ $s_4(t)$ $s_6(t)$ $s_4(t)$

(a)

	T	H	I	N	K

Character coding (6-bit ASCII):
0 0 1 0 1 0 0 0 0 1 0 0 1 0 0 1 0 0 0 1 1 1 0 0 1 1 0 1 0 0

32-ary digits (symbols):   5   1   4   17   25   20

32-ary waveforms: $s_5(t)$ $s_1(t)$ $s_4(t)$ $s_{17}(t)$ $s_{25}(t)$ $s_{20}(t)$

(b)

Figure 2.5 Messages, characters, and symbols. (a) 8-ary example. (b) 32-ary example.

represent the 10 octal symbols to be transmitted. The transmitter must have a repertoire of eight waveforms, $s_i(t)$, where $i = 1, \ldots, 8$, to represent the possible symbols, any one of which may be transmitted during a symbol time. The final row of Figure 2.5a lists the 10 waveforms that an 8-ary modulating system transmits to represent the textual message "THINK."

In Figure 2.5b, the symbol set size, M, has been chosen to be 32 (each symbol represents a 32-ary digit). The bits are therefore taken five at a time, and the resulting group of six numbers represent the six 32-ary symbols to be transmitted. Notice that there is no need for the symbol boundaries and the character boundaries to coincide. The first symbol represents $\frac{5}{6}$ of the first character, "T." The second symbol represents the remaining $\frac{1}{6}$ of the character "T" and $\frac{4}{6}$ of the next character, "H," and so on. It is not necessary that the characters be partitioned more aesthetically. The system sees the characters as a string of digits to be transmitted; only the end user (or the user's teleprinter machine) ascribes textual meaning to the final delivered sequence of bits. In this 32-ary case, a transmitter needs a repertoire of 32 waveforms, $s_i(t)$, where $i = 1, \ldots, 32$, one for each possible symbol that may be transmitted. The final row of the figure lists the six waveforms that a 32-ary modulating system transmits to represent the textual message "THINK."

2.4 FORMATTING ANALOG INFORMATION

If the information is analog, it cannot be character encoded as in the case of textual data; the information must first be transformed into a digital format. The process of transforming an analog waveform into a form that is compatible with a digital communication system starts with sampling the waveform to produce a discrete pulse-amplitude-modulated waveform, as described below.

2.4.1 The Sampling Theorem

The link between an analog waveform and its sampled version is provided by what is known as the *sampling process*. This process can be implemented in several ways, the most popular being the *sample-and-hold* operation. In this operation, a switch and storage mechanism (such as a transistor and a capacitor, or a shutter and a filmstrip) form a sequence of samples of the continuous input waveform. The output of the sampling process is called *pulse amplitude modulation* (PAM) because the successive output intervals can be described as a sequence of pulses with amplitudes derived from the input waveform samples. The analog waveform can be approximately retrieved from a PAM waveform by simple low-pass filtering. An important question is: How closely can a filtered PAM waveform approximate the original input waveform? This question can be answered by reviewing the *sampling theorem*, which states [1]: A bandlimited signal having no spectral components above f_m hertz can be determined uniquely by values sampled at uniform intervals of T_s seconds, where

$$T_s \le \frac{1}{2f_m} \tag{2.1}$$

This particular statement is also known as the *uniform sampling theorem*. Stated another way, the upper limit on T_s can be expressed in terms of the sampling rate, denoted $f_s = 1/T_s$. The restriction, stated in terms of the sampling rate, is known as the *Nyquist criterion*. The statement is

$$f_s \ge 2f_m \tag{2.2}$$

The sampling rate $f_s = 2f_m$ is also called the *Nyquist rate*. The Nyquist criterion is a theoretically sufficient condition to allow an analog signal to be *reconstructed completely* from a set of uniformly spaced discrete-time samples. In the sections that follow, the validity of the sampling theorem is demonstrated using different sampling approaches.

2.4.1.1 Impulse Sampling

Here we demonstrate the validity of the sampling theorem using the frequency convolution property of the Fourier transform. Let us first examine the case of *ideal sampling* with a sequence of unit impulse functions. Assume an analog waveform, $x(t)$, as shown in Figure 2.6a, with a Fourier transform, $X(f)$, which is zero outside the interval $(-f_m < f < f_m)$, as shown in Figure 2.6b. The sampling of $x(t)$ can be viewed as the product of $x(t)$ with a periodic train of unit impulse functions, $x_\delta(t)$, shown in Figure 2.6c and defined as follows:

$$x_\delta(t) = \sum_{n=-\infty}^{\infty} \delta(t - nT_s) \tag{2.3}$$

where T_s is the sampling period and $\delta(t)$ is the unit impulse or Dirac delta function defined in Section 1.2.5. Let us choose $T_s = 1/2f_m$, so that the Nyquist criterion is just satisfied.

The *sifting property* of the impulse function (see Section A.4.1) states that

$$x(t)\delta(t - t_0) = x(t_0)\delta(t - t_0) \tag{2.4}$$

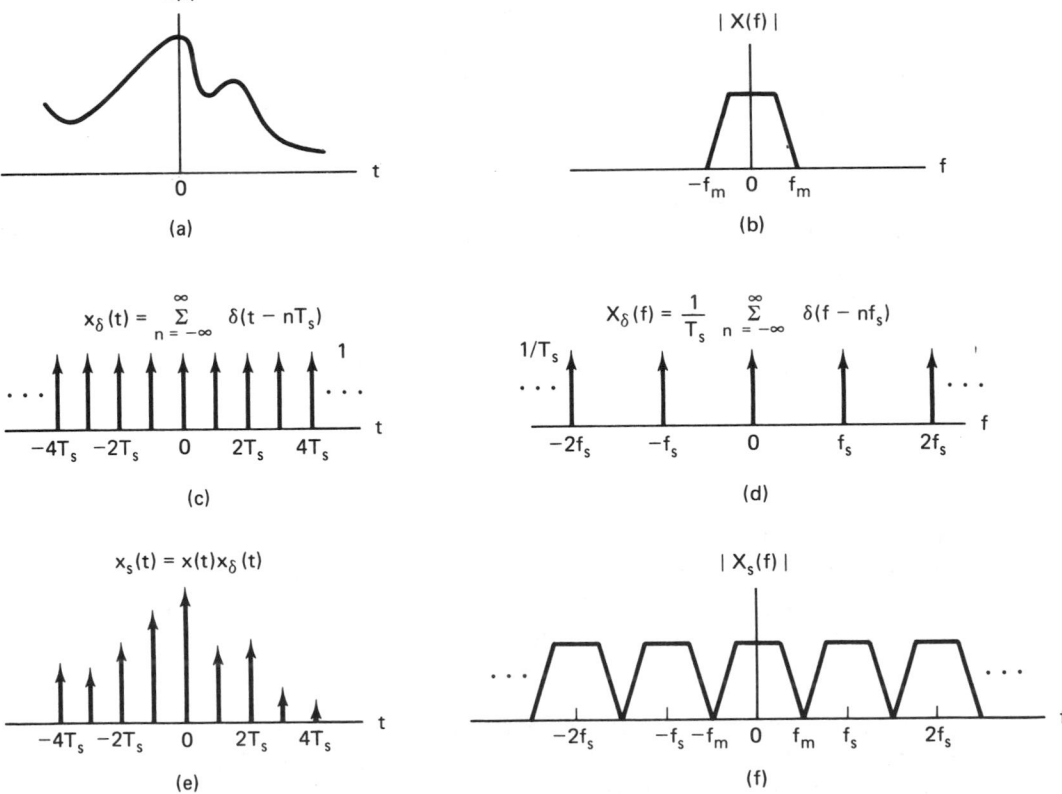

Figure 2.6 Sampling theorem using the frequency convolution property of the Fourier transform.

Using this property, we can see that $x_s(t)$, the sampled version of $x(t)$, shown in Figure 2.6e, is given by

$$x_s(t) = x(t)x_\delta(t) = \sum_{n=-\infty}^{\infty} x(t)\delta(t - nT_s)$$

$$= \sum_{n=-\infty}^{\infty} x(nT_s)\delta(t - nT_s)$$

(2.5)

Using the *frequency convolution property* of the Fourier transform (see Section A.5.3), the time-domain product $x(t)x_\delta(t)$ of Equation (2.5) transforms to the frequency-domain convolution $X(f) * X_\delta(f)$, where $X_\delta(f)$ is the Fourier transform of the impulse train $x_\delta(t)$,

$$X_\delta(f) = \frac{1}{T_s} \sum_{n=-\infty}^{\infty} \delta(f - nf_s)$$

(2.6)

and where $f_s = 1/T_s$ is the sampling frequency. Notice that the Fourier transform of an impulse train is another impulse train; the values of the periods of the two trains are reciprocally related to one another. Figures 2.6c and d illustrate the impulse train $x_\delta(t)$ and its Fourier transform $X_\delta(f)$, respectively.

Convolution with an impulse function simply shifts the original function, as follows:

$$X(f) * \delta(f - nf_s) = X(f - nf_s)$$

(2.7)

We can solve for the transform, $X_s(f)$, of the sampled waveform as follows:

$$X_s(f) = X(f) * X_\delta(f) = X(f) * \left[\frac{1}{T_s} \sum_{n=-\infty}^{\infty} \delta(f - nf_s) \right]$$

$$= \frac{1}{T_s} \sum_{n=-\infty}^{\infty} X(f - nf_s)$$

(2.8)

We therefore conclude that within the original bandwidth, the spectrum $X_s(f)$ of the sampled signal $x_s(t)$ is, to within a constant factor $(1/T_s)$, exactly the same as that of $x(t)$. In addition, the spectrum repeats itself periodically in frequency every f_s hertz. The sifting property of an impulse function makes the convolving of an impulse train with another function easy to visualize. The impulses act as sampling functions. Hence, convolution can be performed graphically by sweeping the impulse train, $X_\delta(f)$, in Figure 2.6d past the transform, $|X(f)|$, in Figure 2.6b. This sampling of $|X(f)|$ at each step in the sweep replicates $|X(f)|$ at each of the frequency positions of the impulse train, resulting in $|X_s(f)|$, shown in Figure 2.6f.

When the sampling rate is chosen, as it has been here, such that $f_s = 2f_m$, each spectral replicate is separated from each of its neighbors by a frequency band exactly equal to f_s hertz, and the analog waveform can theoretically be completely recovered from the samples, by the use of filtering. However, a filter with infinitely steep sides would be required. It should be clear that if $f_s > 2f_m$, the replications will move farther apart in frequency, as shown in Figure 2.7a, making it easier to perform the filtering operation. A typical low-pass filter characteristic that might be used to separate the baseband spectrum from those at higher frequencies is shown in the figure. When the sampling rate is reduced, such that $f_s < 2f_m$, the replications will overlap, as shown in Figure 2.7b, and some information will be lost. This phenomenon, the result of undersampling (sampling at too low a rate), is called *aliasing*. The Nyquist rate, $f_s = 2f_m$, is the sampling rate below which aliasing occurs; to avoid aliasing, the Nyquist criterion, $f_s \geq 2f_m$, must be satisfied.

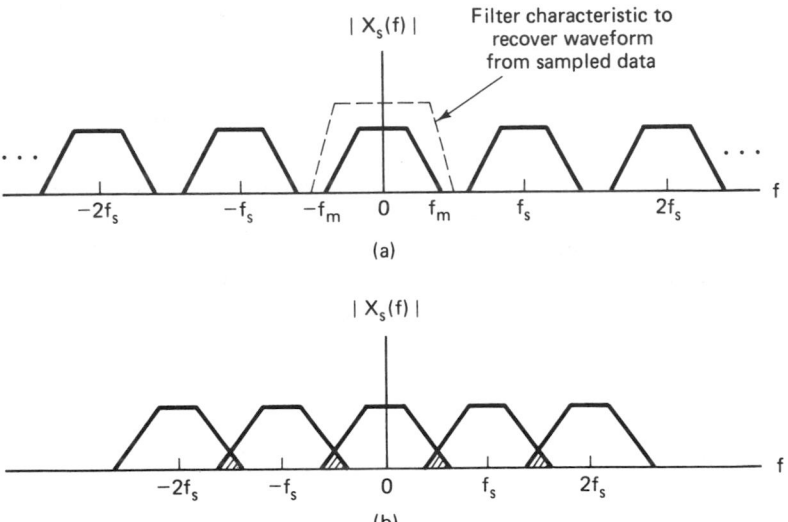

Figure 2.7 Spectra for various sampling rates. (a) Sampled spectrum ($f_s > 2f_m$). (b) Sampled spectrum ($f_s < 2f_m$).

As a matter of practical consideration, neither waveforms of engineering interest nor realizable bandlimiting filters are strictly bandlimited. These signals and filters can, however, be considered to be "essentially" bandlimited. By this we mean that a bandwidth can be determined beyond which the spectral components are attenuated to a level that is considered negligible.

2.4.1.2 Natural Sampling

Here we demonstrate the validity of the sampling theorem using the frequency shifting property of the Fourier transform. Although instantaneous sampling is a convenient model, a more practical way of accomplishing the sampling of a bandlimited analog signal, $x(t)$, is to multiply $x(t)$, shown in Figure 2.8a, by the pulse train or switching waveform, $x_p(t)$, shown in Figure 2.8c. Each pulse in $x_p(t)$ has width T and amplitude $1/T$. Multiplication by $x_p(t)$ can be viewed as the opening and closing of a switch. As before, the sampling frequency is designated f_s, and its reciprocal, the time period between samples, is designated T_s.

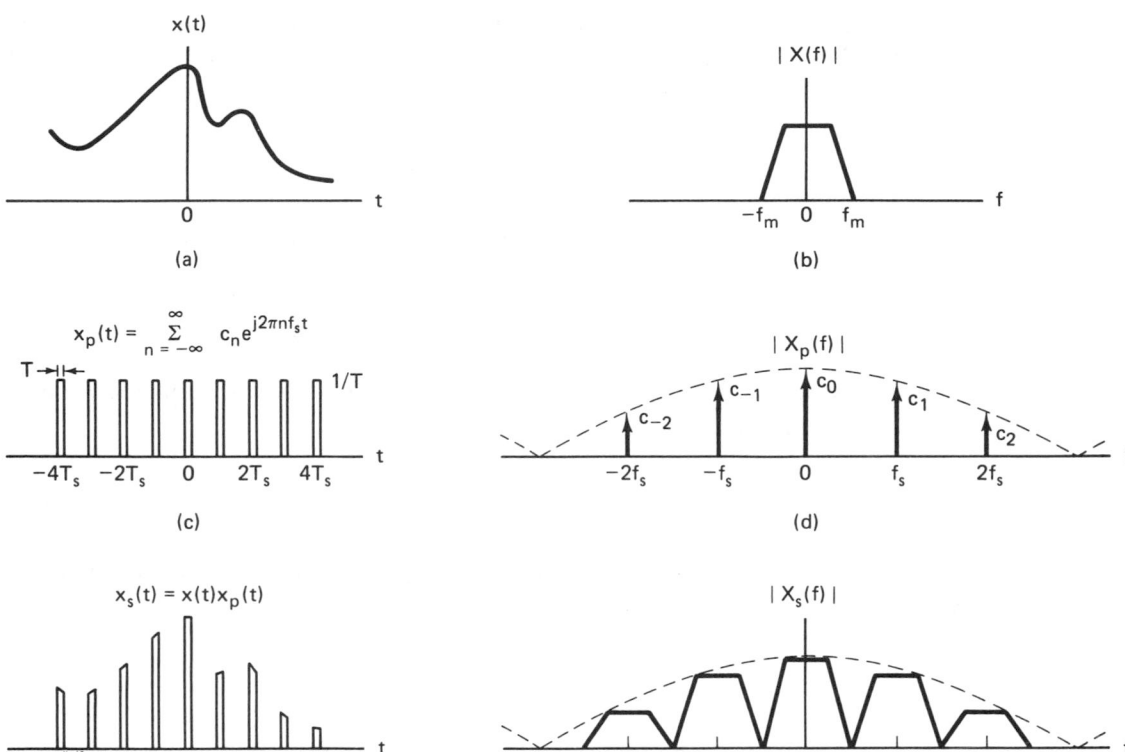

Figure 2.8 Sampling theorem using the frequency shifting property of the Fourier transform.

The resulting sampled-data sequence, $x_s(t)$, is illustrated in Figure 2.8e and is expressed as

$$x_s(t) = x(t)x_p(t) \qquad (2.9)$$

The sampling here is termed *natural sampling*, since the top of each pulse in the $x_s(t)$ sequence retains the shape of its corresponding analog segment during the pulse interval. Using Equation (A.13), we can express the periodic pulse train $x_p(t)$ as a Fourier series in the form

$$x_p(t) = \sum_{n=-\infty}^{\infty} c_n e^{j2\pi n f_s t} \qquad (2.10)$$

where the sampling rate, $f_s = 1/T_s$, is chosen equal to $2f_m$, so that the Nyquist criterion is just satisfied. From Equation (A.24), $c_n = (1/T_s)$ sinc (nT/T_s), where T is the pulse width, $1/T$ is the pulse amplitude, and

$$\text{sinc } y = \frac{\sin \pi y}{\pi y}$$

The envelope of the magnitude spectrum of the pulse train, seen as a dashed line in Figure 2.8d, has the characteristic sinc shape. Combining Equations (2.9) and (2.10), we can express $x_s(t)$ as

$$x_s(t) = x(t) \sum_{n=-\infty}^{\infty} c_n e^{j2\pi n f_s t} \tag{2.11}$$

The transform, $X_s(f)$, of the sampled waveform is found as follows:

$$X_s(f) = \mathcal{F}\left\{ x(t) \sum_{n=-\infty}^{\infty} c_n e^{j2\pi n f_s t} \right\} \tag{2.12}$$

For linear systems, we can interchange the operations of summation and Fourier transformation. Therefore, we can write

$$X_s(f) = \sum_{n=-\infty}^{\infty} c_n \mathcal{F}\{x(t) e^{j2\pi n f_s t}\} \tag{2.13}$$

Using the *frequency translation* property of the Fourier transform (see Section A.3.2), we solve for $X_s(f)$ as follows:

$$X_s(f) = \sum_{n=-\infty}^{\infty} c_n X(f - n f_s) \tag{2.14}$$

Similar to the unit impulse sampling case, Equation (2.14) and Figure 2.8f illustrate that $X_s(f)$ is a replication of $X(f)$, periodically repeated in frequency every f_s hertz. In this natural-sampled case, however, we see that $X_s(f)$ is weighted by the Fourier series coefficients of the pulse train, compared to a constant value in the impulse-sampled case. It is satisfying to note that *in the limit*, as the pulse width, T, approaches zero, c_n approaches $1/T_s$ for all n (see the example that follows), and Equation (2.14) converges to Equation (2.8).

Example 2.1 Comparison of Impulse Sampling and Natural Sampling

Consider a given waveform, $x(t)$, with Fourier transform, $X(f)$. Let $X_{s1}(f)$ be the spectrum of $x_{s1}(t)$, which is the result of sampling $x(t)$ with a unit impulse train $x_\delta(t)$. Let $X_{s2}(f)$ be the spectrum of $x_{s2}(t)$, the result of sampling $x(t)$ with a pulse train, $x_p(t)$, with pulse width, T, amplitude $1/T$ and period, T_s. Show that in the limit, as T approaches zero, $X_{s1}(f) = X_{s2}(f)$.

Solution

From Equation (2.8),

$$X_{s1}(f) = \frac{1}{T_s} \sum_{n=-\infty}^{\infty} X(f - n f_s)$$

and from Equation (2.14),

$$X_{s2}(f) = \sum_{n=-\infty}^{\infty} c_n X(f - n f_s)$$

As the pulse width $T \to 0$, and the pulse amplitude approaches infinity (the area of the pulse remains unity), $x_p(t) \to x_\delta(t)$. Using Equation (A.14), we can solve for c_n in the limit as follows:

$$c_n = \lim_{T \to 0} \frac{1}{T_s} \int_{-T_s/2}^{T_s/2} x_p(t)e^{-j2\pi nf_s t}\, dt$$

$$= \frac{1}{T_s} \int_{-T_s/2}^{T_s/2} x_\delta(t)e^{-j2\pi nf_s t}\, dt$$

Since, within the range of integration, $-T_s/2$ to $T_s/2$, the only contribution of $x_\delta(t)$ is that due to the impulse at the origin, we can write

$$c_n = \frac{1}{T_s} \int_{-T_s/2}^{T_s/2} \delta(t)e^{-j2\pi nf_s t}\, dt = \frac{1}{T_s}$$

Therefore, in the limit, $X_{s1}(f) = X_{s2}(f)$ for all n.

2.4.1.3 Sample-and-Hold Operation

The simplest and thus most popular sampling method, *sample and hold*, can be described by the convolution of the sampled pulse train, $[x(t)x_\delta(t)]$, shown in Figure 2.6e, with a unity amplitude rectangular pulse, $p(t)$, of pulse width T_s. This time convolution results in the *flat-top* sampled sequence, $x_s(t)$:

$$x_s(t) = p(t) * [x(t)x_\delta(t)]$$

$$= p(t) * \left[x(t) \sum_{n=-\infty}^{\infty} \delta(t - nT_s) \right] \tag{2.15}$$

The Fourier transform, $X_s(f)$, of the time convolution in Equation (2.15) is the frequency-domain product between the transform $P(f)$ of the rectangular pulse and the periodic spectrum, shown in Figure 2.6f, of the impulse-sampled data:

$$X_s(f) = P(f)\mathscr{F}\left\{ x(t) \sum_{n=-\infty}^{\infty} \delta(t - nT_s) \right\}$$

$$= P(f) \left\{ X(f) * \left[\frac{1}{T_s} \sum_{n=-\infty}^{\infty} \delta(f - nf_s) \right] \right\} \tag{2.16}$$

$$= P(f) \frac{1}{T_s} \sum_{n=-\infty}^{\infty} X(f - nf_s)$$

where $P(f)$ is of the form T_s sinc fT_s. The effect of this product operation results in a spectrum similar in appearance to the natural-sampled example presented in Figure 2.8f. The most obvious effect of the hold operation is the significant attenuation of the higher-frequency spectral replicates (compare Figure 2.8f to Figure 2.6f), which is a desired effect. Additional analog postfiltering is usually required to finish the filtering process by further attenuating the residual spectral components located at the multiples of the sample rate. A secondary effect of the hold operation is the nonuniform spectral gain, $P(f)$, applied to the desired baseband spectrum shown in Equation (2.16). The postfiltering operation can compensate for this attenuation by incorporating the inverse of $P(f)$ over the signal passband.

2.4.2 Aliasing

Figure 2.9 is a detailed view of the positive half of the baseband spectrum and one of the replicates from Figure 2.7b. It illustrates aliasing in the frequency domain. The overlapped region, shown in Figure 2.9b, contains that part of the spectrum which is aliased due to *undersampling*. The aliased spectral components represent ambiguous data that can be retrieved only under special conditions (see Section 11.4.4, on subband coding). In general, the ambiguity is not resolved and the ambiguous data appear in the frequency band between $(f_s - f_m)$ and f_m. Figure 2.10 illustrates that a higher sampling rate, f'_s, can eliminate the aliasing by separating the spectral replicates; the resulting spectrum in Figure 2.10b corresponds to the case in Figure 2.7a. Figures 2.11 and 2.12 illustrate two ways of eliminating aliasing using *antialiasing filters*. In Figure 2.11 the analog signal is *prefiltered* so that the new maximum frequency, f'_m, is reduced to $f_s/2$ or less. Thus there are no aliased components seen in Figure 2.11b, since $f_s > 2f'_m$. Eliminating the aliasing terms prior to sampling is good engineering practice. When the signal structure is well known, the aliased terms can be eliminated after sampling, with a low-pass filter operating on the sampled data [2]. In Figure 2.12 the aliased components are removed by *postfiltering* after sampling; the filter cutoff frequency, f''_m, removes the aliased components; f''_m needs to be less than

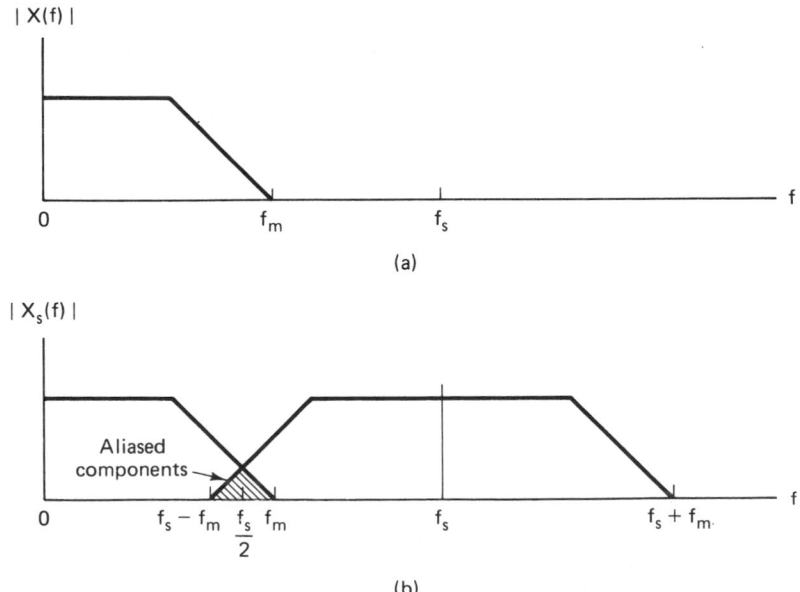

Figure 2.9 Aliasing in the frequency domain. (a) Continuous signal spectrum. (b) Sampled signal spectrum.

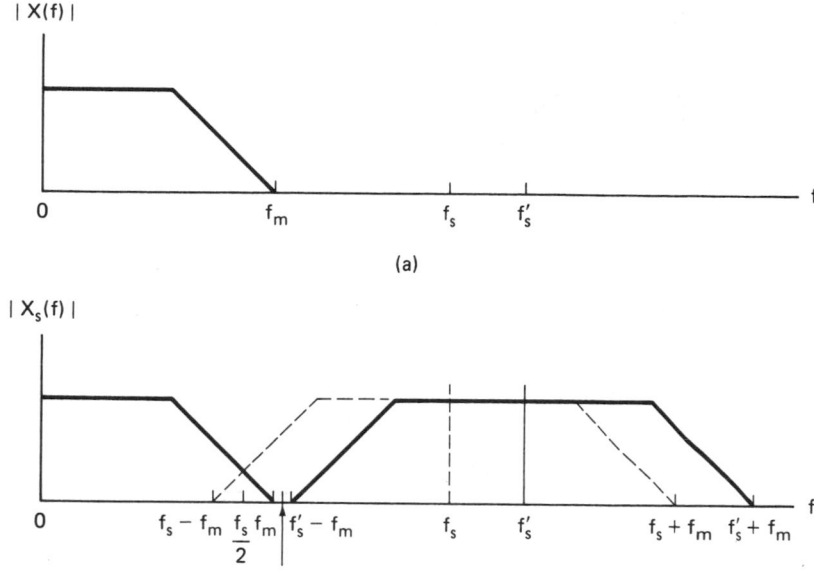

Figure 2.10 Higher sampling rate eliminates aliasing. (a) Continuous signal spectrum. (b) Sampled signal spectrum.

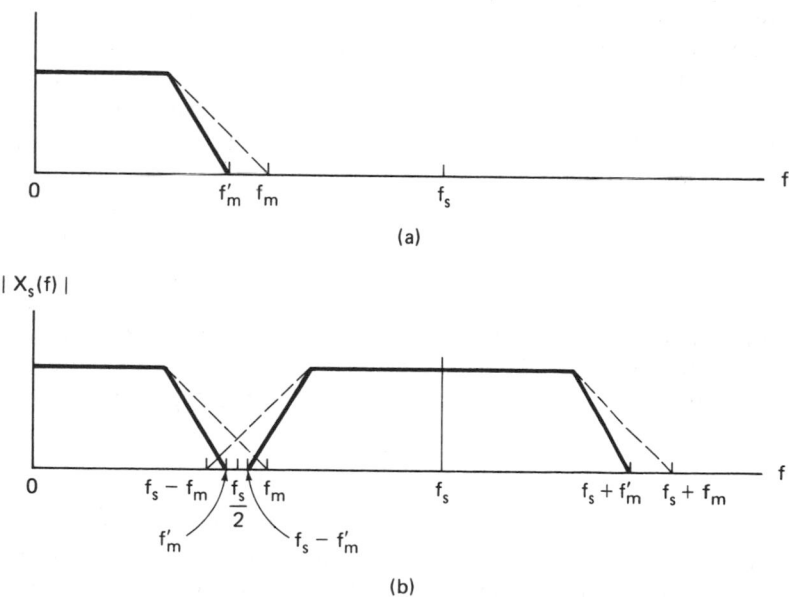

Figure 2.11 Sharper-cutoff filters eliminate aliasing. (a) Continous signal spectrum. (b) Sampled signal spectrum.

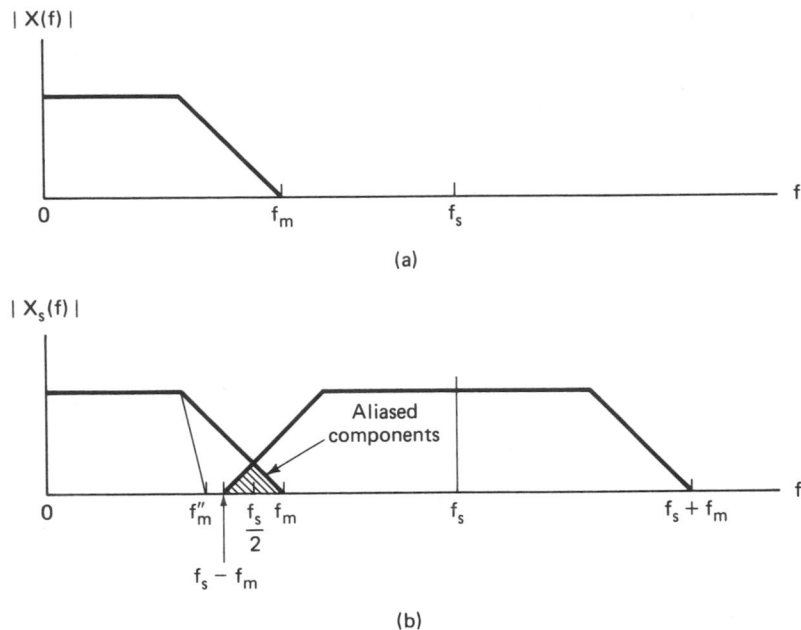

Figure 2.12 Postfilter eliminates aliased portion of spectrum. (a) Continuous signal spectrum. (b) Sampled signal spectrum.

$(f_s - f_m)$. Notice that the filtering techniques for eliminating the aliased portion of the spectrum in Figures 2.11 and 2.12 *will result in a loss* of some of the signal information. For this reason, the sample rate, cutoff bandwidth, and filter type selected for a particular signal bandwidth are all interrelated.

Realizable filters require a nonzero bandwidth for the transition between the passband and the required out-of-band attenuation. This is called the *transition bandwidth*. To minimize the system sample rate, we desire that the antialiasing filter have a small transition bandwidth. Filter complexity and cost rise sharply with narrower transition bandwidth, so a trade-off is required between the cost of a small transition bandwidth and the costs of the higher sampling rate, which are those of more storage and higher transmission rates. In many systems the answer has been to make the transition bandwidth between 10 and 20% of the signal bandwidth. If we account for the 20% transition bandwidth of the antialiasing filter, we have an *engineer's version* of the Nyquist sampling rate:

$$f_s \geq 2.2 f_m \tag{2.17}$$

Figure 2.13 provides some insight into aliasing as seen in the time domain. The sampling instants of the solid-line sinusoid have been chosen so that the sinusoidal signal is undersampled. Notice that the resulting ambiguity allows one to draw a totally different (dashed-line) sinusoid, following the undersampled points.

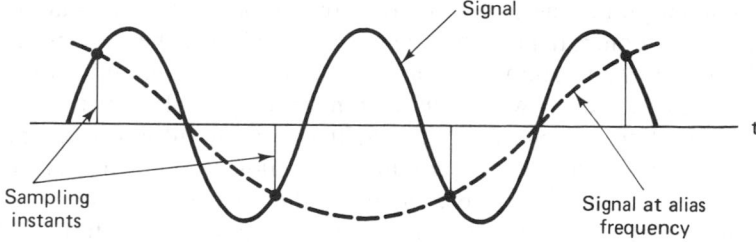

Figure 2.13 Alias frequency generated by sub-Nyquist sampling rate.

Example 2.2 Sampling Rate for a High-Quality Music System

We wish to produce a high-quality digitization of a 20-kHz bandwidth music source. We are to determine a reasonable sample rate for this source. By the engineer's version of the Nyquist rate, in Equation (2.17), the sampling rate should be greater than 44.0 ksamples/s. As a matter of comparison, the standard sampling rate for the compact disc digital audio player is 44.1 ksamples/s, and the standard sampling rate for studio-quality audio is 48.0 ksamples/s.

2.4.3 Signal Interface for a Digital System

Let us examine four ways in which analog source information can be described. Figure 2.14 illustrates the choices. Let us refer to the waveform in Figure 2.14a as the *original analog waveform*. Figure 2.14b represents a sampled version of the original waveform, typically referred to as *natural-sampled data* or PAM

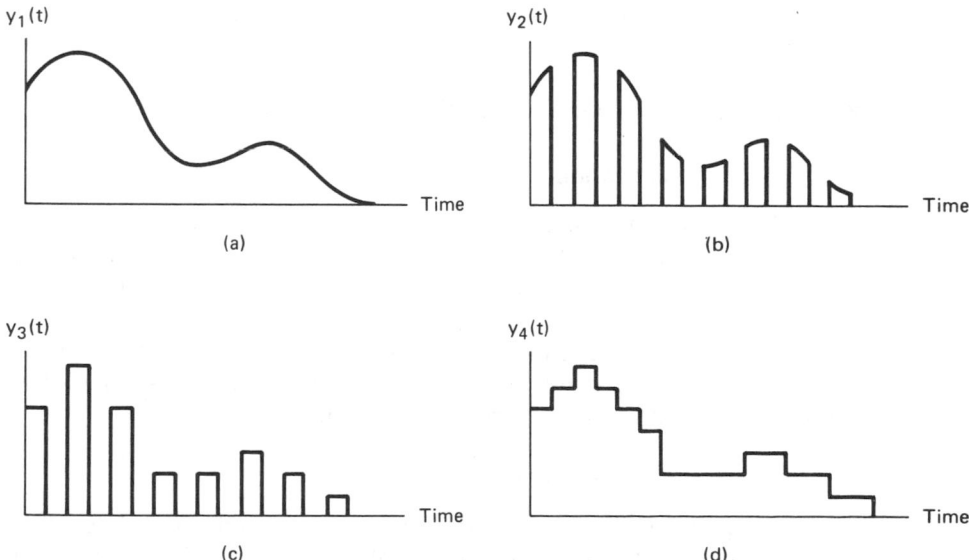

Figure 2.14 Amplitude and time coordinates of source data. (a) Original analog waveform. (b) Natural-sampled data. (c) Quantized samples. (d) Sample and hold.

(pulse amplitude modulation). Do you suppose that the sampled data in Figure 2.14b are compatible with a digital system? No, they are not, because the amplitude of each natural sample still has an infinite number of possible values; a digital system deals with a finite number of symbols. Even if the sampling is flat-top sampling, the possible pulse values form an infinite set, since they reflect all the possible values of the continuous analog waveform. Figure 2.14c illustrates the original waveform represented by discrete pulses. Here the pulses have flat tops *and* the pulse amplitude values are limited to a finite set. Each pulse is expressed as a level from a finite number of predetermined levels; each such level can be represented by a symbol from a finite alphabet. The pulses in Figure 2.14c are referred to as *quantized samples*; such a format is the obvious choice for interfacing with a digital system. The format in Figure 2.14d may be construed as the output of a sample-and-hold circuit. When the sample values are quantized to a finite set, this format can also interface with a digital system. After quantization, the analog waveform can still be recovered, but not precisely; improved reconstruction fidelity of the analog waveform can be achieved by increasing the number of quantization levels (requiring increased system bandwidth). Signal distortion due to quantization is treated in the following sections (and in Chapter 11).

2.5 SOURCES OF CORRUPTION

The analog signal recovered from the sampled, quantized, and transmitted pulses will contain corruption from several sources. The sources of corruption are related to (1) sampling and quantizing effects, and (2) channel effects. These effects are considered in the sections that follow.

2.5.1 Sampling and Quantizing Effects

2.5.1.1 Quantization Noise

The distortion inherent in quantization is a round-off or truncation error. The process of encoding the PAM waveform into a quantized waveform involves discarding some of the original analog information. This distortion, introduced by the need to approximate the analog waveform with quantized samples, is referred to as *quantization noise*; the amount of such noise is inversely proportional to the number of levels employed in the quantization process. The signal-to-noise ratio of quantized pulses is treated in Section 2.5.3.

2.5.1.2 Quantizer Saturation

The quantizer (or analog-to-digital converter) allocates L levels to the task of approximating the continuous range of inputs with a finite set of outputs. The range of inputs for which the difference between the input and output is small is called the *operating range* of the converter. If the input exceeds this range, the difference between the input and the output becomes large, and we say that the converter is operating in *saturation*. Saturation errors, being large, are more objectionable than quantizing noise. Generally, saturation is avoided by the use of automatic gain control (AGC), which effectively extends the operating range of the converter. Chapter 11 covers quantizer saturation in greater detail.

2.5.1.3 Timing Jitter

Our analysis of the sampling theorem predicted precise reconstruction of the signal based on uniformly spaced samples of the signal. If there is a slight jitter in the position of the sample, the sampling is no longer uniform. Although exact reconstruction is still possible if the sample positions are accurately known, the jitter is usually a random process and thus the sample positions are not accurately known. The effect of the jitter is equivalent to frequency modulation (FM) of the baseband signal. If the jitter is random, a low-level wideband spectral contribution is induced whose properties are very close to those of the quantizing noise. If the jitter exhibits periodic components, as might be found in data extracted from a tape recorder, the periodic FM will induce low-level spectral lines in the data. Timing jitter can be controlled with very good power supply isolation and stable clock references.

2.5.2 Channel Effects

2.5.2.1 Channel Noise

Thermal noise, interference from other users, and interference from circuit switching transients can cause errors in detecting the pulses carrying the digitized samples. Channel-induced errors can degrade the reconstructed signal quality quite quickly. This rapid degradation of output signal quality with channel-induced errors is called a *threshold effect*. If the channel noise is small, there will be no problem detecting the presence of the waveforms. Thus small noise does not corrupt the reconstructed signals. In this case, the only noise present in the reconstruction is the quantization noise. On the other hand, if the channel noise is large enough to affect our ability to detect the waveforms, the resultant detection error causes reconstruction errors. A large difference in behavior can occur for very small changes in channel noise level.

2.5.2.2 Intersymbol Interference

The channel is always bandlimited. A bandlimited channel disperses or spreads a pulse waveform passing through it (see Section 1.6.4). When the channel bandwidth is much greater than the pulse bandwidth, the spreading of the pulse will be slight. When the channel bandwidth is close to the signal bandwidth, the spreading will exceed a symbol duration and cause signal pulses to overlap. This overlapping is called *intersymbol interference* (ISI). Like any other source of interference, ISI causes system degradation (higher error rates); it is a particularly insidious form of interference because raising the signal power to overcome the interference will not improve the error performance. Details of how ISI is handled are presented in Section 2.11.

2.5.3 Signal-to-Noise Ratio for Quantized Pulses

Figure 2.15 illustrates an L-level linear quantizer for an analog signal with a peak-to-peak voltage range of $V_{pp} = V_p - (-V_p) = 2V_p$ volts. The quantized pulses assume positive and negative values, as shown in the figure. The step size between quantization levels, called the *quantile interval*, is denoted q volts. When the quantization levels are uniformly distributed over the full range, the quantizer is called a *uniform or linear quantizer*. Each sample value of the analog waveform is approximated with a quantized pulse; the approximation will result in an error no larger than $q/2$ in the positive direction or $-q/2$ in the negative direction. The degradation of the signal due to quantization is therefore limited to half a quantile interval, $\pm q/2$ volts.

A useful figure of merit for the uniform quantizer is the quantizer variance (mean-square error assuming zero mean). If we assume that the quantization error, e, is uniformly distributed over a single quantile interval q-wide (i.e., the analog input takes on all values with equal probability), the quantizer error variance is found to be

$$\sigma^2 = \int_{-q/2}^{+q/2} e^2 p(e) \, de \qquad (2.18a)$$

$$= \int_{-q/2}^{+q/2} e^2 \frac{1}{q} \, de = \frac{q^2}{12} \qquad (2.18b)$$

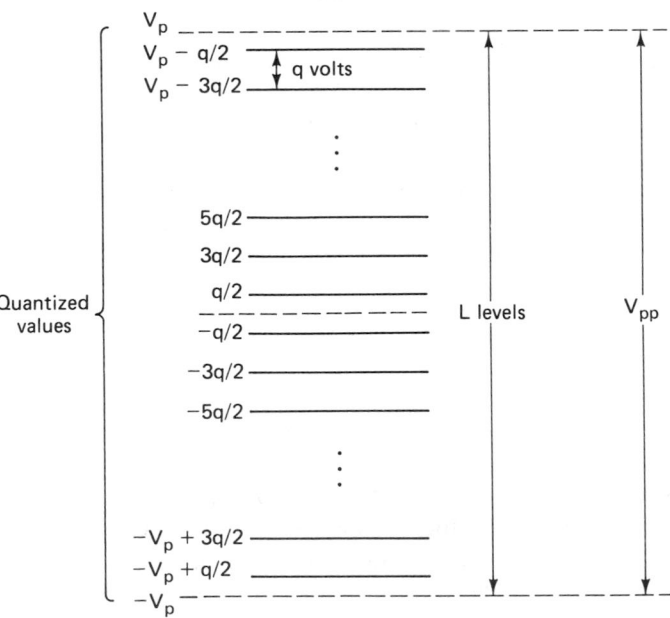

Figure 2.15 Quantization levels.

Formatting and Baseband Transmission   Chap. 2

where $p(e) = 1/q$ is the (uniform) probability density function of the quantization error. The variance, σ^2, corresponds to the *average quantization noise power*. The peak power of the analog signal (normalized to 1 Ω) can be expressed as

$$V_p^2 = \left(\frac{V_{pp}}{2}\right)^2 = \left(\frac{Lq}{2}\right)^2 = \frac{L^2 q^2}{4} \tag{2.19}$$

where L is the number of quantization levels. Equations (2.18) and (2.19) combined yield the ratio of peak signal power to average quantization noise power $(S/N)_q$, assuming that there are no errors due to ISI or channel noise:

$$\left(\frac{S}{N}\right)_q = \frac{L^2 q^2/4}{q^2/12} = 3L^2 \tag{2.20}$$

It is intuitively satisfying to see that $(S/N)_q$ improves as a function of the number of quantization levels squared. In the limit (as $L \to \infty$), the signal approaches the PAM format (with no quantization), and the signal-to-quantization noise ratio is infinite; in other words, with an infinite number of quantization levels, there is zero quantization noise.

2.6 PULSE CODE MODULATION

Pulse code modulation (PCM) is the name given to the class of baseband signals obtained from the quantized PAM signals by encoding each quantized sample into a *digital word* [3]. The source information is sampled and quantized to one of L levels; then each quantized sample is digitally encoded into an ℓ-bit ($\ell = \log_2 L$) codeword. For baseband transmission, the codeword bits will then be transformed to pulse waveforms. The essential features of binary PCM are shown in Figure 2.16. Assume that an analog signal, $x(t)$, is limited in its excursions to the range -4 to $+4$ V. The step size between quantization levels has been set at 1 V. Thus eight quantization levels are employed; these are located at -3.5, -2.5, . . . , $+3.5$ V. We assign the code number 0 to the level at -3.5 V, the code number 1 to the level at -2.5 V, and so on, until the level at 3.5 V, which is assigned the code number 7. Each code number has its representation in binary arithmetic, ranging from 000 for code number 0 to 111 for code number 7.

The ordinate in Figure 2.16 is labeled with quantization levels and their code numbers. Each sample of the analog signal is assigned to the quantization level closest to the value of the sample. Beneath the analog waveform, $x(t)$, are seen four representations of $x(t)$, as follows: the natural sample values, the quantized sample values, the code numbers, and the PCM sequence.

Note that in the example of Figure 2.16, each sample is represented by a 3-bit codeword. If the signal, $x(t)$, had been quantized to 16 levels, a 4-bit codeword would be needed to characterize each sample, or if $x(t)$ had been quantized to four levels, a 2-bit codeword would be needed. From Equation (2.20) it can be seen that the greater the number of quantization levels, the lower will be the quantization noise. Hence quantization noise performance can be traded off versus data rate.

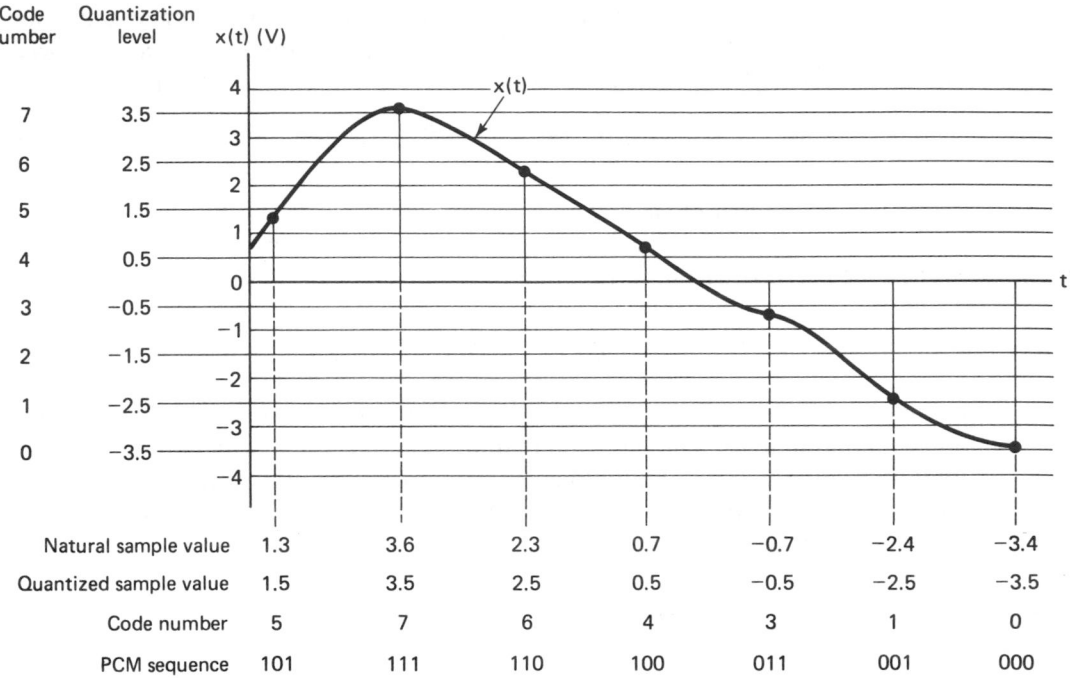

Natural sample value	1.3	3.6	2.3	0.7	−0.7	−2.4	−3.4
Quantized sample value	1.5	3.5	2.5	0.5	−0.5	−2.5	−3.5
Code number	5	7	6	4	3	1	0
PCM sequence	101	111	110	100	011	001	000

Figure 2.16 Natural samples, quantized samples, and pulse code modulation. (Reprinted with permission from Taub and Schilling, *Principles of Communication Systems*, McGraw-Hill Book Company, New York, 1971, Fig. 6.5-1, p. 205.)

2.7 UNIFORM AND NONUNIFORM QUANTIZATION

2.7.1 Statistics of Speech Amplitudes

Speech communication is a very important and specialized area of digital communications. Human speech is characterized by unique statistical properties; one such property is illustrated in Figure 2.17. The abscissa represents speech signal magnitudes, normalized to the root-mean-square (rms) value of such magnitudes through a typical communication channel, and the ordinate is probability. For most voice communication channels, very low speech volumes predominate; 50% of the time, the voltage characterizing detected speech energy is less than one-fourth of the rms value. Large amplitude values are relatively rare; only 15% of the time does the voltage exceed the rms value. We see from Equation (2.18b) that the quantization noise depends on the step size (size of the quantile interval). When the steps are uniform in size the quantization is known as *uniform quantization*. Such a system would be wasteful for speech signals; many of the quantizing steps would rarely be used. In a system that uses equally spaced quantization levels, the quantization noise is the same for all signal magnitudes. Therefore, with uniform quantization, the signal-to-noise ratio (SNR) is worse for low-level signals than for high-level signals. *Nonuniform quantization* can provide fine quantization of the weak signals and coarse quantization of the strong signals. Thus in the case of nonuniform quantization, quantization noise can be made

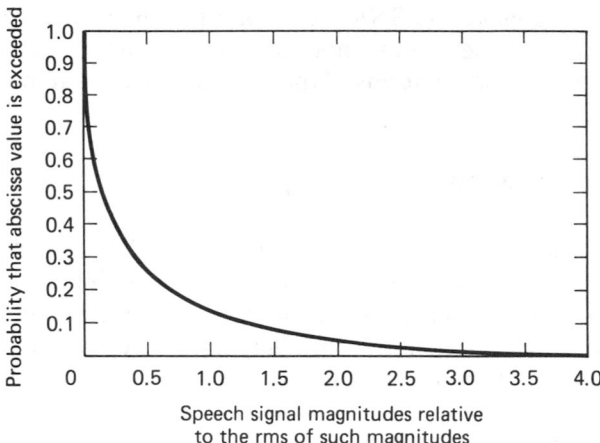

Figure 2.17 Statistical distribution of single-talker speech signal magnitudes.

proportional to signal size. The effect is to improve the overall SNR by reducing the noise for the predominant weak signals, at the expense of an increase in noise for the rarely occurring strong signals. Figure 2.18 compares the quantization of a strong versus a weak signal for uniform and nonuniform quantization. The staircase-like waveforms represent the approximations to the analog waveforms (after quantization distortion has been introduced). The SNR improvement that nonuniform quantization provides for the weak signal should be apparent. Nonuni-

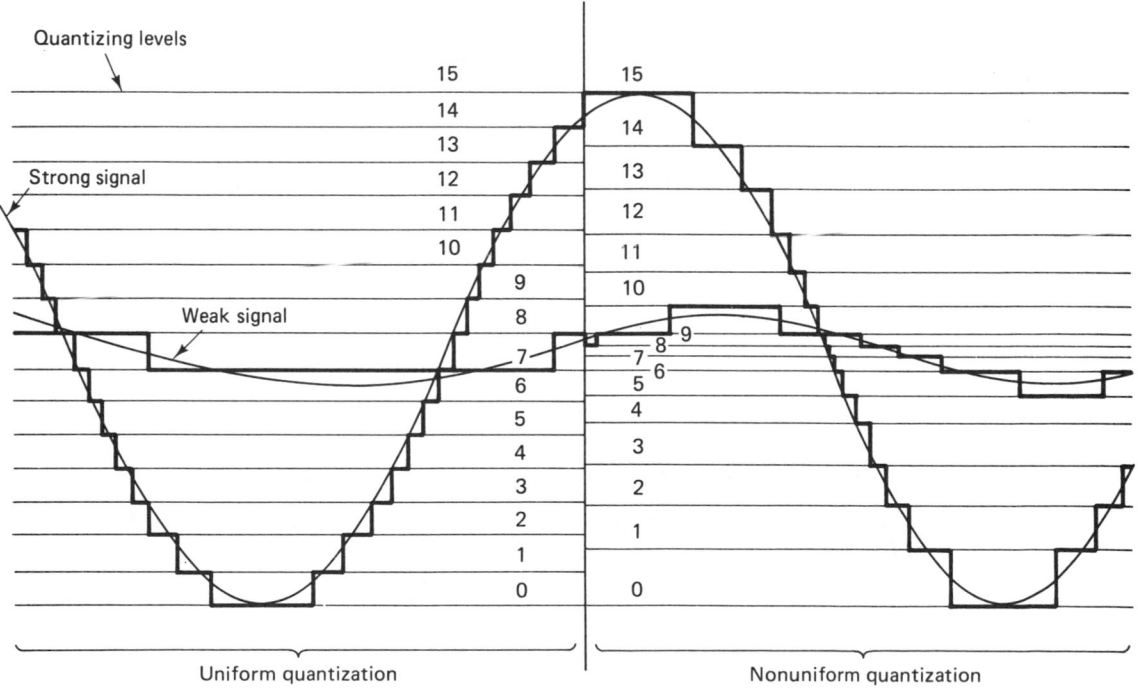

Figure 2.18 Uniform and nonuniform quantization of signals.

form quantization can be used to make the SNR a constant for all signals within the input range. For voice signals, the typical input signal dynamic range is 40 decibels (dB), where a decibel is defined in terms of the ratio of power P_2 to power P_1:

$$\text{number of dB} = 10 \log_{10} \frac{P_2}{P_1} \tag{2.21}$$

With a uniform quantizer, weak signals would experience a 40-dB-poorer SNR than that of strong signals. The standard telephone technique of handling the large range of possible input signal levels is to use a *logarithmic-compressed* quantizer instead of a uniform one. With such a nonuniform compressor the output SNR is independent of the distribution of input signal levels.

2.7.2 Nonuniform Quantization

One way of achieving nonuniform quantization is to use a nonuniform quantizer characteristic, shown in Figure 2.19a. More often, nonuniform quantization is achieved by first distorting the original signal with a logarithmic compression characteristic, as shown in Figure 2.19b, and then using a uniform quantizer. For small magnitude signals the compression characteristic has a much steeper slope than for large magnitude signals. Thus a given signal change at small magnitudes will carry the uniform quantizer through more steps than the same change at large

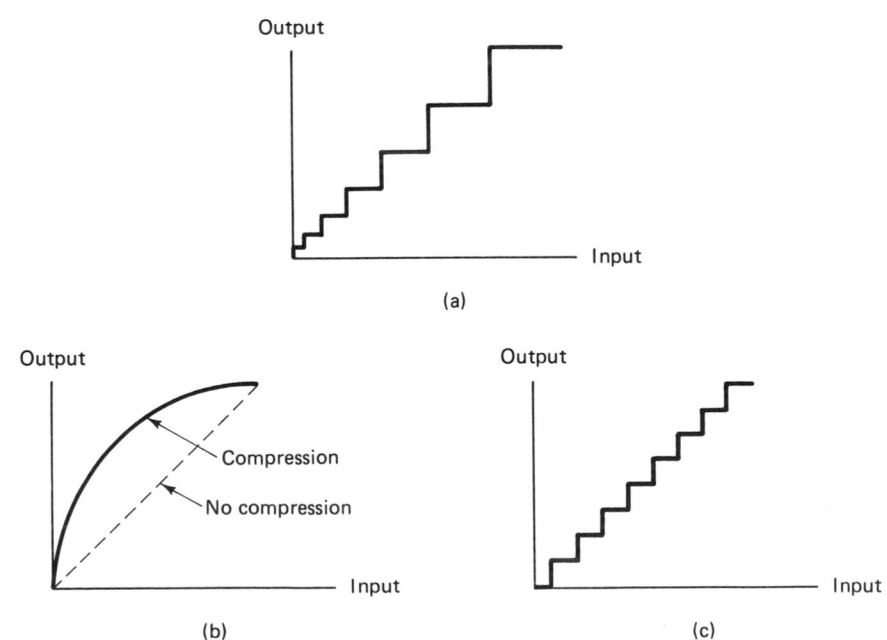

Figure 2.19 (a) Nonuniform quantizer characteristic. (b) Compression characteristic. (c) Uniform quantizer characteristic.

magnitudes. The compression characteristic effectively changes the distribution of the input signal magnitudes so that there is not a preponderance of *low* magnitude signals at the output of the compressor. After compression, the distorted signal is used as the input to a uniform (linear) quantizer characteristic, shown in Figure 2.19c. At the receiver, an inverse compression characteristic, called *expansion*, is applied so that the overall transmission is not distorted. The processing pair (compression and expansion) is usually referred to as *companding*.

2.7.3 Companding Characteristics

The early PCM systems implemented a smooth logarithmic compression function. Today, most PCM systems use a piecewise linear approximation to the logarithmic compression characteristic. In North America a μ-law compression characteristic is used:

$$y = y_{max} \frac{\log_e[1 + \mu(|x|/x_{max})]}{\log_e(1 + \mu)} \text{ sgn } x \qquad (2.22)$$

where

$$\text{sgn } x = \begin{cases} +1 & \text{for } x \geq 0 \\ -1 & \text{for } x < 0 \end{cases}$$

and where μ is a positive constant, x and y represent input and output voltages, and x_{max} and y_{max} are the maximum positive excursions of the input and output voltages, respectively. The compression characteristic is shown in Figure 2.20a for several values of μ. The standard value for μ is 255. Notice that $\mu = 0$ corresponds to linear amplification (uniform quantization).

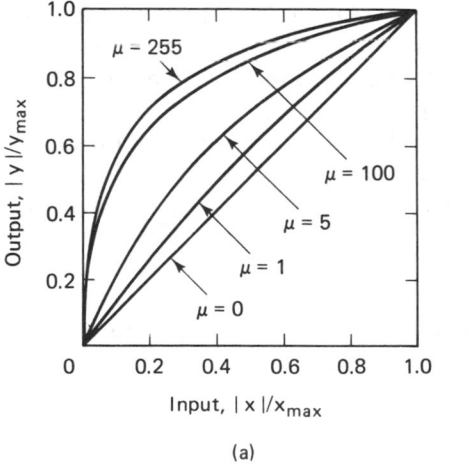

(a)

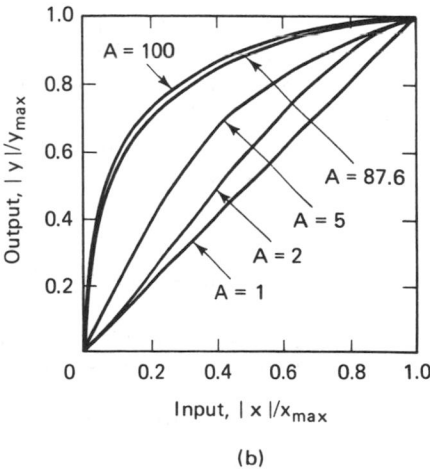

(b)

Figure 2.20 Compression characteristics. (a) μ-law characteristic. (b) A-law characteristic.

Another compression characteristic, used mainly in Europe, is the A-law characteristic, defined as

$$
y = \begin{cases}
y_{\max} \dfrac{A(\, |x|\, /x_{\max})}{1 + \log_e A} \, \mathrm{sgn}\, x & 0 < \dfrac{|x|}{x_{\max}} \leq \dfrac{1}{A} \\[4mm]
y_{\max} \dfrac{1 + \log_e[A(\, |x|\, /x_{\max})]}{1 + \log_e A} \, \mathrm{sgn}\, x & \dfrac{1}{A} < \dfrac{|x|}{x_{\max}} < 1
\end{cases}
\tag{2.23}
$$

where A is a positive constant and x and y are as defined in Equation (2.22). The A-law compression characteristic is shown in Figure 2.20b for several values of A. A standard value for A is 87.6. See Chapter 11 for a more detailed treatment of μ-law and A-law companding characteristics.

2.8 BASEBAND TRANSMISSION

2.8.1 Waveform Representation of Binary Digits

We need to represent PCM binary digits by electrical pulses in order to transmit them through a baseband channel. Such a representation is shown in Figure 2.21. Codeword time slots are shown in Figure 2.21a, where the codeword is a 4-bit representation of each quantized sample. In Figure 2.21b, each binary one is represented by a pulse and each binary zero is represented by the absence of a pulse. Thus a sequence of electrical pulses having the pattern shown in Figure 2.21b can be used to transmit the information in the PCM bit stream, and hence the information in the quantized samples of a message.

At the receiver, a determination must be made as to the presence or absence of a pulse in each bit time slot. It will be shown in Section 2.9 that the likelihood of correctly detecting the presence of a pulse is a function of the pulse energy (or area under the pulse). Thus there is an advantage in making the pulse width, T', in Figure 2.21b as wide as possible. If we increase the pulse width to the maximum possible (equal to the bit time duration, T), we have the waveform shown in Figure 2.21c. Rather than describe this waveform as a sequence of present or absent pulses, we can describe it as a sequence of transitions between two levels. When the waveform occupies the upper voltage level it represents a binary one; when it occupies the lower voltage level it represents a binary zero.

2.8.2 PCM Waveform Types

Figure 2.22 illustrates the most commonly used PCM waveforms. The various waveforms are classified into the following groups:

1. Nonreturn-to-zero (NRZ)
2. Return-to-zero (RZ)
3. Phase encoded
4. Multilevel binary

Formatting and Baseband Transmission   Chap. 2

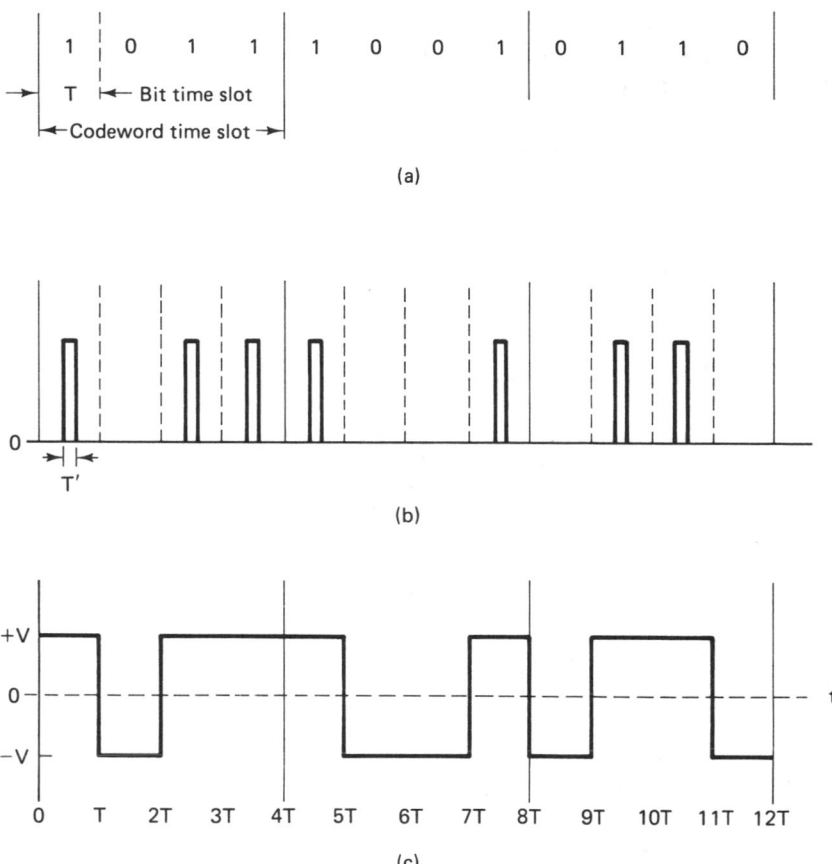

Figure 2.21 Example of waveform representation of binary digits. (a) PCM sequence. (b) Pulse representation of PCM. (c) Pulse waveform (transition between two levels).

The NRZ group is probably the most commonly used PCM waveform. It can be partitioned into the following subgroups: NRZ-L (L for level), NRZ-M (M for mark), and NRZ-S (S for space). NRZ-L is used extensively in digital logic. A binary one is represented by one level and a binary zero is represented by another level. There is a change in level whenever the data change from a one to a zero or from a zero to a one. With NRZ-M, the one, or *mark*, is represented by a change in level, and the zero, or *space*, is represented by no change in level. This is often referred to as *differential encoding*. NRZ-M is used primarily in magnetic tape recording. NRZ-S is the complement of NRZ-M: A one is represented by no change in level, and a zero is represented by a change in level.

The RZ waveforms consist of unipolar-RZ, bipolar-RZ, and RZ-AMI. These codes find application in baseband data transmission and in magnetic recording. With unipolar-RZ, a one is represented by a half-bit-wide pulse, and a zero is represented by the absence of a pulse. With bipolar-RZ, the ones and zeros are represented by opposite-level pulses that are one-half-bit wide. There is a pulse

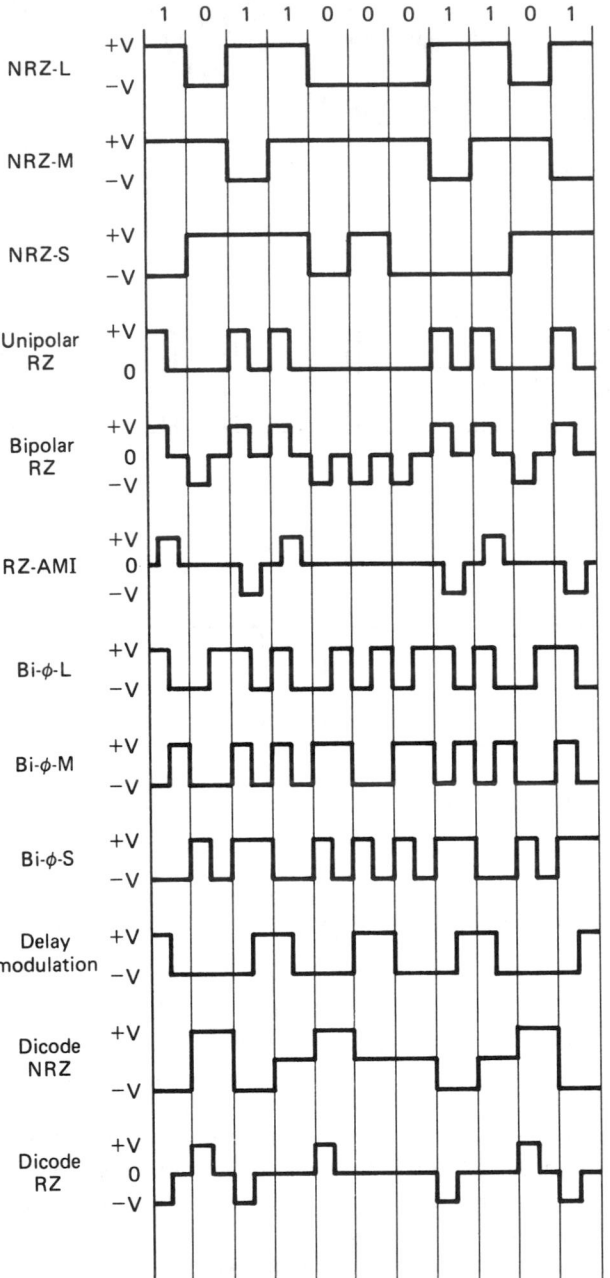

Figure 2.22 Various PCM waveforms.

present in each bit interval. RZ-AMI (AMI for "alternate mark inversion") is the coding scheme most often used in telemetry systems. The ones are represented by equal-amplitude alternating pulses. The zeros are represented by the absence of pulses.

The phase-encoded group consists of bi-ϕ-L (bi-phase-level), better known as *Manchester coding*; bi-ϕ-M (bi-phase-mark); bi-ϕ-S (bi-phase-space); and *delay modulation* (DM), or *Miller coding*. The phase-encoding schemes are used in magnetic recording systems and optical communications and in some satellite telemetry links. With bi-ϕ-L, a one is represented by a half-bit-wide pulse positioned during the first half of the bit interval; a zero is represented by a half-bit-wide pulse positioned during the second half of the bit interval. With bi-ϕ-M, a transition occurs at the beginning of every bit interval. A one is represented by a second transition one-half bit interval later; a zero is represented by no second transition. With bi-ϕ-S, a transition also occurs at the beginning of every bit interval. A one is represented by no second transition; a zero is represented by a second transition one-half bit interval later. With delay modulation [4], a one is represented by a transition at the midpoint of the bit interval. A zero is represented by no transition, unless it is followed by another zero. In this case, a transition is placed at the end of the bit interval of the first zero. Reference to the illustration in Figure 2.22 should help to make these descriptions clear.

Many binary waveforms use three levels, instead of two, to encode the binary data. Bipolar RZ and RZ-AMI belong to this group. The group also contains formats called *dicode* and *duobinary*. With dicode-NRZ, the one-to-zero or zero-to-one data transition changes the pulse polarity; without a data transition, the zero level is sent. With dicode-RZ, the one-to-zero or zero-to-one transition produces a half-duration polarity change; otherwise, a zero level is sent. The three-level duobinary signaling scheme is treated in Section 2.12.

One might ask why there are so many PCM waveforms. Are there really so many unique applications necessitating such a variety of waveforms to represent digits? The reason for the large selection relates to the differences in performance that characterize each waveform [5]. In choosing a coding scheme for a particular application, some of the parameters worth examining are the following:

1. *Dc component*. Eliminating the dc energy from the signal's power spectrum enables the system to be ac coupled. Magnetic recording systems, or systems using transformer coupling, have little sensitivity to very low frequency signal components. Thus low-frequency information could be lost.

2. *Self-Clocking*. Symbol or bit synchronization is required for any digital communication system. Some PCM coding schemes have inherent synchronizing or clocking features that aid in the recovery of the clock signal. For example, the Manchester code has a transition in the middle of every bit interval whether a one or a zero is being sent. This guaranteed transition provides a clocking signal.

3. *Error detection*. Some schemes, such as duobinary, provide the means of detecting data errors without introducing additional error-detection bits into the data sequence.

4. *Bandwidth compression*. Some schemes, such as multilevel codes, increase the efficiency of bandwidth utilization by allowing a reduction in required bandwidth for a given data rate; thus there is more information transmitted per unit bandwidth.

5. *Differential encoding.* This technique is useful because it allows the polarity of differentially encoded waveforms to be inverted without affecting the data detection. In communication systems where waveforms sometimes experience inversion, this is a great advantage. Differential encoding is treated in greater detail in Section 3.6.2.

6. *Noise immunity.* The various PCM waveform types can be further characterized by probability of bit error versus signal-to-noise ratio. Some of the schemes are more immune than others to noise. For example, the NRZ waveforms have better error performance than does the unipolar RZ waveform.

2.8.3 Spectral Attributes of PCM Waveforms

The most common criteria used for comparing PCM waveforms and for selecting one waveform type from the many available are: spectral characteristics, bit synchronization capabilities, error-detecting capabilities, interference and noise immunity, and cost and complexity of implementation. Figure 2.23 shows the spectral characteristics of some of the most popular PCM waveforms. The figure plots power spectral density in watts/hertz versus normalized bandwidth (frequency times pulse width). The spectral characteristic of a PCM waveform establishes the required system bandwidth and indicates how efficiently the bandwidth is being used. Bandwidth efficiency is addressed in detail in Chapter 7. The features that are easily observed in Figure 2.23 are the energy content at low frequency and the bandwidth requirements. Notice that the NRZ and duobinary schemes

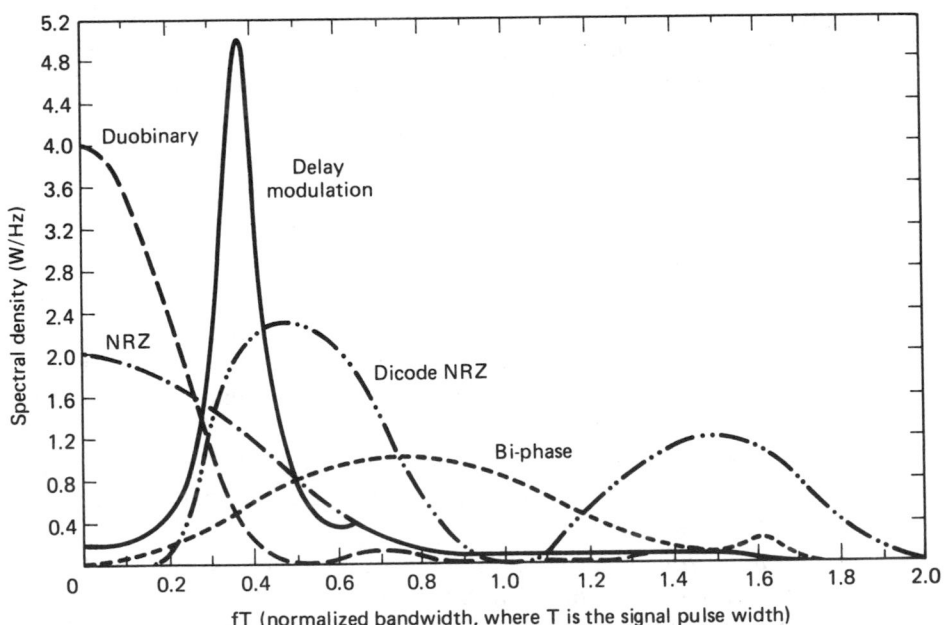

Figure 2.23 Spectral densities of various PCM waveforms.

have large spectral components at low frequency. Notice also that the bi-phase schemes have no energy at dc. However, bi-phase requires a relatively large system bandwidth, as does the dicode scheme. The methods that are particularly bandwidth efficient are the duobinary and delay modulation. Duobinary signaling is treated in Section 2.12.

2.9 DETECTION OF BINARY SIGNALS IN GAUSSIAN NOISE

Once the digital symbols are transformed into electrical waveforms, they can then be transmitted through the channel. During a given signaling interval, T, a binary system will transmit one of two waveforms, denoted $s_1(t)$ and $s_2(t)$. The transmitted signal over a symbol interval $(0, T)$ is represented by

$$s_i(t) = \begin{cases} s_1(t) & 0 \leq t \leq T & \text{for a binary 1} \\ s_2(t) & 0 \leq t \leq T & \text{for a binary 0} \end{cases}$$

The signal, $r(t)$, received by the receiver is represented by

$$r(t) = s_i(t) + n(t) \qquad i = 1, 2; \qquad 0 \leq t \leq T \tag{2.24}$$

where $n(t)$ is a zero-mean additive white Gaussian noise (AWGN) process.

Figure 2.24 highlights the *two separate* steps involved in signal detection. The *first step* consists of reducing the received waveform, $r(t)$ (whether baseband or bandpass), to a *single number*, $z(t = T)$. This operation can be performed by a linear filter followed by a sampler, as shown in block 1 of Figure 2.24, or optimally by a matched filter or correlator, which will be treated in later sections. The initial conditions of the filter or correlator are set to zero just prior to the arrival of each new symbol. At the end of a symbol duration, T, the output of block 1 yields the sample, $z(T)$, sometimes called the *test statistic*. We have assumed that the input noise is a Gaussian random process, and we have stated that the input filter is linear. A linear operation on a Gaussian random process will produce a second Gaussian random process [6]. Thus the filter output noise is Gaussian. If a nonlinear detector is used, the output noise will not be Gaussian

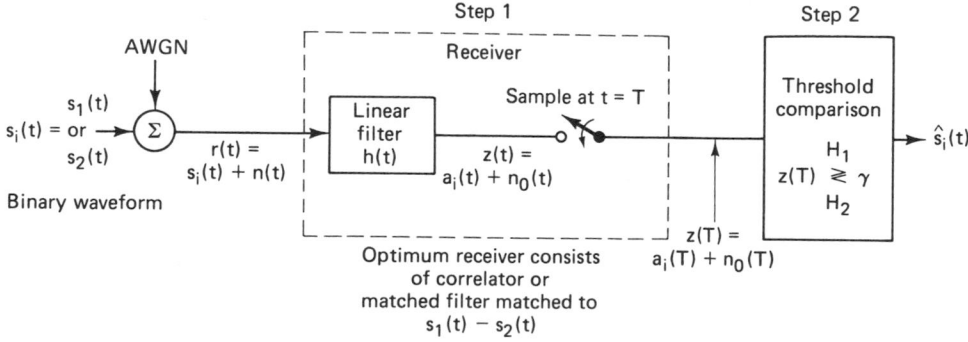

Figure 2.24   Two basic steps in digital signal detection.

and the following analysis will not apply. The output of block 1, sampled at $t = T$, yields

$$z(T) = a_i(T) + n_0(T) \qquad i = 1, 2 \qquad (2.25)$$

where $a_i(T)$ is the signal component of $z(T)$ and $n_0(T)$ is the noise component. To shorten the notation, we sometimes write Equation (2.25) as $z = a_i + n_0$. The noise component, n_0, is a zero-mean *Gaussian random variable*, and thus $z(T)$ is a *Gaussian random variable* with a mean of either a_1 or a_2 depending on whether a binary one or binary zero was sent. The probability density function (pdf) of the Gaussian random noise, n_0, can be expressed as

$$p(n_0) = \frac{1}{\sigma_0 \sqrt{2\pi}} \exp\left[-\frac{1}{2} \left(\frac{n_0}{\sigma_0} \right)^2 \right] \qquad (2.26)$$

where σ_0^2 is the noise variance. Thus it follows from Equations (2.25) and (2.26) that the conditional probability density functions (pdfs), $p(z|s_1)$ and $p(z|s_2)$ can be expressed as

$$p(z|s_1) = \frac{1}{\sigma_0 \sqrt{2\pi}} \exp\left[-\frac{1}{2} \left(\frac{z - a_1}{\sigma_0} \right)^2 \right] \qquad (2.27)$$

$$p(z|s_2) = \frac{1}{\sigma_0 \sqrt{2\pi}} \exp\left[-\frac{1}{2} \left(\frac{z - a_2}{\sigma_0} \right)^2 \right] \qquad (2.28)$$

These conditional pdfs are illustrated in Figure 2.25. The rightmost conditional pdf, $p(z|s_1)$, illustrates the probability density of the detector output, $z(T)$, given that $s_1(t)$ was transmitted. Similarly, the leftmost conditional pdf, $p(z|s_2)$, illustrates the probability density of $z(T)$ given that $s_2(t)$ was transmitted. The abscissa, $z(T)$, represents the full range of possible sample output values from block 1 of Figure 2.24.

The *second step* of the signal detection process consists of comparing the test statistic, $z(T)$, to a threshold level, γ, in block 2 of Figure 2.24, in order to estimate which signal, $s_1(t)$ or $s_2(t)$, has been transmitted. The filtering operation in block 1 does not depend on the decision criterion in block 2. Thus the choice of how best to implement block 1 can be independent of the particular decision strategy (choice of the threshold setting, γ).

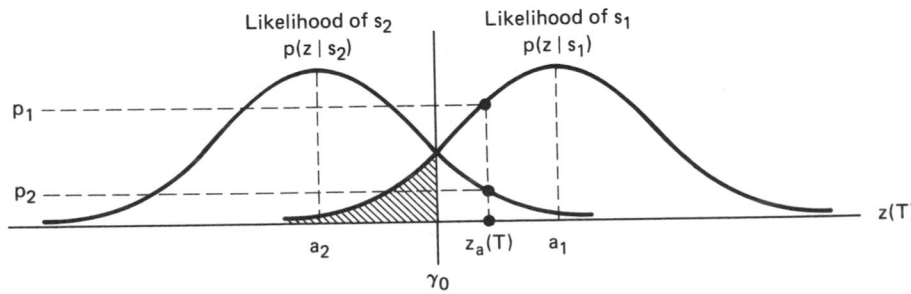

Figure 2.25 Conditional probability density functions: $p(z|s_1)$ and $p(z|s_2)$.

Formatting and Baseband Transmission   Chap. 2

Once a received waveform, $r(t)$, is transformed to a number $z(T)$, the actual shape of the waveform is no longer important; all waveform types that are transformed to the same value of $z(T)$ are identical for detection purposes. We will see in Section 2.9.2 that a *matched filter* receiver in block 1 of Figure 2.24 is one that maps all signals of equal energy into the same point, $z(T)$. Therefore, the *signal energy* (not its shape) is the important parameter in the detection process. Thus the detection analysis for *baseband signals* is the same as that for *bandpass signals*. The final step in block 2 is to make the decision

$$z(T) \underset{H_2}{\overset{H_1}{\gtrless}} \gamma \qquad (2.29)$$

where H_1 and H_2 are the two possible (binary) hypotheses. Choosing H_1 is equivalent to deciding that signal $s_1(t)$ was sent, and choosing H_2 is equivalent to deciding that signal $s_2(t)$ was sent. The inequality relationship indicates that hypothesis H_1 is chosen if $z(T) > \gamma$, and hypothesis H_2 is chosen if $z(T) < \gamma$. If $z(T) = \gamma$, the decision can be an arbitrary one.

2.9.1 Maximum Likelihood Receiver Structure

A popular criterion for choosing the threshold level, γ, for the binary decision is based on minimizing the probability of error. The computation for this *minimum error* value of $\gamma = \gamma_0$ starts with forming an inequality expression between the ratio of conditional probability density functions and the signal a priori probabilities. The conditional density function, $p(z|s_i)$, is also called the *likelihood* of s_i. Thus the formulation as shown below is called the *likelihood ratio test* (see Appendix B).

$$\frac{p(z|s_1)}{p(z|s_2)} \underset{H_2}{\overset{H_1}{\gtrless}} \frac{P(s_2)}{P(s_1)} \qquad (2.30)$$

where $P(s_1)$ and $P(s_2)$ are the a priori probabilities that $s_1(t)$ and $s_2(t)$, respectively, are transmitted, and H_1 and H_2 are the two possible hypotheses. The rule for minimizing the error probability in Equation (2.30) states that we should choose hypothesis H_1 if the ratio of likelihoods is greater than the ratio of a priori probabilities.

It is shown in Section B.3.1 that if $P(s_1) = P(s_2)$, and if the likelihoods, $p(z|s_i)$ $(i = 1, 2)$, are symmetrical, the substitution of Equations (2.27) and (2.28) into (2.30) yields

$$z(T) \underset{H_2}{\overset{H_1}{\gtrless}} \frac{a_1 + a_2}{2} = \gamma_0 \qquad (2.31)$$

where a_1 is the signal component of $z(T)$ when $s_1(t)$ is transmitted, and a_2 is the signal component of $z(T)$ when $s_2(t)$ is transmitted. The threshold level, γ_0, represented by $(a_1 + a_2)/2$, is the *optimum threshold* for minimizing the probability of making an incorrect decision for this important special case. This strategy is known as the *minimum error criterion*.

For equally likely signals, the optimum threshold, γ_0, passes through the intersection of the likelihood functions, as shown in Figure 2.25. Thus by following Equation (2.31), the decision stage effectively selects the hypothesis that corresponds to the signal with the *maximum likelihood*. For example, given an arbitrary detector output value, $z_a(T)$, for which there is a nonzero likelihood that $z_a(T)$ belongs to either signal class $s_1(t)$ or $s_2(t)$, one can think of the likelihood test as a comparison of the likelihood values $p(z_a|s_1)$ and $p(z_a|s_2)$. The signal corresponding to the maximum pdf is chosen as the most likely to have been transmitted. In other words, the detector chooses $s_1(t)$ if

$$p(z_a|s_1) > p(z_a|s_2) \tag{2.32}$$

Otherwise, the detector chooses $s_2(t)$. A detector that minimizes the error probability (for the case where the signal classes are equally likely) is also known as a *maximum likelihood detector*.

Figure 2.25 illustrates that Equation (2.32) is just a "common sense" way to make a decision when there exists statistical knowledge of the classes. Given the detector output value, $z_a(T)$, we see in Figure 2.25 that $z_a(T)$ intersects the likelihood of $s_1(t)$ at a value p_1, and it intersects the likelihood of $s_2(t)$ at a value p_2. What is the most reasonable decision for the detector to make? For this example, choosing class $s_1(t)$, which has the greater likelihood, is the most sensible choice. If this was an M-ary instead of a binary example, there would be a total of M likelihood functions representing the M signal classes to which a received signal might belong. The maximum likelihood decision would then be to choose the class that had the greatest likelihood of all M likelihoods. Refer to Appendix B for a review of decision theory fundamentals.

2.9.1.1 Error Probability

For the binary example in Figure 2.25, there are two ways in which errors can occur. An error, e, will occur when $s_1(t)$ is sent, and channel noise results in the receiver output signal, $z(T)$, being less than γ_0. The probability of such an occurrence is

$$P(e|s_1) = P(H_2|s_1) = \int_{-\infty}^{\gamma_0} p(z|s_1) \, dz \tag{2.33}$$

This is illustrated by the shaded area to the left of γ_0 in Figure 2.25. Similarly, an error occurs when $s_2(t)$ is sent, and the channel noise results in $z(T)$ being greater than γ_0. The probability of this occurrence is

$$P(e|s_2) = P(H_1|s_2) = \int_{\gamma_0}^{\infty} p(z|s_2) \, dz \tag{2.34}$$

The probability of an error is the sum of the probabilities of all the ways that an error can occur. For the binary case, we can express the probability of bit error, P_B, as follows:

$$P_B = \sum_{i=1}^{2} P(e, s_i) \tag{2.35}$$

Formatting and Baseband Transmission   Chap. 2

Combining Equations (2.33) to (2.35), we can write

$$P_B = P(e|s_1)P(s_1) + P(e|s_2)P(s_2) \tag{2.36a}$$

or equivalently,

$$P_B = P(H_2|s_1)P(s_1) + P(H_1|s_2)P(s_2) \tag{2.36b}$$

That is, given that signal $s_1(t)$ was transmitted, an error results if hypothesis H_2 is chosen; or given that signal $s_2(t)$ was transmitted, an error results if hypothesis H_1 is chosen. For the case where the a priori probabilities are equal, that is, $P(s_1) = P(s_2) = \frac{1}{2}$,

$$P_B = \frac{1}{2}P(H_2|s_1) + \frac{1}{2}P(H_1|s_2) \tag{2.37}$$

and because of the symmetry of the probability density functions

$$P_B = P(H_2|s_1) = P(H_1|s_2) \tag{2.38}$$

The probability of a bit error, P_B, is numerically equal to the area under the "tail" of either likelihood function, $p(z|s_1)$ or $p(z|s_2)$, falling on the "incorrect" side of the threshold. We can therefore compute P_B by integrating $p(z|s_1)$ between the limits $-\infty$ and γ_0, or as shown below, by integrating $p(z|s_2)$ between the limits γ_0 and ∞:

$$P_B = \int_{\gamma_0 = (a_1 + a_2)/2}^{\infty} p(z|s_2) \, dz \tag{2.39}$$

where $\gamma_0 = (a_1 + a_2)/2$ is the optimum threshold from Equation (2.31). Replacing the likelihood $p(z|s_2)$ with its Gaussian equivalent from Equation (2.28), we have

$$P_B = \int_{\gamma_0 = (a_1 + a_2)/2}^{\infty} \frac{1}{\sigma_0 \sqrt{2\pi}} \exp\left[-\frac{1}{2} \left(\frac{z - a_2}{\sigma_0} \right)^2 \right] dz \tag{2.40}$$

where σ_0^2 is the variance of the noise out of the correlator.

Let $u = (z - a_2)/\sigma_0$. Then $\sigma_0 \, du = dz$ and

$$P_B = \int_{u = (a_1 - a_2)/2\sigma_0}^{u = \infty} \frac{1}{\sqrt{2\pi}} \exp\left(-\frac{u^2}{2} \right) du = Q\left(\frac{a_1 - a_2}{2\sigma_0} \right) \tag{2.41}$$

where $Q(x)$, called the *complementary error function* or *co-error function*, is a commonly used symbol for the probability under the tail of the Gaussian distribution. It is defined as

$$Q(x) = \frac{1}{\sqrt{2\pi}} \int_x^{\infty} \exp\left(-\frac{u^2}{2} \right) du \tag{2.42}$$

Note that the co-error function is defined in several ways (see Appendix B); however, all definitions are essentially equivalent. $Q(x)$ cannot be evaluated in closed form. It is presented in tabular form in Table B.1. Good approximations

Sec. 2.9   Detection of Binary Signals in Gaussian Noise   **87** to $Q(x)$ by simpler functions can be found in Reference [7]. One such approximation, valid for $x > 3$, is

$$Q(x) \simeq \frac{1}{x\sqrt{2\pi}} \exp\left(-\frac{x^2}{2}\right)$$ (2.43)

We have optimized (in the sense of minimizing P_B) the threshold level, γ, but have not optimized the filter in block 1 of Figure 2.24; we next consider optimizing this filter by maximizing the argument of $Q(x)$ in Equation (2.41).

2.9.2 The Matched Filter

A matched filter is a linear filter designed to provide the maximum signal-to-noise power ratio at its output for a given transmitted symbol waveform. Consider that a known signal $s(t)$ plus AWGN, $n(t)$, is the input to a linear, time-invariant filter followed by a sampler, as shown in Figure 2.24. At time $t = T$, the receiver output, $z(T)$, consists of a signal component, a_i, and a noise component, n_0. The variance of the output noise (average noise power) is denoted by σ_0^2, so that the ratio of the instantaneous signal power to average noise power, $(S/N)_T$, at time $t = T$, out of the receiver in block 1, is

$$\left(\frac{S}{N}\right)_T = \frac{a_i^2}{\sigma_0^2}$$ (2.44)

We wish to find the filter transfer function, $H_0(f)$, that *maximizes* Equation (2.44). We can express the signal, $a(t)$, at the filter output, in terms of the filter transfer function, $H(f)$ (before optimization), and the Fourier transform of the input signal, as follows:

$$a(t) = \int_{-\infty}^{\infty} H(f)S(f)e^{j2\pi ft}\, df$$ (2.45)

where $S(f)$ is the Fourier transform of the input signal, $s(t)$. If the two-sided power spectral density of the input noise is $N_0/2$ watts/hertz, then using Equations (1.19) and (1.53), we can express the output noise power, σ_0^2, as

$$\sigma_0^2 = \frac{N_0}{2} \int_{-\infty}^{\infty} |H(f)|^2\, df$$ (2.46)

We then combine Equations (2.44) to (2.46) to express $(S/N)_T$, as follows:

$$\left(\frac{S}{N}\right)_T = \frac{\left| \int_{-\infty}^{\infty} H(f)S(f)e^{j2\pi fT}\, df \right|^2}{N_0/2 \int_{-\infty}^{\infty} |H(f)|^2\, df}$$ (2.47)

We next find that value of $H(f) = H_0(f)$ for which the maximum $(S/N)_T$ is achieved, by using *Schwarz's inequality*. One form of the inequality can be stated as

$$\left| \int_{-\infty}^{\infty} f_1(x) f_2(x) \, dx \right|^2 \leq \int_{-\infty}^{\infty} |f_1(x)|^2 \, dx \int_{-\infty}^{\infty} |f_2(x)|^2 \, dx \qquad (2.48)$$

The equality holds if $f_1(x) = kf_2^*(x)$, where k is an arbitrary constant and * indicates complex conjugate. If we identify $H(f)$ with $f_1(x)$ and $S(f) e^{j2\pi fT}$ with $f_2(x)$, we can write

$$\left| \int_{-\infty}^{\infty} H(f) S(f) e^{j2\pi fT} \, df \right|^2 \leq \int_{-\infty}^{\infty} |H(f)|^2 \, df \int_{-\infty}^{\infty} |S(f)|^2 \, df \qquad (2.49)$$

Substituting into Equation (2.47) yields

$$\left(\frac{S}{N}\right)_T \leq \frac{2}{N_0} \int_{-\infty}^{\infty} |S(f)|^2 \, df \qquad (2.50)$$

or

$$\max \left(\frac{S}{N}\right)_T = \frac{2E}{N_0} \qquad (2.51)$$

where the energy, E, of the input signal $s(t)$ is

$$E = \int_{-\infty}^{\infty} |S(f)|^2 \, df \qquad (2.52)$$

Thus the maximum output $(S/N)_T$ depends on the input *signal energy* and the power spectral density of the noise, *not on the particular shape* of the waveform that is used.

The equality in Equation (2.51) holds only if the optimum filter transfer function, $H_0(f)$, is employed, such that

$$H(f) = H_0(f) = kS^*(f)e^{-j2\pi fT} \qquad (2.53)$$

or

$$h(t) = \mathcal{F}^{-1}\{kS^*(f)e^{-j2\pi fT}\} \qquad (2.54)$$

Since $s(t)$ is a real-valued signal, we can write from Equations (A.29) and (A.31),

$$h(t) = \begin{cases} ks(T - t) & 0 \leq t \leq T \\ 0 & \text{elsewhere} \end{cases} \qquad (2.55)$$

Thus the impulse response of a filter that produces the maximum output signal-to-noise ratio is the mirror image of the message signal, $s(t)$, *delayed* by the symbol time duration, T. Note that the delay of T seconds makes Equation (2.55) *causal*; that is, the delay of T seconds makes $h(t)$ a function of positive time in the interval $0 \leq t \leq T$. Without the delay of T seconds, the response, $s(-t)$, is unrealizable because it describes a response as a function of negative time.

2.9.3 Correlation Realization of the Matched Filter

The term *matched filter* is often used synonymously with *product integrator* or *correlator*. Equation (2.55) and Figure 2.26a illustrate the matched filter's basic property: The impulse response of the filter is a delayed version of the mirror image (rotated on the $t = 0$ axis) of the signal waveform. Therefore, if the signal waveform is $s(t)$, its mirror image is $s(-t)$, and the mirror image delayed by T seconds is $s(T - t)$. The output, $z(t)$, of a causal filter can be described in the time domain as the convolution of a received input waveform, $r(t)$, with the impulse response of the filter (see Section A.5):

$$z(t) = r(t) * h(t) = \int_0^t r(\tau)h(t - \tau)\, d\tau \tag{2.56}$$

Substituting $h(t)$ of Equation (2.55) into $h(t - \tau)$ of Equation (2.56) and arbitrarily setting the constant k equal to unity, we get

$$z(t) = \int_0^t r(\tau)s[T - (t - \tau)]\, d\tau$$

$$= \int_0^t r(\tau)s(T - t + \tau)\, d\tau \tag{2.57}$$

When $t = T$, we can write Equation (2.57) as

$$z(T) = \int_0^T r(\tau)s(\tau)\, d\tau \tag{2.58}$$

The operation of Equation (2.58), the product integration of the received signal, $r(t)$, with a replica of the transmitted waveform, $s(t)$, over one symbol interval is known as the *correlation* of $r(t)$ with $s(t)$. Consider that a received signal, $r(t)$, is correlated with each prototype signal, $s_i(t)$ $(i = 1, \ldots, M)$, using a bank of M correlators. The signal $s_i(t)$ whose product integration or correlation with $r(t)$ yields the maximum output $z_i(T)$ is the signal that matches $r(t)$ better than all the other $s_j(t)$, $j \neq i$. We will subsequently use this correlation characteristic for the optimum detection of signals.

2.9.3.1 Comparison of Convolution and Correlation

It is important to note that the correlator output and the matched filter output are the same *only at time* $t = T$. For a sine-wave input, the output of the correlator, $z(t)$, is approximately a linear ramp for $0 \leq t \leq T$. However, the matched filter output is approximately a sine-wave amplitude modulated by a linear ramp for $0 \leq t \leq T$. The comparison is shown in Figure 2.26b. To understand the similarities and differences between a matched filter and a product integrator, one might first ask: What are the similarities between *convolution* as expressed in Equation (2.56) and *correlation* as expressed in Equation (2.58)? With correlation, we simply multiply two functions together and integrate (compute the area under their product curve). We are calculating how closely two waveforms *match each other* in a given time period. With convolution, we sweep (step) two functions past one

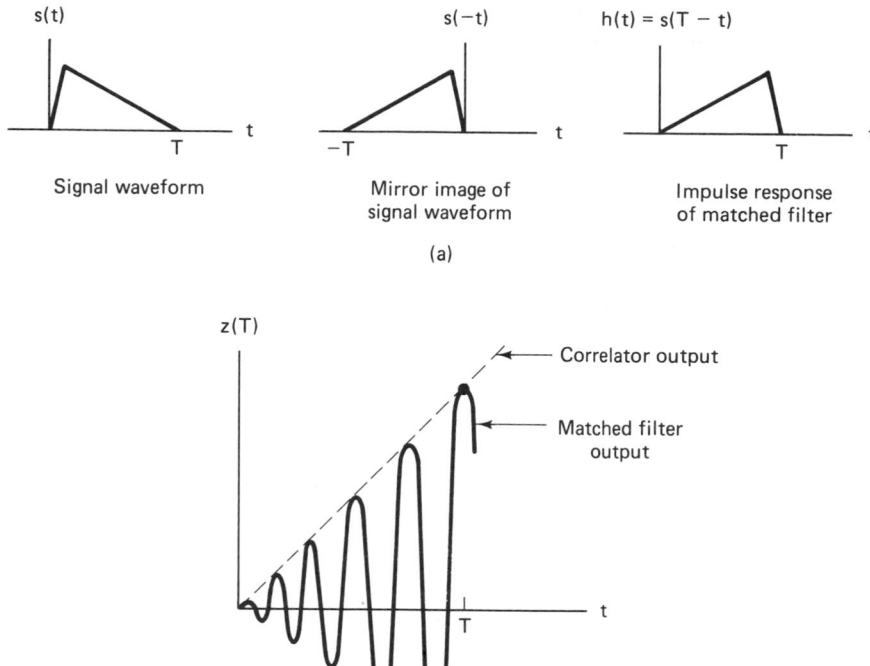

Figure 2.26 Correlator and matched filter. (a) Matched filter characteristic. (b) Comparison of correlator and matched filter outputs.

another and calculate a sequence of correlations (one for each step). The matched filter, used as a demodulator, only utilizes the correlation made at the symbol duration, T. Since the matched filter output and the correlator output are identical at the sampling time $t = T$, the matched filter and correlator functions, pictured in Figure 2.27, are used interchangeably.

2.9.4 Application of the Matched Filter

In Equation (2.41) we found that the optimum decision threshold resulted in $P_B = Q[(a_1 - a_2)/2\sigma_0]$. Finding the optimum threshold alone is not sufficient to optimize the detection process. To minimize P_B, we also need to select an optimum filter to maximize the argument of $Q(x)$. Thus we need to determine the linear filter that maximizes $(a_1 - a_2)/2\sigma_0$, or equivalently, that maximizes

$$\frac{(a_1 - a_2)^2}{\sigma_0^2} \tag{2.59}$$

where $(a_1 - a_2)$ is the difference of the signal components at the filter output, at time $t = T$, and the square of this difference signal is the instantaneous power

Sec. 2.9   Detection of Binary Signals in Gaussian Noise

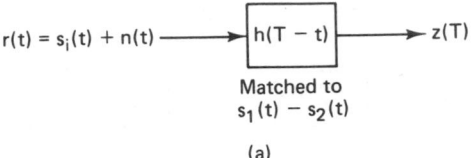

Matched to
$s_1(t) - s_2(t)$

(a)

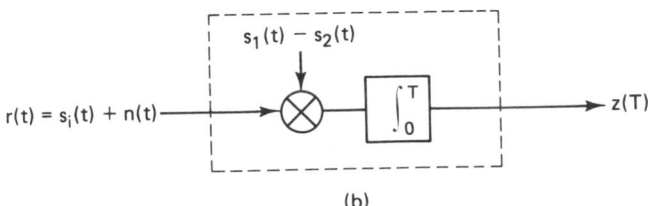

(b)

Figure 2.27 Equivalence of matched filter and correlator. (a) Matched filter. (b) Correlator.

of the difference signal. In Section 2.9.2 we described a filter that maximizes the output signal-to-noise ratio—the matched filter. Consider a filter that is *matched* to the input difference signal $[s_1(t) - s_2(t)]$. From Equations (2.44) and (2.51), the ratio of the instantaneous signal power to average noise power, $(S/N)_T$, at time $t = T$ out of this matched filter can be expressed as

$$\left(\frac{S}{N}\right)_T = \frac{(a_1 - a_2)^2}{\sigma_0^2} = \frac{2E_d}{N_0} \tag{2.60}$$

where $N_0/2$ is the two-sided power spectral density of the noise at the filter input, and E_d is the energy of the difference signal at the filter input:

$$E_d = \int_0^T [s_1(t) - s_2(t)]^2 \, dt \tag{2.61}$$

Thus, using Equations (2.41) and (2.60), we have

$$P_B = Q\left(\sqrt{\frac{E_d}{2N_0}}\right) \tag{2.62}$$

2.9.5 Error Probability Performance of Binary Signaling

2.9.5.1 Unipolar Signaling

Figure 2.28a illustrates an example of a baseband waveform used for unipolar signaling where

$$s_1(t) = A \quad 0 \le t \le T \quad \text{for binary 1} \tag{2.63}$$
$$s_2(t) = 0 \quad 0 \le t \le T \quad \text{for binary 0}$$

where $A > 0$ is the amplitude of signal $s_1(t)$. Assume that the unipolar signal plus white Gaussian noise is present at the input of a matched filter, with sampling time $t = T$. The correlator detector for such a signal type is shown in Figure 2.28b. The correlator multiplies and integrates the incoming signal, $r(t)$, with the

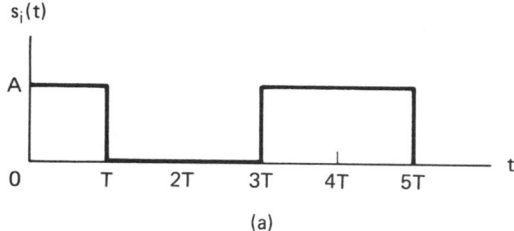

(a)

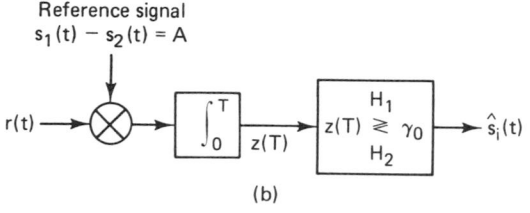

Reference signal
$s_1(t) - s_2(t) = A$

(b)

Figure 2.28 Detection of unipolar baseband signaling. (a) Unipolar signaling example. (b) Correlator detector.

difference of the prototype signals $[s_1(t) - s_2(t)] = A$, and after a symbol duration, T, compares the result, $z(T)$, with the threshold, γ_0. When $r(t) = s_1(t) + n(t)$, the signal component, $a_1(T)$, of $z(T)$ is found, using Equation (2.58), to be

$$a_1(T) = \mathbf{E}\{z(T)\} = \mathbf{E}\left\{ \int_0^T A^2 + An(t)\, dt \right\} = A^2T$$

where $\mathbf{E}\{\cdot\}$ is the *expected value operator*. This follows since $\mathbf{E}\{n(t)\} = 0$. Similarly, when $r(t) = s_2(t) + n(t)$, then $a_2(T) = 0$. Thus the optimum threshold is $\gamma_0 = (a_1 + a_2)/2 = \frac{1}{2}A^2T$. If the correlator output, $z(T)$, is greater than γ_0, the signal is declared to be $s_1(t)$; otherwise, it is declared to be $s_2(t)$.

The energy difference signal, from Equation (2.61), is $E_d = A^2T$. Then the bit error performance at the output is obtained from Equation (2.62) as follows:

$$P_B = Q\left(\sqrt{\frac{E_d}{2N_0}} \right) = Q\left(\sqrt{\frac{A^2T}{2N_0}} \right) = Q\left(\sqrt{\frac{E_b}{N_0}} \right) \tag{2.64}$$

where the average energy per bit is $E_b = A^2T/2$.

2.9.5.2 Bipolar Signaling

Figure (2.29a) illustrates an example of a bipolar baseband waveform, where

$$
\begin{aligned}
s_1(t) &= +A \quad 0 \le t \le T \quad \text{for binary 1} \\
s_2(t) &= -A \quad 0 \le t \le T \quad \text{for binary 0}
\end{aligned}
\tag{2.65}
$$

Binary waveforms that are the negative of one another, such as the bipolar pair above, where $s_1(t) = -s_2(t)$, are called *antipodal signals*. A correlator receiver for this antipodal type of waveform can be configured as shown in Figure 2.29b. One correlator multiplies and integrates the incoming signal $r(t)$ with the prototype signal, $s_1(t)$; the second correlator multiplies and integrates $r(t)$ with $s_2(t)$. The

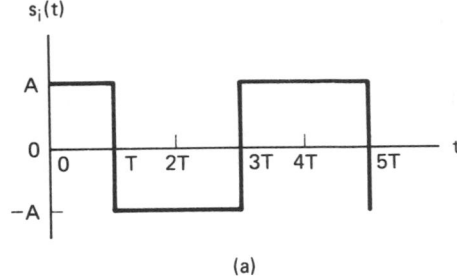

(a)

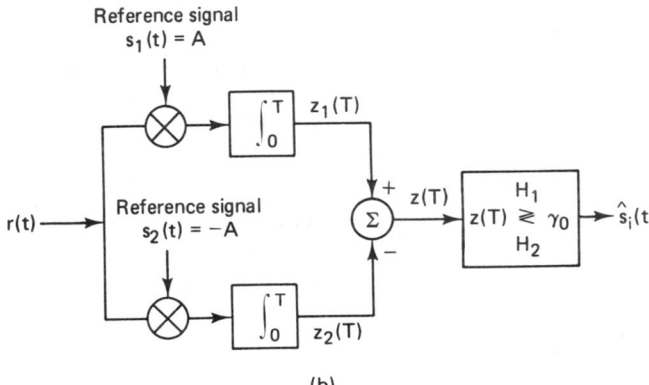

(b)

Figure 2.29 Detection of bipolar baseband signaling. (a) Bipolar signaling example. (b) Correlator detector.

correlator outputs are designated $z_i(T)$ ($i = 1, 2$). The point in the decision space, $z(T)$, is formed from the difference of the correlator outputs, as follows:

$$z(T) = z_1(T) - z_2(T) \tag{2.66}$$

and the decision is made according to Equation (2.31). For antipodal signals, $a_1 = -a_2$; therefore, $\gamma_0 = 0$. Thus if the *test statistic*, $z(T)$, is positive, the signal is declared to be $s_1(t)$, and if it is negative, it is declared to be $s_2(t)$.

The energy difference signal, $E_d = (2A)^2 T$. Then the bit error performance from Equation (2.62) is

$$P_B = Q\left(\sqrt{\frac{2A^2 T}{N_0}} \right) = Q\left(\sqrt{\frac{2E_b}{N_0}} \right) \tag{2.67}$$

where the average energy per bit is $E_b = A^2 T$. Figure 2.30 illustrates curves of P_B versus E_b/N_0 for unipolar and bipolar signaling. In examining the two curves, we can see a 3-dB error performance improvement for bipolar compared to unipolar signaling. This difference could have been predicted by the factor-of-2 difference in the coefficient of E_b in Equation (2.67) compared with Equation (2.64). In Chapter 3 we shall see that the error performance of *bandpass antipodal signaling* (e.g., coherently detected binary phase shift keying) is the same as that for *baseband antipodal signaling* (matched filter reception). Also, we shall see that the error performance of *bandpass orthogonal signaling* (e.g., coherently detected

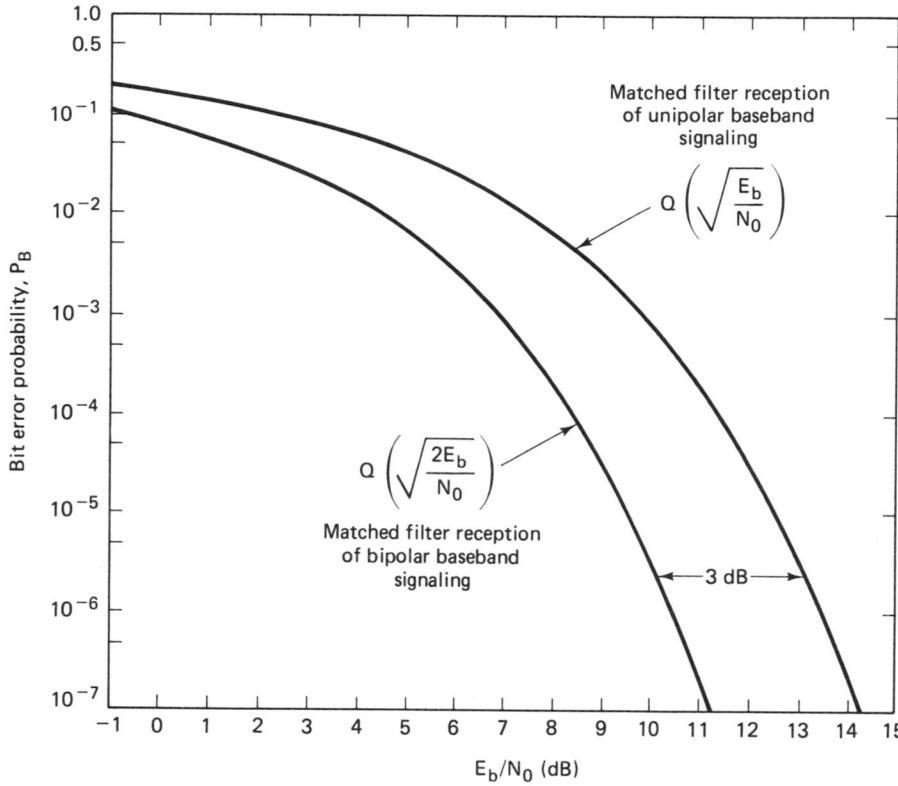

Figure 2.30 Bit error performance of unipolar and bipolar signaling.

frequency shift keying) is the same as that for *baseband unipolar signaling* (matched filter reception).

2.10 MULTILEVEL BASEBAND TRANSMISSION

The system bandwidth required for binary PCM signaling may be very large. What might we do to reduce the required bandwidth? One possibility is to use *multilevel signaling*. Consider a binary PCM bit stream with data rate R bits per second. Instead of transmitting a pulse waveform for each bit, we first partition the data into k-bit groups. We then use $M = 2^k$-level pulses for transmission. Each pulse waveform can now represent a k-bit symbol in a symbol stream of rate R/k symbols per second. Thus multilevel signaling, where $M > 2$, can be used to reduce the number of symbols transmitted per second, or thus to reduce the bandwidth requirements of the channel. Is there a price to be paid for such bandwidth reduction? Of course there is; it is discussed below.

Consider the task that the pulse receiver must perform; it needs to distinguish between the possible levels of each pulse. Can the receiver distinguish among the eight possible levels of each octal pulse in Figure 2.31a as easily as it can distin-

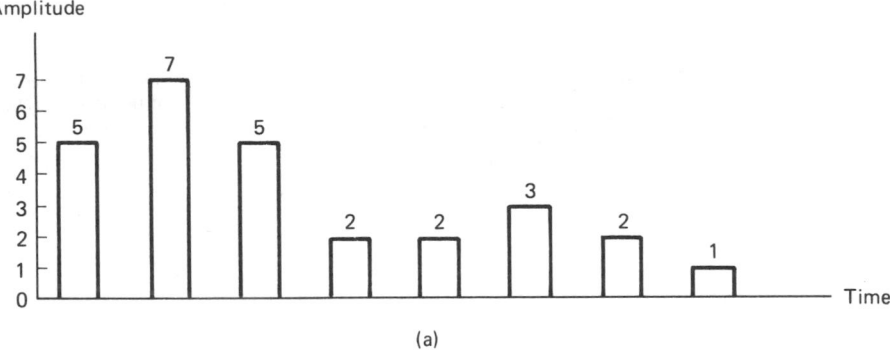

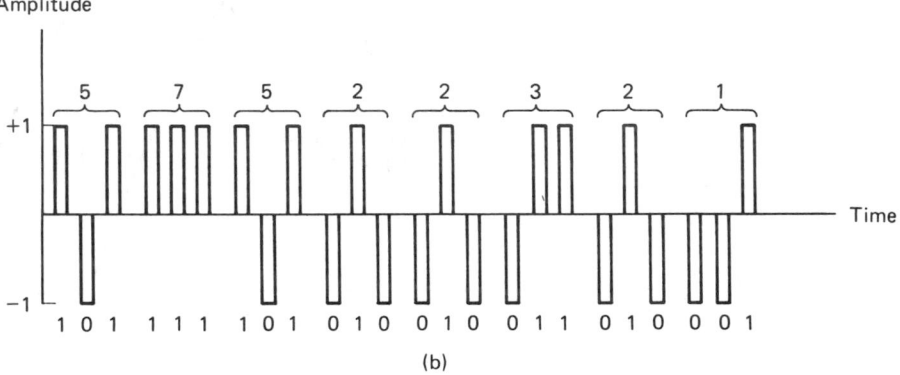

Figure 2.31 Pulse code modulation signaling. (a) Eight-level signaling. (b) Two-level signaling.

guish between the two possible levels of each binary pulse in Figure 2.31b? The transmission of an 8-level (compared to a 2-level) pulse requires a greater amount of energy for equivalent detection performance. (It is the amount of signal energy that determines how reliably a signal will be detected.) For equal average power in the binary and the octal pulses, it is easier to detect the binary pulses because the detector has more signal energy per level for making a binary decision than an 8-level decision. What price does a system designer pay if he or she chooses the transmission waveform to be the easier-to-detect binary PCM, rather than eight-level PCM? The engineer pays the price of needing three times as much system bandwidth for a given data rate, compared to the octal pulses, since each octal pulse must be replaced with three binary pulses (each one-third as wide as the octal pulses). One might ask: Why not use binary pulses with the same pulse duration as the original octal pulses, and suffer the information delay? For some cases this might be appropriate, but for most communication systems, such an increase in delay cannot be tolerated; the six o'clock news *must* be received at six o'clock. In Chapter 7 we examine in detail the trade-off between signal power and system bandwidth.

2.10.1 PCM Word Size

How many bits shall we assign to each analog sample? For digital telephone channels, each speech sample is PCM encoded using 8 bits, yielding 2^8 or 256 levels per sample. The choice of the number of levels, or bits per sample, depends on how much distortion we are willing to tolerate with the PCM format. It is useful to develop a general relationship between the required number of bits per analog sample (the PCM word size) and the allowable quantization distortion. Let the magnitude of the quantization distortion error, $|e|$, be specified not to exceed a fraction, p, of the peak-to-peak analog voltage, V_{pp}, as follows:

$$|e| \leq pV_{pp} \tag{2.68}$$

Since the quantization error can be no larger than $q/2$, where q is the quantile interval, we can write

$$|e|_{max} = \frac{q}{2} = \frac{V_{pp}}{2L} \tag{2.69}$$

where L is the number of quantization levels. Then

$$\frac{V_{pp}}{2L} \leq pV_{pp} \tag{2.70}$$

$$2^\ell = L \geq \frac{1}{2p} \text{ levels} \tag{2.71}$$

$$\ell \geq \log_2 \frac{1}{2p} \text{ bits} \tag{2.72}$$

It is important that we do not confuse the idea of bits per PCM word, denoted by ℓ in Equation (2.72), with the M-level transmission concept of k data bits per symbol. The following example should clarify the distinction.

Example 2.3 Quantization Levels and Multilevel Signaling

The information in an analog waveform, with maximum frequency $f_m = 3$ kHz, is to be transmitted over an M-level PCM system, where the number of pulse levels is $M = 16$. The quantization distortion is specified not to exceed $\pm 1\%$ of the peak-to-peak analog signal.

(a) What is the minimum number of bits/sample, or bits/PCM word, that should be used in this PCM system?

(b) What is the minimum required sampling rate, and what is the resulting bit transmission rate?

(c) What is the PCM pulse or symbol transmission rate?

In this example we are concerned with two types of *levels*: the number of quantization levels for fulfilling the distortion requirement, and the 16 levels of the multilevel PCM pulses.

Solution

(a) Using Equation (2.72), we calculate

$$\ell \geq \log_2 \frac{1}{0.02} = \log_2 50 \approx 5.6$$

Therefore, use $\ell = 6$ bits/sample to meet the distortion requirement.

(b) Using the Nyquist sampling criterion, the minimum sampling rate $f_s = 2f_m = 6000$ samples/second (samples/s). From part (a), each sample will give rise to a PCM word composed of 6 bits. Therefore, the bit transmission rate $R = \ell f_s = 36,000$ bits/s.

(c) Since multilevel pulses are to be used with $M = 2^k = 16$ levels, $k = \log_2 16 = 4$ bits/symbol. Therefore, the bit stream will be partitioned into groups of 4 bits to form the new 16-level PCM digits, and the resulting symbol transmission rate R_s is $R/k = 36,000/4 = 9000$ symbols/s.

2.11 INTERSYMBOL INTERFERENCE

Figure 2.32a highlights the major filtering aspects of a typical baseband digital system; there are circuit reactances throughout the system—in the transmitter, in the receiver, and in the channel. The pulses at the input might be impulse-like samples, or flat-top samples. In either case, they are low-pass filtered at the transmitter to confine them to some desired bandwidth. Channel reactances can cause amplitude and phase variations that distort the pulses. The receiving filter, called the *equalizing filter*, should be configured to compensate for the distortion

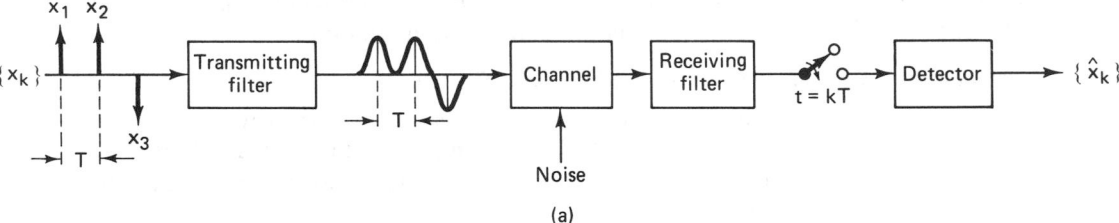

(a)

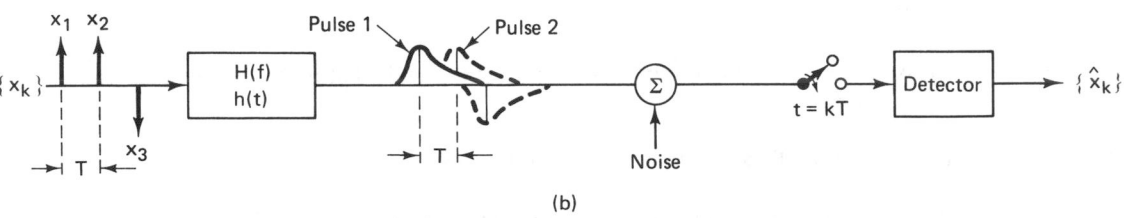

(b)

Figure 2.32 Intersymbol interference in the detection process. (a) Typical baseband digital system. (b) Equivalent model.

caused by the transmitter and the channel [8]. In a binary system with a commonly used PCM format, such as NRZ-L, the detector makes symbol decisions by comparing the received bipolar pulses to a threshold; for example, the detector decides that a binary one was sent if the received pulse is positive, and that a binary zero was sent if the received pulse is negative. Figure 2.32b illustrates a convenient model for the system, lumping all the filtering effects into one overall equivalent system transfer function, $H(f)$:

$$H(f) = H_t(f)H_c(f)H_r(f) \qquad (2.73)$$

where $H_t(f)$ characterizes the transmitting filter, $H_c(f)$ the filtering within the channel, and $H_r(f)$ the receiving or equalizing filter. The characteristic $H(f)$, then, represents the composite system transfer function due to all of the filtering at various locations throughout the transmitter/channel/receiver chain. Due to the effects of system filtering, the received pulses overlap one another as shown in Figure 2.32b; the tail of one pulse "smears" into adjacent symbol intervals so as to interfere with the detection process; such interference is termed *intersymbol interference* (ISI). Even in the absence of noise, imperfect filtering and system bandwidth constraints lead to ISI. In practice, $H_c(f)$ is usually specified, and the problem remains to determine $H_t(f)$ and $H_r(f)$ such that the ISI of the pulses are minimized at the output of $H_r(f)$.

Nyquist [9] investigated the problem of specifying a received pulse shape so that no ISI occurs at the detector. He showed that the theoretical minimum system bandwidth needed to detect R_s symbols/s, without ISI, is $R_s/2$ hertz. This occurs when the system transfer function, $H(f)$, is made rectangular, as shown in Figure 2.33a. When $H(f)$ is such an ideal filter with bandwidth $1/2T$, its impulse response, the inverse Fourier transform of $H(f)$ (from Table A.1) is $h(t) = $ sinc (t/T), shown in Figure 2.33b. Thus $h(t)$ is the received pulse shape resulting from the application of an impulse at the input of such an ideal system. Nyquist established that if each pulse of a received sequence is of the form $h(t)$, the pulses can be detected without ISI. The bandwidth required to detect $1/T$ such pulses (symbols) per second is equal to $1/2T$; in other words, a system with bandwidth $W = 1/2T = R_s/2$ hertz can support a maximum transmission rate of $2W = 1/T = R_s$ symbols/s (*Nyquist bandwidth constraint*) without ISI. Figure

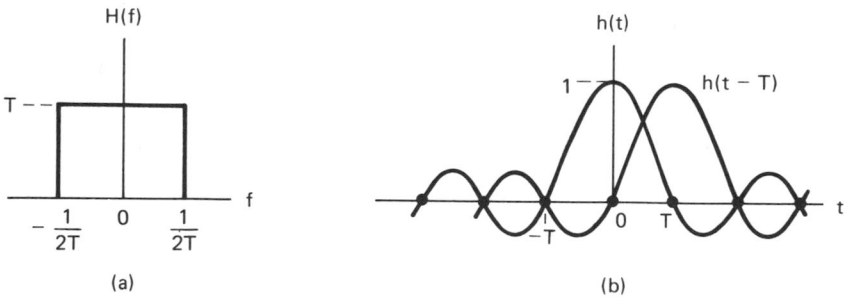

(a) (b)

Figure 2.33 Nyquist channels for zero ISI. (a) Rectangular system transfer function $H(f)$. (b) Received pulse shape $h(t) = $ sinc (t/T).

Sec. 2.11 Intersymbol Interference

2.33b illustrates how ISI is avoided. The figure shows two successive received pulses, $h(t)$ and $h(t - T)$. Even though $h(t)$ has a long tail, it passes through zero at the instant that $h(t - T)$ is sampled (at $t = T$) and therefore causes no degradation to the detection process. With such an ideal received pulse shape, the maximum possible symbol transmission rate per hertz, called the *symbol-rate packing*, is 2 symbols/s/Hz, without ISI.

What does the Nyquist bandwidth constraint say about the maximum number of *bits*/s/Hz that can be received without ISI? It says nothing about bits, directly. The constraint deals only with pulses or symbols, and the ability to detect their amplitude values without distortion from other pulses. The assignment of how many bits each symbol represents is a separate issue. In theory, each symbol can represent M levels or k bits ($M = 2^k$); as k or M increases in value, so does the complexity of the system. For example, when $k = 6$ bits/symbol, each symbol represents $M = 64$ levels. The number of bits/s/Hz that a system can support is referred to as the *bandwidth efficiency* of the system; this subject is treated separately in Chapter 7.

For most communication systems (with the exception of spread-spectrum systems, covered in Chapter 10), our goal is to reduce the required system bandwidth as much as possible; Nyquist has provided us with a basic limitation to such bandwidth reduction. What would happen if we tried to force a system to operate at smaller bandwidths than the constraint dictates? We would find that restricting the bandwidth would spread the pulses in time; this would degrade the system's error performance, due to the increase in ISI.

2.11.1 Pulse Shaping to Reduce ISI

The Nyquist requirement for a sinc (t/T) received pulse shape is not physically realizable since it dictates a rectangular bandwidth characteristic and an infinite time delay. Also, with such a characteristic, the detection process would be very sensitive to small timing errors. In Figure 2.33b the pulse $h(t)$ has zero value in adjacent pulse times *only* when the sampling is performed at exactly the correct sampling time; timing errors will produce ISI. Therefore, we cannot implement systems using the Nyquist bandwidth; we need to provide some "excess bandwidth" beyond the theoretical minimum. One frequently used system transfer function, $H(f)$, is called the *raised cosine filter*. It can be expressed as

$$H(f) = \begin{cases} 1 & \text{for } |f| < 2W_0 - W \\ \cos^2\left(\frac{\pi}{4} \frac{|f| + W - 2W_0}{W - W_0}\right) & \text{for } 2W_0 - W < |f| < W \quad (2.74) \\ 0 & \text{for } |f| > W \end{cases}$$

where W is the absolute bandwidth, and $W_0 = 1/2T$ represents the minimum Nyquist bandwidth for the rectangular spectrum and the -6-dB bandwidth (or half-amplitude point) for the raised cosine spectrum. The difference ($W - W_0$) is termed the *excess bandwidth*; notice that $W = W_0$ for the rectangular spectrum.

The *roll-off factor* is defined to be $r = (W - W_0)/W_0$. It represents the excess bandwidth divided by the filter -6-dB bandwidth (i.e., the fractional excess bandwidth). For a given W_0, r specifies the required excess bandwidth (as a fraction of W_0) and characterizes the steepness of the filter roll-off. The raised cosine characteristic is illustrated in Figure 2.34a for roll-off values of $r = 0$, $r = 0.5$, and $r = 1.0$. The $r = 0$ roll-off is the Nyquist minimum-bandwidth case. Notice that when $r = 1.0$, the required excess bandwidth is 100%; a system with such an overall spectral characteristic can provide a symbol rate of R_s symbols/s using a bandwidth of R_s hertz (twice the Nyquist bandwidth), thus yielding a symbol-

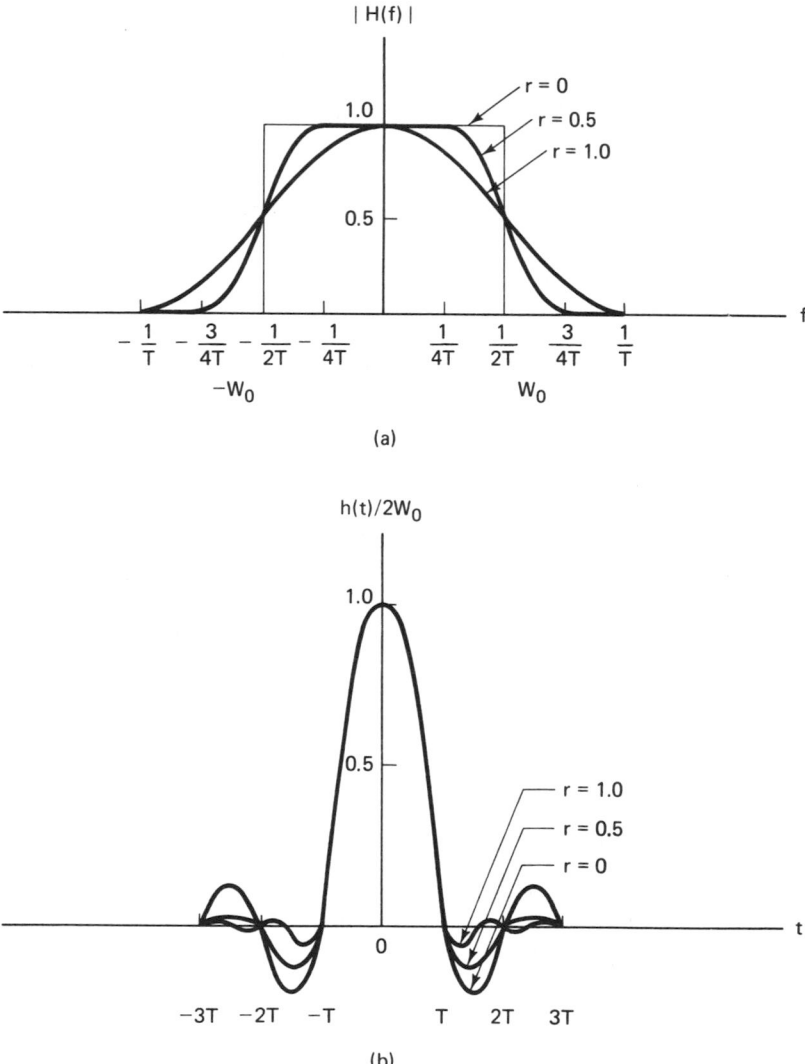

Figure 2.34 Raised cosine filter characteristics. (a) System transfer function. (b) System impulse response.

rate packing of 1 symbol/s/Hz. The corresponding impulse response for the $H(f)$ of Equation (2.74) is

$$h(t) = 2W_0(\text{sinc } 2W_0 t) \frac{\cos [2\pi(W - W_0)t]}{1 - 4(W - W_0)t^2} \tag{2.75}$$

The impulse response is shown in Figure 2.34b for $r = 0$, $r = 0.5$, and $r = 1.0$.

Recall that for zero ISI, we shall choose the system received pulse shape to be equal to $h(t)$; we can only do this approximately, since strictly speaking, the raised cosine pulse spectrum is not precisely physically realizable. A realizable frequency characteristic must have a time response that is zero prior to the pulse turn-on time, which is not the case for the family of raised cosine characteristics. These unrealizable filters are *noncausal* (the filter impulse response begins at time $t = -\infty$). However, a delayed version of $h(t)$, say $h(t - t_0)$, may be approximately generated by real filters if the delay t_0 is chosen such that $h(t - t_0) \simeq 0$, for $t < 0$. Notice in Figure 2.34b that timing errors will still result in some ISI degradation for $r = 1$. However, the problem is not as serious as it is for $r = 0$, because the tails of the $h(t)$ waveform are of much smaller amplitude for $r = 1$ than they are for $r = 0$.

The Nyquist bandwidth constraint states that the theoretical minimum required system bandwidth, W, for a symbol rate of R_s symbols/s without ISI, is $R_s/2$ hertz. A more general relationship between required bandwidth and symbol transmission rate involves the filter roll-off factor r, and can be stated as

$$W = \tfrac{1}{2}(1 + r)R_s \tag{2.76}$$

Thus with $r = 0$, Equation (2.76) describes the required bandwidth for ideal rectangular filtering, also referred to as *Nyquist filtering*. Bandpass-modulated signals (baseband signals that have been shifted in frequency), such as amplitude shift keying (ASK) and phase shift keying (PSK), require twice the transmission bandwidth of the equivalent baseband signals (see Section 1.7.1). Such frequency-translated signals, occupying twice their baseband bandwidth, are often called double-sideband (DSB) signals. Therefore, for ASK- and PSK-modulated signals, the relationship between the required DSB bandwidth, W_{DSB}, and the symbol transmission rate, R_s, is

$$W_{\text{DSB}} = (1 + r)R_s \tag{2.77}$$

Example 2.4 Bandwidth Requirements

(a) Find the minimum required bandwidth for the baseband transmission of a four-level PCM pulse sequence having a data rate of $R = 2400$ bits/s if the system transfer characteristic consists of a raised cosine spectrum with 100% excess bandwidth ($r = 1$).

(b) The same PCM sequence is modulated onto a carrier wave, so that the baseband spectrum is shifted and centered at frequency f_0. Find the minimum required DSB bandwidth for transmitting the modulated PCM sequence. Assume that the system transfer characteristic is the same as in part (a).

Solution

(a) $M = 2^k$; since $M = 4$ levels, then $k = 2$.

$$\text{Symbol or pulse rate } R_s = \frac{R}{k} = \frac{2400}{2} = 1200 \text{ symbols/s}$$

$$\text{Minimum bandwidth } W = \tfrac{1}{2}(1 + r)R_s = \tfrac{1}{2}(2)(1200) = 1200 \text{ Hz}$$

Figure 2.35a illustrates the baseband PCM received pulse in the time domain—an approximation to the $h(t)$ in Equation (2.75). Figure 2.35b illustrates the Fourier transform of $h(t)$—the raised cosine spectrum. Notice that the required bandwidth, W, starts at zero frequency and extends to $f = 1/T$; it is twice the size of the Nyquist theoretical minimum bandwidth.

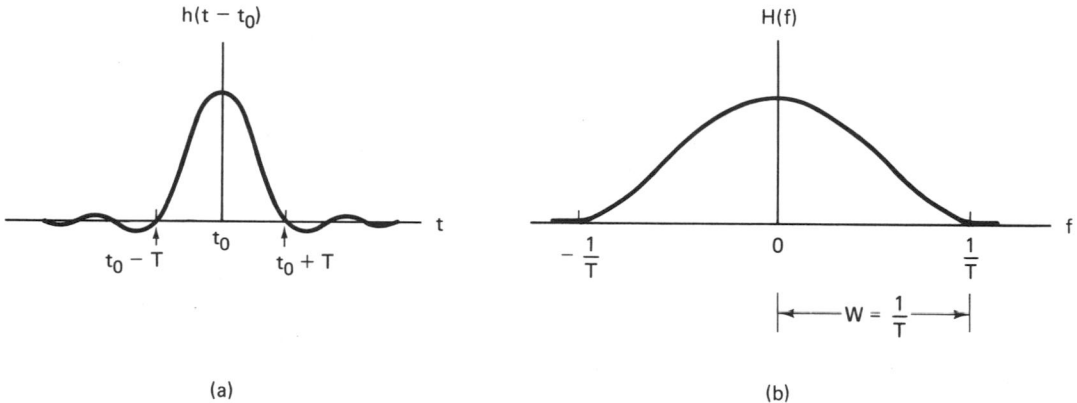

(a)   (b)

Figure 2.35 (a) Shaped pulse. (b) Baseband raised cosine spectrum.

(b) As in part (a),

$$R_s = 1200 \text{ symbols/s}$$

$$W_{\text{DSB}} = (1 + r)R_s = 2(1200) = 2400 \text{ Hz}$$

Figure 2.36a illustrates the modulated PCM received pulse. This waveform can be viewed as the product of a high-frequency sinusoidal carrier wave and a waveform with the pulse shape of Figure 2.35a. The single-sided spectral plot in Figure 2.36b illustrates that the modulated bandwidth, W_{DSB}, is

$$W_{\text{DSB}} = \left(f_0 + \frac{1}{T}\right) - \left(f_0 - \frac{1}{T}\right) = \frac{2}{T}$$

When the spectrum of Figure 2.35b is shifted up in frequency, the negative and positive halves of the baseband spectrum are shifted up in frequency, thereby doubling the required transmission bandwidth. As the name implies, the DSB signal has two sidebands: the upper sideband (USB), derived from the baseband positive half, and the lower sideband (LSB), derived from the baseband negative half.

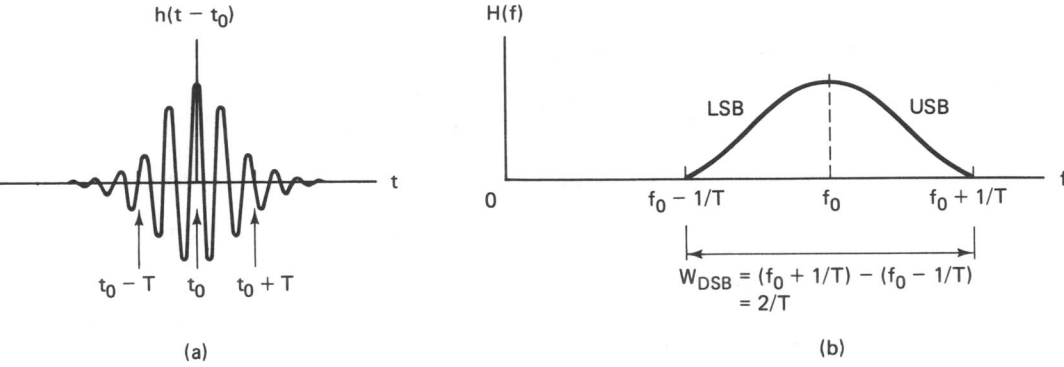

Figure 2.36 (a) Modulated shaped pulse. (b) DSB-modulated raised cosine spectrum.

Example 2.5 Digital Telephone Circuits

Compare the system bandwidth requirements for a 3-kHz analog telephone voice circuit versus a PCM voice circuit. Assume that the sampling rate for the analog-to-digital (A/D) conversion is 8000 samples/s. Also, assume that each voice sample is quantized to one of 256 levels (8-bit quantization).

Solution

The result of the sampling and quantization process yields a PAM signal such that each pulse (symbol) has one of 256 different levels. From Equation (2.76) we can write that the required system bandwidth (without ISI) for R_s symbols/s is

$$W \geq \frac{R_s}{2} \qquad \text{hertz}$$

where the equality sign holds true only for Nyquist filtering. For binary PCM, having $L = 256$ levels, each sample is converted to $\ell = \log_2 L = 8$ bits. Therefore, the system bandwidth required to transmit voice using PCM with 8-bit words is

$$W_{\text{PCM}} \geq (\log_2 L) \frac{R_s}{2} \qquad \text{hertz}$$

$$\geq \tfrac{1}{2}(8 \text{ bits/symbol})(8000 \text{ symbols/s}) = 32 \text{ kHz}$$

The 3-kHz analog voice circuit will generally require approximately 4 kHz of bandwidth (including some bandwidth separation between channels, called *guard bands*). Therefore, the PCM format using 8-bit quantization requires *at least* eight times the bandwidth required by the analog format.

2.11.2 Equalization

In practical systems, the frequency response of the channel is not known with sufficient precision to allow for a receiver design that will compensate for the intersymbol interference (ISI) for all time. In practice, the filter for handling ISI at the receiver contains various parameters that are adjusted on the basis of measurements of the channel characteristics. The process of thus correcting the channel-induced distortion is called *equalization*. A *transversal filter*—a delay line with T-second taps (where T is the symbol duration)—is a common choice for the *equalizer filter*. The outputs of the taps are amplified, summed, and fed to a decision device. The tap coefficients, c_n, are set to subtract the effects of interference from symbols that are adjacent in time to the desired symbol. Consider that there are $(2N + 1)$ taps with coefficients $c_{-N}, c_{-N+1}, \ldots, c_N$ as shown in Figure 2.37. Output samples, $\{y_k\}$, of the equalizer are then expressed in terms of the input samples, $\{x_k\}$, and tap coefficients as

$$y_k = \sum_{n=-N}^{N} c_n x_{k-n} \qquad k = -2N, \ldots, 2N \qquad (2.78)$$

By defining the matrices **y**, **c**, and **x** as

$$\mathbf{y} = \begin{bmatrix} y_{-2N} \\ \vdots \\ y_0 \\ \vdots \\ y_{2N} \end{bmatrix} \qquad \mathbf{c} = \begin{bmatrix} c_{-N} \\ \vdots \\ c_0 \\ \vdots \\ c_N \end{bmatrix} \qquad (2.79)$$

$$\mathbf{x} = \begin{bmatrix} x_{-N} & 0 & 0 & \cdots & 0 & 0 \\ x_{-N+1} & x_{-N} & 0 & \cdots & \cdots & \cdots \\ \vdots & & & & \vdots & \vdots \\ x_N & x_{N-1} & x_{N-2} & \cdots & x_{-N+1} & x_{-N} \\ \vdots & & & & \vdots & \vdots \\ 0 & 0 & 0 & \cdots & x_N & x_{N-1} \\ 0 & 0 & 0 & \cdots & 0 & x_N \end{bmatrix} \qquad (2.80)$$

we can simplify the computation for $\{y_k\}$ as follows:

$$\mathbf{y} = \mathbf{x}\mathbf{c} \qquad (2.81)$$

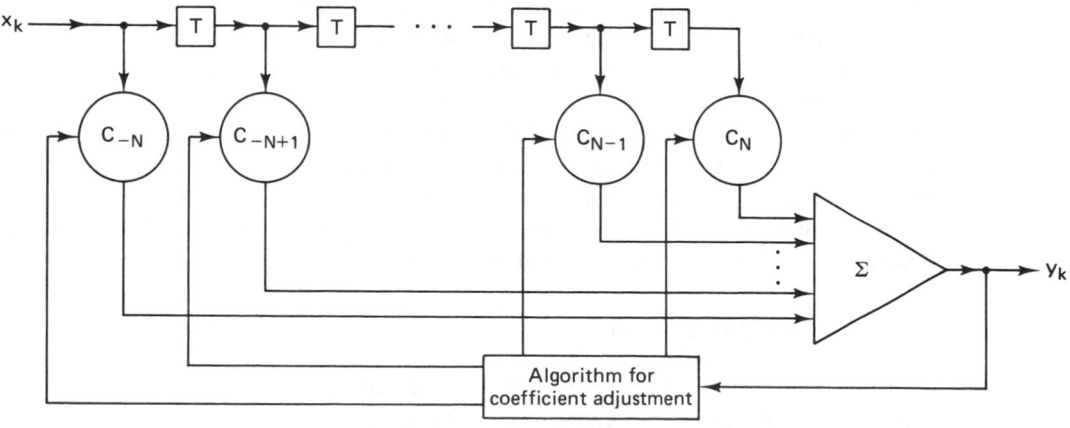

Figure 2.37 Transversal filter.

The criterion for selecting the c_n coefficients is typically based on the minimization of either peak distortion or mean-square distortion. Minimizing peak distortion can be accomplished by selecting the c_n coefficients so that the equalizer output is forced to zero at N sample points on either side of the desired pulse. That is,

$$y_k = \begin{cases} 1 & \text{for } k = 0 \\ 0 & \text{for } k = \pm 1, \pm 2, \ldots, \pm N \end{cases} \tag{2.82}$$

We then solve for c_n by combining Equations (2.79) to (2.81) and solving $2N + 1$ simultaneous equations. Minimizing the mean-square distortion similarly results in $2N + 1$ simultaneous equations.

There are two general types of automatic equalization. The first, *preset equalization*, transmits a training sequence that is compared at the receiver with a locally generated sequence. The differences between the two sequences are used to set the coefficients c_n. With the second method, *adaptive equalization*, the coefficients are continually and automatically adjusted directly from the transmitted data. A disadvantage of preset equalization is that it requires an initial training session, which must be repeated after any break in transmission. Also, a time-varying channel can degrade in ISI since the coefficients are fixed. Adaptive equalization can perform well if the channel error performance is satisfactory. However, if the error performance is poor, received channel errors may not allow the algorithm to converge. A common solution employs preset equalization initially to provide good channel error performance; once normal transmission begins, the system switches to an adaptive algorithm. A significant amount of research and development has taken place in the area of equalization during the past two decades [8, 10, 11].

2.12 PARTIAL RESPONSE SIGNALING

In 1963, Adam Lender [12, 13] showed that it is possible to transmit $2W$ symbols/s with zero ISI, using the theoretical minimum bandwidth of W hertz, without infinitely sharp filters. Lender used a technique called *duobinary signaling*, also referred to by the names *partial response signaling* and *correlative coding*. The basic idea behind the duobinary technique is to introduce some controlled amount of ISI into the data stream rather than trying to eliminate it completely. By introducing correlated interference between the pulses, and by changing the detection procedure, Lender, in effect, "canceled out" the interference at the detector, and thereby achieved the ideal symbol-rate packing of 2 symbols/s/Hz, an amount that had been considered unrealizable.

2.12.1 Duobinary Signaling

To understand how duobinary signaling introduces controlled ISI, let us look at a model of the process. We can think of the duobinary coding operation as if it were implemented as shown in Figure 2.38. Assume that a sequence of binary

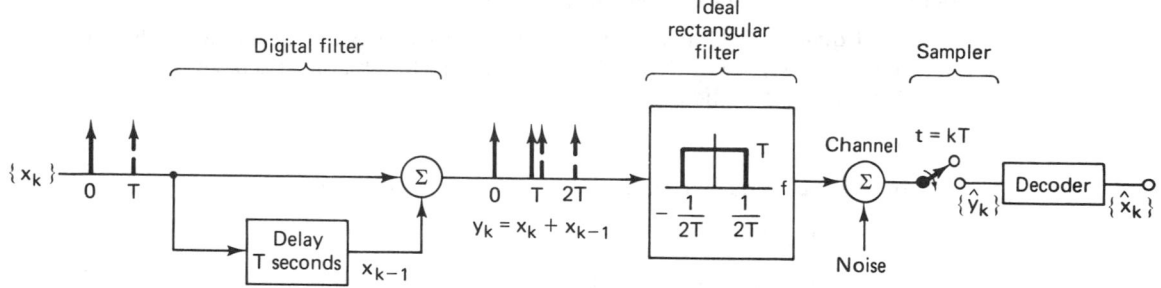

Figure 2.38 Duobinary signaling.

symbols $\{x_k\}$ is to be transmitted at the rate of R symbols/s over a system having an ideal rectangular spectrum of bandwidth $W = R/2 = 1/2T$ hertz. You might ask: How is this rectangular spectrum, in Figure 2.38, different from the unrealizable Nyquist characteristic? It has the same ideal characteristic; but we are not trying to implement the ideal rectangular filter. It is only the part of our equivalent model that is used for developing a filter that is easier to approximate. Before being shaped by the ideal filter, the pulses pass through a simple digital filter, as shown in the figure. The digital filter incorporates a one-digit delay; to each incoming pulse, the filter adds the value of the previous pulse. In other words, for every pulse into the digital filter, we get the summation of two pulses out. Each pulse of the sequence $\{y_k\}$ out of the digital filter can be expressed as

$$y_k = x_k + x_{k-1} \tag{2.83}$$

Hence the $\{y_k\}$ amplitudes are not independent; each y_k digit carries with it the *memory* of the prior digit. The ISI introduced to each y_k digit comes only from the preceding x_{k-1} digit. This correlation between the pulse amplitudes of $\{y_k\}$ can be thought of as the controlled ISI introduced by the duobinary coding. Controlled interference is the essence of this novel technique, because at the detector, such controlled interference can be removed as easily as it was added. The $\{y_k\}$ sequence is followed by the ideal Nyquist filter that does not introduce any ISI. At the receiver sampler, in Figure 2.38, we would expect to recover the sequence $\{y_k\}$, exactly in the absence of noise. Since all systems experience noise contamination, we shall refer to the *received* $\{y_k\}$ as the estimate of $\{y_k\}$ and denote it $\{\hat{y}_k\}$. Removing the controlled interference with the duobinary decoder yields an estimate of $\{x_k\}$ which we shall denote as $\{\hat{x}_k\}$.

2.12.2 Duobinary Decoding

If the binary digit x_k is equal to ± 1, then using Equation (2.83), y_k has one of three possible values: $+2$, 0, or -2. The duobinary code results in a three-level output: in general for M-ary transmission, partial response signaling results in $2M - 1$ output levels. The decoding procedure involves the inverse of the coding procedure, namely, subtracting the x_{k-1} decision from the y_k digit. Consider the following coding/decoding example.

Example 2.6 Duobinary Coding and Decoding

Use Equation (2.83) to demonstrate duobinary coding and decoding for the following sequence: $\{x_k\} = 0\ 0\ 1\ 0\ 1\ 1\ 0$. Consider the first bit of the sequence to be a startup digit, not part of the data.

Solution

Binary digit sequence $\{x_k\}$:	0	0	1	0	1	1	0
Bipolar amplitudes $\{x_k\}$:	-1	-1	$+1$	-1	$+1$	$+1$	-1
Coding rule: $y_k = x_k + x_{k-1}$:		-2	0	0	0	2	0

Decoding decision rule: If $\hat{y}_k = \ \ 2$, decide that $\hat{x}_k = \ +1$ (or binary one)

If $\hat{y}_k = \ -2$, decide that $\hat{x}_k = \ -1$ (or binary zero).

If $\hat{y}_k = \ \ 0$, decide opposite of the previous decision.

Decoded bipolar sequence $\{\hat{x}_k\}$:	-1	$+1$	-1	$+1$	$+1$	-1
Decoded binary sequence $\{\hat{x}_k\}$:	0	1	0	1	1	0

The decision rule simply implements the subtraction of each $\hat{x}_{k-1}$ decision from each $\hat{y}_k$. One drawback of this detection technique is that once an error is made, it tends to propagate, causing further errors, since present decisions depend on prior decisions. A means of avoiding this error propagation is known as *precoding*.

2.12.3 Precoding

Precoding is accomplished by first differentially encoding the $\{x_k\}$ binary sequence into a new $\{w_k\}$ binary sequence as follows:

$$w_k = x_k \oplus w_{k-1} \tag{2.84}$$

where the symbol $\oplus$ represents modulo-2 addition (equivalent to the logical *exclusive-or* operation) of the binary digits. The rules of modulo-2 addition are as follows:

$$0 \oplus 0 = 0$$
$$0 \oplus 1 = 1$$
$$1 \oplus 0 = 1$$
$$1 \oplus 1 = 0 \qquad .$$

The $\{w_k\}$ binary sequence is then converted to a bipolar pulse sequence, and the coding operation proceeds in the same way as it did in Example 2.6. However, with precoding, the detection process is quite different from the detection of ordinary duobinary, as shown below in Example 2.7. The precoding model is shown in Figure 2.39; in this figure it is implicit that the modulo-2 addition producing the precoded $\{w_k\}$ sequence is performed on the *binary* digits, while the digital filtering producing the $\{y_k\}$ sequence is performed on the *bipolar* pulses.

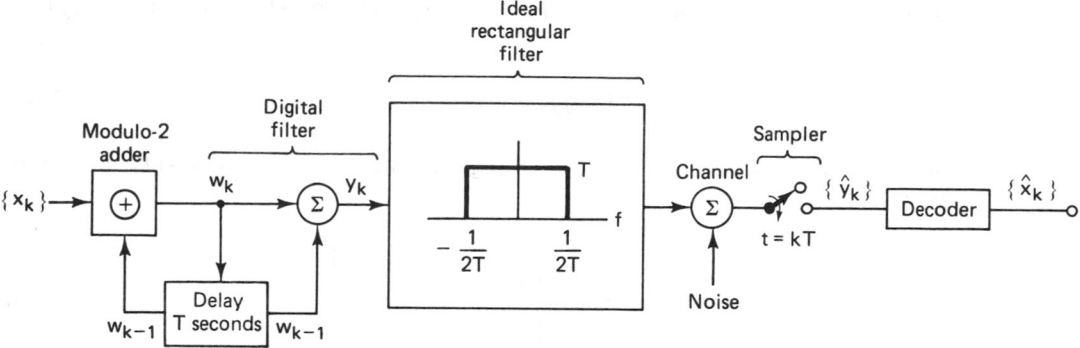

Figure 2.39 Precoded duobinary signaling.

Example 2.7 Duobinary Precoding

Illustrate the duobinary coding and decoding rules when using the differential precoding of Equation (2.84). Assume the same $\{x_k\}$ sequence as that given in Example 2.6.

Solution

Binary digit sequence $\{x_k\}$:	0	0	1	0	1	1	0
Precoded sequence $w_k = x_k \oplus w_{k-1}$:	0	0	1	1	0	1	1
Bipolar sequence $\{w_k\}$:	-1	-1	$+1$	$+1$	-1	$+1$	$+1$
Coding rule: $y_k = w_k + w_{k-1}$:		-2	0	$+2$	0	0	$+2$

Decoding decision rule: If $\hat{y}_k = \pm 2$, decide that $\hat{x}_k$ = binary zero.

If $\hat{y}_k = \ \ 0$, decide that $\hat{x}_k$ = binary one.

Decoded binary sequence $\{\hat{x}_k\}$: 0 1 0 1 1 0

The differential precoding enables us to decode the $\{\hat{y}_k\}$ sequence by making a decision on each received sample singly, without resorting to prior decisions which could be in error. The major advantage is that in the event of a digit error due to noise, such an error does not propagate to other digits. Notice that the first bit in the differentially precoded binary sequence $\{w_k\}$ is an arbitrary choice. If the startup bit in $\{w_k\}$ had been chosen to be a binary one instead of a binary zero, the decoded result would have been the same.

2.12.4 Duobinary Equivalent Transfer Function

In Section 2.12.1 we described the duobinary transfer function as a digital filter incorporating a one-digit delay, followed by an ideal rectangular transfer function. Let us now examine an equivalent model. The Fourier transform of a delay can be described as $e^{-j2\pi fT}$ (see Section A.3.1); therefore, the input digital filter of

Figure 2.38 can be characterized with the frequency transfer function, $H_1(f)$, as follows:

$$H_1(f) = 1 + e^{-j2\pi fT} \qquad (2.85)$$

The transfer function of the ideal rectangular filter, designated $H_2(f)$, is shown below.

$$H_2(f) = \begin{cases} T & \text{for } |f| < \dfrac{1}{2T} \\ 0 & \text{elsewhere} \end{cases} \qquad (2.86)$$

The overall equivalent transfer function $H_e(f)$, of the digital filter cascaded with the ideal rectangular filter is then given by

$$H_e(f) = H_1(f)H_2(f) \qquad \text{for } |f| < \frac{1}{2T}$$

$$= (1 + e^{-j2\pi fT})T \qquad (2.87)$$

$$= T(e^{j\pi fT} + e^{-j\pi fT})e^{-j\pi fT}$$

$$|H_e(f)| = \begin{cases} 2T \cos \pi fT & \text{for } |f| < \dfrac{1}{2T} \\ 0 & \text{elsewhere} \end{cases} \qquad (2.88)$$

Thus $H_e(f)$, the composite transfer function for the cascaded digital and rectangular filters, has a gradual roll-off to the band edge, as can be seen in Figure 2.40a. The transfer function can be approximated by using realizable analog filtering; a separate digital filter is not needed. The duobinary equivalent $H_e(f)$ is called a *cosine filter* [14] (not to be confused with the raised cosine filter described in Section 2.11.1). The corresponding impulse response, $h_e(t)$, found by taking the inverse Fourier transform of $H_e(f)$ in Equation (2.87), is

$$h_e(t) = \text{sinc}\left(\frac{t}{T}\right) + \text{sinc}\left(\frac{t - T}{T}\right) \qquad (2.89)$$

and is plotted in Figure 2.40b. For every impulse, $\delta(t)$, at the input of Figure 2.38, the output is $h_e(t)$ with an appropriate polarity. Notice that there are only two nonzero samples, at T-second intervals, giving rise to controlled ISI from the adjacent bit. The introduced ISI is eliminated by use of the decoding procedure discussed in Section 2.12.2. Although the cosine filter is noncausal and therefore nonrealizable, it can be easily approximated. The implementation of the precoded duobinary technique described in Section 2.12.3 can be accomplished by first differentially encoding the binary sequence $\{x_k\}$ into the sequence $\{w_k\}$ (see Example 2.7). The pulse sequence $\{w_k\}$ is then filtered by the equivalent cosine characteristic described in Equation (2.88).

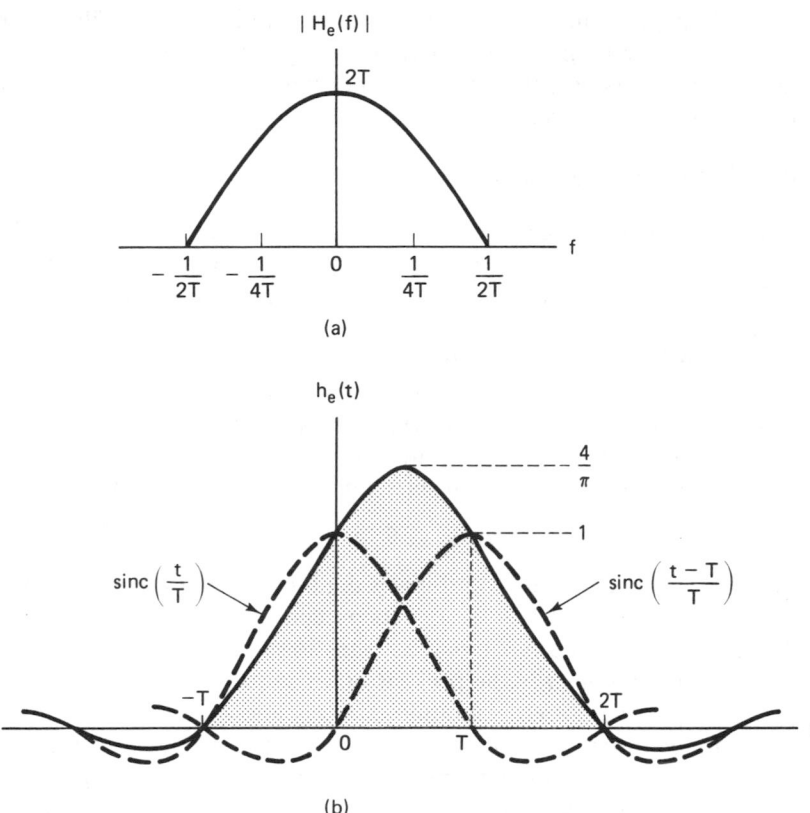

Figure 2.40 Duobinary transfer function and pulse shape. (a) Cosine filter. (b) Impulse response of the cosine filter.

2.12.5 Comparison of Binary with Duobinary Signaling

The duobinary technique introduces correlation between pulse amplitudes, whereas the more restrictive Nyquist criterion assumes that the transmitted pulse amplitudes are independent of one another. We have shown that duobinary signaling can exploit this introduced correlation to achieve zero ISI signal transmission, using a smaller system bandwidth than is otherwise possible. Do we get this performance improvement without paying a price? Such is hardly ever the case with engineering design options; there is almost always a trade-off involved. We saw that duobinary coding requires three levels, compared to the usual two levels for binary coding. Recall our discussion in Section 2.10, where we compared the performance and the required signal power for making eight-level PCM decisions versus two-level PCM decisions. For a fixed amount of signal power, the ease of making reliable decisions is inversely related to the number of levels that must be distinguished in each waveform. Therefore, it should be no surprise that although duobinary signaling accomplishes the zero ISI requirement with minimum bandwidth, duobinary also requires more power than binary signaling, for equivalent performance against noise. For a given probability of bit error (P_B), duobinary signaling requires approximately 2.5 dB greater SNR than binary signaling, while using only $1/(1 + r)$ the bandwidth that binary signaling requires [13], where r is the filter roll-off.

2.12.6 Polybinary Signaling

Duobinary signaling can be extended to more than three digits or levels, resulting in greater bandwidth efficiency; such systems are called *polybinary* [13, 15]. Consider that a binary message with two signaling levels is transformed into a signal with j signaling levels, numbered consecutively from zero to $(j - 1)$. The transformation from binary to polybinary takes place in two steps. First, the original sequence $\{x_k\}$, consisting of binary ones and zeros, is converted into another binary sequence $\{y_k\}$, as follows: The present binary digit of sequence $\{y_k\}$ is formed from the modulo-2 addition of the $(j - 2)$ immediately preceding digits of sequence $\{y_k\}$ and the present digit x_k. For example, let

$$y_k = x_k \oplus y_{k-1} \oplus y_{k-2} \oplus y_{k-3} \qquad (2.90)$$

Here x_k represents the input binary digit and y_k the kth encoded digit. Since the expression involves $(j - 2) = 3$ bits preceding y_k, there are $j = 5$ signaling levels. Next, the binary sequence $\{y_k\}$ is transformed into a polybinary pulse train $\{z_k\}$ by adding *algebraically* the present bit of sequence $\{y_k\}$ to the $(j - 2)$ preceding bits of $\{y_k\}$. Therefore, z_k modulo-2 $= x_k$, and the binary elements one and zero are mapped into even- and odd-valued pulses in the sequence $\{z_k\}$. Note that each digit in $\{z_k\}$ can be independently detected despite the strong correlation between bits. The primary advantage of such a signaling scheme is the redistribution of the spectral density of the original sequence $\{x_k\}$, so as to favor the low frequencies, thus improving system bandwidth efficiency.

2.13 CONCLUSION

In this chapter we have considered the first important step in any digital communication system, transforming the source information (both textual and analog) to a form that is compatible with a digital system. We treated various aspects of sampling, quantization (both uniform and nonuniform), and pulse code modulation (PCM). We also considered the selection of PCM waveforms for the transmission of baseband signals through the channel.

We described the detection of binary signals plus Gaussian noise in terms of two basic steps. In the first step the received waveform is reduced to a single number, $z(T)$, and in the second step a decision is made as to which signal was transmitted, on the basis of comparing $z(T)$ to a threshold. We discussed how to best choose this threshold. We also showed that a linear filter known as a matched filter or correlator is the optimum choice for maximizing the output signal-to-noise ratio, and thus minimizing the probability of error.

We defined intersymbol interference (ISI) and explained the importance of Nyquist's work in establishing a theoretical minimum bandwidth for symbol detection without ISI. We also introduced the duobinary concept of adding a controlled amount of ISI to achieve an improvement in bandwidth efficiency at the expense of an increase in power.

REFERENCES

1. Black, H. S., *Modulation Theory*, D. Van Nostrand Company, Princeton, N.J., 1953.

2. Oppenheim, A. V., *Applications of Digital Signal Processing*, Prentice-Hall, Inc., Englewood Cliffs, N.J., 1978.

3. Stiltz, H., ed., *Aerospace Telemetry*, Vol. 1, Prentice-Hall, Inc., Englewood Cliffs, N.J., 1961. p. 179.

4. Hecht, M., and Guida, A., "Delay Modulation," *Proc. IEEE*, vol. 57, no. 7, July 1969, pp. 1314–1316.

5. Deffebach, H. L., and Frost, W. O., "A Survey of Digital Baseband Signaling Techniques," *NASA Technical Memorandum NASATM X-64615*, June 30, 1971.

6. Van Trees, H. L., *Detection, Estimation, and Modulation Theory*, Part 1, John Wiley & Sons, Inc., New York, 1968.

7. Borjesson, P. O., and Sundberg, C. E., "Simple Approximations of the Error Function $Q(x)$ for Communications Applications," *IEEE Trans. Commun.*, vol. COM27, Mar. 1979, pp. 639–642.

8. Proakis, J. G., *Digital Communications*, McGraw-Hill Book Company, New York, 1983.

9. Nyquist, H., "Certain Topics of Telegraph Transmission Theory," *Trans. Am. Inst. Electr. Eng.*, vol. 47, Apr. 1928, pp. 617–644.

10. Korn, I., *Digital Communications*, Van Nostrand Reinhold Company, Inc., New York, 1985.

11. Wu, W. W., *Elements of Digital Satellite Communication*, Computer Science Press, Inc., Rockville, Md., 1984.

12. Lender, A., "The Duobinary Technique for High Speed Data Transmission," *IEEE Trans. Commun. Electron.*, vol. 82, May 1963, pp. 214–218.

13. Lender, A., "Correlative (Partial Response) Techniques and Applications to Digital Radio Systems," in K. Feher, *Digital Communications: Microwave Applications*, Prentice-Hall, Inc., Englewood Cliffs, N.J., 1981, Chap. 7.

14. Couch, L. W., II, *Digital and Analog Communication Systems*, Macmillan Publishing Company, New York, 1982.

15. Lender, A., "Correlative Digital Communication Techniques," *IEEE Trans. Commun. Technol.*, Dec. 1964, pp. 128–135.

PROBLEMS

2.1. You want to transmit the word "HOW" using an 8-ary system.
  (a) Encode the word "HOW" into a sequence of bits, using 7-bit ASCII coding, followed by an eighth bit for error detection, per character. The eighth bit is chosen so that the number of ones in the 8 bits is an even number. How many total bits are there in the message?

(b) Partition the bit stream into $k = 3$ bit segments. Represent each of the 3-bit segments as an octal number (symbol). How many octal symbols are there in the message?

(c) If the system were designed with 16-ary modulation, how many symbols would be used to represent the word "HOW"?

(d) If the system were designed with 256-ary modulation, how many symbols would be used to represent the word "HOW"?

2.2. We want to transmit 800 characters/s, where each character is represented by its 7-bit ASCII codeword, followed by an eighth bit for error detection, per character, as in Problem 2.1. A multilevel PCM format with $M = 16$ levels is used.

(a) What is the effective transmitted bit rate?

(b) What is the PCM symbol rate?

2.3. We wish to transmit a 100-character alphanumeric message in 2 s, using 7-bit ASCII coding, followed by an eighth bit for error detection, per character, as in Problem 2.1. A multilevel PCM format with $M = 32$ levels is used.

(a) Calculate the effective transmitted bit rate and the PCM symbol rate.

(b) Repeat part (a) for 16-level PCM, eight-level PCM, four-level PCM, and binary PCM.

2.4. Given an analog waveform that has been sampled at its Nyquist rate, f_s, using natural sampling, prove that a waveform (proportional to the original waveform) can be recovered from the samples, using the recovery techniques shown in Figure P2.1. The parameter mf_s is the frequency of the local oscillator, where m is an integer.

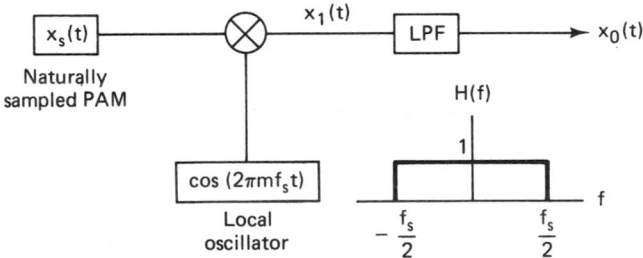

Figure P2.1

2.5. An analog signal is sampled at its Nyquist rate $1/T_s$, and quantized using L quantization levels. The derived digital signal is then transmitted on some channel.

(a) Show that the time duration, T, of one bit of the transmitted binary encoded signal must satisfy $T \leq T_s/(\log_2 L)$.

(b) When is the equality sign valid?

2.6. Determine the number of quantization levels that are implied if the number of bits per sample in a given PCM code is (a) 5; (b) 8; (c) x.

2.7. Determine the minimum sampling rate necessary to sample and perfectly reconstruct the signal $x(t) = \sin (6280t)/(6280t)$.

2.8. Consider an audio signal with spectral components limited to the frequency band 300 to 3300 Hz. Assume that a sampling rate of 8000 samples/s will be used to generate a PCM signal. Assume that the ratio of peak signal power to average quantization noise power at the output needs to be 30 dB.

(a) What is the minimum number of uniform quantization levels needed, and what is the minimum number of bits per sample needed?

(b) Calculate the system bandwidth (as specified by the main spectral lobe of the signal) required for the detection of such a PCM signal.

2.9. A waveform, $x(t) = 10 \cos (1000t + \pi/3) + 20 \cos (2000t + \pi/6)$ is to be uniformly sampled for digital transmission.

    (a) What is the maximum allowable time interval between sample values that will ensure perfect signal reproduction?

    (b) If we want to reproduce 1 hour of this waveform, how many sample values need to be stored?

2.10. **(a)** A waveform that is bandlimited to 50 kHz is sampled every 10 μs. Show graphically that these samples uniquely characterize the waveform. (Use a sinusoidal example for simplicity. Avoid sampling at points where the waveform equals zero.)

    (b) If samples are taken 30 μs apart instead of 10 μs, show graphically that waveforms other than the original can be characterized by the samples.

2.11. Use the method of convolution to illustrate the effect of undersampling the waveform $x(t) = \cos 2\pi f_0 t$ for a sampling rate of $f_s = \frac{3}{2}f_0$.

2.12. **(a)** Sketch the complete $\mu = 10$ compression characteristic that will handle input voltages in the range -5 to $+5$ V.

    (b) Plot the corresponding expansion characteristic.

    (c) Draw a 16-level nonuniform quantizer characteristic that corresponds to the $\mu = 10$ compression characteristic.

2.13. Assume a binary sequence with equally likely binary levels. The sequence can be represented by either a bipolar or a unipolar signal set. Show that if the corresponding bipolar signal and unipolar signal have the same peak-to-peak amplitude separation, the bipolar signal uses less average power than the unipolar signal.

2.14. Assume that in a binary digital communication system, the signal component out of the correlator receiver is $a_i(T) = +1$ or -1 V with equal probability. If the Gaussian noise at the correlator output has unit variance, find the probability of a bit error.

2.15. A bipolar binary signal, $s_i(t)$, is a $+1$- or -1-V pulse during the interval $(0, T)$. Additive white Gaussian noise having two-sided power spectral density of 10^{-3} W/Hz is added to the signal. If the received signal is detected with a matched filter, determine the maximum bit rate that can be sent with a bit error probability of $P_B \leq 10^{-3}$.

2.16. Bipolar pulse signals, $s_i(t)$ $(i = 1, 2)$, of amplitude ± 1 V are received in the presence of Gaussian noise with $\sigma^2 = 0.1$ V². Determine the optimum (minimum probability of error) detection threshold, γ_0, for matched filter detection if the a priori probabilities are: **(a)** $P(s_1) = 0.5$; **(b)** $P(s_1) = 0.7$; **(c)** $P(s_1) = 0.2$. **(d)** Explain the effect of the a priori probabilities on the value of γ_0. [*Hint:* Refer to Equations (B.10) to (B.12).]

2.17. A binary communication system transmits signals $s_i(t)$ $(i = 1, 2)$. The receiver test statistic, $z(T) = a_i + n_0$, where the signal component, a_i, is either $a_1 = +1$ or $a_2 = -1$, and the noise component, n_0 is uniformly distributed, yielding the conditional density functions $p(z|s_i)$ given by

$$p(z|s_1) = \begin{cases} \frac{1}{2} & \text{for } -0.2 \leq z \leq 1.8 \\ 0 & \text{otherwise} \end{cases}$$

$$p(z|s_2) = \begin{cases} \frac{1}{2} & \text{for } -1.8 \leq z \leq 0.2 \\ 0 & \text{otherwise} \end{cases}$$

Find the probability of a bit error, P_B, for the case of equally likely signaling and the use of an optimum decision threshold.

2.18. The information in an analog waveform, whose maximum frequency $f_m = 4000$ Hz, is to be transmitted using a 16-level PCM system. The quantization distortion must not exceed $\pm 1\%$ of the peak-to-peak analog signal.

(a) What is the minimum number of bits per sample or bits per PCM word that should be used in this PCM system?

(b) What is the minimum required sampling rate, and what is the resulting bit rate?

(c) What is the PCM pulse or symbol transmission rate?

2.19. (a) What is the theoretical minimum system bandwidth needed for a 10-Mbits/s signal using 16-level PCM without ISI?

(b) How large can the filter roll-off factor be if the allowable system bandwidth is 1.375 MHz?

2.20. A voice signal (300 to 3300 Hz) is digitized such that the quantization distortion $\leq \pm 0.1\%$ of the peak-to-peak signal voltage. Assume a sampling rate of 8000 samples/s and a multilevel PCM format with $M = 32$ levels. Find the theoretical minimum system bandwidth that avoids ISI.

2.21. A binary waveform of 9600 bits/s is converted to an octal waveform that is transmitted over a system having a raised cosine roll-off filter characteristic. The system has a conditioned (equalized) response out to 2.4 kHz.

(a) What is the octal symbol rate?

(b) What is the roll-off factor of the filter characteristic?

2.22. A voice signal in the range 300 to 3300 Hz is sampled at 8000 samples/s. We may transmit these samples directly as PAM, or we may first convert them into codewords using PCM.

(a) What is the minimum system bandwidth required for the detection of PAM with no ISI and with a filter roll-off characteristic of $r = 1$?

(b) Using the same filter roll-off characteristic, what is the minimum bandwidth required for the detection of binary PCM if the samples are quantized to eight levels?

(c) Repeat part (b) using 128 quantization levels.

2.23. A signal in the frequency range 300 to 3300 Hz is limited to a peak-to-peak swing of 10 V. It is sampled at 8000 samples/s and the samples are quantized to 64 evenly spaced levels. Calculate and compare the bandwidths and ratio of peak signal power to rms quantization noise if the quantized samples are transmitted either as binary pulses or as four-level pulses. Assume that the system bandwidth is defined by the main spectral lobe of the signal.

2.24. An analog signal is to be converted to a binary PCM signal and transmitted over a channel that is bandlimited to 100 kHz. Assume that 32 quantization levels are used and that the overall equivalent transfer function is of the raised cosine type with roll-off $r = 0.6$.

(a) Find the maximum PCM bit rate that can be used by this system without introducing ISI.

(b) Find the maximum signal bandwidth that can be accommodated for the analog signal.

(c) Repeat parts (a) and (b) for an eight-level PCM signal.

CHAPTER 3

Bandpass Modulation
and
Demodulation

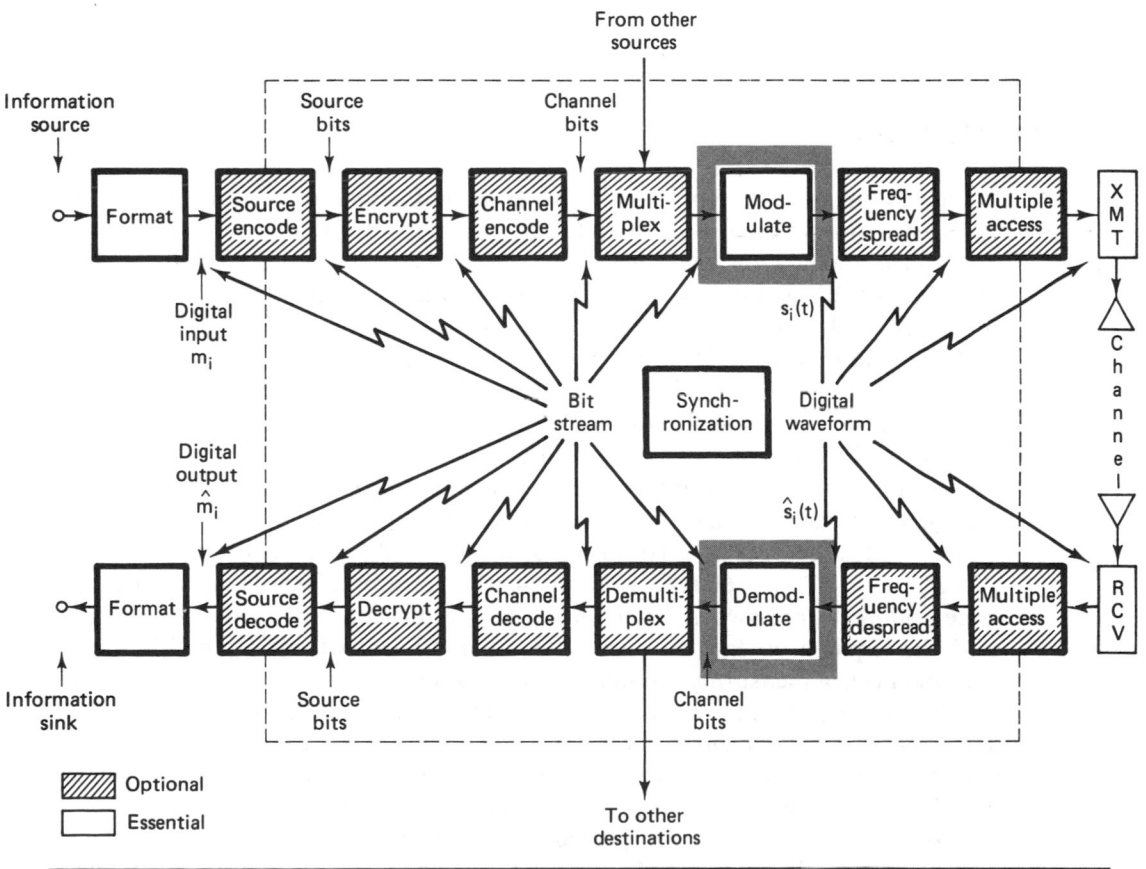

3.1 WHY MODULATE?

Digital modulation is the process by which digital symbols are transformed into waveforms that are compatible with the characteristics of the channel. In the case of baseband modulation, these waveforms are pulses, but in the case of *bandpass modulation* the desired information signal modulates a sinusoid called a *carrier wave,* or simply a *carrier*; for radio transmission the carrier is converted to an electromagnetic (EM) field for propagation to the desired destination. One might ask why it is necessary to use a carrier for the radio transmission of baseband signals. The answer is as follows. The transmission of EM fields through space is accomplished with the use of antennas. To efficiently couple the transmitted EM energy into space, the dimensions of the antenna aperture should be at least as large as the wavelength being transmitted. Wavelength, λ, is equal to c/f, where c, the speed of light, is 3×10^8 m/s. For a baseband signal with frequency $f = 3000$ Hz, $\lambda = 10^5$ m $\simeq 60$ miles. To efficiently transmit a 3000-Hz signal through space *without carrier-wave modulation*, an antenna that spans at least 60 miles would be required. Even if we were willing to inefficiently transmit the EM energy with an antenna measuring one-tenth of a wavelength, we are faced with an impossible antenna size. However, if the information to be transmitted is first modulated on a higher frequency carrier, for example a 30-GHz carrier, the equivalent antenna diameter is then less than $\frac{1}{2}$ in. For this reason, carrier-wave or bandpass modulation is an essential step for all systems involving radio transmission.

Bandpass modulation can provide other important benefits in signal transmission. If more than one signal utilizes a single channel, modulation may be used to separate the different signals. Such a technique, known as *frequency-division multiplexing,* is discussed in Chapter 9. Modulation can be used to minimize the effects of interference. A class of such modulation schemes, known as *spread-spectrum modulation*, requires a system bandwidth much larger than the minimum bandwidth that would be required by the message. The trade-off of bandwidth for interference rejection is considered in Chapter 10. Modulation can also be used to place a signal in a frequency band where design requirements, such as filtering and amplification, can be easily met. This is the case when radio-frequency (RF) signals are converted to an intermediate frequency (IF) in a receiver.

3.2 SIGNALS AND NOISE

3.2.1 Noise in Radio Communication Systems

The task of the demodulator or detector is to retrieve the bit stream from the received waveform, as nearly error free as possible, notwithstanding the distortion to which the signal may have been subjected. There are two primary causes for signal distortion. The first is the filtering effects of the transmitter, channel, and receiver discussed in Section 2.11. As described there, a nonideal system transfer function causes symbol "smearing," which can produce *intersymbol interference*.

The second cause for signal distortion is the noise that is produced by a variety of sources, such as galaxy noise, terrestrial noise, amplifier noise, and unwanted signals from other sources. An unavoidable cause of noise is the thermal motion of electrons in any conducting media. This motion produces *thermal noise* in amplifiers and circuits which corrupts the signal in an additive fashion; that is, the received signal, $r(t)$, is the sum of the transmitted signal, $s(t)$, and the thermal noise, $n(t)$. The statistics of thermal noise have been developed using quantum mechanics and are well known [1].

The primary statistical characteristic of thermal noise is that the noise amplitudes are distributed according to a normal or Gaussian distribution, discussed in Section 1.5.5 and shown in Figure 1.7. The probability density function (pdf), $p(n)$, of the zero-mean noise voltage is expressed as

$$p(n) = \frac{1}{\sigma\sqrt{2\pi}} \exp\left[-\frac{1}{2}\left(\frac{n}{\sigma}\right)^2 \right] \qquad (3.1)$$

where σ^2 is the noise variance. In Figure 1.7 it can be seen that the most probable noise amplitudes are those with small positive or negative values. In theory, the noise can be infinitely large, but very large noise amplitudes are rare.

The primary spectral characteristic of thermal noise is that its two-sided power spectral density, $G_n(f) = N_0/2$, is flat for all frequencies of interest for radio communication systems. In other words, thermal noise, on the average, has just as much power per hertz in low-frequency fluctuations as in high-frequency fluctuations—up to a frequency of about 10^{12} hertz. When the noise power is characterized by a constant power spectral density, as shown in Figure 1.8a, we refer to it as *white noise*. Since thermal noise is present in all communication systems and is the predominant noise source for most systems, the thermal noise characteristics (additive, white, and Gaussian) are most often used to model the noise in the detection process and in the design of optimum receivers.

3.2.2 A Geometric View of Signals and Noise

Let us define an *N*-dimensional *orthogonal space* as one characterized by a set of *N* linearly independent functions, $\{\psi_j(t)\}$, called *basis functions*. Any arbitrary function in the space can be generated by a linear combination of these basis functions. The basis functions must satisfy the following conditions:

$$\int_0^T \psi_j(t)\psi_k(t)\,dt = K_j\delta_{jk} \qquad 0 \le t \le T; \ j, k = 1, \ldots, N \qquad (3.2a)$$

$$\delta_{jk} = \begin{cases} 1 & \text{for } j = k \\ 0 & \text{otherwise} \end{cases} \qquad (3.2b)$$

where the operator δ_{jk} is called the *Kronecker delta function* and is defined by Equation (3.2b). When the K_j constants are nonzero, the signal space is called *orthogonal*. When the basis functions are normalized so that each $K_j = 1$, the space is called an *orthonormal* space. The principal requirement for orthogonality can be stated as follows: Each $\psi_j(t)$ function of the set of basis functions must be independent of the other members of the set. Each $\psi_j(t)$ must not interfere with any other members of the set in the detection process. From a geometric point of view, each $\psi_j(t)$ is mutually perpendicular to each of the other $\psi_k(t)$ for $j \ne k$. An example of such a space with $N = 3$ is shown in Figure 3.1, where the mutually perpendicular axes are designated $\psi_1(t)$, $\psi_2(t)$, and $\psi_3(t)$. If $\psi_j(t)$ corresponds to a real-valued voltage or current waveform component, associated with a 1-Ω

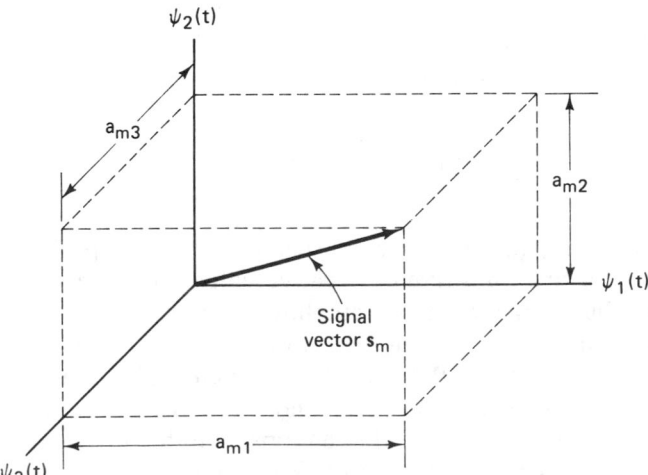

Figure 3.1 Vectorial representation of the signal waveform $s_m(t)$.

Bandpass Modulation and Demodulation   Chap. 3

resistive load, then using Equations (1.5) and (3.2), the normalized energy in joules dissipated in the load in T seconds, due to ψ_j, is

$$E_j = \int_0^T \psi_j^2(t)\, dt = K_j \tag{3.3}$$

One reason we focus on an *orthogonal signal space* is that Euclidean distance measurements, fundamental to the detection process, are easily formulated in such a space. However, even if the signaling waveforms do not comprise such an orthogonal set, they can be transformed into linear combinations of orthogonal waveforms. It can be shown [2] that *any arbitrary* finite set of waveforms $\{s_i(t)\}$ $(i = 1, \ldots, M)$, where each member of the set is physically realizable and of duration T, can be expressed as a linear combination of N orthogonal waveforms $\psi_1(t), \psi_2(t), \ldots, \psi_N(t)$, where $N \le M$, such that

$$s_1(t) = a_{11}\psi_1(t) + a_{12}\psi_2(t) + \cdots + a_{1N}\psi_N(t)$$

$$s_2(t) = a_{21}\psi_1(t) + a_{22}\psi_2(t) + \cdots + a_{2N}\psi_N(t)$$

$$\vdots \qquad\qquad\qquad\qquad\qquad \vdots$$

$$s_M(t) = a_{M1}\psi_1(t) + a_{M2}\psi_2(t) + \cdots + a_{MN}\psi_N(t)$$

These relationships are expressed in more compact notation as follows:

$$s_i(t) = \sum_{j=1}^{N} a_{ij}\psi_j(t) \qquad \begin{matrix} i = 1, \ldots, M \\ N \le M \end{matrix} \tag{3.4}$$

where

$$a_{ij} = \frac{1}{K_j} \int_0^T s_i(t)\psi_j(t)\, dt \qquad \begin{matrix} i = 1, \ldots, M; \quad 0 \le t \le T \\ j = 1, \ldots, N \end{matrix} \tag{3.5}$$

The coefficient a_{ij} is the value of the $\psi_j(t)$ component of signal, $s_i(t)$. The form of the $\{\psi_j(t)\}$ is not specified; it is chosen for convenience and will depend on the form of the signal waveforms. The set of signal waveforms, $\{s_i(t)\}$, can be viewed as a set of vectors, $\{s_i\} = \{a_{i1}, a_{i2}, \ldots, a_{iN}\}$. If, for example, $N = 3$, we may plot the vector, s_m, corresponding to the waveform

$$s_m(t) = a_{m1}\psi_1(t) + a_{m2}\psi_2(t) + a_{m3}\psi_3(t)$$

as a point in a three-dimensional Euclidean space with coordinates (a_{m1}, a_{m2}, a_{m3}), as shown in Figure 3.1. The orientation among the signal vectors describes the relation of the signals to one another (with respect to phase or frequency), and the amplitude of each vector in the set $\{s_i\}$ is a measure of the signal energy transmitted during a symbol duration. In general, once a set of N orthogonal functions has been adopted, each of the transmitted signal waveforms, $s_i(t)$, is completely determined by the vector of its coefficients

$$s_i = (a_{i1}, a_{i2}, \ldots, a_{iN}) \qquad i = 1, \ldots, M \tag{3.6}$$

We shall employ the notation of signal vectors, {s}, or signal waveforms, {s(t)}, as best suits the discussion. A typical detection problem, conveniently viewed in terms of signal vectors, is illustrated in Figure 3.2. Vectors s_j and s_k represent *prototype* or *reference signals* belonging to the set of M waveforms, {$s_i(t)$}. The receiver knows, a priori, the location in the signal space of each prototype vector belonging to the M-ary set. During the transmission of any signal, the signal is perturbed by noise so that the resultant vector that is actually received is a perturbed version (e.g., $s_j + n$ or $s_k + n$) of the original one, where n represents a noise vector. The noise is additive and has a Gaussian distribution; therefore, the resulting distribution of possible received signals is a cluster or cloud of points around s_j and s_k. The cluster is dense in the center and becomes sparse with increasing distance from the prototype. The arrow marked r represents a signal vector that might arrive at the receiver during some symbol interval. The task of the receiver is to decide whether r has a close "resemblance" to the prototype s_j, whether it more closely resembles s_k, or whether it is closer to some other prototype signal in the M-ary set. The measurement can be thought of as a *distance* measurement. The question that the receiver or detector must resolve is: Which of the prototypes within the signal space is *closest* in distance to the received vector, r? The analysis of all demodulation or detection schemes involves this concept of *distance* between a received waveform and a set of possible transmitted waveforms. A simple rule for the detector to follow is to decide that r belongs to the same class as its nearest neighbor (nearest prototype vector).

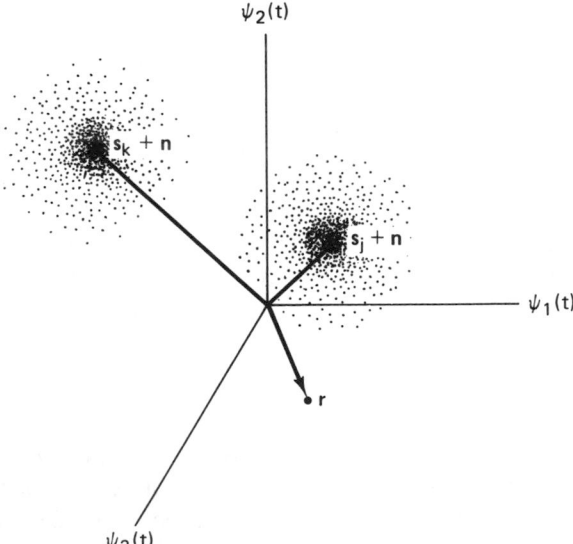

Figure 3.2 Signals and noise in a three-dimensional vector space.

3.2.2.1 Waveform Energy

Using Equations (1.5), (3.4), and (3.2), the normalized energy, E_i, associated with the waveform, $s_i(t)$, over a symbol interval, T, can be expressed in terms of the orthogonal components of $s_i(t)$ as follows:

$$E_i = \int_0^T s_i^2(t)\, dt = \int_0^T \left[\sum_j a_{ij}\psi_j(t)\right]^2 dt \tag{3.7}$$

$$= \int_0^T \sum_j a_{ij}\psi_j(t) \sum_k a_{ik}\psi_k(t)\, dt \tag{3.8}$$

$$= \sum_j \sum_k a_{ij}a_{ik} \int_0^T \psi_j(t)\psi_k(t)\, dt \tag{3.9}$$

$$= \sum_j \sum_k a_{ij}a_{ik}K_j\delta_{jk} \tag{3.10}$$

$$= \sum_{j=1}^N a_{ij}^2 K_j \qquad i = 1, \ldots, M \tag{3.11}$$

Equation (3.11) is a special case of Parseval's theorem relating the integral of the square of the waveform, $s_i(t)$, to the sum of the square of the orthogonal series coefficients. If orthonormal functions are used (i.e., $K_j = 1$), the normalized energy over a symbol duration T is given by

$$E_i = \sum_{j=1}^N a_{ij}^2 \tag{3.12}$$

If there is equal energy, E, in each of the $s_i(t)$ waveforms, we can write Equation (3.12) in the form

$$E = \sum_{j=1}^N a_{ij}^2 \qquad \text{for all } i \tag{3.13}$$

3.2.2.2 Generalized Fourier Transforms

The transformation described by Equations (3.2), (3.4), and (3.5) is referred to as the *generalized Fourier transformation*. In the case of ordinary Fourier transforms, the $\{\psi_j(t)\}$ set is comprised of sine and cosine harmonic functions. But in the case of generalized Fourier transforms, the $\{\psi_j(t)\}$ set is not constrained to any specific form; it must only satisfy the orthogonality statement of Equation (3.2). *Any* arbitrary integrable waveform set, as well as noise, can be represented as a linear combination of orthogonal waveforms through such a generalized Fourier transformation [2]. Therefore, in such an orthogonal space, we are justified in using distance (Euclidean distance) as a decision criterion for the detection of *any* signal set in the presence of AWGN. The most important application of this orthogonal transformation has to do with the way in which signals are actually transmitted and received. The transmission of a nonorthogonal signal set is generally accomplished by the appropriate weighting of the orthogonal carrier components. For example, in Section 3.5.3 we show that multiple phase shift keying (MPSK) signals are fully characterized by weighted sine and cosine components of the carrier.

Example 3.1 Orthogonal Representation of Waveforms

Figure 3.3 illustrates the statement that any arbitrary integrable waveform set can be represented as a linear combination of orthogonal waveforms. Figure 3.3a shows a set of three waveforms, $s_1(t)$, $s_2(t)$, $s_3(t)$.

(a) Demonstrate that these waveforms *do not* form an orthogonal set.

(b) Figure 3.3b shows a set of two waveforms, $\psi_1(t)$ and $\psi_2(t)$. Verify that these waveforms form an orthogonal set.

(c) Show how the nonorthogonal waveform set in part (a) can be expressed as a linear combination of the orthogonal set in part (b).

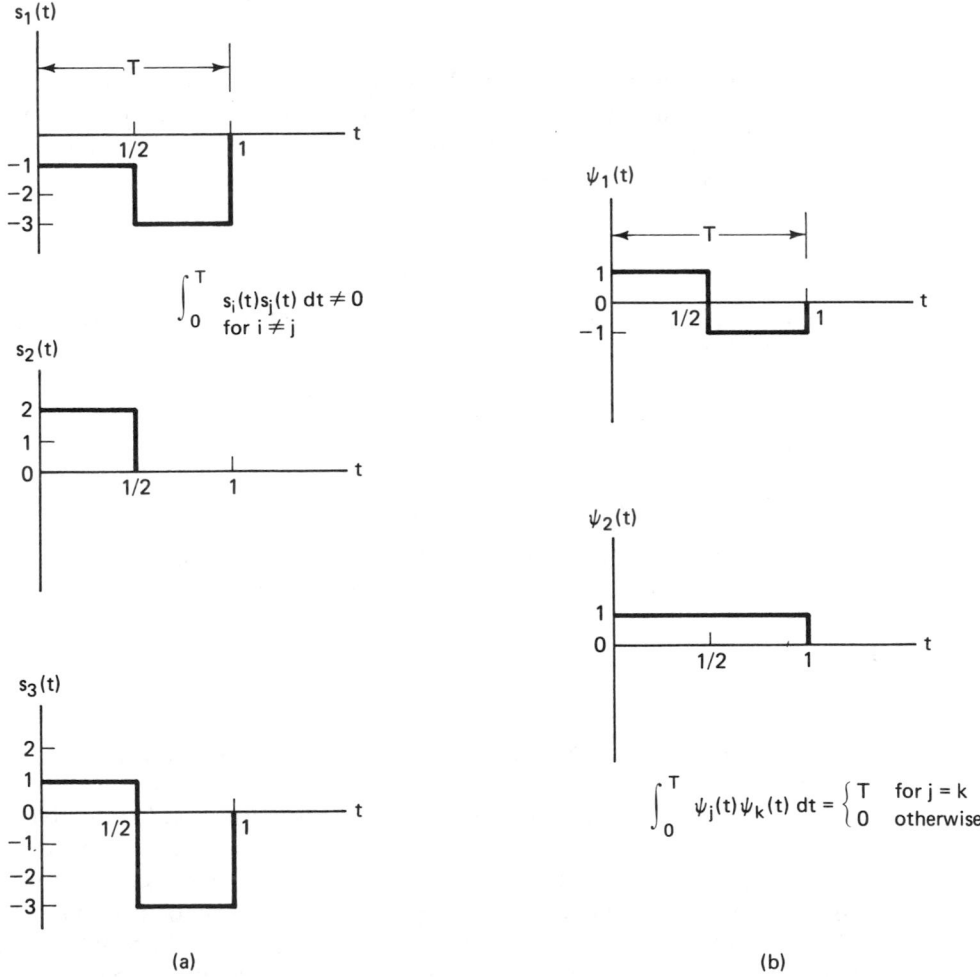

(a)

(b)

Figure 3.3 Example of an arbitrary signal set in terms of an orthogonal set. (a) Arbitrary signal set. (b) Orthogonal basis functions.

Bandpass Modulation and Demodulation   Chap. 3

Solution

(a) $s_1(t)$, $s_2(t)$, and $s_3(t)$ are clearly not orthogonal, since they do not meet the requirements of Equation (3.2); that is, the time integrated value (over a symbol duration) of the cross-product of any two of the three waveforms is not zero. Let us verify this for $s_1(t)$ and $s_2(t)$.

$$\int_0^T s_1(t)s_2(t)\ dt = \int_0^{T/2} s_1(t)s_2(t)\ dt + \int_{T/2}^T s_1(t)s_2(t)\ dt$$

$$= \int_0^{T/2} (-1)(2)\ dt + \int_{T/2}^T (-3)(0)\ dt = -T$$

Similarly, the integral over the interval, T, of each of the cross-products $s_1(t)s_3(t)$ and $s_2(t)s_3(t)$ results in nonzero values. Hence the waveform set $\{s_i(t)\}$ ($i = 1, 2, 3$) in Figure 3.3a is not an orthogonal set.

(b) Using Equation (3.2), we verify that $\psi_1(t)$ and $\psi_2(t)$ form an orthogonal set as follows:

$$\int_0^T \psi_1(t)\psi_2(t)\ dt = \int_0^{T/2} (1)(1)\ dt + \int_{T/2}^T (-1)(1)\ dt = 0$$

(c) We can express the nonorthogonal set $\{s_i(t)\}$ ($i = 1, 2, 3$) as a linear combination of the orthogonal basis waveforms $\{\psi_j(t)\}$ ($j = 1, 2$), as follows, by using Equation (3.5), where $K_j = T$:

$$s_1(t) = \psi_1(t) - 2\psi_2(t)$$

$$s_2(t) = \psi_1(t) + \psi_2(t)$$

$$s_3(t) = 2\psi_1(t) - \psi_2(t)$$

These relationships illustrate how an arbitrary waveform set $\{s_i(t)\}$ can be expressed as a linear combination of an orthogonal set $\{\psi_j(t)\}$, as described in Equations (3.4) and (3.5). What are the practical applications for being able to describe $s_1(t)$, $s_2(t)$, and $s_3(t)$, in terms of $\psi_1(t)$, $\psi_2(t)$, and the appropriate coefficients? If we want a system for transmitting waveforms $s_1(t)$, $s_2(t)$, and $s_3(t)$, the transmitter and the receiver need only be implemented using the two basis functions $\psi_1(t)$ and $\psi_2(t)$ instead of the three original waveforms. A convenient way in which an appropriate choice of a basis function set, $\{\psi_j(t)\}$, can be obtained for any given signal set, $\{s_i(t)\}$, is called the *Gram–Schmidt orthogonalization procedure*. It is described in Appendix 4A of Reference [3].

3.2.2.3 Representing White Noise with Orthogonal Waveforms

Additive white Gaussian noise (AWGN) can be expressed as a linear combination of orthogonal waveforms in the same way as signals. For the signal detection problem, the noise can be partitioned into two components,

$$n(t) = \hat{n}(t) + \tilde{n}(t) \tag{3.14}$$

where

$$\hat{n}(t) = \sum_{j=1}^N n_j \psi_j(t) \tag{3.15}$$

is taken to be the noise within the signal space, or the projection of the noise components on the signal coordinates $\psi_1(t), \ldots, \psi_N(t)$, and

$$\tilde{n}(t) = n(t) - \hat{n}(t) \tag{3.16}$$

is defined as the noise outside the signal space. In other words, $\tilde{n}(t)$ may be thought of as the noise that is effectively tuned out by the detector. The symbol $\hat{n}(t)$ represents the noise that will interfere with the detection process. We can express the noise waveform, $n(t)$, as follows:

$$n(t) = \sum_{j=1}^{N} n_j \psi_j(t) + \tilde{n}(t) \tag{3.17}$$

where

$$n_j = \frac{1}{K_j} \int_0^T n(t)\psi_j(t)\, dt \qquad \text{for all } j \tag{3.18}$$

and

$$0 = \int_0^T \tilde{n}(t)\psi_j(t)\, dt \tag{3.19}$$

The interfering portion of the noise, $\hat{n}(t)$, expressed in Equation (3.15) will henceforth be referred to simply as $n(t)$. We can express $n(t)$ by a vector of its coefficients similar to the way we did for signals in Equation (3.6).

$$\mathbf{n} = (n_1, n_2, \ldots, n_N) \tag{3.20}$$

where $\mathbf{n}$ is a random vector with zero mean and Gaussian distribution, and where the noise components n_i ($i = 1, \ldots, N$) are independent.

3.2.2.4 Variance of White Noise

White noise is an *idealized process* with two-sided power spectral density equal to a constant, $N_0/2$, for all frequencies from $-\infty$ to $+\infty$. Hence the noise variance (average noise power, since the noise has zero mean) is

$$\sigma^2 = \text{var}\,[n(t)] = \int_{-\infty}^{\infty} \left(\frac{N_0}{2}\right) df = \infty \tag{3.21}$$

Although the variance for AWGN is infinite, the variance for *filtered* AWGN is finite. For example, if AWGN is correlated with one of a set of orthonormal functions $\psi_j(t)$, the variance of the correlator output is given by

$$\sigma^2 = \text{var}\,(n_j) = \mathbf{E}\left\{\left[\int_0^T n(t)\psi_j(t)\, dt\right]^2\right\} = \frac{N_0}{2} \tag{3.22}$$

The proof of Equation (3.22) is given in Appendix C. Henceforth we shall assume that the noise of interest in the detection process is the output noise of a correlator or matched filter with variance $\sigma^2 = N_0/2$ as expressed in Equation (3.22).

3.3 DIGITAL BANDPASS MODULATION TECHNIQUES

Bandpass modulation (either analog or digital) is the process by which an information signal is converted to a sinusoidal waveform; for digital modulation, such a sinusoid of duration T is referred to as a digital symbol. The sinusoid has just three features that can be used to distinguish it from other sinusoids: amplitude, frequency, and phase. Thus bandpass modulation can be defined as the process whereby the amplitude, frequency, or phase of an RF carrier, or a combination of them, is varied in accordance with the information to be transmitted. The general form of the carrier wave, $s(t)$, is as follows:

$$s(t) = A(t) \cos \theta(t) \qquad (3.23)$$

where $A(t)$ is the time-varying amplitude and $\theta(t)$ is the time-varying angle. It is convenient to write

$$\theta(t) = \omega_0 t + \phi(t) \qquad (3.24)$$

so that

$$s(t) = A(t) \cos [\omega_0 t + \phi(t)] \qquad (3.25)$$

where ω_0 is the *radian frequency* of the carrier and $\phi(t)$ is the *phase*. The terms f and ω will each be used to denote frequency. When f is used, frequency in hertz is intended; when ω is used, frequency in radians per second is intended. The two frequency parameters are related by $\omega = 2\pi f$.

The basic *digital modulation/demodulation* types are listed in Figure 3.4. When the receiver exploits knowledge of the carrier's phase to detect the signals, the process is called *coherent detection*; when the receiver does not utilize such phase reference information, the process is called *noncoherent detection*. In digital communications, the terms *demodulation* and *detection* are used somewhat interchangeably, although demodulation emphasizes removal of the carrier, and detection includes the process of symbol decision. In ideal coherent detection, there is available at the receiver a prototype of each possible arriving signal. These prototype waveforms attempt to duplicate the transmitted signal set in every respect, even RF phase. The receiver is then said to be *phase locked* to the incoming signal. During detection, the receiver multiplies and integrates (correlates) the incoming signal with each of its prototype replicas. Under the heading of coherent modulation/demodulation in Figure 3.4 are listed phase shift keying (PSK), frequency shift keying (FSK), amplitude shift keying (ASK), continuous phase modulation (CPM), and hybrid combinations. The basic bandpass modulation formats are discussed in this chapter. Some specialized formats, such as offset quadrature PSK (OQPSK), minimum shift keying (MSK) belonging to the CPM class, and quadrature amplitude modulation (QAM), are treated in Chapter 7.

Noncoherent demodulation refers to systems employing demodulators that are designed to operate without knowledge of the absolute value of the incoming signal's phase; therefore, phase estimation is not required. Thus the advantage of noncoherent over coherent systems is reduced complexity, and the price paid is increased probability of error (P_E). In Figure 3.4 the modulation/demodulation

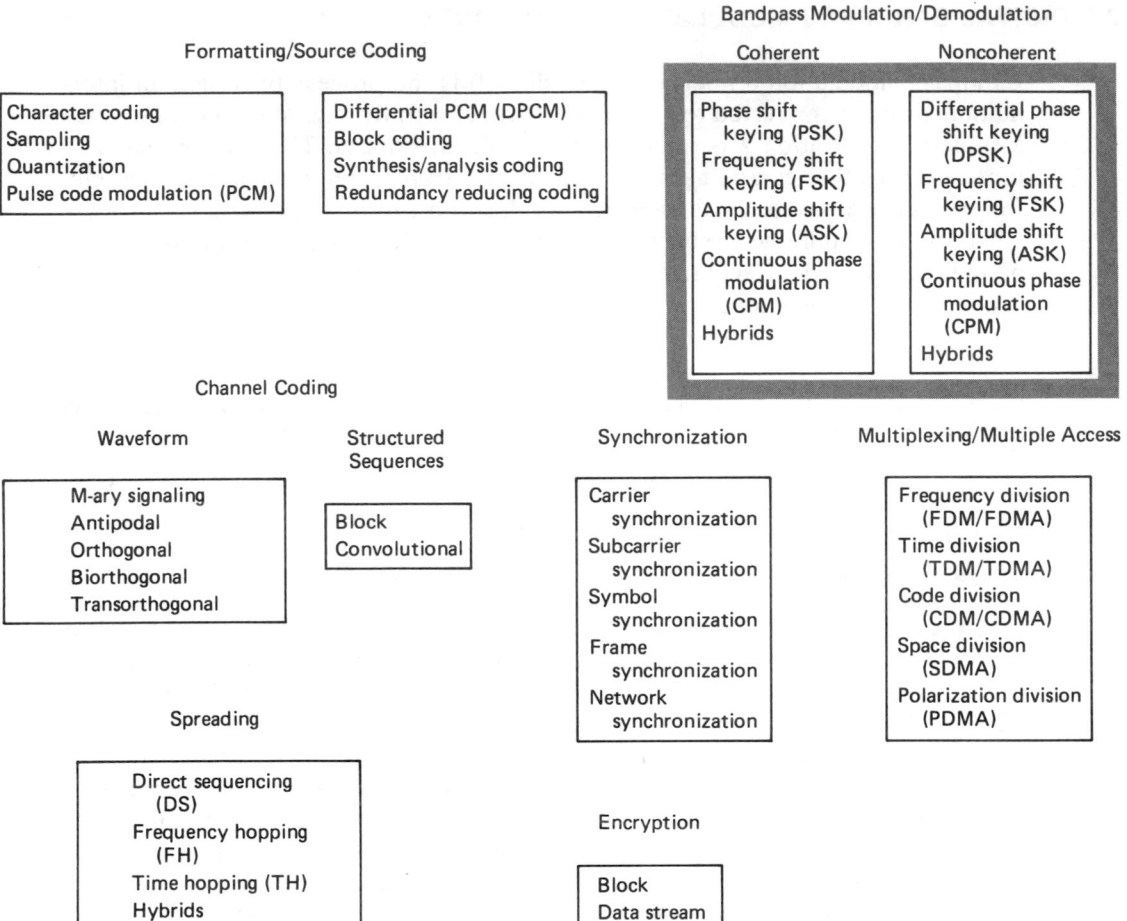

Figure 3.4 Basic digital communication transformations.

types that are listed in the noncoherent column, DPSK, FSK, ASK, CPM, and hybrids, are similar to those listed in the coherent column. We had implied that phase information is not used for noncoherent reception; how do you account for the fact that there is a form of phase shift keying under the noncoherent heading? It turns out that an important form of PSK can be classified as noncoherent (or differentially coherent) since it does not require a reference in phase with the received carrier. This "pseudo-PSK," termed *differential PSK* (DPSK), utilizes phase information of the prior symbol as a phase reference for detecting the current symbol. This is described in Sections 3.6.1 and 3.6.2.

Figure 3.5 illustrates examples of the most common digital modulation formats: PSK, FSK, ASK, and a hybrid combination of ASK and PSK (ASK/PSK or APK). The first column lists the analytic expression, the second is a typical pictorial of the waveform versus time, and the third is a vectorial schematic, with the orthogonal axes labeled $\{\psi_j(t)\}$. In the general *M*-ary signaling case, the pro-

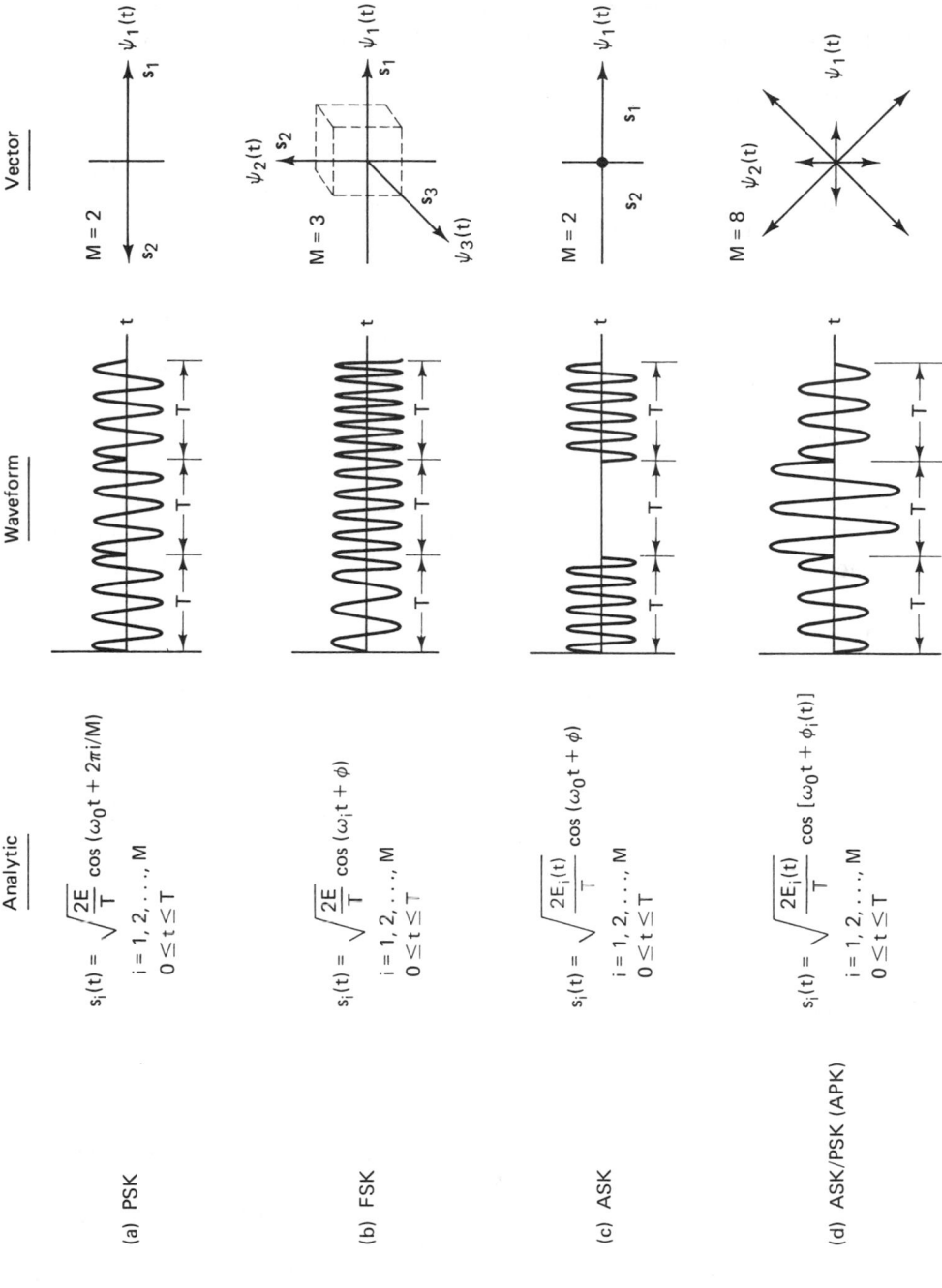

Figure 3.5 Digital modulations. (a) PSK. (b) FSK. (c) ASK. (d) ASK/PSK (APK).

Analytic

(a) PSK

$$s_i(t) = \sqrt{\frac{2E}{T}} \cos(\omega_0 t + 2\pi i/M)$$

$$i = 1, 2, \ldots, M$$
$$0 \le t \le T$$

(b) FSK

$$s_i(t) = \sqrt{\frac{2E}{T}} \cos(\omega_i t + \phi)$$

$$i = 1, 2, \ldots, M$$
$$0 \le t \le T$$

(c) ASK

$$s_i(t) = \sqrt{\frac{2E_i(t)}{T}} \cos(\omega_0 t + \phi)$$

$$i = 1, 2, \ldots, M$$
$$0 \le t \le T$$

(d) ASK/PSK (APK)

$$s_i(t) = \sqrt{\frac{2E_i(t)}{T}} \cos[\omega_0 t + \phi_i(t)]$$

$$i = 1, 2, \ldots, M$$
$$0 \le t \le T$$

cessor accepts k source bits at a time and instructs the modulator to produce one of an available set of $M = 2^k$ waveform types. Binary modulation, where $k = 1$, is just a special case of M-ary modulation. Each example shown in Figure 3.5 illustrates the set of signal waveforms with a particular value chosen for M.

3.3.1 Phase Shift Keying

Phase shift keying (PSK) was developed during the early days of the deep-space program; PSK is now widely used in both military and commercial communications systems. The general analytic expression for PSK is

$$s_i(t) = \sqrt{\frac{2E}{T}} \cos \left[\omega_0 t + \phi_i(t) \right] \quad \begin{array}{l} 0 \leq t \leq T \\ i = 1, \ldots, M \end{array} \qquad (3.26)$$

where the phase term, $\phi_i(t)$, will have M discrete values, typically given by

$$\phi_i(t) = \frac{2\pi i}{M} \quad i = 1, \ldots, M$$

For the binary PSK (BPSK) example in Figure 3.5a, M is 2. The parameter E is symbol energy, T is symbol time duration, and $0 \leq t \leq T$. In BPSK modulation, the modulating data signal shifts the phase of the waveform, $s_i(t)$, to one of two states, either zero or π (180°). The waveform sketch in Figure 3.5a shows a typical BPSK waveform with its abrupt phase changes at the symbol transitions; if the modulating data stream were to consist of alternating ones and zeros, there would be such an abrupt change at each transition. The signal waveforms can be represented as vectors on a polar plot; the vector length corresponds to the signal amplitude, and the vector direction, for the general M-ary case, corresponds to the signal phase relative to the other $M - 1$ signals in the set. For the BPSK example, the vectorial picture illustrates the two 180° opposing vectors. Signal sets that can be depicted with such opposing vectors are called *antipodal signal sets*.

3.3.2 Frequency Shift Keying

The general analytic expression for FSK modulation is

$$s_i(t) = \sqrt{\frac{2E}{T}} \cos \left(\omega_i t + \phi \right) \quad \begin{array}{l} 0 \leq t \leq T \\ i = 1, \ldots, M \end{array} \qquad (3.27)$$

where the frequency term, ω_i, will have M discrete values, and the phase term, ϕ, is an arbitrary constant. The FSK waveform sketch in Figure 3.5b illustrates the typical abrupt frequency changes at the symbol transitions. In this example, M has been chosen equal to 3, corresponding to the same number of waveform types (3-ary); note that this $M = 3$ choice for FSK has been selected to emphasize the mutually perpendicular axes. In practice, M is usually a nonzero power of 2 (2, 4, 8, 16, . . .). The signal set is characterized by Cartesian coordinates, such that each of the mutually perpendicular axes represents a sinusoid with a different frequency. As described earlier, signal sets that can be characterized with such mutually perpendicular vectors are called *orthogonal* signals. The required frequency spacing between the orthogonal tones is discussed in Section 3.6.4.

3.3.3 Amplitude Shift Keying

For the ASK example in Figure 3.5c, the general analytic expression is

$$s_i(t) = \sqrt{\frac{2E_i(t)}{T}} \cos(\omega_0 t + \phi) \qquad \begin{array}{l} 0 \leq t \leq T \\ i = 1, \ldots, M \end{array} \qquad (3.28)$$

where the amplitude term, $\sqrt{2E_i(t)/T}$, will have M discrete values, and the phase term, ϕ, is an arbitrary constant. In Figure 3.5c, M has been chosen equal to 2, corresponding to two waveform types. The ASK waveform sketch in the figure can describe a radar transmission example, where the two signal amplitude states would be $\sqrt{2E/T}$ and zero. The vectorial picture utilizes the same phase–amplitude polar coordinates as the PSK example. Here we see a vector corresponding to the maximum-amplitude state, and a point at the origin corresponding to the zero-amplitude state. Binary ASK signaling (also called on–off keying) was one of the earliest forms of digital modulation used in radio telegraphy at the beginning of this century. Simple ASK is no longer widely used in digital communication systems; therefore, it will not be treated in detail.

3.3.4 Amplitude Phase Keying

For the combination of ASK and PSK (APK) example in Figure 3.5d, the general analytic expression

$$s_i(t) = \sqrt{\frac{2E_i(t)}{T}} \cos[\omega_0 t + \phi_i(t)] \qquad \begin{array}{l} 0 \leq t \leq T \\ i = 1, \ldots, M \end{array} \qquad (3.29)$$

illustrates the indexing of both the signal amplitude term and the phase term. The APK waveform picture in Figure 3.5d illustrates some typical simultaneous phase and amplitude changes at the symbol transition times. For this example, M has been chosen equal to 8, corresponding to eight waveforms (8-ary). The figure illustrates a hypothetical eight-vector signal set on the phase–amplitude plane. Four of the vectors are at one amplitude; the other four vectors are at a different amplitude; and each of the vectors is separated by 45°. When the set of M symbols in the two-dimensional signal space are arranged in a rectangular constellation, the signaling is referred to as quadrature amplitude modulation (QAM); examples of QAM are considered in Chapter 7.

The vectorial picture for each of the modulation types described in Figure 3.5 (except the FSK case) is characterized on a plane whose *polar* coordinates represent signal *amplitude* and *phase*. The FSK case is characterized in a *Cartesian* coordinate space, with each axis representing a *frequency tone* ($\cos \omega_i t$) from the M-ary set of orthogonal tones.

3.3.5 Waveform Amplitude Coefficient

The waveform amplitude coefficient appearing in Equations (3.26) to (3.29) has the same general form, $\sqrt{2E/T}$, for all modulation formats. This expression is derived as follows:

$$s(t) = A \cos \omega t \tag{3.30}$$

where A is the peak value of the waveform. Since the peak value of a sinusoidal waveform equals $\sqrt{2}$ times the root-mean-square (rms) value, we can write

$$s(t) = \sqrt{2} A_{rms} \cos \omega t$$
$$= \sqrt{2 A_{rms}^2} \cos \omega t$$

Assuming the signal to be a voltage or a current waveform, A_{rms}^2 represents average power P (normalized to 1 Ω). Therefore, we can write

$$s(t) = \sqrt{2P} \cos \omega t \tag{3.31}$$

Replacing P watts by E joules/T seconds, we get

$$s(t) = \sqrt{\frac{2E}{T}} \cos \omega t \tag{3.32}$$

We shall use either the amplitude notation, A, in Equation (3.30) or the designation $\sqrt{2E/T}$ in Equation (3.32). Since the *energy* in a signal is the key parameter in determining the error performance of the detection process, it is often more convenient to use the amplitude notation in Equation (3.32) because it facilitates solving directly for the probability of error, P_E, as a function of signal energy.

3.4 DETECTION OF SIGNALS IN GAUSSIAN NOISE

3.4.1 Decision Regions

Consider that the two-dimensional signal space in Figure 3.6 is the locus of the noise-perturbed prototype binary vectors $(\mathbf{s}_1 + \mathbf{n})$ and $(\mathbf{s}_2 + \mathbf{n})$. The noise vector, $\mathbf{n}$, is a zero-mean random vector; hence the received signal vector, $\mathbf{r}$, is a random vector with mean $\mathbf{s}_1$ or $\mathbf{s}_2$. The detector's task after receiving $\mathbf{r}$ is to decide which of the signals, $\mathbf{s}_1$ or $\mathbf{s}_2$, was actually transmitted. The method is usually to decide on the signal classification that yields the minimum expected P_E, although other strategies are possible [4]. For the case where M equals 2, with $\mathbf{s}_1$ and $\mathbf{s}_2$ being equally likely and with the noise being an additive white Gaussian noise (AWGN) process, we will see that the minimum-error decision rule is equivalent to choosing the signal class such that the distance $d(\mathbf{r}, \mathbf{s}_i) = \| \mathbf{r} - \mathbf{s}_i \|$ is minimized, where $\| \mathbf{x} \|$ is called the *norm* or *magnitude* of vector $\mathbf{x}$. This rule is often stated in terms of decision regions. In Figure 3.6, let us construct decision regions in the following

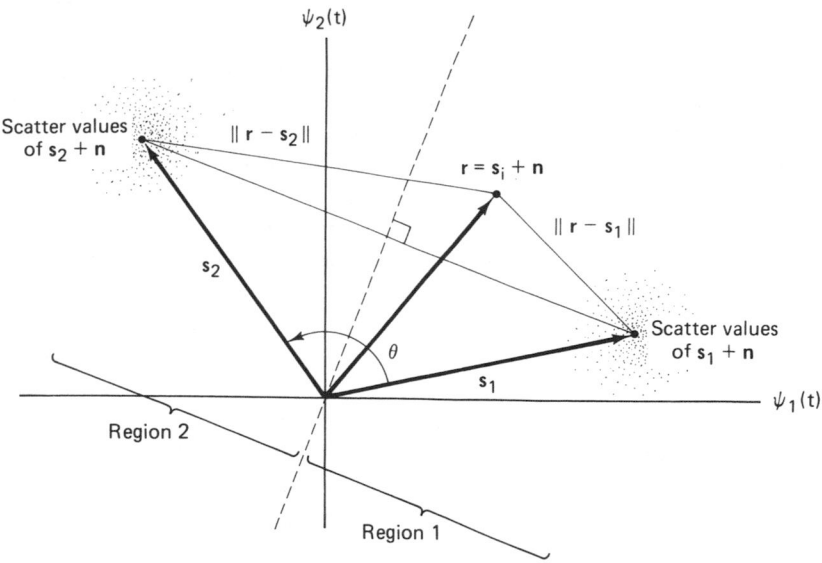

Figure 3.6 Two-dimensional signal space, with arbitrary equal-amplitude vectors s_1 and s_2.

way. Draw a line connecting the tips of the prototype vectors, s_1 and s_2. Next, construct the perpendicular bisector of the connecting line. Notice that this bisector passes through the origin of the space if s_1 and s_2 are equal in amplitude. For this $M = 2$ example in Figure 3.6, the constructed perpendicular bisector represents the locus of points equidistant between s_1 and s_2; hence the bisector describes the boundary between decision region 1 and decision region 2. The *decision rule* for the detector, stated in terms of *decision regions,* is: Whenever the received signal r is located in region 1, choose signal s_1; when it is located in region 2, choose signal s_2.

3.4.2 Correlation Receiver

In Section 2.9 we treated the detection of *baseband* binary signals in Gaussian noise. Since the detection of *bandpass* signals employs the same concepts, we shall summarize the key findings of that section. We focus particularly on that realization of a matched filter known as a *correlator*. In addition to binary detection, we also consider the more general case of *M*-ary detection. We assume that the only performance degradation is due to AWGN. The received signal, $r(t)$, is the sum of the transmitted prototype signal plus the random noise:

$$r(t) = s_i(t) + n(t) \quad \begin{array}{c} 0 \leq t \leq T \\ i = 1, \ldots, M \end{array} \tag{3.33}$$

Given such a received signal, the detection process consists of *two basic steps.* In the first step, the received waveform, $r(t)$, is reduced to a *single random variable, $z(T)$,* or a *set of random variables, $z_i(T)$ ($i = 1, \ldots, M$),* formed at the output of the correlator(s) at time $t = T$, where T is the symbol duration. In the second step, a symbol decision is made, on the basis of comparing $z(T)$ to a threshold or on the basis of choosing the maximum $z_i(T)$. Step 1 can be thought of as transforming the waveform into a point in the decision space. Step 2 can be thought of as determining *in which decision region* the point is located. For the detector to be optimized (in the sense of minimizing the error probability), it is necessary to optimize the waveform-to-random-variable transformation, by using matched filters or correlators in step 1, and by also optimizing the decision criterion in step 2.

In Sections 2.9.2 and 2.9.3 we found that the matched filter provides the maximum signal-to-noise ratio at the filter output at time $t = T$. We described a correlator as one realization of a matched filter. We can define a *correlation receiver* comprised of M correlators, as shown in Figure 3.7a, that transforms a received waveform, $r(t)$, to a sequence of M numbers or correlator outputs, $z_i(T)$ ($i = 1, \ldots, M$). Each correlator output is characterized by the following product integration or correlation with the received signal.

$$z_i(T) = \int_0^T r(t)s_i(t) \, dt \qquad i = 1, \ldots, M \qquad (3.34)$$

The verb "to correlate" means "to match." The correlators attempt to match the incoming received signal, $r(t)$, with each of the candidate prototype waveforms, $s_i(t)$, known a priori to the receiver. A reasonable decision rule is to choose the waveform, $s_i(t)$, that *matches best* or has the *largest correlation* with $r(t)$. In other words, the decision rule is:

$$\text{Choose the } s_i(t) \text{ whose index} \atop \text{corresponds to the max } z_i(T) \qquad (3.35)$$

Following Equation (3.4), any signal set, $\{s_i(t)\}$ ($i = 1, \ldots, M$), can be expressed in terms of some set of basis functions, $\{\psi_j(t)\}$ ($j = 1, \ldots, N$), where $N \le M$. Then the bank of M correlators in Figure 3.7a may be replaced with a bank of N correlators, shown in Figure 3.7b, where the set of basis functions $\{\psi_j(t)\}$ form *reference signals*. The decision stage of this receiver consists of logic circuitry for choosing the signal, $s_i(t)$. The choice of $s_i(t)$ is made according to the best match of the coefficients, a_{ij}, seen in Equation (3.4), with the set of outputs $\{z_j(T)\}$. When the prototype waveform set, $\{s_i(t)\}$, is an orthogonal set, the receiver implementation in Figure 3.7a is identical to that in Figure 3.7b (differing perhaps by a scale factor). However, when $\{s_i(t)\}$ is *not* an orthogonal set, the receiver in Figure 3.7b, using N correlators instead of M, with reference signals $\{\psi_j(t)\}$, can represent a cost-effective implementation. We examine such an application for the detection of multiple phase shift keying (MPSK) in Section 3.5.3. For the other applications in this chapter, we shall assume a correlator receiver with reference signals $\{s_i(t)\}$.

In the case of *binary detection,* the correlation receiver can be configured as a single matched filter or product integrator, as shown in Figure 3.8a, with the reference signal being the difference between the binary prototype signals, $s_1(t) - s_2(t)$. The output of the correlator, $z(T)$, is fed directly to the decision stage.

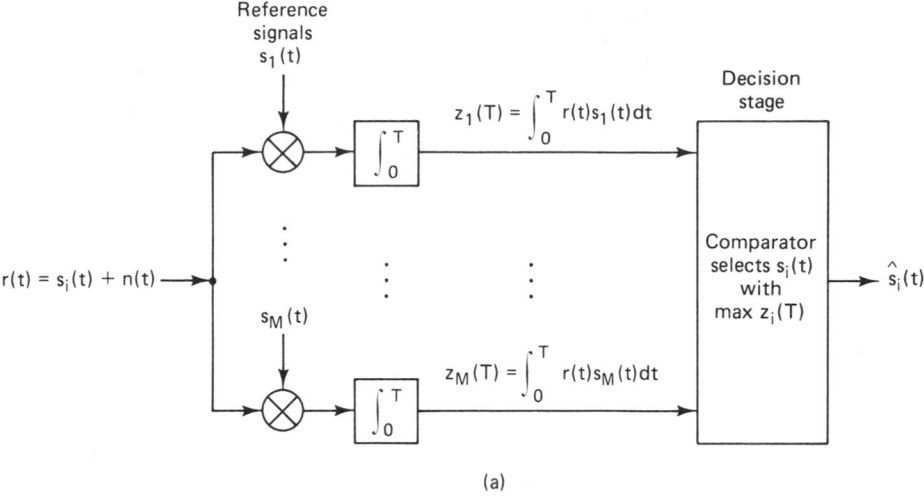

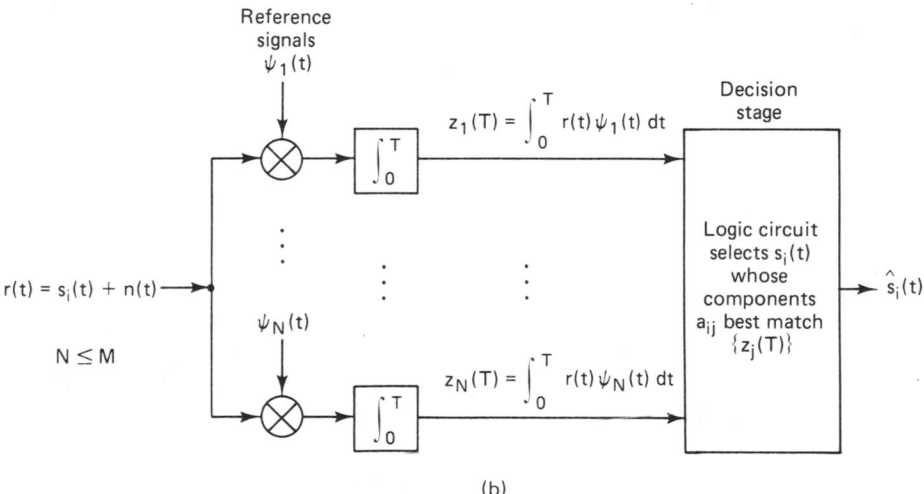

Figure 3.7 (a) Correlator receiver with reference signals $\{s_i(t)\}$. (b) Correlator receiver with reference signals $\{\psi_j(t)\}$.

For binary detection, the correlation receiver can also be drawn, as shown in Figure 3.8b, as two matched filters or product integrators, each of which is matched to one of the prototype reference signals, $s_1(t)$ or $s_2(t)$. The decision stage can then be configured to follow the rule in Equation (3.35), or the correlator outputs, $z_i(T)$ ($i = 1, 2$), can be differenced to form

$$z(T) = z_1(T) - z_2(T) \tag{3.36}$$

as shown in Figure 3.8b. Then, $z(T)$, called the *test statistic,* is fed to the decision stage, as in the case of the single correlator. In the *absence of noise,* an input

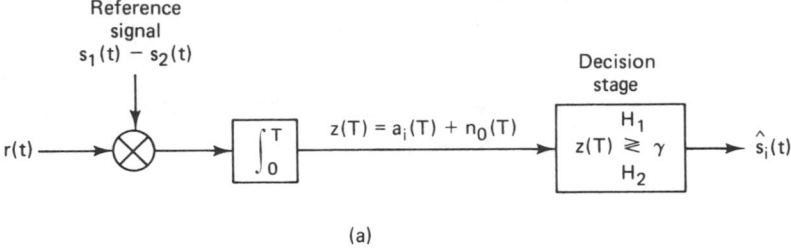

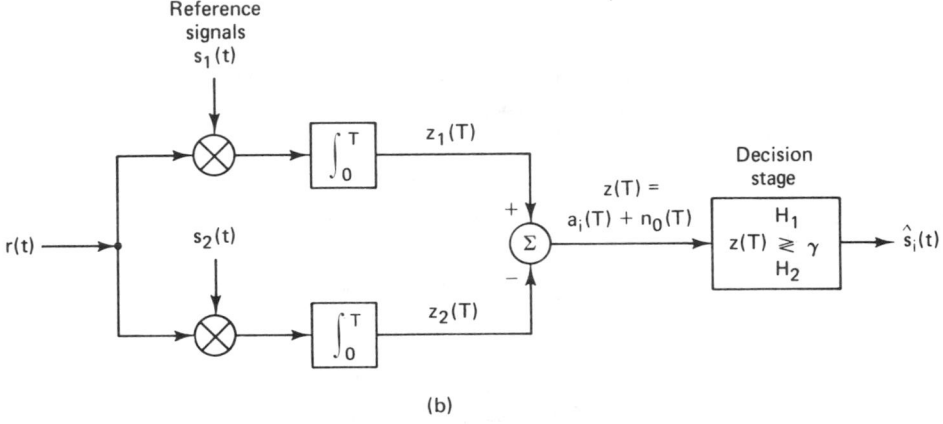

Figure 3.8 Binary correlator receiver. (a) Using a single correlator. (b) Using two correlators.

waveform, $s_i(t)$, yields the output, $z(T) = a_i(T)$, a signal-only component. The input noise, $n(t)$, is a Gaussian random process. Since the correlator is a *linear* device, the output noise is also a Gaussian random process [4]. Thus the output of the correlator, sampled at $t = T$, yields

$$z(T) = a_i(T) + n_0(T) \qquad i = 1, 2$$

where $n_0(T)$ is the noise component. To shorten the notation we sometimes express $z(T)$ as $a_i + n_0$. The noise component, n_0, is a zero-mean *Gaussian random variable*, and thus $z(T)$ is a *Gaussian random variable* with a mean of either a_1 or a_2 depending on whether a binary one or binary zero was sent.

3.4.2.1 Binary Decision Threshold

For the random variable, $z(T)$, Figure 3.9 illustrates the two conditional probability density functions (pdfs), $p(z|s_1)$ and $p(z|s_2)$, with mean value of a_1 and a_2, respectively (these pdfs are also called the *likelihood* of s_1 and the *likelihood*

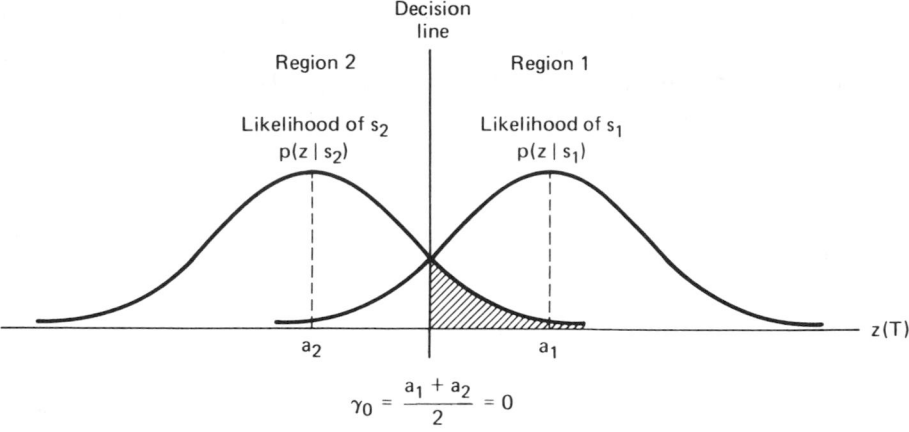

Figure 3.9 Conditional probability density functions: $p(z|s_1)$, $p(z|s_2)$.

of s_2, respectively):

$$p(z|s_1) = \frac{1}{\sigma_0\sqrt{2\pi}} \exp\left[-\frac{1}{2}\left(\frac{z - a_1}{\sigma_0}\right)^2 \right] \qquad (3.37a)$$

$$p(z|s_2) = \frac{1}{\sigma_0\sqrt{2\pi}} \exp\left[-\frac{1}{2}\left(\frac{z - a_2}{\sigma_0}\right)^2 \right] \qquad (3.37b)$$

where σ_0^2 is the noise variance. In Figure 3.9 the rightmost likelihood, $p(z|s_1)$, illustrates the probability density of the detector output, $z(T)$, given that $s_1(t)$ was transmitted. Similarly, the leftmost likelihood $p(z|s_2)$, illustrates the probability density of $z(T)$ given that $s_2(t)$ was transmitted. The abscissa, $z(T)$, represents the full range of possible sample output values from the correlation receiver in Figure 3.8.

With regard to optimizing the binary decision threshold for deciding in which region a received signal is located, we found in Section 2.9.1 that the *minimum error* criterion for equally likely binary signals corrupted by Gaussian noise can be stated as follows:

$$z(T) \underset{H_2}{\overset{H_1}{\gtrless}} \frac{a_1 + a_2}{2} = \gamma_0 \qquad (3.38)$$

where a_1 is the signal component of $z(T)$ when $s_1(t)$ is transmitted, and a_2 is the signal component of $z(T)$ when $s_2(t)$ is transmitted. The threshold level, γ_0, represented by $(a_1 + a_2)/2$, is the *optimum threshold* for minimizing the probability of making an incorrect decision given equally likely signals and symmetrical likelihoods. The decision rule in Equation (3.38) states that hypothesis H_1 should be selected [equivalent to deciding that signal $s_1(t)$ was sent] if $z(T) > \gamma_0$, and hypothesis H_2 should be selected [equivalent to deciding that $s_2(t)$ was sent] if $z(T)$ $< \gamma_0$. If $z(T) = \gamma_0$, the decision can be an arbitrary one. For equal-energy, equally likely antipodal signals, where $s_1(t) = -s_2(t)$ and $a_1 = -a_2$, the optimum decision rule becomes

$$z(T) \underset{H_2}{\overset{H_1}{\gtrless}} \gamma_0 = 0 \tag{3.39a}$$

or

$$
\begin{array}{ll}
\text{decide } s_1(t) & \text{if } z_1(T) > z_2(T) \\
\text{decide } s_2(t) & \text{otherwise}
\end{array}
\tag{3.39b}
$$

In the next section we illustrate the use of correlators and matched filters for the coherent detection of PSK and FSK modulation. In later sections we consider noncoherent detection, and we treat the error performance of various modulation types.

3.5 COHERENT DETECTION

3.5.1 Coherent Detection of PSK

The detector shown in Figure 3.7 can be used for the coherent detection of any digital waveforms. Such a correlating detector is often referred to as a *maximum likelihood detector*. Consider the following binary PSK (BPSK) example. Let

$$s_1(t) = \sqrt{\frac{2E}{T}} \cos(\omega_0 t + \phi) \qquad 0 \le t \le T \tag{3.40a}$$

$$s_2(t) = \sqrt{\frac{2E}{T}} \cos(\omega_0 t + \phi + \pi)$$

$$= -\sqrt{\frac{2E}{T}} \cos(\omega_0 t + \phi) \qquad 0 \le t \le T \tag{3.40b}$$

$$n(t) = \text{zero-mean white Gaussian random process}$$

where the phase term, ϕ, is an arbitrary constant, so that the analysis is unaffected by setting $\phi = 0$. The parameter, E, is the signal energy per symbol, and T is the symbol duration. For this antipodal case, only a single basis function is needed. If an orthonormal signal space is assumed in Equations (3.4) and (3.5) (i.e., $K_j = 1$), we can express a basis function, $\psi_1(t)$, as follows:

$$\psi_1(t) = \sqrt{\frac{2}{T}} \cos \omega_0 t \qquad \text{for } 0 \le t \le T \tag{3.41}$$

Thus we may express the transmitted signals $s_i(t)$ in terms of $\psi_1(t)$ and the coefficients $a_{i1}(t)$:

$$s_i(t) = a_{i1}\psi_1(t) \tag{3.42a}$$

$$s_1(t) = a_{11}\psi_1(t) = \sqrt{E}\psi_1(t) \tag{3.42b}$$

$$s_2(t) = a_{21}\psi_1(t) = -\sqrt{E}\psi_1(t) \tag{3.42c}$$

Assume that $s_1(t)$ was transmitted. Then the expected values of the product integrators in Figure 3.7b, with reference signals $\psi_1(t)$ and $-\psi_1(t)$, are found as follows:

$$E\{z_1|s_1\} = E\left\{ \int_0^T \sqrt{E}\psi_1^2(t) + n(t)\psi_1(t)\, dt \right\} \tag{3.43a}$$

$$E\{z_2|s_1\} = E\left\{ \int_0^T -\sqrt{E}\psi_1^2(t) - n(t)\psi_1(t)\, dt \right\} \tag{3.43b}$$

$$E\{z_1|s_1\} = E\left\{ \int_0^T \frac{2}{T}\sqrt{E}\cos^2\omega_0 t + n(t)\sqrt{\frac{2}{T}}\cos\omega_0 t\, dt \right\} = \sqrt{E} \tag{3.44a}$$

$$E\{z_2|s_1\} = E\left\{ \int_0^T -\frac{2}{T}\sqrt{E}\cos^2\omega_0 t - n(t)\sqrt{\frac{2}{T}}\cos\omega_0 t\, dt \right\} = -\sqrt{E} \tag{3.44b}$$

where $E\{\cdot\}$ denotes the ensemble average, referred to as the *expected value*. Equation (3.44) follows because $E\{n(t)\} = 0$. The decision stage must decide which signal was transmitted by determining its location within the signal space. For this example, the choice of $\psi_1(t) = \sqrt{2/T}\cos\omega_0 t$ normalizes $E\{z_i(T)\}$ to be $\pm\sqrt{E}$. The prototype signals $\{s_i(t)\}$ are the same as the reference signals $\{\psi_j(t)\}$ except for the normalizing scale factor. The decision stage chooses the signal with the largest value of $z_i(T)$. Thus, the received signal in this example is judged to be $s_1(t)$. The error performance for such coherently detected BPSK systems is treated in Section 3.7.1.

3.5.2 Sampled Matched Filter

In Section 2.9.2 we discussed the basic characteristic of the matched filter— namely, that its impulse response is a delayed version of the mirror image (rotated on the $t = 0$ axis) of the input signal waveform. Therefore, if the signal waveform is $s(t)$, its mirror image is $s(-t)$ and the mirror image delayed by T seconds is $s(T - t)$. The impulse response, $h(t)$, of a filter matched to $s(t)$ is then described by

$$h(t) = \begin{cases} s(T - t) & 0 \leq t \leq T \\ 0 & \text{elsewhere} \end{cases} \tag{3.45}$$

Figure 3.10a illustrates how a matched filter can be implemented using digital hardware. The input signal, $r(t)$, is comprised of the prototype signal, $s(t)$, plus noise, $n(t)$. The bandwidth of the signal is $W = 1/2T_s$, where the Nyquist sampling

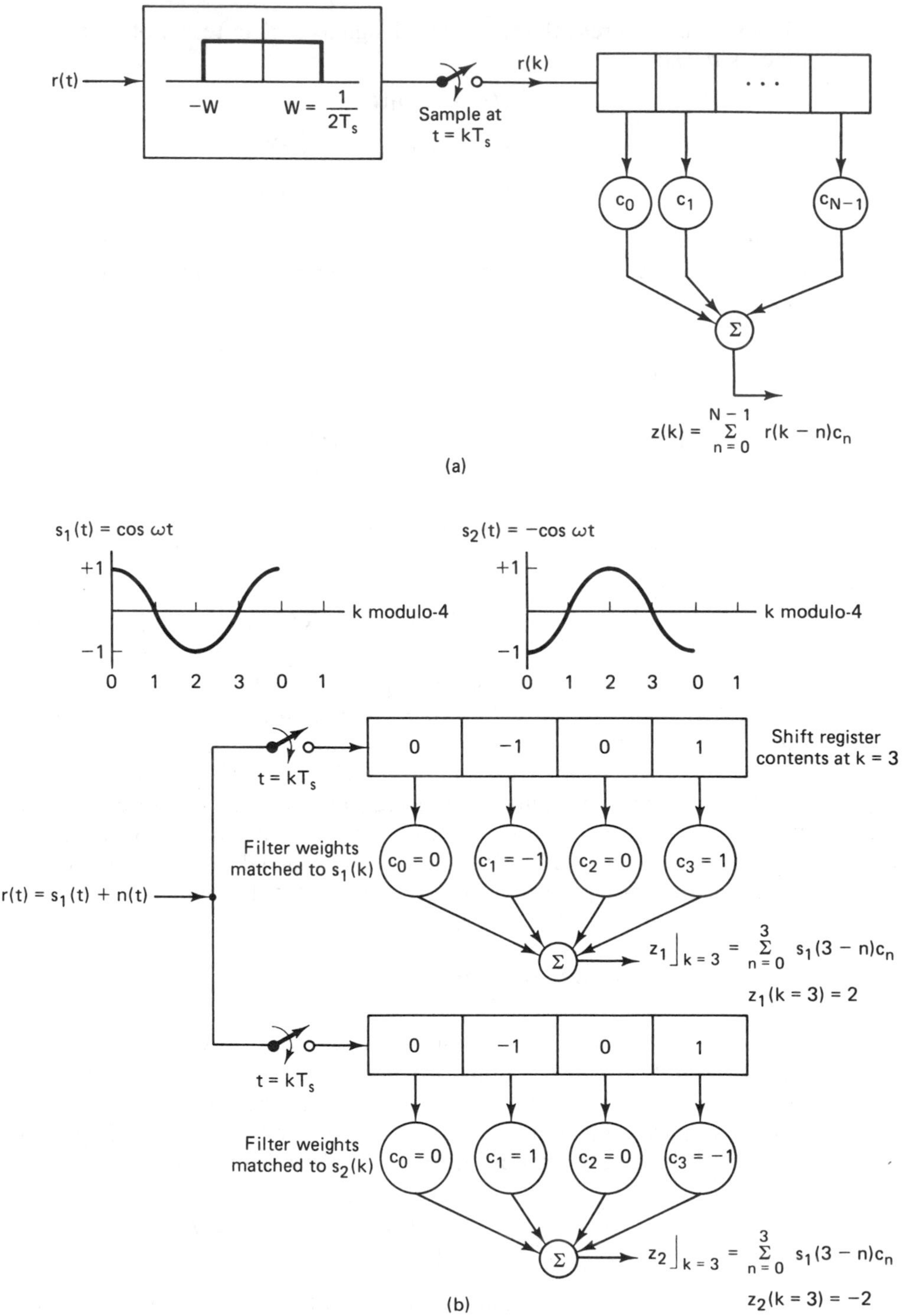

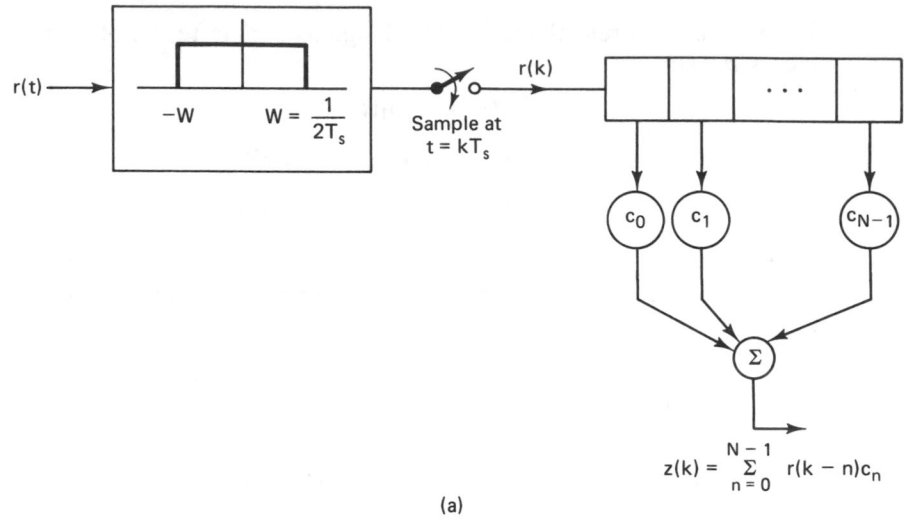

(a)

$$z(k) = \sum_{n=0}^{N-1} r(k-n)c_n$$

$s_1(t) = \cos \omega t$

$s_2(t) = -\cos \omega t$

Shift register contents at $k = 3$

| 0 | −1 | 0 | 1 |

Filter weights matched to $s_1(k)$

$c_0 = 0$ $c_1 = -1$ $c_2 = 0$ $c_3 = 1$

$$z_1\Big]_{k=3} = \sum_{n=0}^{3} s_1(3-n)c_n$$

$z_1(k=3) = 2$

| 0 | −1 | 0 | 1 |

Filter weights matched to $s_2(k)$

$c_0 = 0$ $c_1 = 1$ $c_2 = 0$ $c_3 = -1$

$$z_2\Big]_{k=3} = \sum_{n=0}^{3} s_1(3-n)c_n$$

$z_2(k=3) = -2$

$r(t) = s_1(t) + n(t)$

(b)

Figure 3.10 (a) Sampled matched filter. (b) Sampled matched filter detection example, in the absence of noise.

rate $f_s = 2W = 1/T_s$; hence the sampling interval is equal to T_s. At the clock times of $t = kT_s$, the analog signal is sampled and the samples are shifted into the register of Figure 3.10a from left to right. The shift register with its coefficients c_0 to c_{N-1} approximate a matched filter. Once the received signal has been sampled, the continuous time notation t is changed to kT_s or simply k to reflect the sampled notation

$$r(k) = s(k) + n(k) \qquad k = 0, 1, \ldots$$

where k represents a sample index. The output, $z(k)$, of the sampled matched filter, at a time corresponding to the kth sample is

$$z(k) = \sum_{n=0}^{N-1} r(k - n)c_n \qquad k = 0, 1, \ldots, \text{modulo-}N \qquad (3.46)$$

where x modulo-y is defined as the remainder of dividing x by y. For the binary demodulation application, $z_i(k)(i = 1, 2)$ outputs are compared to a threshold at each value of $k = N - 1$ corresponding to the end of a symbol. The c_n values are the filter weights constituting the filter impulse response that is matched to the signal, where n is the index of the weights and the register stages (from left to right) and k is the index of the samples as they are produced by the sampler. One can see the similarity between the convolution integral of Equation (2.56) and the summation of Equation (3.46), especially with regard to the mirror-image rotation of one of the functions prior to multiplication. Since we assume the noise to have zero mean, the expected value of a received sample for the binary case is expressed as

$$\mathbf{E}\{r(k)\} = s_i(k) \qquad i = 1, 2 \qquad (3.47)$$

If $s_1(t)$ had been transmitted, the expected matched filter outputs would be

$$\mathbf{E}\{z_i(k)\} = \sum_{n=0}^{N-1} s_1(k - n)c_n \qquad (3.48)$$

where the filter weights, c_n, are matched to the corresponding $s_i(k)$ for each branch.

Example 3.2   Sampled Matched Filter

Consider the BPSK waveform set

$$s_1(t) = \cos \omega t$$

and

$$s_2(t) = -\cos \omega t$$

Illustrate how a *sampled* matched filter or correlator, as shown in Figure 3.10a, can be used to detect a received signal, say $s_1(t)$, from the BPSK waveform set, in the absence of noise.

Solution

First, the waveform is sampled so that $s_1(t)$ is transformed into the set of samples, $\{s_1(k)\}$. The sampled matched filter receiver will be shown with two branches, following the analog implementation in Figure 3.8b. The top branch is made up of shift registers and coefficients matched to the $\{s_1(k)\}$ sample points. The bottom branch is similarly matched to the $\{s_2(k)\}$ sample points. The four equally spaced sample points ($k = 0, 1, 2, 3$) for each of the $\{s_i(k)\}$ are as follows (see Figure 3.10b):

$$s_1(k = 0) = 1, \qquad s_1(k = 1) = 0, \qquad s_1(k = 2) = -1, \qquad s_1(k = 3) = 0$$

$$s_2(k = 0) = -1, \qquad s_2(k = 1) = 0, \qquad s_2(k = 2) = 1, \qquad s_2(k = 3) = 0$$

The c_n coefficients represent the delayed mirror-image rotation of the signal to which the filter is matched. Therefore, $c_n = s_i(N - 1 - n)$, where $n = 0, \ldots, N - 1$, and we can write $c_0 = s_i(3)$, $c_1 = s_i(2)$, $c_2 = s_i(1)$, $c_3 = s_i(0)$. It is here that the reader can gain some insight as to why the convolution operation (with its mirror-image rotation) results in the appropriate lining up of the received signal samples with the weights (reference signal).

Consider the top branch in Figure 3.10b. At the $k = 0$ clock time, the first sample, $s_1(k = 0) = 1$, enters the leftmost stage of each register. At the next clock time, the second sample, $s_1(k = 1) = 0$, enters the leftmost stage of each register; at this same time the first sample, $s_1(k = 0) = 1$, has been shifted to the next right stage in each register, and so on. At the $k = 3$ clock time the sample, $s_1(k = 3) = 0$, enters the leftmost stage; by this time the first sample, $s_1(k = 0) = 1$, has been shifted into the rightmost stage. The four signal samples are now located in the registers in mirror-image arrangement compared to the way the prototype waveform, $s_1(t)$, is drawn in Figure 3.10b. The task of the demodulator is to find the best match to the incoming signal; the demodulator matches the reference coefficients of each branch with the incoming signal samples, in the order in which the samples arrive. Hence the convolution operation is an appropriate expression for describing the alignment of the incoming waveform samples with the reference coefficients, to maximize the correlation in the proper branch.

3.5.3 Coherent Detection of Multiple Phase Shift Keying

Figure 3.11 illustrates the signal space for a multiple phase shift keying (MPSK) signal set; the figure describes a four-level (4-ary) PSK or quadriphase shift keying (QPSK) example ($M = 4$). Binary source digits are collected two at a time, and for each symbol interval the two sequential digits instruct the modulator as to which of the four waveforms to produce. For typical coherent M-ary PSK (MPSK) systems, $s_i(t)$ can be expressed as

$$s_i(t) = \sqrt{\frac{2E}{T}} \cos\left(\omega_0 t - \frac{2\pi i}{M}\right) \qquad \begin{matrix} 0 \le t \le T \\ i = 1, \ldots, M \end{matrix} \qquad (3.49)$$

where E is the energy content of $s_i(t)$ over each symbol duration T, and ω_0 is the carrier frequency. If an orthonormal signal space is assumed in Equations (3.4) and (3.5), we can choose a convenient set of axes, as follows:

$$\psi_1(t) = \sqrt{\frac{2}{T}} \cos \omega_0 t \qquad (3.50a)$$

$$\psi_2(t) = \sqrt{\frac{2}{T}} \sin \omega_0 t \qquad (3.50b)$$

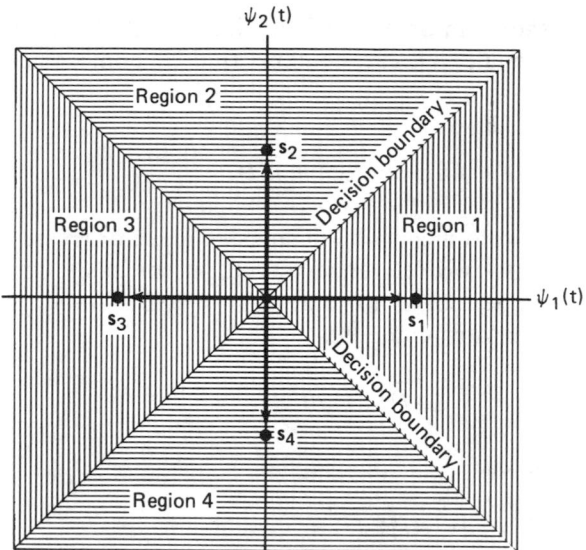

Figure 3.11 Signal space and decision regions for a QPSK system.

where the amplitude $\sqrt{2/T}$ has been chosen to normalize the expected output of the detector, as was done in Section 3.5.1. Now $s_i(t)$ can be written in terms of these orthonormal coordinates, giving

$$s_i(t) = a_{i1}\psi_1(t) + a_{i2}\psi_2(t) \qquad \begin{matrix} 0 \le t \le T \\ i = 1, \ldots, M \end{matrix} \qquad (3.51a)$$

$$= \sqrt{E}\cos\left(\frac{2\pi i}{M}\right)\psi_1(t) + \sqrt{E}\sin\left(\frac{2\pi i}{M}\right)\psi_2(t) \qquad (3.51b)$$

Notice that Equation (3.51) describes a set of M multiple phase waveforms (intrinsically nonorthogonal) in terms of only two orthogonal carrier-wave components. The $M = 4$ (QPSK) case is unique among MPSK signal sets in the sense that the QPSK waveform set is represented by a combination of antipodal and orthogonal members. The decision boundaries partition the signal space into $M = 4$ regions; the construction is similar to the procedure outlined in Section 3.4.1 and Figure 3.6 for $M = 2$. The decision rule for the detector (see Figure 3.11) is to decide that $s_1(t)$ was transmitted if the received signal vector falls in region 1, that $s_2(t)$ was transmitted if the received signal vector falls in region 2, and so on. In other words, the decision rule is to choose the ith waveform if $z_i(T)$ is the largest of the correlator outputs (seen in Figure 3.7).

The form of the correlator shown in Figure 3.7a implies that there are always M product correlators used for the demodulation of MPSK signals. The figure infers that for each of the M branches, a reference signal with the appropriate phase shift is configured. In practice, the implementation of an MPSK demodulator follows Figure 3.7b, requiring only $N = 2$ product integrators regardless of the size of the signal set M. The savings in implementation is possible because any arbitrary integrable waveform set can be expressed as a linear combination of orthogonal waveforms, as shown in Section 3.2.2. Figure 3.12 illustrates such a demodulator. The received signal, $r(t)$, can be expressed by combining Equations (3.50) and (3.51) as follows:

$$r(t) = \sqrt{\frac{2E}{T}}(\cos \phi_i \cos \omega_0 t + \sin \phi_i \sin \omega_0 t) + n(t) \quad \begin{array}{l} 0 \leq t \leq T \\ i = 1, \ldots, M \end{array} \quad (3.52)$$

where $\phi_i = 2\pi i/M$, and $n(t)$ is a zero-mean white Gaussian noise process. Notice in Figure 3.12 that there are only two reference waveforms or basis functions, $\psi_1(t) = \sqrt{2/T} \cos \omega_0 t$ for the upper correlator and $\psi_2(t) = \sqrt{2/T} \sin \omega_0 t$ for the lower correlator. The upper correlator computes

$$X = \int_0^T r(t)\psi_1(t) \, dt \qquad (3.53)$$

and the lower correlator computes

$$Y = \int_0^T r(t)\psi_2(t) \, dt \qquad (3.54)$$

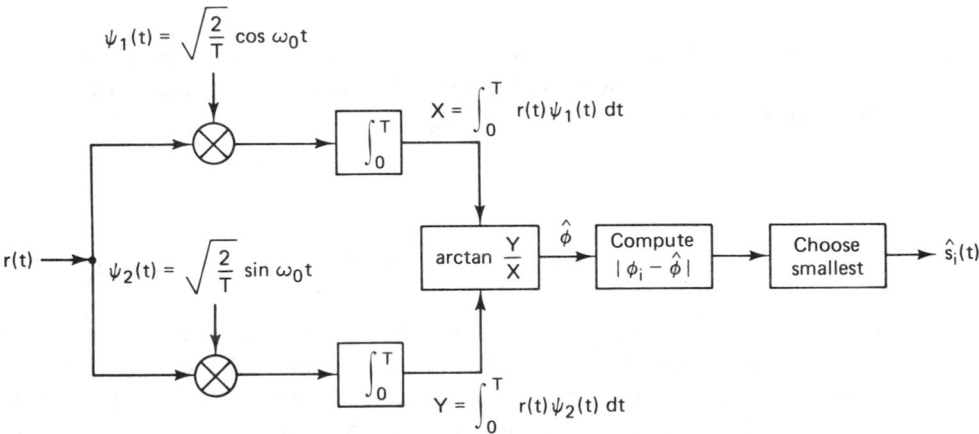

Figure 3.12 Demodulator for MPSK signals.

Figure 3.13 illustrates that the computation of the received phase angle $\hat{\phi}$ can be accomplished by computing the arctan of Y/X, where X can be thought of as the in-phase component of the received signal, Y is the quadrature component, and $\hat{\phi}$ is a noisy estimate of the transmitted ϕ_i. In other words, the upper correlator of Figure 3.12 produces an output X, the magnitude of the in-phase projection of the vector $\mathbf{r}$, and the lower correlator produces an output Y, the magnitude of the quadrature projection of the vector $\mathbf{r}$. The X and Y outputs of the correlators feed into the block marked arctan (Y/X). The resulting value of the angle $\hat{\phi}$ is compared with each of the stored prototype phase angles, ϕ_i. The demodulator selects the ϕ_i that is closest to the angle $\hat{\phi}$. In other words, the demodulator computes $|\phi_i - \hat{\phi}|$ for each of the ϕ_i prototypes and chooses the ϕ_i yielding the smallest output.

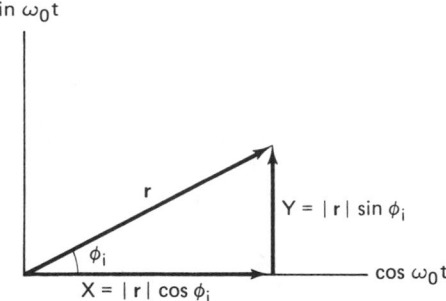

$$\hat{\phi} = \arctan(Y/X) \begin{cases} \text{Noisy estimate} \\ \text{of transmitted } \phi_i \end{cases}$$

Figure 3.13 In-phase and quadrature components of the received signal vector **r**.

3.5.4 Coherent Detection of FSK

FSK modulation is characterized by the information being contained in the frequency of the carrier. A typical set of FSK signal waveforms was described in Equation (3.27) as

$$s_i(t) = \sqrt{\frac{2E}{T}} \cos(\omega_i t + \phi) \qquad \begin{array}{l} 0 \le t \le T \\ i = 1, \ldots, M \end{array}$$

where E is the energy content of $s_i(t)$ over each symbol duration T, and $(\omega_{i+1} - \omega_i)$ is typically assumed to be an integral multiple of π/T. The phase term, ϕ, is an arbitrary constant and can be set equal to zero. Assuming that the basis functions $\psi_1(t), \psi_2(t), \ldots, \psi_N(t)$ form an orthonormal set, the most useful form for $\{\psi_j(t)\}$ is shown below.

$$\psi_j(t) = \sqrt{\frac{2}{T}} \cos \omega_j t \qquad j = 1, \ldots, N \tag{3.55}$$

where, as before, the amplitude $\sqrt{2/T}$ normalizes the expected output of the detector. From Equation (3.5) we can write

$$a_{ij} = \int_0^T \sqrt{\frac{2E}{T}} \cos(\omega_i t) \sqrt{\frac{2}{T}} \cos \omega_j t \, dt \tag{3.56}$$

Therefore,

$$a_{ij} = \begin{cases} \sqrt{E} & \text{for } i = j \\ 0 & \text{otherwise} \end{cases} \tag{3.57}$$

In other words, the ith prototype signal vector is located on the ith coordinate axis at a displacement $\sqrt{E}$ from the origin of the signal space. In this scheme, for the general M-ary case, the distance between any two prototype signal vectors s_i and s_j is constant:

$$d(s_i, s_j) = \| s_i - s_j \| = \sqrt{2E} \qquad \text{for } i \ne j \tag{3.58}$$

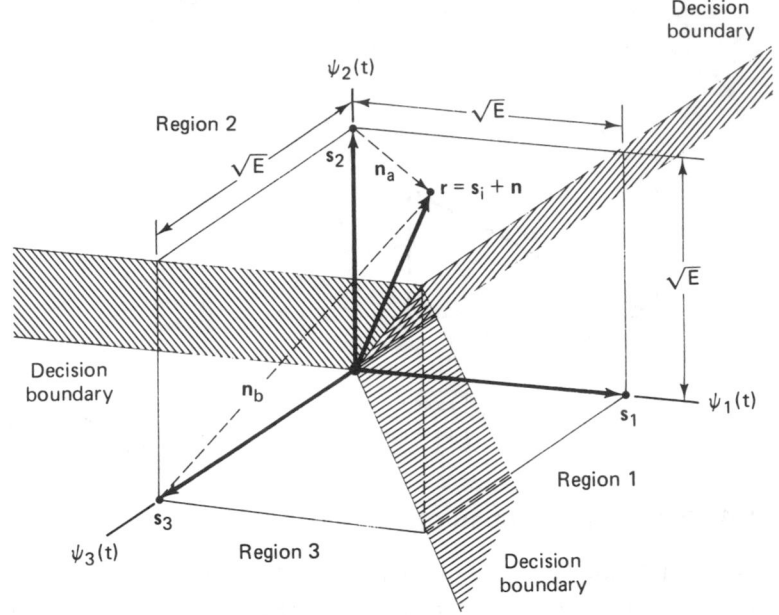

Figure 3.14 Partitioning the signal space for a 3-ary FSK signal.

Figure 3.14 illustrates the prototype signal vectors and the decision regions for a 3-ary ($M = 3$) coherently detected FSK system. As in the PSK case, the signal space is partitioned into M distinct regions, each containing one prototype signal vector; here, because the decision region is three-dimensional, the decision boundaries are planes instead of lines. The optimum decision rule is to decide that the transmitted signal belongs to the class whose index corresponds to the region where the received signal is found. In Figure 3.14, a received signal vector **r** is shown in region 2. Using the decision rule stated above, the detector classifies **r** as signal s_2. Since the noise is a Gaussian random vector, there is a probability greater than zero that **r** could have been produced by some signal other than s_2. For example, if the transmitter had sent s_2, then **r** would be the sum of signal plus noise, $s_2 + \mathbf{n}_a$, and the decision to choose s_2 is correct; however, if the transmitter had actually sent s_3, then **r** would be the sum of signal plus noise, $s_3 + \mathbf{n}_b$ and the decision to select s_2 is an error. The error performance of coherently detected FSK systems is treated in Section 3.7.3.

3.6 NONCOHERENT DETECTION

3.6.1 Detection of Differential PSK

The name *differential PSK* (DPSK) sometimes needs clarification because two separate aspects of the modulation/demodulation format are being referred to: the encoding procedure and the detection procedure. The term *differential encoding* refers to the procedure of encoding the data differentially; that is, the presence

146   Bandpass Modulation and Demodulation   Chap. 3

of a binary one or zero is manifested by the symbol's similarity or difference when compared to the preceding symbol. The term *differentially coherent detection* of differentially encoded PSK, the usual meaning of DPSK, refers to a detection scheme often classified as noncoherent because it does not require a reference in phase with the received carrier. Sometimes, differentially encoded PSK is *coherently* detected. This will be discussed in Section 3.7.2.

With noncoherent systems, no attempt is made to determine the actual value of the phase of the incoming signal. Therefore, if the transmitted waveform is

$$s_i(t) = \sqrt{\frac{2E}{T}} \cos\left[\omega_0 t + \theta_i(t)\right] \qquad \begin{array}{l} 0 \le t \le T \\ i = 1, \ldots, M \end{array}$$

the received signal can be characterized by

$$r(t) = \sqrt{\frac{2E}{T}} \cos\left[\omega_0 t + \theta_i(t) + \alpha\right] + n(t) \qquad \begin{array}{l} 0 \le t \le T \\ i = 1, \ldots, M \end{array} \qquad (3.59)$$

where α is an arbitrary constant and is typically assumed to be a random variable uniformly distributed between zero and 2π, and $n(t)$ is an AWGN process.

For coherent detection, matched filters (or their equivalents) are used; for noncoherent detection, this is not possible because the matched filter output is a function of the unknown angle α. However, if we assume that α varies slowly relative to two period times ($2T$), the phase difference between two successive waveforms, $\theta_j(T_1)$ and $\theta_k(T_2)$ is independent of α, that is,

$$[\theta_k(T_2) + \alpha] - [\theta_j(T_1) + \alpha] = \theta_k(T_2) - \theta_j(T_1) = \phi_i(T_2) \qquad (3.60)$$

The basis for *differentially coherent detection* of differentially encoded PSK (DPSK) is as follows. The carrier phase of the previous signaling interval can be used as a phase reference for demodulation. Its use requires *differential encoding* of the message sequence at the transmitter since the information is carried by the difference in phase between two successive waveforms. To send the ith message ($i = 1, 2, \ldots, M$), the present signal waveform must have its phase advanced by $\phi_i = 2\pi i/M$ radians over the previous waveform. The detector, in general, calculates the coordinates of the incoming signal by correlating it with locally generated waveforms such as $\sqrt{2/T} \cos \omega_0 t$ and $\sqrt{2/T} \sin \omega_0 t$. The detector then measures the angle between the currently received signal vector and the previously received signal vector, as illustrated in Figure 3.15.

In general, DPSK signaling performs less efficiently than PSK, because the errors in DPSK tend to propagate (to adjacent symbol times) due to the correlation between signaling waveforms. One way of viewing the difference between PSK and DPSK is that the former compares the received signal with a clean reference; in the latter, however, two noisy signals are compared with each other. We might say that there is twice as much noise associated with DPSK signaling compared to PSK signaling. Consequently, as a first guess, we might estimate that DPSK manifests a degradation of approximately 3 dB when compared with PSK; this degradation decreases rapidly with increasing signal-to-noise ratio. The trade-off

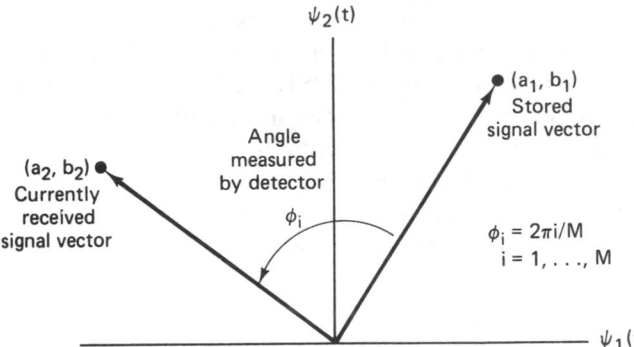

$\phi_i = 2\pi i/M$
$i = 1, \ldots, M$

Figure 3.15 Signal space for DPSK.

for this performance loss is reduced system complexity. The error performance for the detection of DPSK is treated in Section 3.7.5.

3.6.2 Binary Differential PSK Example

The essence of differentially coherent detection in DPSK is that the identity of the data is inferred from the changes in phase from symbol to symbol. Therefore, since the data are detected by differentially examining the waveform, the transmitted waveform would first be encoded in a differential fashion. Figure 3.16a illustrates a differential encoding of a binary message data stream, $m(k)$, where k is the sample time index. The differential encoding starts (third row in the figure) with the first bit of the code bit sequence, $c(k = 0)$, chosen arbitrarily (here taken to be a one). Then the sequence of encoded bits, $c(k)$, can, in general, be encoded in one of two ways:

$$c(k) = c(k - 1) \oplus m(k) \tag{3.61}$$

or

$$c(k) = \overline{c(k - 1) \oplus m(k)} \tag{3.62}$$

where the symbol $\oplus$ represents modulo-2 addition (defined in Section 2.12.3) and the overbar denotes complement. In Figure 3.16a the differentially encoded message was obtained by using Equation (3.62). In other words, the present code bit, $c(k)$, is a one if the message bit, $m(k)$, and the prior coded bit, $c(k - 1)$, are the same, otherwise, $c(k)$ is a zero. The fourth row translates the coded bit sequence, $c(k)$, into the phase shift sequence, $\theta(k)$, where a one is characterized by a 180° phase shift, and a zero is characterized by a 0° phase shift.

Figure 3.16b illustrates the binary DPSK detection scheme in block diagram form. Notice that the basic product integrator of Figure 3.7 is the essence of this detection process; as with coherent PSK, we are still attempting to correlate a received signal with a reference. The interesting difference here is that the reference signal is simply a delayed version of the received signal. In other words, during each symbol time, we are matching a received symbol with the prior symbol and looking for a correlation or an anticorrelation (180° out of phase).

Consider the received signal with phase shift sequence, $\theta(k)$, entering the

Bandpass Modulation and Demodulation   Chap. 3

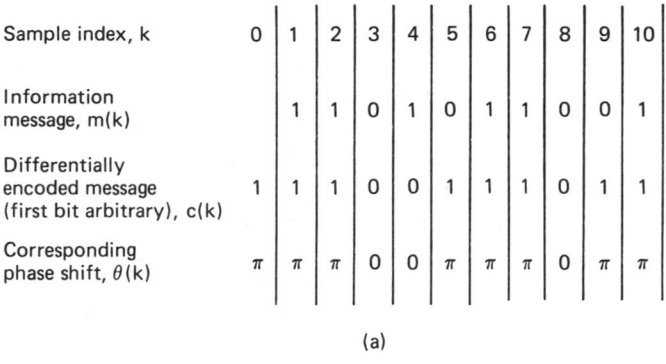

(a)

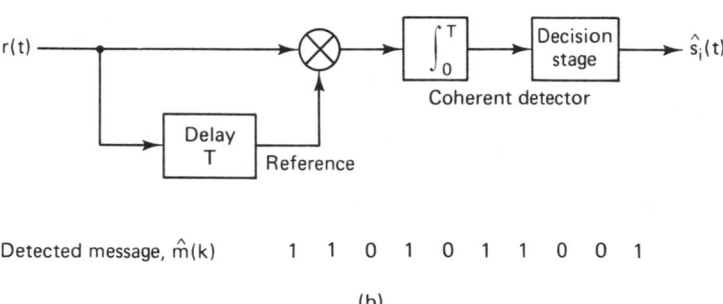

(b)

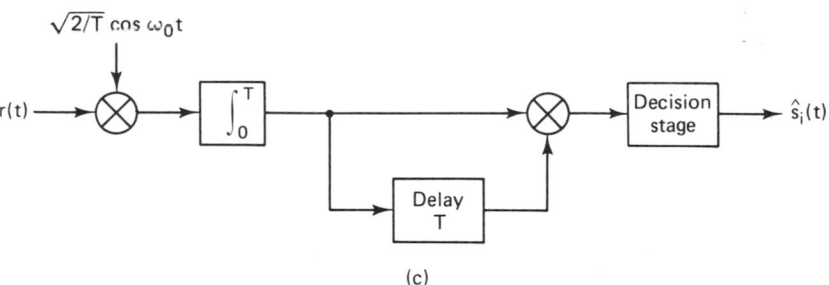

(c)

Figure 3.16 Differential PSK (DPSK). (a) Differential encoding. (b) Differentially coherent detection. (c) Optimum differentially coherent detection.

detector of Figure 3.16b, in the absence of noise. The phase, $\theta(k = 1)$, is matched with $\theta(k = 0)$; they have the same value, π; hence the first bit of the detected output is $\hat{m}(k = 1) = 1$. Then $\theta(k = 2)$ is matched with $\theta(k = 1)$; again they have the same value, and $\hat{m}(k = 2) = 1$. Then $\theta(k = 3)$ is matched with $\theta(k = 2)$; they are different, so that $\hat{m}(k = 3) = 0$, and so on.

It must be pointed out that the detector in Figure 3.16b is suboptimum [5] in the sense of error performance. The optimum differential detector for DPSK requires a reference carrier in frequency but not necessarily in phase with the received carrier. Hence the optimum differential detector is shown in Figure 3.16c [6]. Its performance is treated in Section 3.7.5.

3.6.3 Noncoherent Detection of FSK

A detector for the noncoherent detection of FSK waveforms described by Equation (3.27) can be implemented with correlators similar to those shown in Figure 3.7. However, the hardware must be configured as an *energy detector*, without exploiting phase measurements. For this reason, the noncoherent detector typically requires twice as many channel branches as the coherent detector. Figure 3.17 illustrates the in-phase (I) and quadrature (Q) channels used to detect a binary FSK (BFSK) signal set noncoherently. Notice that the upper two branches are configured to detect the signal with frequency ω_1; the reference signals are $\sqrt{2/T} \cos \omega_1 t$ for the I branch and $\sqrt{2/T} \sin \omega_1 t$ for the Q branch. Similarly, the lower two branches are configured to detect the signal with frequency ω_2; the reference signals are $\sqrt{2/T} \cos \omega_2 t$ for the I branch and $\sqrt{2/T} \sin \omega_2 t$ for the Q branch. Imagine that the received signal $r(t)$, by chance alone, is exactly of the form $\cos \omega_1 t + n(t)$; that is, the phase is exactly zero, and thus the signal component of the received signal exactly matches the top-branch reference signal with regard to frequency and phase. In that event, the product integrator of the top branch should yield the maximum output. The second branch should yield a near-zero output (integrated zero-mean noise) since its reference signal $\sqrt{2/T} \sin \omega_1 t$

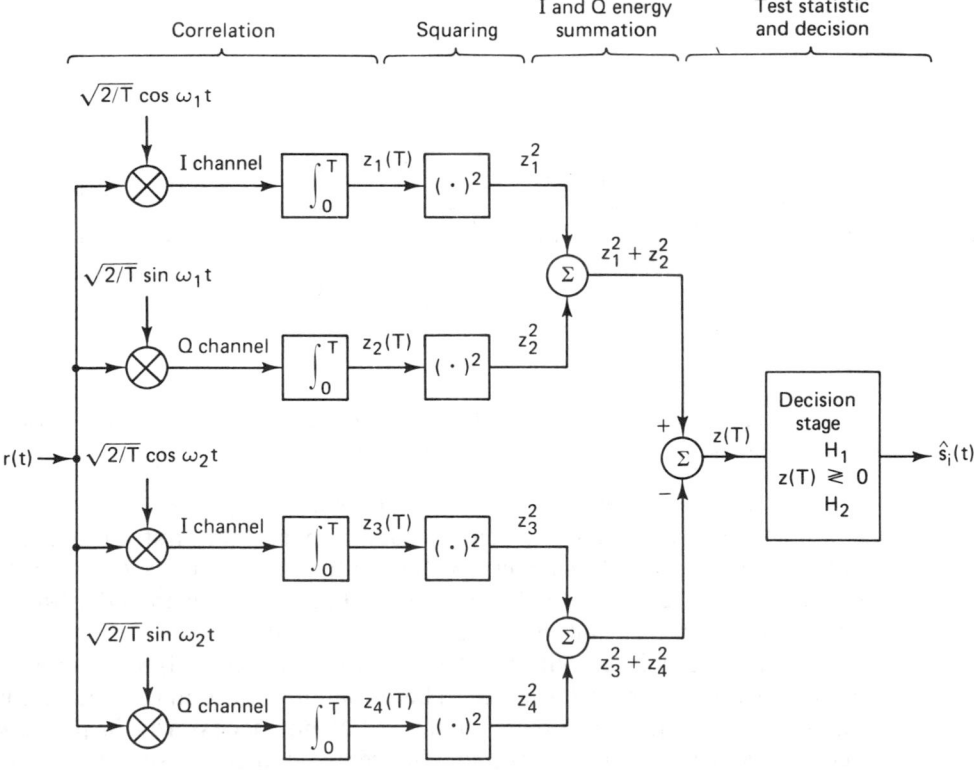

Figure 3.17 Quadrature receiver.

is orthogonal to the signal component of $r(t)$. The third and fourth branches should also yield near-zero outputs since their ω_2 reference signals are also orthogonal to the signal component of $r(t)$.

Now, imagine a different scenario; suppose that by chance alone, the received signal, $r(t)$, is of the form $\sin \omega_1 t + n(t)$. In that event, the second branch in Figure 3.17 should yield the maximum output, while the others should yield near-zero outputs. In actual practice, the most likely scenario is that $r(t)$ is of the form $\cos (\omega_1 t + \phi) + n(t)$; that is, the incoming signal will *partially* correlate with the $\cos \omega_1 t$ reference and *partially* correlate with the $\sin \omega_1 t$ reference. Now it should be obvious why a noncoherent quadrature receiver uses twice as many branches as a coherent one; the receiver knows nothing about the incoming signal's phase. The receiver essentially resolves the signal into an I component and a Q (90° out of phase) component. In Figure 3.17 the blocks following the product integrators perform a squaring operation to prevent the appearance of any negative values. Then for each of the signal types in the set (two in this binary example) the energy from the I and Q channels is added. The final stage forms the test statistic, $z(T)$, and chooses the signal with frequency ω_1 or the signal with frequency ω_2 depending on which pair of energy detectors yielded the maximum output.

Another possible implementation for noncoherent FSK detection uses bandpass filters, centered at $f_i = \omega_i/2\pi$, with bandwidth, $W_f = 1/T$, followed by *envelope detectors*, as shown in Figure 3.18. An envelope detector consists of a rectifier and a low-pass filter. The detectors are matched to the *signal envelopes* and not to the signals themselves. The phase of the carrier is of no importance in defining the envelope; hence no phase information is used. In the case of binary FSK, the decision as to whether a one or a zero was transmitted is made on the basis of which of two envelope detectors has the largest amplitude at the moment of measurement. Similarly, for a multiple frequency shift keying (MFSK) system, the decision as to which of M signals was transmitted is made on the basis of which of the M envelope detectors has the maximum output.

Even though the envelope detector block diagram of Figure 3.18 looks func-

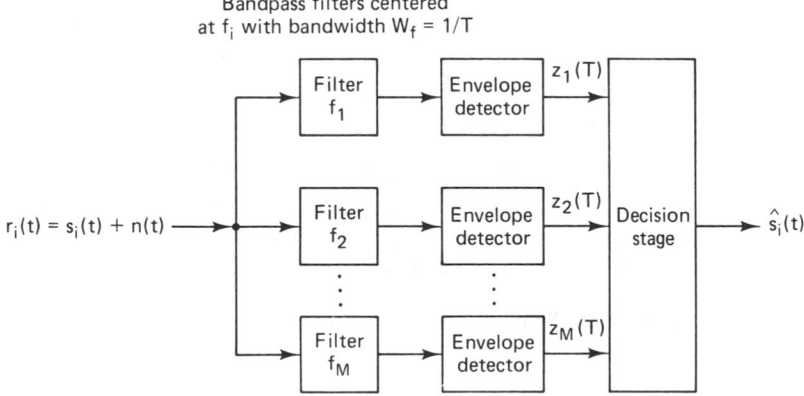

Figure 3.18 Noncoherent detection of FSK using envelope detectors.

tionally simpler than the quadrature receiver of Figure 3.17, the use of filters usually results in the envelope detector design having greater weight and cost than the quadrature receiver. Quadrature receivers can be implemented digitally; thus, with the advent of large-scale integrated (LSI) circuits, they are often the preferred choice for noncoherent detectors. The detector in Figure 3.18 can also be implemented digitally by performing discrete Fourier transformations instead of using analog filters, but such a design is usually more complex than a digital implementation of the quadrature receiver.

3.6.4 Minimum Required Tone Spacing for Noncoherent Orthogonal FSK Signaling

Frequency shift keying is usually implemented as orthogonal signaling where each tone (sinusoid) in the signal set cannot interfere with any of the other tones. In order for the signal set to be orthogonal, any pair of adjacent tones must have a frequency separation of a multiple of $1/T$ hertz. A tone with frequency f_i, that is switched on for a symbol duration of T seconds and then switched off, such as the FSK tone described in Equation (3.27), can be analytically described by

$$s_i(t) = (\cos 2\pi f_i t) \, \text{rect} \, (t/T)$$

$$\text{where rect} \, (t/T) = \begin{cases} 1 & \text{for } -T/2 \le t \le T/2 \\ 0 & \text{for } |t| > T/2 \end{cases}$$

The Fourier transform of $s_i(t)$, from Table A.1, is

$$\mathcal{F} \{s_i(t)\} = T \, \text{sinc} \, (f - f_i)T$$

where the sinc function is as defined in Equation (1.39). The spectra of two such adjacent tones, tone 1 with frequency f_1 and tone 2 with frequency f_2, are plotted in Figure 3.19.

In order that the two tones not interfere with each other during detection, the peak of the spectrum of tone 1 must coincide with one of the zero crossings of the spectrum of tone 2 and similarly, the peak of the tone 2 spectrum must coincide with one of the zero crossings of the tone 1 spectrum. The frequency difference between the center of the spectral main lobe and the first zero crossing represents the *minimum required spacing*. This corresponds to a minimum tone separation of $1/T$ hertz.

Example 3.3 Minimum Tone Spacing for Noncoherent Orthogonal FSK

Consider two waveforms $\cos (2\pi f_1 t + \phi)$ and $\cos 2\pi f_2 t$ to be used for *noncoherent* FSK signaling, where $f_1 > f_2$. The symbol rate is equal to $1/T$ symbols/s, where T is the symbol duration and ϕ is a constant arbitrary angle from 0 to 2π.

(a) Prove that the minimum tone spacing for *noncoherently detected* orthogonal FSK signaling is $1/T$.

(b) What is the minimum tone spacing for *coherently detected* orthogonal FSK signaling?

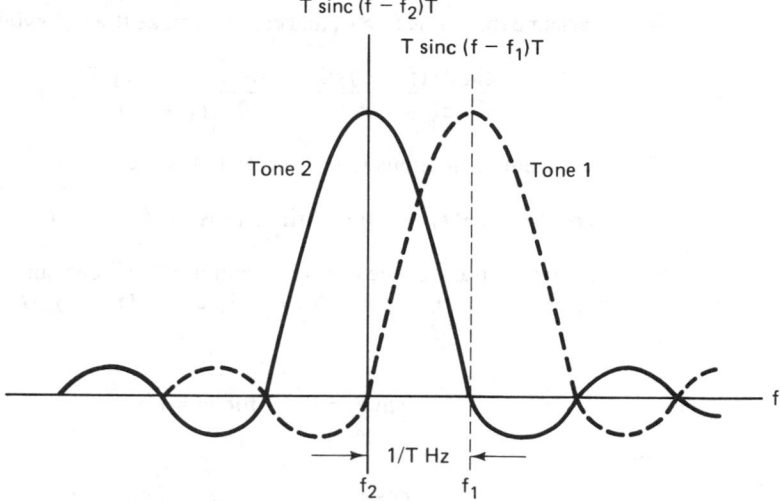

T sinc (f − f₂)T

T sinc (f − f₁)T

Tone 2

Tone 1

1/T Hz

f₂   f₁

f

Figure 3.19 Minimum tone spacing for noncoherently detected orthogonal FSK signaling.

Solution

(a) For the two waveforms to be orthogonal, they must fulfill the orthogonality constraint of Equation (3.2):

$$\int_0^T \cos(2\pi f_1 t + \phi) \cos 2\pi f_2 t \, dt = 0 \qquad (3.63)$$

Using the basic trigonometric identities shown in Equations (D.6) and (D.1) to (D.3), we can write Equation (3.63) as

$$\cos \phi \int_0^T \cos 2\pi f_1 t \cos 2\pi f_2 t \, dt$$

$$- \sin \phi \int_0^T \sin 2\pi f_1 t \cos 2\pi f_2 t \, dt = 0 \qquad (3.64)$$

$$\cos \phi \int_0^T [\cos 2\pi(f_1 + f_2)t + \cos 2\pi(f_1 - f_2)t] \, dt$$

$$- \sin \phi \int_0^T [\sin 2\pi(f_1 + f_2)t + \sin 2\pi(f_1 - f_2)t] \, dt = 0 \qquad (3.65)$$

$$\cos \phi \left[\frac{\sin 2\pi(f_1 + f_2)t}{2\pi(f_1 + f_2)} + \frac{\sin 2\pi(f_1 - f_2)t}{2\pi(f_1 - f_2)} \right]_0^T$$

$$+ \sin \phi \left[\frac{\cos 2\pi(f_1 + f_2)t}{2\pi(f_1 + f_2)} + \frac{\cos 2\pi(f_1 - f_2)t}{2\pi(f_1 - f_2)} \right]_0^T = 0 \qquad (3.66)$$

$$\cos \phi \left[\frac{\sin 2\pi(f_1 + f_2)T}{2\pi(f_1 + f_2)} + \frac{\sin 2\pi(f_1 - f_2)T}{2\pi(f_1 - f_2)} \right]$$

$$+ \sin \phi \left[\frac{\cos 2\pi(f_1 + f_2)T - 1}{2\pi(f_1 + f_2)} + \frac{\cos 2\pi(f_1 - f_2)T - 1}{2\pi(f_1 - f_2)} \right] = 0 \qquad (3.67)$$

Sec. 3.6   Noncoherent Detection

153

We can assume that $f_1 + f_2 \gg 1$ and can thus make the following approximation:

$$\frac{\sin 2\pi(f_1 + f_2)T}{2\pi(f_1 + f_2)} \cong \frac{\cos 2\pi(f_1 + f_2)T}{2\pi(f_1 + f_2)} \cong 0 \qquad (3.68)$$

Then, combining Equations (3.67) and (3.68), we can write

$$\cos \phi \sin 2\pi(f_1 - f_2)T + \sin \phi \, [\cos 2\pi(f_1 - f_2)T - 1] \approx 0 \qquad (3.69)$$

Note that for arbitrary ϕ, the terms in Equation (3.69) can sum to zero only when $\sin 2\pi(f_1 - f_2)T = 0$, and simultaneously $\cos 2\pi(f_1 - f_2)T = 1$.

Since

$$\sin x = 0 \qquad \text{for } x = n\pi$$

and

$$\cos x = 1 \qquad \text{for } x = 2k\pi$$

where n and k are integers, then both $\sin x = 0$ and $\cos x = 1$ occur simultaneously when $n = 2k$. From Equation (3.69), for arbitrary ϕ, we can therefore write:

$$2\pi(f_1 - f_2)T = 2k\pi$$
$$f_1 - f_2 = \frac{k}{T} \qquad (3.70)$$

Thus the minimum tone spacing for *noncoherent* FSK signaling occurs for $k = 1$:

$$f_1 - f_2 = \frac{1}{T} \qquad (3.71)$$

(b) To find the minimum tone spacing for *coherent* FSK, where the angle ϕ is zero, we simply rewrite Equation (3.69) with $\phi = 0$, which gives

$$\sin 2\pi(f_1 - f_2)T = 0 \qquad (3.72)$$

$$f_1 - f_2 = \frac{n}{2T} \qquad (3.73)$$

Thus the minimum tone spacing for *coherent* FSK signaling occurs for $n = 1$ as follows:

$$f_1 - f_2 = \frac{1}{2T} \qquad (3.74)$$

Therefore, for the same symbol rate, coherently detected FSK can occupy less bandwidth than noncoherently detected FSK and still retain orthogonal signaling. We can say that coherent FSK is more *bandwidth efficient*. The subject of bandwidth efficiency is addressed in greater detail in Chapter 7.

3.7 ERROR PERFORMANCE FOR BINARY SYSTEMS

3.7.1 Probability of Bit Error
for Coherently Detected BPSK

An important measure of performance used for comparing digital modulation schemes is the probability of error, P_E. For the correlator or matched filter detector, the calculations for obtaining P_E can be viewed geometrically (see Figure 3.6). They involve finding the probability that given a particular transmitted signal vector, say s_1, the noise vector, n, will give rise to a received signal falling outside region 1. The probability of the detector making an incorrect decision is termed the *probability of symbol error* (P_E). It is often convenient to specify system performance by the probability of bit error (P_B), even when decisions are made on the basis of symbols for which $M > 2$. The relationship between P_B and P_E is treated in Section 3.9.3 for orthogonal signaling and in Section 3.9.4 for multiple phase signaling.

For convenience, this section is restricted to the coherent detection of BPSK modulation. For this case the symbol error probability is the bit error probability. Assume that the signals are equally likely. Also assume that when signal, $s_i(t)$ ($i = 1, 2$), is transmitted, the received signal, $r(t)$, is equal to $s_i(t) + n(t)$, where $n(t)$ is an AWGN process. The antipodal signals, $s_1(t)$ and $s_2(t)$, can be characterized in a one-dimensional signal space as described in Section 3.5.1, where

$$s_1(t) = \sqrt{E}\psi_1(t)$$
$$0 \le t \le T \qquad (3.75)$$
$$s_2(t) = -\sqrt{E}\psi_1(t)$$

The decision stage of the detector will choose the $s_i(t)$ with the largest correlator output $z_i(T)$, or in this case of equal-energy antipodal signals, the detector, using the decision rule in Equation (3.39a), decides

$$
\begin{array}{ll}
s_1(t) & \text{if } z(T) > \gamma_0 = 0 \\
s_2(t) & \text{otherwise}
\end{array}
\qquad (3.76)
$$

Two types of errors can be made, as shown in Figure 3.9: The first type of error takes place if signal $s_1(t)$ is transmitted but the noise is such that the detector measures a negative value for $z(T)$ and chooses hypothesis H_2 [the hypothesis that signal $s_2(t)$ was sent]. The second type of error takes place if signal $s_2(t)$ is transmitted but the detector measures a positive value for $z(T)$ and chooses hypothesis H_1 [the hypothesis that signal $s_1(t)$ was sent].

To calculate the probability of a bit error, P_B, for this binary *minimum error* detector, we use the relationships developed in Section 2.9, starting with Equation (2.36b):

$$P_B = P(H_2|s_1)P(s_1) + P(H_1|s_2)P(s_2) \qquad (3.77)$$

For the case when the a priori probabilities are equal, that is, $P(s_1) = P(s_2) = \frac{1}{2}$, we can write

$$P_B = \tfrac{1}{2}P(H_2|s_1) + \tfrac{1}{2}P(H_1|s_2) \qquad (3.78)$$

Because of the symmetry of the probability density functions in Figure 3.9, we can also write

$$P_B = P(H_2|s_1) = P(H_1|s_2) \qquad (3.79)$$

Thus the probability of a bit error, P_B, is numerically equal to the area under the "tail" of either pdf, $p(z|s_1)$ or $p(z|s_2)$, that falls on the "incorrect" side of the threshold. We can therefore compute P_B by integrating $p(z|s_1)$ between the limits $-\infty$ and γ_0, or as shown below, by integrating $p(z|s_2)$ between the limits γ_0 and ∞.

$$P_B = \int_{\gamma_0 = (a_1 + a_2)/2}^{\infty} p(z|s_2) \, dz \qquad (3.80)$$

where the likelihoods, $p(z|s_i)$ $(i = 1, 2)$, are Gaussian functions with mean value, a_i, and the optimum threshold, γ_0, as shown in Section B.3.1, is equal to $(a_1 + a_2)/2$. The area-related probability of bit error, P_B, is seen to be the shaded area in Figure 3.9. It is shown in Section B.3.2 that Equation (3.80) reduces to

$$P_B = \int_{(a_1 - a_2)/2\sigma_0}^{\infty} \frac{1}{\sqrt{2\pi}} \exp\left(-\frac{u^2}{2}\right) du = Q\left(\frac{a_1 - a_2}{2\sigma_0}\right) \qquad (3.81)$$

where σ_0 is the standard deviation of the noise out of the correlator. The function, $Q(x)$, called the *complementary error function* or *co-error function*, is defined as

$$Q(X) = \frac{1}{\sqrt{2\pi}} \int_x^{\infty} \exp\left(-\frac{u^2}{2}\right) du \qquad (3.82)$$

and is described in greater detail in Sections 2.9 and B.3.2.

For equal-energy antipodal signaling, such as the BPSK format in Equation (3.75), the receiver output signal components are $a_1 = \sqrt{E_b}$ when $s_1(t)$ is sent and $a_2 = -\sqrt{E_b}$ when $s_2(t)$ is sent, where E_b is the signal energy per binary symbol. For AWGN we can replace the noise variance, σ_0^2, out of the correlator with $N_0/2$ (see Appendix C), so that we can rewrite Equation (3.81) as follows:

$$P_B = \int_{\sqrt{2E_b/N_0}}^{\infty} \frac{1}{\sqrt{2\pi}} \exp\left(-\frac{u^2}{2}\right) du \qquad (3.83)$$

$$= Q\left(\sqrt{\frac{2E_b}{N_0}}\right) \qquad (3.84)$$

This result could also have been obtained by noting that the energy difference, E_d, between the *antipodal signal vectors*, s_1 and s_2, with amplitudes of $\pm\sqrt{E_b}$, as seen in Figure 3.20a, can be computed as the square of the distance between

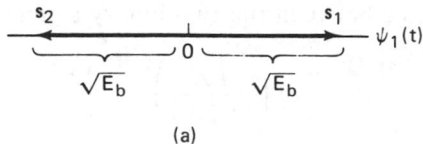

(a)

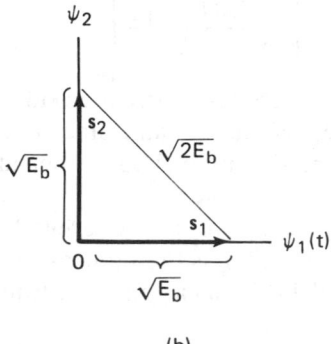

(b)

Figure 3.20 Binary signal vectors. (a) Antipodal. (b) Orthogonal.

the heads of the antipodal vectors, or in terms of the waveforms

$$E_d = \int_0^T [s_1(t) - s_2(t)]^2 \, dt \tag{3.85}$$

$$= \int_0^T s_1^2(t) \, dt + \int_0^T s_2^2(t) \, dt - 2 \int_0^T s_1(t)s_2(t) \, dt \tag{3.86}$$

Assuming equal energy signals,

$$E_b = \int_0^T s_1^2(t) \, dt = \int_0^T s_2^2(t) \, dt \tag{3.87}$$

$$E_d = 2E_b - 2E_b\rho = 2E_b(1 - \rho) \tag{3.88}$$

where

$$\rho = \frac{1}{E_b} \int_0^T s_1(t)s_2(t) \, dt \tag{3.89}$$

is the time cross-correlation coefficient and E_b is the average energy of the binary signals, $s_1(t)$ and $s_2(t)$. The correlation coefficient, ρ, is a measure of similarity between the two signals, $s_1(t)$ and $s_2(t)$, such that

$$-1 \le \rho \le 1 \tag{3.90}$$

In terms of signal vectors, the cross-correlation coefficient can be written

$$\rho = \cos \theta \tag{3.91}$$

where θ is the angle between the two signal vectors s_1 and s_2 (see Figure 3.6). In Equation (2.62), we developed an expression for the probability of bit error in terms of the energy difference between the two binary signals, as follows:

$$P_B = Q\left(\sqrt{\frac{E_d}{2N_0}}\right) \qquad (3.92)$$

Substituting Equation (3.88) into Equation (3.92), we get

$$P_B = Q\left[\sqrt{\frac{E_b(1 - \rho)}{N_0}}\right] \qquad (3.93)$$

For $\rho = 1$ (or $\theta = 0$), the signals are perfectly correlated (identical). For $\rho = -1$ (or $\theta = \pi$), the signals are anticorrelated (antipodal). Since the binary PSK signals are antipodal, we can set $\rho = -1$, and Equation (3.93) is then identical to Equation (3.84).

Note that the bit error probability, P_B, for the coherent detection of bandpass antipodal signals, as seen in Equation (3.84), is the same as the P_B for the matched filter detection of baseband antipodal (bipolar) signals in Equation (2.67).

3.7.1.1 The Basic SNR Parameter for Digital Communication Systems

The parameter E_b/N_0 in Equation (3.84) can be expressed as the ratio of average signal power to average noise power, S/N (or SNR). By introducing the signal bandwidth W, we can write the following identities, showing the relationship between E_b/N_0 and SNR for binary signals.

$$\frac{E_b}{N_0} = \frac{ST}{N_0} = \frac{S}{RN_0} = \frac{SW}{RN_0W} = \frac{S}{N}\left(\frac{W}{R}\right) \qquad (3.94)$$

where

$$S = \text{average modulating signal power}$$

$$T = \text{bit time duration}$$

$$R = 1/T = \text{bit rate}$$

$$N = N_0W$$

Analysis similar to that used for developing P_B in Equations (3.84) and (3.93) is used in finding the P_B expressions for other types of modulation. Figure 3.21 illustrates the "waterfall-like" shape of most probability of error curves in the field of digital communications. The curve describes a system's error probability performance in terms of available E_b/N_0. For $E_b/N_0 \geq x_0$, $P_E \leq P_0$. The dimensionless ratio E_b/N_0 is a standard quality measure for digital communications system performance. Note that optimum digital signal detection implies a correlator (or matched filter) implementation, in which case the signal bandwidth is equal to the noise bandwidth. Often we are faced with a system model for which this is not the case; in practice, we include a factor in the required E_b/N_0 that accounts for such suboptimal detection performance. Required E_b/N_0 can be considered a metric that characterizes the performance of one system versus another;

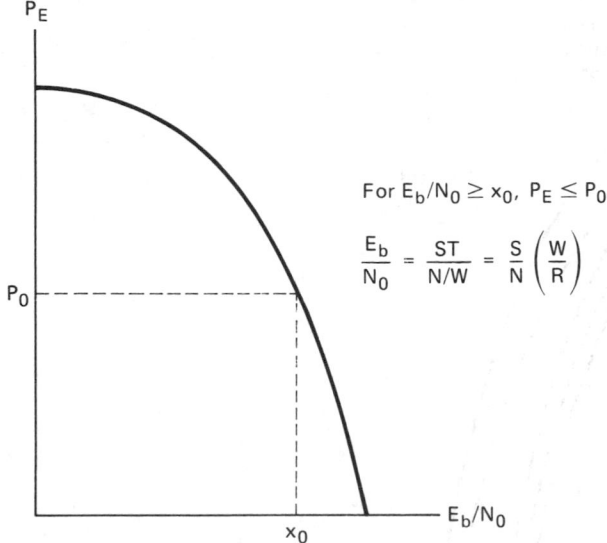

For $E_b/N_0 \geq x_0$, $P_E \leq P_0$

$$\frac{E_b}{N_0} = \frac{ST}{N/W} = \frac{S}{N}\left(\frac{W}{R}\right)$$

Figure 3.21 General shape of the P_E versus E_b/N_0 curve.

the *smaller* the required E_b/N_0, the *more efficient* is the system modulation and detection process for a given probability of error. Figure 3.22 is a plot comparing the bit error probability, P_B, of several binary modulation/demodulation types. The P_B for coherent detection of PSK, as shown in Equation (3.84), is plotted as the leftmost P_B curve.

Example 3.4 Bit Error Probability for BPSK Signaling

Find the bit error probability for a BPSK system with a bit rate of 1 Mbit/s. The received waveforms, $s_1(t) = A \cos \omega_0 t$ and $s_2(t) = -A \cos \omega_0 t$, are coherently detected with a matched filter. The value of A is 10 mV. Assume that the single-sided noise power spectral density is $N_0 = 10^{-11}$ W/Hz and that signal power and energy per bit are normalized relative to a 1-Ω load.

Solution

$$A = \sqrt{\frac{2E_b}{T}} = 10^{-2} \text{ V} \qquad T = \frac{1}{R} = 10^{-6} \text{ s}$$

Thus

$$E_b = \frac{A^2}{2} T = 5 \times 10^{-11} \text{ J} \qquad \text{and} \qquad \sqrt{\frac{2E_b}{N_0}} = 3.16$$

$$P_B = Q\left(\sqrt{\frac{2E_b}{N_0}}\right) = Q(3.16)$$

Using Table B.1 or Equation (2.43), we obtain

$$P_B = 8 \times 10^{-4}$$

Sec. 3.7   Error Performance for Binary Systems

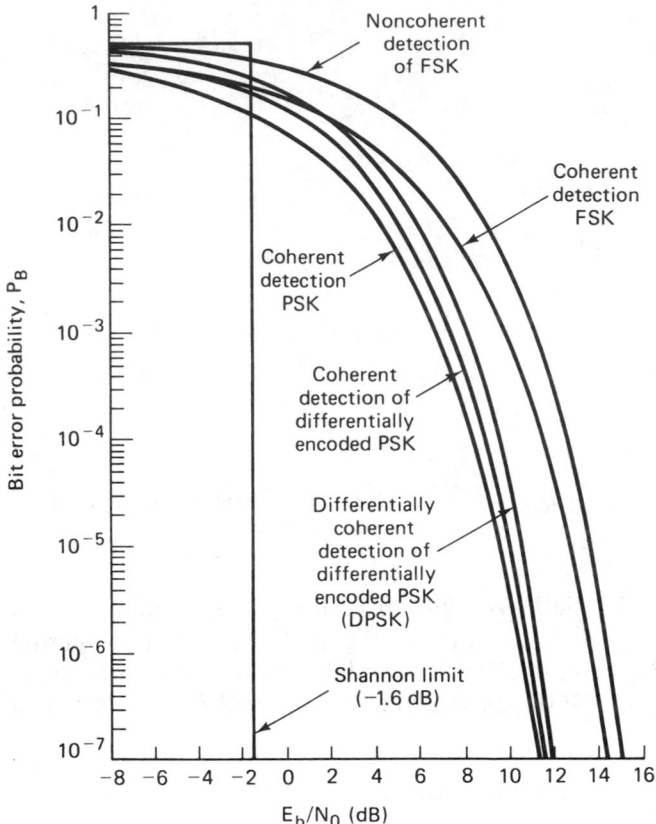

Figure 3.22 Bit error probability for several types of binary systems.

3.7.2 Probability of Bit Error for Coherently Detected Differentially Encoded PSK

Channel waveforms sometimes experience inversion; for example, when using a coherent reference generated by a phase-locked loop (see Chapter 8), one may have phase ambiguity. If the carrier phase were reversed in a DPSK modulation application, what would be the effect on the message? The only effect would be an error in the bit during which inversion occurred or the bit just after inversion, since the message information is encoded in the similarity or difference between adjacent symbols. The similarity or difference quality remains unchanged if the carrier is inverted. Sometimes, systems are *differentially encoded* and *coherently detected*, simply to avoid these phase ambiguities.

The probability of bit error for coherently detected, differentially encoded PSK is given by [7]

$$P_B = 2Q\left(\sqrt{\frac{2E_b}{N_0}}\right)\left[1 - Q\left(\sqrt{\frac{2E_b}{N_0}}\right)\right] \qquad (3.95)$$

This relationship is plotted in Figure 3.22. Notice that there is a slight degradation of error performance compared to the coherent detection of PSK. This is due to the differential encoding since any single detection error results in two decision errors. Error performance for the more popular differentially coherent detection (DPSK) is covered in Section 3.7.5.

3.7.3 Probability of Bit Error for Coherently Detected FSK

Equations (3.83) and (3.84) describe the probability of bit error for coherent antipodal signals. A more general treatment for binary coherent signals (not limited to antipodal signals) yields the following equation for P_B [8]:

$$P_B = \frac{1}{\sqrt{2\pi}} \int_{\sqrt{(1-\rho)E_b/N_0}}^{\infty} \exp\left(-\frac{u^2}{2}\right) du \tag{3.96}$$

From Equation (3.91), $\rho = \cos\theta$ is the time cross-correlation coefficient between signal $s_1(t)$ and $s_2(t)$, where θ is the angle between signal vectors s_1 and s_2 (see Figure 3.6). For antipodal signals such as BPSK, $\theta = \pi$, thus $\rho = -1$.

For orthogonal signals such as binary FSK (BFSK), $\theta = \pi/2$, since the s_1 and s_2 vectors are perpendicular to each other; thus $\rho = 0$, as can be verified with Equation (3.89), and Equation (3.96) can then be written

$$P_B = \frac{1}{\sqrt{2\pi}} \int_{\sqrt{E_b/N_0}}^{\infty} \exp\left(-\frac{u^2}{2}\right) du = Q\left(\sqrt{\frac{E_b}{N_0}}\right) \tag{3.97}$$

where the *co-error function, $Q(x)$*, is defined in Equation (3.82). The result could also have been obtained by noting that the energy difference between the orthogonal signal vectors, s_1 and s_2, with amplitudes of $\sqrt{E_b}$, as shown in Figure 3.20b, can be computed as the square of the distance between the heads of the orthogonal vectors, to be $E_d = 2E_b$. Using this result in Equation (3.92) yields the same result as in Equation (3.97). Equation (3.97) is plotted in Figure 3.22 (coherent detection of FSK). If we compare Equation (3.97) with Equation (3.84), we can see that 3 dB (a factor of 2) more E_b/N_0 is required for BFSK to provide the same performance as BPSK. It should not be surprising that the performance of BFSK signaling is worse than BPSK signaling, since for a given signal power, orthogonal vectors are spaced closer to one another than antipodal vectors.

The bit error probability, P_B, for the coherent detection of orthogonal bandpass signals as seen in Equation (3.97) is the same as the P_B for the matched filter detection of baseband unipolar signals in Equation (2.64). As mentioned earlier, the details of on–off keying (OOK) are not treated in this book. However, it is worth noting that the P_B, described in Equation (3.97), is also identical to the error performance for the coherent detection of OOK signaling (matched filter reception).

3.7.4 Probability of Bit Error for
Noncoherently Detected FSK

Consider the equally likely binary FSK signal set, $\{s_1(t)\}$, defined in Equation (3.27) as follows:

$$s_i(t) = \sqrt{\frac{2E}{T}} \cos(\omega_i t + \phi) \qquad 0 \le t \le T, \qquad i = 1, 2$$

The phase term, ϕ, is unknown and assumed constant. The detector is characterized by $M = 2$ channels of bandpass filters and envelope detectors, as shown in Figure 3.18. The input to the detector consists of the received signal, $r(t) = s_i(t) + n(t)$, where $n(t)$ is a white Gaussian noise process with two-sided power spectral density, $N_0/2$. Assume that $s_1(t)$ and $s_2(t)$ are separated in frequency sufficiently that they have negligible overlap. We start the probability of error, P_B, computation the same way that we did for coherently detected PSK, with Equation (3.78).

$$P_B = \tfrac{1}{2}P(H_2|s_1) + \tfrac{1}{2}P(H_1|s_2) \qquad (3.98)$$

$$= \frac{1}{2} \int_0^T p(z|s_1) \, dz + \frac{1}{2} \int_0^T p(z|s_2) \, dz$$

For the binary case, the *test statistic*, $z(T)$, is defined by $z_1(T) - z_2(T)$. Assume that the bandwidth of the filter, W_f, is $1/T$, so that the envelope of the FSK signal is (approximately) preserved at the filter output. If there was no noise at the receiver, the value of $z(T) = \sqrt{2E/T}$ when $s_1(t)$ is sent, and $z(T) = -\sqrt{2E/T}$ when $s_2(t)$ is sent. Because of this symmetry, the optimum threshold is $\gamma_0 = 0$. The pdf $p(z|s_1)$ is similar to $p(z|s_2)$; that is,

$$p(z|s_1) = p(-z|s_2) \qquad (3.99)$$

Therefore, we can write

$$P_B = \int_0^T p(z|s_2) \, dz \qquad (3.100)$$

or

$$P_B = P(z_1 > z_2|s_2) \qquad (3.101)$$

where z_1 and z_2 denote the outputs $z_1(T)$ and $z_2(T)$ from the envelope detectors shown in Figure 3.18. For the case where the tone $s_2(t) = \cos \omega_2 t$ is sent, such that $r(t) = s_2(t) + n(t)$, the output, $z_1(T)$, is a *Gaussian noise random variable only*; it has no signal component. A Gaussian distribution into the *nonlinear envelope detector* yields a Rayleigh distribution at the output [8], so that

$$p(z_1|s_2) = \begin{cases} \dfrac{z_1}{\sigma_0^2} \exp\left(-\dfrac{z_1^2}{2\sigma_0^2}\right) & z_1 \ge 0 \\[2mm] 0 & z_1 < 0 \end{cases} \qquad (3.102)$$

where σ_0^2 is the noise at the filter output. On the other hand, $z_2(T)$ has a Rician distribution, since the input to the lower envelope detector is a sinusoid plus noise [8]. The pdf, $p(z_2|s_2)$, is written as

$$p(z_2|s_2) = \begin{cases} \dfrac{z_2}{\sigma_0^2} \exp\left[-\dfrac{(z_2^2 + A^2)}{2\sigma_0^2}\right] I_0\left(\dfrac{z_2 A}{\sigma_0^2}\right) & z_2 \geq 0 \\ \\ 0 & z_2 < 0 \end{cases} \tag{3.103}$$

where $A = \sqrt{2E/T}$, and as before, σ_0^2 is the noise at the filter output. The function $I_0(x)$, known as the modified zero-order Bessel function of the first kind [9], is defined as

$$I_0(x) = \frac{1}{2\pi} \int_0^{2\pi} \exp\left(x \cos \theta\right) d\theta \tag{3.104}$$

When $s_2(t)$ is transmitted, the receiver makes an error whenever the envelope sample $z_1(T)$ obtained from the upper channel (due to noise alone) exceeds the envelope sample $z_2(T)$ obtained from the lower channel (due to signal plus noise). Thus the probability of this error can be obtained by integrating $p(z_1|s_2)$ with respect to z_1 from z_2 to infinity, and then averaging over all possible values of z_2. That is,

$$P_B = P(z_1 > z_2|s_2)$$

$$= \int_0^\infty p(z_2|s_2) \left[\int_{z_2}^\infty p(z_1|s_2)\, dz_1\right] dz_2 \tag{3.105}$$

$$= \int_0^\infty \frac{z_2}{\sigma_0^2} \exp\left[-\frac{(z_2^2 + A^2)}{2\sigma_0^2}\right] I_0\left(\frac{z_2 A}{\sigma_0^2}\right) \left[\int_{z_2}^\infty \frac{z_1}{\sigma_0^2} \exp\left(-\frac{z_1^2}{2\sigma_0^2}\right) dz_1\right] dz_2 \tag{3.106}$$

where $A = \sqrt{2E/T}$ and where the inner integral is the conditional probability of an error for a fixed value of z_2, given that $s_2(t)$ was sent, and the outer integral averages this conditional probability over all possible values of z_2. This integral can be evaluated [10], to yield

$$P_B = \frac{1}{2} \exp\left(-\frac{A^2}{4\sigma_0^2}\right) \tag{3.107}$$

Using Equation (1.19), we can express the filter output noise, σ_0^2, as

$$\sigma_0^2 = 2\left(\frac{N_0}{2}\right) W_f \tag{3.108}$$

where $G_n(f) = N_0/2$ and W_f is the filter bandwidth. Thus Equation (3.107) becomes

$$P_B = \frac{1}{2} \exp\left(-\frac{A^2}{4 N_0 W_f}\right) \tag{3.109}$$

Equation (3.109) indicates that the error performance depends on the bandpass filter bandwidth, and that P_B becomes smaller as W_f is decreased. The result is valid only when the intersymbol interference (ISI) is negligible. The minimum W_f allowed (i.e., for no ISI) is obtained from Equation (2.77) with the filter roll-off factor $r = 0$. Thus $W_f = R$ bits/s $= 1/T$, and we can write Equation (3.109) as

$$P_B = \frac{1}{2} \exp \left(-\frac{A^2 T}{4N_0} \right) \tag{3.110}$$

$$= \frac{1}{2} \exp \left(-\frac{E_b}{2N_0} \right) \tag{3.111}$$

where $E_b = (1/2)A^2 T$ is the energy per bit. When comparing the error performance of noncoherent FSK with coherent FSK (see Figure 3.22), it is seen that for the same P_B, noncoherent FSK requires approximately 1 dB more E_b/N_0 than that for coherent FSK (for $P_B \leq 10^{-4}$). The noncoherent receiver is easier to implement, since coherent reference signals need not be generated. Therefore, almost all FSK receivers use noncoherent detection. It can be seen in the following section that when comparing noncoherent FSK to noncoherent DPSK, the same 3-dB difference occurs as for the comparison between coherent FSK and coherent PSK.

As mentioned earlier, the details of on–off keying (OOK) are not treated in this book. However, it is worth noting that the bit error probability, P_B, described in Equation (3.111) is identical to the P_B for the noncoherent detection of OOK signaling.

3.7.5 Probability of Bit Error for DPSK

Let us define a BPSK signal set

$$x_1(t) = \sqrt{\frac{2E}{T}} \cos (\omega_0 t + \phi) \qquad 0 \leq t \leq T$$

$$\tag{3.112}$$

$$x_2(t) = \sqrt{\frac{2E}{T}} \cos (\omega_0 t + \phi \pm \pi) \qquad 0 \leq t \leq T$$

A characteristic of DPSK is that there are no fixed decision regions in the signal space. Instead, the decision is based on the phase difference between successively received signals. Then for DPSK signaling we are really transmitting each bit with the binary signal pair

$$s_1(t) = (x_1, x_1) \quad \text{or} \quad (x_2, x_2) \qquad 0 \leq t \leq 2T$$
$$s_2(t) = (x_1, x_2) \quad \text{or} \quad (x_2, x_1) \qquad 0 \leq t \leq 2T \tag{3.113}$$

where (x_i, x_j) $(i, j = 1, 2)$ denotes $x_i(t)$ followed by $x_j(t)$ defined in Equation (3.112). The first T seconds of each waveform are actually the last T seconds of the previous waveform. Note that $s_1(t)$ and $s_2(t)$ can each have either of two possible forms and that $x_1(t)$ and $x_2(t)$ are antipodal signals. Thus the correlation between $s_1(t)$ and $s_2(t)$ for *any combination* of forms can be written as

$$z(2T) = \int_0^{2T} s_1(t)s_2(t) \, dt$$

$$= \int_0^T [x_1(t)]^2 \, dt - \int_0^T [x_1(t)]^2 \, dt = 0$$

(3.114)

Therefore, pairs of DPSK signals can be represented as orthogonal signals $2T$ seconds long. Detection could correspond to noncoherent envelope detection with four channels matched to each of the possible envelope outputs, as shown in Figure 3.23a. Since the two envelope detectors representing each symbol are negatives of each other, the envelope sample of each will be the same. Hence we can implement the detector as a single channel for $s_1(t)$ matched to either (x_1, x_1) or (x_2, x_2), and a single channel for $s_2(t)$ matched to either (x_1, x_2) or (x_2, x_1), as shown in Figure 3.23b. The DPSK detector is therefore reduced to a standard two-channel noncoherent detector. In reality, the filter can be matched to the difference signal so that only one channel is necessary. For orthogonal signals, this operates with the bit error probability in Equation (3.111). Since the DPSK signals have a bit interval of $2T$, the $s_i(t)$ signals defined in Equation (3.113) have

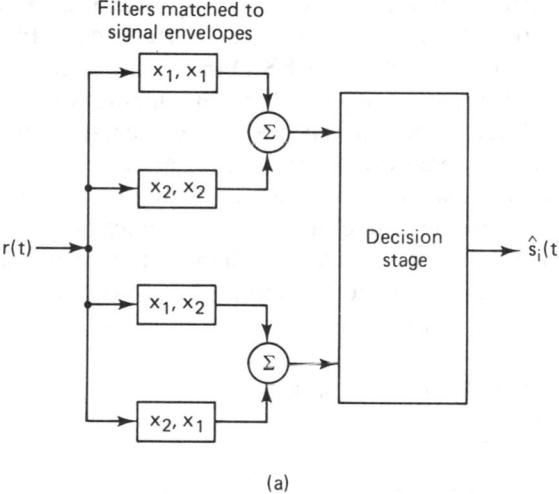

(a)

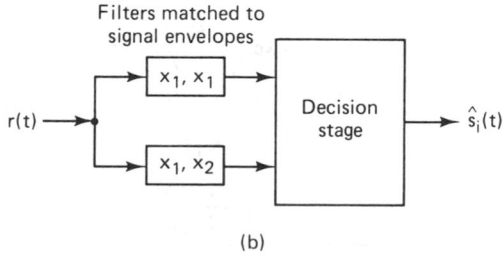

(b)

Figure 3.23 DPSK detection. (a) Four-channel differentially coherent detection of binary DPSK. (b) Equivalent two-channel detector for binary DPSK.

twice the energy of a signal defined over a single-symbol duration. Thus we may write P_B as

$$P_B = \frac{1}{2} \exp \left(- \frac{E_b}{N_0} \right) \qquad (3.115)$$

Equation (3.115) is seen plotted in Figure 3.22, designated as differentially coherent detection of differentially encoded PSK, or simply DPSK. This expression is valid for the optimum DPSK detector shown in Figure 3.16c. For the detector shown in Figure 3.16b, the error probability will be slightly inferior to that given in Equation (3.115) [5]. When comparing the error performance of Equation (3.115) with that of coherent PSK (see Figure 3.22), it is seen that for the same P_B, DPSK requires approximately 1 dB more E_b/N_0 than does BPSK (for $P_B \leq 10^{-4}$). It is easier to implement a DPSK system than a PSK system, since the DPSK receiver does not need phase synchronization. For this reason, DPSK, although less efficient than PSK, is sometimes the preferred choice between the two.

3.7.6 Comparison of Bit Error Performance for Various Modulation Types

The P_B expressions for the best known of the binary modulation schemes discussed above are listed in Table 3.1 and are illustrated in Figure 3.22. For $P_B = 10^{-4}$, it can be seen that there is approximately a 4-dB difference between the best (coherent PSK) and the worst (noncoherent FSK) that were discussed here. In some cases, 4 dB is a small price to pay for the implementation simplicity gained in going from coherent PSK to noncoherent FSK; however, for other cases, even a 1-dB saving is worthwhile. There are other considerations besides P_B and system complexity; for example, in some cases (such as a randomly fading channel), a noncoherent system is more desirable because there may be difficulty in establishing and maintaining a coherent reference. Signals that can withstand significant degradation before their ability to be detected is affected are clearly desirable in military and space applications.

TABLE 3.1 Probability of Error for Selected Binary Modulation Schemes

Modulation	P_B
Coherent PSK	$Q\left(\sqrt{\dfrac{2E_b}{N_0}} \right)$
Noncoherent DPSK	$\dfrac{1}{2} \exp \left(-\dfrac{E_b}{N_0} \right)$
Coherent FSK	$Q\left(\sqrt{\dfrac{E_b}{N_0}} \right)$
Noncoherent FSK	$\dfrac{1}{2} \exp \left(-\dfrac{1}{2}\dfrac{E_b}{N_0} \right)$

3.8 *M*-ARY SIGNALING AND PERFORMANCE

3.8.1 Ideal Probability of Bit Error Performance

The typical probability of error versus E_b/N_0 curve was shown to have a waterfall-like shape in Figure 3.21. The probability of bit error (P_B) characteristics of various binary modulation schemes in AWGN also display this shape, as shown in Figure 3.22. What should an *ideal* P_B versus E_b/N_0 curve look like? Figure 3.24 displays the ideal characteristic as the *Shannon limit*. The limit represents the threshold E_b/N_0 below which reliable communication cannot be maintained. Shannon's work is described in greater detail in Chapter 7.

We can describe the ideal curve in Figure 3.24 as follows. For all values of E_b/N_0 above the Shannon limit, P_B is zero. Once E_b/N_0 is reduced below the Shannon limit, P_B degrades to the worst-case value of $\frac{1}{2}$. (Note that $P_B = 1$ is not the worst case for binary signaling, since that value is just as good as $P_B = 0$; if the probability of making a bit error is 100%, the bit stream could simply be inverted to retrieve the correct data.) It should be clear, by comparing the typical P_B curve with the ideal one in Figure 3.24 that the large arrow in the figure describes the desired direction of movement to achieve improved P_B performance.

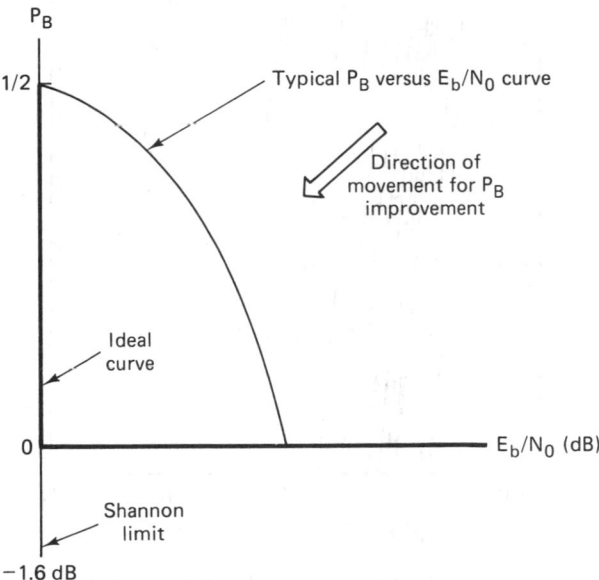

Figure 3.24 Ideal P_B versus E_b/N_0 curve.

3.8.2 *M*-ary Signaling

Let us review *M*-ary signaling. The processor considers k bits at a time. It instructs the modulator to produce one of $M = 2^k$ waveforms; binary signaling is the special case where $k = 1$. Does *M*-ary signaling improve or degrade performance? Be careful with your answer—the question is a loaded one. Figure 3.25 illustrates the probability of bit error, $P_B(M)$, versus E_b/N_0 for coherently detected *orthog-*

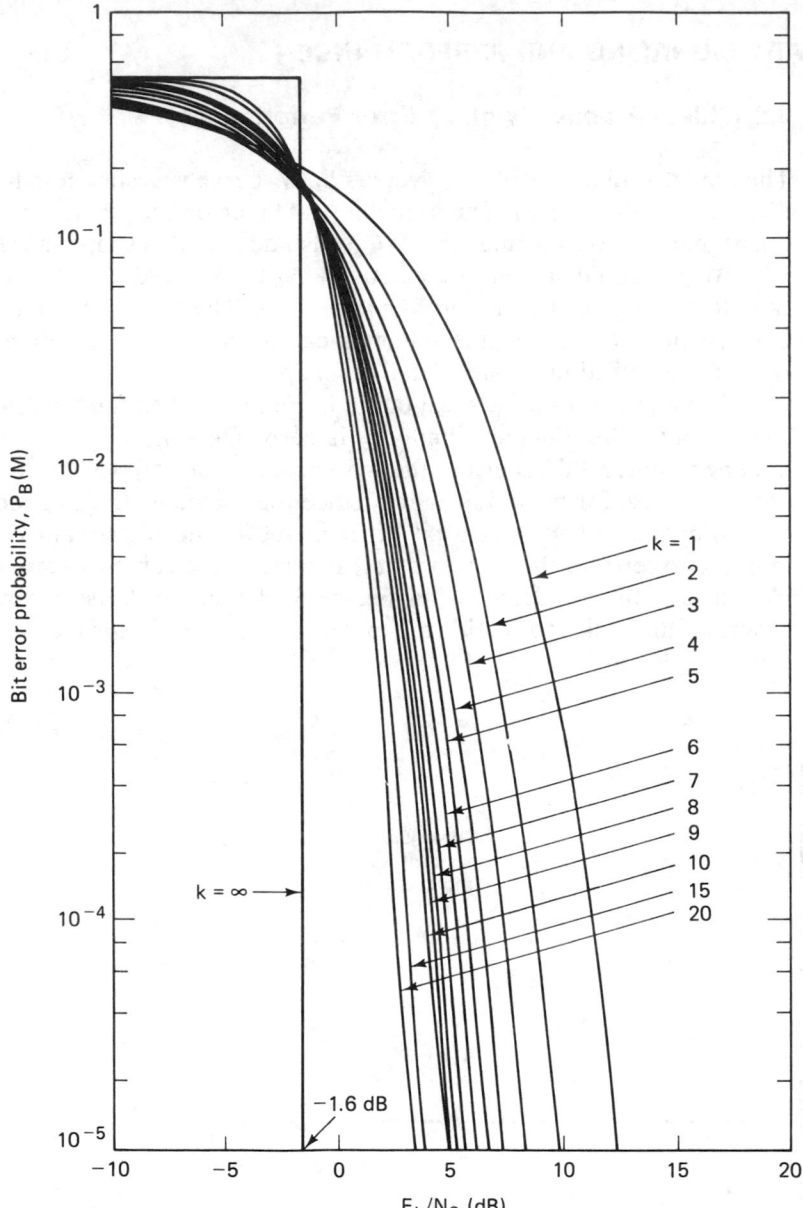

Figure 3.25 Bit error probability for coherently detected *M*-ary orthogonal signaling. (Reprinted from W. C. Lindsey and M. K. Simon, *Telecommunication Systems Engineering*, Prentice-Hall, Inc., Englewood Cliffs, N.J., 1973, courtesy of W. C. Lindsey and Marvin K. Simon.)

onal M-ary signaling over a Gaussian channel. Figure 3.26 similarly illustrates $P_B(M)$ versus E_b/N_0 for coherently detected *multiple phase M*-ary signaling over a Gaussian channel. In which direction do the curves move as the value of k (or M) increases? From Figure 3.24 we know the directions of curve movement for

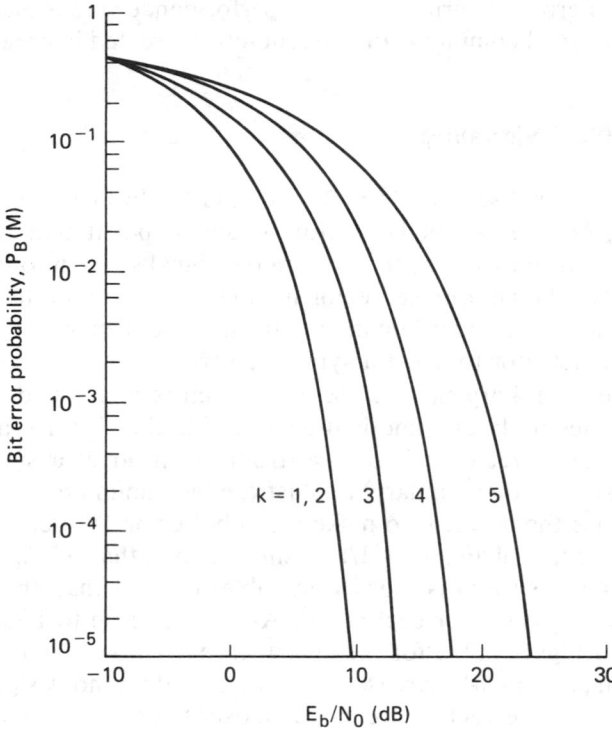

The figure shows: Bit error probability, $P_B(M)$ (vertical axis) versus E_b/N_0 (dB) (horizontal axis). Curves labeled $k = 1, 2$; 3; 4; 5.

Figure 3.26 Bit error probability for coherently detected multiple phase signaling.

improved and degraded error performance. In Figure 3.25, as k increases, the curves move in the direction of improved error performance. In Figure 3.26, as k increases, the curves move in the direction of degraded error performance. Such movement tells us that M-ary signaling produces improved error performance with orthogonal signaling and degraded error performance with multiple phase signaling. Can that be true? Why would anyone ever use multiple phase PSK signaling if it provides degraded error performance compared to binary PSK signaling? It *is* true, and many systems do use multiple phase signaling. The question, as stated, is loaded because it implies that error probability versus E_b/N_0 is the *only* performance criterion; there are many others (e.g., bandwidth, power, throughput, complexity), but in Figures 3.25 and 3.26, error performance is the characteristic that stands out explicitly.

A performance characteristic that is not explicitly seen in Figures 3.25 and 3.26 is the required system bandwidth. For the curves characterizing M-ary orthogonal signals in Figure 3.25, as k increases, the required bandwidth also increases. For the M-ary multiple phase curves in Figure 3.26, as k increases, a larger bit rate can be transmitted within the same bandwidth. In other words, for a fixed data rate, the required bandwidth is decreased. Therefore, *both* the orthogonal and multiple phase error performance curves tell us that M-ary signaling represents a vehicle for performing a system trade-off. In the case of orthogonal signaling, error performance improvement can be achieved at the expense of bandwidth. In the case of multiple phase signaling, bandwidth performance can be achieved at the expense of error performance. Error performance versus bandwidth performance, a fundamental communications trade-off, is treated in greater detail in Chapter 7.

3.8.3 Vectorial View of MPSK Signaling

Figure 3.27 illustrates MPSK signal sets for $M = 2, 4, 8,$ and 16. In Figure 3.27a we see the binary ($k = 1, M = 2$) antipodal vectors s_1 and s_2 positioned 180° apart. The decision boundary is drawn so as to partition the signal space into two regions. On the figure is also shown a noise vector n equal in magnitude to s_1. The figure establishes the magnitude and orientation of the minimum energy noise vector that would cause the detector to make a symbol error.

In Figure 3.27b we see the 4-ary ($k = 2, M = 4$) vectors positioned 90° apart. The decision boundaries (only one line is drawn) divide the signal space into four regions. Again a noise vector n is drawn (from the head of a signal vector, normal to the closest decision boundary) to illustrate the minimum energy noise vector that would cause the detector to make a symbol error. Notice that the minimum energy noise vector of Figure 3.27b is smaller than that of Figure 3.27a, illustrating that the 4-ary system is more vulnerable to noise than the 2-ary system (signal energy being equal for each case). As we move on to Figure 3.27c for the 8-ary case and Figure 3.27d for the 16-ary case, it should be clear that for multiple phase signaling, as M increases, we are crowding more signal vectors into the signal plane. As the vectors are moved closer together, a smaller amount of noise energy is required to cause an error.

Figure 3.27 adds some insight as to why the curves of Figure 3.26 behave as they do as k is increased. Figure 3.27 also provides some insight into a basic trade-off in multiple phase signaling. Crowding more signal vectors into the signal space is tantamount to increasing the data rate without increasing the system bandwidth (the vectors are all confined to the same plane). In other words, we have increased the bandwidth utilization at the expense of error performance.

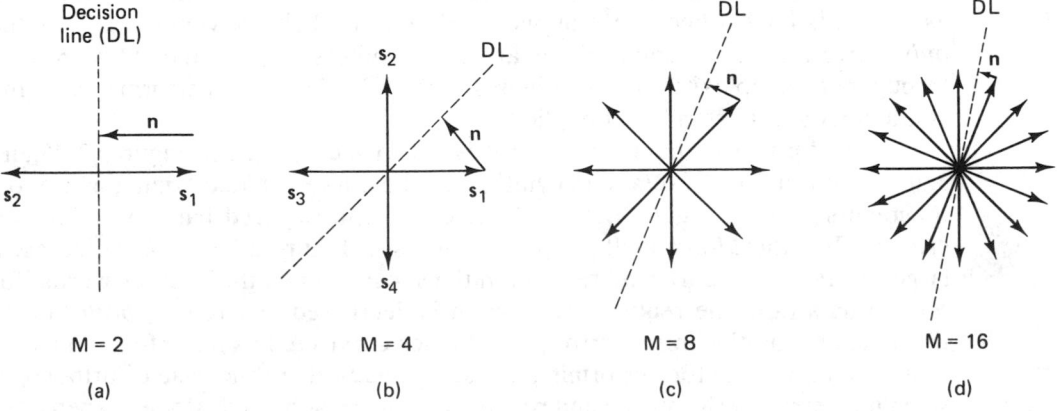

Figure 3.27   MPSK signal sets for $M = 2, 4, 8, 16$.

Look at Figure 3.27d, where the error performance is worse than any of the other examples in Figure 3.27. How might we "buy back" the degraded error performance; that is, what can we trade-off so that the distance between neighboring signal vectors in Figure 3.27d is increased to that in Figure 3.27a? We can increase the signal strength (make the signal vectors larger) until the minimum distance from the head of a signal vector to a decision line equals the length of the noise vector in Figure 3.27a. Therefore, in a multiple phase system, as M is increased, we can either achieve improved bandwidth performance at the expense of error performance, or if we increase the E_b/N_0 so that the error probability is not degraded, we can achieve improved bandwidth performance at the expense of increasing E_b/N_0.

3.8.4 BPSK and QPSK Have the Same Bit Error Probability

In Equation (3.94) we stated the general relationship between E_b/N_0 and S/N_0 for binary transmission, as follows:

$$\frac{E_b}{N_0} = \frac{S}{N_0} \cdot \left(\frac{1}{R}\right) \tag{3.116}$$

where S is the average signal power and R is the bit rate. A BPSK signal with the available E_b/N_0 found from Equation (3.116) will perform with a P_B that can be read from the $k = 1$ curve in Figure 3.26. QPSK can be characterized as two orthogonal BPSK channels. The QPSK bit stream is usually partitioned into an even and odd (I and Q) stream; each new stream modulates an orthogonal component of the carrier at half the bit rate of the original stream. The I stream modulates the cos $\omega_0 t$ term and the Q stream modulates the sin $\omega_0 t$ term. If the magnitude of the original QPSK vector has the value A, the magnitude of the I and Q component vectors each has a value of $A/\sqrt{2}$, as shown in Figure 3.28. Thus, each of the quadrature BPSK signals has half of the average power of the original QPSK signal. Hence if the original QPSK waveform has a bit rate of R bits/s and an average power of S watts, the quadrature partitioning results in each

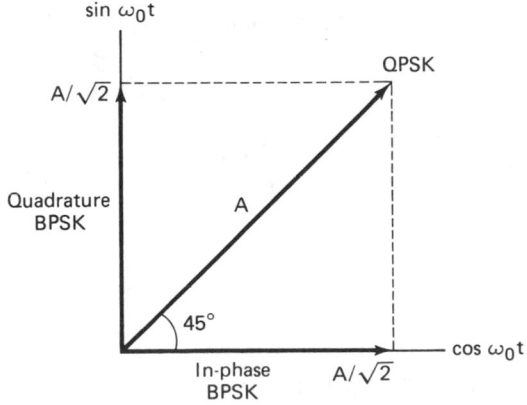

Figure 3.28   In-phase and quadrature BPSK components of QPSK signaling.

of the BPSK waveforms having a bit rate of $R/2$ bits/s and an average power of $S/2$ watts.

Therefore, the E_b/N_0 characterizing each of the orthogonal BPSK channels, comprising the QPSK signal, is equivalent to the E_b/N_0 in Equation (3.116) since it can be written as

$$\frac{E_b}{N_0} = \frac{S/2}{N_0}\left(\frac{W}{R/2}\right) = \frac{S}{N_0}\left(\frac{1}{R}\right) \tag{3.117}$$

Thus each of the orthogonal BPSK channels, and hence the composite QPSK signal, is characterized by the same E_b/N_0 and hence the same P_B performance as a BPSK signal. The natural orthogonality of the 90° phase shifts between adjacent QPSK symbols results in the *bit error probabilities* being equal for both BPSK and QPSK signaling. It is important to note that the *symbol error probabilities* are *not* equal for BPSK and QPSK signaling. The relationship between bit error probability and symbol error probability is treated in Sections 3.9.3 and 3.9.4.

3.8.5 Vectorial View of MFSK Signaling

In Section 3.8.3, Figure 3.27 provides some insight as to why the error performance of MPSK signaling degrades as k (or M) increases. It would be useful to have a similar vectorial illustration for the error performance of MFSK signaling as seen in the curves of Figure 3.25. Since the MFSK signal space is characterized by M mutually perpendicular axes, we can only conveniently illustrate the cases, $M = 2$ and $M = 3$. In Figure 3.29a we see the binary orthogonal vectors s_1 and s_2 positioned 90° apart. The decision boundary is drawn so as to partition the signal space into two regions. On the figure is also shown a noise vector n, which represents the minimum noise vector that would cause the detector to make an error.

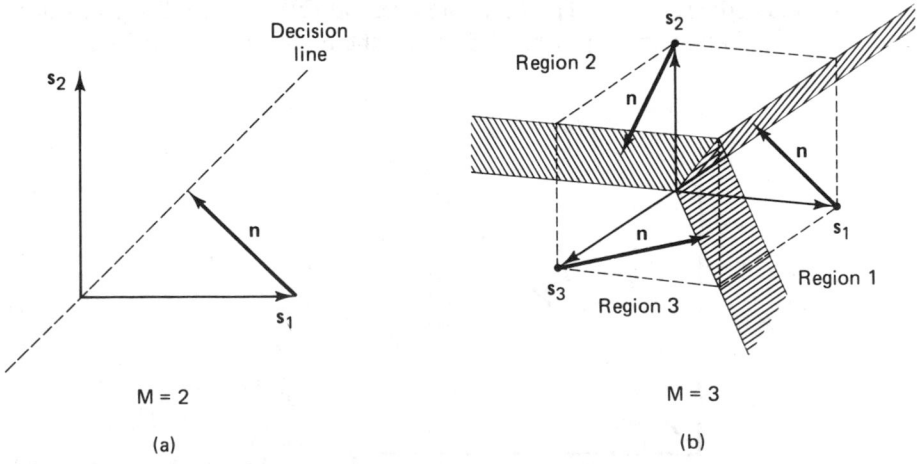

Figure 3.29 MFSK signal sets for $M = 2, 3$.

In Figure 3.29b we see a 3-ary signal space with axes positioned 90° apart. Here decision planes partition the signal space into three regions. Noise vectors **n** are shown added to each of the prototype signal vectors s_1, s_2, and s_3; each noise vector illustrates an example of the minimum noise energy that would cause the detector to make a symbol error. The minimum noise energy vectors in Figure 3.29b are the same length as the noise vector in Figure 3.29a. In Section 3.5.4 we stated that the distance between any two prototype signal vectors s_i and s_j in an M-ary orthogonal space is constant. It follows that the minimum distance between a prototype signal vector and any of the decision boundaries remains fixed as M increases. Unlike the case of MPSK signaling, where adding new signals to the signal set makes the signals vulnerable to smaller noise vectors, here in the case of MFSK signaling, adding new signals to the signal set does *not* make the signals vulnerable to smaller noise vectors.

It would be convenient to illustrate the point by drawing higher-dimensional orthogonal spaces, but of course this is not possible. We can only use our "mind's eye" to understand that increasing the signal set, M, by adding additional axes, where each new axis is mutually perpendicular to all the others, does not crowd the signal set more closely together; thus a transmitted signal from an orthogonal set is *not* more vulnerable to a noise vector when the set is increased in size. In fact, we see from Figure 3.25 that as k increases, the bit error performance improves.

Understanding the error performance improvement of orthogonal signaling, as illustrated in Figure 3.25, is facilitated by comparing the probability of symbol error (P_E) versus unnormalized SNR, with P_E versus E_b/N_0. Figure 3.30 represents a set of P_E performance curves plotted against unnormalized SNR for coherent FSK signaling. Here we see that P_E degrades as M is increased. Didn't we say that an orthogonal signal is *not* made more vulnerable to a given noise vector, as the orthogonal set is increased in size? It is correct that for orthogonal signaling, with a given SNR it takes a fixed-size noise vector to perturb a transmitted signal into an error region; the signals do not become vulnerable to smaller noise vectors as M increases. However, as M increases, more neighboring decision regions are introduced; thus the number of ways in which a symbol error can be made increases. Figure 3.30 reflects the degradation in P_E versus unnormalized SNR as M is increased; there are $(M - 1)$ ways to make an error. Examining performance under the condition of a fixed SNR (as M increases) is not very useful for digital communications. A fixed SNR means a fixed amount of energy per symbol; thus as M increases, there is a fixed amount of energy to be apportioned over a larger number of bits, or there is less energy per bit. The most useful way of comparing one digital system with another is on the basis of *bit-normalized SNR* or E_b/N_0. The error performance improvement with increasing M, seen in Figure 3.25, manifests itself only when error probability is plotted against E_b/N_0. For this case, as M increases, the required E_b/N_0 (to meet a given error probability) is reduced for a fixed SNR; therefore, we need to map the Figure 3.30 plot into a new plot, similar to Figure 3.25, where the abscissa represents E_b/N_0 instead of SNR. Figure 3.31 illustrates such a mapping; it demonstrates that curves manifesting degraded P_E with increasing M (such as Figure 3.30) are

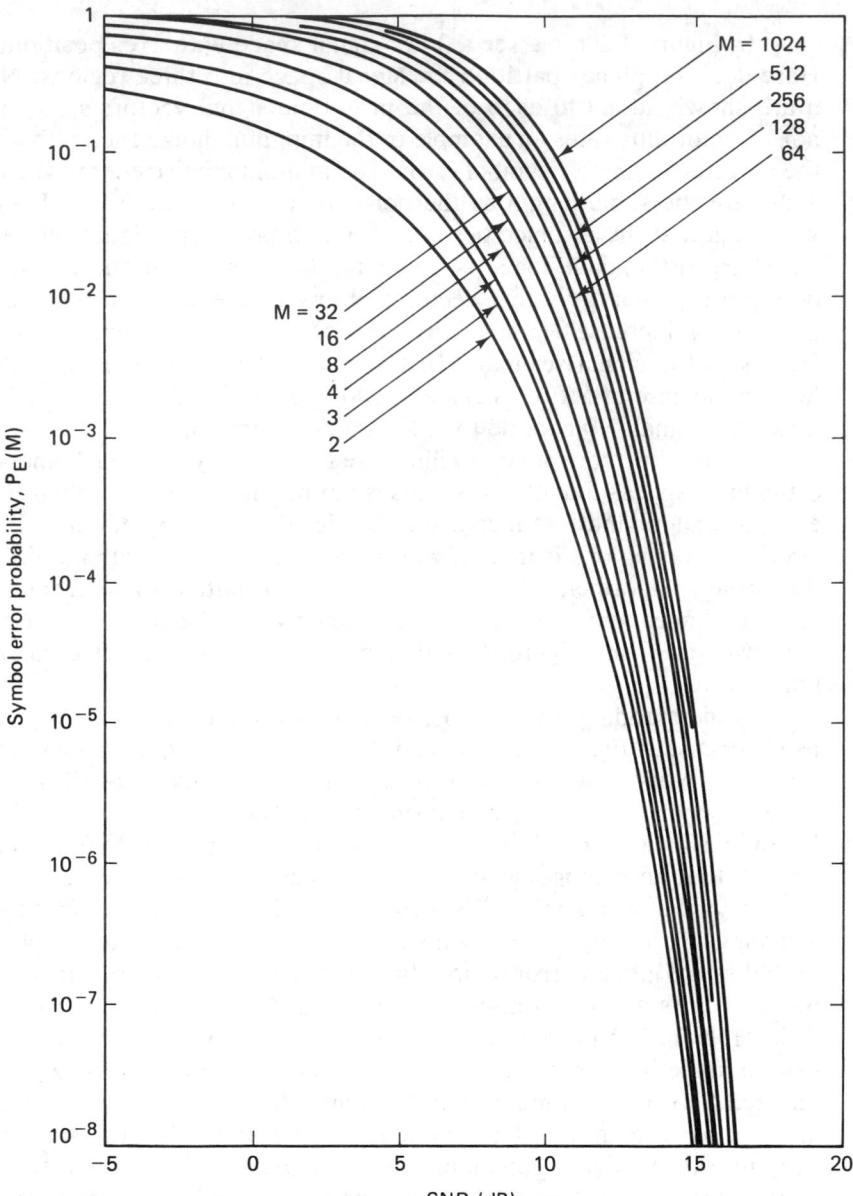

Figure 3.30 Symbol error probability versus SNR for coherent FSK signaling. (From Bureau of Standards, *Technical Note 167*, March 1963.) (Reprinted from *Central Radio Propagation Laboratory Technical Note 167*, March 25, 1963, Fig. 1, p. 5, courtesy of National Bureau of Standards.)

transformed into curves manifesting improved P_E with increasing M. The basic mapping relationship is expressed in Equation (3.94):

$$\frac{E_b}{N_0} = \frac{S}{N}\left(\frac{W}{R}\right)$$

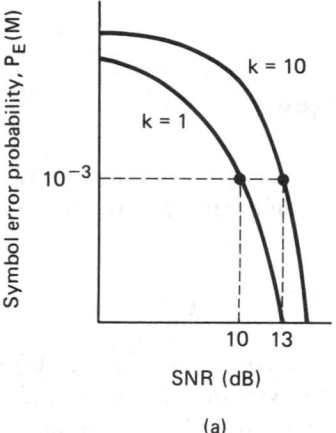
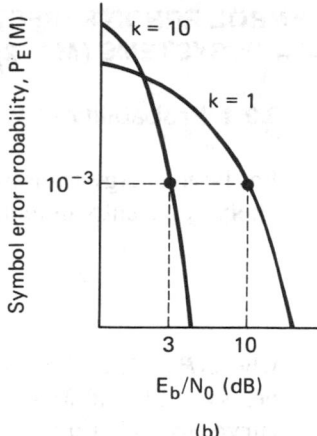

Figure 3.31 Mapping P_E versus SNR into P_E versus E_b/N_0 for orthogonal signaling. (a) Unnormalized. (b) Normalized.

where W is the detection bandwidth. Since

$$R = \frac{\log_2 M}{T} = \frac{k}{T}$$

where T is the symbol duration, we can then write

$$\frac{E_b}{N_0} = \frac{S}{N}\left(\frac{WT}{\log_2 M}\right) = \frac{S}{N}\left(\frac{WT}{k}\right) \tag{3.118}$$

For FSK signaling the detection bandwidth, W in hertz, is typically equal in value to the symbol rate $1/T$, in other words, $WT \simeq 1$. Therefore,

$$\frac{E_b}{N_0} \simeq \frac{S}{N}\left(\frac{1}{k}\right) \tag{3.119}$$

Figure 3.31 illustrates the mapping from P_E versus SNR to P_E versus E_b/N_0 for coherently detected M-ary orthogonal signaling. In Figure 3.31a, on the $k = 1$ curve is shown an operating point corresponding to $P_E = 10^{-3}$ and SNR = 10 dB. On the $k = 10$ curve is shown an operating point at the same $P_E = 10^{-3}$ but with SNR = 13 dB (approximate values taken from Figure 3.30). Here we clearly see the degradation in error performance as k increases. Consider the same $k = 1$ and $k = 10$ cases mapped onto the Figure 3.31b plane, where the abscissa is E_b/N_0. The $k = 1$ case looks exactly the same as it does in Figure 3.31a. But for the $k = 10$ case, the required E_b/N_0 is obtained from Equation (3.119) as follows: $E_b/N_0 = 20(\frac{1}{10}) = 2$ (3 dB), thus showing the error performance improvement as k is increased. In digital communication systems, error performance is almost always considered in terms of E_b/N_0, since such a measurement makes for a meaningful comparison between one system's performance and another. Therefore, the curves of Figures 3.30 and 3.31a are hardly ever seen.

3.9 SYMBOL ERROR PERFORMANCE FOR M-ARY SYSTEMS ($M > 2$)

3.9.1 Probability of Symbol Error for MPSK

For large energy-to-noise ratios, the symbol error performance, $P_E(M)$, for equally likely coherently detected M-ary PSK signaling can be expressed [9] as follows:

$$P_E(M) \simeq 2Q \left(\sqrt{\frac{2E_s}{N_0}} \sin \frac{\pi}{M} \right) \qquad (3.120)$$

where $P_E(M)$ is the probability of symbol error, $E_s = E_b(\log_2 M)$ is the energy per symbol, and $M = 2^k$ is the size of the symbol set. The $P_E(M)$ performance curves for coherently detected MPSK signaling are plotted versus E_b/N_0 in Figure 3.32.

The symbol error performance for differentially coherent detection of M-

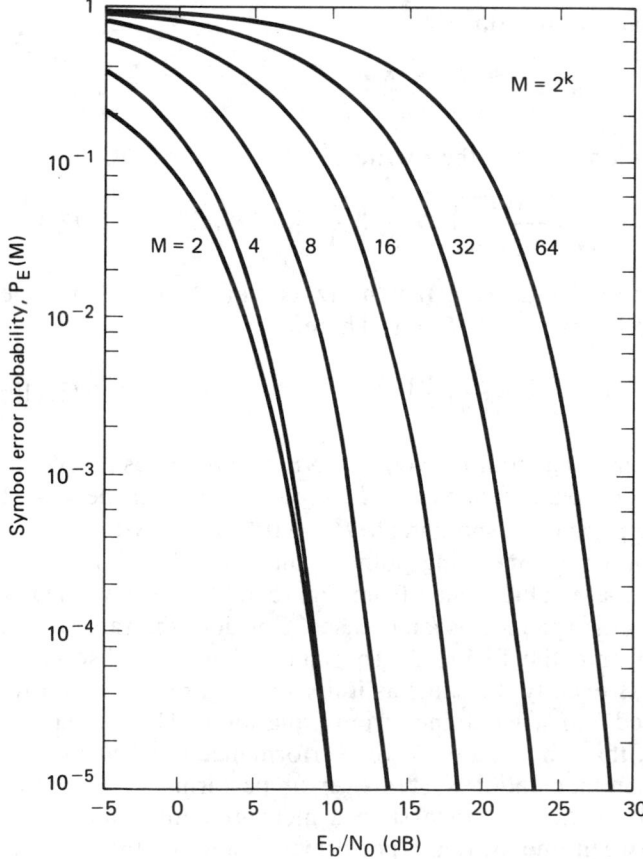

Figure 3.32 Symbol error probability for coherently detected multiple phase signaling. (Reprinted from W. C. Lindsey and M. K. Simon, *Telecommunication Systems Engineering*, Prentice-Hall, Inc., Englewood Cliffs, N.J., 1973, courtesy of W. C. Lindsey and Marvin K. Simon.)

ary DPSK (for large E_s/N_0) is similarly expressed [9] as

$$P_E(M) \simeq 2Q\left(\sqrt{\frac{2E_s}{N_0}}\, \sin \frac{\pi}{\sqrt{2M}}\right) \tag{3.121}$$

3.9.2 Probability of Symbol Error for MFSK

The symbol error performance $P_E(M)$, for equally likely *coherently* detected M-ary orthogonal signaling can be upper bounded [7] as follows:

$$P_E(M) \le (M - 1)Q\left(\sqrt{\frac{E_s}{N_0}}\right) \tag{3.122}$$

where $E_s = E_b(\log_2 M)$ is the energy per symbol and M is the size of the symbol set. The $P_E(M)$ performance curves for coherently detected M-ary orthogonal signaling are plotted versus E_b/N_0 in Figure 3.33.

The symbol error performance for equally likely *noncoherently* detected M-ary orthogonal signaling is [11]

$$P_E(M) = \frac{1}{M} \exp\left(-\frac{E_s}{N_0}\right) \sum_{j=2}^{M} (-1)^j \binom{M}{j} \exp\left(\frac{E_s}{jN_0}\right) \tag{3.123}$$

where

$$\binom{M}{j} = \frac{M!}{j!\,(M - j)!} \tag{3.124}$$

is the standard binomial coefficient yielding the number of ways in which j symbols out of M may be in error. Note that for the binary case, Equation (3.123) reduces to

$$P_B = \frac{1}{2} \exp\left(-\frac{E_b}{2N_0}\right) \tag{3.125}$$

which is the same result as that described by Equation (3.111). The $P_E(M)$ performance curves for noncoherently detected M-ary orthogonal signaling are plotted versus E_b/N_0 in Figure 3.34. If we compare this noncoherent orthogonal $P_E(M)$ performance with the corresponding $P_E(M)$ results for the coherent detection of orthogonal signals in Figure 3.33, it can be seen that for $k > 7$, there is a negligible difference. An upper bound for coherent as well as noncoherent reception of orthogonal signals is [11]

$$P_E(M) < \frac{M - 1}{2} \exp\left(-\frac{E_s}{2N_0}\right) \tag{3.126}$$

where E_s is the energy per symbol and M is the size of the symbol set.

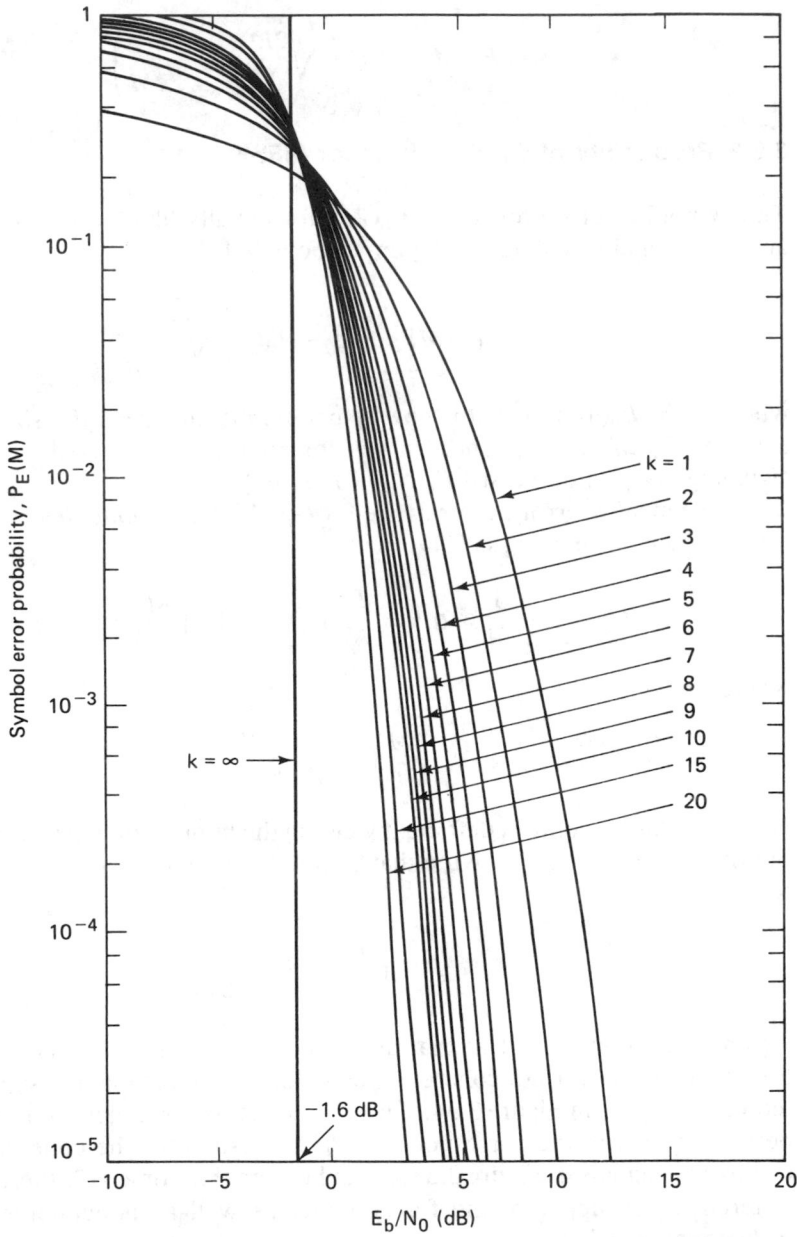

Figure 3.33 Symbol error probability for coherently detected *M*-ary orthogonal signaling. (Reprinted from W. C. Lindsey and M. K. Simon, *Telecommunication Systems Engineering*, Prentice-Hall, Inc., Englewood Cliffs, N.J., 1973, courtesy of W. C. Lindsey and Marvin K. Simon.)

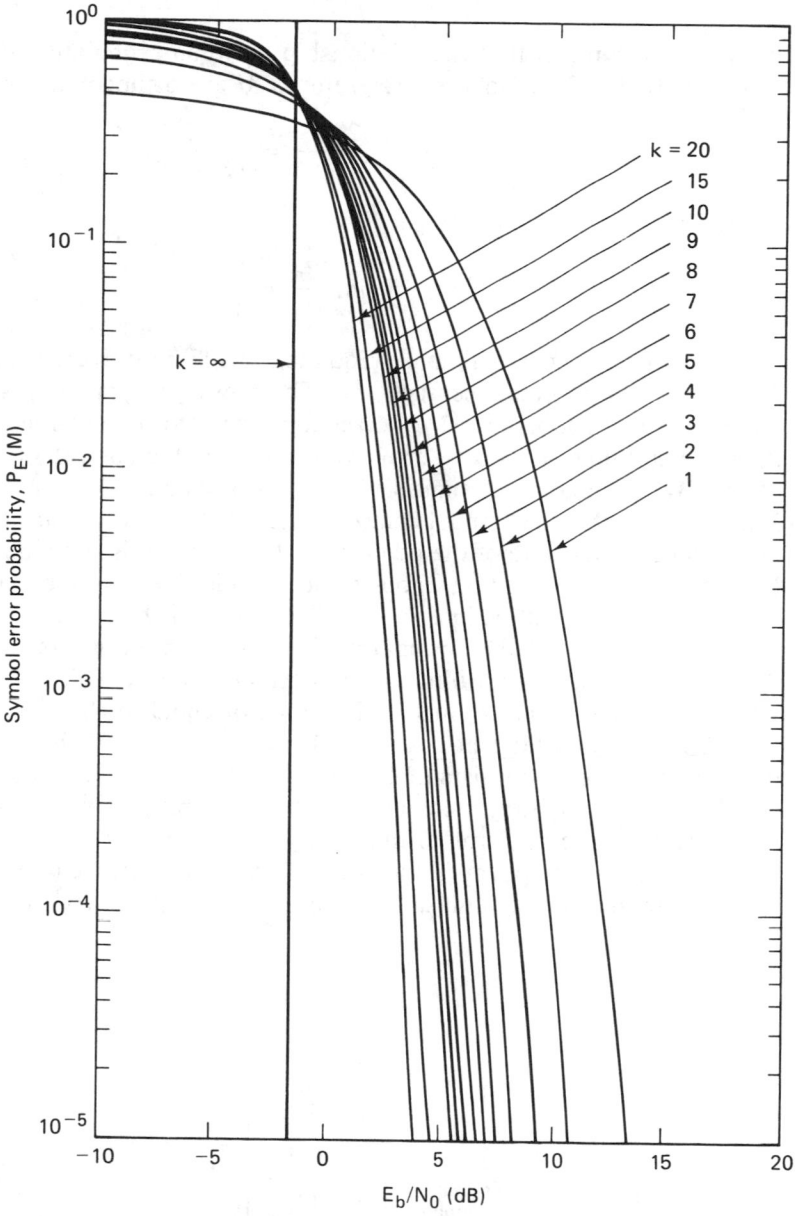

Figure 3.34 Symbol error probability for noncoherently detected *M*-ary orthogonal signaling. (Reprinted from W. C. Lindsey and M. K. Simon, *Telecommunication Systems Engineering*, Prentice-Hall, Inc., Englewood Cliffs, N.J., 1973, courtesy of W. C. Lindsey and Marvin K. Simon.)

3.9.3 Bit Error Probability versus Symbol Error Probability for Orthogonal Signals

It can be shown [11] that the relationship between probability of bit error (P_B) and probability of symbol error (P_E) for an M-ary orthogonal signal set is

$$\frac{P_B}{P_E} = \frac{2^{k-1}}{2^k - 1} = \frac{M/2}{M - 1} \tag{3.127}$$

In the limit as k increases we get

$$\lim_{k \to \infty} \frac{P_B}{P_E} = \frac{1}{2}$$

A simple example will make Equation (3.127) intuitively acceptable. Figure 3.35 describes an octal message set. The message symbols (assumed equally likely) are to be transmitted on orthogonal waveforms such as FSK. With orthogonal signaling, a decision error willl transform the correct signal into any one of the ($M - 1$) incorrect signals with equal probability. The example in Figure 3.35 indicates that the symbol comprised of bits 0 1 1 was transmitted. An error might occur in any one of the other $2^k - 1 = 7$ symbols, with equal probability. Notice that just because a symbol error is made does not mean that all the bits within the symbol will be in error. In Figure 3.35, if the receiver decides that the transmitted symbol is the bottom one listed, comprised of bits 1 1 1, two of the three transmitted symbol bits will be correct; only one bit will be in error. It should be apparent that P_B will be less than or equal to P_E.

Consider any of the bit-position columns in Figure 3.35. For each bit position, the digit occupancy consists of 50% ones and 50% zeros. In the context of the first bit position (rightmost column) and the transmitted symbol, how many ways are there to cause an error to the binary one? There are $2^{k-1} = 4$ ways (four places where zeros appear in the column) that a bit error can be made; it is the same for each of the columns. The final relationship, P_B/P_E, for orthogonal sig-

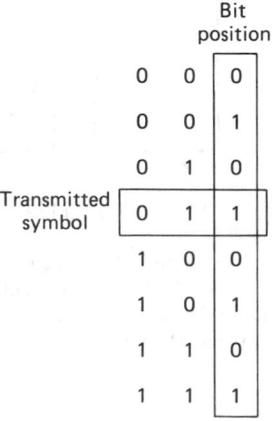

Figure 3.35   Example of P_B versus P_E.

naling, in Equation (3.127), is obtained by forming the following ratio: the number of ways that a bit error can be made (2^{k-1}) divided by the number of ways that a symbol error can be made ($2^k - 1$). For the Figure 3.35 example, $P_B/P_E = 4/7$.

3.9.4 Bit Error Probability versus Symbol Error Probability for Multiple Phase Signaling

For the case of MPSK signaling, P_B is less than or equal to P_E, just as in the case of MFSK signaling. However, there is an important difference. For orthogonal signaling, selecting any one of the ($M - 1$) erroneous symbols is equally likely. In the case of MPSK signaling, each signal vector is not equidistant from all of the others. Figure 3.36a illustrates an 8-ary decision space with the pie-shaped regions denoted by the 8-ary symbols in binary notation. If symbol (0 1 1) is transmitted, it is clear that should an error occur, the transmitted signal will most likely be mistaken for one of its closest neighbors, (0 1 0) or (1 0 0). The likelihood that (0 1 1) would get mistaken for (1 1 1) is relatively remote. If the assignment of bits to symbols follows the binary sequence shown in the symbol decision regions of Figure 3.36a, some symbol errors will usually result in two or more bit errors, even with a large signal-to-noise ratio.

For nonorthogonal schemes, such as MPSK signaling, one often uses a binary-to M-ary code such that binary sequences corresponding to adjacent symbols (phase shifts) differ in only one bit position; thus when an M-ary symbol error occurs, it is more likely that only one of the k input bits will be in error. A code that provides this desirable feature is the Gray code [9]; Figure 3.36b illustrates the bit-to-symbol assignment using a Gray code for 8-ary PSK. Here it can be seen that neighboring symbols differ from one another in only one bit position. Therefore, the occurrence of a multibit error, for a given symbol error, is much

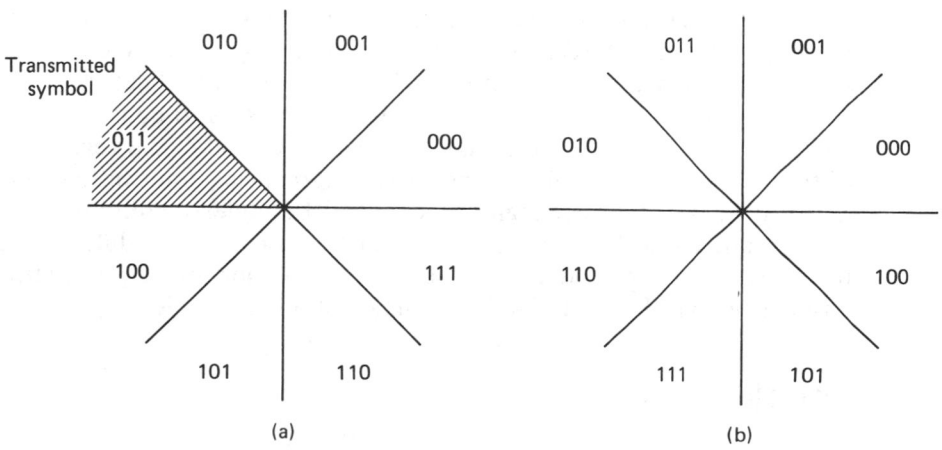

(a)   (b)

Figure 3.36 Binary-coded versus Gray-coded decision regions in an MPSK signal space. (a) Binary coded. (b) Gray coded.

reduced compared to the uncoded binary assignment seen in Figure 3.36a. Utilizing the Gray code assignment, it can be shown [7] that

$$P_B \simeq \frac{P_E}{\log_2 M} = \frac{P_E}{k} \qquad \text{(for } P_E \ll 1) \qquad (3.128)$$

Recall from Section 3.8.4 that BPSK and QPSK signaling have the same bit error probability. Here in Equation (3.128) we verify that they do not have the same symbol error probability. For BPSK, $P_E = P_B$. However, for QPSK, $P_E \simeq 2P_B$.

An exact closed-form expression for the bit error probability, P_B, of 8-ary PSK, together with tight upper and lower bounds on P_B for M-ary PSK with larger M, may be found in Lee [12].

3.9.5 Effects of Intersymbol Interference

In the previous sections and in Chapter 2 we have treated the detection of signals in the presence of AWGN under the assumption that there is no intersymbol interference (ISI). Thus the analysis has been straightforward, since the zero-mean AWGN process is characterized by its variance alone. In practice we find that ISI is often a second source of interference which must be accounted for. As explained in Section 2.11, ISI can be generated by the use of bandlimiting filters at the transmitter output, in the channel, or at the receiver input. The result of this additional interference is to degrade the error probabilities for coherent as well as for noncoherent reception. Analysis involving ISI in addition to AWGN is much more complicated since it involves the impulse response of the channel. The subject will not be treated here; however, for those readers interested in the details of the analysis, References [13–18] should prove interesting.

3.10 CONCLUSION

We have catalogued some basic bandpass digital modulation formats, particularly phase shift keying (PSK) and frequency shift keying (FSK). We have considered a geometric view of signal vectors and noise vectors, particularly antipodal and orthogonal signal sets. This geometric view allows us to consider the detection problem in the light of an orthogonal signal space and signal regions. This view of the space, and the effect of noise vectors causing transmitted signals to be received in the incorrect region, facilitates the understanding of the detection problem and the performance of various modulation and demodulation techniques. In Chapter 7 we reconsider the subjects of modulation and demodulation, and we investigate some bandwidth-efficient modulation techniques.

REFERENCES

1. Nyquist, H., "Thermal Agitation of Electric Charge in Conductors," *Phys. Rev.,* vol. 32, July 1928, pp. 110–113.

2. Arthurs, E., and Dym, H., "On the Optimum Detection of Digital Signals in the Presence of White Gaussian Noise—A Geometric Interpretation of Three Basic Data Transmission Systems," *IRE Trans. Commun. Syst.*, December 1962.

3. Wozencraft, J. M., and Jacobs, I. M., *Principles of Communication Engineering*, John Wiley & Sons, Inc., New York, 1965.

4. Van Trees, H. L., *Detection, Estimation, and Modulation Theory*, Part 1, John Wiley & Sons, Inc., New York, 1968.

5. Park, J. H., Jr., "On Binary DPSK Detection," *IEEE Trans. Commun.*, vol. COM26, no. 4, Apr. 1978, pp. 484–486.

6. Ziemer, R. E., and Peterson, R. L., *Digital Communications and Spread Spectrum Systems*, Macmillan Publishing Company, Inc., New York, 1985.

7. Lindsey, W. C., and Simon, M. K., *Telecommunication Systems Engineering*, Prentice-Hall, Inc., Englewood Cliffs, N.J., 1973.

8. Whalen, A. D., *Detection of Signals in Noise*, Academic Press, Inc., New York, 1971.

9. Korn, I., *Digital Communications*, Van Nostrand Reinhold Company, Inc., New York, 1985.

10. Couch, L. W. II, *Digital and Analog Communication Systems*, Macmillan Publishing Company, New York, 1983.

11. Viterbi, A. J., *Principles of Coherent Communications*, McGraw-Hill Book Company, New York, 1966.

12. Lee, P. J., "Computation of the Bit Error Rate of Coherent M-ary PSK with Gray Code Bit Mapping," *IEEE Trans. Commun.*, vol. COM34, no. 5, May 1986, pp. 488–491.

13. Hoo, E. Y., and Yeh, Y. S., "A New Approach for Evaluating the Error Probability in the Presence of Intersymbol Interference and Additive Gaussian Noise," *Bell Syst. Tech. J.*, vol. 49, Nov. 1970, pp. 2249–2266.

14. Shimbo, O., Fang, R. J., and Celebiler, M., "Performance of M-ary PSK Systems in Gaussian Noise and Intersymbol Interference," *IEEE Trans. Inf. Theory*, vol. IT19, Jan. 1973, pp. 44–58.

15. Prabhu, V. K., "Error Probability Performance of M-ary CPSK Systems with Intersymbol Interference," *IEEE Trans. Commun.*, vol. COM21, Feb. 1973, pp. 97–109.

16. Yao, K., and Tobin, R. M., "Moment Space Upper and Lower Error Bounds for Digital Systems with Intersymbol Interference," *IEEE Trans. Inf. Theory*, vol. IT22, Jan. 1976, pp. 65–74.

17. King, M. A., Jr., "Three Dimensional Geometric Moment Bounding Techniques," *J. Franklin Inst.*, vol. 309, no. 4, Apr. 1980, pp. 195–213.

18. Prabhu, V. K., and Salz, J., "On the Performance of Phase-Shift Keying Systems," *Bell Syst. Tech. J.*, vol. 60, Dec. 1981, pp. 2307–2343.

PROBLEMS

3.1. Determine whether or not $s_1(t)$ and $s_2(t)$ are orthogonal over the interval $(-1.5T_2 < t < 1.5T_2)$, where $s_1(t) = \cos(2\pi f_1 t + \phi_1)$, $s_2(t) = \cos(2\pi f_2 t + \phi_2)$, and $f_2 = 1/T_2$ for the following cases.
   (a) $f_1 = f_2$ and $\phi_1 = \phi_2$
   (b) $f_1 = \frac{1}{3}f_2$ and $\phi_1 = \phi_2$

(c) $f_1 = 2f_2$ and $\phi_1 = \phi_2$

(d) $f_1 = \pi f_2$ and $\phi_1 = \phi_2$

(e) $f_1 = f_2$ and $\phi_1 = \phi_2 + \pi/2$

(f) $f_1 = f_2$ and $\phi_1 = \phi_2 + \pi$

3.2. (a) Show that the three functions illustrated in Figure P3.1 are pairwise orthogonal over the interval $(-2, 2)$.

(b) Determine the value of the constant, A, that makes the set of functions in part (a) an orthonormal set.

(c) Express the following waveform, $x(t)$, in terms of the orthonormal set of part (b).

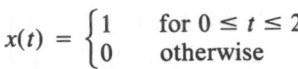

$$x(t) = \begin{cases} 1 & \text{for } 0 \le t \le 2 \\ 0 & \text{otherwise} \end{cases}$$

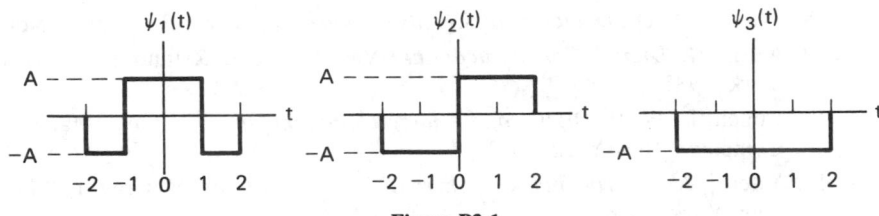

Figure P3.1

3.3. Consider the functions

$$\psi_1(t) = \exp(-|t|) \quad \text{and} \quad \psi_2 = 1 - A \exp(-2|t|)$$

Determine the constant, A, such that $\psi_1(t)$ and $\psi_2(t)$ are orthogonal over the interval $(-\infty, \infty)$.

3.4. Find the expected number of bit errors made in one day by the following continuously operating coherent BPSK receiver. The data rate is 5000 bits/s. The input digital waveforms are $s_1(t) = A \cos \omega_0 t$ and $s_2(t) = -A \cos \omega_0 t$, where $A = 1$ mV and the single-sided noise power spectral density is $N_0 = 10^{-11}$ W/Hz. Assume that signal power and energy per bit are normalized relative to a 1-Ω resistive load.

3.5. A continuously operating coherent BPSK system makes errors at the average rate of 100 errors per day. The data rate is 1000 bits/s. The single-sided noise power spectral density is $N_0 = 10^{-10}$ W/Hz.

(a) If the system is ergodic, what is the average bit error probability?

(b) If the value of received average signal power per bit is adjusted to be 10^{-6} W, will this received power be adequate to maintain the error probability found in part (a)?

3.6. If a system's main performance criterion is bit error probability, which of the following two modulation schemes would be selected for an AWGN channel? Show computations.

Binary coherent orthogonal FSK with $E_b/N_0 = 12$ dB

Binary noncoherent orthogonal FSK with $E_b/N_0 = 14$ dB

3.7. If a system's main performance criterion is bit error probability, which of the following two modulation schemes would be selected for an AWGN channel? Show computations.

$$\text{Binary noncoherent orthogonal FSK with } E_b/N_0 = 13 \text{ dB}$$

$$\text{Binary coherent PSK with } E_b/N_0 = 8 \text{ dB}$$

3.8. The bit stream

$$1\ 0\ 1\ 0\ 1\ 0\ 1\ 1\ 1\ 0\ 1\ 0\ 1\ 0\ 1\ 0\ 0\ 0\ 0\ 1\ 1\ 1\ 1$$

is to be transmitted using DPSK modulation. Show four different differentially encoded sequences that can represent the data sequence above, and explain the algorithm that generated each.

3.9. (a) Calculate the minimum required bandwidth for a noncoherently detected orthogonal binary FSK system. The higher-frequency signaling tone is 1 MHz and the symbol duration is 1 ms.

(b) What is the minimum required bandwidth for a noncoherent MFSK system having the same symbol duration?

3.10. Consider a BPSK system with equally likely waveforms $s_1(t) = \cos \omega_0 t$ and $s_2(t) = -\cos \omega_0 t$. At the matched filter detector, the $s_1(t)$ reference is $\cos (\omega_0 t + \phi)$, where ϕ is a phase error. Calculate the value of the phase error that would increase the probability of bit error from 2.0×10^{-3} to 2.5×10^{-3} relative to no phase error for an AWGN channel.

3.11. Find the probability of bit error, P_B, for the coherent matched filter detection of the equally likely binary FSK signals

$$s_1(t) = 0.5 \cos 2000\pi t$$

$$s_2(t) = 0.5 \cos 2020\pi t$$

where the two-sided AWGN power spectral density is $N_0/2 = 0.0001$. Assume that the symbol duration is $T = 0.01$ s.

3.12. Find the optimum (minimum probability of error) threshold, γ_0, for detecting the equally likely signals $s_1(t) = \sqrt{2E/T} \cos \omega_0 t$ and $s_2(t) = \sqrt{\frac{1}{2}E/T} \cos (\omega_0 t + \pi)$ in AWGN, using a correlator receiver as shown in Figure 3.7b. Assume a reference signal of $\psi_1(t) = \sqrt{2/T} \cos \omega_0 t$.

3.13. A system using matched filter detection of equally likely BPSK signals, $s_1(t) = \sqrt{2E/T} \cos \omega_0 t$ and $s_2(t) = \sqrt{2E/T} \cos (\omega_0 t + \pi)$, operates in AWGN with a received E_b/N_0 of 6.8 dB. Assume that $\mathbf{E}\{z(T)\} = \pm\sqrt{E}$.

(a) Find the minimum probability of bit error, P_B, for this signal set and E_b/N_0.

(b) If the decision threshold is $\gamma = 0.1\sqrt{E}$, find P_B.

(c) The threshold of $\gamma = 0.1\sqrt{E}$ is optimum for a particular set of a priori probabilities, $P(s_1)$ and $P(s_2)$. Find the values of these probabilities (refer to Section B.2).

3.14. A binary source with equally likely symbols controls the switch position in a transmitter operating over an AWGN channel, as shown in Figure P3.2. The noise has two-sided spectral density $N_0/2$. Assume antipodal signals of time duration T seconds and energy E joules. The system clock produces a clock pulse every T seconds, and the binary source rate is $1/T$ bits/s. Under *normal* operation, the switch is up when the source produces a binary zero, and it is down when the source produces a binary one. However, the switch is *faulty*. With probability, p, it will be thrown in the wrong direction during a given T-second interval. The presence of a switch error during any interval is independent of the presence of a switch error at any other time. Assume that $\mathbf{E}\{z(T)\} = \pm\sqrt{E}$.

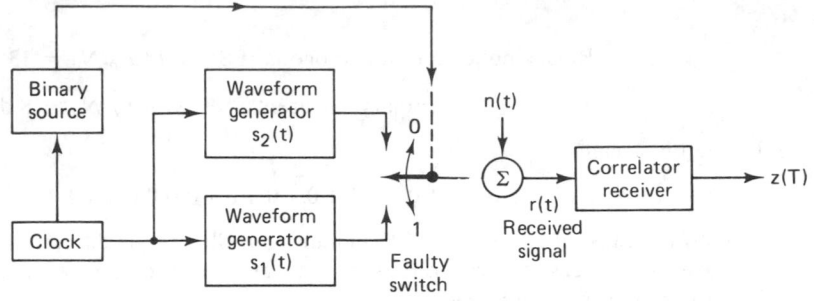

Figure P3.2

(a) Sketch the conditional probability functions, $p(z|s_1)$ and $p(z|s_2)$.

(b) The correlator receiver observes $r(t)$ in the interval $(0, T)$. Sketch the block diagram of an optimum receiver for minimizing the bit error probability when it is known that the switch is faulty with probability, p.

(c) Which one of the following two systems would you prefer to have?

$$p = 0.1 \quad \text{and} \quad \frac{E_b}{N_0} = \infty$$

$$p = 0 \quad \text{and} \quad \frac{E_b}{N_0} = 7 \text{ dB}$$

3.15. (a) Consider a 16-ary PSK system with symbol error probability, $P_E = 10^{-5}$. A Gray code is used for the symbol to bit assignment. What is the approximate bit error probability?

(b) Repeat part (a) for a 16-ary orthogonal FSK system.

3.16. Consider a coherent orthogonal MFSK system with $M = 8$ having the equally likely waveforms $s_i(t) = A \cos 2\pi f_i t$, $i = 1, \ldots, M$, $0 \le t \le T$, where $T = 0.2$ ms. The received carrier amplitude, A, is 1 mV, and the two-sided AWGN spectral density, $N_0/2$, is 10^{-11} W/Hz. Calculate the probability of bit error, P_B.

3.17. A bit error probability of $P_B = 10^{-3}$ is required for a system with a data rate of 100 kbits/s to be transmitted over an AWGN channel using coherently detected MPSK modulation. The system handwidth is 50 kHz. Assume that the filter has a roll-off characteristic of $r = 1$ and that a Gray code is used for the symbol to bit assignment.

(a) What E_s/N_0 is required for the specified P_B?

(b) What E_b/N_0 is required?

3.18. A differentially coherent MPSK system operates over an AWGN channel with an E_b/N_0 of 10 dB. What is the symbol error probability for $M = 8$ and equally likely symbols?

3.19. If a system's main performance criterion is bit error probability, which of the following two modulation schemes would be selected for transmission over an AWGN channel? Show computations.

coherent 8-ary orthogonal FSK with $\dfrac{E_b}{N_0} = 8$ dB

coherent 8-ary PSK with $\dfrac{E_b}{N_0} = 13$ dB

(Assume that a Gray code is used for the MPSK symbol-to-bit assignment.)

CHAPTER 4

Communications Link Analysis

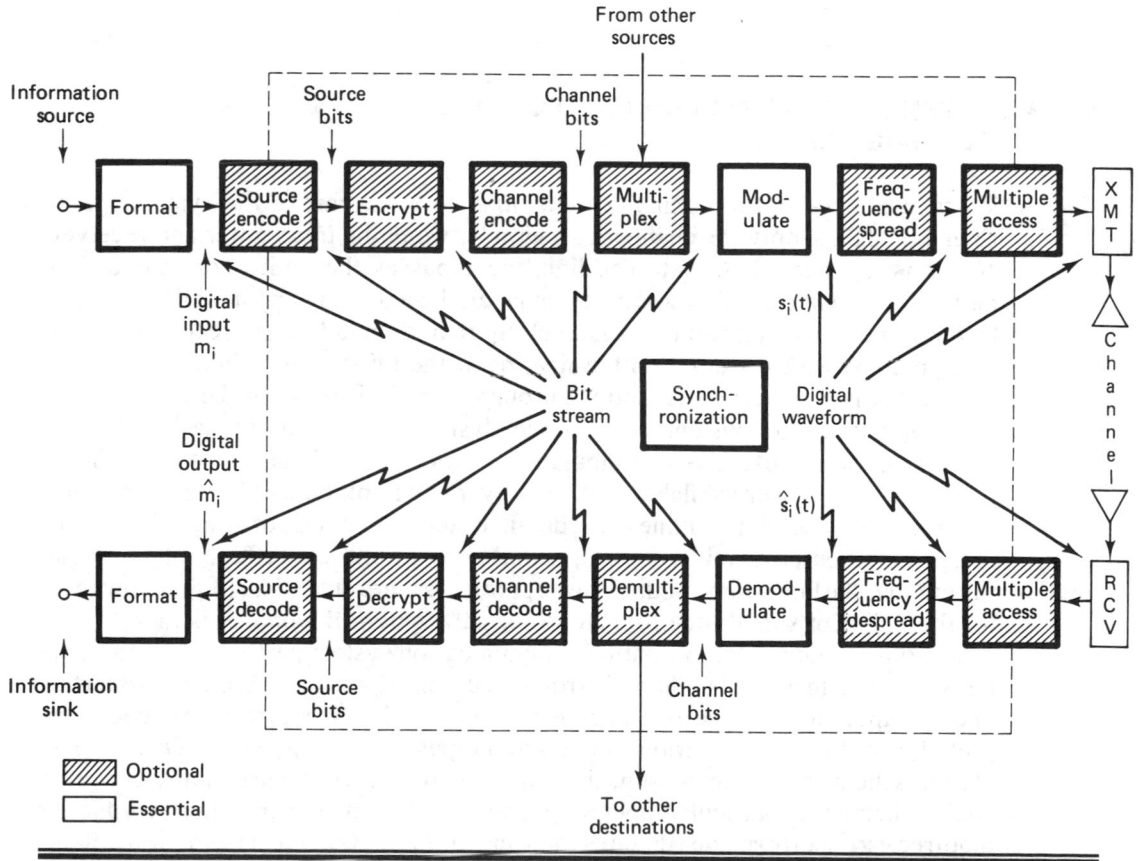

4.1 WHAT THE SYSTEM LINK BUDGET TELLS THE SYSTEM ENGINEER

When we talk about a communications *link,* to what part of the system are we referring? Is it simply the channel or region between the transmitter and receiver? No, it is far more than that. The link encompasses the entire communications path, from the information source, through all the encoding and modulation steps, through the transmitter and the channel, up to and including the receiver with all its signal processing steps, and terminating at the information sink.

What is a link analysis, and what purpose does it serve in the development of a communication system? The link analysis, and its output, the *link budget,* consist of the calculations and tabulation of the useful signal power and the interfering noise power available at the receiver. The link budget is a balance sheet of gains and losses; it outlines the detailed apportionment of transmission and reception resources, noise sources, signal attenuators, and effects of processes throughout the link. Some of the budget parameters are statistical (e.g., allowances for the fading of signals due to meteorological events); the budget is therefore an *estimation* technique for evaluating communication system performance. In Chapter 3 we examined probability of error versus E_b/N_0 curves having a "waterfall-like" shape, such as the one shown in Figure 3.21. We thereby related error probability to E_b/N_0 for various modulation types in Gaussian noise. Once a modulation scheme has been chosen, the requirement to meet a particular error probability dictates a particular operating point on the curve; in other words, the required error performance dictates the value of E_b/N_0 that must be made available at the receiver in order to meet that performance. The primary purpose of a link analysis is to determine the *actual* system operating point in Figure 3.21 and to establish that the error probability associated with that point is less than or equal to the system requirement. Of the many specifications, analyses, and tabulations that are used in the development of a communication system, the link budget stands out as a basic tool for providing the system engineer with overall system insight.

By examining the link budget, one can learn many things about overall system design and performance. For example, from the link margin, one learns whether the system will meet its requirements comfortably, marginally, or not at all. It will be evident if there are any hardware constraints, and whether such constraints can be compensated for in other parts of the link. The link budget is often used as a "score sheet" in considering system trade-offs and configuration changes, and in understanding subsystem nuances and interdependencies. From a quick examination of the link budget and its supporting documentation, one can judge whether the analysis was done precisely or if it represents a rough estimate. Together with other modeling techniques, the link budget can help predict equipment weight, size, prime power requirements, technical risk, and cost. The link budget is one of the system manager's most useful documents; it represents the "bottom-line" tally in the search for optimimum system performance.

4.2 THE CHANNEL

The propagating medium or electromagnetic path connecting the transmitter and receiver is called the *channel*. In general, a communications channel might consist of wires, coaxial cables, fiber optic cables, and in the case of radio-frequency (RF) links, waveguides, the atmosphere, or empty space. For most terrestrial communication links, the channel space is occupied by the atmosphere and partially bounded by the earth's surface. For satellite links, the channel is occupied mostly by empty space. Although some atmospheric effects occur at altitudes up to 100 km, the *bulk* of the atmosphere extends to an altitude of 20 km. Therefore, only a small part (0.05%) of the total synchronous altitude (35,800 km) path is occupied by significant amounts of atmosphere. Most of this chapter is presented in the context of such a satellite communications link.

4.2.1 The Concept of Free Space

The concept of *free space* assumes a channel free of all hindrances to RF propagation, such as absorption, reflection, refraction, or diffraction. If there is any atmosphere in the channel, it must be perfectly uniform and meet all these conditions. Also, we assume that the earth is infinitely far away or that its reflection coefficient is negligible. The RF energy arriving at the receiver is assumed to be a function only of distance from the transmitter (following the inverse-square law of optics). A free-space channel characterizes an ideal RF propagation path; in practice, propagation through the atmosphere and near the ground results in absorption, reflection, refraction, and diffraction, which modify the free-space transmission. Atmospheric absorption is treated in later sections. Reflection, refraction, and diffraction, which play an important role in determining terrestrial communications performance, are not treated here; Panter [1] provides a comprehensive treatment of these mechanisms.

4.2.2 Signal-to-Noise Ratio Degradation

The signal-to-noise power ratio (SNR) defined below is a convenient measure of performance at various points in the link.

$$SNR = \frac{\text{signal power}}{\text{noise power}}$$

Unless otherwise stated, SNR refers to *average* signal power and *average* noise power. The signal can be an information signal, a baseband waveform, or a modulated carrier. The SNR can degrade in two ways: (1) through the decrease of the desired signal power, and (2) through the increase of noise power, or the increase of interfering signal power. Let us refer to these degradations as *loss* and *noise* (or *interference*), respectively. Losses occur when a portion of the signal is absorbed, diverted, scattered, or reflected along its route to the intended receiver; thus a portion of the transmitted energy does not arrive at the receiver. There are four primary noise sources: (1) thermal noise can be generated within the link, (2) sky noise (e.g., galaxy noise, atmospheric noise) can be introduced into the link, (3) system nonlinearities can cause spurious signals to be created within the link, and (4) interfering signals from other users of the same frequency can be introduced into the link. Industry usage of the terms *loss* and *noise* frequently confuses the underlying degradation mechanism; however, the net effect on the SNR is the same.

4.2.3 Sources of Signal Loss and Noise

Figure 4.1 is a block diagram of a satellite communications link, emphasizing the sources of signal loss and noise. In the figure a signal loss is distinguished from a noise source by a dot pattern or line pattern, respectively. The contributors of *both* signal loss *and* noise are identified by a crosshatched line pattern. The following list of 21 sources of degradation represents a partial catalog of the major contributors to SNR degradation. The numbers correspond to the numbered circles in Figure 4.1.

1. *Bandlimiting loss.* All systems use filters in the transmitter to ensure that the transmitted energy is confined to the allocated or assigned bandwidth. This is to avoid interfering with other channels or users and to meet the requirements of regulatory agencies. Such filtering reduces the total amount of energy that would otherwise have been transmitted; the result is a *loss* in signal.

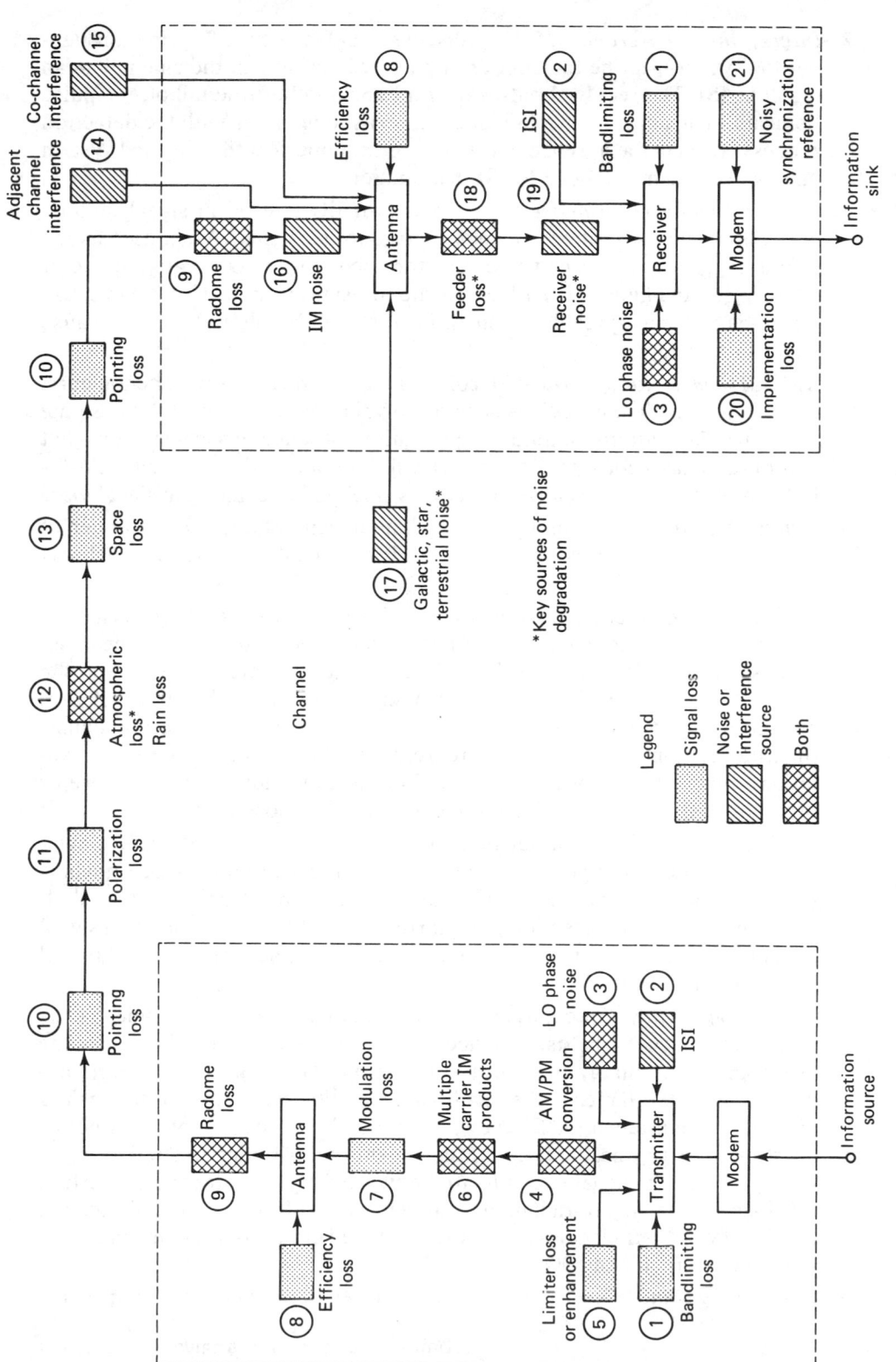

Figure 4.1 Satellite transmitter-to-receiver link with typical loss and noise sources.

2. *Intersymbol interference (ISI)*. As discussed in Chapter 2, filtering through-out the system—in the transmitter, in the receiver, and in the channel—can result in ISI. The received pulses overlap one another; the tail of one pulse "smears" into adjacent symbol intervals so as to *interfere* with the detection process. Even in the absence of thermal noise, imperfect filtering and system bandwidth constraints lead to ISI degradation.

3. *Local oscillator (LO) phase noise*. When an LO is used in signal mixing, phase fluctuations or jitter adds phase *noise* to the signal. When used as the reference signal in a receiver correlator, phase jitter can cause detector degradation and hence signal *loss*. At the transmitter, phase jitter can cause out-of-band signal spreading, which, in turn, will be filtered out and cause a *loss* in signal.

4. *AM/PM conversion*. AM-to-PM conversion is a phase *noise* phenomenon occurring in nonlinear devices such as traveling-wave tubes (TWT). Signal amplitude fluctuations (amplitude modulation) produce phase variations that contribute phase *noise* to signals that will be coherently detected. AM-to-PM conversion can also cause extraneous sidebands, resulting in signal *loss*.

5. *Limiter loss or enhancement*. A hard limiter can enhance the stronger of two signals, and suppress the weaker; this can result in either a signal *loss* or a signal *gain* [2].

6. *Multiple-carrier intermodulation (IM) products*. When several signals having different carrier frequencies are simultaneously present in a nonlinear de-vice, such as a TWT, the result is a multiplicative interaction between the carrier frequencies which can produce signals at all combinations of sum and difference frequencies. The energy apportioned to these spurious signals (intermodulation or IM products) represents a *loss* in signal energy. In ad-dition, if these IM products appear within the bandwidth region of these or other signals, the effect is that of added *noise* for those signals.

7. *Modulation loss*. The link budget is a calculation of received useful power (or energy). Only the power associated with information-bearing signals is useful. Error performance is a function of energy per transmitted symbol. Any power used for transmitting the carrier rather than the modulating signal (symbols) is a modulation *loss*. (However, energy in the carrier may be useful in aiding synchronization.)

8. *Antenna efficiency*. Antennas are transducers that convert electronic signals into electromagnetic fields, and vice versa. They are also used to focus the electromagnetic energy in a desired direction. The larger the antenna ap-erture (area), the larger is the resulting signal power density in the desired direction. An antenna's efficiency is described by the ratio of its effective aperture to its physical aperture. Mechanisms contributing to a reduction in efficiency (*loss* in signal strength) are known as amplitude tapering, aperture blockage, scattering, re-radiation, spillover, edge diffraction, and dissipative loss [3]. Typical efficiencies due to the combined effects of these mechanisms range between 50 and 80%.

9. *Radome loss and noise*. A radome is a protective cover, used with some

antennas, for shielding against weather effects. The radome, being in the path of the signal, will scatter and absorb some of the signal energy, thus resulting in a signal *loss*. A basic law of physics holds that a body capable of absorbing energy also radiates energy (at temperatures above 0 K). Some of this energy falls in the bandwidth of the receiver and constitutes injected *noise*.

10. *Pointing loss.* There is a *loss* of signal when either the transmitting antenna or the receiving antenna is imperfectly pointed.

11. *Polarization loss.* The polarization of an electromagnetic (EM) field is defined as the direction in space along which the field lines point, and the polarization of an antenna is described by the polarization of its radiated field. There is a *loss* of signal due to any polarization mismatch between the transmitting and receiving antennas.

12. *Atmospheric loss and noise.* The atmosphere is responsible for signal loss and is also a contributor of unwanted noise. The bulk of the atmosphere extends to an altitude of approximately 20 km; yet within that relatively short path, important loss and noise mechanisms are at work. Figure 4.2 is a plot of the theoretical one-way attenuation from a specified height to the top of the atmosphere. The calculations were made for several heights (0 km is sea level) and for a water vapor content of 7.5 g/m^3 at the earth's surface. The plot qualifies the magnitude of signal *loss* by indicating where it occurs, due to oxygen (O_2) and water vapor absorption as a function of carrier frequency. Local maxima of attenuation occur in the vicinities of 22 GIIz (water vapor), and 60 and 120 GHz (O_2). The atmosphere also contributes *noise* energy into the link. As in the case of the radome, molecules that absorb energy also radiate energy. The oxygen and water vapor molecules radiate noise throughout the RF spectrum. The portion of this noise that falls within the bandwidth of a given communication system will degrade its SNR. A primary atmospheric cause of signal *loss* and contributor of *noise* is rainfall. The more intense the rainfall, the more signal energy it will absorb. Also, on a day when rain passes through the antenna beam, there is a larger amount of atmospheric noise radiated into the system receiver than there is on a clear day. More will be said about atmospheric noise in later sections.

13. *Space loss.* There is a decrease in the electric field strength, and thus in signal strength (power density or flux density), as a function of distance. For a satellite communications link, the space loss is the largest single *loss* in the system. It is a loss in the sense that all the radiated energy is not focused on the intended receiving antenna.

14. *Adjacent channel interference.* This *interference* is characterized by unwanted signals from other frequency channels "spilling over" or injecting energy into the channel of interest. The proximity with which channels can be located in frequency is determined by the modulation spectral roll-off and the width and shape of the main spectral lobe.

15. *Co-channel interference.* This *interference* refers to the degradation caused by an interfering waveform appearing within the signal bandwidth. It can be

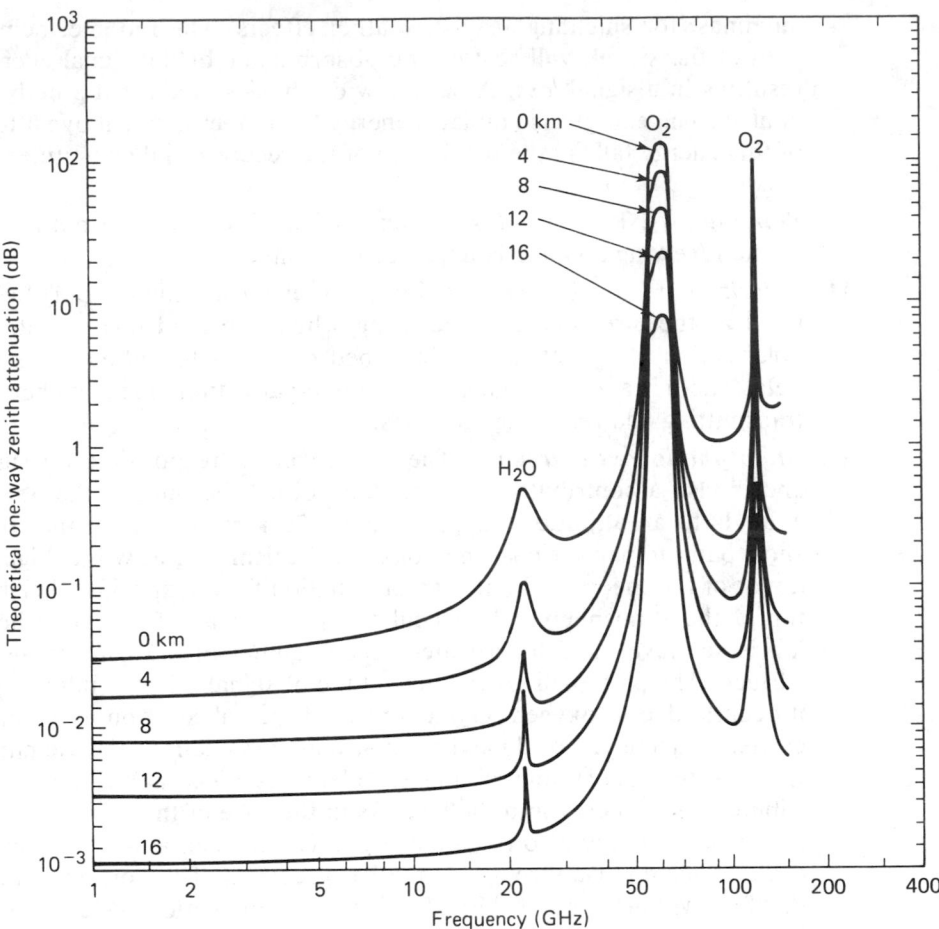

Figure 4.2 Theoretical vertical one-way attenuation from specified height to top of atmosphere for 7.5 g/m^3 of water vapor at the surface. (Does not include effect of rain or cloud attenuation.) (Reprinted from NASA Reference Publication 1082(03), "Propagation Effects Handbook for Satellite Systems Design," June 1983, Fig. 6.2-1, p. 218, courtesy of the National Aeronautics and Space Administration.)

introduced by a variety of ways, such as accidental transmissions, insufficient vertical and horizontal polarization discrimination, or by radiation spillover from an antenna sidelobe (low-energy beam surrounding the main antenna beam). It can be brought about by other authorized users of the same spectrum.

16. *Intermodulation (IM) noise.* The IM products described in item 6 result from multiple-carrier signals interacting in a nonlinear device. Such IM products are sometimes called *active intermods*; as described in item 6, they can either cause a loss in signal energy or be responsible for noise injected into a link. Here we consider *passive intermods*; these are caused by multiple-carrier

transmission signals interacting with nonlinear components at the transmitter output. These nonlinearities generally occur at the junction of waveguide coupling joints, at corroded surfaces, and at surfaces having poor electrical contact. When large EM fields impinge on surfaces that have a diode-like transfer function (work potential), they cause multiplicative products, and hence *noise*. If such noise radiates into a closely located receiving antenna, it can seriously degrade the receiver performance.

17. *Galactic or cosmic, star, and terrestrial noise.* All the celestial bodies, such as the stars and the planets, radiate energy. Such *noise* energy in the field of view of the antenna will degrade the SNR.

18. *Feeder line loss.* The level of the received signal might be very small (e.g., 10^{-12} W), and thus will be particularly susceptible to noise degradation. The receiver front end, therefore, is a region where great care is taken to keep the noise as small as possible until the signal has been suitably amplified. The waveguide or cable (feeder line) between the receiving antenna and the receiver front end contributes both signal *attenuation* and thermal *noise*; the details are treated in Section 4.5.3.

19. *Receiver noise.* This is the thermal *noise* generated within the receiver; the details are treated in Sections 4.5.1 to 4.5.4.

20. *Implementation loss.* This *loss* in performance is the difference between theoretical detection performance and the actual performance due to imperfections such as timing errors, frequency offsets, finite rise and fall times of waveforms, and finite-value arithmetic.

21. *Imperfect synchronization reference.* When the carrier phase, the subcarrier phase, and the symbol timing references are all derived perfectly, the error probability is a well-defined function of E_b/N_0 discussed in Chapter 3. In general, they are not derived perfectly, resulting in a system *loss*.

4.3 RECEIVED SIGNAL POWER AND NOISE POWER

4.3.1 The Range Equation

In radio communication systems, the carrier wave is propagated from the transmitter by the use of a transmitting antenna. The transmitting antenna is a transducer that converts electronic signals into electromagnetic (EM) fields. At the receiver, a receiving antenna performs the reverse function; it converts EM fields into electronic signals. The development of the fundamental power relationship between the receiver and transmitter usually begins with the assumption of an omnidirectional RF source, transmitting uniformly over 4π steradians. Such an ideal source, called an *isotropic radiator,* is illustrated in Figure 4.3. The power density, $p(d)$, on a hypothetical sphere at a distance, d, from the source is related to the transmitted power, P_t, by

$$p(d) = \frac{P_t}{4\pi d^2} \qquad \text{watts/m}^2 \qquad (4.1)$$

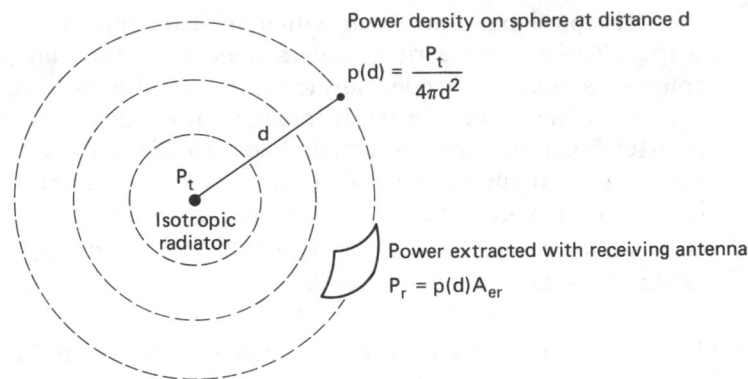

Power density on sphere at distance d

$$p(d) = \frac{P_t}{4\pi d^2}$$

Isotropic radiator

Power extracted with receiving antenna

$$P_r = p(d)A_{er}$$

Figure 4.3 Range equation. Expresses received power in terms of distance.

since $4\pi d^2$ is the area of the sphere. The power, P_r, extracted with the receiving antenna can be written

$$P_r = p(d)A_{er} = \frac{P_t A_{er}}{4\pi d^2} \tag{4.2}$$

where the parameter, A_{er}, is the absorption cross section (effective area) of the receiving antenna, defined by

$$A_{er} = \frac{\text{total power extracted}}{\text{incident power flux density}} \tag{4.3}$$

If the antenna under consideration is a transmitting antenna, its effective area is designated by A_{et}. If the antenna in question is unspecified as to its receiving or transmitting function, its effective area is designated simply by A_e.

An antenna's effective area, A_e, and physical area, A_p, are related by an efficiency parameter, η, as

$$A_e = \eta A_p \tag{4.4}$$

which accounts for the fact that the total incident power is not extracted; it is lost through various mechanisms [3]. Nominal values for η are 0.55 for a dish (parabolic-shaped reflector) and 0.75 for a horn-shaped antenna.

The antenna parameter that relates the power output (or input) to that of an isotropic radiator as a purely geometric ratio is the antenna directivity or *directive gain, G,* where

$$G = \frac{\text{maximum power intensity}}{\text{average power intensity over } 4\pi \text{ steradians}} \tag{4.5}$$

In the absence of any dissipative loss or impedance mismatch loss, the antenna *gain* (in the direction of maximum intensity) is defined simply as the directive gain in Equation (4.5). However, in the event that there exists some dissipative or impedance mismatch loss, the antenna gain is then equal to the directive gain times a loss factor to account for these losses [4]. In this chapter we shall assume that the dissipative loss is zero and that the impedances are perfectly matched. Therefore, Equation (4.5) describes the *peak antenna gain*; it can be viewed as the result of concentrating the RF flux in some restricted region less than 4π steradians, as shown in Figure 4.4. Now we can define an *effective radiated power*, with respect to an isotropic radiator (EIRP), as the product of the transmitted power, P_t, and the gain of the transmitting antenna, G_t, as follows:

$$\text{EIRP} = P_t G_t \tag{4.6}$$

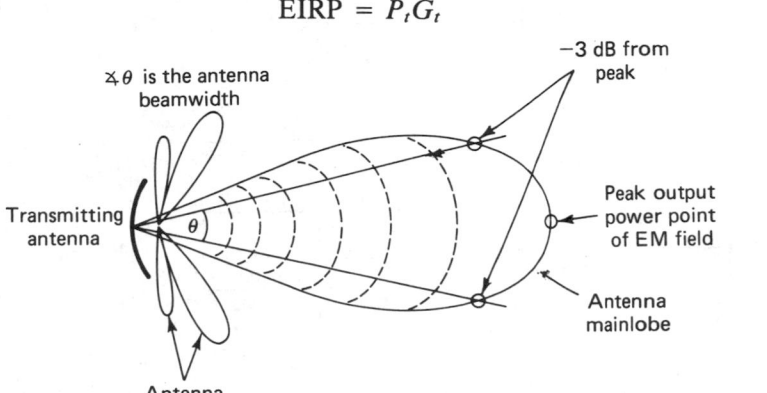

Figure 4.4 Antenna gain is the result of concentrating the isotropic RF flux.

Example 4.1 Effective Isotropic Radiated Power

Show that the same value of EIRP can be produced equally well by using a transmitter with $P_t = 100$ W or with $P_t = 0.1$ W, by employing the appropriate antenna in each case.

Solution

Figure 4.5a depicts a 100-W transmitter coupled to an isotropic antenna; the EIRP $= P_t G_t = 100 \times 1 = 100$ W. Figure 4.5b depicts a 0.1-W transmitter coupled to an antenna with gain $G_t = 1000$; the EIRP $= P_t G_t = 0.1 \times 1000 = 100$ W. If field-strength meters were positioned, as shown, to measure the effective power, the measurements could not distinguish between the two cases.

4.3.1.1 Back to the Range Equation

For the more general case in which the transmitter has some antenna gain relative to an isotropic antenna, we replace P_t with EIRP in Equation (4.2) to yield

$$P_r = \text{EIRP}\,\frac{A_{er}}{4\pi d^2} \tag{4.7}$$

The relationship between antenna gain, G, and antenna effective area, A_e, is [4]

$$G = \frac{4\pi A_e}{\lambda^2} \quad \text{(for } A_e \gg \lambda^2) \tag{4.8}$$

where λ is the wavelength of the carrier. Wavelength λ and frequency f are

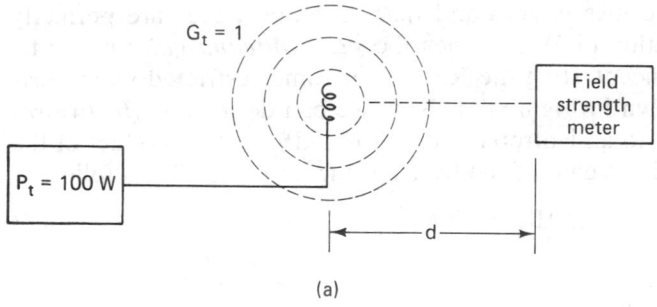

(a)

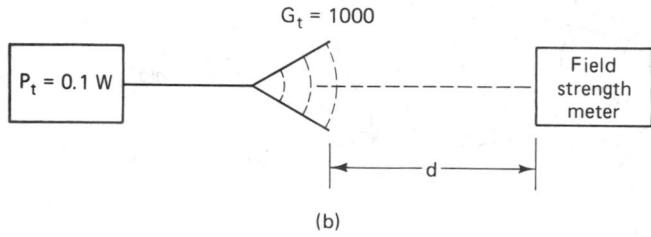

(b)

Figure 4.5 The same value of EIRP produced two different ways.

reciprocally related by $\lambda = c/f$, where c is the speed of light ($\approx 3 \times 10^8$ m/s). Similar expressions apply for both the transmitting and receiving antennas. The *reciprocity theorem* states that for a given antenna and carrier wavelength, the transmitting and receiving gains are identical [4].

The antenna field of view is a measure of the solid angle into which most of the field power is concentrated. Field of view is a measure of the directional properties of the antenna; it is inversely related to antenna gain—high-gain antennas are commensurate with narrow fields of view. Instead of using the solid-angle field of view, we often deal with the planar angle *beamwidth* measured in radians or degrees. Figure 4.4 pictures a directive antenna pattern and illustrates the common definition of the antenna beamwidth. The beamwidth is the angle that subtends the points at which the peak field power is reduced by 3 dB. How does the antenna beamwidth vary with signal frequency? How does the beamwidth vary with antenna size? As can be seen from Equation (4.8), the antenna gain increases with decreased wavelength (increased frequency); antenna gain also increases with increased effective area. Increasing antenna gain is tantamount to focusing the flux density into a more restricted cone angle; hence, increasing either the signal frequency or the antenna size results in a *narrower beamwidth*.

We can calculate the effective area of an isotropic antenna by setting $G = 1$ in Equation (4.8), and solving for A_e as follows:

$$A_e = \frac{\lambda^2}{4\pi} \qquad (4.9)$$

Then to find the power received, P_r, when the receiving antenna is isotropic, we substitute Equation (4.9) into Equation (4.7) to get

$$P_r = \frac{\text{EIRP}}{(4\pi d/\lambda)^2} = \frac{\text{EIRP}}{L_s} \tag{4.10}$$

where the collection of terms $(4\pi d/\lambda)^2$, called the *path loss* or *free-space loss,* is designated by L_s. Notice that Equation (4.10) states that the power received by an isotropic antenna is equal to the effective transmitted power, reduced only by the path loss. When the receiving antenna is not isotropic, replacing A_{er} in Equation (4.7) with $G_r\lambda^2/4\pi$ from Equation (4.8) yields the more general expresion for P_r.

$$P_r = \frac{\text{EIRP}\ G_r\lambda^2}{(4\pi d)^2} = \frac{\text{EIRP}\ G_r}{L_s} \tag{4.11}$$

where G_r is the receiving antenna gain.

4.3.2 Received Signal Power as a Function of Frequency

Since the transmitting antenna and the receiving antenna can each be expressed as a gain or an area, P_r can be expressed four different ways:

$$P_r = \frac{P_t G_t A_{er}}{4\pi d^2} \tag{4.12}$$

$$P_r = \frac{P_t A_{et} A_{er}}{\lambda^2 d^2} \tag{4.13}$$

$$P_r = \frac{P_t A_{et} G_r}{4\pi d^2} \tag{4.14}$$

$$P_r = \frac{P_t G_t G_r \lambda^2}{(4\pi d)^2} \tag{4.15}$$

where A_{er} and A_{et} are the effective areas of the receiving and transmitting antennas, respectively.

In Equations (4.12) to (4.15) the dependent variable is received signal power, P_r, and the independent variables involve parameters such as transmitted power, antenna gain, antenna area, wavelength, and range. Suppose that we ask the question: How does received power vary as wavelength is decreased (or as frequency is increased), all other independent variables remaining constant? From Equations (4.12) and (4.14) it appears that P_r and wavelength are not related at all. From Equation (4.13), P_r appears to be inversely proportional to wavelength squared, and from Equation (4.15), P_r appears to be directly proportional to wavelength squared. Is there a paradox here? Of course there is not; Equations (4.12) to (4.15) seem to conflict only because antenna gain and antenna area are wavelength related, as stated in Equation (4.8). When should one use each of the Equations (4.12) to (4.15) for determining P_r as a function of wavelength? Consider a system that is already configured; that is, the antennas have already been built or their dimensions are fixed (A_{et} and A_{er} are fixed). Then Equation (4.13) is the appropriate choice for calculating the P_r performance. Equation (4.13) states that for fixed-size antennas, the received power increases as the wavelength is decreased.

Consider the use of Equation (4.12), where G_t and A_{er} are independent variables. We want G_t and A_{er} held fixed over the range of P_r versus wavelength calculations. What happens to the gain of a fixed-dimension transmitting antenna as the independent variable, λ, is decreased? G_t increases [see Equation (4.8)]. But we cannot have G_t increasing in Equation (4.12)—we want G_t held fixed. In other words, to ensure that G_t remains fixed, we would need to reduce the transmitting antenna size as wavelength decreases. It should be apparent that Equation (4.12) is the appropriate equation when starting with a *fixed transmitting antenna gain* (or beamwidth) requirement and the parameter, A_{et}, is not fixed. For similar reasons, Equation (4.14) is used when A_{et} and G_r are fixed, and Equation (4.15) is used when both the transmitting and receiving antenna gains (or beamwidths) are fixed.

Figure 4.6 illustrates a satellite application where the downlink antenna beam is required to provide earth coverage (a beamwidth of approximately 17° from synchronous altitude). Since the satellite antenna gain, G_t, must be fixed, the resulting P_r is independent of wavelength, as shown in Equation (4.12). If the transmission at some frequency f_1 ($= c/\lambda_1$) provides earth coverage, then a frequency change to f_2, where $f_2 > f_1$, will result in reduced coverage (since for a given antenna, G_t will increase), hence the antenna size must be reduced to maintain the required earth coverage or beamwidth. Thus earth coverage antennas become smaller as the carrier frequency is increased.

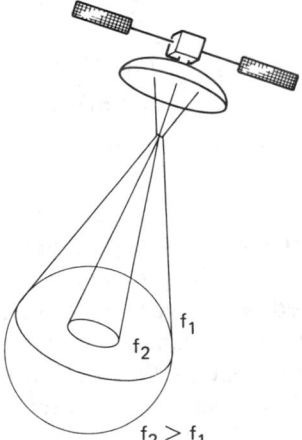

Figure 4.6 Received power as a function of frequency.

4.3.3 Path Loss Is Frequency Dependent

From Equation (4.10) it can be seen that path loss, L_s, is wavelength (frequency) dependent. The question is often asked: Why should path loss, which is just a geometric inverse-square loss, be a function of frequency? The answer is that path loss, as characterized in Equation (4.10), is a *definition* predicated on the use of an isotropic receiving antenna ($G_r = 1$). Hence path loss is a convenient tool; it represents a hypothetical received-power loss that *would occur if the receiving antenna were isotropic*. Figure 4.3 and Equation (4.1) have established that power density, $p(d)$, is a function of distance—a purely geometric consideration; $p(d)$ is *not* a function of frequency. However, since path loss is predicated on $G_r = 1$, when we attempt to collect some P_r with an *isotropic antenna,* the result is characterized by Equation (4.10). Again let us emphasize that L_s can be viewed as a convenient collection of terms that have been assigned the unfortunate name *path loss*. The name conjures up an image of a purely geometric effect and fails to emphasize the requirement that $G_r = 1$. A better choice of a name would have been *unity-gain propagation loss*. In a radio communication system, path loss accounts for the largest loss in signal power. In satellite systems, the path loss for a C-band (6-GHz) link to a synchronous altitude satellite is typically 200 dB.

Example 4.2   Antenna Design for Measuring Path Loss

Design a hypothetical experiment to measure path loss L_s, at frequencies $f_1 = 30$ MHz and $f_2 = 60$ MHz, when the distance between the transmitter and receiver is 100 km. Find the effective area of the receiving antenna, and calculate the path loss in decibels for each case.

Solution

Figure 4.7 illustrates the two links for measuring L_s at frequencies f_1 and f_2, respectively. The power density, $p(d)$, at each receiver is identical and equal to

$$p(d) = \frac{\text{EIRP}}{4\pi d^2}$$

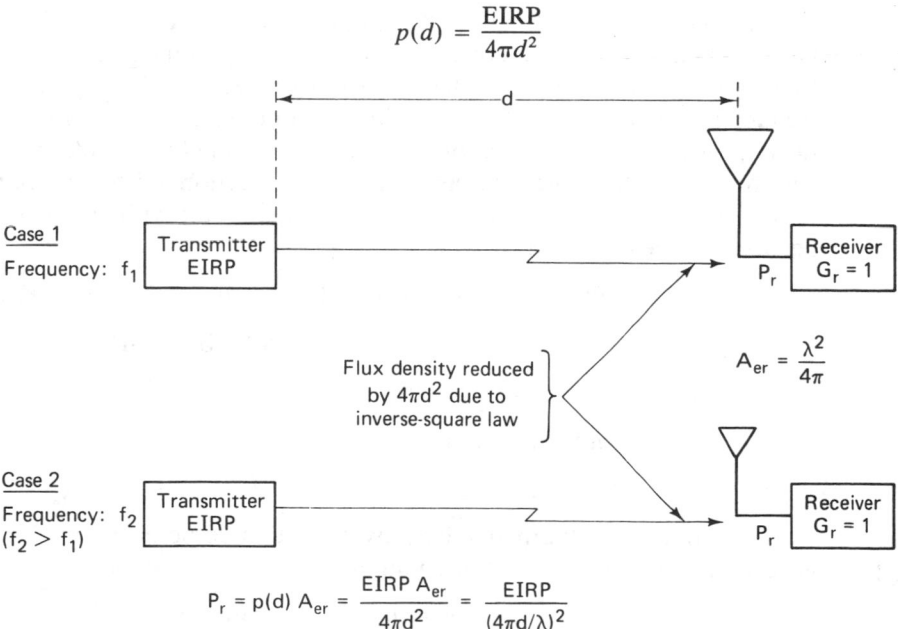

$$P_r = p(d) A_{er} = \frac{\text{EIRP } A_{er}}{4\pi d^2} = \frac{\text{EIRP}}{(4\pi d/\lambda)^2}$$

Figure 4.7   Path loss versus frequency. Hypothetical experiment to measure path loss at two different frequencies.

This reduction in power density is due *only* to the inverse-square law. The actual power received at each receiver is found by multiplying the power density $p(d)$ at the receiver by the effective area, A_{er}, of the collecting antenna, as shown in Equation (4.7). Since path loss is predicated on $G_r = 1$, we compute the effective area, A_{er1}, at frequency f_1, and A_{er2} at frequency f_2, using Equation (4.9):

$$A_{er} = \frac{\lambda^2}{4\pi} = \frac{(c/f)^2}{4\pi}$$

$$A_{er1} = \frac{(3 \times 10^8/30 \times 10^6)^2}{4\pi} \simeq 8 \text{ m}^2$$

$$A_{er2} = \frac{(3 \times 10^8/60 \times 10^6)^2}{4\pi} \simeq 2 \text{ m}^2$$

The path loss for each case in decibels is

$$L_{s1} = 10 \times \log_{10}\left(\frac{4\pi d}{\lambda_1}\right)^2 = 10 \times \log_{10}\left(\frac{4\pi \times 10^5}{3 \times 10^8/30 \times 10^6}\right)^2$$

$$\simeq 102 \text{ dB}$$

$$L_{s2} = 10 \times \log_{10}\left(\frac{4\pi d}{\lambda_2}\right)^2 = 10 \times \log_{10}\left(\frac{4\pi \times 10^5}{3 \times 10^8/60 \times 10^6}\right)^2$$

$$= 108 \text{ dB}$$

4.3.4 Thermal Noise Power

Thermal noise is caused by the thermal motion of electrons in all conductors. It is generated in the lossy coupling between an antenna and receiver and in the first stages of the receiver. The noise power spectral density is constant at all frequencies up to about 10^{12} Hz, giving rise to the name *white noise*. The thermal noise process in communication receivers is modeled as an additive white Gaussian noise (AWGN) process, as described in Section 1.5.5. The physical model [5, 6] for thermal or Johnson noise is a noise generator with an open-circuit mean-square voltage of $4\kappa T^\circ W \mathcal{R}$, where

$$\kappa = \text{Boltzmann's constant} = 1.38 \times 10^{-23} \text{ J/K or W/K-Hz}$$

$$= -228.6 \text{ dBW/K-Hz}$$

T° = temperature, kelvin

W = bandwidth, hertz

$\mathcal{R}$ = resistance, ohms

The maximum thermal noise power, N, that could be coupled from the noise generator into the front end of an amplifier, is

$$N = \kappa T^\circ W \qquad \text{watts} \qquad (4.16)$$

Thus the maximum single-sided noise power spectral density, N_0 (noise power in a 1-Hz bandwidth), available at the amplifier input is

$$N_0 = \frac{N}{W} = \kappa T° \qquad \text{watts/hertz} \qquad (4.17)$$

It might seem that the noise power should depend on the magnitude of the resistance—but it does not. Consider an intuitive argument to verify this. Electrically connect a large resistance to a small one, such that they form a closed path and such that their physical temperatures are the same. If noise power were a function of resistance, there would be a net power flow from the large resistance to the small one; the large resistance would become cooler and the small one would become warmer. This violates our experience, not to mention the second law of thermodynamics. Therefore, the power delivered from the large resistance to the small one must be equal to the power it receives.

The available power from a thermal noise source is dependent on the ambient temperature of the source (the *noise temperature*), as is seen in Equation (4.16). This leads to the useful concept of an *effective noise temperature* for noise sources that are not necessarily thermal in origin (e.g., galactic, atmospheric, interfering signals) that can be introduced into the receiving antenna. The effective noise temperature of such a noise source is defined as the temperature of a hypothetical thermal noise source that would give rise to an equivalent amount of interfering power. The subject of noise temperature is treated in greater detail in Section 4.5.

Example 4.3 Maximum Available Noise Power

Using a noise generator with mean-square voltage equal to $4\kappa T° W\mathcal{R}$, demonstrate that the maximum amount of noise power that can be coupled from this source into an amplifier is $N_i = \kappa T° W$.

Solution

A theorem from network theory states that maximum power is delivered to a load when the value of the load impedance is made equal to the complex conjugate of the generator impedance [7]. In this case the generator is a pure resistance, $\mathcal{R}$; therefore, the condition for maximum power transfer is fulfilled when the input resistance of the amplifier equals $\mathcal{R}$. Figure 4.8 illustrates such a network. The input thermal noise source is represented by an electrically equivalent model consisting of a noise-

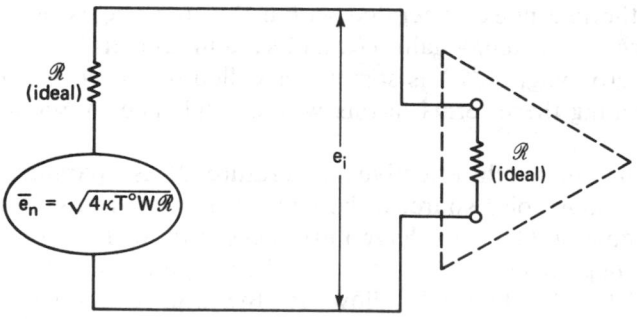

Figure 4.8 Electrical model of maximum available thermal noise power at amplifier input.

less source resistor in series with an ideal voltage generator whose rms noise voltage is $\sqrt{4\kappa T^\circ W\mathcal{R}}$. The input resistance of the amplifier is made equal to $\mathcal{R}$. The noise voltage delivered to the amplifier input is just one-half the generator voltage, following basic circuit principles. The noise power, N_i, delivered to the amplifier input can accordingly be expressed as follows:

$$N_i = \frac{(\sqrt{4\kappa T^\circ W\mathcal{R}}/2)^2}{\mathcal{R}} = \frac{4\kappa T^\circ W\mathcal{R}}{4\mathcal{R}}$$

$$= \kappa T^\circ W$$

4.4 LINK BUDGET ANALYSIS

In evaluating system performance, the quantity of greatest interest is the signal-to-noise ratio (SNR). This is because the basic system design centers on our ability to detect the signal, with an acceptable error probability, in the presence of noise. Since the desired signal here is a modulated carrier waveform, we often speak of the average carrier power-to-noise ratio (C/N) or (P_r/N) as the SNR of particular interest. We obtain P_r/N by dividing Equation (4.11) by noise power, N:

$$\frac{P_r}{N} = \frac{\text{EIRP } G_r/N}{L_s} \tag{4.18}$$

Equation (4.18) applies to any one-way RF link. With *analog receivers,* the noise bandwidth (generally referred to as the effective or equivalent noise bandwidth) seen by the demodulator is usually greater than the signal bandwidth, and P_r/N is the main parameter for measuring signal detectability and performance quality. With *digital receivers,* however, correlators or matched filters are usually implemented, and signal bandwidth is taken to be equal to noise bandwidth. Rather than consider input noise power, a common formulation for digital links is to replace noise power with *noise power spectral density*. We can use Equation (4.17) to rewrite Equation (4.18) as

$$\frac{P_r}{N_0} = \frac{\text{EIRP } G_r/T^\circ}{\kappa L_s L_o} \tag{4.19}$$

where the system effective temperature, T°, is a function of the noise radiated into the antenna and the thermal noise generated within the first stages of the receiver. Note that the receiving antenna gain, G_r, and system temperature, T°, are grouped together. The grouping, G_r/T°, is sometimes called the *receiver sensitivity*. The reason for treating these terms in this way is explained in Section 4.6.2.

It is important to emphasize that the effective temperature, T°, is a parameter that *models* the effect of various noise sources; the subject is treated in greater detail in Section 4.5. In Equation (4.19) we have introduced a term, L_o, to represent all other losses and degradation factors not specifically addressed by the other terms of Equation (4.18). The factor L_o allows for the large assortment of different losses and noise sources cataloged earlier. Equation (4.19) summarizes the key parameters of any link analysis; they are the received signal power-to-noise power spectral density (P_r/N_0), the effective transmitted power (EIRP), the receiver sensitivity $(G_r/T°)$, and the losses (L_s, L_o).

If we assume that all the received power is in the modulating (information bearing) signal, we can write from Equation (3.94),

$$\frac{P_r}{N_0} = \frac{S}{N_0} = \frac{E_b}{N_0} R \qquad (4.20)$$

recalling that S is the average modulating signal power, E_b/N_0 the bit energy per noise power spectral density, and R the bit rate. If some of the received power is unmodulated carrier power (a signal power loss), we can still employ Equation (4.20), except that the carrier power is accounted for as a loss factor [within the parameter L_o of Equation (4.19)].

4.4.1 Two E_b/N_0 Values of Interest

We have referred to E_b/N_0 as that value of bit energy per noise power spectral density required to yield a specified error probability. To facilitate calculating a margin or safety factor M, we need to differentiate between the *required* E_b/N_0 and the actual or *received* E_b/N_0. From this point on we will refer to the former as $(E_b/N_0)_{reqd}$ and to the latter as $(E_b/N_0)_r$. Figure 4.9 depicts an example with two operating points. The first is associated with $P_B = 10^{-3}$; let us call this operating point the system required error performance. Let us assume that an $(E_b/N_0)_{reqd}$ value of 10 dB will yield this required performance. Do you suppose we would build this system so that the demodulator received this 10-dB value *exactly*? Of course not; we would specify and design the system to have a safety margin, so that the $(E_b/N_0)_r$ actually received would be somewhat larger than the $(E_b/N_0)_{reqd}$. Thus we might design the system to operate at the second operating point on Figure 4.9; here $(E_b/N_0)_r = 12$ dB and $P_B = 10^{-5}$. For this example we can describe the safety margin or *link margin*, as providing a two-order-of-magnitude improved P_B, or as is more usual, we can describe the link margin in terms of providing 2 dB more E_b/N_0 than is required. We can rewrite Equation (4.20), introducing the link margin parameter, M, as

$$\frac{P_r}{N_0} = \left(\frac{E_b}{N_0}\right)_r R = M \left(\frac{E_b}{N_0}\right)_{reqd} R \qquad (4.21)$$

The difference in decibels between $(E_b/N_0)_r$ and $(E_b/N_0)_{reqd}$ yields the link margin:

$$M \text{ (dB)} = \left(\frac{E_b}{N_0}\right)_r \text{ (dB)} - \left(\frac{E_b}{N_0}\right)_{reqd} \text{ (dB)} \qquad (4.22)$$

The parameter $(E_b/N_0)_{reqd}$ reflects the differences from one system design to another; these might be due to differences in modulation or coding schemes. A larger than expected $(E_b/N_0)_{reqd}$ may be due to a suboptimal RF system, which manifests large timing errors or which allows more noise into the detection process than does an ideal matched filter.

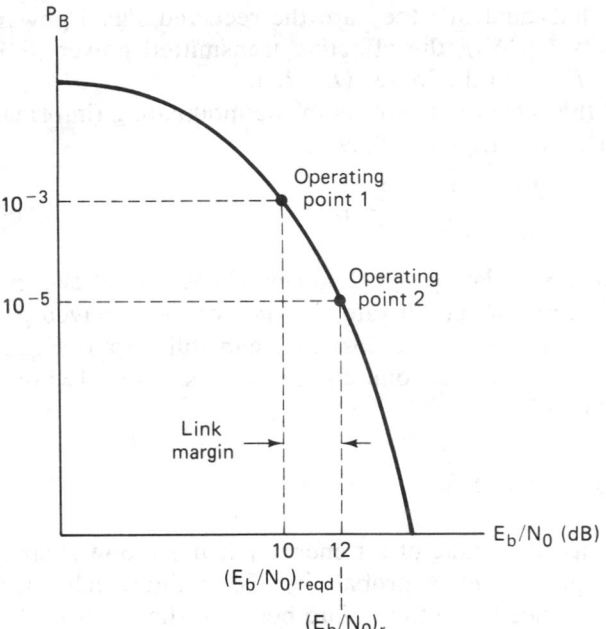

Figure 4.9 Two E_b/N_0 values of interest.

Combining Equations (4.19) and (4.21) and solving for the link margin M yields

$$M = \frac{\text{EIRP } G_r/T^\circ}{(E_b/N_0)_{\text{reqd}}R\kappa L_s L_o} \tag{4.23}$$

4.4.2 Link Budgets Are Typically Calculated in Decibels

Since link budget analysis is typically calculated in decibels, we can express Equation (4.23) as

$$M \text{ (dB)} = \text{EIRP (dBW)} + G_r \text{ (dBi)} - \left(\frac{E_b}{N_0}\right)_{\text{reqd}} \text{ (dB)} - R \text{ (dB-bit/s)}$$

$$- \kappa T^\circ \text{ (dBW/Hz)} - L_s \text{ (dB)} - L_o \text{ (dB)} \tag{4.24}$$

Transmitted signal power, EIRP, is expressed in decibel-watts (dBW); noise power spectral density, N_0, is in decibel-watts per hertz (dBW/Hz); antenna gain, G_r, is in decibels referenced to isotropic gain (dBi); data rate, R, is in decibels referenced to 1 bit/s (dB-bit/s); and all other terms are in decibels (dB). The numerical values of the Equation (4.24) parameters constitute the link budget, a useful tool for allocating communications resources. In an effort to maintain a positive margin, we might trade off any parameter with any other parameter; we might choose to reduce transmitter power by giving up excess margin, or we might elect to increase the data rate by reducing $(E_b/N_0)_{\text{reqd}}$ (through the selection of improved modulation and coding). Any one of the Equation (4.24) decibels, regardless of the parameter from which it stems, is just as good as any other decibel. It should be noted, however, that as requirements become more constrained, it may not be possible to trade or yield on some items. For example, even though binary PSK modulation outperforms binary FSK (in the P_B sense), requirements to operate on a fading channel may dictate the avoidance of PSK and the choice of the more robust FSK. Also, certain beamwidth requirements may constrain antenna dimensions, so that one might *not* have the freedom of trading off or selecting any antenna gain that one desires.

4.4.3 How Much Link Margin Is Enough?

The question of how much link margin should be designed into a system is asked frequently. The answer is that if all sources of gain, loss, and noise have been rigorously detailed (worst case), and if the link parameters with large variances (e.g., fades due to weather) match the statistical requirements for link availability, very little additional margin is needed. The margin needed depends on how much confidence one has in each of the link budget entries. For systems employing new technology or new operating frequencies, one needs more margin than for systems that have been repeatedly built and tested. Sometimes the link budget provides an allowance for fades due to weather directly, as a line item. Other times, however, the required value of margin reflects the link requirements for a given rain degradation. For satellite communications at C-band (uplink at 6 GHz, downlink at 4 GHz), where the parameters are well known and fairly well behaved, it should be possible to design a system with only 1 dB of link margin. Receive-only television stations operating with 16-ft-diameter dishes at C-band are frequently designed with only a fraction of a decibel of margin. However, telephone communications via satellite using standards of 99.9% availability require considerably more margin; some of the INTELSAT systems have 4 to 5 dB of margin. When nominal rather than worst-case computations are performed, allowances are usually made for unit-to-unit equipment variations over the operating temperature range, line voltage variations, and mission duration. Also, for space communications, there may be an allowance for errors in tracking a satellite's location.

Designs using higher frequencies (e.g., 14/12 GHz) generally call for larger (weather) margins because atmospheric losses increase with frequency and are highly variable. It should be noted that a by-product of the attenuation due to atmospheric loss is greater antenna noise. With low-noise amplifiers, small weather changes can result in increases of 40 to 50 K in antenna temperature. Table 4.1 represents a link analysis proposed to the Federal Communications Commission (FCC) by Satellite Television Corporation for the Direct Broadcast Satellite (DBS) service. Notice that the downlink budget is tabulated for two alternative weather conditions: clear, and 5-dB loss due to rain. The signal loss due to atmospheric attenuation is only a small fraction of a decibel for clear weather and is the full 5 dB during rain. The next item in the downlink tabulation, home receiver $G/T°$, illustrates the additional degradation caused by the rain; additional thermal noise irradiates the receiving antenna, making the effective system noise temperature, $T°$, increase, and the home receiver $G/T°$ decrease

TABLE 4.1 Proposed Direct Broadcast Satellite (DBS) from Satellite Television Corp.

Uplink	
Earth station EIRP	86.6 dBW
Free-space loss (17.6 GHz, 48° elevation)	208.9 dB
Assumed rain attenuation	12.0 dB
Satellite $G/T°$	7.7 dB/K
Uplink $C/\kappa T°$	102.0 dB-Hz

	Atmospheric condition	
Downlink	Clear	5-dB rain attenuation
Satellite EIRP	57.0 dBW	57.0 dBW
Free-space loss (12.5 GHz, 30° elevation)	206.1 dB	206.1 dB
Atmospheric attenuation	0.14 dB	5.0 dB
Home receiver $G/T°$ (0.75 m)	9.4 dB/K	8.1 dB/K
Receiver pointing loss (0.5° error)	0.6 dB	0.6 dB
Polarization mismatch loss (average)	0.04 dB	0.04 dB
Downlink $C/\kappa T°$	88.1 dB-Hz	82.0 dB-Hz
Overall $C/\kappa T°$	87.9 dB-Hz	82.0 dB-Hz
Overall C/N (in 16 MHz)	15.9 dB	10.0 dB
Reference threshold C/N	10.0 dB	10.0 dB
Margin over threshold	5.9 dB	0.0 dB

(from 9.4 dB/K to 8.1 dB/K). Therefore, when extra margin is allowed for weather loss, additional margin should simultaneously be added to compensate for the increase in antenna noise temperature.

With regard to satellite links, in industry one often hears such expressions as "the link *can* be closed," meaning that the margin, in decibels, has a positive value and the required error performance will be satisfied, or "the link *cannot* be closed," meaning that the margin has a negative value and the required error performance will *not* be satisfied. Even though the words "the link closes" or "the link does not close" give the impression of an "on–off" condition, it is worth emphasizing that lack of link closure, or a negative margin, means that the error performance falls short of the system requirement; it does not necessarily mean that communications cease. For example, consider a system whose $(E_b/N_0)_{\text{reqd}} = 10$ dB, as shown in Figure 4.9, but whose $(E_b/N_0)_r = 8$ dB. Assume that 8 dB corresponds to $P_B = 10^{-2}$. Thus there is a margin of -2 dB, and a bit error probability of 10 times the specified error probability. The link may still be useful, though degraded.

4.4.4 Link Availability

Link availability is usually a measure of long-term link utility stated on an average annual basis; for a given geographical location, the link availability measures the percentage of time the link can be closed. For example, for a particular link between Washington, D.C., and a satellite repeater, the long-term weather pattern may be such that a 10-dB weather margin is adequate for link closure 98% of the time; for 2% of the time, heavy rains result in greater than 10 dB SNR degradation, so that the link does not close. Since the effect of rain on SNR degradation is a function of signal frequency, link availability and required margin must be examined in the context of a particular transmission frequency.

Figure 4.10 summarizes worldwide satellite link availability at a frequency of 44 GHz. The plot illustrates percentage of the earth visible (the link closes, and a prescribed probability of error is met) as a function of margin for the case of three equispaced geostationary satellites. A *geostationary satellite* is located in a circular orbit in the same plane as the earth's equatorial plane and at the synchronous altitude of 35,800 km. The satellite's orbital period is identical with that of the earth's rotational period, and therefore the satellite appears stationary when viewed from the earth. Figure 4.10 shows a family of visibility curves with different required link availabilities, ranging from benign (95% availability) to fairly stringent (99% availability). In general, for a fixed link margin, visibility is inversely proportional to required availability, and for a fixed availability, visibility increases monotonically with margin [8]. Figures 4.11 to 4.13 illustrate, by unshaded and shaded areas, the parts of the earth from which the 44-GHz link can and cannot be closed 99% of the time for three different values of link margin. Figure 4.11 illustrates the link coverage of such locations for a margin value of 14 dB. Notice that this figure can be used to pinpoint the regions of heaviest

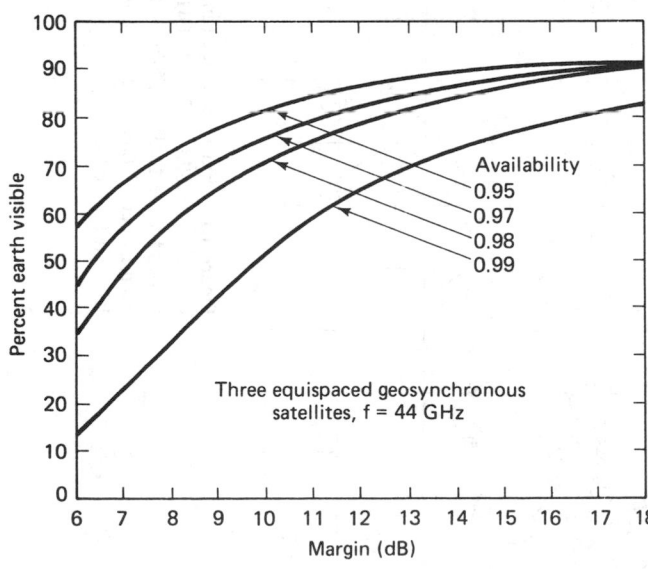

Figure 4.10 Earth coverage versus link margin for various values of link availability. (Reprinted from L. M. Schwab, "World-Wide Link Availability for Geostationary and Critically Inclined Orbits Including Rain Effects," *Lincoln Laboratory*, *Rep. DCA-9*, Jan. 27, 1981, Fig. 14, p. 38, courtesy of Lincoln Laboratory.)

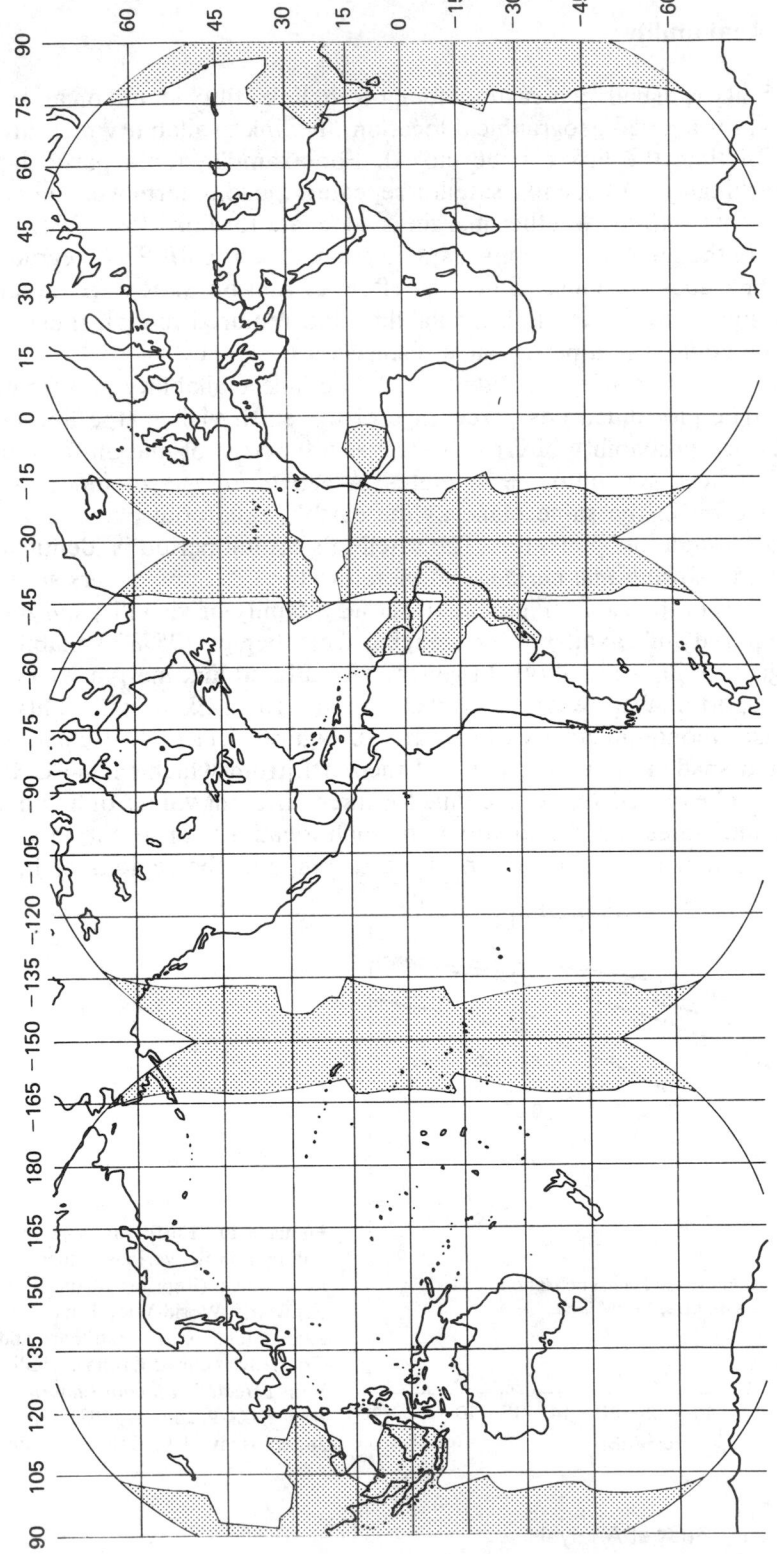

Figure 4.11 Earth coverage (unshaded) for 0.99 link availability for three equispaced geostationary satellites, f = 44 GHz, link margin = 14 dB. (Reprinted from L. M. Schwab, "World-Wide Link Availability for Geostationary and Critically Inclined Orbits Including Rain Effects," *Lincoln Laboratory, Rep. DCA-9,* Jan. 27, 1981, Fig. 17, p. 42, courtesy of Lincoln Laboratory.)

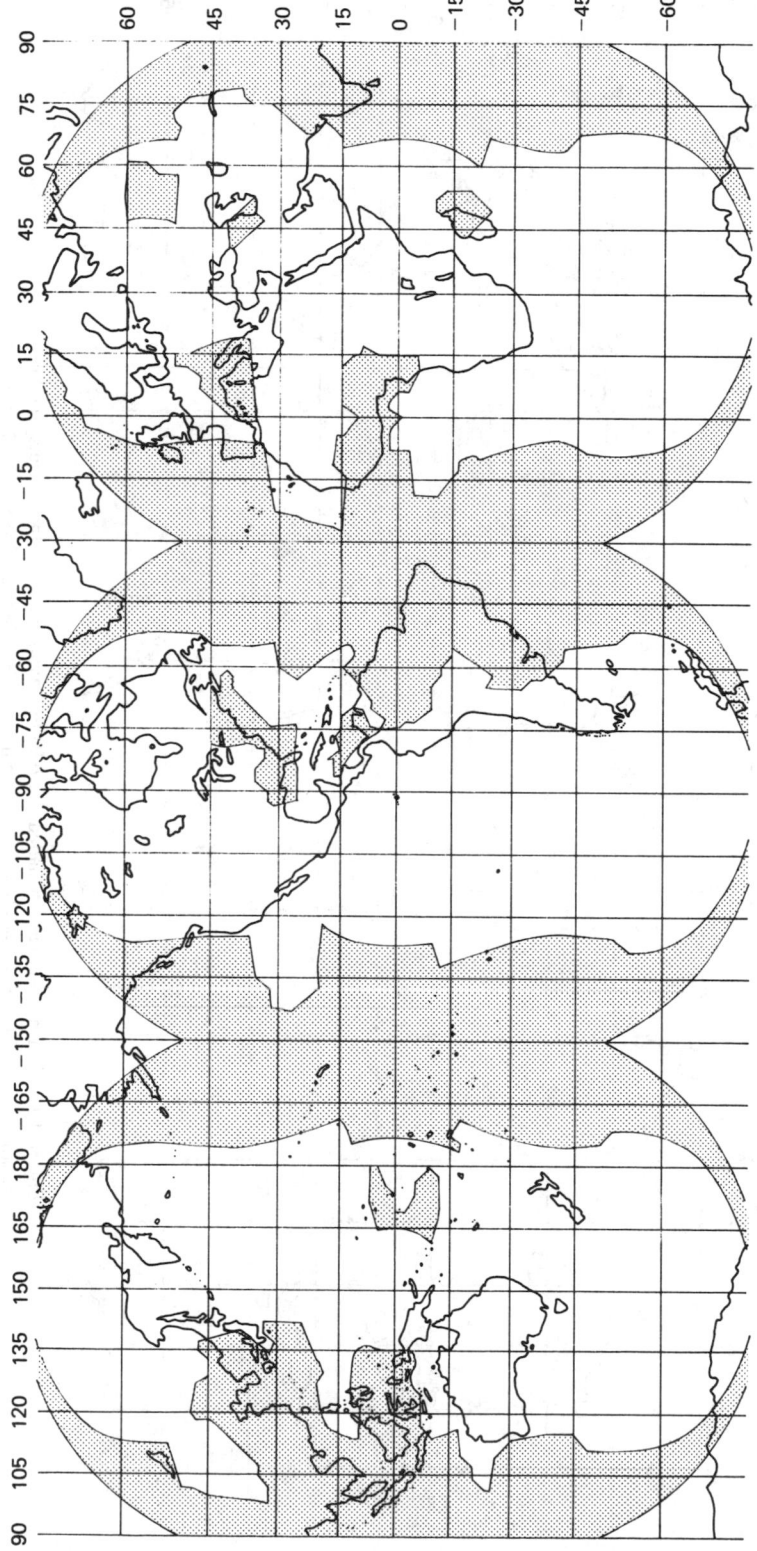

Figure 4.12 Earth coverage (unshaded) for 0.99 link availability for three equispaced geostationary satellites, $f = 44$ GHz, link margin $= 10$ dB. (Reprinted from L. M. Schwab, "World-Wide Link Availability for Geostationary and Critically Inclined Orbits Including Rain Effects," *Lincoln Laboratory, Rep. DCA-9*, Jan. 27, 1981, Fig. 18, p. 43, courtesy of Lincoln Laboratory.)

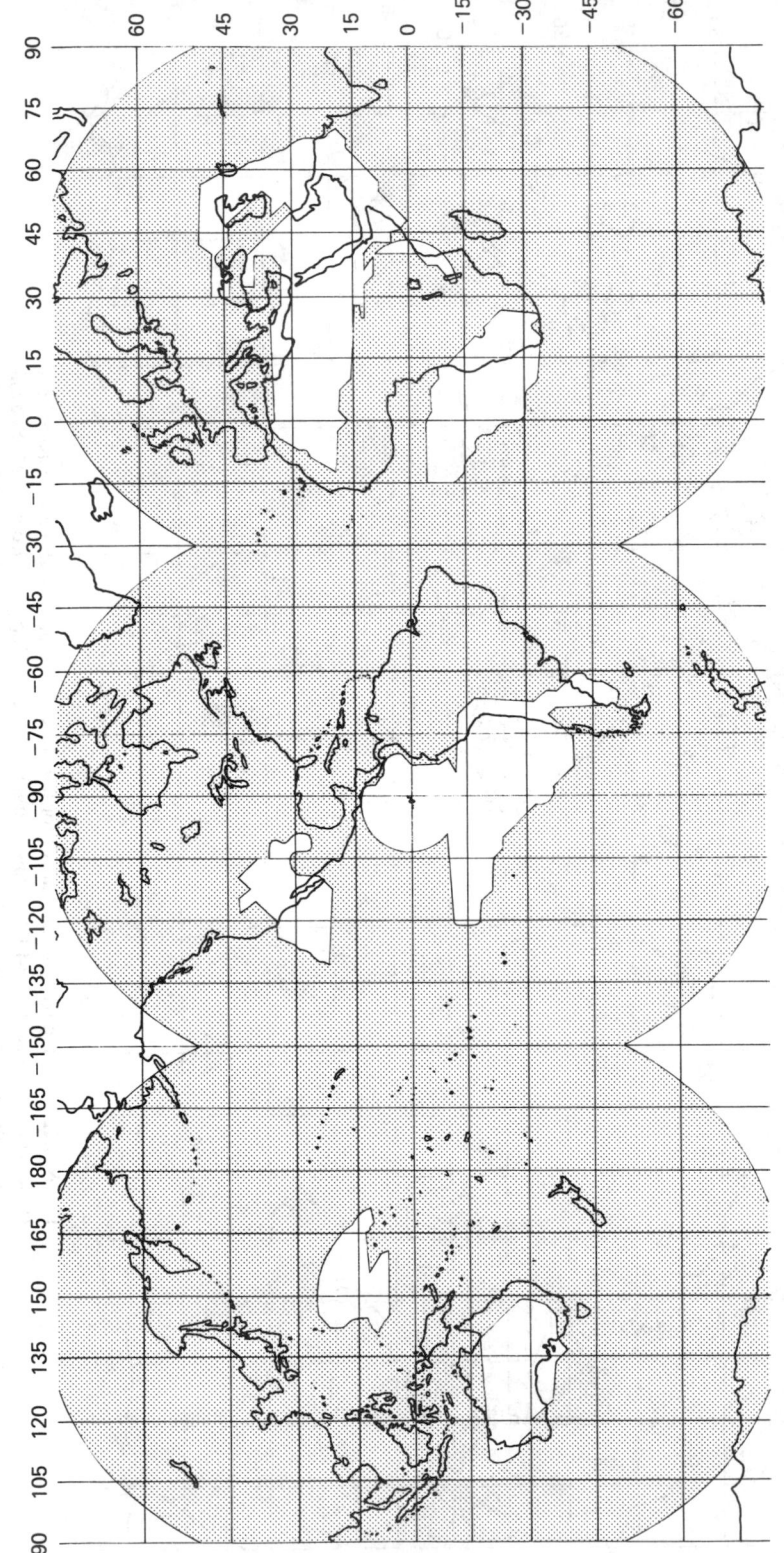

Figure 4.13 Earth coverage (unshaded) for 0.99 link availability for three equispaced geostationary satellites, f = 44 GHz, link margin = 6 dB. (Reprinted from L. M. Schwab, "World-Wide Link Availability for Geostationary and Critically Inclined Orbits Including Rain Effects," *Lincoln Laboratory, Rep. DCA-9*, Jan. 27, 1981, Fig. 19, p. 44, courtesy of Lincoln Laboratory.)

rainfall, such as Brazil and Indonesia. The figure represents the result of a link calculation performed in concert with a weather model of the earth.

In Figure 4.11 there are shaded strips on the east and west boundaries of each satellite's field of view. Why do you suppose the link availability is not met in these regions? At the edge of the earth the propagation path between the satellite and ground is longer than the path directly beneath the satellite. Degradation occurs in three ways: (1) the longer path results in reduced power density at the receiving antenna; (2) the edge of coverage sites will experience reduced satellite antenna gain, unless the satellite antenna pattern is designed to be uniform over its entire field of view (typically the pattern is − 3 dB at the beam edge compared to the peak gain at the beam center); and (3) propagation to the edge of the earth traverses a thicker atmospheric layer because of the oblique path and the earth's curvature. The third item is of prime importance at those signal frequencies that are most attenuated by the atmosphere. Why do you suppose you do not see the same shaded areas near the north and south poles in Figure 4.11? Snowfall does not have the same deleterious effect on signal propagation as does rainfall; the phenomenon is known as the *freeze effect*.

Figure 4.12 illustrates the parts of the earth that can and cannot close the 44-GHz link 99% of the time with 10-dB link margin. Notice that the shaded areas have grown considerably compared to the 14-dB margin case; now, the east coast of the United States, the Mediterranean, and most of Japan cannot close the link 99% of the time. Figure 4.13 illustrates similar link performance for a margin of 6 dB. Whereas Figure 4.11 could be used to locate the regions of greatest rainfall, Figure 4.13 can be used to locate the driest weather regions on the earth; such areas are seen to be the southwestern part of the United States, most of Australia, the coast of Peru and Chile, and the Sahara desert in Africa.

4.5 NOISE FIGURE, NOISE TEMPERATURE, AND SYSTEM TEMPERATURE

4.5.1 Noise Figure

Noise figure, F, relates the SNR at the input of a network to the SNR at the output of the network. Thus noise figure measures the SNR degradation caused by the network. Figure 4.14 illustrates such an example. Figure 4.14a depicts the SNR at an *amplifier input,* $(SNR)_{in}$, as a function of frequency. At its peak, the signal is 40 dB above the noise floor. Figure 4.14b depicts the SNR at the *amplifier output* $(SNR)_{out}$. The amplifier gain has increased the signal by 20 dB; however, the amplifier has added its own additional noise. The output signal, at its peak, is only 30 dB above the noise floor. Since the SNR degradation from input to output is 10 dB, this is tantamount to describing the amplifier as having a 10-dB noise figure. Noise figure is a parameter that expresses the noisiness of a two-port network or device, such as an amplifier, compared to a reference noise source at the input port. It can be written as follows:

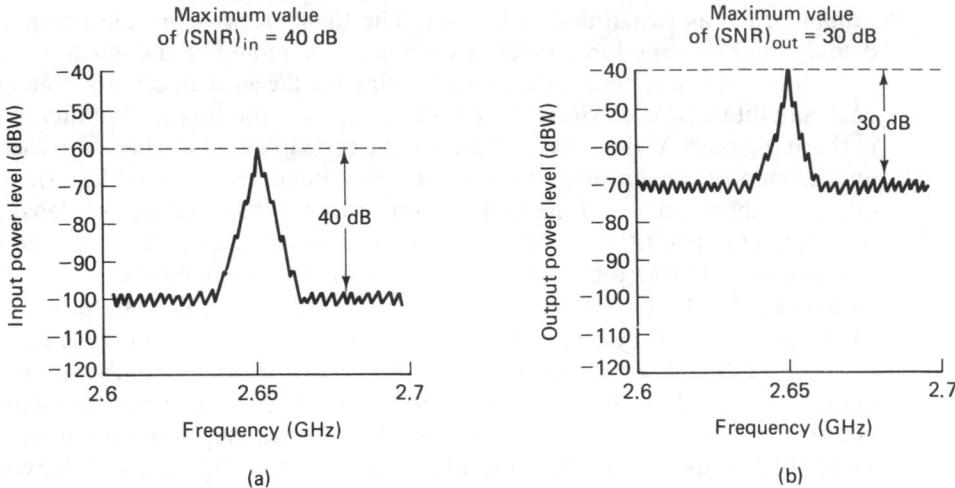

Figure 4.14 Amplifier signal and noise levels as a function of frequency. (a) Amplifier input. (b) Amplifier output.

$$F = \frac{(\text{SNR})_{in}}{(\text{SNR})_{out}} = \frac{S_i/N_i}{GS_i/G(N_i + N_{ai})} \qquad (4.25)$$

where

$$S_i = \text{signal power at the amplifier input port}$$

$$N_i = \text{noise power at the amplifier input port}$$

$$N_{ai} = \text{amplifier noise referred to the input port}$$

$$G = \text{amplifier gain}$$

Figure 4.15 is an example illustrating Equation (4.25). Figure 4.15a represents a *realizable amplifier* example with a gain, $G = 100$, and internal noise power, $N_a = 10 \ \mu\text{W}$. The source noise, external to the amplifier, is $N_i = 1 \ \mu\text{W}$. In Figure 4.15b we assume that the *amplifier is ideal,* and we ascribe the noisiness of the real amplifier, from part (a) of the figure, to an external source, N_{ai}, in series with the original source, N_i. The value of N_{ai} is obtained by reducing N_a by the amplifier gain. As shown in Figure 4.15b, Equation (4.25) references all noise to the amplifier input, whether the noise is actually present at the input or is internal to the device. As can be seen in Figure 4.15, the noise power output from the real amplifier is identical to that of its electrically equivalent model.

Equation (4.25) reduces to the following:

$$F = \frac{N_i + N_{ai}}{N_i} = 1 + \frac{N_{ai}}{N_i} \qquad (4.26)$$

Notice from Equation (4.26) that the noise figure expresses the noisiness of a

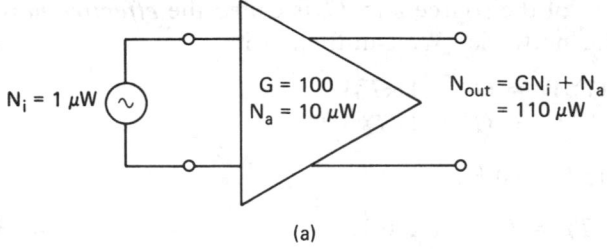

$N_i = 1 \ \mu W$

$G = 100$
$N_a = 10 \ \mu W$

$N_{out} = GN_i + N_a$
$= 110 \ \mu W$

(a)

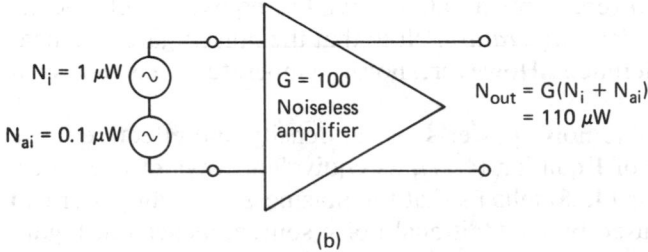

$N_i = 1 \ \mu W$

$N_{ai} = 0.1 \ \mu W$

$G = 100$
Noiseless
amplifier

$N_{out} = G(N_i + N_{ai})$
$= 110 \ \mu W$

(b)

Figure 4.15 Example of noise treatment in amplifiers.

network relative to an input source noise; noise figure is *not* an absolute measure of noise. An ideal amplifier or network, one that contributes no noise ($N_{ai} = 0$), has a noise figure equal to unity (0 dB).

For the concept of noise figure to have utility, we need to be able to make equitable comparisons among devices on the basis of Equation (4.26). We must, therefore, choose a value of N_i as a *reference*. The noise figure of any device will then represent a measure of how much noisier the device is than the reference. In 1944, Friis [9] suggested that noise figure be defined for a noise source at a reference temperature of $T_0^\circ = 290$ K. That suggestion was subsequently adopted by the IEEE as part of its standard definition for noise figure [10]. From Equation (4.17) we see that the maximum available noise power spectral density from any source resistance is established by specifying its temperature. The value of 290 K was selected as the reference because it is a reasonable approximation of the source temperature for many links. Also with T_0° chosen to be 290 K, the value of noise spectral density N_0 at T_0° results in an aesthetically pleasing number, as shown below:

$$N_0 = \kappa T_0^\circ = 1.38 \times 10^{-23} \times 290 = 4.00 \times 10^{-21} \ \text{W/Hz}$$

Expressed in decibels,

$$N_0 = -204 \ \text{dBW/Hz}$$

4.5.2 Noise Temperature

Rearranging Equation (4.26), we can write

$$N_{ai} = (F - 1)N_i \qquad (4.27)$$

From Equation (4.16) we can replace N_i with $\kappa T_0^\circ W$ and N_{ai} with $\kappa T_R^\circ W$, where

T_0° is the reference temperature of the source and T_R° is called the *effective noise temperature* of the receiver (or network). We can then write

$$\kappa T_R^\circ W = (F - 1)\kappa T_0^\circ W$$
$$T_R^\circ = (F - 1)\, T_0^\circ$$

or since T_0° has been chosen to be 290 K,

$$T_R^\circ = (F - 1)290 \text{ K} \tag{4.28}$$

Equation (4.26) uses the concept of noise figure to characterize the noisiness of an amplifier. Equation (4.28) represents an alternative but equivalent characterization known as *effective noise temperature*. Note that the noise figure is a measurement relative to a reference. However, noise temperature has no such constraint.

We can think of available noise power spectral density and effective noise temperature, in the context of Equation (4.17), as equivalent ways of characterizing noise sources. Equation (4.28) tells us that the noisiness of an amplifier can be modeled as if it were caused by an additional noise source, as seen in Figure 4.15b, operating at some effective temperature called T_R°. For a purely resistive termination, T_R° is never less than ambient temperature unless it is cooled. It is important to note that for reactive terminations, such as uncooled parametric amplifiers or other low-noise devices, T_R° can be much less than 290 K, even though the ambient temperature is higher [11]. For the output noise of an amplifier, N_{out}, as a function of its effective temperature, we can use Equations (4.25) and (4.16) to write

$$N_{\text{out}} = GN_i + GN_{ai}$$
$$= G\kappa T_g^\circ W + G\kappa T_R^\circ W = G\kappa(T_g^\circ + T_R^\circ)W \tag{4.29}$$

where T_g° is the temperature of the source.

4.5.3 Line Loss

The difference between amplifier networks and lossy line networks can be viewed in the context of the degradation mechanisms *loss* and *noise*, described earlier. Noisy networks in Sections 4.5.1 and 4.5.2 were discussed with amplifiers in mind. We saw that SNR degradation resulted from injecting additional (amplifier) noise into the link, as shown in Figure 4.15. However, in the case of a lossy line, we shall show that the SNR degradation results from the signal being attenuated while the noise remains fixed (for the case where the line temperature is equal to or less than the source temperature). The degradation effect will nonetheless be measured as an increase in noise figure or effective noise temperature.

Consider the lossy line or network shown in Figure 4.16. Assume the line is matched with its characteristic impedance at the source and at the load. We shall define power loss, L, as

$$L = \frac{\text{input power}}{\text{output power}}$$

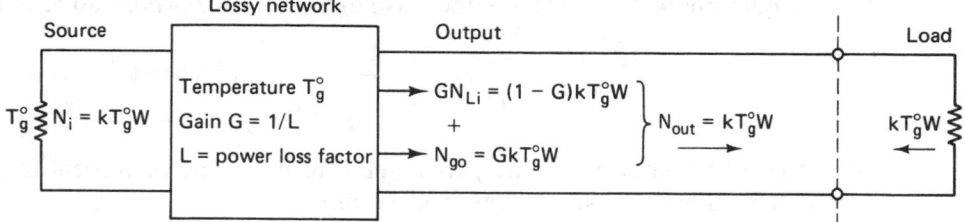

Figure 4.16 Lossy line: impedance matched and temperature matched at both ends.

Then the network gain, G, equals $1/L$ (less than unity for a lossy line). Let all components be at temperature T_g°. The total output noise power flowing from the network into the load is

$$N_{\text{out}} = \kappa T_g^\circ W$$

since the network output appears as a pure resistance at the temperature T_g°. The total power flowing from the load back into the network must also equal N_{out}, to ensure thermal equilibrium. Recall that available noise power, $\kappa T^\circ W$, is dependent only on temperature, bandwidth, and impedance matching; it is not dependent on the resistance value. N_{out} can be considered as being comprised of two components, N_{go} and GN_{Li}, such that

$$N_{\text{out}} = \kappa T_g^\circ W = N_{go} + GN_{Li} \tag{4.30}$$

where

$$N_{go} = G\kappa T_g^\circ W \tag{4.31}$$

is the component of output noise power due to the source and GN_{Li} is the component of output noise power due to the lossy network, where N_{Li} is the network noise relative to its input. Combining Equations (4.30) and (4.31), we can write

$$\kappa T_g^\circ W = G\kappa T_g^\circ W + GN_{Li} \tag{4.32}$$

Solving for N_{Li} yields

$$N_{Li} = \frac{1 - G}{G} \kappa T_g^\circ W = \kappa T_L^\circ W \tag{4.33}$$

Therefore, the effective noise temperature of the line is

$$T_L^\circ = \frac{1 - G}{G} T_g^\circ \tag{4.34}$$

and since $G = 1/L$,

$$T_L^\circ = (L - 1)T_g^\circ \tag{4.35}$$

Choosing $T_g^\circ = 290$ K as the reference temperature, we can write

$$T_L^\circ = (L - 1)290 \text{ K} \tag{4.36}$$

Using Equations (4.28) and (4.36), the *noise figure for a lossy line* can be expressed as

$$F = 1 + \frac{T_L^\circ}{290} = L \qquad (4.37)$$

Note that some authors use the parameter L to mean the reciprocal of the loss factor defined here. In such cases, noise figure $F = 1/L$.

Example 4.4 Lossy Line

A line at temperature $T^\circ = 290$ K is fed from a source whose noise temperature is $T_g^\circ = 290$ K. The input signal power, S_i, is 100 picowatts (pW) and the signal bandwidth is 1 GHz. The line has a loss factor $L = 2$. Calculate the $(SNR)_{in}$, the effective line temperature, T_L°, the output signal power, S_{out}, the $(SNR)_{out}$, and the noise figure, F.

Solution

$$N_i = \kappa T_g^\circ W$$

$$= 1.38 \times 10^{-23} \text{ W/K-Hz} \times 290 \text{ K} \times 10^9 \text{ Hz}$$

$$= 4 \times 10^{-12} \text{ W} = 4 \text{ pW}$$

$$(SNR)_{in} = \frac{100 \text{ pW}}{4 \text{ pW}} = 25 \text{ (14 dB)}$$

$$T_L^\circ = (L - 1)290 \text{ K} = 290 \text{ K}$$

$$S_{out} = \frac{S_i}{L} = \frac{100 \text{ pW}}{2} = 50 \text{ pW}$$

$$N_{out} = \kappa T^\circ W = 4 \text{ pW}$$

$$(SNR)_{out} = \frac{50 \text{ pW}}{4 \text{ pW}} = 12.5 \text{ (11 dB)}$$

$$F = L = 2 \text{ (3 dB)}$$

4.5.4 Composite Noise Figure and Composite Noise Temperature

When two networks are connected in series, as shown in Figure 4.17a, their composite noise figure can be written as

$$F_{comp} = F_1 + \frac{F_2 - 1}{G_1} \qquad (4.38)$$

where G_1 is the gain associated with network 1. When n networks are connected in series the relationship between stages expressed in Equation (4.38) continues, so that the *composite noise figure* for a sequence of n stages is written as

$$F_{comp} = F_1 + \frac{F_2 - 1}{G_1} + \frac{F_3 - 1}{G_1 G_2} + \cdots + \frac{F_n - 1}{G_1 G_2 \cdots G_{n-1}} \qquad (4.39)$$

Communications Link Analysis Chap. 4

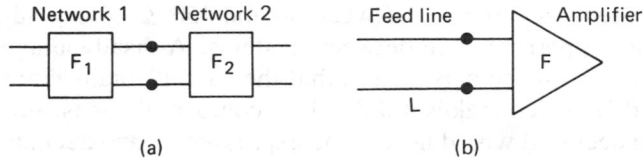

Network 1   Network 2   Feed line   Amplifier

F_1 F_2 F

                                L

   (a)   (b)

Figure 4.17 Networks connected in series.

Can you guess from Equation (4.39) what the design goals for the front end of the receiver (especially the first stage or the first couple of stages) should be? At the front end of the receiver, the signal is most susceptible to added noise; therefore, the first stage should have as low a noise figure, F_1, as possible. Also, because the noise figure of each subsequent stage is reduced by the gains of the prior stages, it behooves us to strive for as high a gain, G_1, as possible. Simultaneously achieving the lowest F_1 and the highest G_1 represents conflicting goals; therefore, compromises are always necessary.

Equations (4.39) and (4.28) can be combined to express the composite effective noise temperature of a sequence of n stages:

$$T_{comp}^{\circ} = T_1^{\circ} + \frac{T_2^{\circ}}{G_1} + \frac{T_3^{\circ}}{G_1 G_2} + \cdots + \frac{T_n^{\circ}}{G_1 G_2 \cdots G_{n-1}} \qquad (4.40)$$

Figure 4.17b illustrates a feed line in series with an amplifier; this is a typical arrangement following a receiving antenna. Using Equation (4.38) to find F_{comp} for such a lossy line and amplifier arrangement, we can write

$$F_{comp} = L + L(F - 1) = LF \qquad (4.41)$$

since the noise figure of the lossy line is L and the gain of the line is $1/L$. By analogy with Equation (4.36), we can write the composite temperature as

$$T_{comp}^{\circ} = (LF - 1)290 \text{ K} \qquad (4.42)$$

We can also write the composite temperature of line and amplifier as follows:

$$T_{comp}^{\circ} = (LF - 1 + L - L)290 \text{ K}$$
$$= [(L - 1) + L(F - 1)]290 \text{ K} \qquad (4.43)$$
$$= T_L^{\circ} + LT_R^{\circ}$$

4.5.4.1 Comparison of Noise Figure and Noise Temperature

Since noise figure, F, and effective noise temperature, T°, characterize the noise performance of devices, some engineers feel compelled to select one of these measures as the more useful. However, they each have their place. For terrestrial applications, F is almost universally used; the concept of SNR degradation for a 290 K source temperature makes sense, because terrestrial source temperatures are typically close to 290 K. Terrestrial noise figure values typically fall in the convenient range 1 to 10 dB.

For space applications, T° is the more common figure of merit. The range of values for commercial systems is typically between 30 and 150 K, giving adequate resolution for comparing performance between systems. A disadvantage of using noise figures for such low-noise networks is that the values obtained are all close to unity (0.5 to 1.5 dB), which makes it difficult to compare devices. For low-noise applications, F (in decibels) would need to be expressed to two decimal places to provide the same resolution or precision as does $T°$. For space applications, a reference temperature of 290 K is not as appropriate as it is for terrestrial applications. When using effective temperature, no reference temperature (other than absolute zero K) is needed for judging degradation. The effective input noise temperature is simply compared to the effective source noise temperature. In general, applications involving very low noise devices seem to favor the effective temperature measure over the noise figure.

4.5.5 System Effective Temperature

Figure 4.18 represents a simplified schematic of a receiving system, identifying those areas—the antenna, the line, and the preamplifier—that play a primary role in SNR degradation. We have already discussed the degradation role of the preamplifier—additional noise is injected into the link. And we have discussed line loss—the signal is attenuated, while the noise is held fixed (for the case where the line temperature is less than or equal to the source temperature). The remaining source of degradation is via the receiving antenna. An antenna is like a lens. Its noise contributions are dictated by what the antenna is "looking at." If the antenna is pointed at a cool portion of the sky, very little thermal noise is introduced, but if the antenna is pointed at a warm body, a greater amount of thermal noise is introduced. The *antenna temperature* is a measure of the effective temperature integrated over the entire antenna pattern.

We now find *system temperature*, $T_S°$, by adding together all the system noise contributors (in terms of effective temperature). The summation is expressed as

$$T_S° = T_A° + T_{comp}° \tag{4.44}$$

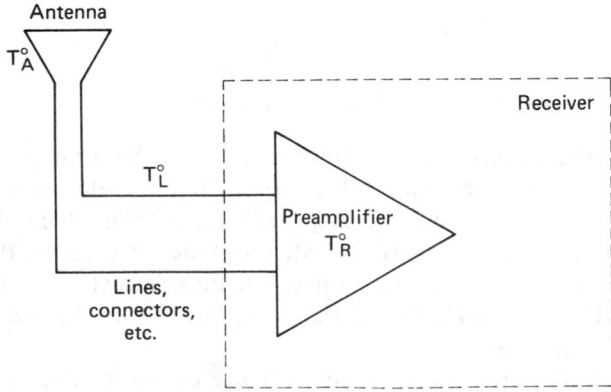

Figure 4.18 Major noise contributors of a receiving system.

where T_A° is the antenna temperature and T_{comp}° is the composite temperature of the line and the preamplifier. Since the system temperature, T_S°, is a new composite, made up of T_A° and the composite effective temperature of the line and preamp, one might ask: Why doesn't Equation (4.44) appear to have the same sequential gain reduction factors as those in Equation (4.40)? We have assumed that the antenna has *no dissipative parts*; its gain, unlike an amplifier or attenuator, can be thought of as a processing gain. Whatever effective temperature is introduced at the antenna comes through, unaltered by the antenna; the antenna represents the source noise, or source temperature, at the input to the line.

Using Equation (4.43), we can modify Equation (4.44) as follows:

$$T_S^\circ = T_A^\circ + T_L^\circ + LT_R^\circ \tag{4.45}$$

$$= T_A^\circ + (L - 1)290 \text{ K} + L(F - 1)290 \text{ K}$$

$$= T_A^\circ + (LF - 1)290 \text{ K} \tag{4.46}$$

If LF is provided in units of decibels, we must first convert LF to a ratio, so that T_S° takes the form

$$T_S^\circ = T_A^\circ + (10^{LF/10} - 1)290 \text{ K} \tag{4.47}$$

Example 4.5 Noise Figure and Noise Temperature

A receiver front end, shown in Figure 4.19a, has a noise figure of 10 dB, a gain of 80 dB, and a bandwidth of 6 MHz. The input signal power, S_i, is 10^{-11} W. Assume that the line loss is zero and the antenna temperature is 150 K. Find T_R°, T_S°, N_{out}, $(SNR)_{in}$, and $(SNR)_{out}$.

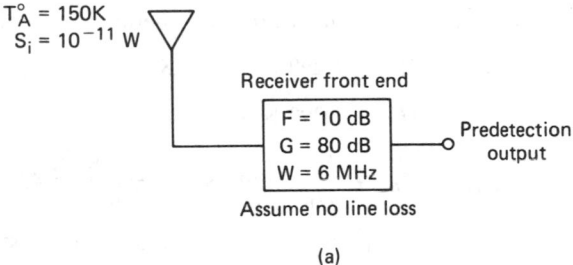

(a)

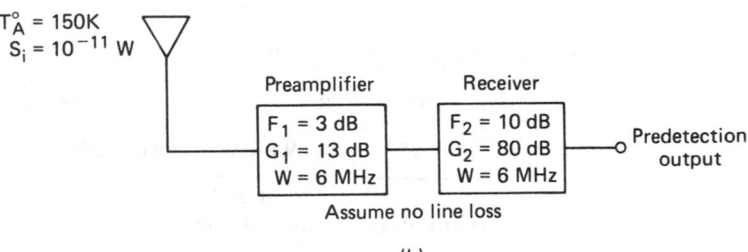

(b)

Figure 4.19 Improving a receiver front end with a low-noise preamplifier.

Solution

First convert all decibel values to ratios.

$$T_R^\circ = (F - 1)290 \text{ K} = 2610 \text{ K}$$

$$T_S^\circ = T_A^\circ + T_R^\circ = 150 \text{ K} + 2610 \text{ K} = 2760 \text{ K}$$

$$N_{\text{out}} = G\kappa T_A^\circ W + G\kappa T_R^\circ W = G\kappa T_s^\circ W$$

$$= 10^8 \times 1.38 \times 10^{-23} \times 6 \times 10^6 (150 \text{ K} + 2610 \text{ K})$$

$$= \underbrace{1.2 \text{ } \mu\text{W}}_{\substack{\text{source} \\ \text{contribution}}} + \underbrace{21.6 \text{ } \mu\text{W}}_{\substack{\text{front-end} \\ \text{contribution}}} = 22.8 \text{ } \mu\text{W}$$

$$(\text{SNR})_{\text{in}} = \frac{S_i}{\kappa T_A^\circ W} = \frac{10^{-11}}{1.24 \times 10^{-14}} = 806.5 \text{ } (29.1 \text{ dB})$$

$$(\text{SNR})_{\text{out}} = \frac{S_{\text{out}}}{N_{\text{out}}} = \frac{10^8 \times 10^{-11}}{22.8 \times 10^{-6}} = 43.9 \text{ } (16.4 \text{ dB})$$

Notice in this example that the amplifier noise is significantly larger than the source noise and represents the major cause of SNR degradation.

Example 4.6   Improving SNR with a Low-Noise Preamplifier

Use a preamplifier, as shown in Figure 4.19b, with a noise figure of 3 dB, a gain of 13 dB, and a bandwidth of 6 MHz to improve the SNR of the receiver in Example 4.5. Find T_{comp}° for the composite preamplifier and receiver. Find T_S°, F_{comp}, N_{out}, and $(\text{SNR})_{\text{out}}$. Assume zero line loss.

Solution

Again, convert all decibel values to ratios before proceeding.

$$T_{R1}^\circ = (F_1 - 1)290 \text{ K} = 290 \text{ K}$$

$$T_{R2}^\circ = (F_2 - 1)290 \text{ K} = 2610 \text{ K}$$

$$T_{\text{comp}}^\circ = T_{R1}^\circ + \frac{T_{R2}^\circ}{G_1} = 290 \text{ K} + \frac{2610 \text{ K}}{20} = 420.5 \text{ K}$$

$$T_S^\circ = T_A^\circ + T_{\text{comp}}^\circ = 150 \text{ K} + 420.5 \text{ K} = 570.5 \text{ K}$$

$$F_{\text{comp}} = F_1 + \frac{F_2 - 1}{G_1} = 2 + \frac{9}{20} = 2.5 \text{ } (4 \text{ dB})$$

$$N_{\text{out}} = G\kappa T_A^\circ W + G\kappa T_{\text{comp}}^\circ W = G\kappa T_S^\circ W$$

$$= 20 \times 10^8 \times 1.38 \times 10^{-23} \times 6 \times 10^6 (150 \text{ K} + 420.5 \text{ K})$$

$$= \underbrace{24.8 \text{ } \mu\text{W}}_{\substack{\text{source} \\ \text{contribution}}} + \underbrace{69.6 \text{ } \mu\text{W}}_{\substack{\text{front-end} \\ \text{contribution}}} = 94.4 \text{ } \mu\text{W}$$

$$(\text{SNR})_{\text{out}} = \frac{S_{\text{out}}}{N_{\text{out}}} = \frac{10^{-11} \times 20 \times 10^8}{94.4 \times 10^{-6}} = 212.0 \text{ } (23.3 \text{ dB})$$

With the added preamplifier the (predetection) output noise has increased (from 22.8 μW in Example 4.5) to 94.4 μW. Even though the noise power has increased, the lower system temperature has resulted in a 6.9-dB improvement in SNR (from 16.4 dB in Example 4.5, to 23.3 dB here). The price we pay for this improvement is the need to provide an F_{comp} improvement of 6 dB (from 10 dB in Example 4.5, to 4 dB in this example).

The unwanted noise is, in part, *injected via the antenna* $(\kappa T_A^\circ W)$, and in part, *generated internally* in the receiver front end $(\kappa T_{comp}^\circ W)$. The amount of system improvement that can be rendered via front-end design depends on what portion of the total noise the front end contributes. We saw in Example 4.5 that the front end contributed the major portion of the noise. Therefore, in Example 4.6, providing a low-noise preamplifier improved the system SNR significantly. In the next example, we show the case where the major portion of the noise is injected via the antenna; we shall see that introducing a low-noise preamplifier in such a case will not help the SNR appreciably.

Example 4.7 Attempting SNR Improvement When the Value of T_A° Is Large

Repeat Examples 4.5 and 4.6 with one change: let $T_A^\circ = 8000$ K. In other words, the preponderant amount of noise is being injected via the antenna; the antenna might have a very hot body (the sun) fully occupying its field of view. Calculate the SNR improvement that would be provided by the preamplifier used in Example 4.6 and Figure 4.19b, and compare the result with that of Example 4.6.

Solution

Without preamplifier

$$N_{out} = G\kappa W(T_A^\circ + T_R^\circ)$$

$$= 10^8 \times 1.38 \times 10^{-23} \times 6 \times 10^6 (8000 \text{ K} + 2610 \text{ K})$$

$$= \underbrace{66.2 \text{ μW}}_{\substack{\text{source} \\ \text{contribution}}} + \underbrace{21.6 \text{ μW}}_{\substack{\text{front-end} \\ \text{contribution}}} = 87.8 \text{ μW}$$

$$(\text{SNR})_{out} = \frac{S_{out}}{N_{out}} = \frac{10^8 \times 10^{-11}}{87.8 \times 10^{-6}} = 11.4 \ (10.6 \text{ dB})$$

With preamplifier

$$N_{out} = 20 \times 10^8 \times 1.38 \times 10^{-23} \times 6 \times 10^6 (8000 \text{ K} + 420.5 \text{ K})$$

$$= \underbrace{1324.8 \text{ μW}}_{\substack{\text{source} \\ \text{contribution}}} + \underbrace{69.6 \text{ μW}}_{\substack{\text{front-end} \\ \text{contribution}}} = 1394.4 \text{ μW}$$

$$(\text{SNR})_{out} = \frac{20 \times 10^8 \times 10^{-11}}{1.39 \times 10^{-3}} = 14.4 \ (11.6 \text{ dB})$$

Therefore, for this case, the SNR improvement is only 1 dB, a far cry from the 6.9 dB accomplished earlier. When the noise is mostly due to devices within the receiver, it is possible to improve the SNR by introducing low-noise devices. However, when the noise is mostly due to external causes, improving the receiver front end will not help much.

Noise figure is a definition, predicated on a reference temperature of 290 K. When the source temperature is other than 290 K, as is the case in Examples 4.5, 4.6, and 4.7, it is necessary to define a *working* or *effective noise figure* that describes the actual $(SNR)_{in}$ versus $(SNR)_{out}$ relationship shown in Equation (4.25). Such an operational noise figure can be found as follows:

$$F_{op} = \frac{T_S^\circ}{T_A^\circ} = \frac{T_A^\circ + T_R^\circ}{T_A^\circ}$$

since

$$F = 1 + \frac{T_R^\circ}{T_0^\circ}$$

then

$$F_{op} = (F-1)\,\frac{T_0^\circ}{T_A^\circ} + 1 \tag{4.48}$$

4.5.6 Sky Noise Temperature

The receiving antenna collects random noise emissions from galactic, solar, and terrestrial sources, constituting the sky background noise. The sky background appears as a combination of galactic effects that decrease with frequency, and atmospheric effects that start becoming significant at 10 GHz and increase with frequency. Figure 4.20 illustrates the sky temperature, as measured from the earth, due to both these effects. Notice that there is a region, between 1 and 10 GHz, where the temperature is lowest; the galaxy noise has become quite small at 1 GHz, and for satellite communications the blackbody radiation noise due to the absorbing atmosphere is not significant below 10 GHz. (For other applications, e.g., passive radiometry, it is still a problem.) This region, known as the *microwave window* or *space window,* is particularly useful for satellite or deep-space communication. The low sky noise is the principal reason that such systems primarily use carrier frequencies in this part of the spectrum. The galaxy and atmospheric noise curves in Figure 4.20 are comprised of a family of curves each at a different elevation angle θ. When θ is zero, the receiving antenna points at the horizon and the propagation path encompasses the longest possible atmospheric layer; when θ is 90°, the receiving antenna points to the zenith, and the resulting propagation path contains the shortest possible atmospheric layer. Thus the upper curve of the family represents the near worst-case (clear-weather) sky noise temperature versus frequency, and the lower curve represents the most benign case. Also shown in Figure 4.20 is a plot of noise temperature versus frequency *due to rain.* Since the intensity of any rainstorm can only be expressed

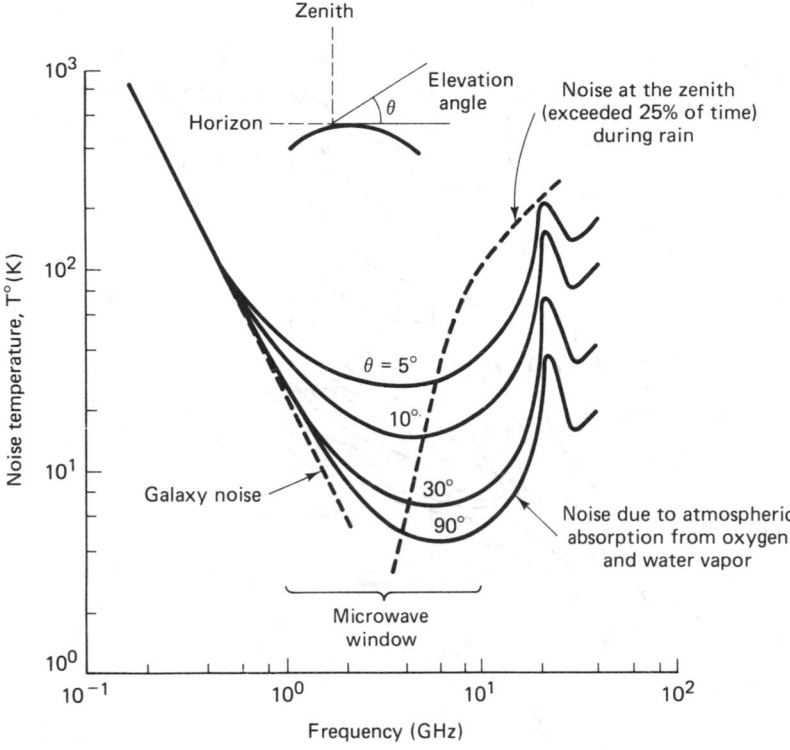

Figure 4.20 Sky noise temperature.

statistically, the noise temperatures shown are values that are exceeded 25% of the time (at the zenith). Which spectral region appears the most benign for space communications when rainfall is taken into account? It is the region at the low end of the space window. For this reason, systems such as the Space Ground Link Subsystem or SGLS (military) and the Unified S-Band Telemetry, Tracking, and Control System (NASA) are located in the 1.8 to 2.4-GHz band.

4.5.6.1 Radio Maps of the Sky

Various researchers have mapped the galactic noise radiation as a function of frequency. Figure 4.21 is such a radio temperature map, after Ko and Kraus [12], indicating the temperature contours of the sky in the region of 250 MHz when viewed from the earth. In general, the sky is composed of localized galactic sources (sun, moon, planets, etc.), each having its own temperature. The map is effectively a weighted sum of the individual galactic source temperatures plus a constant sky background. The coordinates of the map, *declination and right ascension,* can be thought of as celestial latitude and longitude with an earth reference (right ascension is calibrated in units of hour-angle, where 24 hours corresponds to a complete rotation of the earth). On Figure 4.21, the temperature contours range from a low of 90 K to a high of 1000 K. The measurements were

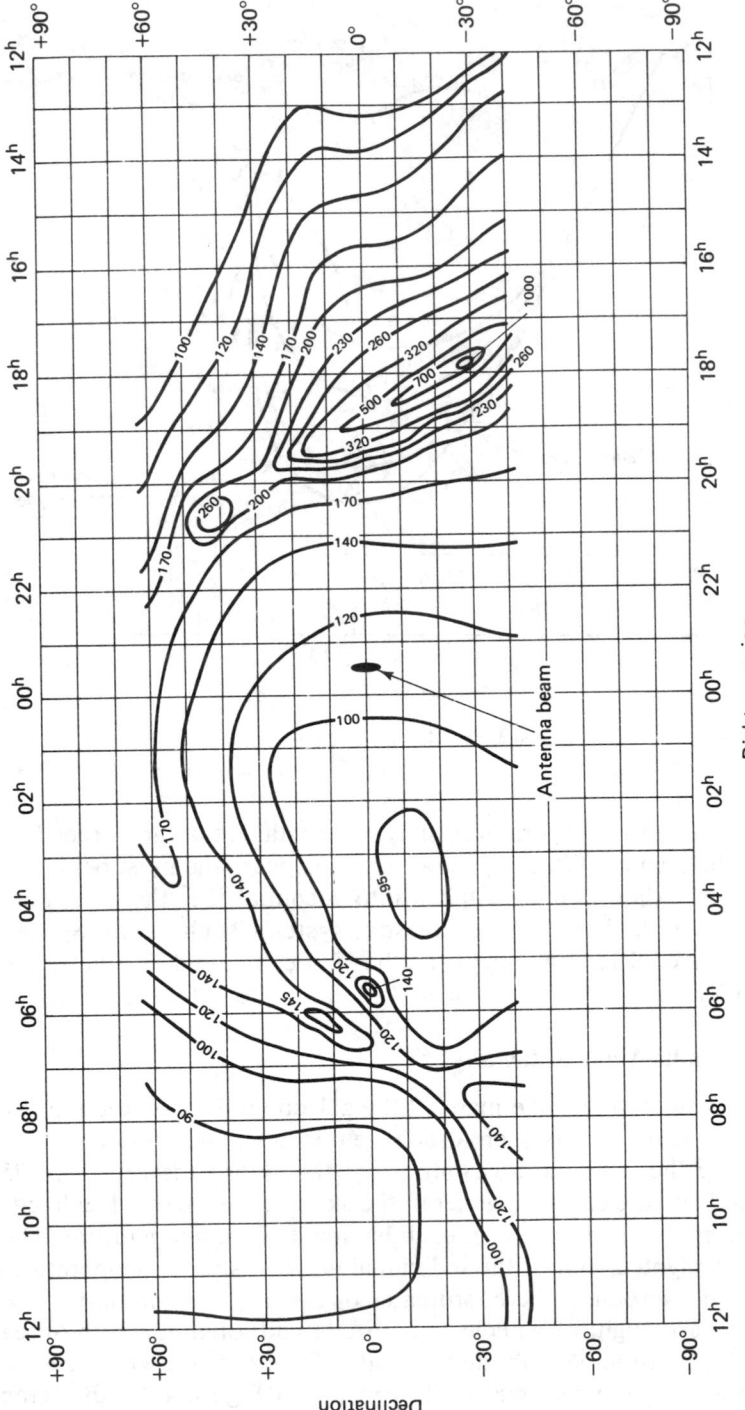

Figure 4.21 Radio map of the sky background at 250 MHz. (Reprinted from H. C. Ko and J. D. Kraus, "A Radio Map of the Sky at 1.2 Meters," *Sky Telesc.*, vol. 16, Feb. 1957, p. 160, with permission from *Sky and Telescope* astronomy magazine, Cambridge, Mass.)

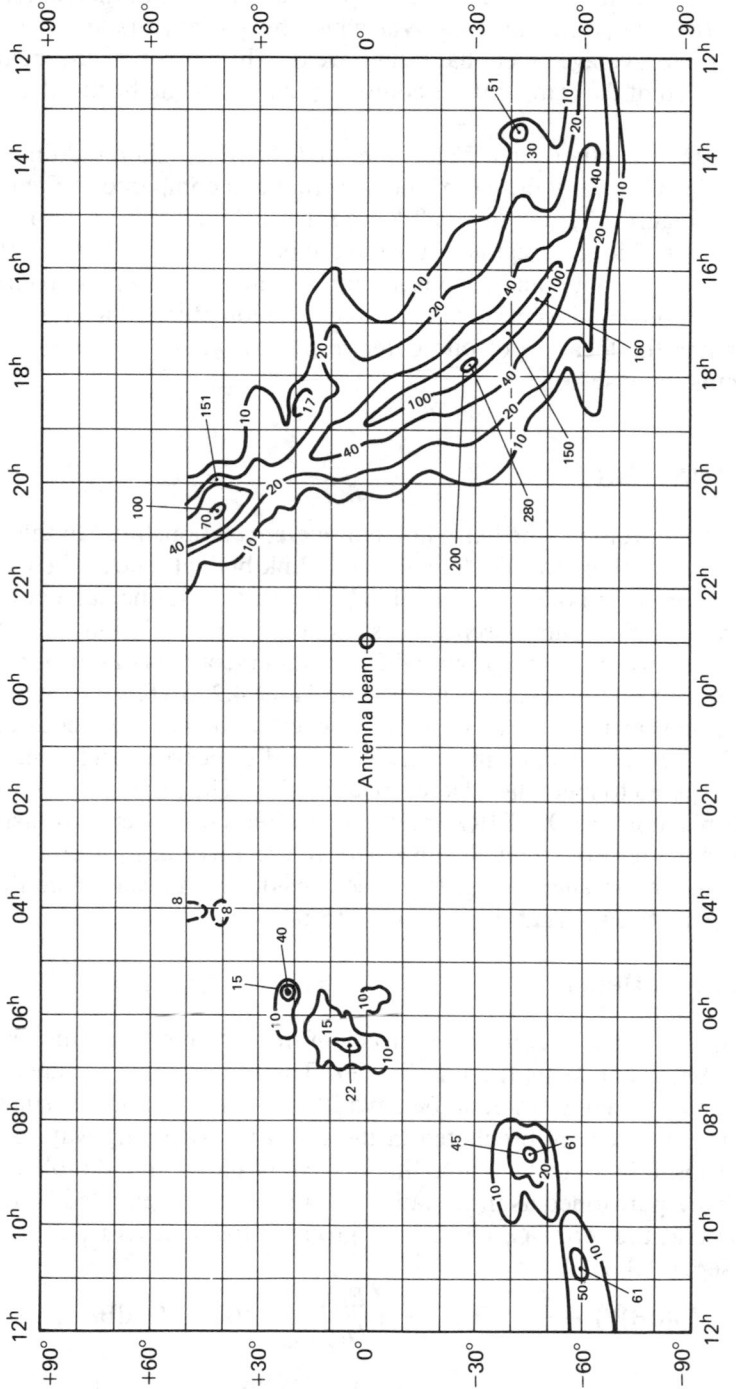

Figure 4.22 Radio map of the sky background at 600 MHz. (Reprinted with permission from J. H. Piddington and G. H. Trent, "A Survey of Cosmic Radio Emission at 600 Mc/s," *Aust. J. Phys.*, vol. 9, Dec. 1956, Fig. 1, pp. 483–486.)

made so as to exclude the sun (night sky). The antenna beam in the center of the map indicates the size of the sky area over which measurements are made (each measurement is an average over that beam area). The narrower the beam, the finer the resolution of the temperature contours; the wider the beam, the coarser the resolution.

Figure 4.22 is another such radio map at 600 MHz, after Piddington and Trent [13]. At this frequency the galaxy noise is reduced compared to Figure 4.21, as predicted in Figure 4.20; the low is 8 K and the high is 280 K. If you examine Figures 4.21 and 4.22 for the region of greatest noise radiation, where on the map do you see the most activity, and what is its significance? It is seen as an elongated region in the right-hand midsection of each map; the longitudinal axis of the elongation designates the location of our *galactic plane,* where such cosmic noise radiation is most intense.

4.6 SAMPLE LINK ANALYSIS

In Section 4.4 we developed the basic link parameter relationships. In this section we use these relationships to calculate a sample link budget, shown in Table 4.2. The table may appear to house a formidable listing of terms; one can get the false impression that the link budget represents a complex compilation. Just the opposite is true, and we introduce Figure 4.23 to underscore this assertion. In this figure we have reduced the set of line items from Table 4.2 to a few key parameters. The goal of a link analysis is to determine whether or not the required error performance is met, by examining the E_b/N_0 actually received and comparing it to the E_b/N_0 required to meet the system specification. The principal items needed for this determination are the EIRP (how much effective power is transmitted), the G/T° figure of merit (how much sensitivity the receiver has for collecting this power), L_s (the largest single loss, the space loss), and L_o (other contributing losses and degradations). That is *all* there is to it!

4.6.1 Link Budget Details

The link budget example in Table 4.2 consists of three columns of numbers. Only the middle column represents the link budget. The other columns consist of ancillary information, such as antenna beamwidth, or computations to support the main tabulation. Losses are bracketed in the usual bookkeeping way. Subtotals are shown enclosed in a box; the final link margin is shown in a double box. The computations are performed as in Equation (4.24), which is repeated below, with the exception that the terms G_r and T° are grouped together as G_r/T° instead of being listed separately.

$$M \text{ (dB)} = \text{EIRP (dBW)} + \frac{G_r}{T^\circ} \text{ (dB/K)} - \left(\frac{E_b}{N_0}\right)_{\text{reqd}} \text{ (dB)} - R \text{ (dB-bits/s)}$$
$$- \kappa \text{ (dBW/K-Hz)} - L_s \text{ (dB)} - L_o \text{ (dB)}$$

Let us examine the 21 line items listed in Table 4.2.

TABLE 4.2 Earth Terminal to Satellite Link Budget Example: Frequency = 8 GHz, Range = 21,915 Nautical Miles.

1.	Transmitter power (dBW)	(100.00 W)	20.0	P_t
2.	Transmitter circuit loss (dB)		⟨2.0⟩	L_o
3.	Transmitter antenna gain (peak dBi)		51.6	G_t
	Dish diameter (ft)	20.00		
	Half-power beamwidth (degrees)	0.45		
4.	Terminal EIRP (dBW)		⌑ 69.6 ⌑	EIRP
5.	Path loss (dB)	(10° elev.)	⟨202.7⟩	L_s
6.	Fade allowance (dB)		⟨4.0⟩	L_o
7.	Other losses (dB)		⟨6.0⟩	L_o
8.	Received isotropic power (dBW)		⌑ −143.1 ⌑	
9.	Receiver antenna gain (peak dBi)		35.1	G_r
	Dish diameter (ft)	3.00		
	Half-power beamwidth (degrees)	2.99		
10.	Edge of coverage loss (dB)		⟨2.0⟩	L_o
11.	Received signal power (dBW)		⌑ −110.0 ⌑	P_r
	Receiver noise figure at antenna port (dB)			11.5
	Receiver temperature (dB-K)			35.8 (3806 K)
	Receiver antenna temperature (dB-K)			24.8 (300 K)
12.	System temperature (dB-K)			36.1 (4106 K)
13.	System $G/T°$ (dB-K)		⌑ −1.0 ⌑	$G/T°$
14.	Boltzmann's constant (dBW/K-Hz)			−228.60
15.	Noise spectral density (dBW/Hz)		⟨−192.5⟩	$N_0 = kT°$
16.	Received P_r/N_0 (dB-Hz)		⌑ 82.5 ⌑	$(P_r/N_0)_r$
17.	Data rate (dB-bit/s)	(2 Mbits/s)	⟨63.0⟩	R
18.	Received E_b/N_0 (dB)		⌑ 19.5 ⌑	$(E_b/N_0)_r$
19.	Implementation loss (dB)		⟨1.5⟩	L_o
20.	Required E_b/N_0 (dB)		⟨10.0⟩	$(E_b/N_0)_{reqd}$
21.	Margin (dB)		⌑ 8.0 ⌑	M

1. Transmitter power is 100 W (20 dBW).
2. Circuit loss between the transmitter and antenna is 2 dB.
3. Transmitting antenna gain is 51.6 dBi.
4. The net tally of items 1 to 3 yields the EIRP = 69.6 dBW.
5. The path loss has been calculated for the range shown in the table title, corresponding to a 10° elevation angle at the earth terminal.
6 **and 7.** Here are allowances made for weather fades and a variety of other, unspecified losses.

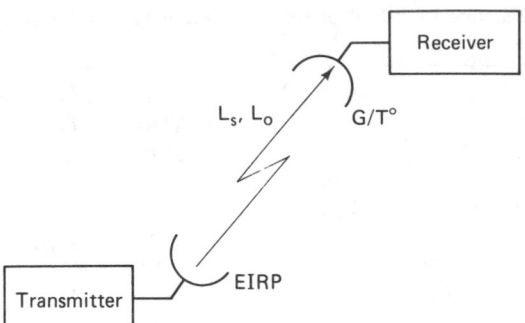

Figure 4.23 Key parameters of a link analysis.

8. Received isotropic power refers to the power that would be received, -143.1 dBW, if the receiving antenna were isotropic.

9. The peak gain of the receiving antenna is 35.1 dBi.

10. Edge of coverage loss is due to the off-axis antenna gain (compared to peak gain) and to the increased range for users at the extreme edge of communication coverage (a nominal 2-dB loss is shown here.)

11. The input power to the receiver, tallied from items 8, 9, and 10, is -110 dBW.

12. Off to the side, in column 3, we compute the system temperature T_S°, from the receiver noise figure and the antenna temperature.

13. We then form G/T° by combining the G_r in item 9, with T_S°.

14. Boltzmann's constant is -228.6 dBW/K-Hz.

15. Boltzmann's constant in decibels (item 14), plus system temperature in decibels (item 12), yields noise power spectral density.

16. Finally, we can form the received signal-to-noise spectral density, 82.5 dB-Hz, by subtracting noise spectral density in decibels (item 15), from received signal power in decibels (item 11).

17. The data rate is listed in dB-bit/s.

18. Since $E_b/N_0 = (1/R)(P_r/N_0)$, we need to subtract R in decibels (item 17), from P_r/N_0 in decibels (item 16), yielding $(E_b/N_0)_r = 19.5$ dB.

19. An implementation loss, here taken to be 1.5 dB, accounts for the difference between theoretically predicted detection performance, and the performance of the actual detector.

20. This is our required E_b/N_0, a result of the modulation and coding chosen, and the probability of error specified.

21. The difference between the received and the required E_b/N_0 in decibels (taking implementation loss into account), yields the final margin.

4.6.2 Receiver Figure of Merit

An explanation of why receiving antenna gain and system temperature have been grouped together as G/T° rather than kept separately, is as follows. In the early days of satellite communications development, the G_r and the T_S° were specified separately. A contractor who agreed to meet these specifications would need to allow himself some safety margin for meeting each specification. Even though the user was generally only interested in the "bottom-line" performance, and not in the explicit value of G_r or T_s°, the contractor would not be able to exploit potential trade-offs. The net result was an overspecified (more costly) system than was necessary. Recognition of such overspecification resulted in specifying the antenna and receiver front end as a single figure-of-merit parameter, G/T°, sometimes called the *receiver sensitivity,* such that cost-effective trade-offs between the antenna design and the receiver design might be employed.

4.6.3 Received Isotropic Power

Another recognized area of overspecification in receiver design is in the separate specification of the required P_R/N_0 (or E_b/N_0) and receiver G/T°. If P_r/N_0 and G/T° are specified separately, the system contractor is forced to meet each value. The contractor will plan for a margin in both places. As in the G/T° case of the preceding section, there are advantages in specifying P_R/N_0 and G/T° as one parameter; this new parameter, called the *received isotropic power* (RIP), can be written as follows:

$$\text{RIP (dBW)} = \frac{P_r}{N_0} \text{ (dB-Hz)} - \frac{G}{T^\circ} \text{ (dB/K)} + \kappa \text{ (dBW/K-Hz)} \qquad (4.49)$$

or, in terms of ratios,

$$\text{RIP} = \frac{P_r}{\kappa T^\circ}\left(\frac{\kappa T^\circ}{G_r}\right) = \frac{P_r}{G_r} \qquad (4.50)$$

It is important to note that P_r/N_0 refers to the predetection signal-to-noise spectral density ratio (SNR) *required* for a particular error probability when using a particular modulation scheme (it usually includes an allowance for *detector implementation losses*). Let us designate the theoretically required SNR to yield a particular P_B as $(P_r/N_0)_{\text{th-rq}}$. We can therefore write

$$\frac{P_r}{N_0} = L_o'\left(\frac{P_r}{N_0}\right)_{\text{th-rq}} \qquad (4.51)$$

where L_o' is called the implementation loss and accounts for the hardware and operational losses in the detection process. Combining Equations (4.50) and (4.51), we can write

$$\text{RIP} = L_o'\left(\frac{P_r}{\kappa T^\circ}\right)_{\text{th-rq}} \frac{\kappa T^\circ}{G_r} \qquad (4.52)$$

Specifying the RIP required to meet the system error performance allows the contractor to commit to meeting a single parameter value. The contractor is allowed to trade off P_r/N_0 versus G/T° and L_o' performance. As G/T° is improved, the detector performance can be degraded, and vice versa.

4.7 SATELLITE REPEATERS

Satellite repeaters retransmit the messages they receive (with a translation in carrier frequency). A *regenerative* (digital) repeater regenerates, that is, demodulates and reconstitutes the digital information embedded in the received waveforms before retransmission; however, a *nonregenerative* repeater only amplifies and retransmits. A nonregenerative repeater, therefore, can be used with many different modulation formats (simultaneously or sequentially without any switching), but a regenerative repeater is usually designed to operate with only one, or a very few, modulation formats. A link analysis for a regenerative satellite repeater treats the uplink and downlink as two separate point-to-point analyses. To calculate the overall bit error performance of a regenerative repeater link, it is necessary to determine separately the bit error probability on the uplink and downlink. Let P_u and P_d be the probability of a bit being in error on the uplink and downlink, respectively. A bit will be correct in the end-to-end link if either the bit is correct on both the up- and downlink, or if it is in error on both the up- and downlink. Therefore, the overall probability that a bit is correct, P_c, is

$$P_c = (1 - P_u)(1 - P_d) + P_u P_d \qquad (4.53)$$

and the overall probability that a bit is in error, P_B, is

$$P_B = 1 - P_c = P_u + P_d - 2P_u P_d \qquad (4.54)$$

For low values of P_u and P_d, the overall bit error performance is approximated simply by summing the individual uplink and downlink bit error probabilities:

$$P_B \simeq P_u + P_d \qquad (4.55)$$

4.7.1 Nonregenerative Repeaters

Link analysis for a nonregenerative repeater treats the entire "round trip" (uplink transmission to the satellite and downlink retransmission to an earth terminal) as a single analysis. Features that are unique to nonregenerative repeaters, are the dependence of the overall SNR on the uplink SNR and the sharing of the repeater downlink power in proportion to the uplink power from each of the various uplink signals and noise. Henceforth, reference to a repeater or transponder will mean a *nonregenerative repeater,* and for simplicity, we will assume that the transponder is operating in its linear range.

A satellite transponder is limited in transmission capability by its downlink power, the earth terminal's uplink power, satellite and earth terminal noise, and channel bandwidth. One of these usually is a dominant performance constraint; most often the downlink power or the channel bandwidth proves to be the major system limitation. Figure 4.24 illustrates the important link parameters of a linear satellite repeater channel. The repeater transmits all uplink signals (or noise, in the absence of signal) without any processing beyond amplification and frequency translation. Let us assume that there are multiple simultaneous uplinks within the receiver's bandwidth W and that they are separated from one another through

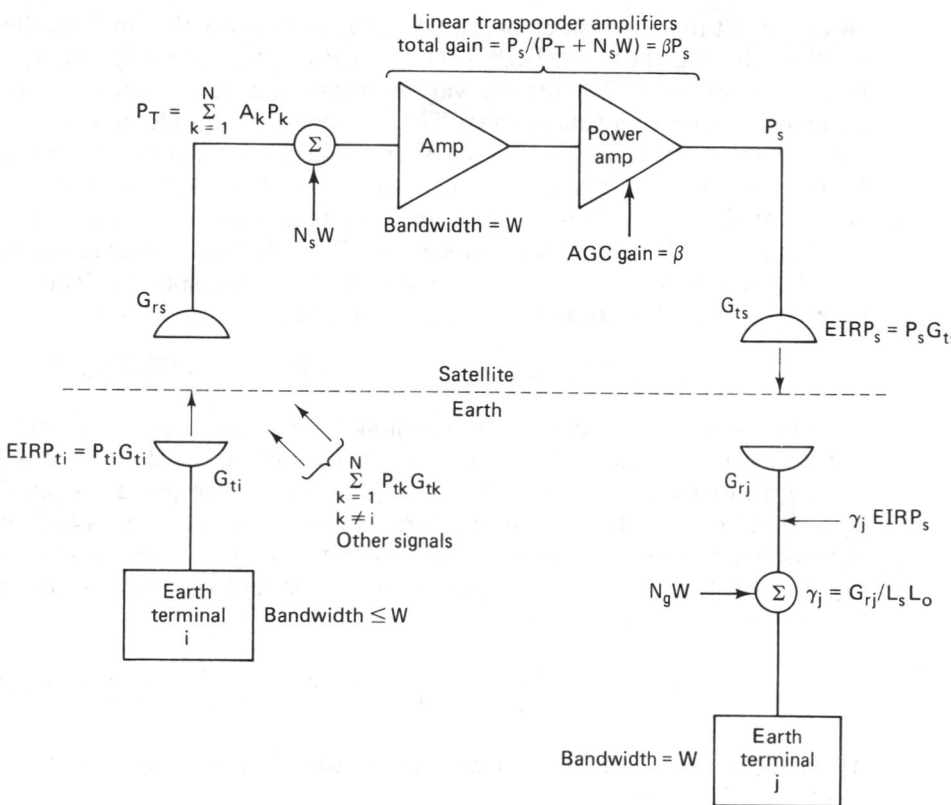

Figure 4.24 Nonregenerative satellite repeater.

the use of a technique called *frequency-division multiple access* (FDMA). FDMA is a communications resource-sharing technique whereby different users occupy disjoint portions of the transponder bandwidth; FDMA is treated in Chapter 9. The satellite effective downlink power $EIRP_s$ is constant and since we are assuming a linear transponder, $EIRP_s$ is shared among the multiple uplink signals (and noise) in proportion to their respective input power levels.

The transmission starts from a ground station (bandwidth $\leq W$), say terminal i, with a terminal $EIRP_{ti} = P_{ti}G_{ti}$. Simultaneously, other signals are being transmitted to the satellite (from other terminals). The EIRP from the kth terminal will henceforth be referred to simply as P_k. At the satellite, a total signal power $P_T = \sum A_k P_k$ is received, where A_k reflects the uplink propagation loss and the satellite receive antenna gain for each terminal. $N_s W$ is the satellite uplink noise power, where N_s is the composite noise power spectral density due to noise radiated into the satellite antenna *and* generated in the satellite receiver. The total satellite down-link $EIRP_s = P_s G_{ts}$, where P_s is the satellite transponder output power and G_{ts} is the satellite transmitting antenna gain, can be expressed by the following identity [14]:

$$EIRP_s = EIRP_s \beta [A_i P_i + (P_T - A_i P_i) + N_s W] \tag{4.56}$$

Both the left and right sides of Equation (4.56) express the total satellite EIRP. On the right side, the term $\beta[A_iP_i + (P_T - A_iP_i) + N_sW]$ constitutes the fractional apportionment of EIRP_s for the various users and uplink noise, such that the composite expression equals unity. The usefulness of this identity should become clear shortly. The total power gain of the transponder can be expressed as βP_s. Since P_s is fixed and the input signals can vary, $\beta = 1/(P_T + N_sW)$ represents an automatic gain control (AGC) term. The total received uplink signal power, P_T, has purposely been written as $A_iP_i + (P_T - A_iP_i)$ to separate signal i power from the remainder of the simultaneous signals in the transponder. The total power received at the jth earth terminal, with bandwidth W, can be written as

$$P_{rj} = \text{EIRP}_s\gamma_j\beta[A_iP_i + (P_T - A_iP_i) + N_sW] + N_gW \qquad (4.57)$$

where $\gamma_j = G_{rj}/L_sL_o$ accounts for downlink losses and receiving antenna gain for the jth earth terminal. $\text{EIRP}_s\gamma_j$ represents the portion of EIRP_s that is received by the jth earth terminal, and N_g is the downlink noise power spectral density generated and introduced into that terminal receiver. Equation (4.57) describes the essence of downlink power apportionment among the various users and noise in a repeater. Let us rewrite Equation (4.57) by replacing β with its equivalent $1/(P_T + N_sW)$, as follows:

$$P_{rj} = \text{EIRP}_s\gamma_j \left(\frac{A_iP_i}{P_T + N_sW} + \frac{P_T - A_iP_i}{P_T + N_sW} + \frac{N_sW}{P_T + N_sW} \right) + N_gW \qquad (4.58)$$

To facilitate our discussion, let us amplify Equation (4.58) with words

$$P_{rj} = \text{EIRP}_s\gamma_j \left(\frac{S_i \text{ U/L power}}{\text{total } (S + N) \text{ U/L power}} + \frac{\text{balance } S \text{ U/L power}}{\text{total } (S + N) \text{ U/L power}} \right.$$
$$\left. + \frac{\text{U/L noise power}}{\text{total } (S + N) \text{ U/L power}} \right) + N_gW$$

where S stands for signal power, N for noise power, and U/L for uplink.

From Equation (4.58), can you recognize an important relationship that must exist among multiple users sharing a nonregenerative transponder? The users must *cooperate with one another,* by not exceeding agreed-upon uplink transmission power levels. Equation (4.58) states that the portion of the downlink EIRP dedicated to any one user (or to uplink noise) is determined by the ratio of that user's uplink power to the total uplink signal plus noise power. Hence if one of the sharing users should choose to "cheat" by increasing his or her uplink power, the effect would be an enhancement of this user's downlink signal level, at the expense of the other users' downlink signal levels. Notice from Equation (4.58) that the uplink noise shares the downlink resource along with the other users. This coupling of uplink noise onto the downlink is a feature unique to nonregenerative repeaters.

From Equation (4.58) we can express the P_r/N for signal i received at the jth terminal as follows:

$$\left(\frac{P_r}{N}\right)_{ij} = \frac{\text{EIRP}_s\gamma_j[A_iP_i/(P_T + N_sW)]}{\text{EIRP}_s\gamma_j[N_sW/(P_T + N_sW)] + N_gW} \tag{4.59}$$

and we can write the overall P_r/N_0 for signal i received at the jth terminal as [14]

$$\left(\frac{P_r}{N_0}\right)_{ij} = \frac{\text{EIRP}_s\gamma_j\beta A_iP_i}{\text{EIRP}_s\gamma_j\beta N_s + N_g} \tag{4.60}$$

Equations (4.58) to (4.60) illustrate that the uplink repeater noise degrades the overall SNR in two ways—it "steals" downlink EIRP, and it contributes to the total system noise. When the satellite uplink noise dominates, that is, when $P_T \ll N_sW$, the link is said to be *uplink limited,* and most of the downlink EIRP$_s$ is wastefully allocated to uplink noise power. When this is the case and when EIRP$_s\gamma_j \gg N_gW$, we can rewrite Equation (4.60) as

$$\left(\frac{P_r}{N_0}\right)_{ij} \simeq \frac{\text{EIRP}_s\gamma_jA_iP_i/N_sW}{(\text{EIRP}_s\gamma_j/W) + N_g} \cong \frac{A_iP_i}{N_s} \tag{4.61}$$

Equation (4.61) illustrates that in the case of an uplink limited channel, the overall P_r/N_0 ratio essentially follows the uplink SNR. The more common situation is the *downlink limited* channel, in which case $P_T \gg N_sW$, and the satellite EIRP is limited. In this case Equation (4.60) can be rewritten as

$$\left(\frac{P_r}{N_0}\right)_{ij} \simeq \frac{\text{EIRP}_s\gamma_jA_iP_i/P_T}{N_g} \tag{4.62}$$

The power of the transponder is then shared primarily among the various uplink transmitted signals; very little uplink noise is transmitted on the downlink. The performance of the repeater, in this case, is constrained only by its downlink parameters.

Table 4.3 illustrates a link analysis example (full round trip) for a nonregenerative repeater. The uplink portion by itself does not constitute a link budget since the transmission is not demodulated at the satellite. Without demodulation, *there are no bits* and therefore there is no way to measure the bit-error performance. After the full round trip, the signal is demodulated at the earth terminal; only then does the link analysis yield the margin. The example in Table 4.3 represents a case where the satellite transponder is servicing 10 simultaneous users. In the block marked "A" is shown the ratio $P_r/(P_T + N_sW)$, which dictates the apportionment of the downlink EIRP for the signal of interest. In this example, with all users transmitting the same power level, each of the signals is allocated 9.8% of the downlink EIRP. In the block marked "B" we see the apportionment of the downlink EIRP. The total is 1514.7 W; the user of interest gets 148.5 W; the other nine users get a total of 1336.1 W; and the uplink noise is apportioned 30.1 W.

An estimate of the performance described in Equation (4.60) can be obtained by using the uplink and downlink values of E_b/N_0 (or P_r/N_0), combined as follows, in the *absence of intermodulation noise* [15]:

$$\left(\frac{E_b}{N_0}\right)_{ov}^{-1} = \left(\frac{E_b}{N_0}\right)_{u}^{-1} + \left(\frac{E_b}{N_0}\right)_{d}^{-1} \qquad (4.63)$$

where the subscripts ov, u, and d, indicate overall, uplink, and downlink values of E_b/N_0, respectively.

Most commercial satellite transponder designs are nonregenerative. However, it seems clear that future commercial systems will require on-board processing, switching, or selective message addressing, and will use regenerative repeaters to transform the received waveforms to message bits. Besides the potential for sophisticated data processing, one of the principal advantages of regenerative compared to nonregenerative repeaters is that the uplink is decoupled from the downlink so that the uplink noise is not retransmitted on the downlink. There are significant performance improvements possible with regenerative satellite repeaters in terms of the E_b/N_0 values needed on the uplinks and downlinks, relative to the values needed for the conventional nonregenerative designs in use today. Improvements of as much as 5 dB on the uplink and 6.8 dB on the downlink (using coherent QPSK modulation, with $P_B = 10^{-4}$) have been demonstrated [16].

4.7.2 Nonlinear Repeater Amplifiers

Power is severely limited in most satellite communication systems, and the inefficiencies associated with linear power amplification stages are expensive to bear. For this reason, many satellite repeaters employ nonlinear power amplifiers. Efficient power amplification is obtained at the cost of signal distortion due to nonlinear operation. The major undesirable effects of the repeater nonlinearities are:

1. Intermodulation (IM) noise due to the interaction of different carriers. The harm is twofold; useful power can be lost from the channel as IM energy (typically 1 to 2 dB), and spurious IM products can be introduced into the channel as interference. The latter problem can be quite serious.
2. AM-to-PM conversion is a phenomenon common to nonlinear devices such as traveling-wave tubes (TWT). Fluctuations in the signal envelope (amplitude modulation) produce phase variations that can affect the error performance, for systems using a coherent or differentially coherent modulation format.
3. In hard limiters, weak signals can be suppressed, relative to stronger signals, by as much as 6 dB [2]. In saturated TWTs, the suppression of weak signals is due not only to limiting, but also to the fact that the signal coupling mechanism of the tube is optimized in favor of the stronger signals. The effect can cause weak signals to be suppressed by as much as 18 dB [17].

Conventional nonregenerative repeaters are generally operated *backed-off* from their highly nonlinear saturated region; this is done to avoid appreciable IM noise and thus to allow efficient utilization of the system's entire bandwidth. However, backing off to the linear region is a compromise; some level of IM noise must be accepted to achieve a useful level of output power.

TABLE 4.3 Link Budget Example For a Nonregenerative Satellite Repeater with 10 Users: Uplink Frequency = 375 MHz, Downlink Frequency = 275 MHz, Range = 22,000 Nautical Miles

	Uplink		Downlink	
Transmitter power (dBW)	27.0	(500.0 W)	13.0	(20.0 W)
Transmitter circuit losses (dB)	1.0		1.0	
Transmitter antenna gain (peak-dBi)	19.0		19.8	
Dish diameter (ft)	10.00		15.00	
Half-power beamwidth (degrees)	19.16		17.42	
EIRP (dBW)	45.0		31.8	(1514.7 W)
Path loss (dB)	176.1		173.4	
Transmitted signal power (dBW)			21.7	(148.5 W)
Transmitted other signal power (dBW)			31.3	(1336.1 W)
Transmitted U/L noise power (dBW)			14.8	(30.1 W)
Other losses (dB)	2.0		2.0	
Received isotropic signal power (dBW)	−133.1		−153.7	
Received isotropic U/L noise power (dBW)			−160.6	
Receiver antenna gain (peak dBi)	22.5		16.3	
Dish diameter (ft)	15.00		10.00	
Half-power beamwidth (degrees)	12.77		26.13	
Received signal power (dBW)	−110.6		−137.4	
Received U/L noise power (dBW)			−144.3	
Receiver antenna temperature (dB/K)	24.6	(290 K)	20.0	(100 K)
Receiver noise figure at antenna port (dB)	10.8		2.0	
Receiver temperature (dB/K)	35.1	(3197 K)	22.3	(170 K)
System temperature (dB/K)	35.4	(3487 K)	24.3	(270 K)
System $G/T°$ (dB/K)	−12.9		−8.0	
Boltzmann's constant (dBW/K-Hz)	−228.6		−228.6	
Noise spectral density (dBW/Hz)	−193.2		−204.3	
System bandwidth (dB/Hz)	75.6	(36.0 MHz)	75.6	(36.0 MHz)
Noise power (dBW)	−117.6		−128.7	
U/L noise + D/L noise power (dBW)			−128.6	
Simultaneous accesses	10			
Received other signal power (dBW)	−101.1			
Other signals + noise (dBW)	−101.0			
$P_r/(P_T + N_sW)$ (dB)	−10.1	(0.098) A		
P_r/N (dB)	7.0		−8.7	
Overall P_r/N (dB)			−8.8	
P_r/N_0 (dB-Hz)	82.6		66.9	
Overall P_r/N_0 (dB-Hz)			66.8	
Data rate (dB-bit/s)			50.0	(100,000 bit/s)
Available E_b/N_0 (dB)			16.8	
Required E_b/N_0 (dB)			10.0	
Margin (dB)			6.8	

B

4.8 SYSTEM TRADE-OFFS

The link budget example in Table 4.3 is a resource allocation document. With such a link tabulation, one can examine potential system trade-offs and attempt to optimize system performance. A decibel is a decibel is a decibel; any decibel, from wherever it comes, is just as good as any other decibel in contributing to the overall performance requirements. The link budget is a natural starting point for considering all sorts of potential trade-offs: margin versus noise figure, antenna size versus transmitter power, and so on. Table 4.4 represents an example of a computer exercise for examining a possible trade-off between the earth station transmitting power and the system noise margin at the receiving terminal. The first row in the table is taken from the Table 4.3 link budget. Suppose a system engineer is concerned that a 500-W transmitter is not practical because of some physical constraints within the transmitting earth terminal; the engineer might then consider a trade-off of transmitter power versus thermal noise margin. The listing of candidate trade-offs is a trivial task for a computer. Table 4.4 was generated by repeating the link budget computation multiple times, and at each iteration, reducing P_t by one-half.

The result is a selection of transmitters (in steps of 3 dB) and uplink, downlink, and overall SNRs, and margin, associated with each transmitter value. The system engineer need only peruse the list to find a likely candidate. For example, if the engineer were satisfied with a margin of 3 to 4 dB, it appears he could reduce the transmitter from 500 W to 20 to 30 W. Or, he might be willing to provide a transmitter with, say, $P_t = 100$ W, since he may want to consider additional trade-offs (perhaps because of having misgivings about one of the other subsystems, say the antenna size). The engineer would then start a new tabulation with $P_t = 100$ W, and again perform a succession of link budget computations, to produce a similar enumeration of other possible trade-offs.

Notice from Table 4.4 that one can recognize the uplink-limited and downlink-limited regions, discussed earlier. In the first few rows, where the uplink SNR is high, a 3-dB degradation in uplink SNR results in only a few tenths of a

TABLE 4.4 Potential Trade-Off: P_t versus Margin

P_t (W)	$(P_r/N_0)_u$ (dB-Hz)	$(P_r/N_0)_d$ (dB-Hz)	$(P_r/N_0)_{ov}$ (dB-Hz)	Margin (dB)
500.0	82.6	66.9	66.8	6.8
250.0	79.6	66.8	66.6	6.6
125.0	76.6	66.6	66.2	6.2
62.5	73.6	66.3	65.5	5.5
31.3	70.5	65.7	64.5	4.5
15.6	67.5	64.8	62.9	2.9
7.8	64.5	63.3	60.8	0.8
3.9	61.5	61.4	58.4	−1.6
2.0	58.4	59.0	55.7	−4.3
1.0	55.4	56.4	52.9	−7.2
0.5	52.4	53.6	49.9	−10.1

decibel degradation to the overall SNR. Here the system is *downlink limited*; that is, the system is constrained primarily by its downlink parameters and is hardly affected by the uplink parameters. In the bottom few rows of the table, we see that a 3-dB degradation to the uplink affects the overall SNR by almost 3 dB. Here the system is *uplink limited*; that is, the system is constrained primarily by the uplink parameters.

4.9 CONCLUSION

Of the many analyses that support a developing communication system, the link budget stands out in its ability to provide overall system insight. By examining the link budget, one can learn many things about the overall system design and performance. For example, from the link margin, one learns whether the system will meet its requirements comfortably, marginally, or not at all. It will be evident if there are any hardware constraints, and whether such constraints can be compensated for in other parts of the link. The link budget is often used for considering system trade-offs and configuration changes, and in understanding subsystem nuances and interdependencies. Together with other modeling techniques, the link budget can help predict weight, size, and cost. We have considered how to formulate this budget and how it might be used for system trade-offs. The link budget is one of the system manager's most useful documents; it represents a "bottom-line" tally in the search for optimum system performance.

REFERENCES

1. Panter, P. F., *Communication Systems Design: Line-of-Sight and Tropo-Scatter Systems,* R. E. Krieger Publishing Co., Inc., Melbourne, Fla., 1982.
2. Jones, J. J., "Hard Limiting of Two Signals in Random Noise," *IEEE Trans. Inf. Theory,* vol. IT9, January 1963.
3. Silver, S., *Microwave Antenna Theory and Design,* MIT Radiation Laboratory Series, Vol. 12, McGraw-Hill Book Company, New York, 1949.
4. Kraus, J. D., *Antennas,* McGraw-Hill Book Company, New York, 1950.
5. Johnson, J. B., "Thermal Agitation of Electricity in Conductors," *Phys. Rev.,* vol. 32, July 1928, pp. 97–109.
6. Nyquist, H., "Thermal Agitation of Electric Charge in Conductors," *Phys. Rev.,* vol. 32, July 1928, pp. 110–113.
7. Desoer, C. A., and Kuh, E. S., *Basic Circuit Theory,* McGraw-Hill Book Company, New York, 1969.
8. Schwab, L. M., "World-Wide Link Availability for Geostationary and Critically Inclined Orbits Including Rain Attenuation Effects," *Lincoln Laboratory, Rep. DCA-9,* Jan. 27, 1981.
9. Friis, H. T., "Noise Figure of Radio Receivers," *Proc. IRE,* July 1944, pp. 419–422.

10. IRE Subcommittee 7.9 on Noise, "Description of the Noise Performance of Amplifiers and Receiving Systems," *Proc. IEEE,* Mar. 1963, pp. 436–442.

11. Blackwell, L. A., and Kotzebue, K. L., *Semiconductor Diode Parametric Amplifiers,* Prentice-Hall, Inc., Englewood Cliffs, N.J., 1961.

12. Ko, H. C., and Kraus, J. D., "A Radio Map of the Sky at 1.2 Meters," *Sky Telesc.,* vol. 16, Feb. 1957, pp. 160–161.

13. Piddington, J. H., and Trent, G. H., "A Survey of Cosmic Radio Emission at 600 Mc/s," *Aust. J. Phys.,* vol. 9, Dec. 1956, pp. 481–493.

14. Spilker, J. J., *Digital Communications by Satellite,* Prentice-Hall, Inc., Englewood Cliffs, N.J., 1977.

15. Pritchard, W. L., and Sciulli, J. A., *Satellite Communication Systems Engineering,* Prentice-Hall, Inc., Englewood Cliffs, N.J., 1986.

16. Campanella, S. J., Assal, F., and Berman, A., "Onboard Regenerative Repeaters," *Int. Conf. Commun.,* Chicago, vol. 1, 1977, pp. 6.2-121–66.2-125.

17. Wolkstein, H. J., "Suppression and Limiting of Undesired Signals in Travelling-Wave-Tube Amplifiers," Publication ST-1583, *RCA Rev.,* vol. 22, no. 2, June 1961, pp. 280–291.

PROBLEMS

4.1. **(a)** What is the value in decibels of the free-space loss for a carrier frequency of 100 MHz and a range of 3 miles?

(b) The transmitter output power is 10 W. Assume that both the transmitting and receiving antennas are isotropic and that there are no other losses. Calculate the received power in dBW.

(c) If in part (b) the EIRP is equal to 20 W, calculate the received power in dBW.

(d) If the diameter of a dish antenna is doubled, calculate the antenna gain increase in decibels.

(e) For the system of part (a), what must the diameter of a dish antenna be in order for the antenna gain to be 10 dB? Assume an antenna efficiency of 0.55.

4.2. A transmitter has an output of 2 W at a carrier frequency of 2 GHz. Assume that the transmitting and receiving antennas are parabolic dishes each 3 ft in diameter. Assume that the efficiency of each antenna is 0.55.

(a) Evaluate the gain of each antenna.

(b) Calculate the EIRP of the transmitted signal in units of dBW.

(c) If the receiving antenna is located 25 miles from the transmitting antenna over a free-space path, find the available signal power out of the receiving antenna in units of dBW.

4.3. From Table 4.1 we see that the proposal from Satellite Television Corporation called for a direct broadcast satellite (DBS) EIRP of 57 dBW and a downlink transmission frequency of 12.5 GHz. Assume that the only loss is the downlink space loss shown. Suppose that the downlink information consists of a digital signal with a data rate of 5×10^7 bits/s. Assume that the required E_b/N_0 is 10 dB, the system temperature at your home receiver is 600 K, and that your rooftop dish has an efficiency of 0.55. What is the minimum dish diameter that you can use in order to close the link? Do you think the neighbors will object?

4.4. An amplifier has an input and output resistance of 50 Ω, a 60-dB gain, and a bandwidth of 10 kHz. When a 50-Ω resistor at 290 K is connected to the input, the output rms noise voltage is 100 μV. Determine the effective noise temperature of the amplifier.

4.5. An amplifier has a noise figure of 4 dB, a bandwidth of 500 kHz, and an input resistance of 50 Ω. Calculate the input signal voltage needed to yield an output SNR = 1 when the amplifier is connected to a signal source of 50 Ω at 290 K.

4.6. Consider a communication system with the following specifications: transmission frequency = 3 GHz, modulation format is BPSK, bit-error probability = 10^{-3}, data rate = 100 bits/s, link margin = 3 dB, EIRP = 100 W, receiver antenna gain = 10 dB, distance between transmitter and receiver = 40,000 km. Assume that the line loss between the receiving antenna and the receiver is negligible.
  (a) Calculate the maximum permissible noise power spectral density in watts/hertz referenced to the receiver input.
  (b) What is the maximum permissible effective noise temperature in kelvin for the receiver if the antenna temperature is 290 K?
  (c) What is the maximum permissible noise figure in dB for the receiver?

4.7. A receiver preamplifier has a noise figure of 13 dB, a gain of 60 dB, and a bandwidth of 2 MHz. The antenna temperature is 490 K, and the input signal power is 10^{-12} W.
  (a) Find the effective temperature, in kelvin, of the preamplifier.
  (b) Find the system temperature in kelvin.
  (c) Find the output SNR in decibels.

4.8. Assume that a receiver has the following parameters: gain = 50 dB, noise figure = 10 dB, bandwidth = 500 MHz, input signal power = 50×10^{-12} W, source temperature, $T_A^\circ = 10$ K, line loss = 0 dB. You are asked to insert a preamplifier between the antenna and the receiver. The preamplifier is to have a gain of 20 dB and a bandwidth of 500 MHz. Find the preamplifier noise figure that would be required to provide a 10-dB improvement in overall system SNR.

4.9. Find the maximum allowable effective system temperature, T_s°, required to *just close* a particular link with a bit error probability of 10^{-5} for a data rate of $R = 10$ kbits/s. The link parameters are as follows: transmission frequency = 12 GHz, EIRP = 10 dBW, receiver antenna gain = 0 dB, modulation type is noncoherently detected BFSK, other losses = 0 dB, and the distance between transmitter and receiver = 100 km.

4.10. Consider a receiver made up of the following three steps: The input stage is a preamplifier with a gain of 20 dB and a noise figure of 6 dB. The second stage is a 3-dB lossy network. The output stage is an amplifier with a gain of 60 dB and a noise figure of 16 dB.
  (a) Find the composite noise figure for the receiver.
  (b) Repeat part (a) with the preamplifier removed.

4.11. (a) Find the effective input noise temperature, T_R°, of a receiver comprised of three amplifier stages connected in series with power gains, from input to output, of 10, 16, and 20 dB, and effective noise temperatures, from input to output, of 1800, 2700, and 4800 K.
  (b) What would the gain of the first stage have to be to reduce the contribution to T_R° of all stages after the first to 10% of the first-stage contribution?

4.12. The effective temperature of a particular multiple-stage receiver is required to be 300 K. Assume that the effective temperatures and gains of stages 2 through 4 are as follows: $T_2^\circ = 600$ K, $T_3^\circ = T_4^\circ = 2000$ K, $G_2 = 13$ dB, and $G_3 = G_4 = 20$ dB.

(a) Compute the required gain, G_1, for the first stage, under the conditions that T_1° = 200, 230, 265, 290, 295, and 300 K.

(b) Plot the G_1 versus T_1° trade-off.

(c) Regarding contributions to the effective temperature of the receiver, why is it reasonable in this case to ignore all stages beyond the fourth stage?

(d) In a practical engineering trade-off between T_1° and G_1, what range of T_1° values do you think should be considered?

4.13. A receiver consists of a preamplifier followd by multiple amplifier stages. The composite effective temperature of all the amplifier stages is 1000 K, referenced to the preamplifier output.

(a) Compute the receiver effective noise temperature, referenced to the preamplifier input, for a single-stage preamplifier with a noise temperature of 400 K and gains of 3, 6, 10, 16, and 20 dB.

(b) Repeat part (a) for a two-stage preamplifier with 400-K noise per stage and gains of 3, 6, 10, and 13 dB per stage.

(c) Plot the receiver effective temperature versus the gain of the first stage for parts (a) and (b).

4.14. Consider increasing the capacity of a working radio link by replacing the antennas with larger antennas. List three reasons why this might not be practical.

4.15. A receiver with 80-dB gain and an effective noise temperature of 3000 K is connected to an antenna that has a noise temperature of 600 K.

(a) Find the noise power that is available from the source over a 40-MHz band.

(b) Find the receiver noise power referenced to the receiver input.

(c) Find the receiver output noise power over a 40-MHz band.

4.16. An antenna is pointed in a direction such that it has a noise temperature of 50 K. It is connected to a preamplifier that has a noise figure of 2 dB and an available gain of 30 dB over an effective bandwidth of 20 MHz. The input signal to the preamplifier has a value of 10^{-12} W.

(a) Find the effective input noise temperature of the preamplifier.

(b) Find the SNR out of the preamplifier.

4.17. A receiver with a noise figure of 13 dB is connected to an antenna through 75 ft of 300-Ω transmission line that has a loss of 3 dB per 100 ft.

(a) Evaluate the composite noise figure of the line and the receiver.

(b) If a 20-dB preamplifier with a 3-dB noise figure is inserted between the line and the receiver, evaluate the composite noise figure of the line, the preamplifier, and the receiver.

(c) Evaluate the composite noise figure if the preamplifier is inserted between the antenna and the transmission line.

4.18. A satellite communication system uses a transmitter that produces 20 W of RF power at a carrier frequency of 8 GHz that is fed into a 2-ft parabolic antenna. The distance to the receiving earth station is 20,000 nautical miles. The receiving system uses an 8-ft parabolic antenna and has a 100-K system noise temperature. Assume that each antenna has an efficiency of 0.55. Also assume that the incidental losses amount to 2 dB.

(a) Calculate the maximum data rate that can be used if the modulation is differentially coherent PSK (DPSK) and the bit error probability is not to exceed 10^{-5}.

(b) Repeat part (a) assuming that the downlink transmission is at a carrier frequency of 2 GHz.

4.19. Consider that an unmanned spacecraft with a carrier frequency of 2 GHz and a 10-W transponder is in the vicinity of the planet Saturn (a distance of 7.9×10^8 miles from the earth). The receiving earth station has a 75-ft antenna and a system noise temperature of 20 K. Calculate the size of the spacecraft antenna that would be required to just close a 100-bits/s data link. Assume that the required E_b/N_0 is 10 dB and that there are incidental losses amounting to 3 dB. Also assume that each antenna has an efficiency of 0.55.

4.20. (a) Assume a receiver front end with the following parameters: gain = 60 dB, bandwidth = 500 MHz, noise figure = 6 dB, input signal power = 6.4×10^{-11} W, source temperature, T_A° = 290 K, line loss = 0 dB. A preamplifier with the following characteristics is inserted between the antenna and the receiver: gain = 10 dB, noise figure = 1 dB. Find the composite receiver noise figure, in decibels. How much noise figure improvement, in decibels, has been realized?

  (b) Find the output SNR improvement, in decibels, as a result of the improved noise figure.

  (c) Repeat part (b) for T_A° = 6000 K. What is the output SNR improvement in decibels?

  (d) Repeat part (b) for T_A° = 15 K. What is the output SNR improvement in decibels?

  (e) What conclusions can you draw from your answers with regard to how the improvement in output SNR tracks the improvement in noise figure? Explain.

4.21. (a) Given the following link parameters, find the maximum allowable receiver noise figure. The modulation is coherent BPSK with a bit-error probability of 10^{-5} for a data rate of 10 Mbits/s. The transmission frequency is 12 GHz. The EIRP is 0 dBW. The receiving antenna diameter is 0.1 m (assume an efficiency of 0.55), and the antenna temperature is 800 K. The distance between the transmitter and receiver is 10 km. The margin is 0 dB and the incidental losses are assumed to be 0 dB.

  (b) If the data rate is doubled, how will that affect the value of the noise figure in part (a)?

  (c) If the antenna diameter is doubled, how will that affect the value of the noise figure in part (a)?

4.22. (a) Ten users simultaneously access a nonregenerative satellite repeater with a 50-MHz bandwidth using an FDMA access scheme. Assume that each user's EIRP is 10 dBW; also assume that each user's coefficient, $A_i = G_{rs}/L_sL_o = -140$ dB. What is the total power, P_T, received by the satellite receiver?

  (b) Assume that the satellite system noise temperature is 2000 K. What is the value of the satellite receiver noise power in watts, referenced to the receiver input?

  (c) What is the uplink SNR at the satellite receiver for each user's signal?

  (d) Assuming the received power at the satellite from each user is the same, what fraction of the satellite EIRP is allocated to each of the 10 users' signals? If the satellite downlink $EIRP_s$ = 1000 W, how many watts per user is downlinked?

  (e) How much of the satellite EIRP is allocated to the transmission of uplink thermal noise?

  (f) Is the satellite uplink limited or downlink limited? Explain.

  (g) At the earth station, the receiver noise temperature is 800 K. What is the resultant (overall) average signal-to-noise power spectral density (P_r/N_0) for a single user's transmission across a 50-MHz band? Assume that the coefficient $\gamma = G_r/L_sL_o$ = -140 dB.

(h) Recalculate P_r/N_0 for a single user's transmission, using an approximation resulting from your answer to part (f).

(i) In the absence of intermodulation noise, the following repeater relationship is often used.

$$\text{overall} \left(\frac{P_r}{N_0}\right)^{-1} = \text{uplink} \left(\frac{P_r}{N_0}\right)^{-1} + \text{downlink} \left(\frac{P_r}{N_0}\right)^{-1}$$

Recalculate P_r/N_0 using this relationship, and compare the result with your answers to parts (g) and (h):

4.23. How many users can simultaneously access a nonregenerative satellite repeater with a 100-MHz bandwidth, such that each user is allocated 50 W of the satellite's EIRP of 5000 W? At the satellite, the effective system temperature, $T_s^\circ = 3500$ K. Assume that each user's uplink EIRP is 10 dBW and that the $G_r/L_s L_o$ term that reduces this EIRP at the satellite receiver is equal to -140 dB for each user.

CHAPTER 5

Channel Coding:
Part 1

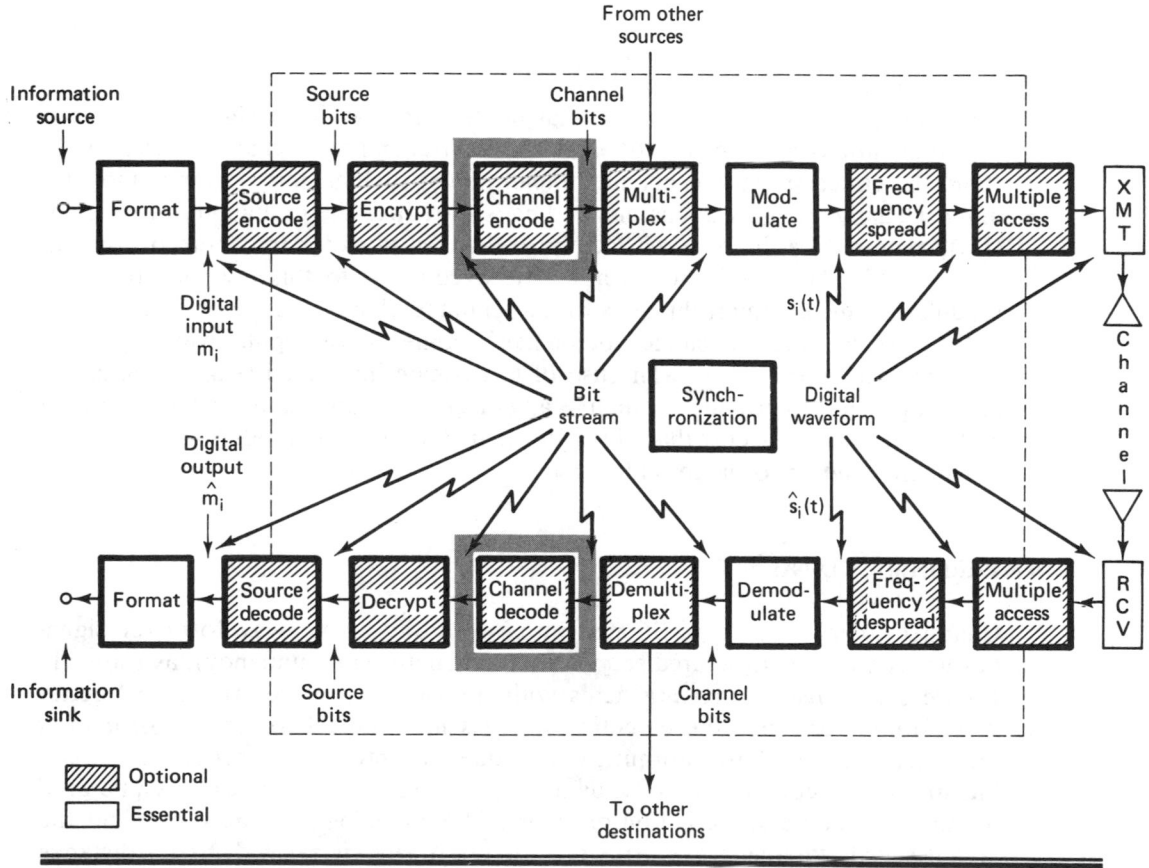

Channel coding refers to the class of signal transformations designed to improve communications performance by enabling the transmitted signals to better withstand the effects of various channel impairments, such as noise, fading, and jamming. Usually, the goal of channel coding is to reduce the probability of bit error (P_B), or to reduce the required E_b/N_0, at the cost of expending more bandwidth than would otherwise be necessary. The exceptions to this are the combined modulation and coding techniques for bandlimited channels described in Chapter 7. Why do you suppose channel coding has become such a popular way to provide performance improvement? The use of large-scale integrated (LSI) circuits has made it possible to provide as much as an 8-dB performance improvement through coding, at much less cost than through the use of other methods such as higher-power transmitters or larger antennas.

5.1 WAVEFORM CODING

Channel coding can be partitioned into two study areas, waveform (or signal design) coding and structured sequences (or structured redundancy), as shown in Figure 5.1. *Waveform coding* deals with transforming waveforms into "better waveforms," to make the detection process less subject to errors. *Structured sequences* deals with transforming data sequences into "better sequences," having structured redundancy (redundant bits). The redundant bits can then be used for the detection and correction of errors. The encoding procedure provides the coded signal (whether waveforms or structured sequences) with better distance

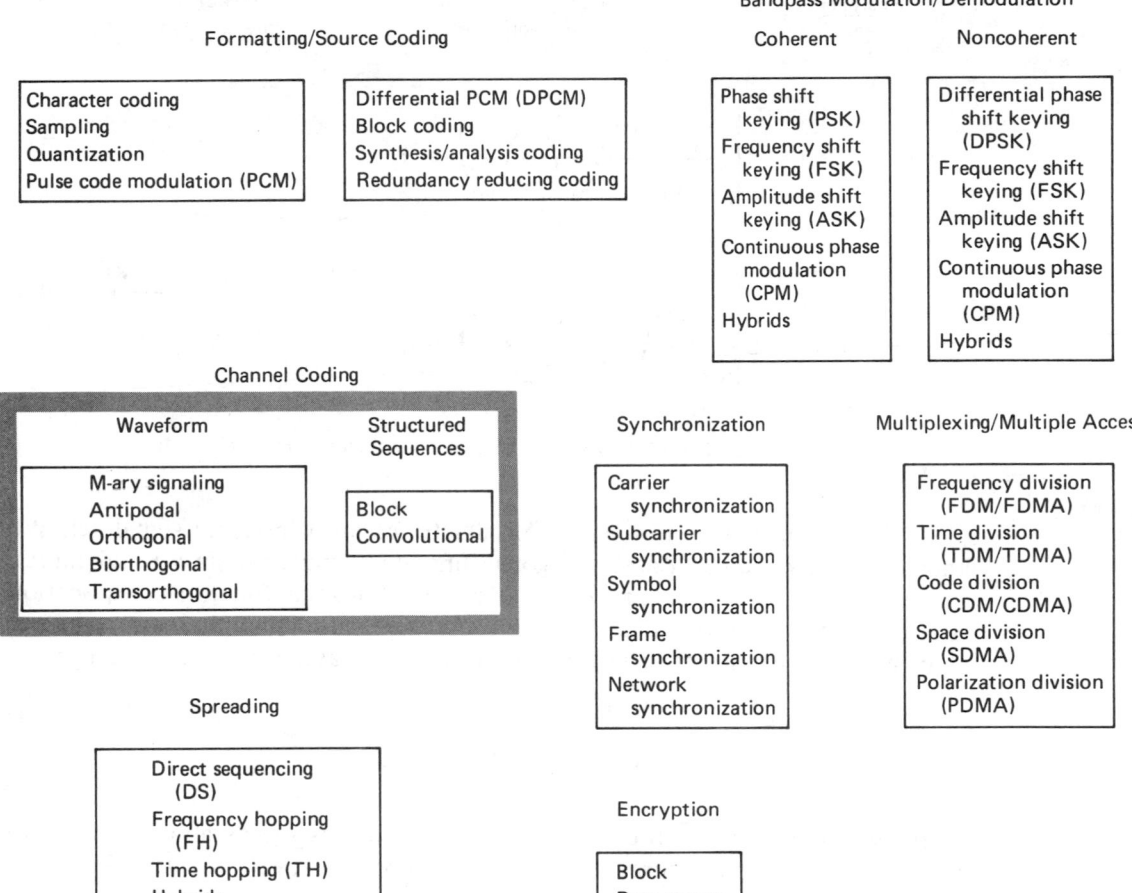

Figure 5.1 Basic digital communication transformations.

properties than those of their uncoded counterparts. First, we consider some waveform coding techniques. Then, starting with Section 5.3, we treat the more popular subject of structured sequences.

5.1.1 Antipodal and Orthogonal Signals

Antipodal and orthogonal signals have been discussed in Chapter 3; we shall repeat the paramount features of these signal classes. The example shown in Figure 5.2 illustrates the analytical representation, $s_1(t) = -s_2(t) = \sin \omega_0 t$, $0 \le t \le T$, of an antipodal signal set, as well as its waveform representation and its vectorial representation. What are some synonyms or analogies that are used to describe *antipodal signals*? We can say that such signals are mirror images, or that one signal is the negative of the other, or that the signals are 180° apart.

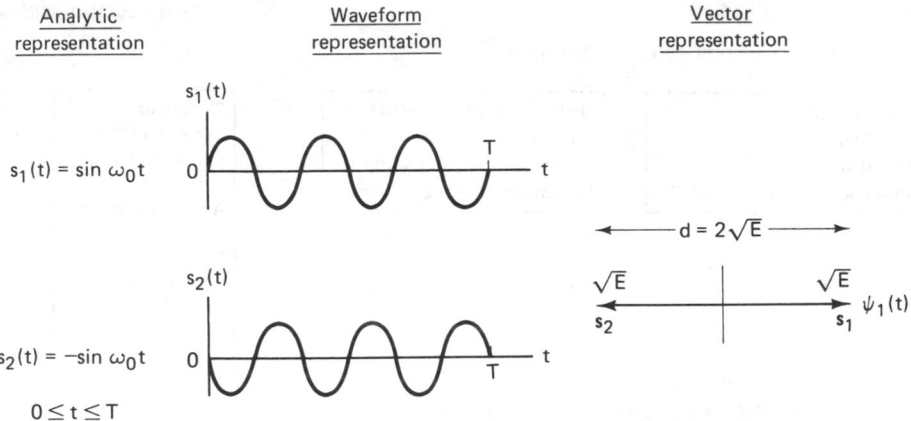

Figure 5.2 Example of an antipodal signal set.

The example shown in Figure 5.3 illustrates an orthogonal signal set. We know that sin x and cos x are orthogonal functions; similarly, sin mx and sin nx, where m and n are integers and $m \neq n$, are also orthogonal functions (see Section A.2.1). In Figure 5.3 we have chosen a pulse waveform example because it provides a clearer picture of orthogonality. The pulse waveform is described by

$$s_1(t) = p(t) \qquad\qquad 0 \leq t \leq T \qquad\qquad (5.1)$$

$$s_2(t) = p\left(t - \frac{T}{2}\right) \qquad 0 \leq t \leq T$$

where $p(t)$ is a pulse with duration $\tau = T/2$, and T is the symbol duration. In general, a set of equal energy signals $s_i(t)$, where $i = 1, 2, \ldots, M$, constitutes

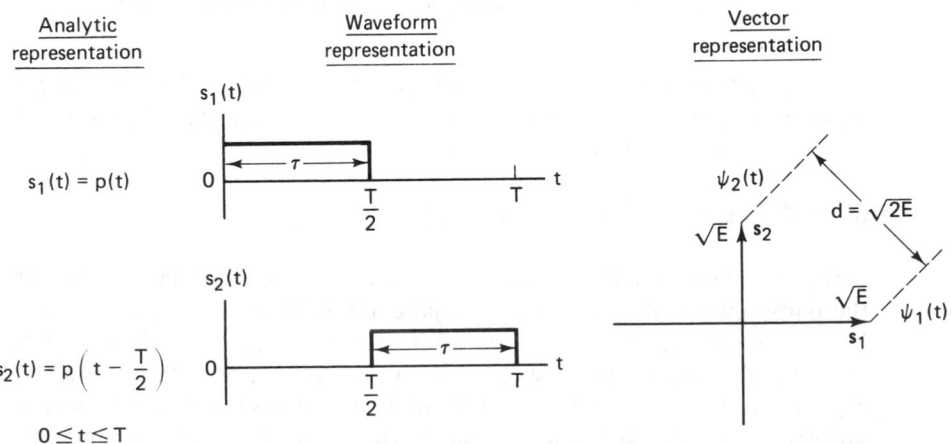

Figure 5.3 Example of a binary orthogonal signal set.

an orthogonal set, if, and only if,

$$z_{ij} = \frac{1}{E} \int_0^T s_i(t)s_j(t)\, dt = \begin{cases} 1 & \text{for } i = j \\ 0 & \text{otherwise} \end{cases} \tag{5.2}$$

where z_{ij} is called the *cross-correlation coefficient*, and where E is the signal energy expressed as

$$E = \int_0^T s_i^2(t)\, dt \tag{5.3}$$

The waveform representation in Figure 5.3 illustrates that $s_1(t)$ and $s_2(t)$ cannot interfere with one another because they are disjoint in time. The vectorial representation illustrates the perpendicular relationship between orthogonal signals. Let us consider some alternative descriptions of orthogonal signals or vectors. We can say that the inner or dot product of two different vectors in the orthogonal set must equal zero. In a two- or three-dimensional Cartesian coordinate space, we can describe the signal vectors, geometrically, as being mutually perpendicular to one another. We can say that one vector has zero projection on the other, or that one signal cannot interfere with the other, since they do not share the same *signal space*.

5.1.2 *M*-ary Signaling

With M-ary signaling, the processor accepts k data bits at a time. It then instructs the modulator to produce one of $M = 2^k$ waveforms; binary signaling is the special case where $k = 1$. For $k > 1$, M-ary signaling, as described in Chapter 3, can be regarded as a *waveform coding* procedure. For orthogonal signaling (e.g., MFSK), as k increases there will be an improved error performance or a reduction in required E_b/N_0, at the expense of bandwidth; nonorthogonal signaling (e.g., MPSK) can manifest improved bandwidth efficiency, at the expense of degraded error performance or an increase in required E_b/N_0. By the appropriate choice of signal waveforms, one can trade off error performance versus E_b/N_0 performance, versus bandwidth efficiency. Such trade-offs are treated in greater detail in Chapter 7.

5.1.3 Waveform Coding with Correlation Detection

Waveform coding procedures transform a waveform set into an improved waveform set. The improved waveform set can then be used to provide improved P_B compared to the original set. The most popular of such *waveform codes* are referred to as *orthogonal* and *biorthogonal* codes. The encoding procedure endeavors to make each of the waveforms in the coded signal set as unalike as possible; the goal is to render the cross-correlation coefficient, z_{ij}, among all pairs of signals, as described in Equation (5.2), as small as possible. The smallest possible value of the cross-correlation coefficient occurs when the signals are anticorrelated ($z_{ij} = -1$); however, this can be achieved only when the number of symbols in the set is two ($M = 2$) and the symbols are *antipodal*. In general, it is possible to make all the cross-correlation coefficients equal to zero [1]. The set is then said to be *orthogonal*. Antipodal signal sets are optimum in the sense that each signal is most distant from the other signal in the set; this is seen in Figure 5.2 where the distance, d, between signal vectors is seen to be $d = 2\sqrt{E}$, where E represents the signal energy during a symbol duration T, as expressed in Equation (5.3). Compared to antipodal signals, the distance properties of orthogonal signal sets can be thought of as "second best" (for a given level of waveform energy). In Figure 5.3 the distance between the orthogonal signal vectors is seen to be $d = \sqrt{2E}$.

The *cross-correlation* between two signals is a measure of the *distance* between the signal vectors. The smaller the cross-correlation, the more distant are the vectors from each other. This can be verified in Figure 5.2, where the antipodal signals (whose $z_{ij} = -1$) are represented by vectors that are most distant from each other, and in Figure 5.3, where the orthogonal signals (whose $z_{ij} = 0$) are represented by vectors that are closer to one another than the antipodal vectors. It should be obvious that the distance between two identical waveforms (whose $z_{ij} = 1$) is zero.

Figure 5.4 illustrates the replacement of a 2-bit data set with an improved (orthogonal) codeword set. Both the original data set and the codeword replacement set are comprised of the binary digits (1, 0). Also shown in the figure is the

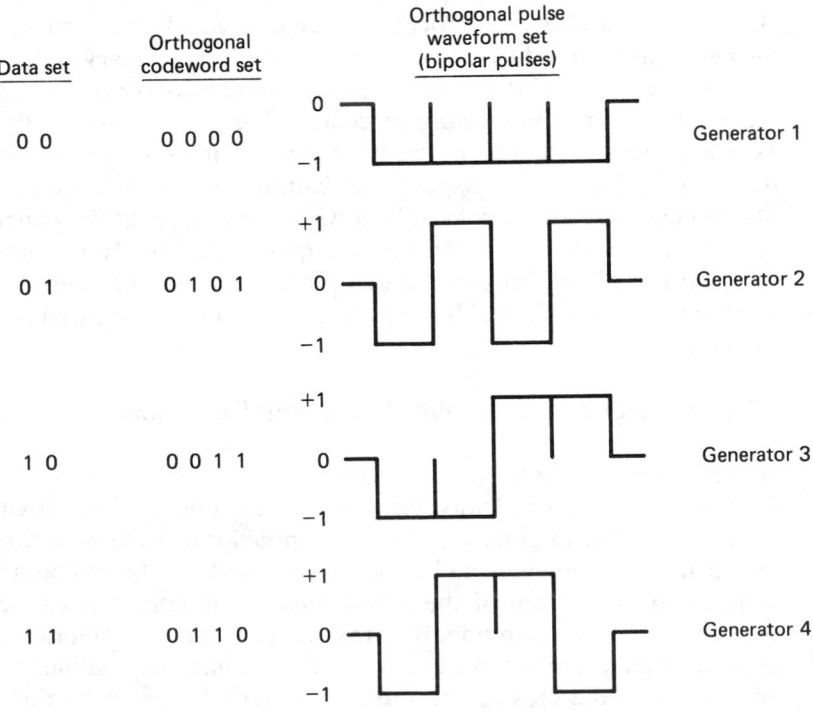

Figure 5.4 Replacement of data set with orthogonal codeword set and waveform set.

waveform set comprised of bipolar pulses $(+1, -1)$ that represents the codeword set. Equation (5.2) is stated in terms of waveforms. However, when the waveform set, $\{s_i(t)\}$, is represented by binary digits, it is easy to show that Equation (5.2) can be simplified as follows:

$$z_{ij} = \frac{\text{number of digit agreements } - \text{ number of digit disagreements}}{\text{total number of digits}} \tag{5.4}$$

$$z_{ij} = \begin{cases} 1 & \text{for } i = j \\ 0 & \text{otherwise} \end{cases}$$

where $i, j = 1, \ldots, M$, and M is the size of the codeword set. Using Equation (5.4), one can quickly verify that the codeword set in Figure 5.4 is orthogonal. Transmitting data with such an orthogonal set in place of the original data set results in larger distances among signaling waveforms, and thus yields better error performance for a given SNR.

Consider a set of $M = 2^k$ messages that are to be transmitted, using PSK modulation, over a channel disturbed by additive white Gaussian noise (AWGN). The transmitter shown in Figure 5.5, stores or generates the M pulse waveforms of the type shown in Figure 5.4. A message is transmitted by selecting one of the M waveform generators to *phase modulate* the carrier, such that the phase ($\phi_j = 0$ or π) of the carrier during each bit time, $0 \le t \le T_b$, corresponds to the amplitudes ($j = -1$ or 1) of the generating pulse waveform. At the receiver in Figure 5.6 the noisy signal is demodulated to baseband and fed to the M correlators (or matched filters). Correlation is performed over a codeword duration, $0 \le t \le T$, where $T = (\log_2 M)T_b = kT_b$. With orthogonally coded waveforms, in the absence of noise, the outputs of all correlators, except the one corresponding to the transmitted codeword, are zero.

5.1.4 Orthogonal Codes

A 1-bit data set can be transformed, using *orthogonal* codewords of two-digits each, described by the matrix $\mathbf{H}_1$ as follows:

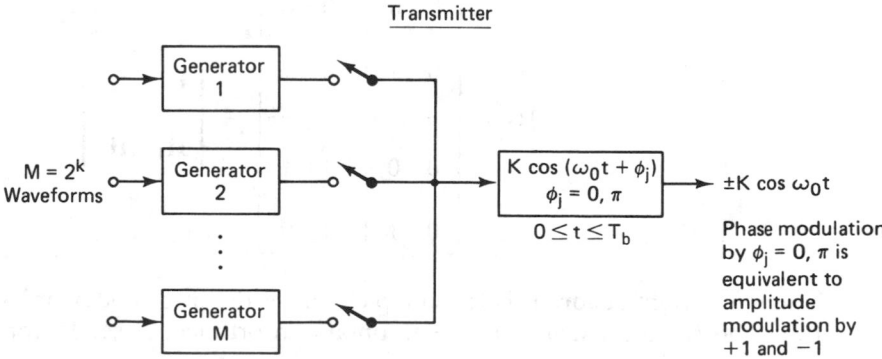

Figure 5.5 Waveform-encoded phase coherent system (transmitter).

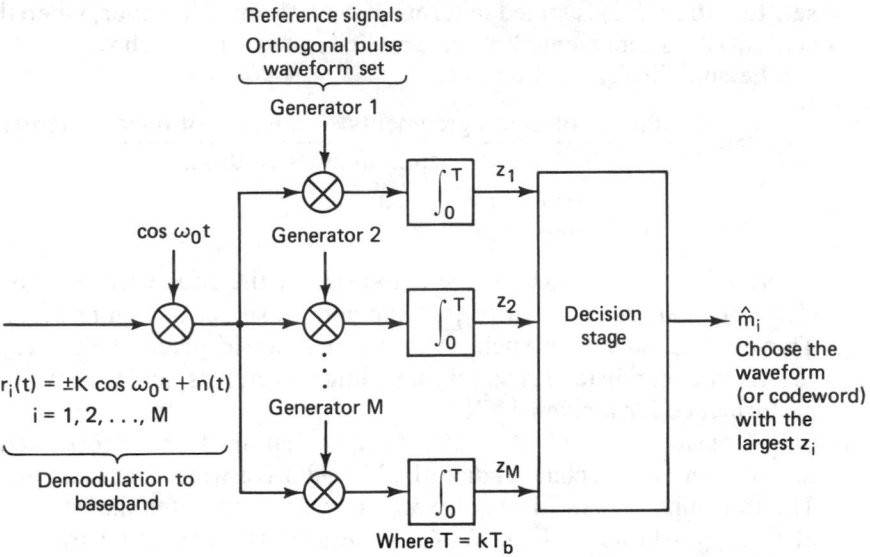

Figure 5.6 Waveform coding with correlation detection.

Data set	Orthogonal codeword set

$$
\begin{array}{c}
0 \\
1
\end{array}
\qquad
\mathbf{H}_1 =
\begin{bmatrix}
0 & 0 \\
0 & 1
\end{bmatrix}
$$

For this, and the following examples, use Equation (5.4) to verify the orthogonality of the codeword set. To encode a 2-bit data set, we extend the foregoing set both horizontally and vertically, creating matrix $\mathbf{H}_2$.

Data set	Orthogonal codeword set

$$
\begin{array}{c}
0\ 0 \\[4pt]
0\ 1 \\[4pt]
1\ 0 \\[4pt]
1\ 1
\end{array}
\qquad
\mathbf{H}_2 =
\left[
\begin{array}{cc:cc}
0 & 0 & 0 & 0 \\
0 & 1 & 0 & 1 \\
\hdashline
0 & 0 & 1 & 1 \\
0 & 1 & 1 & 0
\end{array}
\right]
=
\begin{bmatrix}
\mathbf{H}_1 & \mathbf{H}_1 \\
\mathbf{H}_1 & \overline{\mathbf{H}_1}
\end{bmatrix}
$$

The lower right quadrant is the complement of the prior codeword set. We continue the same construction rule to obtain an orthogonal set $\mathbf{H}_3$ for a 3-bit data set.

Data set	Orthogonal codeword set

$$
\mathbf{H_3} =
\left[
\begin{array}{cccc:cccc}
0 & 0 & 0 & 0 & 0 & 0 & 0 & 0 \\
0 & 1 & 0 & 1 & 0 & 1 & 0 & 1 \\
0 & 0 & 1 & 1 & 0 & 0 & 1 & 1 \\
0 & 1 & 1 & 0 & 0 & 1 & 1 & 0 \\
\hdashline
0 & 0 & 0 & 0 & 1 & 1 & 1 & 1 \\
0 & 1 & 0 & 1 & 1 & 0 & 1 & 0 \\
0 & 0 & 1 & 1 & 1 & 1 & 0 & 0 \\
0 & 1 & 1 & 0 & 1 & 0 & 0 & 1
\end{array}
\right]
=
\left[
\begin{array}{cc}
\mathbf{H_2} & \mathbf{H_2} \\
\mathbf{H_2} & \overline{\mathbf{H_2}}
\end{array}
\right]
$$

Data set rows (left column): 0 0 0, 0 0 1, 0 1 0, 0 1 1, 1 0 0, 1 0 1, 1 1 0, 1 1 1

In general, we can construct a codeword set, $\mathbf{H}_k$, of dimension $2^k \times 2^k$, called a *Hadamard matrix*, for a k-bit data set from the $\mathbf{H}_{k-1}$ matrix, as follows:

$$
\mathbf{H}_k =
\left[
\begin{array}{cc}
\mathbf{H}_{k-1} & \mathbf{H}_{k-1} \\
\mathbf{H}_{k-1} & \overline{\mathbf{H}_{k-1}}
\end{array}
\right]
$$

Each pair of words in each codeword set, $\mathbf{H}_1$, $\mathbf{H}_2$, $\mathbf{H}_3$, . . . , $\mathbf{H}_k$, . . . , has as many digit agreements as disagreements [2]. Hence, in accordance with Equation (5.4), $z_{ij} = 0$ (for $i \neq j$), and each of the sets is orthogonal.

Just as M-ary signaling with an orthogonal modulation format (such as MFSK) improves the P_B performance, waveform coding with an orthogonally constructed signal set, in combination with correlation detection, produces *exactly the same* improvement. For equally likely, equal-energy orthogonal signals, the probability of codeword (symbol) error can be upper bounded, as follows [2]:

$$
P_E(k) \leq (2^k - 1)Q\left(\sqrt{\frac{kE_b}{N_0}} \right) \tag{5.5}
$$

where $Q(x)$ is defined in Equation (2.42). For fixed k, as E_b/N_0 is increased, the bound becomes increasingly tight. For $P_E(k) \leq 10^{-3}$, Equation (5.5) is a good approximation of the error probability. The relationship between $P_B(k)$ and $P_E(k)$ given in Equation (3.127) is repeated here:

$$\frac{P_B(k)}{P_E(k)} = \frac{2^{k-1}}{2^k - 1} \qquad (5.6)$$

Combining Equations (5.5) and (5.6), the probability of bit error can be bounded as follows:

$$P_B(k) \leq (2^{k-1})Q\left(\sqrt{\frac{kE_b}{N_0}}\right) \qquad (5.7)$$

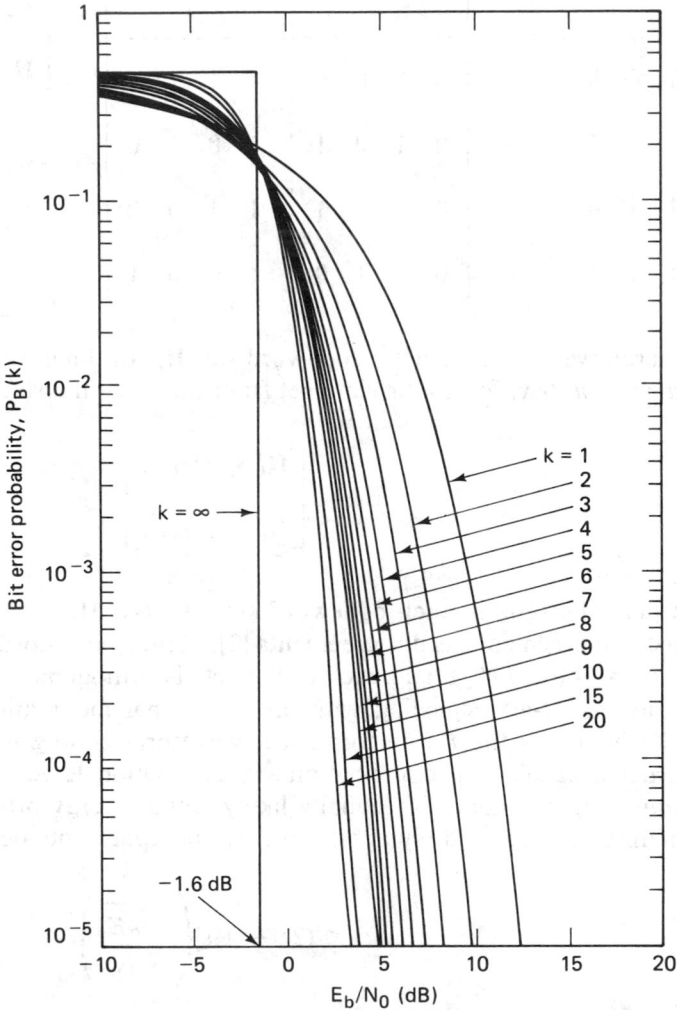

Figure 5.7 Coherent detection of orthogonally coded transmission. (Reprinted from W. C. Lindsey and M. K. Simon, *Telecommunication Systems Engineering*, Prentice-Hall, Inc., Englewood Cliffs, N.J., 1973, courtesy of W. C. Lindsey and Marvin K. Simon.)

$P_B(k)$ is plotted in Figure 5.7 for various values of k; the uncoded case corresponds to the $k = 1$ curve. The performance improvement for $k > 1$ should be obvious. The curves are identical to the orthogonal signaling performance (such as FSK) of Figure 3.25. What price do we pay for this improvement? We need to expend more transmission bandwidth. The orthogonal codes can be described as having $(2^k - k)$ redundant digits. For example, the orthogonal $\mathbf{H}_3$ matrix above reassigns 3-bit messages into 8-bit codewords, resulting in five redundant digits. Therefore, the bandwidth is increased by $\frac{8}{3}$ or, in general, by $2^k/k$. For orthogonal codes, the required transmission bandwidth increases exponentially with k. Compared to structured sequences, this type of coding *does not utilize bandwidth efficiently*.

5.1.5 Biorthogonal Codes

A *biorthogonal* signal set of M total signals or codewords can be obtained from an orthogonal set of $M/2$ signals by augmenting it with the negative of each signal, as follows:

$$\mathbf{B}_k = \begin{bmatrix} \mathbf{H}_{k-1} \\ \hline \mathbf{H}_{k-1} \end{bmatrix}$$

For example, a 3-bit data set can be transformed into a biorthogonal codeword set as follows:

Data set	Biorthogonal codeword set

$$
\begin{array}{ccc}
0 & 0 & 0 \\
\\
0 & 0 & 1 \\
\\
0 & 1 & 0 \\
\\
0 & 1 & 1 \\
\\
1 & 0 & 0 \\
\\
1 & 0 & 1 \\
\\
1 & 1 & 0 \\
\\
1 & 1 & 1 \\
\end{array}
\qquad
\mathbf{B}_3 =
\begin{bmatrix}
0 & 0 & 0 & 0 \\
0 & 1 & 0 & 1 \\
0 & 0 & 1 & 1 \\
0 & 1 & 1 & 0 \\
\hdashline
1 & 1 & 1 & 1 \\
1 & 0 & 1 & 0 \\
1 & 1 & 0 & 0 \\
1 & 0 & 0 & 1 \\
\end{bmatrix}
$$

The biorthogonal set is really two sets of orthogonal codes such that each codeword in one set has its antipodal codeword in the other set. The biorthogonal set consists of a *combination of orthogonal and antipodal signals*. With respect to z_{ij} of Equations (5.2) or (5.4), biorthogonal codes can be characterized as

$$
z_{ij} = \begin{cases} 1 & \text{for } i = j \\ -1 & \text{for } i \neq j, \, |\,i - j\,| = \dfrac{M}{2} \\ 0 & \text{for } i \neq j, \, |\,i - j\,| \neq \dfrac{M}{2} \end{cases} \tag{5.8}
$$

One advantage of a biorthogonal code over an orthogonal one for the same data set, is that the biorthogonal code requires *one-half* as many bits per codeword (compare the $\mathbf{B}_3$ matrix with the $\mathbf{H}_3$ matrix). Thus the bandwidth requirements for

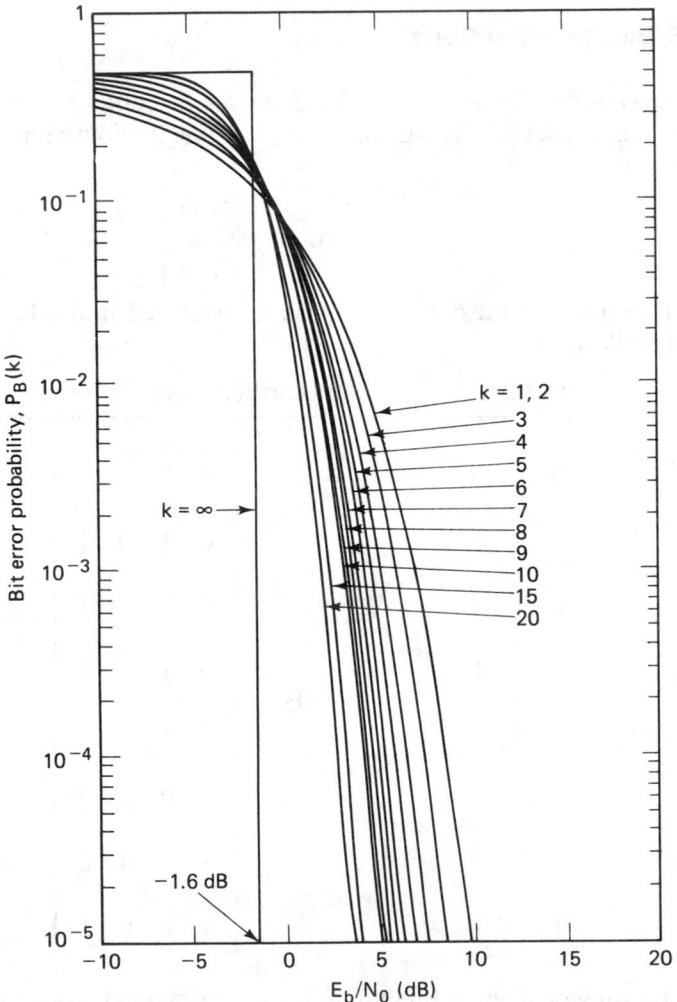

Figure 5.8 Coherent detection of biorthogonally coded transmission. (Reprinted from W. C. Lindsey and M. K. Simon, *Telecommunication Systems Engineering*, Prentice-Hall, Inc., Englewood Cliffs, N.J., 1973, courtesy of W. C. Lindsey and Marvin K. Simon.)

Channel Coding: Part 1   Chap. 5

biorthogonal codes are one-half the requirements for comparable orthogonal ones. Since antipodal signal vectors have better distance properties than orthogonal ones, it should come as no surprise that biorthogonal codes perform slightly better than orthogonal ones. For equally likely, equal-energy biorthogonal signals, the probability of codeword (symbol) error can be upper bounded, as follows [2]:

$$P_E(k) \le (2^k - 2)Q\left(\sqrt{\frac{kE_b}{N_0}}\right) + Q\left(\sqrt{\frac{2kE_b}{N_0}}\right) \qquad (5.9)$$

which becomes increasingly tight for fixed k as E_b/N_0 is increased. $P_B(k)$ is a complicated function of $P_E(k)$; we can approximate it with the relationship [2]

$$P_B(k) \simeq \frac{P_E(k)}{2}$$

The approximation is quite good for $k > 3$. Therefore, we can write

$$P_B(k) \lesssim \tfrac{1}{2}\left[(2^k - 2)Q\left(\sqrt{\frac{kE_b}{N_0}}\right) + Q\left(\sqrt{\frac{2kE_b}{N_0}}\right)\right] \qquad (5.10)$$

The P_B performance of these biorthogonal codes, shown in Figure 5.8, offers improved performance, compared to the performance of the orthogonal codes shown in Figure 5.7, and requires only *half the bandwidth* of orthogonal codes.

5.1.6 Transorthogonal (Simplex) Codes

A code generated from an orthogonal set by deleting the first digit of each codeword is called a *transorthogonal* or *simplex code*. Such a code is characterized by

$$z_{ij} = \begin{cases} 1 & \text{for } i = j \\ \dfrac{-1}{M-1} & \text{for } i \ne j \end{cases} \qquad (5.11)$$

A simplex code represents the *minimum energy* equivalent (in the error probability sense) of the equally likely orthogonal set. In comparing the error performance of orthogonal, biorthogonal, and simplex codes, we can state that simplex coding requires the minimum E_b/N_0 for a specified symbol error rate. However, for a *large value of k*, all three schemes are *essentially identical* in error performance. Biorthogonal coding requires half the bandwidth of the others. However, for each of these codes, bandwidth requirements (and system complexity) grow exponentially with the value of k; therefore, such coding schemes are attractive only when large bandwidths are available. When bandwidth is not plentiful, the structured redundancy techniques (see Section 5.3 through Chapter 6) are more attractive [3]. When bandwidth is *very scarce*, the so-called combined modulation and coding techniques for bandlimited channels are most promising (see Sections 7.10.6 and 7.10.7).

5.2 TYPES OF ERROR CONTROL

Before we discuss the details of structured redundancy, let us describe the two basic ways such redundancy is used for controlling errors. The first, *error detection and retransmission*, utilizes *parity bits* (redundant bits added to the data) to detect that an error has been made. The receiving terminal does not attempt to correct the error; it simply requests the transmitter to retransmit the data. Notice that a two-way link is required for such dialogue between the transmitter and receiver. The second type of error control, *forward error correction* (FEC), requires a one-way link only, since in this case the parity bits are designed for both the detection and correction of errors. We shall see that not all error patterns can be corrected; error-correcting codes are classified according to their error-correcting capabilities.

5.2.1 Terminal Connectivity

Communication terminals are often classified according to their connectivity with other terminals. The possible connections, shown in Figure 5.9, are termed *simplex* (not to be confused with the simplex or transorthogonal codes), *half-duplex*, and *full-duplex*. The simplex connection, in Figure 5.9a, is a one-way link. Transmissions are made from terminal A to terminal B only, never in the reverse direction. The half-duplex connection, in Figure 5.9b, is a link whereby transmissions may be made in either direction but not simultaneously. Finally, the full-duplex connection, in Figure 5.9c, is a two-way link, where transmissions may proceed in both directions simultaneously.

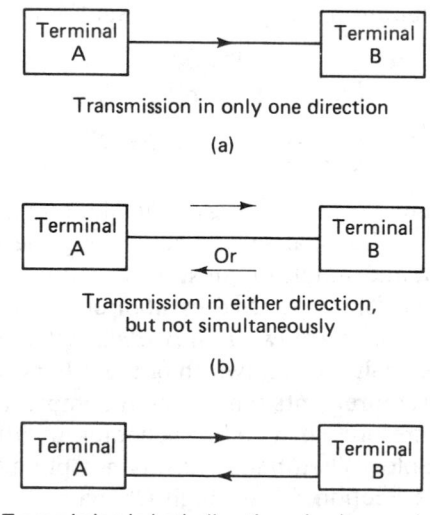

Figure 5.9   Terminal connectivity classifications. (a) Simplex. (b) Half-duplex. (c) Full-duplex.

5.2.2 Automatic Repeat Request

When the error control consists of error detection only, the communication system generally needs to provide a means of alerting the transmitter that an error has been detected and that a retransmission is necessary. Such error control procedures are known as *automatic repeat request* or automatic retransmission query (ARQ) methods. Figure 5.10 illustrates three of the most popular ARQ procedures. In each of the diagrams, time is advancing from left to right. The first procedure, called *stop-and-wait ARQ*, is shown in Figure 5.10a. It requires a half-duplex connection only, since the transmitter waits for an acknowledgment (ACK) of each transmission before it proceeds with the next transmission. In the figure, the third transmission block is received in error; therefore, the receiver responds with a negative acknowledgment (NAK), and the transmitter retransmits this third

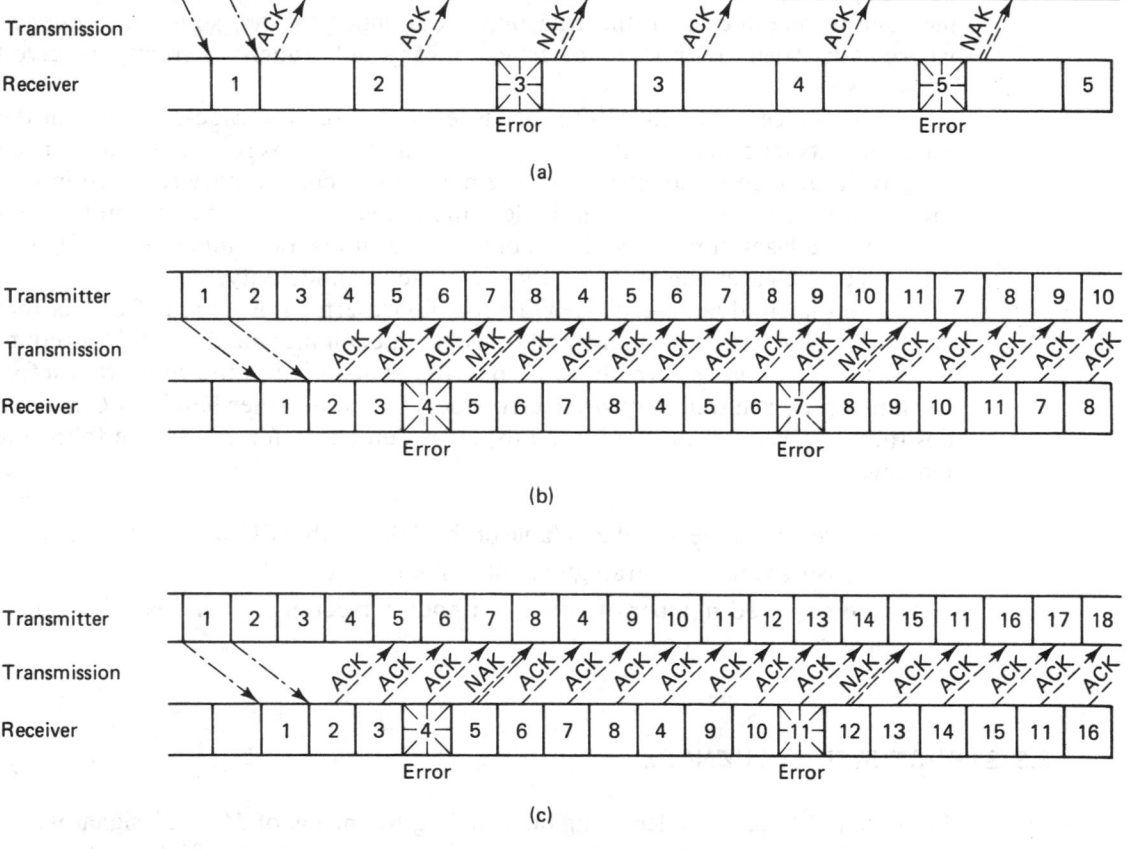

Figure 5.10 Automatic repeat request (ARQ). (a) Stop-and-wait ARQ (half-duplex). (b) Continuous ARQ with pullback (full-duplex). (c) Continuous ARQ with selective repeat (full-duplex).

message block before transmitting the next in the sequence. The second ARQ procedure, called *continuous ARQ with pullback*, is shown in Figure 5.10b. Here a full-duplex connection is necessary. Both terminals are transmitting simultaneously; the transmitter is sending message data and the receiver is sending acknowledgment data. Notice that a sequence number has to be assigned to each block of data. Also, the ACKs and NAKs need to reference such numbers, or else there needs to be a priori knowledge of the propagation delays so that the transmitter knows which messages are associated with which acknowledgments. In the example of Figure 5.10b there is a fixed separation of four blocks between the message being transmitted and the acknowledgment being simultaneously received. For example, when message 8 is being sent, a NAK corresponding to the corrupted message 4 is being received. In this ARQ procedure, the transmitter "pulls back" to the message in error and retransmits all message data, starting with the corrupted message. The final method, called *continuous ARQ with selective repeat*, is shown in Figure 5.10c. Here, as with the second ARQ procedure, a full-duplex connection is needed. However, in this procedure, only the corrupted message is repeated; then the transmitter continues the transmission sequence where it had left off instead of repeating any subsequent correctly received messages.

The choice of which ARQ procedure to choose is a trade-off between the requirements for efficient utilization of the communications resource and the need to provide full-duplex connectivity. The half-duplex connectivity required in Figure 5.10a is less costly than full-duplex; the associated inefficiency can be measured by the blank time slots. The more efficient utilization illustrated in Figures 5.10b and c requires the more costly full-duplex connectivity.

The major advantage of ARQ over forward error correction (FEC) is that error detection requires much simpler decoding equipment and much less redundancy than does error correction. Also, ARQ is adaptive in the sense that information is retransmitted only when errors occur. On the other hand, FEC may be desirable in place of, or in addition to, error detection, for any of the following reasons:

1. A reverse channel is not available or the delay with ARQ would be excessive.
2. The retransmission strategy is not conveniently implemented.
3. The expected number of errors, without corrections, would require excessive retransmissions.

5.3 STRUCTURED SEQUENCES

In Section 3.8 we considered digital signaling by means of $M = 2^k$ signal waveforms (M-ary signaling), where each waveform contains k bits of information. We saw that in the case of orthogonal M-ary signaling, we can decrease P_B by increasing M (expanding the bandwidth). Similarly, in Section 5.1 we showed that it is possible to decrease P_B by encoding k binary digits into one of M orthogonal codewords. The major disadvantage with such orthogonal coding techniques is the associated inefficient use of bandwidth. The required transmission bandwidth grows exponentially with k for an orthogonal set of $M = 2^k$ waveforms. In this and subsequent sections we abandon the need for antipodal or orthogonal properties and focus on a class of encoding procedures known as *parity-check codes*. Such channel coding procedures are classified as *structured sequences* because they represent methods of inserting structured redundancy into the source data so that the presence of errors can be detected or the errors corrected. Structured sequences are partitioned into two important subcategories as shown in Figure 5.1: *block coding* and *convolutional coding*. Block coding (primarily) is treated in this chapter, and convolutional coding is treated in Chapter 6. These techniques allow us to attain a P_B performance comparable to waveform encoding techniques but with lower bandwidth requirements. The codewords of these codes (structured sequences) are usually *nonorthogonal* [3].

5.3.1 Channel Models

5.3.1.1 Discrete Memoryless Channel

A *discrete memoryless channel* (DMC) is characterized by a discrete input alphabet, a discrete output alphabet, and a set of conditional probabilities, $P(j|i)$ ($1 \le i \le M$, $1 \le j \le Q$), where i represents a modulator M-ary input symbol, j represents a demodulator Q-ary output symbol, and $P(j|i)$ is the probability of receiving j given that i was transmitted. Each output symbol of the channel depends only on the corresponding input, so that for a given input sequence $\mathbf{U} = u_1, u_2, \ldots, u_m, \ldots, u_N$ the conditional probability of a corresponding output sequence $\mathbf{Z} = z_1, z_2, \ldots, z_m, \ldots, z_N$ may be expressed as

$$P(\mathbf{Z}|\mathbf{U}) = \prod_{m=1}^{N} P(z_m|u_m) \tag{5.12}$$

In the event that the channel *has memory* (i.e., noise or fading that occurs in bursts), the conditional probability of the sequence $\mathbf{Z}$ would need to be expressed as the *joint* probability of all the elements of the sequence. Equation (5.12) expresses the *memoryless* condition of the channel. Since the channel noise in a memoryless channel is defined to affect each symbol independently of all the other symbols, the conditional probability of $\mathbf{Z}$ is seen as the product of the independent element probabilities.

5.3.1.2 Binary Symmetric Channel

A *binary symmetric channel* (BSC) is a special case of a DMC; the input and output alphabet sets consist of the binary elements (0 and 1). The conditional probabilities are symmetric:

$$
\begin{aligned}
P(0|1) &= P(1|0) = p \\
P(1|1) &= P(0|0) = 1 - p
\end{aligned}
\tag{5.13}
$$

Equation (5.13) states the channel *transition probabilities*. That is, given that a channel symbol was transmitted, the probability that it is received in error is p (related to the symbol energy), and the probability that it is received correctly is $(1 - p)$. Since the demodulator output consists of the discrete elements 0 and 1, the demodulator is said to make a firm or *hard decision* on each symbol. A commonly used code system consists of BPSK modulated coded data, hard decision demodulated. Then the channel symbol error probability is found using the methods discussed in Section 3.7.1 and Equation (3.84) to be

$$p = Q\left(\sqrt{\frac{2E_c}{N_0}}\right)$$

where E_c/N_0 is the channel symbol energy per noise density, and $Q(x)$ is defined in Equation (2.42).

When such hard decisions are used in a binary coded system, the demodulator feeds the two-valued *code symbols* or *channel bits* to the decoder. Since the decoder then operates on the hard decisions made by the demodulator, decoding with a BSC channel is called *hard-decision decoding*.

5.3.1.3 Gaussian Channel

We can generalize our definition of the DMC to channels with alphabets that are not discrete. An example is the *Gaussian channel* with a discrete input alphabet and a continuous output alphabet over the range $(-\infty, \infty)$. The channel adds noise to the symbols. Since the noise is a Gaussian random variable, with zero mean and variance σ^2, the resulting probability density function (pdf) of the received random variable z, conditioned on the symbol u_k (the likelihood of u_k), can be written

$$p(z|u_k) = \frac{1}{\sigma\sqrt{2\pi}} \exp\left[\frac{-(z - u_k)^2}{2\sigma^2}\right] \tag{5.14}$$

for all z, where $k = 1, 2, \ldots, M$. For this case, *memoryless* has the same meaning as it does in Section 5.3.1.1, and Equation (5.12) can be used to obtain the conditional probability for the sequence, **Z**.

When the demodulator output consists of a continuous alphabet or its quantized approximation (with greater than two quantization levels), the demodulator is said to make *soft decisions*. In the case of a coded system, the demodulator feeds such quantized code symbols to the decoder. Since the decoder then operates on the soft decisions made by the demodulator, decoding with a Gaussian channel is called *soft-decision decoding*.

In the case of a hard-decision channel, we are able to characterize the detection process with a channel symbol error probability. However, in the case of a soft-decision channel, the detector makes the kind of decisions (soft decisions) that cannot be labeled as correct or incorrect. Thus, since there are no firm decisions, there cannot be a probability of making an error; the detector can only formulate a family of conditional probabilities or likelihoods of the different symbol types.

It is possible to design decoders using soft decisions, but block code soft-decision decorders are substantially more complex than hard-decision decoders; therefore, block codes are usually implemented with hard-decision decoders. For convolutional codes, both hard- and soft-decision implementations are equally popular. In this chapter we consider that the channel is a binary symmetric channel (BSC), and hence the decoder employs hard decisions. In Chapter 6 we further discuss channel models, as well as hard- versus soft-decision decoding for convolutional codes.

5.3.2 Code Rate and Redundancy

In the case of block codes, the source data are segmented into blocks of k data bits, also called information bits or message bits; each block can represent any one of 2^k distinct messages. The encoder transforms each k-bit data block into a larger block of n bits, called code bits or channel symbols. The $(n - k)$ bits, which the encoder adds to each data block, are called *redundant bits*, *parity bits*, or *check bits*; they carry no new information. The code is referred to as an (n, k) code. The ratio of redundant bits to data bits, $(n - k)/k$, within a block is called the *redundancy* of the code, and the ratio of data bits to total bits, k/n, is called the *code rate*. The code rate can be thought of as the portion of a code bit that constitutes information. For example, in a rate $\frac{1}{2}$ code, each code bit carries $\frac{1}{2}$ bit of information.

In this chapter and Chapter 6 we consider those coding techniques that provide redundancy by increasing the required transmission bandwidth. For example, an error control technique that employs a rate 1/2 code (100% redundancy) will require double the bandwidth of an uncoded system. However, if a rate 3/4 code is used, the redundancy is 33% and the bandwidth expansion is only 4/3. In Chapter 7 we consider modulation/coding techniques for bandlimited channels where complexity, instead of bandwidth, is traded for error performance improvement.

5.3.3 Parity-Check Codes

5.3.3.1 Single-Parity-Check Code

Parity-check codes use linear sums of the information bits, called *parity symbols* or *parity bits*, for error detection or correction. A single-parity check code is constructed by adding a single-parity bit to a block of data bits. The parity bit takes on the value of one or zero as needed to ensure that the summation of all the bits in the codeword yields an even (or odd) result. The summation operation is performed using modulo-2 arithmetic (exclusive-or logic), as described in Section 2.12.3. If the added parity is designed to yield an even result, the method is termed *even parity*, and if designed to yield an odd result, it is termed *odd*

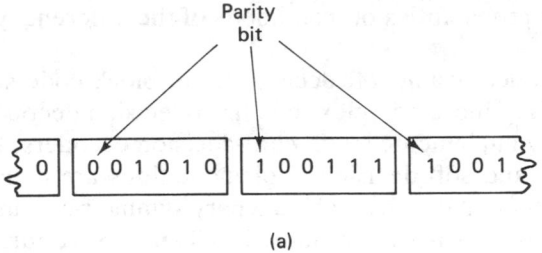

(a)

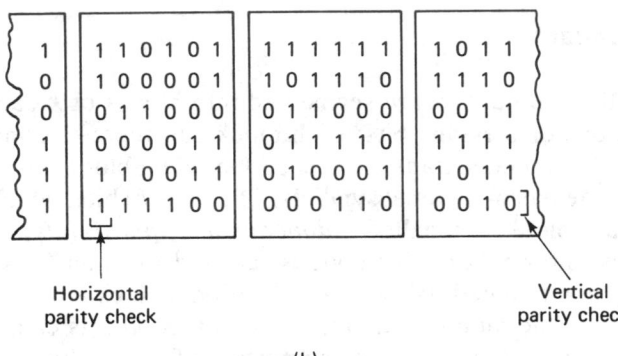

Horizontal
parity check

Vertical
parity check

(b)

Figure 5.11 Parity checks for serial and parallel transmission. (a) Serial transmission. (b) Parallel transmission.

parity. Figure 5.11a illustrates a serial data transmission (the rightmost bit is the earliest bit). A single-parity bit is added (the leftmost bit in each block) to yield even parity.

At the receiving terminal, the decoding procedure consists of testing that the modulo-2 sum of the codeword bits yields a zero result (even parity). If the result is found to be one instead of zero, the codeword is known to contain errors. The rate of the code can be expressed as $k/(k + 1)$. Do you suppose the decoder can automatically *correct* a digit that is received in error? No, it cannot. It can only *detect* the presence of an odd number of bit errors (if an even number of bits are inverted, the parity test will appear correct; this represents the case of an *undetected error*). Assuming that all bit errors are equally likely and occur independently, we can write the probability of j errors occurring in a block of n symbols as

$$P(j, n) = \binom{n}{j} p^j (1 - p)^{n-j} \tag{5.15}$$

where p is the probability that a *channel symbol* is received in error, and where

$$\binom{n}{j} = \frac{n!}{j!(n - j)!}$$

is the number of various ways in which j bits out of n may be in error. Thus for a single-parity error-detection code, the probability of an undetected error, P_{nd}, within a block of n bits is computed, as follows:

$$P_{nd} = \sum_{j=1}^{\substack{n/2 \text{ (for } n \text{ even)} \\ (n-1)/2 \text{ (for } n \text{ odd)}}} \binom{n}{2j} p^{2j}(1 - p)^{n-2j} \tag{5.16}$$

Example 5.1 Even-Parity Code

Configure a (4, 3) even-parity error-detection code such that the parity symbol appears as the leftmost symbol of the codeword. Which error patterns can the code detect? Compute the probability of an undetected message error, assuming that all symbol errors are independent events and that the probability of a channel symbol error is $p = 10^{-3}$.

Solution

Message	Parity	Codeword
000	0	0 000
100	1	1 100
010	1	1 010
110	0	0 110
001	1	1 001
101	0	0 101
011	0	0 011
111	1	1 111

$$\overbrace{\qquad}^{\nearrow} \overbrace{\qquad}^{\searrow}$$
parity   message

The code is capable of detecting all single- and triple-error patterns. The probability of an undetected error is equal to the probability that two or four errors occur anywhere in a codeword.

$$P_{nd} = \binom{4}{2} p^2 (1 - p)^2 + \binom{4}{4} p^4$$

$$= 6p^2 (1 - p)^2 + p^4$$

$$= 6p^2 - 12p^3 + 7p^4$$

$$= 6(10^{-3})^2 - 12(10^{-3})^3 + 7(10^{-3})^4 \approx 6 \times 10^{-6}$$

5.3.3.2 Rectangular Code

A *rectangular code*, also called a *product code*, can be thought of as a parallel data transmission, depicted in Figure 5.11b. First we form a rectangle of message bits comprised of M rows and N columns; then a horizontal parity check is appended to each row and a vertical parity check is appended to each column, resulting in an augmented array of dimensions $(M + 1) \times (N + 1)$. The rate of the rectangular code, k/n, can then be written as

$$\frac{k}{n} = \frac{MN}{(M + 1)(N + 1)} \tag{5.17}$$

How much more powerful is the rectangular code than the single-parity code, which is only capable of error detection? Notice that any single bit error will cause a parity check failure in one of the array columns *and* in one of the array rows. Therefore, the rectangular code can correct a single error pattern since the error is uniquely located at the intersection of the error-detecting row and the error-detecting column. For the example shown in Figure 5.11b, the array dimensions are $M = N = 5$; therefore, the figure depicts a (36, 25) code that can correct a single error located anywhere in the 36 bit positions. For an error-correcting block code, we compute the probability that the decoded block has an uncorrected error by accounting for all the ways in which a *message error* can be made. Starting with the probability of j errors in a block of n symbols, expressed in Equation (5.15), we can write the probability of a message error, also called a *block error* or *word error*, P_M, for a code that can correct all t and fewer error patterns:

$$P_M = \sum_{j=t+1}^{n} \binom{n}{j} p^j (1 - p)^{n-j} \qquad (5.18)$$

where p is the probability that a *channel symbol* is received in error. For the example in Figure 5.11b, the code can correct all single error patterns ($t = 1$) within the rectangular block of $n = 36$ bits. Hence the summation in Equation (5.18) starts with $j = 2$:

$$P_M = \sum_{j=2}^{36} \binom{36}{j} p^j (1 - p)^{36-j} \qquad (5.19)$$

When p is reasonably small, the first term in the summation is the dominant one; we can therefore write for this (36, 25) rectangular code example

$$P_M \simeq \binom{36}{2} p^2 (1 - p)^{34}$$

The *bit error probability*, P_B, depends on the particular code and decoder. An approximation for P_B is given in Section 5.5.3.

5.3.4 Coding Gain

Figure 5.12 illustrates the probability of bit error, P_B, versus E_b/N_0 for coherent binary PSK modulation in combination with examples of various (n, k) codes over a Gaussian channel. The (1, 1) curve illustrates the uncoded PSK performance, while the (24, 12) and (127, 92) curves illustrate coded PSK performance using block codes with $(n - k) = 12$ parity bits and 35 parity bits, respectively. From Figure 3.24 we know in which direction the waterfall-like curves move, corresponding to P_B performance improvement. Look at the various curves in Figure 5.12. Can you explain why the coded curves (to which we attribute P_B performance improvement) appear to be moving in the wrong direction when compared with the uncoded curve? Where does the strength of the code manifest itself? The curves in Figure 5.12 indicate that the strength of a code is seen only after an E_b/N_0 threshold has been exceeded (approximately 5.5 dB in this example). For values of E_b/N_0 less than the threshold, the coding manifests itself only as *over-*

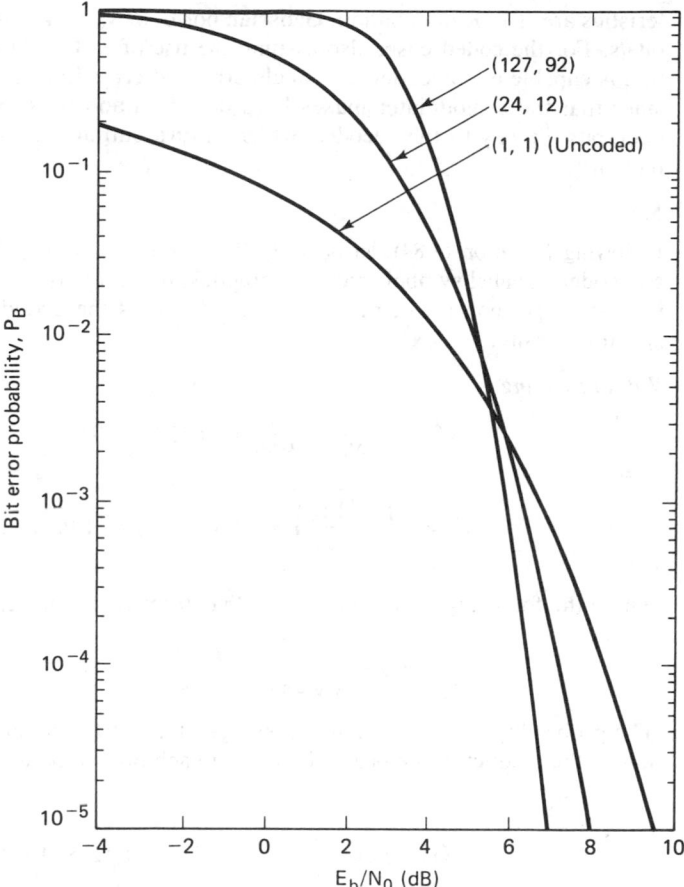

Figure 5.12 Coded versus uncoded bit error performance for coherent PSK with various (n, k) codes.

head bits resulting in *reduced energy per bit*, compared to the uncoded case; before the threshold is exceeded, the redundant bits are simply "excess baggage" without the ability to improve performance. Once the threshold is exceeded, the performance improvement of the code more than compensates for the reduction in energy per coded bit. Therefore, in Figure 5.12, once the threshold value of $E_b/N_0 = 5.5$ dB is exceeded, the relative positions of the curves reverse themselves compared to their positions at less-than-threshold E_b/N_0. *Coding gain* is defined as the reduction, expressed in decibels, in the required E_b/N_0 to achieve a specified error performance of an error-correcting coded system over an uncoded one with the same modulation. For example, in Figure 5.12, for $P_B = 10^{-5}$, the (24, 12) code has a coding gain of about 1.5 dB.

Example 5.2 Coded versus Uncoded Performance

Compare the message error probability for a communications link with and without the use of error-correction coding. Assume that the uncoded transmission characteristics are: BPSK modulation, Gaussian noise, $S/N_0 = 43{,}776$, data rate $R = 4800$ bits/s. For the coded case, also assume the use of a (15, 11) error-correcting code that is capable of correcting any single-error pattern within a block of 15 bits. Consider that the demodulator makes hard decisions and thus feeds the demodulated code bits directly to the decoder, which in turn outputs an estimate of the original message.

Solution

Following Equation (3.84), let $p_u = Q\sqrt{2E_b/N_0}$ and $p_c = Q\sqrt{2E_c/N_0}$ be the uncoded and coded channel symbol error probabilities, respectively, where E_b/N_0 is uncoded bit energy per noise spectral density and E_c/N_0 is the coded bit energy per noise spectral density.

Without coding

$$\frac{E_b}{N_0} = \frac{S}{RN_0} = 9.12 \ (9.6 \text{ dB})$$

$$p_u = Q\left(\sqrt{\frac{2E_b}{N_0}}\right) = Q(\sqrt{18.24}) = 1.02 \times 10^{-5} \tag{5.20}$$

where the following approximation of $Q(x)$ from Equation (2.43) was used:

$$Q(x) \simeq \frac{1}{x\sqrt{2\pi}} \exp\left(\frac{-x^2}{2}\right) \quad \text{for } x > 3$$

The probability that the uncoded message block, P_M^u, will be received in error is 1 minus the product of the probabilities that each bit will be detected correctly. Thus

$$P_M^u = 1 - (1 - p_u)^k$$
$$= 1 - \underbrace{(1 - p_u)^{11}}_{\substack{\text{probability that all} \\ \text{11 bits in uncoded} \\ \text{block are correct}}} \qquad \underbrace{= 1.12 \times 10^{-4}}_{\substack{\text{probability that at} \\ \text{least 1 bit out of} \\ \text{11 is in error}}} \tag{5.21}$$

With coding:

The channel symbol rate, sometimes called the coded bit rate, R_c is 15/11 times the data bit rate.

$$R_c = 4800 \times \tfrac{15}{11} \approx 6545 \text{ bps}$$

$$\frac{E_c}{N_0} = \frac{S}{R_c N_0} = 6.688 \ (8.25 \text{ dB})$$

The E_c/N_0 for each code bit is less than that for the uncoded bit because the channel bit rate has increased but the transmitter power is assumed to be fixed.

$$p_c = Q\left(\sqrt{\frac{2E_c}{N_0}}\right) = Q(\sqrt{13.38}) = 1.36 \times 10^{-4} \tag{5.22}$$

It can be seen by comparing the results of Equation (5.20) with (5.22) that the channel bit error probability has degraded. More bits must be detected during the same time interval, and with the same available power; the performance improvement due to the coding *is not yet apparent*. We now compute the coded message error rate, P_M^c, using Equation (5.18).

$$P_M^c = \sum_{j=2}^{n=15} \binom{15}{j}(p_c)^j(1 - p_c)^{15-j}$$

The summation is started with $j = 2$ since the code corrects all single errors within a block of $n = 15$ bits. A good approximation is obtained by using only the first term of the summation. For p_c we use the value calculated in Equation (5.22):

$$P_M^c = \binom{15}{2}(p_c)^2(1 - p_c)^{13} = 1.94 \times 10^{-6} \tag{5.23}$$

By comparing the results of Equation (5.21) with (5.23), it is seen that the probability of message error has improved by a factor of 58 due to the error-correcting code used in this example.

5.4 LINEAR BLOCK CODES

Linear block codes (such as the one in Example 5.2) are a class of parity check codes that can be characterized by the (n, k) notation described earlier. The encoder transforms a block of k message digits (a message vector) into a longer block of n codeword digits (a code vector), constructed from a given alphabet of elements. When the alphabet consists of two elements (0 and 1), the code is a binary code comprised of binary digits (bits). Our discussion of linear block codes is restricted to binary codes, unless otherwise noted.

The k-bit messages form 2^k distinct message sequences referred to as *k-tuples* (sequences of k digits). The n-bit blocks can form as many as 2^n distinct sequences, referred to as *n-tuples*. The encoding procedure assigns to each of the 2^k message k-tuples *one* of the 2^n n-tuples. A block code represents a one-to-one assignment, whereby the 2^k message k-tuples are *uniquely* mapped into a new set of 2^k codeword n-tuples; the mapping can be accomplished via a look-up table. For *linear codes*, the mapping transformation is, of course, *linear*.

5.4.1 Vector Spaces

The set of all binary n-tuples, V_n, is called a *vector space* over the binary field of two elements (0 and 1). The binary field has two operations, addition and multiplication, such that the results of all operations are in the same set of two elements. The arithmetic operations of addition and multiplication are defined by the conventions of the algebraic field [4]. For example, in a binary field, the rules of addition and multiplication are as follows:

Addition	Multiplication
$0 \oplus 0 = 0$	$0 \cdot 0 = 0$
$0 \oplus 1 = 1$	$0 \cdot 1 = 0$
$1 \oplus 0 = 1$	$1 \cdot 0 = 0$
$1 \oplus 1 = 0$	$1 \cdot 1 = 1$

The addition operation, designated with the symbol $\oplus$, is the same modulo-2 operation described in Section 2.12.3.

5.4.2 Vector Subspaces

A subset S of the vector space V_n is called a *subspace* if the following two conditions are met:

1. The all-zeros vector is in S.
2. The sum of any two vectors in S is also in S (known as the *closure property*).

These properties are fundamental for the algebraic characterization of *linear block codes*. Suppose that $\mathbf{V}_i$ and $\mathbf{V}_j$ are two codewords (also called code vectors) in an

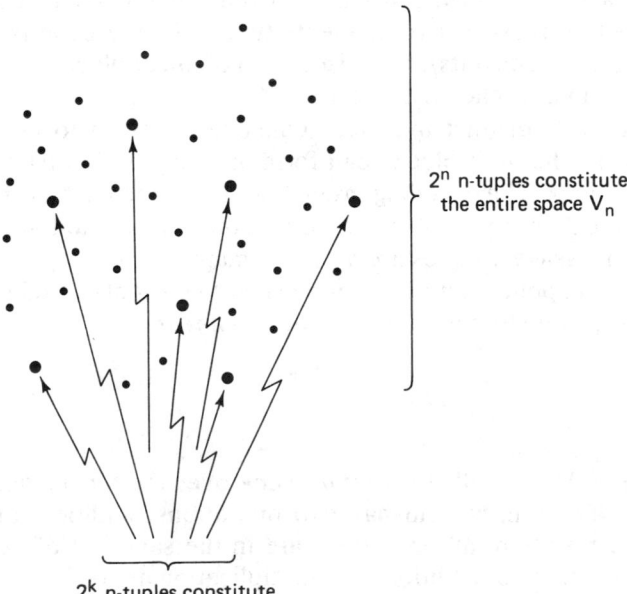

2^n n-tuples constitute the entire space V_n

2^k n-tuples constitute the subspace of codewords

Figure 5.13 Linear block-code structure.

(n, k) binary block code. The code is said to be *linear* if, and only if, $(\mathbf{V}_i \oplus \mathbf{V}_j)$ is also a code vector. A linear block code, then, is one in which vectors outside the subspace cannot be created by the addition of legitimate code vectors (members of the subspace).

For example, the vector space V_4 is totally populated by the following 2^4 = sixteen 4-tuples:

0000 0001 0010 0011 0100 0101 0110 0111

1000 1001 1010 1011 1100 1101 1110 1111

An example of a subset of V_4 that forms a subspace is

0000 0101 1010 1111

It is easy to verify that the addition of any two vectors in the subspace can only yield one of the other members of the subspace. A set of 2^k n-tuples is called a *linear block code* if, and only if, it is a subspace of the vector space V_n of all n-tuples. Figure 5.13 illustrates, with a simple geometric analogy, the structure behind linear block codes. We can imagine the vector space V_n comprised of 2^n n-tuples. Within this vector space there exists a subset of 2^k n-tuples comprising a subspace. These 2^k vectors or points, shown "sprinkled" among the more numerous 2^n points, represent the legitimate or allowable codeword assignments. A message is encoded into one of the 2^k allowable code vectors and then transmitted. Because of noise in the channel, a perturbed version of the code vector (one of the other 2^n vectors in the n-tuple space) may be received. If the perturbed vector is not too unlike (not too distant from) the valid code vector, the decoder can decode the message correctly. The basic goals in choosing a particular code, similar to the goals in selecting a set of modulation waveforms, can be stated in the context of Figure 5.13 as follows:

1. We want to strive for coding efficiency by packing the V_n space with as many code vectors as possible. This is tantamount to saying that we only want to expend a *small amount of redundancy* (excess bandwidth).
2. We want the code vectors to be as *far apart from one another* as possible, so that even if the vectors experience some corruption during transmission, they may still be correctly decoded, with a high probability.

5.4.3 A (6, 3) Linear Block Code Example

Examine the following coding assignment that describes a (6, 3) code. There are $2^k = 2^3 = 8$ message vectors, and therefore eight code vectors. There are $2^n = 2^6 =$ sixty-four 6-tuples in the V_6 vector space.

Message vector	Code vector
0 0 0	0 0 0 0 0 0
1 0 0	1 1 0 1 0 0
0 1 0	0 1 1 0 1 0
1 1 0	1 0 1 1 1 0
0 0 1	1 0 1 0 0 1
1 0 1	0 1 1 1 0 1
0 1 1	1 1 0 0 1 1
1 1 1	0 0 0 1 1 1

It is easy to check that the eight code vectors shown above form a subspace of V_6 (the all-zeros vector is present, and the sum of any two code vectors yields another code vector member of the subspace). Therefore, these code vectors represent a *linear block code*, as defined in Section 5.4.2.

5.4.4 Generator Matrix

If k is large, a *table look-up* implementation of the encoder becomes prohibitive. For a (127, 92) code there are 2^{92} or approximately 5×10^{27} code vectors. If the encoding procedure consists of a simple look-up table, imagine the size of the memory necessary to contain such a large number of code vectors. Fortunately, it is possible to reduce complexity by generating the required code vectors as needed, instead of storing them.

Since a set of code vectors that forms a linear block code is a k-dimensional subspace of the n-dimensional binary vector space ($k < n$), it is always possible to find a set of n-tuples, fewer than 2^k, that can generate all the 2^k member vectors of the subspace. The generating set of vectors is said to *span* the subspace. The smallest *linearly independent* set that spans the subspace is called a *basis* of the subspace, and the number of vectors in this basis set is the dimension of the subspace. Any basis set of k linearly independent n-tuples $\mathbf{V}_1, \mathbf{V}_2, \ldots, \mathbf{V}_k$ can be used to generate the required linear block code vectors, since each code vector is a linear combination of $\mathbf{V}_1, \mathbf{V}_2, \ldots, \mathbf{V}_k$. That is, each of the set of 2^k code vectors $\mathbf{U}$ can be described by

$$\mathbf{U} = m_1\mathbf{V}_1 + m_2\mathbf{V}_2 + \cdots + m_k\mathbf{V}_k$$

where $m_i = $ (0 or 1) are the message digits and $i = 1, \ldots, k$.

In general, we can define a *generator matrix* by the following $k \times n$ array:

$$\mathbf{G} = \begin{bmatrix} \mathbf{V}_1 \\ \mathbf{V}_2 \\ \vdots \\ \mathbf{V}_k \end{bmatrix} = \begin{bmatrix} v_{11} & v_{12} & \cdots & v_{1n} \\ v_{21} & v_{22} & \cdots & v_{2n} \\ \vdots & & & \\ v_{k1} & v_{k2} & \cdots & v_{kn} \end{bmatrix} \qquad (5.24)$$

Code vectors, by convention, are usually designated as row vectors. Thus, the message **m**, a sequence of k message bits, is shown below as a row vector ($1 \times k$ matrix having one row and k columns).

$$\mathbf{m} = m_1, m_2, \ldots, m_k$$

The generation of the code vector, U, is written in matrix notation as the product of **m** and **G**, as follows:

$$\mathbf{U} = \mathbf{mG} \qquad (5.25)$$

where, in general, the matrix multiplication $\mathbf{C} = \mathbf{AB}$ is performed in the usual way by using the rule

$$c_{ij} = \sum_{k}^{n} a_{ik} b_{kj} \qquad i = 1, \ldots, l \quad j = 1, \ldots, m$$

where **A** is an $l \times n$ matrix, **B** is an $n \times m$ matrix, and the result **C** is an $l \times m$ matrix. For the example introduced in the preceding section, we can fashion a generator matrix as follows:

$$\mathbf{G} = \begin{bmatrix} \mathbf{V}_1 \\ \mathbf{V}_2 \\ \mathbf{V}_3 \end{bmatrix} = \begin{bmatrix} 1 & 1 & 0 & 1 & 0 & 0 \\ 0 & 1 & 1 & 0 & 1 & 0 \\ 1 & 0 & 1 & 0 & 0 & 1 \end{bmatrix} \qquad (5.26)$$

where $\mathbf{V}_1$, $\mathbf{V}_2$, and $\mathbf{V}_3$ are three *linearly independent vectors* (a subset of the eight code vectors) that can generate all the code vectors. Notice that the sum of any two generating vectors does not yield any of the other generating vectors (opposite of closure). Let us generate the code vector for the message vector 1 1 0, using the generator matrix of Equation (5.26).

$$\mathbf{U} = [1 \quad 1 \quad 0] \begin{bmatrix} \mathbf{V}_1 \\ \mathbf{V}_2 \\ \mathbf{V}_3 \end{bmatrix} = 1 \cdot \mathbf{V}_1 + 1 \cdot \mathbf{V}_2 + 0 \cdot \mathbf{V}_3$$

$$= 1\ 1\ 0\ 1\ 0\ 0 + 0\ 1\ 1\ 0\ 1\ 0 + 0\ 0\ 0\ 0\ 0\ 0$$

$$= 1\ 0\ 1\ 1\ 1\ 0 \quad \text{(code vector for the message vector 1 1 0)}$$

Thus the code vector corresponding to a message vector is a linear combination of the rows of **G**. Since the code is totally defined by **G**, the encoder need only store the k rows of G instead of the total 2^k vectors of the code. For this example notice that the generator array of dimension 3×6 replaces the original code vector array of dimension 8×6, representing a reduction in system complexity.

5.4.5 Systematic Linear Block Codes

A systematic (n, k) linear block code is a mapping from a k-dimensional message vector to an n-dimensional code vector in such a way that part of the sequence generated coincides with the k message digits. The remaining $(n - k)$ digits are parity digits. A systematic linear block code will have a generator matrix of the form

$$\mathbf{G} = \left[\begin{array}{c|c} \mathbf{P} & \mathbf{I}_k \end{array} \right]$$

$$= \begin{bmatrix} p_{11} & p_{12} & \cdots & p_{1,(n-k)} & 1 & 0 & \cdots & 0 \\ p_{21} & p_{22} & \cdots & p_{2,(n-k)} & 0 & 1 & \cdots & 0 \\ \vdots & & & & & & \vdots & \\ p_{k1} & p_{k2} & \cdots & p_{k,(n-k)} & 0 & 0 & \cdots & 1 \end{bmatrix} \quad (5.27)$$

where $\mathbf{P}$ is the parity array portion of the generator matrix, $p_{ij} = (0 \text{ or } 1)$, and $\mathbf{I}_k$ is the $k \times k$ identity matrix (ones on the main diagonal and zeros elsewhere). Notice that with this systematic generator, the encoding complexity is further reduced since it is not necessary to store the identity matrix portion of the array. By combining Equations (5.25) and (5.27), each code vector is expressed as follows:

$$u_1, u_2, \ldots, u_n = [m_1, m_2, \ldots, m_k]$$

$$\times \begin{bmatrix} p_{11} & p_{12} & \cdots & p_{1,(n-k)} & 1 & 0 & \cdots & 0 \\ p_{21} & p_{22} & \cdots & p_{2,(n-k)} & 0 & 1 & \cdots & 0 \\ \vdots & & & & & & & \vdots \\ p_{k1} & p_{k2} & \cdots & p_{k,(n-k)} & 0 & 0 & \cdots & 1 \end{bmatrix}$$

where

$$u_i = m_1 p_{1i} + m_2 p_{2i} + \cdots + m_k p_{ki} \quad \text{for } i = 1, \ldots, (n-k)$$

$$= m_{i-n+k} \quad \text{for } i = (n-k+1), \ldots, n$$

Given the message k-tuple

$$\mathbf{m} = m_1, m_2, \ldots, m_k$$

and the general code vector n-tuple

$$\mathbf{U} = u_1, u_2, \ldots, u_n$$

the systematic code vector can be expressed as

$$\mathbf{U} = \underbrace{p_1, p_2, \ldots, p_{n-k}}_{\text{parity bits}}, \underbrace{m_1, m_2, \ldots, m_k}_{\text{message bits}} \quad (5.28)$$

where

$$p_1 = m_1 p_{11} + m_2 p_{21} + \cdots + m_k p_{k1}$$

$$p_2 = m_1 p_{12} + m_2 p_{22} + \cdots + m_k p_{k2} \quad (5.29)$$

$$p_{n-k} = m_1 p_{1,(n-k)} + m_2 p_{2,(n-k)} + \cdots + m_k p_{k,(n-k)}$$

Systematic code vectors are sometimes written so that the message bits occupy the left-hand portion of the code vector and the parity bits occupy the right-hand portion. This reordering has no effect on the error detection or error correction properties of the code, and will not be considered further.

For the (6, 3) code example in Section 5.4.3, the code vectors are described as follows:

$$\mathbf{U} = [m_1, m_2, m_3] \underbrace{\begin{bmatrix} 1 & 1 & 0 \\ 0 & 1 & 1 \\ 1 & 0 & 1 \end{bmatrix}}_{\mathbf{P}} \left.\begin{matrix} \vdots \\ \vdots \\ \vdots \end{matrix}\right| \underbrace{\begin{bmatrix} 1 & 0 & 0 \\ 0 & 1 & 0 \\ 0 & 0 & 1 \end{bmatrix}}_{\mathbf{I}_3} \tag{5.30}$$

$$\mathbf{U} = \underbrace{m_1 + m_3}_{u_1}, \underbrace{m_1 + m_2}_{u_2}, \underbrace{m_2 + m_3}_{u_3}, \underbrace{m_1}_{u_4}, \underbrace{m_2}_{u_5}, \underbrace{m_3}_{u_6}, \tag{5.31}$$

Equation (5.31) gives us some insight regarding the structure of linear block codes. We see that the redundant digits are produced in a variety of ways. The first parity bit is the sum of the first and third message bits; the second parity bit is the sum of the first and second message bits, and the third parity bit is the sum of the second and third message bits. Intuition tells us that such structure, compared to single-parity checks or simple digit-repeat procedures, may provide greater ability to detect and correct errors.

5.4.6 Parity-Check Matrix

Let us define a matrix, $\mathbf{H}$, called the *parity-check matrix*, that will enable us to decode the received vectors. For each $(k \times n)$ generator matrix, $\mathbf{G}$, there exists an $(n - k) \times n$ matrix, $\mathbf{H}$, such that the rows of $\mathbf{G}$ are orthogonal to the rows of $\mathbf{H}$; that is $\mathbf{GH}^T = \mathbf{0}$, where $\mathbf{H}^T$ is the *transpose* of $\mathbf{H}$, and $\mathbf{0}$ is a $k \times (n - k)$ all-zeros matrix. $\mathbf{H}^T$ is an $n \times (n - k)$ matrix whose rows are the columns of $\mathbf{H}$ and whose columns are the rows of $\mathbf{H}$. To fulfill the orthogonality requirements, the components of the $\mathbf{H}$ matrix are written

$$\mathbf{H} = [\mathbf{I}_{n-k} \mid \mathbf{P}^T] \tag{5.32}$$

Hence, the $\mathbf{H}^T$ matrix is written

$$\mathbf{H}^T = \begin{bmatrix} \mathbf{I}_{n-k} \\ \text{------} \\ \mathbf{P} \end{bmatrix} \tag{5.33a}$$

$$= \begin{bmatrix} 1 & 0 & \cdots & 0 \\ 0 & 1 & \cdots & 0 \\ \vdots & & & \\ 0 & 0 & \cdots & 1 \\ p_{11} & p_{12} & \cdots & p_{1,(n-k)} \\ p_{21} & p_{22} & \cdots & p_{2,(n-k)} \\ \vdots & & & \\ p_{k1} & p_{k2} & \cdots & p_{k,(n-k)} \end{bmatrix} \tag{5.33b}$$

It is easy to verify that the product $\mathbf{UH}^T$ of each code vector $\mathbf{U}$ generated by $\mathbf{G}$, and the $\mathbf{H}^T$ matrix, yields the zero vector:

$$\mathbf{UH}^T = p_1 + p_1, p_2 + p_2, \ldots, p_{n-k} + p_{n-k} = \mathbf{0}$$

where the parity bits $p_1, p_2, \ldots, p_{n-k}$ are defined in Equation (5.29). Thus once the *parity-check matrix*, $\mathbf{H}$, is constructed to fulfill the foregoing orthogonality requirements, we can use it to test whether a received vector is a valid member of the codeword set. $\mathbf{U}$ is a code vector generated by matrix $\mathbf{G}$ if, and only if, $\mathbf{UH}^T = \mathbf{0}$.

5.4.7 Syndrome Testing

Let $\mathbf{r} = r_1, r_2, \ldots, r_n$ be the received code vector (one of 2^n n-tuples) resulting from the transmission of $\mathbf{U} = u_1, u_2, \ldots, u_n$ (one of 2^k n-tuples). We can therefore describe $\mathbf{r}$ as

$$\mathbf{r} = \mathbf{U} + \mathbf{e} \tag{5.34}$$

where $\mathbf{e} = e_1, e_2, \ldots, e_n$ is the error vector or error pattern introduced by the channel. There are a total of $2^n - 1$ potential nonzero error patterns in the space of 2^n n-tuples. The *syndrome* of $\mathbf{r}$ is defined as

$$\mathbf{S} = \mathbf{rH}^T \tag{5.35}$$

The syndrome is the result of a parity check performed on $\mathbf{r}$ to determine whether $\mathbf{r}$ is a valid member of the codeword set. If, in fact, $\mathbf{r}$ is a member, the syndrome $\mathbf{S}$ has a value $\mathbf{0}$. If $\mathbf{r}$ contains detectable errors, the syndrome has some nonzero value. If $\mathbf{r}$ contains correctable errors, the syndrome (like the symptom of an illness) has some nonzero value which can earmark the particular error pattern. The decoder, depending upon whether it has been implemented to perform FEC or ARQ, will then take actions to locate the errors and correct them (FEC), or will request a retransmission (ARQ). Combining Equations (5.34) and (5.35), the syndrome of $\mathbf{r}$ is seen to be

$$\mathbf{S} = (\mathbf{U} + \mathbf{e})\mathbf{H}^T$$

$$= \mathbf{UH}^T + \mathbf{eH}^T \tag{5.36}$$

However, $\mathbf{UH}^T = \mathbf{0}$ for all members of the codeword set. Therefore,

$$\mathbf{S} = \mathbf{eH}^T \tag{5.37}$$

The foregoing development, starting with Equation (5.34) and terminating with Equation (5.37), is evidence that the syndrome test, whether performed on either a corrupted code vector or on the error pattern that caused it, yields the same syndrome. An important property of linear block codes, fundamental to the decoding process, is that the mapping between correctable error patterns and syndromes is one to one.

It is interesting to note the following two required properties of the parity-check matrix.

1. No column of **H** can be all zeros, or else an error in the corresponding code vector position would not affect the syndrome and would be undetectable.

2. All columns of **H** must be unique. If two columns of **H** were identical, errors in these two corresponding code vector positions would be indistinguishable.

Example 5.3 Syndrome Test

Suppose that code vector $\mathbf{U} = 1\ 0\ 1\ 1\ 1\ 0$ from the example in Section 5.4.3 is transmitted and the vector $\mathbf{r} = 0\ 0\ 1\ 1\ 1\ 0$ is received; that is, the leftmost bit is received in error. Find the syndrome vector value $\mathbf{S} = \mathbf{r}\mathbf{H}^T$ and verify that it is equal to $\mathbf{e}\mathbf{H}^T$.

Solution

$$\mathbf{S} = \mathbf{r}\mathbf{H}^T$$

$$= [0\ \ 0\ \ 1\ \ 1\ \ 1\ \ 0] \begin{bmatrix} 1 & 0 & 0 \\ 0 & 1 & 0 \\ 0 & 0 & 1 \\ 1 & 1 & 0 \\ 0 & 1 & 1 \\ 1 & 0 & 1 \end{bmatrix}$$

$$= [1,\ \ 1+1,\ \ 1+1] = [1\ \ 0\ \ 0] \qquad \text{(syndrome of corrupted code vector)}$$

Next, we verify that the syndrome of the corrupted code vector is the same as the syndrome of the error pattern that caused the error.

$$\mathbf{S} = \mathbf{e}\mathbf{H}^T = [1\ \ 0\ \ 0\ \ 0\ \ 0\ \ 0]\mathbf{H}^T = [1\ \ 0\ \ 0] \qquad \text{(syndrome of error pattern)}$$

5.4.8 Error Correction

We have detected a single error and have shown that the syndrome test performed on either the corrupted code vector, or on the error pattern that caused it, yields the same syndrome. This should be a clue that we not only can detect the error, but since there is a one-to-one correspondence between correctable error patterns and syndromes, we can correct such error patterns. Let us arrange the 2^n n-tuples that represent possible received vectors in an array, called the *standard array*, such that the first row contains all the code vectors, starting with the all-zeros vector, and the first column contains all the correctable error patterns. Recall from the basic properties of linear codes (see Section 5.4.2) that the all-zeros vector must be a member of the codeword set. Each row, called a *coset*, consists of an error pattern in the first column, called the *coset leader*, followed by the code vectors perturbed by that error pattern. The standard array format for an (n, k) code is as follows:

$$
\begin{array}{ccccc}
\mathbf{U}_1 & \mathbf{U}_2 & \cdots\ \mathbf{U}_i & \cdots\ \mathbf{U}_{2^k} \\
\mathbf{e}_2 & \mathbf{U}_2 + \mathbf{e}_2 & \cdots\ \mathbf{U}_i + \mathbf{e}_2 & \cdots\ \mathbf{U}_{2^k} + \mathbf{e}_2 \\
\mathbf{e}_3 & \mathbf{U}_2 + \mathbf{e}_3 & \cdots\ \mathbf{U}_i + \mathbf{e}_3 & \cdots\ \mathbf{U}_{2^k} + \mathbf{e}_3 \\
\vdots & \vdots & \vdots & \\
\mathbf{e}_j & \mathbf{U}_2 + \mathbf{e}_j & \cdots\ \mathbf{U}_i + \mathbf{e}_j & \cdots\ \mathbf{U}_{2^k} + \mathbf{e}_j \\
\vdots & \vdots & \vdots & \\
\mathbf{e}_{2^{n-k}} & \mathbf{U}_2 + \mathbf{e}_{2^{n-k}} & \cdots\ \mathbf{U}_i + \mathbf{e}_{2^{n-k}} & \cdots\ \mathbf{U}_{2^k} + \mathbf{e}_{2^{n-k}}
\end{array}
\qquad (5.38)
$$

The array contains all 2^n n-tuples in the space V_n (each n-tuple appears in *only one* location). Each coset consists of 2^k n-tuples. Therefore, there are $(2^n/2^k) = 2^{n-k}$ cosets. Suppose that a code vector $\mathbf{U}_i$ is transmitted over a noisy channel. If the error pattern caused by the channel is a coset leader, the received vector will be decoded correctly into the transmitted code vector $\mathbf{U}_i$. If the error pattern is not a coset leader, an erroneous decoding will result.

5.4.8.1 The Syndrome of a Coset

If $\mathbf{e}_j$ is the coset leader or error pattern of the jth coset, then $\mathbf{U}_i + \mathbf{e}_j$ is an n-tuple in this coset. The syndrome of this n-tuple can be written

$$\mathbf{S} = (\mathbf{U}_i + \mathbf{e}_j)\mathbf{H}^T = \mathbf{U}_i\mathbf{H}^T + \mathbf{e}_j\mathbf{H}^T$$

Since $\mathbf{U}_i$ is a code vector, $\mathbf{U}_i\mathbf{H}^T = \mathbf{0}$, and we can write, as in Equation (5.37)

$$\mathbf{S} = (\mathbf{U}_i + \mathbf{e}_j)\mathbf{H}^T = \mathbf{e}_j\mathbf{H}^T \tag{5.39}$$

From Equation (5.39) it is clear that all members of a coset have the *same syndrome*, and in fact, the syndrome is used to estimate the error pattern. The syndrome for every coset is different.

5.4.8.2 Error Correction Decoding

The procedure for error correction decoding proceeds as follows:

1. Calculate the syndrome of $\mathbf{r}$ using $\mathbf{S} = \mathbf{r}\mathbf{H}^T$.
2. Locate the coset leader (error pattern), $\mathbf{e}_j$, whose syndrome equals $\mathbf{r}\mathbf{H}^T$.
3. This error pattern is assumed to be the corruption caused by the channel.
4. The corrected received vector, or code vector, is identified as $\mathbf{U} = \mathbf{r} + \mathbf{e}_j$. We can say that we retrieve the valid code vector by subtracting out the identified error; in modulo-2 arithmetic the operation of subtraction is identical to that of addition.

5.4.8.3 Locating the Error Pattern

Returning to the example of Section 5.4.3, we arrange the $2^6 =$ sixty-four 6-tuples in a standard array as shown in Figure 5.14. The valid code vectors are the eight vectors in the first row, and the *correctable error patterns* are the eight *coset leaders* in the first column. Notice that all 1-bit error patterns are correctable. Also notice that after exhausting all 1-bit error patterns, there remains some error-correcting capability since we have not yet accounted for all sixty-four 6-tuples. There is one unassigned coset leader; therefore, there remains the capability of correcting one additional error pattern. We have the flexibility of choosing this error pattern to be any of the n-tuples in the remaining coset. In Figure 5.14 this final correctable error pattern is chosen, somewhat arbitrarily, to be the 2-bit error pattern 0 1 0 0 0 1. Decoding will be correct if, and only if, the error pattern caused by the channel is one of the coset leaders.

We now determine the syndrome corresponding to each of the correctable

000000	110100	011010	101110	101001	011101	110011	000111
000001	110101	011011	101111	101000	011100	110010	000110
000010	110110	011000	101100	101011	011111	110001	000101
000100	110000	011110	101010	101101	011001	110111	000011
001000	111100	010010	100110	100001	010101	111011	001111
010000	100100	001010	111110	111001	001101	100011	010111
100000	010100	111010	001110	001001	111101	010011	100111
010001	100101	001011	111111	111000	001100	100010	010110

Figure 5.14 Example of a standard array for a (6, 3) code.

error sequences by computing $e_j H^T$ for each coset leader, as follows:

$$S = e_j \begin{bmatrix} 1 & 0 & 0 \\ 0 & 1 & 0 \\ 0 & 0 & 1 \\ 1 & 1 & 0 \\ 0 & 1 & 1 \\ 1 & 0 & 1 \end{bmatrix}$$

The results are listed in Table 5.1. Since each syndrome in the table is unique, the decoder can identify the error pattern **e** to which it corresponds.

TABLE 5.1 Syndrome Look-Up Table

Error pattern	Syndrome
0 0 0 0 0 0	0 0 0
0 0 0 0 0 1	1 0 1
0 0 0 0 1 0	0 1 1
0 0 0 1 0 0	1 1 0
0 0 1 0 0 0	0 0 1
0 1 0 0 0 0	0 1 0
1 0 0 0 0 0	1 0 0
0 1 0 0 0 1	1 1 1

5.4.8.4 Error Correction Example

As outlined in Section 5.4.8.2, we receive the vector **r** and calculate its syndrome using $S = r H^T$. We then use the syndrome look-up table (Table 5.1), developed in the preceding section, to find the corresponding error pattern. This error pattern is an estimate of the error, and we denote it $\hat{e}$. The decoder then adds $\hat{e}$ to **r** to obtain an estimate of the transmitted code vector $\hat{U}$.

$$\hat{U} = r + \hat{e} = (U + e) + \hat{e} = U + (e + \hat{e}) \qquad (5.40)$$

If the estimated error pattern is the same as the actual error pattern, that is, if ê = e, then the estimate Û is equal to the transmitted code vector U. On the other hand, if the error estimate is incorrect, the decoder will estimate a code vector that was not transmitted, and we have an *undetectable decoding error*.

Example 5.4   Error Correction

Assume that code vector U = 1 0 1 1 1 0, from the Section 5.4.3 example, is transmitted, and the vector r = 0 0 1 1 1 0 is received. Show how a decoder, using the Table 5.1 syndrome look-up table, can correct the error.

Solution

The syndrome of r is computed:

$$S = [0 \ \ 0 \ \ 1 \ \ 1 \ \ 1 \ \ 0]H^T = [1 \ \ 0 \ \ 0]$$

Using Table 5.1, the error pattern corresponding to the syndrome above is estimated to be

$$ê = 1 \ 0 \ 0 \ 0 \ 0 \ 0$$

The corrected vector is then estimated by

$$Û = r + ê$$

$$= 0 \ 0 \ 1 \ 1 \ 1 \ 0 + 1 \ 0 \ 0 \ 0 \ 0 \ 0$$

$$= 1 \ 0 \ 1 \ 1 \ 1 \ 0$$

Since the estimated error pattern is the actual error pattern in this example, the error correction procedure yields Û = U.

5.5 CODING STRENGTH

5.5.1 Weight and Distance of Binary Vectors

It should be clear that not all error patterns can be correctly decoded. The error correction capability of a code will be investigated by first defining its structure. The *Hamming weight*, $w(U)$, of a vector U is defined to be the number of nonzero elements in U. For a binary vector this is equivalent to the number of ones in the vector. For example, if U = 1 0 0 1 0 1 1 0 1, then $w(U) = 5$. The *Hamming distance* between two code vectors U and V, denoted $d(U, V)$, is defined to be the number of elements in which they differ: for example,

$$U = 1 \ 0 \ 0 \ 1 \ 0 \ 1 \ 1 \ 0 \ 1$$

$$V = 0 \ 1 \ 1 \ 1 \ 1 \ 0 \ 1 \ 0 \ 0$$

$$d(U, V) = 6$$

By the properties of modulo-2 addition, we note that the sum of two binary vectors is another vector whose binary ones are located in those positions in which the two vectors differ: for example,

$$U + V = 1\ 1\ 1\ 0\ 1\ 1\ 0\ 0\ 1$$

Thus we observe that the Hamming distance between two code vectors is equal to the Hamming weight of their sum: that is, $d(U, V) = w(U + V)$. Also, we see that the Hamming weight of a code vector is equal to its Hamming distance from the all-zeros vector.

5.5.2 Minimum Distance of a Linear Code

Consider the set of distances between all pairs of code vectors in the space V_n. The smallest member of the set is the *minimum distance* of the code and is denoted d_{min}. Why do you suppose we have an interest in the minimum distance; why not the maximum distance? The minimum distance, like the weakest link in a chain, gives us a measure of the code's minimum capability and therefore characterizes the code's strength.

As discussed earlier, the sum of any two code vectors yields another code vector member of the subspace. This property of linear codes is stated simply as: If U and V are code vectors, then $W = U + V$ must also be a code vector. Hence the distance between two code vectors is equal to the weight of a third code vector; that is, $d(U, V) = w(U + V) = w(W)$. Thus the minimum distance of a linear code can be ascertained without examining the distance between all combinations of code vector pairs. We only need to examine the weight of each code vector (excluding the all-zeros vector) in the subspace; the minimum weight corresponds to the minimum distance, d_{min}. Equivalently, d_{min} corresponds to the smallest of the set of distances between the all-zeros code vector and all the other code vectors.

5.5.3 Error Detection and Correction

The task of the decoder, having received the vector r, is to estimate the transmitted code vector U_i. The optimal decoder strategy can be expressed in terms of the *maximum likelihood* algorithm (see Appendix B) as follows: Decide in favor of U_i if

$$P(r|U_i) = \max_{\text{over all } U_j} P(r|U_j) \tag{5.41}$$

Since for the binary symmetric channel (BSC), the likelihood of U_i with respect to r is inversely proportional to the distance between r and U_i, we can write: Decide in favor of U_i if

$$d(r, U_i) = \min_{\text{over all } U_j} d(r, U_j) \tag{5.42}$$

In other words, the decoder determines the distance between $\mathbf{r}$ and each of the possible transmitted code vectors $\mathbf{U}_i$, and selects as most likely a $\mathbf{U}_i$ for which

$$d(\mathbf{r}, \mathbf{U}_i) \le d(\mathbf{r}, \mathbf{U}_j) \qquad \text{for } i, j = 1, \ldots, M \quad \text{and} \quad i \ne j \qquad (5.43)$$

where $M = 2^k$ is the size of the code vector set. If the minimum is not unique, the choice between minimum distance codewords is arbitrary. Distance metrics are treated further in Chapter 6.

In Figure 5.15 the distance between two code vectors $\mathbf{U}$ and $\mathbf{V}$ is shown using a number line calibrated in *Hamming distance*. Each black dot represents a corrupted code vector. Figure 5.15a illustrates the reception of vector $\mathbf{r}_1$, which is distance 1 from $\mathbf{U}$ and distance 4 from $\mathbf{V}$. An error-correcting decoder, following the maximum likelihood strategy, will select $\mathbf{U}$ upon receiving $\mathbf{r}_1$. If $\mathbf{r}_1$ had been the result of a 1-bit corruption to the transmitted code vector $\mathbf{U}$, the decoder has successfully corrected the error. But if $\mathbf{r}_1$ had been the result of a 4-bit corruption to the transmitted code vector $\mathbf{V}$, the result is a decoding error. Similarly, a double error in transmission of $\mathbf{U}$ might result in the received vector $\mathbf{r}_2$, which is distance 2 from $\mathbf{U}$ and distance 3 from $\mathbf{V}$, as shown in Figure 5.15b. Here, too, the decoder will select $\mathbf{U}$ upon receiving $\mathbf{r}_2$. A triple error in transmission of $\mathbf{U}$ might result in a received vector $\mathbf{r}_3$ which is distance 3 from $\mathbf{U}$ and distance 2 from $\mathbf{V}$, as shown in Figure 5.15c. Here the decoder will select $\mathbf{V}$ upon receiving $\mathbf{r}_3$, and will have made an error in decoding. From Figure 5.15 it should be clear that if error

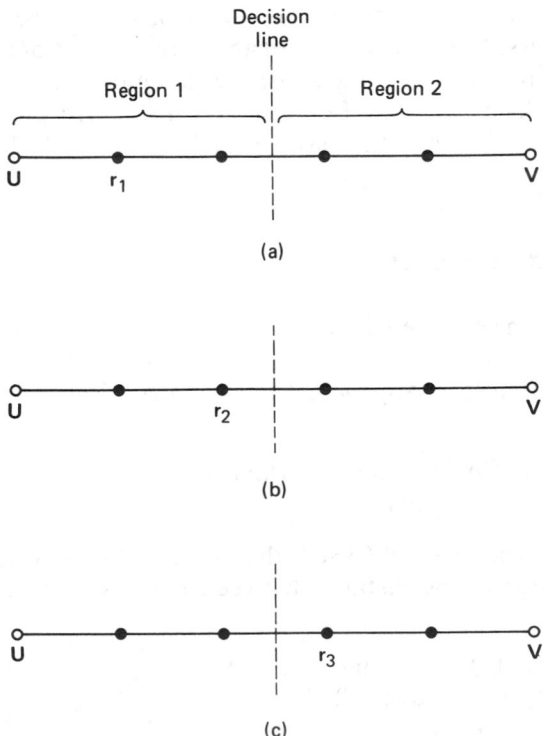

(a)

(b)

(c)

Figure 5.15 Error correction and detection strength. (a) Received vector $\mathbf{r}_1$. (b) Received vector $\mathbf{r}_2$. (c) Received vector $\mathbf{r}_3$.

Channel Coding: Part 1   Chap. 5

detection and not correction is the task, a corrupted vector, characterized by a black dot and representing a 1-bit, 2-bit, 3-bit, or 4-bit error, can be detected. However, five errors in transmission might result in code vector **V** being received when code vector **U** was actually transmitted; such an error would be *undetectable*.

From Figure 5.15 we can see that the error-detecting and error-correcting capabilities of a code are related to the *minimum distance* between code vectors. The decision line in the figure serves the same purpose in the process of decoding as it does in demodulation, to define the decision regions. In the Figure 5.15 example, the decision criterion of choosing **U** if **r** falls in region 1, and choosing **V** if **r** falls in region 2, illustrates that such a code, with $d_{min} = 5$, can correct two errors. In general, the *error-correcting capability*, t, of a code is defined as the maximum number of guaranteed correctable errors per codeword, and is written [4]

$$ t = \left\lfloor \frac{d_{min} - 1}{2} \right\rfloor \tag{5.44} $$

where $\lfloor x \rfloor$ means the largest integer not to exceed x. Often, a code that corrects all possible sequences of t or fewer errors can also correct certain sequences of $t + 1$ errors. This can be seen in Figure 5.14. In this example $d_{min} = 3$, and thus from Equation (5.44), we can see that *all* $t = 1$ bit-error patterns are correctable. Also, *a single* $t + 1$ or 2-bit error pattern is correctable. In general, a t-error-correcting (n, k) linear code is capable of correcting a total of 2^{n-k} error patterns. If a t-error-correcting block code is used strictly for error correction on a binary symmetric channel (BSC) with transition probability p, the probability that the decoder commits an erroneous decoding, and that the n-bit block is in error, can be calculated by using Equation (5.18) as an upper bound:

$$ P_M \leq \sum_{j=t+1}^{n} \binom{n}{j} p^j (1 - p)^{n-j} \tag{5.45} $$

The bound becomes an equality when the decoder corrects all combinations of errors up to and including t errors, but no combinations of errors greater than t. Such decoders are called *bounded distance decoders*. The decoded bit-error probability depends on the particular code and decoder. It can be expressed [5] by the following approximation:

$$ P_B \simeq \frac{1}{n} \sum_{j=t+1}^{n} j \binom{n}{j} p^j (1 - p)^{n-j} \tag{5.46} $$

A code can be used to detect errors prior to, or instead of, correcting them. It should be clear from Figure 5.15 that any received vector characterized by a black dot (a corrupted code vector) can be identified as an error. Therefore, the error-detecting capability, e, is defined in terms of d_{min} as

$$ e = d_{min} - 1 \tag{5.47} $$

A block code with minimum distance d_{min} guarantees that all error patterns of

$d_{min} - 1$ or fewer errors can be detected. Such a code is also capable of detecting a large fraction of error patterns with d_{min} or more errors. In fact, an (n, k) code is capable of detecting $2^n - 2^k$ error patterns of length n. The reasoning is as follows. There are a total of $2^n - 1$ possible nonzero error patterns in the space of 2^n n-tuples. Even the bit pattern of a valid codeword represents a potential error pattern. Thus there are $2^k - 1$ error patterns that are identical to the $2^k - 1$ nonzero codewords. If any of these $2^k - 1$ error patterns occurs, it alters the transmitted codeword $\mathbf{U}_i$ into another codeword $\mathbf{U}_j$. Thus $\mathbf{U}_j$ will be received and its syndrome is zero. The decoder accepts $\mathbf{U}_j$ as the transmitted codeword and thereby commits an incorrect decoding. Therefore, there are $2^k - 1$ undetectable error patterns. If the error pattern is not identical to one of the 2^k codewords, the syndrome test on the received vector $\mathbf{r}$ yields a nonzero syndrome, and the error is detected. Therefore, there are exactly $2^n - 2^k$ detectable error patterns. For large n, where $2^k \ll 2^n$, only a small fraction of error patterns are undetected.

5.5.3.1 Code Vector Weight Distribution

Let A_j be the number of code vectors of weight j within an (n, k) linear code. The numbers $A_0, A_1, \ldots, A_n$ are called the *weight distribution* of the code. If the code is used only for error detection, on a BSC, the probability, P_{nd}, that the decoder does not detect an error can be computed from the weight distribution of the code [5] as follows:

$$P_{nd} = \sum_{j=1}^{n} A_j p^j (1 - p)^{n-j} \tag{5.48}$$

where p is the transition probability of the BSC. If the minimum distance of the code is d_{min}, the values of A_1 to $A_{d_{min}-1}$ are zero.

Example 5.5 Probability of an Undetected Error in an Error Detecting Code

Consider that the (6, 3) code, given in Section 5.4.3, is used only for error detection. Calculate the probability of an undetected error if the channel is a BSC and the transition probability is 10^{-2}.

Solution

The weight distribution of this code is $A_0 = 1$, $A_1 = A_2 = 0$, $A_3 = 4$, $A_4 = 3$, $A_5 = 0$, $A_6 = 0$. Therefore, we can write, using Equation (5.48),

$$P_{nd} = 4p^3(1 - p)^3 + 3p^4(1 - p)^2$$

For $p = 10^{-2}$, the probability of an undetected error is 3.9×10^{-6}.

5.5.3.2 Simultaneous Error Correction and Detection

It is possible to trade correction capability from the maximum guaranteed (t), where t is defined in Equation (5.44), for the ability to simultaneously detect a class of errors. A code can be used for the simultaneous correction of α errors and detection of β errors where $\beta \geq \alpha$, provided that its minimum distance is [4]

$$d_{min} \geq \alpha + \beta + 1 \tag{5.49}$$

When t or fewer errors occur, the code is capable of detecting and correcting them. When more than t but fewer than $e + 1$ errors occur, where e is defined in Equation (5.47), the code is capable of detecting their presence but correcting only a subset of them. For example, a code with $d_{min} = 7$ can be used to simultaneously detect and correct in any one of the following ways:

Detect (β)	Correct (α)
3	3
4	2
5	1
6	0

Note that correction implies prior detection. For the above example, when there are three errors, all of them can be detected and corrected. When there are five errors, all of them can be detected but only a subset of them (one) can be corrected.

5.5.4 Visualization of a 6-Tuple Space

Figure 5.16 is a visualization of the eight codewords from the example of Section 5.4.3. The codewords are generated from linear combinations of the three independent 6-tuples in Equation (5.26); the codewords form a three-dimensional subspace. The figure shows such a subspace completely occupied by the eight codewords (large black circles); the coordinates of the subspace have purposely been drawn to emphasize their nonorthogonality. Figure 5.16 is an attempt to illustrate the entire space, containing sixty-four 6-tuples, even though there is no precise way to draw or construct such a model. Spherical layers or shells are shown around each codeword. Each of the nonintersecting inner layers is a Hamming distance of 1 from its associated codeword; each outer layer is a Hamming distance of 2 from its codeword. Larger distances are not useful in this example. For each codeword, the two layers shown are occupied by perturbed codewords. There are six such points on each inner sphere (a total of 48 points), representing the six possible 1-bit error-perturbed vectors associated with each codeword. These 1-bit perturbed codewords are distinct in the sense that they can best be associated with only one codeword, and therefore can be corrected. As is seen from the standard array of Figure 5.14, there is also one 2-bit error pattern that can be corrected. There is a total of $\binom{6}{2} = 15$ different 2-bit error patterns that can be inflicted on each codeword, but only one of them, in our example the 0 1 0 0 0 1 error pattern, can be corrected. The other fourteen 2-bit error patterns yield vectors that cannot be uniquely identified with just one codeword; these noncorrectable error patterns yield vectors that are equivalent to the error-perturbed vectors of two or more codewords. In the figure, all correctable (fifty-six) 1- and 2-bit error-perturbed codewords are shown as small black circles. Perturbed codewords that cannot be corrected are shown as small clear circles.

Figure 5.16 is useful for visualizing the properties of a class of codes known as *perfect codes*. A t-error-correcting code is called a perfect code if its standard array has all the error patterns of t or fewer errors and no others as coset leaders.

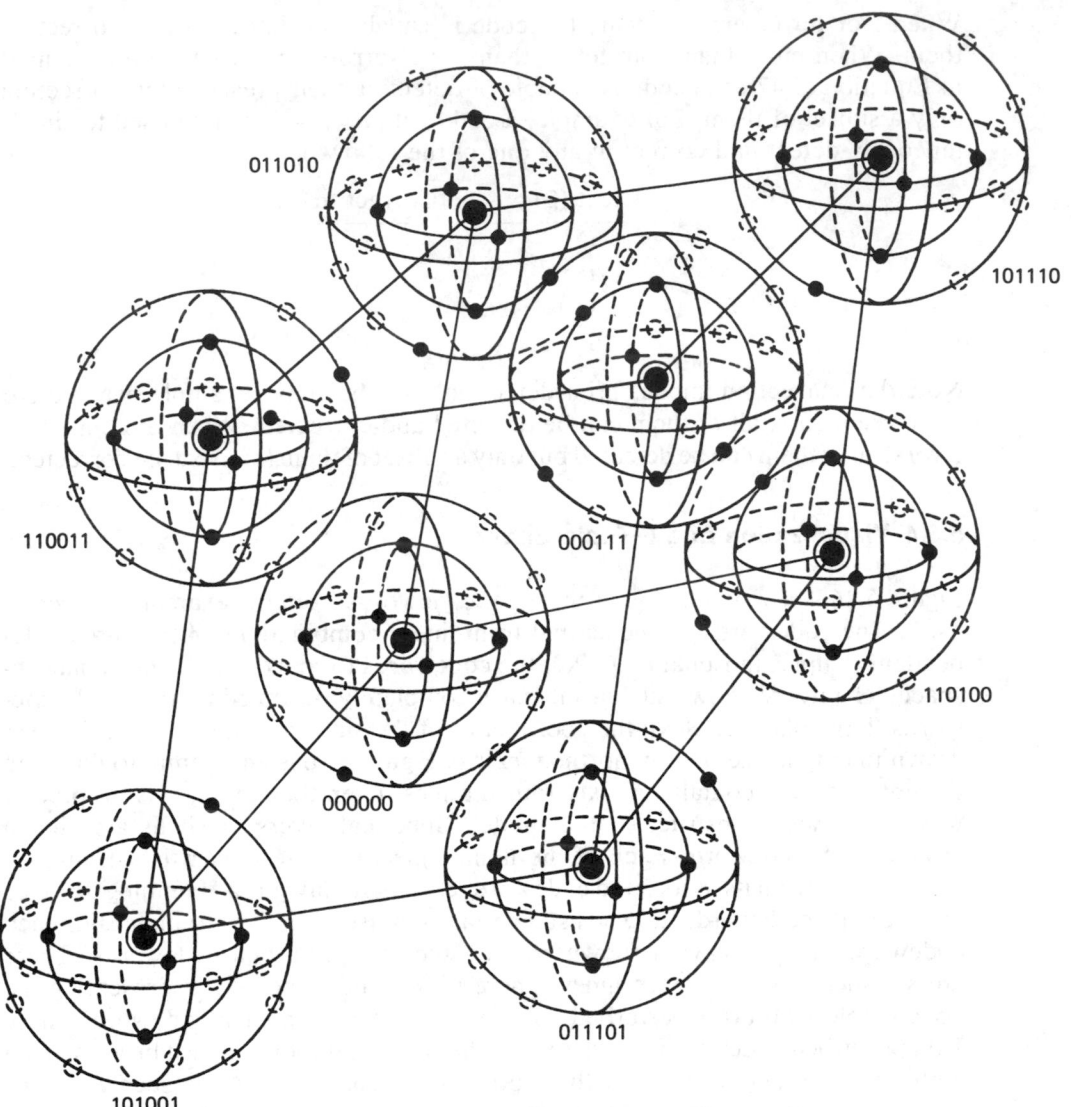

Figure 5.16 Example of eight codewords in a 6-tuple space.

In terms of Figure 5.16, a *t*-error-correcting perfect code is one that can, with maximum likelihood decoding, correct all perturbed code vectors occupying a shell at Hamming distance *t* or less from its originating codeword, and cannot correct any perturbed vectors occupying shells at distances greater than *t*.

Figure 5.16 is also useful for understanding the basic goal in the search for good codes. We would like for the space to be filled with as many codewords as possible (efficient utilization of the added redundancy), and we would also like these codewords to be as far away from one another as possible. Obviously, these goals conflict.

5.5.5 Erasure Correction

A receiver may be designed to declare a symbol *erased* when it is received ambiguously or when the receiver recognizes the presence of interference or a transient malfunction. Such a channel has an input alphabet of size Q and an output alphabet of size $Q + 1$; the extra output symbol is called an *erasure flag*, or simply an *erasure*. When a demodulator makes a symbol error, two parameters are needed to correct that error, its *location* and its *correct* symbol value. In the case of binary symbols, this reduces to needing only the error location. However, if the demodulator declares a symbol *erased*, although the correct symbol value is not known, the symbol location *is* known, and for this reason, the decoding of erased codewords can be simpler than error correcting. An error control code can be used to correct erasures or to correct errors and erasures simultaneously. If the code has minimum distance d_{min}, any pattern of ρ or fewer erasures can be corrected if [6]

$$d_{min} \geq \rho + 1 \tag{5.50}$$

Assume for the moment that no errors occur outside the erasure positions. The advantage of correcting by means of erasures is expressed quantitatively as follows: If a code has a minimum distance d_{min}, then from Equation (5.50), $d_{min} - 1$ erasures can be reconstituted. Since the number of errors that can be corrected without erasure information is $(d_{min} - 1)/2$ at most, from Equation (5.44), the advantage of correcting by means of erasures is clear. Further, any pattern of α errors and γ erasures can be corrected simultaneously if [6]

$$d_{min} \geq 2\alpha + \gamma + 1 \tag{5.51}$$

Simultaneous erasure correction and error correction can be accomplished in the following way. First, the γ-erased positions are replaced with zeros and the resulting codeword is decoded normally. Next, the γ-erased positions are replaced with ones, and the decoding operation is repeated on this version of the codeword. Of the two codewords obtained (one with erasures replaced by zeros, and the other with erasures replaced by ones) the one corresponding to the smallest number of errors corrected outside the γ-erased positions is selected. This technique will always result in correct decoding if Equation (5.51) is satisfied.

Example 5.6   Erasure Correction

Consider the codeword set presented in Section 5.4.3:

000000   110100   011010   101110   101001   011101   110011   000111

Suppose that the codeword 110011 was transmitted and that the two leftmost digits were declared by the receiver to be erasures. Verify that the received flawed sequence xx0011 can be corrected.

Solution

Since $d_{min} = \rho + 1 = 3$, the code can correct as many as $\rho = 2$ erasures. This is easily verified above or with Figure 5.14 by comparing the rightmost four digits of xx0011 with each of the allowable codewords. The codeword that was actually transmitted is closest in Hamming distance to the flawed sequence.

5.6 CYCLIC CODES

Binary cyclic codes are an important subclass of linear block codes. The codes are easily implemented with feedback shift registers; the syndrome calculation is easily accomplished with similar feedback shift registers; and the underlying algebraic structure of a cyclic code lends itself to efficient decoding methods. An (n, k) linear code is called a *cyclic code* if it can be described by the following property. If the n-tuple $\mathbf{U} = (u_0, u_1, u_2, \ldots, u_{n-1})$ is a code vector in the subspace S, then $\mathbf{U}^{(1)} = (u_{n-1}, u_0, u_1, \ldots, u_{n-2})$ obtained by an end-around shift, is also a code vector in S. Or in general, $\mathbf{U}^{(i)} = (u_{n-i}, u_{n-i+1}, \ldots, u_{n-1}, u_0, u_1, \ldots, u_{n-i-1})$, obtained by i end-around or cyclic shifts, is also a code vector in S.

The components of a code vector $\mathbf{U} = (u_0, u_1, u_2, \ldots, u_{n-1})$ can be treated as the coefficients of a polynomial $\mathbf{U}(X)$ as follows:

$$\mathbf{U}(X) = u_0 + u_1 X + u_2 X^2 + \cdots + u_{n-1} X^{n-1} \tag{5.52}$$

The polynomial function $\mathbf{U}(X)$ can be thought of as a "placeholder" for the digits of the code vector $\mathbf{U}$; that is, an n-tuple vector is described by a polynomial of degree $n - 1$ or less. The presence or absence of each term in the polynomial indicates the presence of a 1 or 0 in the corresponding location of the n-tuple. If the u_{n-1} component is nonzero, the polynomial is of degree $n - 1$. The usefulness of this polynomial description of a codeword will become clear as we discuss the algebraic structure of the cyclic codes.

5.6.1 Algebraic Structure of Cyclic Codes

Expressing the code vectors in polynomial form, the cyclic nature of the code manifests itself in the following way. If $\mathbf{U}(X)$ is an $(n - 1)$-degree codeword polynomial, then $\mathbf{U}^{(i)}(X)$, the remainder resulting from dividing $X^i \mathbf{U}(X)$ by $X^n + 1$, is also a codeword; that is,

$$\frac{X^i \mathbf{U}(X)}{X^n + 1} = \mathbf{q}(X) + \frac{\mathbf{U}^{(i)}(X)}{X^n + 1} \tag{5.53}$$

or, multiplying through by $X^n + 1$,

$$X^i \mathbf{U}(X) = \mathbf{q}(X)(X^n + 1) \underbrace{+ \ \mathbf{U}^{(i)}(X)}_{\text{remainder}} \tag{5.54}$$

which can also be described in terms of modulo arithmetic as follows:

$$\mathbf{U}^{(i)}(X) = X^i \mathbf{U}(X) \text{ modulo } (X^n + 1) \tag{5.55}$$

where x modulo y is defined as the remainder obtained from dividing x by y. Let us demonstrate the validity of Equation (5.55) for the case of $i = 1$.

$$\mathbf{U}(X) = u_0 + u_1 X + u_2 X^2 + \cdots + u_{n-2} X^{n-2} + u_{n-1} X^{n-1}$$

$$X\mathbf{U}(X) = u_0 X + u_1 X^2 + u_2 X^3 + \cdots + u_{n-2} X^{n-1} + u_{n-1} X^n$$

We now add and subtract u_{n-1}, or since we are using modulo-2 arithmetic, we add u_{n-1} twice, as follows:

$$X\mathbf{U}(X) = \underbrace{u_{n-1} + u_0 X + u_1 X^2 + u_2 X^3 + \cdots + u_{n-2} X^{n-1}}_{\mathbf{U}^{(1)}(X)} + u_{n-1} X^n + u_{n-1}$$

$$= \mathbf{U}^{(1)}(X) + u_{n-1}(X^n + 1)$$

Since $\mathbf{U}^{(1)}(X)$ is of degree $n - 1$, it cannot be divided by $X^n + 1$. Thus we can write from Equation (5.53)

$$\mathbf{U}^{(1)}(X) = X\mathbf{U}(X) \text{ modulo } (X^n + 1)$$

By extension we can write

$$\mathbf{U}^{(i)}(X) = X^i \mathbf{U}(X) \text{ modulo } (X^n + 1) \tag{5.56}$$

Example 5.7 Cyclic Shift of a Code Vector

Let $\mathbf{U} = 1\ 1\ 0\ 1$, for $n = 4$. Express the code vector in polynomial form, and using Equation (5.54), solve for the third end-around shift of the code vector.

Solution

$$\mathbf{U}(X) = 1 + X + X^3 \qquad \text{(polynomial is written low order to high order)}$$

$$X^i \mathbf{U}(X) = X^3 + X^4 + X^6 \qquad \text{where } i = 3$$

Divide $X^3 \mathbf{U}(X)$ by $X^4 + 1$, and solve for the remainder using polynomial division.

$$
\begin{array}{r}
X^2 + 1 \\
X^4 + 1\ \overline{)\,X^6 + X^4 + X^3} \\
\underline{X^6 + X^2} \\
X^4 + X^3 + X^2 \\
\underline{X^4 + 1} \\
X^3 + X^2 + 1 \qquad \text{remainder } \mathbf{U}^{(3)}(X)
\end{array}
$$

Writing the remainder low order to high order: $1 + X^2 + X^3$, the codeword $\mathbf{U}^{(3)} = 1\ 0\ 1\ 1$ is three cyclic shifts of $\mathbf{U} = 1\ 1\ 0\ 1$. Remember that for binary codes, the addition operation is performed modulo-2, so that $+1 = -1$, and we consequently do not show any minus signs in the computation.

5.6.2 Binary Cyclic Code Properties

We can generate a cyclic code using a *generator polynomial* in much the way that we generated a block code using a generator matrix. The generator polynomial $g(X)$ for an (n, k) cyclic code is unique and is of the form

$$g(X) = g_0 + g_1 X + g_2 X^2 + \cdots + g_r X^r \tag{5.57}$$

where g_0 and g_r must equal 1. Every codeword polynomial in the subspace is of the form $U(X) = m(X)g(X)$, where $U(X)$ is a polynomial of degree $n - 1$ or less. Therefore, the message polynomial $m(X)$ is written

$$m(X) = m_0 + m_1 X + m_2 X^2 + \cdots + m_{n-r-1} X^{n-r-1} \tag{5.58}$$

There are 2^{n-r} codeword polynomials, and there are 2^k code vectors in an (n, k) code. Since there must be one codeword polynomial for each code vector

$$n - r = k$$

or

$$r = n - k$$

Hence $g(X)$, as shown in Equation (5.57), must be of degree $n - k$, and every codeword polynomial in the (n, k) cyclic code can be expressed as

$$U(X) = (m_0 + m_1 X + m_2 X^2 + \cdots + m_{k-1} X^{k-1})g(X) \tag{5.59}$$

U is said to be a valid code vector of the subspace S if, *and only if*, $g(X)$ divides into $U(X)$ without a remainder.

A generator polynomial $g(X)$ of an (n, k) cyclic code is a factor of $X^n + 1$; that is, $X^n + 1 = g(X)h(X)$. For example,

$$X^7 + 1 = (1 + X + X^3)(1 + X + X^2 + X^4)$$

Using $g(X) = 1 + X + X^3$ as a generator polynomial of degree $n - k = 3$, we can generate an $(n, k) = (7, 4)$ cyclic code. Or, using $g(X) = 1 + X + X^2 + X^4$ where $n - k = 4$ we can generate a $(7, 3)$ cyclic code. In summary, if $g(X)$ is a polynomial of degree $n - k$ and is a factor of $X^n + 1$, then $g(X)$ uniquely generates an (n, k) cyclic code.

5.6.3 Encoding in Systematic Form

In Section 5.4.5 we introduced the *systematic* form and discussed the reduction in complexity that makes this encoding form attractive. Let us use some of the algebraic properties of the cyclic code to establish a systematic encoding procedure. We can express the message vector in polynomial form, as follows:

$$m(X) = m_0 + m_1 X + m_2 X^2 + \cdots + m_{k-1} X^{k-1} \tag{5.60}$$

In systematic form, the message digits are utilized as part of the code vector. We can think of shifting the message digits into the rightmost k stages of a codeword register, and then appending the parity digits by placing them in the leftmost $n - k$ stages. Therefore, we want to manipulate the message polynomial algebraically so that it is right-shifted $n - k$ positions. If we multiply $\mathbf{m}(X)$ by X^{n-k} we get the right-shifted message polynomial:

$$X^{n-k}\mathbf{m}(X) = m_0 X^{n-k} + m_1 X^{n-k+1} + \cdots + m_{k-1}X^{n-1} \tag{5.61}$$

If we next divide Equation (5.61) by $\mathbf{g}(X)$, the result can be expressed as

$$X^{n-k}\mathbf{m}(X) = \mathbf{q}(X)\mathbf{g}(X) + \mathbf{r}(X) \tag{5.62}$$

where

$$\mathbf{r}(X) = r_0 + r_1 X + r_2 X^2 + \cdots + r_{n-k-1}X^{n-k-1}$$

We can also say that

$$\mathbf{r}(X) = X^{n-k}\mathbf{m}(X) \text{ modulo } \mathbf{g}(X) \tag{5.63}$$

Adding $\mathbf{r}(X)$ to both sides of Equation (5.62), using modulo-2 arithmetic, we get

$$\mathbf{r}(X) + X^{n-k}\mathbf{m}(X) = \mathbf{q}(X)\mathbf{g}(X) = \mathbf{U}(X) \tag{5.64}$$

The left-hand side of Equation (5.64) is recognized as a valid codeword polynomial, since it is a polynomial of degree $n - 1$ or less, and when divided by $\mathbf{g}(X)$ there is a zero remainder. This codeword can be expanded into its polynomial terms as follows:

$$\mathbf{r}(X) + X^{n-k}\mathbf{m}(X) = r_0 + r_1 X + \cdots + r_{n-k-1}X^{n-k-1}$$
$$+ m_0 X^{n-k} + m_1 X^{n-k+1} + \cdots + m_{k-1}X^{n-1}$$

The codeword polynomial corresponds to the code vector

$$\mathbf{U} = (\underbrace{r_0, r_1, \ldots, r_{n-k-1}}_{(n-k) \text{ parity bits}}, \underbrace{m_0, m_1, \ldots, m_{k-1}}_{k \text{ message bits}}) \tag{5.65}$$

Example 5.8 Cyclic Code in Systematic Form

Using the generator polynomial $\mathbf{g}(X) = 1 + X + X^3$, generate a systematic code vector from the (7, 4) codeword set for the message vector $\mathbf{m} = 1\ 0\ 1\ 1$.

Solution

$$\mathbf{m}(X) = 1 + X^2 + X^3, \quad n = 7, \quad k = 4, \quad n - k = 3$$
$$X^{n-k}\mathbf{m}(X) = X^3(1 + X^2 + X^3) = X^3 + X^5 + X^6$$

Dividing $X^{n-k}\mathbf{m}(X)$ by $\mathbf{g}(X)$ using polynomial division, we can write

$$X^3 + X^5 + X^6 = \underbrace{(1 + X + X^2 + X^3)}_{\substack{\text{quotient} \\ \mathbf{q}(X)}} \underbrace{(1 + X + X^3)}_{\substack{\text{generator} \\ \mathbf{g}(X)}} + \underbrace{1}_{\substack{\text{remainder} \\ \mathbf{r}(X)}}$$

Using Equation (5.64) yields

$$\mathbf{U}(X) = \mathbf{r}(X) + X^3\mathbf{m}(X) = 1 + X^3 + X^5 + X^6$$

$$\mathbf{U} = \underbrace{1\ 0\ 0}_{\substack{\text{parity} \\ \text{bits}}} \quad \underbrace{1\ 0\ 1\ 1}_{\substack{\text{message} \\ \text{bits}}}$$

5.6.4 Circuit for Dividing Polynomials

We have seen that the cyclic shift of a codeword polynomial and that the encoding of a message polynomial involves the division of one polynomial by another. Such an operation is readily accomplished by a *dividing circuit* (feedback shift register). Given two polynomials $\mathbf{V}(X)$ and $\mathbf{g}(X)$, where

$$\mathbf{V}(X) = v_0 + v_1X + v_2X^2 + \cdots + v_mX^m$$

and

$$\mathbf{g}(X) = g_0 + g_1X + g_2X^2 + \cdots + g_rX^r$$

such that $m \geq r$, the divider circuit of Figure 5.17 performs the polynomial division steps of dividing $\mathbf{V}(X)$ by $\mathbf{g}(X)$, thereby determining the quotient and remainder terms:

$$\frac{\mathbf{V}(X)}{\mathbf{g}(X)} = \mathbf{q}(X) + \frac{\mathbf{r}(X)}{\mathbf{g}(X)}$$

The stages of the register are first initialized by being filled with zeros. The first r shifts enter the most significant (higher-order) coefficients of $\mathbf{V}(X)$. After the rth shift, the quotient output is $g_r^{-1}v_m$; this is the highest-order term in the quotient. For each quotient coefficient q_i the polynomial $q_ig(X)$ must be subtracted from the dividend. The feedback connections in Figure 5.17 perform this subtraction. The difference between the leftmost r terms remaining in the dividend

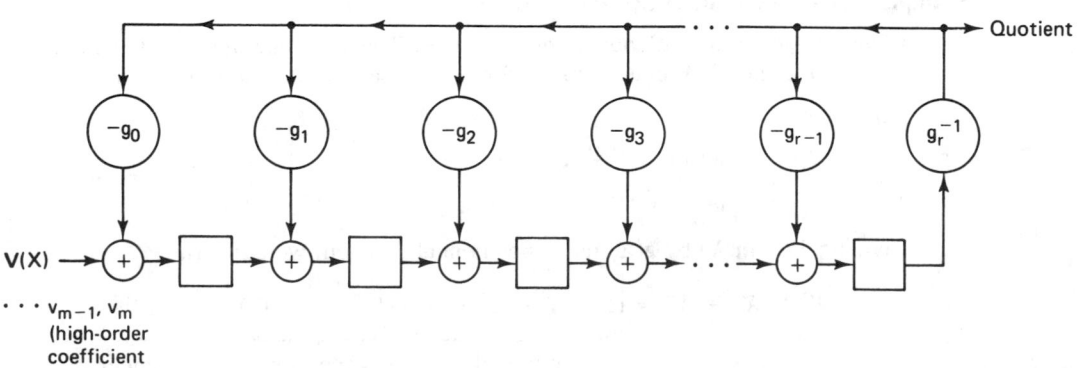

Figure 5.17 Circuit for dividing polynomials.

and the feedback terms $q_i g(X)$ is formed on each shift of the circuit and appears as the contents of the register. At each shift of the register, the difference is shifted one stage; the highest-order term (which by construction is zero) is shifted out, while the next significant coefficient of $V(X)$ is shifted in. After $m + 1$ total shifts into the register, the quotient has been serially presented at the output and the remainder resides in the register.

Example 5.9 Dividing Circuit

Use a dividing circuit of the form shown in Figure 5.17 to divide $V(X) = X^3 + X^5 + X^6$ ($V = 0\ 0\ 0\ 1\ 0\ 1\ 1$) by $g(X) = (1 + X + X^3)$. Find the quotient and remainder terms. Compare the circuit implementation to the polynomial division steps performed by hand.

Solution

The dividing circuit needs to perform the following operation:

$$\frac{X^3 + X^5 + X^6}{1 + X + X^3} = q(X) + \frac{r(X)}{1 + X + X^3}$$

The required feedback shift register, following the general form of Figure 5.17, is shown in Figure 5.18. Assume that the register contents are initially zero. The operational steps of the circuit are as follows:

Input queue	Shift number	Register contents	Output
0 0 0 1 0 1 1	0	0 0 0	–
0 0 0 1 0 1	1	1 0 0	0
0 0 0 1 0	2	1 1 0	0
0 0 0 1	3	0 1 1	0
0 0 0	4	0 1 1	1
0 0	5	1 1 1	1
0	6	1 0 1	1
–	7	1 0 0	1

After the fourth shift, the quotient coefficients $\{q_i\}$ serially presented at the output are seen to be 1 1 1 1, or the quotient polynomial is $q(X) = 1 + X + X^2 + X^3$. The remainder coefficients $\{r_i\}$ are 1 0 0, or the remainder polynomial $r(X) = 1$. In

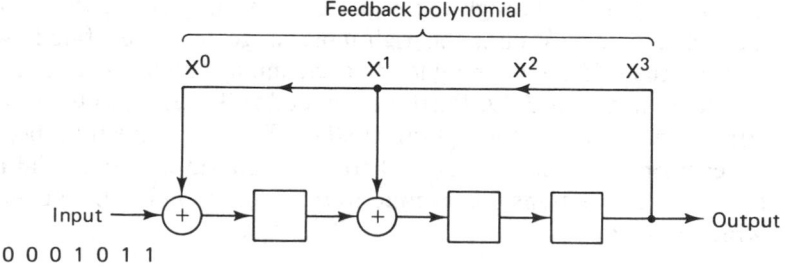

Figure 5.18 Dividing circuit for Example 5.9.

summary, the circuit computation $V(X)/g(X)$ is seen to be

$$\frac{X^3 + X^5 + X^6}{1 + X + X^3} = 1 + X + X^2 + X^3 + \frac{1}{1 + X + X^3}$$

The polynomial division steps are as follows:

Output after shift number:

$$
\begin{array}{cccc}
4 & 5 & 6 & 7 \\
\downarrow & \downarrow & \downarrow & \downarrow
\end{array}
$$

$$X^3 + X^2 + X + 1$$

$$X^3 + X + 1 \overline{) X^6 + X^5 \qquad + X^3 }$$

$$X^6 \qquad + X^4 + X^3 \longleftarrow \text{feedback after 4th shift}$$

$$X^5 + X^4 \longleftarrow \text{register after 4th shift}$$

$$X^5 \qquad + X^3 + X^2 \longleftarrow \text{feedback after 5th shift}$$

$$X^4 + X^3 + X^2 \longleftarrow \text{register after 5th shift}$$

$$X^4 \qquad + X^2 + X \longleftarrow \text{feedback after 6th shift}$$

$$X^3 \qquad + X \longleftarrow \text{register after 6th shift}$$

$$X^3 \qquad + X + 1 \longleftarrow \text{feedback after 7th shift}$$

$$1 \longleftarrow \text{register after 7th shift (remainder)}$$

5.6.5 Systematic Encoding with an (n − k)-Stage Shift Register

The encoding of a cyclic code in systematic form has been shown, in Section 5.6.3, to involve the computation of parity bits as the result of the formation of $X^{n-k} m(X)$ modulo $g(X)$, in other words, the *division* of an *upshifted* (right shifted) message polynomial by a generator polynomial $g(X)$. The need for upshifting is to make room for the parity bits, which are appended to the message bits, yielding the code vector in systematic form. Upshifting the message bits by $n - k$ positions is a trivial operation and is not really performed as part of the dividing circuit. Instead, only the parity bits are computed; they are then placed in the appropriate location alongside the message bits. The parity polynomial is the *remainder* after dividing by the generator polynomial; it is available in the register after n shifts through the $(n - k)$-stage feedback register shown in Figure 5.18. Notice that the first $n - k$ shifts through the register are simply filling the register. We cannot have any feedback until the rightmost stage has been filled; we therefore can shorten the shifting cycle by loading the input data to the output of the last stage, as shown in Figure 5.19. Further, the feedback term into the leftmost stage is the sum of the input and the rightmost stage. We guarantee that this sum is generated by ensuring that $g_0 = g_{n-k} = 1$ for any generator polynomial $g(X)$. The circuit feedback connections correspond to the coefficients of the generator polynomial, which is written

$$g(X) = 1 + g_1 X + g_2 X^2 + \cdots + g_{n-k-1} X^{n-k-1} + X^{n-k} \qquad (5.66)$$

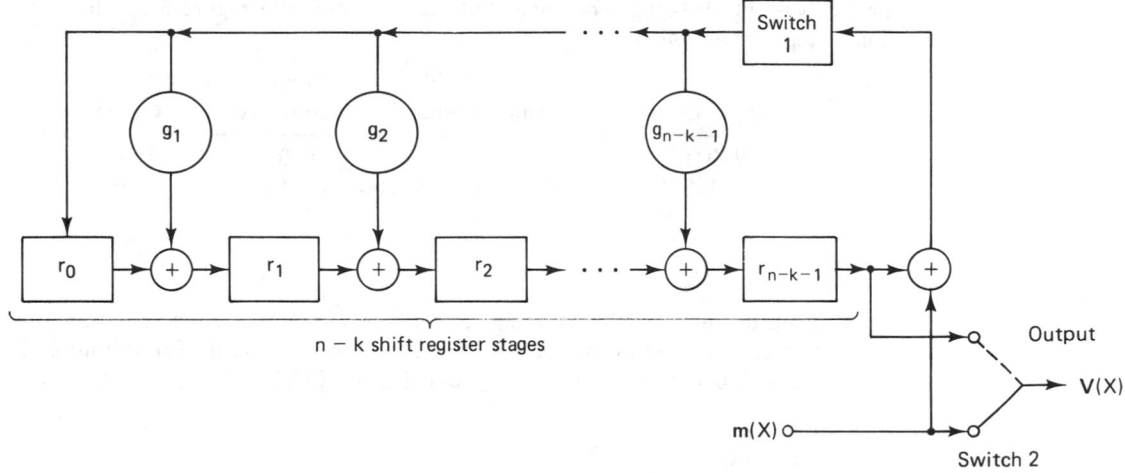

Figure 5.19 Encoding with an $(n - k)$-stage shift register.

The following steps describe the encoding procedure used with the Figure 5.19 encoder.

1. Switch 1 is closed during the first k shifts, to allow transmission of the message bits into the $n - k$ stage encoding shift register.
2. Switch 2 is in the down position to allow transmission of the message bits directly to an output register during the first k shifts.
3. After transmission of the kth message bit, switch 1 is opened and switch 2 is moved to the up position.
4. The remaining $n - k$ shifts clear the encoding register by moving the parity bits to the output register.
5. The total number of shifts is equal to n, and the contents of the output register is the codeword polynomial $r(X) + X^{n-k}m(X)$.

Example 5.10   Systematic Encoding of a Cyclic Code

Use a feedback shift register of the form shown in Figure 5.19 to encode the message vector $\mathbf{m} = 1\ 0\ 1\ 1$ into a (7, 4) code vector using the generator polynomial $g(X) = 1 + X + X^3$.

Solution

$$\mathbf{m} = 1\ 0\ 1\ 1$$

$$\mathbf{m}(X) = 1 + X^2 + X^3$$

$$X^{n-k}\mathbf{m}(X) = X^3\mathbf{m}(X) = X^3 + X^5 + X^6$$

$$X^{n-k}\mathbf{m}(X) = \mathbf{q}(X)\mathbf{g}(X) + \mathbf{r}(X)$$

$$\mathbf{r}(X) = X^3 + X^5 + X^6 \text{ modulo } (1 + X + X^3)$$

For the $(n - k) = 3$-stage encoding shift register shown in Figure 5.20, the operational steps are as follows:

Input queue	Shift number	Register contents	Output
1 0 1 1	0	0 0 0	–
1 0 1	1	1 1 0	1
1 0	2	1 0 1	1
1	3	1 0 0	0
–	4	1 0 0	1

After the fourth shift, switch 1 is opened, switch 2 is moved to the up position, and the parity bits contained in the register are shifted to the output. The output code vector is $U = 1\ 0\ 0\ 1\ 0\ 1\ 1$, or in polynomial form, $U(X) = 1 + X^3 + X^5 + X^6$.

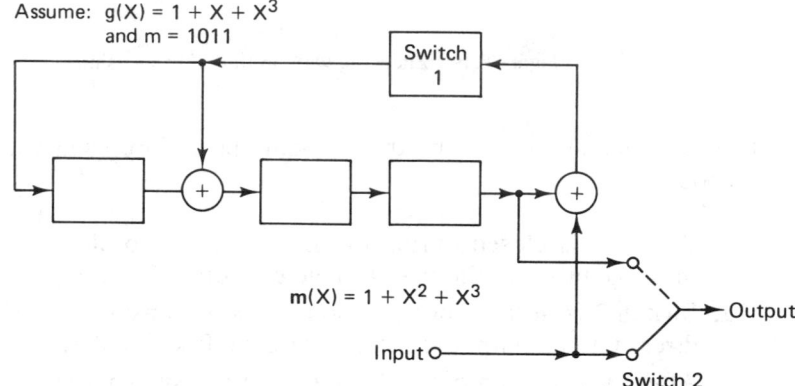

Assume: $g(X) = 1 + X + X^3$ and m = 1011

$m(X) = 1 + X^2 + X^3$

Figure 5.20 Example of encoding a $(7, 4)$ cyclic code with an $(n - k)$-stage shift register.

5.6.6 Error Detection with an $(n - k)$-Stage Shift Register

A transmitted code vector may be perturbed by noise, and hence the vector received may be a corrupted version of the transmitted code vector. Let us assume that a codeword with polynomial representation $U(X)$ is transmitted and that a vector with polynomial representation $Z(X)$ is received. Since $U(X)$ is a code polynomial, it must be a multiple of the generator polynomial $g(X)$, that is,

$$U(X) = m(X)g(X) \tag{5.67}$$

and $Z(X)$, the corrupted version of $U(X)$, can be written

$$Z(X) = U(X) + e(X) \tag{5.68}$$

where $e(X)$ is the error pattern polynomial. The decoder tests whether $Z(X)$ is a codeword polynomial, that is, whether it is divisible by $g(X)$, with a zero remainder. This is accomplished by *calculating the syndrome* of the received polynomial. The syndrome $S(X)$ is equal to the remainder resulting from dividing $Z(X)$ by $g(X)$, that is,

$$Z(X) = q(X)g(X) + S(X) \qquad (5.69)$$

where $S(X)$ is a polynomial of degree $n - k - 1$ or less. Thus the syndrome is an $(n - k)$-tuple. By combining Equations (5.67) to (5.69), we obtain

$$e(X) = [m(X) + q(X)]g(X) + S(X) \qquad (5.70)$$

By comparing Equations (5.69) and (5.70), we see that the syndrome $S(X)$, obtained as the remainder of $Z(X)$ modulo $g(X)$, is exactly the same polynomial obtained as the remainder of $e(X)$ modulo $g(X)$. Thus the syndrome of the received polynomial $Z(X)$ contains the information needed for correction of the error pattern. The syndrome calculation is accomplished by a division circuit, almost identical to the encoding circuit used at the transmitter. An example of syndrome calculation with an $n - k$ shift register is shown in Figure 5.21 using the code vector generated in Example 5.10. Switch 1 is initially closed, and switch 2 is open. The received vector is shifted into the register input, with all stages initially set to zero. After the entire received vector has been entered into the shift register, the contents of the register is the syndrome. Switch 1 is then opened and switch 2 is closed, so that the syndrome vector can be shifted out of the register. The operational steps of the decoder are as follows:

Input queue	Shift number	Register contents
1 0 0 1 0 1 1	0	0 0 0
1 0 0 1 0 1	1	1 0 0
1 0 0 1 0	2	1 1 0
1 0 0 1	3	0 1 1
1 0 0	4	0 1 1
1 0	5	1 1 1
1	6	1 0 1
–	7	$\boxed{0\ 0\ 0}$ Syndrome

If the syndrome is an all-zeros vector, the received vector is assumed to be a valid code vector. If the syndrome is a nonzero vector, the received vector is a

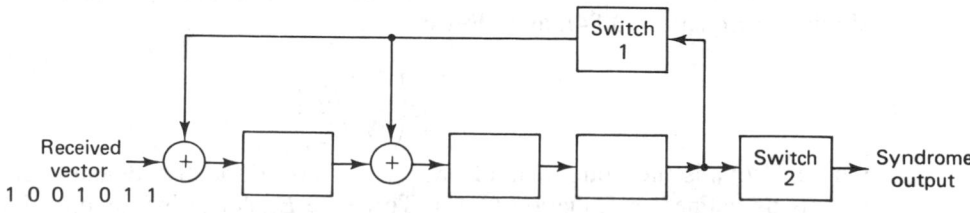

Figure 5.21 Example of syndrome calculation with an $(n - k)$-stage shift register.

perturbed code vector and errors have been detected; such errors can be corrected by adding the error vector (indicated by the syndrome) to the received vector, similar to the procedure described in Section 5.4.8. This method of decoding is useful for simple codes. More complex codes require the use of algebraic techniques to obtain practical decoders [6, 7].

5.7 WELL-KNOWN BLOCK CODES

5.7.1 Hamming Codes

Hamming codes are a simple class of block codes characterized by the following (n, k) structure:

$$(n, k) = (2^m - 1, 2^m - 1 - m) \tag{5.71}$$

where $m = 2, 3, \ldots$. These codes have a minimum distance of 3 and thus, from Equations (5.44) and (5.47), they are capable of correcting all single errors or detecting all combinations of two or fewer errors within a block. Syndrome decoding is especially suited for Hamming codes. In fact, the syndrome can be formed to act as a binary pointer to identify the error location [5]. Although Hamming codes are not very powerful, they belong to a very limited class of block codes known as *perfect* codes, described in Section 5.5.4.

Assuming hard decision decoding the bit error probability can be written, from Equation (5.46), as follows:

$$P_B \simeq \frac{1}{n} \sum_{j=2}^{n} j \binom{n}{j} p^j (1 - p)^{n-j} \tag{5.72}$$

where p is the channel symbol error probability (transition probability on the binary symmetric channel). In place of Equation (5.72) we can use the following equivalent equation. Its identity with Equation (5.72) is proven in Appendix D, Equation (D.16).

$$P_B \cong p - p(1 - p)^{n-1} \tag{5.73}$$

Figure 5.22 is a plot of P_B versus channel symbol error probability, illustrating the comparative performance for different types of block codes. For the Hamming codes, the plots are shown for $m = 3, 4,$ and 5, or $(n, k) = (7, 4), (15, 11),$ and $(31, 26)$. For performance over a Gaussian channel using coherently demodulated BPSK, we can express the channel symbol error probability in terms of E_c/N_0, similar to Equation (3.84), as follows:

$$p = Q\left(\sqrt{\frac{2E_c}{N_0}}\right) \tag{5.74}$$

where E_c/N_0 is the code symbol energy per noise spectral density, and where $Q(x)$ is as defined in Equation (2.42). To relate E_c/N_0 to information bit energy per noise spectral density (E_b/N_0), we use

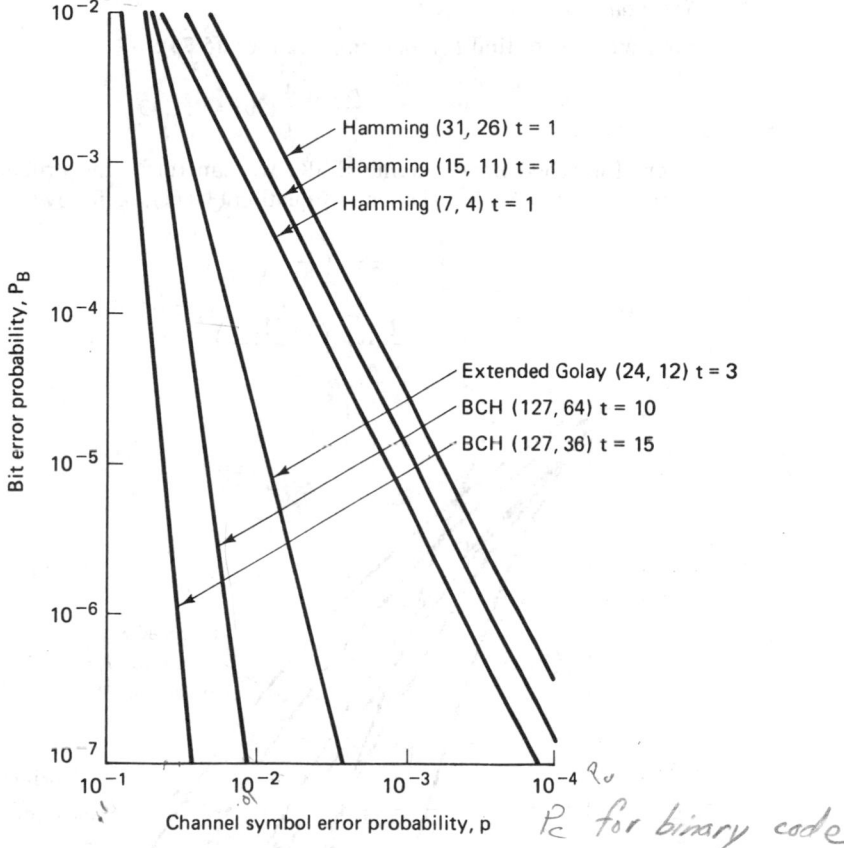

Figure 5.22 Bit error probability versus channel symbol error probability for several block codes.

$$\frac{E_c}{N_0} = \left(\frac{k}{n}\right) \frac{E_b}{N_0} \tag{5.75}$$

For Hamming codes, Equation (5.75) becomes

$$\frac{E_c}{N_0} = \frac{2^m - 1 - m}{2^m - 1} \frac{E_b}{N_0} \tag{5.76}$$

Combining Equations (5.73), (5.74), and (5.76), P_B can be expressed as a function of E_b/N_0 for coherently demodulated BPSK over a Gaussian channel. The results are plotted in Figure 5.23 for different types of block codes. For the Hamming codes, plots are shown for $(n, k) = (7, 4)$, $(15, 11)$, and $(31, 26)$.

Example 5.11 Error Probability for Modulated and Coded Signals

A coded BFSK modulated signal is transmitted over a Gaussian channel. The signal is noncoherently detected and hard-decision decoded. Find the decoded bit error probability if the coding is a Hamming (7, 4) block code and the received E_b/N_0 is equal to 20.

Solution

First we need to find E_c/N_0 using Equation (5.75):

$$\frac{E_c}{N_0} = \frac{4}{7}(20) = 11.43$$

Then, for coded noncoherent BFSK, we can relate the probability of a channel symbol error to E_c/N_0, similar to Equation (3.111), as follows:

$$p = \frac{1}{2}\exp\left(-\frac{E_c}{2N_0}\right)$$

$$= \frac{1}{2}\exp\left(-\frac{11.43}{2}\right) = 1.6 \times 10^{-3}$$

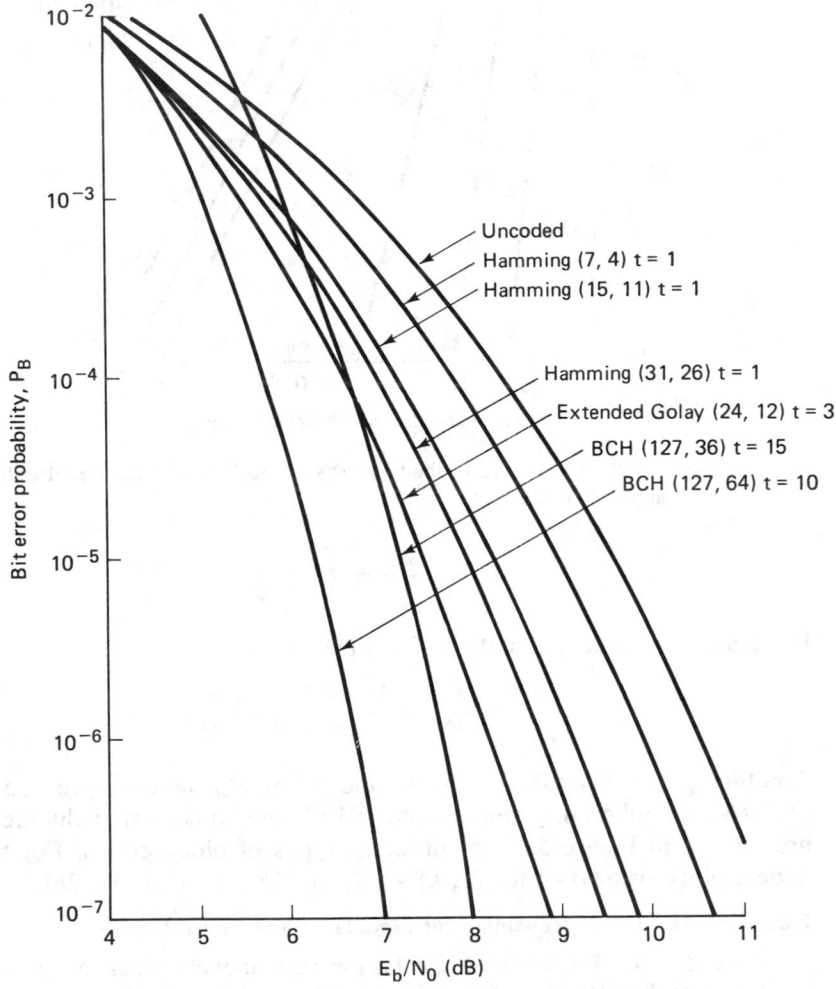

Figure 5.23 P_B versus E_b/N_0 for coherently demodulated BPSK over a Gaussian channel for several block codes.

Using this result in Equation (5.73), we solve for the probability of a decoded bit error, as follows:

$$P_B \simeq p - p(1 - p)^6 \simeq 1.6 \times 10^{-5}$$

5.7.2 Extended Golay Code

One of the more useful block codes is the binary (24, 12) *extended Golay code*, which is formed by adding an overall parity bit to the perfect (23, 12) code, known as the *Golay code*. This added parity bit increases the minimum distance d_{min} from 7 to 8 and produces a rate $\frac{1}{2}$ code, which is easier to implement (with regard to system clocks) than the rate 12/23 original Golay code. Extended Golay codes are considerably more powerful than the Hamming codes described in the preceding section. The price paid for the improved performance is a more complex decoder, a lower code rate, and hence a larger bandwidth expansion.

Since $d_{min} = 8$ for the extended Golay code, we see from Equation (5.44) that the code is guaranteed to correct all triple errors. The decoder can additionally be designed to correct *some but not all* four-error patterns. Since only 19% of the four-error patterns can be corrected, the decoder, for the sake of simplicity, is usually designed to only correct three-error patterns [5]. Assuming hard decision decoding, the bit error probability for the extended Golay code can be written as a function of the channel symbol error probability, p, from Equation (5.46), as follows:

$$P_B \simeq \frac{1}{24} \sum_{j=4}^{24} j \binom{24}{j} p^j (1 - p)^{24-j} \tag{5.77}$$

The plot of Equation (5.77) is shown in Figure 5.22; the error performance of the extended Golay code is seen to be significantly better than that of the Hamming codes. Combining Equations (5.77), (5.74), and (5.75), we can relate P_B versus E_b/N_0 for coherently demodulated BPSK with extended Golay coding over a Gaussian channel. The result is plotted in Figure 5.23.

5.7.3 BCH Codes

Bose–Chadhuri–Hocquenghem (BCH) codes are a generalization of Hamming codes that allow multiple error correction. They are a *powerful class of cyclic codes* that provide a large selection of block lengths, code rates, alphabet sizes, and error-correcting capability. Table 5.2 lists some commonly used code generators, $g(x)$, for the construction of BCH codes [8], for various values of n, k, and t, up to a block length of 255. The coefficients of $g(x)$ are presented as octal numbers arranged so that when they are converted to binary digits the rightmost digit corresponds to the zero-degree coefficient of $g(x)$. BCH codes are important, because at block lengths of a few hundred, the BCH codes outperform all other block codes with the same block length and code rate. The most commonly used BCH codes employ a binary alphabet and a codeword block length of $n = 2^m - 1$, where $m = 3, 4, \ldots$.

TABLE 5.2 Generators of Primitive BCH Codes

n	k	t	g(x)
7	4	1	13
15	11	1	23
	7	2	721
	5	3	2467
31	26	1	45
	21	2	3551
	16	3	107657
	11	5	5423325
	6	7	313365047
63	57	1	103
	51	2	12471
	45	3	1701317
	39	4	166623567
	36	5	1033500423
	30	6	157464165547
	24	7	17323260404441
	18	10	1363026512351725
	16	11	6331141367235453
	10	13	472622305527250155
	7	15	5231045543503271737

n	k	t	g(x)
255	171	11	15416214212342356077061630637
	163	12	7500415510075602551574724514601
	155	13	37575130054076650157225064644677633
	147	14	1642130173537165525304165305441011711
	139	15	461401732060175561570722730247453567445
	131	18	215713331471510151261250277442142024165471
	123	19	120614052242066003717210326516141226272506267
	115	21	6052665572100247263636404600276352556313472737
	107	22	22205772322066256312417300235347420176574750154441
	99	23	1065666725347317422274141620157433225241107643203431
	91	25	675026503027444172723631724732511075550762720724344561
	87	26	110136763141743236435231634307172046206722545273311721317
	79	27	66700035637657500020270344207366174621015326711766541342355

n	k	t	$g(x)$
127	120	1	211
	113	2	41567
	106	3	11554743
	99	4	3447023271
	92	5	624730022327
	85	6	130704476322273
	78	7	26230002166130115
	71	9	6255010713253127753
	64	10	1206534025570773100045
	57	11	335265252505705053517721
	50	13	54446512523314012421501421
	43	14	17721772213651227512220574343
	36	15	3146074666522075044764574721735
	29	21	403114461367670603667530141176155
	22	23	1233760700404722522243544562663647043
	15	27	22057042456045454770523013762217604353
	8	31	7047264052751030651476224271567733130217
	71	29	2402471052064432151555417211233116320544250362557643221706035
	63	30	1075447505516354432515217357707003666111726455267613656702543301
	55	31	7315425203501100133015275306032054325414326755010557044426035473617
	47	42	2533542017062646563033041377406233175123334145446045005066024552543173
	45	43	15202056055234161131101346376423701563670024470762373033202157025051541
	37	45	5136330255067007414177447245437530420735706174323432347644354737403044003
	29	47	30257155366730714655270640123613771153422423242011741140602547574410403565037
	21	55	1256215257060332656001773153607612103227341405653074542521153121614466513473725
	13	59	4641732005052564544426573714250066004330677445476564140317467721357026134460500547
	9	63	1572602521747246320103104325535134614162367212044074545112766115547705561677516057
255	247	1	435
	239	2	267543
	231	3	156720665
	223	4	75626641375
	215	5	23157564726421
	207	6	16176560567636227
	199	7	7633031270407222341
	191	8	266347076115333714567
	187	9	52755313540001322236351
	179	10	22624710717340432416300455

Source: Reprinted with permission from "Table of Generators for BCH Codes," *IEEE Trans. Inf. Theory*, vol. IT10, no. 4, Oct. 1964, p. 391. © 1964 IEEE.

The title of Table 5.2 indicates that the generators shown are for those BCH codes known as *primitive codes*. The term "primitive" is a number-theoretic concept requiring an elaborate algebraic development [9–11], which will not be presented here. In Figures 5.22 and 5.23 are plotted error performance curves of two BCH codes (127, 64) and (127, 36), to illustrate comparative performance. Assuming hard decision decoding, the P_B versus channel error probability is shown in Figure 5.22. The P_B versus E_b/N_0 for coherently demodulated BPSK over a Gaussian channel is shown in Figure 5.23. The curves in Figure 5.23 seem to depart from our expectations. They each have the same block size, yet the more redundant (127, 36) code does not exhibit as much coding gain as does the less redundant (127, 64) code. It has been shown that a relatively broad maximum of coding gain versus code rate for fixed n occurs roughly between coding rates of $\frac{1}{3}$ and $\frac{3}{4}$ for BCH codes [12]. Performance over a Gaussian channel degrades substantially at very high or very low rates [11].

Figure 5.24 represents computed performance of BCH codes [13] using coherently demodulated BPSK with both *hard-* and *soft-decision decoding*. Soft-decision decoding is not usually used with block codes because of its complexity. However, whenever it is implemented, it offers an approximate 2-dB coding gain over hard-decision decoding. For a given code rate, the decoded error probability is known to improve with increasing block length n [4]. Thus for a given code rate, it is interesting to compare the block length that would be required for the hard-decision-decoding performance to be comparable to the soft-decision-decoding performance. In Figure 5.24, the BCH codes shown all have code rates of approximately $\frac{1}{2}$. From the figure [13] it appears that for a fixed code rate, the hard-decision-decoded BCH code of length 8 times n or longer has a better performance than that of a soft-decision-decoded BCH code of length n.

5.7.4 Reed–Solomon Codes

One special subclass of the BCH codes (the discovery of which preceded the BCH codes) is the particularly useful *nonbinary* set called Reed–Solomon codes. Reed–Solomon codes achieve the *largest possible code minimum distance* for any linear code with the same encoder input and output block lengths. For nonbinary codes, the distance between two code words is defined as the number of nonbinary symbols in which the sequences differ. For Reed–Solomon codes the code minimum distance is given by [14]

$$d_{\min} = n - k + 1 \tag{5.78}$$

where k is the number of data symbols being encoded, and n is the total number of code symbols in the encoded block. Following Equation (5.44), the code is capable of correcting any combination of t or fewer symbol errors, as follows:

$$t = \frac{d_{\min} - 1}{2} = \frac{n - k}{2} \tag{5.79}$$

and thus requires no more than $2t$ parity check symbols.

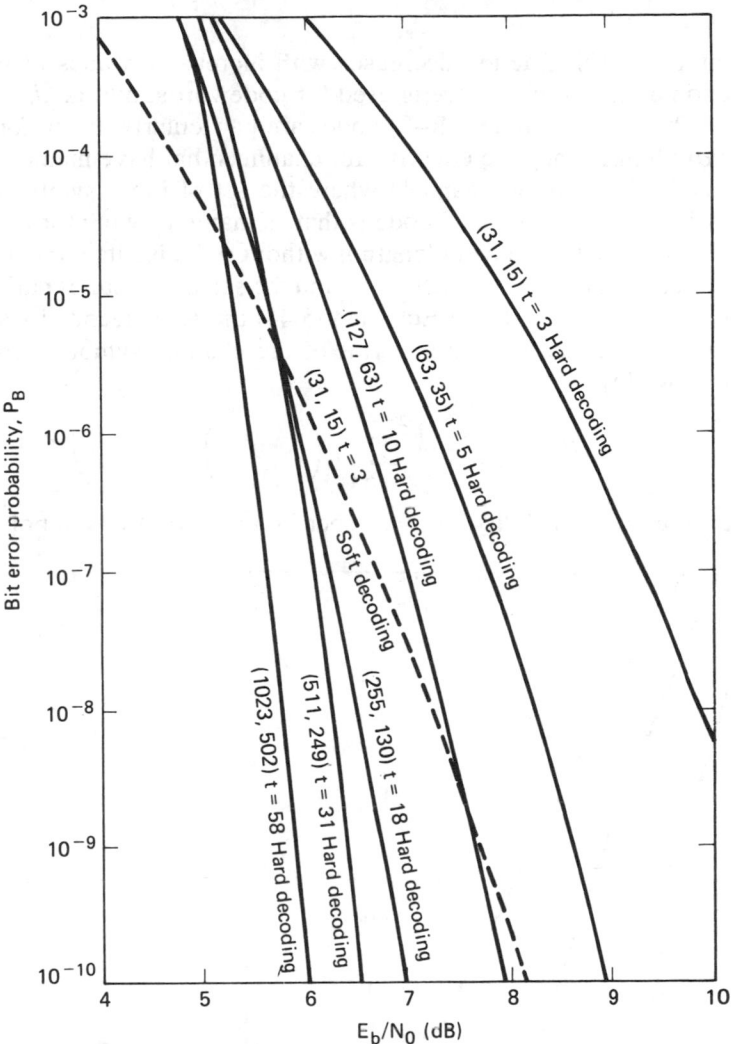

Figure 5.24 P_B versus E_b/N_0 for coherently demodulated BPSK over a Gaussian channel using BCH codes. (Reprinted with permission from L. J. Weng, "Soft and Hard Decoding Performance Comparisons for BCH Codes," *Proc. Int. Conf. Commun.*, 1979, Fig. 3, p. 25.5.5. © 1979 IEEE.)

A *t*-error-correcting Reed–Solomon code with an alphabet of 2^m symbols has $n = 2^m - 1$ and $k = 2^m - 1 - 2t$, where $m = 2, 3, \ldots$. An advantage of nonbinary codes such as a Reed–Solomon code can be seen by the following comparison. Consider a binary $(n, k) = (7, 3)$ code. The entire *n*-tuple space amounts to $2^n = 2^7 = 128$ binary words, of which $2^k = 2^3 = 8$ (or $\frac{1}{16}$ of the *n*-tuples) are codewords. Next, consider a nonbinary $(n, k) = (7, 3)$ code where each symbol is comprised of $m = 3$ bits. The *n*-tuple space amounts to $2^{nm} = 2^{21} = 2,097,152$ binary words, of which $2^{km} = 2^9 = 512$ (or 1/4096 of the *n*-tuples) are codewords. With symbols, each made up of *m* bits, only a small fraction (i.e., 2^{km} of the large number 2^{nm}) of possible different words of *n* symbols become codewords. This fraction decreases with increasing values of m. When a small fraction of the n-tuple space is used for codewords, a large $d_{\min}$ can be created.

The Reed–Solomon (R–S) codes are particularly useful for *burst-error correction*; that is, they are effective for channels that have memory. Also, they can be used efficiently on channels where the set of input symbols is large. An interesting feature of the R–S code is that as many as two information symbols can be added to an R–S code of length n without reducing its minimum distance. This extended R–S code has length $n + 2$ and the same number of parity check symbols as the original code. From Equation (5.46) the R–S decoded symbol error probability, P_E, can be written in terms of the channel symbol error probability, p, as follows [5]:

$$P_E \simeq \frac{1}{2^m - 1} \sum_{j=t+1}^{2^m - 1} j \binom{2^m - 1}{j} p^j (1 - p)^{2^m - 1 - j} \tag{5.80}$$

The bit error probability can be upper bounded by the symbol error probability

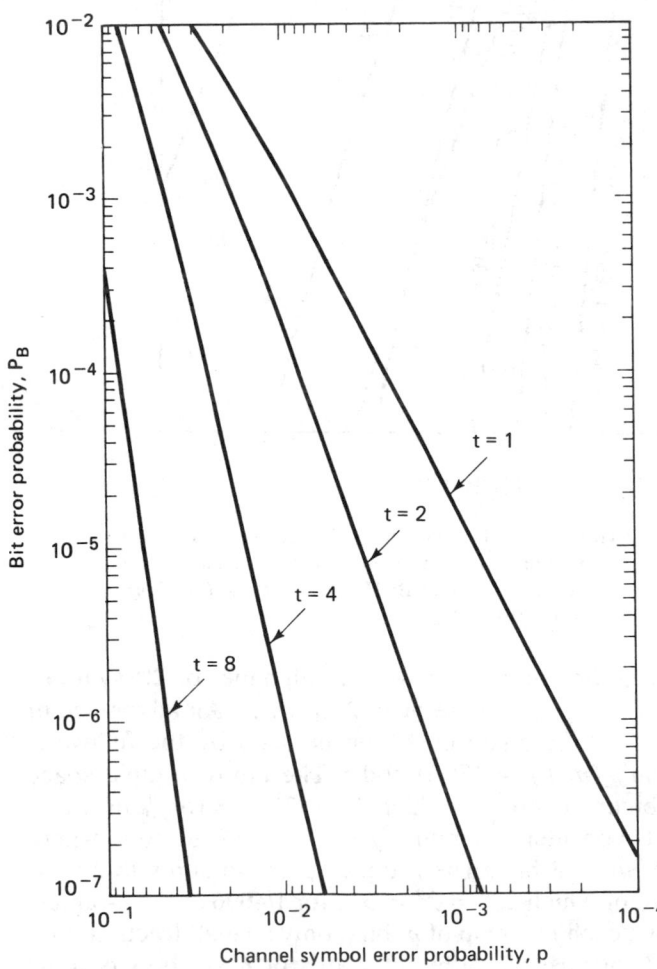

Figure 5.25 P_B versus p for 32-ary orthogonal signaling and $n = 31$, t-error-correcting Reed–Solomon coding. (Reprinted with permission from *Data Communications, Networks, and Systems*, ed. Thomas C. Bartee, Howard W. Sams Company, Indianapolis, Ind., 1985, p. 311. Originally published in J. P. Odenwalder, *Error Control Coding Handbook*, M/A-COM LINKABIT, Inc., San Diego, Calif., July 15, 1976, p. 91.)

for specific modulation types. For MFSK modulation with $M = 2^m$, the relationship between P_B and P_E as given in Equation (3.127) is repeated here:

$$\frac{P_B}{P_E} = \frac{2^{m-1}}{2^m - 1} \tag{5.81}$$

Figure 5.25 shows P_B versus the channel symbol error probability, p, plotted from Equations (5.80) and (5.81) for various t-error-correcting 32-ary orthogonal Reed–

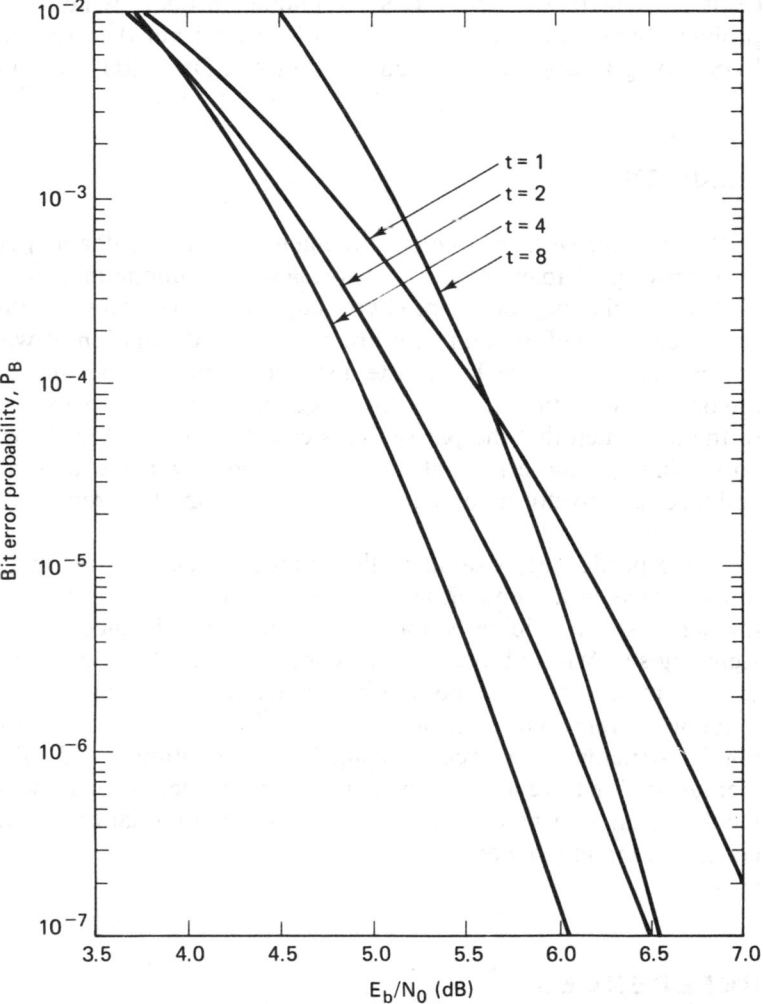

Figure 5.26 Bit error probability versus E_b/N_0 performance of several $n = 31$, t-error-correcting Reed–Solomon coding systems with 32-ary MFSK modulation over an AWGN channel. (Reprinted with permission from *Data Communications, Networks, and Systems*, ed. Thomas C. Bartee, Howard W. Sams Company, Indianapolis, Ind., 1985, p. 312. Originally published in J. P. Odenwalder, *Error Control Coding Handbook*, M/A-COM LINKABIT, Inc., San Diego, Calif., July 15, 1976, p. 92.)

Solomon codes with $n = 31$ (thirty-one 5-bit symbols per code block). Figure 5.26 shows P_B versus E_b/N_0 for such a coded system using 32-ary MFSK modulation and noncoherent demodulation over an AWGN channel [5]. For R–S codes, error probability is an exponentially decreasing function of block length, n, and decoding complexity is proportional to a small power of the block length [14]. The R–S codes are sometimes used in a concatenated arrangement. In such a system, an inner convolutional decoder first provides some error control by operating on soft-decision demodulator outputs; the convolutional decoder then presents hard-decision data to the outer Reed–Solomon decoder, which further reduces the probability of error. In Chapter 6 we discuss further the use of concatenated and R–S coding as applied to the compact disc (CD) digital audio system.

5.8 CONCLUSION

In this chapter we have explored the general goals of channel coding, all leading to improve probability of error performance at a minimum cost in bandwidth. We partitioned channel coding into two study groups: waveform coding and structured sequences. Waveform coding represents a transformation of waveforms into improved waveforms, such that the distance properties are improved compared to the original waveforms. Structured sequences involve the addition of parity digits to the data such that the parity digits can then be employed for detecting and/or correcting specific error patterns. The main advantage of structured sequence coding over waveform coding is that it can accomplish improved P_B performance using less bandwidth.

We particularly examined linear block codes. Geometric analogies can be drawn between the coding and modulation disciplines. They both seek to pack the signal space efficiently and to maximize the distance between signals in the signaling set. Within block codes we looked at cyclic codes, which are relatively easy to implement using modern integrated circuit techniques. We considered the polynomial representation of codes and the correspondence between the polynomial structure, the necessary algebraic operations, and the hardware implementation. We have also looked at performance details of some of the well-known block codes. A large class of codes, the convolutional codes, have been left for consideration in Chapter 6.

REFERENCES

1. Viterbi, A. J., "On Coded Phase-Coherent Communications," *IRE Trans. Space Electron. Telem.,* vol. SET7, Mar. 1961, pp. 3–14.
2. Lindsey, W. C., and Simon, M. K., *Telecommunication Systems Engineering,* Prentice-Hall, Inc., Englewood Cliffs, N.J., 1973.

3. Proakis, J. G., *Digital Communications,* McGraw-Hill Book Company, New York, 1983.

4. Lin, S., and Costello, D. J., Jr., *Error Control Coding: Fundamentals and Applications,* Prentice-Hall, Inc., Englewood Cliffs, N.J., 1983.

5. Odenwalder, J. P., *Error Control Coding Handbook,* Linkabit Corporation, San Diego, Calif., July 15, 1976.

6. Blahut, R. E., *Theory and Practice of Error Control Codes,* Addison-Wesley Publishing Company, Inc., Reading, Mass., 1983.

7. Blahut, R. E., "Algebraic Fields, Signal Processing, and Error Control," *Proc. IEEE,* vol. 73, May 1985, pp. 874–893.

8. Stenbit, J. P., "Tables of Generators for Bose–Chadhuri Codes, *IEEE Trans. Inf. Theory,* vol. IT10, no. 4, Oct. 1964, pp. 390–391.

9. Berlekamp, E. R., *Algebraic Coding Theory,* McGraw-Hill Book Company, New York, 1968.

10. Peterson, W. W., and Weldon, E. J., *Error Correcting Codes,* 2nd ed., The MIT Press, Cambridge, Mass., 1972.

11. Clark, G. C., Jr., and Cain, J. B., *Error-Correction Coding for Digital Communications,* Plenum Press, New York, 1981.

12. Wozencraft, J. M., and Jacobs, I. M., *Principles of Communication Engineering,* John Wiley & Sons, Inc., New York, 1965.

13. Weng, L. J., "Soft and Hard Decoding Performance Comparisons for BCH Codes," *Proc. Int. Conf. Commun.,* 1979, pp. 25.5.1–25.5.5.

14. Gallager, R. G., *Information Theory and Reliable Communication,* John Wiley & Sons, Inc., New York, 1968.

PROBLEMS

5.1. Design an (n, k) single-parity code that will detect all 1-, 3-, 5-, and 7-error patterns in a block. Show the values of n and k, and find the probability of an undetected block error if the probability of channel symbol error is 10^{-2}.

5.2. Calculate the probability of message error for a 12-bit data sequence encoded with a (24, 12) linear block code. Assume that the code corrects all 1-bit and 2-bit error patterns and assume that it corrects no error patterns with more than two errors. Also, assume that the probability of a channel symbol error is 10^{-3}.

5.3. Consider a (127, 92) linear block code capable of triple error corrections.
   (a) What is the probability of message error for an uncoded block of 92 bits if the channel symbol error probability is 10^{-3}?
   (b) What is the probability of message error when using the (127, 92) block code if the channel symbol error probability is 10^{-3}?

5.4. Calculate the improvement in probability of message error relative to an uncoded transmission for a (24, 12) double-error-correcting linear block code. Assume that coherent BPSK modulation is used and that the received $E_b/N_0 = 10$ dB.

5.5. Consider a (24, 12) linear block code capable of double-error corrections. Assume that a noncoherently detected BFSK modulation format is used and that the received $E_b/N_0 = 14$ dB.

(a) Does the code provide any improvement in probability of message error? If it does, how much? If it does not, explain why not.

(b) Repeat part (a) with $E_b/N_0 = 10$ dB.

5.6. The telephone company uses a "best-of-five" encoder for some of its digital data channels. In this system every data bit is repeated five times, and at the receiver, a majority vote decides the value of each data bit. If the uncoded probability of bit error is 10^{-3}, calculate the coded bit error probability when using such a best-of-five code.

5.7. The minimum distance for a particular linear block code is 11. Find the maximum error-correcting capability, the maximum error-detecting capability, and the maximum erasure-correcting capability in a block length.

5.8. Consider a (7, 4) code whose generator matrix is

$$G = \begin{bmatrix} 1 & 1 & 1 & 1 & 0 & 0 & 0 \\ 1 & 0 & 1 & 0 & 1 & 0 & 0 \\ 0 & 1 & 1 & 0 & 0 & 1 & 0 \\ 1 & 1 & 0 & 0 & 0 & 0 & 1 \end{bmatrix}$$

(a) Find all the code vectors of the code.

(b) Find **H**, the parity-check matrix of the code.

(c) Compute the syndrome for the received vector 1 1 0 1 1 0 1. Is this a valid code vector?

(d) What is the error-correcting capability of the code?

(e) What is the error-detecting capability of the code?

5.9. Consider a systematic block code whose parity-check equations are

$$p_1 = m_1 + m_2 + m_4$$

$$p_2 = m_1 + m_3 + m_4$$

$$p_3 = m_1 + m_2 + m_3$$

$$p_4 = m_2 + m_3 + m_4$$

where m_i are message digits and p_i are check digits.

(a) Find the generator matrix and the parity-check matrix for this code.

(b) How many errors can the code correct?

(c) Is the vector 10101010 a codeword?

(d) Is the vector 01011100 a codeword?

5.10. Consider the linear block code with the codeword defined by

$$U = m_1 + m_2 + m_4 + m_5, \; m_1 + m_3 + m_4 + m_5, \; m_1 + m_2 + m_3 + m_5,$$
$$m_1 + m_2 + m_3 + m_4, m_1, m_2, m_3, m_4, m_5$$

(a) Show the generator matrix.

(b) Show the parity-check matrix.

(c) Find n, k, and d_{min}.

5.11. Design a (4, 2) linear block code.

    (a) Choose the codewords to be in systematic form, and choose them with the goal of maximizing d_{min}.

    (b) Find the generator matrix for the codeword set.

    (c) Calculate the parity-check matrix.

    (d) Enter the sixteen 4-tuples into a standard array.

    (e) What are the error-correcting and error-detecting capabilities of the code?

    (f) Make a syndrome table for the correctable error patterns.

5.12. Consider the (5, 1) repetition code, which consists of the two codewords 00000 and 11111, corresponding to messages 0 and 1, respectively. Derive the standard array for this code. Is this a perfect code?

5.13. Design a (3, 1) code that will correct all single-error patterns. Choose the codeword set and show the standard array.

5.14. Is a (7, 3) code a perfect code? Is a (7, 4) code a perfect code? Is a (15, 11) code a perfect code? Justify your answers.

5.15. A (15, 11) linear block code can be defined by the following parity array:

$$
P = \begin{bmatrix}
0 & 0 & 1 & 1 \\
0 & 1 & 0 & 1 \\
1 & 0 & 0 & 1 \\
0 & 1 & 1 & 0 \\
1 & 0 & 1 & 0 \\
1 & 1 & 0 & 0 \\
0 & 1 & 1 & 1 \\
1 & 1 & 1 & 0 \\
1 & 1 & 0 & 1 \\
1 & 0 & 1 & 1 \\
1 & 1 & 1 & 1
\end{bmatrix}
$$

    (a) Show the parity-check matrix for this code.

    (b) List the coset leaders from the standard array. Is this code a perfect code? Justify your answer.

    (c) A received vector is V = 0 1 1 1 1 1 0 0 1 0 1 1 0 1 1. Compute the syndrome. Assuming that a single bit error has been made, find the correct codeword.

    (d) How many erasures can this code correct? Explain.

5.16. Is it possible that a nonzero error pattern can produce a syndrome of $S = 0$? If yes, how many such error patterns can give this result for an (n, k) code? Use Figure 5.14 to justify your answer.

5.17. Determine which, if any, of the following polynomials can generate a cyclic code with codeword length $n \leq 7$. Find the (n, k) values of any such codes that can be generated.

    (a) $1 + X^3 + X^4$

    (b) $1 + X^2 + X^4$

    (c) $1 + X + X^3 + X^4$

    (d) $1 + X + X^2 + X^4$

    (e) $1 + X^3 + X^5$

5.18. Encode the message 1 0 1 in systematic form using polynomial division and the generator $g(X) = 1 + X + X^2 + X^4$.

5.19. Design a feedback shift register encoder for an (8, 5) cyclic code with a generator $g(x) = 1 + X + X^2 + X^3$. Use the encoder to find the codeword for the message 1 0 1 0 1 in systematic form.

5.20. In Figure P5.1 the signal is differentially coherent PSK (DPSK), the encoded symbol rate is 10,000 code symbols per second, and the decoder is a single-error-correcting (7, 4) decoder. Is a predetection signal-to-noise spectral density ratio of $P_r/N_0 = 48$ dBW sufficient to provide a probability of message error of 10^{-3} at the output? Justify your answer. Assume that a message block contains 4 data bits and that any single-error pattern in a block length of 7 bits can be corrected.

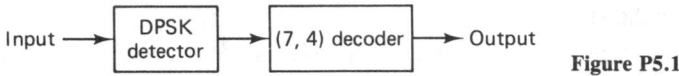

Figure P5.1

5.21. A (15, 5) cyclic code has a generator polynomial as follows:

$$g(X) = 1 + X + X^2 + X^5 + X^8 + X^{10}$$

(a) Draw a diagram of an encoder for this code.
(b) Find the code polynomial (in systematic form) for the message $m(X) = 1 + X^2 + X^4$.
(c) Is $V(X) = 1 + X^4 + X^6 + X^8 + X^{14}$ a code polynomial in this system? Justify your answer.

5.22. Consider the (15, 11) cyclic code generated by $g(X) = 1 + X + X^4$.
(a) Devise a feedback register encoder and decoder for this code.
(b) Illustrate the encoding procedure with the message vector 11001101011 by listing the states of the register (the rightmost bit is the earliest bit).
(c) Repeat part (b) for the decoding procedure.

5.23. For a fixed probability of channel symbol error, the probability of bit error for a Hamming (15, 11) code is worse than that for a Hamming (7, 4) code. Explain why. What, then, is the advantage of the (15, 11) code? What basic trade-off is involved?

5.24. A (63, 36) BCH code can correct five errors. Nine blocks of a (7, 4) code can correct nine errors. Both codes have the same code rate.
(a) The (7, 4) code can correct more errors. Is it more powerful? Explain.
(b) Compare the two codes when five errors occur randomly in 63 bits.

5.25. Information from a source is organized in 36-bit messages that are to be transmitted over an AWGN channel using noncoherently detected BFSK modulation.
(a) If no error control coding is used, compute the E_b/N_0 required to provide a message error probability of 10^{-3}.
(b) Consider the use of a (127, 36) linear block code (minimum distance is 31) in the transmission of these messages. Compute the coding gain for this code for a message error probability of 10^{-3}. (*Hint:* The coding gain is defined as the difference between the E_b/N_0 required without coding and the E_b/N_0 required with coding.)

5.26. (a) Consider a data sequence encoded with a (127, 64) BCH code and then modulated using coherent 16-ary PSK. If the received E_b/N_0 is 10 dB, find the MPSK probability of symbol error, the probability of coded bit error (assuming that a Gray code is used for symbol-to-bit assignment), and the probability of information bit error.

    (b) For the same probability of information bit error found in part (a), determine the value of E_b/N_0 required if the modulation in part (a) is changed to coherent 16-ary FSK. Explain the difference.

5.27. A message consists of English text (assume that each word in the message contains six letters). Each letter is encoded using the 7-bit ASCII character code. Thus, each word of text consists of a 42-bit sequence. The message is to be transmitted over a channel having a symbol error probability of 10^{-3}.

    (a) What is the probability that a word will be received in error?

    (b) If a repetition code is used such that each letter in each word is repeated three times, and at the receiver, majority voting is used to decode the message, what is the probability that a decoded word will be in error?

    (c) If a (126, 42) BCH code with error-correcting capability of $t = 14$ is used to encode each 42-bit word, what is the probability that a decoded word will be in error?

    (d) For a real system, it is not fair to compare uncoded versus coded message error performance on the basis of a fixed probability of channel symbol error, since this implies a fixed level of received E_c/N_0 for all choices of coding (or lack of coding). Therefore, repeat parts (a), (b), and (c) under the condition that the channel symbol error probability is determined by a received E_b/N_0 of 12 dB, where E_b/N_0 is the information bit energy per noise spectral density. Assume that the information rate must be the same for all choices of coding or lack of coding. Also assume that noncoherent binary FSK modulation is used over an AWGN channel.

    (e) Discuss the relative error performance capabilities of the above coding schemes under the two postulated conditions—fixed channel symbol error probability, and fixed E_b/N_0. Under what circumstances can a repetition code offer error performance improvement? When will it cause performance degradation?

CHAPTER 6

Channel Coding:
Part 2

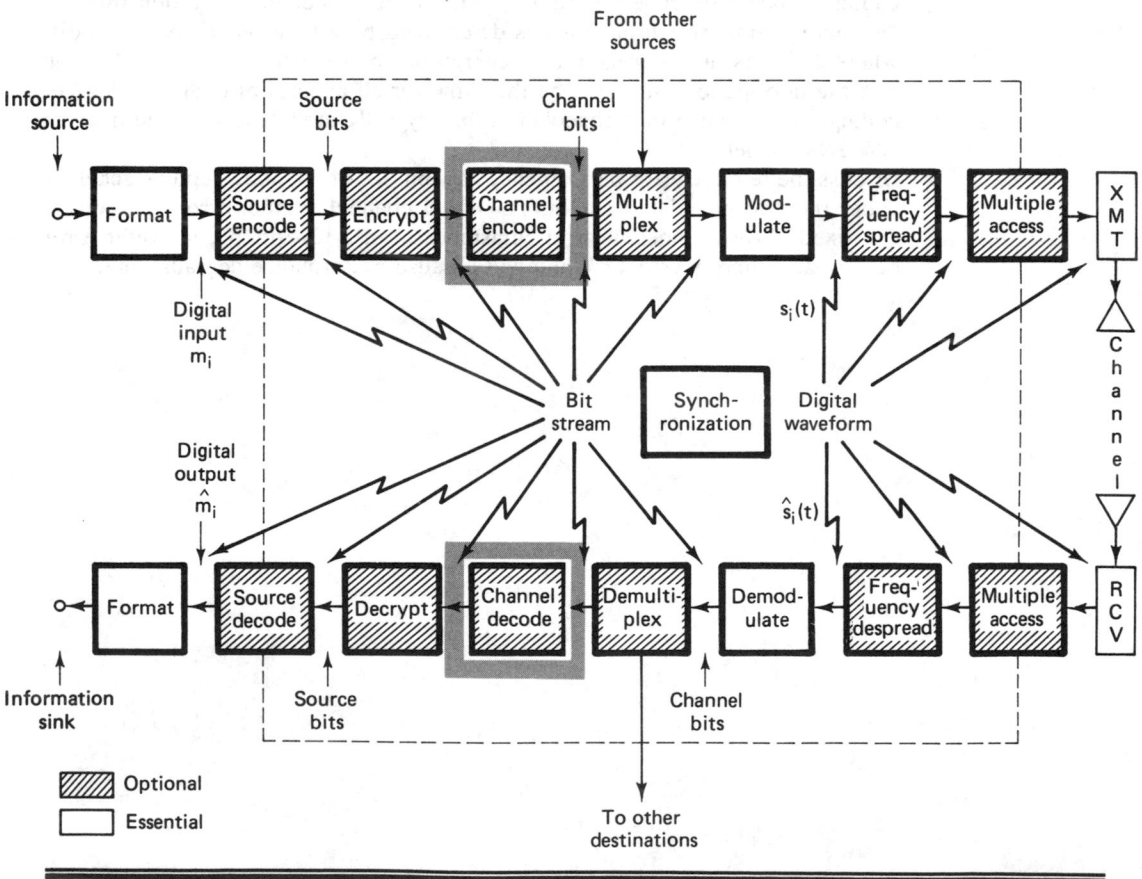

There are two major categories of channel codes: block and convolutional. Chapter 5 deals mainly with block coding. This chapter deals mainly with convolutional coding. A *linear block code* is described by two integers, n and k, and a generator matrix or polynomial. The integer k is the number of data bits that form an input to a block encoder. The integer n is the total number of bits in the associated codeword out of the encoder. A characteristic of linear block codes is that each codeword n-tuple is uniquely determined by the input message k-tuple. The ratio k/n is called the *rate* of the code—a measure of the amount of added redundancy. A *convolutional code* is described by three integers, n, k, and K, where the ratio k/n has the same code rate significance (information per coded bit) that it has for block codes; however, n does *not* define a block or codeword length as it does for block codes. The integer K is a parameter known as the *constraint length*; it represents the number of k-tuple stages in the encoding shift register. An important characteristic of convolutional codes, different from block codes, is that the encoder has memory—the n-tuple emitted by the convolutional encoding procedure is not only a function of an input k-tuple, but is also a function of the previous $K - 1$ input k-tuples. In practice, n and k are small integers and K is varied to control the redundancy.

6.1 CONVOLUTIONAL ENCODING

In Figure 1.2 we presented a typical block diagram of a digital communication system. A version of this functional diagram, focusing primarily on the convolutional encode/decode and modulate/demodulate portions of the communication link, is shown in Figure 6.1. The input message source is denoted by the sequence $\mathbf{m} = m_1, m_2, \ldots, m_i, \ldots$, where each m_i represents a binary digit (bit). We shall assume that each m_i is equally likely to be a one or a zero, and independent from digit to digit. Being independent, the bit sequence lacks any redundancy; that is, knowledge about bit m_i gives no information about m_j $(i \neq j)$. The encoder transforms each sequence $\mathbf{m}$ into a unique codeword sequence $\mathbf{U} = G(\mathbf{m})$. Even though the sequence $\mathbf{m}$ uniquely defines the sequence $\mathbf{U}$, a key feature of convolutional codes is that a given k-tuple within $\mathbf{m}$ does *not* uniquely define its associated n-tuple within $\mathbf{U}$ since the encoding of each k-tuple is *not only* a function of that k-tuple but is also a function of the $K - 1$ input k-tuples that precede it. The sequence $\mathbf{U}$ can be partitioned into a sequence of branch words: $\mathbf{U} = U_1$, $U_2, \ldots, U_i, \ldots$. Each branch word U_i is made up of binary *code symbols*, often called *channel symbols*, *channel bits*, or *coded bits*; unlike the input message bits the code symbols are not independent.

In a typical communication application, the codeword sequence $\mathbf{U}$ modulates a waveform $s(t)$. During transmission, the waveform $s(t)$ is corrupted by noise, resulting in a received waveform $\hat{s}(t)$ and a demodulated sequence $\mathbf{Z} = Z_1, Z_2$, $\ldots, Z_i, \ldots$, as indicated in Figure 6.1. The task of the decoder is to produce an estimate $\hat{\mathbf{m}} = \hat{m}_1, \hat{m}_2, \ldots, \hat{m}_i, \ldots$, of the original message sequence, using the received sequence $\mathbf{Z}$ together with a priori knowledge of the encoding procedure.

A general convolutional encoder, shown in Figure 6.2, is mechanized with a kK-stage shift register and n modulo-2 adders, where K is the constraint length. The constraint length represents the number of k-bit shifts over which a single information bit can influence the encoder output. At each unit of time, k bits are shifted into the first k stages of the register; all bits in the register are shifted k stages to the right, and the outputs of the n adders are sequentially sampled to

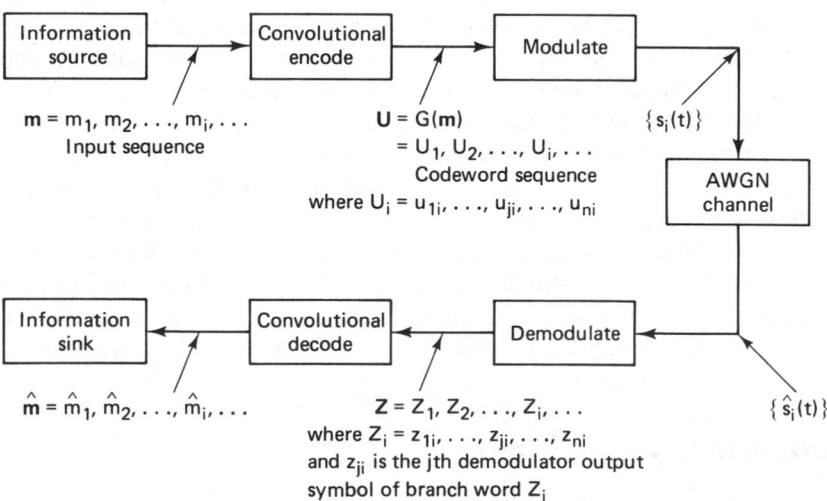

Figure 6.1 Encode/decode and modulate/demodulate portions of a communication link.

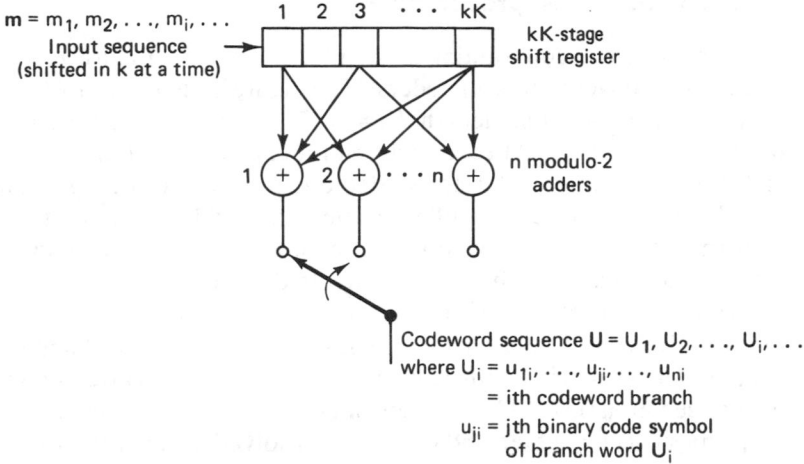

Figure 6.2 Convolutional encoder with constraint length K and rate k/n.

yield the binary code symbols or coded bits. These code symbols are then used by the modulator to specify the waveforms to be transmitted over the channel. Since there are n coded bits for each input group of k message bits, the code rate is k/n message bit per coded bit, where $k < n$.

We shall consider only the most commonly used binary convolutional encoders for which $k = 1$, that is, those encoders in which the message bits are shifted into the encoder one bit at a time, although generalization to higher-order alphabets is straightforward [1, 2]. For the $k = 1$ encoder, at the ith unit of time, message bit m_i is shifted into the first shift register stage; all previous bits in the register are shifted one stage to the right, and as in the more general case, the outputs of the n adders are sequentially sampled and transmitted. Since there are n coded bits for each message bit, the code rate is $1/n$. The n code symbols occurring at time t_i comprise the ith branch word, $U_i = u_{1i}, u_{2i}, \ldots, u_{ni}$, where u_{ji} ($j = 1, 2, \ldots, n$) is the jth code symbol belonging to the ith branch word. Note that for the rate $1/n$ encoder, the kK-stage shift register can be referred to simply as a K-stage register, and the constraint length K, which was expressed in units of k-tuple stages, can be referred to as constraint length in units of bits.

6.2 CONVOLUTIONAL ENCODER REPRESENTATION

To describe a convolutional code, one needs to characterize the encoding function $G(\mathbf{m})$, so that given an input sequence $\mathbf{m}$, one can readily compute the output sequence U. Several methods are used for representing a convolutional encoder, the most popular being the *connection pictorial, connection vectors or polynomials*, the *state diagram*, the *tree diagram*, and the *trellis diagram*. They are each described below.

6.2.1 Connection Representation

We shall use the convolutional encoder, shown in Figure 6.3, as a model for discussing convolutional encoders. The figure illustrates a (2, 1) convolutional encoder with constraint length $K = 3$. There are $n = 2$ modulo-2 adders; thus the code rate k/n is $\frac{1}{2}$. At each input bit time, a bit is shifted into the leftmost stage and the bits in the register are shifted one position to the right. Next, the output switch samples the output of each modulo-2 adder (i.e., first the upper adder, then the lower adder), thus forming the code symbol pair making up the branch word associated with the bit just inputted. The sampling is repeated for each inputted bit. The choice of connections between the adders and the stages of the register gives rise to the characteristics of the code. Any change in the choice of connections results in a different code. The connections are, of course, *not* chosen or changed arbitrarily. The problem of choosing connections to yield good distance properties is complicated and has not been solved in general; however, good codes have been found by computer search for all constraint lengths less than about 20 [3–5].

Unlike a block code that has a fixed word length n, a convolutional code has no particular block size. However, convolutional codes are often forced into a block structure by *periodic truncation*. This requires a number of zero bits to be appended to the end of the input data sequence, for the purpose of clearing or *flushing* the encoding shift register of the data bits. Since the added zeros carry no information, the *effective code rate* falls below k/n. To keep the code rate close to k/n, the truncation period is generally made as long as practical.

One way to represent the encoder is to specify a set of *n connection vectors*, one for each of the n modulo-2 adders. Each vector has dimension K and describes the connection of the encoding shift register to that modulo-2 adder. A one in the *i*th position of the vector indicates that the corresponding stage in the shift register is connected to the modulo-2 adder, and a zero in a given position indicates that no connection exists between the stage and the modulo-2 adder. For the encoder example in Figure 6.3, we can write the connection vector $\mathbf{g}_1$ for the upper connections and $\mathbf{g}_2$ for the lower connections as follows:

$$\mathbf{g}_1 = 1\ 1\ 1$$

$$\mathbf{g}_2 = 1\ 0\ 1$$

Consider that a message vector $\mathbf{m} = 1\ 0\ 1$ is convolutionally encoded with the encoder shown in Figure 6.3. The three message bits are inputted, one at a time, at times t_1, t_2, and t_3, as shown in Figure 6.4. Subsequently, $(K - 1) = 2$ zeros are inputted at times t_4 and t_5 to flush the register and thus ensure that the tail end of the message is shifted the full length of the register. The output sequence is seen to be 1 1 1 0 0 0 1 0 1 1, where the leftmost symbol represents the earliest transmission. The entire output sequence, including the code symbols as a result of flushing, are needed to decode the message. To flush the message from the encoder requires one less zero than the number of stages in the register, or $K - 1$ flush bits. Another zero input is shown at time t_6, for the reader to verify that the corresponding branch word output is then 00.

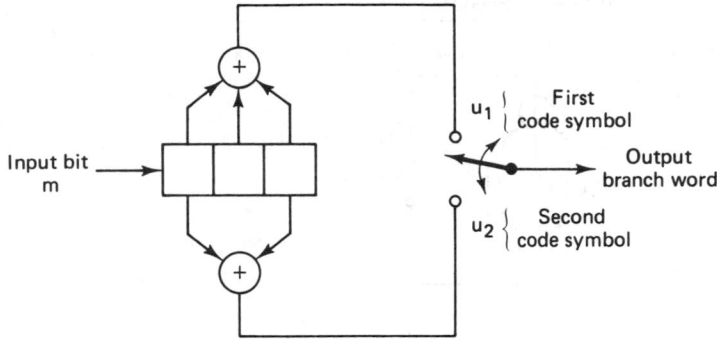

Figure 6.3 Convolutional encoder (rate $\frac{1}{2}$, $K = 3$).

6.2.1.1 Impulse Response of the Encoder

We can approach the encoder in terms of its *impulse response*—that is, the response of the encoder to a single "one" bit that moves through it. Consider the contents of the register in Figure 6.3 as a one moves through it.

Register contents	Branch word u_1	u_2
1 0 0	1	1
0 1 0	1	0
0 0 1	1	1

Input sequence: 1 0 0
Output sequence: 1 1 1 0 1 1

The output sequence for the input "one" is called the impulse response of the encoder. Then for the input sequence $\mathbf{m} = 1\ 0\ 1$, the output may be found by the *superposition* or the *linear addition* of the time-shifted input "impulses" as follows:

Input **m**	Output				
1	1 1	1 0	1 1		
0		0 0	0 0	0 0	
1			1 1	1 0	1 1
Modulo-2 sum:	1 1	1 0	0 0	1 0	1 1

Observe that this is the same output as that obtained in Figure 6.4, demonstrating that *convolutional codes are linear*—just as the linear block codes of Chapter 5. It is from this property of generating the output by the linear addition of time-shifted impulses, or the convolution of the input sequence with the impulse response of the encoder, that we derive the name *convolutional encoder*. Often,

m = 101 ⟶ Encoder ⟶ U

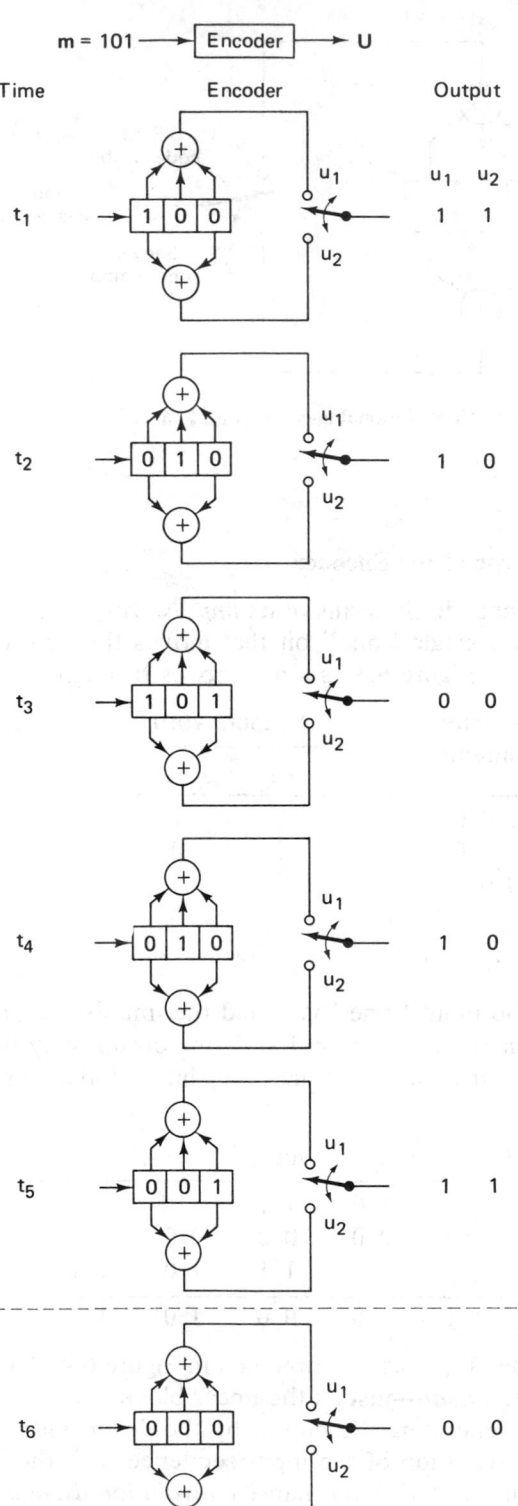

Figure 6.4 Convolutionally encoding a message sequence with a rate $\frac{1}{2}$, $K = 3$ encoder.

this encoder characterization is presented in terms of an infinite-order generator matrix [6].

Notice that the *effective code rate* for the foregoing example with 3-bit input sequence and 10-bit output sequence is $k/n = \frac{3}{10}$—quite a bit less than the rate $\frac{1}{2}$ that might have been expected from the knowledge that each input data bit yields a pair of output channel bits. The reason for the disparity is that the final data bit into the encoder needs to be shifted through the encoder. All of the output channel bits are needed in the decoding process. If the message had been longer, say 300 bits, the output codeword sequence would contain 604 bits, resulting in a code rate of 300/604—much closer to $\frac{1}{2}$.

6.2.1.2 Polynomial Representation

Sometimes, the encoder connections are characterized by *generator polynomials*, similar to those used in Chapter 5 for describing the feedback shift register implementation of cyclic codes. We can represent a convolutional encoder with a set of n generator polynomials, one for each of the n modulo-2 adders. Each polynomial is of degree $K - 1$ or less and describes the connection of the encoding shift register to that modulo-2 adder, much the same way that a connection vector does. The coefficient of each term in the $(K - 1)$-degree polynomial is either 1 or 0, depending on whether a connection exists or does not exist between the shift register and the modulo-2 adder in question. For the encoder example in Figure 6.3, we can write the generator polynomial $g_1(X)$ for the upper connections and $g_2(X)$ for the lower connections as follows:

$$g_1(X) = 1 + X + X^2$$

$$g_2(X) = 1 + X^2$$

where the lowest-order term in the polynomial corresponds to the input stage of the register. The output sequence is found as follows:

$$U(X) = m(X)g_1(X) \text{ interlaced with } m(X)g_2(X)$$

First, express the message vector $m = 1\ 0\ 1$ as a polynomial—that is, $m(X) = 1 + X^2$. We shall again assume the use of zeros following the message bits, to flush the register. Then the output polynomial, $U(X)$, or the output sequence, U, of the Figure 6.3 encoder can be found for the input message m as follows:

$$m(X)g_1(X) = (1 + X^2)(1 + X + X^2) = 1 + X + X^3 + X^4$$

$$m(X)g_2(X) = (1 + X^2)(1 + X^2) = 1 + X^4$$

$$\frac{}{m(X)g_1(X) = 1 + X + 0X^2 + X^3 + X^4}$$

$$m(X)g_2(X) = 1 + 0X + 0X^2 + 0X^3 + X^4$$

$$\overline{U(X) = (1, 1) + (1, 0)X + (0, 0)X^2 + (1, 0)X^3 + (1, 1)X^4}$$

$$U = 1\ 1 \qquad 1\ 0 \qquad 0\ 0 \qquad 1\ 0 \qquad 1\ 1$$

In this example we started with another point of view—that the convolutional encoder can be treated as a set of *cyclic code shift registers*. We represented the encoder with *polynomial generators* as used for describing cyclic codes. However, we arrived at the same output sequence as in Figure 6.4, and the same output sequence as the impulse response treatment of the preceding section. For a good presentation of convolutional code structure in the context of linear sequential circuits, see Reference [7].

6.2.2 State Representation and the State Diagram

The state of a rate $1/n$ convolutional encoder is defined as the contents of the rightmost $K - 1$ stages (see Figure 6.3). Knowledge of the state together with knowledge of the next input is necessary and sufficient to determine the next output. Let the state of the encoder at time, t_i, be defined as $X_i = m_{i-1}, m_{i-2}, \ldots, m_{i-K+1}$. The ith codeword branch, U_i, is completely determined by state X_i and the present input bit m_i; thus the state X_i represents the past history of the encoder in determining the encoder output. The encoder state is said to be *Markov*, in the sense that the probability, $P(X_{i+1}|X_i, X_{i-1}, \ldots, X_0)$, of being in state X_{i+1}, given all previous states, depends only on the most recent state, X_i; that is, the probability is equal to $P(X_{i+1}|X_i)$.

One way to represent simple encoders is with a *state diagram*; such a representation for the encoder in Figure 6.3 is shown in Figure 6.5. The states, shown in the boxes of the diagram, represent the possible contents of the rightmost $K - 1$ stages of the register, and the paths between the states represent the output branch words resulting from such state transitions. The states of the register are designated $a = 00$, $b = 10$, $c = 01$, and $d = 11$; the diagram shown in Figure 6.5 illustrates all the state transitions that are possible for the encoder in Figure

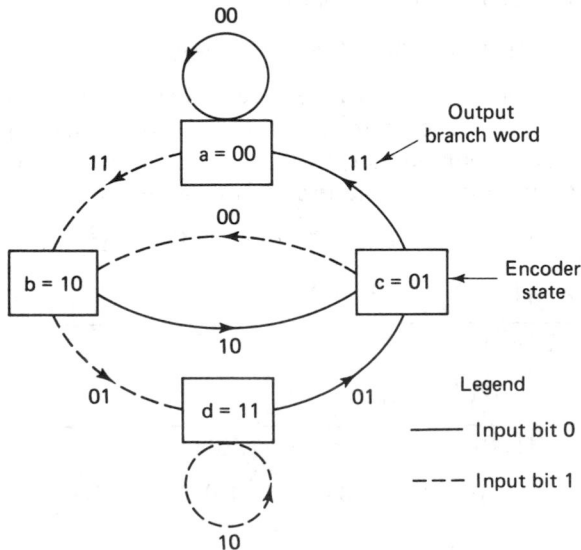

Figure 6.5 Encoder state diagram (rate $\frac{1}{2}$, $K = 3$).

6.3. There are *only two transitions* emanating from each state, corresponding to the two possible input bits. Next to each path between states is written the output branch word associated with the state transition. In drawing the path, we use the convention that a solid line denotes a path associated with an input bit, zero, and a dashed line denotes a path associated with an input bit, one. Notice that it is *not possible* in a single transition to move from a given state to *any arbitrary state*. As a consequence of shifting-in one bit at a time, there are only two possible state transitions that the register can make at each bit time. For example, if the present encoder state is 00, the *only possibilities* for the state at the next shift are 00 or 10.

Example 6.1 Convolutional Encoding

For the encoder shown in Figure 6.3, show the state changes and the resulting output codeword sequence U for the message sequence $\mathbf{m} = 1\ 1\ 0\ 1\ 1$, followed by $K - 1 = 2$ zeros to flush the register. Assume that the initial contents of the register are all zeros.

Solution

Input bit m_i	Register contents	State at time t_i	State at time t_{i+1}	Branch word at time t_i u_1	u_2
–	0 0 0	0 0	0 0	–	
1	1 0 0	0 0	1 0	1	1
1	1 1 0	1 0	1 1	0	1
0	0 1 1	1 1	0 1	0	1
1	1 0 1	0 1	1 0	0	0
1	1 1 0	1 0	1 1	0	1
0	0 1 1	1 1	0 1	0	1
0	0 0 1	0 1	0 0	1	1

state t_i

state t_{i+1}

Output sequence: U = 1 1   0 1   0 1   0 0   0 1   0 1   1 1

Example 6.2 Convolutional Encoding

In Example 6.1 the initial contents of the register are all zeros. This is equivalent to the condition that the given input sequence is preceded by two zero bits (the encoding is a function of the present bit and the $K - 1$ prior bits). Repeat Example 6.1 with the assumption that the given input sequence is preceded by two one bits, and verify that now the codeword sequence U for input sequence $\mathbf{m} = 1\ 1\ 0\ 1\ 1$ is different than the codeword found in Example 6.1.

Solution

The entry "×" signifies "don't know."

Input bit m_i	Register contents	State at time t_i	State at time t_{i+1}	Branch word at time t_i	
				u_1	u_2
−	1 1 ×	1 ×	1 1	−	
1	1 1 1	1 1	1 1	1	0
1	1 1 1	1 1	1 1	1	0
0	0 1 1	1 1	0 1	0	1
1	1 0 1	0 1	1 0	0	0
1	1 1 0	1 0	1 1	0	1
0	0 1 1	1 1	0 1	0	1
0	0 0 1	0 1	0 0	1	1

state t_i

state t_{i+1}

Output sequence: $U = 1\ 0 \quad 1\ 0 \quad 0\ 1 \quad 0\ 0 \quad 0\ 1 \quad 0\ 1 \quad 1\ 1$

By comparing this result with that of Example 6.1, we can see that each branch word of the output sequence U is *not only* a function of the input bit, but is also a function of the $K - 1$ prior bits.

6.2.3 The Tree Diagram

Although the state diagram completely characterizes the encoder, one cannot easily use it for tracking the encoder transitions as a function of time since the diagram cannot represent time history. The tree diagram adds the *dimension of time* to the state diagram. The tree diagram for the covolutional encoder shown in Figure 6.3 is illustrated in Figure 6.6. At each successive input bit time the encoding procedure can be described by traversing the diagram from left to right, each tree branch describing an output branch word. The branching rule for finding a codeword sequence is as follows: If the input bit is a zero, its associated branch word is found by moving to the next rightmost branch in the upward direction. If the input bit is a one, its branch word is found by moving to the next rightmost branch in the downward direction. Assuming that the initial contents of the encoder is all zeros, the diagram shows that if the first input bit is a zero, the output branch word is 00 and, if the first input bit is a one, the output branch word is 11. Similarly, if the first input bit is a one and the second input bit is a zero, the second output branch word is 10. Or, if the first input bit is a one and the second input bit is a one, the second output branch word is 01. Following this procedure we see that the input sequence 1 1 0 1 1 traces the heavy line drawn on the tree

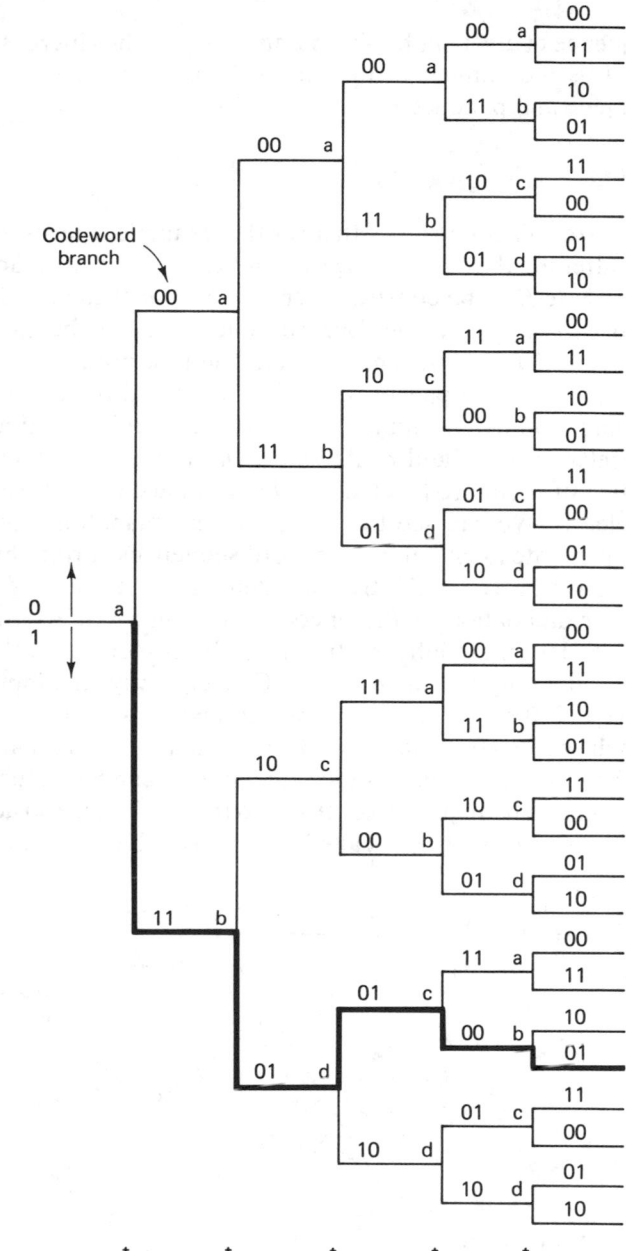

Codeword branch

t_1 t_2 t_3 t_4 t_5

Figure 6.6 Tree representation of encoder (rate $\frac{1}{2}$, $K = 3$).

diagram in Figure 6.6. This path corresponds to the following output codeword sequence: 1 1 0 1 0 1 0 0 0 1.

The added dimension of time in the tree diagram (compared to the state diagram) allows one to dynamically describe the encoder as a function of a particular input sequence. However, can you see one problem in trying to use a tree diagram for describing a sequence of any length? The number of branches increase as a function of 2^L, where L is the number of bits in the input sequence. You would quickly run out of paper, and patience.

6.2.4 The Trellis Diagram

Observation of the Figure 6.6 tree diagram shows that for this example, the structure repeats itself at time t_4, after the third branching (in general, the tree structure *repeats after K branchings*, where K is the constraint length). We label each node in the tree of Figure 6.6 to correspond to the four possible states in the shift register, as follows: $a = 00$, $b = 10$, $c = 01$, and $d = 11$. The first branching of the tree structure, at time t_1, produces a pair of nodes labeled a and b. At each successive branching the number of nodes double. The second branching, at time t_2, results in four nodes labeled a, b, c, and d. After the *third* branching there are a total of eight nodes; two of them are labeled a, two are labeled b, two are labeled c, and two are labeled d. We can see that all branches emanating from two nodes of the same state generate identical branch word sequences. From this point on, the upper and the lower halves of the tree are identical. The reason for this should be obvious from examination of the encoder in Figure 6.3. As the fourth input bit enters the encoder on the left, the first input bit is ejected on the right and no longer influences the output branch words. Consequently, the input sequences 1 0 0 x y . . . and 0 0 0 x y . . . , where the leftmost bit is the earliest bit, generate the same branch words after the ($K = 3$)rd branching. This means that any two nodes having the same state label, at the same time t_i, can be merged since all succeeding paths will be indistinguishable. If we do this to the tree structure of Figure 6.6, we obtain another diagram, called the trellis. The *trellis dia-*

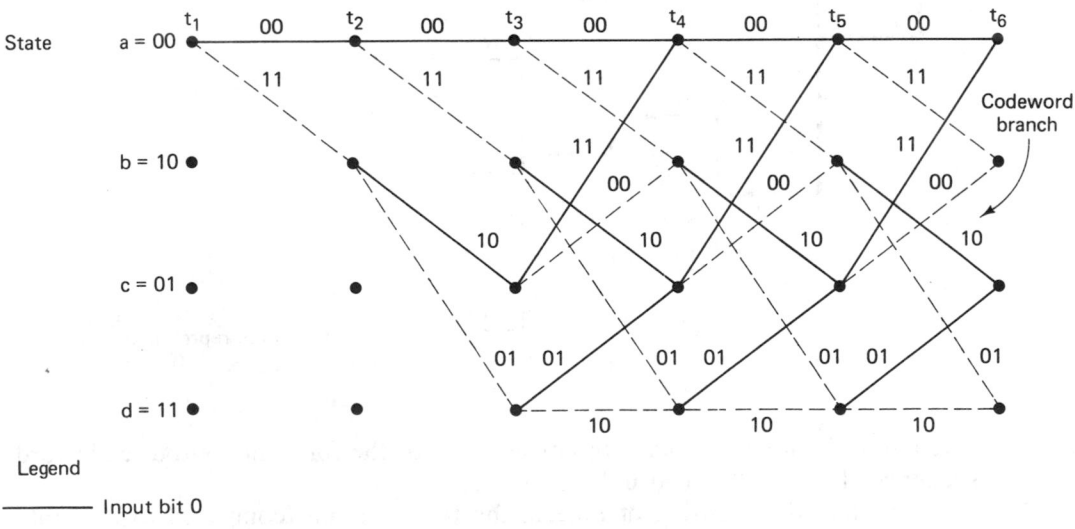

Figure 6.7 Encoder trellis diagram (rate $\frac{1}{2}$, $K = 3$).

gram, by exploiting the repetitive structure, provides a more manageable encoder description than does the tree diagram. The trellis diagram for the convolutional encoder of Figure 6.3 is shown in Figure 6.7.

In drawing the trellis diagram, we use the same convention that we introduced with the state diagram—that a solid line denotes the output generated by an input bit, zero, and a dashed line denotes the output generated by an input bit, one. The nodes of the trellis characterize the encoder states; the first row nodes correspond to the state $a = 00$, the second and subsequent rows correspond to the states $b = 10$, $c = 01$, and $d = 11$. At each unit of time the trellis requires 2^{K-1} nodes to represent the 2^{K-1} possible encoder states. The trellis in our example assumes a fixed periodic structure after trellis depth 3 is reached (at time t_4). In the general case, the fixed structure prevails after depth K is reached. After this point, each of the states can be entered from either of two preceding states. Also, each of the states can transition to one of two states. Of the two outgoing branches, one corresponds to an input bit zero and the other corresponds to an input bit one. On Figure 6.7 the output branch words corresponding to the state transitions appear as labels on the trellis branches.

6.3 FORMULATION OF THE CONVOLUTIONAL DECODING PROBLEM

6.3.1 Maximum Likelihood Decoding

If all input message sequences are equally likely, a decoder that achieves the minimum probability of error is one that compares the conditional probabilities, also called the *likelihood functions*, $P(\mathbf{Z}|\mathbf{U}^{(m)})$, where $\mathbf{Z}$ is the received sequence and $\mathbf{U}^{(m)}$ is one of the possible transmitted sequences, and chooses the maximum. The decoder chooses $\mathbf{U}^{(m')}$ if

$$P(\mathbf{Z}|\mathbf{U}^{(m')}) = \max_{\text{all } \mathbf{U}^{(m)}} P(\mathbf{Z}|\mathbf{U}^{(m)}) \qquad (6.1)$$

The *maximum likelihood* concept, as stated in Equation (6.1), is a fundamental development of decision theory (see Appendix B); it is the formalization of a "common-sense" way to make decisions when there is statistical knowledge of the possibilities. In the binary demodulation treatment in Chapters 2 and 3 there were *only two* equally likely possible signals, $s_1(t)$ or $s_2(t)$, that might have been transmitted. Therefore, to make the binary maximum likelihood decision, given a received signal, meant only to decide that $s_1(t)$ was transmitted if

$$p(z|s_1) > p(z|s_2)$$

otherwise, to decide that $s_2(t)$ was transmitted. The parameter z represents $z(T)$, the receiver output at a symbol duration time $t = T$. However, when applying maximum likelihood to the convolutional decoding problem, there are typically a *multitude* of possible codeword sequences that might have been transmitted. To be specific, an L-bit codeword sequence is a member of a set of 2^L possible sequences. Therefore, in the maximum likelihood context, we can say that the decoder chooses a particular $\mathbf{U}^{(m')}$ as the transmitted sequence if the likelihood $P(\mathbf{Z}|\mathbf{U}^{(m')})$ is greater than the likelihoods of all the other possible transmitted sequences. Such an optimal decoder, which minimizes the error probability (for the case where all transmitted sequences are equally likely), is known as a *maximum likelihood decoder*. The likelihood functions are given or computed from the specifications of the channel.

We will assume that the noise is additive white Gaussian with zero mean and the channel is *memoryless*, which means that the noise affects each code symbol *independently* of all the other symbols. For a convolutional code of rate $1/n$, we can therefore express the likelihood, $P(\mathbf{Z}|\mathbf{U}^{(m)})$ as follows:

$$P(\mathbf{Z}|\mathbf{U}^{(m)}) = \prod_{i=1}^{\infty} P(Z_i|U_i^{(m)}) = \prod_{i=1}^{\infty} \prod_{j=1}^{n} P(z_{ji}|u_{ji}^{(m)}) \qquad (6.2)$$

where Z_i is the ith branch of the received sequence $\mathbf{Z}$, $U_i^{(m)}$ the ith branch of a particular codeword sequence $\mathbf{U}^{(m)}$, z_{ji} the jth code symbol of Z_i, and $u_{ji}^{(m)}$ the jth code symbol of $U_i^{(m)}$, each branch comprising n code symbols. The decoder problem consists of choosing a path through the trellis of Figure 6.7 (each possible path defines a codeword) such that

$$\prod_{i=1}^{\infty} \prod_{j=1}^{n} P(z_{ji}|u_{ji}^{(m)}) \text{ is maximized} \qquad (6.3)$$

Generally, it is computationally more convenient to use the logarithm of the likelihood function since this permits the summation, instead of the multiplication, of terms. We are able to use this transformation because the logarithm is a monotonically increasing function and thus will not alter the final result in our codeword selection. We can define the log-likelihood function $\gamma_U(m)$ as

$$\gamma_U(m) = \log P(\mathbf{Z}|\mathbf{U}^{(m)}) = \sum_{i=1}^{\infty} \log P(Z_i|U_i^{(m)}) = \sum_{i=1}^{\infty} \sum_{j=1}^{n} \log P(z_{ji}|u_{ji}^{(m)}) \qquad (6.4)$$

The decoder problem now consists of choosing a path through the tree of Figure 6.6 or the trellis of Figure 6.7 such that $\gamma_U(m)$ is maximized. For the decoding of convolutional codes, either the tree or the trellis structure can be used. In the tree representation of the code, the fact that the paths remerge is ignored. Since the number of possible sequences for an L-symbol-long sequence is 2^L, maximum likelihood decoding of an L-bit-long received sequence, using a tree diagram, requires the "brute force" or exhaustive comparison of 2^L accumulated log-likelihood metrics, representing all the possible different codewords that could have been transmitted. Hence it is not practical to consider maximum likelihood decoding with a tree structure. It is shown in a later section that with the use of the trellis representation of the code, it is possible to configure a decoder which can discard the paths that could not possibly be candidates for the maximum likelihood sequence. The decoded path is chosen from some reduced set of *surviving paths*. Such a decoder is still optimum in the sense that the decoded path is the same as the decoded path obtained from a "brute force" maximum likelihood decoder, but the early rejection of unlikely paths reduces the decoding complexity.

For an excellent tutorial on the structure of convolutional codes, maximum likelihood decoding, and code performance, see Reference [8]. There are several algorithms that yield *approximate* solutions to the maximum likelihood decoding problem, including sequential [9, 10] and threshold [11]. Each of these algorithms is suited to certain special applications, but are all suboptimal. In contrast, the *Viterbi decoding algorithm* performs maximum likelihood decoding and is therefore optimal. This does not imply that the Viterbi algorithm is best for every application; there are severe constraints imposed by hardware complexity. The Viterbi algorithm is considered in Sections 6.3.3 and 6.3.4.

6.3.2 Channel Models: Hard versus Soft Decisions

Before specifying an algorithm that will determine the maximum likelihood decision, let us describe the channel. The codeword sequence $U^{(m)}$, made up of branch words, with each branch word comprised of n code symbols, can be considered to be an endless stream, as opposed to a block code, in which the source data and their codewords are partitioned into precise block sizes. The codeword sequence shown in Figure 6.1 emanates from the convolutional encoder and enters the modulator, where the code symbols are transformed into signal waveforms. The modulation may be baseband (e.g., pulse waveforms) or bandpass (e.g., PSK or FSK). In general, ℓ symbols at a time, where ℓ is an integer, are mapped into signal waveforms $s_i(t)$, where $i = 1, 2, \ldots, M = 2^\ell$. When $\ell = 1$, the modulator maps each code symbol into a binary waveform. The channel over which the waveform is transmitted is assumed to corrupt the signal with Gaussian noise. When the corrupted signal is received, it is first processed by the demodulator and then by the decoder.

Consider that a binary signal, transmitted over a symbol interval $(0, T)$, is represented by $s_1(t)$ for a binary one and $s_2(t)$ for a binary zero. The received signal is $r(t) = s_i(t) + n(t)$, where $n(t)$ is a zero-mean Gaussian noise process. In Sections 2.9 and 3.4 we described the detection of $r(t)$ in terms of two basic steps. In the first step, the received waveform is reduced to a single number, $z(T) = a_i + n_0$, where a_i is the signal component of $z(T)$ and n_0 is the noise component. The noise component, n_0, is a zero-mean *Gaussian random variable*, and thus $z(T)$ is a *Gaussian random variable* with a mean of either a_1 or a_2 depending on whether a binary one or binary zero was sent. In the second step of the detection process a decision was made as to which signal was transmitted, on the basis of comparing $z(T)$ to a threshold. The conditional probabilities of $z(T)$, $p(z|s_1)$, and $p(z|s_2)$ are shown in Figure 6.8, labeled likelihood of s_1 and likelihood of s_2. The demodulator in Figure 6.1, converts the set of time-ordered random variables, $\{z(T)\}$, into a code sequence, Z, and passes it on to the decoder. The demodulator output can be configured in a variety of ways. It can be implemented to make a *firm or hard decision* as to whether $z(T)$ represents a zero or a one. In this case, the output of the demodulator is quantized to two levels, zero and one, and fed

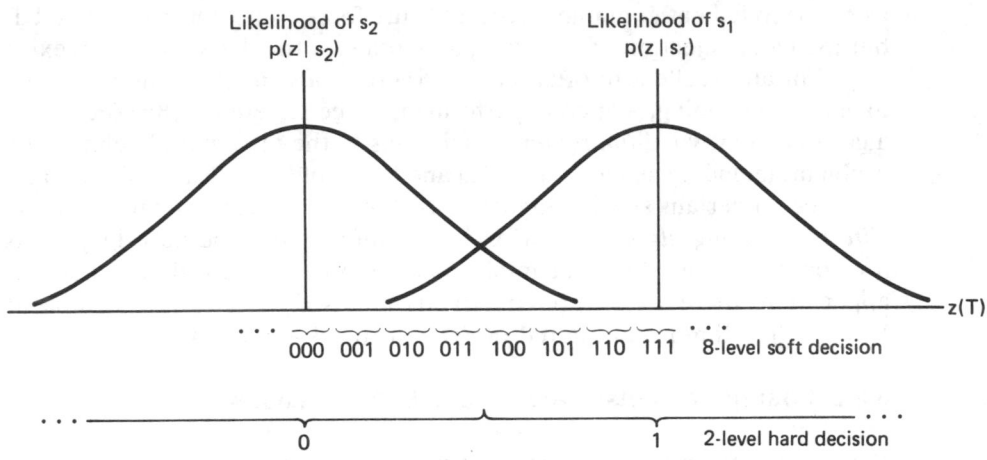

Figure 6.8 Hard and soft decoding decisions.

into the decoder (this is exactly the same threshold decision that was made in Chapters 2 and 3). Since the decoder operates on the hard decisions made by the demodulator, the decoding is called *hard-decision decoding*.

The demodulator can also be configured to feed the decoder with a *quantized value* of $z(T)$ *greater than two levels*, or with an unquantized or analog value of $z(T)$. Such an implementation furnishes the decoder with more information than is provided in the hard-decision case. When the quantization level of the demodulator output is greater than two, the decoding is called *soft-decision decoding*. Eight levels (3-bits) of quantization are illustrated on the abscissa of Figure 6.8. When the demodulator sends a hard binary decision to the decoder, it sends it a single binary symbol. When the demodulator sends a soft binary decision, quantized to eight levels, it sends the decoder a 3-bit word describing an interval along $z(T)$. In effect, sending such a 3-bit word in place of a single binary symbol is equivalent to sending the decoder a *measure of confidence* along with the code symbol. Referring to Figure 6.8, if the demodulator sends 1 1 1 to the decoder, this is tantamount to declaring the code symbol to be a one with very high confidence, while sending a 1 0 0 is tantamount to declaring the code symbol to be a one with very low confidence. It should be clear that ultimately, every message decision out of the decoder must be a hard decision; otherwise, one might see computer printouts that read: "think it's a 1," "think it's a 0," and so on. The idea behind the demodulator *not making hard decisions* and sending more data (soft decisions) to the decoder can be thought of as an interim step to provide the decoder with more information, which the decoder then uses for recovering the message sequence (with better error performance than it could in the case of hard-decision decoding).

For a Gaussian channel, eight-level quantization results in a performance improvement of approximately 2 dB in required signal-to-noise ratio compared to two-level quantization. This means that eight-level soft-decision decoding can provide the same probability of bit error as that of hard-decision decoding, but requires 2 dB *less* E_b/N_0 for the same performance. Analog (or infinite-level quantization) results in a 2.2-dB performance improvement over two-level quantization; therefore, *eight-level quantization* results in a loss of approximately 0.2 dB compared to infinitely fine quantization. For this reason, quantization to more than eight levels can yield little performance improvement [12]. What price is paid for such improved soft-decision-decoder performance? In the case of hard-decision decoding, a single bit is used to describe each code symbol, while for eight-level quantized soft-decision decoding 3 bits are used to describe each code symbol; therefore, three times the amount of data must be handled during the decoding process. Hence the price paid for soft-decision decoding is an increase in required memory size at the decoder (and possibly a speed penalty).

Block decoding algorithms and convolutional decoding algorithms have been devised to operate with hard *or* soft decisions. However, soft-decision decoding is generally not used with block codes because it is considerably more difficult than hard-decision decoding to implement. The most prevalent use of soft-decision decoding is with the *Viterbi convolutional decoding algorithm*, since with Viterbi decoding, soft decisions represent only a trivial increase in computation.

6.3.2.1 Binary Symmetric Channel

A binary symmetric channel (BSC) is a discrete memoryless channel (see Section 5.3.1) that has binary input and output alphabets and symmetric transition probabilities. It can be described by the conditional probabilities

$$P(0|1) = P(1|0) = p \qquad (6.5)$$
$$P(1|1) = P(0|0) = 1 - p$$

as illustrated in Figure 6.9. The probability that an output symbol will differ from the input symbol is p, and the probability that the output symbol will be identical to the input symbol is $(1 - p)$. The BSC is an example of a *hard-decision channel*, which means that, even though continuous-valued signals may be received by the demodulator, a BSC allows only firm decisions such that each demodulator output symbol, z_{ji}, as shown in Figure 6.1, consists of one of two binary values. The indexing of z_{ji} pertains to the jth code symbol of the ith branch word, Z_i. The demodulator then feeds the sequence $\mathbf{Z} = \{Z_i\}$ to the decoder.

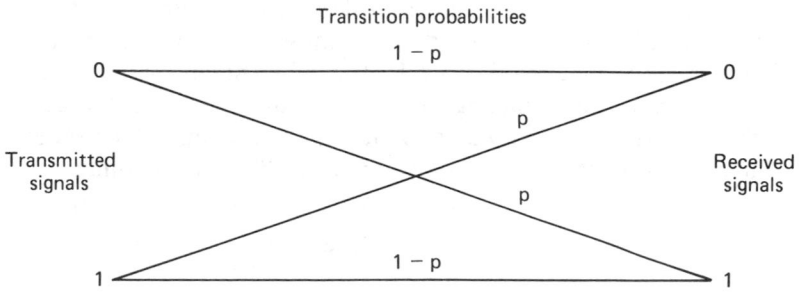

Figure 6.9 Binary symmetric channel (hard-decision channel).

Let $\mathbf{U}^{(m)}$ be a transmitted codeword over a BSC with symbol error probability p, and let $\mathbf{Z}$ be the corresponding received decoder sequence. As noted previously, a maximum likelihood decoder chooses the codeword $\mathbf{U}^{(m')}$ which maximizes the likelihood, $P(\mathbf{Z}|\mathbf{U}^{(m)})$, or its logarithm. For a BSC, this is equivalent to choosing the codeword, $\mathbf{U}^{(m')}$, that is closest in *Hamming distance* to $\mathbf{Z}$ [8]. Thus Hamming distance is an appropriate metric to describe the distance or closeness of fit between $\mathbf{U}^{(m)}$ and $\mathbf{Z}$. From all the possible transmitted sequences, $\mathbf{U}^{(m)}$, the decoder chooses the $\mathbf{U}^{(m')}$ sequence for which the distance to $\mathbf{Z}$ is minimum.

Suppose that $\mathbf{U}^{(m)}$ and $\mathbf{Z}$ are each L-bit-long sequences and that they differ in d_m positions [i.e., the Hamming distance between $\mathbf{U}^{(m)}$ and $\mathbf{Z}$ is d_m]. Then, since the channel is assumed memoryless, the probability that this $\mathbf{U}^{(m)}$ was transformed to the specific received $\mathbf{Z}$ at distance d_m from it can be written

$$P(\mathbf{Z}|\mathbf{U}^{(m)}) = p^{d_m}(1 - p)^{L - d_m} \tag{6.6}$$

and the log-likelihood function is

$$\log P(\mathbf{Z}|\mathbf{U}^{(m)}) = -d_m \log\left(\frac{1 - p}{p}\right) + L \log(1 - p) \tag{6.7}$$

If we compute this quantity for each possible transmitted sequence, the second term will be constant in each case. Assuming that $p < 0.5$, we can express Equation (6.7) as

$$\log P(\mathbf{Z}|\mathbf{U}^{(m)}) = -A d_m - B \tag{6.8}$$

where A and B are positive constants. Therefore, choosing the codeword $\mathbf{U}^{(m')}$ such that the Hamming distance, d_m, to the received sequence $\mathbf{Z}$ is minimized corresponds to *maximizing the likelihood or log-likelihood metric*. Consequently, over a BSC, the log-likelihood metric is conveniently replaced by the Hamming distance, and a maximum likelihood decoder will choose, in the tree or trellis diagram, the path whose corresponding sequence, $\mathbf{U}^{(m')}$, is at the *minimum Hamming distance* to the received sequence $\mathbf{Z}$.

6.3.2.2 Gaussian Channel

For a Gaussian channel, each demodulator output symbol, z_{ji}, as shown in Figure 6.1, is a value from a continuous alphabet. The symbol z_{ji} cannot be labeled as a correct or incorrect detection decision. Sending the decoder such soft decisions can be viewed as sending a family of conditional probabilities of the different symbols (see Section 5.3.1). It can be shown [8] that maximizing $P(\mathbf{Z}|\mathbf{U}^{(m)})$ is equivalent to maximizing the inner product between the codeword sequence, $\mathbf{U}^{(m)}$ (consisting of binary symbols), and the analog-valued received sequence, $\mathbf{Z}$. Thus the decoder chooses the codeword $\mathbf{U}^{(m')}$ if it maximizes

$$\sum_{i=1}^{\infty} \sum_{j=1}^{n} z_{ji} u_{ji}^{(m)} \tag{6.9}$$

This is equivalent to choosing the codeword $\mathbf{U}^{(m')}$ that is closest in *Euclidean distance* to $\mathbf{Z}$. Even though the hard- and soft-decision channels require different metrics, the concept of choosing the codeword $\mathbf{U}^{(m')}$ that is closest to the received sequence, $\mathbf{Z}$, is the same in both cases. To implement the maximization of Equation (6.9) exactly, the decoder would have to be able to handle analog-valued arithmetic operations. This is impractical because the decoder is generally implemented digitally. Thus it is necessary to quantize the received symbols z_{ji}. Does Equation (6.9) remind you of the demodulation treatment in Chapter 3? Equation (6.9) is the discrete version of correlating an input received waveform, $r(t)$, with a reference waveform, $s_i(t)$, as expressed in Equation (3.34). The quantized Gaussian channel, typically referred to as a *soft-decision channel*, is the channel model assumed for the soft-decision decoding described earlier.

6.3.3 The Viterbi Convolutional Decoding Algorithm

The Viterbi decoding algorithm was discovered and analyzed by Viterbi [13] in 1967. The Viterbi algorithm essentially performs maximum likelihood decoding; however, it reduces the computational load by taking advantage of the special structure in the code trellis. The advantage of Viterbi decoding, compared with brute-force decoding, is that the complexity of a Viterbi decoder is not a function of the number of symbols in the codeword sequence. The algorithm involves calculating a *measure of similarity*, *or distance*, between the received signal, at time t_i, and all the trellis paths entering each state at time t_i. The Viterbi algorithm removes from consideration those trellis paths that could not possibly be candidates for the maximum likelihood choice. When two paths enter the same state, the one having the best metric is chosen; this path is called the *surviving path*. This selection of surviving paths is performed for all the states. The decoder continues in this way to advance deeper into the trellis, making decisions by eliminating the least likely paths. The early rejection of the unlikely paths reduces the decoding complexity. In 1969, Omura [14] demonstrated that the Viterbi algorithm is, in fact, maximum likelihood. Note that the goal of selecting the optimum path can be expressed, equivalently, as choosing the codeword with the *maximum likelihood metric*, or as choosing the codeword with the *minimum distance metric*.

6.3.4 An Example of Viterbi Convolutional Decoding

For simplicity, a BSC is assumed; thus Hamming distance is a proper distance measure. The encoder for this example is shown in Figure 6.3, and the encoder trellis diagram is shown in Figure 6.7. A similar trellis can be used to represent the decoder, as shown in Figure 6.10. The basic idea behind the decoding procedure can best be understood by examining the Figure 6.7 encoder trellis in concert with the Figure 6.10 decoder trellis. For the decoder trellis it is convenient to label each trellis branch at time t_i with the *Hamming distance* between the received code symbols and the corresponding branch word from the encoder trellis. The example in Figure 6.10, shows a message sequence, $\mathbf{m}$, the corresponding codeword sequence, $\mathbf{U}$, and a noise corrupted received sequence, $\mathbf{Z} = $ 11 01 01 10 01 . . . . The branch words seen on the *encoder trellis* branches

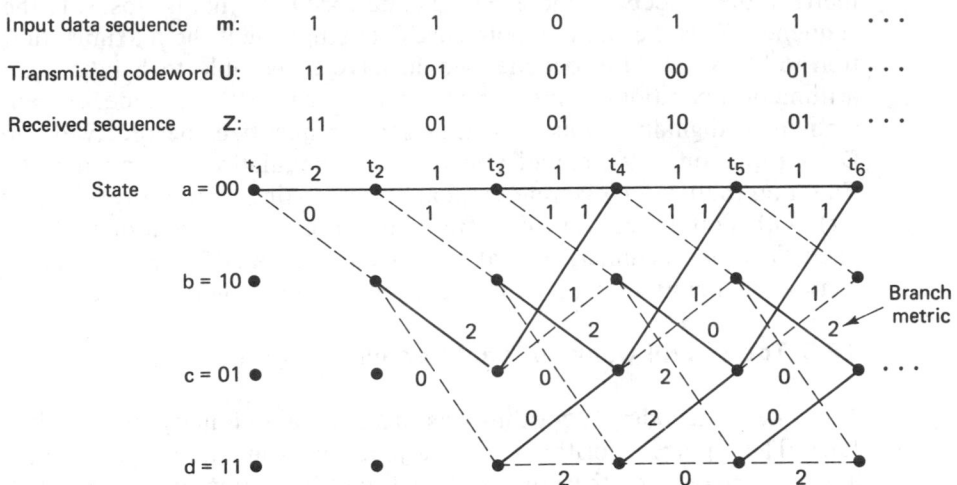

Input data sequence   m: 1   1   0   1   1   · · ·

Transmitted codeword U: 11   01   01   00   01   · · ·

Received sequence   Z: 11   01   01   10   01   · · ·

Figure 6.10   Decoder trellis diagram (rate $\frac{1}{2}$, $K = 3$).

characterize the encoder in Figure 6.3, and are known a priori to both the encoder and the decoder. These encoder branch words are the code symbols that would be expected to come from the encoder output as a result of each of the state transitions. The labels on the *decoder trellis* branches are accumulated by the decoder *on the fly*. That is, as the code symbols are received, each branch of the decoder trellis is labeled with a metric of similarity (Hamming distance) between the received code symbols and each of the branch words for that time interval. From the received sequence, Z, in Figure 6.10, we see that the code symbols received at time t_1 are 11. In order to label the decoder branches at time t_1 with the appropriate Hamming distance metric, we look at the Figure 6.7 encoder trellis. Here we see that a state $00 \rightarrow 00$ transition yields an output branch word of 00. But we received 11. Therefore, on the decoder trellis we label the state $00 \rightarrow 00$ transition with the Hamming distance between them, namely 2. Looking at the encoder trellis again, we see that a state $00 \rightarrow 10$ transition yields an output branch word of 11, which corresponds exactly with the code symbols we received at time t_1. Therefore, on the decoder trellis, we label the state $00 \rightarrow 10$ transition with a Hamming distance of 0. We continue labeling the decoder trellis branches in this way as the symbols are received at each time t_i. The decoding algorithm uses these Hamming distance metrics to find the *most likely* (minimum distance) path through the trellis.

The basis of *Viterbi decoding* is the following observation: If any two paths in the trellis merge to a single state, one of them can always be eliminated in the search for an optimum path. For example, Figure 6.11 shows two paths merging at time t_5 to state 00. Let us define the *cumulative Hamming path metric* of a given path at time t_i as the sum of the branch Hamming distance metrics along that path up to time t_i. In Figure 6.11 the upper path has metric 4; the lower has metric 1. The upper path cannot be a portion of the optimum path because the lower path, which enters the same state, has a lower metric. This observation holds because of the Markov nature of the encoder state: The present state summarizes the encoder history in the sense that previous states cannot affect future states or future output branches.

At each time t_i there are 2^{K-1} states in the trellis, where K is the constraint length, and each state can be *entered by means of two paths*. Viterbi decoding consists of computing the metrics for the two paths entering each state and *eliminating one of them*. This computation is done for each of the 2^{K-1} nodes at time t_i; then the decoder moves to time t_{i+1} and repeats the process. The first few steps in our decoding example are as follows (see Figure 6.12). Assume that the input data sequence $\mathbf{m}$, codeword $\mathbf{U}$, and received sequence $\mathbf{Z}$ are as shown in Figure 6.10. Assume that the decoder knows the correct initial state of the trellis. (This assumption is not necessary in practice, but simplifies the explanation.) At time t_1 the received code symbols are 11. From state 00 the only possible transitions are to state 00 or state 10, as shown in Figure 6.12a. State $00 \rightarrow 00$ transition has branch metric 2; state $00 \rightarrow 10$ transition has branch metric 0. At time t_2 there are two possible branches leaving each state, as shown in Figure 6.12b. The cumulative path metrics of these branches are labeled λ_a, λ_b, λ_c, and λ_d, corresponding to the terminating state. At time t_3 in Figure 6.12c there are again two branches diverging from each state. As a result, there are two paths entering each state at time t_4. As noted previously, one path entering each state can be eliminated, namely, the one having the larger cumulative path metric. Should metrics of the two entering paths be of equal value, one path is chosen for elimination by using an arbitrary rule. The surviving path into each state is shown in Figure 6.12d. At this point in the decoding process, there is only a single surviving path between times t_1 and t_2. Therefore, the decoder can now decide that the state transition which occurred between t_1 and t_2 was $00 \rightarrow 10$. Since this transition is produced by an input bit one, the decoder outputs a one as the first decoded bit. Here we can see how the decoding of the surviving branch is facilitated by having drawn the lattice branches with solid lines for input zeros and dashed lines for input ones. Note that the first bit was not decoded until the path metric computation had proceeded to a much greater depth into the trellis. For a typical

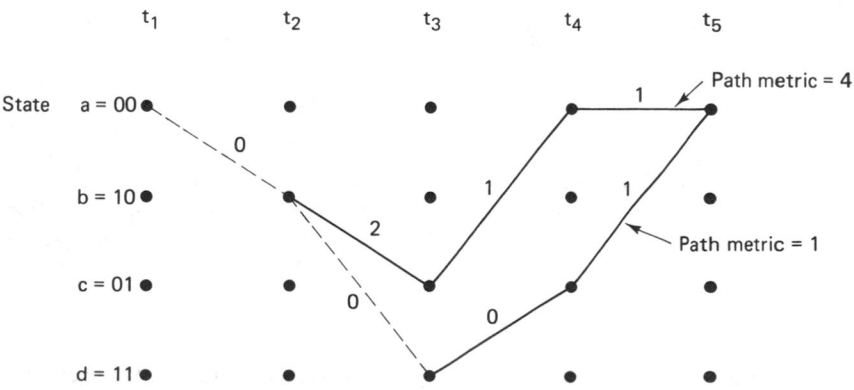

Figure 6.11 Path metrics for two merging paths.

decoder implementation, this represents a decoding delay which can be as much as five times the constraint length in bits.

At each succeeding step in the decoding process, there will always be two possible paths entering each state; one of the two will be eliminated by comparing the path metrics. Figure 6.12e shows the next step in the decoding process. Again, at time t_5 there are two paths entering each state, and one of each pair can be eliminated. Figure 6.12f shows the survivors at time t_5. Notice that in our example we cannot yet make a decision on the second input data bit because there still are two paths leaving the state 10 node at time t_2. At time t_6 in Figure 6.12g we again see the pattern of remerging paths, and in Figure 6.12h we see the survivors at time t_6. Also, in Figure 6.12h the decoder outputs one as the second decoded bit, corresponding to the single surviving path between t_2 and t_3. The decoder continues in this way to advance deeper into the trellis and to make decisions on the input data bits by eliminating all paths but one.

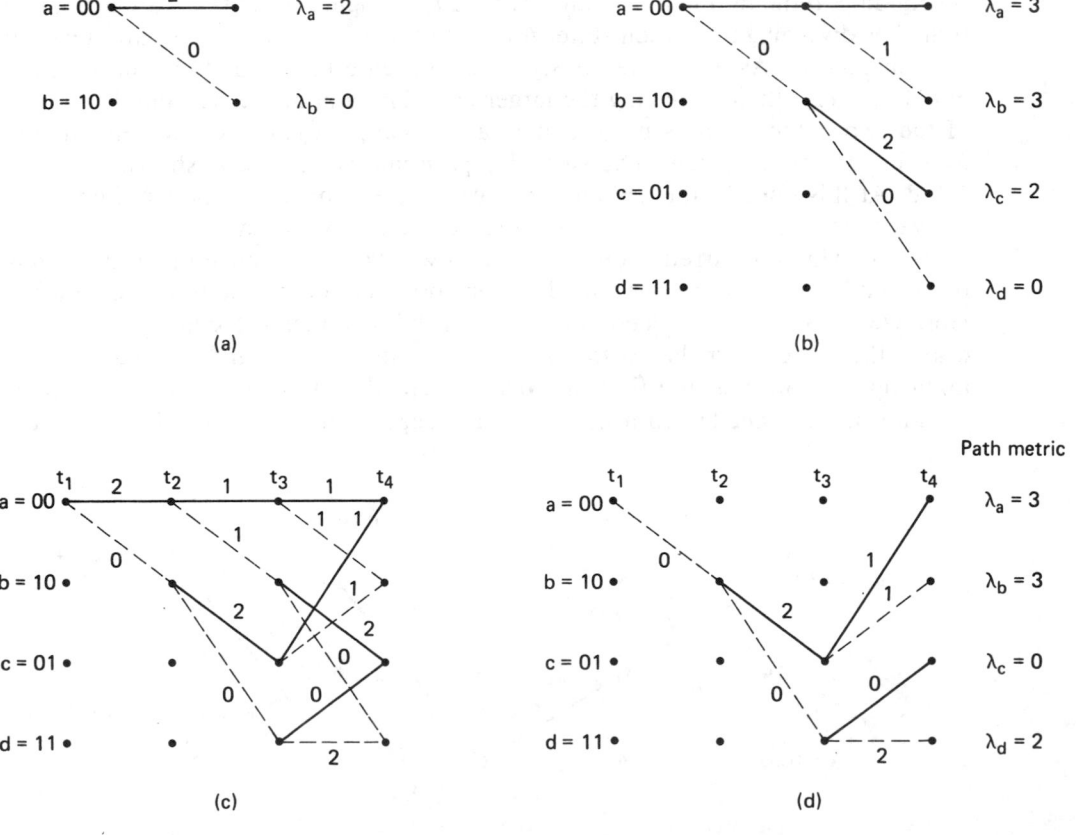

Figure 6.12 Selection of survivor paths. (a) Survivors at t_2. (b) Survivors at t_3. (c) Metric comparisons at t_4. (d) Survivors at t_4. (e) Metric comparisons at t_5. (f) Survivors at t_5. (g) Metric comparisons at t_6. (h) Survivors at t_6.

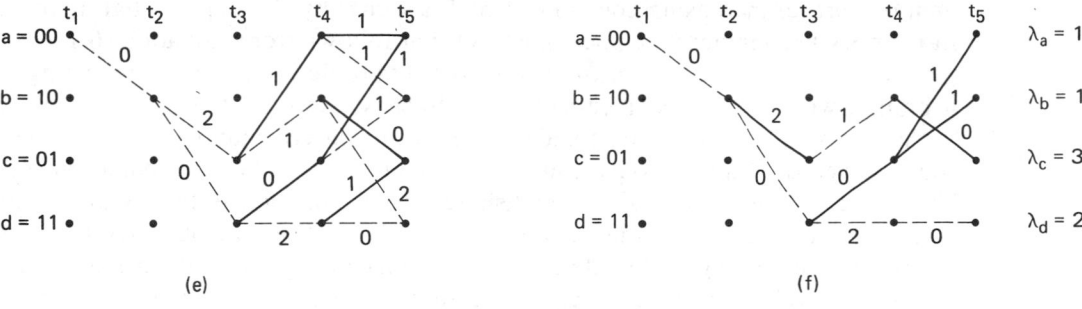

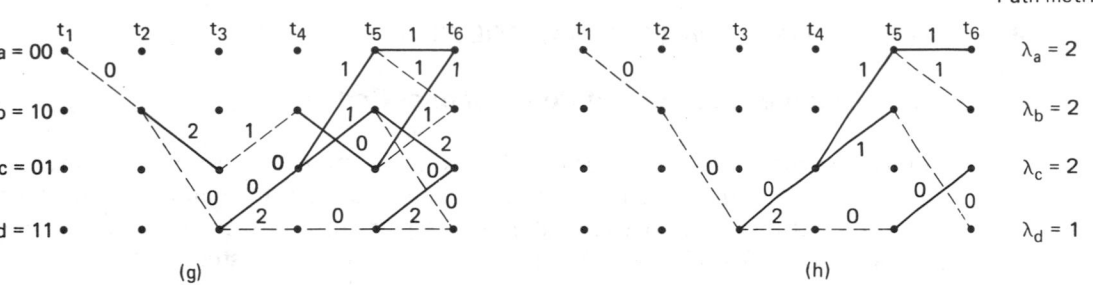

Figure 6.12 (*Continued*)

6.3.5 Path Memory and Synchronization

The storage requirements of the Viterbi decoder grow exponentially with constraint length K. For a code with rate $1/n$, the decoder retains a set of 2^{K-1} paths after each decoding step. With high probability, these paths will not be mutually disjoint very far back from the present decoding depth [12]. All of the 2^{K-1} paths tend to have a common stem which eventually branches to the various states. Thus if the decoder stores enough of the history of the 2^{K-1} paths, the oldest bits on all paths will be the same. A simple decoder implementation, then, contains a *fixed amount of path history* and outputs the oldest bit on an arbitrary path each time it steps one level deeper into the trellis. The amount of path storage required, u, is [12]

$$u = h2^{K-1} \tag{6.10}$$

where h is the length of the information bit path history per state. A refinement, which minimizes the value of h, uses the oldest bit on the most likely path as the decoder output, instead of the oldest bit on an arbitrary path. It has been demonstrated [12] that a value of h of 4 or 5 times the code constraint length is sufficient for near-optimum decoder performance. The storage requirement, u, is the basic limitation on the implementation of Viterbi decoders. The current state of the art limits decoders to a constraint length of about $K = 10$. Efforts to increase coding gain by further increasing constraint length are met by the exponential increase in memory requirements (and complexity) that follows from Equation (6.10).

Branch word synchronization is the process of determining the beginning of a branch word in the received sequence. Such synchronization can take place without new information being added to the transmitted symbol stream because the received data appear to have an excessive error rate when not synchronized. Therefore, a simple way of accomplishing synchronization is to monitor some concomitant indication of this large error rate, that is, the rate at which the path metrics are increasing or the rate at which the surviving paths in the trellis merge. The monitored parameters are compared to a threshold, and synchronization is then adjusted accordingly.

6.4 PROPERTIES OF CONVOLUTIONAL CODES

6.4.1 Distance Properties of Convolutional Codes

Let us consider the distance properties of convolutional codes in the context of our simple encoder in Figure 6.3 and its trellis diagram in Figure 6.7. We want to evaluate the distance between all possible pairs of codeword sequences. As in the case of block codes (see Section 5.5.2), we are interested in the *minimum distance* between all pairs of such codeword sequences in the code, since the minimum distance is related to the error-correcting capability of the code. Because a convolutional code is a group or *linear code* [6], there is no loss in generality in simply finding the minimum distance between each of the codeword sequences and the all-zeros sequence. Assuming that the all-zeros input sequence was transmitted, the paths of interest are those that start and end in the 00 state and do not return to the 00 state anywhere in between. An error will occur whenever the distance of any other path that merges with the $a = 00$ state at time t_i is less than that of the all-zeros path up to time t_i, causing the all-zeros path to be discarded in the decoding process. In other words, given the all-zeros transmission, an error occurs whenever the *all-zeros path does not survive*. The minimum distance for making such an error can be found by exhaustively examining every path from the 00 state to the 00 state. First, let us redraw the trellis diagram, shown in Figure 6.13, labeling each branch with its Hamming distance from the all-zeros codeword instead of with its branch word symbols. The Hamming distance between two unequal-length sequences will be found by first appending the necessary number of zeros to the shorter sequence to make the two sequences equal in length. Consider all the paths that diverge from the all-zeros path and then remerge for the first time at some arbitrary node. From Figure 6.13 we can compute the distances of these paths from the all-zeros path. There is one path at distance 5 from the all-zeros path; this path departs from the all-zeros path at time t_1 and merges with it at time t_4. Similarly, there are two paths at distance 6, one which departs at time t_1 and merges at time t_5, and the other which departs at time t_1 and merges at time t_6, and so on. We can also see from the dashed and solid lines

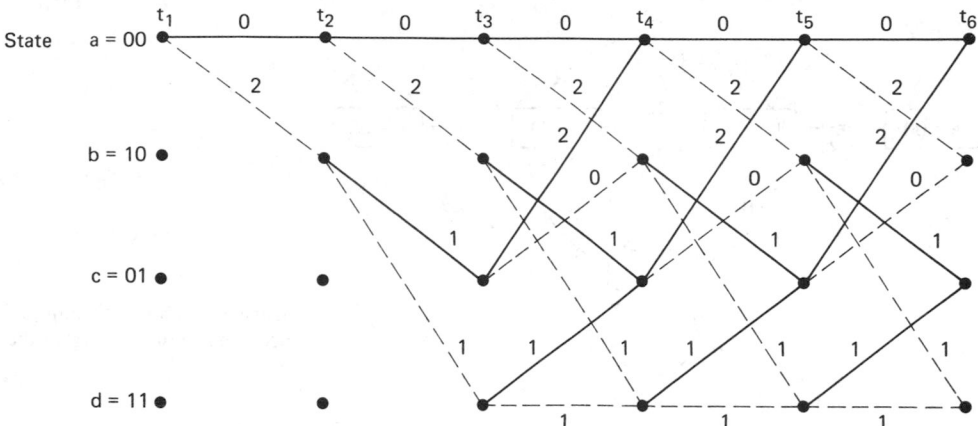

Figure 6.13 Trellis diagram, labeled with distances from the all-zeros path.

of the diagram that the input bits for the distance 5 path are 1 0 0; it differs in only one input bit from the all-zeros input sequence. Similarly, the input bits for the distance 6 paths are 1 1 0 0 and 1 0 1 0 0; each differs in two positions from the all-zeros path. The minimum distance in the set of all arbitrarily long paths that diverge and remerge, called the *minimum free distance* or simply the *free distance*, is seen to be 5 in this example. For calculating the error-correcting capability of the code, we repeat Equation (5.44) with the minimum distance, $d_{\min}$, replaced by the free distance, d_f.

$$t = \left\lfloor \frac{d_f - 1}{2} \right\rfloor \tag{6.11}$$

where $\lfloor x \rfloor$ means the largest integer no greater than x. Setting $d_f = 5$, we see that the code, characterized by the Figure 6.3 encoder, can correct any two channel errors.

Although Figure 6.13 presents the computation of free distance in a straightforward way, a more direct closed-form expression can be obtained by starting with the state diagram in Figure 6.5. First, we label the branches of the state diagram as either $D^0 = 1$, D^1, or D^2, shown in Figure 6.14, where the exponent of D denotes the Hamming distance from the branch word of that branch to the all-zeros branch. The self-loop at node a can be eliminated since it contributes nothing to the distance properties of a codeword sequence relative to the all-zeros sequence. Furthermore, node a can be split into two nodes (labeled a and e), one of which represents the input and the other the output of the state diagram. All paths originating at $a = 00$ and terminating at $e = 00$ can be traced on the modified state diagram of Figure 6.14. We can calculate the transfer function of path $a\ b\ c\ e$ (starting and ending at state 00) in terms of the indeterminate "placeholder" D, as $D^2\ D\ D^2 = D^5$. The exponent of D represents the cumulative tally of the number of ones in the path, and hence the Hamming distance from the all-zeros path. Similarly, the paths $a\ b\ d\ c\ e$ and $a\ b\ c\ b\ c\ e$ each have the transfer

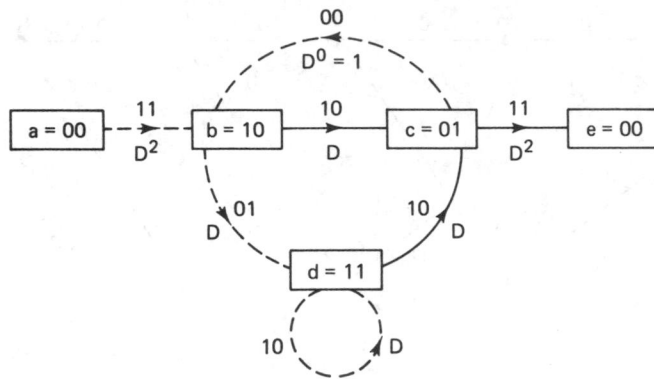

Figure 6.14 State diagram, labeled according to distance from the all-zeros path.

function D^6 and thus a Hamming distance of 6 from the all-zeros path. We now write the following state equations:

$$X_b = D^2 X_a + X_c$$
$$X_c = DX_b + DX_d$$
$$X_d = DX_b + DX_d \tag{6.12}$$
$$X_e = D^2 X_c$$

where $X_a, \ldots, X_e$ are dummy variables for the partial paths to the intermediate nodes. The *transfer function*, $T(D)$, sometimes called the *generating function* of the code can be expressed as $T(D) = X_e/X_a$. By solving the state equations shown in Equation (6.12), we obtain [15, 16]

$$T(D) = \frac{D^5}{1 - 2D} \tag{6.13}$$
$$= D^5 + 2D^6 + 4D^7 + \cdots + 2^\ell D^{\ell+5} + \cdots$$

The transfer function for this code indicates that there is a single path of distance 5 from the all-zeros path, two of distance 6, four of distance 7, and in general, there are 2^ℓ paths of distance $\ell + 5$ from the all-zeros path, where $\ell = 0, 1, 2, \ldots$. The free distance d_f of the code is the Hamming weight of the lowest-order term in the expansion of $T(D)$. In this example $d_f = 5$. In evaluating distance properties, the transfer function, $T(D)$, cannot be used for long constraint lengths since the complexity of $T(D)$ increases exponentially with constraint length.

The transfer function can be used to provide more detailed information than just the distance of the various paths. Let us introduce a factor L into each branch of the state diagram so that the exponent of L can serve as a counter to indicate the number of branches in any given path from state $a = 00$ to state $e = 00$. Furthermore, we can introduce a factor N into all branch transitions caused by the input bit one. Thus, as each branch is traversed, the cumulative exponent on N increases by one, only if that branch transition is due to an input bit one. For the convolutional code characterized in our Figure 6.3 example, the additional factors L and N are shown on the modified state diagram of Figure 6.15. We can now modify Equations (6.12) as follows:

$$X_b = D^2 LN X_a + LN X_c$$
$$X_c = DL X_b + DL X_d$$
$$X_d = DLN X_b + DLN X_d \tag{6.14}$$
$$X_e = D^2 LX_c$$

The transfer function of this augmented state diagram is

$$T(D, L, N) = \frac{D^5 L^3 N}{1 - DL(1 + L)N}$$

$$= D^5 L^3 N + D^6 L^4 (1 + L) N^2 + D^7 L^5 (1 + L)^2 N^3 \tag{6.15}$$

$$+ \cdots + D^{\ell+5} L^{\ell+3} N^{\ell+1} + \cdots$$

Thus we can verify some of the path properties displayed in Figure 6.13. There is one path of distance 5, length 3, which differs in one input bit from the all-zeros path. There are two paths of distance 6, one of which is length 4, the other length 5, and both differ in two input bits from the all-zeros path. Also, of the distance 7 paths, one is of length 5, two are of length 6, and one is of length 7; all four paths correspond to input sequences that differ in three input bits from the all-zeros path. Thus if the all-zeros path is the correct path and the noise causes us to choose one of the incorrect paths of distance 7, three bit errors will be made.

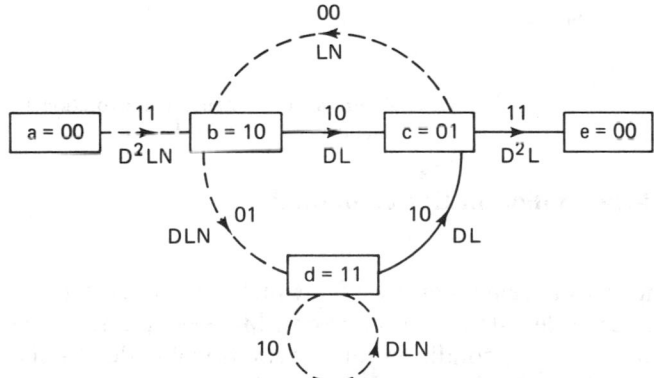

Figure 6.15 State diagram, labeled according to distance, length, and number of input ones.

6.4.1.1 Error-Correcting Capability of Convolutional Codes

In the study of block codes in Chapter 5, we saw that the error-correcting capability, t, represented the number of code symbol errors that could, with maximum likelihood decoding, be corrected in each block length of the code. However, when decoding convolutional codes, the error-correcting capability cannot be stated so succinctly. With regard to Equation (6.11), we can say that the code can, with maximum likelihood decoding, correct t errors within a few constraint lengths, where "few" here means 3 to 5. The exact length depends on how the errors are distributed. For a particular code and error pattern, the length can be bounded using transfer function methods. A computer program for convolutional decoding with the Viterbi algorithm, called VITALG, is provided in Appendix E. The interested reader can use this tool for verifying the capability of Viterbi decoding of convolutional codes with various choices of code generators, code rates, constraint lengths, and path memory lengths.

6.4.2 Systematic and Nonsystematic Convolutional Codes

A *systematic* convolutional code is one in which the input k-tuple appears as part of the output branch word n-tuple associated with that k-tuple. Figure 6.16 shows a binary, rate $\frac{1}{2}$, $K = 3$ systematic encoder. For linear block codes, any nonsystematic code can be transformed into a systematic code with the same block distance properties. This is not the case for convolutional codes. The reason for this is that convolutional codes depend largely on *free distance*; making the convolutional code systematic, in general, *reduces* the maximum possible free distance for a given constraint length and rate.

Table 6.1 shows the maximum free distance for rate $\frac{1}{2}$ systematic and nonsystematic codes for $K = 2$ through 8. For large constraint lengths the results are even more widely separated [17].

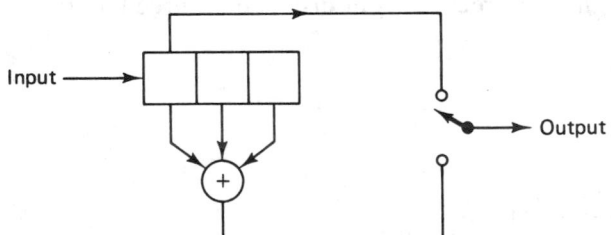

Figure 6.16 Systematic convolutional encoder, rate $\frac{1}{2}$, $K = 3$.

6.4.3 Catastrophic Error Propagation in Convolutional Codes

A *catastrophic error* is defined as an event whereby a finite number of code symbol errors cause an infinite number of decoded data bit errors. Massey and Sain [18] have derived a necessary and sufficient condition for convolutional codes to display catastrophic error propagation. For rate $1/n$ codes with register taps designated by polynomial generators, as described in Section 6.2.1, the condition for catastrophic error propagation is that the generators have a *common polynomial factor* (of degree at least one). For example, Figure 6.17a illustrates a rate $\frac{1}{2}$, $K = 3$ encoder with upper polynomial $g_1(X)$ and lower polynomial $g_2(X)$, as follows:

$$g_1(X) = 1 + X \qquad (6.16)$$
$$g_2(X) = 1 + X^2$$

TABLE 6.1 Comparison of Systematic and Nonsystematic Free Distance, Rate ½

Constraint length	Free distance systematic	Free distance nonsystematic
2	3	3
3	4	5
4	4	6
5	5	7
6	6	8
7	6	10
8	7	10

Source: A. J. Viterbi and J. K. Omura, *Principles of Digital Communication and Coding*, McGraw-Hill Book Company, New York, 1979, p. 251.

The generators $g_1(X)$ and $g_2(X)$ have in common the polynomial factor, $1 + X$, since

$$1 + X^2 = (1 + X)(1 + X)$$

Therefore, the encoder in Figure 6.17a can manifest *catastrophic error propagation*.

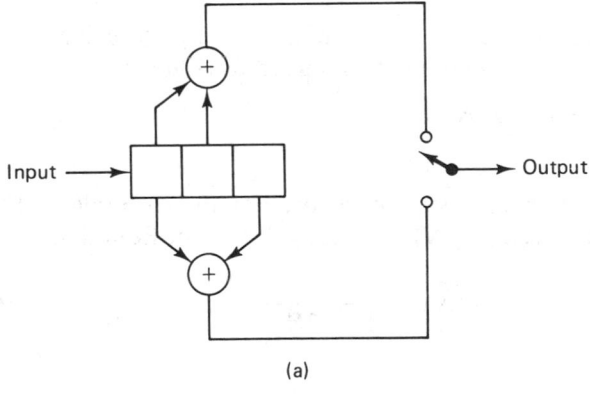

(a)

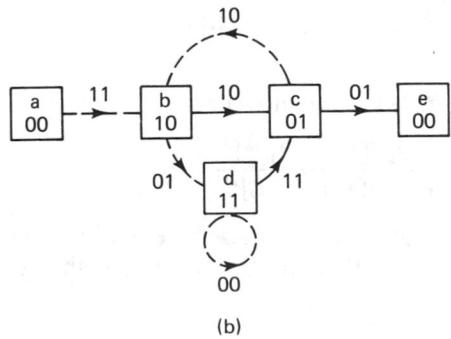

(b)

Figure 6.17 Encoder displaying catastrophic error propagation. (a) Encoder. (b) State diagram.

In terms of the state diagram for any-rate code, catastrophic errors can occur if, and only if, any closed-loop path in the diagram has zero weight (zero distance from the all-zeros path). To illustrate this, consider the example of Figure 6.17. The state diagram in Figure 6.17b is drawn with the state $a = 00$ node split into two nodes, a and e, as before. Assuming that the all-zeros path is the correct path, the incorrect path $a\ b\ d\ d\ \ldots\ d\ c\ e$ has exactly 6 ones, no matter how many times we go around the self-loop at node d. Thus for a BSC, for example, three channel errors may cause us to choose this incorrect path. An arbitrarily large number of errors (two plus the number of times the self-loop is traversed) can be made on such a path. We observe that for rate $1/n$ codes, if each adder in the encoder has an even number of connections, the self-loop corresponding to the all-ones data state will have zero weight, and consequently, *the code will be catastrophic.*

The only advantage of a systematic code, described earlier, is that it can never be catastrophic, since each closed loop must contain at least one branch generated by a nonzero input bit, and thus each closed loop must have a nonzero code symbol. However, it can be shown [19] that only a small fraction of non-systematic codes (excluding those where all adders have an even number of taps) are catastrophic.

6.4.4 Performance Bounds for Convolutional Codes

The probability of bit error, P_B, for a binary convolutional code using hard-decision decoding can be shown [8] to be upper bounded as follows:

$$P_B \le \frac{dT(D, N)}{dN}\bigg|_{N=1, D=2\sqrt{p(1-p)}} \tag{6.17}$$

where p is the probability of channel symbol error. For the example of Figure 6.3, $T(D, N)$ is obtained from $T(D, L, N)$ by setting $L = 1$ in Equation (6.15).

$$T(D, N) = \frac{D^5 N}{1 - 2DN} \tag{6.18}$$

and

$$\frac{dT(D, N)}{dN}\bigg|_{N=1} = \frac{D^5}{(1 - 2D)^2} \tag{6.19}$$

Combining Equations (6.17) and (6.19), we can write

$$P_B \le \frac{\{2[p(1 - p)]^{1/2}\}^5}{\{1 - 4[p(1 - p)]^{1/2}\}^2} \tag{6.20}$$

For coherent BPSK modulation over an additive white Gaussian noise (AWGN) channel, it can be shown [8] that the bit error probability is bounded by

$$P_B \leq Q\left(\sqrt{2d_f \frac{E_c}{N_0}}\right) \exp\left(d_f \frac{E_c}{N_0}\right) \frac{dT(D, N)}{dN}\Bigg|_{N=1, D=\exp(-E_c/N_0)} \qquad (6.21)$$

where

$$E_c/N_0 = rE_b/N_0$$

E_b/N_0 = ratio of information bit energy to noise power spectral density

E_c/N_0 = ratio of channel symbol energy to noise power spectral density

$$r = k/n = \text{rate of the code}$$

and where $Q(x)$ is defined in Equations (2.42) and (2.43) and tabulated in Table B.1. Therefore, for the rate $\frac{1}{2}$ code with free distance $d_f = 5$, in conjunction with coherent BPSK and hard-decision decoding, we can write

$$P_B \leq Q\left(\sqrt{\frac{5E_b}{N_0}}\right) \exp\left(\frac{5E_b}{2N_0}\right) \frac{\exp(-5E_b/2N_0)}{[1 - 2\exp(-E_b/2N_0)]^2}$$

$$\leq \frac{Q(\sqrt{5E_b/N_0})}{[1 - 2\exp(-E_b/2N_0)]^2} \qquad (6.22)$$

6.4.5 Coding Gain

Coding gain is defined as the reduction, usually expressed in decibels, in the required E_b/N_0 to achieve a specified error probability of the coded system over an uncoded system with the same modulation and channel characteristics. Table 6.2 lists an upper bound on the coding gains, compared to uncoded coherent BPSK, for several maximum free distance convolutional codes with constraint lengths varying from 3 to 9 over a Gaussian channel with hard-decision decoding. The table illustrates that it is possible to achieve significant coding gain even with

TABLE 6.2 Coding Gain Upper Bounds for Some Convolutional Codes

Rate $\frac{1}{2}$ codes			Rate $\frac{1}{3}$ codes		
K	d_f	Upper bound (dB)	K	d_f	Upper bound (dB)
3	5	3.97	3	8	4.26
4	6	4.76	4	10	5.23
5	7	5.43	5	12	6.02
6	8	6.00	6	13	6.37
7	10	6.99	7	15	6.99
8	10	6.99	8	16	7.27
9	12	7.78	9	18	7.78

Source: V. K. Bhargava, D. Haccoun, R. Matyas, and P. Nuspl, *Digital Communications by Satellite*, John Wiley & Sons, Inc., New York, 1981.

a simple convolutional code. The actual coding gain will vary with the required bit error probability [20].

Table 6.3 lists the measured coding gains, compared to uncoded coherent BPSK, achieved with hardware implementation or computer simulation over a Gaussian channel with soft-decision decoding [21]. The uncoded E_b/N_0 is given in the leftmost column. From Table 6.3 we can see that coding gain increases as the bit error probability is decreased. However, the coding gain cannot increase indefinitely; it has an upper bound as shown in the table. This bound in decibels can be shown [21] to be

$$\text{coding gain} \le 10 \log_{10} (rd_f) \tag{6.23}$$

where r is the code rate and d_f is the free distance. Examination of Table 6.3 also reveals that at $P_B = 10^{-7}$, for code rates of $\frac{1}{2}$ and $\frac{2}{3}$, the weaker codes tend to be closer to the upper bound than are the more powerful codes.

Typically, Viterbi decoding is used over binary input channels with either hard or 3-bit soft quantized outputs. The constraint lengths vary between 3 and 9, the code rate is rarely smaller than $\frac{1}{3}$, and the path memory is usually a few constraint lengths [12]. The path memory refers to the depth of the input bit history stored by the decoder. From the Viterbi decoding example in Section 6.3.4, one might question the notion of a fixed path memory. It seems from the example that the decoding of a branch word, at any arbitrary node, can take place as soon as there is only a single surviving branch at that node. That is true; however, to actually implement the decoder in this way would entail an extensive amount of processing to continually check when the branch word can be decoded. Instead, *a fixed delay is provided*, after which the branch word is decoded. It has been shown [12, 22] that a fixed amount of path history, namely 4 or 5 times the constraint length, is sufficient to limit the degradation from the optimum decoder performance to about 0.1 dB for the BSC and Gaussian channels. Typical error performance curves are shown in Figure 6.18 for rate $\frac{1}{2}$ codes using coherent BPSK over a soft (8-level) quantized channel, with Viterbi decoding, and a 32-bit path memory. Also plotted are the transfer function bounds for infinitely fine quantized received data [12]. Figure 6.19 gives the simulation results for Viterbi decoding with hard decision quantization [12]. Notice that each increment in constraint

TABLE 6.3 Basic Coding Gain (dB) for Soft Decision Viterbi Decoding

Uncoded E_b/N_0 (dB)	Code rate		$\frac{1}{3}$		$\frac{1}{2}$			$\frac{2}{3}$		$\frac{3}{4}$	
	P_B	K	7	8	5	6	7	6	8	6	9
6.8	10^{-3}		4.2	4.4	3.3	3.5	3.8	2.9	3.1	2.6	2.6
9.6	10^{-5}		5.7	5.9	4.3	4.6	5.1	4.2	4.6	3.6	4.2
11.3	10^{-7}		6.2	6.5	4.9	5.3	5.8	4.7	5.2	3.9	4.8
	Upper bound		7.0	7.3	5.4	6.0	7.0	5.2	6.7	4.8	5.7

Source: I. M. Jacobs, "Practical Applications of Coding," *IEEE Trans. Inf. Theory*, vol. IT20, May 1974, pp. 305–310.

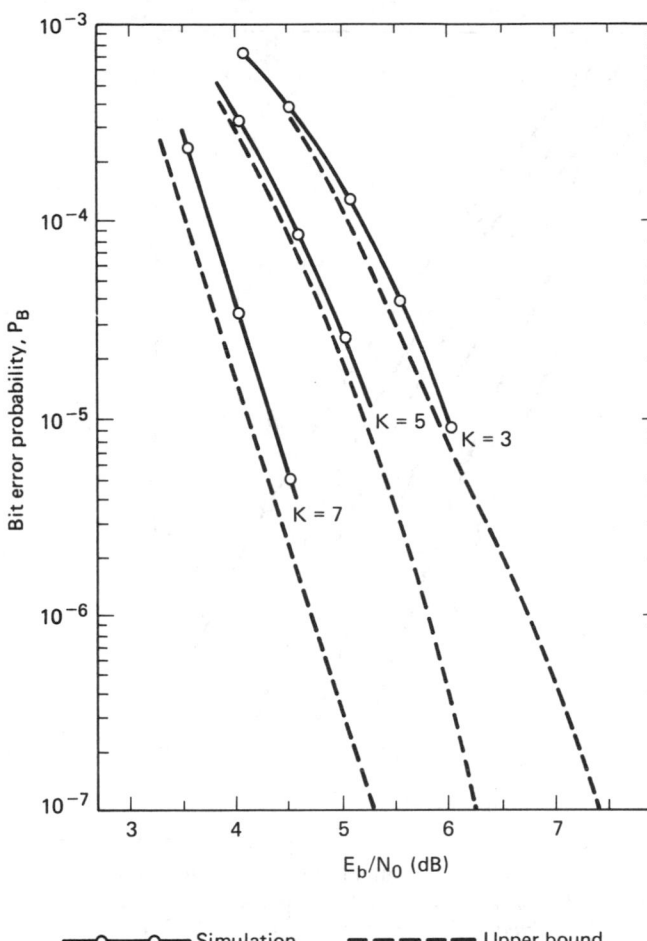

Figure 6.18 Bit error probability versus E_b/N_0 for rate $\frac{1}{2}$ codes using coherent BPSK over a soft quantized channel, Viterbi decoding, and a 32-bit path memory. (Reprinted with permission from J. A. Heller and I. M. Jacobs, "Viterbi Decoding for Satellite and Space Communication," *IEEE Trans. Commun. Technol.*, vol. COM19, no. 5, October 1971, Fig. 5, p. 84. © 1971 IEEE.)

length improves the required E_b/N_0 by a factor of approximately 0.5 dB at $P_B = 10^{-5}$. Also, as expected, the 3-bit soft decisions of the channel output result in approximately a 2-dB gain over the hard quantized BSC.

6.4.6 Best Known Convolutional Codes

The connection vectors or polynomial generators of a convolutional code are usually selected based on the code's free distance properties. The first criterion is to select a code that does not have catastrophic error propagation and that has the maximum free distance for the given rate and constraint length. Then the number of paths at the free distance d_f, or the number of data bit errors the paths represent, should be minimized. The selection procedure can be further refined by considering the number of paths or bit errors at $d_f + 1$, at $d_f + 2$, and so on, until only one code or class of codes remains. A list of the best known codes of rate $\frac{1}{2}$, $K = 3$ to 9, and rate $\frac{1}{3}$, $K = 3$ to 8, based on this criterion was compiled by Odenwalder [3, 23] and is given in Table 6.4. The connection vectors in this

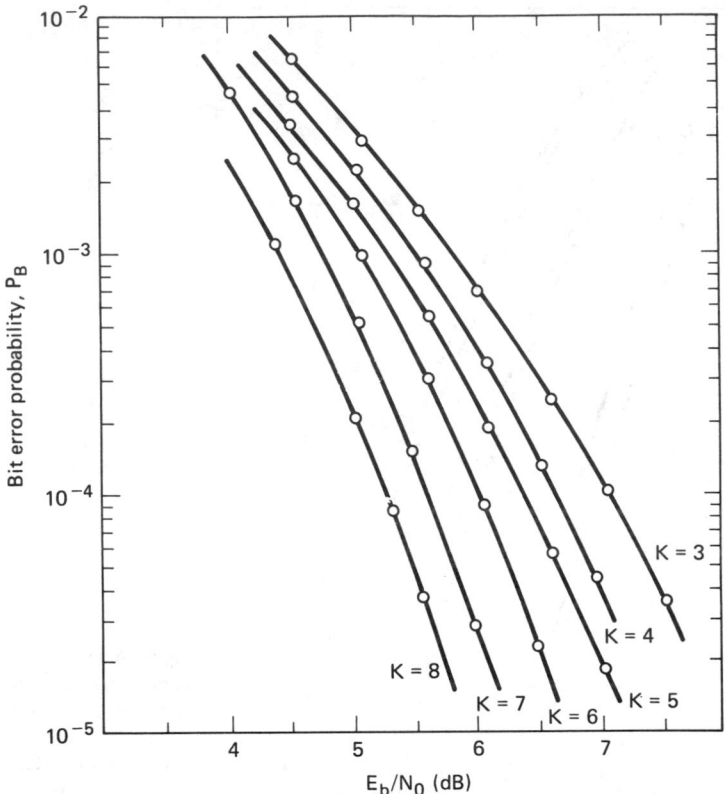

Figure 6.19 Bit error probability versus E_b/N_0 for rate $\frac{1}{2}$ codes using coherent BPSK over a BSC, Viterbi decoding, and a 32-bit path memory. (Reprinted with permission from J. A. Heller and I. M. Jacobs, "Viterbi Decoding for Satellite and Space Communication," *IEEE Trans. Commun. Technol.*, vol. COM19, no. 5, October 1971, Fig. 7, p. 84. © 1971 IEEE.)

table represent the presence or absence (1 or 0) of a tap connection on the corresponding stage of the convolutional encoder. The leftmost term corresponds to the leftmost stage of the encoder register, and the rightmost term corresponds to the rightmost stage, following the notation established in Figure 6.3. It is interesting to note that these connections can be inverted (leftmost and rightmost can be interchanged in the above description). Under the condition of Viterbi decoding, the inverted connections give rise to codes with identical distance properties, and hence identical performance, as those in Table 6.4.

6.4.7 Convolutional Code Rate Trade-Off

6.4.7.1 Performance with Coherent PSK Signaling

The error-correcting capability of a coding scheme increases as the number of channel symbols n per information bit k increases or the rate, k/n, decreases. However, the channel bandwidth and the decoder complexity both increase with n. The advantage of lower code rates when using convolutional codes with co-

TABLE 6.4 Optimum Short Constraint Length Convolutional Codes (Rate $\frac{1}{2}$ and Rate $\frac{1}{3}$)

Rate	Constraint length	Free distance	Code vector
$\frac{1}{2}$	3	5	111 101
$\frac{1}{2}$	4	6	1111 1011
$\frac{1}{2}$	5	7	10111 11001
$\frac{1}{2}$	6	8	101111 110101
$\frac{1}{2}$	7	10	1001111 1101101
$\frac{1}{2}$	8	10	10011111 11100101
$\frac{1}{2}$	9	12	110101111 100011101
$\frac{1}{3}$	3	8	111 111 101
$\frac{1}{3}$	4	10	1111 1011 1101
$\frac{1}{3}$	5	12	11111 11011 10101
$\frac{1}{3}$	6	13	101111 110101 111001
$\frac{1}{3}$	7	15	1001111 1010111 1101101
$\frac{1}{3}$	8	16	11101111 10011011 10101001

Source: J. P. Odenwalder, *Error Control Coding Handbook*, Linkabit Corp., San Diego, Calif., July 15, 1976.

herent PSK, is that the required E_b/N_0 is decreased (for a large range of code rates), permitting the transmission of higher data rates for a given amount of power, or permitting reduced power for a given data rate. Simulation studies have shown [16, 22] that for a fixed constraint length, a decrease in the code rate from $\frac{1}{2}$ to $\frac{1}{3}$ results in a reduction of the required E_b/N_0 of roughly 0.4 dB. However, the corresponding increase in decoder complexity is about 17%. For smaller values of code rate, the improvement in performance relative to the increased decoding complexity diminishes rapidly [22]. Eventually, a point is reached where further decrease in code rate is characterized by a reduction in coding gain.

6.4.7.2 Performance with Noncoherent Orthogonal Signaling

In contrast to PSK, there is an optimum code rate of about $\frac{1}{2}$ for noncoherent orthogonal signaling. Error performance at rates of $\frac{1}{3}$, $\frac{2}{3}$, and $\frac{3}{4}$ are each worse than those for rate $\frac{1}{2}$. For a fixed constraint length, the rate $\frac{1}{3}$, $\frac{2}{3}$, and $\frac{3}{4}$ codes typically degrade by about 0.25, 0.5, and 0.3 dB, respectively, relative to the rate $\frac{1}{2}$ performance [16].

6.5 OTHER CONVOLUTIONAL DECODING ALGORITHMS

6.5.1 Sequential Decoding

Prior to the discovery of an optimum algorithm by Viterbi, other algorithms had been proposed for decoding convolutional codes. The earliest was the *sequential decoding algorithm*, originally proposed by Wozencraft [24, 25] and subsequently modified by Fano [2]. A sequential decoder works by generating hypotheses about the transmitted codeword sequence; it computes a metric between these hypotheses and the received signal. It goes forward as long as the metric indicates that its choices are likely; otherwise, it goes backward, changing hypotheses until, through a systematic trial-and-error search, it finds a likely hypothesis. Sequential decoders can be implemented to work with hard or soft decisions, but soft decisions are usually avoided because they greatly increase the amount of the required storage and the complexity of the computations.

Consider that using the encoder shown in Figure 6.3, a sequence $\mathbf{m}$ = 1 1 0 1 1 is encoded into the codeword sequence $\mathbf{U}$ = 1 1 0 1 0 1 0 0 0 1, as shown in Example 6.1. Assume that the received sequence $\mathbf{Z}$ is, in fact, a *correct* rendition of $\mathbf{U}$. The decoder has available a replica of the encoder code tree, shown in Figure 6.6, and can use the received sequence $\mathbf{Z}$ to penetrate the tree. The decoder starts at the time t_1 node of the tree and generates both paths leaving that node. The decoder follows that path which agrees with the received n code symbols. At the next level in the tree, the decoder again generates both paths leaving that node, and follows the path agreeing with the second group of n code symbols. Proceeding in this manner, the decoder quickly penetrates the tree.

Suppose, however, that the received sequence $\mathbf{Z}$ is a *corrupted* version of $\mathbf{U}$. The decoder starts at the time t_1 node of the code tree and generates both paths leading from that node. If the received n code symbols coincide with one of the generated paths, the decoder follows that path. If there is not agreement, the decoder follows the *most likely path* but keeps a cumulative count on the number of disagreements between the received symbols and the branch words on the path being followed. If two branches appear equally likely, the receiver uses an arbitrary rule, such as following the zero input path. At each new level in the tree, the decoder generates new branches and compares them with the next set of n received code symbols. The search continues to penetrate the tree along the most likely path and maintains the cumulative disagreement count.

If the disagreement count exceeds a certain number (which may increase as we penetrate the tree), the decoder decides that it is on an incorrect path, backs out of the path, and tries another. The decoder keeps track of the discarded pathways to avoid repeating any path excursions. For example, assume that the encoder in Figure 6.3 is used to encode the message sequence $\mathbf{m} = 1\ 1\ 0\ 1\ 1$ into the codeword sequence $\mathbf{U}$ as shown in Example 6.1. Suppose that the fourth and seventh bits of the transmitted sequence $\mathbf{U}$ are received in error, such that:

Time:		t_1	t_2	t_3	t_4	t_5
Message sequence:	$\mathbf{m} =$	1	1	0	1	1
Transmitted sequence:	$\mathbf{U} =$	1 1	0 1	0 1	0 0	0 1
Received sequence:	$\mathbf{Z} =$	1 1	0 0	0 1	1 0	0 1

Let us follow the decoder path trajectory with the aid of Figure 6.20. Assume that a cumulative path disagreement count of 3 is the criterion for backing up and trying an alternative path. On Figure 6.20 the numbers along the path trajectory represent the current disagreement count.

1. At time t_1 we receive symbols 11 and compare them with the branch words leaving the first node.
2. The most likely branch is the one with branch word 11 (corresponding to an input bit one or downward branching), so the decoder decides that input bit one is the correct decoding, and moves to the next level.
3. At time t_2, the decoder receives symbols 00 and compares them with the available branch words 10 and 01 at this second level.
4. There is no "best" path, so the decoder arbitrarily takes the input bit zero (or branch word 10) path, and the disagreement count registers a disagreement of 1.
5. At time t_3, the decoder receives symbols 01 and compares them with the available branch words 11 and 00 at this third level.
6. Again, there is no best path, so the decoder arbitrarily takes the input zero (or branch word 11) path, and the disagreement count is increased to 2.
7. At time t_4, the decoder receives symbols 10 and compares them with the available branch words 00 and 11 at this fourth level.
8. Again, there is no best path, so the decoder takes the input bit zero (or branch word 00) path, and the disagreement count is increased to 3.
9. But a disagreement count of 3 is the turnaround criterion, so the decoder "backs out" and tries the alternative path. The disagreement counter is reset to 2.
10. The alternative path is the input bit one (or branch word 11) path at the t_4

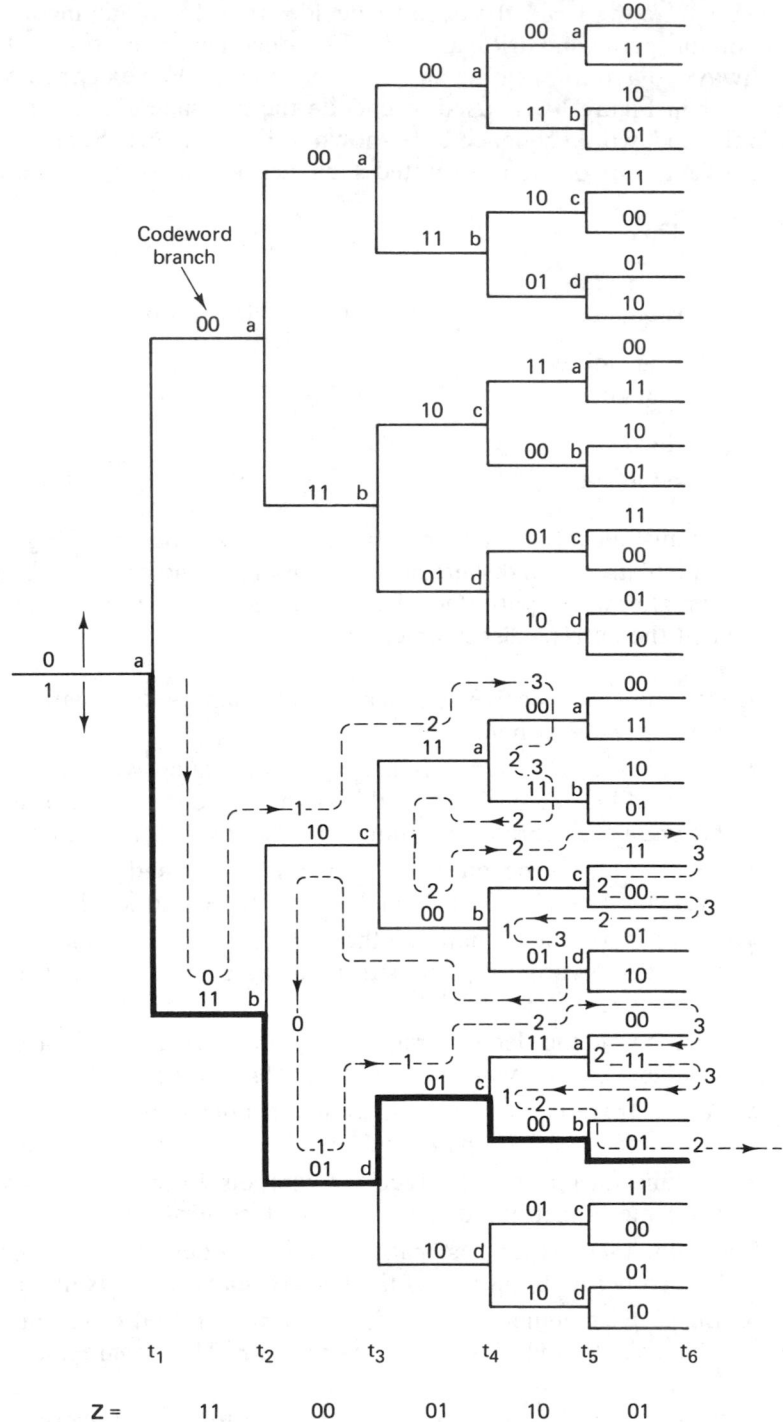

Figure 6.20 Sequential decoding example.

level. The decoder tries this, but compared to the received symbols 10, there is still a disagreement of 1, and the counter is reset to 3.

11. But, 3 being the turnaround criterion, the decoder backs out of this path, and the counter is reset to 2. All of the alternatives have now been traversed at this t_4 level, so the decoder returns to the node at t_3, and resets the counter to 1.

12. At the t_3 node, the decoder compares the symbols received at time t_3, namely 01, with the untried 00 path. There is a disagreement of 1, and the counter is increased to 2.

13. At the t_4 node, the decoder follows the branch word 10 that matches its t_4 code symbols of 10. The counter remains unchanged at 2.

14. At the t_5 node, there is no best path, so the decoder follows the upper branch, as is the rule, and the counter is increased to 3.

15. At this count, the decoder backs up, resets the counter to 2, and tries the alternative path at node t_5. Since the alternate branch word is 00, there is a disagreement of 1 with the received code symbols 01 at time t_5, and the counter is again increased to 3.

16. The decoder backs out of this path, and the counter is reset to 2. All of the alternatives have now been traversed at this t_5 level, so the decoder returns to the node at t_4 and resets the counter to 1.

17. The decoder tries the alternative path at t_4, which raises the metric to 3 since there is a disagreement in two positions of the branch word. This time the decoder must back up all the way to the time t_2 node because all of the other paths at higher levels have been tried. The counter is now decremented to zero.

18. At the t_2 node, the decoder now follows the branch word 01, and because there is a disagreement of 1 with the received code symbols 00 at time t_2, the counter is increased to 1.

The decoder continues in this way. As shown in Figure 6.20, the final path, which has not increased the counter to its turnaround criterion, yields the correctly decoded message sequence, 1 1 0 1 1. Sequential decoding can be viewed as a trial-and-error technique for searching out the correct path in the code tree. It performs the search in a sequential manner, always operating on just a single path at a time. If an incorrect decision is made, subsequent extensions of the path will be wrong. The decoder can eventually recognize its error by monitoring the path metric. The algorithm is similar to the case of an automobile traveler following a road map. As long as the traveler recognizes that the passing landmarks correspond to those on the map, he continues on the path. When he notices strange landmarks (an increase in his dissimilarity metric) the traveler eventually assumes that he is on an incorrect road, and he backs up to a point where he can now recognize the landmarks (his metric returns to an acceptable range). He then tries an alternative road.

6.5.2 Comparisons and Limitations of Viterbi and Sequential Decoding

The major drawback of the Viterbi algorithm is that while error probability decreases exponentially with constraint length, the number of code states, and consequently decoder complexity, *grows exponentially with constraint length*. On the other hand, the computational complexity of the Viterbi algorithm is independent of channel characteristics (compared to hard-decision decoding, soft-decision decoding requires only a trivial increase in the number of computations). Sequential decoding achieves asymptotically the same error probability as maximum likelihood decoding but without searching all possible states. In fact, with sequential decoding the number of states searched is essentially *independent of constraint length*, thus making it possible to use very large ($K = 41$) constraint

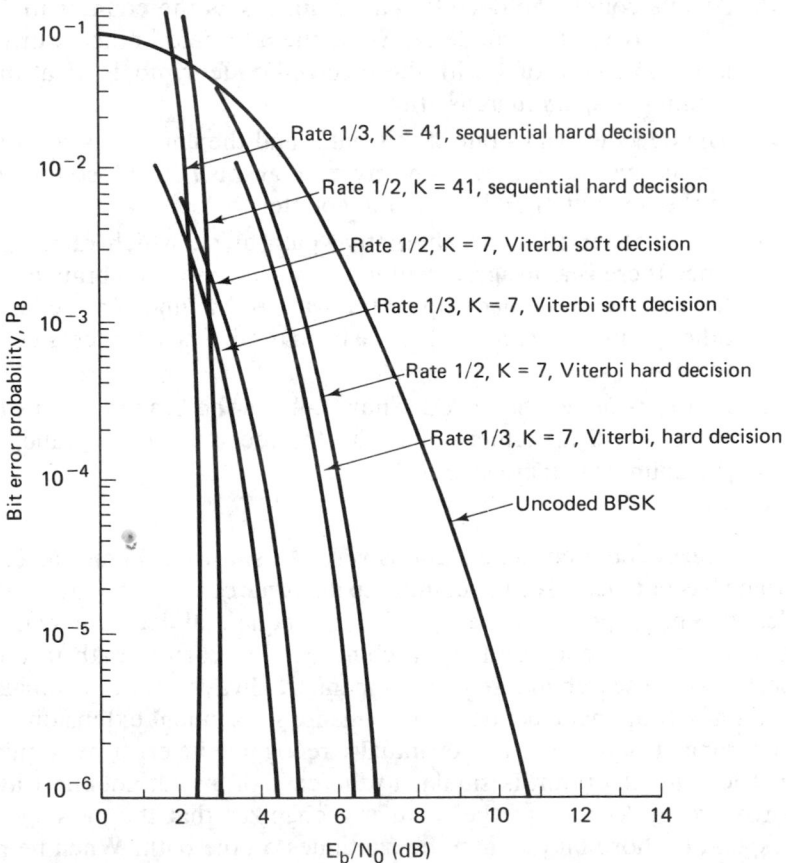

Figure 6.21 Bit error performance for various Viterbi and sequential decoding schemes using coherent BPSK over an AWGN channel. (Reprinted with permission from J. K. Omura and B. K. Levitt, "Coded Error Probability Evaluation for Antijam Communication Systems," *IEEE Trans. Commun.*, vol. COM30, no. 5, May 1982, Fig. 4, p. 900. © 1982 IEEE.)

lengths. This is an important factor in providing such low error probabilities. The major drawback of sequential decoding is that the number of state metrics searched is a random variable. For sequential decoding, the expected number of poor hypotheses and backward searches is a function of the channel SNR. With a low SNR, more hypotheses must be tried than with a high SNR. Because of this variability in computational load, buffers must be provided to store the arriving sequences. Under low SNR, the received sequences must be buffered while the decoder is laboring to find a likely hypothesis. If the average symbol arrival rate exceeds the average symbol decode rate, the buffer will overflow, no matter how large it is, causing a loss of data. The sequential decoder typically puts out error-free data until the buffer overflows, at which time the decoder has to go through a recovery procedure. The buffer overflow threshold is a very sensitive function of SNR. Therefore, an important part of a sequential decoder specification is the *probability of buffer overflow*.

In Figure 6.21, some typical P_B versus E_b/N_0 curves for these two popular solutions to the convolutional decoding problem, Viterbi decoding and sequential decoding, illustrate their comparative performance using coherent BPSK over an AWGN channel. The curves compare Viterbi decoding (rates $\frac{1}{2}$ and $\frac{1}{3}$ hard decision, $K = 7$) versus Viterbi decoding (rates $\frac{1}{2}$ and $\frac{1}{3}$ soft decision, $K = 7$) versus sequential decoding (rates $\frac{1}{2}$ and $\frac{1}{3}$ hard decision, $K = 41$). One can see from Figure 6.21 that coding gains of approximately 8 dB at $P_B = 10^{-6}$ can be achieved with sequential decoders. Since the work of Shannon [26] foretold the potential of approximately 11 dB of coding gain compared to uncoded BPSK, it appears that the major portion of what is theoretically possible can already be accomplished.

6.5.3 Feedback Decoding

A *feedback decoder* makes a hard decision on the data bit at stage j based on metrics computed from stages $j, j + 1, \ldots, j + m$, where m is a preselected positive integer. *Look-ahead length*, L, is defined as $L = m + 1$, the number of received code symbols, expressed in terms of the corresponding number of encoder input bits that are used to decode an information bit. The decision of whether the data bit is zero or one depends on which branch the minimum Hamming distance path traverses in the *look-ahead window* from stage j to stage $j + m$. The detailed operation is best understood in terms of a specific example. Let us consider the use of a feedback decoder for the rate $\frac{1}{2}$ convolutional code shown in Figure 6.3. Figure 6.22 illustrates the tree diagram and the operation of the feedback decoder for $L = 3$. That is, in decoding the bit at branch j, the decoder considers the paths at branches $j, j + 1$, and $j + 2$.

Beginning with the first branch, the decoder computes 2^L or eight cumulative Hamming path metrics and decides that the bit for the first branch is zero if the minimum distance path is contained in the upper part of the tree, and decides one if the minimum distance path is in the lower part of the tree. Assume that the received sequence is $Z = 1\ 1\ 0\ 0\ 0\ 1\ 0\ 0\ 0\ 1$. We now examine the eight paths from time t_1 through time t_3 in the block marked A in Figure 6.22, and compute

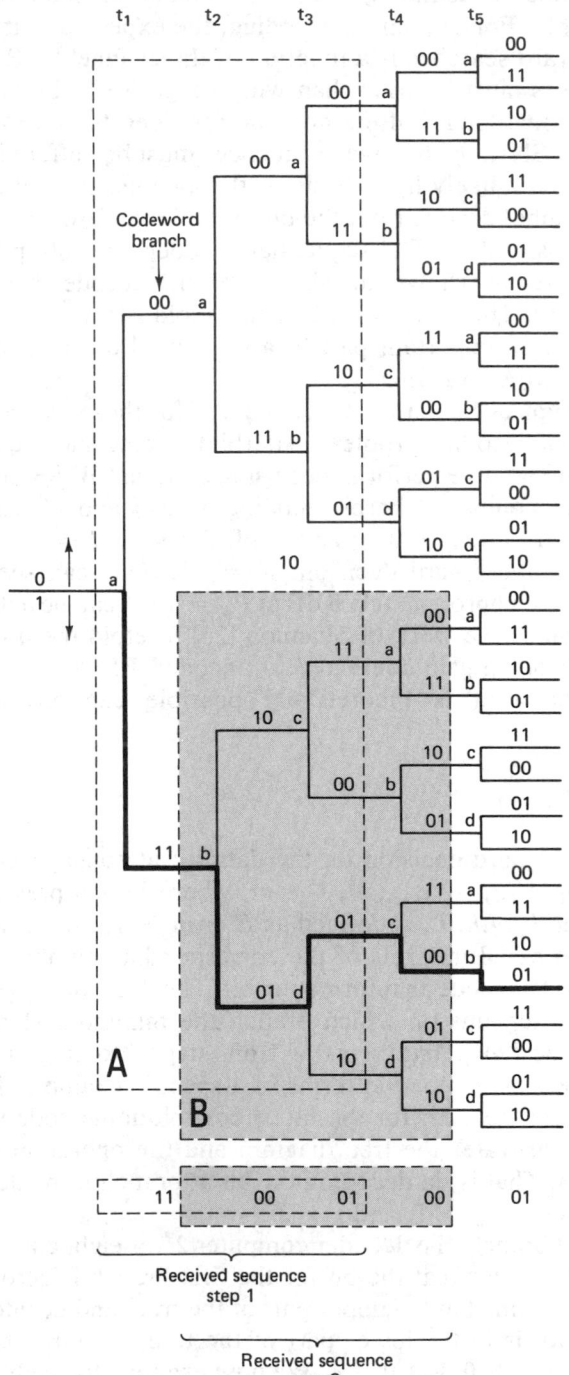

Figure 6.22 Feedback decoding example.

metrics comparing these eight paths with the first six received code symbols (three branches deep times two symbols per branch). Listing the Hamming cumulative path metrics (starting from the top path), they are:

Upper-half metrics: 3, 3, 6, 4

Lower-half metrics: 2, 2, 1, 3

We see that the minimum metric is contained in the lower part of the tree. Therefore, the first decoded bit is one (characterized by a downward movement on the tree). The next step is to extend the lower part of the tree (the part that survived) one stage deeper, and again compute eight metrics, this time from t_2 through t_4. Having decoded the first two code symbols, we now slide over two code symbols to the right and again compute the path metrics for six code symbols. This takes place in the block marked B in Figure 6.22. Again, listing the metrics from top path to bottom path, they are:

Upper-half metrics: 2, 4, 3, 3

Lower-half metrics: 3, 1, 4, 4

For the assumed received sequence, the minimum metric is found in the lower half of block B. Therefore, the second decoded bit is one.

The same procedure continues until the entire message is decoded. The decoder is called a *feedback decoder* because the detection decisions are *fed back* to the decoder in determining the subset of code paths that are to be considered next. On the BSC, the feedback decoder can perform nearly as well as the Viterbi decoder [17] in that it can correct all the more probable error patterns, namely all those of weight $(d_f - 1)/2$ or less, where d_f is the free distance of the code. An important design parameter for feedback convolutional decoders is L, the look-ahead length. Increasing L increases the coding gain but also increases the decoder implementation complexity.

6.6 INTERLEAVING AND CONCATENATED CODES

Throughout this chapter and Chapter 5 we have assumed that the channel is *memoryless,* since we have considered codes that are designed to combat random independent errors. A channel that has *memory* is one that exhibits mutually dependent signal transmission impairments. An example of such a channel is a fading channel, particularly when the fading varies slowly compared to one symbol time. Another type of impairment, called *multipath,* involves signal arrivals at the receiver over two or more paths of different lengths. The effect is that the signals *arrive out of phase* with each other, and the cumulative received signal is distorted. High-frequency (HF) and tropospheric propagation radio channels suffer from such phenomena. Also, some channels suffer from switching noise and other burst noise (e.g., telephone channels or channels disturbed by pulse jamming). All of these time-correlated impairments result in statistical dependence among successive symbol transmissions. That is, the disturbances tend to cause errors that occur in bursts, instead of as isolated events.

Under the assumption that the channel has memory, the errors no longer can be characterized as single randomly distributed bit errors whose occurrence is independent from bit to bit. Most block or convolutional codes are designed to combat random independent errors. The result of a channel having memory on such coded signals is to cause degradation in error performance. Coding techniques for channels with memory have been proposed [27, 28], but the greatest problem with such coding is the difficulty in obtaining accurate models of the often time-varying statistics of such channels. One technique, which only requires a knowledge of the *duration or span* of the channel memory, *not* its exact statistical characterization, is the use of time diversity or *interleaving*.

Interleaving the coded message before transmission and deinterleaving after reception causes bursts of channel errors to be spread out in time and thus to be handled by the decoder as if they were random errors. Since, in all practical cases, the channel memory decreases with time separation, the idea behind interleaving is to separate the codeword symbols in time. The intervening times are similarly filled by the symbols of other codewords. Separating the symbols in time effectively transforms a channel with memory to a *memoryless* one, and thereby enables the random-error-correcting codes to be useful in a burst-noise channel.

The interleaver shuffles the code symbols over a span of several block lengths (for block codes) or several constraint lengths (for convolutional codes). The span required is determined by the burst duration. The details of the bit redistribution pattern must be known to the receiver in order for the symbol stream to be deinterleaved before being decoded. Figure 6.23 illustrates a simple interleaving example. In Figure 6.23a we see seven uninterleaved codewords, A through G. Each codeword is comprised of seven code symbols. Let us assume that the code has a single-error-correcting capability within each seven-symbol sequence. If the memory span of the channel is one codeword in duration, such a seven-symbol-time noise burst could destroy the information contained in one or two codewords. However, suppose that, after having encoded the data, the code symbols were then *interleaved* or shuffled, as shown in Figure 6.23b. That is, each code symbol of each codeword is separated from its preinterleaved neighbors by a span of seven symbol times. The interleaved stream is then used to modulate a waveform that is transmitted over the channel. A contiguous channel noise burst occupying seven symbol times is seen in Figure 6.23b, to affect one code symbol from each of the original seven codewords. Upon reception, the stream is first deinterleaved so that it resembles the original coded sequence in Figure 6.23a. Then the stream is decoded. Since each codeword possesses a single-error-correcting capability, the burst noise has no degrading effect on the final sequence.

Interleaving techniques have proven useful for all the convolutional and block codes described here and in Chapter 5. Two types of interleavers are commonly used, *block interleavers* and *convolutional interleavers*. They are each described below.

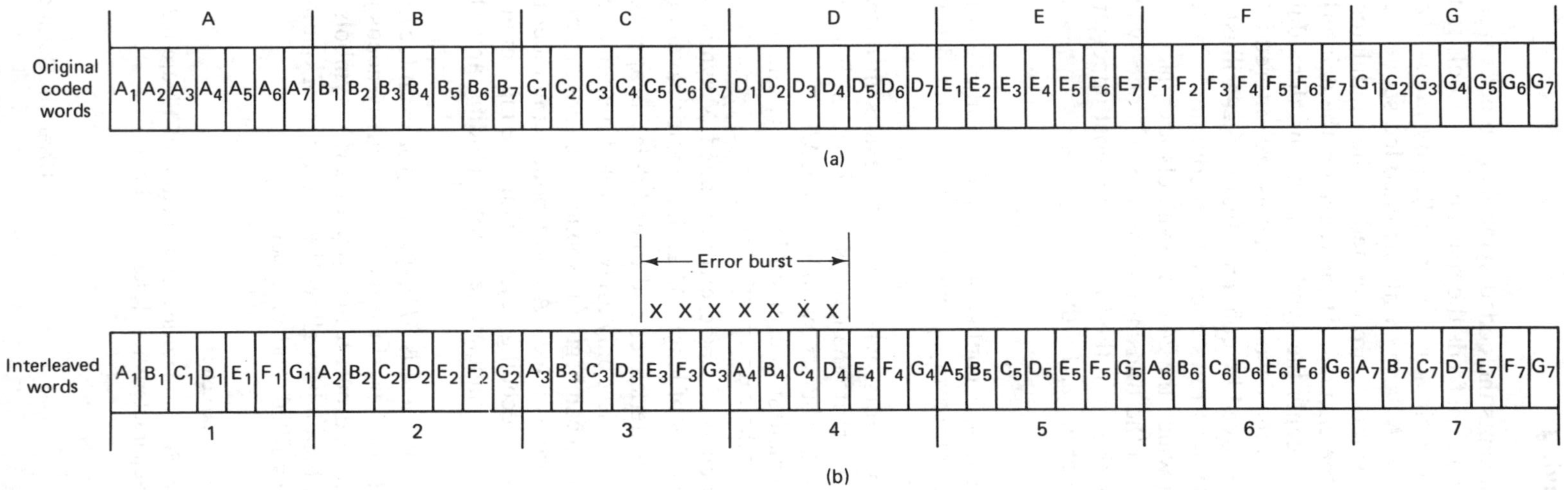

Figure 6.23 Interleaving example. (a) Original uninterleaved codewords, each comprised of seven code symbols. (b) Interleaved code symbols.

6.6.1 Block Interleaving

A block interleaver accepts the coded symbols in blocks from the encoder, permutes the symbols, and then feeds the rearranged symbols to the modulator. The usual permutation of the block is accomplished by *filling the columns* of an *M*-row-by *N*-column ($M \times N$) array with the encoded sequence. After the array is completely filled, the symbols are then fed to the modulator *one row at a time* and transmitted over the channel. At the receiver, the deinterleaver performs the inverse operation; it accepts the symbols from the demodulator, deinterleaves them, and feeds them to the decoder. Symbols are entered into the deinterleaver array by rows, and removed by columns. Figure 6.24a illustrates an example of an interleaver with $M = 4$ rows and $N = 6$ columns. The entries in the array illustrate the order in which the 24 code symbols are placed into the interleaver. The output sequence to the transmitter consists of code symbols removed from the array by rows, as shown in the figure. The most important characteristics of such a block interleaver are as follows:

1. Any burst of less than N contiguous channel symbol errors results in isolated errors at the deinterleaver output that are separated from each other by at least M symbols.

2. Any bN burst of errors, where $b > 1$, results in output bursts from the deinterleaver of no more than $\lceil b \rceil$ symbol errors. Each output burst is separated from the other bursts by no less than $M - \lfloor b \rfloor$ symbols. The notation $\lceil x \rceil$ means the smallest integer no less than x, and $\lfloor x \rfloor$ means the largest integer no greater than x.

3. A periodic sequence of single errors spaced N symbols apart results in a single burst of errors of length M at the deinterleaver output.

4. The interleaver/deinterleaver end-to-end delay is approximately $2MN$ symbol times. To be precise, only $M(N - 1) + 1$ memory cells need to be filled before transmission can begin (as soon as the first symbol of the last column of the $M \times N$ array is filled). A corresponding number needs to be filled at the receiver before decoding begins. Thus the minimum end-to-end delay is $(2MN - 2M + 2)$ symbol times, not including any channel propagation delay.

5. The memory requirement is MN symbols for each location (interleaver and deinterleaver). However, since the $M \times N$ array needs to be (mostly) filled before it can be read out, a memory of $2MN$ symbols is generally implemented at each location to allow the emptying of one $M \times N$ array while the other is being filled, and vice versa.

Example 6.3   Interleaver Characteristics

Using the $M = 4$, $N = 6$ interleaver structure of Figure 6.24a, verify each of the block interleaver characteristics described above.

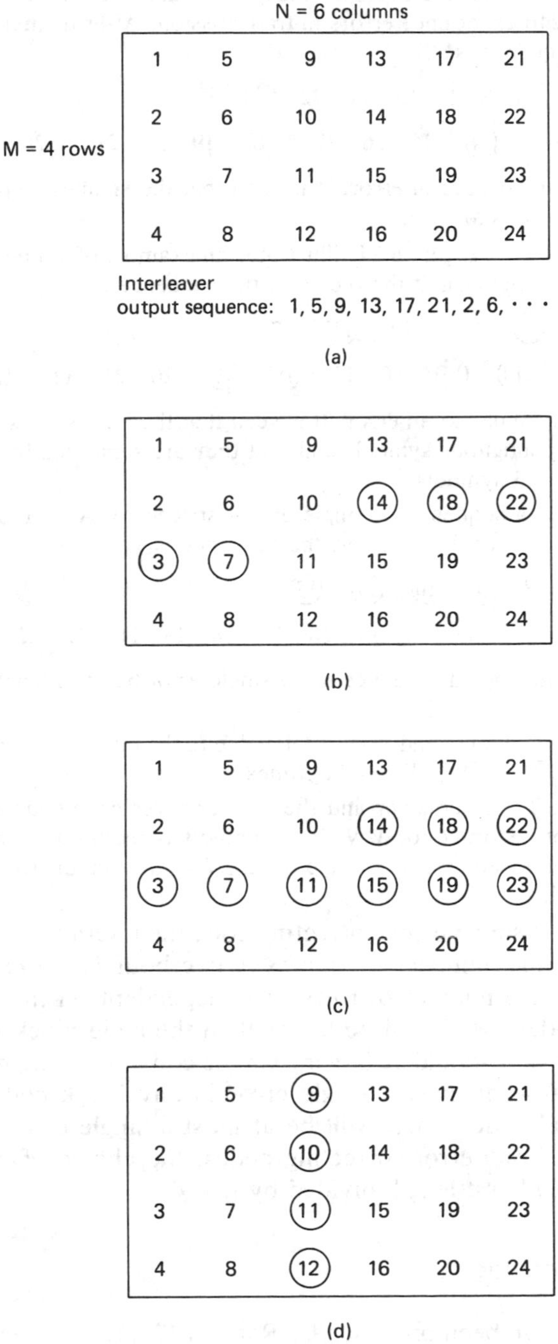

N = 6 columns

M = 4 rows

1	5	9	13	17	21
2	6	10	14	18	22
3	7	11	15	19	23
4	8	12	16	20	24

Interleaver
output sequence: 1, 5, 9, 13, 17, 21, 2, 6, · · ·

(a)

(b)

(c)

(d)

Figure 6.24 Block interleaver example. (a) $M \times N$ block interleaver. (b) Five-symbol error burst. (c) Nine-symbol error burst. (d) Periodic single-error sequence spaced $N = 6$ symbols apart.

Solution

1. Let there be a noise burst of five symbol times, such that the symbols shown encircled in Figure 6.24b experience errors in transmission. After deinterleaving at the receiver, the sequence is

1　2　③　4　5　6　⑦　8　9　10　11　12

13　⑭　15　16　17　⑱　19　20　21　㉒　23　24

where the encircled symbols are in error. It is seen that the smallest separation between symbols in error is $M = 4$.

2. Let $b = 1.5$ so that $bN = 9$. Figure 6.24c illustrates an example of a nine-symbol error burst. After deinterleaving at the receiver, the sequence is

1　2　③　4　5　6　⑦　8　9　10　⑪　12

13　⑭　⑮　16　17　⑱　⑲　20　21　㉒　㉓　24

Again, the encircled symbols are in error. It is seen that the bursts consist of no more than $\lceil 1.5 \rceil = 2$ contiguous symbols and that they are separated by at least $M - \lfloor 1.5 \rfloor = 4 - 1 = 3$ symbols.

3. Figure 6.24d illustrates a sequence of single errors spaced by $N = 6$ symbols apart. After deinterleaving at the receiver, the sequence is

1　2　3　4　5　6　7　8　⑨　⑩　⑪　⑫

13　14　15　16　17　18　19　20　21　22　23　24

It is seen that the deinterleaved sequence has a single error burst of length $M = 4$ symbols.

4. End-to-end delay: The minimum end-to-end delay due to the interleaver and deinterleaver is $(2MN - 2M + 2) = 42$ symbol times.

5. Memory requirement: The interleaver and the deinterleaver arrays are each of size $M \times N$. Therefore, storage for $MN = 24$ symbols is required at each end of the channel. As mentioned earlier, storage for $2MN = 48$ symbols would generally be implemented.

Typically, for use with a single-error-correcting code the interleaver parameters are selected such that the number of columns N overbounds the *expected burst length*. The choice of the number of rows M is dependent on the coding scheme used. For block codes, M should be larger than the code block length, while for convolutional codes, M should be larger than the constraint length. Thus a burst of length N can cause at most a single error in any block codeword; similarly, with convolutional codes, there will be at most a single error in any decoding constraint length. For t-error-correcting codes, the choice of N need only overbound the expected burst length divided by t.

6.6.2 Convolutional Interleaving

Convolutional interleavers have been proposed by Ramsey [29] and Forney [30]. The structure proposed by Forney appears in Figure 6.25. The code symbols are

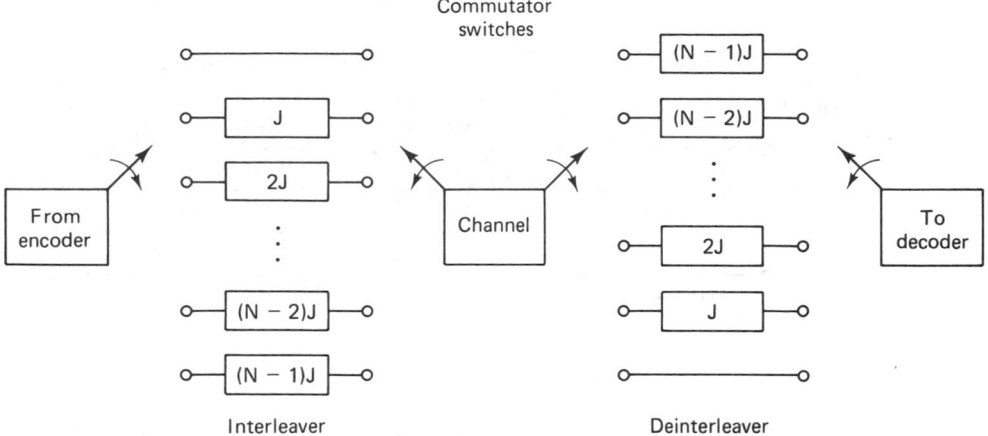

Figure 6.25 Shift register implementation of a convolutional interleaver/deinterleaver.

sequentially shifted into the bank of N registers; each successive register provides J symbols more storage than did the preceding one. The zeroth register provides no storage (the symbol is transmitted immediately). With each new code symbol the commutator switches to a new register, and the new code symbol is shifted in while the oldest code symbol in that register is shifted out to the modulator/transmitter. After the $(N - 1)$th register, the commutator returns to the zeroth register and starts again. The deinterleaver performs the inverse operation, and the input and output commutators for both interleaving and deinterleaving must be synchronized.

Figure 6.26 illustrates an example of a simple convolutional four-register ($J = 1$) interleaver being loaded by a sequence of code symbols. The synchronized deinterleaver is shown simultaneously feeding the deinterleaved symbols to the decoder. Figure 6.26a shows symbols 1 to 4 being loaded; the $\times$s represent unknown states. Figure 6.26b shows the first four symbols shifted within the registers and the entry of symbols 5 to 8 to the interleaver input. Figure 6.6c shows symbols 9 to 12 entering the interleaver. The deinterleaver is now filled with message symbols, but nothing useful is being fed to the decoder yet. Finally, Figure 6.6d shows symbols 13 to 16 entering the interleaver, and at the output of the deinterleaver, symbols 1 to 4 are being passed to the decoder. The process continues in this way until the entire codeword sequence, in its original preinterleaved form, is presented to the decoder.

The performance of a convolutional interleaver is very similar to that of a block interleaver. The important advantage of convolutional over block interleaving is that with convolutional interleaving the end-to-end delay is $M(N - 1)$ symbols, where $M = NJ$, and the memory required is $M(N - 1)/2$ at both ends of the channel. Therefore, there is a reduction of one-half in delay and memory over the block interleaving requirements [16].

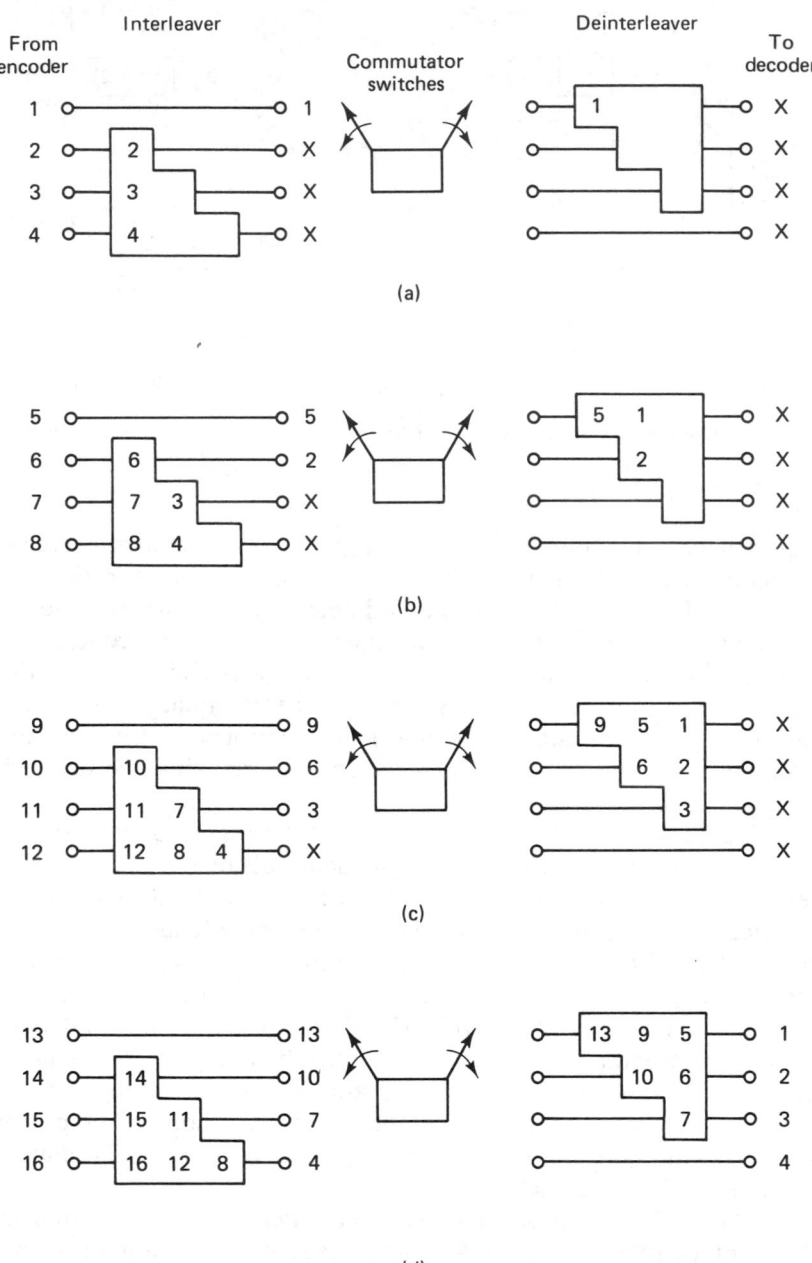

Figure 6.26 Convolutional interleaver/deinterleaver example.

6.6.3 Concatenated Codes

A concatenated code is one that uses two levels of coding, an inner code and an outer code, to achieve the desired error performance. Figure 6.27 illustrates the order of encoding and decoding. The inner code, the one that interfaces with the modulator/demodulator and channel, is usually configured to correct most of the channel errors. The outer code, usually a higher-rate (lower-redundancy) code, then reduces the probability of error to the specified level. The primary reason for using a concatenated code is to achieve a low error rate with an overall implementation complexity which is less than that which would be required by a single coding operation. In Figure 6.27 an interleaver is shown between the two coding steps. This is usually required to spread any error bursts that may appear at the output of the inner coding operation.

One of the most popular concatenated coding systems uses a Viterbi-decoded convolutional inner code and a Reed–Solomon (R–S) outer code, with interleaving between the two coding steps [23]. Operation of such systems with E_b/N_0 in the range 2.0 to 2.5 dB to achieve $P_B = 10^{-5}$ (only about 4 dB away from the Shannon limit) is now feasible with practical hardware [16]. In this system, the demodulator outputs soft quantized code symbols to the inner convolutional decoder, which in turn outputs hard quantized code symbols with bursty errors to the R–S decoder. (In a Viterbi-decoded system, the output errors tend to occur in bursts.) The outer R–S code is formed from m-bit segments of the binary data stream (see Section 5.7.4). The performance of such a (nonbinary) R–S code depends only on the number of *symbol errors* in the block. The code is undisturbed by burst errors within an m-bit symbol. That is, for a given symbol error, the R–S code performance is the same whether the symbol error is due to one bit being in error or m bits being in error. However, the concatenated system performance is severely degraded by correlated errors among successive symbols. Hence the interleaving between codes needs to take place at the symbol level (not at the bit level). In the next section we consider a popular consumer application of such symbol interleaving in a concatenated system.

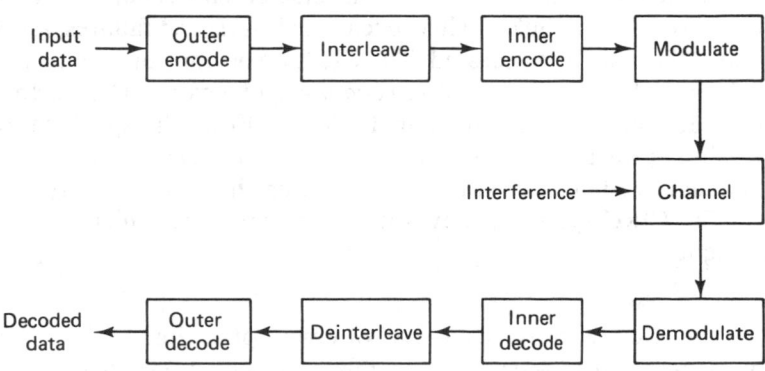

Figure 6.27 Block diagram of a concatenated coding system.

6.7 CODING AND INTERLEAVING APPLIED TO THE COMPACT DISC DIGITAL AUDIO SYSTEM

In 1979, Philips Corp. of the Netherlands and Sony Corp. of Japan defined a standard for the digital storage and reproduction of audio signals, known as the *compact disc (CD) digital audio system*. This CD system has become the world standard for achieving fidelity of sound reproduction that far surpasses any other available technique. A plastic disc 120 mm in diameter is used to store the digitized audio waveform. The waveform is sampled at 44.1 kilosamples/s to provide a recorded bandwidth of 20 kHz; each audio sample is uniformly quantized to one of 2^{16} levels (16 bits/sample), resulting in a dynamic range of 96 dB and a total harmonic distortion of 0.005%. A single disc (playing time approximately 70 minutes) stores about 10^{10} bits in the form of minute *pits* that are optically scanned by a laser.

There are several sources of channel errors: (1) small unwanted particles or air bubbles in the plastic material or pit inaccuracies arising in manufacturing, and (2) fingerprints or scratches during handling. It is difficult to predict how, on the average, a CD will get damaged; but in the absence of an accurate channel model, it is safe to assume that the channel mainly has a *burstlike* error behavior, since a scratch or fingerprint will cause *several* consecutive data samples to be in error. An important aspect of the system design contributing to the high-fidelity performance is a concatenated control scheme called the *cross-interleave Reed–Solomon* code (CIRC). The data are rearranged in time so that digits stemming from contiguous samples of the waveform are *spread out in time*. In this way, error bursts are made to appear as single random events (see the earlier sections on interleaving). The digital information is protected by adding parity bytes derived in two Reed–Solomon (R–S) encoders (see Section 5.7.4). Error control applied to the compact disc depends mostly on R–S coding and multiple layers of interleaving. Material on the CD is treated in this chapter rather than in Chapter 5 with R–S coding because it follows naturally after the subject of interleaving and concatenated codes in the previous sections.

In digital audio applications, an undetected decoding error is very serious since it results in clicks, while occasional *detected* failures are not so serious because they can be concealed. The CIRC error-control scheme in the CD system involves both *correction* and *concealment* of errors. The performance specifications for the CIRC are given in Table 6.5. From the specifications in the table it would appear that the CD can endure much damage (e.g., 8-mm holes punched in the disc) without any noticeable effect on the sound quality.

The CIRC system achieves its error control by a hierarchy of the following techniques:

1. The decoder provides a level of error correction.
2. If the error correction capability is exceeded, the decoder provides a level of erasure correction (see Section 5.5.5).
3. If the erasure correction capability is exceeded, the decoder attempts to

TABLE 6.5 Specifications for the CD Cross-Interleave Reed–Solomon Code

Maximum correctable burst length	$\simeq$ 4000 bits (2.5-mm track length on the disc)
Maximum interpolatable burst length	$\simeq$ 12,000 bits (8 mm)
Sample interpolation rate	One sample every 10 hours at $P_B = 10^{-4}$
	1000 samples/min at $P_B = 10^{-3}$
Undetected error samples (clicks)	Less than one every 750 hours at $P_B = 10^{-3}$
	Negligible at $P_B \leq 10^{-4}$
New discs are characterized by	$P_B \simeq 10^{-4}$

conceal unreliable data samples by *interpolating* between reliable neighboring samples.

4. If the interpolation capability is exceeded, the decoder blanks out or *mutes* the system for the duration of the unreliable samples.

6.7.1 CIRC Encoding

Figure 6.28 illustrates the basic CIRC encoder block diagram (within the CD recording equipment) and the decoder block diagram (within the CD player equipment). Encoding consists of the encoding and interleaving steps designated as: Δ interleave, C_2 encode, D* interleave, C_1 encode, and D interleave. The decoder steps, consisting of deinterleaving and decoding, are performed in the *reverse* order of the encoding steps and are designated as: D deinterleave, C_1 decode, D* deinterleave, C_2 decode, and Δ deinterleave.

Figure 6.29 illustrates the basic system frame time, comprised of six sampling periods, each made up of a stereo sample pair (16-bit left sample and 16-bit right sample). The bits are organized into symbols or bytes of 8 bits each. Therefore, each sample pair contains 4 bytes, and the uncoded frame contains $k = 24$ bytes. Figure 6.29a–e summarizes the *five encoding steps* that characterize the CIRC system. The function of each of these steps will best be understood when we consider the decoding operation.

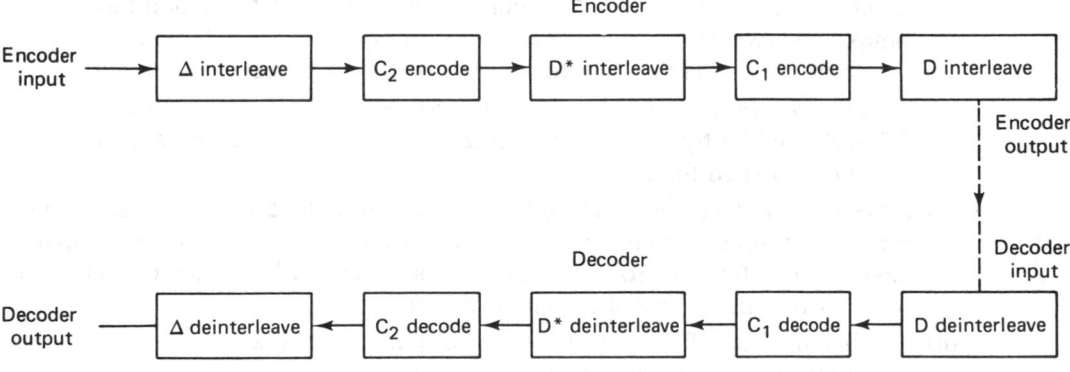

Figure 6.28 CIRC encoder and decoder.

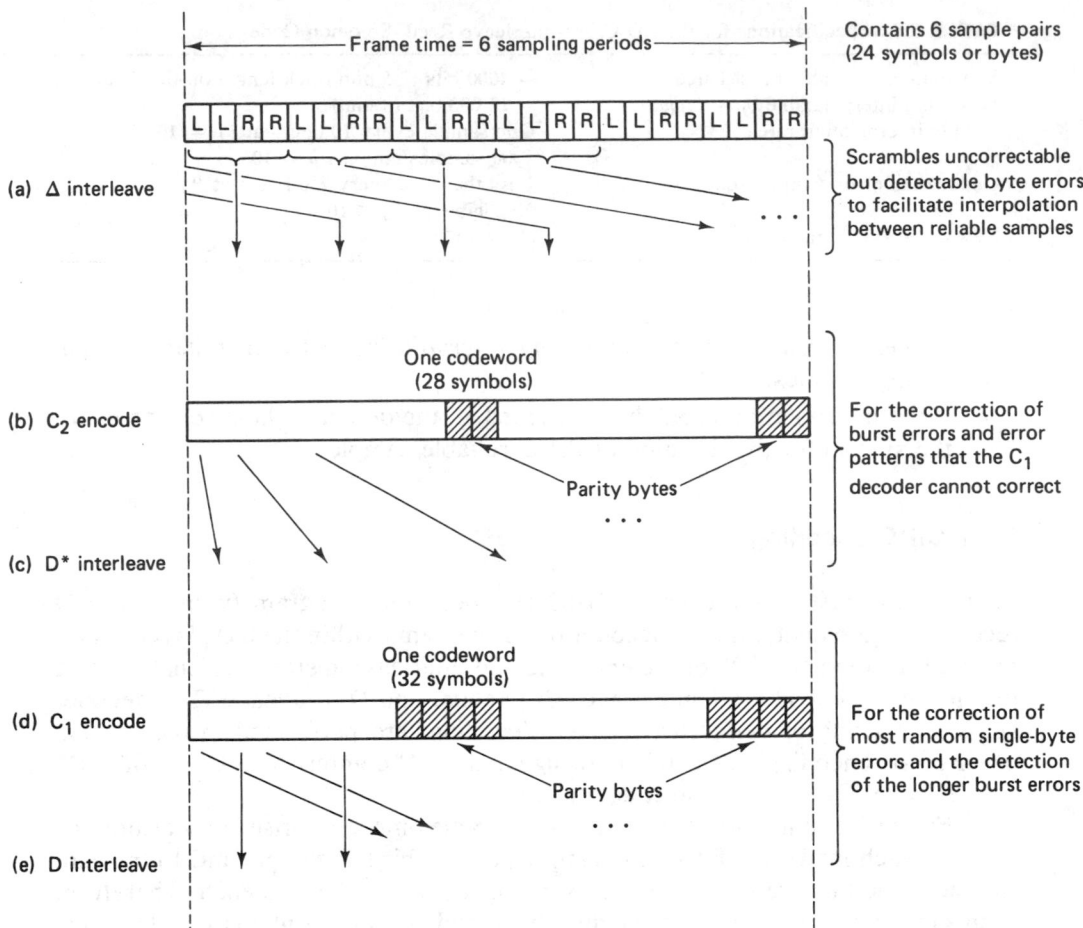

Figure 6.29 Compact disc encoder. (a) Δ interleave. (b) C_2 encode. (c) D^* interleave. (d) C_1 encode. (e) D interleave.

(a) Δ *interleave.* Even-numbered samples are separated from odd-numbered samples by two frame times in order to scramble uncorrectable but detectable byte errors. This facilitates the interpolation process.

(b) C_2 *encode.* Four Reed–Solomon (R–S) parity bytes are added to the Δ-interleaved 24-byte frame, resulting in a total of $n = 28$ bytes. This (28, 24) code is called the *inner code.*

(c) D^* *interleave.* Here each byte is delayed a different length, thereby spreading errors over several codewords. C_2 encoding together with D^* interleaving have the function of providing for the correction of burst errors and error patterns that the C_1 decoder cannot correct.

(d) C_1 *encode.* Four R–S parity bytes are added to the $k = 28$ bytes of the D^*-interleaved frame, resulting in a total of $n = 32$ bytes. This (32, 28) code is called the *outer code.*

(e) *D interleave*. The purpose is to *cross-interleave* the *even bytes* of a frame with the *odd bytes* of the next frame. By this procedure, two consecutive bytes on the disc will always end up in two different codewords. Upon decoding, this interleaving, together with the C_1 decoding, results in the correction of most random single errors and the detection of longer burst errors.

6.7.1.1 Shortening the R–S Code

In Section 5.7.4 an (n, k) R–S code is expressed in terms of $n = 2^m - 1$ total symbols and $k = 2^m - 1 - 2t$ data symbols, where m is the number of bits per symbol and t is the error-correcting capability of the code in symbols. For the CD system, where a symbol is made up of 8 bits, a 2-symbol error-correcting code can be configured as a (255, 251) code. However, the CD system uses a considerably shorter block length. Any block code (in systematic form) can be shortened without affecting the number of errors that can be corrected within a block length. In terms of the (255, 251) R–S code, imagine that 227 of the 251 data symbols are a set of all-zero symbols (which are not actually transmitted and hence are not subject to any errors). Then the code is really a (28, 24) code with the same 2-symbol error-correcting capability. This is what is done in the C_2 encoder of the CD system.

We can think of the 28 total symbols out of the C_2 encoder as the data symbols into the C_1 encoder. Again, we can configure a shortened 2-symbol error-correcting (255, 251) code by throwing away 223 data symbols—the result being a (32, 28) code.

6.7.2 CIRC Decoding

The inner and outer R–S codes with (n, k) values (32, 28) and (28, 24) each use four parity bytes. The code rate of the CIRC is $(k_1/n_1)(k_2/n_2) = 24/32 = 3/4$. From Equation (5.78) the minimum distance of the C_1 and C_2 R–S codes is $d_{min} = n - k + 1 = 5$. From Equations (5.79) and (5.50),

$$t \le \frac{d_{min} - 1}{2} \qquad (6.24)$$

$$\rho \le d_{min} - 1 \qquad (6.25)$$

where t is the error-correcting capability and ρ is the erasure-correcting capability, it is seen that the C_1 or C_2 decoder can correct a maximum of 2 symbol errors or 4 symbol erasures per codeword. Or, as described by Equation (5.51), it is possible to correct any pattern of α errors and γ erasures simultaneously provided that

$$d_{min} \ge 2\alpha + \gamma + 1 \qquad (6.26)$$

There is a trade-off between error correction and erasure correction; the larger the error correcting capability used, the smaller will be the erasure correcting capability.

The benefits of CIRC are best seen at the *decoder*, where the processing steps, shown in Figure 6.30, are in the reverse order of the encoder steps. The decoder steps are as follows:

1. *D deinterleave*. This function is performed by the alternating delay lines marked D. The 32 bytes ($B_{i1}, \ldots, B_{i32}$) of an encoded frame are applied in parallel to the 32 inputs of the D deinterleaver. Each delay is equal to the duration of 1 byte, so that the information of the *even bytes* of a frame is cross-deinterleaved with that of the *odd bytes* of the next frame.

2. C_1 *decode*. The D deinterleaver and the C_1 decoder are designed to correct a single byte error in the block of 32 bytes and to detect larger burst errors. If multiple errors occur, the C_1 decoder passes them on unchanged, attaching to all 28 remaining bytes an erasure flag, sent via the dashed lines (the four parity bytes used in the C_1 decoder are no longer retained).

3. *D* deinterleave*. Due to the different lengths of the deinterleaving delay lines $D^*(1, \ldots, 27)$, errors that occur in one word at the output of the C_1 decoder are *spread over a number of words* at the input of the C_2 decoder. This results in reducing the number of errors per input word of the C_2 decoder, enabling the C_2 decoder to correct these errors.

4. C_2 *decode*. The C_2 decoder is intended for the correction of burst errors that the C_1 decoder could not correct. If the C_2 decoder cannot correct these errors, the 24-byte codeword is passed on unchanged to the Δ deinterleaver

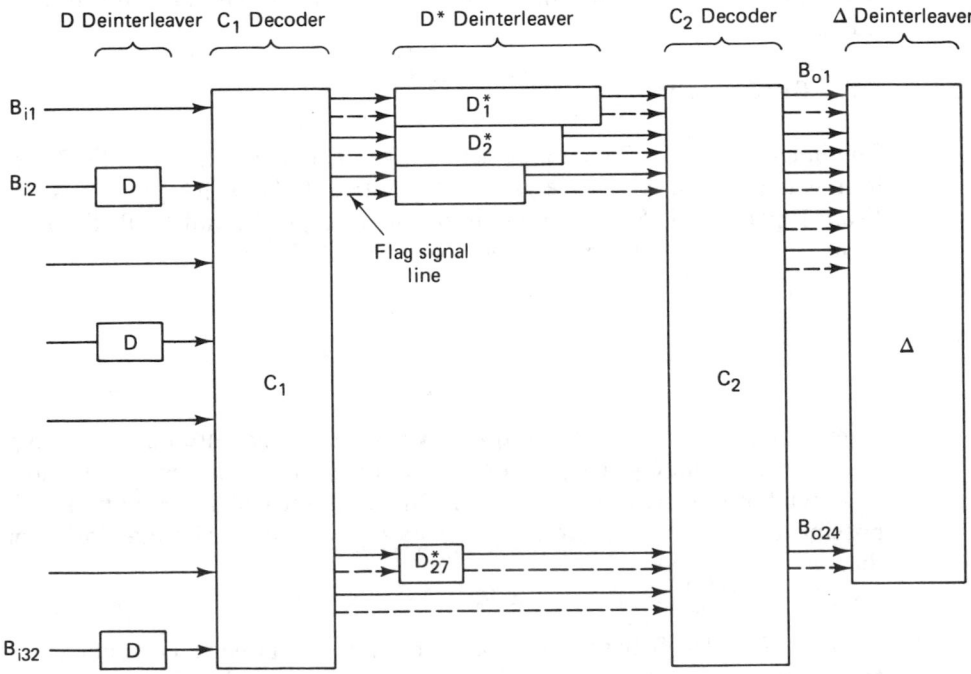

Figure 6.30 Compact disc decoder.

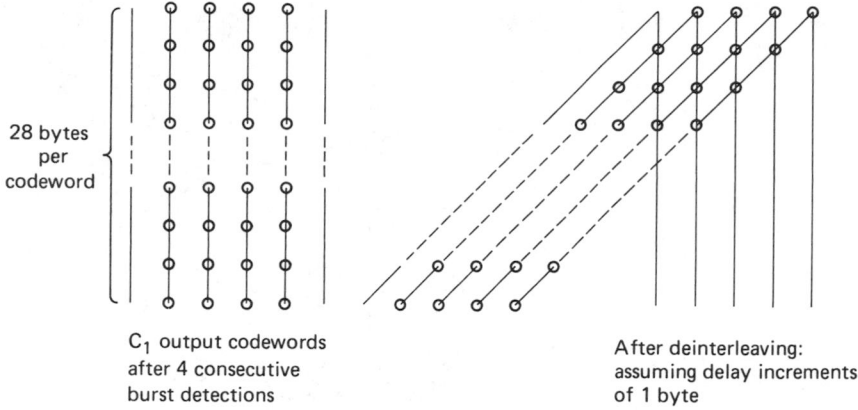

28 bytes per codeword

C_1 output codewords
after 4 consecutive
burst detections

After deinterleaving:
assuming delay increments
of 1 byte

Figure 6.31 Example of 4-byte erasure capability. (Rightmost event is at the earliest time.)

and the associated positions are given an *erasure flag* via the dashed output lines, $B_{o1}, \ldots, B_{o24}$.

5. Δ *deinterleave*. The final operation deinterleaves uncorrectable but detected byte errors in such a way that *interpolation* can be used between reliable neighboring samples.

Figure 6.31 highlights the decoder steps 2, 3, and 4. At the output of the C_1 decoder is seen a sequence of four 28-byte codewords that have exceeded the 1 byte per codeword error correction design. Therefore, each of the symbols in these codewords is tagged with an erasure flag (shown with circles). The D* deinterleaver provides a staggered delay for each byte of a codeword, so that the bytes of a given codeword arrive in different codewords at the input to the C_2 decoder. If we assume that the delay increments of the D* deinterleaver in Figure 6.31 are 1 byte, it would be possible to correct error bursts of as many as four consecutive C_1 codewords (since the C_2 decoder is capable of four erasure corrections per codeword). In the actual CD system, the delay increments are 4 bytes; therefore, the maximum burst error correction capability consists of 16 consecutive uncorrectable C_1 words.

6.7.3 Interpolation and Muting

Samples that cannot be corrected by the C_2 decoder could cause audible disturbances. The function of the *interpolation* process is to insert new samples, estimated from reliable neighbors, in place of the unreliable ones. If an entire C_2 word is detected as unreliable, this would make it impossible to apply interpolation without additional interleaving, since both even- and odd-numbered samples are unreliable. This can happen if the C_1 decoder fails to detect an error but the C_2 decoder detects it. It is the purpose of Δ deinterleaving (over a span of two frame

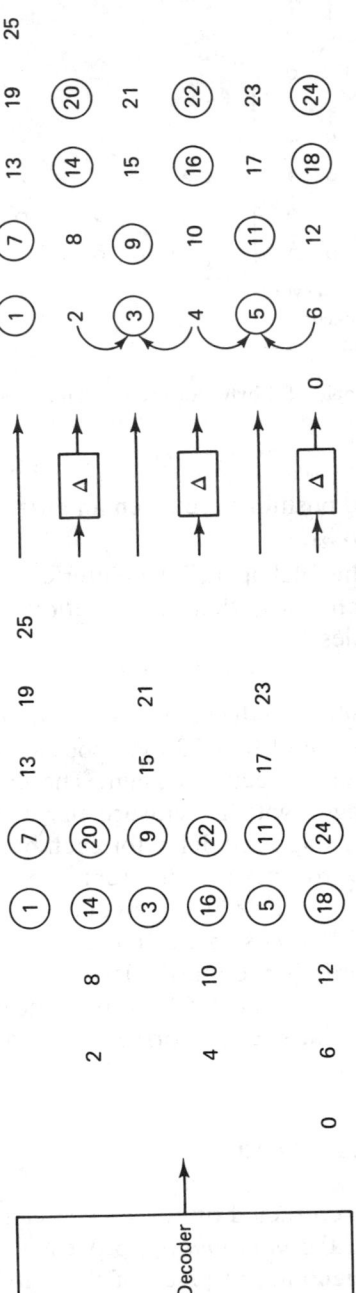

Figure 6.32 Effect of interleaving. (Rightmost event is at the earliest time.)

times) to obtain a pattern where even-numbered samples can be interpolated from reliable odd-numbered samples, or vice versa.

Two successive unreliable words consisting of 12 sample pairs are shown in Figure 6.32. A sample pair consists of a sample (2 bytes) from the right audio channel and a sample from the left audio channel. The numbers indicate the ordering of the sets of samples. An encircled sample set denotes an *erasure flag*. After Δ deinterleaving, the unreliable samples shown in the figure are estimated by a first-order linear interpolation between neighboring samples that stem from a different location on the disc.

In CD players, another level of error control is provided in case a burst length of 48 frames is exceeded and 2 or more consecutive unreliable samples result. In this case the system is *muted* (audio is softly blanked out), which is not discernible to the human ear if the muting time does not exceed a few milliseconds. For a more detailed treatment of the CIRC coding scheme in the CD system, see References [31–34].

6.8 CONCLUSION

In the last decade, coding emphasis has been in the area of convolutional codes since in almost every application, convolutional codes outperform block codes for the same implementation complexity of the encoder–decoder. For satellite communication channels, forward error correction techniques can easily reduce the required SNR for a specified error performance by 5 to 6 dB. This coding gain can translate directly into an equivalent reduction in required satellite effective radiated power (EIRP), with consequently reduced satellite weight and cost.

In this chapter we have outlined the essential structural difference between block codes and convolutional codes—the fact that rate $1/n$ convolutional codes have a memory of the prior $K - 1$ bits, where K is the encoder constraint length. With such memory, the encoding of each input data bit not only depends on the value of that bit but on the values of the $K - 1$ input bits that precede it. We presented the decoding problem in the context of the maximum likelihood algorithm, examining all the candidate codeword sequences which could possibly be created by the encoder, and selecting the one that appears statistically most likely; the decision is based on a distance metric for the received code symbols. The error performance analysis of convolutional codes is more complicated than the simple binomial expansion describing the error performance of many block codes. We laid out the concept of free distance, and we presented the relationship between free distance and error performance in terms of bounds. We also described the basic idea behind sequential decoding and feedback decoding and showed some comparative performance curves and tables for various coding schemes.

Finally, we described a technique, interleaving, that allows the popular block and convolutional coding schemes to be used over channels that exhibit bursty noise or periodic fading, without suffering degradation. We used the CD digital audio system as an example of how interleaving plays an important role in ameliorating the effects of burst noise.

Appendix E consists of a FORTRAN program called VITALG for the convolutional encoding and Viterbi decoding of messages. The messages can be in the form of binary sequences or ASCII characters. The user has a choice of code rate, constraint length, connection vectors, and path memory length. The program can be used to insert errors into a bit stream after it has been encoded. From the program output, the user sees the error correcting that results from the use of hard-decision Viterbi decoding of his chosen message. It should prove interesting to use the VITALG program for verifying the performance of the optimum Odenwalder codes shown in Table 6.4.

REFERENCES

1. Gallager, R. G., *Information Theory and Reliable Communication,* John Wiley & Sons, Inc., New York, 1968.

2. Fano, R. M., "A Heuristic Discussion of Probabilistic Decoding," IRE *Trans. Inf. Theory,* vol. IT9, no. 2, 1963, pp. 64–74.

3. Odenwalder, J. P., *Optimal Decoding of Convolutional Codes,* Ph.D. dissertation, University of California, Los Angeles, 1970.

4. Curry, S. J., *Selection of Convolutional Codes Having Large Free Distance,* Ph.D. dissertation, University of California, Los Angeles, 1971.

5. Larsen, K. J., "Short Convolutional Codes with Maximal Free Distance for Rates $\frac{1}{2}$, $\frac{1}{3}$, and $\frac{1}{4}$," *IEEE Trans. Inf. Theory,* vol. IT19, no. 3, 1973, pp. 371–372.

6. Lin, S., and Costello, D. J., Jr., *Error Control Coding: Fundamentals and Applications,* Prentice-Hall, Inc., Englewood Cliffs, N.J., 1983.

7. Forney, G. D., Jr., "Convolutional Codes: I. Algebraic Structure," *IEEE Trans. Inf. Theory,* vol. IT16, no. 6, Nov. 1970, pp. 720–738.

8. Viterbi, A., "Convolutional Codes and Their Performance in Communication Systems," *IEEE Trans. Commun. Technol.,* vol. COM19, no. 5, Oct. 1971, pp. 751–772.

9. Forney, G. D., Jr., and Bower, E. K., "A High Speed Sequential Decoder: Prototype Design and Test," *IEEE Trans. Commun. Technol.,* vol. COM19, no. 5, Oct. 1971, pp. 821–835.

10. Jelinek, F., "Fast Sequential Decoding Algorithm Using a Stack," *IBM J. Res. Dev.,* vol. 13, Nov. 1969, pp. 675–685.

11. Massey, J. L., *Threshold Decoding,* The MIT Press, Cambridge, Mass., 1963.

12. Heller, J. A., and Jacobs, I. W., "Viterbi Decoding for Satellite and Space Communication," *IEEE Trans. Commun. Technol.,* vol. COM19, no. 5, October 1971, pp. 835–848.

13. Viterbi, A. J., "Error Bounds for Convolutional Codes and an Asymptotically Optimum Decoding Algorithm," *IEEE Trans. Inf. Theory,* vol. IT13, April 1967, pp. 260–269.

14. Omura, J. K., "On the Viterbi Decoding Algorithm" (correspondence), *IEEE Trans. Inf. Theory,* vol. IT15, Jan. 1969, pp. 177–179.

15. Mason, S. J., and Zimmerman, H. J., *Electronic Circuits, Signals, and Systems,* John Wiley & Sons, Inc., New York, 1960.

16. Clark, G. C., Jr., and Cain, J. B., *Error-Correction Coding for Digital Communications,* Plenum Press, New York, 1981.

17. Viterbi, A. J., and Omura, J. K., *Principles of Digital Communication and Coding,* McGraw-Hill Book Company, New York, 1979.

18. Massey, J. L., and Sain, M. K., "Inverse of Linear Sequential Circuits," *IEEE Trans. Comput.,* vol. C17, Apr. 1968, pp. 330–337.

19. Rosenberg, W. J., *Structural Properties of Convolutional Codes,* Ph.D. dissertation, University of California, Los Angeles, 1971.

20. Bhargava, V. K., Haccoun, D., Matyas, R., and Nuspl, P., *Digital Communications by Satellite,* John Wiley & Sons, Inc., New York, 1981.

21. Jacobs, I. M., "Practical Applications of Coding," *IEEE Trans. Inf. Theory,* vol. IT20, May 1974, pp. 305–310.

22. Linkabit Corporation, "Coding Systems Study for High Data Rate Telemetry Links," *NASA Ames Res. Center, Final Rep. CR-114278,* Contract NAS-2-6-24, Moffett Field, Calif., 1970.

23. Odenwalder, J. P., *Error Control Coding Handbook,* Linkabit Corporation, San Diego, Calif., July 15, 1976.

24. Wozencraft, J. M., "Sequential Decoding for Reliable Communication," *IRE Natl. Conv. Rec.,* vol. 5, pt. 2, 1957, pp. 11–25.

25. Wozencraft, J. M., and Reiffen, B., *Sequential Decoding,* The MIT Press, Cambridge, Mass., 1961.

26. Shannon, C. E., "A Mathematical Theory of Communication," *Bell Syst. Tech. J.,* vol. 27, 1948, pp. 379–423, 623–656.

27. Brayer, K., "Error Correcting Code Performance on HF, Troposcatter, and Satellite Channels," *IEEE Trans. Commun. Technol.,* vol. COM19, 1971, pp. 835–848.

28. Kohlenberg, A., and Forney, G. D., "Convolutional Coding for Channels with Memory," *IEEE Trans. Inf. Theory,* vol. IT2, 1968, pp. 618–626.

29. Ramsey, J. L., "Realization of Optimum Interleavers, *IEEE Trans. Inf. Theory,* vol. IT16, no. 3, May 1970, pp. 338–345.

30. Forney, G. D., "Burst-Correcting Codes for the Classic Bursty Channel," *IEEE Trans. Commun. Technol.,* vol. COM19, Oct. 1971, pp. 772–781.

31. Peek, J. B. H., "Communications Aspects of the Compact Disc Digital Audio System," *IEEE Commun. Mag.,* vol. 23, no. 2, Feb. 1985, pp. 7–20.

32. Berkhout, P. J., and Eggermont, L. D. J., "Digital Audio Systems," *IEEE ASSP Mag.,* Oct. 1985, pp. 45–67.

33. Driessen, L. M. H. E., and Vries, L. B., "Performance Calculations of the Compact Disc Error Correcting Code on a Memoryless Channel," *Fourth Int. Conf. Video and Data Record.,* Southampton, England, Apr. 20–23, 1982, *IERE Conf. Proc.,* vol. 54, pp. 385–395.

34. Hoeve, H., Timmermans, J., and Vries, L. B., "Error Correction in the Compact Disc System," *Philips Tech. Rev.,* vol. 40, no. 6, 1982, pp. 166–172.

PROBLEMS

6.1. Draw the state diagram, tree diagram, and trellis diagram for the $K = 3$, rate $\frac{1}{3}$ code generated by

$$g_1(X) = X + X^2$$

$$g_2(X) = 1 + X$$

$$g_3(X) = 1 + X + X^2$$

6.2. Given a $K = 3$, rate $\frac{1}{2}$, binary convolutional code with the partially completed state diagram shown in Figure P6.1, find the complete state diagram and sketch a diagram for the encoder.

6.3. Draw the state diagram, tree diagram, and trellis diagram for the convolutional encoder characterized by the block diagram in Figure P6.2.

6.4. Suppose that you were trying to find the quickest way to get from London to Vienna by boat or train. The diagram in Figure P6.3 was constructed from various schedules. The labels on each path are travel times. Using the Viterbi algorithm, find the fastest route from London to Vienna. In a general sense, explain how the algorithm works, what calculations must be made, and what information must be retained in the memory used by the algorithm.

6.5. Consider the convolutional encoder shown in Figure P6.4.
   (a) Write the connection vectors and polynomials for this encoder.
   (b) Draw the state diagram, tree diagram, and trellis diagram.

6.6. What is the impulse response of the encoder of Problem 6.5? Using the impulse response, determine the output sequence when the input is 1 0 1. Verify by using the generator polynomials.

6.7. Does the encoder of Problem 6.5 allow catastrophic error propagation? Justify your answer with an example.

6.8. Find the free distance of the encoder of Problem 6.3 by the transfer function method.

6.9. Let the codewords of a coding scheme be

$$a = 0\ 0\ 0\ 0\ 0\ 0$$

$$b = 1\ 0\ 1\ 0\ 1\ 0$$

$$c = 0\ 1\ 0\ 1\ 0\ 1$$

$$d = 1\ 1\ 1\ 1\ 1\ 1$$

If the received sequence over a binary symmetric channel is 1 1 1 0 1 0 and a maximum likelihood decoder is used, what will be the decoded symbol?

6.10. Consider that the $K = 3$, rate $\frac{1}{2}$ encoder of Figure 6.3 is used over a binary symmetric channel (BSC). Assume that the initial encoder state is the 00 state. At the output of the BSC, the sequence $\mathbf{Z} = (1\ 1\ 0\ 0\ 0\ 0\ 1\ 0\ 1\ 1$ rest all "0") is received.
   (a) Find the maximum likelihood path through the trellis diagram, and determine the first 5 decoded information bits. If a tie occurs between any two merged paths, choose the upper branch entering the particular state.
   (b) Identify any channel bits in $\mathbf{Z}$ that were inverted by the channel during transmission.

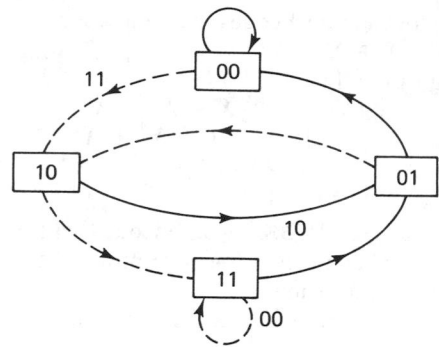

Figure P6.1

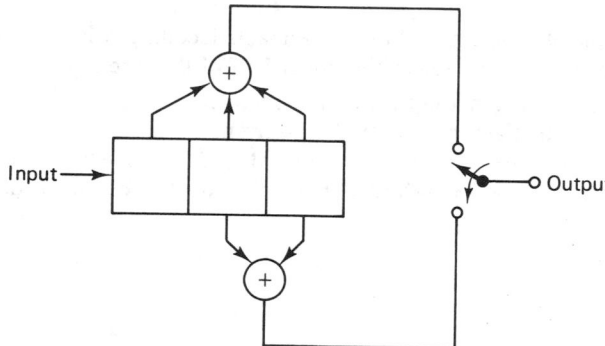

Figure P6.2

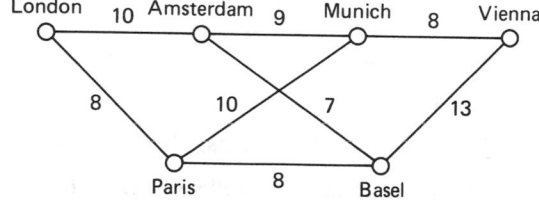

Figure P6.3

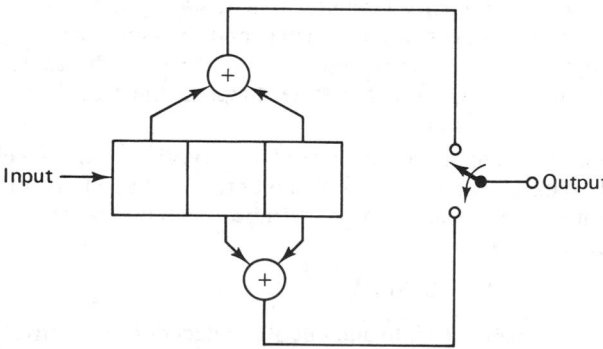

Figure P6.4

6.11. Determine which of the following rate $\frac{1}{2}$ codes are catastrophic.
  (a) $g_1(X) = X^2,\qquad g_2(X) = 1 + X + X^3$
  (b) $g_1(X) = 1 + X^2,\qquad g_2(X) = 1 + X^3$
  (c) $g_1(X) = 1 + X + X^2,\qquad g_2(X) = 1 + X + X^3 + X^4$
  (d) $g_1(X) = 1 + X + X^3 + X^4,\qquad g_2(X) = 1 + X^2 + X^4$
  (e) $g_1(X) = 1 + X^4 + X^6 + X^7,\qquad g_2(X) = 1 + X^3 + X^4$
  (f) $g_1(X) = 1 + X^3 + X^4,\qquad g_2(X) = 1 + X + X^2 + X^4$

6.12. (a) Consider a coherently detected BPSK signal encoded with the encoder shown in Figure 6.3. Find an upper bound on the bit error probability, P_B, if the available E_b/N_0 is 6 dB. Assume hard decision decoding.
  (b) Compare P_B with the uncoded case and calculate the improvement factor.

6.13. Using sequential decoding, illustrate the path along the tree diagram shown in Figure 6.20 when the received sequence is 0 1 1 1 0 0 0 1 1 1. The backup criterion is three disagreements.

6.14. Repeat the decoding example of Problem 6.13 using feedback decoding, with a look-ahead length of 3. In the event of a tie, select the upper half of the tree.

6.15. Figure P6.5 depicts a constraint length 2 convolutional encoder.
  (a) Draw the state diagram, tree diagram, and trellis diagram.
  (b) Assume that a received message from this encoder is 1 1 0 0 1 0. Use a feedback decoding algorithm with a look-ahead length of 2 to decode the coded message sequence.

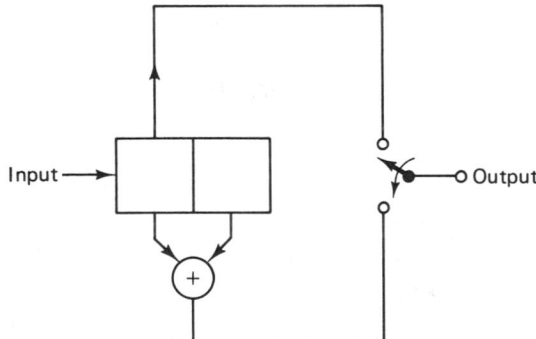

Figure P6.5

6.16. Using the branch word information on the encoder trellis of Figure 6.7, decode the sequence $\mathbf{Z} = (01\ 11\ 00\ 01\ 11$ rest all "0"), using hard-decision Viterbi decoding.

6.17. Consider the rate $\frac{2}{3}$ convolutional encoder shown in Figure P6.6. In this encoder, $k = 2$ bits at a time are shifted into the encoder and $n = 3$ bits are generated at the encoder output. There are $kK = 4$ stages in the register, and the constraint length is $K = 2$ in units of 2-bit bytes. The state of the encoder is defined as the contents of the rightmost $K - 1$ k-tuple stages. Draw the state diagram, the tree diagram, and the trellis diagram.

6.18. Find the ratio of the predetection signal-to-noise spectral density, P_r/N_0, in decibels, required to yield a decoded data rate of 1 Mbit/s with a bit error probability of 10^{-5}. Assume binary noncoherent FSK modulation. Also, assume convolutional encoding with the following decoder relationship:

$$P_b = 2000\,P_B^4$$

where P_B and P_b are bit error probabilities into and out of the decoder, respectively.

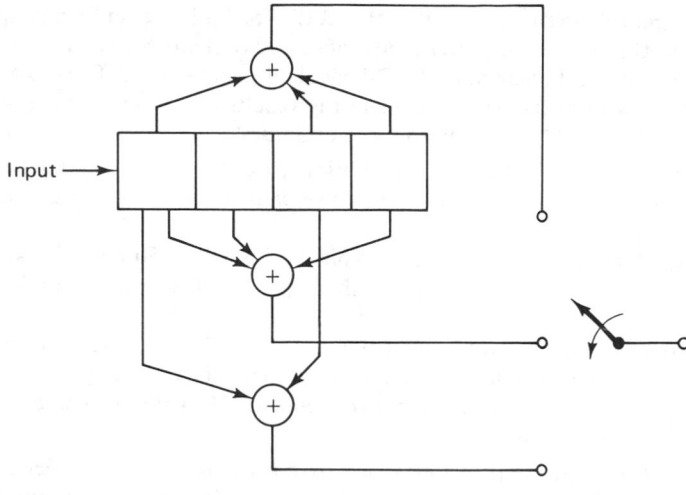

Figure P6.6

6.19. The sequence

$$1\ 0\ 1\ 1\ 0\ 1\ 1\ 0\ 0\ 0\ 1\ 0\ 1\ 1\ 0\ 0$$

is the input to a 4 × 4 block interleaver. What is the output sequence? The same input sequence is applied to the convolutional interleaver of Figure 6.26. What is the output sequence?

6.20. Using the computer program VITALG listed in Appendix E, perform the following calculations. Let the uncoded message consist of a sequence of binary zeros, where the number of zeros is 10 times the constraint length of the code being used. Convolutionally encode the message and emulate a memoryless AWGN channel by inserting random transmission errors into the coded sequence (space the errors at approximately a uniform distance from one another). Decode the corrupted coded message using the Viterbi algorithm with the path memory chosen to be five times the constraint length. Use the code generators described as optimum by Odenwalder in Table 6.4. Record the errors corrected by the code, and tabulate the maximum number of errors that can be corrected using each of the following codes:
(a) Rate $\frac{1}{2}$, constraint length 3
(b) Rate $\frac{1}{3}$, constraint length 3
(c) Rate $\frac{1}{2}$, constraint length 5
(d) Rate $\frac{1}{3}$, constraint length 5
(e) Rate $\frac{1}{2}$, constraint length 7
(f) Rate $\frac{1}{3}$, constraint length 7
Explain the results.

6.21. Repeat Problem 6.20. However, instead of a memoryless AWGN channel, emulate a channel that has memory by inserting error bursts into the coded message. Let a burst consist of an uninterrupted sequence of errors placed approximately in the middle of the message. Tabulate the maximum number of errors that can be corrected using each of the code types (a) through (f) listed in Problem 6.20, and compare the error-correcting capabilities of the codes with these two different channel environments. Explain the results.

6.22. Repeat Problem 6.20, parts (e) and (f), for both the uniformly spaced error pattern and the burst error pattern (described in Problems 6.20 and 6.21). In each case compare the performance of the Odenwalder generators in Table 6.4 to other generators of your own choosing. Tabulate the results. Do your findings support the premise that the Table 6.4 generators are optimum?

6.23. For each of the following conditions, design an interleaver for a communication system operating over a bursty noise channel at a transmission rate of 19,200 coded symbols/s.

   (a) A contiguous noise burst typically lasts for 250 ms. The system code consists of a (127, 36) BCH code with $d_{min} = 31$. The end-to-end delay is not to exceed 5 s.

   (b) A contiguous noise burst typically lasts for 20 ms. The system code consists of a rate $\frac{1}{2}$ convolutional code with a feedback decoding algorithm that corrects an average of 3 symbols in a sequence of 21 symbols. The end-to-end delay is not to exceed 160 ms.

6.24. **(a)** Calculate the probability of a byte (symbol) error after decoding the data stored on a compact disc (CD) as described in Section 6.7. Assume that the probability of a channel symbol error for the disc is 10^{-3}. Also assume that the inner and outer R–S decoders are each configured to correct all 2-symbol errors, and that the interleaving process results in channel symbol errors being uncorrelated from one another.

   (b) Repeat part (a) for a disc that has a probability of channel symbol error equal to 10^{-2}.

CHAPTER 7

Modulation and Coding Trade-Offs

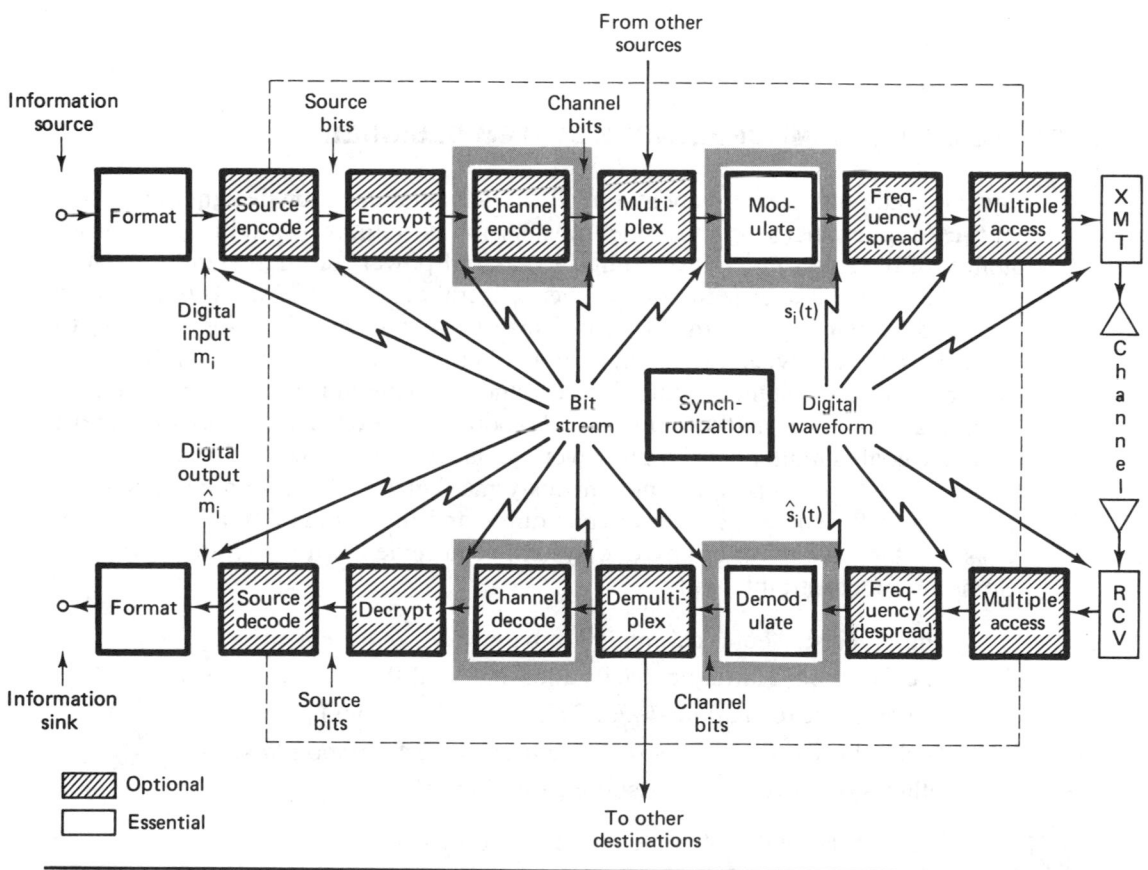

7.1 GOALS OF THE COMMUNICATIONS SYSTEM DESIGNER

System trade-offs are fundamental to all digital communication designs. The goals of the designer are (1) to maximize transmission bit rate, R; (2) to minimize probability of bit error, P_B; (3) to minimize required power, or equivalently, to minimize required bit energy to noise power spectral density, E_b/N_0; (4) to minimize required system bandwidth, W; (5) to maximize system utilization, that is, to provide reliable service for a maximum number of users with minimum delay and with maximum resistance to interference; and (6) to minimize system complexity, computational load, and system cost. A good system designer seeks to achieve all these goals simultaneously. However, goals 1 and 2 are clearly in conflict with goals 3 and 4; they call for simultaneously maximizing R, while minimizing P_B, E_b/N_0, and W. There are several constraints and theoretical limitations that necessitate the trading off of any one system requirement with each of the others. Some of the constraints are:

The Nyquist theoretical minimum bandwidth requirement
The Shannon–Hartley capacity theorem (and the Shannon limit)
Government regulations (e.g., frequency allocations)
Technological limitations (e.g., state-of-the art components)
Other system requirements (e.g., satellite orbits)

Some of the realizable modulation and coding trade-offs can best be viewed as a change in operating point on one of two performance planes. These planes will be referred to as the error probability plane and the bandwidth efficiency plane; they are described in the following sections.

7.2 ERROR PROBABILITY PLANE

Figure 7.1 illustrates the family of P_B versus E_b/N_0 curves for the coherent detection of orthogonal signaling (Figure 7.1a) and multiple phase signaling (Figure 7.1b). For signaling schemes that process k bits at a time, the signaling is called M-ary (see Section 3.8). The modulator uses one of its $M = 2^k$ waveforms to represent each k-bit sequence, where M is the size of the symbol set. Figure 7.1a illustrates the potential bit error improvement with orthogonal signaling as k (or M) is increased. For orthogonal signal sets, such as frequency shift keying (FSK) modulation, increasing the size of the symbol set can provide an improvement in P_B, or a reduction in the E_b/N_0 required, at the cost of increased bandwidth. Figure 7.1b illustrates potential bit error degradation with nonorthogonal signaling as k (or M) increases. For nonorthogonal signal sets, such as multiple phase shift keying (MPSK) modulation, increasing the size of the symbol set can reduce the bandwidth requirement, but at the cost of a degraded P_B, or an increased E_b/N_0 requirement. We shall refer to these families of curves (Figure 7.1a or b) as *error probability performance curves,* and to the plane on which they are plotted as an *error probability plane.* Such a plane describes the locus of operating points available for a particular type of modulation and coding. For a given system information rate, each curve in the plane can be associated with a different fixed minimum required bandwidth; therefore, the set of curves can be termed *equibandwidth curves.* As the curves move in the direction of the ordinate, the required transmission bandwidth increases; as the curves move in the opposite direction, the required bandwidth decreases. Once a modulation and coding scheme and an available E_b/N_0 are determined, system operation is characterized by a particular point in the error probability plane. Possible trade-offs can be viewed as changes in the operating point on one of the curves or as changes in the operating point from one curve to another curve of the family. These trade-offs are seen in Figure 7.1a and b as changes in the system operating point in the direction shown by the arrows. Movement of the operating point along line 1, between points a and b, can be viewed as trading off P_B for E_b/N_0 performance (with W fixed). Similarly, movement along line 2, between points c and d, is seen as trading P_B for W performance (with E_b/N_0 fixed). Finally, movement along line 3, between points e and f, illustrates trading W for E_b/N_0 performance (with P_B fixed). Movement along line 1 is effected by increasing or decreasing the available E_b/N_0. This can be achieved, for example, by increasing transmitter power, which means that the trade-off might be accomplished simply by "turning a knob," even after the system is configured. However, the other trade-offs (movement along line 2 or line 3) involve some change in the system modulation or coding scheme, and therefore need to be accomplished during the system design phase.

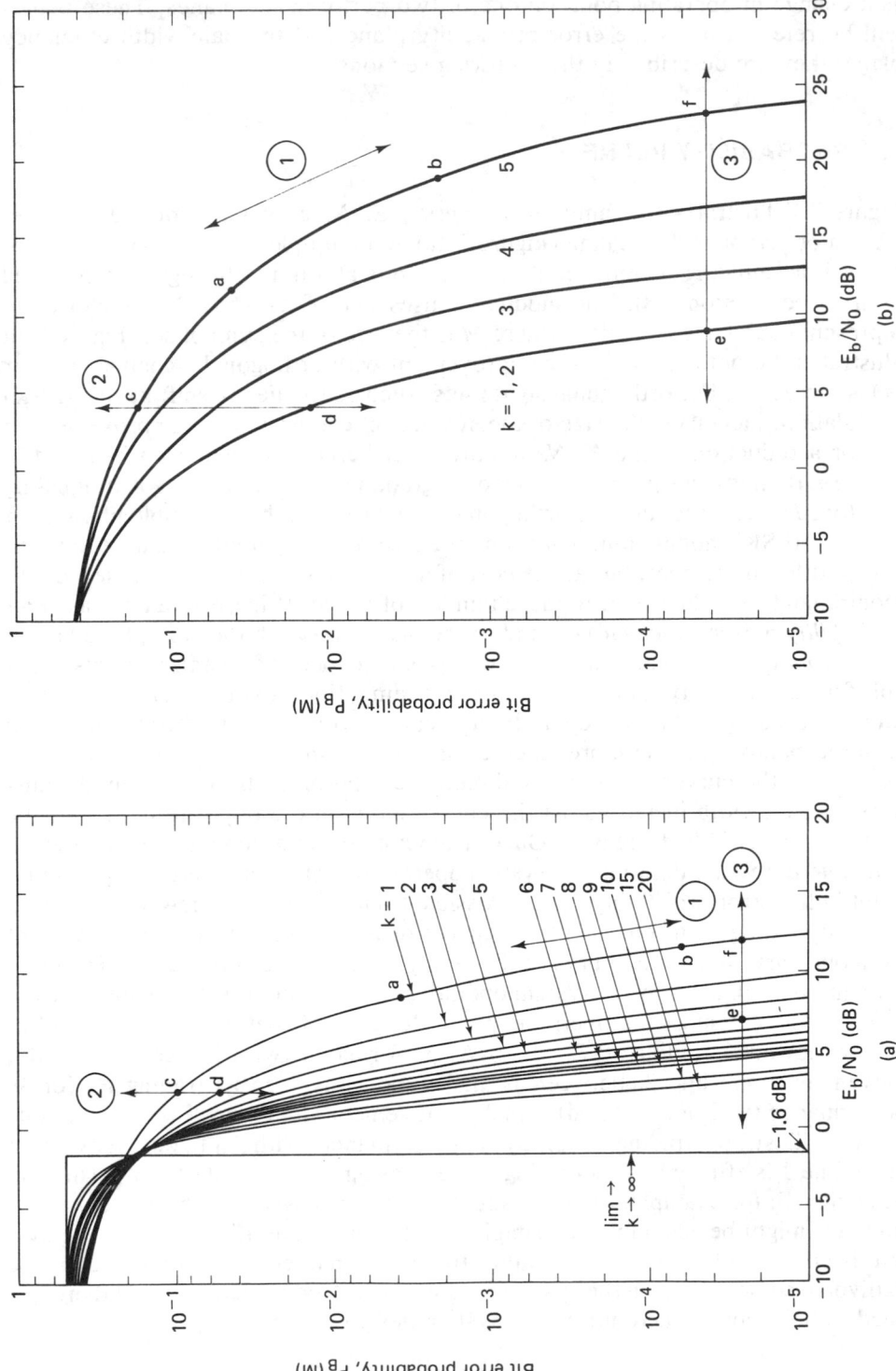

Figure 7.1 Bit error probability versus E_b/N_0 for coherently detected M-ary signaling. (a) Orthogonal signaling. (b) Multiple phase signaling.

7.3 NYQUIST MINIMUM BANDWIDTH

Every realizable system having some nonideal filtering will suffer from intersymbol interference (ISI)—the tail of one pulse spilling over into adjacent symbol intervals so as to interfere with correct detection. Nyquist [1] showed that, in theory, R_s symbols per second could be detected without ISI in an $R_s/2$ hertz minimum bandwidth (Nyquist bandwidth); this is a basic theoretical constraint, limiting the designer's goal to expend as little bandwidth as possible (see Section 2.11). In practice, R_s hertz is typically required for the transmission of R_s symbols per second. In other words, *typical* digital communication throughput, without ISI, is limited to 1 symbol/s per hertz. The modulation or coding system assigns to each symbol, of its set of M symbols, a k-bit meaning, where $M = 2^k$. For a signaling scheme with a fixed bandwidth, such as MPSK, as k increases, the allowable data rate, R, increases, and hence the bandwidth efficiency, R/W, measured in bits per second per hertz, also increases. For example, movement along line 3, from point e to point f in Figure 7.1b represents trading E_b/N_0 for a reduced bandwidth requirement. In other words, with the same system bandwidth one can transmit at an increased data rate, hence at an increased R/W.

7.4 SHANNON–HARTLEY CAPACITY THEOREM

Shannon [2] showed that the system capacity, C, of a channel perturbed by additive white Gaussian noise (AWGN) is a function of the average received signal power, S, the average noise power, N, and the bandwidth, W. The capacity relationship (Shannon–Hartley theorem) can be stated as

$$C = W \log_2 \left(1 + \frac{S}{N} \right)$$ (7.1)

When W is in hertz and the logarithm is taken to the base 2, as shown, the capacity is given in bits/s. It is theoretically possible to transmit information over such a channel at any rate, R, where $R \leq C$, with an *arbitrarily small* error probability by using a sufficiently complicated coding scheme. For an information rate $R > C$, it is not possible to find a code that can achieve an arbitrarily small error probability. Shannon's work showed that the values of S, N, and W *set a limit on transmission rate, not on error probability*. Shannon [3] used Equation (7.1) to graphically exhibit a bound for the achievable performance of practical systems. This plot, shown in Figure 7.2, gives the normalized channel capacity C/W in bits/s/Hz as a function of the channel signal-to-noise ratio (SNR). A related plot, shown in Figure 7.3, indicates the normalized channel bandwidth W/C in Hz/bits/s as a function of SNR in the channel. Figure 7.3 is sometimes used to illustrate the power–bandwidth trade-off inherent in the ideal channel. However, it is not a pure trade-off [4] because the detected noise power is proportional to bandwidth.

$$N = N_0 W$$ (7.2)

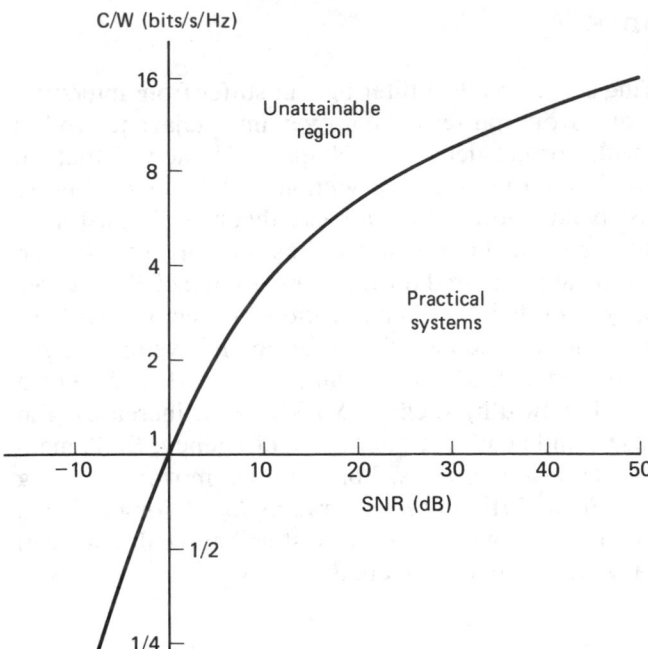

C/W (bits/s/Hz)

Unattainable region

Practical systems

SNR (dB)

Figure 7.2 Normalized channel capacity versus channel SNR.

Substituting Equation (7.2) into Equation (7.1) and rearranging terms yields

$$\frac{C}{W} = \log_2\left(1 + \frac{S}{N_0 W}\right) \tag{7.3}$$

For the case where transmission bit rate is equal to channel capacity, $R = C$, we can use the identity presented in Equation (3.94) to write

$$\frac{S}{N_0 C} = \frac{E_b}{N_0} \tag{7.4}$$

Hence we can modify Equation (7.3) as follows:

$$\frac{C}{W} = \log_2\left[1 + \frac{E_b}{N_0}\left(\frac{C}{W}\right)\right] \tag{7.5}$$

$$2^{C/W} = 1 + \frac{E_b}{N_0}\left(\frac{C}{W}\right)$$

$$\frac{E_b}{N_0} = \frac{W}{C}(2^{C/W} - 1) \tag{7.6}$$

Figure 7.4 is a plot of W/C versus E_b/N_0 in accordance with Equation (7.6).

Modulation and Coding Trade-Offs   Chap. 7

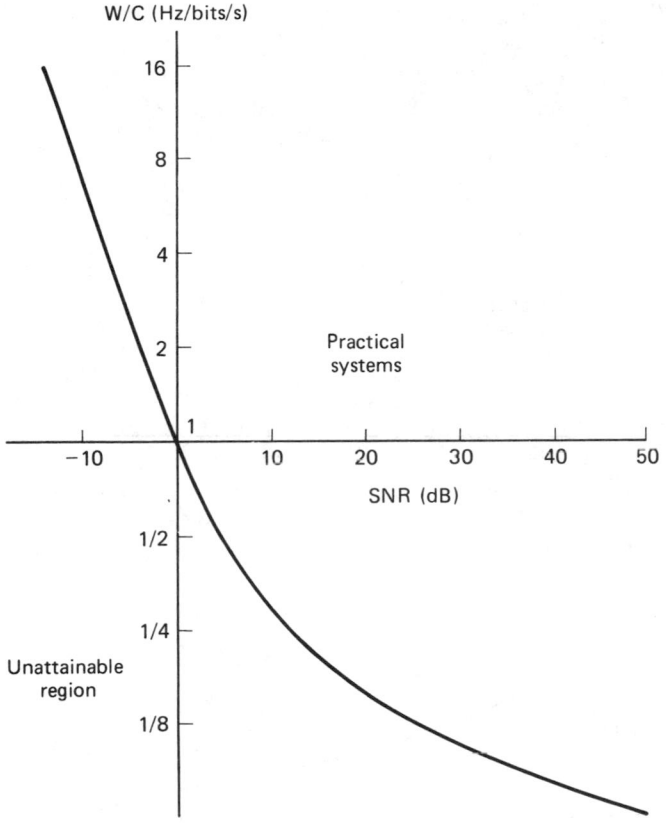

Figure 7.3 Normalized channel bandwidth versus channel SNR.

The asymptotic behavior of this curve as $C/W \to 0$ (or $W/C \to \infty$) is discussed in the next section.

7.4.1 Shannon Limit

There exists a limiting value of E_b/N_0 below which there can be no error-free communication at any information rate. Using the identity

$$\lim_{x \to 0} (1 + x)^{1/x} = e$$

we can calculate the limiting value of E_b/N_0 as follows. Let

$$x = \frac{E_b}{N_0} \left(\frac{C}{W} \right)$$

Then from Equation (7.5),

$$\frac{C}{W} = x \log_2 (1 + x)^{1/x}$$

$$1 = \frac{E_b}{N_0} \log_2 (1 + x)^{1/x}$$

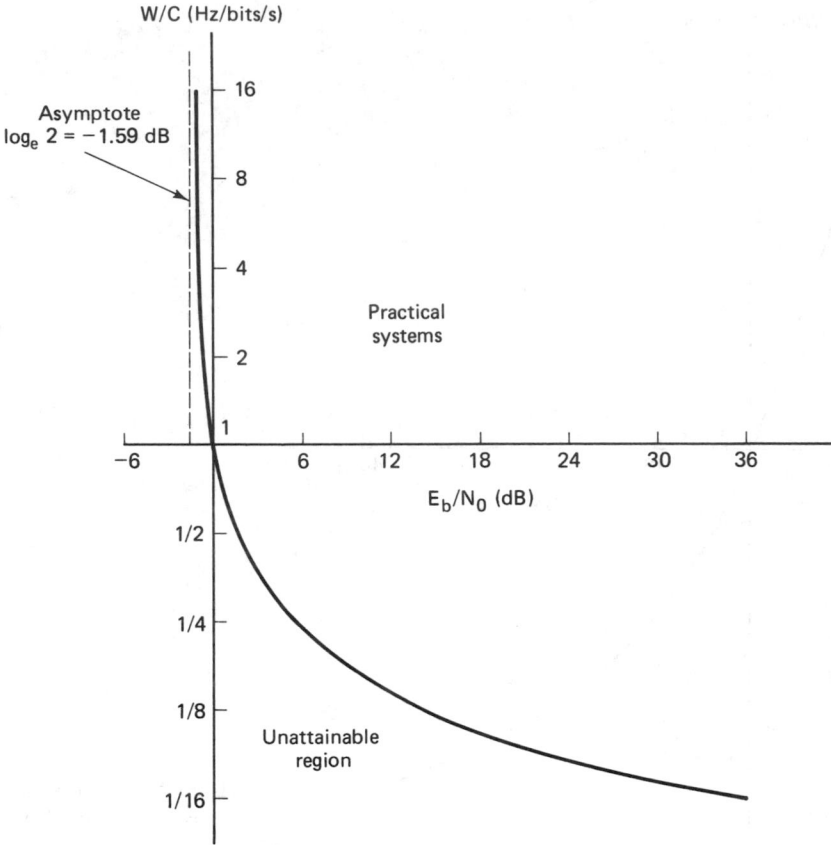

Figure 7.4 Normalized channel bandwidth versus channel E_b/N_0.

In the limit, as $C/W \to 0$, we get

$$\frac{E_b}{N_0} = \frac{1}{\log_2 e} = 0.693$$

or, in decibels $= -1.59$ dB

(7.7)

This value of E_b/N_0 is called the *Shannon limit*. On Figure 7.1a the Shannon limit is the P_B versus E_b/N_0 curve corresponding to $k \to \infty$. The curve is discontinuous, going from a value of $P_B = \frac{1}{2}$ to $P_B = 0$ at $E_b/N_0 = -1.59$ dB. It is not possible in practice to reach the Shannon limit, because as k increases without bound, the bandwidth requirement and the implementation complexity increase without bound. Shannon's work provided a theoretical proof for the existence of codes that could improve the P_B performance, or reduce the E_b/N_0 required, from the levels of the uncoded binary modulation schemes to levels approaching the limiting curve. For a bit error probability of 10^{-5}, binary phase shift keying (BPSK) modulation requires an E_b/N_0 of 9.6 dB (the optimum uncoded binary modulation). Therefore, Shannon's work promised the existence of a theoretical performance

Modulation and Coding Trade-Offs   Chap. 7

improvement of 11.2 dB over the performance of optimum uncoded binary modulation, through the use of coding techniques. Today, most of that promised improvement (approximately 7 dB) is realizable [5]. Optimum system design can best be described as a search for rational compromises or trade-offs among the various constraints and conflicting goals. The modulation and coding trade-off, that is, the selection of modulation and coding techniques to make the best use of transmitter power and channel bandwidth, is important, since there are strong incentives to reduce the cost of generating power and to conserve the radio spectrum.

7.4.2 Entropy

To design a communications system with a specified message handling capability, we need a metric for measuring the information content to be transmitted. Shannon [2] developed such a metric, H, called the entropy of the message source (having n possible outputs). *Entropy* is defined as the average amount of information per source output and is expressed by

$$H = - \sum_{i=1}^{n} p_i \log_2 p_i \qquad \text{bits/source output} \qquad (7.8)$$

where p_i is the probability of the ith output and $\sum p_i = 1$. In the case of a binary message or a source having only two possible outputs, with probabilities p and $q = (1 - p)$, the entropy is written

$$H = -(p \log_2 p + q \log_2 q) \qquad (7.9)$$

and is plotted versus p in Figure 7.5.

The quantity H has a number of interesting properties, including the following:

1. When the logarithm in Equation (7.8) is taken to the base 2, as shown, the unit for H is average bits per event. The unit *bit*, here, is a measure of *information content* and is not to be confused with the term "bit," meaning "binary digit."
2. The term "entropy" has the same uncertainty connotation as it does in certain formulations of statistical mechanics. For the information source with two equally likely possibilities (e.g., the flipping of a fair coin), it can be seen from Figure 7.5 that the uncertainty in the event, and hence the average information content, is maximum. As the probabilities depart from the equally likely case, the average information content decreases. In the limit, when one of the probabilities goes to zero, H also goes to zero. We know the result before the event happens, so the result conveys no additional information.
3. To illustrate that information content is related to a priori probability (if the a priori message probability at the receiver is zero or one, we need not send the message), consider the following example: At the end of her nine-month

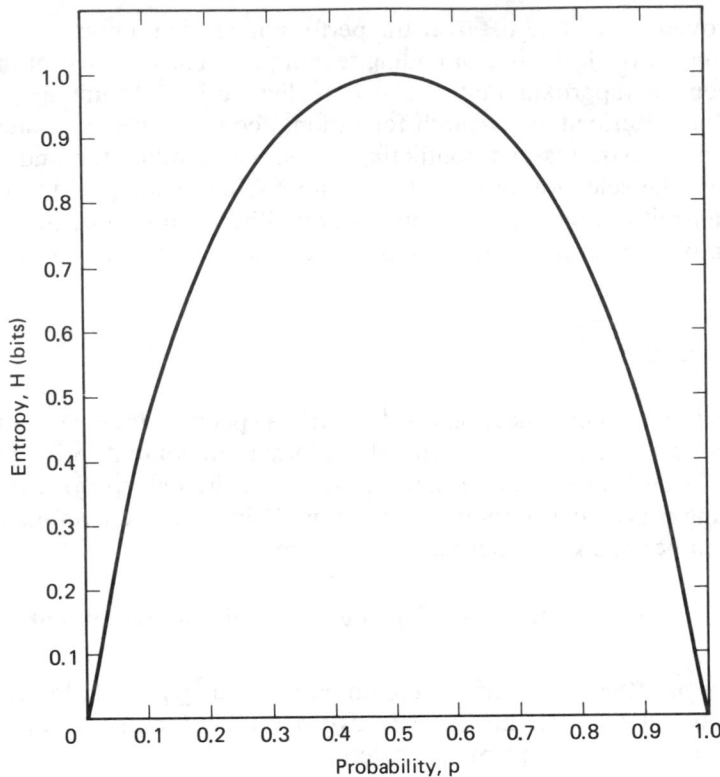

Figure 7.5 Entropy versus probability (two events).

pregnancy, a woman enters the delivery room of a local hospital to give birth. Her husband waits anxiously in the waiting room. After some time, a physician approaches the husband and says: "Congratulations, you are the father of a child." How much information has the physician given the father *beyond the medical outcome*? Almost none; the father has known with virtual certainty that a child was forthcoming. Had the physician said, "you are the father of a boy" or "you are the father of a girl," he would have transmitted 1 bit of information, since there was a 50% chance that the child could have been a boy or a girl.

Example 7.1 Average Information Content in the English Language

    (a) Calculate the average information in bits/character for the English language, assuming that each of the 26 characters in the alphabet occurs with equal likelihood. Neglect spaces and punctuation.

    (b) Since the alphabetic characters do not appear with equal frequency in the English language (or any other language), the answer to part (a) will represent an upper bound on average information content per character. Repeat part (a) under the assumption that the alphabetic characters occur with the following probabilities:

$$p = 0.10: \quad \text{for the letters a, e, o, t}$$

$$p = 0.07: \quad \text{for the letters h, i, n, r, s}$$

$$p = 0.02: \quad \text{for the letters c, d, f, l, m, p, u, y}$$

$$p = 0.01: \quad \text{for the letters b, g, j, k, q, v, w, x, z}$$

Solution

(a) $H = -\sum\limits_{i=1}^{26} \dfrac{1}{26} \log_2 \left(\dfrac{1}{26}\right)$

$\quad = 4.7 \text{ bits/character}$

(b) $H = -(4 \times 0.1 \log_2 0.1 + 5 \times 0.07 \log_2 0.07$

$\quad\quad\quad + 8 \times 0.02 \log_2 0.02 + 9 \times 0.01 \log_2 0.01)$

$\quad = 4.17 \text{ bits/character}$

If we want to express the 26 letters of the alphabet with some binary-digit coding scheme, we generally need five binary digits for each character. Example 7.1 demonstrates that there may be a way to encode the English language with a fewer number of binary digits per character, *on the average,* by exploiting the fact that the average amount of information contained within each character is less than 5 bits. The subject of source coding, which deals with this exploitation, is treated in Chapter 11.

7.4.3 Equivocation and Effective Transmission Rate

Suppose that we are transmitting information at a rate of 1000 binary symbols/s over a binary symmetric channel (defined in Section 5.3.1), and that the a priori probability of transmitting either a one or a zero is equally likely. Suppose also that the noise in the channel is so great that the probability of receiving a one is $\frac{1}{2}$, whatever was transmitted, and similarly for receiving a zero. In such a case, half the received symbols would be correct *due to chance alone,* and the system might appear to be providing 500 bits/s while actually no information is being received at all. Equally "good" reception could be obtained by dispensing with the channel entirely and "flipping a coin" within the receiver. The proper correction to apply to the amount of information transmitted is the amount of information that is lost in the channel. Shannon [2] uses a correction factor called *equivocation* to account for the uncertainty in the received signal. Equivocation is defined as the *conditional entropy* of the message X, given Y, as shown below:

$$H(X|Y) = -\sum_{X,Y} P(X, Y) \log_2 P(X|Y)$$

$$= -\sum_{Y} P(Y) \sum_{X} P(X|Y) \log_2 P(X|Y) \tag{7.10}$$

where X is the transmitted source message, Y is the received signal, $P(X, Y)$ is the joint probability of X and Y, and $P(X|Y)$ is the conditional probability of X given Y. Equivocation can be thought of as the uncertainty that message X was sent, having received Y. For an *error-free channel*, $H(X|Y) = 0$, because having received Y, there is complete certainty about the message X. However, for a channel with a nonzero probability of symbol error, $H(X|Y) > 0$, because the channel introduces uncertainty. Consider a binary sequence, X, where the a priori source probabilities are $P(X = 1) = P(X = 0) = \frac{1}{2}$, and where, on the average, the channel produces one error in a received sequence of 100 bits ($P_B = 0.01$). Using Equation (7.10), the equivocation $H(X|Y)$ is expressed as

$$H(X|Y) = -[(1 - P_B) \log_2 (1 - P_B) + P_B \log_2 P_B]$$

$$= -(0.99 \log_2 0.99 + 0.01 \log_2 0.01)$$

$$= 0.081 \text{ bit/received symbol}$$

Thus, the channel introduces 0.081 bit of uncertainty to each received symbol.

Shannon showed that the average effective information content, H_{eff}, at the receiver, is obtained by subtracting the equivocation from the entropy of the source. Therefore,

$$H_{eff} = H(X) - H(X|Y) \tag{7.11}$$

For a system transmitting equally likely binary symbols, the entropy, $H(X)$, is 1 bit/symbol. When the symbols are received with $P_B = 0.01$ the equivocation is 0.081 bit/received symbol as was calculated above. Then using Equation (7.11), the effective entropy of the received signal, H_{eff}, is

$$H_{eff} = 1 - 0.081 = 0.919 \text{ bit/received symbol}$$

Thus, if $R = 1000$ binary symbols transmitted per second, for example, the effective information bit rate, R_{eff}, can be expressed as

$$R_{eff} = RH_{eff} \tag{7.12}$$

$$= 1000 \text{ symbols/s} \times 0.919 \text{ bit/symbol} = 919 \text{ bits/s}$$

Notice that in the extreme case where $P_B = 0.5$,

$$H(X|Y) = -(0.5 \log_2 0.5 + 0.5 \log_2 0.5)$$

$$= 1 \text{ bit/symbol}$$

and, applying Equations (7.12) and (7.11) to the $R = 1000$ symbols/s example, yields

$$R_{eff} = 1000 \text{ symbols/s } (1 - 1) = 0 \text{ bit/s}$$

as should be expected.

Example 7.2 Apparent Contradiction in the Shannon Limit

Plots of P_B versus E_b/N_0 typically display a smooth increase of P_B as E_b/N_0 is decreased. For example, the bit error probability for the curves in Figure 7.1 shows P_B *tending* to 0.5 in the limit as E_b/N_0 approaches zero. Thus there is apparently always a nonvanishing information rate, regardless of how small E_b/N_0 becomes. This *appears to contradict* the Shannon limit of $E_b/N_0 = -1.59$ dB, below which no error-free information rate can be supported per unit bandwidth, or below which even an infinite bandwidth cannot support a finite information rate (see Figure 7.4).

(a) Suggest a way of resolving the apparent contradiction.
(b) Show how Shannon's equivocation correction can resolve it for a binary PSK system where the source has an entropy of 1 bit/symbol. Consider that the operating point on Figure 7.1b corresponds to $E_b/N_0 = 0.1$ (-10 dB).

Solution

(a) The value of E_b, traditionally used in link calculations for practical systems, is invariably the received signal energy per *transmitted symbol*. However, the meaning of E_b in Equation (7.6) is the signal energy per bit of *received information*. The information loss caused by the noisy channel must be taken into account to resolve the apparent contradiction.
(b) Following Equation (3.84) for BPSK,

$$P_B = Q(\sqrt{2E_b/N_0}) = Q(0.447)$$

where Q is defined in Equation (2.42) and tabulated in Table B.1. From the tabulation, P_B is found to be 0.33. Next, we solve for the equivocation and effective entropy:

$$H(X|Y) = -[(1 - P_B) \log_2 (1 - P_B) + P_B \log_2 P_B]$$
$$= -(0.67 \log_2 0.67 + 0.33 \log_2 0.33)$$
$$= 0.915 \text{ bit/symbol}$$

$$H_{\text{eff}} = H(X) - H(X|Y)$$
$$= 1 - 0.915$$
$$= 0.085 \text{ bit/symbol}$$

Hence

$$\left(\frac{E_b}{N_0}\right)_{\text{eff}} = \frac{(E_b/N_0) \quad \text{joules per symbol/watts per hertz}}{H_{\text{eff}} \quad \text{bits/symbol}}$$

$$= \frac{0.1}{0.085} = 1.176 \frac{\text{joules per bit}}{\text{watts/Hz}}$$

$$= 0.7 \text{ dB}$$

Thus the effective value of E_b/N_0 is equal to 0.7 dB per received information bit, which is well above Shannon's limit of -1.59 dB.

7.5 BANDWIDTH-EFFICIENCY PLANE

Using Equation (7.6), we can plot normalized channel bandwidth W/C in Hz/bits/s versus E_b/N_0, as shown in Figure 7.4. Here, with the abscissa taken as E_b/N_0, we see the *true power–bandwidth trade-off* at work. It can be shown [4] that well-designed systems tend to operate near the "knee" of this power–bandwidth trade-off curve for the ideal ($R = C$) channel. Actual systems are frequently within 10 dB or less of the performance of the ideal. The existence of the knee means that systems seeking to reduce the channel bandwidth they occupy or to reduce the signal power they require must make an increasingly unfavorable exchange in the other parameter. For example, from Figure 7.4, an ideal system operating at an E_b/N_0 of 1.8 dB and using a normalized bandwidth of 0.5 Hz/bits/s would have to increase E_b/N_0 to 20 dB to reduce the bandwidth occupancy to 0.1 Hz/bits/s. Trade-offs in the other direction are similarly inequitable.

Using Equation (7.6), we can also plot C/W versus E_b/N_0. This relationship is shown plotted on the R/W versus E_b/N_0 plane in Figure 7.6. We shall denote

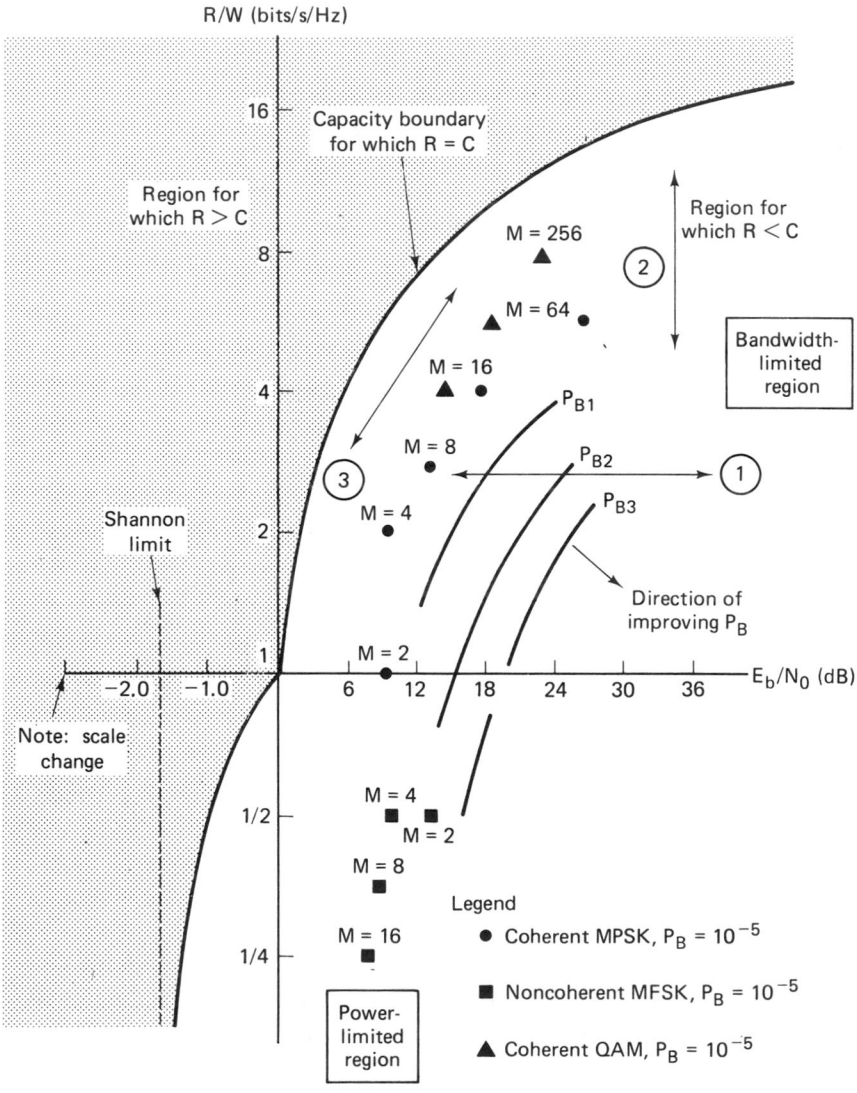

Figure 7.6 Bandwidth-efficiency plane.

this plane as the *bandwidth-efficiency plane*. The ordinate, R/W, is a measure of how much data can be communicated in a specified bandwidth within a given time; it therefore reflects how efficiently the bandwidth resource is utilized. The abscissa is E_b/N_0 in units of decibels. For the case where $R = C$ in Figure 7.6, the curve represents a boundary that separates a region characterizing practical communication systems from a region where such communication systems are not theoretically possible. Like Figure 7.2, the bandwidth-efficiency plane in Figure 7.6 sets the limiting performance that can be achieved by practical systems. Since the abscissa in Figure 7.6 is E_b/N_0 rather than SNR, Figure 7.6 is more useful for comparing digital communication modulation and coding trade-offs than is Figure 7.2.

7.5.1 Bandwidth Efficiency of MPSK and MFSK Modulation

On the bandwidth-efficiency plane of Figure 7.6 are plotted the operating points for coherent MPSK modulation at a bit error probability of 10^{-5}. We assume Nyquist (ideal rectangular) filtering at baseband, so that the minimum double-sideband (DSB) bandwidth at an intermediate frequency (IF) is $W_{IF} = 1/T$, where T is the symbol duration. Thus the bandwidth efficiency is $R/W = \log_2 M$, where M is the symbol set size. For realistic channels and waveforms, the performance must be reduced to account for the bandwidth increase required to implement realizable filters. Notice that for MPSK modulation, R/W increases with increasing M. Notice also that the location of the MPSK points indicates that BPSK ($M = 2$) and quaternary PSK or QPSK ($M = 4$) require the same E_b/N_0. That is, for the same value of E_b/N_0, QPSK has a bandwidth efficiency of 2 bits/s/Hz, compared to 1 bit/s/Hz for BPSK. This unique feature stems from the fact that QPSK is effectively a composite of two BPSK signals transmitted on orthogonal components of the carrier.

Also plotted on the bandwidth-efficiency plane of Figure 7.6 are the operating points for noncoherent MFSK modulation at a bit error probability of 10^{-5}. We assume that the IF transmission bandwidth is $W_{IF} = M/T$, and thus the bandwidth efficiency is $R/W = k/M$. Notice that for MFSK modulation, R/W decreases with increasing M. Notice also that the position of the MFSK points indicates that BFSK ($M = 2$) and quaternary FSK ($M = 4$) have the same bandwidth efficiency, even though the former requires greater E_b/N_0 for the same error probability. The bandwidth efficiency varies with the modulation index (tone spacing in hertz divided by bit rate). Under the assumption that an equal increment of bandwidth is required for each MFSK tone the system uses, it can be seen that for $M = 2$, the bandwidth efficiency is 1 bit/s/2 Hz or $\frac{1}{2}$, and for $M = 4$, similarly, the R/W is 2 bits/s/4 Hz or $\frac{1}{2}$.

Operating points for coherent quadrature amplitude modulation (QAM) are also plotted in Figure 7.6. Of the modulations shown, QAM is clearly the most bandwidth efficient; it is treated in greater detail in Section 7.9.3.

7.5.2 Analogies between Bandwidth-Efficiency and Error Probability Planes

The bandwidth-efficiency plane in Figure 7.6 is analogous to the error probability plane in Figure 7.1. The Shannon limit of the Figure 7.1 plane is analogous to the capacity boundary of the Figure 7.6 plane. The curves in Figure 7.1 were referred to as equibandwidth curves. In Figure 7.6, we can analogously describe equi-error-probability curves for various modulation and coding schemes. The curves, labeled P_{B1}, P_{B2}, and P_{B3}, are hypothetical constructions for some arbitrary modulation and coding scheme; the P_{B1} curve represents the largest error probability of the three curves, and the P_{B3} curve represents the smallest. The general direction in which the curves move for improved P_B is indicated on the figure.

Just as potential trade-offs among P_B, E_b/N_0, and W were considered for the error probability plane, the same trade-offs can be considered on the bandwidth efficiency plane. The potential trade-offs are seen in Figure 7.6 as changes in operating point in the direction shown by the arrows. Movement of the operating point along line 1 can be viewed as trading P_B for E_b/N_0, with R/W fixed. Similarly, movement along line 2 is seen as trading P_B for W (or R/W), with E_b/N_0 fixed. Finally, movement along line 3 illustrates trading W (or R/W) for E_b/N_0, with P_B fixed. In Figure 7.6, as in Figure 7.1, movement along line 1 can be effected by increasing or decreasing the available E_b/N_0. However, movement along line 2 or line 3 requires changes in the system modulation or coding scheme.

The two primary communications resources are the transmitted power and the channel bandwidth. In many communication systems, one of these resources may be more precious than the other, and hence most systems can be classified as either power limited or bandwidth limited. In *power-limited systems,* coding schemes can be used to save power at the expense of bandwidth, whereas in *bandwidth-limited systems,* spectrally efficient modulation techniques can be used to save bandwidth at the expense of power.

7.6 POWER-LIMITED SYSTEMS

For the case of power-limited systems, systems in which power is scarce but system bandwidth is available (e.g., a space communication link), the following trade-offs might be made: (1) improved P_B can be achieved by expending bandwidth (for a given E_b/N_0); or (2) required E_b/N_0 can be reduced by expending bandwidth (for a given P_B). The error probability plane of Figure 7.1a can be very useful for examining these potential trade-offs. It is on such a plane that we can verify whether or not a candidate modulation or code offers improvement in required E_b/N_0 for a particular channel and for a specified P_B (or whether the modulation or code offers improvement in P_B for a given E_b/N_0).

7.7 BANDWIDTH-LIMITED SYSTEMS

Any digital scheme that transmits $\log_2 M$ bits in T seconds using a bandwidth of W hertz operates at a bandwidth efficiency of $R/W = (\log_2 M)/WT$ bits/s/Hz. From this expression it can be seen that the smaller the WT product, the more bandwidth efficient will be the system. Signals with small WT products are more often used with bandwidth-limited systems—systems in which channel bandwidth is constrained but power is available. For this case the usual objective is to design the link so as to maximize the transmitted information rate over the bandlimited channel, at the expense of E_b/N_0 (while maintaining a specified value of P_B). For bandlimited operation, bandwidth efficiency is a useful criterion of system performance, and the bandwidth-efficiency plane of Figure 7.6 is useful for examining potential trade-offs.

Two regions, the bandwidth-limited region and the power-limited region, are shown on the bandwidth efficiency plane of Figure 7.6. Notice that the desirable trade-offs associated with each of these regions are not equitable. For the bandwidth-limited region, large R/W is desired; however, as E_b/N_0 is increased, the capacity boundary curve flattens out and ever-increasing amounts of additional E_b/N_0 are required to achieve improvement in R/W. A similar relationship is at work in the power-limited region. Here a savings in E_b/N_0 is desired, but the capacity boundary curve is steep; to achieve a small reduction in required E_b/N_0 requires a large reduction in R/W.

7.8 MODULATION AND CODING TRADE-OFFS

Figure 7.7 is useful in pointing out analogies between the two performance planes, the error probability plane of Figure 7.1 and the bandwidth efficiency plane of Figure 7.6. Figure 7.7a and b represent the same planes as Figures 7.1 and 7.6, respectively. They have been redrawn as symmetrical, by choosing appropriate scales. In each case the arrows and their labels describe the general effect of moving an operating point in the direction of the arrow by means of appropriate modulation and coding techniques. The notations G, C, and F stand for the trade-off considerations "*G*ained or achieved," "*C*ost or expended," and "*F*ixed or unchanged," respectively. The parameters being traded are P_B, W, R/W, and P (power or S/N). Just as the movement of an operating point toward the Shannon limit in Figure 7.7a can achieve improved P_B or reduced required transmitter power at the cost of bandwidth, so too movement toward the capacity boundary in Figure 7.7b can improve bandwidth efficiency at the cost of increased required power or degraded P_B.

Most often, these trade-offs are examined with a fixed P_B (constrained by the system requirement) in mind. Therefore, the most interesting arrows are those having bit error probability (marked F: P_B). There are four such arrows on Figure

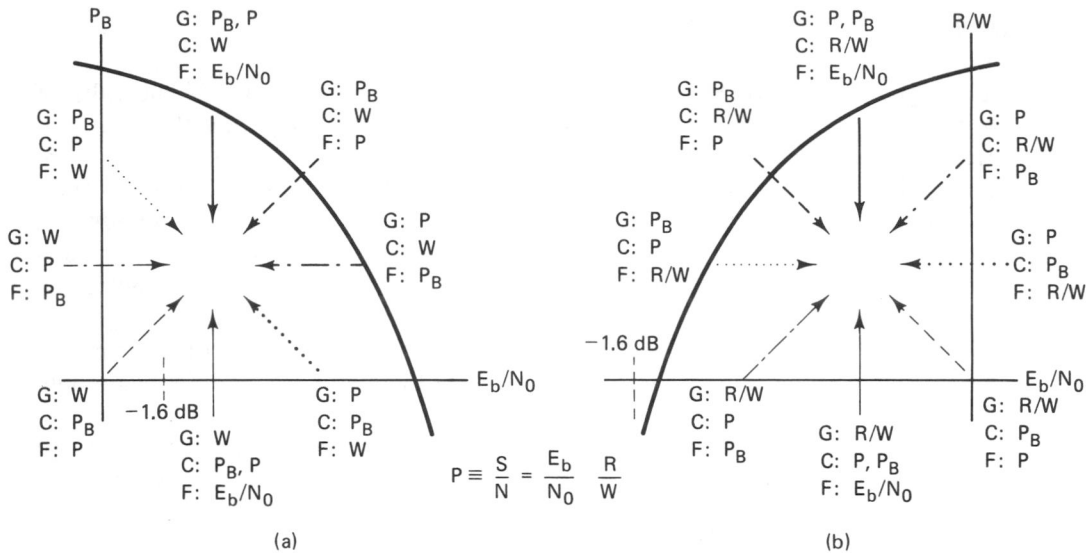

Figure 7.7 Modulation/coding trade-offs. (a) Error probability plane. (b) Bandwidth-efficiency plane.

7.7, two on the error probability plane and two on the bandwidth-efficiency plane. Arrows marked with the same pattern indicate correspondence between the two planes. System operation can be characterized by either of these two planes. The planes represent two ways of looking at some of the key system parameters; each plane highlights slightly different aspects of the overall design problem. The error probability plane tends to be most useful with *power-limited systems,* where as we move from curve to curve, the bandwidth requirements are only inferred, while the bit error probability is clearly displayed. The bandwidth efficiency plane is generally more useful for examining *bandwidth-limited systems*; here as we move from curve to curve, the bit error probability is only inferred, but the bandwidth requirements are explicit.

The two system trade-off planes, error probability and bandwidth efficiency, have been presented *heuristically* with simple examples (orthogonal and multiple phase signaling) to provide some insight into the design issues of trading-off error probability, bandwidth, and power. The ideas are useful for *most modulation and coding schemes,* with the following caveat. For *some* codes or combined modulation and coding schemes, the performance curves *do not move as predictably* as those for the examples chosen here. The reason has to do with the strength and bandwidth expansion features of the particular code. For example, the performance of coherent PSK combined with several codes was illustrated in Figure 5.23. Examine the curves characterizing the two BCH codes, (127, 64) and (127, 36). It should be clear from their relative positions that the (127, 64) code manifests *greater coding gain* than the (127, 36) code. This violates our expectations since, within the same block size, the latter code has greater redundancy (requires more bandwidth expansion) than the former. Also, in the area of trellis-coded modulation covered in Section 7.10.6, we consider codes that provide coding gain without any bandwidth expansion. Performance curves for such coding schemes will also behave differently from the curves of most modulation and coding schemes discussed so far.

7.9 BANDWIDTH-EFFICIENT MODULATION

The primary objective of spectrally efficient modulation techniques is to maximize bandwidth efficiency. The increasing demand for digital transmission channels has led to the investigation of spectrally efficient modulation techniques [6] to maximize bandwidth efficiency and thus help ameliorate the spectral congestion problem.

Some systems have additional modulation requirements besides spectral efficiency. For example, satellite systems with highly nonlinear transponders require a constant envelope modulation. This is because the nonlinear transponder produces extraneous sidebands when passing a signal with amplitude fluctuations (due to a mechanism called AM-to-PM conversion). These sidebands deprive the information signals of some of their portion of transponder power, and also can interfere with nearby channels (adjacent channel interference) or with other communication systems (co-channel interference). *Offset QPSK* (OQPSK) and *Minimum shift keying* (MSK) are two examples of constant envelope modulation schemes that are attractive for systems using nonlinear transponders.

7.9.1 QPSK and Offset QPSK Signaling

Figure 7.8 illustrates the partitioning of a typical pulse stream for QPSK modulation. Figure 7.8a shows the original data stream $d_k(t) = d_0, d_1, d_2, \ldots$ consisting of bipolar pulses; that is, the values of $d_k(t)$ are $+1$ or -1, representing binary one and zero, respectively. This pulse stream is divided into an in-phase stream, $d_I(t)$, and a quadrature stream, $d_Q(t)$, illustrated in Figure 7.8b, as follows:

$$d_I(t) = d_0, d_2, d_4, \ldots \quad \text{(even bits)}$$
$$d_Q(t) = d_1, d_3, d_5, \ldots \quad \text{(odd bits)} \tag{7.13}$$

Note that $d_I(t)$ and $d_Q(t)$ each have half the bit rate of $d_k(t)$. A convenient orthogonal realization of a QPSK waveform, $s(t)$, is achieved by amplitude modulating the in-phase and quadrature data streams onto the cosine and sine functions of a carrier wave, as follows:

$$s(t) = \frac{1}{\sqrt{2}} d_I(t) \cos\left(2\pi f_0 t + \frac{\pi}{4}\right) + \frac{1}{\sqrt{2}} d_Q(t) \sin\left(2\pi f_0 t + \frac{\pi}{4}\right) \tag{7.14}$$

Using the trigonometric identities shown in Equations (D.5) and (D.6), Equation

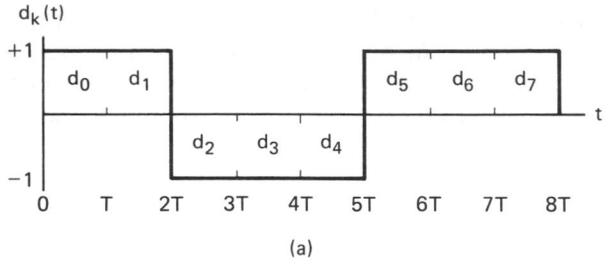

(a)

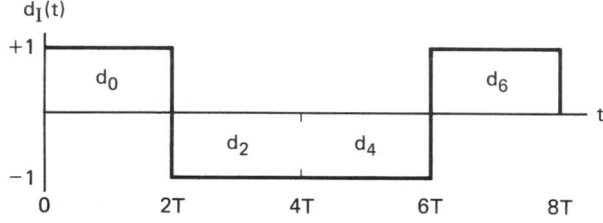

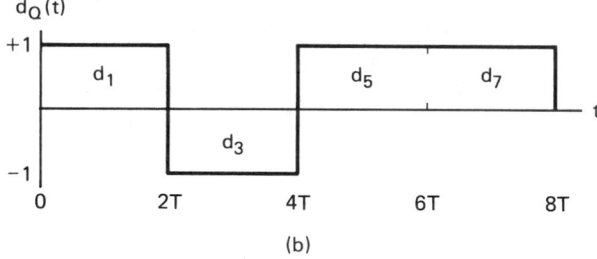

(b)

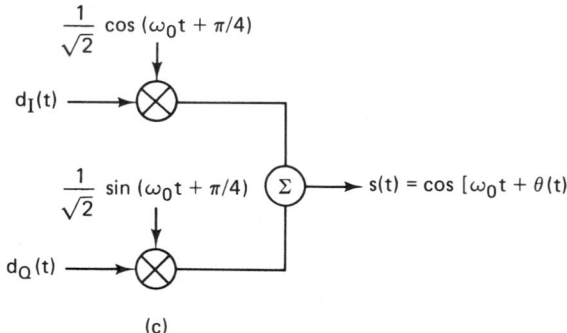

(c)

Figure 7.8 QPSK modulation.

(7.14) can also be written as

$$s(t) = \cos\left[2\pi f_0 t + \theta(t)\right] \qquad (7.15)$$

The QPSK modulator is shown in the block diagram of Figure 7.8c. The pulse stream $d_I(t)$ amplitude-modulates the cosine function with an amplitude of $+1$ or -1. This is equivalent to shifting the phase of the cosine function by 0 or π; consequently, this produces a BPSK waveform. Similarly, the pulse stream $d_Q(t)$ modulates the sine function, yielding a BPSK waveform orthogonal to the cosine function. The summation of these two orthogonal components of the carrier yields the QPSK waveform. The value of $\theta(t)$ will correspond to one of the four possible combinations of $d_I(t)$ and $d_Q(t)$ in Equation (7.14). These values are: $\theta(t) = 0°$, $\pm 90°$, or $180°$, and the resulting signal vectors are seen in the signal space illustrated in Figure 7.9. Because $\cos(2\pi f_0 t + \pi/4)$ and $\sin(2\pi f_0 t + \pi/4)$ are orthogonal, the two BPSK signals can be detected separately.

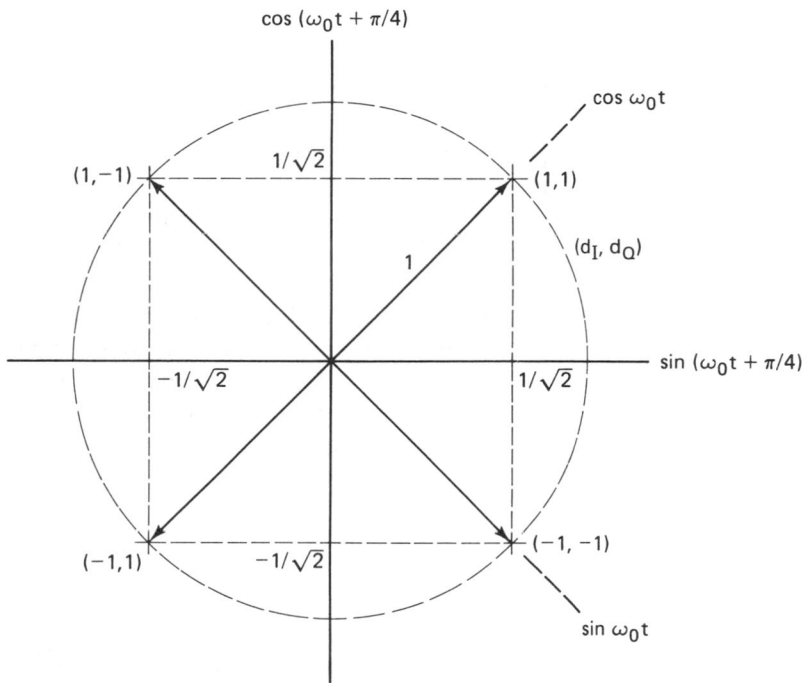

Figure 7.9 Signal space for QPSK and OQPSK.

Offset QPSK (OQPSK) signaling can also be represented by Equations (7.14) and (7.15); the difference between the two modulation schemes, QPSK and OQPSK, is only in the *alignment* of the two baseband waveforms. As shown in Figure 7.8, the duration of each original pulse is T (Figure 7.8a), and hence in the partitioned streams of Figure 7.8b, the duration of each pulse is $2T$. In standard QPSK, the odd and even pulse streams are both transmitted at the rate of $1/2T$ bits/s and are synchronously aligned, such that their transitions coincide, as shown in Figure 7.8b. In OQPSK, sometimes called *staggered QPSK* (SQPSK), there is the same data stream partitioning and orthogonal transmission; the difference is that the timing of the pulse stream $d_I(t)$ and $d_Q(t)$ is shifted such that the alignment of the two streams is offset by T. Figure 7.10 illustrates this offset.

In standard QPSK, due to the coincident alignment of $d_I(t)$ and $d_Q(t)$, the carrier phase can change only once every $2T$. The carrier phase during any $2T$ interval can be any one of the four phases shown in Figure 7.9, depending on the values of $d_I(t)$ and $d_Q(t)$ during that interval. During the next $2T$ interval, if neither pulse stream changes sign, the carrier phase remains the same. If only one of the

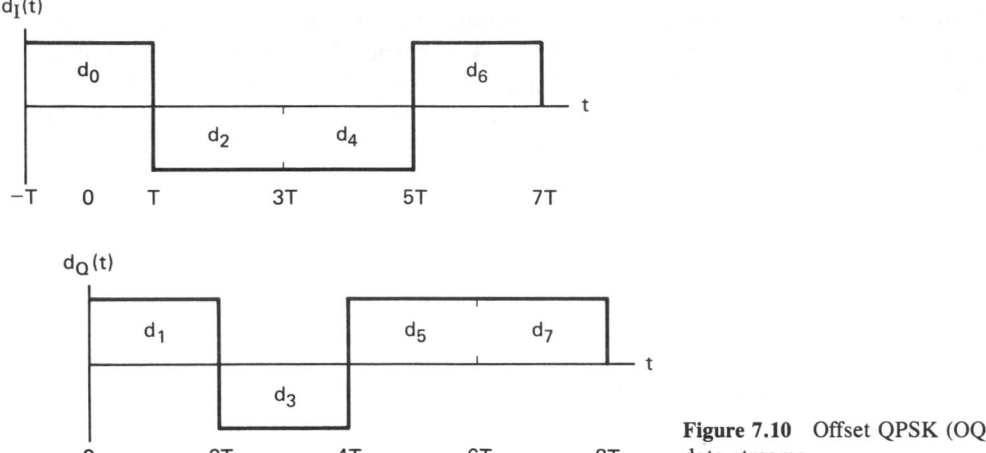

Figure 7.10 Offset QPSK (OQPSK) data streams.

pulse streams changes sign, a phase shift of $\pm 90°$ occurs. A change in both streams results in a carrier phase shift of 180°. Figure 7.11a shows a typical QPSK waveform for the sample sequence $d_I(t)$ and $d_Q(t)$ shown in Figure 7.8.

If a QPSK modulated signal undergoes filtering to reduce the spectral sidelobes, the resulting waveform will no longer have a constant envelope and in fact,

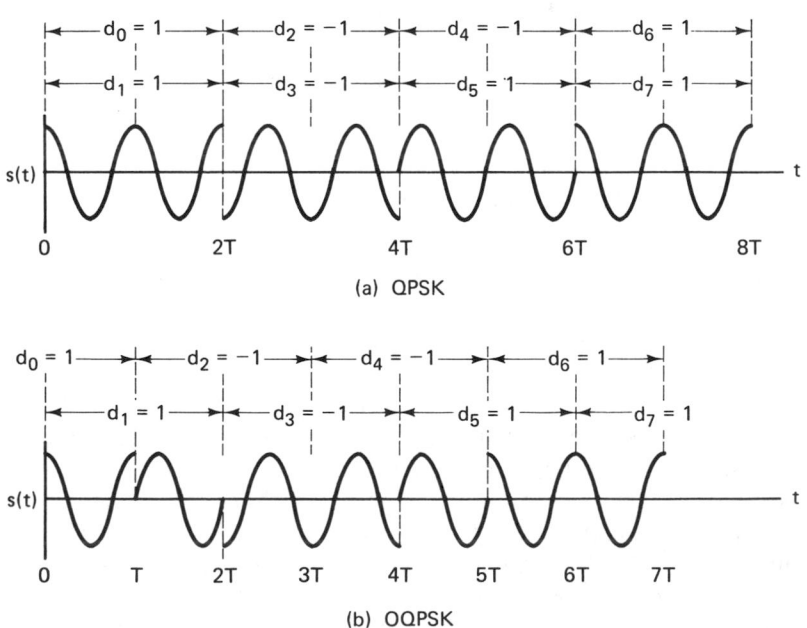

Figure 7.11 (a) QPSK and (b) OQPSK waveforms. (Reprinted with permission from S. Pasupathy, "Minimum Shift Keying: A Spectrally Efficient Modulation," *IEEE Commun. Mag.*, July 1979, Fig. 4, p. 17. © 1979 IEEE.)

the occasional 180° phase shifts will cause the envelope to go to zero momentarily (see Figure 7.11a). When these signals are used in satellite channels employing highly nonlinear amplifiers, the constant envelope will tend to be restored. However, at the same time, all of the *undesirable* frequency side-lobes, which can interfere with nearby channels and other communication systems, are also restored.

In OQPSK, the pulse streams $d_I(t)$ and $d_Q(t)$ are staggered and thus do not change states simultaneously. The possibility of the carrier changing phase by 180° is eliminated, since only one component can make a transition at one time. Changes are limited to 0° and $\pm 90°$ every T seconds. Figure 7.11b shows a typical OQPSK waveform for the sample sequence in Figure 7.10. When an OQPSK signal undergoes bandlimiting, the resulting intersymbol interference causes the envelope to droop slightly to the region of $\pm 90°$ phase transition, but since the phase transitions of 180° have been avoided in OQPSK, the envelope will not go to zero as it does with QPSK. When the bandlimited OQPSK goes through a nonlinear transponder, the envelope droop is removed; however, the high-frequency components associated with the collapse of the envelope are not reinforced. Thus out-of-band interference is avoided [7].

7.9.2 Minimum Shift Keying

The main advantage of OQPSK over QPSK, that of suppressing out-of-band interference, suggests that further improvement is possible if the OQPSK format is modified to avoid discontinuous phase transitions. This was the motivation for designing continuous phase modulation (CPM) schemes. *Minimum shift keying* (MSK) is one such scheme [7–9]. MSK can be viewed as either a special case of *continuous-phase frequency shift keying* (CPFSK), or a special case of OQPSK with sinusoidal symbol weighting. When viewed as CPFSK, the MSK waveform can be expressed as [8]

$$s(t) = \cos\left[2\pi\left(f_0 + \frac{d_k}{4T}\right)t + x_k\right] \qquad kT < t < (k+1)T \qquad (7.16)$$

where f_0 is the carrier frequency, $d_k = \pm 1$ represents the bipolar data being transmitted at a rate $R = 1/T$, and x_k is a phase constant which is valid over the kth binary data interval. Notice that for $d_k = 1$, the frequency transmitted is $f_0 + 1/4T$, and for $d_k = -1$, the frequency transmitted is $f_0 - 1/4T$. The tone spacing in MSK is thus one-half that employed for noncoherently demodulated orthogonal FSK, giving rise to the name *minimum* shift keying. During each T-second data interval, the value of x_k is a constant, that is, $x_k = 0$ or π, determined by the requirement that the phase of the waveform be continuous at $t = kT$. This requirement results in the following recursive phase constraint for x_k:

$$x_k = \left[x_{k-1} + \frac{\pi k}{2}(d_{k-1} - d_k)\right] \text{ modulo } 2\pi \qquad (7.17)$$

Equation (7.16) can be expressed in a quadrature representation, as follows, using the identities in Equations (D.5) and (D.6):

$$s(t) = a_k \cos \frac{\pi t}{2T} \cos 2\pi f_0 t - b_k \sin \frac{\pi t}{2T} \sin 2\pi f_0 t$$

$$kT < t < (k+1)T \quad (7.18)$$

where

$$a_k = \cos x_k = \pm 1$$
$$b_k = d_k \cos x_k = \pm 1$$
$$(7.19)$$

The in-phase (I) component is identified as $a_k \cos (\pi t/2T) \cos 2\pi f_0 t$, where cos $2\pi f_0 t$ is the carrier, $\cos (\pi t/2T)$ can be regarded as a *sinusoidal symbol weighting*, and a_k is a data-dependent term. Similarly, the quadrature (Q) component is identified as $b_k \sin (\pi t/2T) \sin 2\pi f_0 t$, where $\sin 2\pi f_0 t$ is the quadrature carrier term, $\sin (\pi t/2T)$ can be regarded as a sinusoidal symbol weighting, and b_k is a data-dependent term. It might appear that the a_k and b_k terms can change every T seconds, since the source data, d_k, can change every T seconds. However, because of the continuous phase constraint, the a_k term can only change value at the zero crossings of $\cos (\pi t/2T)$ and the b_k term can only change value at the zero crossings of $\sin (\pi t/2T)$. Thus the symbol weighting in either the I- or Q-channel is a half-cycle sinsuoidal pulse of duration $2T$ seconds with alternating sign. As in the case of OQPSK, the I and Q components are offset T seconds with respect to one another.

Notice that x_k in Equation (7.17) is a function of the difference between the prior data bit and the present data bit (differential encoding). Hence the a_k and b_k terms in Equation (7.18) can be viewed as *differentially encoded* components of the d_k source data. However, for bit-to-bit independent data d_k, the signs of successive I- or Q-channel pulses are also random from one $2T$-second pulse interval to the next. Thus when viewed as a special case of OQPSK, Equation (7.18) can be rewritten with more straightforward (nondifferential) data encoding [8] as follows:

$$s(t) = d_I(t) \cos \frac{\pi t}{2T} \cos 2\pi f_0 t + d_Q(t) \sin \frac{\pi t}{2T} \sin 2\pi f_0 t \quad (7.20)$$

where $d_I(t)$ and $d_Q(t)$ have the same in-phase and quadrature data stream interpretation as in Equation (7.13). This MSK format in Equation (7.20) is sometimes referred to as *precoded MSK*. Figure 7.12 illustrates Equation (7.20) pictorially. Figure 7.12a and c show the sinusoidal weighting of the I- and Q-channel pulses. These sequences represent the same data sequences as in Figure 7.10, but here, multiplication by a sinusoid results in more gradual phase transitions compared to those of the original data representation. Figure 7.12b and d illustrate the modulation of the orthogonal components $\cos 2\pi f_0 t$ and $\sin 2\pi f_0 t$, respectively, by the sinusoidally shaped data streams. Figure 7.12e illustrates the summation of the orthogonal components from Figure 7.12b and d. In summary, the following properties of MSK modulation can be deduced from Equation (7.20) and Figure

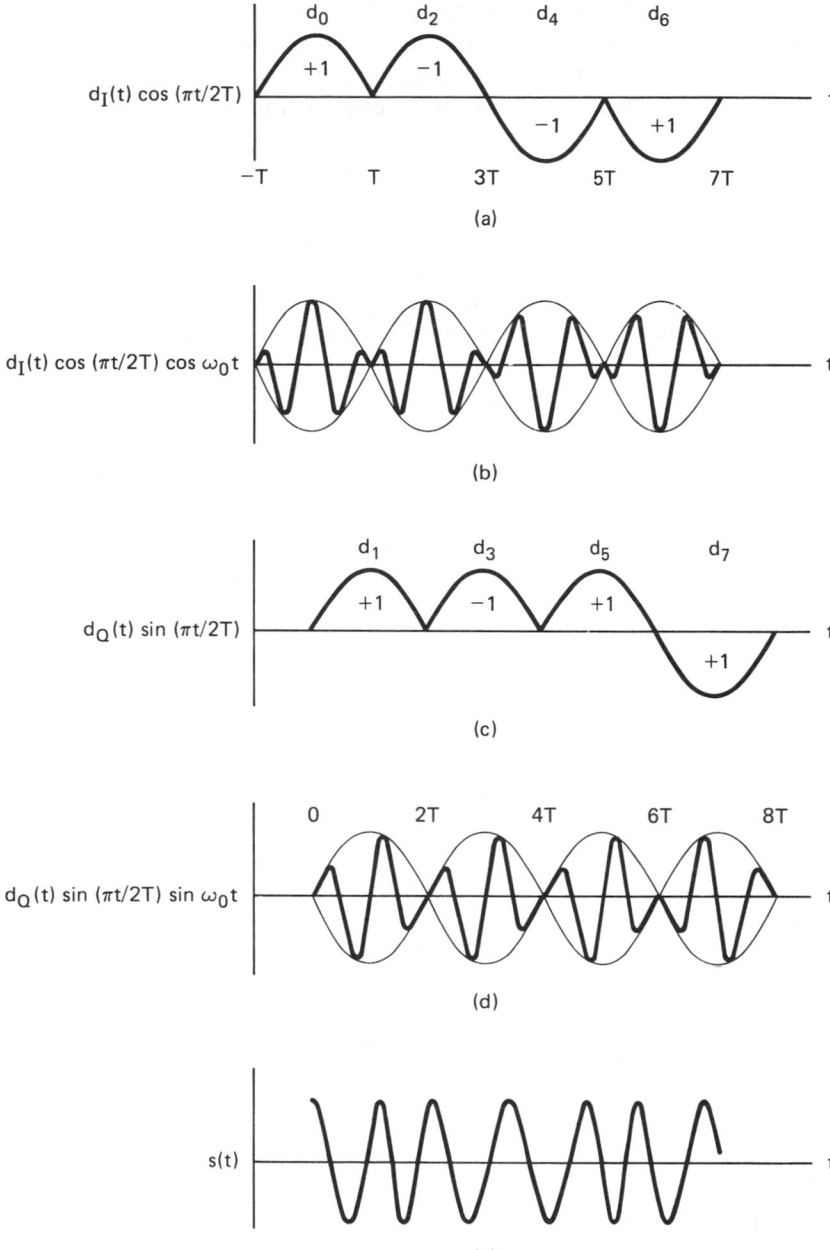

Figure 7.12 Minimum shift keying (MSK). (a) Modified I bit stream. (b) I bit stream times carrier. (c) Modified Q bit stream. (d) Q bit stream times carrier. (e) MSK waveform. (Reprinted with permission from S. Pasupathy, "Minimum Shift Keying: A Spectrally Efficient Modulation," *IEEE Commun. Mag.*, July 1979, Fig. 5, p. 18. © 1979 IEEE.)

7.12: (1) the waveform $s(t)$ has constant envelope; (2) there is phase continuity in the RF carrier at the bit transitions; and (3) the waveform $s(t)$ can be regarded as an FSK waveform with signaling frequencies $f_0 + 1/4T$ and $f_0 - 1/4T$. Therefore, the minimum tone separation required for MSK modulation is

$$\left(f_0 + \frac{1}{4T} \right) - \left(f_0 - \frac{1}{4T} \right) = \frac{1}{2T} \tag{7.21}$$

which is equal to half the bit rate. Notice that the required tone spacing for MSK is one-half the spacing, $1/T$, required for the noncoherent detection of FSK signals (see Section 3.6.4). That is because it is being coherently demodulated.

The power spectral density $G(f)$ for QPSK and OQPSK is given by [8]

$$G(f) = 2PT \left(\frac{\sin 2\pi fT}{2\pi fT} \right)^2 \tag{7.22}$$

where P is the average power in the modulated waveform. For MSK, $G(f)$ is given by [8]

$$G(f) = \frac{16PT}{\pi^2} \left(\frac{\cos 2\pi fT}{1 - 16f^2 T^2} \right)^2 \tag{7.23}$$

The normalized power spectral density ($P = 1$ W) for QPSK, OQPSK, and MSK are sketched in Figure 7.13. A spectral plot of BPSK is included for comparison. The fact that BPSK requires more bandwidth than the others for a given level of spectral density should come as no surprise. In Section 7.5.1 and Figure 7.6 we saw that the theoretical bandwidth efficiency of BPSK is half that of QPSK. It is seen from Figure 7.13 that MSK has lower sidelobes than QPSK or OQPSK. This is a consequence of multiplying the data stream with a sinusoid, yielding more *gradual phase transitions*. The more gradual the transition, the faster the spectral tails drop to zero. MSK is *spectrally more efficient* than QPSK or OQPSK; however, as can be seen from Figure 7.13, the MSK spectrum has a wider mainlobe than QPSK and OQPSK. Therefore, MSK may not be the preferred method for narrowband links.

7.9.2.1 Error Performance of OQPSK and MSK

We have seen that BPSK and QPSK have the same bit error probability because QPSK is configured as two BPSK signals modulating orthogonal components of the carrier. Since staggering the bit streams does not change the orthogonality of the carriers, OQPSK has the same theoretical bit error performance as BPSK and QPSK.

Minimum shift keying uses antipodal symbol shapes, $\pm \cos (\pi t/2T)$ and $\pm \sin (\pi t/2T)$, over $2T$ to modulate the two quadrature components of the carrier. Thus when a matched filter is used to recover the data from each of the quadrature components independently, MSK, as defined in Equation (7.20), has the same error performance properties as BPSK, QPSK, and OQPSK [7]. However, if MSK is coherently detected as an FSK signal over an observation interval of T seconds, it would be poorer than BPSK by 3 dB [7]. MSK, with differentially encoded

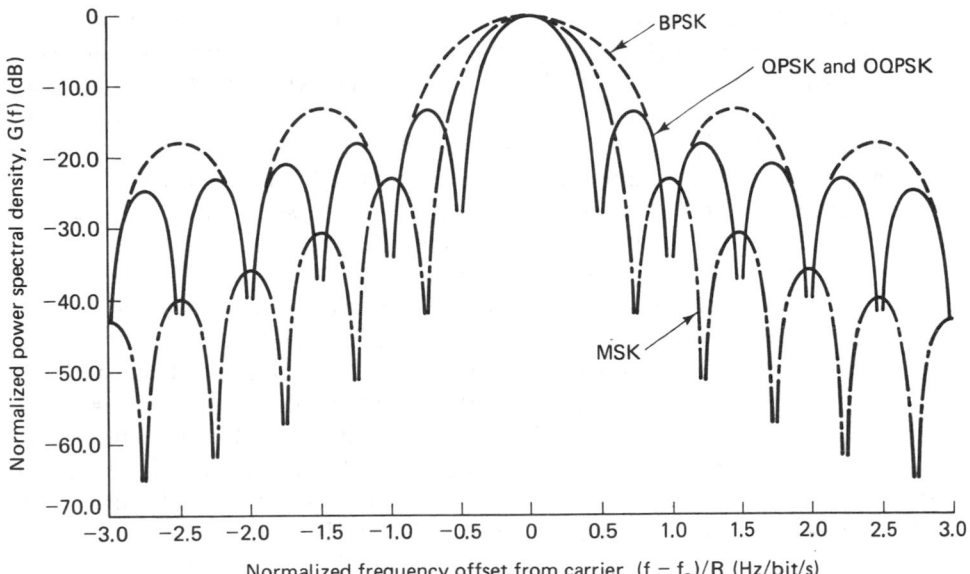

Figure 7.13 Normalized power spectral density for BPSK, QPSK, OQPSK, and MSK. (Reprinted with permission from F. Amoroso, "The Bandwidth of Digital Data Signals," *IEEE Commun. Mag.*, vol. 18, no. 6, Nov. 1980, Fig. 2A, p. 16. © 1980 IEEE.)

data, as defined in Equation (7.16), has the same error probability performance as the coherent detection of differentially encoded PSK.

QPSK systems require a fully coherent or differentially coherent detection scheme. However, since MSK is a type of FSK, it can also be noncoherently detected. This permits inexpensive demodulation of MSK when the value of received E_b/N_0 permits.

7.9.3 Quadrature Amplitude Modulation

Coherent *M*-ary phase shift keying (MPSK) modulation is a well-known technique for achieving bandwidth reduction. Instead of using a binary alphabet with 1 bit of information per channel symbol period, an alphabet with M symbols is used, permitting the transmission of $k = \log_2 M$ bits during each symbol period. Since the use of *M*-ary symbols allows a *k*-fold increase in the data rate within the same bandwidth, then for a fixed data rate, use of *M*-ary PSK reduces the required bandwidth by a factor k (see Section 3.8.3).

From Equation (7.14) it can be seen that QPSK modulation consists of two independent streams. One stream amplitude-modulates the cosine function of a carrier wave with levels $+1$ and -1, and the other stream similarly amplitude-modulates the sine function. The resultant waveform is termed a double-sideband suppressed-carrier (DSB-SC) wave, since the RF bandwidth is twice the baseband bandwidth (see Section 1.7.1) and there is no isolated carrier term. *Quadrature amplitude modulation* (QAM) can be considered a logical extension of QPSK, since QAM also consists of two independently amplitude-modulated carriers in quadrature. Each block of k bits (k assumed even) can be split into two ($k/2$)-bit blocks which use ($k/2$)-bit digital-to-analog (D/A) converters to provide the required modulating voltages for the carriers. At the receiver, each of the two signals is independently detected using matched filters. QAM signaling can also be viewed as a combination of amplitude shift keying (ASK) and phase shift keying (PSK), giving rise to the alternative name, *amplitude phase keying* (APK). Finally, it can also be viewed as amplitude shift keying in two dimensions, giving rise to the name *quadrature amplitude shift keying* (QASK).

Figure 7.14a illustrates a two-dimensional signal space and a set of 16-ary QAM signal vectors or points arranged in a rectangular constellation. A canonical QAM modulator is shown in Figure 7.14b. Assuming that Gaussian noise is the only channel disturbance, the simple channel model of Figure 7.14c applies. Signals are sent in pairs (x, y). The model indicates that the signal point coordinates (x, y) are transmitted over separate channels and independently perturbed by Gaussian noise variables (n_x, n_y), each with zero mean and variance N. Or we can say that the two-dimensional signal point is perturbed by a two-dimensional Gaussian noise variable. If the average signal energy (mean-square value of the signal coordinates) is S, then the signal-to-noise ratio is S/N. The simplest method of digital signaling through such a system is to use one-dimensional pulse amplitude modulation (PAM) independently for each signal coordinate. In PAM, to

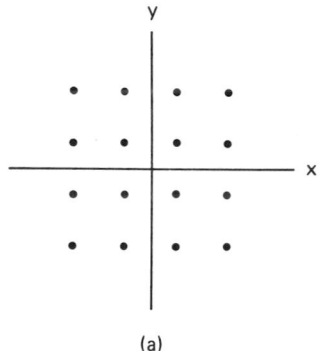

(a)

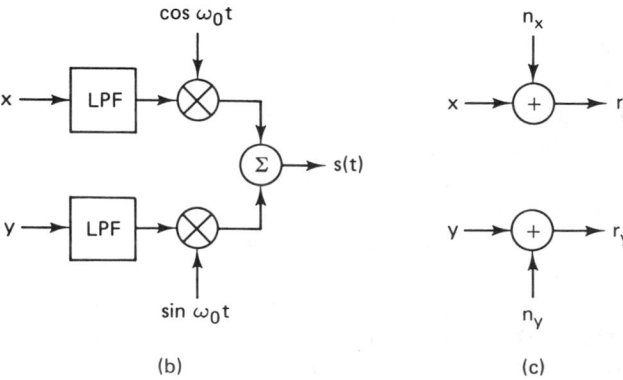

(b) (c)

Figure 7.14 QAM modulation. (a) 16-ary signal space. (b) Canonical QAM modulator. (c) QAM channel model.

send k bits/dimension over a Gaussian channel, each signal point coordinate takes on one of 2^k equally likely equispaced amplitudes. By convention, the signal points are grouped about the center of the space at amplitudes $\pm 1, \pm 3, \ldots, \pm(2^k - 1)$.

7.9.3.1 QAM Probability of Bit Error

For a rectangular constellation, a Gaussian channel, and matched filter reception, the probability of bit error is expressed [10] by

$$P_B \simeq \frac{2(1 - L^{-1})}{\log_2 L} Q \left[\sqrt{\left(\frac{3 \log_2 L}{L^2 - 1}\right) \frac{2E_b}{N_0}} \right] \qquad (7.24)$$

where $Q(x)$ is as defined in Equation (2.42) and L represents the number of amplitude levels in one dimension. We assume that a sequence of $\log_2 L$ bits are assigned to an L-ary symbol using a Gray code (defined in Section 3.9.4).

7.9.3.2 Bandwidth–Power Trade-Off

The bandwidth–power trade-off of M-ary QAM at a bit error probability of 10^{-5} is displayed on the bandwidth-efficiency plane in Figure 7.6, with the abscissa measured in average E_b/N_0. We assume Nyquist filtering of the baseband pulses so that the DSB transmission bandwidth at IF is $W_{IF} = 1/T$, where T is the symbol duration. Thus the bandwidth efficiency is $R/W = \log_2 M$, where M is the symbol set size. For realistic channels and waveforms, the performance must be reduced to account for the increased bandwidth necessary to implement realizable filters. From Figure 7.6 it can be seen that QAM represents a method of reducing the bandwidth required for the transmission of digital data. As with M-ary PSK, bandwidth efficiency can be exchanged for power or E_b/N_0; however, in the case of QAM, a *much more efficient exchange* is possible than in the case of M-ary PSK.

For a comparative treatment of digital modulation techniques in general, Reference [11] contains useful performance data and an extensive list of other references.

Example 7.3 Waveform Design

Assume that a data stream with data rate $R = 144$ Mbits/s is to be transmitted on an RF channel using a DSB modulation scheme. Assume Nyquist filtering and an allowable DSB bandwidth of 36 MHz. Which modulation technique would you choose for this requirement? If the available E_b/N_0 is 20, what would be the resulting probability of bit error?

Solution

The required spectral efficiency is

$$\frac{R}{W} = \frac{144 \text{ Mbits/s}}{36 \text{ MHz}} = 4 \text{ bits/s/Hz}$$

From Figure 7.6 we note that 16-ary QAM, with a theoretical spectral efficiency of 4 bits/s/Hz, requires a lower E_b/N_0 than that of 16-ary PSK for the same P_B. Based on these considerations we choose a 16-ary QAM modem.

With the available E_b/N_0 given as 20, we use Equation (7.24) to calculate the expected bit error probability as

$$P_B \simeq \frac{3}{4} Q \left(\sqrt{\frac{4}{5} \frac{E_b}{N_0}} \right) = 2.5 \times 10^{-5}$$

Example 7.4   Spectral Efficiency

(a) Explain the computation of the QAM spectral efficiency in Example 7.3, considering that QAM is transmitted on orthogonal components of a carrier wave.
(b) Since the DSB bandwidth is 36 MHz in Example 7.3, consider using half that amount at baseband to transmit the 144-Mbits/s data stream, using multilevel PAM. What is the spectral efficiency needed to accomplish this, and how many levels of PAM would be required? Assume Nyquist filtering.

Solution

(a) *Bandpass channel using QAM:* The 144-Mbits/s data stream is partitioned into a 72-Mbits/s in-phase and a 72-Mbits/s quadrature stream; one stream amplitude-modulates the cosine component of a carrier over a bandwidth of 36 MHz, and the other stream amplitude-modulates the sine component of the carrier wave over the same 36-MHz bandwidth. Since each 72-Mbits/s stream modulates an orthogonal component of the carrier, the 36 MHz suffices for both streams, or for the full 144 Mbits/s. Thus the spectral efficiency is (144 Mbits/s)/36 MHz = 4 bits/s/Hz.
(b) *Required spectral efficiency at baseband*

$$\frac{R}{W} = \frac{144 \text{ Mbits/s}}{18 \text{ MHz}} = 8 \text{ bits/s/Hz}$$

Assuming Nyquist filtering, a bandwidth of 18 MHz can support a maximum symbol rate of $R_s = 2W = 36$ megasymbols/s [see Equation (2.76)]. Each PAM pulse must therefore have an ℓ-bit meaning, such that

$$R = \ell R_s$$

Hence

$$\ell = \frac{144 \text{ Mbits/s}}{36 \text{ megapulses/s}} = 4 \text{ bits/pulse}$$

where $\ell = \log_2 L$, and $L = 16$ levels.

7.10 MODULATION AND CODING FOR BANDLIMITED CHANNELS

The channel coding techniques of Chapters 5 and 6 have generally *not* been associated with voice-grade telephone channels (although the first field test of sequential decoding of convolutional codes was on a telephone line). Recently, however, there has been considerable interest in techniques that can provide coding gain for bandlimited channels. The motivation is to enable the reliable transmission of *higher data rates* over voice-grade channels. The potential gain is about 3 bits/symbol (for a given signal-to-noise ratio) [12] or, alternatively, a given error performance could be achieved with a power savings of 9 dB [12].

The greatest interest is in the following three separate coding research areas:

1. Optimum signal constellation boundaries (choosing a closely packed signal subset from any regular array or lattice of candidate points)
2. Higher-density lattice structures (adding improvement to the signal subset choice by starting with the densest possible lattice for the space)
3. Trellis-coded modulation (combined modulation and coding techniques for obtaining coding gain for bandlimited channels)

The first two areas are not "true" error control coding schemes. By "true error control coding" we refer to those techniques that employ some structured redundancy to improve the error performance. Only the third technique, trellis-coded modulation, involves redundancy. Each of these coding research areas and their expected performance improvements are discussed below.

7.10.1 Commercial Telephone Modems

The use of efficient modulation techniques has traditionally been spearheaded by the telecommunications industry, since the telephone company's foremost resource consists of sharply bandlimited voice-grade channels. The typical telephone channel is characterized by a high signal-to-noise ratio (SNR) of approximately 30 dB and a bandwidth of approximately 3 kHz. Table 7.1 lists the evolution of high-speed telephone modems with bandwidth efficiencies (R/W) ranging from 2 to 8 bits/s/Hz. The list starts with the Bell 201, introduced in about 1962, which used QPSK in a nominal 1200-Hz bandwidth to achieve 2400 bits/s on private lines. The first commercially important 4800-bits/s modem was the Milgo 4400/48, introduced in about 1967. It utilized a nominal 1600-Hz bandwidth in conjunction with 8-ary PSK to achieve a bandwidth efficiency of 3 bits/s/Hz. In 1971 the Codex 9600C was introduced. It provided 9600 bits/s in a 2400-Hz

TABLE 7.1 Modem Milestones

Year	Model	Speed (bits/s)	Bandwidth (Hz)	Modulation	R/W (bits/s/Hz)
1962	Bell 201	2,400	1200	4-PSK	2
1967	Milgo 4400/48	4,800	1600	8-PSK	3
1971	Codex 9600C	9,600	2400	16-QAM	4
1980	Paradyne MP14400	14,400	2400	64-QAM	6
1981	Codex SP14.4	14,400	2400	64-QAM	6
1984	Codex 2660	14,400	2400	Trellis-coded QAM	6
1985	Codex 2680	19,200	2400	Trellis-coded QAM	8

bandwidth ($R/W = 4$ bits/s/Hz) using 16-ary QAM. Note that as channel equalization techniques (see Section 2.11.2) improved, a larger bandwidth portion of the voice-grade channel became usable. Whereas in 1962, only 1200 Hz could be reliably employed, that value doubled by 1971.

In 1980, first-generation 14,400-bits/s modems were introduced by Paradyne (MP14400), followed in 1981 by Codex (SP14.4). These modems improved the bandwidth efficiency by utilizing 64-ary QAM with an R/W of 6 bits/s/Hz. In a second generation, appearing in 1984, trellis-coded QAM modulation (treated in Section 7.10.6) was introduced to provide better error performance. In 1985, Codex introduced a modem with $R/W = 8$ bits/s/Hz, thereby achieving a data rate of 19,200 bits/s in a nominal bandwidth of 2400 Hz. Without any major upgrading of the telephone network, 19,200 bits/s is considered the maximum achievable data rate for an unconditioned voice-grade telephone channel [12].

7.10.2 Signal Constellation Boundaries

Several researchers [13–17] have examined large numbers of possible QAM signal constellations in a search for designs that result in the best error performance for a given average signal-to-noise ratio. Figure 7.15 illustrates some examples of symbol constellations for $M = 4$, 8, and 16 that have been considered [13]. The circular sets are designated by the notation $(a, b, . . .)$, where there are a quantity of a signals on the inner circle, b signals on the next circle, and so on. In general, the constellation rule, known as the Campopiano–Glazer construction rule [15], that yields optimum signal set performance can be summarized as follows: From an infinite array of points closely packed in a *regular array or lattice,* select a closely packed subset of 2^k points as a signal constellation. In this case "optimum" means minimum average or peak power for a given error probability. In a two-dimensional signal space the optimum boundary surrounding an array of points tends toward a circle. Figure 7.16 illustrates examples of 64-ary ($k = 6$) and 128-ary ($k = 7$) signal sets from a rectangular array. The cross-shaped boundaries are a compromise to the optimum circle. The $k = 6$ constellation was used in the Paradyne 14.4-kbits/s modem. Compared to a square, the performance improvement resulting from a circular boundary is only a modest 0.2 dB [12].

7.10.3 Higher-Dimensional Signal Constellations

For any particular information rate and a channel noise process that is independent and identically distributed in the two dimensions, signaling in a two-dimensional QAM space can provide the same error performance with less average (or peak) power than signaling in a one-dimensional pulse amplitude (PAM) space. We stated earlier that this is accomplished by choosing signal points on a two-dimensional lattice from within a circular rather than a rectangular boundary. In the same way, by going to a higher number, N, of dimensions and choosing points on an N-dimensional lattice from within an N-sphere rather than an N-cube, further energy savings are possible. Several researchers [18–21] have studied multidimensional signal constellations. Consider the four-dimensional configuration

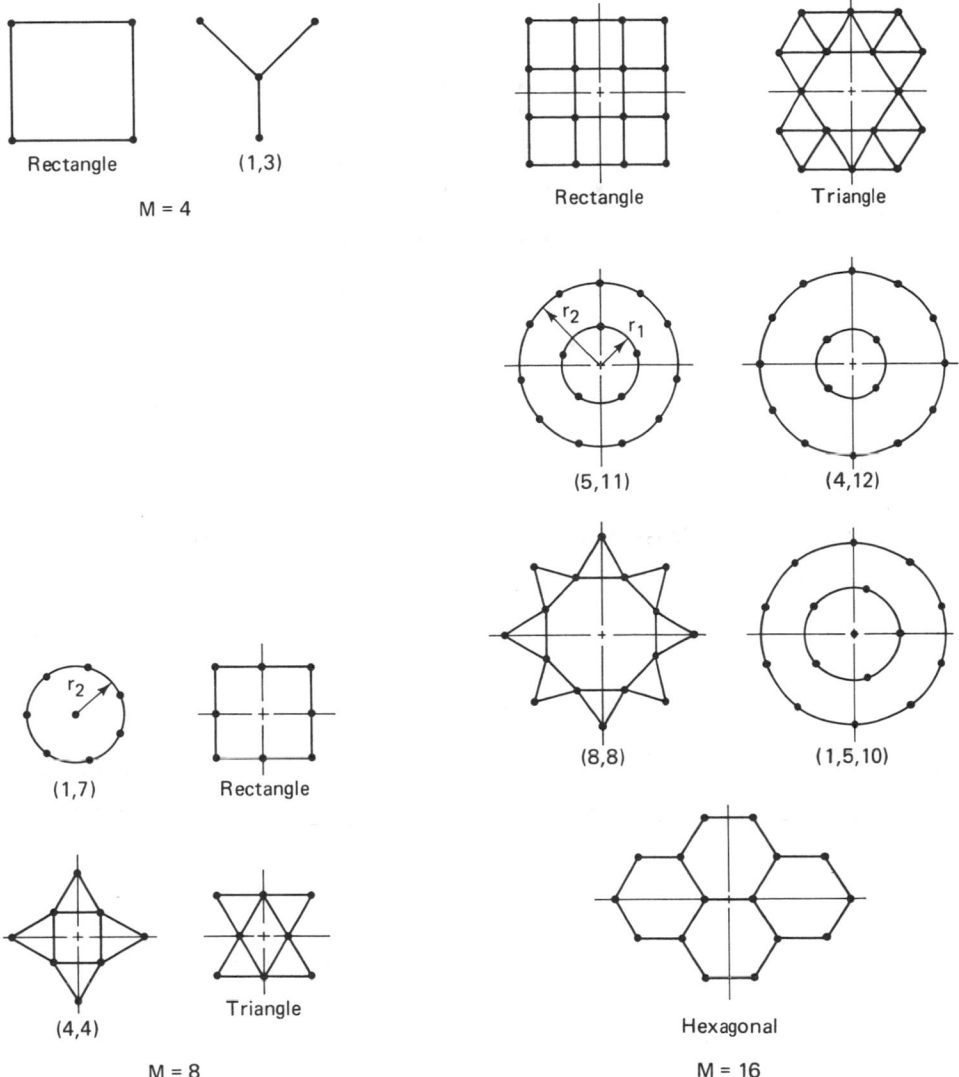

Figure 7.15 *M*-ary symbol constellations. (Reprinted with permission from C. M. Thomas, M. Y. Weidner, and S. H. Durrani, "Digital Amplitude-Phase Keying with *M*-ary Alphabets," *IEEE Trans. Commun.*, vol. COM22, no. 2, Feb. 1974, Figs. 2 and 3, p. 170. © 1974 IEEE.)

illustrated in Figure 7.17. The transmitter transmits four simultaneous sequences of pulses over four bandlimited Gaussian channels. We assume that the source produces one of M symbols, $m_i = 1, 2, \ldots, M$, every T seconds. A given symbol m_i causes four pulses to be emitted—$a_i s(t)$, $b_i s(t)$, $c_i s(t)$, $d_i s(t)$—as shown in Figure 7.17. These are transmitted on separate noninterfering channels. The pulses are distorted by independent AWGN in each channel, and at the receiver they are detected separately with matched filters. The four independent channels can

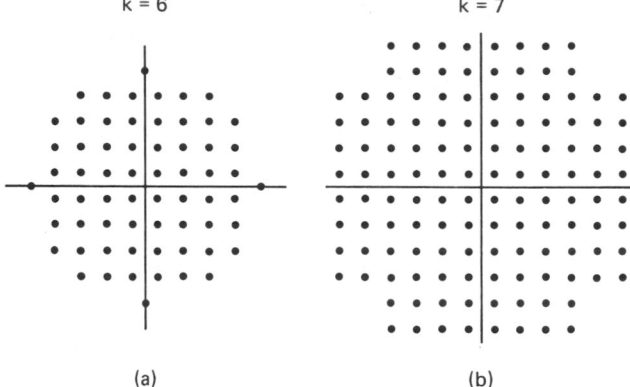

k = 6   k = 7

(a)   (b)

Figure 7.16 Examples of *M*-ary constellations using a rectangular array.

be implemented in a number of ways:

1. Two bandpass channels can be used, each with separately modulated in-phase and quadrature components (QAM or MPSK modulation on each channel).
2. The two bandpass channels can be time- or frequency-division multiplexed and carried on a common transmission line.
3. Orthogonal electromagnetic wave polarization can be used.

Let us compare a two-dimensional 16-ary QAM system with a four-dimensional alternative. In the two-dimensional modulation case, during each *T*-second

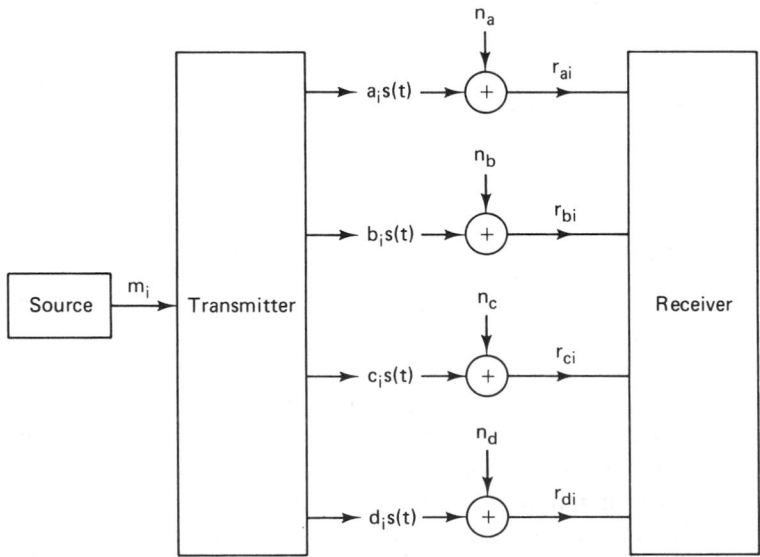

Figure 7.17 Four-dimensional system configuration.

interval, a symbol (4 bits) is transmitted by the modulation of the in-phase and quadrature components of a carrier. In other words, two 4-ary numbers can be transmitted to describe a signal vector in the 16-ary two-dimensional signaling plane. In the case of four-dimensional modulation, two successive symbols (8 bits) are sent each $2T$ seconds by transmitting four 4-ary numbers representing a point in a 256-ary signal space. It can be shown that increasing the dimensionality of the signal space offers a potential savings in average signal energy for a given level of error performance. That is, in going to a higher-dimensional space, one can effect an energy savings based on the selection of signal points from an N-sphere versus an N-cube of the same volume—the average energy of the signal points from the N-sphere is less than that from the N-cube. Table 7.2 gives the energy savings possible in N dimensions. Of course, the implementation of such a scheme involves added complexity. To send n bits per symbol in N dimensions (assuming N even), incoming bits must be grouped in blocks of $nN/2$. A mapping must then be performed into the space of $2^{(nN/2)}$ N-dimensional vectors which have the least energy among all such vectors. A corresponding inverse mapping must be made at the receiver. The added complexity may, of course, outweigh the performance gain. As N goes to infinity, the gain goes to 1.53 dB [12].

TABLE 7.2 Energy Savings from N-Sphere Mapping versus N-Cube Mapping

Dimensions (N)	N-sphere mapping gain (dB)
2	0.20
4	0.45
8	0.73
16	0.98
24	1.10
32	1.17
48	1.26
64	1.31

Source: G. D. Forney, Jr., et al., "Efficient Modulation for Bandlimited Channels," *IEEE J. Sel. Areas Commun.*, vol. SAC2, no. 5, September 1984, pp. 632–647.

7.10.4 Higher-Density Lattice Structures

In Section 7.10.3 we discussed the selection of a closely packed subset of points from any regular array or lattice. Here we consider the added improvement by starting with the *densest possible lattice* in the space. In a two-dimensional signal space, the densest lattice is the hexagonal lattice (try penny packing). The result of employing a hexagonal lattice instead of a rectangular one, such as those shown in Figure 7.16, can be a 0.6-dB savings in average energy. Figure 7.18 illustrates some examples of hexagonal packing. The strange-looking $k = 4$ constellation in Figure 7.18a was discovered by Foschini et al. [17] and is stilll the best 16-ary

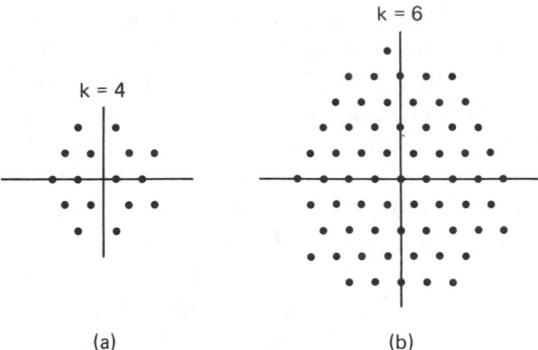

Figure 7.18 Examples of *M*-ary constellations using a hexagonal array.

constellation known. The $k = 6$ constellation in Figure 7.18b is used in the Codex SP14.4 modem.

The hexagonal lattice is optimum for two dimensions. For higher dimensions there are other lattice structures that provide the densest packing. Table 7.3 gives the gain over the rectangular lattice, in decibels, due to the densest packings currently known for various dimensions.

TABLE 7.3 Energy Savings from Dense Lattices versus the Rectangular Lattice

Dimensions (N)	Dense lattice gain (dB)
2	0.62
4	1.51
8	3.01
16	4.52
24	6.02
32	6.02
48	7.78
64	8.09

Source: G. D. Forney, Jr., et al., "Efficient Modulation for Bandlimited Channels," *IEEE J. Sel. Areas Commun.*, vol. SAC2, no. 5, September 1984, pp. 632–647.

7.10.5 Combined Gain: *N*-Sphere Mapping and Dense Lattice

It is possible to combine the benefits of the Campopiano–Glazer boundary construction in N dimensions with the gain from the densest lattice in N-space. The resulting gain is a combination of N-sphere versus N-cube boundary gain of Table 7.2 and the lattice packing density gain of Table 7.3. The combined energy savings are shown in Table 7.4.

TABLE 7.4 Combined Energy Savings
from *N*-Sphere Mapping and Dense Lattices

Dimensions (*N*)	Combined savings gain (dB)
2	0.82
4	1.96
8	3.74
16	5.50
24	7.12
32	7.19
48	9.04
64	9.40

Source: G. D. Forney, Jr., et al., "Efficient Modulation for Bandlimited Channels," *IEEE J. Sel. Areas Commun.*, vol. SAC2, no. 5, September 1984, pp. 632–647.

7.10.6 Trellis-Coded Modulation

The codes described in Chapters 5 and 6 achieve an improvement in bit error probability (P_B) by *bandwidth expansion*. In the case of both block codes and convolutional codes, bandwidth is increased by replacing each *k*-tuple message with an *n*-tuple codeword, where $n > k$. In the case of bandlimited channels, *bandwidth expansion is not possible*. In the past, therefore, coding has never been popular for bandlimited channels such as telephone channels. Recently, however, there has been increasing interest in some types of combined modulation and coding schemes, called *trellis-coded modulation,* that achieve coding gain without any bandwidth expansion. At first it may seem that this statement violates some basic power–bandwidth–error probability trade-off principle. However, there is still a trade-off at work; trellis-coded modulation achieves coding gain at the expense of *decoder complexity*.

Trellis-coded modulation combines a *multilevel/phase* modulation signaling set with a state-oriented trellis coding scheme. Multilevel/phase signal sets are signal constellations having multiple amplitudes, multiple phases, or a combination of multiple amplitudes and multiple phases. A trellis code is one that can be characterized with a trellis diagram. The convolutional codes described in Chapter 6 are linear trellis codes, but trellis codes are *not constrained to be linear*. Coding gains can be realized with block codes or trellis codes, but we shall consider only trellis codes because the availability of the Viterbi decoding algorithm makes trellis decoding attractive. Coding for bandlimited channels still requires controlled introduction of redundancy. However, in this case, the redundancy is due to an increased signal alphabet, achieved through multilevel/phase signaling, so that channel bandwidth is not increased. Ungerboeck [22] investigated the design of multilevel/phase trellis codes that provide *coding gain without band-*

width expansion. He showed that in the presence of AWGN, net coding gains of 3 to 6 dB, relative to the uncoded case, could be achieved.

7.10.6.1 The Idea behind Trellis-Coded Modulation

The error performance of an uncoded nonorthogonal *M*-ary modulation (such as PAM, PSK, or QAM) depends on the distance between the closest pair of signal points. This minimum distance is determined by the average transmitter power and the number and position of the signal points. For a constant average power, the minimum distance between points decreases as the number of points increases. Therefore, assuming a constant channel symbol rate and constant average power, the error performance degrades for systems that attempt to increase the transmission bit rate by increasing the size of the symbol set. The objective of trellis coding is to increase the minimum distance between the signals that are the *most likely to be confused,* without increasing the average power.

Trellis coding may be implemented with a convolutional encoder (see Chapter 6) wherein *k* current bits and $K - 1$ prior bits are used to produce $n = k + p$ coded bits, where *K* is the encoder constraint length and where *p* is the number of parity bits. The $n = k + p$ coded bits require 2^n binary channel symbols for transmission. Notice that encoding increases the signal set size from 2^k to 2^{k+p}. Figure 7.19a illustrates an uncoded 4-ary PAM signal set, before and after being rate $\frac{2}{3}$ encoded into an 8-ary PAM signal set. Similarly, Figure 7.19b illustrates an uncoded 4-ary PSK (QPSK) signal set before and after being rate $\frac{2}{3}$ encoded into an 8-ary PSK signal set. Similarly, Figure 7.19c illustrates an uncoded 16-

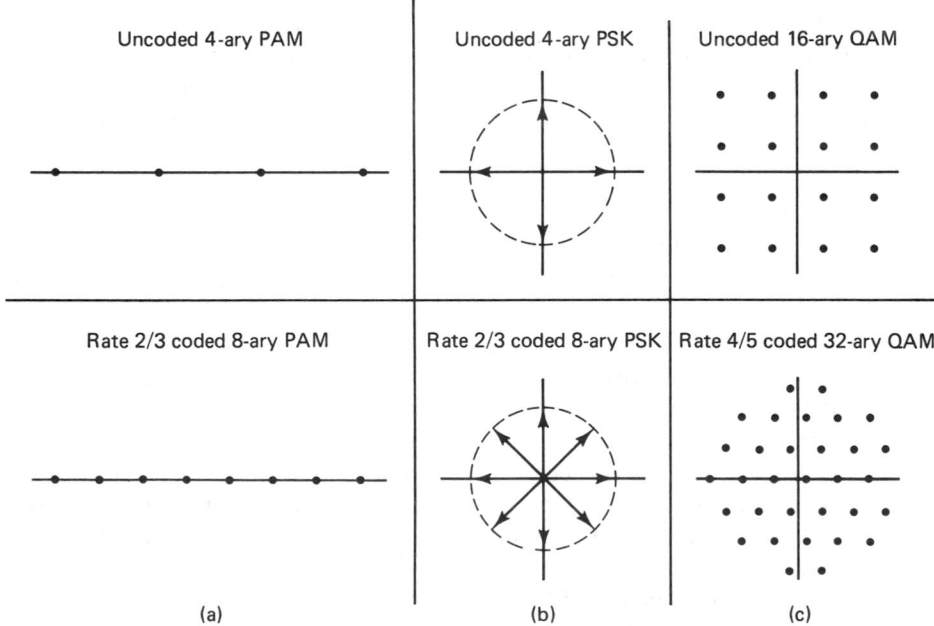

Figure 7.19 Increase of signal set size for trellis-coded modulation.

ary QAM signal set before and after being rate $\frac{4}{5}$ encoded into a 32-ary QAM signal set. In each of these three cases the system is configured to use the same average signal power before and after coding. The examples in Figure 7.19 illustrate the basic idea behind trellis-coded modulation. In each case the symbol set size is increased from 2^k to 2^{k+1} (there are twice as many coded symbols as uncoded ones) to provide the needed coding redundancy; however, in each case, the increase in the number of signals *does not* result in an increase in required bandwidth. The expanded signal set does result in a *reduced distance* between adjacent symbol points for a given average power. However, because of the redundancy introduced by the code, this reduced distance no longer determines the error performance. Instead, the *free distance* (see Section 6.4.1), which is the minimum distance between members of the set of *allowable code symbol sequences,* determines the error performance. Ungerboeck [22] investigated the increase in channel capacity achievable by signal set expansion and concluded that by *doubling* the number of channel signals ($p = 1$), it is possible to gain almost all the channel capacity that can be gained. This can be accomplished by encoding with a rate $k/(k + 1)$ code, and subsequently mapping groups of $k + 1$ bits into the larger set of 2^{k+1} channel symbols.

7.10.6.2 An Error Event

Figure 7.20 illustrates an error event in a trellis code; that is, the figure illustrates a transmitted sequence marked $\mathbf{U} = \ldots, U_1, U_2, U_3, \ldots$ and an alternative sequence marked $\mathbf{V} = \ldots, V_1, V_2, V_3, \ldots$. The alternative sequence is

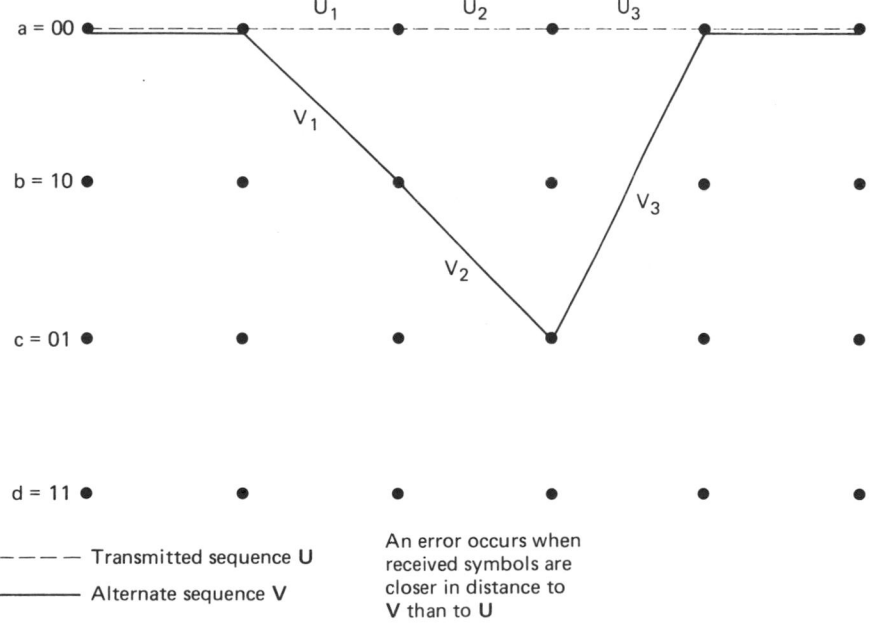

An error occurs when received symbols are closer in distance to V than to U

– – – – Transmitted sequence U

———— Alternate sequence V

Figure 7.20 Illustration of an error event.

seen to diverge and then remerge with the transmitted sequence. Assuming soft-decision decoding, an *error event occurs* whenever the received symbols are closer in Euclidean distance to some alternative sequence, **V**, than to the actually transmitted sequence, **U**. Thus the separation of **U** and **V** describes an error event. This implies that codes for multilevel/phase signals should be designed to achieve *maximum free Euclidean distance* rather than maximum free Hamming distance; for soft-decision decoding, the larger the Euclidean distance, the lower the probability of error. Therefore, assigning signal points to the coded bits in a way that maximizes Euclidean distance is the key to optimizing the trellis codes. Ungerboeck [22–24] investigated this bit-to-symbol mapping problem and devised an assignment procedure, called the *method of set partitioning,* which will always provide coding gain, given an adequate choice of trellis states. The rules for this bit-to-symbol mapping are based on the method of set partitioning, which can be summarized as follows:

1. All parallel transitions in the trellis structure are separated by the maximum possible Euclidean distance. *Parallel transitions* refer to the branch words resulting from the transmission of uncoded bits together with coded bits (see the example in the following section). The reasoning behind this is based on the fact that parallel transitions imply that single signal-error events can occur. This limits the achievable free Euclidean distance to the minimum distance in the subsets of signals assigned to parallel transitions.
2. All transitions diverging from or merging into a trellis state are assigned the next maximum possible Euclidean distance separation.

   In summary, trellis coding for bandlimited channels employs larger signal alphabets achieved through multilevel/phase signaling, such that channel bandwidth is not increased (e.g., M-ary PAM, MPSK, or QAM). Even though the increase in signal set size *reduces* the minimum distance between symbols, the free Euclidean distance between trellis code sequences *more than compensates* for the signal points being crowded together. The result is a net error-performance gain of 3 to 6 dB without any bandwidth expansion [22]. We illustrate these ideas by considering a rate $\frac{2}{3}$ convolutional encoder in the following section.

7.10.7 Trellis-Coding Example

A rate $\frac{2}{3}$ convolutional encoder with constraint length $K = 3$ is shown in Figure 7.21. The rate $\frac{2}{3}$ encoding is accomplished by transmitting one bit from each pair of bits in the input sequence unmodified, and encoding the other bit into two channel bits using a rate $\frac{1}{2}$ encoder. The resulting trellis diagram is shown in Figure 7.22, where the parallel transitions are due to the uncoded bit m_1 shown as the leftmost bit on each trellis branch. The two upper branches emerging from each state represent transitions due to $m_1 m_2$ being 00 and 10, respectively; the two lower branches represent transitions due to $m_1 m_2$ being 01 and 11, respectively. The Viterbi decoding technique for finding the maximum likelihood path through the trellis proceeds in exactly the same way as in the example of Section 6.3.4.

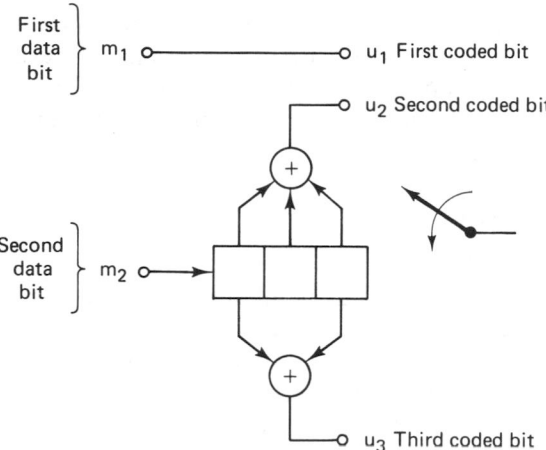

Figure 7.21 Rate $\frac{2}{3}$ convolutional encoder.

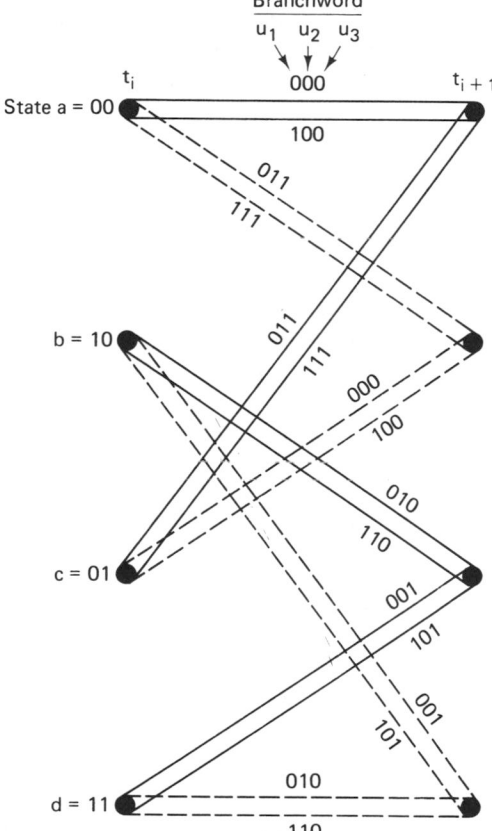

Figure 7.22 Trellis diagram (rate $\frac{2}{3}$ code).

The only (operational) differences are: (1) In this rate $\frac{2}{3}$ example, there are twice as many branches to consider than in the ordinary $K = 3$ convolutional code; and (2) in choosing the two decoded bits for a surviving branch, the first decoded bit of the pair is the same as the first bit u_1 of that branch word since u_1 is the same as the *uncoded* bit m_1. The second decoded bit of the pair corresponds to the input bit m_2 that produced the state transition of the branch being decoded. In Figure 7.22 a branch having a solid line corresponds to $m_2 = 0$, and a branch having a dashed line corresponds to $m_2 = 1$.

7.10.7.1 Coding Gain for Trellis Coding

Consider the coding gain of the rate $\frac{2}{3}$ trellis-coding example described in the preceding section. Let us assume a simple one-dimensional signal space with multilevel pulse amplitude modulation (PAM), as shown in Figure 7.23. In Figure 7.23a is an 8-ary PAM signal set. Since soft decisions are assumed, the appropriate distance metric is the Euclidean distance. The Euclidean distance of each signal, from the center of the signal space, is shown in arbitrary units. Also shown in Figure 7.23a is the bit-to-symbol assignment according to the set partitioning rules outlined earlier. Notice the adherence to these rules, by comparing Figure 7.23a with Figure 7.22. All parallel transitions are separated by a distance of eight units, and all branches diverging from a given state are separated by at least four units.

The average signal power, S_{av}, is computed as follows:

$$S_{av} = \frac{d_1^2 + d_2^2 + \cdots + d_M^2}{M} \tag{7.25}$$

where d_i is the Euclidean distance of the ith signal from the center of the space, and M is the number of codeword symbols in the set. For the signal set shown in Figure 7.23a, where $M = 8$, use of Equation (7.25) yields $S_{av} = 21$. Figure 7.23b illustrates a 4-ary PAM signal set which is the uncoded equivalent of the rate $\frac{2}{3}$ codeword set; the Euclidean distances have been chosen to yield the same average signal power as in the coded case in Figure 7.23a.

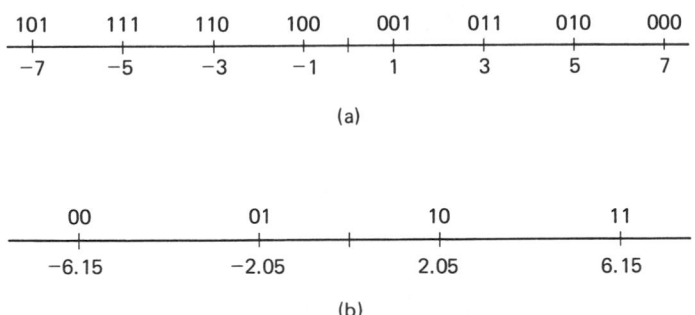

Figure 7.23 8-ary and 4-ary PAM signal sets.

Figure 7.24 illustrates the minimum distance error event for the rate $\frac{2}{3}$ encoder shown in Figure 7.21. The transmitted sequence is assumed to correspond to the all-zeros path. Each of the branch words on this path has a Euclidean distance of 7 units from the center of the space. The error event diverges from the all-zeros path by first transitioning to state 10, then state 01, and finally re-

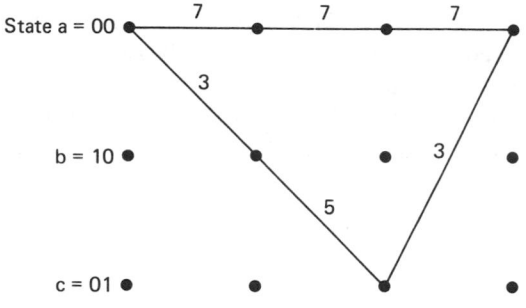

State a = 00

7   7   7

3

b = 10

3

5

c = 01

Figure 7.24 Minimum distance error event for rate $\frac{2}{3}$ convolutional code.

d = 11

merging to state 00. On each branch of the error event is written its Euclidean distance from the center of the space, assuming the uncoded bit $m_1 = 0$; this assumption assures that the distance between the all-zeros transmission and the error event is minimum. To verify the Euclidean distances for the error event in Figure 7.24, first note the state from which and to which each branch transitions. From Figure 7.22 determine the branch word sequence for each transition (with $m_1 = 0$), and then from Figure 7.23a determine the Euclidean distance from the center of the space for each branch in the error event.

Signal amplitude corresponds to the distance of the signal point from the center of the signal space; signal power corresponds to the square of this distance. In comparing the relative performance of the uncoded 4-ary PAM modulation with the trellis-coded 8-ary modulation, it is therefore appropriate to compare the square of the minimum distance d_{min}^2 for an error event in each system, given that the average power is the same in both cases. In general, allowing the average signal power to be different in each case, we can solve for the coding gain G, as follows [25]:

$$G = \frac{(d_{min}^2/S_{av})_{coded}}{(d_{min}^2/S_{av})_{uncoded}} \tag{7.26}$$

For the rate $\frac{2}{3}$ trellis-coding example, the d_{min}^2 value for each error event is calculated as follows:

$$(d_{min}^2)_{coded} = (7 - 3)^2 + (7 - 5)^2 + (7 - 3)^2 = 36$$

$$(d_{min}^2)_{uncoded} = (6.15 - 2.05)^2 = 16.8$$

Then the coding gain is calculated by using Equation (7.26) as follows:

$$G = \frac{36/21}{16.8/21} = 2.14$$

or, in decibels = 3.31 dB

Therefore, even for this simple $K = 3$ example, a significant amount of coding gain has been provided *without any bandwidth expansion*. Larger coding gains can be achieved with an increased number of trellis states (larger constraint length) at the expense of increased decoding complexity. Table 7.5 lists the average power coding gain as a function of number of trellis states [26] for the rate $\frac{2}{3}$ coded 8-ary PAM example discussed here. Gain is computed relative to an uncoded 4-ary PAM signal set.

TABLE 7.5 Coding Gain Obtained for 8-ary PAM with Rate 2/3 Trellis Coding

Number of trellis states	Constraint length (K)	Average power gain (dB)
4	3	3.31
8	4	3.77
16	5	4.18
32	6	4.56
64	7	5.23
128	8	5.23
256	9	5.83

Source: G. C. Clark, Jr. and J. B. Cain, *Error Correction Coding for Digital Communications*, Plenum Press, New York, 1981, p. 388.

At the transmitter, there is only a slight increase in complexity due to the trellis coding. However, the decoding problem at the receiver is made much more complex [12], so that the trade-off consists of evaluating the *coding gain versus the decoding complexity*. The increased availability of large-scale integrated (LSI) circuits and very high speed integrated circuits (VHSIC) can ameliorate this problem and make these coding techniques extremely attractive for achieving coding gain for bandlimited channels. Before concluding, we point out that further coding gain, without bandwidth expansion, is possible by introducing asymmetry into the signal point constellation [27–28].

7.11 CONCLUSION

In this chapter we have integrated some of the ideas in Chapters 3, 5, and 6 dealing with modulation and coding. We have reviewed the basic system design goals: to maximize data rate while simultaneously minimizing error probability, bandwidth, E_b/N_0, and complexity. We examined the trade-offs heuristically on two performance planes: the error probability plane and the bandwidth efficiency plane. The former explicitly illustrates the P_B versus E_b/N_0 trade-offs while only implicitly displaying the bandwidth expenditure. The latter explicitly illustrates the R/W versus E_b/N_0 trade-offs while only implicitly displaying the P_B performance. We discussed some of the basic constraints to improvement without limit. The Nyquist criterion establishes that we cannot continue to reduce system bandwidth indefinitely. There is a theoretical limitation; in order to transmit R_s symbols/second without intersymbol interference, we must utilize a minimum of $R_s/2$ hertz of bandwidth. The Shannon–Hartley theorem relates to the power–bandwidth trade-off and results in another important limitation, the Shannon limit. The Shannon limit of -1.59 dB is the minimum amount of E_b/N_0 that is necessary (in concert with channel coding) to achieve an arbitrarily low error probability over an AWGN channel. The more general limitation is the channel capacity, above which there cannot be error-free signaling. We have also examined some of the bandwidth-efficient modulation schemes, such as minimum shift keying (MSK), quadrature amplitude modulation (QAM), and trellis-coded modulation. The latter technique offers an attractive way to obtain coding gain without paying the price of additional bandwidth.

REFERENCES

1. Nyquist, H., "Certain Topics on Telegraph Transmission Theory," *Trans. Am. Inst. Electr. Eng.*, vol. 47, Apr. 1928, pp. 617–644.

2. Shannon, C. E., "A Mathematical Theory of Communication," *Bell Syst. Tech. J.*, vol. 27, 1948, pp. 379–423, 623–657.

3. Shannon, C. E., "Communication in the Presence of Noise," *Proc. IRE,* vol. 37, no. 1, Jan. 1949, pp. 10–21.

4. Bedrosian, E., "Spectrum Conservation by Efficient Channel Utilization," *Rand Corp., Rep. WN-9275-ARPA,* Contract DAHC-15-73-C-0181, Santa Monica, Calif., Oct. 1975.

5. Odenwalder, J. P., *Error Control Coding Handbook,* Linkabit Corporation, San Diego, Calif., July 15, 1977.

6. Smith, J. G., "Spectrally Efficient Modulation," *Proc. IEEE Int. Conv. Commun. (ICC '77),* June 1977, pp. 3.1-37–3.1-41.

7. Pasupathy, S., "Minimum Shift Keying: A Spectrally Efficient Modulation," *IEEE Commun. Mag.,* July 1979, pp. 14–22.

8. Gronemeyer, S. A., and McBride, A. L., "MSK and Offset QPSK Modulation," *IEEE Trans. Commun.,* vol. COM-24, Aug. 1976, pp. 809–820.

9. M. K. Simon, "A Generalization of Minimum Shift Keying (MSK) Type Signaling Based upon Input Data Symbol Pulse Shaping," *IEEE Trans. Commun.,* vol. COM24, Aug. 1976, pp. 845–857.

10. Korn, I., *Digital Communications,* Van Nostrand Reinhold Company, Inc., New York, 1985.

11. Oetting, J. D., "A Comparison of Modulation Techniques for Digital Radio," *IEEE Trans. Commun.,* vol. COM27, no. 12, Dec. 1979, pp. 1752–1762.

12. Forney, G. D., Jr. et al., "Efficient Modulation for Bandlimited Channels," *IEEE J. Sel. Areas Commun.,* vol. SAC2, no. 5, Sept. 1984, pp. 632–647.

13. Thomas, C. M., Weidner, M. Y., and Durrani, S. H., "Digital Amplitude-Phase Keying with *M*-ary Alphabets," *IEEE Trans. Commun.,* vol. COM22, no. 2, Feb. 1974, pp. 168–180.

14. Lucky, R. W., and Hancock, J. C., "On the Optimum Performance of *N*-ary Systems Having Two Degrees of Freedom," *IRE Trans. Commun. Syst.,* vol. CS10, June 1962, pp. 185–192.

15. Campopiano, C. N., and Glazer, B. G., "A Coherent Digital Amplitude and Phase Modulation Scheme," *IRE Trans. Commun. Syst.,* vol. CS10, June 1962, pp. 90–95.

16. Cahn, C. R., "Combined Digital Phase and Amplitude Modulation Communication Systems," *IRE Trans. Commun. Technol.*, Sept. 1960.

17. Foschini, G. J., Gitlin, R. D., and Weinstein, S. B., "Optimization of Two Dimensional Signal Constellations in the Presence of Gaussian Noise," *IEEE Trans. Commun.*, vol. COM22, no. 1, Jan. 1974, pp. 28–38.

18. Welti, G. R., and Jhong, S. L., "Digital Transmission with Coherent Four-Dimensional Modulation," *IEEE Trans. Inf. Theory*, vol. IT20, no. 4, July 1974, pp. 497–502.

19. Gersho, A., and Lawrence, V. B., "Multidimensional Signal Constellations for Voiceband Data Transmission," *IEEE J. Sel. Areas Commun.*, vol. SAC2, no. 5, Sept. 1984, pp. 687–702.

20. Zetterberg, L. H., and Brandstrom, H., "Codes for Combined Phase and Amplitude Modulated Signals in a Four-Dimensional Space," *IEEE Trans. Commun.*, vol. COM25, no. 9, Sept. 1977, pp. 943–950.

21. Wilson, S. G., Sleeper, H. A., and Srinath, N. K., "Four-Dimensional Modulation and Coding: An Alternative to Frequency Reuse," *IEEE 1984 Intl. Commun. Conf.*, pp. 919–923.

22. Ungerboeck, G., "Channel Coding with Multilevel/Phase Signals," *IEEE Trans. Inf. Theory*, vol. IT28, Jan. 1982, pp. 55–67.

23. Ungerboeck, G., "Trellis-Coded Modulation with Redundant Signal Sets, Part I. Introduction," *IEEE Commun. Mag.*, vol. 25, no. 2, Feb. 1987, pp. 5–11.

24. Ungerboeck, G., "Trellis-Coded Modulation with Redundant Signal Sets, Part II; State of the Art," *IEEE Commun. Mag.*, vol. 25, no. 2, Feb. 1987, pp. 12–21.

25. Thapar, H. K., "Real-Time Application of Trellis Coding to High-Speed Voiceband Data Transmission," *IEEE J. Sel. Areas Commun.*, vol. SAC2, no. 5, Sept. 1984, pp. 648–658.

26. Clark, G. C., Jr., and Cain, J. B., *Error Correction Coding for Digital Communications*, Plenum Press, New York, 1981.

27. Divsalar, D., and Yuen, J. H., "Asymmetric MPSK for Trellis Codes," *GLOBECOM '84*, Nov. 26–29, 1984.

28. Divsalar, D., Simon, M. K., and Yuen, J. H., "Trellis Coding with Asymmetric Modulations," *IEEE Trans. Commun.*, vol. COM35, no. 2, Feb. 1987.

PROBLEMS

7.1. Consider a voice-grade telephone circuit with a bandwidth of 3 kHz. Assume that the circuit can be modeled as an AWGN channel.
   (a) What is the capacity of such a circuit if the SNR is 30 dB?
   (b) What is the minimum SNR required for a data rate of 4800 bits/s on such a voice-grade circuit?
   (c) Repeat part (b) for a data rate of 19,200 bits/s.

7.2. Consider that a 100-kbits/s data stream is to be transmitted on a voice-grade telephone circuit (with a bandwidth of 3 kHz). Is it possible to achieve error-free transmission with a SNR of 10 dB? Justify your answer. If it is not possible, suggest system modifications that might be made.

7.3. Consider a source that produces six messages with probabilities $\frac{1}{2}$, $\frac{1}{4}$, $\frac{1}{8}$, $\frac{1}{16}$, $\frac{1}{32}$, and $\frac{1}{32}$. Determine the average information content in bits, of a message.

7.4. A given source alphabet consists of 300 words, of which 15 occur with probability 0.06 each and the remaining 285 words occur with probability 0.00035 each. If 1000 words are transmitted each second, what is the average rate of information transmission?

7.5. (a) Find the average capacity in bits per second that would be required to transmit a high-resolution black-and-white TV signal at the rate of 32 pictures per second if each picture is made up of 2×10^6 picture elements and 16 different brightness levels. All picture elements are assumed to be independent and all levels have equal likelihood of occurrence.

(b) For color TV, this system additionally provides for 64 different shades of color. How much more system capacity is required for a color system compared to the black and white system?

(c) Find the required capacity if 100 of the possible brightness–color combinations occur with a probability of 0.003 each, 300 of the combinations occur with a probability of 0.001, and 624 of the combinations occur with a probability of 0.00064.

7.6. Prove that entropy is maximized when all source outputs have equal probability.

7.7. Compute the equivocation or message uncertainty in bits per character for a textual transmission using 7-bit ASCII coding. Assume that each character is equally likely and that the noise on the channel results in a bit error probability of 0.01.

7.8. Suppose a binary noncoherent FSK link has a maximum data rate of 2.4 kbits/s without ISI over a channel whose nominal bandwidth is 2.4 kHz. Suggest ways of increasing the data rate under the following system constraints.

(a) The system is power limited.

(b) The system is bandwidth limited.

(c) The system is both power and bandwidth limited.

7.9. Table P7.1 characterizes four different satellite-to-earth-terminal links. For each link assume that the space loss is 196 dB, the margin is 0 dB, and there are no other incidental losses. For each link, plot an operating point on the bandwidth efficiency plane, R/W versus E_b/N_0, and characterize the link according to one of the following descriptions: bandwidth limited, severely bandwidth limited, power limited, and severely power limited. Justify your answers.

7.10. In designing a communication system for a 9600-bits/s data stream with a bit error

TABLE P7.1 Downlink Capacity for Four Satellite Links

Satellite	Receive terminal	Maximum data rate

INTELSAT IV
    EIRP = 22.5 dBW
    Bandwidth = 36 MHz
	Large fixed
	    antenna diameter = 30 m
	    G/T = 40.7 dB/K
		165 Mbits/s

DSCS II
    EIRP = 28 dBW
    Bandwidth = 50 MHz
	Shipboard
	    antenna diameter = 4 ft
	    G/T = 10 dB/K
		100 kbits/s

DSCS II
    EIRP = 28 dBW
    Bandwidth = 50 MHz
	Large fixed
	    antenna diameter = 60 ft
	    G/T = 39 dB/K
		72 Mbits/s

GAPSAT/MARISAT
    EIRP = 28 dBW
    Bandwidth = 500 kHz
	Aircraft
	    antenna gain = 0 dB
	    G/T = −30 dB/K
		500 bits/s

probability of 10^{-5} or better, assume that you are required to choose the modulation, coding, and interleaving from the schemes described below.

8-ary noncoherent FSK

16-ary QAM (matched filter detection)

(127, 92) BCH code, $d_{min} = 11$

Rate $\frac{1}{2}$, feedback-decoded convolutional code, corrects an average of three symbol errors out of a sequence of 21 symbols

Block interleaver (16 × 32)

Convolutional interleaver (150 × 300)

Choose a modulation technique and if deemed necessary, a coding scheme or a coding/interleaving scheme for the following applications. Justify your choices.

(a) Voice-grade telephone channel with 2400 Hz of usable bandwidth and available $E_b/N_0 = 14$ dB.

(b) Satellite channel with 40 kHz of usable bandwidth and with an available E_b/N_0 of 7.3 dB.

(c) Voice-grade link over a bursty noise channel. A noise burst typically lasts for 100 ms. The usable bandwidth is 3400 Hz and the available $E_b/N_0 = 10$ dB.

7.11. (a) For a fixed error probability, show that the relationship between alphabet size, M, and required average power for MPSK versus QAM can be expressed as

$$\frac{\text{average power for MPSK}}{\text{average power for QAM}} \simeq \frac{3M^2}{2(M-1)\pi^2}$$

(b) Discuss the advantage of one type of signaling over the other.

7.12. Telephone modems operating at 19.2 kbits/s are now available using trellis-coded QAM modulation.

(a) Calculate the bandwidth efficiency of such modems, assuming that the usable channel bandwidth is 2400 Hz.

(b) Assuming AWGN and an available $E_b/N_0 = 10$ dB, calculate the theoretically available capacity in the 2400-Hz bandwidth.

(c) What is the required E_b/N_0 that will enable a 2400-Hz bandwidth to have a capacity of 19.2 kbits/s?

7.13. Figure 7.15 shows several 16-ary symbol constellations.

(a) For the (5, 11) circular constellations, compute the minimum radial distances r_1 and r_2 if the minimum distance between each symbol must be 1 unit.

(b) Compute the average signal power for the (5, 11) circular constellation, and compare it to the average signal power for the 4 × 4 ($M = 16$) square constellation (with the same minimum distance between symbols).

(c) Why might the square constellation be more practical?

7.14. Consider that the rate $\frac{2}{3}$ trellis-coded system of Section 7.10.7 is used over a binary symmetric channel (BSC). Assume that the initial encoder state is the 00 state. At the output of the BSC, the sequence **Z** = (1 1 1 0 0 1 1 0 1 0 1 1 rest all "0") is received.

(a) Find the maximum likelihood path through the trellis diagram and determine the first 6 decoded information bits. If a tie occurs between any two merged paths, choose the upper branch entering the particular state.

(b) Determine if any channel bits in **Z** had been inverted by the channel during transmission, and if so, identify them.

(c) Explain how you would proceed with the problem if the channel were specified as a Gaussian channel instead of a BSC.

CHAPTER 8

Synchronization

Maurice A. King, Jr.
The Aerospace Corporation
El Segundo, California

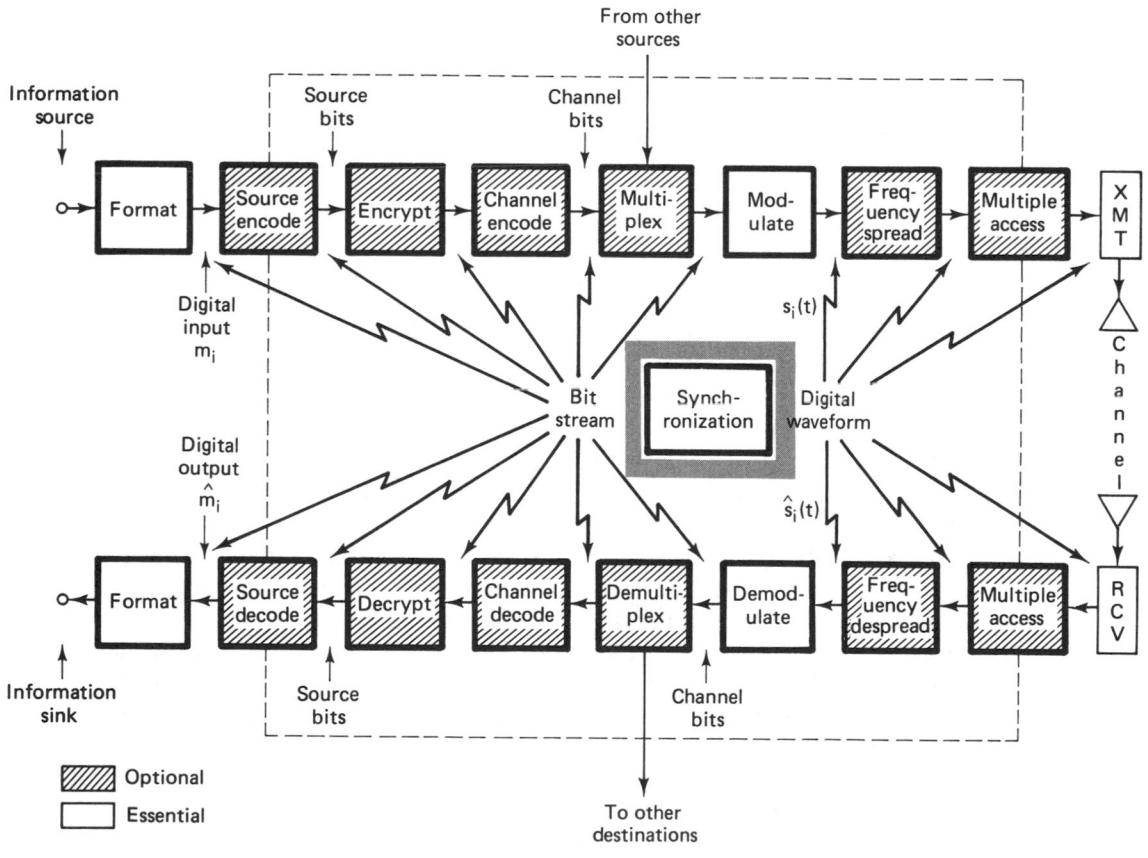

8.1 SYNCHRONIZATION IN THE CONTEXT OF DIGITAL COMMUNICATIONS

8.1.1 What It Means to Be Synchronized

In most of this book some level of signal synchronization is assumed, although perhaps not explicitly. In the discussions of demodulation and channel decoding it is assumed that many things about the incoming signal are accurately known. For example, in the case of coherent phase modulation (PSK), the receiver is assumed to be able to generate a set of reference signals whose phases are identical (except perhaps for a constant offset) to the phases of the signaling alphabet in use at the transmitter. These reference signals are compared to the incoming signals in the process of making maximum likelihood symbol decisions.

In order to be able to generate these reference signals, the receiver has to be in synchronization with the received carrier. This means that there has to be phase concurrence between the incoming carrier sinusoid and a replica of it in the receiver. In other words, if there were no information modulated on the incoming carrier, the incoming sinusoid and the replica in the receiver would pass through zero simultaneously. This is what is known as being in *phase lock* and is a condition that must be closely approximated if coherently modulated signals are going to be accurately demodulated at the receiver. If the information-bearing signal is not modulated directly on the carrier but indirectly through the use of a subcarrier, both the phase of the carrier and that of the subcarrier must be determined. If the carrier and subcarrier are not kept in phase synchronism by the transmitter (they typically are not), this will require the generation of a replica of the subcarrier by the receiver, where the phase of the subcarrier replica is controlled separately from that of the carrier replica. This will enable the receiver to achieve phase lock on both the carrier and subcarrier.

It is also assumed that the receiver has accurate knowledge of when an incoming symbol started and when it is over. This knowledge is required in order to know the proper symbol integration interval—the interval over which energy is integrated prior to making symbol decisions. Clearly, if the receiver integrates over an interval of an inappropriate length, or over an interval that spans two symbols, the ability to make accurate symbol decisions will be degraded.

It can be seen that symbol synchronization and phase synchronization are similar in that they both involve producing in the receiver a replica of a portion of the transmitted signal. For phase synchronization it was the sinusoidal carrier. For symbol synchronization the replica is a square wave at the symbol transition rate. The receiver must be able to produce a square wave that will transition through zero simultaneously with the incoming signal's transitions between symbols. A receiver that is able to do this can be said to have symbol synchronization, or be in *symbol lock*. Since there are typically a very large number of carrier cycles per symbol period, this second level of synchronization is much coarser than phase synchronization and is usually done with different circuitry than that used for phase synchronization.

In many communication systems an even higher level of synchronization is required. This is usually called *frame synchronization*. Frame synchronization is required when the information is organized in blocks, or messages of some uniform number of symbols. This will occur, for example, if a block code is used for forward error control, or if the communications channel is being time shared, on a regular basis, by several users (TDMA). In the case of block coding, the decoder needs to know where the boundaries between codewords are in order to decode the message correctly. In the case of a time-shared channel, it is necessary to know where the boundaries between channel users are, in order to know where the information is coming from and to where it should be routed. Similar to symbol synchronization, frame synchronization is equivalent to being able to generate a square wave at the frame rate, with the zero crossings coincident with the transitions from one frame to the next.

Most digital communications systems using coherent modulation require all three of these levels of synchronization: phase, symbol, and frame. Systems using noncoherent modulation techniques will typically require symbol and frame synchronization, but since the modulation is not coherent, phase lock is not required. Instead, noncoherent systems require *frequency synchronization*. Frequency synchronization differs from phase synchronization in that with frequency synchronization, the replica of the carrier that is generated by the receiver is allowed to have an arbitrary constant phase offset from the received carrier. Receiver designs can be simplified by removing the requirement to determine the exact value of the incoming carrier phase. Unfortunately, as is shown in the discussion of modulation techniques, this simplification carries a penalty in terms of degraded performance versus signal-to-noise ratio. The relative trade-offs of synchronization levels versus performance and system versatility are discussed further in the next section.

All of the discussion thus far has been oriented toward the receiving end of a communications link. There are instances, however, when the transmitter assumes the more active role in synchronization, by varying the timing and frequency of its transmissions to correspond to the expectations of the receiver. An example of this situation is a satellite communication network, where many terrestrial terminals are beaming signals toward a single satellite receiver. In most of these cases the transmitter relies on a return path from the receiver to determine the accuracy of its synchronization. Thus transmitter synchronization often implies two-way communications, or a network, in order to be successful. Thus transmitter synchronization is often called *network synchronization*. Transmitter or network synchronization is discussed later in this chapter.

8.1.2 Costs versus Benefits of Synchronization Levels

The need for a receiver to be synchronized with a communication system implies costs to the system. Each level of synchronization that is added implies more cost. The most obvious cost is in the need for additional hardware or software in the receiver for acquisition and tracking loops. Possibly less obvious costs lie in the extra time required to achieve synchronization before commencing communications, or in the energy expended by the transmitter on signals to be used at the receiver as acquisition or tracking aids. In the face of these costs to the system, one might question why a communications system designer would consider a system design requiring a high degree of synchronization. The answers are improved performance and versatility.

Consider a standard commercial AM radio. This radio may be considered part of a broadcast communication system involving a central transmitter and many receivers. This communication system involves no synchronization. However, the receiver passband must be wide enough to accommodate not only the information-bearing signal, but also any fluctuations in the carrier, due perhaps to Doppler shift* or drift in the transmitter's frequency reference. This requirement on the receiver passband means that additional noise energy is passed to the detector, over and above the amount theoretically required by the bandwidth of the information. A somewhat more complicated receiver that includes a carrier-frequency tracking loop would be able to keep a narrow passband filter centered about the carrier, thereby substantiallly reducing the detected noise energy and improving the received signal-to-noise ratio. Thus although a standard radio may be perfectly adequate for reception of signals from large transmitters a few tens of kilometers distant, it may prove totally inadequate under less benign conditions.

---

* An offset in frequency as perceived by a receiver, from the nominally transmitted frequency, caused by relative motion of the transmitter and receiver. Ignoring second- and higher-order effects, the value of the frequency offset, Δf, is given by $v f_0/c$, where v is the relative velocity (positive when the relative distance between transmitter and receiver is being reduced), f_0 the nominal frequency, and c the speed of light.

For digital communications, another example of a trade-off between performance and receiver complexity is in the choice of modulation. Among the simplest digital receivers are those designed to be used with noncoherently detected binary FSK. The only synchronization requirements are bit timing and frequency tracking. However, the same bit error probability could be achieved with approximately 4 dB less signal-to-noise ratio if the modulation is coherent BPSK. The disadvantage of BPSK is that the receiver requires a phase tracking loop, which can be a complex design problem for signals that may experience high Doppler rates* or deep signal fades.†

A third instance of a cost versus performance trade-off involving synchronization is in the use of error control coding. As was established in Chapters 5 and 6, there are substantial performance advantages in the use of appropriate error control coding techniques. The cost, measured in receiver complexity, however, can be high. For a block decoder to operate properly requires the receiver to achieve block, message, or frame synchronism. This is a procedure over and above the usual decoding procedure, although some error-correcting codes have been designed with block synchronization aids built in [1]. Convolutional codes also require some degree of additional synchronization in order to provide optimum performance. Although the performance analysis of convolutional codes often makes the assumption that the input data sequence is infinitely long, in practice it is not. In order to provide the minimum error probability, the decoder must know the beginning state (usually all zeros) when the data sequence will begin, the eventual ending state, and when the ending state is to be reached. Knowing when the beginning state was left and when the ending state is to be reached, however, is equivalent to having frame synchronization. In addition, the decoder will have to know how to group the channel symbols in order to make branch decisions. This is also a synchronization requirement.

The trade-offs discussed thus far have been in terms of the performance versus complexity of individual links and receivers. The ability to synchronize has a large potential consequence in terms of system efficiency and versatility as well. Frame synchronization allows the use of advanced, versatile multiple access techniques, such as the variety of demand assignment multiple access (DAMA) schemes, which are becoming increasingly popular as communication channel resources become increasingly scarce. In addition, the use of spread-spectrum techniques, both as multiple access schemes and for interference rejection, requires a high level of system synchronization. Spread-spectrum techniques are treated in depth in following chapters. It will be seen that these techniques provide the potential for a great deal of system versatility, which is a very valuable feature if the system will encounter changing or unstable conditions, such as the effects of intentional and unintentional interference from external sources.

---

* The rate of change of the Doppler shift. This rate sets requirements on the tracking ability of the phase tracking loop.

† Variations in the received signal strength caused by random variations in channel parameters.

8.2 RECEIVER SYNCHRONIZATION

All digital communication systems require some degree of synchronization to incoming signals by the receivers. In this section the fundamentals of the various levels of receiver synchronization are discussed. The discussion will begin with the basic level of synchronization required for coherent reception—phase synchronization—and a brief discussion of the principles of phase-locked-loop (PLL) operation and design. The discussion will then broaden into the topics of frequency and symbol synchronization. Frequency and symbol synchronization are required for all digital communications reception, either coherent or noncoherent. The final topics in the section are receiver frame synchronization and techniques for achieving and maintaining it.

8.2.1 Coherent Systems: Phase-Locked Loops

At the heart of all phase synchronization circuits is some version of a phase-locked loop (PLL). A schematic diagram of the basic PLL is given in Figure 8.1. Phase-locked loops are servo-control loops, whose controlled parameter is the phase of a locally generated replica of the incoming carrier signal. Phase-locked loops have three basic components: a phase detector, a loop filter, and a voltage-controlled oscillator (VCO). The phase detector is a device that produces a measure of the difference in phase between an incoming signal and the local replica. As the incoming signal and the local replica change with respect to each other, the phase difference (or phase error) becomes a time-varying signal into the loop filter. The loop filter governs the PLL's response to these variations in the error signal. A well-designed loop should be able to track changes in the incoming signal's phase but not be overly responsive to receiver noise. The VCO is the device that produces the carrier replica. The VCO, as the name implies, is a sinusoidal oscillator whose frequency is controlled by a voltage level at the device input. In Figure 8.1 the phase detector is shown as a multiplier, the loop filter is described by its impulse response function, $f(t)$, with Fourier transform, $F(\omega)$, and the VCO is so indicated.

A VCO is an oscillator whose output frequency is a linear function of its input voltage over some range of input and output. A positive input voltage will cause the VCO output frequency to be greater than its uncontrolled value, ω_0,

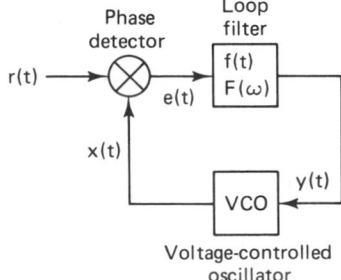

Voltage-controlled oscillator

Figure 8.1 Schematic of the basic phase-locked loop.

while a negative voltage will cause it to be less. Phase lock is achieved by feeding a filtered version of the phase difference (i.e., the phase error) between the incoming signal, $r(t)$, and the output of the VCO, $x(t)$, back to the input of the VCO, $y(t)$.

Consider a normalized input signal of the form

$$r(t) = \sin [\omega_0 t + \theta(t)] \tag{8.1}$$

where ω_0 is the nominal carrier frequency and $\theta(t)$ is a slowly varying phase. Similarly, consider a normalized VCO output of the form

$$x(t) = 2 \cos [\omega_0 t + \hat{\theta}(t)] \tag{8.2}$$

These signals will produce an output error signal at the phase detector output of the form

$$e(t) = x(t)r(t) = 2 \cos [\omega_0 t + \hat{\theta}(t)] \sin [\omega_0 t + \theta(t)]$$

$$= \sin [\theta(t) - \hat{\theta}(t)] + \sin [2\omega_0 t + \theta(t) + \hat{\theta}(t)] \tag{8.3}$$

Assuming that the loop filter is low pass, the second term on the right hand side of Equation (8.3) will be filtered out and can be ignored. This low-pass assumption is a reasonable loop design decision. A low-pass filter provides an error signal that is solely a function of the difference in phases between the input [Equation (8.1)] and the VCO output [Equation (8.2)]. This is exactly the error signal that is needed. The VCO output frequency is the time derivative of the argument of the sine function in Equation (8.2). If we make the assumption that ω_0 is the uncontrolled frequency of the VCO (the output frequency when the input voltage is zero), we can express the difference in the VCO output frequency from ω_0 as the time differential of the phase term $\hat{\theta}(t)$. The output frequency of the VCO is a linear function of the input voltage. Therefore, since an input voltage of zero produces an output frequency of ω_0, the difference in the output frequency from ω_0 will be proportional to the value of the input voltage, $y(t)$, or

$$\Delta\omega(t) = \frac{d}{dt} [\hat{\theta}(t)] = K_0 y(t)$$

$$= K_0 e(t) * f(t)$$

$$\simeq K_0 [\theta(t) - \hat{\theta}(t)] * f(t) \tag{8.4}$$

where $\Delta\omega(t)$ denotes the frequency difference, the notation $*$ indicates the convolution operation (see Appendix A), and the small-angle approximation [i.e., $e(t) = \sin [\theta(t) - \hat{\theta}(t)] \simeq \theta(t) - \hat{\theta}(t)]$ has been used in the last line of Equation (8.4). The small-angle approximation will be accurate when the output phase error is small (the loop is close to phase lock). This will be the situation when the loop is operating normally. The factor K_0 is the gain of the VCO, and $f(t)$ is the loop filter impulse response. This linear differential equation in $\hat{\theta}(t)$ (utilizing the small-angle approximation) is known as the linearized loop equation. It is the single most useful relationship in determining loop behavior during normal operation (where the phase error is small).

Example 8.1 Linearized Loop Equation

Show that for appropriately chosen K_0 and $f(t)$ the linearized loop equation [Equation (8.4)] demonstrates a tendency toward phase lock—that is, the phase difference between the incoming signal and the VCO output tends to decrease.

Solution

Consider the case where the phase of the input signal, $\theta(t)$, is slowly varying with time. It can be seen that if the phase difference on the right-hand side of Equation (8.4) is positive [i.e., $\theta(t) > \hat{\theta}(t)$], then by appropriate choice of K_0 and $f(t)$, the time derivative of $\hat{\theta}(t)$ will be positive, so that $\hat{\theta}(t)$ will increase with time, which will tend to reduce the magnitude of the difference $|\theta(t) - \hat{\theta}(t)|$. On the other hand, if the phase difference is negative, $\hat{\theta}(t)$ will decrease with time, which will also reduce the magnitude of the phase difference. Finally, if $\theta(t) = \hat{\theta}(t)$, then Equation (8.4) indicates that $\hat{\theta}(t)$ will not change with time, and the equality will be maintained.

Consider the Fourier transform of Equation (8.4),

$$j\omega\hat{\Theta}(\omega) = K_0[\Theta(\omega) - \hat{\Theta}(\omega)]F(\omega) \tag{8.5}$$

where the capitalized functions of ω are the Fourier transforms of the lowercase functions of t in Equation (8.4). That is, $\hat{\Theta}(\omega) \leftrightarrow \hat{\theta}(t)$, $\Theta(\omega) \leftrightarrow \theta(t)$, and $F(\omega) \leftrightarrow f(t)$. Reorganizing Equation (8.5) provides

$$\frac{\hat{\Theta}(\omega)}{\Theta(\omega)} = \frac{K_0 F(\omega)}{j\omega + K_0 F(\omega)} = H(\omega) \tag{8.6}$$

The term $H(\omega)$ is known as the closed-loop transfer function of the PLL. This term is very useful in characterizing the transient response of a PLL. The order of a PLL is defined to be the order of the highest-order term in $j\omega$ in the denominator of $H(\omega)$. Equation (8.6) indicates that this is always one more than the order of the loop filter, $F(\omega)$. This is because when $F(\omega)$ is expressed analytically as $F(\omega) = N(\omega)/D(\omega)$, the denominator of $H(\omega)$ when expressed as a polynomial in $j\omega$ will have the term $j\omega D(\omega)$, which must have a term in $j\omega$ that is one order higher than the highest-order term in $D(\omega)$ alone. The order of a PLL is critical for determining the loop's steady-state response to a steady-state input. This is discussed in the next section.

8.2.1.1 Steady-State Errors and Loop Filter Characteristics

By reorganizing Equation (8.6) we can obtain an expression for the Fourier transform of the phase error.

$$\begin{aligned}
E(\omega) &= \mathcal{F}\{e(t)\} \\
&= \Theta(\omega) - \hat{\Theta}(\omega) \\
&= [1 - H(\omega)]\Theta(\omega) \\
&= \frac{j\omega\Theta(\omega)}{j\omega + K_0 F(\omega)}
\end{aligned} \tag{8.7}$$

Equation (8.7) can be used in conjunction with the final value theorem of Fourier transforms to determine the steady-state error response of a loop to a variety of possible input characteristics. The steady-state error is the residual error after all transients have died away, and thus provides a measure of a loop's ability to cope with various types of changes in the input. The final value theorem states that

$$\lim_{t \to \infty} e(t) = \lim_{j\omega \to 0} j\omega E(\omega) \qquad (8.8)$$

Combining Equations (8.7) and (8.8) yields

$$\lim_{t \to \infty} e(t) = \lim_{j\omega \to 0} \frac{(j\omega)^2 \Theta(\omega)}{j\omega + K_0 F(\omega)} \qquad (8.9)$$

Example 8.2 Response to a Phase Step

Consider a loop's steady-state response to a phase step at the loop input.

Solution

Assuming that the PLL was originally in phase lock, a phase step will throw the loop out of lock. Having abruptly changed, however, the input phase again becomes stable. This should be the easiest type of phase disturbance for a PLL to deal with. The Fourier transform of a phase step will be taken to be

$$\Theta(\omega) = \mathcal{F}\{\Delta\phi \, u(t)\}$$

$$= \frac{\Delta\phi}{j\omega} \qquad (8.10)$$

where $\Delta\phi$ is the magnitude of the step and $u(t)$ is the unit step function,

$$u(t) = \begin{cases} 1 & \text{for } t > 0 \\ 0 & \text{for } t < 0 \end{cases}$$

$$= \int_{-\infty}^{t} \delta(\tau) \, d\tau$$

where $\delta(\tau)$ is the Dirac delta function. From Equations (8.9) and (8.10),

$$\lim_{t \to \infty} e(t) = \lim_{j\omega \to 0} \frac{j\omega \, \Delta\phi}{j\omega + K_0 F(\omega)} = 0$$

assuming that $F(0) \neq 0$. Thus the loop will eventually track out any phase step that appears at the input if the loop filter has a nonzero dc response. This means that for any loop filter with the property that $F(\omega) = N(\omega)/D(\omega)$ and $N(0) \neq 0$, the PLL will automatically tend to recover phase lock if the input is displaced by a constant phase. This is clearly a very desirable loop characteristic.

Example 8.3 Response to a Frequency Step

Next, consider a loop's steady-state response to a frequency step at the input.

Solution

A frequency step can approximate the effect of a Doppler shift in the incoming signal frequency due to relative motion between the transmitter and the receiver. Thus, this is an important example for systems with mobile terminals. Since phase is the integral of frequency, the input phase will change linearly as a function of time for a constant input frequency offset. The Fourier transform of the phase characteristic will be the transform of the integral of the frequency characteristic. Since the frequency characteristic is a step, and the transform of an integral is the transform of the integrand divided by the parameter $j\omega$,

$$\Theta(\omega) = \frac{\Delta\omega}{(j\omega)^2} \tag{8.11}$$

where $\Delta\omega$ is the magnitude of the frequency step. Substituting Equation (8.11) into Equation (8.9) yields

$$\lim_{t\to\infty} e(t) = \lim_{j\omega\to 0} \frac{\Delta\omega}{j\omega + K_0 F(\omega)} = \frac{\Delta\omega}{K_0 F(0)} \tag{8.12}$$

The steady-state result in this case depends on more properties of the loop filter than merely a nonzero dc response. If the filter is an all-pass,

$$F_{\mathrm{ap}}(\omega) = 1 \tag{8.13}$$

a low-pass,

$$F_{\ell p}(\omega) = \frac{\omega_1}{j\omega + \omega_1} \tag{8.14}$$

or a lead-lag,

$$F_{\ell\ell}(\omega) = \left(\frac{\omega_1}{\omega_2}\right) \frac{j\omega + \omega_2}{j\omega + \omega_1} \tag{8.15}$$

Equation (8.12) indicates that the loop will track the input phase ramp with a constant steady-state error whose value will depend on the gain term, K_0, and the magnitude of the frequency step. Using any of $F_{\mathrm{ap}}(\omega)$, $F_{\ell p}(\omega)$, or $F_{\ell\ell}(\omega)$ for $F(\omega)$ in Equation (8.12) yields

$$\lim_{t\to\infty} e(t) = \frac{\Delta\omega}{K_0}$$

Notice that a product of several filters with filter characteristics of the form of Equation (8.13), (8.14), or (8.15) would still produce this result. This steady-state error, which is called the *velocity error*, will exist regardless of the order of the filter, unless the denominator of $F(\omega)$, contains $j\omega$ as a factor [$\omega_1 = 0$ in the denominator of Equation (8.14) or (8.15) with the appropriate renormalization in the numerators]. Having $j\omega$ as a factor of $D(\omega)$ is equivalent to having a perfect integrator in the loop filter. It is not possible to build a perfect integrator, but one may be closely approximated either digitally or by using active integrated circuits [2]. Thus if the system design requires the tracking of Doppler shifts with zero steady-state error, the loop filter design must contain an approximation to a perfect integrator. It should be noted that even with a nonzero velocity error, the frequency is still being tracked; there are important applications where tracking to zero phase error is not important. Noncoherent signaling, such as the standard use of FSK modulation, is an example. For noncoherent signaling it is actually frequency tracking that is required, and the absolute value of phase is unimportant.

Example 8.4   Response to a Frequency Ramp

Consider a loop's steady-state response when the input frequency is changing linearly with time (a frequency ramp function).

Solution

This example corresponds to the effect of a step change in the time derivative of the input frequency. This would approximate a change in the Doppler rate, which could model acceleration in the motion between a satellite or an aircraft and a ground receiver. In this case the Fourier transform of the phase characteristic is given by

$$\Theta(\omega) = \frac{\Delta\dot{\omega}}{(j\omega)^3} \tag{8.16}$$

where $\Delta\dot{\omega}$ is the magnitude of the rate of frequency change. In this case, Equation (8.9) yields

$$\lim_{t\to\infty} e(t) = \lim_{j\omega\to 0} \frac{\Delta\dot{\omega}/j\omega}{j\omega + K_0 F(\omega)} = \lim_{j\omega\to 0} \frac{\Delta\dot{\omega}}{j\omega K_0 F(\omega)} \tag{8.17}$$

If the loop has a nonzero velocity error, that is, if the right-hand side of Equation (8.12) is not equal to zero, Equation (8.17) shows the steady-state phase error due to a frequency ramp to be unbounded. This says that a PLL with loop filters given by any of Equations (8.13) to (8.15) will not be able to track a frequency ramp. In order to track a frequency ramp, the denominator of the loop filter transform, $D(\omega)$, must have $j\omega$ as a factor. From Equation (8.17) it can be seen that a loop filter with this type of transfer function, $F(\omega) = N(\omega)/j\omega D_1(\omega)$, will allow the PLL to track a frequency ramp with a constant phase error. This implies that in order to track a signal with a linearly changing Doppler shift (constant relative acceleration), the receiver must have a PLL that is second order or higher. To track a frequency ramp with zero phase error, the loop filter would be required to have a transfer function with $(j\omega)^2$ as a factor of the denominator, $F(\omega) = N(\omega)/(j\omega)^2 D_2(\omega)$. This implies a PLL that is third order or higher. Thus high-performance aircraft that need to track phase accurately through violent maneuvers may require third- or higher-order PLLs. In all cases, frequency lock is available with a loop of one order less than that required for phase lock. Steady-state error analysis is therefore a useful indicator of the required complexity of the loop filters.

In practice, the vast majority of PLL designs are second order. This is because a second-order loop can be made to be unconditionally stable [2] (i.e., the loop will always try to track the input—no set of input conditions, regardless of how extreme, will cause the loop to become unable to respond in the appropriate direction to changes in the input). Second-order loops will track out the effect of a frequency step (Doppler shift), and they are relatively easy to analyze, since the closed-form results obtained for first-order loops are good approximations for second-order loop performance. Third-order loops are used for some special applications [e.g., some Global Positioning System (GPS) navigation receivers have a third-order PLL], but loop performance for third-order loops is relatively difficult to determine, and third- and higher-order loops are only conditionally stable. A typical communication system design decision is that if the signal dynamics are expected to be such that high-order loops would be required for coherent demodulation, noncoherent demodulation is used instead.

8.2.1.2 Performance in Noise

The steady-state analysis of the preceding section tacitly assumed that the input signal was noise free. In some situations this may be approximately correct, but as in other parts of communication analysis, the more general case would include the effects of noise.

Reconsider the normalized loop input signal of Equation (8.1) and Figure 8.1. With the inclusion of normalized narrowband additive Gaussian noise, $n(t)$, the expression for the input becomes

$$r(t) = \sin (\omega_0 t + \theta) + n(t) \tag{8.18}$$

where, for the moment, we consider the input phase offset, θ, to be a constant. The noise process, $n(t)$, assumed to be a zero-mean narrowband Gaussian process, can be expanded into quadrature components about the carrier frequency as [3]

$$n(t) = n_c(t) \cos \omega_0 t + n_s(t) \sin \omega_0 t \tag{8.19}$$

where both $n_c(t)$ and $n_s(t)$ are zero-mean Gaussian random processes and are statistically independent. Now the output of the phase detector can be written as [see Equation (8.3)]

$$
\begin{aligned}
e(t) &= x(t)r(t) \\
&= \sin (\theta - \hat{\theta}) + n_c(t) \cos \hat{\theta} - n_s(t) \sin \hat{\theta} \\
&\quad + \text{(terms at twice the carrier frequency)}
\end{aligned} \tag{8.20}
$$

As before, the twice-carrier-frequency terms are assumed to be eliminated by the loop filter. Denoting the second and third terms of Equation (8.20) as

$$n'(t) = n_c(t) \cos \hat{\theta} - n_s(t) \sin \hat{\theta} \tag{8.21}$$

it is easy to verify that the variance of $n'(t)$ is identical to the variance of $n(t)$. This variance will be denoted by σ_n^2.

Consider the autocorrelation function of $n'(t)$,

$$
\begin{aligned}
R(t_1, t_2) &= E\{n'(t_1)n'(t_2)\} \\
&= E\{n_c(t_1)n_c(t_2)\} \cos^2 \hat{\theta} + E\{n_s(t_1)n_s(t_2)\} \sin^2 \hat{\theta} \\
&\quad - [E\{n_c(t_1)n_s(t_2)\} + E\{n_s(t_1)n_c(t_2)\}] \sin \hat{\theta} \cos \hat{\theta}
\end{aligned} \tag{8.22}
$$

where $E\{\cdot\}$ denotes the expected value. The cross-terms on the right-hand side of Equation (8.22) are equal to zero because n_c and n_s are mutually independent and have zero means [3]. With the assumption of wide-sense stationarity [4] we have

$$R(\tau) = R_c(\tau) \cos^2 \hat{\theta} + R_s(\tau) \sin^2 \hat{\theta} \tag{8.23}$$

where $\tau = t_1 - t_2$. Taking Fourier transforms, the power spectral density of $n'(t)$ is seen to be

$$G(\omega) = \mathcal{F}[R(\tau)]$$

$$= G_c(\omega) \cos^2 \hat{\theta} + G_s(\omega) \sin^2 \hat{\theta} \qquad (8.24)$$

where G_c and G_s are the Fourier transforms of R_c and R_s, respectively. But from Equation (8.19), it can be seen that the spectra G_c and G_s are made of shifted versions of the spectra of the original noise process $n(t)$. Therefore, because of our construction [5],

$$G_s(\omega) = G_c(\omega) = G_n(\omega_0 - \omega) + G_n(\omega_0 + \omega)$$

where $G_n(\omega)$ is the spectral density of the original bandpass noise process $n(t)$. Equation (8.24) can be rewritten as

$$G(\omega) = G_n(\omega_0 - \omega) + G_n(\omega_0 + \omega) \qquad (8.25)$$

For the special case of white noise, we have $G_n(\omega) = N_0/2$ watts/hertz, where N_0 is the single-sided spectral density of the white noise. Thus, from Equation (8.25), for this important special case,

$$G(\omega) = N_0 \qquad (8.26)$$

The value in this development is that for the same small-angle approximations that were made in the preceding section, the spectral density of the VCO phase, $G_{\hat{\theta}}$, is related to the spectral density of the noise process through the loop transfer function [Equation (8.6)]. That is,

$$G_{\hat{\theta}}(\omega) = G(\omega) |H(\omega)|^2 \qquad (8.27)$$

where $G(\omega)$ is as given in Equation (8.25) and $H(\omega)$ as defined in Equation (8.6). The variance of the output phase is then

$$\sigma_{\hat{\theta}}^2 = \frac{1}{2\pi} \int_{-\infty}^{\infty} G(\omega) |H(\omega)|^2 \, d\omega \qquad (8.28)$$

For the special case of white noise,

$$\sigma_{\hat{\theta}}^2 = \frac{N_0}{2\pi} \int_{-\infty}^{\infty} |H(\omega)|^2 \, d\omega \qquad (8.29)$$

The integral in Equation (8.29) (renormalized to natural frequency) is called the *two-sided loop bandwidth*, W_L. The *single-sided loop bandwidth* is termed B_L. The definitions of these terms are

$$W_L = 2B_L = \frac{1}{2\pi} \int_{-\infty}^{\infty} |H(\omega)|^2 \, d\omega \qquad \text{hertz} \qquad (8.30)$$

Thus, if the noise process is white and the small-angle approximation holds (in other words, the loop is successfully tracking the input phase), the phase variance is given by

$$\sigma_{\hat{\theta}}^2 = 2N_0 B_L \tag{8.31}$$

The phase variance is a measure of the amount of jitter or wobble in the VCO output due to noise at the input. Equations (8.31) and (8.7) highlight one of the many trade-offs in communication theory. Clearly, one would wish $\sigma_{\hat{\theta}}^2$ to be small, which for a given noise level implies a small loop bandwidth, B_L, which from Equation (8.30) implies a narrow $H(\omega)$. However, it can be inferred from Equation (8.7) that the narrower the effective bandwidth of $H(\omega)$, the poorer will be the loop's ability to track incoming signal phase changes, $\Theta(\omega)$. Thus a loop design must balance noise response with desired input phase response. The designer's dilemma is to design a loop that responds appropriately to the changes in the input signal, while not being overly responsive to the apparent changes, which are actually only artifacts of the noise process.

8.2.1.3 Nonlinear Loop Analysis

All of the PLL discussion in the previous sections has utilized what is called the linearized PLL model. This model is shown schematically in Figure 8.2. The model makes use of the small-angle approximation

$$\sin(\theta - \hat{\theta}) \simeq \theta - \hat{\theta} \tag{8.32}$$

which is accurate when the loop is "in lock" and performing as desired (i.e., with small phase errors). Clearly, these conditions form only part of the picture. A complete analysis of PLL performance must allow for the times when Equation (8.32) is not accurate. When the small-angle approximation is inaccurate, an appropriate model is the one shown schematically in Figure 8.3. From Equations (8.4), (8.20), and (8.21) and Figure 8.3, the model can be described by the differential equation

$$\frac{d}{dt}[\hat{\theta}(t)] = K_0 f(t) * \sin[\theta(t) - \hat{\theta}(t)] + f(t) * n'(t) \tag{8.33}$$

where, as before, $*$ denotes the convolution operation. In spite of the best efforts of many researchers, this differential equation has resisted general solution for many years. However, Viterbi [5] has derived a closed-form solution for an important special case.

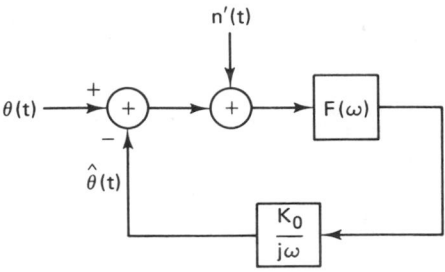

Figure 8.2 Schematic of linearized PLL model.

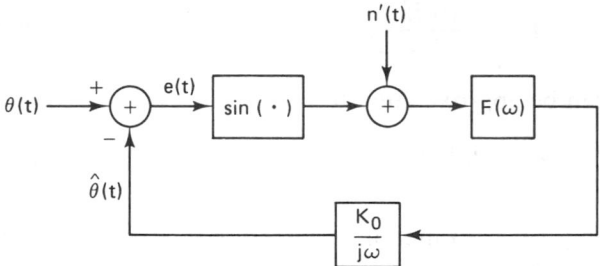

Figure 8.3 Schematic of nonlinearized PLL model.

Consider the case where $\theta(t)$, the input phase as a function of time, is a constant, θ. We can now define a new phase variable

$$\phi(t) = [\theta - \hat{\theta}(t)] \text{ modulo } 2\pi \tag{8.34}$$

Since θ is constant, Equation (8.33) can be rewritten as

$$\frac{d}{dt}[\phi(t)] = K_0 f(t) * \sin \phi(t) + f(t) * n'(t) \tag{8.35}$$

Since from Equation (8.35), $\phi(t)$ is a function of the random process $n'(t)$, $\phi(t)$ itself is a random process. Because $\phi(t)$ is defined modulo 2π, it can be shown [2] that $\phi(t)$ is stationary in the limit when all transient effects have died down (i.e., θ is a constant). Viterbi [5] has determined that for a first-order PLL [i.e., the loop filter is a short circuit, or equivalently $f(t) = \delta(t)$], the probability density function of ϕ is of the form

$$p(\phi) = \frac{\exp(\rho \cos \phi)}{2\pi I_0(\rho)} \quad \text{for } |\phi| \leq \pi \tag{8.36}$$

where $\rho = 1/\sigma_{\hat{\theta}}^2$ [see Equation (8.31)] is the normalized (to unit signal energy) loop signal-to-noise ratio, and $I_0(\rho)$ is the zeroth-order modified Bessel function of the first kind, evaluated at ρ. The phase variance, modulo 2π, can now be computed using Equation (8.36). The resulting value of the phase variance will be exact for first-order loops, and is an extremely useful approximation for the behavior of many second-order loops [2]. It has also been shown to be an exact form for higher-order loops under a modified definition of ρ [6].

The change of variable from a phase that can take any real value to a phase that is modulo 2π results in the concept of loop cycle slips. A cycle slip occurs when the magnitude of the original phase error, $|\theta - \hat{\theta}(t)|$, exceeds 2π radians. This will cause the value of ϕ [Equation (8.34)] to abruptly change from about 2π to about 0. This event can be thought of as a momentary loss of lock with an almost immediate reacquisition. The statistics of cycle slips can be as important an indicator of PLL performance as phase variance—especially at low loop signal-to-noise ratios when cycle slips may occur frequently.

By manipulating his phase distribution results, Viterbi [5] derived an expression for the mean time to the first cycle slip, T_m, beginning at some arbitrary reference time,

$$T_m = \frac{\pi^2 \rho I_0^2(\rho)}{2B_L} \tag{8.37}$$

For large ρ this expression can be approximated by

$$T_m \simeq \frac{\pi \exp (2\rho)}{4B_L} \tag{8.38}$$

As was true with the probability density function of Equation (8.36), these results were derived for first-order loops, but they are useful approximations for the behavior of second-order loops, and provide an upper bound to second-order loop performance at medium and large loop signal-to-noise ratios. In addition, computer simulations and laboratory measurements [2] indicate that the time between cycle slips, T, is exponentially distributed,

$$P(T) = 1 - \exp \left(-\frac{T}{T_m} \right) \tag{8.39}$$

This is to say that the probability that a loop will cycle-slip within time T, starting from zero phase error, is given by Equation (8.39).

8.2.1.4 Suppressed Carrier Loops

The discussion of PLLs to this point has presumed that the carrier input is a fairly stable sinusoid with some known positive average energy. In the case of a phase modulated communication system, if the carrier phase variation due to the modulation is less than $\pi/2$ radians, there will be positive energy at the carrier frequency. This is called a system design that has a residual carrier component, and all of the discussion of PLL development to this point would apply directly to this residual component. A diagram of the signal space for a binary phase modulated system with a residual carrier component is given in Figure 8.4, for a modulating angle of $\gamma \leq \pi/2$. At one time, most phase modulated systems were designed in this way. However, the residual carrier component is, in a sense, wasted energy—in the sense that the energy in the residual carrier is not being used to transmit the information, only to transmit the carrier. Thus most modern phase modulated systems are suppressed carrier systems. This means that there is no average energy transmitted at the carrier frequency. All of the transmitted energy goes into the modulation. Unfortunately, this means that there is no longer any signal for the basic PLL of Figure 8.1 to track.

Consider, as an example, a BPSK signal

$$r(t) = m(t) \sin (\omega_0 t + \theta) + n(t) \tag{8.40}$$

where $m(t) = \pm 1$ with equal probability. This is a suppressed carrier transmission—the average energy at radian frequency ω_0 is zero. In Figure 8.4 this signal is the situation when $\gamma = \pi/2$. The figure indicates that for this case the vertical carrier component will vanish. To acquire and track the phase of the carrier, the effects of the modulation must be eliminated. One way to eliminate the modulation

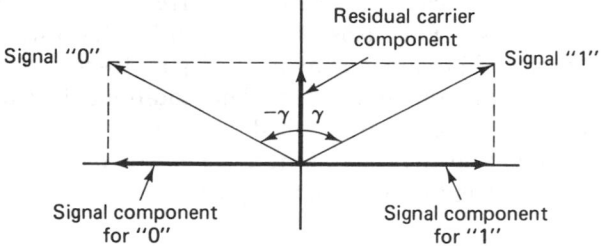

Signal "0" — Residual carrier component — Signal "1"

$-\gamma$ | γ

Signal component for "0" | Signal component for "1"

Figure 8.4 Residual carrier binary phase modulation.

is to square the signal;

$$r^2(t) = m^2(t) \sin^2 (\omega_0 t + \theta) + n^2(t) + 2n(t)m(t) \sin (\omega_0 t + \theta)$$

$$= \tfrac{1}{2} - \tfrac{1}{2} \cos (2\omega_0 t + 2\theta) + n^2(t) + 2n(t)m(t) \sin (\omega_0 t + \theta) \quad (8.41)$$

where use has been made of the fact that $m^2(t) = 1$. The second term on the right-hand side of Equation (8.41) is a carrier-related term (at twice the original carrier frequency) that can be acquired and tracked with a basic PLL of the type illustrated in Figure 8.1. Such an arrangement is illustrated in Figure 8.5. When the incoming suppressed-carrier waveform is squared, the resulting twice-carrier component can be acquired and tracked by a PLL of standard design. Some of the problems with this procedure can be inferred from Equation (8.41). The first problem is simply that all phase angles have been doubled. Thus the phase noise and phase jitter has been doubled, and the phase error variance (related to the phase noise squared) is larger by a factor of 4 than that of the original signal. This angle doubling is offset by the divide-by-2 circuit at the VCO output, and therefore, does not directly affect the accuracy of the loop's output signal that is used by the data demodulator, but this larger internal variation will cause the PLL to require a 6-dB-larger carrier signal-to-noise ratio than a residual carrier system in order to maintain phase lock. In addition, there are now two effective noise

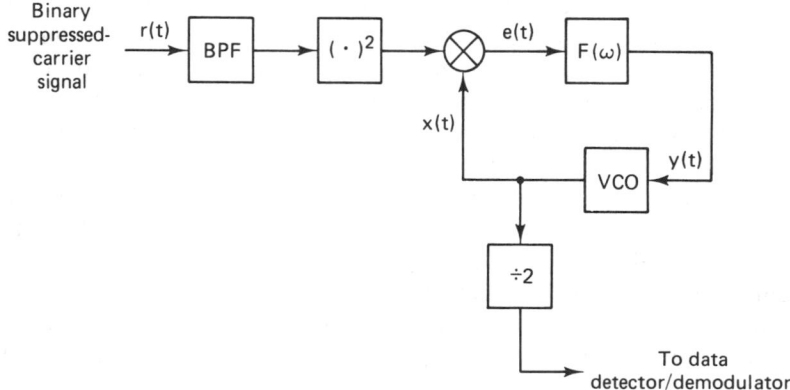

Binary suppressed-carrier signal → $r(t)$ → BPF → $(\cdot)^2$ → ⊗ → $e(t)$ → $F(\omega)$

$x(t)$

VCO → $y(t)$

÷2

To data detector/demodulator

Figure 8.5 Basic squaring loop schematic.

terms interfering with loop operation, because of the cross-correlation term between noise and signal in Equation (8.41). For cases of medium or low loop signal-to-noise ratio, these two noise terms will reduce the available signal-to-noise ratio even further, relative to the original unmodulated carrier. This additional loss due to signal-times-noise and noise-times-noise terms is called the *loop squaring loss*, S_L. Gardner [2] shows that if the input noise process, $n(t)$, is a narrowband Gaussian noise of bandwidth B_i, the squaring loss is upper bounded by

$$S_L = 1 + N_0 B_i \qquad (8.42)$$

where, as before, N_0 is the single-sided power spectral density of the prefiltered, normalized white Gaussian noise process. Equation (8.42) is an upper bound because the filter bandwidth B_i is tacitly assumed to be wide enough to pass the signal undistorted. In an actual design, signal distortion can be traded for squaring loss, as is shown in [7].

Since the normalization in Equation (8.42) is with respect to the signal power, the second term is proportional to a signal-to-noise ratio

$$\rho_i = \frac{1}{2N_0 B_i} \qquad (8.43)$$

where ρ_i is the signal-to-noise ratio in the input filter bandwidth. For large loop signal-to-noise ratios, the output phase variance can now be expressed as

$$\sigma_\theta^2 = 2N_0 B_L S_L = 2N_0 B_L \left(1 + \frac{1}{2\rho_i}\right) \qquad (8.44)$$

The leading term on the right-hand side of Equation (8.44) can be seen to be identical to that of Equation (8.31), the phase variance of the standard PLL. It can also be seen that for large input signal-to-noise ratios, the second term in the squaring loss will vanish, and we are left with the phase variance of the standard PLL.

Another potentially serious problem, associated mainly with suppressed carrier loops, is that of *false lock* [2, 8–10]. This can be a problem especially during acquisition or reacquisition of carrier phase. The interaction of the data stream with the loop nonlinearities (especially the squaring circuit) and loop filters will produce sidebands in the spectrum that is input to the phase detector. These sidebands can contain stable frequency components. Care must be taken that these stable components are not allowed to capture the tracking loop. If the loop is captured, it will appear to be operating correctly; the VCO control signal, $y(t)$, will be small but the VCO output will be offset in frequency from the correct carrier component. This is false lock. The loop is tracking a sideband frequency component, and the real carrier is being filtered out by the loop filter. False lock is a hardware implementation problem that typically sets an effective lower limit on the bandwidth of the loop filters. Residual carrier loops, having fewer nonlinear elements, are not usually bothered by false locking.

8.2.1.5 Costas Loops

An important form of a suppressed carrier loop is the Costas loop, shown schematically in Figure 8.6. This loop design is important because it eliminates the square-law device, which can be difficult to implement at carrier frequencies, and replaces it with a multiplier and relatively simple low-pass filters. Although the appearance of the circuits in Figures 8.5 and 8.6 is quite different, their theoretical performance can be shown to be the same [2]. The main remaining implementation problem with Costas loops is that to achieve the theoretically optimum performance, the two low-pass arm filters must be perfectly matched. This can only be approximated in any hardware implementation. Thus the decision as to whether to implement a Costas loop or the classical design of Figure 8.5 amounts to a design decision between the difficulty of implementing the squaring device and the difficulty of keeping the arm filters nearly matched. This design decision will depend on the parameters and requirements of the particular receiving system, and cannot be generalized here.

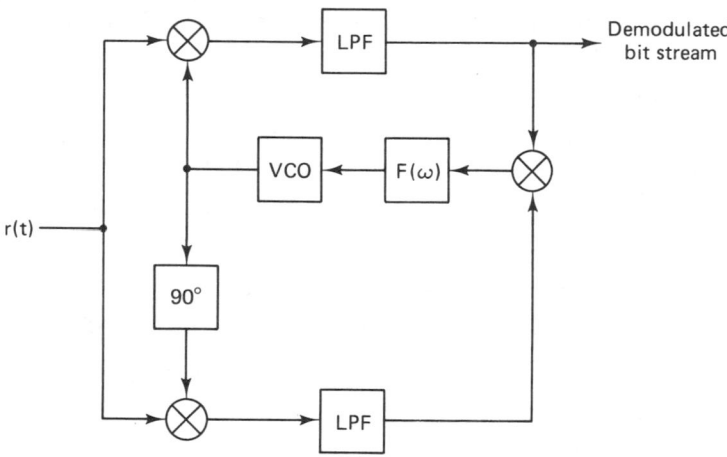

Figure 8.6 Costas loop.

8.2.1.6 High-Order Suppressed-Carrier Loops

Binary phase shift keying is not the only type of suppressed-carrier modulation. In fact, assuming that all signals are equally likely a priori, any modulation scheme whose average amplitude, averaged over the signal set, is zero will have no average energy in the transmitted carrier. Perhaps the most common nonbinary suppressed-carrier modulation is quadrature phase shift keying, or QPSK (4-ary PSK). If a QPSK signal is squared, the result "looks like" a BPSK signal. Thus for equally likely QPSK signals, the carrier is still suppressed. However, squaring the signal a second time, equivalent to taking the original signal to the fourth power, can be seen to produce a term with a carrier component, at four times the transmitted carrier's frequency. As in the binary case, operating on the incoming signal with a power-law device produces cross products among the noise terms and signal terms, and introduces the equivalent of a "squaring loss." Under the assumption that the noise bandwidth will pass the signal undistorted, the loss for fourth-power loops is upper bounded by [2]

$$S_L = 1 + \frac{9}{\rho_i} + \frac{6}{\rho_i^2} + \frac{3}{2\rho_i^3} \qquad (8.45)$$

As was the case with the squaring loop, for sufficiently high input signal-to-noise ratios, ρ_i, Equation (8.45) indicates that the additional loss terms vanish, and the loop performance approaches that of the basic loop. As was also the case for the squaring loop, there are Costas loop designs equivalent to fourth-order loops [2, 11, 12] that may exhibit hardware implementation advantages. Their theoretical performance, however, is the same as that of the straightforward fourth-power design.

Example 8.5   Squaring Loss Bounds

Compare the upper bounds on squaring loss, S_L, given by Equations (8.42) and (8.45) for second- and fourth-power loops, respectively, for an input loop signal-to-noise ratio, ρ_i, of 10 dB.

Solution

A 10 dB signal-to-noise ratio is also 10 in terms of its power ratio. Therefore, from Equations (8.42)) to (8.44), for the squaring loop,

$$S_L = 1 + \frac{1}{2\rho_i} = 1.05 = 0.2 \text{ dB}$$

From Equation (8.45), for the fourth-power loop

$$S_L = 1 + 0.9 + 0.06 + 0.0015 = 1.9615 = 2.9 \text{ dB}$$

Thus, while an input signal-to-noise ratio of 10 dB is adequate to keep losses small for the squaring loop, the same signal-to-noise ratio may allow significant losses for the fourth-power loop.

8.2.1.7 Acquisition

In most of the discussion thus far, the assumption has been that the PLL is in lock. This was the justification for assuming that the phase error, $|\theta - \hat{\theta}|$, was small. At one time or another, however, every loop must acquire lock; that is, it must be brought into lock. Acquisition can be accomplished with the aid of external circuits or signals (aided acquisition) or in some cases by an unaided PLL (self-acquisition) [2].

Acqusition is an inherently nonlinear operation, and therefore is difficult to analyze in general. However, some intuition may be obtained by considering a noise-free first-order loop. Such a loop is shown schematically in Figure 8.3, where $n'(t) = 0$ (noise-free) and $F(\omega) = 1$ (first-order). Denote the input phase as

$$\theta(t) = \omega_i t$$

and the output phase as

$$\hat{\theta}(t) = \omega_0 t + \int_0^t K_0 \sin e(\tau) \, d\tau + \hat{\theta}(0) \tag{8.46}$$

where ω_i and ω_0 are the radian frequencies of the input and output signals, respectively. Thus the phase error is given by

$$e(t) = \theta(t) - \hat{\theta}(t)$$

$$= (\omega_i - \omega_0)t - K_0 \int_0^t \sin e(\tau) \, d\tau - \hat{\theta}(0) \tag{8.47}$$

Differentiating both sides and letting $\Delta\omega = \omega_i - \omega_0$ provides

$$\frac{de}{dt} = \Delta\omega - K_0 \sin e \tag{8.48}$$

where the time dependence of the function $e(t)$ has been suppressed to ease notation. This differential equation describes the behavior of the first-order noise-free PLL. The loop being in lock requires that

$$\frac{de}{dt} = 0 \tag{8.49}$$

Equation (8.49) is a necessary but not a sufficient condition for phase lock. This can be verified by observing the phase plane diagram of Figure 8.7. This figure is obtained by dividing both sides of Equation (8.48) by the gain term K_0, and plotting the results. First observe point a. If the phase error is displaced a little to the left or right of point a, the sign of the derivative term is such that the phase

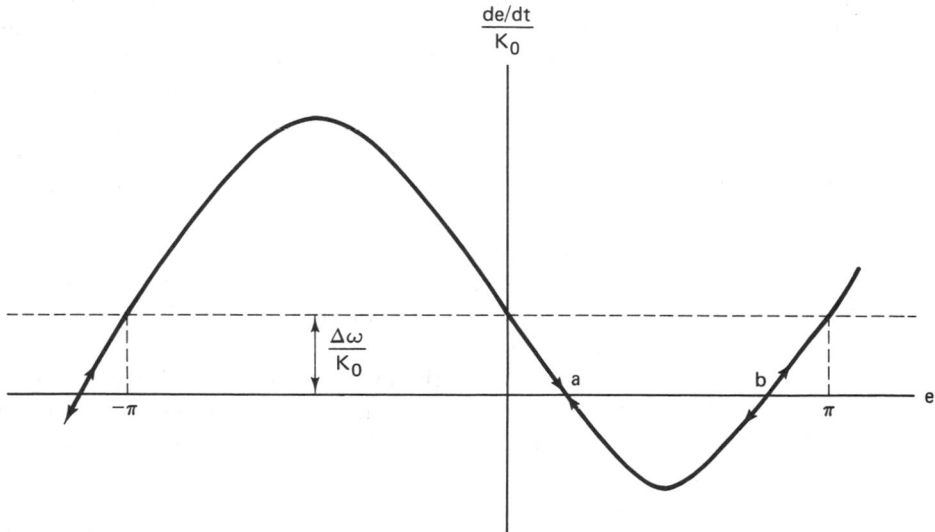

Figure 8.7 Phase-plane plot of first-order loop.

error e, will be driven back toward a. Thus point a is a stable point of the system, a point where phase lock can be obtained and will be maintained. For the case of point b, however, if the phase error is exactly at b, Equation (8.49) will be satisfied, but if there is any slight offset from b, the sign of the derivative term will be such that the error will be driven away from b. Thus b is a point of marginal stability for the loop, a point where Equation (8.49) is satisfied, but not a stable lock point.

The amount of time required for a loop to come into lock can be a very important system design consideration. By observing Equation (8.48) we can see that the requirement of Equation (8.49) for phase lock cannot be met unless

$$\frac{|\Delta\omega|}{|K_0|} \le 1 \qquad (8.50)$$

This is because sinusoidal functions have a maximum amplitude of unity. This range of frequency difference, $-K_0 < \Delta\omega < K_0$, is sometimes called the lock-in range of the loop. Assuming that Equation (8.50) holds, Gardner [2] gives a rule of thumb of $3/K_0$ seconds for the time required for loop acquisition. Actual values can be obtained from Equation (8.47) for well-defined sets of initial conditions. It can be seen from the phase plot of Figure 8.7 that the time required will vary widely as a function of the initial phase error. For phase errors very close to point b, the driving force $(de/dt)/K_0$ will be very small. Thus, for this worst-case phase error, the error could "linger" in the vicinity of b for a long time. This phenomenon is called *terminal loop hang-up* [13] and can be a serious problem for system designs that depend on self-acquisition.

Perhaps the most important operational difference between first-order and higher-order loops is the higher-order loop's ability to "pull in" from frequency differences that are larger than the lock-in range. A first-order loop with a frequency error larger than the lock-in range will drift toward lock but never quite lock in (why?). Second- and higher-order loops can pull in and achieve phase lock because of their more complicated phase-plane characteristics. Interested readers should consult Viterbi [5] and other texts on PLLs for more details [2, 6, 14–16].

The study of self-acquisition for phase-locked loops is mostly of academic interest. Gardner [2] states that loops using self-acquisition can be guaranteed to acquire in reasonable time only under very benign circumstances. This, unfortunately, is rarely the case in practice.

Acquisition aiding drives the loop through the region of phase space expected to contain the lock-in region by means of some external driving signal. This is the most common means of achieving acquisition. Aiding can be implemented by simply applying a voltage ramp to the input of the VCO. This driving signal will cause the VCO output frequency to vary linearly with time. As was shown earlier [Equation (8.17)], loops with loop filters that do not contain $j\omega$ as a factor of their transfer function's denominator cannot track a frequency ramp with finite phase error. Therefore, if frequency sweeping is to be employed with a first-order loop, or a second-order loop without this transfer function characteristic, the rate of frequency sweep must be slow enough so that when the loop achieves lock, the presence of phase lock can be detected and the sweeping signal removed before it drives the loop back out of lock. With loops that contain $j\omega$ as a factor of $D(\omega)$, it may not be necessary to remove the sweeping signal at all, because, at least in theory, the loop will be able to track out the frequency ramp. In any case, the sweep rate must not be too large, or the loop will be driven through the lock point so fast that it will fail to acquire. For a second-order loop with loop transfer function [see Equation (8.6)]

$$H(\omega) = \frac{1}{-(j\omega/\omega_n)^2 + 2\zeta(j\omega/\omega_n) + 1} \tag{8.51}$$

Gardner [2] indicates that the maximum sweep rate, $\Delta\dot\omega$, must be in the vicinity of

$$\Delta\dot\omega \simeq \tfrac{1}{2}\omega_n^2(1 - 2\sigma_{\hat\theta}) \tag{8.52}$$

where $\sigma_{\hat\theta}$ is as defined in Equation (8.31), and ω_n, implicitly defined in Equation (8.51), is called the *natural frequency* of a second-order PLL and is related to the loop bandwidth, B_L, and loop damping factor, ζ [6], by

$$\omega_n = \frac{8\zeta}{4\zeta^2 + 1} B_L$$

Blanchard [14] gives more detailed results for aided phase acquisition.

8.2.1.8 Phase Tracking Errors and Link Performance

If a loop is unable to track out all phase errors, the received symbol error probability will be degraded relative to what is theoretically achievable. The analysis required to determine the amount of the degradation is very involved, but for most of the standard coherent signaling systems, curves are available [11, 12, 17]. Figure 8.8 is an example of such a performance curve for a residual carrier-phase tracking loop operating on a signal with BPSK modulation in additive Gaussian noise. It can be seen that for signal-to-noise ratios of moderate value, small phase errors produce very little degradation. It is only when the standard deviation of phase error exceeds 0.3 that the degradations become significant. This means that the inherent degradation in performance caused by a well-designed loop operating in benign conditions can generally be ignored. The curve also indicates that if conditions are such that the phase variance is large, increasing the data signal-to-Gaussian-noise ratio may not be effective in reducing the detected error probability. It should be noted that the presence of an irreducible error in these situations is a characteristic of residual carrier designs with constant loop signal-to-noise ratios, ρ_i. Suppressed carrier tracking loops tend not to have irreducible errors, because an increase in the data signal-to-noise ratio will increase the signal-to-noise ratio of the suppressed carrier tracking loop, reducing the tracking error.

Example 8.6   PLL Signal-to-Noise Ratio

Develop an integral expression for the effect on link bit error probability of slowly varying phase tracking errors for a residual carrier BPSK link. Compare the effect

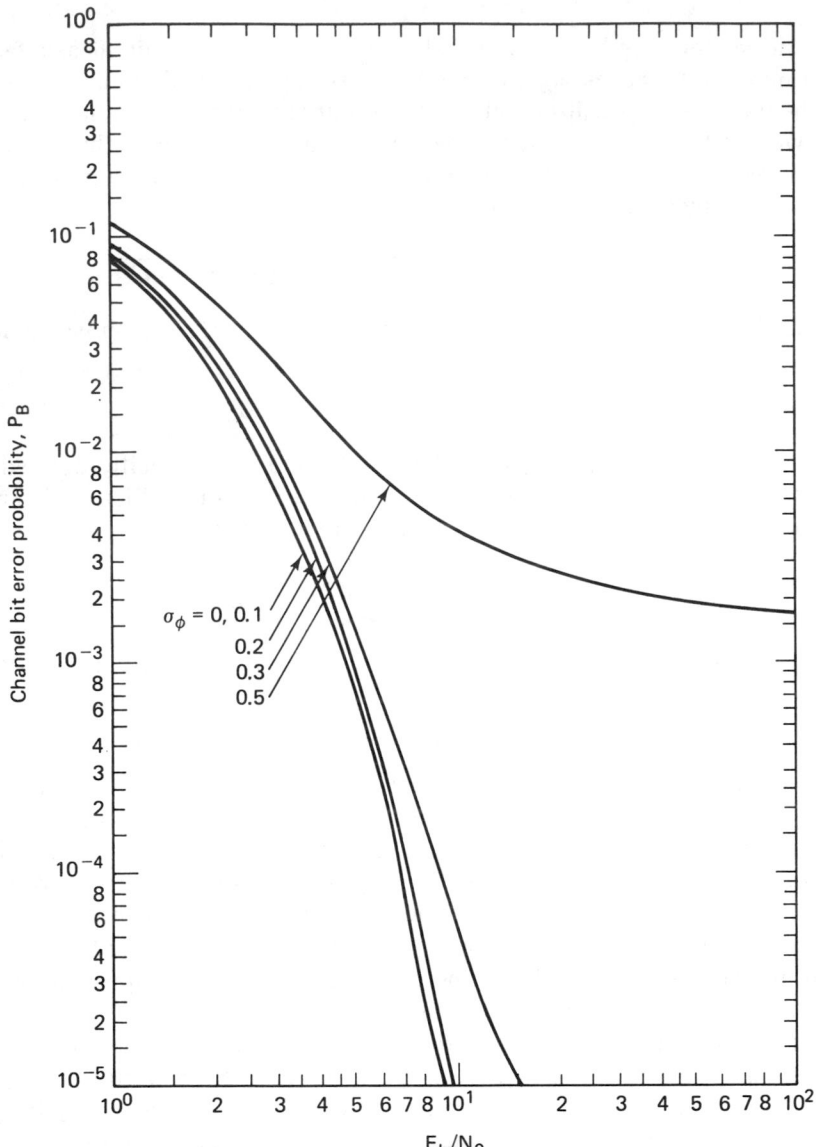

Figure 8.8 Channel bit error probability versus E_b/N_0 for BPSK with imperfect carrier synchronization. (Reprinted with permission from J. J. Stiffler, *Theory of Synchronous Communications*, Prentice-Hall, Inc., Englewood Cliffs, N.J., 1971, Fig. 9.1, p. 270.)

of a normalized loop signal-to-noise ratio ($\rho = 1/\sigma_{\hat{\theta}}^2$) of 20 dB with one of 10 dB on error performance at a desired bit error probability of 10^{-5} using Figure 8.8.

Solution

From Chapter 3 we have that the theoretically possible bit error probability for a BPSK link in additive white Gaussian noise of single-sided spectral density N_0

watts/hertz is given by

$$P_B = Q\left(\sqrt{\frac{2E_b}{N_0}}\right)$$

where E_b is the received energy per bit time. From the derivation of this expression for error probability, it can be shown that if there is a slowly varying (with respect to the data rate) phase tracking error of β radians, the resulting probability of error will be given by

$$P_B(\beta) = Q\left(\sqrt{\frac{2E_b \cos \beta}{N_0}}\right)$$

Now if the phase error β is the result of tracking errors caused by system noise, β will be described stochastically by some probability density function $p(\beta)$. Then the expected bit error probability is given by

$$P_B = \int_0^{2\pi} P_B(\beta)p(\beta)\, d\beta$$

For the special case of a first-order loop, the probability density function is given by Equation (8.36). Then the final expression for expected bit error probability is given by

$$P_B = \int_0^{2\pi} Q\left(\sqrt{\frac{2E_b \cos \beta}{N_0}}\right) \frac{\exp(\rho \cos \beta)}{2\pi I_0(\rho)}\, d\beta$$

A loop signal-to-noise ratio (ρ_i) of 20 dB will correspond to a standard deviation of phase noise of $\sigma_{\hat{\theta}} = 0.1$ rad. From Figure 8.8, this small amount of phase noise produces no appreciable degradation in the bit error probability. A loop ρ_i of 10 dB, however, corresponds to a phase noise standard deviation of $\sigma_{\hat{\theta}} = 0.32$ rad. It can be seen from Figure 8.8 that for a bit error probability of 10^{-5}, this phase noise standard deviation will require a data SNR of somewhat more than 11 (10.4 dB) rather than a data SNR of 9.1 (9.6 dB) for perfect phase tracking. Thus this loop signal-to-noise will cause an error performance degradation of somewhat more than 0.8 dB at an error probability of 10^{-5}. It should be noted that for loop SNRs less than about 10 dB, the degradation in performance increases very rapidly. Thus 10 dB is something of a threshold for reasonable system performance for residual carrier designs. Suppressed carrier designs, having no problem with irreducible error, may do better.

8.2.2 Symbol Synchronization

All digital receivers need to have their demodulators synchronized to the incoming digital symbol transitions in order to achieve optimum demodulation. In the discussion that follows, we will consider several of the basic types of designs of symbol or data synchronizers. The discussion will center on a random binary baseband signal, for ease of terminology and notation, but it is expected that the extension to nonbinary baseband signals will be apparent.

The symbol synchronizers that will be considered here can be classified into two basic groups. The first group consists of the *open-loop synchronizers*. These circuits recover a replica of the transmitter data clock output directly from operations on the incoming data stream. The second group comprises the *closed-loop synchronizers*. Closed-loop data synchronizers attempt to lock a local data clock to the incoming signal by use of comparative measurements on the local and incoming signals. Closed-loop methods tend to be more accurate, but are much more costly and complex.

8.2.2.1 Open-Loop Symbol Synchronizers

Open-loop symbol synchronizers are also occasionally called *nonlinear filter synchronizers* [17], a very descriptive title. This class of synchronizers generate a frequency component at the symbol rate by operating on the incoming baseband sequence with a combination of filtering and a nonlinear device. The operation is analogous to carrier recovery in a suppressed carrier tracking loop. In the present case, the desired frequency component, at the data symbol rate, is isolated with a bandpass filter, and "shaped" with a high-gain saturating amplifier. The shaping recovers the square-wave appearance of the data clock signal.

Three examples of open-loop bit synchronizers are shown in Figure 8.9. In the first example (Figure 8.9a), the incoming signal, $s(t)$, is filtered with a matched filter. The output of this filter will be the autocorrelation function of the input signal shape. For square-wave signaling, for example, the output will be the familiar isosceles-triangular waveshape. The sequence of bit autocorrelation wave-

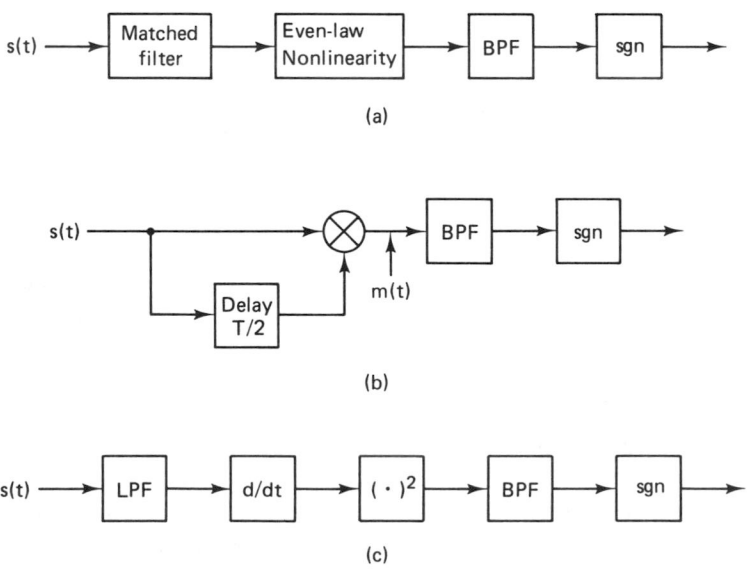

Figure 8.9 Three types of open-loop bit synchronizers.

shapes is then "rectified" by some type of memoryless even-law nonlinearity, a square-law device, for example. The resulting waveform will have positive amplitude peaks that correspond, to within a time delay, with the input symbol transitions. This sequence of processes is illustrated in Figure 8.10. Thus the output waveform from the even-law device will contain a Fourier component at the fundamental frequency of the data clock. This frequency component is isolated from its harmonics with a bandpass filter (BPF), and shaped with an ideal saturating amplifier, with transfer function sgn x, defined as follows:

$$\text{sgn } x = \begin{cases} 1 & \text{for } x > 0 \\ -1 & \text{otherwise} \end{cases} \tag{8.53}$$

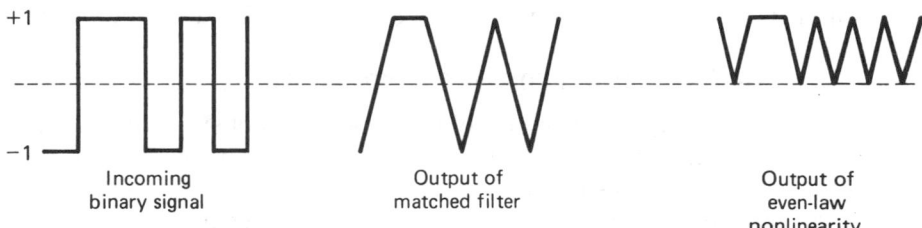

Figure 8.10 Open-loop bit synchronizer illustration.

The second example in Figure 8.9 produces a Fourier component at the data clock frequency by means of a delay and multiply. The delay shown in Figure 8.9b is half a bit period, which is the best value because it provides the strongest Fourier component [17]. The waveform $m(t)$ will always be positive in the second half of every bit period, but will have a negative first half if there has been a state change in the incoming bit stream, $s(t)$. This produces a square-wave signal with spectral components at the data rate and all harmonics, as in Figure 8.9a. As before, the appropriate spectral component can be isolated with a BPF and shaped.

The final example (Figure 8.9c) amounts to an edge detector. The main operations are those of differentiation and rectification (by use of a square-law device). For a square-wave input, the differentiator will produce positive or negative spikes at all symbol transitions. When rectified, the resulting sequence of positive spikes will have a Fourier component at the data symbol rate. A potential problem with this particular scheme is that differentiators are typically very sensitive to wideband noise. This necessitates the low-pass filter (LPF) that precedes the differentiator in Figure 8.9c. The LPF, however, will also remove the high-frequency components of the data symbols, causing them to lose their original rectangular waveshape. This will cause the resulting differential signal to have some finite rise and fall time, rather than being a set of impulses.

Clearly, there will be some hardware delay associated with the signal processing steps illustrated in Figure 8.9. Wintz and Luecke [18] have shown that for a BPF that effectively averages K input symbols (bandwidth $\approx 1/KT$), the magnitude of the fractional mean time error (delay) is approximated by

$$\frac{|\bar{\epsilon}|}{T} \simeq \frac{0.33}{\sqrt{KE_b/N_0}} \qquad \text{for } \frac{E_b}{N_0} > 5, \quad K \geq 18 \tag{8.54}$$

where T is the bit period, E_b the detected energy per bit, and N_0 the single-sided received noise spectral density. Wintz and Luecke have also shown that at high signal-to-noise ratios the fractional standard deviation of the fractional timing error is given by

$$\frac{\sigma_\epsilon}{T} \simeq \frac{0.411}{\sqrt{KE_b/N_0}} \qquad \text{for } \frac{E_b}{N_0} > 1 \tag{8.55}$$

Thus, for a given BPF, when the received signal-to-noise ratio is sufficiently large, all of the techniques shown in Figure 8.9 will provide accurate bit timing.

8.2.2.2 Closed-Loop Symbol Synchronizers

The primary disadvantage of open-loop symbol synchronization methods is that there is an unavoidable non-zero-mean tracking error. This error can be made small for large signal-to-noise ratios, but since the synchronization signal waveform depends directly on the incoming signal, the error will never vanish.

Closed-loop symbol data synchronizers use comparative measurements on the incoming signal and a locally generated data clock signal to bring the locally generated signal into synchronism with the incoming data transitions. The procedure is essentially the same as that used for closed-loop carrier tracking.

Among the most popular of the closed-loop symbol synchronizers is the early/late-gate synchronizer. An example of such a synchronizer is shown schematically in Figure 8.11. The synchronizer operates by performing two separate integrations of the incoming signal energy over two different $(T - d)$ second portions of a symbol interval. The first integration (the early gate) begins integration at the loop's best estimate of the beginning of a symbol period (the nominal time zero), and integrates for the next $(T - d)$ seconds. The second integral (the late gate) delays the start of its integration for d seconds, and then integrates to the end of the symbol period (the nominal time T). The difference in the absolute values of the outputs of these two integrations, y_1 and y_2, is a measure of the receiver's symbol timing error, and can be fed back to the loop's timing reference to correct loop timing.

The action of the early/late-gate synchronizer can be understood by referring to Figure 8.12. In the case of perfect synchronization, Figure 8.12a shows that both gates are entirely within a single symbol interval. In this case, both integrators will accumulate the same amount of signal, and their difference (the error signal, e, in Figure 8.11) is zero. Thus, when the device is synchronized, it is stable—there is no tendency to drive itself away from synchronization. The case shown in Figure 8.12b is for a receiver whose data clock is early relative to the incoming data. In this case the first portion of the early gate falls in the previous bit interval, while the late gate is still entirely inside the current symbol. The late-gate integrator will accumulate signal over its entire $(T - d)$ integration interval, as in the

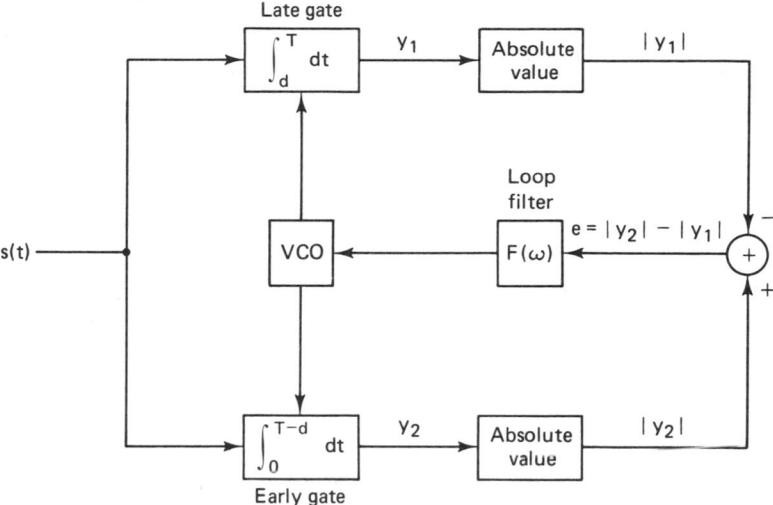

Figure 8.11 Early/late-gate data synchronizer.

case in Figure 8.12a, but the early-gate integrator will end up with energy accumulated only over $[(T - d) - 2\Delta)]$, where Δ is the portion of the early-gate interval falling in the previous bit interval. Thus, for this case the error signal will be $e = -2\Delta$, which will lower the input voltage to the VCO in Figure 8.11. This will reduce the VCO output frequency and retard the receiver's timing to bring it back toward the incoming signal's bit timing. Using Figure 8.12 as a guide, it can be seen that if the receiver's timing had been late, the amounts of energy integrated in the early gate and late gate would be reversed, as would the sign of the error signal. Thus late receiver timing produces an increase in the VCO input voltage, increasing the output frequency and advancing the receiver's timing toward that of the incoming signal.

The example illustrated in Figure 8.12 tacitly assumes that there will be data state changes before and after the channel symbol of interest. If there are no transitions, it can be seen that the early gate and late gate will have the same integrated energy. Thus there will be no error signal generated for cases where there is no data state change. This is a practical implementation consideration in the use of all symbol synchronizers. Reconsider Figure 8.11. It is not possible to build two integrators that are exactly the same. Thus the signals from the two arms of the early/late-gate loop will contain an offset with respect to each other, even when they should be identical. This offset will be small for well-designed integrators but will cause the loop to drift out of synchronism if there are long sequences of identical data symbols. There are two common responses to this problem. The first, and perhaps most obvious, is to format the data in a manner which ensures that there will be no transitionless intervals that are long enough to allow the loop to break lock. The second response is to modify the loop design so that it contains a single integrator. An example of this type of modified design

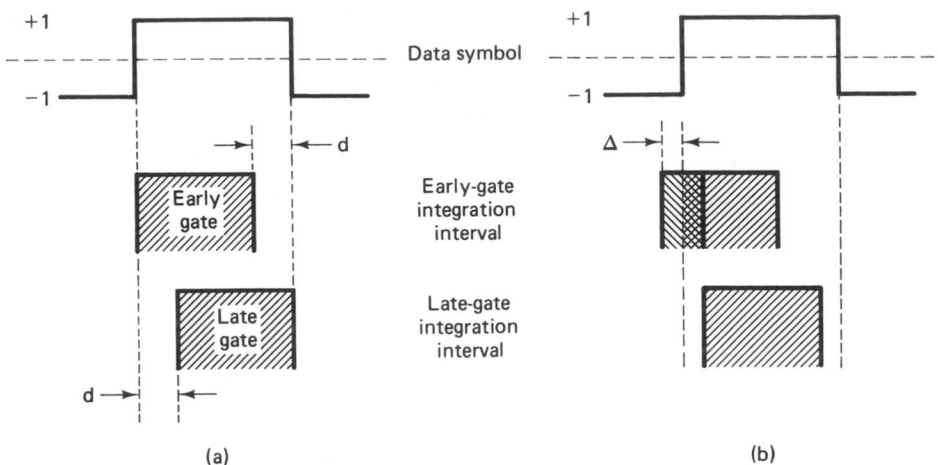

Figure 8.12 (a) Correct receiver timing. (b) Early receiver timing.

is the tau-dither loop, considered in conjunction with the synchronization of spread-spectrum systems in Chapter 10.

Another loop design issue is the integration interval of the two gates. The example illustrated in Figure 8.12 shows the gates to occupy about three-fourths of a symbol period. Actually, this interval can vary from half a symbol interval to nearly a whole symbol interval (why not less than half?). The trade-off is between the amount of integrated noise and interference in a gate versus the amount of signal. As was true with the nonlinear model of phase-locked loops, loops of this type are difficult to analyze; the determination of the best design may require computer simulation. This will be especially true for overlapping gates, as in Figure 8.12, because the noise samples in the two gates will be correlated. Gardner [2] has shown that for a normalized incoming signal of one volt, additive white Gaussian noise, random data (the probability of a transition is $\frac{1}{2}$), and early and late gates that are half a bit interval in duration, for large loop signal-to-noise ratios the fractional timing jitter is approximated by

$$\frac{\sigma_e^2}{T^2} = 2N_0 B_L \tag{8.56}$$

where N_0 is the (normalized) noise power spectral density, T the symbol interval, and B_L the loop bandwidth.

8.2.2.3 Symbol Synchronization Errors and Symbol Error Performance

The effect of symbol synchronization error on bit error probability for a BPSK signal in additive white Gaussian noise is shown in Figure 8.13. It can be seen from the figure that the degradation is less than about 1 dB in signal-to-noise ratio for a fractional timing jitter of less than 5%. Comparing symbol timing error effects to the effect of phase noise (see Figure 8.8), it is seen that the symbol

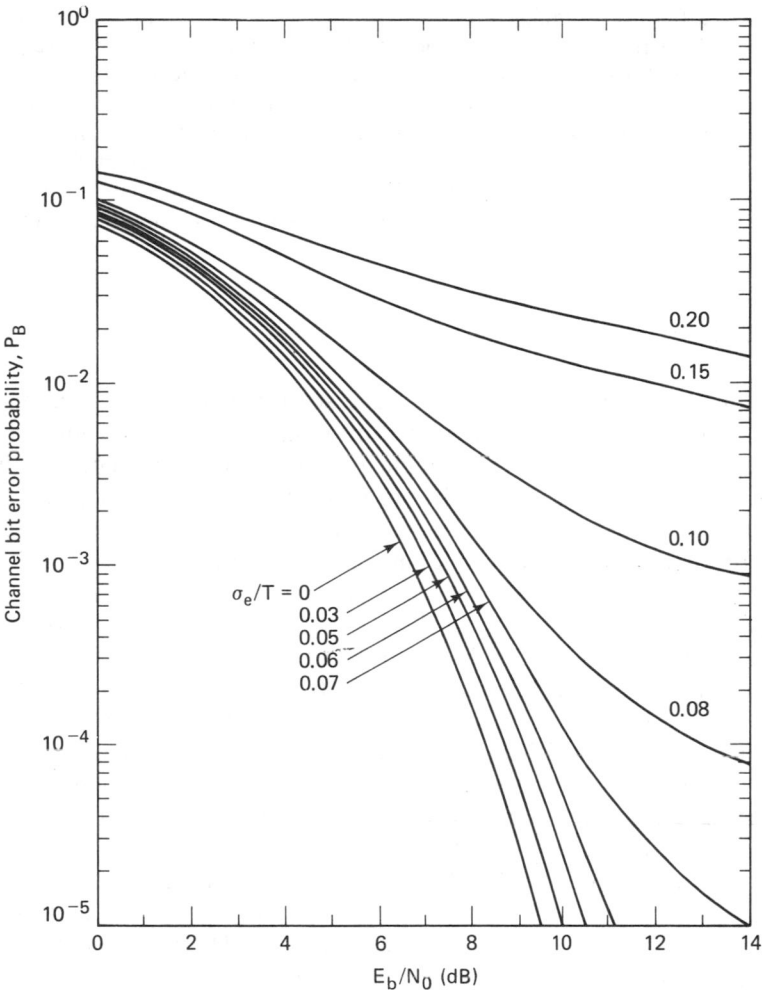

Figure 8.13 Channel bit error probability versus E_b/N_0 with the standard deviation of the symbol sync error σ_e as a parameter. (Reprinted from W. C. Lindsey and M. K. Simon, *Telecommunication Systems Engineering*, Prentice-Hall, Inc., Englewood Cliffs, N.J., 1973, courtesy of W. C. Lindsey and Marvin K. Simon.)

synchronization error, taken as a fraction of the symbol interval, does not affect system performance as strongly as does phase noise taken as a fraction of a cycle. In both cases, however, the degradation increases with increases in error.

Example 8.7 Effect of Timing Jitter

Through the use of Figure 8.13, determine the effect of a 10% symbol fractional timing jitter on a system required to maintain a 10^{-3} bit error probability.

Solution

It can be seen from Figure 8.13 that a 10^{-3}-bit error probability will require a SNR of about 6.7 dB in the absence of all timing jitter. The same figure indicates that for a fractional timing jitter of 10% ($\sigma_e/T = 0.1$) a SNR of about 12.9 dB is required. Thus the ability to accommodate this large timing jitter would require a 6.2-dB higher signal-to-noise ratio than that needed to maintain a 10^{-3}-bit error probability without jitter. This illustrates a use to which Figure 8.13 can be put; however, this example is clearly extreme. No communication system would be designed with over four times the nominally required power level in order to accommodate a large symbol synchronization error. Some other answer would be found, such as redesigning the system filtering to increase the value of K in Equation (8.55), which will reduce the symbol timing jitter.

8.2.3 Frame Synchronization

Almost all digital data streams have some sort of frame structure. This is to say that the data stream is organized into uniformly sized groups of bits. If the data stream is digitized TV, example groups might correspond to horizontal raster scans, which would be further organized in terms of vertical raster scans. Computer data are typically organized into words in one of the several standard digital formats (ASCII, for example) and these, in turn, are organized into card images or files. Any system that uses block error control coding must be organized around the codeword length. Even digital speech is typically transmitted in packets of bits with a constant number of bits in a packet.

For a receiver to make sense of the incoming data stream, the receiver needs to be synchronized with the data stream's frame structure. Frame synchronization is usually accomplished with the aid of some special signaling procedure from the transmitter. This procedure may be very simple, or fairly involved, depending on the environment in which the system is required to operate.

Probably the simplest frame synchronization aid is the *frame marker,* illustrated in Figure 8.14. The frame marker is a single bit, or a short pattern of bits that the transmitter injects periodically into the data stream. The receiver must know the pattern and the injection interval. The receiver, having achieved data synchronization, correlates the known pattern with the incoming data stream at the known injection interval. If the receiver is not in synchronization with the framing pattern, the accumulated correlation will be low. When the receiver comes into frame synch, however, the correlation should be nearly perfect, blemished only by an occasional detection error.

The advantage of the frame marker is its simplicity. Even a single bit can suffice as a frame marker if a sufficient number of correlations are accumulated before deciding whether or not the system has achieved synchronization. The major drawback is that the sufficient number may be very large, and thus the expected time required to acquire synchronization would be long. Therefore, frame markers are most useful in systems that transmit data continuously, like some computer links, and would be inappropriate for systems that transmit in isolated bursts or systems that require rapid frame acquisition.

An approach for systems with inconsistent or bursty transmissions, or systems with rapid acquisition requirements, is a *synchronization codeword*. A synchronization codeword would typically be sent as part of a message header. The

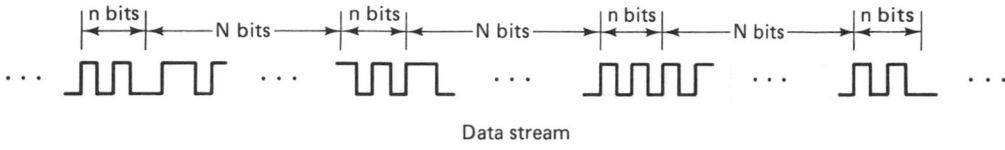

Data stream

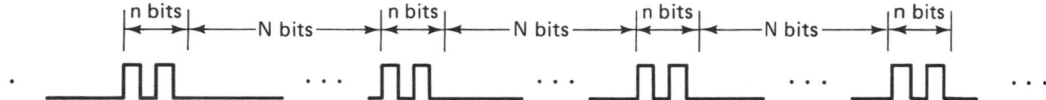

Receiver generated frame marker replica

Figure 8.14 Frame marker illustration.

receiver must know the codeword and be constantly searching for it in the data stream, possibly with a matched filter correlator. Detection of the codeword would indicate a known position (typically the beginning) in the data frame. The advantage of this system is that frame acquisition can be essentially immediate. The only delay would be that required to process the incoming codeword. The disadvantage is that the codeword must be relatively long, relative to the frame marker, to keep the probability of false detections low. The complexity of the correlation operation is proportional to the length of the sequence, so the correlator must be relatively complicated.

A good synchronization codeword is one that has the property that the absolute value of its "correlation sidelobes" is small. A correlation sidelobe is the value of the correlation of a codeword with a time-shifted version of itself. Thus the correlation sidelobe value, C_k, for a k-symbol shift of an N-bit code sequence, $\{X_j\}$, is given by

$$C_k = \sum_{j=1}^{N-k} X_j X_{j+k} \tag{8.57}$$

where X_i $(1 \le i \le N)$ is an individual code symbol taking values ± 1, and the adjacent data symbols (associated with index values $i > N$) are assumed to be zero. An example of correlation sidelobe computation is shown in Figure 8.15. The 5-bit sequence in the example is seen to have good correlation properties, in that the largest sidelobe is one-fifth of the main lobe, C_0. Sequences like the example in Figure 8.15, with the property that their largest sidelobe has a magnitude of unity, are known as Barker sequences or Barker words [19]. There is no known constructive method for finding Barker words, and only 10 unique words are known, the longest of which has 13 symbols. The known unique Barker words are given in Table 8.1. Some thought should make it clear that a completely exhaustive list of known Barker sequences would include those sequences produced by inverting the sign of the symbols and those produced by reversing the time ordering of the symbols in the sequences of Table 8.1.

The sidelobe correlation properties of Barker codes are based on the as-

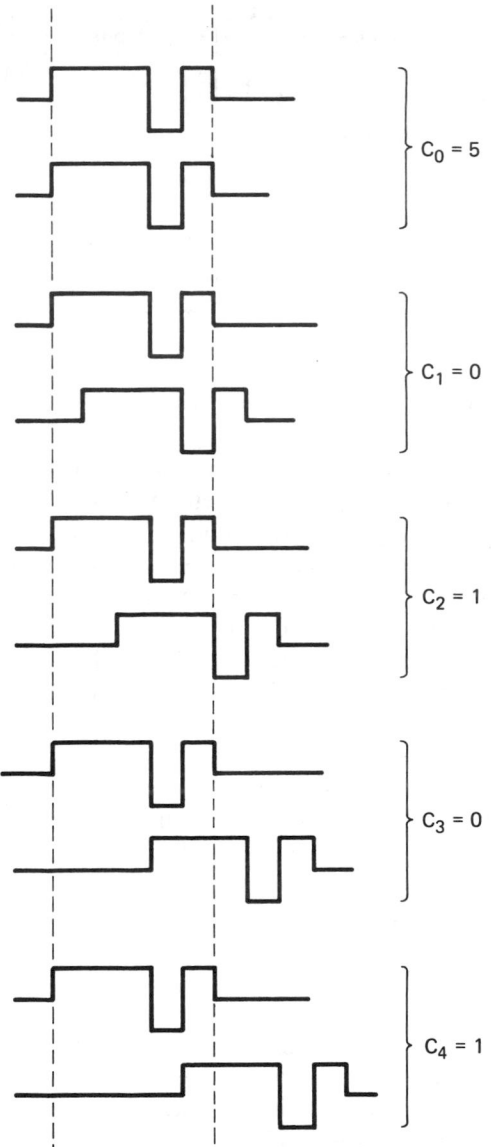

$C_0 = 5$

$C_1 = 0$

$C_2 = 1$

$C_3 = 0$

$C_4 = 1$

Figure 8.15 Correlation sidelobe example.

sumption that the adjacent symbols have zero value. This is an approximation to the effect of equally likely random binary data adjacent to the Barker word, taking the values ± 1. Unfortunately, the Barker sequences are too short for this approximation to provide the best codeword in random binary data in all cases. Willard [20] found the best sequences, in terms of the minimum probability of false synchronization, by the use of computer simulation, for random adjacent symbols for the Barker word lengths. The Willard sequences are shown in Table 8.2.

TABLE 8.1 Barker Synchronization Codewords

N	Barker sequence
1	+
2	+ + or + −
3	+ + −
4	+ + + − or + + − +
5	+ + + − +
7	+ + + − − + −
11	+ + + − − − + − − + −
13	+ + + + + − − + + − + − +

Two probabilities characterize the performance of a system using a synchronization word. These are the probability of a missed detection and the probability of false alarm. Clearly, the system designer would wish both probabilities to be as small as possible. These are conflicting desires. In order to decrease the probability of a miss, the system design may allow less than perfect correlation of an incoming synchronization word. That is, a word may be accepted even if it contains a small number of errors. This, however, enlarges the number of symbol patterns that will be accepted and thereby increases the probability of a false alarm. The probability of a miss for an N-bit word where k or fewer errors are accepted is given by

$$P_m = \sum_{j=k+1}^{N} \binom{N}{j} p^j (1 - p)^{N-j} \qquad (8.58)$$

where p is the probability of a detector bit error. The probability of a false alarm generated by N bits of random data is given by

$$P_{FA} = \sum_{j=0}^{k} \frac{\binom{N}{j}}{2^N} \qquad (8.59)$$

It can be seen that for small p, P_m will decrease roughly exponentially with in-

TABLE 8.2 Willard Synchronization Codewords

N	Willard sequences
1	+
2	+ −
3	+ + −
4	+ + − −
5	+ + − + −
7	+ + + − + − −
11	+ + + − + + − + − − −
13	+ + + + + − − + − + − − −

creasing k. Unfortunately, P_{FA} increases roughly exponentially with increasing k. To obtain acceptable values of both P_m and P_{FA} for a given value of p, the system designer often needs values of N larger than those provided by the Barker and Willard sequences. Fortunately, there is a fairly large and growing body of literature dealing with long sequences. Most of these sequences have been discovered through exhaustive computer searches. Spilker [17] lists sequences of up to $N = 24$ found by Newman and Hofman [21] and mentions that their original paper has sequences to $N = 100$. Wu [22] provides a list of Maury-Styles sequences to length $N = 30$, a list of Linder sequences to length 40, and a fairly complete discussion of the topic of synchronization sequences, including constructive techniques for reasonable but nonoptimum sequences, and insight into the frame synchronization procedures of some operational satellite digital communication systems.

8.3 NETWORK SYNCHRONIZATION

For systems using coherent modulation techniques, one-direction communications such as broadcast channels, or single-link communications, such as most microwave links, land-line links, or fiber optics links, the synchronization architecture that makes the most sense is to make synchronization totally a receiver function. For communications systems using noncoherent modulation techniques, or that involve many users accessing a central communication node, such as many satellite communication systems, it often makes sense for synchronization to be mostly or entirely a terminal function. This means that the terminal transmitter parameters are modified to achieve synchronization, rather than modifying the central node's receiver parameters. This must be the approach if the system uses time-division multiple access (TDMA). In TDMA each user is allotted a segment of time in which to transmit its information. The terminal transmitter must be synchronized with the system in order for its transmitted burst of data to arrive at the central node at the time when the node is prepared to receive the data. Synchronization of the terminal transmitter also makes sense with systems that combine signal processing at the central node with frequency-division multiple access (FDMA). If the terminals precorrect their transmission to be synchronized with the central node, the node can use a fixed set of channel filters and a single timing reference for the processing of all channels. Otherwise, the node would require a separate time and frequency acquisition and tracking capability for each incoming channel, and would need to deal with the possibility of varying amounts of adjacent channel interference. It seems clear that terminal transmitter synchronization is often the cleaner, more reasonable system approach to synchronizing a network.

Transmitter synchronization procedures may be classified as being either open loop or closed loop. Open-loop techniques do not rely on any measurement of the arriving signal parameters at the central node. The terminal precorrects its transmission based on stored knowledge of link parameters that have been provided by some external authority but may possibly be modified by observations of a return signal from the central node. Open-loop techniques rely on link parameters being accurately known and predictable. They work best when link geometry is nominally fixed, and the links themselves operate continually for relatively long periods, once established. They tend to be difficult to use efficiently when the link geometry is not static or when the terminals access the system sporadically. The main advantages of the open-loop methods are that acquisition is fast, the procedure can work without a return link, and the amount of real-time computation that is required is small. The disadvantages of open-loop methods are that they require the existence of the external authority that provides knowledge of the required link parameters and that they are relatively inflexible. The lack of any direct real-time measure of system characteristics means that the system cannot adjust quickly to any unplanned change in conditions.

Closed-loop techniques, on the other hand, require little in the way of a priori knowledge of link parameters. Knowledge would be useful in reducing the time required for acquisition, but need not be precise as is required by open-loop methods. Closed-loop methods involve measurements of the synchronization accuracy of the incoming transmissions from the terminal upon their arrival at the central node, and the return of the results of these measurements to the terminal via a return path. Thus closed-loop methods require a return path that provides a response to the terminal's transmission, the ability in the terminal to recognize the response for what it is, and the ability in the terminal to modify the transmitter characteristics appropriately, based on the response. This amounts to a requirement for a relatively large amount of real-time processing in the terminal, and two-way links between every terminal and the central node. The disadvantages of closed-loop methods are that they require a relatively large amount of real-time processing, require two-way links to every terminal, and that acquisition can take a relatively long time. The advantages are that no external source of knowledge is required for the system to work, and the responses on the return link allow the system to adapt easily and quickly to changing geometries and link conditions.

8.3.1 Open-Loop Transmitter Synchronization

Open-loop systems can be further subdivided into systems that employ information gained by observing a return link, and those that do not. Those that do not are the simplest of all, in terms of real-time processing requirements, but communication performance for these simple terminals is clearly very dependent on stable link characteristics.

All transmitter synchronization schemes attempt to precorrect the timing and transmission frequency of the signal in such a manner that the signal will arrive at a receiver with the expected frequency and at the expected time. Thus, to precorrect time, a transmitter would divide the distance between itself and the receiver by the speed of light to get the transmission transit time, and then shift the message transmission timing that much ahead. By transmitting the signal early, it will arrive at the receiver at the appropriate time. The time of arrival at the node, T_A, is given by

$$T_A = T_t + \frac{d}{c} \qquad (8.60)$$

where T_t is the actual transmission start time, d the transmit distance, and c the speed of light. Similarly, to precorrect the transmission frequency, the transmitter must allow for the Doppler shift caused by relative motion between the transmitter and the intended receiver. To be received correctly, the required *transmission radian frequency* is

$$\omega \simeq \left(1 - \frac{v}{c}\right)\omega_0 \qquad (8.61)$$

where c is the speed of light, v the relative velocity (positive for decreasing transmission distance), and ω_0 the nominal transmission radian frequency.

Unfortunately, in practice neither the time nor the frequency precorrection can be done exactly. Even satellites in nominally geostationary orbits move slightly with respect to a point on the earth, and the behavior of the time and frequency references in the terminal and the central node are never entirely predictable. Thus there will always be some time and frequency precorrection error. The time error may be expressed as

$$T_e = \frac{r_e}{c} + \Delta t \qquad (8.62)$$

where r_e is the error in the range estimate and Δt is the difference between the time reference at the terminal and the reference at the receiver. The frequency error may be expressed as

$$\omega_e = \frac{v_e \omega_0}{c} + \Delta\omega \qquad (8.63)$$

where v_e is the error in the measured or predicted relative velocity of the transmitter and receiver—the Doppler error, and $\Delta\omega$ is the frequency difference between the transmitter and the receiver frequency references. There are many other sources of time and frequency error in addition to those mentioned here, but they are typically much less important. Spilker [17] gives a reasonably complete accounting of time and frequency error sources for satellite systems.

The error terms Δt and $\Delta\omega$ are typically due to random fluctuations in frequency references. The time reference for a transmitter or receiver is generally obtained by counting cycles of the frequency reference, so errors in the accuracy of the time and frequency references are related. The fluctuation in a frequency reference are very difficult to characterize statistically, although the power spectral density of the fluctuations is approximated by a sequence of power-law segments [12]. Frequency references are often specified in terms of a maximum allowable fractional frequency change per day:

$$\delta = \frac{\Delta\omega}{\omega_0} \qquad \text{hertz/hertz/day} \qquad (8.64)$$

Typical values for δ range from 10^{-5} to 10^{-6} for inexpensive crystal oscillators, to 10^{-9} to 10^{-11} for high-quality crystal oscillators, to 10^{-12} for rubidium standards, to 10^{-13} for cesium standards. An effect of specifying system frequency references by the maximum fractional frequency is that if there is no intervention, the offset from the nominal frequency, ω_0, can grow linearly with time.

$$\Delta\omega(T) = \omega_0 \int_0^T \delta \, dt + \Delta\omega(0) = \omega_0 \, \delta T + \Delta\omega(0) \qquad \text{hertz} \qquad (8.65)$$

For a cycle-counting time reference, however, the cumulative time offset is related to the cumulative phase error of the reference:

$$\Delta t(T) = \int_0^T \frac{\Delta\omega(t)}{\omega_0} \, dt + \Delta t(0)$$

$$= \int_0^T \delta t \, dt + \int_0^T \frac{\Delta\omega(0)}{\omega} \, dt + \Delta t(0)$$

$$= \frac{1}{2} \delta T^2 + \frac{\Delta\omega(0) \, T}{\omega_0} + \Delta t(0) \qquad (8.66)$$

Thus, without intervention, a time reference error can grow quadratically with time. For open-loop transmitter synchronization systems this quadratic growth in time error often sets limits on how often the exernal authority must intervene, either to update the terminal's knowledge of receiver timing, or to reset both the receiver's and the transmitter's time references to nominal. The quadratic error growth usually means that timing errors are more of an operational problem than are frequency errors, although this will depend on the system design.

If the transmitter does not have information from measurements on a return link, the time and frequency offsets as modeled by Equations (8.62) to (8.66) will allow a system designer the ability to determine the maximum interval between interventions, based on a probability-of-error criterion. Time and frequency reference recalibration is often a burdensome procedure, to be done as rarely as possible.

If a terminal has access to a return link from the central node and the ability to make comparative measurements between the local reference and incoming signal parameters, the interval between recalibrations can be made much longer. Large satellite control stations can measure and model the orbital parameters of nominally geostationary satellites to an accuracy of a few tens of feet in range and a few feet/second in velocity relative to the ground terminal. Thus, for the important special case of a synchronous satellite as the central node, the first terms on the right-hand side of Equations (8.62) and (8.63) are usually negligible. When this is true, the differences between the incoming signal parameters and those generated by the terminal's time and frequency references will approximate the error terms Δt and $\Delta\omega$. These error terms measured on the downlink can be used to compute appropriate corrections to the uplink transmissions. On the other hand, if the time and frequency references are known to be accurate but the link geometry is somewhat in question, perhaps because the terminal is mobile or the satellite is nongeostationary, the same sort of return link measurement could be used to resolve range or velocity uncertainties. These measures of range or relative velocity can then be used to precorrect uplink timing and frequency.

The case where a terminal is able to utilize measurements made on a return link signal is sometimes called quasi-closed-loop transmitter synchronization. The quasi-closed technique is clearly more adaptable to uncertainties in the communication system than is the purely open-loop system. The purely open-loop system requires complete a priori knowledge of all important link parameters in order to operate successfully. Unanticipated changes in the links cannot be tolerated. The quasi-closed-loop system, on the other hand, requires a priori knowledge of all but one of the important parameters in each of time and frequency, but the remaining term can be determined from observations of the return link. This adds complexity to the terminal, but also adds the ability to adapt to certain types of unplanned link changes. This degree of adaptability can greatly reduce the frequency of required system calibration.

8.3.2 Closed-Loop Transmitter Synchronization

Closed-loop transmitter synchronization involves the transmission of special synchronization signals that are used to determine the signal's time or frequency error relative to the desired timing or frequency when the signal arrives at the receiver. The results of this determination are then fed back to the transmitter on a return link. The determination of synchronization errors can be either implicit or explicit. If the central node has sufficient processing capacity, the central node may make an actual error measurement. Such a measurement might be the amount and direction of offset, or perhaps simply the direction alone. This information would be formatted and returned to the transmitter on a return link. If the central node has little processing capability, the special synchronization signal may simply be turned around and returned to the transmitter on the return link. In this case it becomes part of the transmitter's task to interpret the returned signal for itself. The design of a special synchronization signal that lends itself to easy unambiguous interpretation can be a challenge.

The relative advantages and disadvantages of the two types of closed-loop systems have to do with the location of the signal processing capability and the efficiency of channel usage. A major advantage of having the processing at the central node is that results of the error measurements that are transmitted on the return link can be a short digital sequence. This efficient use of the return link can be important if a single return link is time-division multiplexed between a large number of terminals. A second potential advantage is that the error-measuring capability in the central node can be shared by all terminals communicating through the node. This can amount to a large savings in system processing capability. The principal potential advantage in having the processing at the terminal is that the central node may not be easily accessible, and reliability considerations may dictate a simple design. This has typically been the case when the central node is a space satellite. With continuing improvements in satellite technology, simplicity requirements can be expected to be less dominant in the future than in the past. Another potential advantage to having the processing in the terminal is that the response can be quicker because there is little processing delay in the central node. This may be important if link parameters are changing very rapidly. The primary disadvantages are the inefficient use of the return channel and that the return signals may be difficult to interpret. This difficulty would arise when the central node is not just a simple repeater, but makes symbol decisions and transmits these decisions on the return link. This symbol decision capability can greatly improve the terminal-to-terminal error performance, but complicates the synchronization procedure. This is because the effects of a time or frequency offset are resident in the return signal indirectly—only as they have affected the symbol decisions. Consider the example of a BFSK transmission to a central node that makes noncoherent bit decisions. The decisions will be dependent on the detected signal energy in the mark and space detectors. If the transmitted signal is an alternating sequence of marks and spaces, the signal at the central node can be modeled as

$$r(t) = \begin{cases} \sin\left[(\omega_0 + \omega_s + \Delta\omega)t + \theta\right] & 0 \le t \le \Delta t \\ \sin\left[(\omega_0 + \Delta\omega)t + \theta\right] & \Delta t < t \le T \end{cases} \tag{8.67}$$

where T is the symbol interval, ω_0 one symbol frequency, $(\omega_0 + \omega_s)$ the other symbol frequency, $\Delta\omega$ the frequency error at the central node, Δt the signal arrival time error at the central node, and θ an arbitrary phase angle. Now if x and y represent the detector quadrature components,

$$x = \frac{1}{T}\int_0^T r(t)\cos\omega_0 t\, dt \tag{8.68}$$

$$y = \frac{1}{T}\int_0^T r(t)\sin\omega_0 t\, dt \tag{8.69}$$

the detected signal energy can be expressed as

$$z^2 = x^2 + y^2$$

$$= \left(\frac{\sin\left[(\omega_s + \Delta\omega)\,\Delta t/2\right]}{(\omega_s + \Delta\omega)T}\right)^2 + \left(\frac{\sin\left[\Delta\omega(T - \Delta t)/2\right]}{\Delta\omega T}\right)^2 \tag{8.70}$$

$$+ \frac{\cos(\Delta\omega\,\Delta t) + \cos\left[\Delta\omega T - (\omega_s + \Delta\omega)\,\Delta t\right] - \cos(\Delta\omega\, T) - \cos(\omega_s\,\Delta t)}{2\Delta\omega(\omega_s + \Delta\omega)T^2}$$

For the special case where the time error, Δt, is zero, Equation (8.70) simplifies to

$$z^2 = \left[\frac{\sin(\Delta\omega\, T/2)}{\Delta\omega\, T}\right]^2 \tag{8.71}$$

For the case where the frequency offset is zero,

$$z^2 = \left(\frac{T - \Delta t}{2T}\right)^2 + \left[\frac{\sin(\omega_s\,\Delta t/2)}{\omega_s T}\right]^2 \tag{8.72}$$

The important thing to notice in Equations (8.70) to (8.72) is that any time error or frequency offset or combination of both will decrease the detected signal energy in the correct symbol detector and introduce signal energy into the incorrect signal detector. This will reduce the effective distance between signals in signal space and degrade error performance. A measurement of error performance, however, which is all that is available on the return link, gives no insight into whether the problem is a frequency offset, a time error, or a combination of both. Thus the transmission of standard signals is not likely to provide a useful response for synchronization.

A useful technique for determining the correct frequency precorrection for our example of BFSK signaling is to transmit a constant tone whose frequency is the average of the two symbol frequencies. Such a tone should produce a random binary sequence on the return link with equal numbers of marks and spaces. A frequency offset from the average would produce predominately marks or spaces. Finding the center frequency in this way allows accurate frequency precorrection of the signals. Once the correct frequency is found, the transmitter can transmit an alternating sequence of marks and spaces in order to discover correct timing. By varying the timing of the transmission through a range of half a symbol interval, the transmitter can look for the timing that provides the worst error performance. When the transmission arrival at the central node is displaced from correct timing by half a symbol interval, the two detectors will detect equal amounts of energy, and the binary sequence on the return link will be random. Determining the time when the transmitted and return signal are decorrelated will allow the transmitter to compute the correct transmission timing. Notice that this procedure works better than attempting to find the point at which error performance is the best. Any well-designed system will have sufficient transmission energy to allow for slight timing offsets, so an error-free return signal could be achieved with less than perfect timing. In fact, the larger the signal-to-noise ratio, the worse a best-finding procedure works. A worst-finding system, however, will work well for any well-designed system and will improve in potential accuracy with increasing signal-to-noise ratio. This can be seen intuitively, because increased signal-to-noise ratios will allow the system to tolerate larger timing errors, so the improvement in error performance as the timing error decreases from half a symbol time will be more rapid in the large signal-to-noise case than in the smaller signal-to-noise ratio case. This will allow a more precise determination of the half-symbol timing position.

8.4 CONCLUSION

This chapter has outlined the fundamental problems and issues associated with synchronization in digital communications. The trade-offs are generally between expense and complexity, on the one hand, and error performance on the other. We have discussed receiver synchronization and phase-locked loops (PLL) in particular. Typically, it is the receiver that takes the most active role in the synchronization of a communications link. Even in cases where a terminal's transmitter assumes the more active role, as in some satellite links, the process is often aided by a return path that has been acquired by the terminal's receiver. Thus receiver synchronization is more fundamental. Phase-locked loops, and their variations, are the primary control circuits used to track variations in phase of an incoming signal. The mathematics needed to describe the response of a PLL to a given input involves the solution to a nonlinear differential equation. It was shown, however, that under steady-state conditions, a linearized model provides a useful approximation to system performance. In circumstances where the linearized model cannot be accurately applied, results by Viterbi [5] for first-order loops were introduced. Although exact for first-order loops only, these results have been shown to be useful approximations to the performance of higher-order loops as well [2].

The extremely important special case of suppressed-carrier loops was discussed. Suppressed-carrier loops are required to track the phase of an incoming signal that has no average energy at the carrier frequency. The common example of such a signal is one that has been modulated with standard antipodal BPSK. In this situation, a harmonic of the suppressed carrier is produced through the use of a nonlinearity, and the harmonic is tracked.

The next higher level of synchronization treated here was symbol synchronization. Two primary classes of symbol synchronization were discussed. Open-loop synchronizers operate directly on the modulated signal to produce a symbol transition indication. Closed-loop synchronizers use a closed-cycle control loop to acquire and track the symbol transitions.

The highest level of synchronization considered was frame synchronization. To receive the data in a useful form, the receiver must determine which symbols belong to which frames. This knowledge is equivalent to having frame sync, which is usually accomplished by including some recognizable pattern, known to the receiver, with the data symbols. The receiver scans the incoming data until it recognizes the pattern. Synchronization can be checked by looking for periodic repetitions of the pattern.

This chapter has necessarily been only an outline of the important problems, issues, and results relating to the synchronization of digital communication systems. The interested reader will find that the references listed are worthy works that will provide much greater depth of coverage than space has allowed here.

REFERENCES

1. Peterson, W. W., and Weldon, E. J., *Error-Correcting Codes,* The MIT Press, Cambridge, Mass., 1972.

2. Gardner, F. M., *Phaselock Techniques,* 2nd ed., John Wiley & Sons, Inc., New York, 1979.

3. Davenport, W. B., and Root, W. L., *Random Signals and Noise,* McGraw-Hill Book Company, New York, 1958.

4. Papoulis, A., *Probability, Random Variables, and Stochastic Processes,* McGraw-Hill Book Company, New York, 1965.

5. Viterbi, A. J., *Principles of Coherent Communications,* McGraw-Hill Book Company, New York, 1966.

6. Lindsey, W. C., *Synchronization Systems in Communication and Control,* Prentice-Hall, Inc., Englewood Cliffs, N.J., 1972.

7. Lindsey, W. C., and Simon, M. K., "Detection of Digital FSK and PSK Using a First-Order Phase-Locked Loop," *IEEE Trans. Commun.,* vol. COM25, no. 2, Feb. 1977, pp. 200–214.

8. Develet, J. A., Jr., "The Influence of Time Delay on Second-Order Phase Lock Loop Acquisition Range," *Int. Telem. Conf.,* London, 1963.

9. Johnson, W. A., "A General Analysis of the False-Lock Problem Associated with the Phase-Lock Loop," *The Aerospace Corp., Rep. TOR-269(4250-45)-1,* NASA Accession N64-13776, 1963.

10. Tausworthe, R. C., "Acquisition and False-Lock Behavior of Phase-Locked Loops with Noisy Inputs," *Jet Propulsion Laboratory, JPL SPS 37-46,* vol. 4, 1967.

11. Franks, L. E., "Synchronization Subsystems: Analysis and Design," in K. Feher, *Digital Communications, Satellite/Earth Station Engineering,* Prentice-Hall, Inc., Englewood Cliffs, N.J., 1981, Chap. 7.

12. Simon, M. K., and Yuen, J. H., "Receiver Design and Performance Characteristics," in J. H. Yuen, ed., *Deep Space Telecommunications Systems Engineering,* Plenum Press, New York, 1983.

13. Gardner, F. M., "Hangup in Phase-Lock Loops," *IEEE Trans. Commun.,* COM25, October 1977.

14. Blanchard, A., *Phase-Locked Loops,* John Wiley & Sons, Inc., New York, 1976.

15. Holmes, J. K., *Coherent Spread Spectrum Systems,* John Wiley & Sons, Inc., New York, 1982.

16. Lindsey, W. C., and Simon, M. K., eds., *Phase Locked Loops and Their Applications,* IEEE Press, New York, 1977.

17. Spilker, J. J., Jr., *Digital Communications by Satellite,* Prentice-Hall, Inc., Englewood Cliffs, N.J., 1977.

18. Wintz, P. A., and Luecke, E. J., "Performance of Optimum and Suboptimum Synchronizers," *IEEE Trans. Commun. Technol.,* June 1969, pp. 380–389.

19. Barker, R. H., "Group Synchronization of Binary Digital Systems," in W. Jackson, ed., *Communication Theory,* Academic Press, Inc., New York, 1953.

20. Willard, M. W., "Optimum Code Patterns for PCM Synchronization," *Proc. Natl. Telem. Conf.,* 1962, paper 5-5.

21. Newman, F., and Hofman, L., "New Pulse Sequences with Desirable Correlation Properties," *Proc. Natl. Telem. Conf.,* 1971, pp. 272–282.

22. Wu, W. W., *Elements of Digital Satellite Communication,* Vol. 1, Computer Science Press, Inc., Rockville, Md., 1984.

PROBLEMS

8.1. A transmitter is sending an unmodulated tone of constant energy (a beacon) to a distant receiver. The receiver and transmitter are in motion with respect to each other such that $d(t) = D[1 - \sin(mt)] + D_0$, where $d(t)$ is the distance between the transmitter and receiver (possibly this represents an aircraft doing "figure-eight" maneuvers over a ground station), and D, m, and D_0 are constants. This relative motion will cause a Doppler shift in the received transmitter frequency of

$$\Delta\omega_D(t) = \frac{\omega_0 v(t)}{c}$$

where $\Delta\omega_D$ is the Doppler shift, ω_0 the nominal carrier frequency, $v(t) = \dot{d}(t)$ the relative velocity between the transmitter and receiver, and c the speed of light. Assuming that the linearized loop equations hold and that the receiver's PLL is in lock (zero phase error) at $t = 0$, show that an appropriately designed first-order loop can maintain frequency lock.

8.2. Consider a transmitter and receiver that are in relative motion as in Problem 8.1. Once again assume that the linearized loop equations hold. Under this assumption determine the PLL phase error as a function of time for the all-pass and low-pass loop filters of Equations (8.13) and (8.14). Demonstrate that the validity of the assumption of the linearized loop equations depends on the value of the gain K_0.

8.3. A high-performance aircraft is transmitting an unmodulated carrier signal to a ground terminal. The ground terminal is initially in phase lock with the signal. The aircraft performs a maneuver whose dynamics are described by the equation for acceleration, $a(t) = At^2$, where A is a constant. Assuming that the linearized equations apply, determine the minimum order of the phase-locked loop required to track the signal from this aircraft.

8.4. Show that the loop bandwidth of a first-order phase-locked loop is given by $B_L = K_0/4$, where K_0 is the loop gain.

8.5. A second-order phase-locked loop has a low-pass loop filter given by

$$F(\omega) = \frac{\omega_1}{j\omega + \omega_1}$$

and a loop gain of K_0. Under the assumption that $K_0 \geq \omega_1/4$, show that the loop bandwidth of this phase-locked loop is given by $B_L = K_0/8$. [*Hint:**

$$\int_{-\infty}^{\infty} \frac{dx}{R} = \frac{\pi \cos (h/2)}{2cq^3 \sin h} \qquad \text{for } 4ac > b^2$$

where $R = a + bx^2 + cx^4$, $q = \sqrt[4]{a/c}$, and $\cos h = -b/2\sqrt{ac}$.]

8.6. A first-order phase-locked loop with loop gain K_0 is disturbed by additive white Gaussian noise of normalized (to unit signal energy) two-sided power spectral density of $N_0/2$ watts/hertz. Determine the necessary relationship between noise power spectral density and loop gain if the loop is designed to cycle slip no more often than once per day.

8.7. Viterbi [5] determined that the probability density function of the output phase of a first-order phase-locked loop disturbed by white Gaussian noise is given by

$$p(\phi) = \frac{\exp (\rho \cos \phi)}{2\pi I_0(\rho)}, \qquad |\phi| \leq \pi, \ \rho \geq 0$$

Demonstrate that $p(\phi)$ given above is in fact a probability density function, and compute the mean and variance of ϕ.

8.8. Computer simulations and laboratory measurements have indicated that the time

---

* I. S. Gradshteyn and I. M. Ryzhik, *Table of Integrals, Series and Products* (New York: Academic Press, 1965), 2.161.1.

between cycle slips is exponentially distributed; that is, the distribution function of the time between cycle slips, T, is given by

$$p(T) = 1 - \exp\left(-\frac{T}{T_m}\right)$$

Given this distribution function, find the mean time between cycle slips and the variance about this mean as functions of T_m. If the mean time between cycle slips is 1 day, what is the probability of cycle slips less than 1 hour apart? More than 3 days apart?

8.9. Consider a second-order phase-locked loop with a low-pass loop filter

$$F(\omega) = \frac{\omega_1}{j\omega + \omega_1}$$

During aided acquisition it is desired that this loop be scanned throughout a 1000-radian uncertainty region in 1 s. If the relationship between loop gain and filter constant is $K_0 = 2\omega_1$, what is the required relationship between loop gain and the single-sided additive white Gaussian noise power spectral density, N_0? Determine the largest value of N_0 that can be accommodated.

8.10. Consider the operation of an open-loop symbol synchronizer whose band pass filter (BPF) has a bandwidth of $0.1/T$ hertz, where T is the symbol period. For a bit energy-to-noise power spectral density ratio (E_b/N_0) of 10 dB, determine the magnitude of the approximate mean and variance of the fractional tracking error, and compute an upper bound on the probability that the tracking error exceeds three times its approximate fractional mean. (*Hint:* Consider the Chebyshev inequality [4].)

8.11. A communication system is used to transmit commands to a payload at a data rate of 100 bits/s. Each command is preceded by an N-bit header that identifies it in the data stream. Assuming that except possibly for the header, the bits appear to be random [$P(1) = P(0) = \frac{1}{2}$], what is the minimum-length header that would provide an expected frequency of false alarms of one per year? For a channel bit error probability of 10^{-5}, what is the probability of missing this header? What is the miss probability if the channel error probability is 2×10^{-2}? If the system is redesigned to accept the header with up to two errors, what is the minimum required length for an expected false alarm rate of one per year? What is the miss probability with this new system and a channel error probability of 2×10^{-2}?

8.12. A deep-space probe is moving away from the earth at a nominal velocity of 15,000 m/s with a velocity uncertainty of ± 3 m/s. The probe frequency reference is specified to have a drift rate of no more than 10^{-9} Hz/Hz/day. The nominal downlink transmission frequency is 8 GHz. After a 1-month (30-day) silence the probe begins a scheduled transmission toward an earth terminal. The earth terminal contains a cesium standard. What center frequency and frequency search bandwidth should be used by the ground station? Assuming that the range to the probe was accurately known at the beginning of the month, and that the uncertainty in the probe's time and frequency references were zero [$\Delta t(0) = 0$, $\Delta \omega(0) = 0$], what is the uncertainty in the time of arrival of the downlink transmission?

8.13. A communications link operates at a nominal center frequency of 10 GHz for a single brief period once a day. The receiver operates with a second-order PLL that has a pull-in range of ± 1 kHz. Assuming that the loop acquires using self-acquisition and that both the transmitter and receiver use the same type of frequency reference, what type of frequency reference must this be?

CHAPTER 9

Multiplexing
and
Multiple Access

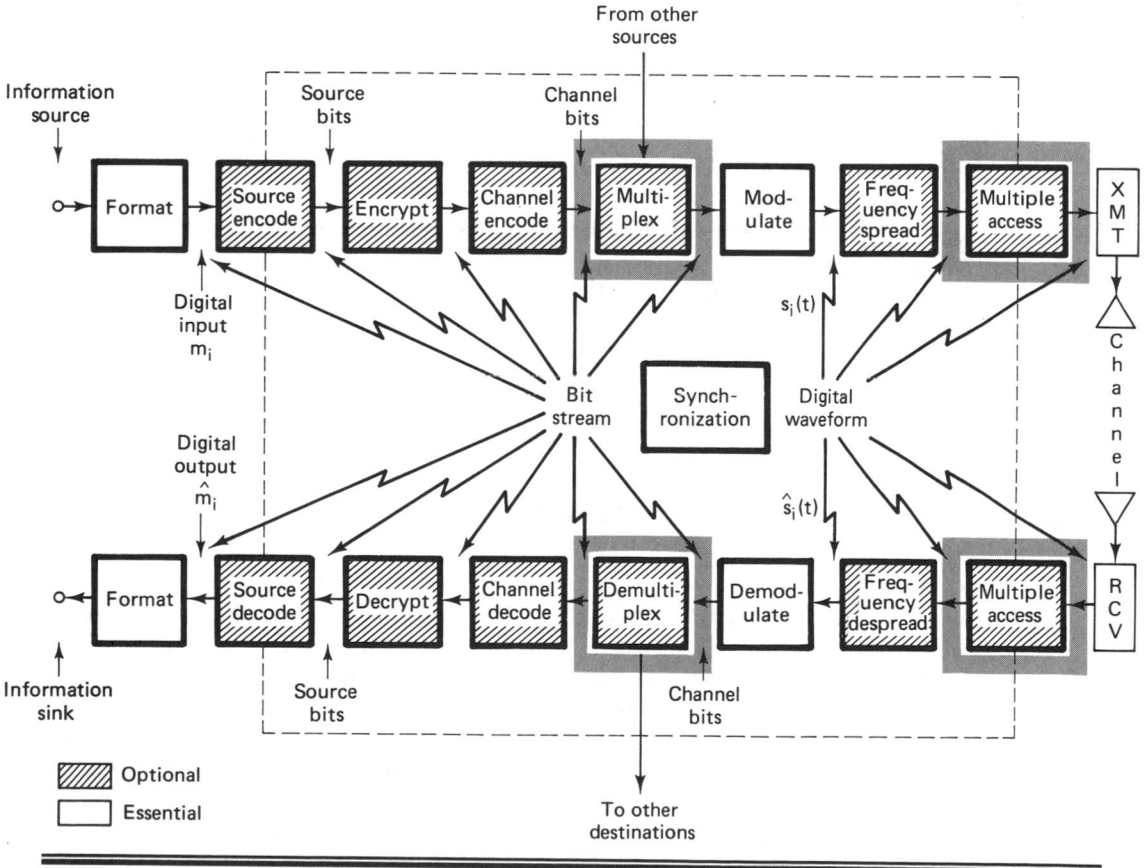

The terms "multiplexing" and "multiple access" refer to the sharing of a fixed communications resource (CR). There is a subtle difference between multiplexing and multiple access. With *multiplexing*, users' requirements or plans for CR sharing are fixed, or at most, slowly changing. The resource allocation is assigned a priori, and the sharing is usually a process that takes place within the confines of a *local site* (e.g., a circuit board). *Multiple access*, however, usually involves the *remote sharing* of a resource, such as a satellite. With a dynamically changing multiple access scheme, a system controller must become aware of each user's CR needs; the amount of time required for this information transfer constitutes an overhead and sets an upper limit on the efficiency of the utilization of the CR.

9.1 ALLOCATION OF THE COMMUNICATIONS RESOURCE

There are three basic ways to increase the throughput (total data rate) of a communications resource (CR). The first way is either to increase the transmitter's effective isotropic radiated power (EIRP) or to reduce system losses so that the received E_b/N_0 is increased. The second way is to provide more channel bandwidth. The third approach is to make the allocation of the CR more efficient. This third approach is the domain of communications multiple access. The problem, in the context of a satellite transponder, is to efficiently allocate portions of the transponder's fixed CR to a large number of users who seek to communicate digital information to each other at a variety of bit rates and duty cycles. The

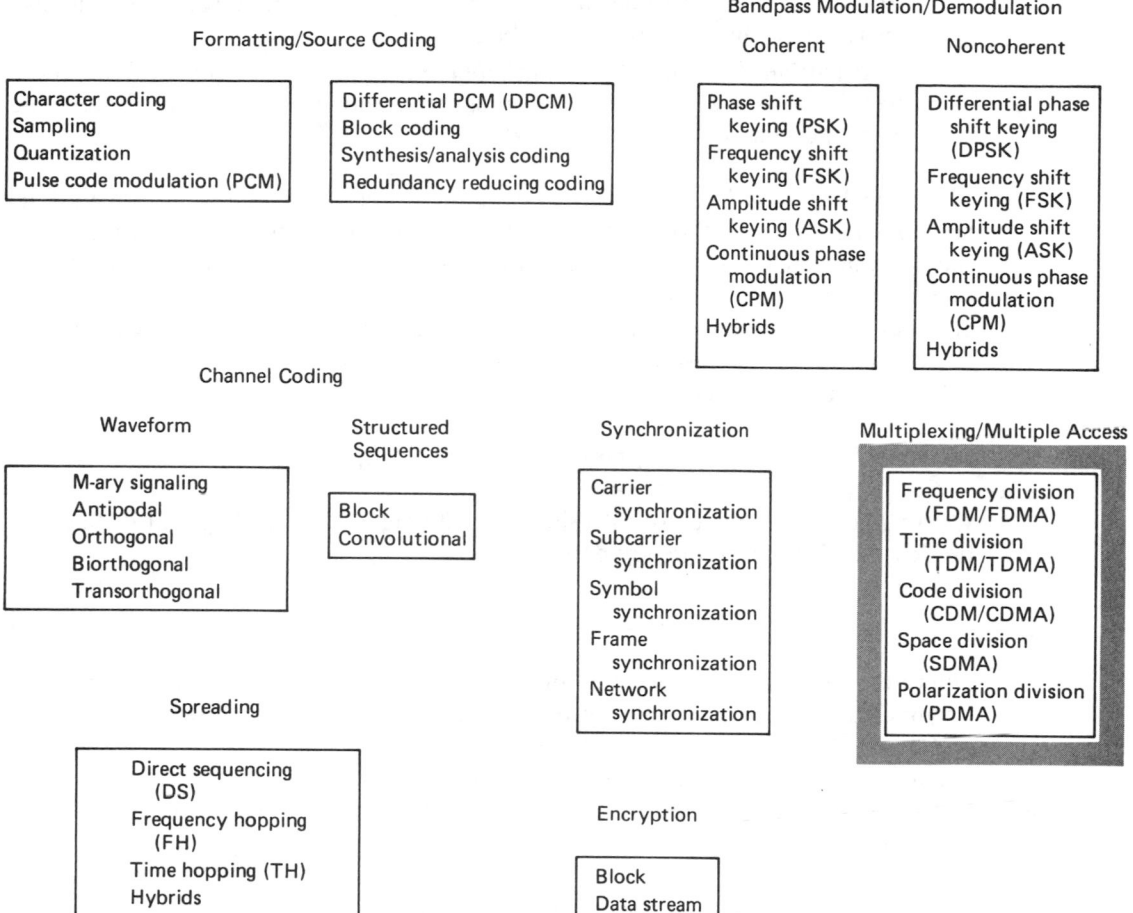

Figure 9.1 Basic digital communication transformations.

basic ways of distributing the communications resource. listed under the heading "multiplexing/multiple access" in Figure 9.1, are:

1. *Frequency division (FD).* Specified subbands of frequency are allocated.
2. *Time division (TD).* Periodically recurring time slots are identified. With some systems, users are provided a fixed assignment in time. With others, users may access the resource at random times.
3. *Code division (CD).* Specified members of a set of orthogonal or nearly orthogonal spread spectrum codes (each using the full channel bandwidth) are allocated.
4. *Space division (SD) or multiple beam frequency reuse.* Spot beam antennas

are used to separate radio signals by pointing in different directions. It allows for reuse of the same frequency band.

5. *Polarization division (PD) or dual polarization frequency reuse.* Orthogonal polarizations are used to separate signals, allowing for reuse of the same frequency band.

The key to *all* multiplexing and multiple access schemes is that various signals share a CR without creating unmanageable interference to each other in the detection process. The allowable limit of such interference is that signals on one CR channel should not significantly increase the probability of error in another channel. Orthogonal signals on separate channels will avoid interference between users. Signal waveforms $x_i(t)$, where $i = 1, 2, \ldots$, are defined to be orthogonal if they can be described in the time domain by

$$\int_{-\infty}^{\infty} x_i(t)x_j(t) \, dt = \begin{cases} K & \text{for } i = j \\ 0 & \text{otherwise} \end{cases} \tag{9.1}$$

where K is a nonzero constant. Similarly, the signals are orthogonal if they can be described in the frequency domain by

$$\int_{-\infty}^{\infty} X_i(f)X_j(f) \, df = \begin{cases} K & \text{for } i = j \\ 0 & \text{otherwise} \end{cases} \tag{9.2}$$

where the functions $X_i(f)$ are the Fourier transforms of the signal waveforms $x_i(t)$. Channelization characterized by orthogonal waveforms, as shown in Equation (9.1), is called time-division multiplexing or time-division multiple access (TDM/TDMA), and that characterized by orthogonal spectra, as shown in Equation (9.2), is called frequency-division multiplexing or frequency-division multiple access (FDM/FDMA).

9.1.1 Frequency-Division Multiplexing/Multiple Access

9.1.1.1 Frequency-Division Multiplex Telephony

In the early days of telephony, a separate pair of wires was needed for each telephone trunk circuit (trunk circuits interconnect intercity switching centers). As illustrated in Figure 9.2, the skies of all the major cities in the world grew dark with overhead wires as the demand for telephone service grew. A major development in the early 1900s, frequency-division multiplex (FDM) telephony, made it possible to transmit several telephone signals simultaneously on a single wire, and thereby transformed the methods of telephone transmission.

The communications resource (CR) is illustrated in Figure 9.3 as the frequency–time plane. The channelized spectrum shown here is an example of FDM or FDMA. The assignment of a signal or user to a frequency band is *long term* or *permanent*; the CR can simultaneously contain several spectrally separate signals. The first frequency band contains signals that operate between frequencies f_0 and f_1, the second between frequencies f_2 and f_3, and so on. The spectral regions between assignments, called *guard bands*, act as buffer zones to reduce

Figure 9.2 In the early days of telephony a pair of wires was needed for each trunk circuit.

interference between adjacent frequency channels. We might ask: How does one transform a baseband signal so that it occupies a higher frequency band? The answer is, by *heterodyning* or *mixing*, also called *modulating* the signal with a fixed frequency from a sine-wave oscillator.

If two input signals to a mixer are sinusoids with frequencies f_A and f_B, the mixing or multiplication will yield new sum and difference frequencies at f_{A+B} and f_{A-B}. The trigonometric identity

$$\cos A \cos B = \tfrac{1}{2}[\cos (A + B) + \cos (A - B)] \qquad (9.3)$$

Sec. 9.1   Allocation of the Communications Resource **479**

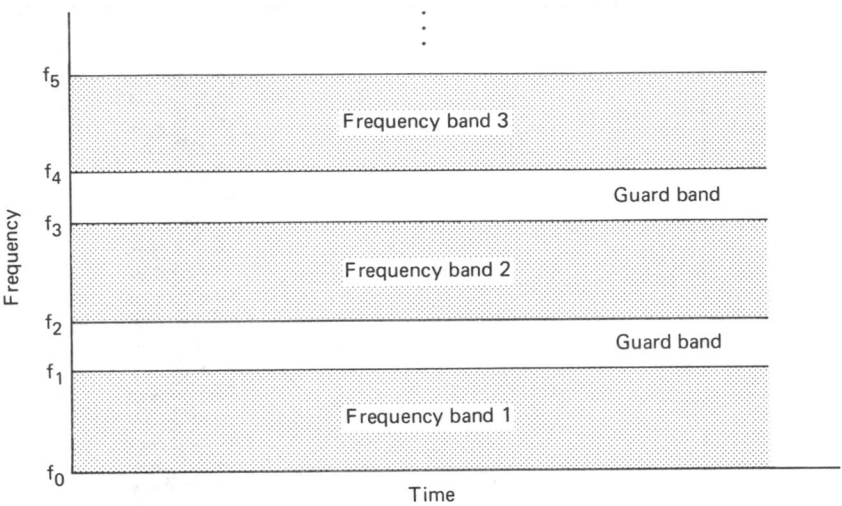

Figure 9.3 Frequency-division multiplexing.

describes the effect of the mixer. Figure 9.4a illustrates the mixing of a typical voice-grade telephone signal, $x(t)$ (baseband frequency range is 300 to 3400 Hz) with a sinusoid from a 20-kHz oscillator. The baseband two-sided magnitude spectrum, $|X(f)|$, is shown in Figure 9.4a. Can the mixer be a linear device? *No*. The output signal of a linear device will only consist of the *same* component frequencies as the input signal, differing only in amplitude and/or phase.

Figure 9.4b illustrates the one-sided magnitude spectrum, $|X(f - f_0)|$, at the mixer output. As a result of the mixing described by Equation (9.3), the output spectrum is a frequency-upshifted version of the baseband spectrum, centered at the oscillator frequency of 20 kHz. This spectrum is called a *double-sideband* (DSB) *spectrum* because the information appears in two different bands of the positive frequency domain. Figure 9.4c shows the lower sideband (LSB), whose frequency range is 16,600 to 19,700 Hz, the result of filtering the DSB spectrum. This sideband is sometimes referred to as the *inverted sideband* because the order of low-to-high frequency components is the reverse of that of the baseband components. Filtering can similarly be used to separate the upper sideband (USB), whose frequency range is 20,300 to 23,400 Hz, as shown in Figure 9.4d. This sideband is sometimes referred to as the *erect sideband* because the order of the low-to-high frequency components corresponds to that of the baseband components. Each sideband of the DSB spectrum contains the same information. Thus, only one sideband, either the USB or the LSB, is needed in order to retrieve the original baseband data.

A simple FDM example with three translated voice channels is seen in Figure 9.5. In channel 1, the 300- to 3400-Hz voice signal is mixed with a 20-kHz oscillator. In channels 2 and 3, a similar type of voice signal is mixed with a 16-kHz and 12-kHz oscillator, respectively. Only the lower sidebands are retained; the result of the mixing and filtering (to remove the upper sidebands) yields the frequency-shifted voice channels shown in Figure 9.5. The composite output waveform is just the sum of the three signals, having a total bandwidth in the range 8.6 to 19.7 kHz.

Figure 9.6 illustrates the two lowest levels of the FDM multiplex hierarchy for telephone channels. The first level consists of a *group* of 12 channels modulated onto subcarriers shown in the range 60 to 108 kHz. The second level is made up of five groups (60 channels) called a *supergroup* modulated onto the subcarriers shown in the range 312 to 552 kHz. The multiplexed channels are now treated as a composite signal that can be transmitted over cables or can be further modulated onto a carrier wave for radio transmission.

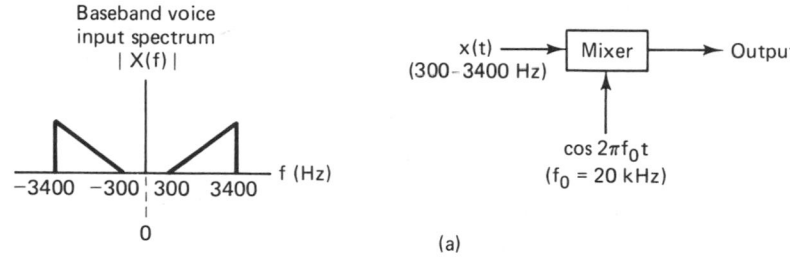

(a)

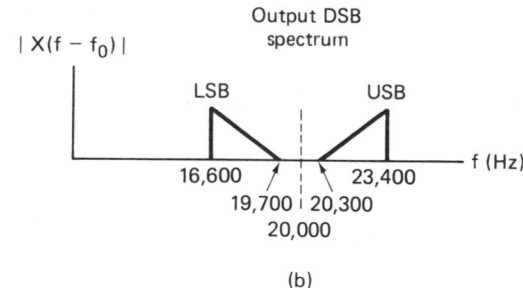

(b)

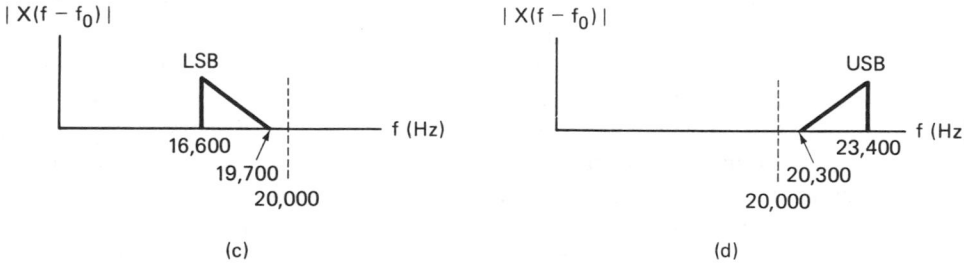

(c)   (d)

Figure 9.4 Heterodyning (mixing). (a) Mixing operation. (b) Mixer output spectrum. (c) Lower sideband. (d) Upper sideband.

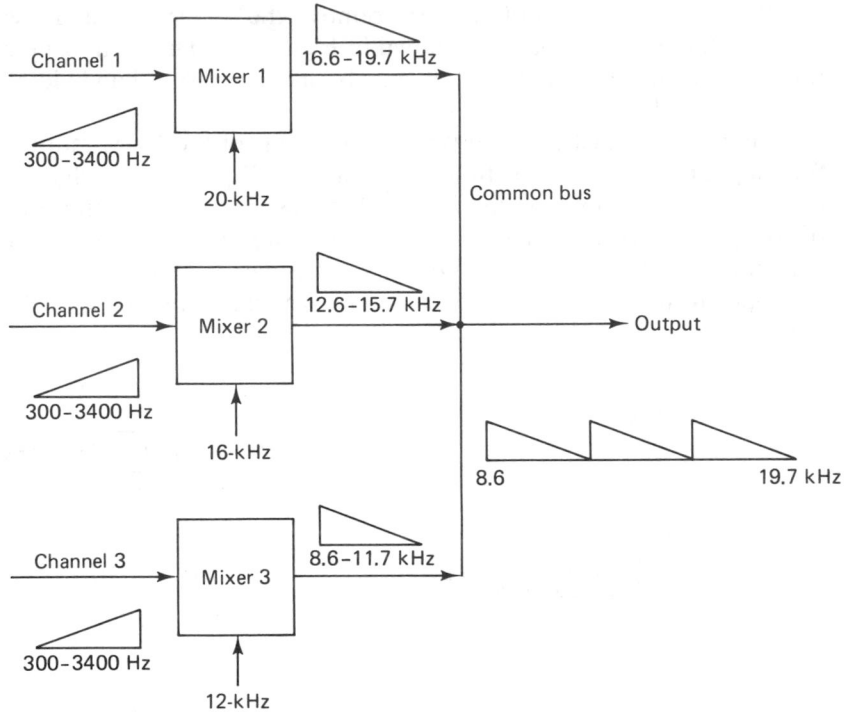

Figure 9.5 Simple FDM example. Three frequency-shifted voice channels.

9.1.1.2 Frequency-Division Multiple Access of Satellite Systems

Most of the free world's communication satellites are positioned in a *geostationary* or *geosynchronous* orbit. This means that the satellite is in a circular orbit, in the same plane as the earth's equatorial plane, and at such an altitude (approximately 19,330 nautical miles) that the orbital period is identical with the earth's rotational period. Since such satellites appear stationary when viewed from the earth, three of them spaced 120° apart can provide worldwide coverage (except for the polar regions). Most communication satellite systems are made up of non-regenerative repeaters or transponders. *Nonregenerative* means that the uplink (earth-to-satellite) transmissions are simply amplified, frequency shifted, and re-transmitted on the downlink (satellite-to-earth) without any demodulation/re-modulation or signal processing (see Section 4.7.1). The most popular frequency band for commercial satellite communications, called *C-band*, uses a 6-GHz carrier for the uplink and a 4-GHz carrier for the downlink. For C-band satellite systems, *each satellite* is permitted, by international agreement, to use a 500-MHz-wide spectral assignment. Typically, each satellite has 12 transponders with a bandwidth of 36 MHz each. The most common 36-MHz transponders operate in an FDM/FM/FDMA (frequency-division multiplex, frequency-modulated, frequency-division multiple access) multidestination mode. Let us consider each component of this name:

1. *FDM.* Signals such as telephone signals, each one having a single-sideband 4-kHz spectrum (including guard bands) are FDM'd to form a multichannel composite signal.
2. *FM.* The composite signal is frequency-modulated (FM) onto a carrier and transmitted to the satellite.
3. *FDMA.* Subdivisions of the 36-MHz transponder bandwidth may be assigned to different users. Each user receives a specific bandwidth allocation whereby he or she can access the transponder.

Thus, composite FDM channels are FM modulated and transmitted to the satellite within the bandwidth allocation of an FDMA plan. The major advantage of FDMA (compared to TDMA) is its simplicity. The FDMA channels require no

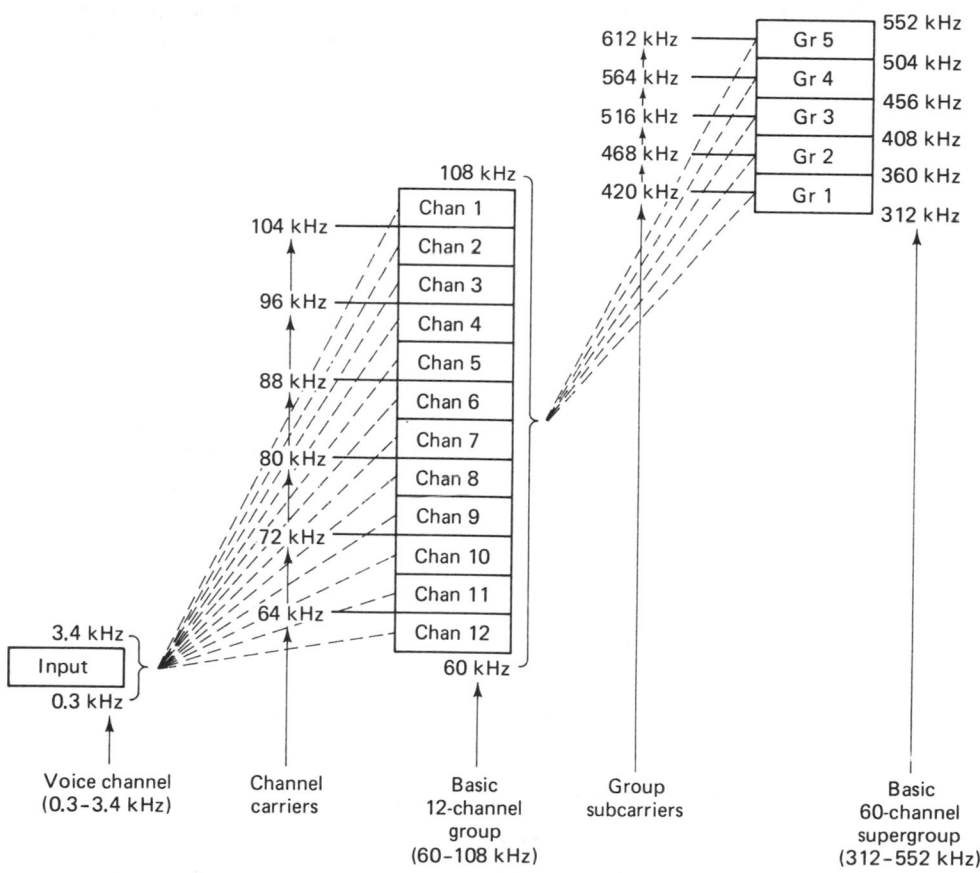

Figure 9.6 Modulation plan of a typical frequency-division multiplex system.

synchronization or central timing; each channel is almost independent of all other channels. Later we discuss some advantages of TDMA compared to FDMA.

9.1.2 Time-Division Multiplexing/Multiple Access

In Figure 9.3, sharing of the communications resource (CR) is accomplished by allocating frequency bands. In Figure 9.7, the same CR is shared by assigning each of M signals or users the full spectral occupancy of the system for a short duration of time called a *time slot*. The unused time regions between slot assignments, called *guard times*, allow for some time uncertainty between signals in adjacent time slots, and thus act as buffer zones to reduce interference. Figure 9.8 is an illustration of a typical TDMA satellite application. Time is segmented into intervals called frames. Each frame is further partitioned into assignable user time slots. The frame structure repeats, so that a fixed TDMA assignment constitutes one or more slots that periodically appear during each frame time. Each earth station transmits its data in bursts, timed so as to arrive at the satellite coincident with its designated time slot(s). When the bursts are received by the satellite transponder, they are retransmitted on the downlink, together with the bursts from other stations. A receiving station detects and demultiplexes the appropriate bursts and feeds the information to the intended user.

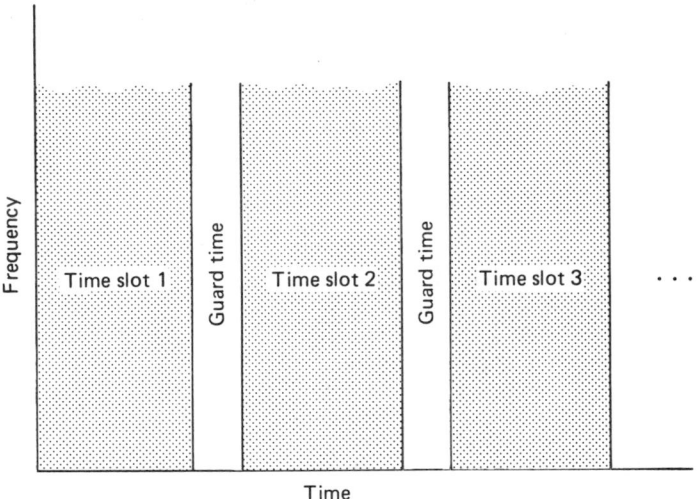

Figure 9.7 Time-division multiplexing.

9.1.2.1 Fixed-Assignment TDM/TDMA

The simplest TDM/TDMA scheme, called *fixed-assignment TDM/TDMA*, is so named because the M time slots that make up each frame are preassigned to signal sources, long term. Figure 9.9 illustrates, in block diagram form, the operation of such a system. The multiplexing operation consists of providing each

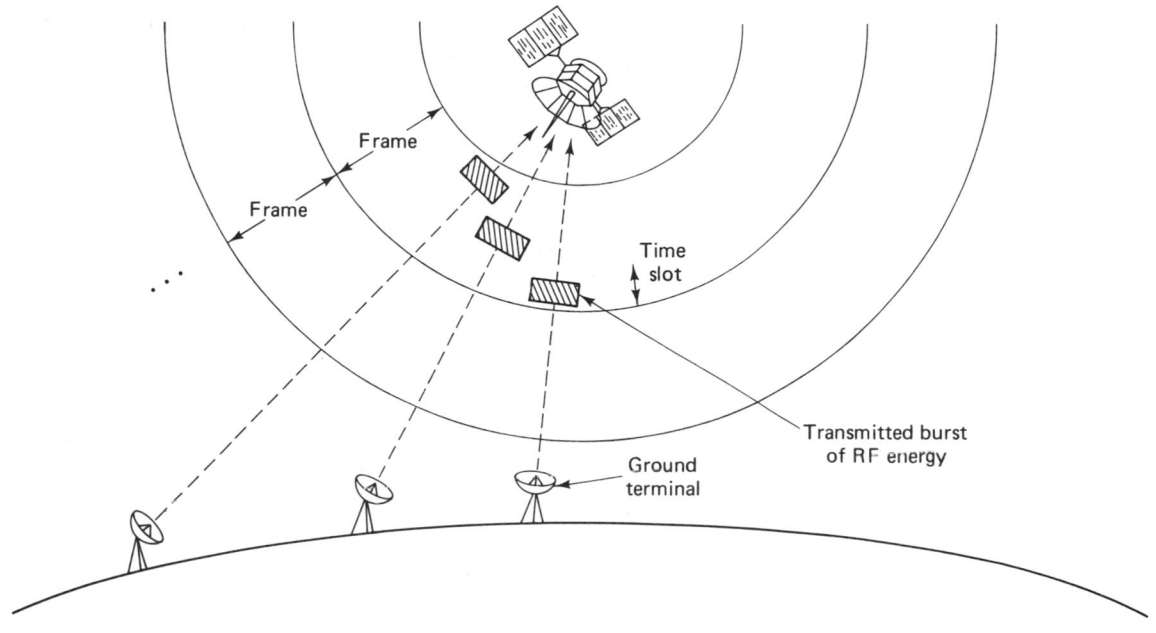

Figure 9.8 Typical TDMA configuration.

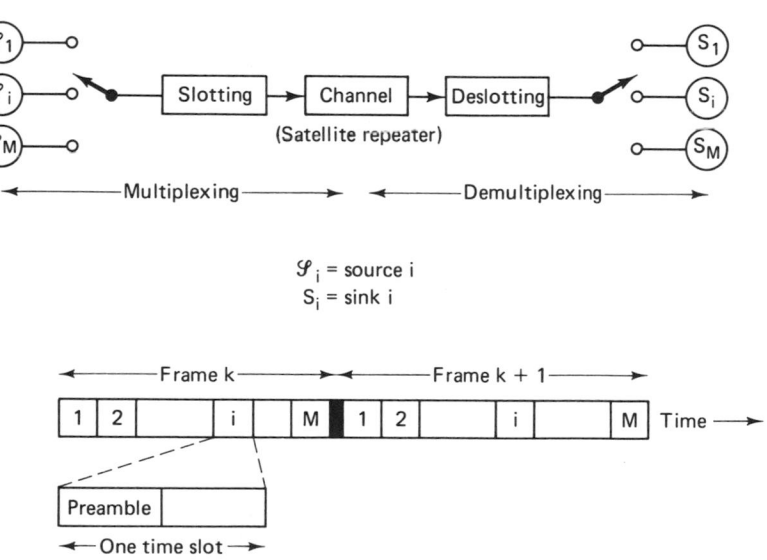

$\mathscr{S}_i$ = source i
S_i = sink i

Figure 9.9 Fixed-assignment TDM.

source with an opportunity to occupy one or more slots. The demultiplexing operation consists of deslotting the information and delivering the data to the intended sink. The two commutating switches in Figure 9.9 have to be synchronized so that the message corresponding to source 1, for example, appears on the channel 1 output, and so on. The message itself is generally comprised of a preamble portion and a data portion. The preamble portion usually contains synchronization, addressing, and error control sequences.

A fixed-assignment TDM/TDMA scheme is extremely efficient when the source requirements are predictable, and the traffic is heavy (the time slots are most always filled). However, for bursty or sporadic traffic, the fixed-assignment scheme is wasteful. Consider the simple example shown in Figure 9.10. In this example there are four time slots per frame; each slot is preassigned to users A, B, C, and D, respectively. In Figure 9.10a we see a typical activity profile of the four users. During the first frame time, user C has no data to transmit; during the second frame time, user B has none, and during the third frame time, user A has none. In a fixed-assignment TDMA scheme, all of the slots within a frame are preassigned. If the "owner" of a slot has *no* data to send during a particular frame, that slot is wasted. The data stream, shown in Figure 9.10b, illustrates the wasted time slots in this example. When source requirements are unpredictable, as in this example, there can be more efficient schemes, involving the dynamic assignment of the slots rather than a fixed assignment. Such schemes are variously known as packet-switched systems, statistical multiplexers, or concentrators; the effect, shown in Figure 9.10c, is to use all the slots in a frame in such a way that capacity is conserved. In later sections we discuss the TDMA systems used in INTELSAT V and VI.

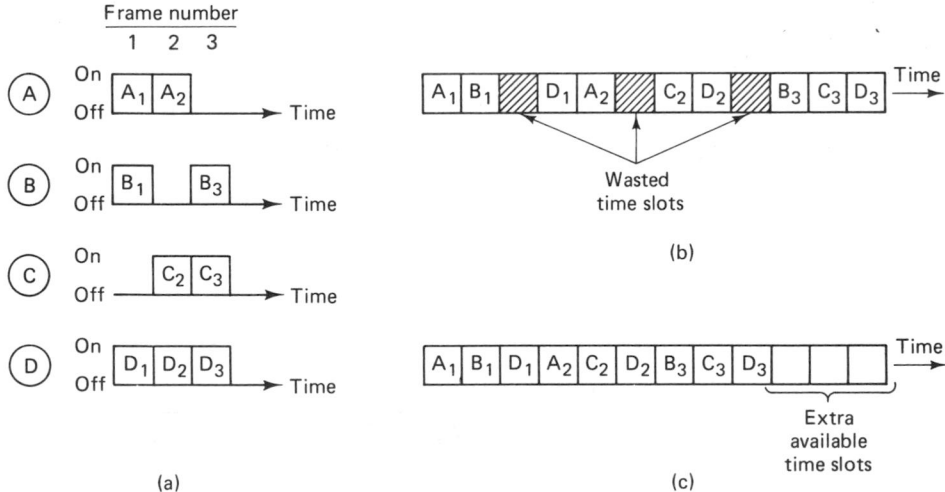

Figure 9.10 Fixed-assignment TDM versus packet switching. (a) Data source activity profiles. (b) Fixed-assignment time-division multiplexing. (c) Time-division packet switching (concentration).

9.1.3 Communications Resource Channelization

In Figure 9.3 we considered that the CR is partitioned into spectral bands, and in Figure 9.7 we viewed the same CR as being partitioned into time slots. Figure 9.11 represents a more general organization of the CR allowing for the assignment of a frequency band for a prescribed period of time. Such a multiple access scheme is referred to as *combined FDMA/TDMA*. For the assignments of frequency bands, let us assume an equal apportionment of the total bandwidth, W, among M user groups or classes, so that M disjoint frequency bands of width W/M hertz are continuously available to their assigned group. Similarly, for the assignment of time slots, the time axis is partitioned into time frames, each of duration T, and the frames are partitioned into N slot times, each of duration T/N. We assume that the users are time synchronized and that the assigned slots are located periodically within the frames. Each user in each frequency band is permitted to transmit during each periodic appearance of the user's assigned slot, and is permitted to use the assigned channel bandwidth for the slot duration. A slot is uniquely determined as the mth slot within the nth frame. Referring to Figure 9.11, we can describe the time of a particular slot (n, m) with reference to time zero as follows:

$$\text{time of slot } (n, m) = nT + \frac{(m - 1)T}{N} \le t \le nT + \frac{mT}{N}$$

$$n = 0, 1, \ldots ; m = 1, 2, \ldots, N. \quad (9.4)$$

The nth frame time, T, is denoted by the time interval $[nT, (n + 1)T]$. As can be seen in Figure 9.11, the domain of the unit signal is the intersection of the time slot (n, m) and the frequency band (j). Assume that a modulation/coding system is chosen so that the full bandwidth W of the CR can support R bits/s. In any

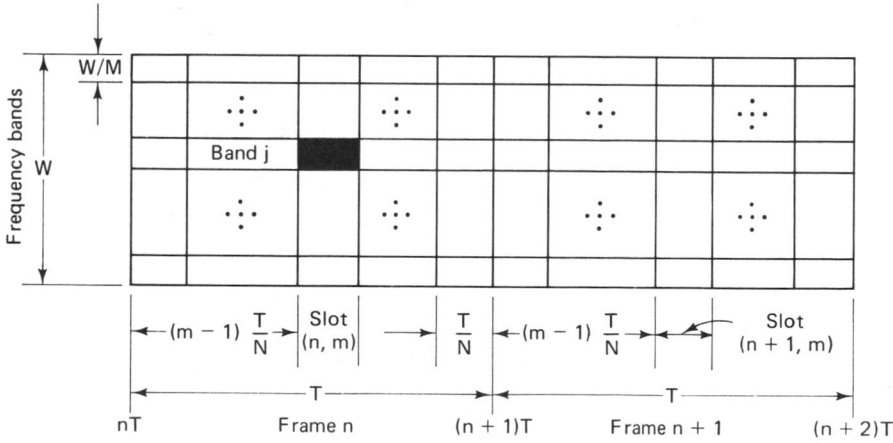

Figure 9.11 Communications resource: time/frequency channelization.

frequency band having a bandwidth of W/M hertz, the associated bit rate will be R/M bits/s. FDMA alone would provide M bands each with a bandwidth of $1/M$ of the full bandwidth of the CR. TDMA alone would provide the full system bandwidth for each of the N slots, where the duration of each slot is $1/N$ of the frame time.

9.1.4 Performance Comparison of FDMA and TDMA

9.1.4.1 Bit Rate Equivalence of FDMA and TDMA

Figure 9.12 highlights the basic differences between an FDMA and TDMA system in a communications resource capable of supporting a total of R bits/s. In Figure 9.12a the system bandwidth is divided into M orthogonal frequency bands. Hence each of the M sources, $\mathcal{S}_m(1 \leq m \leq M)$ can simultaneously transmit at a bit rate of R/M bits/s. In Figure 9.12b the frame is divided into M orthogonal time slots. Hence each of the M sources bursts its transmission at R bits/s, M times faster than the equivalent FDMA user for $(1/M)$th the time. In both cases, the source $\mathcal{S}_m$ transmits information at an average rate of R/M bits/s.

Let the information generated by each of the sources in Figure 9.12 be organized into b-bit groups, or *packets*. In the case of FDMA, the b-bit packets are transmitted in T seconds over each of the M disjoint channels. Therefore, the

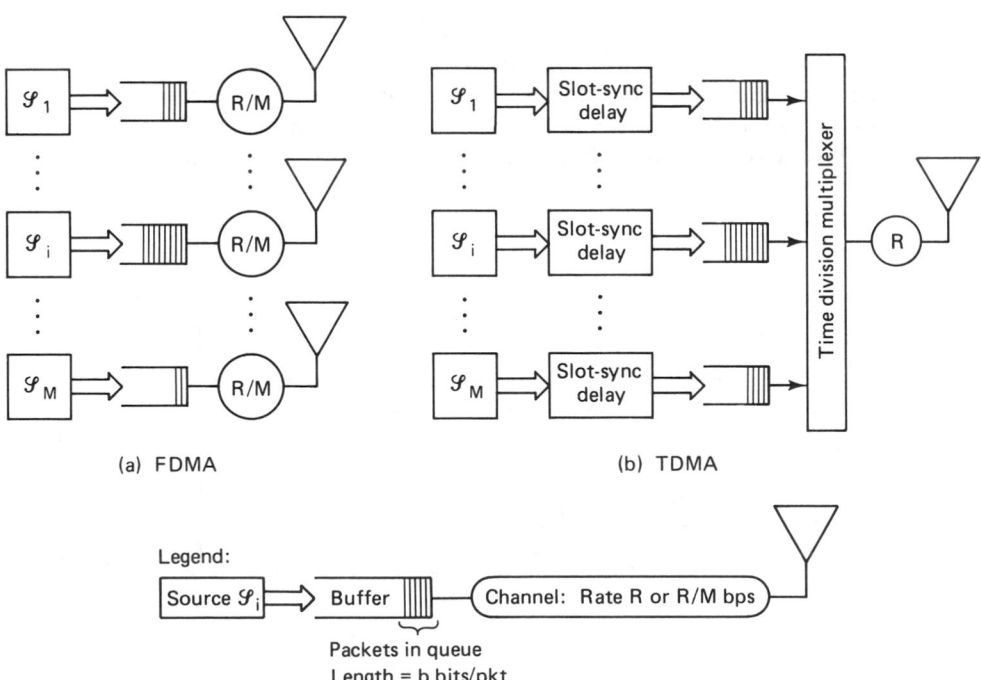

Figure 9.12 (a) FDMA: frequency divided into M orthogonal frequency bands. (b) TDMA: time divided into M orthogonal time slots (one packet per time slot).

total bit rate, R_{FD}, required is

$$R_{FD} = M \frac{b}{T} \quad \text{bits/s} \tag{9.5}$$

In the case of TDMA, the b bits are transmitted in T/M seconds from each source. Therefore, the bit rate, R_{TD}, required is

$$R_{TD} = \frac{b}{T/M} \quad \text{bits/s} \tag{9.6}$$

Since Equations (9.5) and (9.6) yield identical results, we can conclude that

$$R_{FD} = R_{TD} = R = \frac{Mb}{T} \quad \text{bits/s} \tag{9.7}$$

Thus both systems require the same full CR data rate, R bits/s.

9.1.4.2 Message Delays in FDMA and TDMA

From the previous sections it might appear that the duality between FDMA and TDMA will result in equivalent performance. This is not the case when the metric of performance is the average packet *delay*. It can be shown [1, 2] that TDMA is inherently superior to FDMA in the sense that the average packet delay using TDMA is less than the delay using FDMA.

As before, we assume that in the case of FDMA the system bandwidth is divided into M orthogonal frequency bands, and in the case of TDMA the frame is divided into M orthogonal time slots. For the analysis of message delay, the simplest case is that of deterministic data sources. It is assumed that the CR is 100% utilized, so that all frequency bands in the case of FDMA, and all time slots in the case of TDMA, are filled with data packets. For simplicity, it is also assumed that there are *no* overhead costs such as guard bands or guard times.

The message delay, D, can be defined as

$$D = w + \tau \tag{9.8}$$

where w is the average packet waiting time (prior to transmission) and τ is the packet transmission time. In the FDMA case, each packet is sent over a T-second interval, so the packet transmission time for FDMA, τ_{FD}, is simply

$$\tau_{FD} = T \tag{9.9}$$

In the TDMA case, each packet is sent in slots of T/M seconds. We can thus write the TDMA packet transmission time, τ_{TD}, with the use of Equation (9.7), as

$$\tau_{TD} = \frac{T}{M} = \frac{b}{R} \tag{9.10}$$

Since the FDMA channel is continuously available and packets are sent as soon as they are generated, the waiting time, w_{FD}, for FDMA is

$$w_{FD} = 0 \qquad (9.11)$$

FDMA and TDMA bit streams are compared in Figure 9.13. For TDMA, Figure 9.13a illustrates that each user's slot begins at a different point in the T-second frame; that is, packet S_{mk} will start at $(m - 1)T/M$ seconds ($1 \le m \le M$) after the packet generation instant. Therefore, the average waiting time, w_{TD}, that a TDMA packet sustains before transmission begins is

$$w_{TD} = \frac{1}{M} \sum_{m=1}^{M} (m - 1) \frac{T}{M} = \frac{T}{M^2} \sum_{n=0}^{M-1} n = \frac{T}{M^2} \frac{(M-1)(M)}{2}$$

$$= \frac{T}{2} \left(1 - \frac{1}{M} \right) \qquad (9.12)$$

The maximum waiting time before transmission of a packet is $(M - 1)T/M$ seconds, and on the average a packet will wait $\frac{1}{2}(M - 1)(T/M) = (T/2)(1 - 1/M)$ seconds, as given by Equation (9.12).

To compare the average delay times, D_{FD} and D_{TD}, for FDMA and TDMA, respectively, we combine Equations (9.9) and (9.11) into Equation (9.8), and sim-

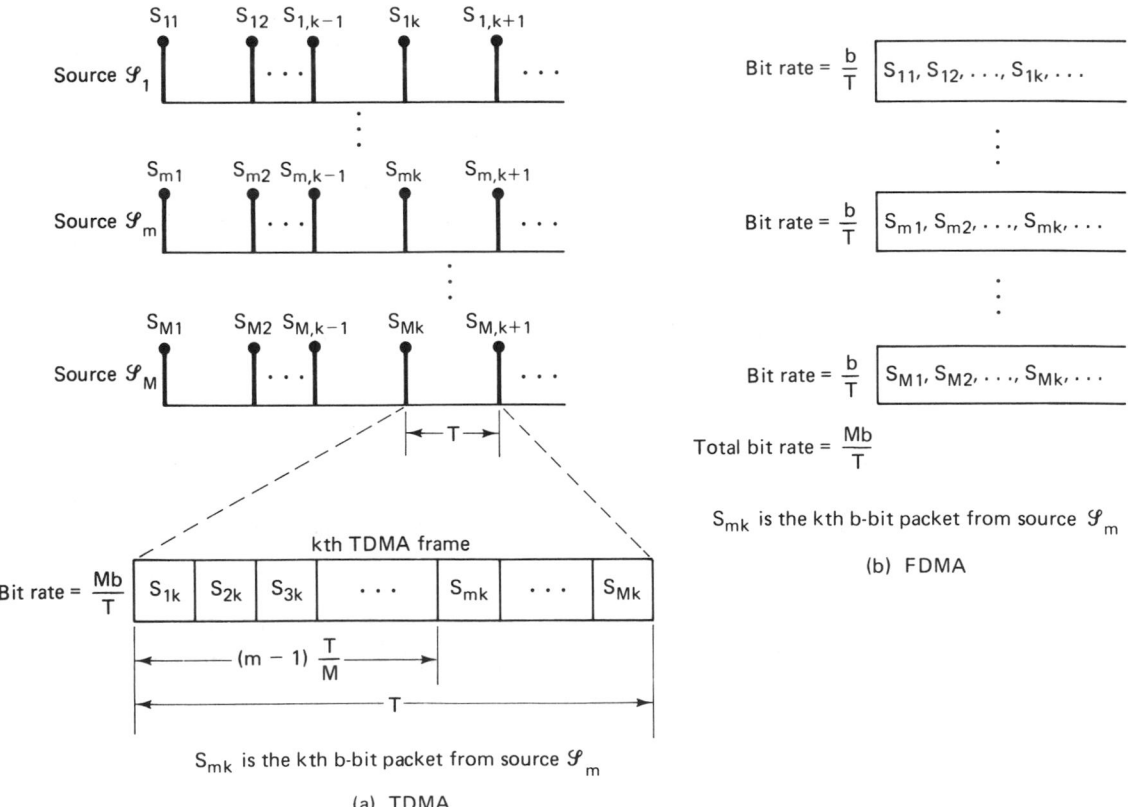

Figure 9.13 (a) TDMA and (b) FDMA channelization.

ilarly combine Equations (9.10) and (9.12) into Equation (9.8), yielding

$$D_{FD} = T \tag{9.13}$$

$$D_{TD} = \frac{T}{2}\left(1 - \frac{1}{M}\right) + \frac{T}{M} = D_{FD} - \frac{T}{2}\left(1 - \frac{1}{M}\right) \tag{9.14}$$

Using Equation (9.7), Equation (9.14) can be written as

$$D_{TD} = D_{FD} - \frac{b}{2R}(M - 1) \tag{9.15}$$

The result indicates that TDMA is inherently superior to FDMA, from a message delay point of view. Although Equation (9.15) assumed that the data source is deterministic, the smaller average message delays for TDMA schemes hold up for any independent message arrival process [1, 2].

9.1.5 Code-Division Multiple Access

In Figure 9.3 the CR plane was illustrated as being shared by slicing it horizontally to form FDMA frequency bands, and in Figure 9.7 the same CR plane was illustrated as being shared by slicing it vertically to form TDMA time slots. These two techniques are the most common choices for multiple access applications. Figure 9.14 illustrates the CR being partitioned by the use of a hybrid combination of FDMA and TDMA known as *code-division multiple access* (CDMA). CDMA

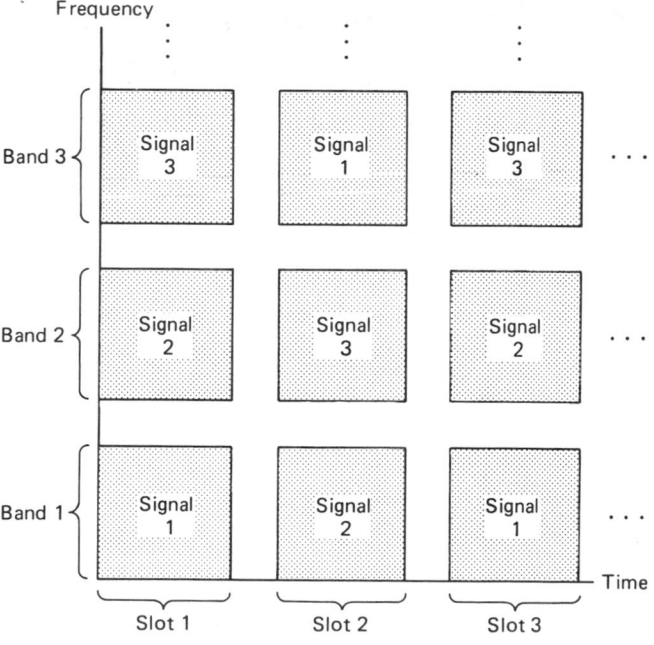

Figure 9.14 Code-division multiplexing.

is an application of spread-spectrum (SS) techniques. Spread-spectrum techniques can be classified into two major categories: *direct-sequence* SS and *frequency hopping* SS. We introduce frequency hopping CDMA (FH-CDMA) in this chapter, and we treat direct-sequence CDMA together with the overall subject of spread-spectrum techniques in Chapter 10.

It is easiest to visualize *frequency hopping* CDMA, illustrated in Figure 9.14, as the short-term assignment of a frequency band to various signal sources. At each successive time slot, whose duration is usually brief, the frequency band assignments are reordered. In Figure 9.14, during time slot 1, signal 1 occupies band 1, signal 2 occupies band 2, and signal 3 occupies band 3. During time slot 2, signal 1 hops to band 3, signal 2 hops to band 1, and signal 3 hops to band 2, and so on. The CR can thus be fully utilized, but the participants, having their frequency bands reassigned at each time slot, appear to be playing "musical chairs." Each user employs a pseudonoise (PN) code, orthogonal (or nearly orthogonal) to all the other user codes, that dictates the frequency hopping band assignments. Details of PN code sequences are treated in Section 10.2. Figure 9.14 is an oversimplified view of the way the CR is shared in frequency hopping CDMA, since the symmetry implies that each frequency hopping signal is in time synchronism with each of the other signals. This is *not the case*. In fact, one of the attractions of CDMA compared to TDMA is that there is no need for synchronization among user groups (only between a transmitter and a receiver within a group).

The block diagram in Figure 9.15 illustrates the frequency hopping modulation process. At each frequency hop time the PN generator feeds a code sequence to a device called a *frequency hopper*. The frequency hopper synthesizes one of the allowable hop frequencies. Assume that the data modulation has an *M*-ary frequency shift keying (MFSK) format. The essential difference between a conventional MFSK system and a frequency hopping (FH) MFSK system is that in the conventional system, a data symbol modulates a carrier wave that is *fixed* in frequency, but in the hopping system, the data symbol modulates a carrier

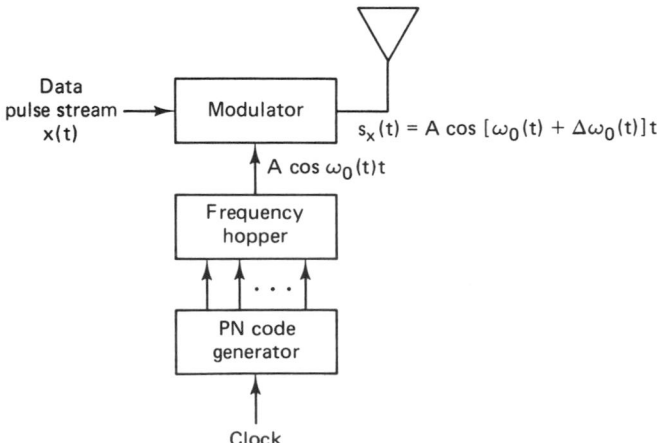

$$s_x(t) = A \cos [\omega_0(t) + \Delta\omega_0(t)]t$$

$$A \cos \omega_0(t)t$$

Figure 9.15 CDMA frequency hopping modulation process.

Multiplexing and Multiple Access   Chap. 9

wave that *hops* across the total CR bandwidth. The FH modulation in Figure 9.15 can be thought of as a two-step process—data modulation and frequency hopping modulation—even though it can be implemented in a single step, where the modulator produces a transmission tone based on the simultaneous dictates of the PN code and the data. Frequency hopping systems are covered in detail in Section 10.4.

One might ask: Don't the FDMA and TDMA options provide sufficient multiple access flexibility? FDMA and TDMA methods can surely be relied on to apportion the communications resource equitably. Of what use is this hybrid technique? CDMA offers some unique advantages, as follows:

1. *Privacy*. When the code for a particular user group is only distributed among authorized users, the CDMA process provides communications privacy, since the transmissions cannot easily be intercepted by unauthorized users without the code.

2. *Fading channels*. If a particular portion of the spectrum is characterized by fading, signals in that frequency range are attenuated. In an FDMA scheme, a user who was unfortunate enough to be assigned to the fading position of the spectrum might experience highly degraded communications for as long as the fading persists. However, in a FH-CDMA scheme, only during the time a user hops into the affected portion of the spectrum will the user experience degradation. Therefore, with CDMA, such degradation is shared among all the users.

3. *Jam resistance*. During a given CDMA hop, the signal bandwidth is identical to the bandwidth of conventional MFSK, which is typically equal to the minimum bandwidth necessary to transmit the MFSK symbol. However, over a duration of many time slots, the system will hop over a frequency band which is much wider than the data bandwidth. We refer to this utilization of bandwidth as spread spectrum. In Chapter 10 we develop, in detail, the resistance to jamming that spread spectrum affords a user.

4. *Flexibility*. The most important advantage of CDMA schemes, compared to TDMA, is that there need be no precise time coordination among the various simultaneous transmitters. The orthogonality between user transmissions on different codes is not affected by transmission-time variations. This will become clear upon closer examination of the autocorrelation and cross-correlation properties of the codes, considered in Chapter 10.

9.1.6 Space-Division and Polarization-Division Multiple Access

Figure 9.16a depicts the INTELSAT IVA application of space-division multiple access (SDMA), also called *multiple-beam frequency reuse*. INTELSAT IVA used a dual-beam receive antenna feeding two receivers to allow simultaneous access of the satellite from two different regions of the earth. The frequency band

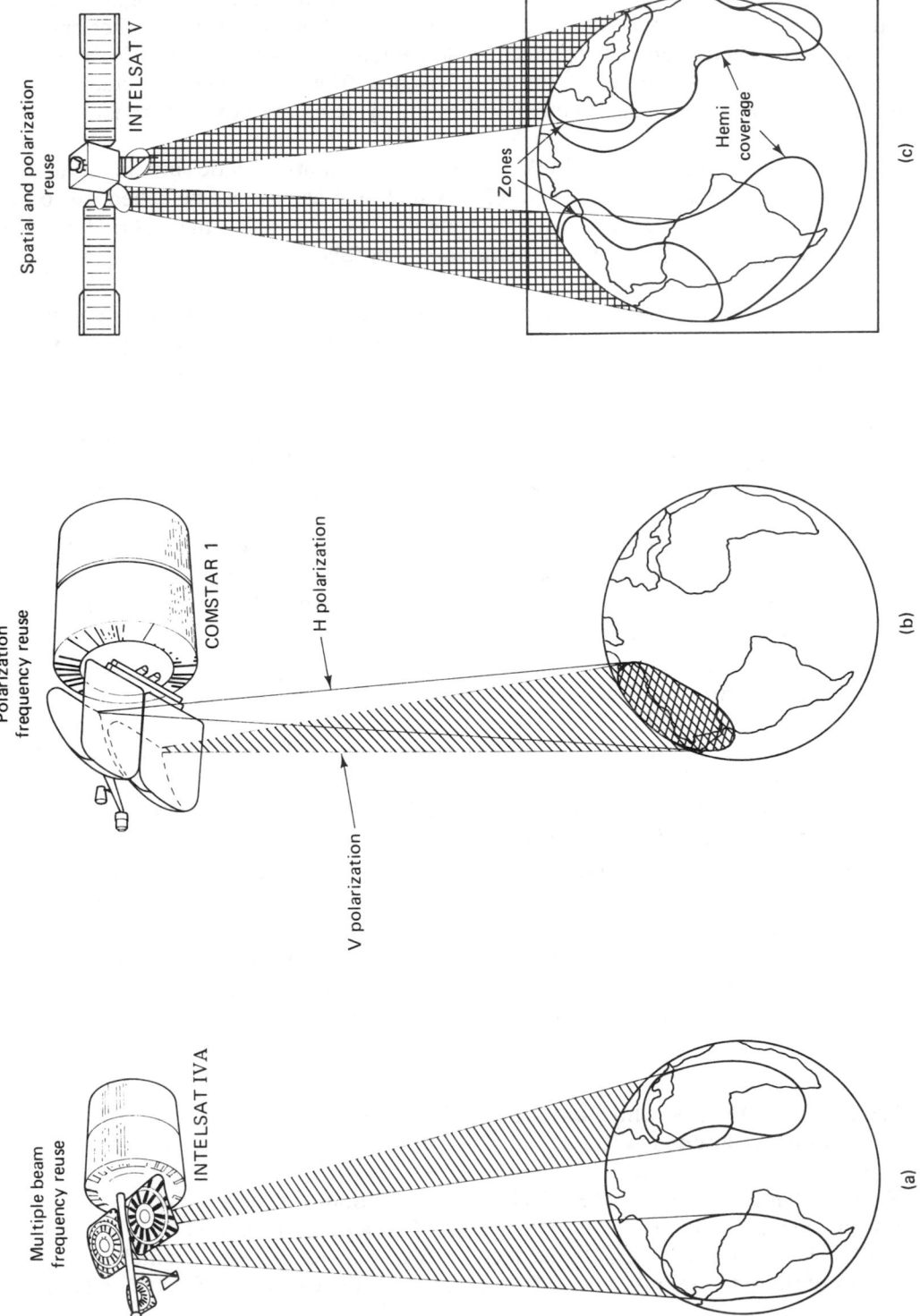

Figure 9.16 SDMA and PDMA. (a) INTELSAT IVA. (b) COMSTAR 1. (c) INTELSAT V (Atlantic coverage).

allocated to each receive beam was identical because the uplink signals were spatially separated. In such cases, the frequency band is said to be *reused*.

Figure 9.16b depicts an application of polarization-division multiple access (PDMA), also called *dual-polarization frequency reuse*, from COMSTAR 1. Here separate antennas are used, each with different polarization and followed by separate receivers, allowing simultaneous access of the satellite from the same region of the earth. Each corresponding earth station antenna needs to be polarized in the same way as its counterpart in the satellite. (This is generally accomplished by providing each participating earth station with an antenna that has dual polarization.) The frequency band allocated to each antenna beam can be identical because the uplink signals are orthogonal in polarization. As with SDMA, the frequency band in PDMA is said to be reused. Figure 9.16c depicts an application of the simultaneous use of SDMA and PDMA in INTELSAT V. There are two separate hemispheric coverages, west and east. There are also two smaller zone beams; each zone beam overlaps a portion of one of the hemispheric beams and is separated from it by orthogonal polarization. Thus there is a fourfold reuse of the spectrum.

9.2 MULTIPLE ACCESS COMMUNICATIONS SYSTEM AND ARCHITECTURE

A *multiple access protocol* or *multiple access algorithm* (MAA) is that rule by which a user knows how to use time, frequency, and code functions to communicate through a satellite to other users. A multiple access system is a com-

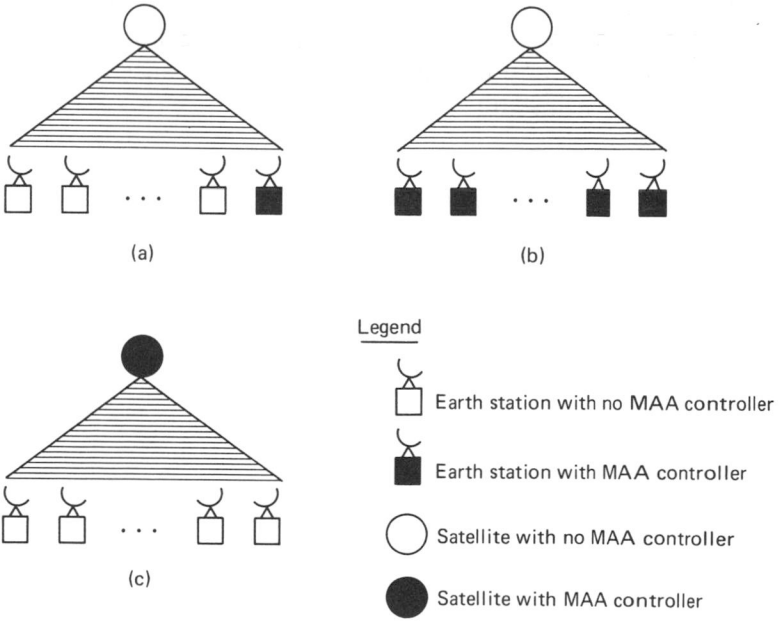

Figure 9.17 Satellite multiple access architecture. (a) Single earth station control. (b) Distributed earth station control. (c) Satellite control.

bination of hardware and software that supports the MAA. The general goal of a multiple access system is to provide communications service in a timely, orderly, and efficient way.

Figure 9.17 illustrates some basic choices for the architecture of a satellite multiple access system. The legend indicates the symbols used for an earth station with and without an MAA controller, and a satellite with and without an MAA controller. Figure 9.17a illustrates the case where one earth station is designated as the master, or the controller. This earth station possesses an MAA computer and responds to the service requests of all other users. Notice that a user's request entails a transmission through the satellite and back down to the controller. The controller's response entails another transmission through the satellite; hence there are two up- and downlink transmissions required for each service assignment. Figure 9.17b illustrates the case where the MAA control is distributed among all the earth stations; there is no single controller. Each earth station uses the same algorithm and they each have identical knowledge regarding access requests and assignments; therefore, only one round trip is required for each service assignment. Figure 9.17c illlustrates the case where the MAA controller is in the satellite. A service request goes from user to satellite, and the response from the satellite can follow immediately; therefore, only one round trip is required for each service assignment.

9.2.1 Multiple Access Information Flow

Figure 9.18 is a flow diagram describing the basic flow of information between the multiple access algorithm (MAA) or controller and an earth station; the numbers below correspond to those on the figure. Recall from the preceding section that the control may be lodged in the satellite, in a master station, or distributed among all the earth stations.

1. *Channelization*. This term refers to the most general allocation information [e.g., channels 1 to N may be allocated for the Army and channels $(N + 1)$

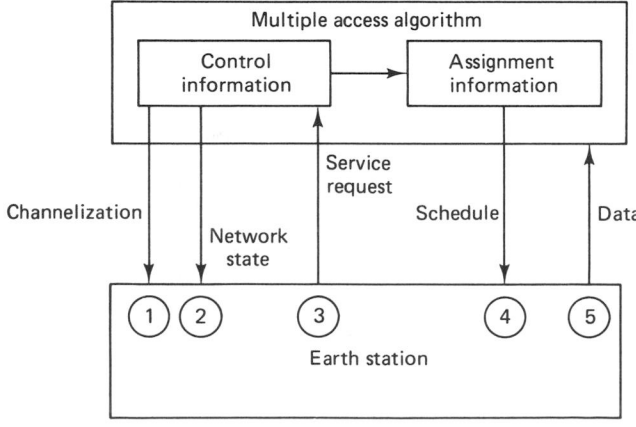

Figure 9.18 Multiple access information flow.

to M for the Navy]. This information seldom changes, and may be distributed to the earth stations by the use of a newsletter rather than via the communication system.

2. *Network state* (*NS*). This term refers to the state of the CR. A station is advised regarding the availability of the communications resource and where in the resource (e.g., time, frequency, code position) to transmit its service request(s).

3. *Service request.* Then the station makes its request(s) for service (e.g., allocation for m message slots).

4. Upon receipt of the service request(s), the controller sends the station a schedule regarding where and when to position its data in the CR.

5. The station transmits its data according to its assigned schedule.

9.2.2 Demand-Assignment Multiple Access

Multiple access schemes are termed *fixed assignment* when a station has periodic access to the channel independent of its actual need. By comparison, dynamic assignment schemes, sometimes called *demand-assignment multiple access* (DAMA), give the station access to the channel only when it requests access. If the traffic from a station tends to be burst-like or intermittent, DAMA procedures can be much more efficient than fixed-assignment procedures. A DAMA scheme capitalizes on the fact that actual demand *rarely* equals the peak demand. If a system's capacity is equal to the total peak demand and if the traffic is bursty, the system will be underutilized most of the time. However, by using buffers and DAMA, a system with reduced average capacity can handle bursty traffic, at the cost of some queueing delay. Figure 9.19 summarizes the difference between a fixed system, whose capacity is equal to the sum of the user requirements, and a dynamic system, whose capacity is equal to the average of the user requirements.

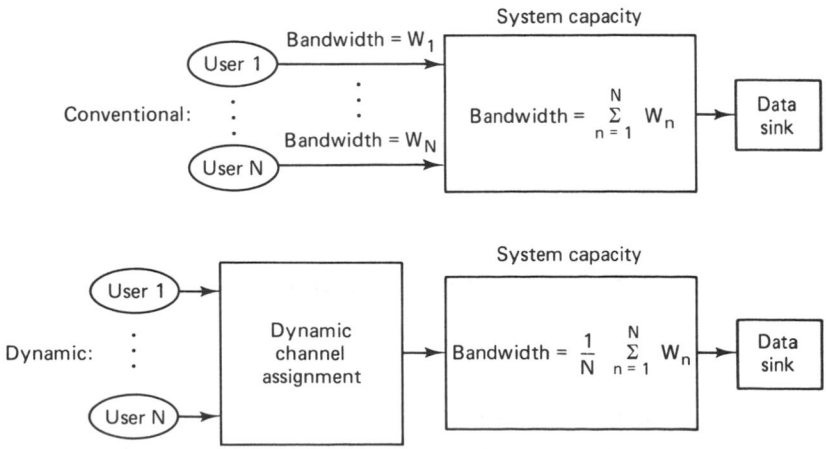

Figure 9.19 Bandwidth reduction for systems using dynamic channel assignment.

9.3 ACCESS ALGORITHMS

9.3.1 ALOHA

In 1971, the University of Hawaii began operation of its ALOHA system. A communication satellite was used to interconnect the several university computers by use of a random access protocol [3–7]. The system concept was extremely simple, consisting of the following modes:

1. *Transmission mode.* Users transmit at any time they desire, encoding their transmissions with an error detection code.
2. *Listening mode.* After a message transmission, a user listens for an acknowledgment (ACK) from the receiver. Transmissions from different users will sometimes overlap in time, causing reception errors in the data in each of the contending messages. We say that the messages have *collided*. In such cases, the errors are detected, and the users receive a negative acknowledgment (NAK).
3. *Retransmission mode.* When a NAK is received, the messages are simply retransmitted. Of course, if the colloding users were to retransmit immediately, they would collide again. Therefore, the users retransmit after a *random* delay.
4. *Timeout mode.* If, after a transmission, the user does not receive either an ACK or NAK within a specified time, the user retransmits the message.

9.3.1.1 Message Arrival Statistics

Assume that the total system demand requires an average message or packet arrival rate of λ successful or accepted messages per second. Because of the presence of collisions, some of the messages will be unsuccessful or rejected. Therefore, we define the total traffic arrival rate, λ_t, as the acceptance rate, λ, plus the rejection rate, λ_r, as follows:

$$\lambda_t = \lambda + \lambda_r \qquad (9.16)$$

Let us denote the length of each message or packet as b bits. Then we can define the average amount of successful traffic or *throughput*, ρ', on the channel in units of bits per second, as

$$\rho' = b\lambda \qquad (9.17)$$

We can also define the *total traffic*, G', on the channel in units of bits per second, as

$$G' = b\lambda_t \qquad (9.18)$$

With the channel capacity (maximum bit rate) designated as R bits per second, let us further define a *normalized throughput*, ρ, and a *normalized total traffic*, G, as

$$\rho = \frac{b\lambda}{R} \qquad (9.19)$$

$$G = \frac{b\lambda_t}{R} \qquad (9.20)$$

Normalized throughput, ρ, expresses throughput as a fraction $(0 \leq \rho \leq 1)$ of channel capacity. Normalized total traffic, G, expresses total traffic as a fraction $(0 \leq G \leq \infty)$ of the channel capacity. Notice that G can take on values greater than unity.

We can also define the transmission time of each packet as folllows:

$$\tau = \frac{b}{R} \qquad \text{seconds/packet} \qquad (9.21)$$

By substituting equation (9.21) into Equations (9.19) and (9.20), we can write

$$\rho = \lambda\tau \qquad (9.22)$$

$$G = \lambda_t\tau \qquad (9.23)$$

A user can successfully transmit a message as long as no other user began one within the previous τ seconds or starts one within the next τ seconds. If another user began a message within the previous τ seconds, its tail end will collide with the current message. If another user begins a message within the next τ seconds, it will collide with the tail end of the current message. Thus a space of 2τ seconds is needed for each message.

The message arrival statistics for unrelated users of a communication system is often modeled as a Poisson process. The probability of having K new messages arrive during a time interval of τ seconds is given by the Poisson distribution [8] to be

$$P(K) = \frac{(\lambda\tau)^K e^{-\lambda\tau}}{K!} \qquad K \geq 0 \qquad (9.24)$$

where λ is the average message arrival rate. Because the users transmit without regard for each other in the ALOHA system, this expression is useful for calculating the probability that exactly $K = 0$ other messages are transmitted during a time interval 2τ. This is the probability, P_s, that a user's message transmission was successful (experienced no collisions). To compute P_s, assuming that all traffic is Poisson, we use λ_t and 2τ in Equation (9.24). Thus

$$P_s = P(K = 0) = \frac{(2\tau\lambda_t)^0 e^{-2\tau\lambda_t}}{0!} = e^{-2\tau\lambda_t} \qquad (9.25)$$

In Equation (9.16) we defined total traffic arrival rate λ_t, in terms of the successful portion, λ, and the repetition or unsuccessful portion, λ_r; then, by definition, the probability of a successful packet can be expressed as

$$P_s = \frac{\lambda}{\lambda_t} \qquad (9.26)$$

By combining Equations (9.25) and (9.26), we have

$$\lambda = \lambda_t e^{-2\tau\lambda_t} \tag{9.27}$$

By combining Equation (9.27) with Equations (9.22) and (9.23), we can write

$$\rho = Ge^{-2G} \tag{9.28}$$

Equation (9.28) relates the normalized throughput, ρ, to the normalized total traffic, G, on the channel for the ALOHA system. A plot of this relationship labeled "pure ALOHA" is shown in Figure 9.20. As G increases, ρ increases until a point is reached where further traffic increases create a large enough collision rate to cause a reduction in the throughput. The maximum ρ, equal to $1/2e$ = 0.18, occurs at a value of $G = 0.5$. Therefore, for a pure ALOHA channel, only 18% of the CR can be utilized. Simplicity of control is achieved at the expense of channel capacity [7, 9].

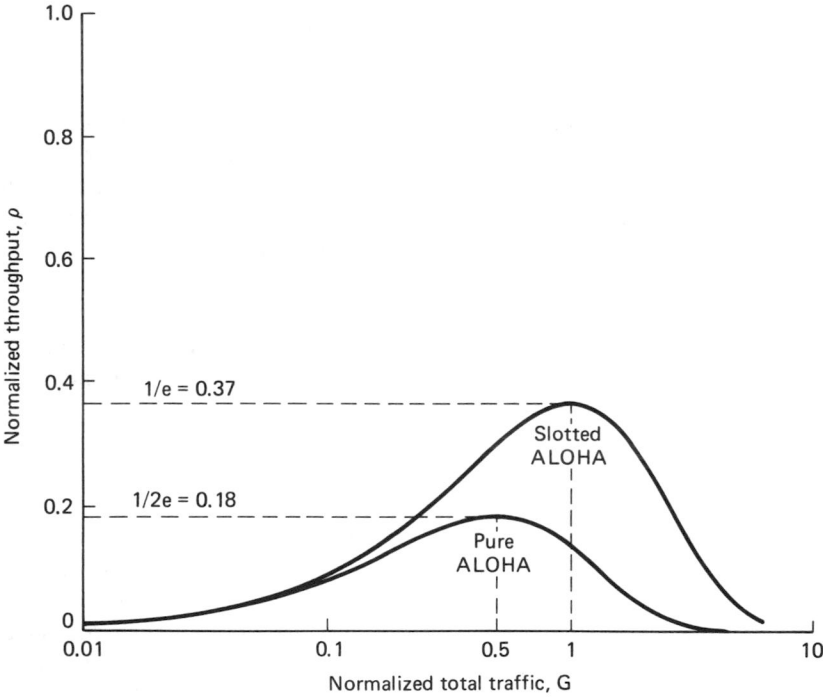

Figure 9.20 Throughput in ALOHA channels (successful transmissions versus total transmissions).

9.3.2 Slotted ALOHA

The pure ALOHA scheme can be improved by requiring a small amount of co-ordination among the stations. The slotted ALOHA (S-ALOHA) is such a system. A sequence of synchronization pulses is broadcast to all stations. As with pure

ALOHA, packet lengths are constant. Messages are required to be sent in the slot time between synchronization pulses, and can be started only at the *beginning* of a time slot. This simple change reduces the rate of collisions by half, since only messages transmitted in the same slot can interfere with one another. It can be shown [9, 10] that for S-ALOHA, the reduction in the *collision window* from 2τ to τ results in the following relationship between normalized throughput, ρ, and normalized total traffic, G.

$$\rho = Ge^{-G} \tag{9.29}$$

The plot of Equation (9.29) is shown in Figure 9.20 labeled "slotted ALOHA." Here the maximum value of ρ is $1/e = 0.37$, or an improvement of two times the pure ALOHA protocol.

The retransmission mode described for the pure ALOHA system was modified for S-ALOHA so that if a negative acknowledgment (NAK) occurs, the user retransmits after a *random* delay of an integer number of slot times. Figure 9.21 illustrates the S-ALOHA operation. A packet of data bits is shown transmitted by user k followed by the satellite acknowledgment (ACK). Also shown are users m and n simultaneously transmitting packets, which results in a collision; a NAK is returned. Each using station employs a random-number generator to select its retransmission time. The figure illustrates an example of the m and n retransmission at their respective randomly selected times. Of course, there is some probability that users m and n will recollide. However, in that case, they simply repeat the retransmission, using another random delay.

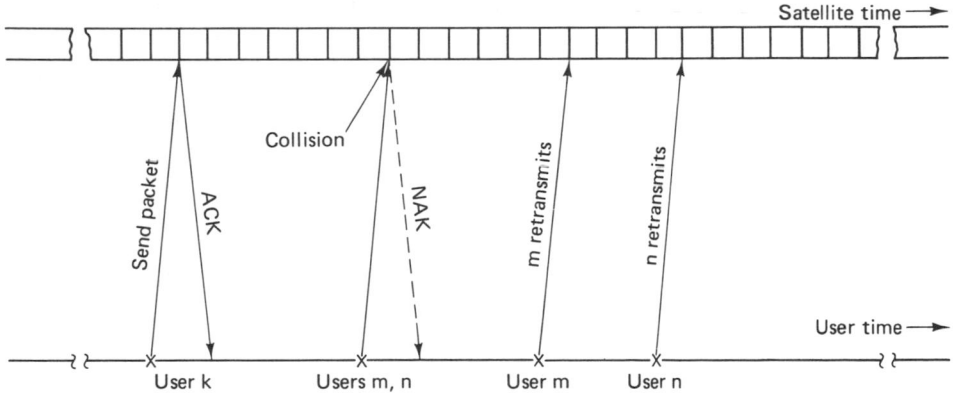

Figure 9.21 Random access scheme: slotted ALOHA operation.

Example 9.1   Poisson Process

Assuming that packet transmissions and retransmission can both be described as a Poisson process, calculate the *probability* that a data packet transmission in an S-ALOHA system will experience a collision with *one other user*. Assume that the total traffic rate $\lambda_t = 10$ packets/s and the packet duration $\tau = 10$ ms.

Solution

$$P(K = 1) = \left.\frac{(\tau\lambda_t)^K e^{-\tau\lambda_t}}{K!}\right|_{K=1}$$

$$= (10 \times 0.01)^1 e^{-0.1} = 0.1e^{-0.1}$$

$$= 0.09$$

9.3.3 Reservation-ALOHA

A significant improvement was made to the ALOHA system with the introduction of the reservation-ALOHA (R-ALOHA) [11] scheme. The R-ALOHA system has two basic modes: an unreserved mode and a reserved mode; each is described below.

Unreserved Mode (Quiescent State)

1. A time frame is established and divided into a number of small reservation subslots.
2. Users use these small subslots to reserve message slots.
3. After requesting a reservation, the user listens for an acknowledgment and a slot assignment.

Reserved Mode

1. The time frame is divided into $M + 1$ slots whenever a reservation is made.
2. The first M slots are used for message transmissions.
3. The last slot is subdivided into subslots to be used for reservation/requests.
4. Users send message packets only in their assigned portions of the M slots.

Consider the R-ALOHA example shown in Figure 9.22. In the quiescent state, with no reservations, time is partitioned into short subslots for making reservations. Once a reservation is made, the system is configured so that $M = 5$ message slots followed by $V = 6$ reservation subslots becomes the timing format. The figure illustrates a request and an acknowledgment in progress. In this example the station seeks to reserve three message slots. The reservation acknowledgment advises the using station where to locate its first data packet. Since the control is distributed so that all participants receive the downlink transmissions and are thus aware of the reservations and time format, the acknowledgment need not disclose any more than the location of the first slot. As shown in Figure 9.22, the station sends its second packet in the slot following the first packet. The user further knows that the next slot is comprised of six subslots for reservations, so *no* packets are transmitted during this time. The third and final packet is sent in the following slot. When there are no reservations taking place, the system reverts back to its quiescent format of subslots only. Since the control is distributed, all the participants are made aware of the quiescent format by receiving appropriate

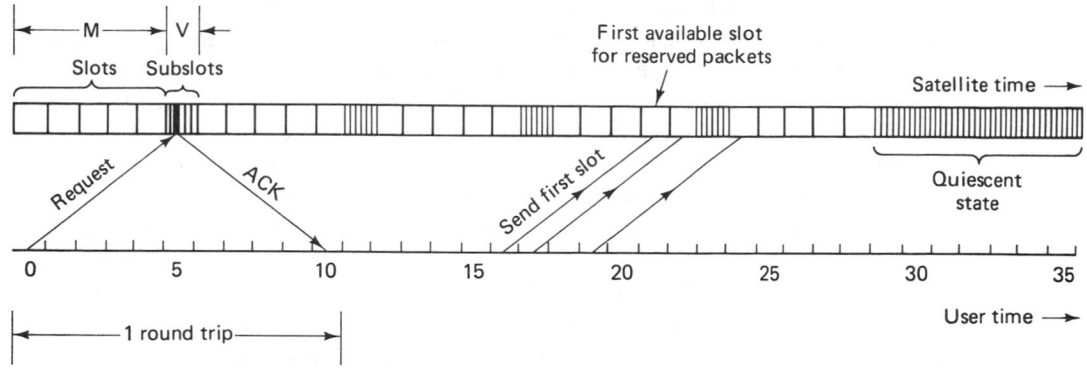

Figure 9.22 Example of reservation ALOHA. Station seeks to reserve three slots ($M = 5$ slots, $V = 6$ subslots).

synchronizing pulses on the downlink. Other interesting reservation schemes are discussed in References [12, 13].

9.3.4 Performance Comparison of S-ALOHA and R-ALOHA

From Chapter 3 the basic quality measure of a digital modulation scheme is its P_B versus E_b/N_0 curve. This measure is particularly useful because E_b/N_0 is a *normalized signal-to-noise ratio*; being normalized, the curves allow us to compare the performance of various modulation schemes. There is a similar performance measure for multiple access schemes. Here we are interested in the average delay versus normalized throughput. What would an *ideal delay–throughput curve* look like? Figure 9.23 illustrates such a curve. For normalized throughput values

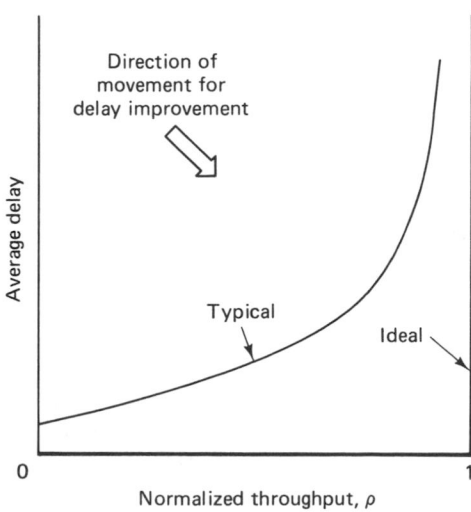

Figure 9.23 Delay–throughput characteristic.

of, $0 \leq \rho < 1$, the delay equals zero until $\rho = 1$; then the delay increases without bound. Figure 9.23 also shows a *typical* delay–throughput curve and the direction in which the curve will move as delay performance improves.

Figure 9.24 compares the delay–throughput performance of S-ALOHA with that of R-ALOHA (formatted with two message slots and six reservation subslots). Knowing the location of the *ideal* curve it is easy to compare the delay perform-ance of these two systems. For a throughput of less than approximately 0.20, the S-ALOHA manifests less average delay than does R-ALOHA. But for values of ρ between 0.20 and 0.67, it is apparent that R-ALOHA is superior, since the average delay is less. Why does the S-ALOHA perform better at low traffic in-tensity? The S-ALOHA algorithm does not require the overhead of the reservation subslots as does R-ALOHA. Therefore, at low values of ρ, R-ALOHA pays the price of greater delay due to the greater overhead. For $\rho > 0.2$, the collisions and retransmissions inherent in the S-ALOHA system cause it to incur greater delay (unbounded at $\rho = 0.37$), more quickly than the R-ALOHA system. At higher throughput ($0.2 < \rho < 0.67$), the overhead structure of R-ALOHA ensures that its delay degradation grows in a more orderly manner than S-ALOHA. For R-ALOHA, an unbounded delay is not reached until $\rho = 0.67$.

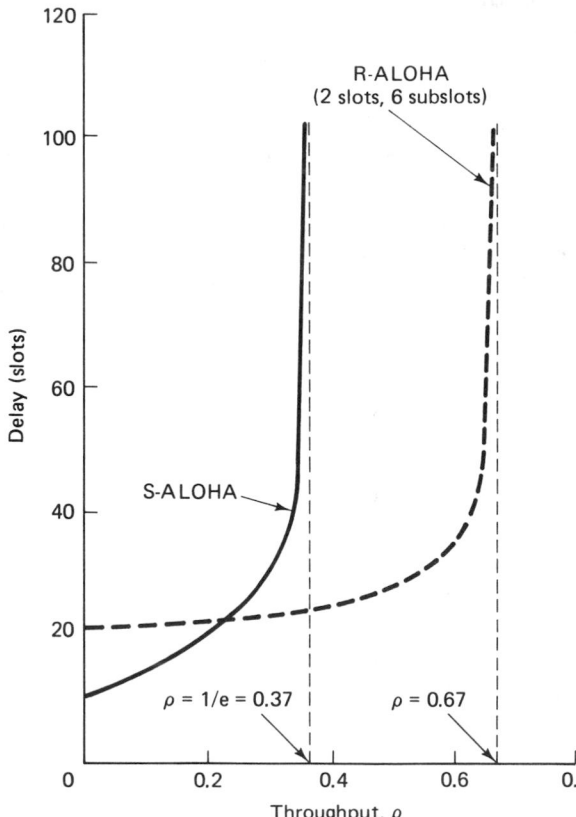

Figure 9.24 Delay–throughput comparison: S-ALOHA versus R-ALOHA on a satellite channel.

Example 9.2   Channel Utilization

(a) Normalized throughput, ρ, is a measure of channel utilization. It can be found by forming the ratio of the successfully transmitted message traffic, in bits per second to the total message traffic, including rejected messages, in bits per second. Calculate the normalized throughput of a channel that has a maximum data rate $R = 50$ kbits/s and operates with $M = 10$ ground stations, each station transmitting at the average rate of $\lambda = 2$ packets/second. The system format provides for $b = 1350$ bits/packet.

(b) Which of the three ALOHA schemes discussed—pure, slotted, and reservation—could be successfully used with this channel?

Solution

(a) Generalizing Equation (9.19) to allow for traffic from multiple stations, we have

$$\rho = \frac{Mb\lambda}{R}$$

$$= \frac{10(1350)(2)}{50,000}$$

$$= 0.54$$

(b) Only the R-ALOHA scheme could be used for this system, since with each of the other schemes, 54% of the resource cannot be utilized.

9.3.5 Polling Techniques

One way to impose order on a system with multiple users having random access requirements is to institute a controller that periodically polls the user population to determine their service requests. If the user population is large (e.g., thousands of terminals) and the traffic is bursty, the time required to poll the population can be an excessive overhead burden. One technique for rapidly polling a user population [4, 14] is called a *binary tree search*. Figure 9.25 illustrates a satellite example of such a tree search to resolve contention among users. In this example, assume that the total user population is eight terminals; let them be identified by the binary numbers 000 to 111 as shown in Figure 9.25. Assume that terminals 001, 100, and 110 are contending for the service of a single channel. The tree search operates by continually partitioning the population until there is just a single branch remaining. The terminal corresponding to that branch is the "winner" and hence the first terminal to access the channel. The operation is repeated and again yields a single terminal that may next use the channel. The algorithm proceeds according to the following steps (see Figure 9.25):

1. The satellite requests the transmission of the contending terminals' first (leftmost) bit of their identification (ID) numbers.
2. Terminal 001 transmits a zero, and terminals 100 and 110 each transmit a one. The satellite, on the basis of received signal strength, selects one or zero as the bit it "heard." In this example the satellite chooses binary one

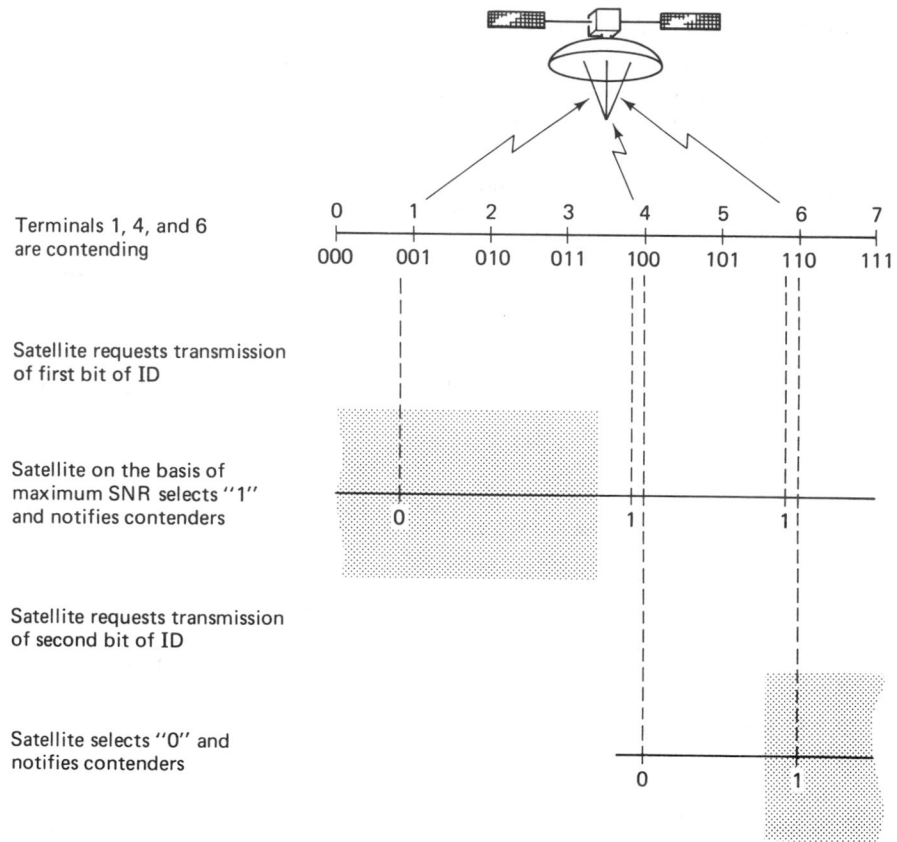

Figure 9.25 Tree search to resolve contention (eight terminal example).

and informs the users accordingly. Half the user population now knows that it has not been selected. The terminals in the "losing" half "bow out" of contention during this pass through the tree. In this example terminal 001 bows out.

3. The satellite requests the transmission of the second identifying bit from the remaining contending terminals.

4. Terminal 100 transmits a zero, and terminal 110 transmits a one.

5. Assume that the satellite selects the zero and notifies the contenders accordingly. Terminal 110 bows out. The process continues until it is clear that terminal 100 is free to access the satellite.

6. When the channel becomes available, steps 1–5 are repeated.

**Example 9.3 Comparison between Binary Tree Search and
Straight Polling**

(a) A binary tree search requires $n = \log_2 Q$ decisions for each pass through a population of Q terminals. A savings in time is possible with a tree search if the population is large and the average demand for service is small. Calculate the time needed for the straight polling of a population of 4096 terminals, to provide channel availability to 100 terminals requesting service. Compare the result with the time needed to perform a binary tree search 100 times, over the same population. Assume that the time required to poll one terminal and the time required for one decision of a binary tree search are each equal to 1 s.

(b) Develop an expression for Q', the largest number of terminals that results in the same (or less) time expended for binary tree searching as compared to straight polling.

(c) Compute Q' for part (a).

Solution

(a) Straight polling of 4096 terminals:

$$T = 4096 \times 1 \text{ s} = 4096 \text{ s}$$

Binary tree search for 100 terminals requires 100 passes through the binary tree:

$$T' = (100 \times \log_2 4096) \times 1 \text{ s} = 1200 \text{ s}$$

(b) Q' is the maximum number of terminals that will result in $T' \leq T$ in part (a). This will occur when

$$Q'' \log_2 Q \times 1 \text{ s/decision} = Q \times 1 \text{ s/poll}$$

$$Q' = \lfloor Q'' \rfloor = \left\lfloor \frac{Q}{\log_2 Q} \right\rfloor \qquad (9.30)$$

where $\lfloor x \rfloor$ is the largest integer no greater than x.

(c) Q' for part (a)

$$Q' = \left\lfloor \frac{4096}{\log_2 4096} \right\rfloor = 341 \text{ terminals}$$

A binary tree search for 341 terminals entails a search time of 4092 s.

9.4 MULTIPLE ACCESS TECHNIQUES EMPLOYED WITH INTELSAT

The first commercial, geostationary communication satellite (INTELSAT I, or Early Bird) launched in 1965, represented the start of a new telecommunications era. Its 240 voice circuits provided more capacity than the undersea cables laid between the United States and Europe during the previous 10 years [15].

Early Bird featured a hard-limiting nonlinear transponder using FDMA. When several signals having different carrier frequencies simultaneously occupy a nonlinear device, the result is the production of intermodulation products which are signals at all combinations of sum and difference frequencies [16–18]. The energy apportioned to these intermodulation or IM products represents a *loss* in the useful signal energy. In addition, if these IM products appear within the bandwidth of other signals, the effect is that of added *noise* for the other signals.

The nonlinear transponder in Early Bird allowed for only two earth stations (one in the United States and one in Europe) to simultaneously access the satellite.

Figure 9.26 illustrates this satellite's operation between the United States and Europe. Three European earth stations were interconnected via a terrestrial network. Each month a different European station accessed the satellite and distributed the traffic to the other two stations.

9.4.1 Preassigned FDM/FM/FDMA or MCPC Operation

INTELSAT II and III improved multiple access capability by operating their travelling-wave tube amplifiers (TWTA) in the linear region instead of the hard-limiting region. This kept the IM products at an acceptable level, allowing more than two simultaneous accesses. (The price paid was a reduction in power amplifier efficiency.) Thus many FM carriers from various earth stations could simultaneously access these satellites. The operation is designated preassigned multidestination FDM/FM/FDMA or simply FDM/FM, or multichannel per carrier (MCPC), and is illustrated in Figure 9.27. Long-distance calls originating in country A enter the telephone exchange and are multiplexed into a supergroup (five groups of 12 voice circuits each). Country A transmits the supergroup on a single FM carrier at frequency f_A. Each group within the supergroup has been preassigned to an earth station in country A for telephone traffic destined to countries

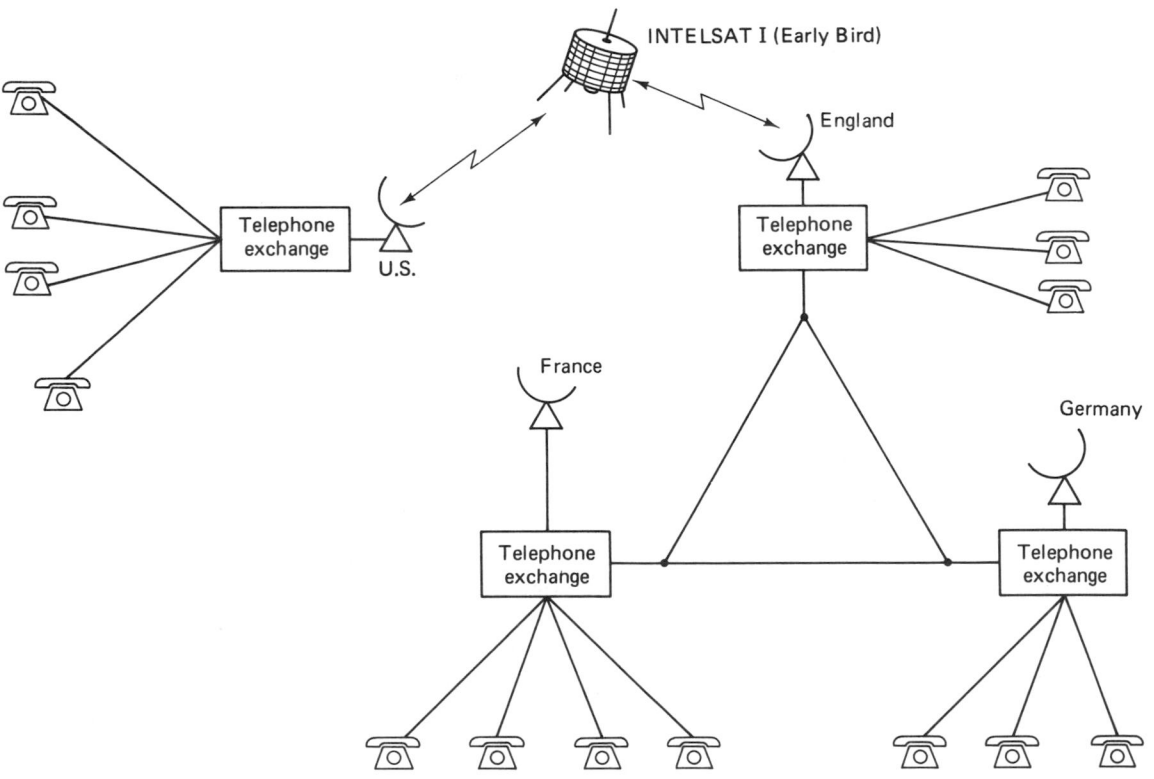

Figure 9.26 Early satellite operation.

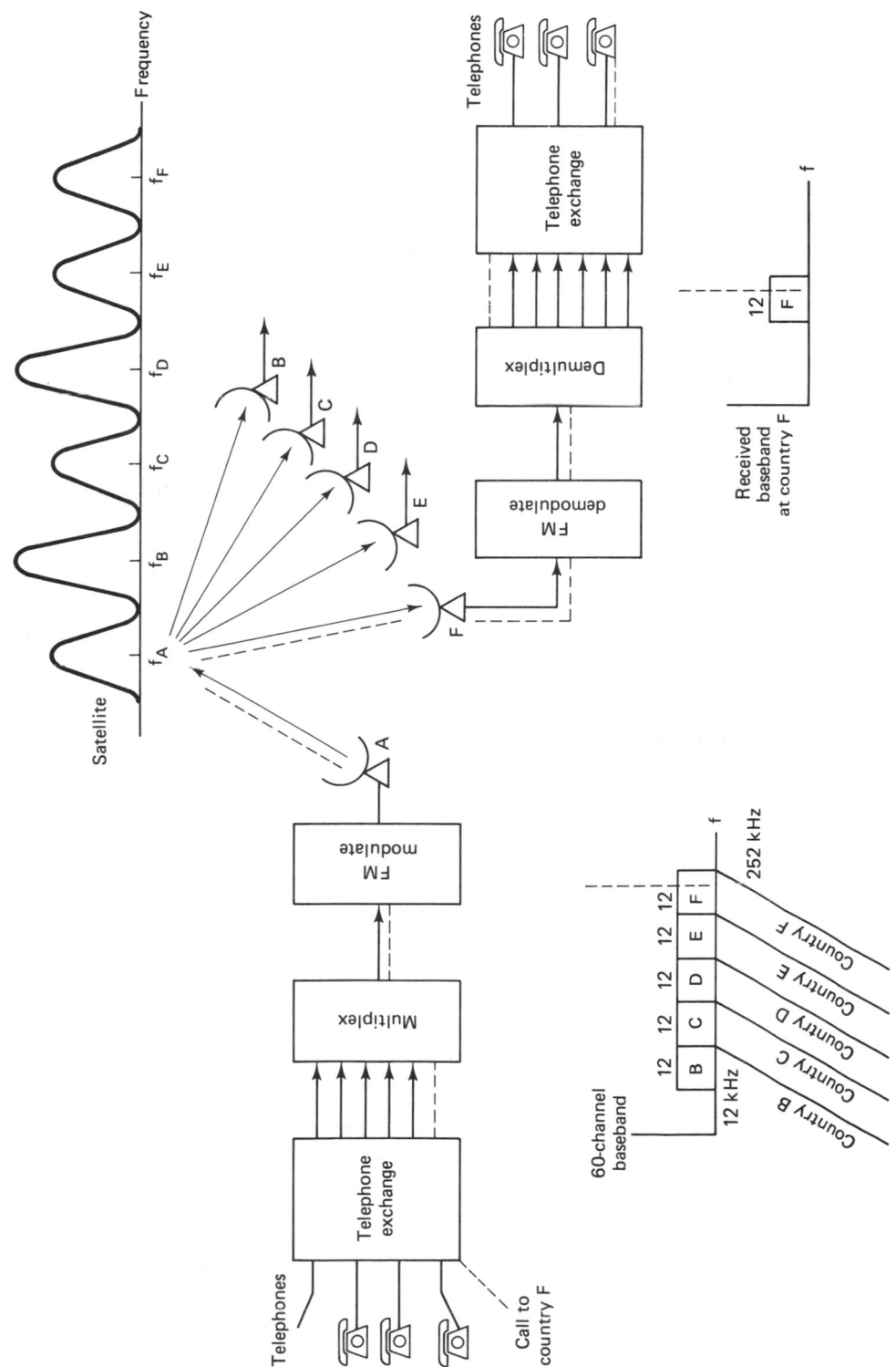

Figure 9.27 Preassigned multidestination FDM/FM carriers. (Reprinted with permission from J. G. Puente and A. M. Werth, "Demand-Assigned Service for the INTELSAT Global Network," *IEEE Spectrum*, Jan. 1971. © 1971 IEEE.)

509

B through *F*. These countries each receive the signal on frequency f_A. The received signal is demodulated and demultiplexed at the destination country, selecting only those 12 channels preassigned to it.

9.4.2 MCPC Modes of Accessing an INTELSAT Satellite

INTELSAT has standardized the ways in which each 36-MHz transponder may be shared by specifying the occupied RF bandwidth and the number of 4-kHz channels per user. Some of these standard channels are shown in Table 9.1. Notice that the capacity of the transponder (last column in Table 9.1) drops as the number of carriers increases. The reasons are as follows:

1. Guard bands are needed between carrier bands; the more carriers there are, the more guard bands are needed. Hence capacity is reduced.
2. Multiple carriers in the nonlinear TWTA cause intermodulation (IM) products. If the TWTA is backed off into the linear region to reduce interference, the TWTA can provide less overall power. The channel becomes power limited and can service fewer carriers.

TABLE 9.1 Standard INTELSAT MCPC Accessing Modes

Number of carriers per transponder	Carrier bandwidth	Number of 4-kHz channels per carrier	Number of 4-kHz channels per transponder
1	36 MHz	900	900
4	3 at 10 MHz	132	456
	1 at 5 MHz	60	
7	5 MHz	60	420
14	2.5 MHz	24	336

Table 9.1 indicates that a single carrier provides the most efficient use of the transponder. Why doesn't INTELSAT always operate its transponders in this mode? The answer is that not all earth stations have enough traffic to justify the assignment of an entire 36-MHz transponder. The other modes are needed so that various combinations of stations having less traffic will be able to share a transponder.

9.4.2.1 Bandwidth-Limited versus Power-Limited Conditions

In the preceding section it was stated that the backed-off transponder cannot support as many channels as the fully saturated transponder. It is useful to examine the two extreme transponder conditions, bandwidth limited and power limited, in the context of a satellite transponder. In Figure 9.28 we assume a 36-MHz transponder with a maximum power output of 20 W. Figure 9.28a illustrates an MCPC mode of operation whereby four carriers share the 36-MHz bandwidth. Assume that each carrier requires 4 W. The total output power is 16 W (less than the maximum capability of the amplifier); therefore, there is still power to spare.

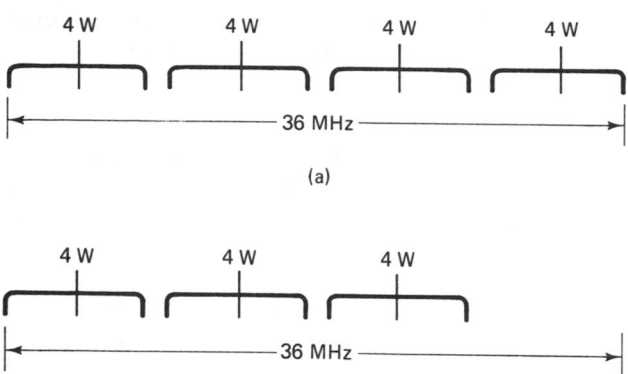

Figure 9.28 Bandwidth-limited versus power-limited configurations. (a) Bandwidth-limited example. (b) Power-limited example.

However, should another user want to access the transponder, the total 36-MHz bandwidth has already been allocated to the existing four carriers; there is no additional bandwidth to spare. Figure 9.28a illustrates this bandwidth-limited case.

Suppose that the previous example results in the production of serious IM products at the transponder. Assume that it is necessary to linearize the transponder by operating it at a reduced maximum power output of 12 W. With only a 12-W capability, the transponder can no longer support four users with 4 W each. One of the users must be "thrown off," as illustrated in Figure 9.28b. Therefore, we have bandwidth to support another user, but not sufficient power. Figure 9.28b illustrates this power-limited case.

9.4.3 SPADE Operation

The preassigned MCPC multiple access scheme is very efficient when the traffic is heavy enough so that the channels are most always filled. However, if out of a 12-channel group, only one channel is active, the other 11 cannot be turned off. The FDM/FM transmission is made with or without actual telephone traffic on the channels. Therefore, the long-term preassignment of carriers to stations having light traffic is wasteful. Since there are many light traffic links, a flexible method to service them was needed. Also, an efficient way to handle overflow traffic from medium-capacity preassigned links was needed. Such was the motivation for a novel DAMA scheme known as SPADE, first used with INTELSAT IV. The acronym SPADE stands for "single-channel-per-carrier PCM multiple access demand assignment equipment." The principal features characterizing SPADE operation [15] are:

1. A single voice-grade channel is analog-to-digital (A/D) converted at a bit rate of 64 kbits/s.
2. This baseband digital signal modulates a carrier using quadrature phase shift keying (QPSK). Unlike the MCPC case, there is *only one* voice channel per carrier.

3. The channel spacing is 45 kHz. Within a transponder, there is bandwidth available for 800 channel carriers. Six carrier positions are vacant by design; thus there are 794 usable carriers.

4. The carrier is dynamically assigned, *upon demand.*

5. The dynamic assignment is accomplished over a 160-kHz common signaling channel (CSC) used as an "order-wire" or control circuit. The bit rate on the CSC is 128 kbits/s, and the modulation is binary phase shift keying (BPSK).

Figure 9.29 illustrates the frequency allocations for the CSC and the 800 carriers in the SPADE system. The SPADE operation can best be understood with the aid of Figure 9.30. The CSC operates in a fixed-assignment TDMA broadcast mode; that is, all earth stations monitor the CSC and are aware of the current state of channel assignments. Each earth station has a 1-ms time slot on the CSC (once every 50 ms) for requesting or releasing a channel. When an earth station needs a channel, it "seizes" a free one by requesting a frequency pair at random and transmitting its selection on the CSC. Random selection makes it unlikely that two stations will simultaneously request the same channel unless there are very few remaining. As soon as the channel is allocated, each of the other earth station processors deletes it from its list of available channels. The list is continually kept updated via the CSC. Thus control of the SPADE access scheme is *distributed* among all the participating earth stations.

When the station finishes with the channel, the station indicates the channel's release by transmitting a signal in its time slot on the CSC. Each station receives this signal and designates the released channel as available. If two stations simultaneously seize the same channel, they each get a "busy" indication. They try again, selecting at random from the pool of available channels.

9.4.3.1 Transponder Capacity Utilization with SPADE

Table 9.2 is a continuation of Table 9.1. We see that the transponder bandwidth utilization with SPADE results in a total capacity of 800 voice channels per transponder. Compare Table 9.2 with Table 9.1. In Table 9.1, as the number of carriers increases from 1 to 14, the total number of channels decreases from 900 to 336. Why doesn't the SPADE system in Table 9.2 exhibit less capacity than the 336 channels associated with 14 carriers? The improved utilization comes about as follows. When there is only one voice channel per carrier, the carrier can be switched off when no speech is detected. Even with all channels operating, they can be switched off approximately 60% of the time. The transponders are power limited; power savings means that more channels can be transmitted. Also, SPADE uses digital voice transmission (QPSK); the bandwidth efficiency of the system is commensurate with the single-carrier FDM/FM case.

9.4.3.2 SPADE Efficiency

With MCPC, capacity is preassigned; a station's unused channels cannot be reallocated to other stations. SPADE is a DAMA system where all channels are shared. The channels are allocated to users as needed. An important telephone

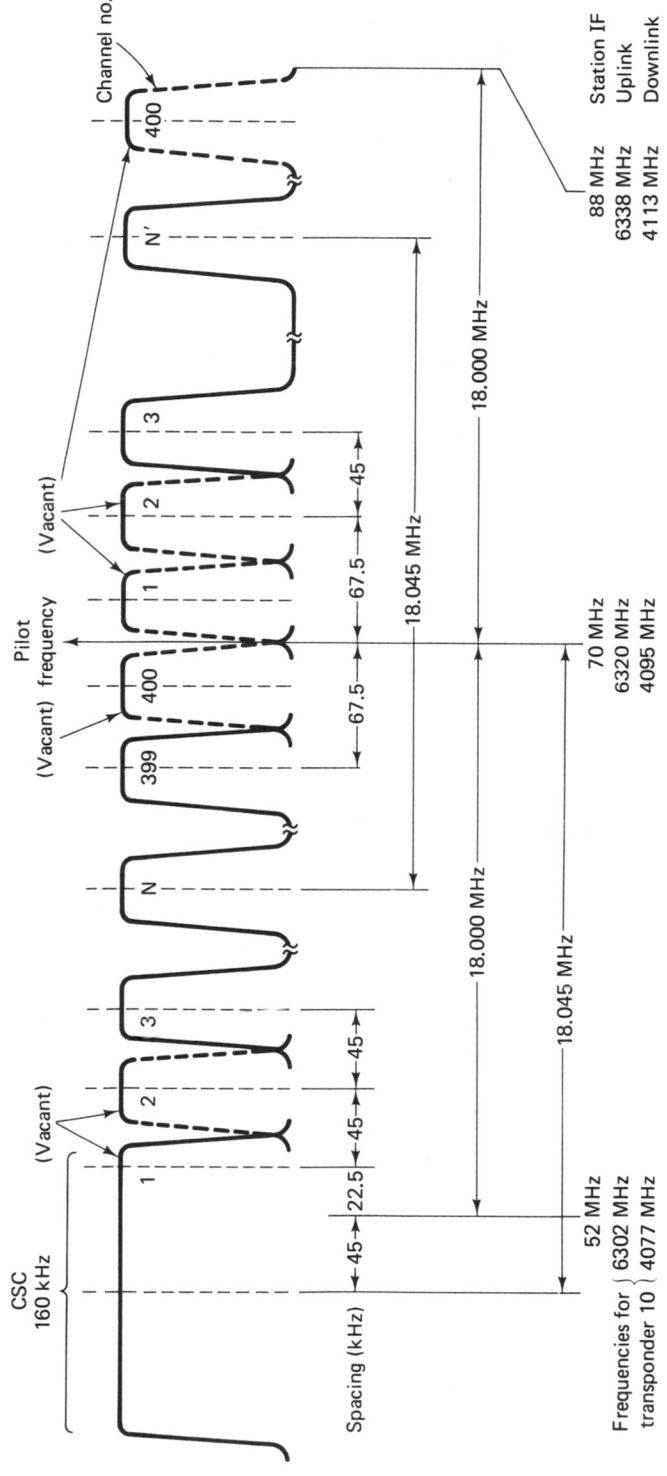

Figure 9.29 SPADE frequency allocations. (Reprinted with permission from J. G. Puente and A. M. Werth, "Demand-Assigned Service for the INTELSAT Global Network," *IEEE Spectrum*, Jan. 1971. © 1971 IEEE.)

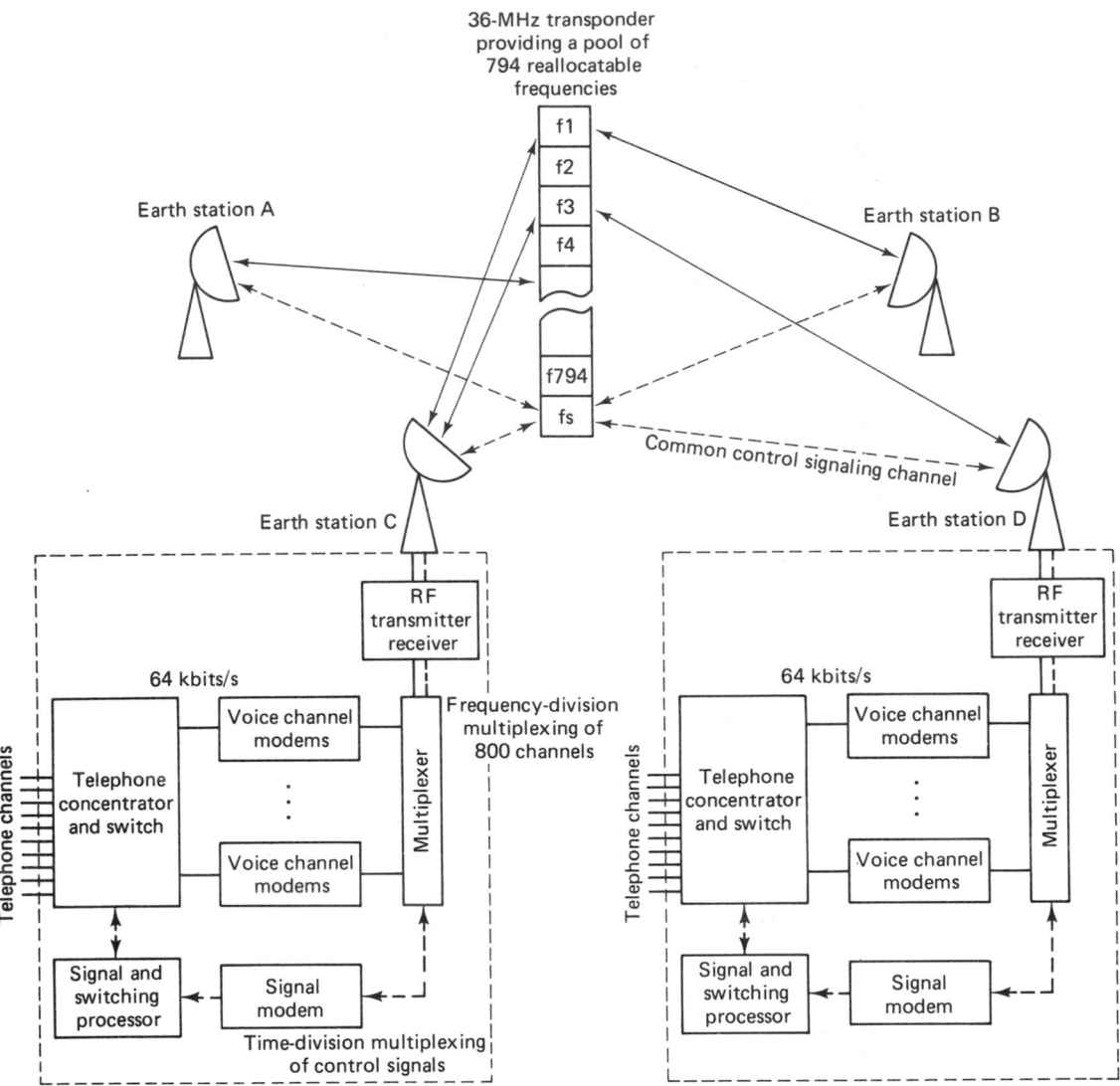

Figure 9.30 SPADE operation. (From James Martin, *Communications Satellite Systems*, © 1978, Fig. 15.2, p. 236. Reprinted by permission of Prentice-Hall, Englewood Cliffs, N.J.)

TABLE 9.2 SPADE Accessing

Number of carriers per transponder	Carrier bandwidth	Number of 4-kHz channels per carrier	Number of 4-kHz channels per transponder
800	45 kHz	1	800

system quality measure, called the probability of blocking, is the probability that a requested circuit is not available. To achieve 1% probability of blocking requires four times as many MCPC channels as SPADE channels. A SPADE transponder with 800 channels is equivalent to 3200 MCPC channels [15].

9.4.3.3 Mixed-Size Earth Station Network Using SPADE

A standard-size INTELSAT earth station has a receiver sensitivity $G/T° = 40.7$ dB/K, whereas the smaller size stations have a $G/T° = 35$ dB/K. If 125 SPADE channels are destined for small stations, the total transponder capacity of 800 standard channels is reduced to 525 channels. This is the point at which half the available power is used to service the standard stations. The relationship between transponder capacity and channels allocated to small stations is shown in Figure 9.31. An explanation of this relationship can best be seen in Figure 9.32. When the total TWTA power provides service to large stations, Figure 9.32a illustrates that the 36-MHz bandwidth transponder is occupied by approximately 800 carriers each at a power level of x dBW (the bandwidth-limited case). When half the power is required to service small stations, Figure 9.32b illustrates that 400 carriers (half of the original 800) each at a power level of x dBW are reserved for the standard stations. Consider what happens to the remaining 400 carrier positions. From Chapter 4 we know that the error performance of a link is directly related to the product of EIRP and $G/T°$. For any link, one can trade off these two parameters, thereby maintaining a fixed level of performance. Since the small station has a $G/T°$ of 5.7 dB less than that of the standard station, it is necessary to supply the small station with 5.7 dB more EIRP for equivalent performance. The carrier power is *increased* by approximately 5.7 dB for each small station, thus the quan-

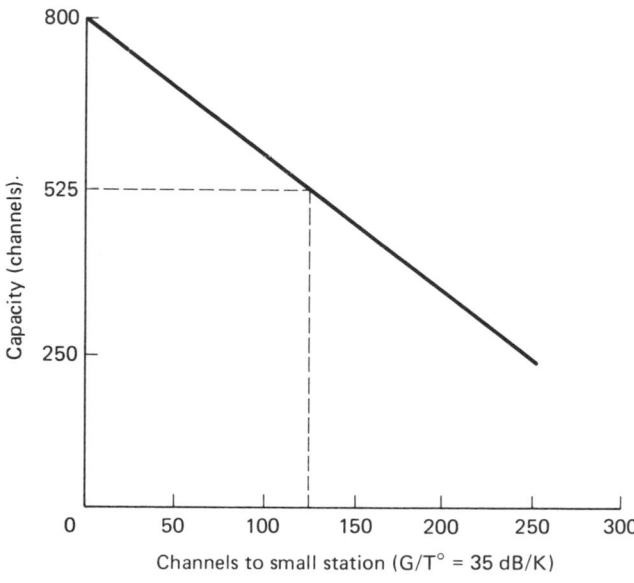

Figure 9.31 SPADE transponder capacity in a mixed-size earth station network.

Channels to small station (G/T° = 35 dB/K)

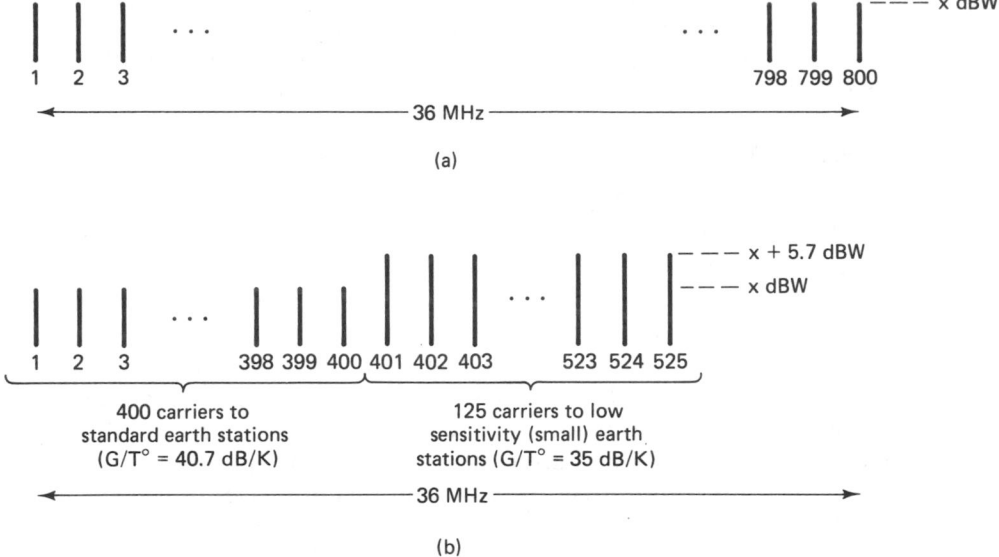

Figure 9.32 Mixed-size earth station network. (a) When the total TWTA power services large stations: bandwidth limited (800 carriers). (b) When half the TWTA power services small stations: power limited (525 carriers).

tity of the remaining carriers serving these small stations is *decreased* by a similar amount. Therefore instead of 400 carriers, 125 (a reduction of 5.1 dB) are used to serve the small stations; the transponder is now power limited.

At the time a channel is assigned to a call, the transmitting station is apprised of the size of the destination station. Recall that these satellites are non-regenerative so that the apportionment of downlink EIRP is established by the transmitting station (see Section 4.7.1). The transmitting station sets its power level according to the needs of the receiving station.

9.4.4 TDMA in INTELSAT

The first generation of multiple access communication systems has been dominated by FDMA systems. The trend, however, is now in favor of TDMA systems, made possible by the availability of precise clocks and high-speed switching elements [19–24]. INTELSAT IV used a 128-kbits/s TDMA scheme for the common signaling channel that controls the SPADE network. Intelsat V introduced a 120-Mbits/s TDMA scheme for multiple-beam international digital service. One disadvantage or cost in implementing a TDMA scheme is the need for providing precise *synchronization* among the participating earth stations and the satellite. FDMA systems, not having such requirements, are less complex from a networking point of view. Comparisons of TDMA versus FDMA operation are summarized as follows:

1. FDMA can cause IM products. This can be avoided by operating the TWTA in its linear region, thereby reducing the available power output.

2. With TDMA, there is only one carrier present at a time in the TWTA. Thus IM distortion cannot occur.

3. TDMA earth station equipment is more sophisticated and hence more costly than FDMA equipment. However, for earth stations providing multiple point-to-point channels, FDMA stations require separate radio-frequency (RF) up-conversion and down-conversion signal processing stages. Thus with FDMA, the amount of equipment grows with the amount of simultaneous connectivity. With TDMA, such growth does not take place since channel selectivity is accomplished in time rather than frequency. Therefore, for a large multiply connected earth station, TDMA can be more cost-effective than FDMA.

4. In multiple-beam systems, each beam may need to communicate with every other beam. TDMA lends itself to conveniently forming connections sequentially as in satellite-switched TDMA (SS/TDMA). INTELSAT VI uses such satellite-switched TDMA (SS/TDMA), described in Section 9.4.5.

An example of the comparative performance of TDMA, FDM/FM, and SPADE is shown for an INTELSAT IV transponder as a plot of channel capacity versus earth station $G/T°$ in Figure 9.33. Figure 9.33a is for an earth coverage antenna, and Figure 9.33b is for a spot-beam antenna. From synchronous altitudes these antennas have half-power beamwidths of 17° and 4.5°, respectively. From these plots it is seen that single-carrier FDM/FM is as efficient as TDMA when the system is operated with standard earth stations ($G/T° = 40.7$ dB/K). For smaller earth stations ($G/T° \leq 31$ dB/K) working through earth-coverage transponders, SPADE is more efficient than TDMA and multicarrier FDM/FM (MCPC); only the four-carrier case is plotted. For earth stations having $G/T°$ in the range 19 to 40.7 dB/K working through a spot-beam transponder, TDMA is superior to SPADE and MCPC. For smaller earth stations having $G/T°$ in the range 6 to 19 dB/K working through a spot-beam transponder, SPADE is superior to TDMA and MCPC. In general, when working through *standard* earth stations it is seen [19] that TDMA is the most efficient multiple access scheme for INTELSAT IV.

9.4.4.1 PCM Multiplex Frame Structures

There are two digital telephony standards for PCM frame structures in operation. The North American standard is called *T-Carrier*; it is built around the 193-bit frame shown in Figure 9.34a. There are 24 channels; each channel contains an 8-bit voice sample. Also, there is one bit per frame with alternating value 1 0 1 0 . . . from frame to frame, used for frame alignment. Since a voice-grade telephone channel has a bandwidth of $W = 4$ kHz (including guard bands), the Nyquist sampling rate for recovering the analog information within 4 kHz is $f_s = 2W = 8000$ samples/s. Therefore, the basic PCM frame, called the *Nyquist*

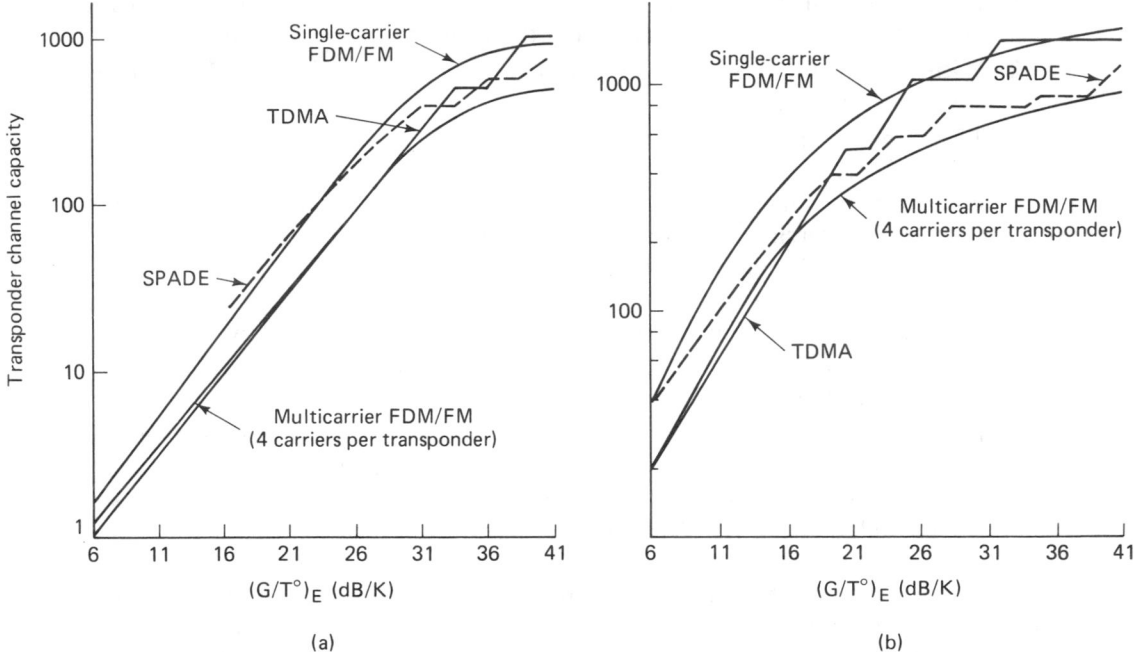

Figure 9.33 Channel capacity versus earth station $G/T°$ for FDMA, TDMA, and SPADE. (a) Global-beam transponder channel capacity as function of $(G/T°)_E$, where $(G/T°)_E$ means earth station $G/T°$. (b) Spot-beam transponder channel capacity as function of $(G/T°)_E$. [From D. Chakraborty, "INTELSAT IV Satellite System (Voice) Channel Capacity versus Earth Station Performance," *IEEE Trans. Commun. Tech.,* vol. COM19, no. 3, June 1971, pp. 355–362. © 1971 IEEE.]

frame, which contains 24 voice samples from 24 different message sources, has a frame rate of 8000 frames/s (duration of 125 μs). Thus the basic T-Carrier bit rate is 193 bits/frame × 8000 frames/s = 1.544 Mbits/s.

The European standard is built around a 256-bit frame shown in Figure 9.34b. There are 30 message channels, each containing an 8-bit voice sample. Also, one 8-bit time slot is used for frame alignment and another 8-bit time slot is used for signaling (addressing) information. The European frame rate is the same as that of the T-Carrier. Therefore, the basic European bit rate is 256 bits/frame × 8000 frames/s = 2.048 Mbits/s.

9.4.4.2 The High-Rate TDMA Frame for Europe

Sixteen Nyquist frames of the European PCM Multiplex format are shown in Figure 9.35a. Each frame contains an 8-bit sample from each of 30 terrestrial channels, plus 8 bits of framing and 8 bits of signaling information. The TDMA frame duration is

$$16 \text{ Nyquist frames} \times 125 \text{ μs/Nyquist frame} = 2 \text{ ms}$$

Within this 2-ms frame are contained

16 Nyquist frames × 256 bits/Nyquist frame = 4096 bits

The basic idea behind TDMA is that a user's low-rate data stream can share the CR with similar streams from other users by *bursting* the transmission at a much faster rate than the rate at which it is generated. Figure 9.35b illustrates a 2-ms high-rate TDMA frame. The frame begins with a reference burst, RB1, emitted by a reference station. The burst contains information necessary to enable other stations to precisely position their message traffic bursts in the frame. There may be a second burst, RB2, for reliability, followed by a sequence of traffic slots. The traffic slots may be preassigned, or they may be assigned according to a DAMA protocol [20].

The PCM multiplex signal with a bit rate of $R_0 = 2.048$ Mbits/s and a frame duration of $T = 2$ ms is compressed (by a factor of 59) and transmitted using QPSK modulation at a burst rate of $R_T = 120.832$ Mbits/s (symbol rate of 60.416 megasymbols/s). The duration of the traffic data field T_{tr} in the high rate TDMA

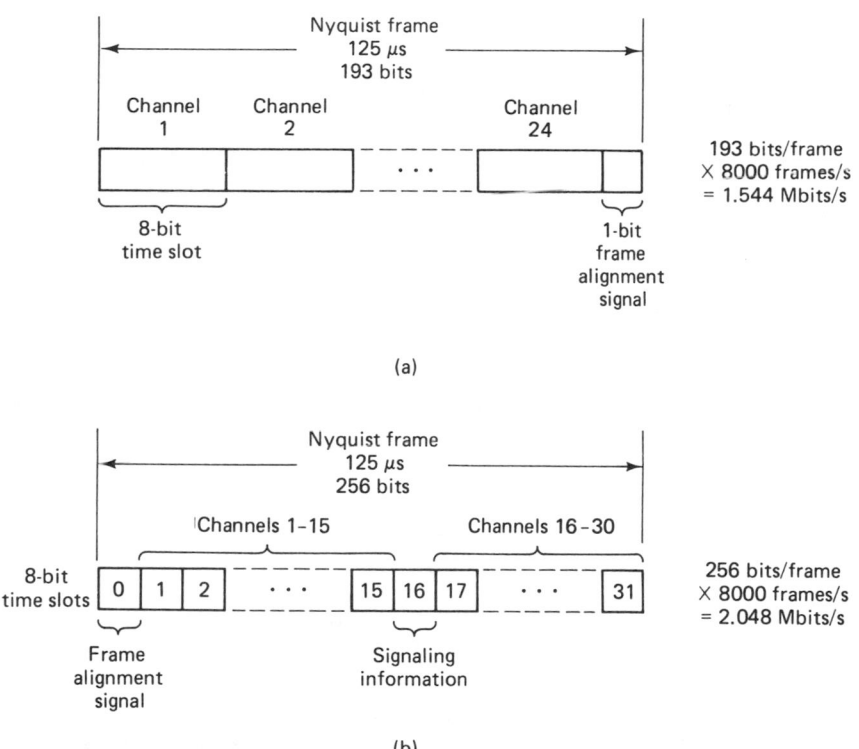

(a)

(b)

Figure 9.34 PCM multiplex frame structure. (a) Frame structure for T-Carrier (North American) PCM multiplex. (b) Frame structure for the European PCM multiplex.

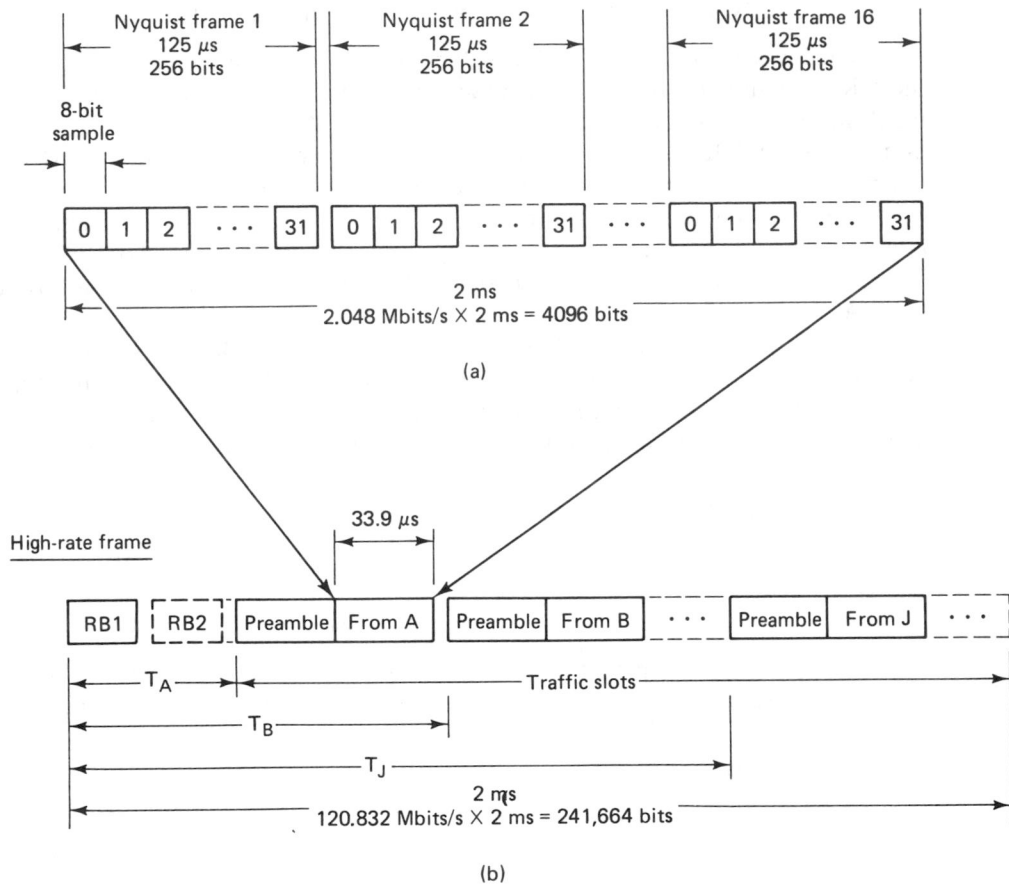

Figure 9.35 INTELSAT digital transmission standards for Europe. (a) Terrestrial PCM multiplex. (b) High-rate frame.

frame is calculated as follows:

$$T_{tr} = \frac{R_0 T}{R_T} \tag{9.31}$$

$$= \frac{2.048 \times 10^6 \times 2 \times 10^{-3}}{120.832 \times 10^6}$$

$$= 33.9 \ \mu s$$

To obtain the total duration of a traffic burst, the time used for the preamble must be added. If the preamble contains S_P symbols, then assuming QPSK modulation, the total length of the traffic burst measured in number of symbols, S_T, is

$$S_T = \frac{R_0 T}{2} + S_P \tag{9.32}$$

and the burst-time duration is

$$T_T = \frac{2S_T}{R_T} \qquad (9.33)$$

If the preamble contains 300 symbols, then

$$S_T = \frac{2.048 \times 10^6 \times 2 \times 10^{-3}}{2} + 300$$

$$= 2348 \text{ symbols}$$

Using this in Equation (9.33), we obtain

$$T_T = \frac{2 \times 2348}{120.832 \times 10^6} = 38.9 \ \mu s$$

9.4.4.3 The High-Rate TDMA Frame for North America

The INTELSAT TDMA burst (bit) rate of $R_T = 120.832$ Mbits/s was chosen to be compatible with both the European and North American standards. Figure 9.36 is similar to Figure 9.35 except that the PCM multiplex signal is the 24-channel T-Carrier instead of the 30-channel European standard. The essential T-Carrier features that are different from the European standard are listed below and are shown on the figure.

1. Each Nyquist frame is comprised of 24 channels or samples $\times$ 8 bits $+$ 1 frame alignment bit $=$ 193 bits.
2. The 16 Nyquist frames contain $16 \times 193 = 3088$ bits.
3. The T-Carrier data rate is 1.544 Mbits/s.
4. The duration of the traffic data field in the high-rate TDMA frame is calculated from Equation (9.31).

$$T_{tr} = \frac{1.544 \times 10^6 \times 2 \times 10^{-3}}{120.832 \times 10^6}$$

$$= 25.6 \ \mu s$$

9.4.4.4 INTELSAT TDMA Operation

At the transmitting earth station, the continuous low-rate data stream enters one of a pair of buffers illustrated in Figure 9.37a. When one buffer is filling at the low rate (1.544 Mbits/s or 2.048 Mbits/s), the other is emptying at the burst rate (120.832 Mbits/s). The buffers alternate functions at each TDMA frame. The time of application of the high-rate clock is controlled so that the traffic burst is transmitted in the proper interval to arrive at the satellite in its assigned position in the TDMA frame.

At the receiving station, the received traffic burst is routed to one of a pair of expansion buffers, shown in Figure 9.37b, that have the inverse function of

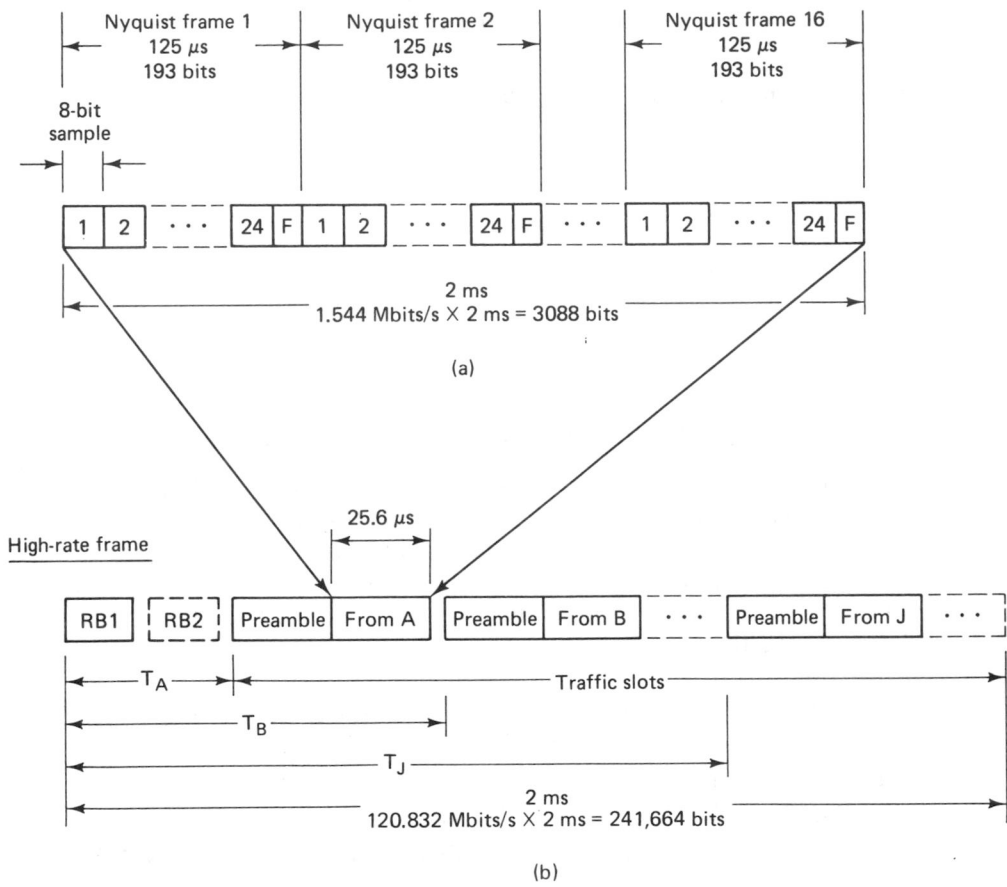

Figure 9.36 INTELSAT digital transmissions standards for T-Carrier. (a) Terrestrial PCM multiplex. (b) High-rate frame.

the compression buffers in Figure 9.37a. When one buffer is filling at the high rate, the other is emptying at the desired output rate.

The most critical aspect of TDMA operation is the precise synchronization needed to assure orthogonality of the time slots [20]. Figure 9.38 illustrates the general idea behind most commercial satellite synchronization schemes. One station is designated as the master or control station. This station transmits periodic bursts of reference timing pulses. User stations also transmit their timing pulses, designated as slave pulses in Figure 9.38. On the downlink, the using station receives the master or reference pulses in addition to its own slave pulses. The time difference between the master and slave pulses corresponds to the timing error. The station adjusts its clock so as to reduce this timing error.

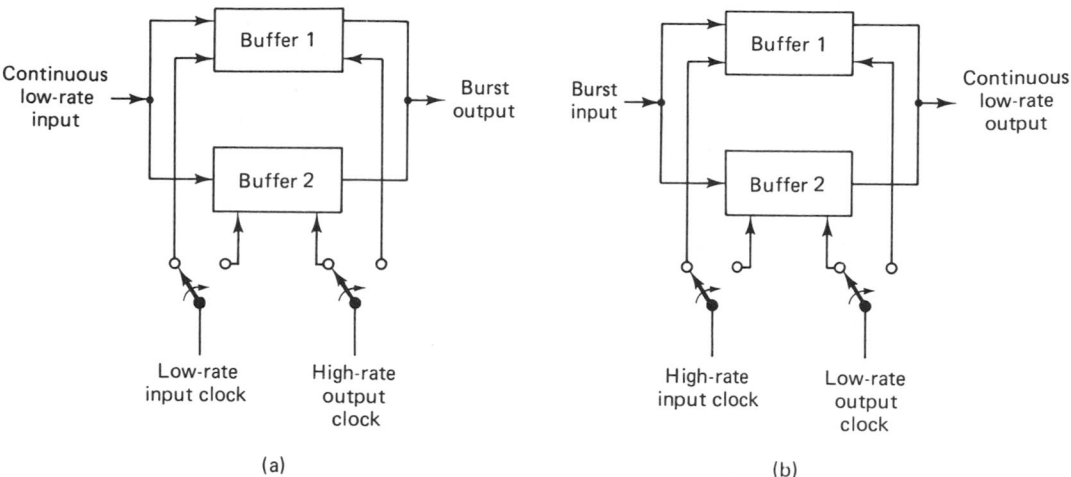

Figure 9.37 Burst compression and expansion buffers. (a) Compression buffers at transmitter. (b) Expansion buffers at receiver.

9.4.5 Satellite-Switched TDMA in INTELSAT

Modern communication satellites often employ several regional antenna beams. For a satellite based over the Atlantic Ocean, separate beams might be aimed at North America, Europe, South America, and Africa. Switches are used to allow the interconnection of stations in one region to communicate with stations in another region. The basic goal of a satellite-switched TDMA (SS/TDMA) scheme is to provide an efficient way of cyclically providing interconnection of TDMA data among various coverage regions.

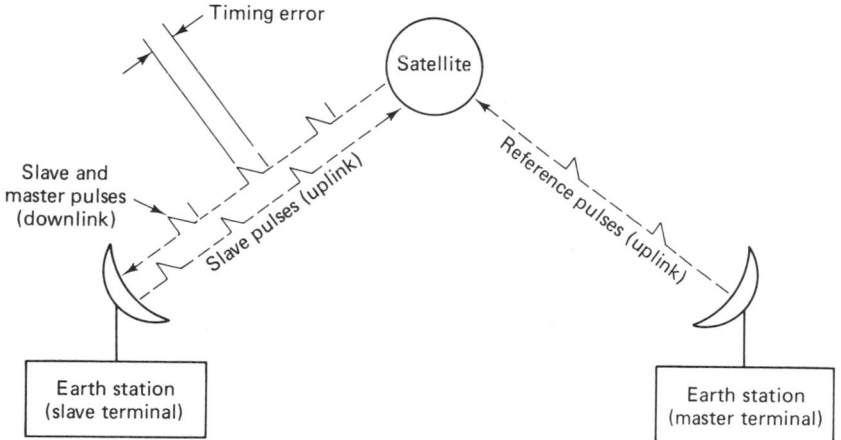

Figure 9.38 TDMA synchronization concept.

The heart of the system consists of a microwave switch matrix, located in the satellite, that is programmed via ground control to change states cyclically in rapid sequence, thus interconnecting distinct uplink beams to distinct downlink beams at each switching time. An earth station in the network communicates with those in other beams by transmitting TDMA bursts in the proper timing positions in the sequence. The pattern of switch states is selected so as to maximize the usable system capacity under the constraints of the traffic demands [21]. For complete interconnectivity between N beams, a total of $N!$ different satellite switch states or *modes* are required. Table 9.3 illustrates the six modes required for the full interconnectivity of a three-beam system.

In mode 1, the satellite receivers in beams A, B, and C are connected to the satellite transmitters for beams A, B, and C, respectively. An earth station in one of these beams can then communicate with other earth stations in the same beam. The beam is said to be *looped back* on itself.

Figure 9.39 illustrates a three-beam (beams A, B, and C) example of a SS/ TDMA system. The satellite microwave switch matrix is configured in a *crossbar* design. This design can be thought of as being made up of row and column lines; when one row and one column are energized, contact is made at the intersection. A crossbar design only permits a single row to communicate with a single column at a time. If uplink A_U is connected to downlink B_D, *neither A_U nor B_D can be* simultaneously connected to any other beam.

In Figure 9.39, three different traffic patterns during time slot intervals T_1, T_2, and T_3, with three different switch states S_1, S_2, and S_3 are shown. During interval T_1, switch state S_1 interconnects the beams in a loop-back fashion which permits the uplink messages in slot T_1 to be delivered to their correct destinations. During time interval T_2, switch state S_2 interconnects uplink beam A_U to downlink beam B_D, uplink beam B_U to downlink beam C_D, and uplink beam C_U to downlink beam A_D. This connection pattern assures that the uplink messages in slot T_2 are delivered to their correct destinations. During time interval T_3, switch state S_3 similarly connects uplink transmissions to downlink beams to assure correct delivery of the data.

The traffic patterns and their durations are programmed to optimize the resource capacity and to serve the users as efficiently as possible. The cyclic pattern can be reprogrammed by ground command to meet changing traffic requirements.

9.4.5.1 Traffic Matrix

Figure 9.40 is a matrix describing the communication traffic among N spot-beam coverages. In this figure, t_{ij} is the traffic volume from the ith beam to the jth beam. The subtotal S_i is the total traffic originating from the ith uplink beam, expressed as

$$S_i = \sum_{j=1}^{N} t_{ij} \tag{9.34}$$

TABLE 9.3 Three-Beam Satellite Switch Modes

Input	Output					
	Mode 1	Mode 2	Mode 3	Mode 4	Mode 5	Mode 6
A	A	A	B	B	C	C
B	B	C	A	C	A	B
C	C	B	C	A	B	A

and R_j is the total traffic received in the *j*th downlink beam, expressed as

$$R_j = \sum_{i=1}^{N} t_{ij} \tag{9.35}$$

When the traffic in a SS/TDMA system is controlled by a nonblocking switch (one that allows for the transmission of *all* messages, without any "busy" signals)

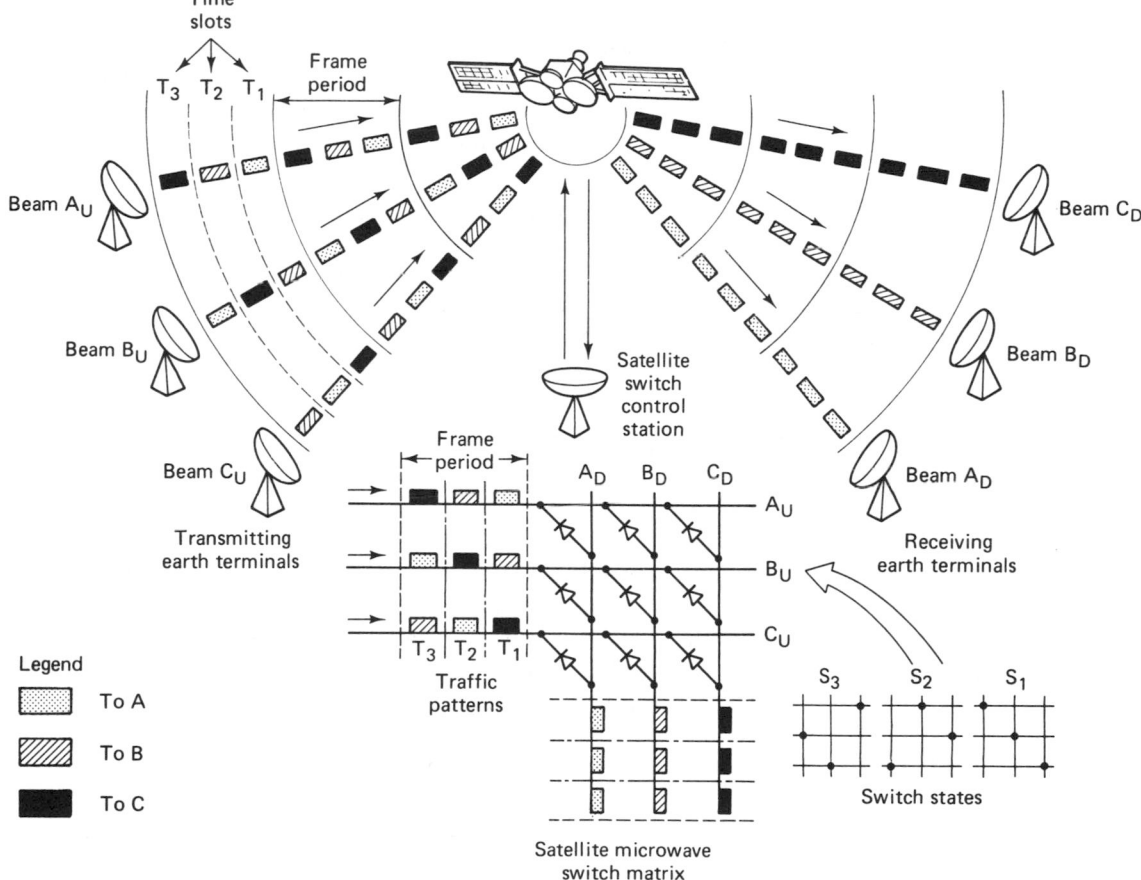

Figure 9.39 Satellite-switched TDMA (SS/TDMA).

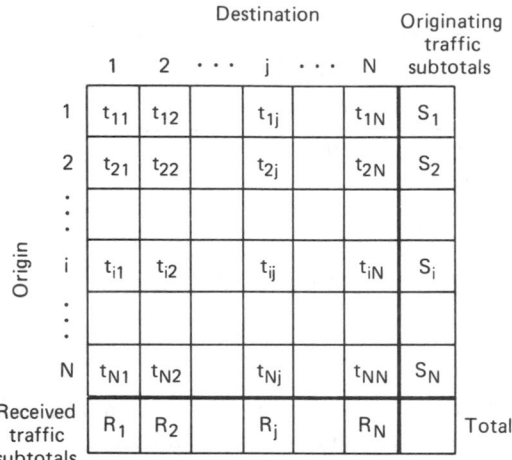

Figure 9.40 Traffic matrix.

a k-second time slot will be assigned to each channel in the TDMA frame. For efficient utilization of the CR, the total traffic in Figure 9.40 should be transmitted within a frame time T which should be made as short as possible. The minimum frame time, $T_{\min}$, for providing such nonblocking connectivity can be expressed [22] as follows:

$$T_{\min} = k \max (\{S_i\}, \{R_j\}) \tag{9.36}$$

where max $(\{S_i\}, \{R_j\})$ is the maximum value taken over the set of all $\{S_i\}$ and $\{R_j\}$. Equation (9.36) describes the minimum time to communicate *all* of the traffic in the traffic matrix, for equal bandwidth per channel.

9.5 MULTIPLE ACCESS TECHNIQUES FOR LOCAL AREA NETWORKS

A local area network (LAN) can be used to interconnect computers, terminals, printers, and so on, located within a building or a small set of buildings. While long-haul networks use the public telephone network for economic reasons, LAN designers usually lay their own high-bandwidth cables. Bandwidth is not as scarce as it is in the long-haul cases. Not being forced to optimize bandwidth, a LAN can use simple access algorithms [6, 25–27].

9.5.1 Carrier-Sense Multiple Access Networks

Ethernet is a LAN access scheme developed by the Xerox Corporation. The Ethernet scheme is based on the assumption that each local machine can sense the state of a common broadcast channel before attempting to use it. The technique is known as *carrier-sense multiple access with collision detection* (CSMA/CD). The word ''carrier,'' here, means *any* electrical activity on the cable. Figure 9.41a

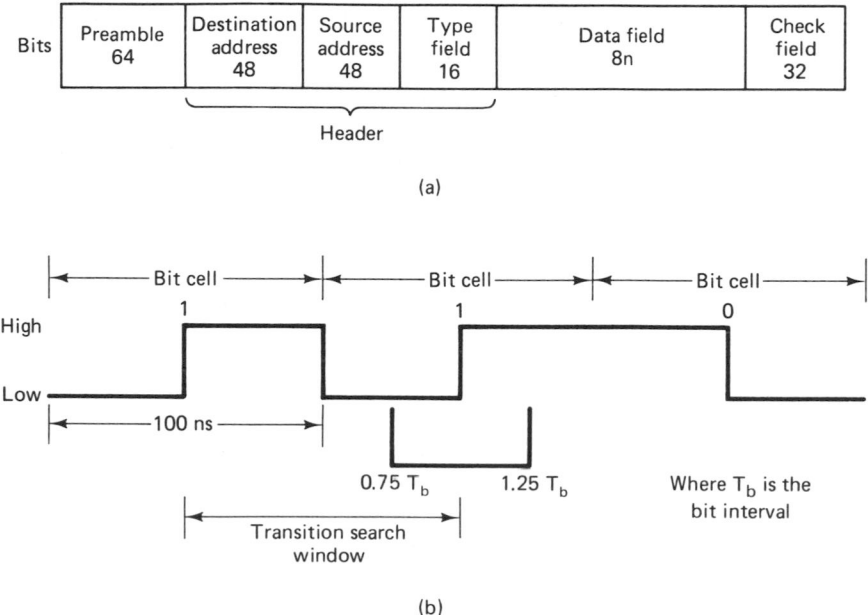

Figure 9.41 Ethernet bit field and PCM format. (a) Ethernet specification. (b) Manchester PCM format.

illustrates the bit field format for the Ethernet specification; the details are listed below.

1. The maximum packet size is 1526 bytes, where a byte is 8 bits. The packet breakdown is 8-byte preamble + 14-byte header + 1500-byte data + 4-byte parity.

2. The minimum packet size is 72 bytes, consisting of an 8-byte preamble + 14-byte header + 46-byte data + 4-byte parity.

3. The minimum spacing between packets is 9.6 μs.

4. The preamble contains a 64-bit synchronization pattern of alternating ones and zeros, ending with two consecutive ones: (1 0 1 0 1 0 ... 1 0 1 0 1 1).

5. The receiving station examines a destination address field in the header to see if it should accept a particular packet. The first bit indicates the type of address (0 = single address, 1 = group address); an entire field of ones means an all-station broadcast.

6. The source address is the unique address of the transmitting machine.

7. The type field determines how the data field is to be interpreted. For example, bits in the type field can be used to describe such things as data encoding, encryption, message priority, and so on.

8. The data field is an integer number of bytes from a minimum of 46 to a maximum of 1500.

9. The parity check field houses the parity bits which are generated by the following generating polynomial (see Section 5.6):

$$X^{32} + X^{26} + X^{23} + X^{22} + X^{14} + X^{12} + X^{11} + X^{10}$$

$$+ X^8 + X^7 + X^5 + X^4 + X^2 + X + 1$$

The Ethernet multiple access algorithm defines the following user action or response:

1. *Defer*. The user must not transmit when the carrier is present or within the minimum packet spacing time.
2. *Transmit*. The user may transmit if not deferring until the end of the packet or until a collision is detected.
3. *Abort*. If a collision is detected, the user must terminate packet transmission and transmit a short jamming signal to ensure that all collision participants are aware of the collision.
4. *Retransmit*. The user must wait a random delay time (similar to the ALOHA system) and then attempt retransmission.
5. *Backoff*. The delay before the nth attempt is a uniformly distributed random number from 0 to $2^n - 1$, for ($0 < n \leq 10$). For $n > 10$, the interval remains 0 to 1023. The unit of time for the retransmission delay is 512 bits (51.2 μs).

Figure 9.41b illustrates a 10-Mbits/s data stream with Manchester PCM formatting from the Ethernet specification. Notice that with such formatting, each bit cell or bit position contains a transition. A binary one is characterized by transitioning from a low level to a high level, while a binary zero has the opposite transition. Therefore, the presence of data transitions denotes to all "listeners" that the carrier is present. If a transition is not seen between 0.75 and 1.25 bit times since the last transition, the carrier has been lost, indicating the end of a packet.

9.5.2 Token-Ring Networks

A carrier-sense network consists of a cable onto which all stations are passively connected. A *ring network,* by comparison, consists of a series of point-to-point cables between consecutive stations. The interfaces between the ring and the stations are active rather than passive. Figure 9.42a illustrates a typical unidirectional ring with interface connections to several stations. Figure 9.42b illustrates the state of the interface for the listen mode and the transmit mode. In the *listen mode* the input bits are copied to the output with a delay of one bit time. In the *transmit mode,* the connection is broken so that the station can enter its own data onto the ring. The token is defined as a special bit pattern (e.g., 1 1 1 1 1 1 1 1) which circulates on the ring whenever all stations are idle. How does the system ensure that message data do not contain a tokenlike sequence? *Bit stuffing* is used to prevent this pattern from occurring in the data. For the

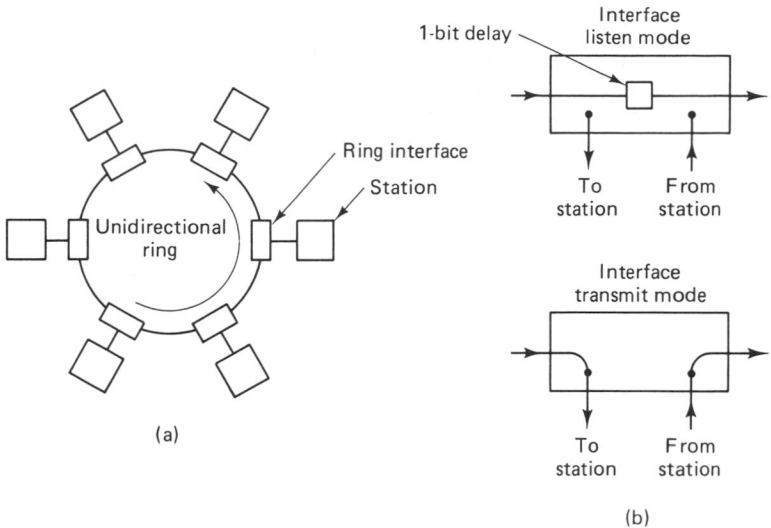

Figure 9.42 Token-ring network. (a) Network. (b) Listen and transmit modes.

8-bit token example shown, a bit-stuffing algorithm would insert a zero into the data stream after each sequence of seven consecutive ones. The data receiver would use a similar algorithm to dispose of the inserted bit following any sequence of seven consecutive ones. The token-ring access scheme works as follows:

1. A station wanting to send a message monitors the token appearing at the interface. When the last bit of the token appears, the station inverts it (e.g., 1 1 1 1 1 1 1 0). The station then breaks the interface connection and enters its own data onto the ring.

2. As bits come back around the ring, they are removed by the sender. There is no limit on the size of the packets, because the entire packet never appears on the ring at one instant.

3. After transmitting the last bit of its message, the station must regenerate the token. After the last data bit has circled the ring and has been removed, the interface is switched back to the listen mode.

4. Contention is not possible with a token-ring system. During heavy traffic, as soon as a token is regenerated, the next downstream station requiring service will see and remove the token. Thereby, permission to transmit rotates smoothly around the ring. Since there is only one token, there is no contention.

The ring itself must have sufficient delay to enable a complete token to circulate when all stations are idle. A major issue in ring network design is the propagation distance or "length" of a bit. If the data rate is R Mbits/s, a bit is emitted every $(1/R)$ microseconds. Since the propagation rate along a typical coaxial cable is 200 m/μs, each bit occupies $200/R$ meters on the ring.

Example 9.4 Minimum Ring Size

If an 8-bit token is to be used on a 5-Mbits/s token-ring network, calculate the minimum *propagation distance*, d_p, needed for the ring circumference. Assume that the propagation velocity v_p is 200 m/μs.

Solution

$$R = 5 \text{ Mbits/s}$$

Time to emit one bit, t_b:

$$t_b = \frac{1}{5 \times 10^6} \text{ s}$$

Time to emit the 8-bit token, t_t:

$$t_t = \frac{8}{5 \times 10^6} \text{ s}$$

Propagation distance for the 8-bit token:

$$d_p = t_t \times v_p$$

$$= \tfrac{8}{5} \text{ μs} \times 200 \text{ m/μs}$$

$$= 320 \text{ m}$$

9.5.3 Performance Comparison of CSMA/CD and Token-Ring Networks

Figure 9.43 compares the delay-throughput characteristics of a CSMA/CD network with a token-ring network. In each case, the cable length is 2 km, there are 50 stations on the network, the average packet length is 1000 bits, and the header size is 24 bits. Figure 9.43a, the case where the transmission rate is 1 Mbits/s, illustrates that under these assumptions, CSMA/CD and token ring perform almost equally well. In Figure 9.43b, only one parameter has been changed as compared to Figure 9.43a; the transmission rate was increased to 10 Mbits/s. The difference for CSMA/CD is considerable; for normalized throughput, $\rho < 0.22$, CSMA/CD performs better than token ring. However, for $\rho > 0.22$, token ring clearly manifests better delay-throughput characteristics. To understand the reason for the poor CSMA/CD performance in Figure 9.43b, let us review the definition of ρ, described in Equations (9.17) and (9.19) and shown as

$$\rho = \frac{b\lambda}{R} = \frac{\rho'}{R}$$

where $\rho' = b\lambda$ is channel throughput in bits per second and R is the channel capacity (maximum transmission bit rate). As R increases, channel throughput must increase accordingly for a given value of ρ. At higher channel throughput rates a significant portion of the CSMA/CD transmission attempts ends in collision [26].

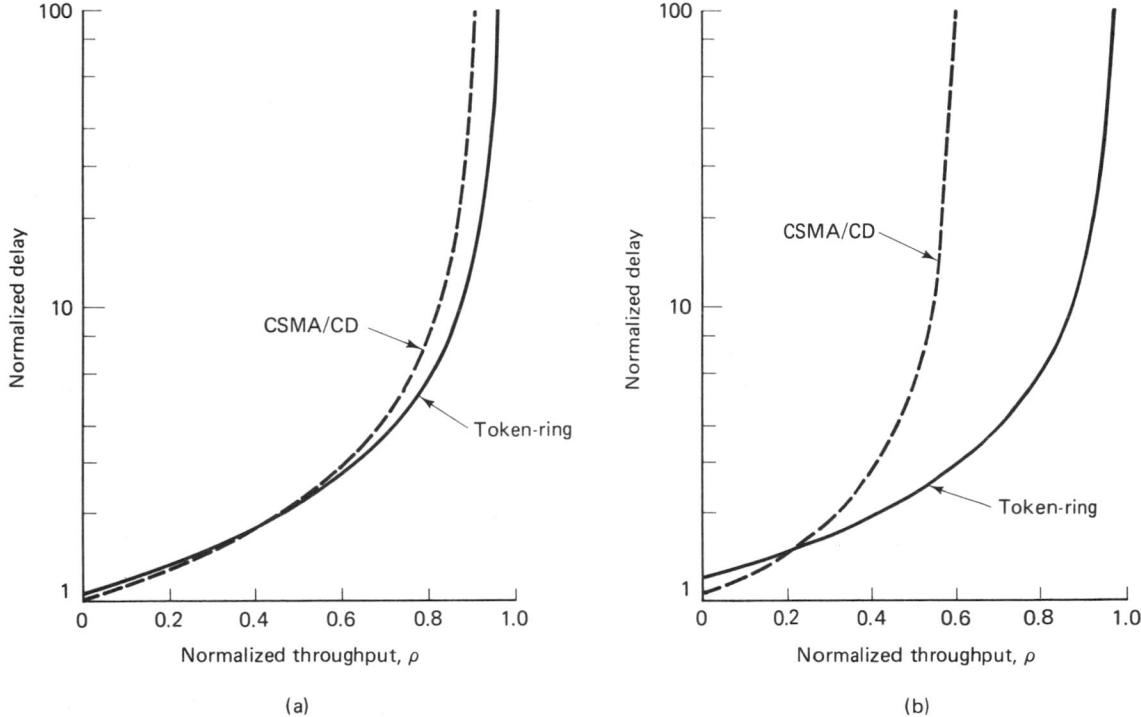

Figure 9.43 Delay versus throughput performance for CSMA/CD and token-ring networks. (a) Transmission rate = 1 Mbits/s. (b) Transmission rate = 10 Mbits/s. (Reprinted with permission from W. Bux, "Local-Area Subnetworks: A Performance Comparison," *IEEE Trans. Commun.*, vol. COM29, no. 10, Oct. 1981, pp. 1465–1473. © 1981 IEEE.)

9.6 CONCLUSION

In this chapter we have outlined the concepts of resource sharing. The classical approaches of FDM/FDMA and TDM/TDMA were discussed in some detail. We also described a hybrid multiple access technique called CDMA, and introduced some of the satellite multiple access techniques that became popular in the 1970s and 1980s, known as multiple-beam frequency reuse and dual-polarization frequency reuse.

We described the demand-assignment (DAMA) techniques in the context of several versions of the ALOHA algorithm, and we considered several of the multiple-access techniques employed with INTELSAT, such as FDM/FM, SPADE, TDMA, and SS/TDMA. Finally, we examined two popular algorithms used for local area networks: carrier-sense multiple access with collision detection (CSMA/CD) and a token-ring network. The goals of the chapter were to introduce an assortment of multiple access techniques rather than attempting a rigorous treatment of any of them.

REFERENCES

1. Rubin, I., "Message Delays in FDMA and TDMA Communication Channels," *IEEE Trans. Commun.,* vol. COM27, no. 5, May 1979, pp. 769–777.

2. Nirenberg, L. M., and Rubin, I., "Multiple Access System Engineering—A Tutorial," *IEEE WESCON/78 Professional Program,* Modern Communication Techniques and Applications, session 21, Los Angeles, Sept. 13, 1978.

3. Abramson, N., "The ALOHA System—Another Alternative for Computer Communications," *Proc. Fall Joint Comput. Conf. AFIPS,* vol. 37, 1970, pp. 281–285.

4. Hayes, J. F., "Local Distribution in Computer Communications," *IEEE Commun. Mag.,* Mar. 1981, pp. 6–14.

5. Schwartz, M., *Computer-Communication Network Design and Analysis,* Prentice-Hall, Inc., Englewood Cliffs, N.J., 1977.

6. Tanenbaum, A. S., *Computer Networks,* Prentice-Hall, Inc., Englewood Cliffs, N.J., 1981.

7. Abramson, N., "The ALOHA System," in N. Abramson and F. F. Kuo, eds., *Computer Communication Networks,* Prentice-Hall, Inc., Englewood Cliffs, N.J., 1973.

8. Kleinrock, L., *Queueing Systems,* Vol. 1, *Theory,* John Wiley & Sons, Inc., New York, 1975.

9. Abramson, N., "Packet Switching with Satellites," *AFIPS Conf. Proc.,* vol. 42, June 1973, pp. 695–702.

10. Rosner, R. D., *Packet Switching,* Lifelong Learning Publications, Wadsworth Publishing Company, Inc., Belmont, Calif., 1982.

11. Crowther, W., Rettberg, R., Walden, D., Ornstein, S., and Heart, F., "A System for Broadcast Communication: Reservation ALOHA," *Proc. Sixth Hawaii Int. Conf. Syst. Sci.,* Jan. 1973, pp. 371–374.

12. Roberts, L., "Dynamic Allocation of Satellite Capacity through Packet Reservation," *AFIPS Conf. Proc.,* vol. 42, June 1973, p. 711.

13. Binder, R., "A Dynamic Packet-Switching System for Satellite Broadcast Channels," *Proc. Int. Conf. Commun.,* June 1975, pp. 41-1–41-5.

14. Capetanakis, J., "Tree Algorithms for Packet Broadcast Channels," *IEEE Trans. Inf. Theory,* vol. IT25, Sept. 1979, pp. 505–515.

15. Puente, J. G., and Werth, A. M., "Demand-Assigned Service for the INTELSAT Global Network," *IEEE Spectrum,* Jan. 1971, pp. 59–69.

16. Jones, J. J., "Hard Limiting of Two Signals in Random Noise," *IEEE Trans. Inf. Theory,* vol. IT9, Jan. 1963, pp. 34–42.

17. Bond, F. E., and Meyer, H. F., "Intermodulation Effects in Limiter Amplifier Repeaters," *IEEE Trans. Commun. Technol.,* vol. COM18, no. 2, Apr. 1970, pp. 127–135.

18. Shimbo, O., "Effects of Intermodulation, AM-PM Conversion, and Additive Noise in Multicarrier TWT Systems," *Proc. IEEE,* vol. 59, Feb. 1971, pp. 230–238.

19. Chakraborty, D. "INTELSAT IV Satellite System (Voice) Channel Capacity versus Earth-Station Performance," *IEEE Trans. Commun. Technol.,* vol. COM19, no. 3, June 1971, 355–362.

20. Campanella, S., and Schaefer, D., "Time Division Multiple Access Systems

(TDMA)," in K. Feher, *Digital Communications, Satellite/Earth Station Engineering,* Prentice-Hall, Inc., Englewood Cliffs, N.J., 1983.

21. Scarcella, T., and Abbott, R. V., "Orbital Efficiency Through Satellite Digital Switching," *IEEE Commun. Mag.,* May 1983, pp. 38–46.

22. Muratani, T., Satellite-Switched Time-Domain Multiple Access," *Proc. IEEE Electron. and Aerosp. Conf. (EASCON),* 1974, pp. 189–196.

23. Dill, G. D., "TDMA, The State-of-the-Art," *Rec. IEEE Electron. Aerosp. Syst. Conv. (EASCON),* Sept. 26–28, 1977, pp. 31-5A–31-5I.

24. Jarett, K., "Operational Aspects of Intelsat VI Satellite-Switched TDMA Communication System," *AIAA Tenth Commun. Satell. Syst. Conf.* Mar. 1984, pp. 107–111.

25. Stallings, W., "Local Network Performance," *IEEE Commun. Mag.,* vol. 22, No. 2, Feb. 1984, pp. 27–36.

26. Bux, W., Local-Area Subnetworks: A Performance Comparison," *IEEE Trans. Commun.,* vol. COM29, no. 10, Oct. 1981, pp. 1465–1473.

27. Dixon, R. C., Strole, N. C., and Markov, J. D., "A Token-Ring Network for Local Data Communications," *IBM Syst. J.,* vol. 22, no. 1–2, 1983, pp. 47–62.

PROBLEMS

9.1. Design an FDM signal set consisting of five voice channels, each in the frequency range 300 to 3400 Hz. The multiplexed composite is to be made up of inverted sidebands and is to occupy the spectral region from 30 to 50 kHz.
   (a) Draw the composite spectrum, indicating individual spectrum and guard band frequency locations.
   (b) Draw a block diagram showing the heterodyning and filtering details and the required local oscillator values.

9.2. A receiver is tuned to receive the lower sideband (LSB) of a radio-frequency (RF) carrier wave with frequency, $f_c = 8$ MHz. The bandwidth of the LSB signal is 100 kHz. The receiver employs a local oscillator (LO) with frequency, f_{LO}, for heterodyning the received signal down to a lower intermediate frequency (IF). Assume that $f_{LO} > f_c$, and that the IF amplifier is centered at 2 MHz. Draw a block diagram of the heterodyning conversion, including the RF filter, the LO, and the IF filter. Indicate the center frequency of each filter and typical spectra of the signals at various points in the diagram.

9.3. Equations (9.13) to (9.15) demonstrate that the average message delay time for TDMA is less than that for FDMA. Discuss the practical benefits of such reduced delay in TDMA, as a function of frame time, for a satellite link with a one-way range of 36,000 km. For what values of frame time can there be a significant advantage of TDMA over FDMA?

9.4. A group of stations share a 56-kbits/s pure ALOHA channel. Each station outputs a packet on the average of once every 10 s, even if the previous one has not yet been sent (i.e., the stations buffer the packets). Each packet is comprised of 3000 bits. What is the maximum number of stations that can share this channel, assuming that the arrival process is Poisson?

9.5. A group of three stations share a 56-kbits/s pure ALOHA channel. The average bit rate transmitted from each of the three stations is $R_1 = 7.5$ kbits/s, $R_2 = 10$ kbits/s, and $R_3 = 20$ kbits/s. The size of each packet is 100 bits/packet. Find the normalized total traffic on the channel, the normalized throughput, the probability of successful transmission, and the arrival rate of successful packets. Assume that the arrival process is Poisson.

9.6. Verify that for a pure ALOHA access scheme, the normalized throughput is bounded by $1/2e$ and that this maximum occurs when the normalized total traffic is equal to 0.5.

9.7. (a) Verify that Equation (9.24) is a valid probability density function (pdf) for a discrete random variable.
   (b) Calculate the mean of a discrete random variable having a pdf like the one given in Equation (9.24).
   (c) Show that your result in part (b) is consistent with the claim that λ is the average packet arrival rate.

9.8. Consider the pure ALOHA arrival scenario shown in Figure P9.1. The vertical arrows indicate packet arrival times. N_n is the number of arriving packets in the time interval $(T_{n-1}, T_n]$, where $(t_x, t_y]$ indicates the interval $t_x < t \le t_y$. N_{n+1} is the number of arriving packets in $(T_n, T_{n+1}]$, and τ is the time duration per packet in seconds. The average arrival rate is λ_t. Assume the arrivals are independent of each other.
   (a) Write an expression for the joint pdf of N_n and N_{n+1}.
   (b) Let T_n define the time at which user A's packet arrives. Express, in terms of the joint pdf of N_n and N_{n+1}, the probability that user A's transmission will be successful.

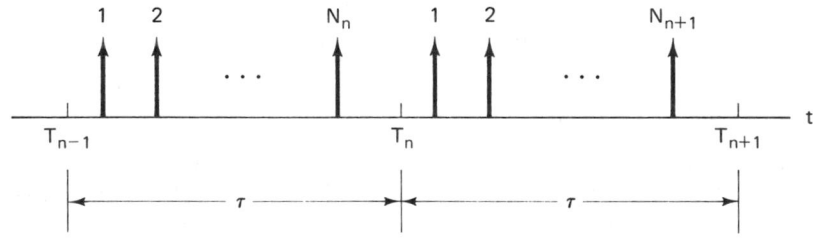

Figure P9.1

9.9. Let $N = N_n + N_{n+1}$, where N_n and N_{n+1} are as defined in Problem 9.8. Write an expression for the pdf of N, and give an interpretation for N.

9.10. Six thousand stations are competing for the use of a single slotted ALOHA channel. The average station makes 30 requests per hour, where each request is for one slot of 500-μs duration. Calculate the normalized total traffic on the channel.

9.11. Consider the arrival scenario of Figure P9.1; the location of the packet arrival times are permissible as shown under pure ALOHA, but not under slotted ALOHA, where arrivals are permitted only at the discrete times T_i, where $i = 0, 1, \ldots$. Assume that the average arrival rate is λ_t.
   (a) How would Figure P9.1 need to be modified if slotted ALOHA is used? How would the pdfs of N_n and N_{n+1} change?
   (b) If user A's packet arrives at time T_n, what is the probability of successful transmission?

9.12. A group of slotted-ALOHA stations generate a total of 120 requests per second, including both original and retransmissions. Each request is for a 12.5-ms duration slot.

(a) What is the normalized total traffic on the channel?
(b) What is the probability of a successful transmission on the first attempt?
(c) What is the probability of exactly two collisions before a successful transmission?

9.13. Measurements of a slotted-ALOHA channel show that 20% of the slots are idle.

(a) What is the normalized total traffic on the channel?
(b) What is the normalized throughput?
(c) Is the channel underloaded or overloaded?

9.14. Show that the sum of two Poisson processes, with rates λ_1 and λ_2, is also a Poisson process, with rate $\lambda_t = \lambda_1 + \lambda_2$. Generalize your result for the sum of n Poisson processes.

9.15. A 10-MHz transponder is occupied by 200 identical carriers, half servicing stations with $G/T = 40$ dB/K, the other half servicing stations with $G/T = 37$ dB/K. All stations have a requirement to operate with a bit error probability of 10^{-5}. The transponder is power limited under this configuration.

(a) What is the maximum possible bandwidth for each carrier?
(b) Suppose that each carrier has a bandwidth of 40 kHz, and the transponder is required to service a group of larger ($G/T = 40$ dB/K) stations only. How many stations can the transponder handle? Will the transponder be power or bandwidth limited?
(c) Repeat part (b) for the case where the transponder is to service a group of small ($G/T = 37$ dB/K) stations only.

9.16. A TDMA system operates at 100 Mbits/s with a 2-ms frame time. Assume that all slots are of equal length and that a guard time of 1 μs is required between slots.

(a) Compute the efficiency of the communications resource (CR) for the case of 1, 2, 5, 10, 20, 50, and 100 slots per frame.
(b) Repeat part (a) assuming that a 100-bit preamble is required at the start of each slot. Compute the efficiency of the CR in terms of the desired information transmission.
(c) Graph the results of parts (a) and (b).

9.17. With reference to Equation (9.36):

(a) Discuss the efficiency of the CR use if all S_i and R_j are equal.
(b) Discuss the effect of a few S_i or R_j being much larger than the majority. How can the efficiency of the CR be improved?
(c) When are the distributions of S_i and R_j likely to be similar? Dissimilar?

9.18. (a) Consider a token-ring network operating at a transmission rate of 10 Mbits/s over a cable having a propagation velocity of 200 m/μs. How many meters of cable is equal to a delay of 1 bit at each ring interface?
(b) If the token is 10 bits long, and all but three stations are switched off during evening hours, what is the minimum cable length needed for the ring?

CHAPTER 10

Spread-Spectrum Techniques

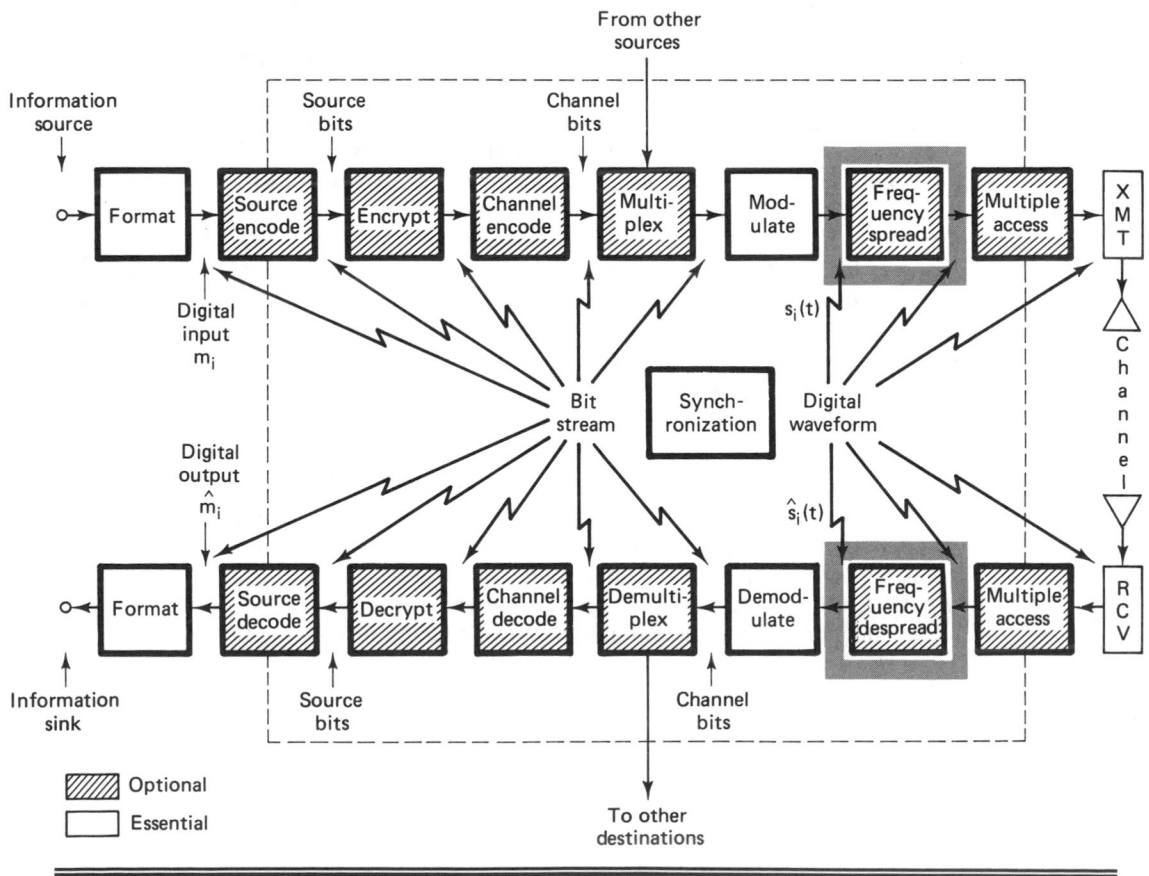

10.1 SPREAD-SPECTRUM OVERVIEW

The initial application of spread-spectrum (SS) techniques was in the development of military guidance and communication systems. By the end of World War II, spectrum spreading for jamming resistance was already a familiar concept to radar engineers [1], and during subsequent years, SS investigation was motivated primarily by the desire to achieve highly jam-resistant communication systems. As a result of this research, there emerged an assortment of other applications in such areas as energy density reduction, high-resolution ranging, and multiple access, which will be discussed in later sections. The techniques considered in this chapter are called *spread spectrum* because the transmission bandwidth employed is much greater than the minimum bandwidth required to transmit the information. A system is defined to be a spread-spectrum system if it fulfills the following requirements:

1. The signal occupies a bandwidth much in excess of the minimum bandwidth necessary to send the information.
2. Spreading is accomplished by means of a *spreading signal*, often called a *code signal*, which is independent of the data. The details of some spreading signals are described in later sections.
3. At the receiver, despreading (recovering the original data) is accomplished by the correlation of the received spread signal with a synchronized replica of the spreading signal used to spread the information.

Standard modulation schemes such as frequency modulation and pulse code modulation also spread the spectrum of an information signal, but they do not qualify as spread-spectrum systems since they do not satisfy all the conditions outlined above.

10.1.1 The Beneficial Attributes of Spread-Spectrum Systems

10.1.1.1 Interference Suppression Benefits

White Gaussian noise is a mathematical model that, by definition, has infinite power spread uniformly over all frequencies. Effective communication is possible with this interfering noise of infinite power because only the finite-power noise components that are present within the signal space (in other words, share the *same coordinates* as the signal components) can interfere with the signal. The balance of the noise power may be thought of as noise that is effectively tuned out by the detector (see Section 3.2.2). For a typical narrowband signal, this means that only the noise in the signal bandwidth can degrade performance. The idea behind a spread-spectrum anti-jam (AJ) system is as follows. Consider that many orthogonal signal coordinates or dimensions are available to a communication link and that only a small subset of these signal coordinates are used at any time. We assume that the jammer cannot determine the signal subset that is currently in use. For signals of bandwidth W and duration T, the number of signaling dimensions can be shown [2] to be approximately $2WT$. Given a specific signal design, the error performance of such a system is only a function of E_b/N_0. Against white Gaussian noise, with *infinite* power, the use of spreading (large $2WT$) offers no performance improvement. However, when the noise stems from a jammer with a *fixed finite* power and with uncertainty as to where in the signal space the signal coordinates are located, the jammer's choices are limited to those shown below.

1. Jam *all* the signal coordinates of the system, with an *equal* amount of power in each one, with the result that *little* power is available for each coordinate.
2. Jam a *few* signal coordinates with *increased* power in each of the jammed coordinates (or more generally, jam all the coordinates with various amounts of power in each).

Figure 10.1 compares the effect of spectrum spreading in the presence of white noise with spreading in the presence of an intentional jammer. The power spectral density of the signal is denoted $G(f)$ before spreading, and $G_{ss}(f)$ after spreading. For simplicity, the figure treats the frequency dimension only. In Figure 10.1a it can be seen that the single-sided power spectral density of white noise, N_0, is unchanged as a result of expanding the signal bandwidth from W to W_{ss}. The average power of white noise (area under the spectral density curve) is infinite. Hence, the use of spreading offers no performance improvement here. Figure 10.1b (upper diagram) illustrates the case of received (fixed finite) jammer power, J, and power spectral density, $J_0' = J/W$, where W is the unspread band-

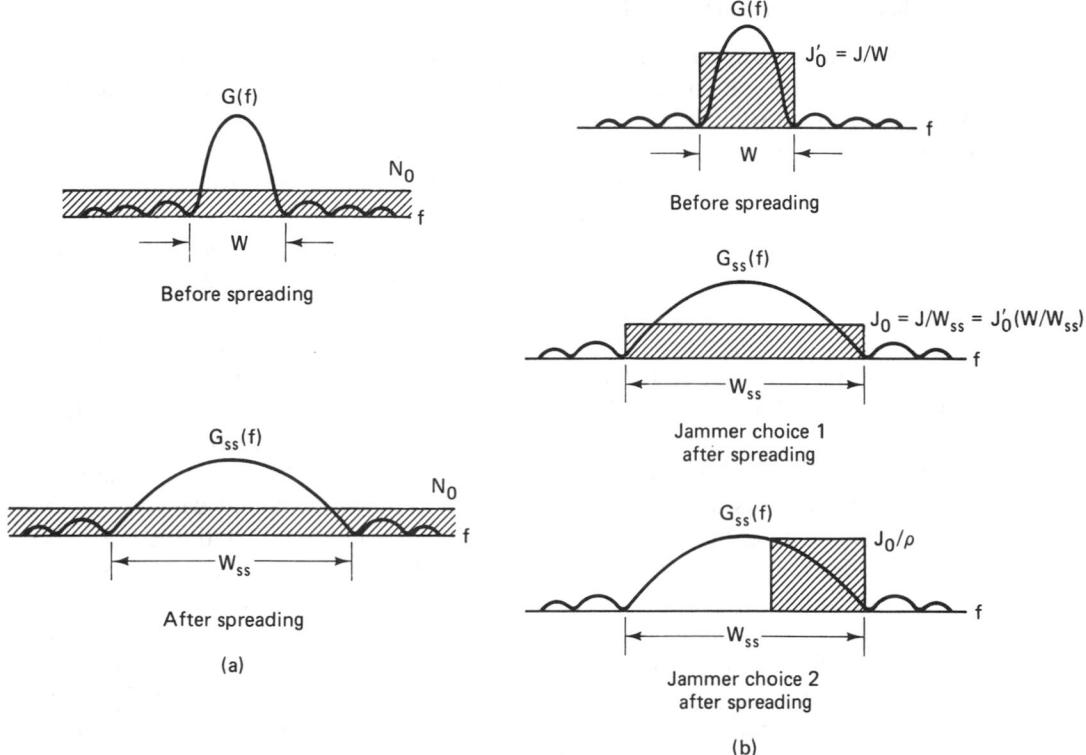

Figure 10.1 Effect of spectrum spreading. (a) Spectrum spreading in the presence of white noise. (b) Spectrum spreading in the presence of an intentional jammer.

width being jammed. Once the signal bandwidth is spread, the jammer can make one of the two choices listed earlier—choice 1 results in a reduction in jammer noise spectral density, J_0', by a factor (W/W_{ss}) across the spread spectrum. The resulting noise spectral density, $J_0 = J/W_{ss}$, is referred to as the *broadband jammer noise spectral density*. Choice 2 results in a reduction in the number of signal coordinates that the jammer occupies. However, with choice 2 the jammer can increase its noise spectral density from J_0 to J_0/ρ $(0 < \rho \leq 1)$, where ρ is the portion of the spread-spectrum band the jammer elects to jam. If the jammer makes a poor choice in the coordinates to be jammed, the average effect of jamming will be less than if it makes a good choice. The larger the dimensionality of the signal set or the more signal coordinates the communicator can choose from, the greater is the jammer's uncertainty regarding the effectiveness of the jamming technique, and the better will be the protection against jamming.

Jamming is not always the result of an intentional act. Sometimes, the jamming signal is caused by natural phenomena, and sometimes it is the result of self-interference caused by *multipath*, in which delayed versions of the signal, arriving via alternative paths, interfere with the direct path transmission.

10.1.1.2 Energy Density Reduction

One can imagine situations where it is desired that a communications link be operated without being detected by anyone other than the intended receiver. Systems designed for this special task are known as *low probability of detection* (LPD) or *low probability of intercept* (LPI) communication systems. These systems are designed to make the detection of their signals as difficult as possible by anyone but the intended receiver. The goal of such a system is to use the minimum signal power and the optimum signaling scheme that results in the minimum probability of being detected. Since, in spread-spectrum systems, the signal is spread over many more signaling coordinates than in conventional modulation schemes, the resulting signal power is, on the average, spread thinly and uniformly in the spread domain. Therefore, not only can the spread-spectrum signal be made difficult to jam, but additionally, the signal's very existence may be rendered difficult to perceive. To anyone who does not possess a synchronized replica of the spreading signal, the spread-spectrum signal will seem "buried in the noise."

A *radiometer* is a simple power measuring instrument that can be used by an adversary to detect the presence of spread-spectrum signals within some bandwidth W. The radiometer, illustrated in Figure 10.2, consists of a bandpass filter (BPF) with bandwidth W, a squaring circuit to ensure a positive output value, since the presence of *signal energy* is being detected, and an integrating circuit. At time $t = T$, the output of the integrator is compared to a preset threshold. If the output of the integrator is larger than the threshold, a signal is declared present; otherwise, the signal is declared absent. References [3, 4] provide details on the detectability of spread-spectrum signals, using radiometers and other more complicated instruments that make use of the features of the SS signal itself.

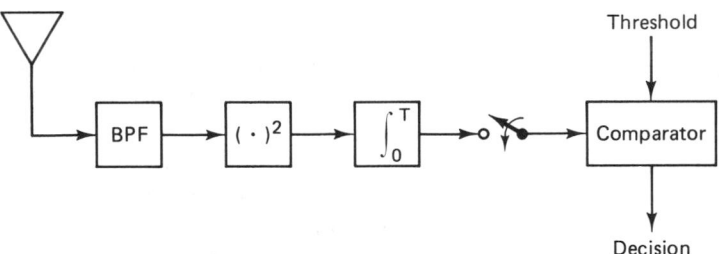

Figure 10.2 Radiometer.

Spread-spectrum systems that are designed to exhibit LPI may also exhibit a *low probability of position fix* (LPPF), which means that even if the presence of the signal is perceived, the direction of the transmitter is difficult to pinpoint. Some spread-spectrum systems also exhibit a *low probability of signal exploitation* (LPSE), which means that the identification of the source is difficult to ascertain.

Another, unrelated application of spread-spectrum signaling deals with the fact that in some cases energy density reduction may be required to meet national allocation regulations. Downlink transmissions from satellites must meet international regulations on the spectral density that impinges on the earth. By spreading the downlink energy over a wider bandwidth, the total transmitted power can be increased and hence performance improved, while the energy density regulations are followed.

10.1.1.3 Fine Time Resolution

Spread-spectrum signals can be used for ranging or determination of position location. Distance can be determined by measuring the time delay of a pulse as it traverses the channel. Uncertainty in the delay measurement is inversely proportional to the bandwidth of the signal pulse. This can be seen by the illustration in Figure 10.3. The uncertainty of the measurement, Δt, is proportional to the rise time of the pulse, which is inversely proportional to the bandwidth of the pulse signal; that is,

$$\Delta t \simeq \frac{1}{W} \tag{10.1}$$

The larger the bandwidth, the more precisely one can measure range. Over a Gaussian channel, a one-shot measurement on a single pulse is not very reliable. The spread-spectrum technique, however, uses a code signal consisting of a long sequence of polarity changes (e.g., a binary PSK-modulated signal) in place of the single pulse. Upon reception, the received sequence is correlated against a local replica and the results of the correlation are used to perform an accurate time-delay or range measurement.

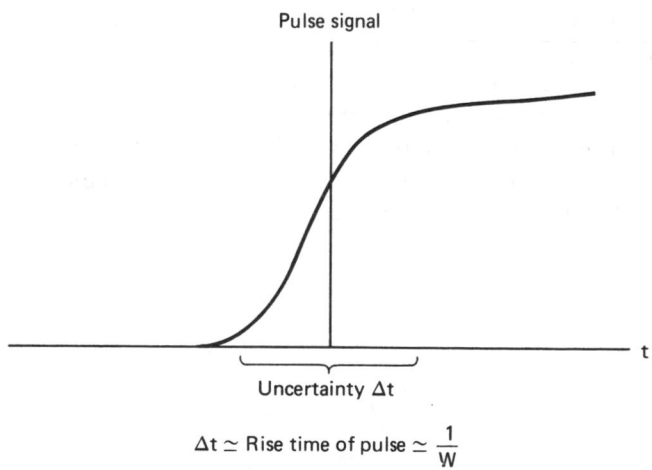

$$\Delta t \simeq \text{Rise time of pulse} \simeq \frac{1}{W}$$

Figure 10.3 Time-delay measurement.

10.1.1.4 Multiple Access

Spread-spectrum methods can be used as a multiple access technique, in order to share a communications resource among numerous users in a coordinated manner. The technique, termed *code-division multiple access* (CDMA), since each simultaneous user employs a unique spread-spectrum signaling code, was discussed briefly in Chapter 9. One of the by-products of this type of multiple access is the ability to provide communication privacy between users with different spreading signals. An unauthorized user (a user not having access to a spreading signal) cannot easily monitor the communications of the authorized users. A more detailed treatment is presented in a later section.

10.1.2 Model for Spread-Spectrum Interference Rejection

Figure 10.4 illustrates a model for spread-spectrum interference rejection. At the modulator, the information signal $x(t)$, with a data rate of R bits/s, is multiplied by a spreading code signal, $g(t)$, having a code symbol rate, usually called the code *chip rate*, R_p chips/second. Assume that the transmission bandwidths for $x(t)$ and $g(t)$ are R hertz and R_p hertz, respectively. Multiplication in the time domain transforms to convolution in the frequency domain:

$$x(t)g(t) \leftrightarrow X(\omega) * G(\omega) \tag{10.2}$$

Therefore, if the data signal is narrowband compared to the spreading signal, the resulting product signal $x(t)g(t)$ will have approximately the bandwidth of the spreading signal (see Section A.5).

At the demodulator, the received signal is ideally multiplied by a synchronized replica of the spreading code signal, $g(t)$, which results in the despreading of the signal. A filter with bandwidth R is used to remove any spurious higher-frequency components. If there is any undesired signal at the receiver, the multiplication by $g(t)$ will spread this undesired signal, in the same way that the multiplication by $g(t)$ at the transmitter spread the desired signal originally. Consider the effect on a jammer that attempts to position a narrowband jamming signal within the information bandwidth. The first operation at the receiver input is multiplication by the spreading signal. Hence the jamming tone is spread to the bandwidth of the spreading signal.

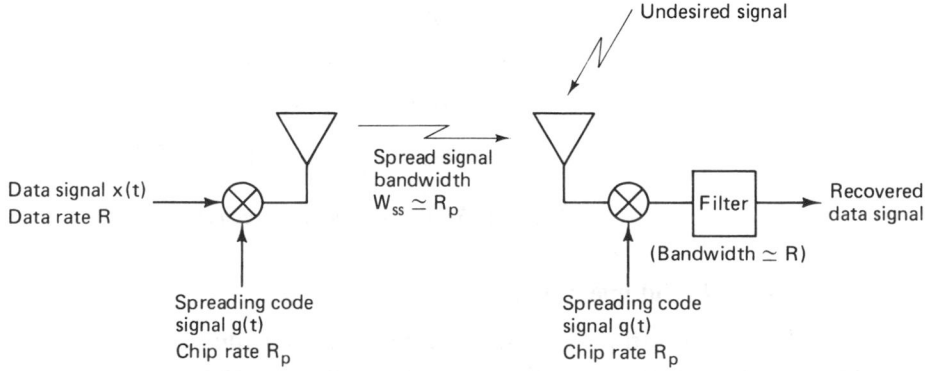

Figure 10.4 Basic spread-spectrum technique.

The essence behind the interference rejection capability of a spread-spectrum system can be summarized as follows:

1. Multiplication by the spreading signal *once* spreads the signal bandwidth.
2. Multiplication by the spreading signal *twice*, followed by filtering, recovers the original signal.
3. The desired signal gets multiplied *twice*, but the interference signal gets multiplied only *once*.

10.1.3 A Catalog of Spreading Techniques

Figure 10.5 highlights the popular techniques for spreading the information signal over a large number of signal coordinates or dimensions. For signals of bandwidth W and duration T, the dimensionality of the signaling space is approximately $2WT$.

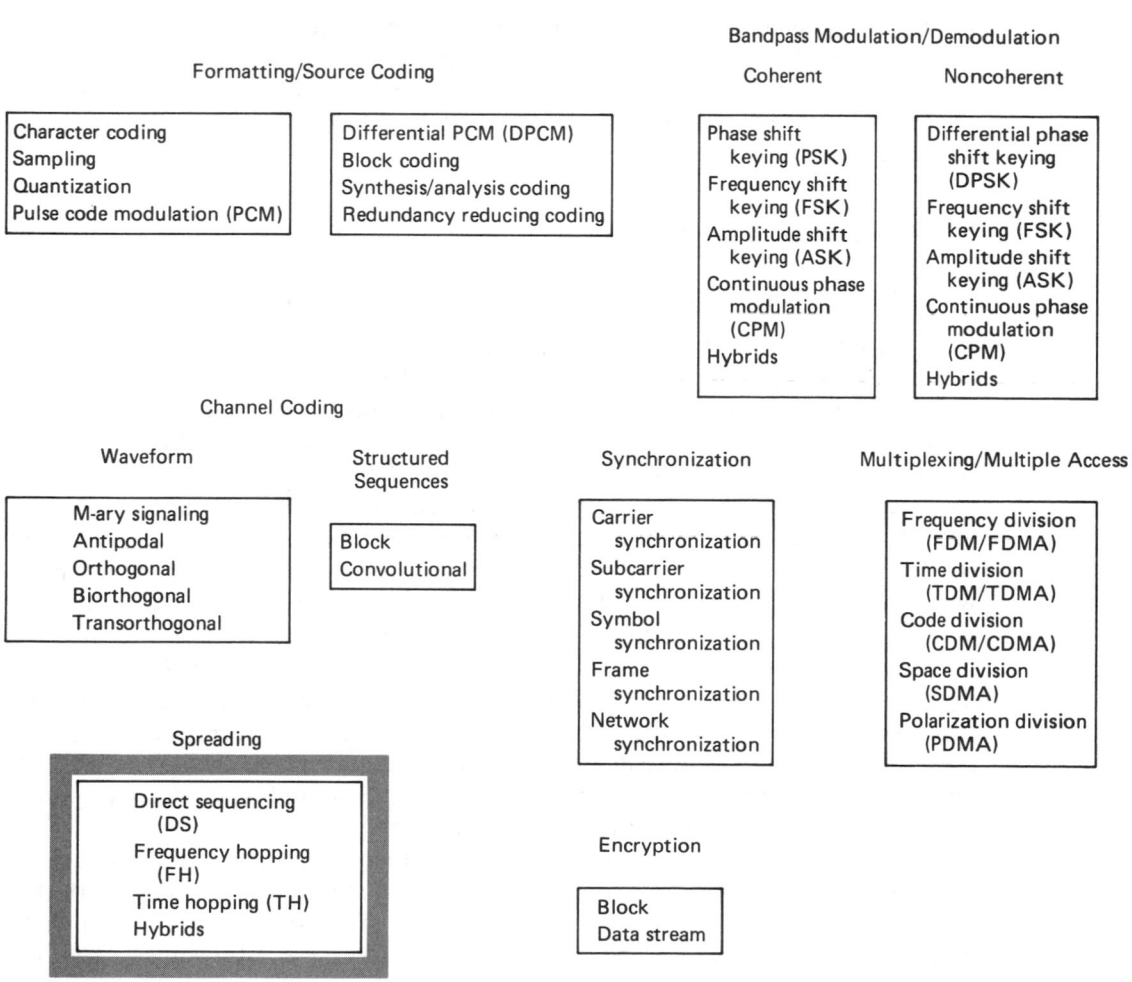

Figure 10.5 Basic digital communication transformations.

To increase the dimensionality, we can either increase W by spectrum spreading, or increase T by time spreading or time hopping (TH). With spectrum spreading the signal is spread in the frequency domain. With time hopping, a message with data rate R is allocated a longer transmission-time duration than would be used with a conventional modulation scheme. During this longer time the data are sent in bursts according to the dictates of a code. We can say that with time hopping the signal is spread in the time domain. For both cases, frequency spreading and time spreading, a jammer will be uncertain regarding the signaling subset that is currently in use.

In Figure 10.5, the first two items listed under the category of spreading, *direct sequencing* (DS) and *frequency hopping* (FH), are the most commonly used techniques for spectrum spreading. As a jamming-rejection technique, *time hopping* (TH), the third item in the list, is similar to spread spectrum, in that the location of the signal coordinates is hidden from potential adversaries. Also, there are hybrid combinations of the spreading techniques, for example, DS/FH, FH/TH, and DS/FH/TH; however, these techniques can be viewed as simple extensions of the material presented here and will not be elaborated on. In this chapter we focus only on the two major spread-spectrum techniques, direct sequencing and frequency hopping.

10.1.4 Historical Background

10.1.4.1 Transmitted Reference versus Stored Reference

During the early years of spread-spectrum investigation one technique that was considered for operating a transmitter and receiver synchronously with a *truly random* spreading signal such as wideband noise, was called a *transmitted reference* (TR) system. In a TR system the transmitter would send two versions of an unpredictable wideband carrier, one modulated by data and the other unmodulated. These two signals were transmitted on separate channels. The receiver used the unmodulated carrier as the reference signal for despreading (correlating) the data-modulated carrier. The principal advantage of a TR system was that there were no significant synchronization problems at the receiver, since the data-modulated signal and the spreading signal used for despreading were transmitted simultaneously. The principal disadvantages of TR systems were that (1) the spreading code was sent in the clear and thus was available to any listener; (2) the system could be easily spoofed by a jammer sending a pair of waveforms acceptable to the receiver; (3) performance degraded at low signal levels since noise was present on both signals; and (4) twice the bandwidth and transmitted power were required because of the need to transmit the reference.

Modern spread-spectrum systems all use a technique called *stored reference* (SR), whereby the spreading code signal is independently generated at both the transmitter and the receiver. The main advantage of an SR system is that a well-designed code signal cannot be predicted by monitoring the transmission. Note that the noiselike code signal in an SR system cannot be truly random as it could in the case of a TR system. Since the same code must be generated independently at two or more sites, the code sequence must be deterministic, even though it should appear random to unauthorized listeners. Such random-appearing deterministic signals are called pseudonoise (PN) or pseudorandom signals; their generation is treated later in greater detail.

10.1.4.2 Noise Wheels

In the late 1940s and early 1950s, Mortimer Rogoff, working at ITT, demonstrated the fundamental operation of spectrum spreading systems with a novel experiment [5]. Using photographic techniques, Rogoff built a "noise wheel" for storing a noiselike signal. He randomly selected 1440 numbers not ending in 00 from the Manhattan telephone directory, and radially plotted the middle two of the last four digits so that the radius at every $\frac{1}{4}°$ represented a new random number. The drawing was transferred to the wheel-shaped film shown in Figure 10.6. When the wheel was rotated past a slit of light, the resulting intensity-modulated light beam provided a stored noiselike spreading signal to be sensed by a photocell.

Rogoff mounted two such identical wheels on a single axis driven by a 900-rpm synchronous motor. One wheel's noiselike spreading signal was modulated

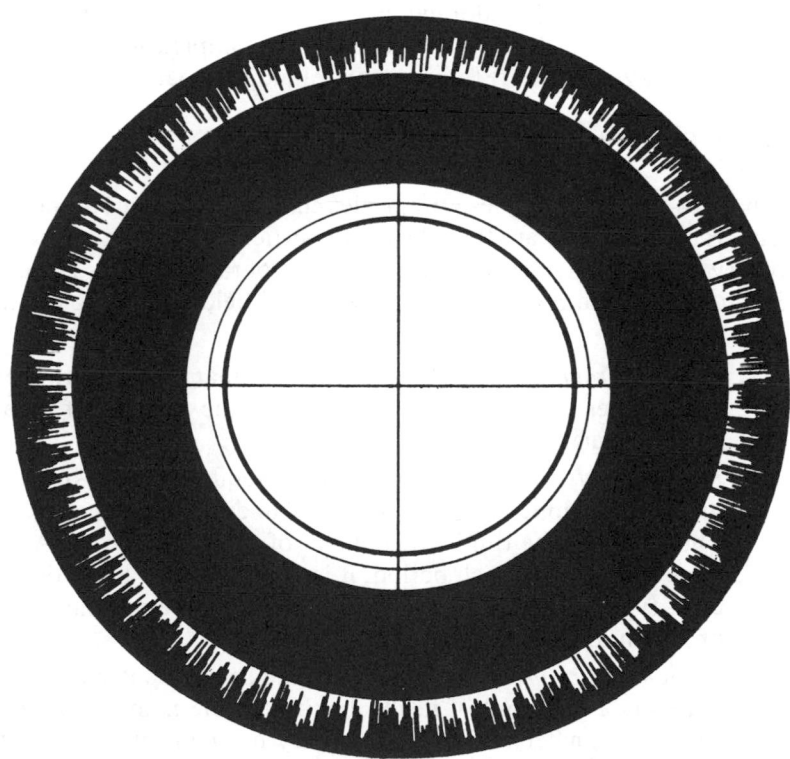

Figure 10.6 Rogoff's noise wheel. [Reprinted from Section I (Communications) of "Application of Statistical Methods to Secrecy Communication Systems," Proposal 946, Fed. Telecomm. Lab., August 28, 1950, Fig. 6, courtesy of ITT.]

with data (and interference) to provide one input to the receiving correlator, while the other wheel's unmodulated spreading signal provided the other input to the correlator. These baseband experiments, performed with data rates of 1 bit/s, demonstrated the feasibility of conveying information hidden in noiselike signals [6].

10.2 PSEUDONOISE SEQUENCES

The spread-spectrum approach called *transmitted reference* (TR) can utilize a *truly* random code signal for spreading and despreading, since the code signal and the data-modulated code signal are simultaneously transmitted over different regions of the spectrum. The *stored reference* (SR) approach *cannot* use a truly random code signal since the code needs to be stored or generated at the receiver. For the SR system a *pseudonoise* or *pseudorandom* code signal must be used.

How does a pseudorandom signal differ from a random one? A random signal *cannot* be predicted; its future variations can only be described in a statistical sense. However, a pseudorandom signal is not random at all; it is a deterministic, periodic signal that is known to both the transmitter and receiver. Why the name "pseudonoise" or "pseudorandom"? Even though the signal is deterministic, it appears to have the statistical properties of sampled white noise. It appears, to an unauthorized listener, to be a truly random signal.

10.2.1 Randomness Properties

What are these randomness properties that make a pseudorandom signal appear truly random? There are three basic properties that can be applied to any periodic binary sequence as a test for the appearance of randomness. The properties, called *balance, run,* and *correlation*, are described below for binary signals:

1. *Balance property*. Good balance requires that in each period of the sequence, the number of binary ones differs from the number of binary zeros by at most one digit.
2. *Run property*. A *run* is defined as a sequence of a single type of binary digit(s). The appearance of the alternate digit in a sequence starts a new run. The length of the run is the number of digits in the run. Among the runs of ones and zeros in each period, it is desirable that about one-half the runs of each type are of length 1, about one-fourth are of length 2, one-eighth are of length 3, and so on.
3. *Correlation property*. If a period of the sequence is compared term by term with any cyclic shift of itself, it is best if the number of agreements differs from the number of disagreements by not more than one count.

In the next section, a PN sequence is generated to test these properties.

10.2.2 Shift Register Sequences

Consider the linear feedback shift register illustrated in Figure 10.7. It is made up of a four-stage register for storage and shifting, a modulo-2 adder, and a feedback path from the adder to the input of the register (modulo-2 addition has been defined in Section 2.12.3). The shift register operation is controlled by a sequence of clock pulses (not shown). At each clock pulse the contents of each stage in the register is shifted one stage to the right. Also, at each clock pulse the contents of stages X_3 and X_4 are modulo-2 added (a linear operation), and the result is fed back to stage X_1. The shift register sequence is defined to be the output of the last stage—stage X_4 in this example.

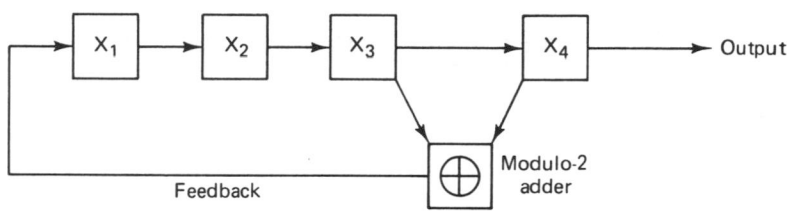

Figure 10.7 Linear feedback shift register example.

Assume that stage X_1 is initially filled with a one and the remaining stages are filled with zeros, that is, the initial state of the register is 1 0 0 0. From Figure 10.7 we can see that the succession of register states will be as follows:

1 0 0 0   0 1 0 0   0 0 1 0   1 0 0 1   1 1 0 0   0 1 1 0   1 0 1 1   0 1 0 1

1 0 1 0   1 1 0 1   1 1 1 0   1 1 1 1   0 1 1 1   0 0 1 1   0 0 0 1   1 0 0 0

Since the last state, 1 0 0 0, corresponds to the initial state, we see that the register repeats the foregoing sequence after 15 clock pulses. The output sequence is obtained by noting the contents of stage X_4 at each clock pulse. The output sequence is seen to be

0 0 0 1 0 0 1 1 0 1 0 1 1 1 1

where the leftmost bit is the earliest bit. Let us test the sequence above for the randomness properties outlined in the preceding section. First, the balance property; there are seven zeros and eight ones in the sequence—therefore, the sequence meets the balance condition. Next, the run property; consider the zero runs—there are four of them. One-half are of length 1, and one-fourth are of length 2. The same is true for the one runs. The sequence is too short to go further, but we can see that the run condition is met. The correlation property is treated in Section 10.2.3.

The shift register generator produces sequences that depend on the number of stages, the feedback tap connections, and initial conditions. The output sequences can be classified as either *maximal length* or *nonmaximal length*. Maximal length sequences have the property that for an n-stage linear feedback shift register the sequence repetition period in clock pulses p is

$$p = 2^n - 1 \tag{10.3}$$

Thus it can be seen that the sequence generated by the shift register generator of Figure 10.7 is an example of a maximal length sequence. If the sequence length is less than $(2^n - 1)$, the sequence is classified as a nonmaximal length sequence.

10.2.3 PN Autocorrelation Function

The autocorrelation function $R_x(\tau)$ of a periodic waveform $x(t)$, with period T_0, was given in Equation (1.23) and is shown below in normalized form.

$$R_x(\tau) = \frac{1}{K} \left(\frac{1}{T_0} \right) \int_{-T_0/2}^{T_0/2} x(t)x(t + \tau)\, dt \qquad \text{for } -\infty < \tau < \infty \tag{10.4}$$

where

$$K = \frac{1}{T_0} \int_{-T_0/2}^{T_0/2} x^2(t)\, dt \tag{10.5}$$

When $x(t)$ is a periodic pulse waveform representing a PN code, we refer to each fundamental pulse as a *PN code symbol* or a *chip*. For such a PN waveform of unit chip duration and period p chips, the normalized autocorrelation function may be expressed as

$$R_x(\tau) = \frac{1}{p} \cdot \left(\begin{array}{l} \text{number of agreements less number of disagreements} \\ \text{in a comparison of one full period of the sequence} \\ \text{with a } \tau \text{ position cyclic shift of the sequence} \end{array} \right) \tag{10.6}$$

The normalized autocorrelation function for a maximal length sequence, $R_x(\tau)$, is shown plotted in Figure 10.8. It is clear that for $\tau = 0$, that is, when $x(t)$ and its replica are perfectly matched, $R(\tau) = 1$. However, for any cyclic shift between $x(t)$ and $x(t + \tau)$ with $(1 \leq \tau < p)$, the autocorrelation function is equal to $-1/p$ (for large p, the sequences are virtually decorrelated for a shift of a *single chip*).

It is now easy to test the output PN sequence of the shift register in Figure 10.7 for the third randomness property—correlation. Below is shown the output sequence; also shown is the same sequence with a single end-around shift:

$$0\ 0\ 0\ 1\ 0\ 0\ 1\ 1\ 0\ 1\ 0\ 1\ 1\ 1\ 1$$
$$1\ 0\ 0\ 0\ 1\ 0\ 0\ 1\ 1\ 0\ 1\ 0\ 1\ 1\ 1$$

$$d\ a\ a\ d\ d\ a\ d\ a\ d\ d\ d\ d\ a\ a\ a$$

The digits that agree are labeled a and those that disagree are labeled d. Following Equation (10.6), the value of the autocorrelation function for this single one-chip shift is seen to be

$$R(\tau = 1) = \tfrac{1}{15}(7 - 8) = -\tfrac{1}{15}$$

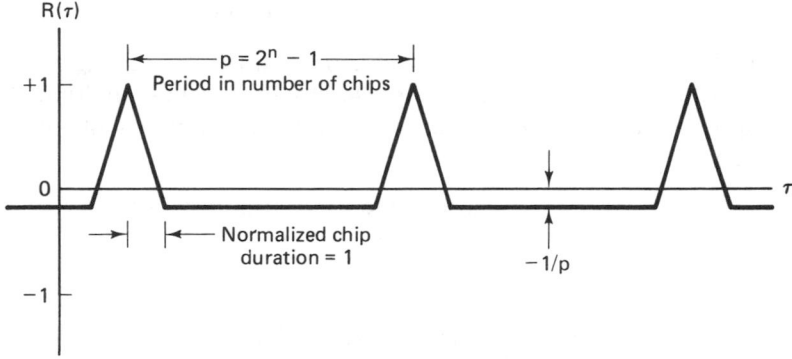

Figure 10.8 PN autocorrelation function.

Any cyclic shift yielding a mismatch from perfect synchronization results in the same autocorrelation value, $-1/p$. Hence the sequence meets the third randomness property.

10.3 DIRECT-SEQUENCE SPREAD-SPECTRUM SYSTEMS

The block diagram in Figure 10.9a depicts a *direct-sequence* (DS) modulator. "Direct sequence" is the name given to the spectrum spreading technique whereby a carrier wave is first modulated with a data signal $x(t)$, then the data-modulated signal is again modulated with a high-speed (wideband) spreading signal $g(t)$. Consider a constant-envelope data-modulated carrier having power P, radian frequency ω_0, and data phase modulation $\theta_x(t)$, given by

$$s_x(t) = \sqrt{2P} \cos [\omega_0 t + \theta_x(t)] \tag{10.7}$$

Upon further constant-envelope modulation by the spreading signal, $g(t)$, the transmitted waveform can be expressed as

$$s(t) = \sqrt{2P} \cos [\omega_0(t) + \theta_x(t) + \theta_g(t)] \tag{10.8}$$

where the phase of the carrier is now seen to have two components: $\theta_x(t)$ due to the data and $\theta_g(t)$ due to the spreading sequence.

In Chapter 3 it was shown that ideal suppressed carrier binary phase shift keying (BPSK) modulation results in instantaneous changes of π radians to the phase of the carrier, according to the dictates of the data. We can equivalently express Equation (10.7) as the multiplication of the carrier wave by $x(t)$, an antipodal pulse stream with pulse values of $+1$ or -1:

$$s_x(t) = \sqrt{2P}\, x(t) \cos \omega_0 t \tag{10.9}$$

If, like the data, the spreading sequence modulation is also BPSK, and $g(t)$ is an antipodal pulse stream with pulse values of $+1$ or -1, Equation (10.8) can be written as

$$s(t) = \sqrt{2P}\, x(t)g(t) \cos \omega_0 t \qquad (10.10)$$

A modulator based on Equation (10.10) is illustrated in Figure 10.9b. The data pulse stream and the spreading pulse stream are first multiplied, and then the composite $x(t)g(t)$ modulates the carrier. If the assignment of pulse value to binary value is

Pulse value	Binary value
1	0
−1	1

then the initial step in the DS/BPSK modulation can be accomplished by the modulo-2 addition of the binary data sequence with the binary spreading sequence.

Demodulation of the DS/BPSK signal is accomplished by correlating or re-modulating the received signal with a synchronized replica of the spreading signal $g(t - \hat{T}_d)$ as seen in Figure 10.9c, where $\hat{T}_d$ is the receiver's estimate of the propagation delay T_d from the transmitter to the receiver. In the absence of noise and interference, the output signal from the correlator can be written as

$$A\sqrt{2P}\, x(t - T_d)g(t - T_d)g(t - \hat{T}_d) \cos [\omega_0(t - T_d) + \phi] \qquad (10.11)$$

where the constant A is a system gain parameter and ϕ is a random phase angle in the range $(0, 2\pi)$. Since $g(t) = \pm 1$, the product $g(t - T_d)g(t - \hat{T}_d)$ will be unity if $\hat{T}_d = T_d$, that is, if the code signal at the receiver is exactly synchronized with the code signal at the transmitter. When it is synchronized, the output of the receiver correlator is the despread data-modulated signal (except for a random phase ϕ and delay T_d). The despreading correlator is then followed by a conventional demodulator for recovering the data.

10.3.1 Example of Direct Sequencing

Figure 10.10 is an example of DS/BPSK modulation and demodulation following the block diagrams of Figure 10.9b and c. In Figure 10.10a are shown the binary data sequence (1, 0) and its bipolar pulse waveform equivalent $x(t)$, where the binary to pulse value assignments are the same as those described in the preceding section. Examples of a binary spreading sequence and its bipolar pulse waveform equivalent $g(t)$ are shown in Figure 10.10b. The modulo-2 addition of the data sequence and the code sequence, and the equivalent waveform of the product $x(t)g(t)$, is shown in Figure 10.10c.

For the BPSK modulation described by Equations (10.8) and (10.10), it is shown in Figure 10.10d that the phase of the carrier, $\theta_x(t) + \theta_g(t)$, equals π when the value of the product waveform $x(t)g(t)$ equals -1 (or the modulo-2 sum of data and code is binary 1). Similarly, the phase of the carrier is zero when the value of $x(t)g(t)$ equals $+1$ (or the modulo-2 sum of data and code is binary 0). One can appreciate the *signal hiding* property of spread-spectrum signals by comparing the code waveform in Figure 10.10b with the composite waveform in Figure 10.10c. The latter has the signal $x(t)$ "hidden" within it. Just as your eye has

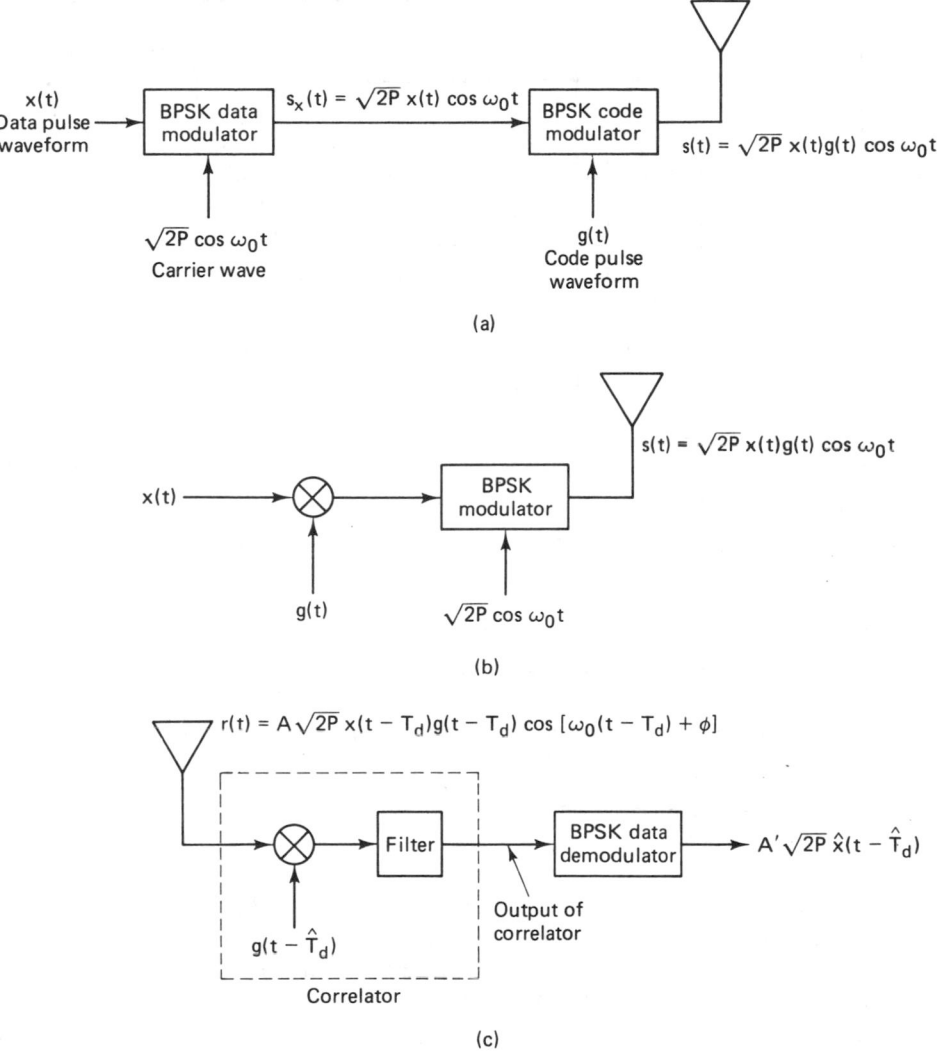

Figure 10.9 Direct-sequence spread-spectrum system. (a) BPSK direct-sequence transmitter. (b) Simplified BPSK direct-sequence transmitter. (c) BPSK direct-sequence receiver.

difficulty finding the slowly moving data signal in the rapidly moving code signal, it is similarly difficult for a receiver to recover a slowly moving signal from a rapidly moving code without having an exact replica of the code.

As shown in Figure 10.9c, DS/BPSK demodulation is a two-step process. The first step, despreading, is accomplished by correlating the received signal with a synchronized replica of the code. The second step, data demodulation, is accomplished with a conventional demodulator. In the example of Figure 10.10

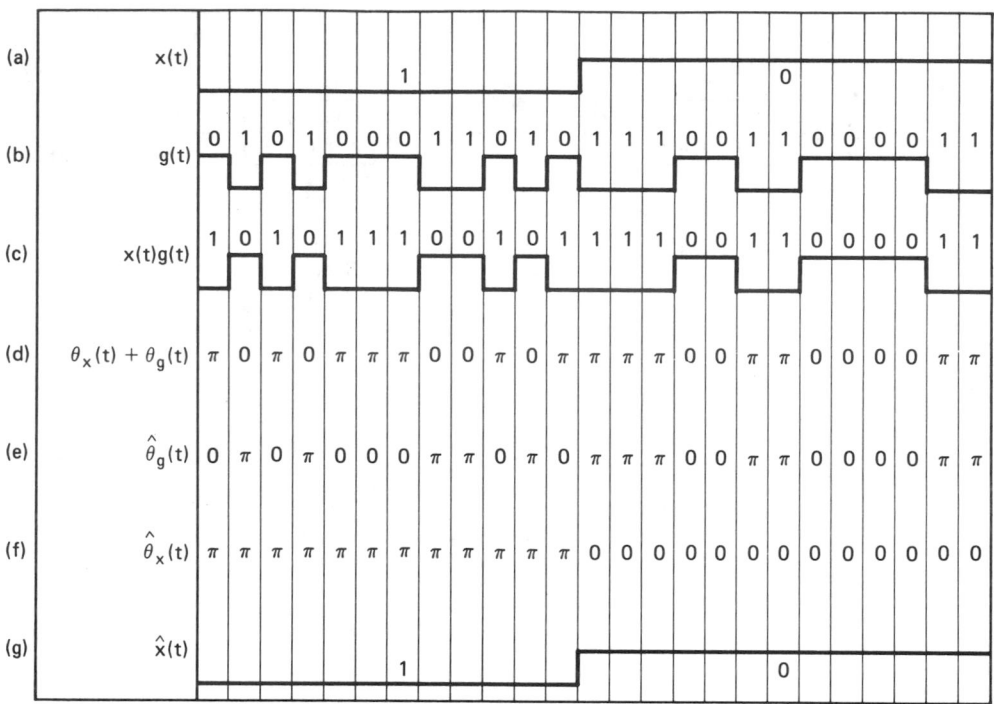

Figure 10.10 Spread-spectrum example using direct sequencing. (a) Binary data waveform to be transmitted. (b) Code sequence. (c) Transmitted sequence. (d) Phase of transmitted carrier. (e) Phase shift produced by receiver code. (f) Phase of received carrier after phase shift by receiver code. (g) Demodulated data waveform.

we see the code replica $\hat{\theta}_g(t)$, in Figure 10.10e, as the phase shift (either 0 or π) that is produced at the receiver by the despreading code. Figure 10.10f illustrates the resulting estimate of the carrier phase, $\hat{\theta}_x(t)$, after despreading or after $\hat{\theta}_g(t)$ has been added to $\theta_x(t) + \theta_g(t)$. At this point one can recognize the original data pattern in the phase terms of the carrier wave. The final step, shown in Figure 10.10g, is to recover an estimate of the data waveform, $\hat{x}(t)$, by the use of a BPSK demodulator.

10.3.2 Processing Gain and Performance

A fundamental issue in spread-spectrum systems is *how much* protection spreading can provide against interfering signals with finite power. Spread-spectrum techniques distribute a relatively low-dimensional signal in a large-dimensional signal space. The signal is "hidden" within the signal space, since we assume that a jammer does not know which signal coordinates are being transmitted at any time. The only recourse for the jammer, intent upon communication disruption, is to jam the entire space with its fixed total power, thus inducing a limited amount of interference in each signal coordinate, or to jam a portion of the signal space with its total power, thus leaving the remainder of the signal space free of interference.

Consider a set of D orthogonal signals, $s_i(t)$, $1 \le i \le D$, in an N-dimensional space, where in general, $D \ll N$. Following the development in Section 3.2.2, we can write

$$s_i(t) = \sum_{j=1}^{N} a_{ij}\psi_j(t) \qquad \begin{matrix} i = 1, 2, \ldots, D; \quad 0 \le t \le T \\ D \ll N \end{matrix} \qquad (10.12)$$

where

$$a_{ij} = \int_{0}^{T} s_i(t)\psi_j(t) \, dt \qquad (10.13)$$

and

$$\int_{0}^{T} \psi_j(t)\psi_k(t) \, dt = \begin{cases} 1 & \text{for } j = k \\ 0 & \text{otherwise} \end{cases} \qquad (10.14)$$

The $\{\psi_j(t)\}$ are linearly independent functions that *span* or characterize the N-dimensional orthonormal space and are called *basis* functions of the space. For every information symbol that is transmitted, a set of coefficients $\{a_{ij}\}$ is chosen independently, using a pseudorandom spreading code, in order to hide the D-dimensional signal set in the larger N-dimensional space. The set of random variables $\{a_{ij}\}$ assume the values $\pm a$, each with a probability of $\frac{1}{2}$. The receiver, of course, has access to each set of coefficients chosen in order to perform the necessary correlation despreading. Even if the same ith symbol is sent repeatedly, the set $\{a_{ij}\}$ used to transmit it is newly selected from symbol to symbol. The energy in each signal waveform of the D signal set will be assumed equal, so that we can write the average energy for each signal as follows:

$$E_s = \int_{0}^{T} \overline{s_i^2(t)} \, dt = \sum_{j=1}^{N} \overline{a_{ij}^2} \qquad i = 1, 2, \ldots, D \qquad (10.15)$$

where the overbar means the expected value over the ensemble of many symbol transmissions. The independent coefficients have zero mean and correlation:

$$\overline{a_{ij}a_{ik}} = \begin{cases} \dfrac{E_s}{N} & \text{for } j = k \\ 0 & \text{otherwise} \end{cases} \qquad (10.16)$$

The standard assumption is that the jammer has no a priori knowledge regarding the selection of the signaling coefficients $\{a_{ij}\}$. As far as the jammer is concerned, the coefficients are uniformly distributed over the N basis coordinates. If the jammer chooses to distribute its power uniformly over the total signal space, the jammer waveform $w(t)$ can be written

$$w(t) = \sum_{j=1}^{N} b_j\psi_j(t) \qquad (10.17)$$

with total energy

$$E_w = \int_0^T w^2(t) \, dt = \sum_{j=1}^{N} b_j^2 \qquad (10.18)$$

A reasonable goal for a jammer would be to devise a strategy for selecting the portions b_j^2, of its fixed total energy E_w so as to minimize the desired signal-to-noise ratio (SNR) at the receiver after demodulation.

At the receiver, the detector output (ignoring receiver noise),

$$r(t) = s_i(t) + w(t) \qquad (10.19)$$

is correlated with the set of possible transmitted signals, so that the output of the ith correlator z_i is

$$z_i = \int_0^T r(t)s_i(t) \, dt = \sum_{j=1}^{N} \left(a_{ij}^2 + b_j a_{ij}\right) \qquad (10.20)$$

The second term on the right side of Equation (10.20) averages to zero over the ensemble of all possible pseudorandom code sequences, since the set of random variables $\{a_{ij}\}$ assume the values $\pm a$, each with probability $\frac{1}{2}$. Therefore, given that $s_m(t)$ was transmitted, the expected value of the output of the ith correlator, $\mathbf{E}(z_i|s_m)$, can be written, following the development in References [7, 8],

$$\mathbf{E}(z_i|s_m) = \sum_{j=1}^{N} \overline{a_{ij}^2} = \begin{cases} E_s & \text{for } i = m \\ 0 & \text{otherwise} \end{cases} \qquad (10.21)$$

In Equation (10.21), the term $\mathbf{E}(z_i|s_m)$ for $i = m$ is to be interpreted as follows. Given that $s_i(t)$ is to be transmitted, N coefficients a_{ij} ($1 \le j \le N$) are chosen pseudorandomly (the receiver is assumed to have access to each choice of the a_{ij} for correlation despreading). Hence, in computing $\mathbf{E}(z_i|s_i)$, even though the ith information symbol is specified at the transmitter, the pattern of coefficients used to send it appears random (to the unauthorized receiver) for each transmission. Equation (10.21) presumes that the jammer has not been successful in its attempt to employ some clever tactics (described in Section 10.7).

Let us assume that all D signals are equally likely. Then the expected value at the output of any of the D correlators is

$$\mathbf{E}(z_i) = \frac{E_s}{D} \qquad (10.22)$$

Similarly, using Equations (10.15) to (10.21), we compute var $(z_i|s_i)$, the variance at the output of the ith correlator, given that the ith signal was transmitted.

$$\text{var } (z_i|s_i) = \sum_{j,k} b_j b_k \overline{a_{ij} a_{ik}}$$

$$= \sum_{j=1}^{N} b_j^2 \overline{a_{ij}^2} \qquad (10.23)$$

$$= \sum_{j=1}^{N} b_j^2 \frac{E_s}{N}$$

$$= \frac{E_w E_s}{N} \tag{10.24}$$

For completeness, the variance at the output of the ith correlator, var $(z_i|s_m)$, given that the mth signal was transmitted, where $i \neq m$, can similarly be computed to be

$$\text{var } (z_i|s_m) = \frac{E_w E_s}{N} + \frac{E_s^2}{N} \tag{10.25}$$

The signal-to-jammer ratio (SJR) at the output of the ith correlator can be defined as

$$\text{SJR} = \sum_{m=1}^{D} \frac{E^2(z_i|s_m)}{\text{var } (z_i|s_m)} P(s_m) = \frac{E_s^2/D}{E_w E_s/N} = \frac{E_s N}{E_w D} \tag{10.26}$$

where the probability of the mth signal $P(s_m) = 1/D$, since the signals are assumed to occur with equal probability, and where the signal energy and the jammer energy in the ith correlator are denoted by $E^2(z_i)$ and var (z_i), respectively. Because of Equation (10.21), the only terms in the summation of Equation (10.26) not equal to zero are those for which $i = m$. The result is independent of the way in which the jammer chooses to distribute its energy. Therefore, regardless of how b_j is chosen, subject to $\sum_j b_j^2 = E_w$, the SJR in Equation (10.26) indicates that spreading gives the signal an advantage of a factor of N/D over the jammer. The ratio N/D is known as the *processing gain* G_p.

Since the approximate dimensionality of a signal with bandwidth W and duration T is $2WT$, we can express the processing gain as

$$G_p = \frac{N}{D} \simeq \frac{2W_{ss}T}{2W_{\min}T} = \frac{W_{ss}}{R} \tag{10.27}$$

where W_{ss} is the spread-spectrum bandwidth (the total bandwidth used by the spreading technique) and $W_{\min}$ is the minimum bandwidth of the data (taken to be the data rate, R). For direct sequence systems, W_{ss} is approximately the code chip rate R_p, and $W_{\min}$ is similarly the data rate R, giving

$$G_p = \frac{R_p}{R} \tag{10.28}$$

10.4 FREQUENCY HOPPING SYSTEMS

We now consider a spread-spectrum technique called frequency hopping (FH). The modulation most commonly used with this technique is M-ary frequency shift keying (MFSK), where $k = \log_2 M$ information bits are used to determine which one of M frequencies is to be transmitted. The position of the M-ary signal set is shifted pseudorandomly by the frequency synthesizer over a hopping bandwidth W_{ss}. A typical FH/MFSK system block diagram is shown in Figure 10.11. In a conventional MFSK system, the data symbol modulates a *fixed frequency* carrier; in an FH/MFSK system, the data symbol modulates a carrier whose frequency is *pseudorandomly* determined. In either case, a single tone is transmitted. The FH system in Figure 10.11 can be thought of as a two-step modulation process—data modulation and frequency hopping modulation—even though it can be implemented as a single step whereby the frequency synthesizer produces a transmission tone based on the simultaneous dictates of the PN code and the data. At each frequency hop time a PN generator feeds the frequency synthesizer a frequency word (a sequence of ℓ chips) which dictates one of 2^{ℓ} symbol-set positions. The frequency hopping bandwidth, W_{ss}, and the minimum frequency spacing between consecutive hop positions, Δf, dictate the minimum number of chips necessary in the frequency word.

For a given hop, the occupied transmission bandwidth is identical to the bandwidth of conventional MFSK, which is typically much smaller than W_{ss}. However, averaged over many hops, the FH/MFSK spectrum occupies the entire spread-spectrum bandwidth. Current technology permits FH bandwidths of the order of several gigahertz, which is an order of magnitude larger than implementable DS bandwidths [9], thus allowing for larger processing gains in FH compared to DS systems. Since frequency hopping techniques operate over such wide bandwidths, it is difficult to maintain phase coherence from hop to hop. Therefore, such schemes are usually configured using noncoherent demodulation. Nevertheless, consideration has been given to coherent FH in Reference [10].

In Figure 10.11 we see that the receiver reverses the signal processing steps of the transmitter. The received signal is first FH demodulated (dehopped) by mixing it with the same sequence of pseudorandomly selected frequency tones that was used for hopping. Then the dehopped signal is applied to a conventional bank of M noncoherent energy detectors to select the most likely symbol.

Example 10.1 Frequency Word Size

A hopping bandwidth W_{ss} of 400 MHz and a frequency step size Δf of 100 Hz are specified. What is the minimum number of PN chips that are required for each frequency word?

Solution

$$\text{Number of tones contained in } W_{ss} = \frac{W_{ss}}{\Delta f} = \frac{400 \text{ MHz}}{100 \text{ Hz}}$$

$$= 4 \times 10^{6}$$

$$\text{Minimum number of chips} = \lceil \log_2 (4 \times 10^6) \rceil$$

$$= 22 \text{ chips}$$

where $\lceil x \rceil$ indicates the smallest integer value not less than x.

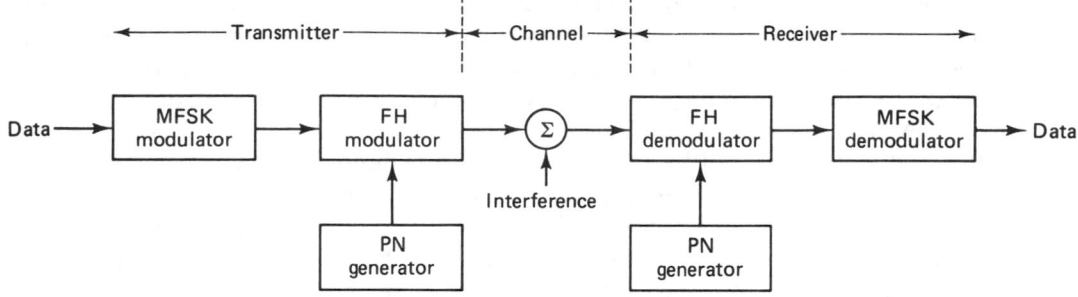

Figure 10.11 FH/MFSK system.

10.4.1 Frequency Hopping Example

Consider the frequency hopping example illustrated in Figure 10.12. The input data consist of a binary sequence with a data rate of $R = 150$ bits/s. The modulation is 8-ary FSK. Therefore, the symbol rate is $R_s = R/(\log_2 8) = 50$ symbols/s (the symbol duration $T = 1/50 = 20$ ms). The frequency is hopped once per symbol, and the hopping is time-synchronous with the symbol boundaries. Thus the hopping rate is 50 hops/s. Figure 10.12 depicts the time–bandwidth plane of the communication resource; the abscissa represents time, and the ordinate represents

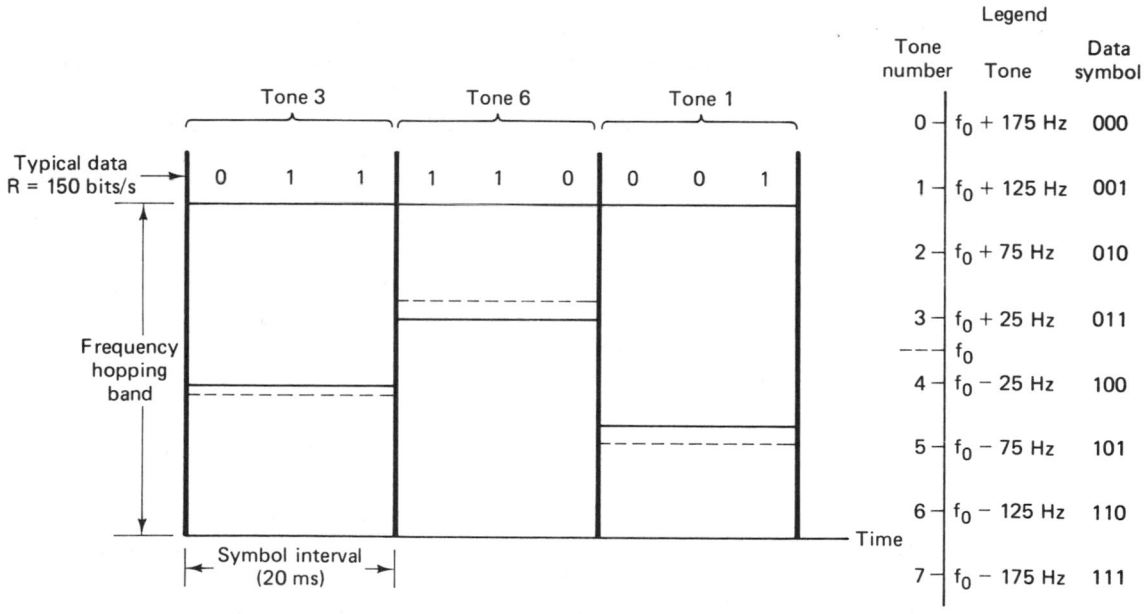

Figure 10.12 Frequency hopping example using 8-ary FSK modulation.

the hopping bandwidth, W_{ss}. The legend on the right side of the figure illustrates a set of 8-ary FSK symbol-to-tone assignments. Notice that the tone separation specified is $1/T = 50$ Hz, which corresponds to the minimum required tone spacing for the orthogonal signaling of this noncoherent FSK example (see Section 3.6.4).

A typical binary data sequence is shown at the top of Figure 10.12. Since the modulation is 8-ary FSK, the bits are grouped three at a time to form symbols. In a *conventional* 8-ary FSK scheme, a single-sideband tone, offset from f_0, the *fixed* center frequency of the data band, would be transmitted (according to an assignment like the one shown in the legend). The only difference in this FH/MFSK example is that the center frequency of the data band, f_0, is *not fixed*. For each new symbol, f_0 hops to a new position in the hop bandwidth, and the entire data-band structure moves with it. In the example of Figure 10.12, the first symbol in the data sequence, 0 1 1, yields a tone 25 Hz above f_0. The diagram depicts f_0 with a dashed line and the symbol tone with a solid line. During the second symbol interval, f_0 has hopped to a new spectral location, as indicated by the dashed line. The second symbol, 1 1 0, dictates that a tone indicated by the solid line, 125 Hz below f_0, shall be transmitted. Similarly, the final symbol in this example, 0 0 1, calls for a tone 125 Hz above f_0. Again, the center frequency has moved, but the relative positions of the symbol tones remain fixed.

10.4.2 Robustness

A common dictionary definition describes the term *robustness* as the state of being strong and healthy; full of vigor; hardy. In the context of communications, the usage is not too different. Robustness characterizes a signal's ability to withstand impairments from the channel, such as noise, jamming, fading, and so on. A signal configured with multiple replicate copies, each transmitted on a different frequency, has a greater likelihood of survival than does a single such signal with equal total power. The greater the diversity (multiple transmissions, at different frequencies, spread in time), the more robust the signal against random interference.

The following example should clarify the concept. Consider a message consisting of four symbols s_1, s_2, s_3, s_4. The introduction of diversity starts by repeating the message N times. Let us choose $N = 8$. Then, the repeated symbols called *chips* can be written

$$s_1 s_1 s_1 s_1 s_1 s_1 s_1 s_1 s_2 s_2 s_2 s_2 s_2 s_2 s_2 s_2 s_3 s_3 s_3 s_3 s_3 s_3 s_3 s_3 s_4 s_4 s_4 s_4 s_4 s_4 s_4 s_4$$

Each chip is transmitted at a different hopping frequency (the center of the data bandwidth is changed for each chip). The resulting transmissions at frequencies f_i, f_j, f_k, . . . yield a more robust signal than without such divesity. A target-shooting analogy is that a pellet from a barrage of shotgun pellets has a better chance of hitting a target, compared to the action of a single bullet.

10.4.3 Frequency Hopping with Diversity

In Figure 10.13 we extend the example illustrated in Figure 10.12, with the additional feature of a chip repeat factor of $N = 4$. During each 20-ms symbol interval, there are now four columns, corresponding to the four separate chips to be transmitted for each symbol. At the top of the figure we see the same data sequence, with $R = 150$ bps, as in the earlier example; and we see the same 3-bit partitioning to form the 8-ary symbols. Each symbol is transmitted four times, and for each transmission the center frequency of the data band is hopped to a new region of the hopping band, under the control of a PN code generator. Therefore, for this example, each chip interval, T_c, is equal to $T/N = 20$ ms/4 = 5 ms in duration, and the hopping rate is now

$$\frac{NR}{\log_2 8} = 200 \text{ hops/s}$$

Notice that the spacing between frequency tones must change to meet the changed requirement for orthogonality. Since the duration of each FSK tone is now equal to the chip duration, that is, $T_c = T/N$, the minimum separation between tones is $1/T_c = N/T = 200$ Hz. As in the earlier example, Figure 10.13 illustrates that the center of the data band (plus the modulation structure) is shifted at each new

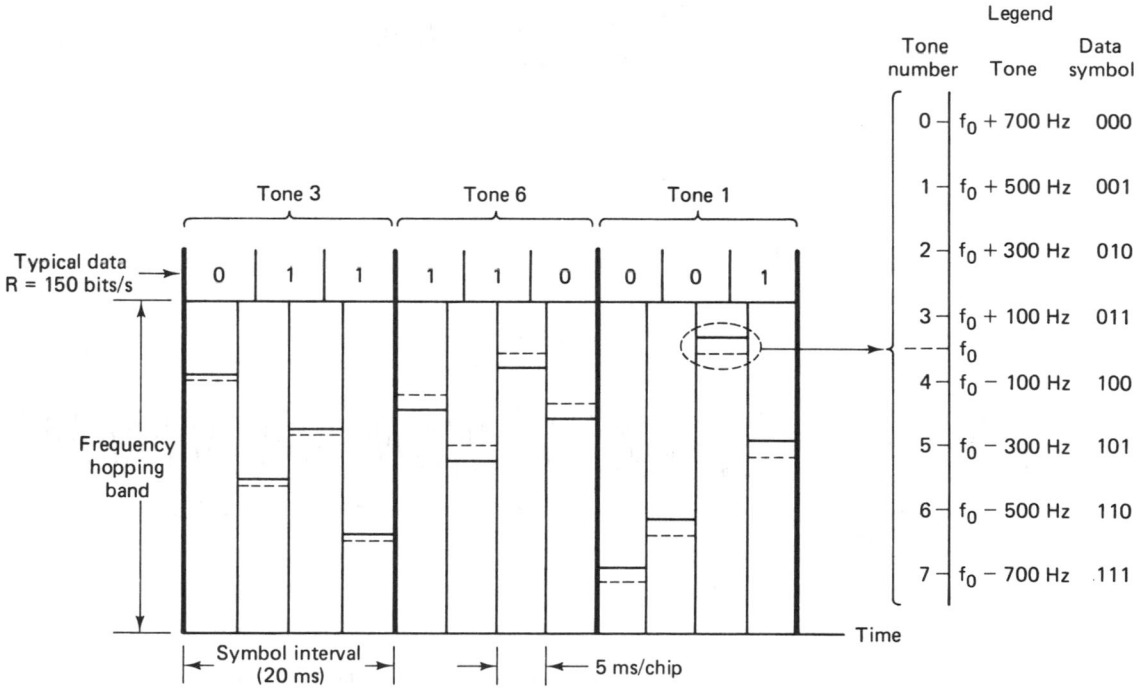

Figure 10.13 Frequency hopping example with diversity ($N = 4$).

chip time. The position of the solid line (transmission frequency) has the same relationship to the dashed line (center of the data band) for each of the chips associated with a given symbol.

10.4.4 Fast Hopping versus Slow Hopping

In the case of direct-sequence spread-spectrum systems, the term "chip" refers to the PN code symbol (the symbol of shortest duration in a DS system). In a similar sense for frequency hopping systems, the term "chip" is used to characterize the shortest uninterrupted waveform in the system. Frequency hopping systems are classified as *slow frequency hopping* (SFH), which means there are several modulation symbols per hop, or as *fast frequency hopping* (FFH), which means that there are several frequency hops per modulation symbol. For SFH, the shortest uninterrupted waveform in the system is that of the data symbol; however, for FFH, the shortest uninterrupted waveform is that of the hop. Figure 10.14a illustrates an example of FFH; the data symbol rate is 30 symbols/s and the frequency hopping rate is 60 hops/s. The figure illustrates the waveform $s(t)$ over one symbol duration ($\frac{1}{30}$ s). The waveform change in (the middle of) $s(t)$ is

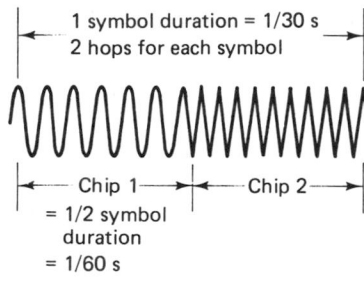

(a)

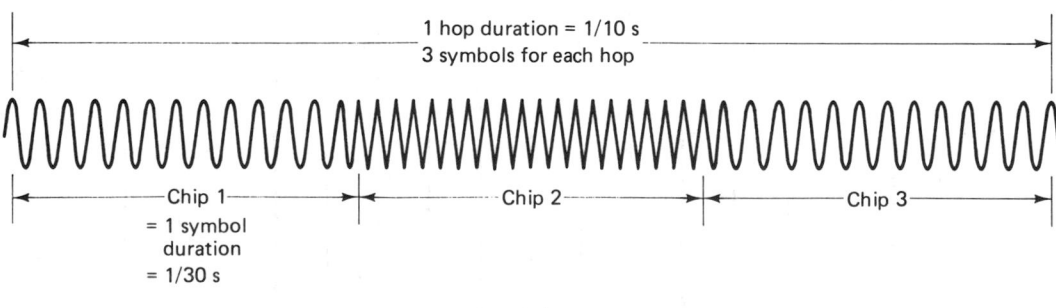

(b)

Figure 10.14 Chip—in the context of an FH/MFSK system. (a) Example 1: Frequency hopping MFSK system with symbol rate = 30 symbols/s and hopping rate = 60 hops/s. 1 chip = 1 hop. (b) Example 2: Same as part (a) except hopping rate = 10 hops/s. 1 chip = 1 symbol.

due to a new frequency hop. In this example, a chip corresponds to a hop since the hop duration is shorter than the symbol duration. Each chip corresponds to half a symbol. Figure 10.14b illustrates an example of SFH; the data symbol rate is still 30 symbols/s, but the frequency hopping rate has been reduced to 10 hops/s. The waveform $s(t)$ is shown over a duration of three symbols ($\frac{1}{10}$ s). In this example, the hopping boundaries appear only at the beginning and end of the three-symbol duration. Here, the changes in the waveform are due to the modulation state changes; therefore, in this example a chip corresponds to a data symbol, since the data symbol is shorter than the hop duration.

Figure 10.15a illustrates an FFH example of a binary FSK system. The diversity is $N = 4$. There are 4 chips transmitted per bit. As in Figure 10.13, the dashed line in each column corresponds to the center of the data band and the solid line corresponds to the symbol frequency. Here, for FFH, the chip duration is the hop duration. Figure 10.15b illustrates an example of an SFH binary FSK system. In this case, there are 3 bits transmitted during the time duration of a single hop. Here, for SFH, the chip duration is the bit duration. If this SFH example were changed from a binary system to an 8-ary system, what would the

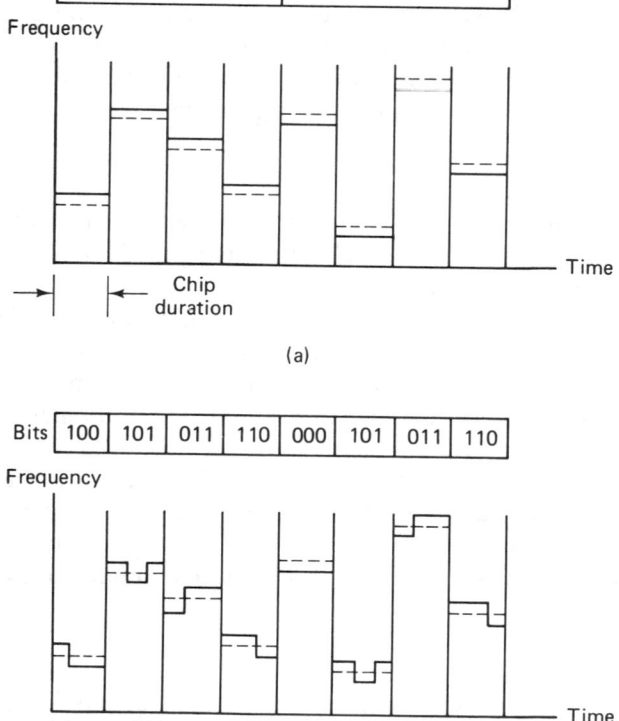

(a)

(b)

Figure 10.15 Fast hopping versus slow hopping in a binary system. (a) Fast-hopping example: 4 hops/bit. (b) Slow-hopping example: 3 bits/hop.

Sec. 10.4   Frequency Hopping Systems

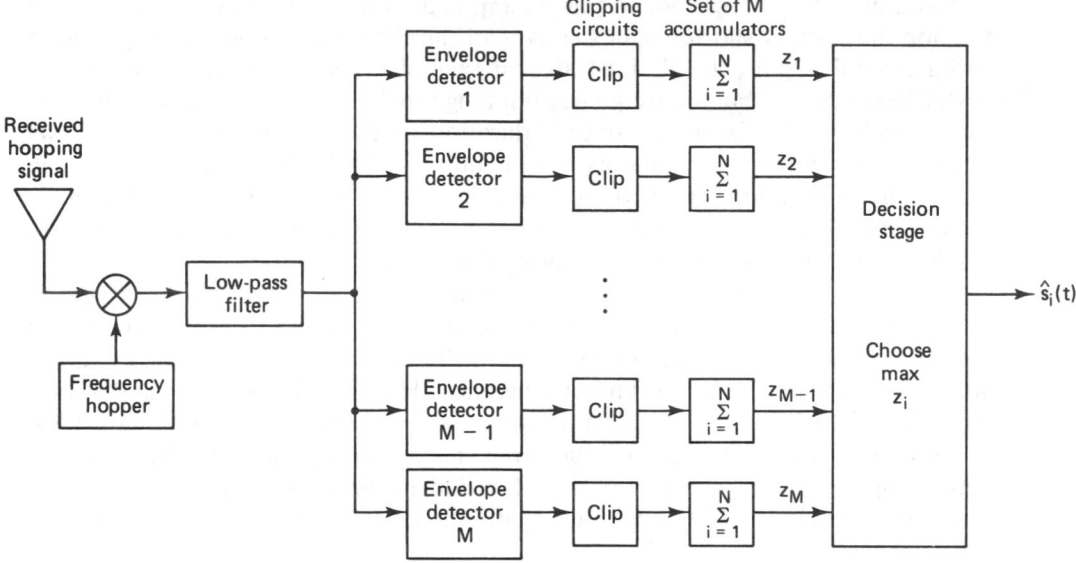

Figure 10.16 FFH/MFSK demodulator.

chip duration then correspond to? If the system were implemented as an 8-ary scheme, each 3 bits would be transmitted as a single data symbol. The symbol boundaries and the hop boundaries would then be the same, and the chip duration, the hop duration, and the symbol duration would all be the same.

10.4.5 FFH/MFSK Demodulator

Figure 10.16 illustrates the schematic for a typical fast frequency hopping MFSK (FFH/MFSK) demodulator. First, the signal is dehopped using a PN generator identical to the one used for hopping. Then, after filtering with a low-pass filter that has a bandwidth equal to the data bandwidth, the signal is demodulated using a bank of M envelope or energy detectors. Each envelope detector is followed by a clipping circuit and an accumulator. The clipping circuit serves an important function in the presence of an intentional jammer or other strong unpredictable interference; it is treated in a later section. The demodulator does *not* make symbol decisions on a chip-by-chip basis. Instead, the energy from the N chips are accumulated, and after the energy from the Nth chip is added to the $N - 1$ earlier ones, the demodulator makes a symbol decision by choosing the symbol that corresponds to the accumulator, z_i ($i = 1, 2, \ldots, M$), with maximum energy.

10.5 SYNCHRONIZATION

For both DS and FH spread-spectrum systems, a receiver must employ a *synchronized* replica of the spreading or code signal to demodulate the received signal successfully. The process of synchronizing the locally generated spreading signal with the received spread-spectrum signal is usually accomplished in two steps.

The first step, called *acquisition,* consists of bringing the two spreading signals into *coarse* alignment with one another. Once the received spread-spectrum signal has been acquired, the second step, called *tracking,* takes over and continuously maintains the best possible waveform *fine* alignment by means of a feedback loop.

10.5.1 Acquisition

The acquisition problem is one of searching throughout a region of time and frequency uncertainty in order to synchronize the received spread-spectrum signal with the locally generated spreading signal. Acquisition schemes can be classified as coherent or noncoherent. Since the despreading process typically takes place before carrier synchronization, and therefore the carrier phase is unknown at this point, most acquisition schemes utilize noncoherent detection. When determining the limits of the uncertainty in time and frequency, the following items must be considered:

1. Uncertainty in the distance between the transmitter and the receiver translates into uncertainty in the amount of propagation delay.
2. Relative clock instabilities between the transmitter and the receiver result in phase differences between the transmitter and receiver spreading signals that will tend to grow as a function of elapsed time between synchronization.
3. Uncertainty of the receiver's relative velocity with respect to the transmitter translates into uncertainty in the value of Doppler frequency offset of the incoming signal.
4. Relative oscillator instabilities between the transmitter and the receiver result in frequency offsets between the two signals.

10.5.1.1 Correlator Structures

A common feature of all acquisition methods is that the received signal and the locally generated signal are first correlated to produce a measure of similarity between the two. This measure is then compared to a threshold to decide if the two signals are in synchronism. If they are, the tracking loop takes over.* If they are not, the acquisition procedure provides for a phase or frequency change in the locally generated code as a part of a systematic search through the receiver's phase and frequency uncertainty region, and another correlation is attempted.

Consider the direct-sequence *parallel-search* acquisition system shown in Figure 10.17. The locally generated code $g(t)$ is available with delays that are spaced one-half chip ($T_c/2$) apart. If the time uncertainty between the local code and the received code is N_c chips and a complete parallel search of the entire time uncertainty region is to be accomplished in a single search time, $2N_c$ correlators are used. Each correlator simultaneously examines a sequence of λ chips, after which the $2N_c$ correlator outputs are compared. The locally generated code, corresponding to the correlator with the largest output is chosen. Conceptually, this is the simplest of the search techniques; it considers all possible code positions

---

* Quite often to maintain a small false alarm probability, the threshold crossing must be further verified by a suitable verification algorithm before the tracking loop takes over [4].

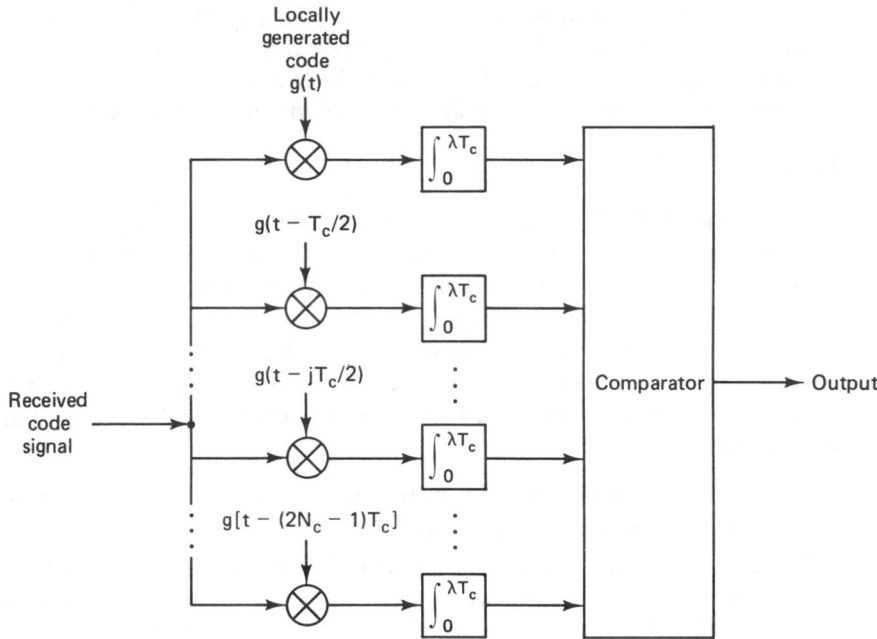

Figure 10.17 Direct-sequence parallel search acquisition.

(or fractional code positions) in parallel and uses a maximum likelihood algorithm for acquiring the code. Each detector output pertains to the identical observation of received signal plus noise. As λ increases, the synchronization error probability (i.e., the probability of choosing the incorrect code alignment) decreases. Thus λ is chosen as a compromise between minimizing the probability of a synchronization error and minimizing the time to acquire.

Figure 10.18 illustrates a simple acquisition scheme for a frequency hopping system. Assume that a sequence of N consecutive frequencies from the hop sequence is chosen as a synchronization pattern (without data modulation). The N noncoherent matched filters each consists of a mixer followed by a bandpass filter (BPF) and a square-law envelope detector (an envelope detector followed by a square-law device). If the frequency hopping sequence is $f_1, f_2, \ldots, f_N$, delays are inserted into the matched filters so that when the correct frequency hopping sequence appears, the system produces a large output, indicating detection of the synchronization sequence. Acquisition can be accomplished rapidly because all possible code offsets are examined simultaneously.

If, during each correlation, λ chips are examined, the maximum time required, $(T_{\text{acq}})_{\text{max}}$, for a fully parallel search is

$$(T_{\text{acq}})_{\text{max}} = \lambda T_c \qquad (10.29)$$

The mean acquisition time of a parallel search system can be approximated by noting that after integrating over λ chips, a correct decision will be made with probability P_D, called the *probability of detection*. If an incorrect output is chosen,

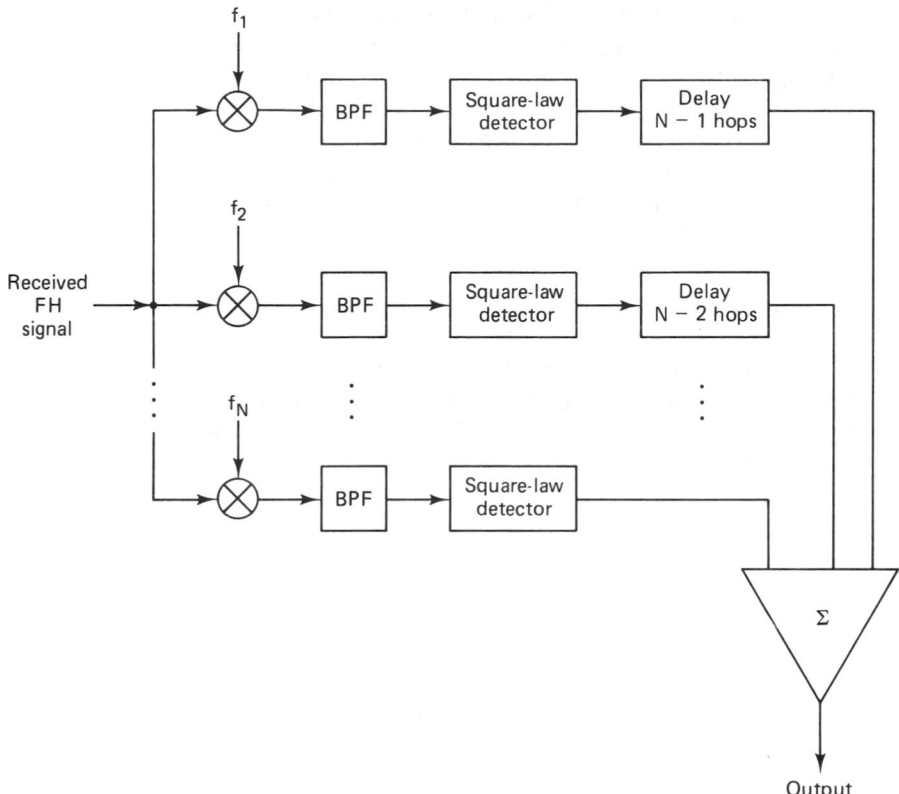

Figure 10.18 Frequency hopping acquisition scheme.

an additional λ chips are again examined to make a determination of the correct output. Therefore, on the average, the acquisition time is [6]

$$\overline{T}_{acq} = \lambda T_c P_D + 2\lambda T_c P_D(1 - P_D) + 3\lambda T_c P_D(1 - P_D)^2 + \cdots$$

$$= \frac{\lambda T_c}{P_D} \tag{10.30}$$

Since the required number of correlators or matched filters can be prohibitively large, fully parallel acquisition techniques are not usually used. In place of Figures 10.17 and 10.18, a single correlator or matched filter can be implemented that will *serially search* until synchronization is achieved. Naturally, trade-offs between fully parallel, fully serial, and combinations of the two involve hardware complexity versus time to acquire for the same uncertainty and chip rate.

10.5.1.2 Serial Search

A popular strategy for the acquisition of spread-spectrum signals is to use a single correlator or matched filter to serially search for the correct phase of the DS code signal or the correct hopping pattern of the FH signal. A considerable reduction in complexity, size, and cost can be achieved by a serial implementation that repeats the correlation procedure for each possible sequence shift. Figures 10.19 and 10.20 illustrate the basic configuration for DS and FH spread-spectrum schemes, respectively. In a stepped serial acquisition scheme for a DS system, the timing epoch of the local PN code is set, and the locally generated PN signal is correlated with the incoming PN signal. At fixed examination intervals of λT_c (search dwell time), where $\lambda \gg 1$, the output signal is compared to a preset threshold. If the output is below the threshold, the phase of the locally generated code signal is incremented by a fraction (usually one-half) of a chip and the correlation is reexamined. When the threshold is exceeded, the PN code is assumed to have been acquired, the phase-incrementing process of the local code is inhibited, and the code tracking procedure will be initiated. In a similar scheme for FH systems, shown in Figure 10.20, the PN code generator controls the frequency hopper. Acquisition is accomplished when the local hopping is aligned with that of the received signal.

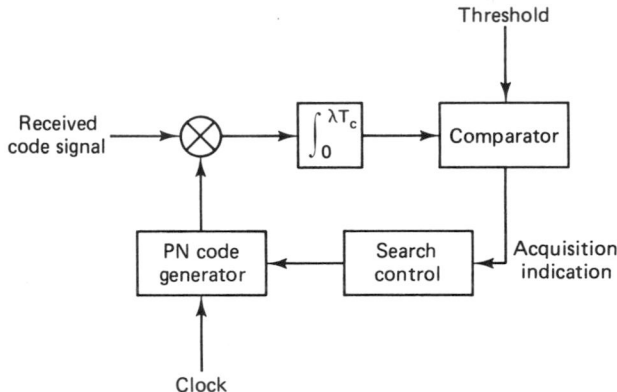

Figure 10.19 Direct-sequence serial search acquisition.

The maximum time required for a fully serial DS search, assuming that the search proceeds in half-chip increments, is

$$(T_{\text{acq}})_{\text{max}} = 2N_c\lambda T_c \qquad (10.31)$$

where the uncertainty region to be searched is N_c chips long. The mean acquisition time of a serial DS search system can be shown, for $N_c \gg \frac{1}{2}$ chip, to be [4]

$$\overline{T}_{\text{acq}} = \frac{(2 - P_D)(1 + KP_{\text{FA}})}{P_D}(N_c\lambda T_c) \qquad (10.32)$$

where λT_c is the search dwell time, P_D the probability of correct detection, and P_{FA} the probability of false alarm. We can regard the time interval $K\lambda T_c$, where $K \gg 1$, as the time needed to verify a detection. Therefore, in the event of a false alarm, $K\lambda T_c$ seconds is the time penalty incurred. For $N_c \gg \frac{1}{2}$ chip and $K \ll 2N_c$, the variance of the acquisition time is

$$(\text{var})_{\text{acq}} = (2N_c\lambda T_c)^2(1 + KP_{\text{FA}})\left(\frac{1}{12} + \frac{1}{P_D^2} - \frac{1}{P_D}\right) \qquad (10.33)$$

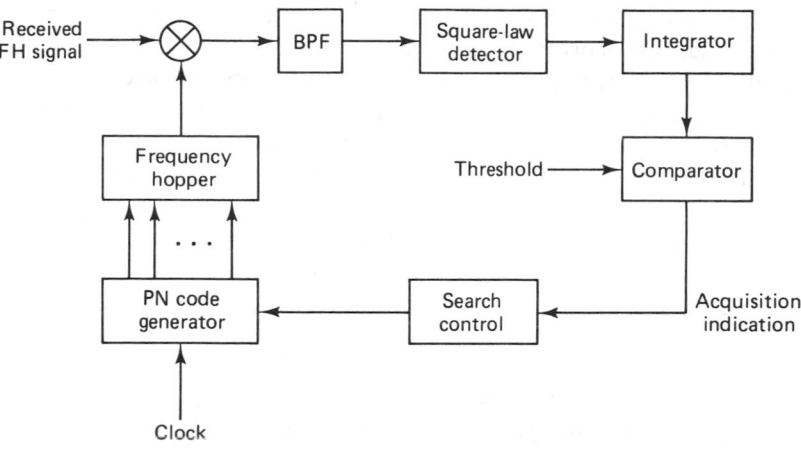

Figure 10.20 Frequency hopping serial search acquisition.

10.5.1.3 Sequential Estimation

Another search technique, called *rapid acquisition by sequential estimation* (RASE), proposed by Ward [11], is illustrated in Figure 10.21. The switch is initially in position 1. The RASE system enters its best estimate of the first n received code chips into the n stages of its local PN generator. The fully loaded register defines a starting state from which the generator begins its operation. A PN sequence has the property that the next combination of register states depends only on the present combination of states. Therefore, if the first n received chips are correctly estimated, all the following chips from the local PN generator will

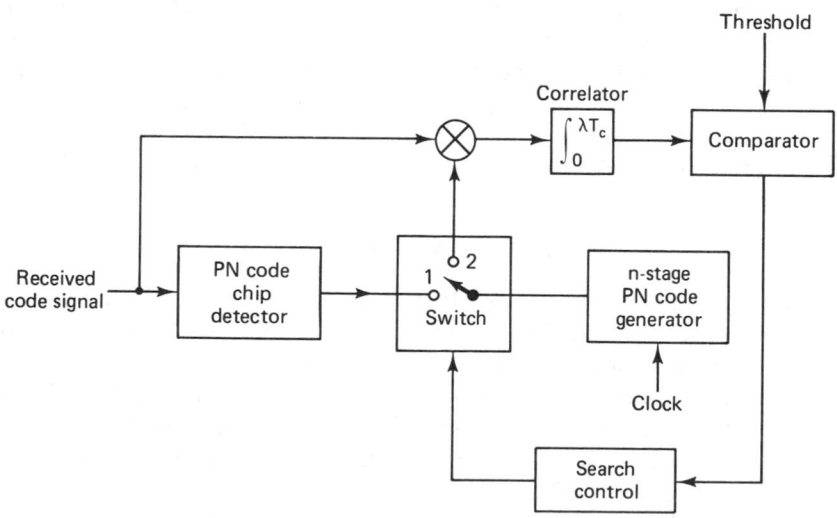

Figure 10.21 Rapid acquisition by sequential estimation.

be correctly generated. The switch is next thrown to position 2. If the starting state had been correctly estimated, the local generator generates the same sequence as the incoming waveform, in the absence of noise. If the correlator output after λT_c exceeds a preset threshold level, we assume that synchronization has occurred. If the output is less than the threshold, the switch is returned to position 1, the register is reloaded with estimates of the next n received chips, and the procedure is repeated. Once synchronization has occurred, the system no longer needs estimates of the input code chips. We can calculate the *minimum* acquisition time for the case when no noise is present. The first n chips will be correctly loaded into the register, and therefore the acquisition time is

$$T_{\text{acq}} = nT_c \qquad (10.34)$$

While the RASE system has a rapid acquisition capability it has the drawback of being highly vulnerable to noise and interference signals. The reason for this is that the estimation process consists of a simple chip-by-chip hard-decision demodulation, without using the interference rejection benefits of the PN code.

For an extensive treatment of sequential estimation, see Reference [4].

10.5.2 Tracking

Once acquisition or coarse synchronization is completed, tracking or fine synchronization takes place. Tracking code loops can be classified as coherent or noncoherent. A coherent loop is one in which the carrier frequency and phase are known exactly so that the loop can operate on a baseband signal. A noncoherent loop is one in which the carrier frequency is not known exactly (due to Doppler effects, for example), nor is the phase. In most instances, since the carrier frequency and phase are not known exactly, a priori, a noncoherent code loop is used to track the received PN code. Tracking loops are further classified as a *full-time* early-late tracking loop, often referred to as a *delay-locked loop* (DLL), or as a *time-shared* early-late tracking loop, frequently referred to as a *tau-dither loop* (TDL). A basic noncoherent DLL loop for a direct-sequence spread-spectrum system using binary phase shift keying (BPSK) is shown in Figure 10.22. The data $x(t)$ and the code $g(t)$ each modulate the carrier wave using BPSK, and as before in the absence of noise and interference, the received waveform can be expressed as

$$r(t) = A\sqrt{2P}\, x(t)g(t) \cos (\omega_0 t + \phi) \qquad (10.35)$$

where the constant A is a system gain parameter and ϕ is a random phase angle in the range $(0, 2\pi)$. The locally generated code of the tracking loop is offset in phase from the incoming $g(t)$ by a time τ, where $\tau < T_c/2$. The loop provides *fine* synchronization by first generating two PN sequences $g(t + T_c/2 + \tau)$ and $g(t - T_c/2 + \tau)$ delayed from each other by one chip. The two bandpass filters are designed to pass the data and to average the product of $g(t)$ and the two PN sequences $g(t \pm T_c/2 + \tau)$. (See Reference [4] for the optimum filter bandwidth for a given filter type.) The square-law envelope detector eliminates the data since $|x(t)| = 1$. The output of each envelope detector is given approximately by

$$E_D \simeq \mathbf{E}\left\{\left| g(t)g\left(t \pm \frac{T_c}{2} + \tau\right)\right|\right\} = \left|R_g\left(\tau \pm \frac{T_c}{2}\right)\right| \qquad (10.36)$$

where the operator $\mathbf{E}\{\cdot\}$ means *expected value* and $R_g(x)$ is the autocorrelation function of the PN waveform as shown in Figure 10.8. The feedback signal $Y(\tau)$ is shown in Figure 10.23. When τ is positive, the feedback signal $Y(\tau)$ instructs the voltage-controlled oscillator (VCO) to increase its frequency, thereby forcing τ to decrease, and when τ is negative, $Y(\tau)$ instructs the VCO to decrease, thereby forcing τ to increase. When τ is a suitably small number, $g(t)g(t + \tau) \simeq 1$, yielding the despread signal $Z(t)$, which is then applied to the input of a conventional data demodulator. Detailed analysis of the DLL can be found in References [4, 12–14].

A problem with the DLL is that the early and late arms must be precisely gain balanced or else the feedback signal $Y(\tau)$ will be offset and will not produce a zero signal when the error is zero. This problem is solved by using a time-shared tracking loop in place of the full-time delay-locked loop. The time-shared loop time shares the use of the early-late correlators. The main advantages are that only one correlator need be used in the design of the loop, and further, that dc offset problems are reduced.

An offshoot of the time-shared tracking loop is called the *tau-dither loop* (TDL), shown in Figure 10.24. This design has the advantage that only one correlator is needed to provide the code *tracking* function *and* the *despreading* function. Just as in the case of a DLL, the received signal is correlated with an early and a late version of the locally generated PN code. As shown in Figure 10.24, the PN code generator is driven by a clock signal whose phase is *dithered* back

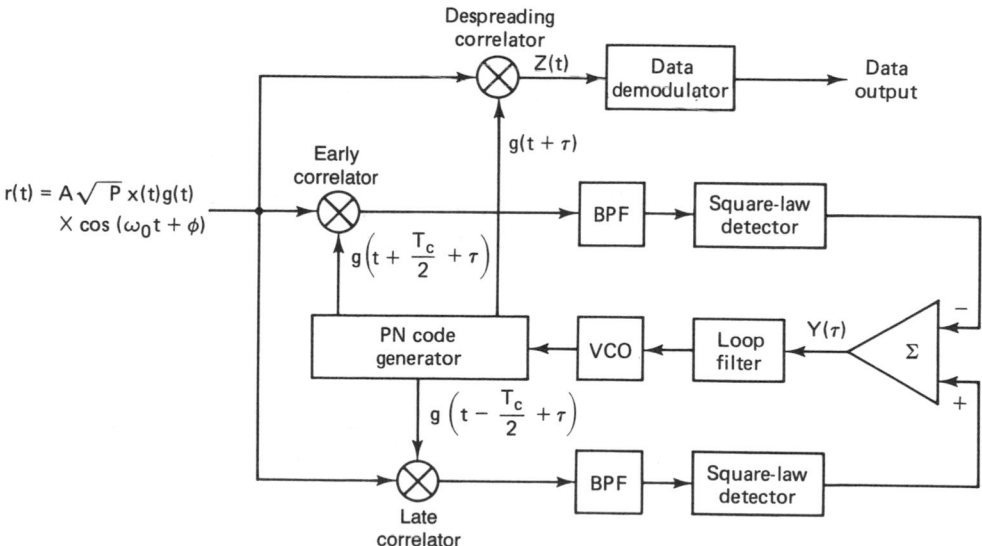

Figure 10.22 Delay-locked loop for tracking direct-sequence signals.

Sec. 10.5   Synchronization

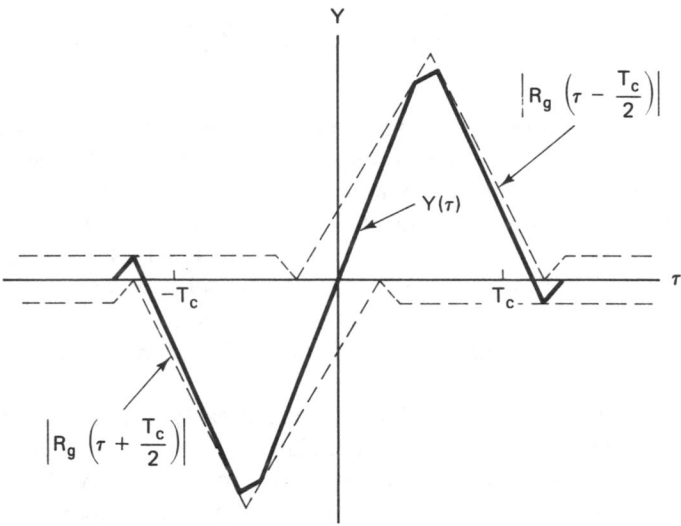

Figure 10.23 DLL feedback signal $Y(\tau)$.

and forth with a square-wave switching function; this eliminates the necessity of ensuring identical transfer functions of the early and late paths. The signal-to-noise performance of the TDL is only about 1.1 dB worse than that of the DLL if the arm filters are designed properly [4]. For a comprehensive treatment of synchronization of PN codes, see References [4, 15,16].

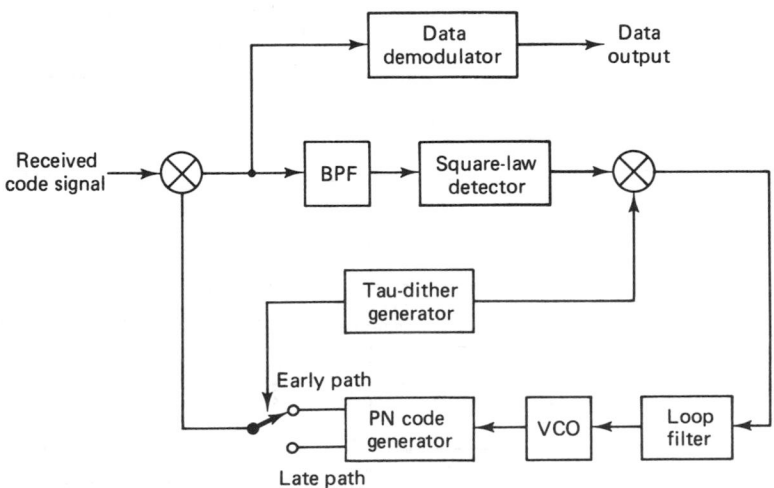

Figure 10.24 Tau-dither tracking loop.

10.6 SPREAD-SPECTRUM APPLICATIONS

10.6.1 Code-Division Multiple Access

Spread-spectrum multiple access techniques allow multiple signals occupying the same RF bandwidth to be transmitted simultaneously without interfering with one another. The application of spread-spectrum techniques to the problem of multiple access was discussed in Chapter 9 for a frequency hopped code-division multiple access (FH/CDMA) scheme. Here we consider CDMA using direct sequence (DS/CDMA). In these schemes, each of N user groups is given its own code, $g_i(t)$, where $i = 1, 2, \ldots, N$. The user codes are approximately orthogonal, so that the cross-correlation of two different codes is near zero. The main advantage of a CDMA system is that all the participants can share the full spectrum of the resource asynchronously; that is, the transition times of the different users' symbols do not have to coincide.

A typical DS/CDMA block diagram is shown in Figure 10.25. The first block illustrates the data modulation of a carrier, $A \cos \omega_0 t$. The output of the data modulator belonging to a user from group 1, $s_1(t)$, is shown below. The waveform is very general in form; no restriction has been placed on the type of modulation that can be used.

$$s_1(t) = A_1(t) \cos [\omega_0 t + \phi_1(t)] \tag{10.37}$$

Next, the data-modulated signal is multiplied by the spreading signal $g_1(t)$ belonging to user group 1, and the resulting signal, $g_1(t)s_1(t)$, is transmitted over the channel. Simultaneously, users from group 2 through N multiply their signals by their own code functions. Frequently, each code function is kept secret, and its use is restricted to the community of authorized users. The signal present at the receiver is the linear combination of the emanations from each of the users. Neglecting signal delays, we show this linear combination below.

$$g_1(t)s_1(t) + g_2(t)s_2(t) + \cdots + g_N(t)s_N(t) \tag{10.38}$$

As mentioned earlier, multiplication of $s_1(t)$ by $g_1(t)$ produces a signal whose spectrum is the convolution of the spectrum of $s_1(t)$ with the spectrum of $g_1(t)$. Thus, assuming that the signal $s_1(t)$ is relatively narrowband compared with the code or spreading signal $g_1(t)$, the product signal $g_1(t)s_1(t)$ will have approximately the bandwidth of $g_1(t)$. Assume that the receiver is configured to receive messages from user group 1. Assume, too, that the $g_1(t)$ code, generated at the receiver, is perfectly synchronized with the received signal from a group 1 user. The first stage of the receiver multiplies the incoming signal of Equation (10.38) by $g_1(t)$. The output of the multiplier will yield the following terms:

Desired signal: $\qquad g_1^2(t)s_1(t)$

Plus a composite of

undesired signals: $\qquad g_1(t)g_2(t)s_2(t) + g_1(t)g_3(t)s_3(t)$

$$+ \cdots + g_1(t)g_N(t)s_N(t) \tag{10.39}$$

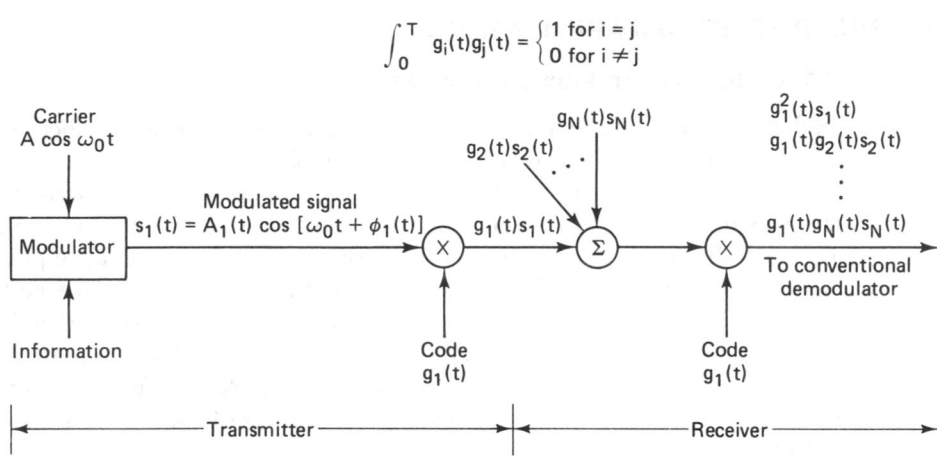

$$\int_0^T g_i(t)g_j(t) = \begin{cases} 1 \text{ for } i = j \\ 0 \text{ for } i \neq j \end{cases}$$

Figure 10.25 Code-division multiple access.

If the code functions, $\{g_i(t)\}$, are chosen with orthogonal properties, similar to Equation (10.14), the desired signal can be extracted perfectly in the absence of noise since $\int_0^T g_i^2(t) = 1$, and the undesired signals are easily rejected, since $\int_0^T g_i(t)g_j(t)\,dt = 0$ for $i \neq j$. In practice, the codes are not perfectly orthogonal; hence the cross-correlation between user codes introduces performance degradation, which limits the maximum number of simultaneous users.

Consider the frequency-domain view of the DS/CDMA receiver. Figure 10.26a illustrates the wideband input to the receiver; it consists of wanted and unwanted signals, each spread by its own code with code rate R_p, and each having a power spectral density of the form sinc^2 (f/R_p). Receiver thermal noise is also shown as having a flat spectrum across the band. The combined waveform of Equation (10.39) (desired plus undesired signals) is applied to the input of the receiver correlator driven by a synchronous replica of $g_1(t)$. Figure 10.26b illustrates the spectrum after correlation with the code $g_1(t)$ (despreading). The desired

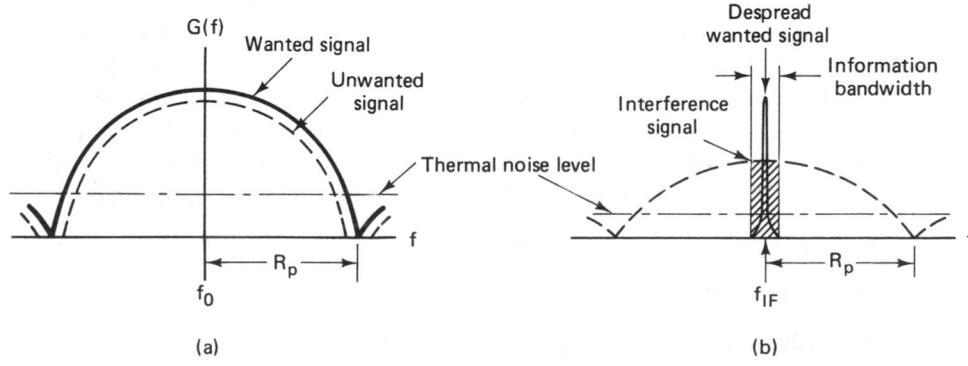

Figure 10.26 Spread-spectrum signal detection. (a) Spectrum at the input to receiver. (b) Spectrum after correlation with the correct and synchronized PN code.

signal, occupying the information bandwidth centered at an intermediate frequency (IF), is then applied to a conventional demodulator, with bandwidth just wide enough to accommodate the despread signal. The undesired signals of Equation (10.39) remain effectively spread by $g_1(t)g_i(t)$. Only that portion of the spectrum of the unwanted signals falling in the information bandwidth of the receiver will cause interference with the desired signal.

Pursley [17] presents an excellent treatment on the performance of SSMA using DS, taking correlation properties of the code sequences into account. Also, Geraniotis [18] and Geraniotis and Pursley [19, 20] evaluate the performance of FH and DS multiple access systems subject to interference.

10.6.2 Multipath Channels

Consider a DS binary PSK communication system operating over a multipath channel that has more than one path from the transmitter to the receiver. Such multiple paths may be due to atmospheric reflection or refraction, or reflections from buildings or other objects, and may result in fluctuations in the received signal level. The different paths may consist of several discrete paths each with a different attentuation and time delay, or they might consist of a continuum of paths. Figure 10.27 illustrates a communication link with two discrete paths. The multipath wave is delayed by some time, τ, compared to the direct wave. In television receivers, signals such as these cause "ghosts," or under extreme conditions, complete loss of picture synchronization.

In a direct-sequence spread-spectrum system, if we assume that the receiver is synchronized to the time delay and RF phase of the direct path, the received signal can be expressed as

$$r(t) = Ax(t)g(t) \cos \omega_0 t + \alpha Ax(t - \tau)g(t - \tau) \cos (\omega_0 t + \theta) + n(t) \quad (10.40)$$

where $x(t)$ is the data signal, $g(t)$ the code signal, $n(t)$ a zero-mean Gaussian noise process, and τ the differential time delay between the two paths, assumed to be in the interval $0 < \tau < T$. The angle θ is a random phase, assumed to be uniformly distributed in the range $(0, 2\pi)$, and α is the attenuation of the multipath signal

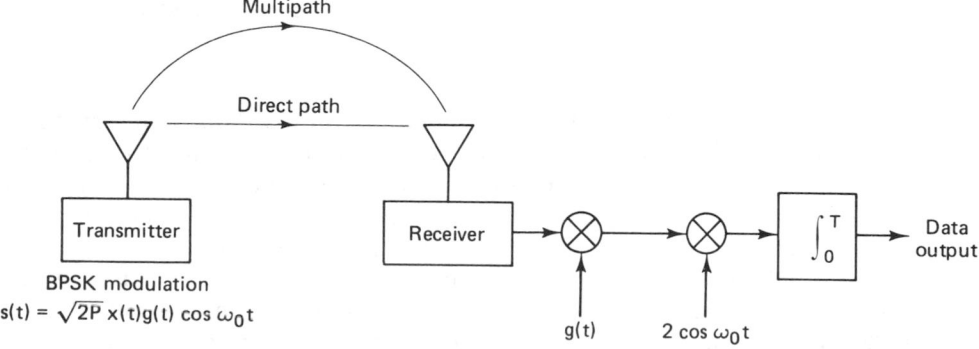

Figure 10.27 Direct-sequence BPSK system operating over a multipath channel.

relative to the direct path signal. For the receiver, synchronized to the direct path signal, the output of the correlator, $z(t = T)$, can be written as

$$z(t = T) = \int_0^T [Ax(t)g^2(t) \cos \omega_0 t$$

$$+ \alpha Ax(t - \tau)g(t)g(t - \tau) \cos (\omega_0 t + \theta) + n(t)g(t)]2 \cos \omega_0 t \, dt \quad (10.41)$$

where $g^2(t) = 1$. Also, for $\tau > T_c$, $g(t)g(t - \tau) \approx 0$ (for codes with long periods), where T_c is the chip duration. Therefore, if T_c is less than the differential time delay between the multipath and direct path signals, we can write

$$z(t = T) = \int_0^T 2Ax(t) \cos^2 \omega_0 t + 2n(t)g(t) \cos \omega_0 t \, dt = Ax(T) + n_0(T) \quad (10.42)$$

where $n_0(T)$ is a zero-mean Gaussian random variable. We see that the spread-spectrum system, similar to the case of CDMA, effectively eliminates the multipath interference by virtue of its code-correlation receiver.

If frequency hopping (FH) is used against the multipath problem, improvement in system performance is also possible but through a different mechanism. FH receivers avoid multipath losses by rapid changes in the transmitter frequency band, thus avoiding the interference by changing the receiver band position before the arrival of the multipath signal.

10.6.3 The Jamming Game

The goals of a jammer are to deny reliable communications to his adversary and to accomplish this at minimum cost. The goals of the communicator are to develop a jam-resistant communication system under the following assumptions: (1) complete invulnerability is not possible; (2) the jammer has a priori knowledge of most system parameters, such as frequency bands, timing, traffic, and so on; (3) the jammer has *no* a priori knowledge of the PN spreading or hopping codes. The signaling waveform should be designed so that the jammer cannot gain any appreciable jamming advantage by choosing a jammer waveform and strategy other than wideband Gaussian noise (i.e., being clever should gain nothing for the jammer). The fundamental design rule in specifying a jam-resistant system is to make it as costly as possible for the jammer to succeed in jamming the system.

10.6.3.1 Jammer Waveforms

There are many different waveforms that can be used for jamming communication systems. The most appropriate choice depends on the targeted system. Figure 10.28 shows power spectral density plots of examples of jammer waveforms versus a communicator's frequency hopped M-ary FSK (FH/MFSK) tone. The range of the abscissa represents the spread-spectrum bandwidth W_{ss}. The three columns in the figure represent three instances in time (three hop times) when symbols having spectra G_1, G_2, and G_3, respectively, are being transmitted. Figure 10.28a illustrates a relatively low-level noise jammer occupying the full spread-spectrum bandwidth. In Figure 10.28b the jammer strategy is to trade bandwidth

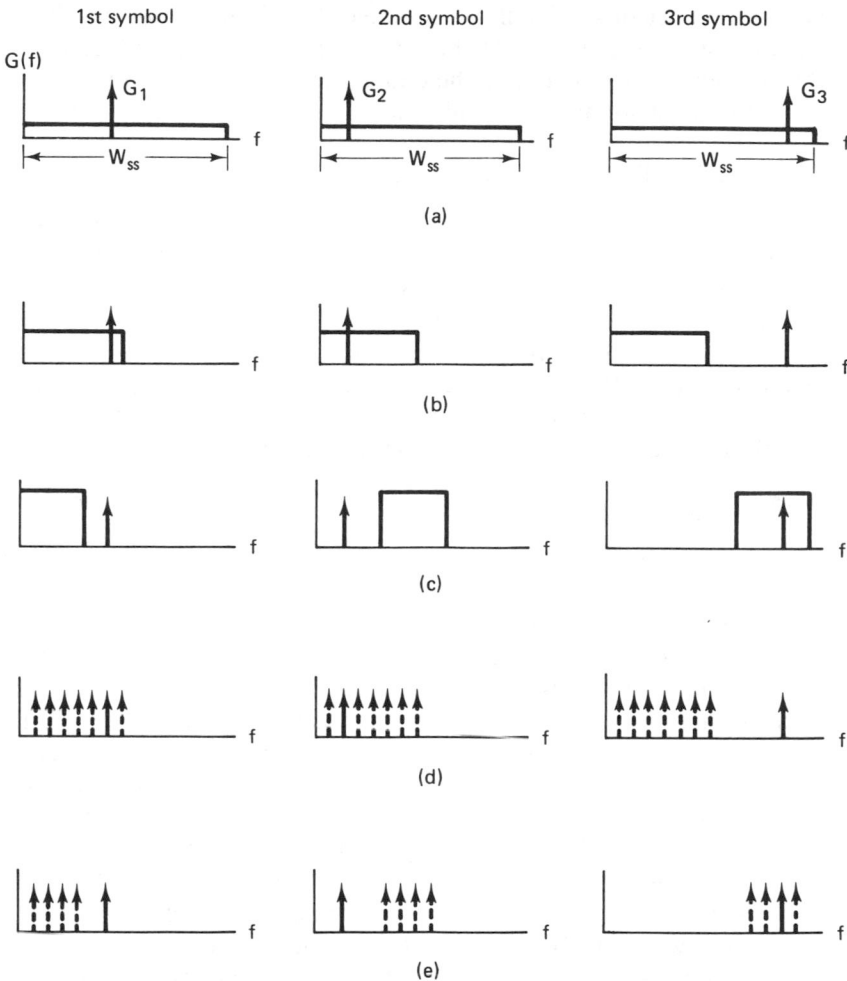

Figure 10.28 Jammer waveforms. (a) Full-band noise. (b) Partial-band noise. (c) Stepped noise. (d) Partial-band tones. (e) Stepped tones.

occupancy for greater power spectral density (the total power, or area under the curve, remains the same). The figure indicates that in this case, the jammer noise does not always share the same bandwidth region as the signal, but when it does, the effect can be destructive. In Figure 10.28c the noise jammer strategy is again to jam only part of the band, so that the jammer power spectral density can be increased, but in this case the jammer steps through different regions of the band at random times, thus preventing the communicator from using adaptive techniques to avoid the jamming. In Figure 10.28d and e the jammer uses a group of tones, instead of a continuous frequency band, in partial-band (Figure 10.28d) and stepped fashion (Figure 10.28e). This is a technique most often used against FH systems. Another jamming technique, not shown in Figure 10.28, is a pulse jammer, consisting of pulse-modulated bandlimited noise. Unless otherwise stated, we shall assume that the jammer waveform is wideband noise and that the jammer strategy is to jam the entire bandwidth W_{ss} continuously. The effects of partial band jamming and pulse jamming are considered later.

10.6.3.2 Tools of the Communicator

The usual design goal for an anti-jam (AJ) communication system is to force a jammer to expend its resources over (1) a wide-frequency band, (2) for a maximum time, and (3) from a diversity of sites. The most prevalent design options are (1) frequency diversity, by the use of direct-sequence and frequency hopping spread-spectrum techniques; (2) time diversity, by the use of time hopping; (3) spatial discrimination, by the use of a narrow-beam antenna which forces a jammer to enter the receiver via an antenna sidelobe and hence suffer, typically, a 20- to 25-dB disadvantage, and (4) combinations of the above.

10.6.3.3 *J/S* Ratio

In Chapter 4 we were concerned primarily with link error performance as a function of thermal noise interference. Emphasis was placed on the signal-to-noise ratio parameters—required E_b/N_0 and available E_b/N_0 for meeting a specified error performance. In this section we are similarly concerned with link error performance as a function of interference. However, here the source of interference is the noise power of a jammer in addition to thermal noise. Therefore, the SNR of interest is $E_b/(N_0 + J_0)$, where J_0 is the noise power spectral density due to the jammer. Unless otherwise specified, J_0 is assumed equal to J/W_{ss}, where J is the average received jammer power (jammer power referred to the receiver front end) and W_{ss} is the spread-spectrum bandwidth. Since the jammer power is generally much greater than the thermal noise power, the SNR of interest in a jammed environment is usually taken to be E_b/J_0. Therefore, similar to the thermal noise case, we define $(E_b/J_0)_{\text{reqd}}$ as the bit energy per jammer noise power spectral density *required* for maintaining the link at a specified error probability. The parameter E_b can be written as

$$E_b = ST_b = \frac{S}{R}$$

where S is the received signal power, T_b the bit duration, and R the data rate in bits/s. Then we can express $(E_b/J_0)_{\text{reqd}}$ as

$$\left(\frac{E_b}{J_0}\right)_{\text{reqd}} = \left(\frac{S/R}{J/W_{ss}}\right)_{\text{reqd}} = \frac{W_{ss}/R}{(J/S)_{\text{reqd}}} = \frac{G_p}{(J/S)_{\text{reqd}}} \tag{10.43}$$

where $G_p = W_{ss}/R$ is denoted the *processing gain*, and $(J/S)_{\text{reqd}}$ can be written

$$\left(\frac{J}{S}\right)_{\text{reqd}} = \frac{G_p}{(E_b/J_0)_{\text{reqd}}} \tag{10.44}$$

The ratio $(J/S)_{\text{reqd}}$ is a figure of merit that provides a measure of how *invulnerable* a system is to interference. Which system has better jammer-rejection capability: one with a larger $(J/S)_{reqd}$ or a smaller $(J/S)_{reqd}$? The *larger* the $(J/S)_{reqd}$, the *greater* is the system's noise rejection capability, since this figure of merit describes how much noise power relative to signal power is *required* in order to degrade the system's specified error performance. Of course, the communicator would like the communication system *not* to degrade at all.

Another way of describing the relationship in Equation (10.44) is as follows. An adversary would like to employ a jamming strategy that forces the effective $(E_b/J_0)_{reqd}$ to be as large as possible. The adversary may employ pulse, tone, or partial-band jamming rather than wideband noise jamming. A large $(E_b/J_0)_{reqd}$ implies a small $(J/S)_{reqd}$ ratio for a fixed processing gain. This may force the communicator to employ a larger processing gain to increase the $(J/S)_{reqd}$. The system designer strives to choose a signaling waveform such that the jammer can gain no special advantage by using a jamming strategy other than wideband Gaussian noise.

10.6.3.4 Anti-Jam Margin

Sometimes the $(J/S)_{reqd}$ ratio is referred to as the *anti-jam* (AJ) *margin* since it characterizes the system jammer-rejection capability. But this is not really a good use of the phrase since AJ margin usually means the safety margin against a *particular threat*. Using the same approach as in Chapter 4 (for calculating the margin against thermal noise), we can define the AJ margin M_{AJ}, as follows:

$$M_{AJ}(dB) = \left(\frac{E_b}{J_0}\right)_r (dB) - \left(\frac{E_b}{J_0}\right)_{reqd} (dB) \qquad (10.45)$$

where $(E_b/J_0)_r$ is the E_b/J_0 *actually received*. Following the same format as Equation (10.43), we can express $(E_b/J_0)_r$ as

$$\left(\frac{E_b}{J_0}\right)_r = \frac{G_p}{(J/S)_r} \qquad (10.46)$$

where $(J/S)_r$, or simply J/S, is the ratio of the actually received jammer power to signal power. We can now combine Equations (10.43), (10.45), and (10.46), as follows:

$$M_{AJ} (dB) = \frac{G_p}{(J/S)_r} (dB) - \frac{G_p}{(J/S)_{reqd}} (dB) \qquad (10.47)$$

$$= \left(\frac{J}{S}\right)_{reqd} (dB) - \left(\frac{J}{S}\right)_r (dB) \qquad (10.48)$$

Example 10.2   Satellite Jamming

Figure 10.29 illustrates a satellite jamming scenario. The airplane terminal is equipped with a frequency hopping (FH) spread-spectrum system transmitting with an $EIRP_T = 20$ dBW. The data rate is $R = 100$ bits/s. The jammer is transmitting wideband Gaussian noise, continually, with an $EIRP_J = 60$ dBW. Assume that $(E_b/J_0)_{reqd} = 10$ dB and that the path loss is identical for both the airplane terminal and the jammer.

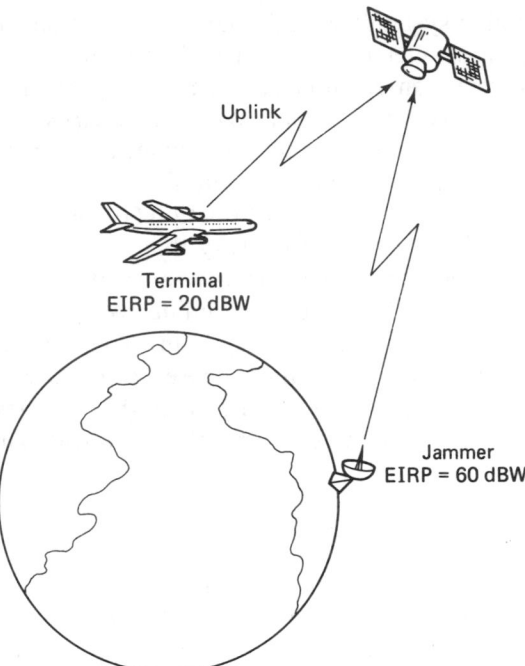

Uplink

Terminal
EIRP = 20 dBW

Jammer
EIRP = 60 dBW

Figure 10.29 Satellite jamming scenario.

(a) Should the communicators be concerned more with the jamming of the uplink or with that of the downlink?

(b) If it is desired to have an AJ margin of 20 dB, what should be the value of the hopping bandwidth W_{ss}?

Solution

(a) Jamming the uplink is of much greater concern, since such single-point interference could degrade the communications of a multitude of terminals that are simultaneously using the satellite transponder. To achieve an equivalent degradation by jamming the downlink, the jammer would have to jam each of the receiving terminals. Downlink jamming is of some concern for critical military missions, but of less concern than uplink jamming.

(b) With the assumption that the path loss is the same for both the communicator and the jammer, we can replace (J/S), in Equation (10.48) with the ratio of *transmitted* jammer-to-signal power, $EIRP_J/EIRP_T$. Therefore, we can write

$$M_{AJ} \text{ (dB)} = (J/S)_{\text{reqd}} \text{ (dB)} + EIRP_T \text{ (dBW)} - EIRP_J \text{ (dBW)}$$

$$= G_p \text{ (dB)} - \left(\frac{E_b}{J_0}\right)_{\text{reqd}} \text{ (dB)} + EIRP_T \text{ (dBW)} - EIRP_J \text{ (dBW)}$$

$$G_p = 20 \text{ dB} + 10 \text{ dB} - 20 \text{ dBW} + 60 \text{ dBW} = 70 \text{ dB}$$

$$W_{ss} = G_p \text{ (dB)} + R \text{ (dB-Hz)} = 70 \text{ dB} + 20 \text{ dB-Hz}$$

$$= 90 \text{ dB-Hz} = 1 \text{ GHz}$$

Example 10.3   Satellite Downlink Jamming

In Example 10.2 the distance from the transmitting airplane to the receiving satellite and the distance from the jammer to the satellite were assumed identical. Certainly, the closer the jammer gets to the receiver, the greater will be the jamming interference. Consider a downlink jamming scenario where the satellite $\text{EIRP}_s = 35$ dBW, the jammer $\text{EIRP}_J = 60$ dBW, the space loss from the satellite to the receiving terminal is $L_s = 200$ dB, and the space loss from the jammer to the receiving terminal is $L_s' = 160$ dB. How much processing gain is needed to close the link with an AJ margin of 0 dB? Assume that $(E_b/J_0)_{\text{reqd}} = 10$ dB.

Solution

For the downlink jamming scenario the proximity of the jammer to the receiving airplane is much closer than that of the satellite to the airplane. These distances show up as the space losses in the $(J/S)_r$ term of Equation (10.48), as follows:

$$M_{\text{AJ}} \text{ (dB)} = \left(\frac{J}{S}\right)_{\text{reqd}} \text{(dB)} - \left(\frac{J}{S}\right)_r \text{(dB)}$$

where

$$\left(\frac{J}{S}\right)_r \text{(dB)} = \text{EIRP}_J \text{ (dBW)} - L_s' \text{ (dB)} - \text{EIRP}_s \text{ (dBW)} + L_s \text{ (dB)}$$

and

$$\left(\frac{J}{S}\right)_{\text{reqd}} \text{(dB)} = \frac{W_{ss}}{R} \text{ (dB)} - \left(\frac{E_b}{J_0}\right)_{\text{reqd}} \text{ (dB)}$$

Combining the above equations, and solving for processing gain, $G_p = W_{ss}/R$, yields

$$G_p \text{ (dB)} = 75 \text{ dB}$$

10.7 FURTHER JAMMING CONSIDERATIONS

10.7.1 Broadband Noise Jamming

If the jamming signal is modeled as a zero-mean wide-sense-stationary Gaussian noise process with a flat power spectral density over the frequency range of interest, then for a fixed jammer received power, J, the jammer power spectral density J_0' is equal to J/W, where W is the bandwidth that the jammer chooses to occupy. If the jammer strategy is to jam the entire spread-spectrum bandwidth, W_{ss}, with its fixed power, the jammer is referred to as a wideband or *broadband jammer,* and the jammer power spectral density is

$$J_0 = \frac{J}{W_{ss}} \tag{10.49}$$

In Chapter 3 it was shown that the bit error probability P_B for a coherently demodulated BPSK system (without channel coding) is

$$P_B = Q\left(\sqrt{\frac{2E_b}{N_0}}\right) \tag{10.50}$$

where $Q(x)$ is defined in Equations (2.42) and (2.43) and tabulated in Table B.1. The single-sided noise power spectral density N_0 represents thermal noise at the front end of the receiver. The presence of the jammer increases this noise power spectral density from N_0 to $(N_0 + J_0)$. Thus the average bit error probability for a coherent BPSK system in the presence of broadband jamming is

$$P_B = Q\left(\sqrt{\frac{2E_b}{N_0 + J_0}}\right) = Q\left[\sqrt{\frac{2E_b/N_0}{1 + (E_b/N_0)(J/S)/G_p}}\right] \tag{10.51}$$

When P_B is plotted versus E_b/N_0 for a given J/S ratio, the resulting curves are such as those in Figure 10.30, [7, 21]. The curves in Figure 10.30, shown for two different values of processing gain, *tend to flatten out* as E_b/N_0 increases, indicating that for a given ratio of jammer power to signal power, the jammer will cause some irreducible error probability. The only way to reduce this error probability is to increase the processing gain.

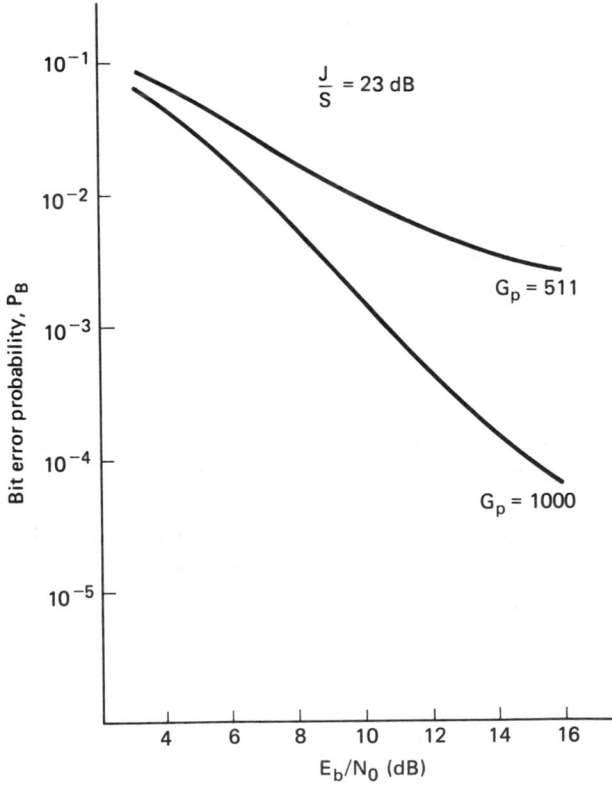

Figure 10.30 Bit error probability versus E_b/N_0 for a given J/S ratio. (Reprinted with permission from R. L. Pickholtz, D. L. Schilling, and L. B. Milstein, "Theory of Spread-Spectrum Communications—A Tutorial," *IEEE Trans. Commun.*, vol. COM30, no. 5, May 1982, Fig. 11, p. 866. © 1982 IEEE.)

10.7.2 Partial-Band Noise Jamming

A jammer can often increase the degradation to a FH system by employing *partial-band* jamming. Assuming that the frequency hopped modulation format is non-coherently detected binary FSK, the probability of a bit error, from Equation (3.111) is

$$P_B = \frac{1}{2} \exp\left(-\frac{E_b}{2N_0}\right) \tag{10.52}$$

Let us define a parameter, ρ, where $0 < \rho \le 1$, representing the fraction of the band being jammed. The jammer can trade bandwidth jammed for in-band jammer power, such that by jamming a band $W = \rho W_{ss}$, the jammer noise power spectral density can be concentrated to a level J_0/ρ, thus maintaining a constant average jamming received power J where $J = J_0 W_{ss}$.

In the case of partial-band jamming, a specific transmitted symbol will be received unjammed, with probability $(1 - \rho)$, and will be perturbed by jammer power with spectral density J_0/ρ, with probability ρ. Therefore, the average bit error probability can be written from Equation (10.52), as follows:

$$P_B = \frac{1 - \rho}{2} \exp\left(-\frac{E_b}{2N_0}\right) + \frac{\rho}{2} \exp\left[-\frac{E_b}{2(N_0 + J_0/\rho)}\right] \tag{10.53}$$

Since, in a jamming environment, it is often the case that $J_0 \gg N_0$, we can simplify Equation (10.53) to the form

$$P_B = \frac{\rho}{2} \exp\left(-\frac{\rho E_b}{2J_0}\right) \tag{10.54}$$

Figure 10.31 illustrates the probability of bit error versus E_b/J_0 for various values of the fraction, ρ. Clearly, the jammer would choose the fraction $\rho = \rho_0$ that maximizes P_B. Notice that ρ_0 decreases with increasing values of E_b/J_0 (see the ρ_0 locus in Figure 10.31). An expression for ρ_0 is easily found by differentiation (setting $dP_B/d\rho = 0$ and solving for ρ). This yields

$$\rho_0 = \begin{cases} \dfrac{2}{E_b/J_0} & \text{for } \dfrac{E_b}{J_0} > 2 \\[4mm] 1 & \text{for } \dfrac{E_b}{J_0} \le 2 \end{cases} \tag{10.55}$$

In this case, $(P_B)_{max}$ is given by

$$(P_B)_{max} = \begin{cases} \dfrac{e^{-1}}{E_b/J_0} & \text{for } \dfrac{E_b}{J_0} > 2 \\[4mm] \dfrac{1}{2} \exp\left(-\dfrac{E_b}{2J_0}\right) & \text{for } \dfrac{E_b}{J_0} \le 2 \end{cases} \tag{10.56}$$

where e is the base of the natural logarithm ($e = 2.7183$). This result is dramatic; the effect of a worst-case partial-band jammer on a system with spread spectrum *but without coding* changes the exponential relationship of Equation (10.54) into the inverse linear one of Equation (10.56). The ρ_0 locus in Figure 10.31 illustrates the P_B versus E_b/J_0 performance for the worst-case partial-band jammer. Here at 10^{-6} bit error probability there is over 40-dB difference between broadband noise jamming and the worst-case partial-band jamming for the same jamming

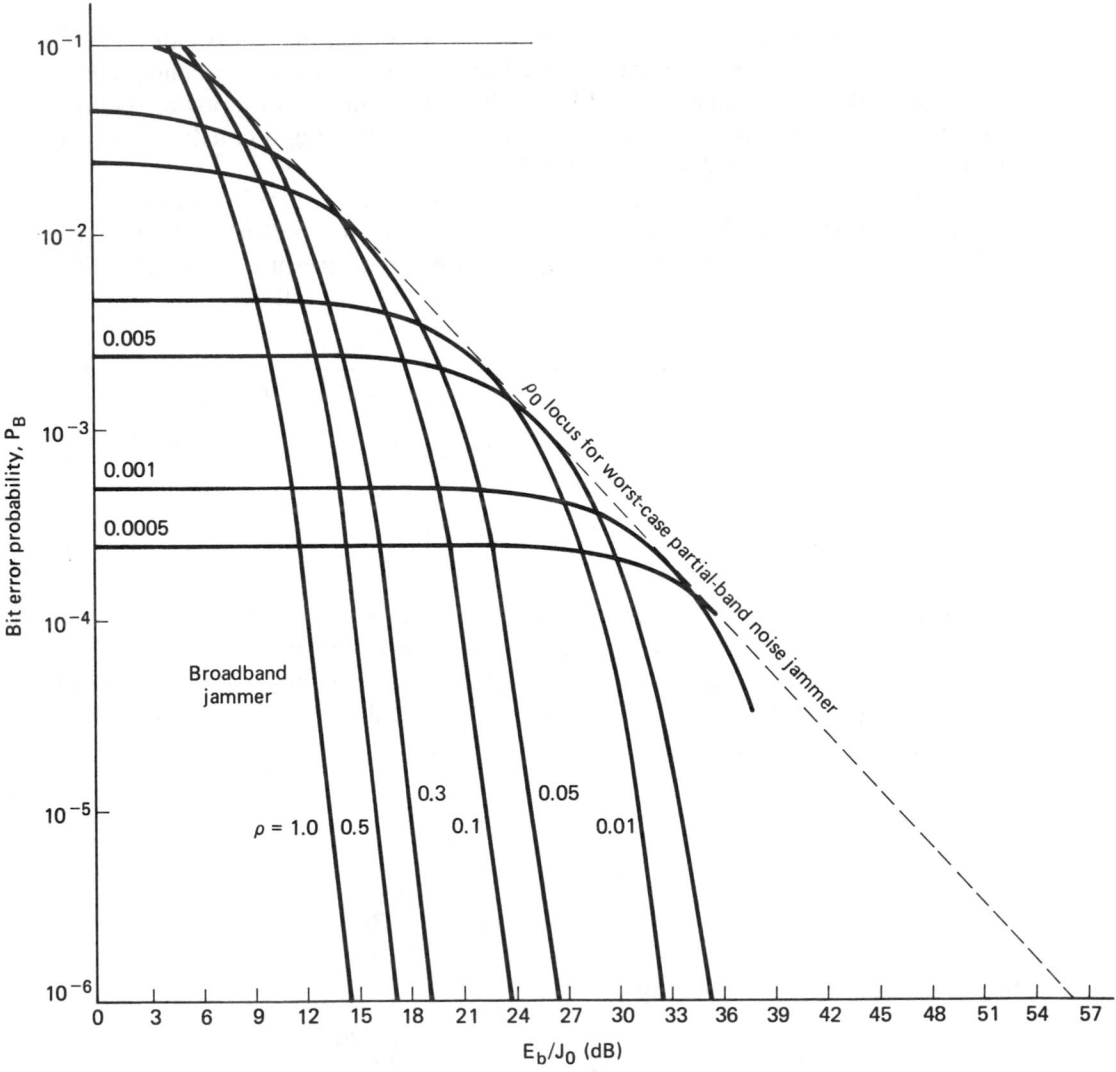

Figure 10.31 Partial-band noise jammer (FH/BFSK signaling). (Reprinted from M. K. Simon, J. K. Omura, R. A. Scholtz, and B. K. Levitt, *Spread Spectrum Communications*, Vol. 1, Fig. 3.24, p. 173. © 1985, with permission of the publisher, Computer Science Press, Inc., 1803 Research Blvd., Rockville, Md. 20850 USA.)

Spread-Spectrum Techniques   Chap. 10

power [6, 22]. Hence, an intelligent jammer, with fixed finite power, can produce significantly greater degradation with partial-band jamming than is possible with broadband jamming. Forward error correction (FEC) coding with appropriate interleaving can mitigate this degradation [9]. In fact, for codes with low-enough rates, FEC can *force* a partial-band jammer to be a worst-case jammer only when operating as a broadband jammer [23, 24].

10.7.3 Multiple-Tone Jamming

In the case of *multiple-tone jamming,* the jammer divides its total received power, J, into distinct, equal-power, random-phase CW tones. These are distributed over the spread-spectrum bandwidth, W_{ss}, according to some strategy [9]. The analysis of the effects of tone jamming is more complicated than that of noise jamming, especially for DS systems. Therefore, the effect of a despread tone is often approximated as Gaussian noise. Reference [25] provides analysis of the performance of DS systems in the presence of multiple-tone interference. For a noncoherent FH/FSK system operating in the presence of partial-band tone jamming, the performance is often assumed the same as that of partial-band noise jamming [26]. However, multiple-CW-tone jamming can be more effective than partial-

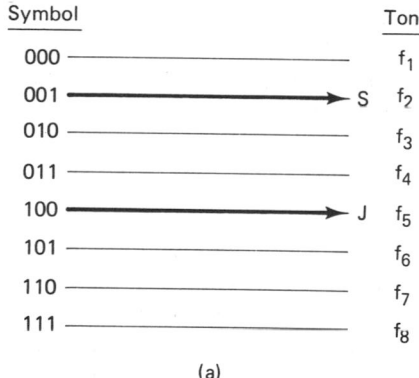

(a)

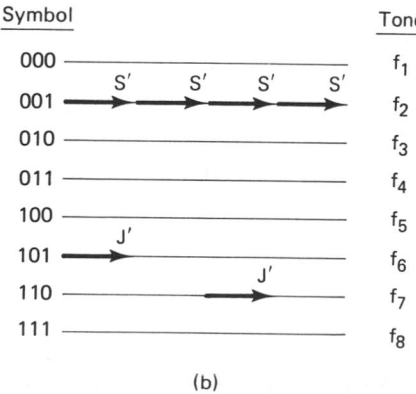

(b)

Figure 10.32 Fast hopping symbol repeat versus tone jamming. (a) One frequency hop. (b) Four frequency hops.

band noise against FH/MFSK signals because CW tones are the most efficient way for a jammer to inject energy into noncoherent detectors [9]. References [9, 10, 26, 27] provide extensive treatment and analysis of the performance of various communication systems in the presence of various types of jammers.

In the FFH/MFSK demodulator of Figure 10.16, a chip clipping circuit is shown between each envelope detector and accumulator. The function of such a circuit in a tone jamming environment can best be understood with the aid of the example shown in Figure 10.32. An 8-ary FSK frequency hopping system with no diversity, indicated in Figure 10.32a, is compared with a *fast*-frequency hopping system that combines chip repeating ($N = 4$ in this example) with the clipping of each chip, indicated in Figure 10.32b. Each row in the figures reprsents one of the $M = 8$ accumulators shown in Figure 10.16. The presence of a signal in the accumulator is indicated by a vector. In Figure 10.32a we see that, for a particular frequency hop, the data band is occupied by a received message symbol with received signal power, S. If, by chance, a jamming tone with received power J, where $J \geq S$, falls on a different tone within this data band during the same hop, the detector would not be able to decide reliably on the correct symbol.

In Figure 10.32b, the communicator's four chips (the length of each vector is a measure of the clipped signal power, S') sum to the maximum capacity of the accumulator. If the jammer tones, by chance, fall in the same spectral region as that of the signal, they will not confuse the detector, since the jamming tones are also clipped to the same level, $J' = S'$, as the signal chips. In Figure 10.32b, two of the jamming tones fall in the data band, but because they are clipped, there is no confusion about the correct symbol decision.

10.7.4 Pulse Jamming

Consider a spread-spectrum DS/BPSK communication system in the presence of a pulse-noise jammer. A pulse-noise jammer transmits pulses of bandlimited white Gaussian noise having a time-averaged received power, J, although the actual power during a jamming pulse duration is larger. Assume that the jammer can choose the center frequency and bandwidth of the noise to be the same as the receiver's center frequency and bandwidth. Assume also that the jammer can trade duty cycle for increased (concentrated) jammer power, such that if the jamming is present for a fraction $0 < \rho < 1$ of the time, then during this time, the jammer power spectral density is increased to a level J_0/ρ, thus maintaining a constant time-averaged power J (where $J = J_0 W_{ss}$ and W_{ss} is the system spread-spectrum bandwidth).

The bit error probability P_B for a coherently demodulated BPSK system (without channel coding) was given in Equation (10.50):

$$P_B = Q\left(\sqrt{\frac{2E_b}{N_0}}\right)$$

The single-sided noise power spectral density N_0 represents thermal noise at the front end of the receiver. The presence of the jammer increases this noise power spectral density from N_0 to $(N_0 + J_0/\rho)$. Since the jammer transmits with duty cycle ρ, the average bit error probability is

$$P_B = (1 - \rho)Q\left(\sqrt{\frac{2E_b}{N_0}}\right) + \rho Q\left(\sqrt{\frac{2E_b}{N_0 + J_0/\rho}}\right) \qquad (10.57)$$

We can generally assume that in a jamming environment, N_0 can be neglected. Therefore, we can write

$$P_B \simeq \rho Q\left(\sqrt{\frac{2E_b\rho}{J_0}}\right) \qquad (10.58)$$

The jammer will, of course, attempt to choose the duty cycle ρ that maximizes P_B. Figure 10.33 illustrates P_B for various values of ρ. The value of $\rho = \rho_0$ that maximizes P_B decreases with increasing values of E_b/J_0, as was the case with partial-band jamming. This is seen by differentiating Equation (10.58) to obtain [6]

$$\rho_0 = \begin{cases} \dfrac{0.709}{E_b/J_0} & \text{for } \dfrac{E_b}{J_0} > 0.709 \\[2ex] 1 & \text{for } \dfrac{E_b}{J_0} \leq 0.709 \end{cases} \qquad (10.59)$$

which results in the maximum bit error probability

$$(P_B)_{\max} = \begin{cases} \dfrac{0.083}{E_b/J_0} & \text{for } \dfrac{E_b}{J_0} > 0.709 \\[2ex] Q\left(\sqrt{\dfrac{2E_b}{J_0}}\right) & \text{for } \dfrac{E_b}{J_0} \leq 0.709 \end{cases} \qquad (10.60)$$

The effect of a worst-case pulse jammer upon a system with spread spectrum *but without coding* changes the complementary error function relationship of Equation (10.58) into the inverse linear one of Equation (10.60). As a result, at an error probability of 10^{-6}, there is almost a 40-dB difference in E_b/J_0 between the broadband jammer and the worst-case pulse jammer (see Figure 10.33). For the same jammer power, the jammer can do considerably more harm to an uncoded DS/BPSK system with pulse jamming than with constant power jamming. The effect of a pulse-noise jammer on uncoded DS/BPSK is similar to the effect of a partial-band noise jammer on uncoded FH/BFSK, treated in Section 10.7.2. In both cases considerable degradation is brought about by concentrating more jammer power on a fraction of the transmitted uncoded symbols. Forward error correction coding with appropriate interleaving can almost fully restore this degraded performance [9, 23–25, 28].

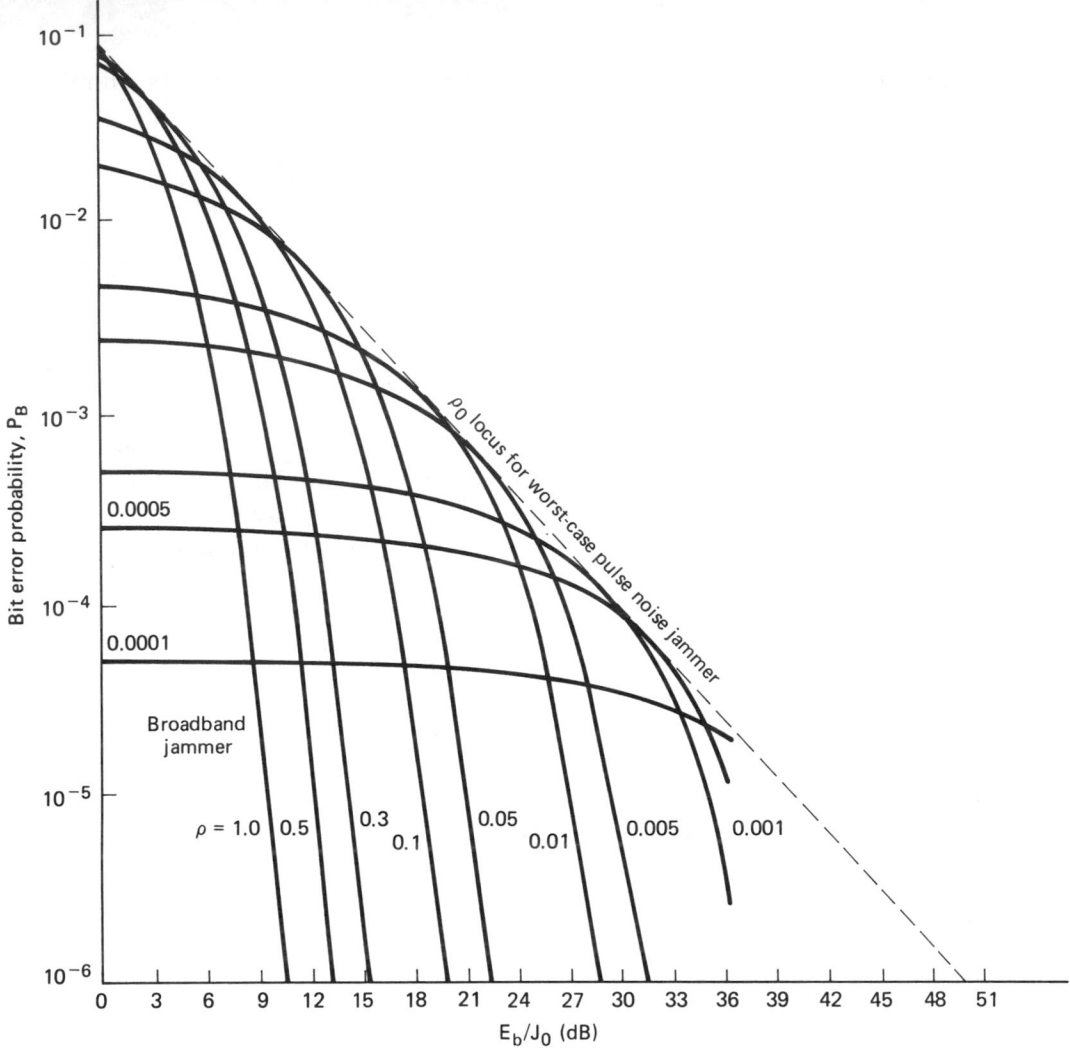

Figure 10.33 Pulse noise jammer (DS/BPSK signaling). (Reprinted from M. K. Simon, J. K. Omura, R. A. Scholtz, and B. K. Levitt, *Spread Spectrum Communications*, Vol. 1, Fig. 3.7, p. 150. © 1985, with permission of the publisher, Computer Science Press, Inc., 1803 Research Blvd., Rockville, Md. 20850 USA.)

10.7.5 Repeat-Back Jamming

In Examples 10.2 and 10.3 we considered an FH spread-spectrum system performance against a broadband Gaussian noise jammer. Notice that the frequency hopping rate did not enter into the margin computations. Isn't this disturbing? Intuitively, it would seem that the faster the frequency hops, the easier it is to "hide" the signal from the jammer. If the hopping rate truly does not enter into the computations, why not hop only once a day or once a week? The answer is that the meaure of jammer-rejection capability, namely processing gain, G_p, is based on the assumption that the jammer is a "dumb" jammer; that is, the jammer knows the extent of the spread-spectrum bandwidth, W_{ss}, but does *not* know the exact spectral location of the signal at any moment in time. We assume that the hopping rate is *fast enough* to preclude the jammer from monitoring the transmitted signal so as to usefully change this jamming strategy. Under what condition is this assumption questionable? There are "smart" jammers that are known as *repeat-back jammers* or *frequency-follower (FF) jammers*. These jammers monitor a communicator's signal (usually via a sidelobe beam from the transmitting antenna). They possess wideband receivers and high-speed signal processing capability that enable them to rapidly concentrate their jamming signal power in the spectral vicinity of a communicator's FH/FSK signal. By so doing, the smart jammer can increase the jamming power in the communicator's instantaneous bandwidth, thereby gaining an advantage over a wideband jammer. Notice that this strategy is useful only against frequency hopping signals. In direct-sequence systems, there is no instantaneous narrowband signal for the jammer to detect.

What can be done to defeat the repeat-back jammer? One method is to simply hop so fast that by the time the jammer receives, detects, and transmits the jamming signal, the communicator is already transmitting at a *new* hop (which of course will be unaffected by jamming at the frequency of the prior hop). The following example should make this point clear.

Example 10.4   Fast Hopping to Evade the Repeat-Back Jammer

Assume that a repeat-back jammer is located $d = 30$ km away from the communicator. Assume further that the jammer can monitor any uplink transmission from the communicator to a nearby satellite, as shown in Figure 10.34. How fast must the communicator hop his frequency to evade the repeat-back jammer? Assume that the jammer can change its jamming frequency in zero time, and that the only differential delay between the communicator's uplink signal and the jamming uplink signal is the propagation delay from the communicator to the jammer.

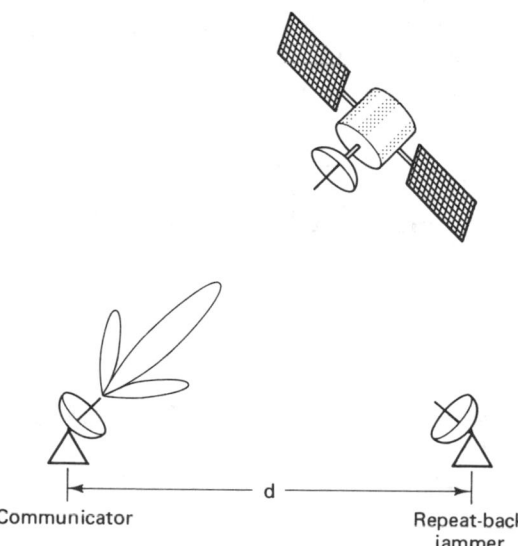

Communicator   Repeat-back   **Figure 10.34**   Example of fast hopping
                                      jammer   to evade the repeat-back jammer.

Solution

To ensure that the communicator's tone transmission and the jammer's attempt to disrupt that tone do not overlap in time, it is necessary that the duration of each hop, T_{hop}, have the value

$$T_{hop} \leq \frac{d}{c} = \frac{3 \times 10^4 \text{ m}}{3 \times 10^8 \text{ m/s}} = 10^{-4} \text{ s}$$

where c is the speed of light. Then $R_{hop} \geq 10,000$ hops/s.

10.7.6 BLADES System

Another technique capable of defeating the repeat-back jammer dates back to the mid-1950s when Sylvania engineers developed a system named the Buffalo Laboratories Application of Digitally Exact Spectra, or BLADES. The system used its code generator to independently select two new frequencies for each bit; the *final choice* of the frequency tone actually transmitted was dictated by the data bit about to be transmitted. Figure 10.35 illustrates a typical data stream of binary ones and zeros, called *marks* and *spaces,* respectively, and a sequence of frequency pairs f_1 and f'_1, f_2 and f'_2, ... The appearance of a mark dictates the choice of frequency f_i, while the appearance of a space dictates the choice of frequency f'_i. As shown in the figure, the data stream in this example gives rise to the sequence of transmitted tones, $f'_1, f_2, f'_3, f'_4, f_5$, and so on. How can such a system defeat a repeat-back jammer? The jammer monitors the transmissions and sends up energy in the neighborhood of the frequencies it perceives. The modulation of the BLADES system has no structure in the usual sense; either there *is* energy present or there is *no* energy present at a given frequency. The jammer sending narrowband energy in the same spectral neighborhood as the signal, is not destroying any modulation structure. For a noncoherent system, the jammer is only enhancing the communicator's signal. The only recourse for the repeat-back jammer is to change strategy by becoming a broadband jammer, and to jam the entire spread-spectrum bandwidth.

Notice that it is not really necessary to have a *pair* of frequencies for each

Mark amplitude

Space amplitude

Marks and spaces generated by independent synthesizers and PN codes:

f_1 f_2 f_3 f_4 f_5 f_6 f_7 f_8 f_9 f_{10} f_{11} f_{12} f_{13}

Mark waveforms:

f'_1 f'_2 f'_3 f'_4 f'_5 f'_6 f'_7 f'_8 f'_9 f'_{10} f'_{11} f'_{12} f'_{13}

Space waveforms:

Figure 10.35   BLADES system.

bit. A *single* frequency will do. The communicator then transmits the pseudo-random frequency for a binary one and sends nothing for a binary zero. The receiver has the same code generator and therefore monitors the same pseudorandom frequencies. A binary one is detected by virtue of energy at the monitored frequency, and a binary zero is known by a lack of energy at the monitored frequency. Of course, the system is not as robust as when the marks and spaces are each transmitted on independently selected frequencies.

10.8 CONCLUSION

Spread-spectrum (SS) technology has only emerged since the 1950s. Yet this novel approach to applications such as multiple access, ranging, and interference rejection has rendered SS techniques extremely important to most current NASA and military communication systems. In this chapter we presented an overview enumerating the benefits and types of spread-spectrum techniques, as well as some historical background.

Pseudorandom sequences are at the heart of all present-day SS systems; we therefore treated PN generation and properties. Emphasis was placed on the two major spread-spectrum techniques: direct sequence and frequency hopping. Also, consideration was given to synchronization, a crucial aspect of spread-spectrum operation. Some application examples were considered, such as code-division multiple access and communications with multipath conditions. Also, attention was devoted to the subject of jamming and jam-resistant systems, since this area represents one of the primary uses for spread-spectrum systems.

REFERENCES

1. Scholtz, R. A., "The Origins of Spread Spectrum Communications," *IEEE Trans. Commun.,* vol. COM30, no. 5, May 1982, pp. 822–854.

2. Shannon, C. E., "Communication in the Presence of Noise," *Proc. IRE,* Jan. 1949, pp. 10–21.

3. Dillard, R. A., "Detectability of Spread Spectrum Signals," *IEEE Trans. Aerosp. Electron. Syst.,* July 1979.

4. Simon, M. K., Omura, J. K., Scholtz, R. A., and Levitt, B. K., *Spread Spectrum Communications,* Vol. 3, Computer Science Press, Inc., Rockville, Md., 1985.

5. de Rosa, L. A., and Rogoff, M., Sec. I (Communications) of *Application of Statistical Methods to Secrecy Communication Systems,* Proposal 946, Fed. Telecommun. Lab., Nutley, N.J., Aug. 28, 1950.

6. Simon, M. K., Omura, J. K., Scholtz, R. A., and Levitt, B. K., *Spread Spectrum Communications,* Vol. 1, Computer Science Press, Inc., Rockville, Md., 1985.

7. Pickholtz, R. L., Schilling, D. L., and Milstein, L. B., "Theory of Spread-Spectrum Communications—A Tutorial," *IEEE Trans. Commun.,* vol. COM30, no. 5, May 1982, pp. 855–884.

8. Pickholtz, R. L., Schilling, D. L., and Milstein, L. B., Revisions to "Theory of Spread-Spectrum Communications—A Tutorial," *IEEE Trans. Commun.*, vol. COM32, no. 2, Feb. 1984, pp. 211–212.

9. Simon, M. K., Omura, J. K., Scholtz, R. A., and Levitt, B. K., *Spread Spectrum Communications,* Vol. 2, Computer Science Press, Inc., Rockville, Md., 1985.

10. Simon, M. K., and Polydoros, A., "Coherent Detection of Frequency-Hopped Quadrature Modulations in the Presence of Jamming: Part I. QPSK and QASK; Part II. QPR class I Modulation," *IEEE Trans. Commun.*, vol. COM29, Nov. 1981, pp. 1644–1668.

11. Ward, R. B., "Acquisition of Pseudonoise Signals by Sequential Estimation," *IEEE Trans. Commun.*, COM13, Dec. 1965, pp. 475–483.

12. Spilker, J. J., and Magill, D. T., "The Delay-Lock Discriminator—An Optimum Tracking Device," *Proc. IRE,* Sept. 1961.

13. Spilker, J. J., "Delay-Lock Tracking of Binary Signals," *IEEE Trans. Space Electron. Telem.,* Mar. 1963.

14. Simon, M. K., "Noncoherent Psuedonoise Code Tracking Performance of Spread Spectrum Receivers," *Commun.,* vol. COM25, Mar. 1977.

15. Ziemer, R. E., and Peterson, R. L., *Digital Communications and Spread Spectrum Systems,* Macmillan Publishing Company, New York, 1985.

16. Holmes, J. K., *Coherent Spread Spectrum Systems,* John Wiley & Sons, Inc., New York, 1982.

17. Pursley, M. B., "Performance Evaluation for Phase-Coded Spread-Spectrum Multiple-Access Communication: Part I. System Analysis," *IEEE Trans. Commun.*, vol. COM25, no. 8, Aug. 1977, pp. 795–799.

18. Geraniotis, E., "Noncoherent Hybrid DS-SFH Spread-Spectrum Multiple-Access Communications," *IEEE Trans. Commun.*, vol. COM34, no. 9, Sept. 1986, pp. 862–872.

19. Geraniotis, E., and Pursley, M. B., "Error Probability for Direct-Sequence Spread-Spectrum Multiple-Access Communications: Part I. Upper and Lower Bounds," *IEEE Trans. Commun.*, vol. COM30, no. 5, May 1982, pp. 985–995.

20. Geraniotis, E., and Pursley, M. B., "Error Probabilities for Direct-Sequence Spread-Spectrum Multiple-Access Communications: Part II. Approximations," *IEEE Trans. Commun.*, vol. COM30, no. 5, May 1982, pp. 996–1009.

21. Schilling, D. L., Milstein, L. B., Pickholtz, R. L., and Brown, R. W., "Optimization of the Processing Gain of an *M*-ary Direct Sequence Spread Spectrum Communication System," *IEEE Trans. Commun.*, vol. COM28, no. 8, Aug. 1980, pp. 1389–1398.

22. Viterbi, A. J., and Jacobs, I. M., "Advances in Coding and Modulation for Noncoherent Channels Affected by Fading, Partial Band, and Multiple Access Interference," in A. S. Viterbi, ed., *Advances in Communication Systems,* Vol. 4, Academic Press, Inc., New York, 1975.

23. Stark, W. E., "Coding for Frequency-Hopped Spread-Spectrum Communication with Partial-Band Interference: Part I. Capacity and Cutoff Rate," *IEEE Trans. Commun.*, vol. COM33, no. 10, Oct. 1985, pp. 1036–1044.

24. Stark, W. E., "Coding for Frequency-Hopped Spread-Spectrum Communication with Partial-Band Interference: Part II. Coded Performance," *IEEE Trans. Commun.*, vol. COM33, no. 10, Oct. 1985, pp. 1045–1057.

25. Milstein, L. B., Davidovici, S., and Schilling, D. L., "The Effect of Multiple-Tone

Interfering Signals on a Direct Sequence Spread Spectrum Communication System," *IEEE Trans. Commun.,* vol. COM30, Mar. 1982, pp. 436–446.

26. Milstein, L. B., Pickholtz, R. L., and Schilling, D. L., "Optimization of the Processing Gain of an FSK-FH system," *IEEE Trans. Commun.,* vol. COM28, July 1980, pp. 1062–1079.

27. Huth, G. K., "Optimization of Coded Spread Spectrum Systems Performance," *IEEE Trans. Commun.,* vol. COM25, Aug. 1977, pp. 763–770.

28. Viterbi, A. J., "Spread Spectrum Communications—Myths and Realities," *IEEE Commun. Mag.,* May 1979, pp. 11–18.

PROBLEMS

10.1. Explain why a maximal-length n-stage linear feedback shift register can produce a sequence with a period no greater than $2^n - 1$.

10.2. Show that in a maximal-length n-stage linear feedback shift register the output stage must always be an input to the feedback network.

10.3. Consider the DS/BPSK spread-spectrum transmitter of Figure 10.9a or b. Let $x(t)$ be the sequence 1 0 0 1 1 0 0 0 1, arriving at a rate of 75 bits/s, where the leftmost bit is the earliest bit. Let $g(t)$ be generated by the shift register of Figure 10.7, with an initial state of 1 1 1 1 and a clock rate of 225 Hz.
  (a) Sketch the final transmitted sequence $x(t)g(t)$.
  (b) What is the bandwidth of the transmitted (spread) signal?
  (c) What is the processing gain?
  (d) Suppose that the estimated delay, $\hat{T}_d$, of Figure 10.9c is too large by one chip time. Sketch the despread chip sequence.
  (e) Choose a decision rule for deciding on $\hat{x}(t)$ and identify the errors.

10.4. A total of 24 equal-power terminals are to share a frequency band through a code-division multiple access (CDMA) system. Each terminal transmits information at 9.6 kbits/s with a direct-sequence spread-spectrum BPSK modulated signal. Calculate the minimum chip rate of the PN code in order to maintain a bit error probability of 10^{-3}. Assume that the receiver noise is negligible with respect to the interference from the other users.

10.5. A feedback shift register PN generator produces a 31-bit PN sequence at a clock rate of 10 MHz. What are the equation and graphical form of the autocorrelation function and power spectral density of the sequence? Assume that the pulses have values of ± 1.

10.6. Consider an FH/MFSK system such as the one shown in Figure 10.11. Let the PN generator be defined by a 20-stage linear feedback shift register with a maximal length sequence. Each state of the register dictates a new center frequency within the hopping band. The minimum step size between center frequencies (hop to hop) is 200 Hz. The register clock rate is 2 kHz. Assume that 8-ary FSK modulation is used and that the data rate is 1.2 kbits/s.
  (a) What is the hopping bandwidth?
  (b) What is the chip rate?
  (c) How many chips are there in each data symbol?
  (d) What is the processing gain?

10.7. The block diagram of Figure 10.16 is described in Section 10.4.5 for a fast frequency hopping (FFH) demodulator. Draw a similar block diagram for a slow frequency hopping (SFH) demodulator, and explain how it would work.

10.8. Find the mean and the standard deviation of the time needed to acquire a 10-megachip/s BPSK modulated PN code sequence using a serial search where 100 chips are examined at a time. Assume that a correct detection results when all 100 received chips match the locally generated ones. The ratio of received chip energy to noise power spectral density is 9.6 dB, and the uncertainty time between the received and local code sequences is 1 ms. Assume that the probability of false lock (false alarm) is negligible.

10.9. There are 11 equal-power terminals in a CDMA communication system, transmitting signals toward a central node. Each terminal transmits information at 1 kbit/s on a 100-kbits/s direct-sequence spreading signal using BPSK modulation.
   (a) If receiver noise is negligible with respect to the interference from other users, what is the received ratio of bit energy to interference power spectral density (E_b/I_0) experienced by each user?
   (b) What is the effect on E_b/I_0 if all users double their output power?
   (c) If the users wish to expand their service to 101 equal-power users, what must be done to the spreading codes to maintain the original E_b/I_0 ratio?

10.10. A CDMA system uses direct-sequence modulation with a data bandwidth of 10 kHz and a spread bandwidth of 10 MHz. With only one signal being transmitted, the received E_b/N_0 is 16 dB.
   (a) If the required E_b/N_0 is 10 dB, how many equal-power users can share the band?
   (b) If each user's transmitted power is reduced by 3 dB, how many equal-power users can share the band?
   (c) What is the maximum number of users that can share the band?
   (d) How many equal-power users could share the band if they switch to TDMA with 98% efficient use of the communications resource?
   (e) Why is the answer to part (d) so much greater than the answer to part (c)? What is the disadvantage of TDMA compared to CDMA (i.e., what penalty is paid to accommodate more users)?

10.11. A DS/SS system is used to combat multipath. If the path length of the multipath wave is 100 m longer than that of the direct wave, what is the minimum chip rate necessary to reject the multipath interference?

10.12. A ground-to-synchronous satellite link must be closed in a jamming environment. The data rate is 1 kbit/s and the ground station has a 60-ft antenna. Antijam protection is provided by a 10-Mbits/s direct-sequence spread-spectrum code. The jammer has a 150-ft antenna and a transmitter with 400 kW of power. Assume equal space and propagation losses. How much power is required of the earth station transmitter to achieve an E_b/J_0 of 16 dB at the satellite receiver? Assume that the receiver noise is negligible.

10.13. Input data at 75 bits/s are channel encoded using a rate $\frac{1}{2}$ encoder. The coded bits are then modulated using 8-ary FSK. The FSK symbols are then spread by frequency hopping at a rate of 2000 hops/s.
   (a) What is the chip rate?
   (b) What is the order of diversity?
   (c) If there are two such signals, time-division multiplexed (TDM'd) on the channel

at the same hopping rate, how would this affect the chip rate, symbol rate, and order of diversity?

(d) If there are 80 such signals TDM'd on the channel, how would this effect the chip rate, symbol rate, and order of diversity?

10.14. A frequency hopping noncoherent binary FSK system operates at an E_b/N_0 of 30 dB with a hopping bandwidth of 2 GHz. Assume that no channel coding is used. A jammer operating over the same broadband bandwidth yields a received $J_0 = 100N_0$.

(a) What is the bit error probability, P_B?

(b) If the jammer becomes a partial-band jammer, what bandwidth should it occupy to be most effective?

(c) What is P_B as a result of such optimum partial-band jamming?

(d) What is the unjammed P_B?

10.15. A noncoherent frequency hopping 8-ary FSK system hops at 12,000 hops/s over a bandwidth of 1 MHz. The symbol rate is 3000 symbols/s. Assume that channel coding is not used. The signal power at the input of the receiver is 10^{-12} W. A partial-band noise jammer occupies 50 kHz (assumed to be entirely within the hopping bandwidth of the signal). The received jammer power is 10^{-11} W. Assume that the system temperature is 290 K. What is the probability of bit error?

10.16. A coherent DS/BPSK system is transmitting at a data rate of 10 kbits/s in the presence of a broadband jammer. Assume that the system does not use channel coding. Also assume that the propagation losses are the same for the system and the jammer.

(a) If the EIRP of the communicator is 20 kW and the EIRP of the jammer is 60 kW, calculate the required spread-spectrum bandwidth to achieve a bit error probability of $P_B = 10^{-5}$.

(b) If the jammer is a pulse jammer, calculate the pulse duty cycle that results in worst-case jamming. What is the value of P_B at this duty cycle?

10.17. A communicator intends to use frequency hopping at a hop rate of 10,000 hops/s to avoid a threat of repeat-back jamming.

(a) Ignoring the curvature of the earth, and assuming that the communicator is transmitting to a satellite at geosynchronous altitude (approximately 36,000 km) that is directly overhead, compute the *radius of vulnerability*, which is the radius outside of which the communicator is unconditionally safe from repeat-back jamming by a ground-based jammer.

(b) If the communicator knows that the jammer requires a minimum of 10 μs to identify the transmission frequency and tune the jammer output, compute the radius of vulnerability conditioned on this information.

10.18. Consider an airborne repeat-back jammer as shown in Figure P10.1. The communicator is using a FH/SS system. What is the minimum hop rate required in order that the repeat-back jamming does not degrade the message? What would be the minimum required hopping rate if the communicator and jammer switched positions (i.e., fixed land jammer and airborne communicator).

10.19. Spread-spectrum techniques can be used to meet government regulations regarding flux (power) density radiating the surface of the earth. If a satellite at synchronous altitude (36,000 km) transmits 4-kbits/s data using 100 W of EIRP, what spreading bandwidth is required to maintain a flux density on the earth's surface no greater than -151 dBW/m^2 in any 4-kHz band?

10.20. A communicator uses noncoherent BFSK modulation and frequency hopping to combat the effects of a jammer. The power of the communicator's signal at the receiver input is 10 μW. The SNR in the absence of jamming is assumed to be very large. The power of the jamming signal at the receiver input is 1 W.

(a) If the jammer jams the entire hopping bandwidth with equal amounts of Gaussian noise (the noise will be white within the band), what bandwidth expansion factor will allow the communicator to maintain a bit error probability of 10^{-4}?

(b) Assume that the jammer decides to "color" its jamming noise by reducing its energy by a fraction, α ($0 \leq \alpha \leq 1$), in half the hop bandwidth, and increasing it by a like amount in the other half (thereby keeping its transmitted energy constant). Assuming that the communicator does not modify his hopping pattern to avoid the jammer strategy, develop an expression for the bit error probability for this case of colored jamming.

(c) Determine the fraction, α, that is optimum from the jammer standpoint for each of the limiting cases (i) when the effective SNR is large and, (ii) when it is small.

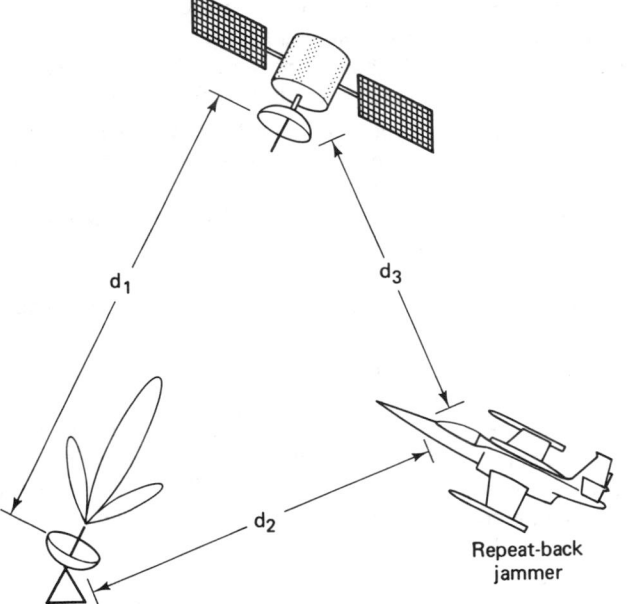

Communicator

d_1

d_3

d_2

Repeat-back jammer

Figure P10.1

CHAPTER 11

Source Coding

Fredric J. Harris
San Diego State University
San Diego, California

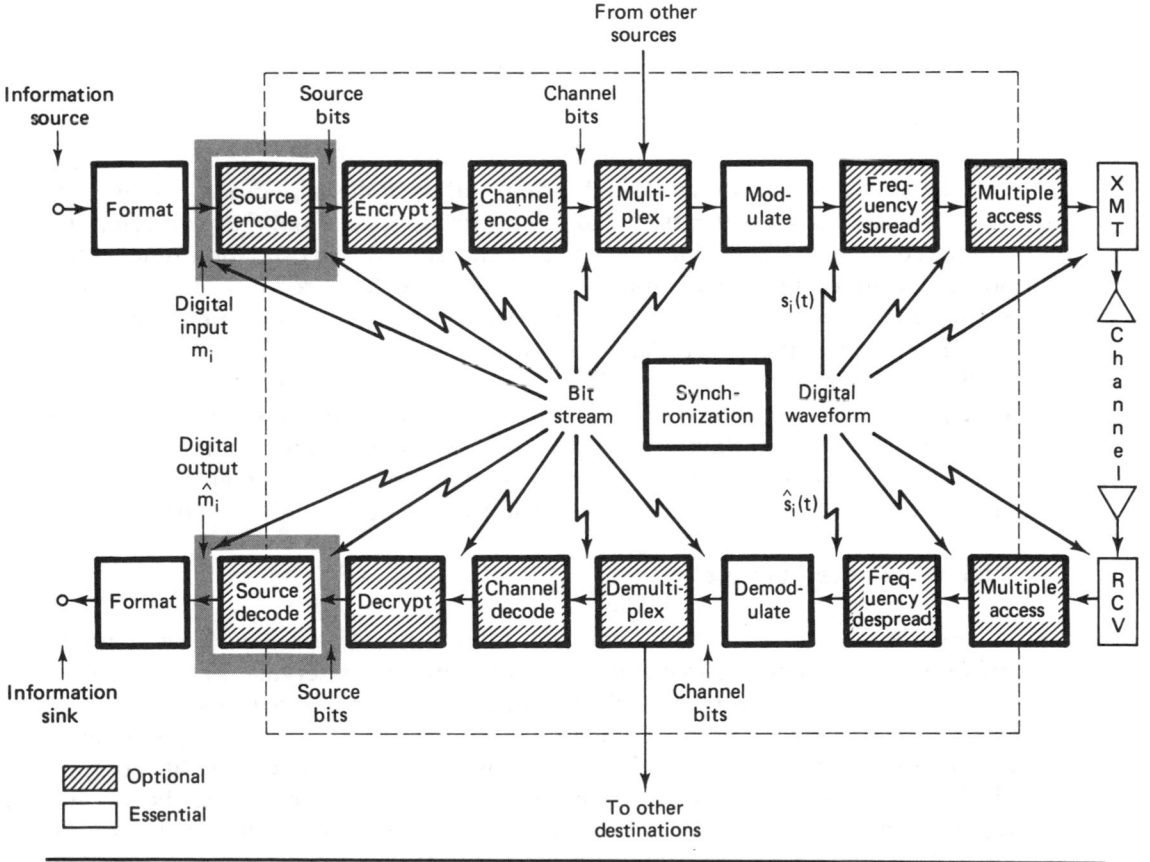

11.1 SOURCES

Source coding deals with the task of forming efficient descriptions of information sources. For discrete sources, the ability to form reduced data rate descriptions is related to the information content and the statistical correlation among the source symbols. For analog sources, the ability to form reduced data rate descriptions, subject to a fixed fidelity criterion, is related to the amplitude distribution and the temporal correlation of the source waveforms. The goal of source coding is either to improve the SNR for a given bit rate or to reduce the bit rate for a given SNR. To understand where the tools and techniques of source coding are effective, it is important to have common measures of source parameters. For this reason, in this section we examine simple models of discrete and analog sources and then describe how well real sources match the ideal models.

11.1.1 Discrete Sources

A discrete source generates (or emits) at a uniform rate a sequence of symbols $x(kT)$, selected from a source alphabet at discrete time intervals, kT, where $k = 1, 2, 3, \ldots$ is an index. If the alphabet contains a finite number of symbols, say N symbols, the source is said to be a *finite discrete source*. An example of such a source is the output of a 12-bit digital-to-analog converter (one of 4096 output levels) or the output of a 10-bit analog-to-digital converter (one of 1024 binary 10-tuples). Another example of a discrete source is the succession of 8-bit ASCII characters emitted by a remote computer terminal.

A finite discrete source is defined by the list of source symbols (sometimes called the alphabet) and the probability assignment to these symbols (or letters). We will assume that the source is short-term stationary, that is, that the probability assignment is fixed over the observation interval. An example in which the alphabet is fixed but the probability assignment changes is found in the sequence of symbols emitted by a terminal for someone typing English text followed by typing Spanish text and then again French text.

If we know that the probability of each symbol, X_j, is $P(X_j) = p_j$, we know the *self-information*, $I(X_j)$, for each symbol in the alphabet set.

$$I(X_j) = -\log_2 p_j \tag{11.1}$$

The average self-information for the symbols in an alphabet is denoted $H(X)$ and is called the *source entropy*.

$$H(X) = E\{I(X_j)\} = -\sum_{j=1}^{N} p_j \log_2 p_j \tag{11.2}$$

where $E\{X\}$ is the expected value of X. The source entropy is defined as the average amount of information per source output; it can be considered to be the average amount of uncertainty that is resolved by use of the alphabet. It is thus the average amount of information that must be moved through the communication channel to resolve that uncertainty. It can be shown that this amount of information in bits per symbol is bounded below by zero if there is no uncertainty, and is bounded above by $\log_2 N$ if there is maximum uncertainty.

$$0 \leq H(X) \leq \log_2 N \tag{11.3}$$

Example 11.1 Entropy of a Binary Source

Consider the binary source that generates independent symbols 0 and 1 with probabilities equal to p and $(1 - p)$, respectively. We described this source in Section 7.4.2 and presented its entropy function in Figure 7.4. If $p = 0.1$ and $(1 - p) = 0.9$, the source entropy is

$$H(X) = -[p \log p + (1 - p) \log (1 - p)] \tag{11.4}$$
$$= 0.47 \text{ bit/symbol}$$

Thus this source can be described (with the use of appropriate coding) with less than half a bit per symbol rather than with one bit per symbol in its present form.

We note that the first reason that source coding works is because the information content of an N-symbol alphabet used in real communication systems is always less than or equal to the upper bound of Equation (11.3). We know from experience, as we have noted in Example 7.1, that the symbols of English text are not all equally likely. For instance, we use the high probability of certain letters in text as part of the strategy to initialize the game of Hangman. (In this game a player must guess the letters, but not the positions, of a hidden word of known length. Penalties accrue to false guesses, and the letters of the entire word must be found prior to the occurrence of six false guesses.)

A discrete source is said to be *memoryless* if the sequence of symbols emitted by the source are statistically independent. In particular this means that for symbols taken two at a time, the joint probability of two elements is simply the product of their respective probabilities.

$$P(X_j, X_k) = P(X_j|X_k)P(X_k) = P(X_j)P(X_k) \qquad (11.5)$$

A result of statistical independence is that the information required to transmit a sequence of M symbols (called an M-tuple) from a given alphabet is precisely M times the average information required to transmit a single symbol. This happens because the probability of a statistically independent M-tuple is given by

$$P(X_1, X_2, \ldots, X_M) = \prod_{m=1}^{M} P(X_m) \qquad (11.6)$$

so that the average entropy per symbol, $H_M(X)$, of a statistically independent M-tuple is given by

$$H_M(X) = \frac{1}{M} E\{-\log_2 P(X_1, X_2, \ldots, X_M)\}$$

$$= \frac{1}{M} \sum_{X_m} [-P(X_m) \log_2 P(X_m)] \qquad (11.7)$$

$$= H(X)$$

A discrete source is said to have memory if the sequence of source elements are not independent. The dependency between symbols means that in a sequence of M symbols there is reduced uncertainty about the Mth symbol when we know the previous $(M - 1)$ symbols. For instance, is there much uncertainty in the next symbol for the 10-tuple CALIFORNI_? Thus the M-tuple with dependent symbols contains less information, or resolves less uncertainty, than does one with independent symbols. The entropy of a source with memory is the limit

$$H(X) = \lim_{M \to \infty} H_M(X) \qquad (11.8)$$

We observe that the entropy of an M-tuple from a source with memory is always less than the entropy of a source with the same alphabet and symbol probability but without memory.

$$H_M(X)_{\text{with memory}} < H_M(X)_{\text{with no memory}} \qquad (11.9)$$

For example, given a symbol (or letter) "q" in English text, we know that the next symbol will probably be a "u." Hence in a communication task, being told that the letter "u" follows a letter "q" adds little information to our knowledge of the word being transmitted. As another example, given the letters "th," the most likely symbols to follow are: a, e, i, o, u, r, and space. Thus adding the next symbol to the given set resolves some uncertainty but not much. A formal statement of this awareness is that the average entropy per symbol of an M-tuple from a source with memory *decreases as the length M increases*. Hence it is more efficient to encode symbols from a source with memory three at a time than it would be to encode them two at a time or one a time. Encoder complexity, memory constraints, and delay considerations require that practical source encoding be performed on finite-length sequences.

To help us understand the gains to be had in coding sources with memory, we form simple models of these sources. One such model is called a *first-order Markov source* [1]. This model identifies a number of states (or symbols in the context of information theory) and the conditional probabilities of transitioning to each next state. In the first-order model, the transition probabilities depend only on the present state. That is, $P(X_{i+1}|X_i, X_{i-1}, \ldots) = P(X_{i+1}|X_i)$. In the context of a binary sequence, this expression gives the probability of the state of the next digit conditioned on the state of the current digit.

Example 11.2 Entropy of a Binary Source with Memory

Consider the binary (i.e., two-symbol) first-order Markov source described by the state transition diagram shown in Figure 11.1. The source is defined by the state transition probabilities $P(0|1)$ and $P(1|0)$ of 0.45 and 0.05, respectively. The entropy of the source, X, is the weighted sum of the conditional entropies that correspond to the transition probabilities of the model.

$$H(X) = P(0)\ H(X|0) + P(1)\ H(X|1) \tag{11.10}$$

where

$$H(X|0) = -[P(0|0) \log_2 P(0|0) + P(1|0) \log_2 P(1|0)]$$

and

$$H(X|1) = -[P(1|1) \log_2 P(1|1) + P(0|1) \log_2 P(0|1)]$$

The a priori probability of each state is found by the total probability equations

$$P(0) = P(0)P(0|0) + P(1)P(0|1)$$

$$P(1) = P(1)P(1|1) + P(0)P(1|0)$$

$$P(0) + P(1) = 1$$

Solving for the a priori probabilities using the transition probabilities, we have

$$P(0) = 0.9 \quad \text{and} \quad P(1) = 0.1$$

Solving for the source entropy using Equation (11.10), we have

$$H(X) = P(0)H(X|0) + P(1)H(X|1)$$

$$= (0.9)(0.286) + (0.1)(0.993)$$

$$= 0.357 \text{ bit/symbol}$$

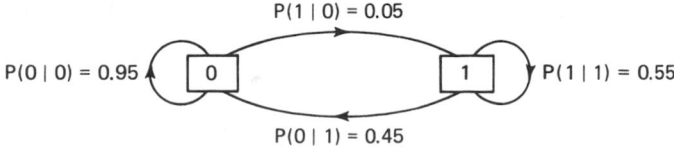

Figure 11.1 State transition diagram for a first-order Markov model.

Comparing this result to the result of Example 11.1, we see that the source with memory has a *lower entropy* than the source without memory even though the a priori symbol probabilities are the same.

Example 11.3 Extension Codes

The source alphabet for the binary Markov source of Example 11.2 consists of 0 and 1 and occur with probabilities of 0.9 and 0.1, respectively. Successive symbols are not independent and we can define a new set of code symbols as binary 2-tuples (*an extension code*) to take advantage of this dependency.

Binary 2-tuple	Extension symbol	Extension symbol probability
00	a	$P(a) = P(0\|0)P(0) = (0.95)(0.9) = 0.855$
11	b	$P(b) = P(1\|1)P(1) = (0.55)(0.1) = 0.055$
01	c	$P(c) = P(0\|1)P(1) = (0.45)(0.1) = 0.045$
10	d	$P(d) = P(1\|0)P(0) = (0.05)(0.9) = 0.045$

where the rightmost digit of the 2-tuple is the earliest digit. The entropy for this code extension alphabet is found by using an extension of Equation (11.10)) as follows:

$$H(\mathbf{X}_2) = P(a)H(\mathbf{X}_2|a) + P(b)H(\mathbf{X}_2|b) + P(c)H(\mathbf{X}_2|c) + P(d)H(\mathbf{X}_2|d)$$

$$H(\mathbf{X}_2) = 0.825 \text{ bit/output symbol}$$

$$= 0.412 \text{ bit/input symbol}$$

where $\mathbf{X}_k$ is the kth-order extension of the source, X. A longer extension code, which takes advantage of the adjacent symbol dependency, is of the form

Binary 3-tuple	Extension symbol	Extension symbol probability
000	a	$P(0\|00)P(00) = (0.95)(0.855) = 0.8123$
100	b	$P(1\|00)P(00) = (0.05)(0.855) = 0.0428$
001	c	$P(0\|01)P(01) = (0.95)(0.045) = 0.0428$
111	d	$P(1\|11)P(11) = (0.55)(0.055) = 0.0303$
110	e	$P(1\|10)P(10) = (0.55)(0.045) = 0.0303$
011	f	$P(0\|11)P(11) = (0.45)(0.055) = 0.0248$
010	g	$P(0\|10)P(10) = (0.45)(0.045) = 0.0248$
101	h	$P(1\|01)P(01) = (0.05)(0.045) = 0.0023$

Again, using an extension of Equation (11.10), the entropy for this extension code is found to be

$$H(\mathbf{X}_3) = 1.223 \text{ bits/output symbol}$$

$$= 0.408 \text{ bit/input symbol}$$

We note that the entropy of the one-symbol, two-symbol, and three-symbol descriptions of the source (0.470, 0.412, and 0.408 bit, respectively) are decreasing asymptotically toward the source entropy of 0.357 bit/input symbol. Remember that the source entropy is the lower bound in bits per input symbol for this (infinite memory) alphabet and this bound can only be approached asymptotically with finite-length coding.

11.1.2 Waveform Sources

A waveform source is a random process of some independent variable. We classically consider this variable to be time, so that the waveform of interest is a time-varying waveform. Important examples of time-varying waveforms are the outputs of transducers used in process control, such as temperature, pressure, velocity, displacement, and flow rates. Examples of particularly high interest include speech and music waveforms. The waveform can also be a function of one or more spatial variables (e.g., displacement). Important examples of spatial waveforms include single images such as a photograph, or moving images such as the successive images (at 24 frames/s) of moving picture film. Spatial images are often converted to time-varying functions by a simple scanning operation. This, for example, is done for facsimile transmission and with a slight modification (called interlacing) for standard broadcast television.

11.1.2.1 Amplitude Density Functions

Discrete sources were described by a list of their possible elements (called letters of an alphabet) and their multidimensional probability density functions (pdfs) of all orders. By analogy, waveform sources are similarly described in terms of their probability density functions as well as parameters and functions derived from these functions. We model many waveforms as random processes with classical probability distribution functions and with simple correlation properties. In the modeling process we distinguish between short-term or local (time) characteristics and long-term or global characteristics. This partition is necessary because many waveforms are nonstationary.

The probability density function of the actual process may not be available to the system designer. Sample density functions can, of course, be rapidly formed in real time during a short preceding interval and used as reasonable estimates over the present interval. A less ambitious task is simply to make estimates of short-term waveform-related averages. These include the sample mean (or time-average value), the sample variance (or mean-square value assuming zero mean), and correlation coefficients formed over the previous sample interval. In many applications of waveform analysis, the input waveform is converted to a zero-mean waveform by subtracting the estimates of the mean. This happens, for example, in a digital panel meter in which an auxiliary circuit measures the effects of the internal dc offset voltages and subtracts them in a process known as *auto-zero*. Further, the variance estimate is often used to scale the range of the input waveform to match the dynamic amplitude range of subsequent waveform-handling equipment. This process, performed in the digital panel meter, is called *autoranging* or *automatic gain control* (AGC). The function of these signal conditioning operations, mean removal and variance control (gain adjustment) shown in Figure 11.2, is to normalize the probability density functions of the input waveform. This normalization assures optimal utility of the limited dynamic range of subsequent recording, transmission, or processing subsystems.

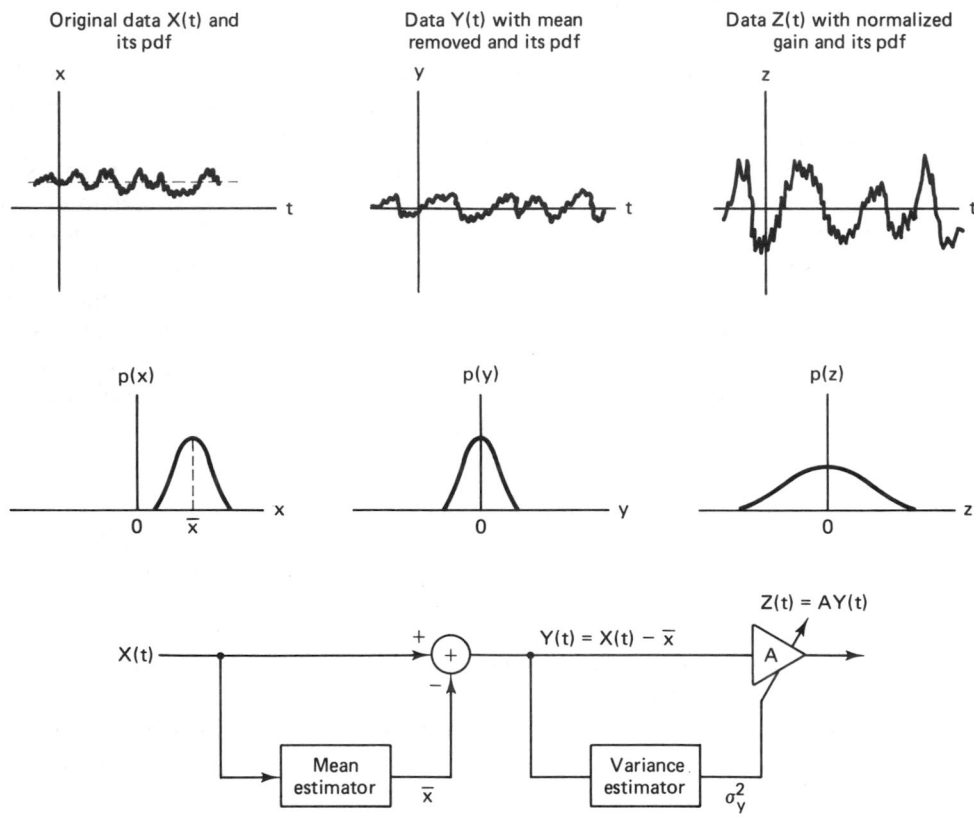

Figure 11.2 Mean removal and variance control (gain adjustment) for a data-dependent signal conditioning system.

11.1.2.2 Autocorrelation Function, Power Spectrum, and Models

There is significant correlation between the amplitudes of many waveform sources in successive time intervals. This correlation means that successive time samples are not independent. If the time sequence is truly independent, the autocorrelation function of the sequence would be an impulse function. The width of the autocorrelation function (in seconds) is called the correlation time of the process and is akin to the time constant of a filter. This time interval is an indication of how much shift along the time axis is necessary to find uncorrelated data samples. If the correlation time is large, we interpret this to mean that the waveform makes significant amplitude changes slowly. Conversely, if the correlation time is small, we infer that the waveform makes significant amplitude changes very quickly.

The Fourier transform of the autocorrelation function is the power spectral density of the waveform process. Thus an alternative description of the autocorrelation function, which reflects the amount of intersample dependence, is the degree of flatness in the waveform power spectrum. A flat spectrum, sometimes called a *white spectrum*, corresponds to source waveforms with independent values sample to sample. A power spectrum with a wide bandwidth implies a time function capable of rapid changes in envelope, while a power spectrum with a narrow bandwidth suggests a time function capable of only slow changes. In general, the larger the deviation from flatness, the more correlation will be found in the waveform samples. Very large changes from flatness in the power spectrum may warrant source descriptions which partition the spectrum, via filters, into subbands each of which is described and quantized separately.

11.2 AMPLITUDE QUANTIZING

Amplitude quantizing is the task of mapping samples of a continuous amplitude waveform to a finite set of amplitudes. The hardware that performs the mapping is the analog-to-digital converter (ADC or A-D). The amplitude quantizing occurs after the sample-and-hold operation. The simplest quantizer to visualize performs an instantaneous mapping from each continuous input sample level to one of the preassigned equally spaced output levels. Quantizers that exhibit equally spaced increments between possible quantized output levels are called *uniform quantizers* or sometimes *linear quantizers*. Possible instantaneous input–output characteristics are easily visualized by a simple staircase graph consisting of risers and treads of the types shown in Figure 11.3. Figure 11.3a, b, and d show quantizers with uniform quantizing steps, while Figure 11.3c is a quantizer with nonuniform quantizing steps. Figure 11.3a depicts a quantizer with *midtread* at the origin, while Figure 11.3b and d present quantizers with *midrisers* at the origin. A distinguishing property of midriser and midtread converters is related to the presence or absence, respectively, of output level changes when the input to the converter is idle noise. Further, Figure 11.3d presents a *biased* (i.e., truncation) quantizer, while the remaining quantizers in the figure are unbiased and are referred to as *rounding quantizers*. Most quantizers are truncation quantizers due to implementation considerations. The terms "midtread" and "midriser" are staircase terms used to describe whether the horizontal or vertical member of the staircase is at the origin. The unity-slope line passing through the origin represents the ideal nonquantized input–output characteristic we are trying to approximate with the staircase. The difference between the staircase and the unity-slope-line segment represents the approximation error made by the quantizer at each input level. Figure 11.4 illustrates the approximation error amplitude versus input amplitude function for each quantizer characteristic in Figure 11.3. Parts (a) through (d) of Figure 11.4 correspond to the same parts in Figure 11.3. This error is often modeled as quantizing noise because the error sequence obtained when quantizing a

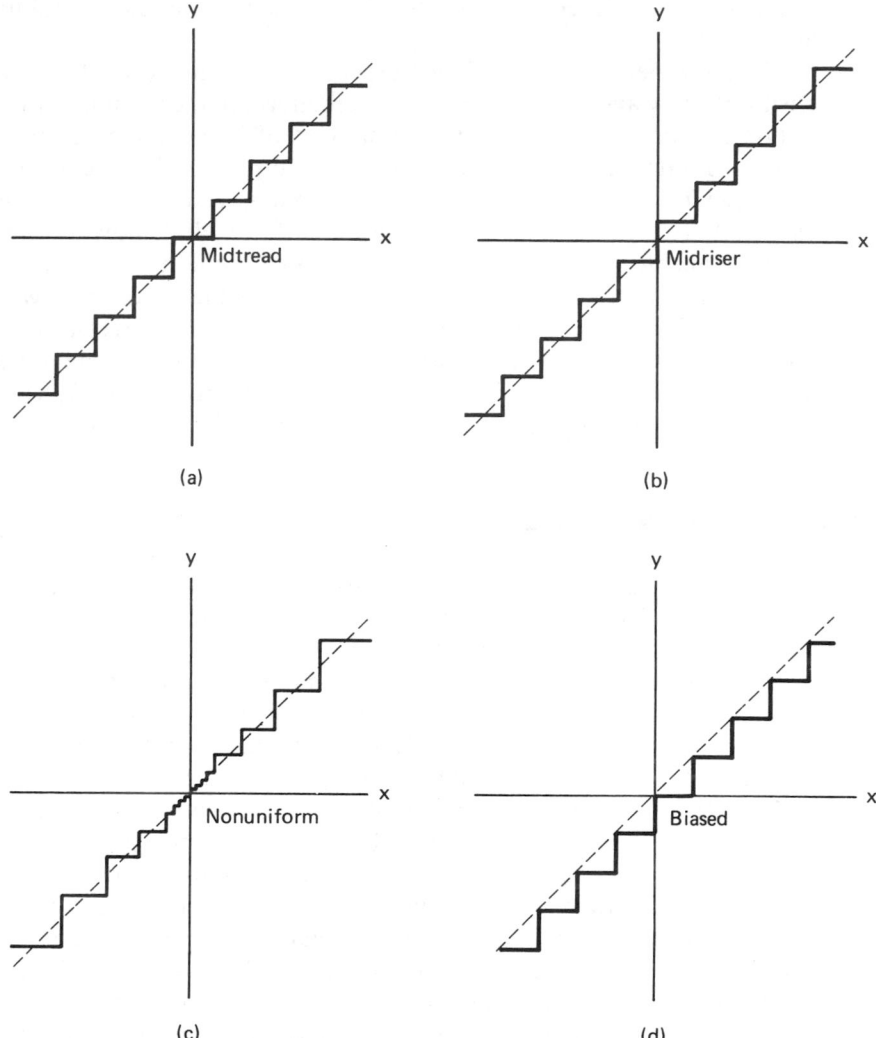

Figure 11.3 Various quantizer transfer functions.

wideband random process is reminiscent of an additive noise sequence. Unlike true additive noise sources, the quantizing errors are signal dependent and are highly structured. It is desirable to break up this structure; this can be accomplished by introducing an independent noise perturbation, known as *dither*, prior to the quantization step. This is discussed in Section 11.2.4.

The linear quantizer is simple to implement and is particularly easy to understand. It is the universal form of the quantizer in the sense that it makes no assumptions about the amplitude statistics and correlation properties of the input waveform, nor does it take advantage of user-related fidelity specifications. Quantizers that take advantage of these considerations are more efficient as source

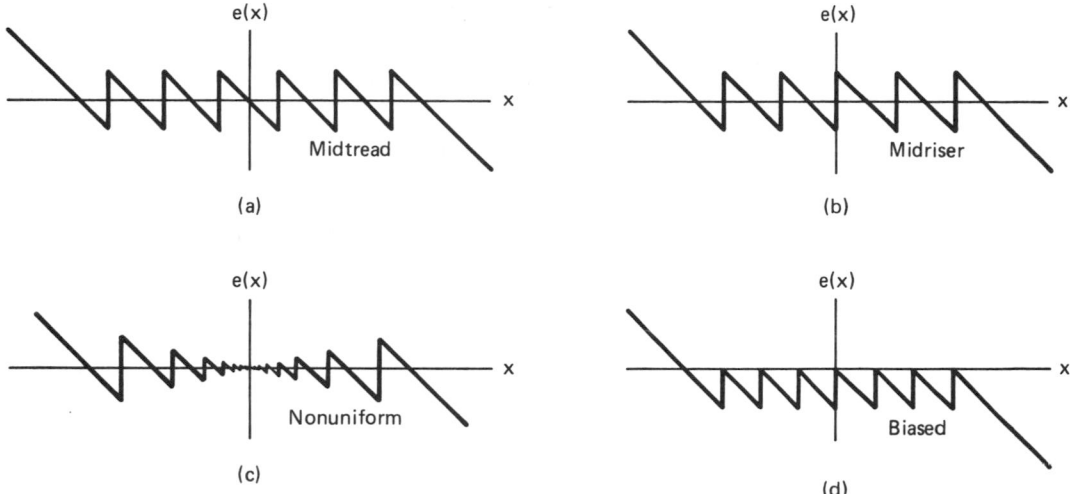

Figure 11.4 Instantaneous error voltage for various quantizer transfer functions.

coders and are more task specific then the general linear quantizer; these quantizers are often more complex and more expensive but are sometimes justified in terms of improved system performance. There are applications for which the uniform quantizer is the most desirable amplitude quantizer. These include signal processing applications, graphics and display applications, and process control applications. There are other applications for which nonuniform, adaptive quantizers are more desirable amplitude quantizers. These include waveform encoders for efficient storage and communication, contour encoders for images, vector encoders for speech, and analysis/synthesis encoders (such as the vocoder) for speech.

11.2.1 Quantizing Noise

The difference between the input and output of a quantizer is called the *quantizing error*. In Figure 11.5 we demonstrate the process of mapping the input sequence $X(t)$ to the quantized output sequence $\hat{X}(t)$. We can visualize forming $\hat{X}(t)$ by adding to each $X(t)$ an error sequence, $e(t)$.

$$\hat{X}(t) = X(t) + e(t)$$

The error sequence, $e(t)$, is deterministically defined by the input amplitude through the instantaneous error versus amplitude characteristic of the form in Figure 11.4. We note that the error sequence exhibits two distinct characteristics over different input operating regions.

The first operating interval is the *granular error region* corresponding to the sawtooth-shaped error characteristic. Within this interval the quantizer errors are confined by the size of the nearby staircase risers. The errors that occur in this region are called the granular errors or sometimes the quantizing errors. The input

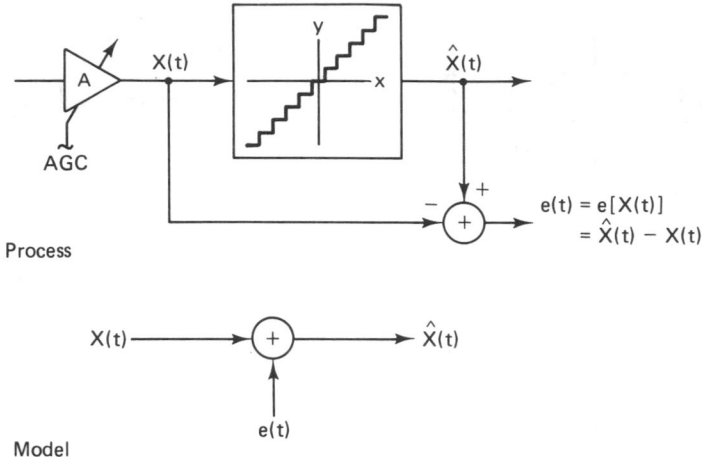

Process

Model

e(t)

Figure 11.5 Process and model of quantizing noise corruption of input signal.

interval for which the quantizing errors are granular defines the *dynamic range of the quantizer*. This interval is sometimes called the *region of linear operation*. Proper use of the quantizer requires that the input signal conditioning somehow match the dynamic range of the input signal to the dynamic range of the quantizer. This is a function of the signal-dependent gain control system called automatic gain control (AGC), indicated in Figure 11.5.

The second operating interval is the nongranular error region corresponding to the linearly increasing (or decreasing) error characteristic. The errors that occur in this interval are called *saturation or overload errors*. When the quantizer operates in this region, we say that the quantizer is saturated. Saturation errors are larger than the granular errors and may have a more objectionable effect on reconstruction fidelity.

The quantization error corresponding to each input amplitude represents an error or noise term associated with that input amplitude. Under the assumptions that the quantization interval is small compared to the dynamic range of the input signal and that the input signal has a smooth probability density function over the quantization interval, we can assume that the quantization errors are uniformly distributed over that interval, as illustrated in Figure 11.6. The pdf with zero mean corresponds to a rounding quantizer, while the pdf with a mean of $-q/2$ corresponds to a truncation quantizer.

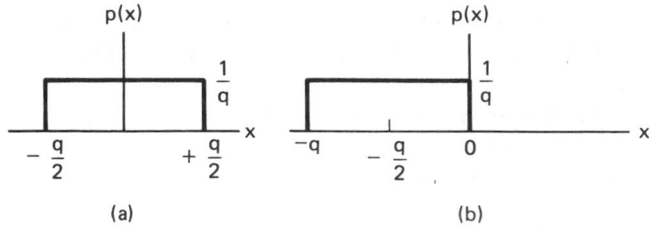

Figure 11.6 Probability density functions for quantizing error uniformly distributed over one quantile, q. (a) Probability density function for a rounding quantizer. (b) Probability density function for a truncating quantizer.

A quantizer or analog-to-digital converter (ADC) is defined by the number, size, and location of its quantizing step boundaries, and the corresponding step sizes. In a uniform quantizer, the step sizes are all equal and are equally spaced. The number of step sizes, N, is usually a power of 2 of the form $N = 2^b$, where b is the number of bits used in the conversion process. This number of levels is equally distributed over the dynamic range of the possible input levels. Normally, this range is defined as $\pm E_{max}$, such as ± 1.0 V or ± 5.0 V. Thus, accounting for the full range of $2E_{max}$, the size of a quantization step q is

$$q = \frac{2E_{max}}{2^b} \tag{11.11}$$

As an example, using Equation (11.11), the quantizing step hereafter called a *quantile*, for a 10-bit converter operating over the ± 1.0 V range is 1.953 mV. Occasionally, the operating range of a converter is altered so that the quantile is a "whole" number. For example, changing the operating range of the converter to ± 1.024 V results in a quantizing step size of 2.0 mV.

A useful figure of merit for the uniform quantizer is the quantizer output variance. If we assume that the quantization error is uniformly distributed over a single quantile interval q-wide, the quantizer variance, assuming zero-mean error, is found to be

$$\sigma^2 = \int_{-q/2}^{+q/2} e^2 p(e)\, de = \int_{-q/2}^{+q//2} e^2 \frac{1}{q}\, de = \frac{q^2}{12} \tag{11.12}$$

where $p(e) = 1/q$ is the probability density function of the quantization error. Thus the rms quantizer noise in a quantile interval of width q is found to be $q/\sqrt{12} = 0.29$ quantile. Equation (11.12) determines the quantizing noise power over one quantile, assuming that the errors are equiprobable over one quantile. In general, the quantiles are not equally sized over the range of input variables. We can account for this amplitude-dependent error by averaging the squared error over the amplitude variable and weighting by the probability of that amplitude. This is expressed by

$$\sigma_q^2 = E\{(x - q(x)]\}^2 = \int_{-\infty}^{+\infty} [e(x)]^2 p(x)\, dx \tag{11.13}$$

where x is the input variable, $q(x)$ is its quantized version, and $p(x)$ is the amplitude probability density function. We can partition the interval of integration in Equation (11.13) into two main intervals, one accounting for errors in the granular region of the quantizer and the second accounting for errors in the saturation region. We shall define the saturation amplitude of the quantizer as E_{max}. Also, we assume an odd symmetric transfer function for the quantizer, a symmetric pdf for the input signal, and redefine the error power to be the total error power, σ_D^2. We now write

$$\sigma_D^2 = 2\int_0^{+\infty} [e(x)]^2 p(x)\, dx \tag{11.14a}$$

$$= 2\int_0^{+E_{max}} [e(x)]^2 p(x)\, dx + 2\int_{+E_{max}}^{+\infty} [e(x)]^2 p(x)\, dx \tag{11.14b}$$

or
$$\sigma_D^2 = \sigma_q^2 + \sigma_s^2 \qquad (11.14c)$$

where σ_q^2 is the power in the linear region and σ_s^2 is the power in the saturation region. The noise power σ_q^2 can be further divided into subintervals corresponding to the successive discrete quantizer output levels (i.e., quantiles). If we assume that there are N such quantile levels, the integral becomes

$$\sigma_q^2 = \sum_{n=1}^{N} \int_{x_n}^{x_{n+1}} [e(x)]^2 p(x) \, dx \qquad (11.15)$$

If we now assume that the density function is approximately uniform over each quantile interval, Equation (11.15) simplifies to

$$\sigma_q^2 = \frac{1}{12} \sum_{n=1}^{N} [x_{n+1} - x_n]^2 \int_{x_n}^{x_{n+1}} p(x) \, dx \qquad (11.16)$$

11.2.2 Uniform Quantizing

If the quantizer has uniform quantiles equal to q, Equation (11.16) simplifies further to

$$\sigma_q^2 = \frac{1}{12} q^2 \, 2 \int_0^{Nq/2} p(x) \, dx \qquad (11.17)$$

where the limits of integration account for the $\pm N/2$ quantiles.

Noise power alone will not fully describe the noise performance of the quantizer. A more meaningful measure of quality is the ratio of output quantizing noise variance to input signal variance. Assuming that the input signal has zero mean, the signal variance is

$$\sigma_X^2 = \int_{-\infty}^{\infty} x^2 p(x) \, dx \qquad (11.18)$$

Further insight into the average quantizer noise requires that we examine a specific density function and a specific quantizer.

Example 11.4   Uniform Quantizer

Determine the quantizer variance for a signal that is uniformly distributed over the full dynamic range of a uniform quantizer with 2^b equally spaced quantile levels. In this case there is no saturation noise and only the granular noise term must be computed. Each quantile interval, denoted by q is

$$q = (2E_{max})2^{-b} \qquad (11.19)$$

where $2E_{max}$ is the input interval between the positive and negative boundaries of the linear quantizing range.

Solution

Substituting Equation (11.19) into Equation (11.12) or (11.17), we have the quantizing noise power (in the linear region):

$$\sigma_q^2 = \tfrac{1}{12}(2E_{\max}2^{-b})^2 = \tfrac{1}{3}(E_{\max})^2 2^{-2b} \tag{11.20}$$

The input signal power is found by performing the integration of Equation (11.18) for a uniform probability density function in the zero-mean interval spanning $2E_{\max}$. In this interval the density is $1/(2E_{\max})$, so the signal variance is found to be

$$\sigma_X^2 = \int_{-E_{\max}}^{+E_{\max}} \frac{1}{2E_{\max}} x^2 \, dx = \frac{1}{3}(E_{\max})^2 \tag{11.21}$$

Taking the ratio of noise power to signal power (NSR), we have

$$\text{NSR} = \frac{\sigma_q^2}{\sigma_X^2} = 2^{-2b} \tag{11.22}$$

Now converting the NSR to decibels, we have

$$\text{NSR}_{dB} = 10 \log_{10} \text{NSR} = 10 \log_{10} 2^{-2b} \tag{11.23a}$$

$$= -20b \log_{10} 2 = -6.02b \text{ (dB)} \tag{11.23b}$$

For our example, Equation (11.23b) suggests that each bit used in the conversion process is worth -6.02 dB in noise-to-signal ratio. In fact, the NSR for any uniform quantizer, not operating in saturation, is of the form

$$\text{NSR}_{dB} = -6.02b + C \tag{11.24}$$

where the term C depends on the signal density function and the ratio of signal standard deviation to quantizer saturation level.

Figure 11.7 presents the discrete Fourier transform of two sinusoids that have been sampled by a linear 10-bit ADC. The two sinusoids have relative amplitudes of 1.0 and 0.01 (i.e., one is reduced 40 dB with respect to the other). In Figure 11.7a the input signal is scaled to full dynamic range of the 10-bit converter. With the full-scale sinusoid and quantizer range set to 1.0, the 10-bit quantizer exhibits a quantile of amplitude 0.001953 and an rms quantization noise amplitude of 0.000564 for a peak signal-to-noise ratio (SNR) of 65 dB. The discrete Fourier transforms performed for Figure 11.7 were of length 256, and since the SNR of a transform increases proportional to integration time, there is an improvement in SNR of 21 dB due to the transform [2]. Thus, at the transform output, the average SNR due to quantizing is 86 dB. There are significant variations about the expected noise power level, typically on the order of 9 dB. Thus the minimum SNR at the transform is on the order of 77 dB, which is the range we observe in Figure 11.7. In Figure 11.7b and c, the input signal is attenuated by 20 and 40 dB, respectively, relative to the full-scale levels of Figure 11.7a. Note that the higher-frequency sinusoidal input signal in Figure 11.7c, now attenuated 80 dB relative to full scale, is at the noise level of the converter and is lost. The low-frequency sinusoid in Figure 11.7c is now attenuated 40 dB relative to full scale, and therefore exhibits a SNR that is 40 dB lower than the same signal in Figure 11.7a.

Given the task of minimizing the average quantizing noise-to-signal ratio in a quantizer, we are faced with a conflict of requirements. On one hand, we wish to keep the signals large with respect to the quantizing level, q, in order to achieve

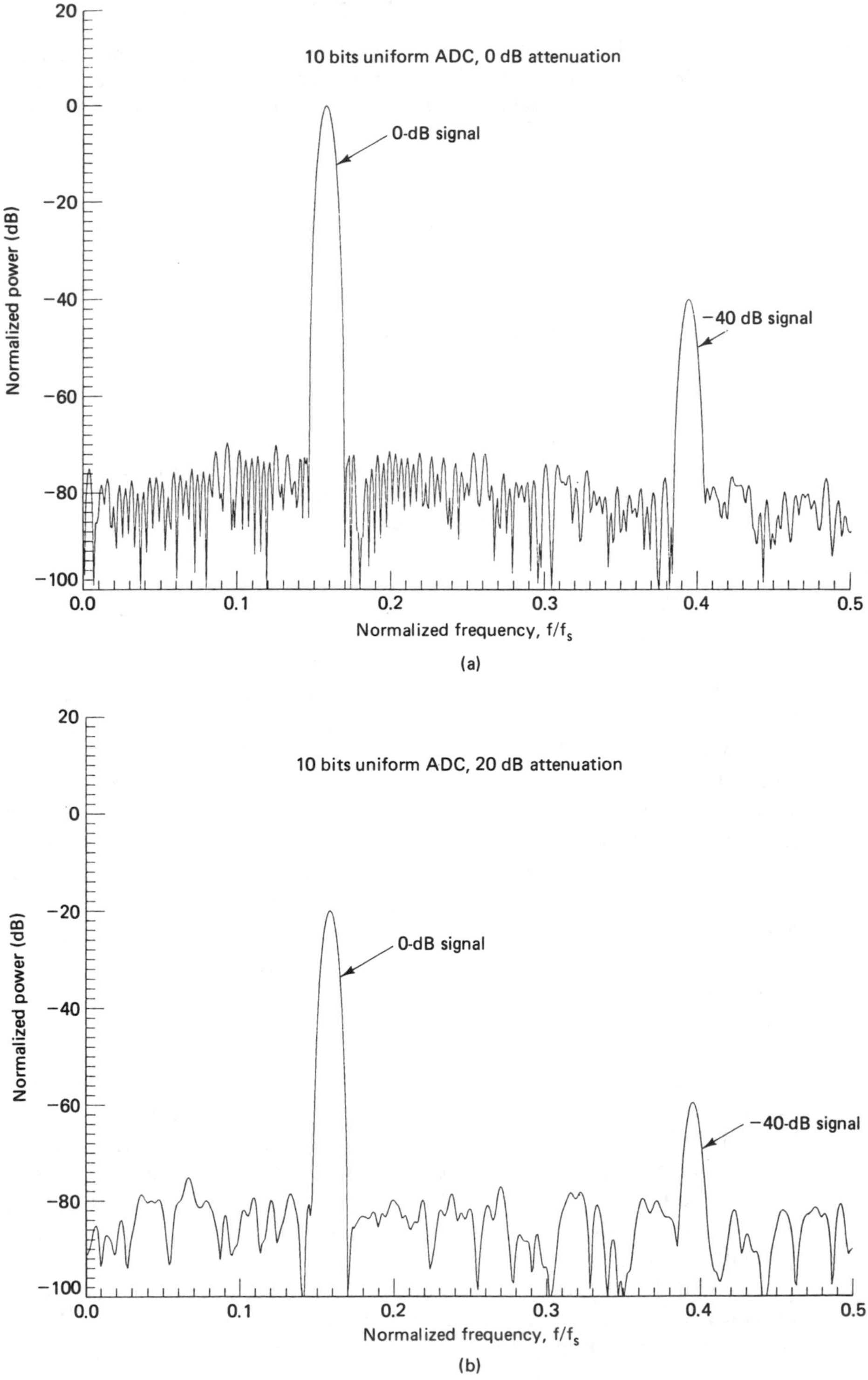

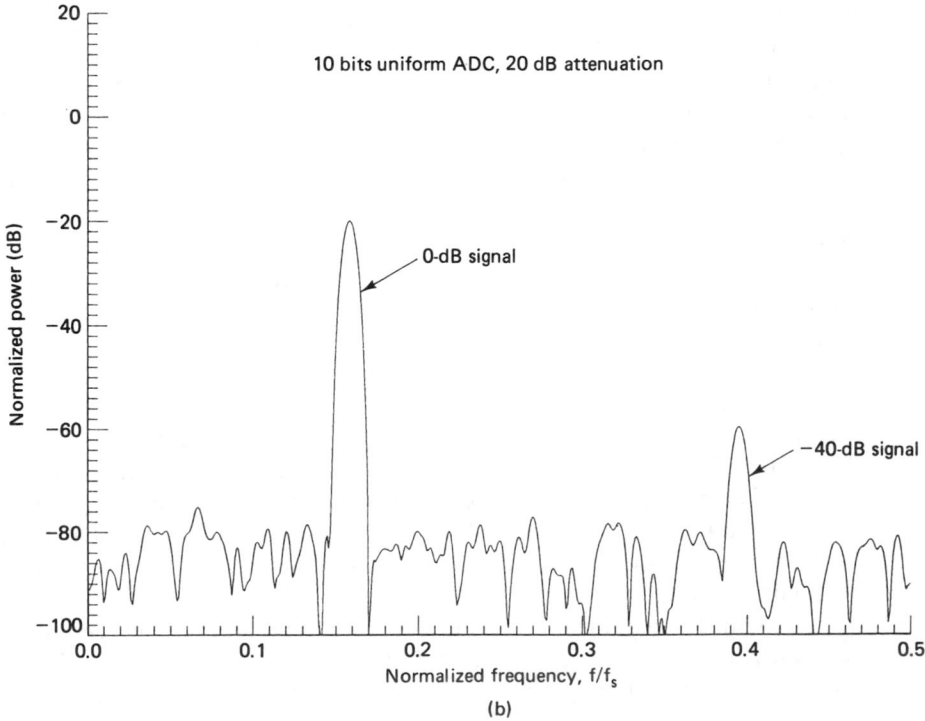

Figure 11.7 Power spectrum of uniformly quantized signals.

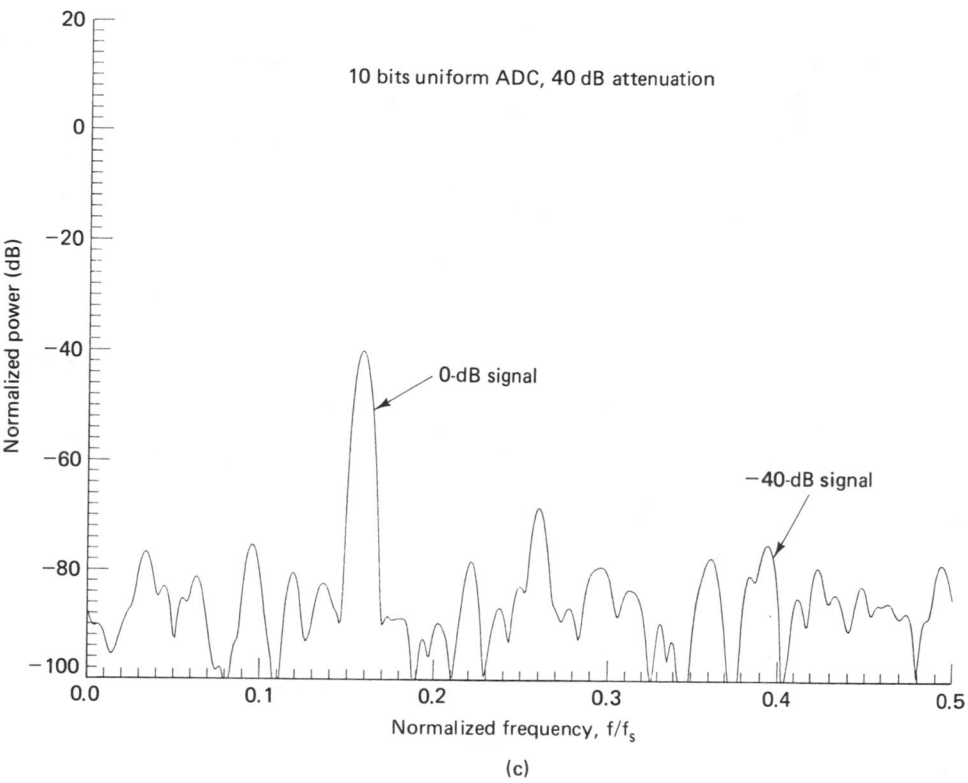

Figure 11.7 (*Continued*)

a high average signal-to-noise ratio. We also find it necessary to keep the signal small to avoid saturating the quantizer. We resolve the opposing requirements by scaling the input signal so that its rms value is a specified fraction of the full-scale quantizer range. The specified fraction is chosen to balance the saturation errors (weighted by their probability of occurrence) against the quantizing errors (which are similarly weighted) and thus achieve a minimum noise-to-signal ratio.

11.2.3 Saturation

Figure 11.8 presents the average NSR of a uniform quantizer as a function of the ratio of quantizer saturation level to rms value of input signal. The figure dramatically demonstrates that saturation noise is more severe than is quantizing noise. This can be simply explained by examining the instantaneous error characteristic, as shown in Figure 11.4, and noting that saturation errors are very large relative to the quantizing errors. Thus a small amount of saturation, even if it occurs infrequently, will make a large contribution to the average noise levels of the quantizer.

Saturation noise and quantization noise differ in another important way. *Quantization noise* tends to be white noise. Dither signals may be intentionally

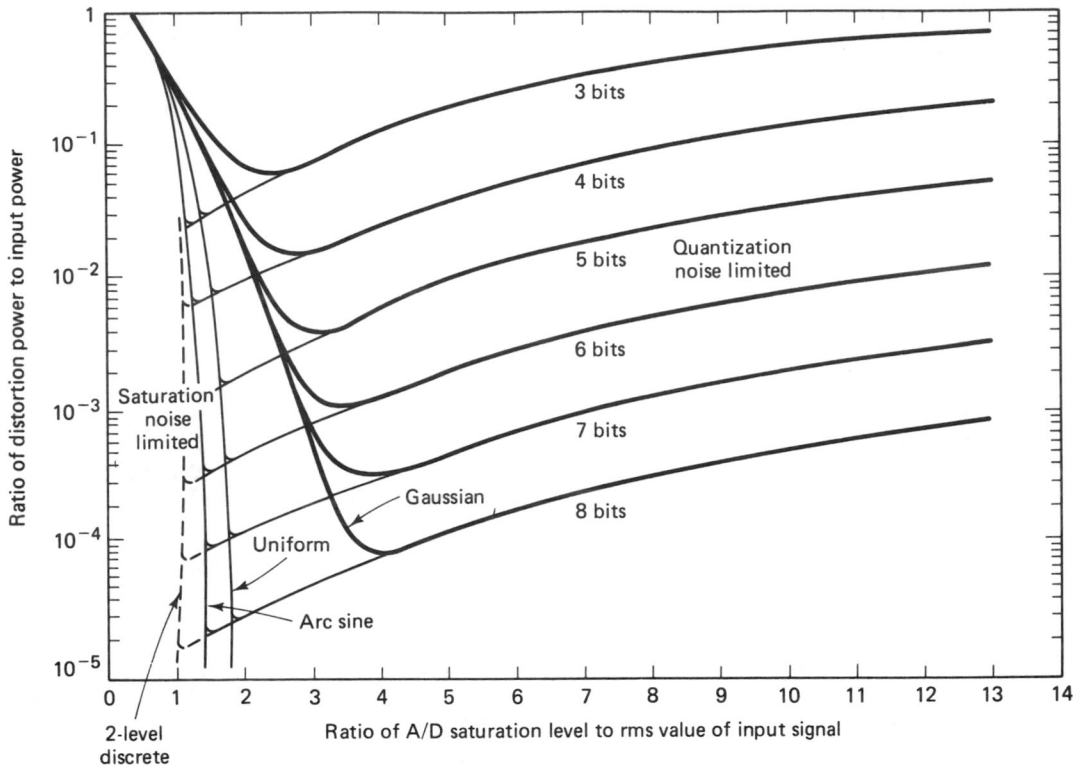

Figure 11.8 Quantizer NSR parameterized on number of bits in ADC for signals with different input probability densities.

added to the analog signal prior to the quantizer to assure this property. *Saturation noise*, on the other hand, tends to be white only when the input signal has a broad bandwidth and tends to be harmonically related to the input signal if it has a narrow bandwidth. Thus the effects of quantizing noise can be filtered or averaged because it truly has the characteristics of noise. Saturation noise, on the other hand, is indistinguishable from signal content and generally cannot be reduced by subsequent averaging or filtering techniques.

Figure 11.9 presents the discrete Fourier transforms of the same signal set presented in Section 11.2.2—two sinusoids of relative amplitude 1.0 and 0.01, quantized with a 10-bit ADC. In Figure 11.9a, b, and c, the peak signal amplitudes were adjusted to 5%, 10%, and 20% (0.42, 0.83, and 1.58 dB, respectively) above the ADC saturation level. Note the very many spectral artifacts caused by the saturation. The saturation noise grows larger as the signal excursion increases into saturation. Also note that some of these artifacts are only down 20 to 40 dB relative to full-scale signal. Compare these figures to Figure 11.7 to see the dramatic difference that too little signal attenuation, hence saturation, makes in the noise output of an ADC.

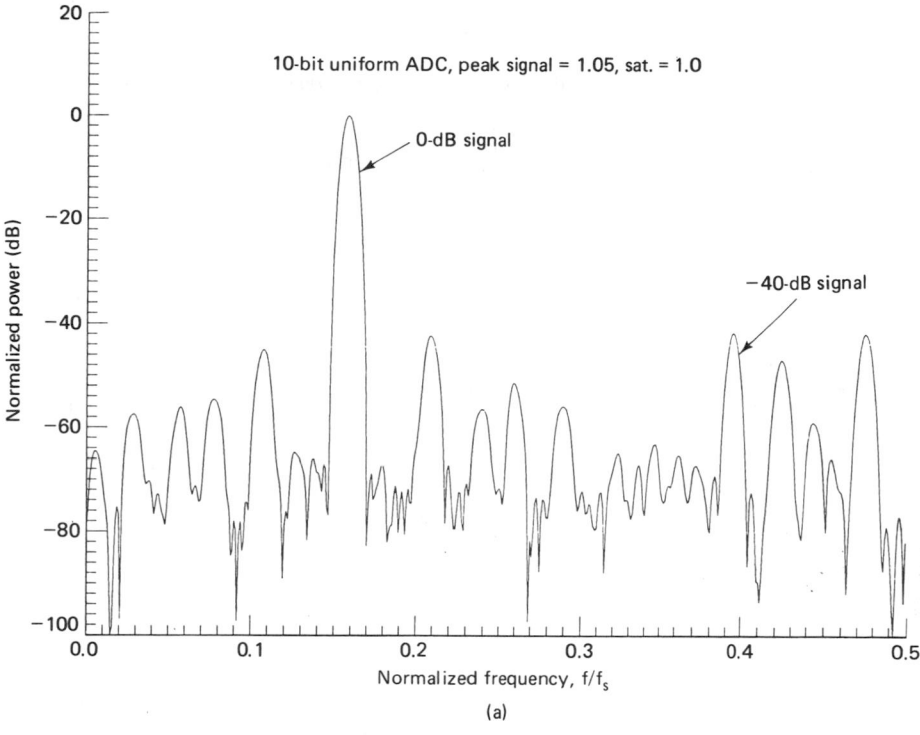

(a)

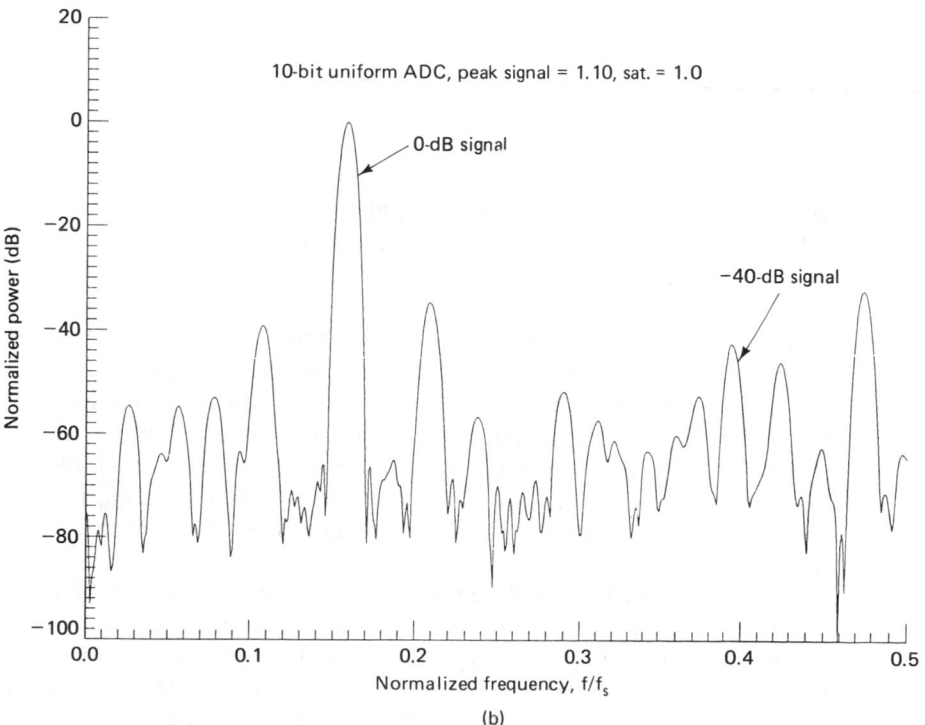

(b)

Figure 11.9 Power spectrum of uniformly quantized signals, with the quantizer saturating on the signal peaks.

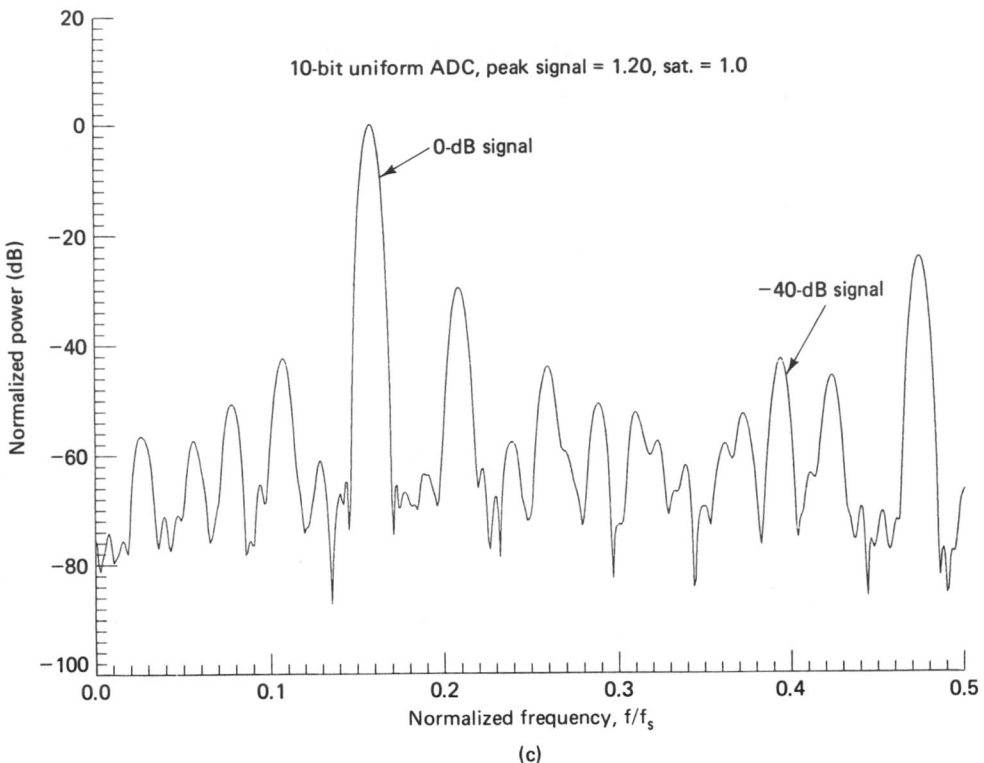

Figure 11.9 (*Continued*)

11.2.4 Dithering

Dithering is one of the most clever applications of noise as a useful engineering tool. To *dither*, according to *Webster's Ninth New Collegiate Dictionary*, is to "act indecisively, to vacillate". A *dither signal* is a small perturbation or disturbance added to a measurement process to reduce the effect of small local nonlinearities. The most familiar form of dither is the slight tapping we apply to the side of a d'Arsonval meter movement prior to taking the reading (before the days of digital meters). The tapping is a sequence of little impulses for displacing the needle movement beyond the local region which exhibits a nonlinear coefficient of friction at low velocities. A more sophisticated example of this same effect is the mechanical dither applied to the counterrotating laser beams of a laser beam gyro to break up low-level frequency entrapment known as deadband [3].

    In the analog-to-digital converter application, the effect of the dither is to reduce or eliminate the local discontinuities (i.e., the risers and flats) of the instantaneous input–output transfer function. We can best visualize the effect of these discontinuities by listing the desired properties of the error sequence formed by the quantizer process and then examining the actual properties of the same sequence. The quantizer error sequence is modeled as additive noise. The desired properties of such a noise sequence, $e(n)$, are

1. *Zero mean:* $\qquad\qquad\qquad E[e(n)] = 0$
2. *White:* $\qquad\qquad\qquad\quad E[e(n)e(n + m)] = \sigma^2\delta(m)$
3. *Uncorrelated with data $X(n)$:* $\quad E[e(n)X(n + m)] = 0$

where n and m are sample indices, and $\delta(m)$ is a Dirac delta function. In Figure 11.10, we examine a sequence of samples formed by a truncating ADC and make the following observations:

1. The error sequence is all of the same polarity; therefore, it is not zero mean.
2. The error sequence is not independent, sample to sample; therefore, it is not white.
3. The error sequence is correlated with the input; therefore, it is not independent.

Repeated measurements of the same signal would result in the same noise, and thus no amount of averaging could reduce the deviation from the true input signal. Paradoxically, we would like this noise to be "noisier." If the noise were independent on successive measurements, averaging would reduce the deviation from the true values. Thus, faced with the problem that the noise we get is not

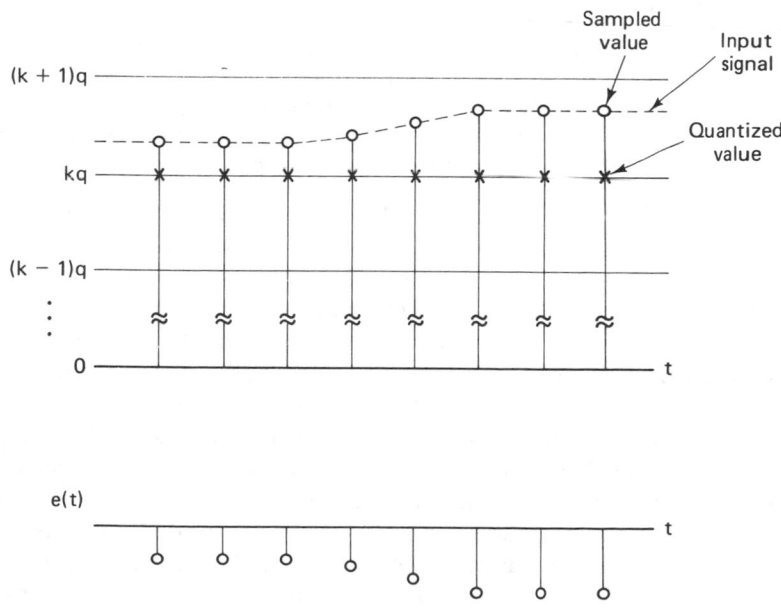

Figure 11.10 Sampled data sequence quantized to next-lowest quantile level and associated error sequence.

the noise we want, we choose to alter that noise by adding our own. We add a perturbation to the measurement to override the undesired low-level structure of the quantizer noise. The added perturbation, in a sense, converts *bad noise* to *good noise* [4].

Example 11.5 Dither Linearization

We hypothesize a quantizer that can only measure integers and converts input data into the next lowest integer (a process called truncation). We make 10 measurements of a signal, say of amplitude 3.7. In the absence of a dither we have readings all equal to 3.0. Now add a uniformly distributed (over 0 to 1) random number sequence to the input prior to performing the reading. The sequence of data has the form

Reading	Raw	Quantized	Dithered	Quantized
1	3.7	3.0	4.0485	4.0
2	3.7	3.0	4.5685	4.0
3	3.7	3.0	3.9789	3.0
4	3.7	3.0	4.0615	4.0
5	3.7	3.0	3.8074	3.0
6	3.7	3.0	3.9629	3.0
7	3.7	3.0	4.6252	4.0
8	3.7	3.0	4.2599	4.0
9	3.7	3.0	4.0408	4.0
10	3.7	3.0	4.2228	4.0
Averages =		3.0	4.1576	3.7
Dither mean =			0.5000	
(Average − Dither mean) =			3.6576	

Note that the average value of the dithered signal is the sum of the signal plus the mean value (0.5) of the dither. Here we have used a biased dither to remove the quantizer bias. The average of the dithered and quantized readings is (for this example) a correct reading and, in general, will be closer to the true signal than will the nondithered and quantized measurements [5, 6].

To help us understand the effect that dithering has on the quantization process, consider the following experiment. Let us apply 60 dB of attenuation to a sinusoidal signal of amplitude 1.0. The attenuated signal, then, has a full-scale amplitude of 0.001, which is approximately one half of the quantization interval 0.001953 of a ten-bit uniform quantizer. The sampled output of a rounding quantizer, with this signal as input, will be essentially all zeros except for an occasional count of ± 1, which occurs when the input crosses the $\pm q/2$ level of 0.000976 (corresponding to the least significant bit of the ADC). If the input signal were attenuated another 0.23 dB, the threshold levels of the least significant bit would never be crossed and the output sequence would be all zeros. Now, let us add a *dither signal* of amplitude 0.001 rms to the attenuated signal of 0.001 so that the signal plus dither regularly crosses the $\pm q/2$ levels of the ADC. Figure 11.11 shows the power spectra obtained by transforming and averaging 32 realizations of this dithered signal. Lo and behold, the 60-dB attenuated signal, at the edge of the ADC's resolvability, is indeed present and has been accurately measured.

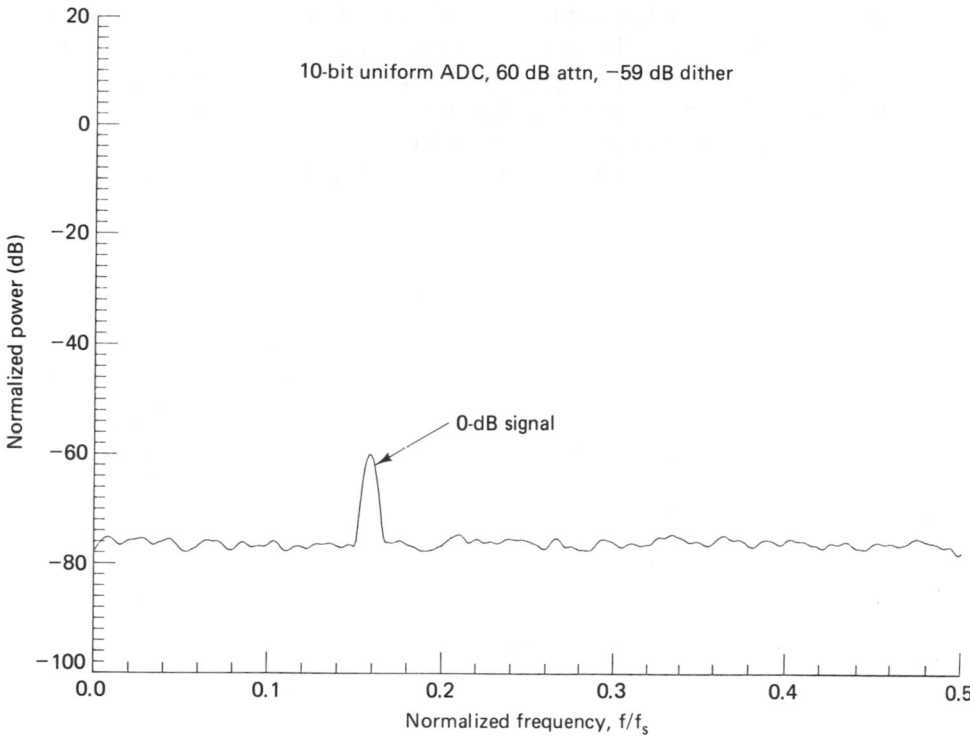

Figure 11.11 Power spectrum of uniformly quantized dithered signal.

The dither signal has had the effect of extending the dynamic range of the ADC (typically by 9 dB or 1.5 bits) and has improved the effective linearity of the low-level ADC staircase approximation.

11.2.5 Nonuniform Quantizing

Uniform quantizers are the common type of analog-to-digital converters because they are the most *robust*. By "robust" we mean that they are relatively insensitive to small changes in the input statistics. They achieve this robustness by not being finely tuned to one specific set of input parameters. This allows them to perform well even in the face of uncertain input parameters, and it means that small changes in input statistics will result in only small changes in output statistics.

When there is small uncertainty in the input signal statistics, it is possible to design a nonuniform quantizer which exhibits a smaller quantizer NSR than a uniform quantizer using the same number of bits. This is accomplished by partitioning the input dynamic range into nonuniform intervals such that the noise power, weighted by the probability of occurrence in each interval, is the same. Iterative solutions for the decision boundaries and step sizes for an optimal quantizer can be found for specific density functions and for a small number of bits. This task is simplified by modeling the nonuniform quantizer as a sequence of operators, as depicted in Figure 11.12. The input levels are first mapped, via a nonlinear function called a compressor, to an alternative range of levels. These levels are uniformly quantized and the quantized signal levels are then mapped, via a complementary nonlinear function called an expander, to the output range of levels. Borrowing part of the name from each of the operations COMpress and exPAND, we form the acronym by which this process is commonly identified, *companding*.

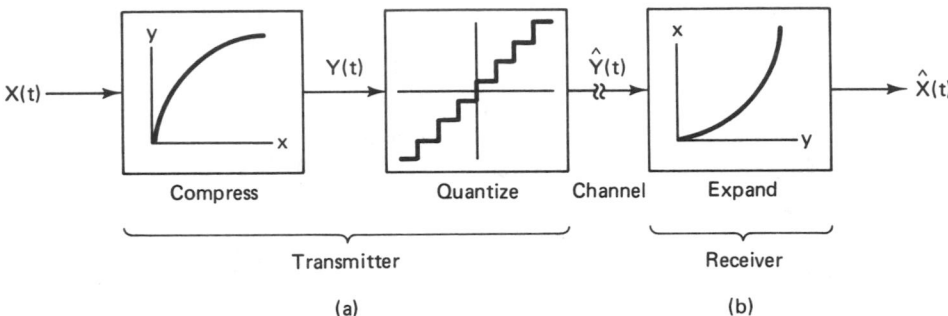

Figure 11.12 Nonuniform quantizer as a sequence of compression, uniform quantization, and expansion.

11.2.5.1 (Near) Optimal Nonuniform Quantizing

Examining the compressor characteristics, $y = C(x)$, of Figure 11.13, we note that the quantizing step sizes for the output variable, y, are related to the quantizing step sizes for the input variable, x, through the slope $\dot{C}(x)$ [i.e., $\Delta y = \Delta x \dot{C}(x)$]. Under reasonable conditions, such as a large number of quantizing levels and a smooth pdf for the input variable, we can arrive at the output quantizing noise variance [7]

$$\sigma_q^2 = \frac{q^2}{12} \int_{-x_{max}}^{+x_{max}} \frac{p(x)}{|\dot{C}(x)|^2} \, dx \tag{11.25}$$

For a specific pdf, the compression characteristic, $C(x)$, can be found which minimizes σ_q^2. The optimal compressor law for a given pdf is [8]

$$C(x) = \int_0^x \sqrt[3]{Kp(z)} \, dz \tag{11.26}$$

We find that the optimal compressor characteristic is proportional to the integral of the cube root of the input probability density function. This is called *fine tuning*. If the compressor is designed to operate with one density function and it is used with some other density function (including scaled versions), the quantizer is said to be *mismatched* and there may be severe performance degradation due to the mismatch [6].

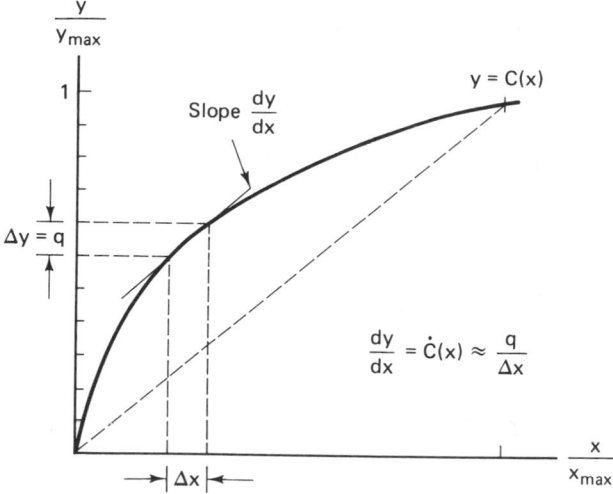

Figure 11.13 Compressor characteristics $y = C(x)$ and estimates to local slope $\dot{C}(x)$.

Within the figure:

$$\text{Slope } \frac{dy}{dx}$$

$$y = C(x)$$

$$\frac{dy}{dx} = \dot{C}(x) \approx \frac{q}{\Delta x}$$

$\Delta y = q$

11.2.5.2 Logarithmic Compression

In the preceding section we presented the compression law for the case in which the input probability density function is well defined. We now address the case for which little is known about the density function. This case occurs, for instance, when the average power of the input signal is a *random variable*. As an example, the voice level of a randomly chosen telephone user may vary from one extreme of a barely audible whisper to the other extreme of a bellowing shout.

For the case of an *unknown density function*, the compressor characteristics of the nonuniform quantizer must be selected such that the resultant noise performance is independent of the specific density function. Although this is a worthy undertaking, it may not be possible to achieve this independence. We are willing to compromise, however, and we will settle for virtual independence over a large range of input variance and input density functions. An example of a quantizer that exhibits a SNR independent of the input density can be visualized with the aid of Figure 2.18. There we saw the very large difference in NSR ratio for different amplitude input signals when quantized with a uniform quantizer. By comparison, we saw that the nonuniform quantizer permits large errors for large signals. This makes intuitive sense. If the SNR is to be independent of the amplitude distribution, the quantizing noise must be proportional to the input level.

Equation (11.25) presented the quantizer noise variance for an arbitrary density function and compressor characteristics. The signal variance for any density function is

$$\sigma_X^2 = \int_{-\infty}^{\infty} x^2 p(x)\, dx \tag{11.27}$$

In the absence of saturation, the quantizer signal-to-noise ratio is of the form

$$\frac{\sigma_x^2}{\sigma_q^2} = \frac{\int_{-x_{max}}^{+x_{max}} x^2 p(x)\, d(x)}{(q^2/12) \int_{-x_{max}}^{+x_{max}} [p(x)/|\dot{C}(x)|^2]\, dx} \tag{11.28}$$

To have the SNR be independent of the specific density function, we require that the numerator be a scaled version of the denominator. This happens if the following is true:

$$|\dot{C}(x)|^2 = \left|\frac{K}{x}\right|^2 \tag{11.29}$$

or

$$\dot{C}(x) = \frac{K}{x} \tag{11.30}$$

from which we obtain by integration,

$$C(x) = \int_0^x \frac{K}{z}\, dz \tag{11.31}$$

or

$$C(x) = \log_e x + \text{constant} \tag{11.32}$$

This result is intuitively appealing. A *logarithmic compressor* allows large signals to have larger errors because of the log scale—equal distances (or errors) are, in fact, equal ratios, which is what we were seeking, a *constant signal-to-noise ratio*. Here the constant is present to match the boundary conditions between x_{max} and y_{max}. Accounting for this boundary condition, we have the logarithmic converter of the form

$$\frac{y}{y_{max}} = \frac{C(x)}{y_{max}} = \log_e \frac{x}{x_{max}} \tag{11.33}$$

The form of the compression suggested by the logarithm function is shown in Figure 11.14a. The first problem with this function is that it does not map the negative input signals. We account for the negative signals by adding a reflected version of the log to the negative axis. This modification results in Figure 11.14b and is of the form

$$\frac{y}{y_{max}} = \frac{C(x)}{y_{max}} = \log_e \frac{|x|}{x_{max}}\, \text{sgn } x \tag{11.34}$$

where

$$\text{sgn } x = \begin{cases} +1 & \text{for } x > 0 \\ -1 & \text{for } x < 0 \end{cases}$$

The remaining problem we face is that the resultant compression is not continuous through the origin; in fact, it completely misses the origin. We need

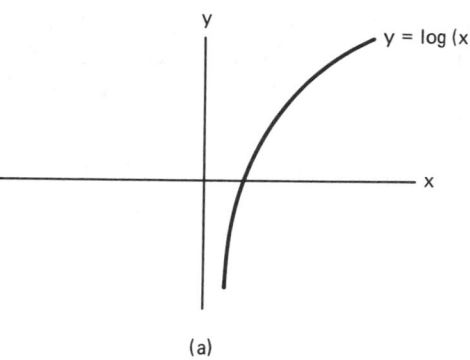

(a)

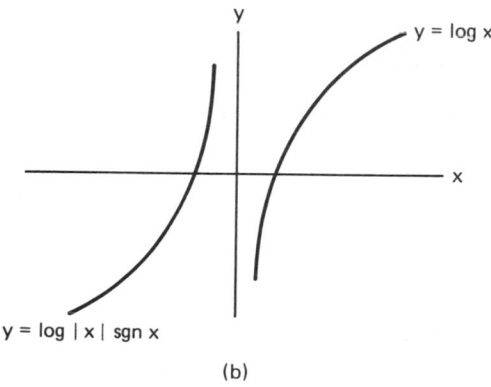

(b)

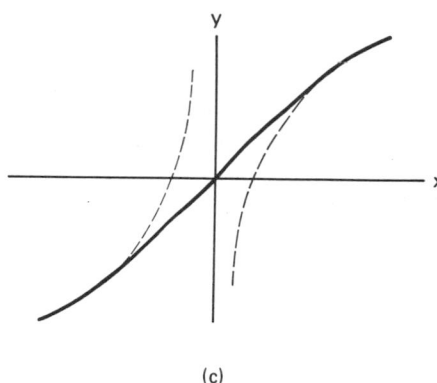

(c)

Figure 11.14 (a) Log function prototype for compression law. (b) Log $|x|$ sgn x function prototype for compression law. (c) Log $|x|$ sgn x function with a smooth transition between segments.

to make a smooth transition between the logarithmic function and a linear segment passing through the origin. There are two standard compression functions that perform this transition: the μ-law and A-law companders.

μ-Law Compander. The μ-law compander is the Bell System (hence the North American standard) compression law. It is of the form

$$y = C(x) = y_{max} \frac{\log_e [1 + \mu(|x|/x_{max})]}{\log_e (1 + \mu)} \text{ sgn } x \qquad (11.35)$$

The approximate behavior of this compressor in the regions corresponding to small and large values of the argument are

$$y = C(x) = \begin{cases} y_{max} \dfrac{\mu(|x|/x_{max})}{\log_e \mu} & \mu\left(\dfrac{|x|}{x_{max}}\right) \ll 1 \\[4mm] y_{max} \dfrac{\log_e [\mu(|x|/x_{max})]}{\log_e \mu} & \mu\left(\dfrac{|x|}{x_{max}}\right) \gg 1 \end{cases} \qquad (11.36)$$

Sec. 11.2   Amplitude Quantizing

621

The parameter μ in the μ-law compander had originally been set to 100 for use with a 7-bit converter. It was later changed to 255 for use with an 8-bit converter. The 8-bit $\mu = 255$ μ-law converter has become the standard North American conversion law.

Example 11.6 Average SNR for μ-Law Compressor

The SNR for the μ-law compressor can be estimated by substituting the μ-law expression into Equation (11.28). For positive values of the input variable x, the compression law is

$$y = C(x) = y_{max} \frac{\log_e [1 + \mu(|x| /x_{max})]}{\log_e (1 + \mu)} \tag{11.37}$$

Then the derivative $\dot{C}(x)$ is

$$\dot{y} = \dot{C}(x) = y_{max} \frac{1}{\log_e (1 + \mu)} \frac{\mu(1/x_{max})}{1 + \mu(|x| /x_{max})} \tag{11.38}$$

For values of the input variable for which $\mu(x/x_{max})$ is large compared to unity, the derivative becomes

$$\dot{y} = \dot{C}(x) = \frac{1}{x} \frac{y_{max}}{\log_e \mu} \tag{11.39}$$

Subsituting for $1/\dot{C}(x)$ in Equation (11.28), we find

$$\text{SNR} = \frac{\sigma_s^2}{\sigma_q^2} = \frac{1}{[(q^2/12)(\log_e \mu/y_{max})]^2} \tag{11.40}$$

$$= 3 \left(\frac{2y_{max}}{q} \right)^2 \left(\frac{1}{\log_e \mu} \right)^2 \tag{11.41}$$

The ratio $2y_{max}/q$ is the number of quantizing levels (2^b) of the compressed quantizer. For the 8-bit converter with $\mu = 255$, the SNR is found to be

$$\text{SNR} = 3 \left[\frac{2^8}{\log_e(255)} \right]^2 = 3(46.166)^2$$

$$= 38.1 \text{ dB} \tag{11.42}$$

For a comparison, the SNR of an actual μ-law quantizer is presented in Figure 11.15. There the SNR is plotted for input sinusoids of different amplitudes. The serration of the performance curve is due to the piecewise linear approximation to the continuous μ-law curve. This is described shortly.

Figure 11.16 presents the discrete Fourier transform of the pair of input sinusoids of relative amplitude 1.0 and 0.01. Here the input signal is quantized with a 10 bit μ-law ($\mu = 40$) converter and the signal levels are attenuated by 0, 20, and 40 dB, respectively, relative to full-scale input. Note that the quantizing noise levels for the full-scale signal are equivalent to that of the uniform quantizer (ADC) (see Figure 11.7). The difference in performance between the log compressed ADC and the uniform ADC is to be seen for the attenuated signals. We see that as the input signal levels are attenuated, the quantizing noise is also reduced. Thus the log compressed ADC has no problem "seeing" the low-level input signal even with 40-dB

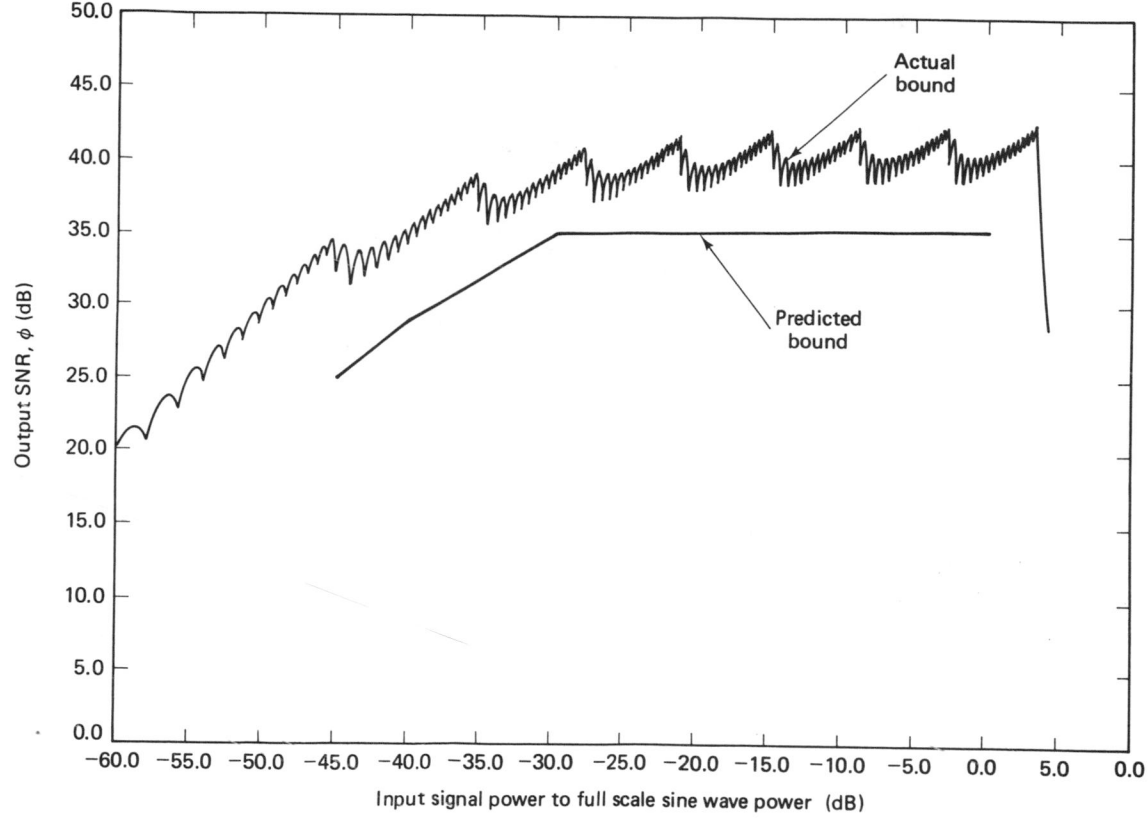

Figure 11.15 Predicted and measured SNR for a μ-law quantizer.

attenuation in Figure 11.16a, while the same signal is lost in the noise of the uniform converter (see Figure 11.7c).

The μ-law compressor realization differs from the expression presented in Equation (11.35) in a minor way. The functional expression is approximated by 16 linear chord segments over the possible 256 output levels, as shown in Figure 11.17. Eight of these segments are in the first quadrant, eight are in the third quadrant, and the "0" segment has the same slope in both quadrants. Over each chord segment the quantization is uniform in the four lower-order conversion bits. Thus the 8-bit compressed conversion format is of the form

$$\underbrace{b_7}_{\substack{\text{sign bit}\\(\text{quadrant})}} \quad \underbrace{b_6\ b_5\ b_4}_{\text{segment}} \quad \underbrace{b_3\ b_2\ b_1\ b_0}_{\substack{\text{position on}\\\text{segment}}}$$

It is the piecewise chord approximation to the smooth function and a staircase

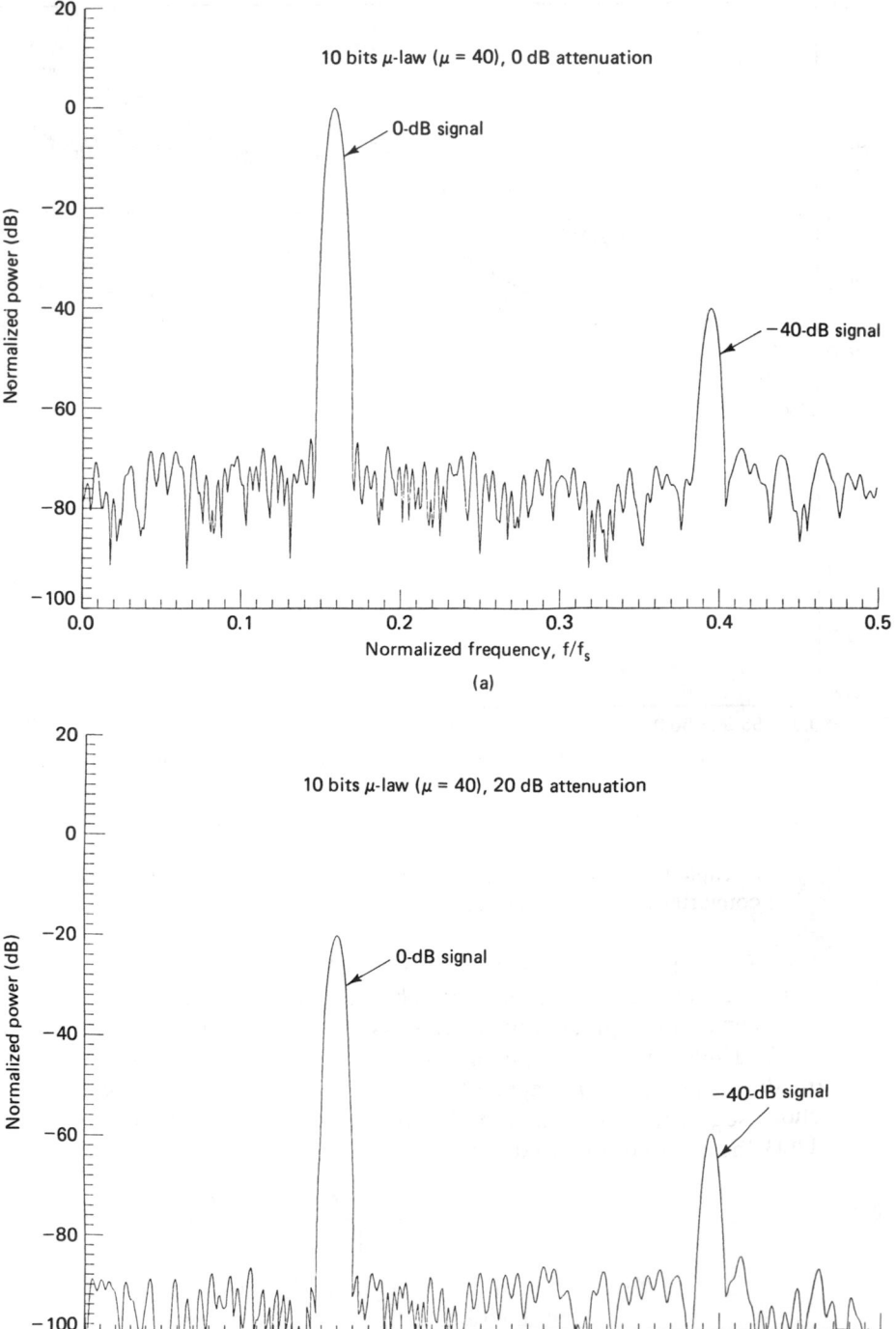

Figure 11.16 Power spectrum of μ-law quantized signals.

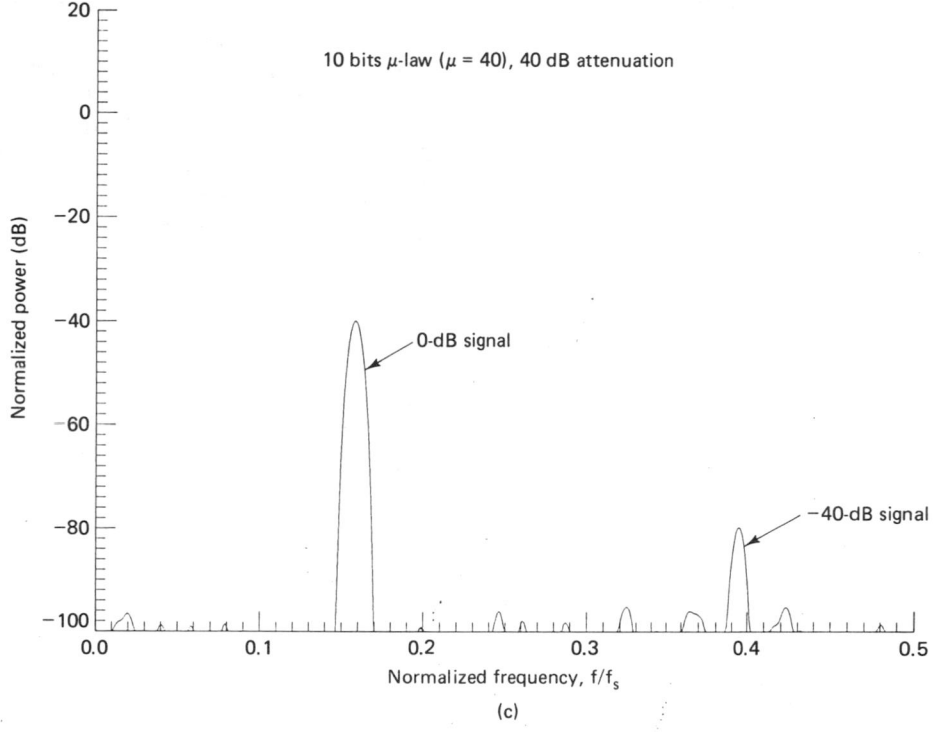

0-dB signal

−40-dB signal

Normalized power (dB)

Normalized frequency, f/f$_s$

(c)

Figure 11.16 (*Continued*)

approximation of each chord that accounts for the cusps in the SNR shown in Figure 11.15.

A-Law Compander. The *A*-law compander is the CCITT (hence the European) standard approximation to the logarithmic compression. The form of the compressor is

$$
y = C(x) = \begin{cases}
y_{max} \dfrac{A(\,|x|\,/x_{max})}{1 + \log_e A} \, \text{sgn } x & 0 < \dfrac{|x|}{x_{max}} < \dfrac{1}{A} \\[4mm]
y_{max} \dfrac{1 + \log_e[A(\,|x|\,/x_{max})]}{1 + \log A} \, \text{sgn } (x) & \dfrac{1}{A} < \dfrac{|x|}{x_{max}} < 1
\end{cases}
\tag{11.43}
$$

The standard value of the parameter *A* is 87.56, and for this value, using an 8-bit conversion, the average SNR is 38.0 dB. The *A*-law compression characteristic is approximated, in a manner similar to the μ-law compressor, by a sequence of 16 linear chords spanning the output range. The lower two chords in each quadrant are in fact a single chord corresponding to the linear segment of the *A*-law compressor. One important difference between the *A*-law and the μ-law compression characteristics is that the *A*-law standard has a midriser at the origin, while the μ-law standard has a midtread at the origin. Thus the *A*-law compressor has no

Sec. 11.2   Amplitude Quantizing

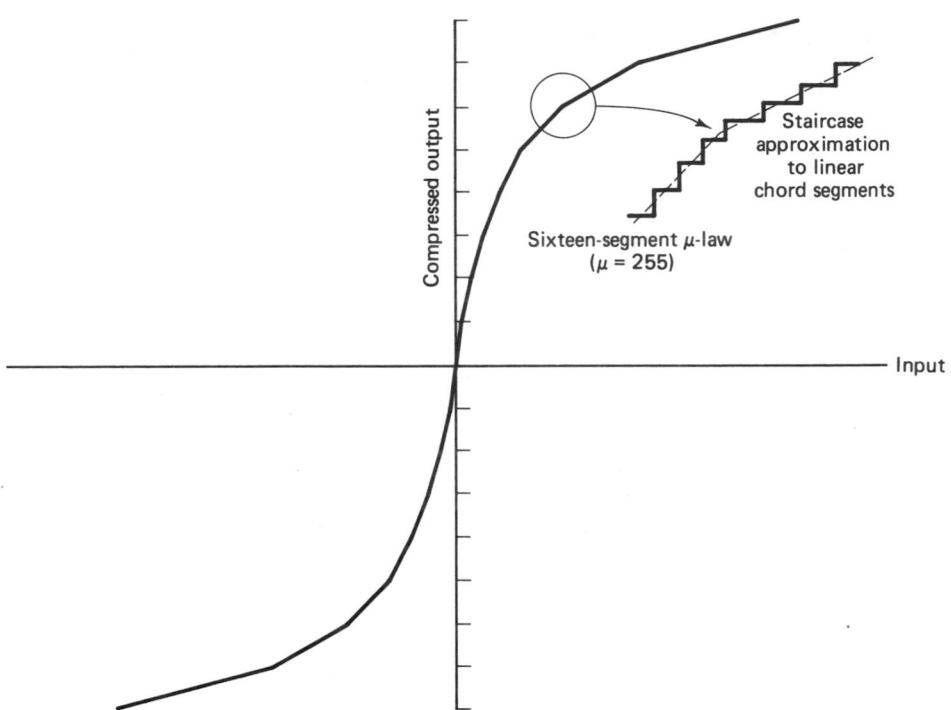

Figure 11.17 Seven-bit compressed quantization with 16-segment approximation to μ-law.

zero value and hence exhibits no interval for which data are not being transmitted for zero input.

There are direct mappings from the *A*-law 8-bit compressed conversion format to a 12-bit linear binary code and from the μ-law 8-bit compressed format to a 13-bit linear code [8]. This operation permits the A/D conversion to be performed with a uniform quantizer and then be mapped to the smaller number of bits in a code converter. This also permits the inverse mapping at the receiver (i.e., the expansion) to be performed on the digital sample.

Pulse Code Modulation.   One of the tasks performed by a pulse code modulation (PCM) process is the conversion of a waveform source to a binary sequence discrete source. This task is performed in a three-step process—sampling, quantizing, and encoding—as indicated in Figure 2.2. We have addressed the sampling process in Chapter 2, and have addressed the quantizing process in this chapter as well as in Chapter 2. We note that the encoding process, which was shown in Figure 2.2 to follow quantization, is often embedded in the hardware that performs the quantization. It can be described as follows: successive approximation (SA) analog-to-digital (A/D) converters form the successive bits of the encoded data by a feedback, comparison, and decision process. In the feedback process, a binary search is conducted over the range of possible input levels by repeatedly asking: Is the input signal above or below the midpoint of the remaining uncertainty interval? By this technique the uncertainty interval is reduced by one-half for each comparison and decision step until the uncertainty range matches the allowable quantizing interval.

In the SA conversion, the results of each previous decision reduce the uncertainty to be resolved during the next decision. In a similar manner, the results of the previous A/D conversions can be used to reduce the uncertainty to be resolved during the next conversion. This reduction in uncertainty is achieved by carrying forward to the next sample auxiliary information from earlier samples. This information is called the redundant part of the signal, and by carrying it forward, we reduce the interval of uncertainty over which the quantizer and encoder must search for the next signal sample. Carrying data forward is one method of achieving *redundancy reduction*.

11.3 DIFFERENTIAL PULSE CODE MODULATION

By the use of past data to assist in measuring (i.e., quantizing) new data, we leave ordinary PCM and enter the realm of differential PCM (DPCM). In DPCM, a prediction of the next sample value is formed from past values. This prediction can be thought of as instructions for the quantizer to conduct its search for the next sample value in a particular interval. By using the redundancy in the signal to form a prediction, the region of uncertainty is reduced and the quantization can be performed with a reduced number of decisions (or bits) for a given quantization level or with reduced quantization levels for a given number of decisions (or bits). The reduction in redundancy is realized by subtracting the prediction from the next sample value. This difference is called the *prediction error*.

The quantizing methods described in Section 11.2 are called *instantaneous or zero memory quantizers* because the digital conversion is based on the single (current) input sample. In Section 11.1 we identified the properties of sources that permitted source rate reductions. These properties were non-equiprobable source levels and nonindependent sample values. Instantaneous quantizers achieve source coding gains by taking into account the probability density assignment for each sample. The quantizing methods that take account of sample-to-sample correlation are noninstantaneous quantizers. These quantizers reduce source redundancy by first converting the correlated input sequence into a related sequence with reduced correlation, reduced variance, or reduced bandwidth. This new sequence is then quantized with fewer bits.

The correlation characteristics of a source can be visualized in the time domain by samples of its autocorrelation function and in the frequency domain by its power spectrum. If we examine a power spectrum, $G_X(f)$, of a short-term speech signal, as shown in Figure 11.18, we find that the spectrum has a local peak in the neighborhood of 300 to 800 Hz and falls off at a rate of 6 to 12 dB/octave. By interpreting this power spectrum, we can infer certain properties of the time function from which it was derived. We observe that large changes in the signal occur slowly (low frequency) and that rapid changes in the signal (high frequency) must be of low amplitude. An equivalent interpretation can be found

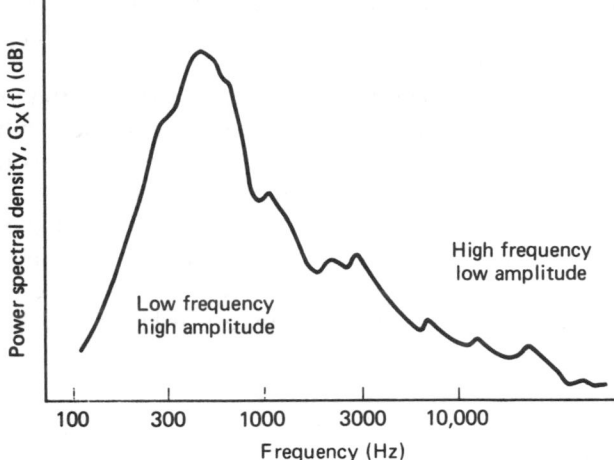

Figure 11.18 Typical power spectrum for speech signals.

in the autocorrelation function, $R_X(\tau)$, of the signal, as shown in Figure 11.19. Here a broad, slowly changing autocorrelation function suggests that there will be only slight change on a sample-to-sample basis, and that a time interval exceeding the correlation distance is required for a full amplitude change. In particular, correlation values for typical single-sample delay is on the order of 0.79 to 0.87 and the correlation distance to the first zero crossing is on the order of 4 to 6 sample intervals of T seconds per interval.

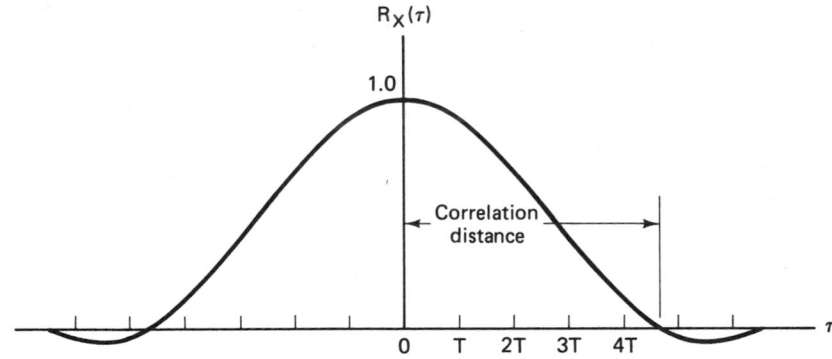

Figure 11.19 Autocorrelation function for typical speech signals.

Since the difference between adjacent time samples for speech is small, sampling techniques have evolved based on transmitting sample-to-sample differences rather than actual sample values. Successive differences are in fact a special case of a class of noninstantaneous converters called one-tap linear predictive coders. These are sometimes called *predictor-corrector coders*. This structure is shown in Figure 11.20. In this type of converter, the transmitter and the receiver have the same prediction model, which is derived from the signal's correlation characteristics. Each use the model to predict the next sample value based

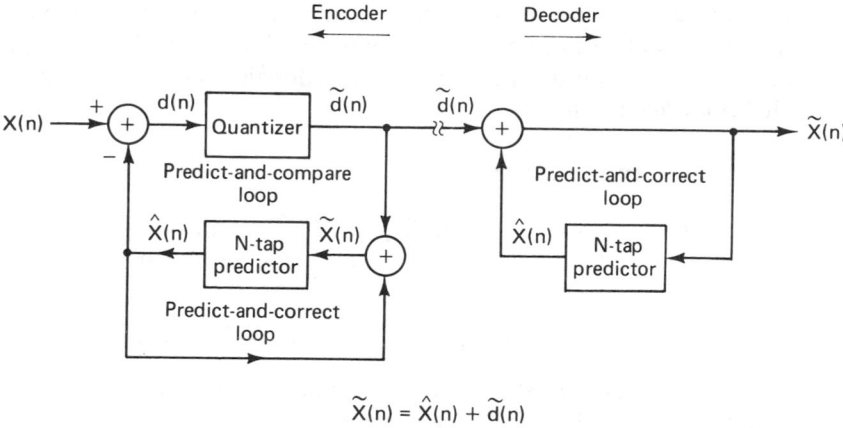

$$\tilde{X}(n) = \hat{X}(n) + \tilde{d}(n)$$

Figure 11.20 One-tap predictor differential pulse code modulator (DPCM).

on recent past sample values. The transmitter forms the *prediction error* (or the *residue*) as the difference between the next measured sample value and the predicted sample value. The equation for the prediction loop is

$$d(n) = X(n) - \hat{X}(n)$$

where $X(n)$ is the nth input sample, $\hat{X}(n)$ the predicted value of that sample, and $d(n)$ the associated prediction error. This is performed in the predict and compare loop, the upper loop of the transmitter in Figure 11.20. The transmitter corrects its prediction by forming the sum of its prediction and the prediction error. The equations for the correction loop are

$$\tilde{d}(n) = \text{quant}[d(n)]$$

$$\tilde{X}(n) = \hat{X}(n) + \tilde{d}(n)$$

where quant(·) represents the quantization operation, and $\tilde{d}(n)$ is the quantized version of the prediction error and $\tilde{X}(n)$ is the corrected (and quantized) version of the input sample. This is performed in the predict and correct loop, the lower loop of the encoder and the only loop of the decoder in Figure 11.20. The decoder must also be informed of the prediction error so that it can use its correction loop to correct its prediction. Thus the communication task is that of transmitting the difference (the error signal) between the predicted and the actual data sample. For this reason, this class of coder is often called a *differential pulse code modulator (DPCM)*. If the prediction model forms predictions which are close to the actual sample values, the residues will exhibit reduced variance (relative to the original signal). From Section 11.2 we know that the number of bits required to move data through the channel with a given fidelity is related to the signal variance. Hence the reduced variance sequence of residues can be moved through the channel with a reduced data rate.

The predictive converters must have a short-term memory that supports the fast-time operations required for the prediction algorithm. In addition, they will often have a long-term memory that supports the slow time, often data-dependent operations, such as automatic gain control and filter coefficient adjustments. Predictors that incorporate the slower, data-dependent, adjustment algorithms are called *adaptive predictors*.

11.3.1 One-Tap Prediction

The one-tap linear prediction coding (LPC) filter in the DPCM process predicts the next input sample value based on the previous input sample value. The prediction equation is of the form

$$X(n \mid n - 1) = aX(n - 1 \mid n - 1) \tag{11.44}$$

where $X(n \mid m)$ is the estimate of X at time n given all the samples collected up through time m, and where a is a parameter used to minimize the prediction error. The prediction error available after the measurement is of the form

$$d(n) = [X(n) - X(n \mid n - 1)] \tag{11.45a}$$

$$= [X(n) - aX(n - 1 \mid n - 1)] \tag{11.45b}$$

The mean-squared error is of the form

$$\mathrm{E}\{d^2(n)\} = \mathrm{E}\{X(n)X(n) - 2aX(n)X(n - 1 \mid n - 1)$$

$$+ a^2 X(n - 1 \mid n - 1)X(n - 1 \mid n - 1)\} \tag{11.46}$$

If $X(n - 1 \mid n - 1)$ is an unbiased estimate of $X(n - 1)$, Equation (11.46) can be written as

$$R_d(0) = R_X(0) - 2aR_X(1) + a^2 R_X(0) \tag{11.47a}$$

$$= R_X(0)[1 + a^2 - 2aC_X(1)] \tag{11.47b}$$

where $R_d(n)$ and $R_X(n)$ are the autocorrelation functions of the prediction error and the input signal, respectively, $R_d(0)$ is the power in the error, $R_X(0)$ the power in the signal, and $C_X(n) = R_X(n)/R_X(0)$ the normalized autocorrelation function. We can select the parameter a to minimize the prediction error power of Equation (11.47) by setting to zero the partial of $R_d(0)$ with respect to a.

$$\frac{\partial R_d(0)}{\partial a} = R_X(0)[2a - 2C_X(1)] \tag{11.48}$$

Setting to zero and solving for a^*, where the * implies the optimal solution, we have

$$a^* = C_X(1) \tag{11.49}$$

Substituting a^* back into Equation (11.47), we have

$$R_d^*(0) = R_X(0)[1 + a^* C_X(1) - 2a^* C_X(1)] \tag{11.50a}$$

$$= R_X(0)[1 - a^* C_X(1)] \tag{11.50b}$$

$$= R_X(0)[1 - C_X^2(1)] \tag{11.50c}$$

We can define the *prediction gain* of the encoder as the ratio of input to output variances, $R_X(0)/R_d(0)$. For a fixed bit rate this gain represents an increase in output SNR, while for a fixed output SNR this gain represents a reduced bit rate description. We note that the prediction gain for the optimal predictor is always greater than one for any value of signal correlation [Equation (11.50b)]. On the other hand, the prediction gain is greater than one for the unity-gain, one-tap predictor, only if the signal correlation exceeds 0.5 [Equation (11.47b)].

Example 11.7   Prediction Gain of a One-Tap LPC Filter

A signal with correlation coefficient $C_X(1)$ equal to 0.8 is to be quantized with a one-tap LPC filter. Determine the prediction gain when the prediction coefficient is (a) optimized with respect to the minimum prediction error, or (b) set to unity.

Solution

(a) From Equation (11.50c),

$$R_d^*(0) = R_X(0)(1 - 0.64) = 0.36R_X(0) \qquad (11.51a)$$

$$\text{Prediction gain} = \frac{1}{0.36} = 2.78 \text{ or } 4.44 \text{ dB} \qquad (11.51b)$$

(b) From Equation (11.47b),

$$R_d(0) = 2R_X(0)(1 - 0.8) = 0.40R_X(0) \qquad (11.51c)$$

$$\text{Prediction gain} = \frac{1}{0.40} = 2.50 \text{ or } 3.98 \text{ dB} \qquad (11.51d)$$

11.3.2 *N*-Tap Prediction

The N-tap LPC filter predicts the next sample value based on a linear combination of the previous N sample values. We will assume that the quantized estimates used by the prediction filters are unbiased and error free. With this assumption, we can drop the double indices (we used in Section 11.3.1) from the data in the filter but still use them for the predictions. Then the N-tap prediction equation takes the form

$$X(n \mid n - 1) = a_1 X(n - 1) + a_2 X(n - 2) + \cdots + a_N X(n - N) \qquad (11.52)$$

The prediction error takes the form

$$d(n) = X(n) - X(n \mid n - 1) \qquad (11.53a)$$

$$= X(n) - a_1 X(n - 1) - a_2 X(n - 2) - \cdots - a_N X(n - N) \qquad (11.53b)$$

The mean-square prediction error is of the form

$$\mathbf{E}\{d(n)d(n)\} = \mathbf{E}\{[X(n) - X(n \mid n - 1)]^2\} \qquad (11.54)$$

Clearly, the mean-square prediction error is quadratic in the filter coefficients a_i. As we did in Section 11.3.1, we can take the partial of the mean-squared error with respect to each coefficient and set those partials to zero. Formally, taking the partial with respect to the jth coefficient prior to expanding $X(n \mid n - 1)$, we have

$$\frac{\partial R_d(0)}{\partial a_j} = \mathbf{E}\left\{2[X(n) - X(n \mid n - 1)] \frac{\partial X(n \mid n - 1)}{\partial a_j} X(n \mid n - 1)\right\} \qquad (11.55a)$$

$$= \mathbf{E}\{2[X(n) - X(n \mid n - 1)] [-X(n - j)]\} \qquad (11.55b)$$

$$= 2\mathbf{E}\{[X(n) - a_1 X(n - 1) - a_2 X(n - 2)$$

$$- \cdots - a_N X(n - N)] X(n - j)\} \qquad (11.55c)$$

$$= 2[R_X(j) - a_1 R_X(j - 1) - a_2 R_X(j - 2)$$

$$- \cdots - a_N R_X(j - N)] \qquad (11.55d)$$

This collection of equations (one for each j) can be arranged in matrix form known as the *normal equations*. This form is

$$\begin{bmatrix} R_X(1) \\ R_X(2) \\ R_X(3) \\ \vdots \\ R_X(N) \end{bmatrix}$$

$$= \begin{bmatrix} R_X(0) & R_X(-1) & R_X(-2) & \cdots & R_X(1 - N) \\ R_X(1) & R_X(0) & R_X(-1) & \cdots & R_X(2 - N) \\ R_X(2) & R_X(1) & R_X(0) & \cdots & R_X(3 - N) \\ & & & & \vdots \\ R_X(N - 1) & R_X(N - 2) & R_X(N - 3) & \cdots & R_X(0) \end{bmatrix} \begin{bmatrix} a_1 \\ a_2 \\ a_3 \\ \vdots \\ a_N \end{bmatrix}^*$$

$$(11.56a)$$

The normal equations can be written more compactly as

$$\mathbf{R}_X(1, N) = \mathbf{R}_X \mathbf{a}^* \qquad (11.56b)$$

where the $\mathbf{R}_X(1, N)$ is the correlation vector of delays from 1 through N, $\mathbf{R}_X$ is the correlation matrix (assuming a zero-mean process), and $\mathbf{a}^*$ is the optimum filter weight vector.

To gain insight into the solution of the normal equations, we now recast the mean-square-error equation (11.54) in matrix form.

$$R_d(0) = \mathbf{E}\{[X(n) - \mathbf{a}^T \mathbf{X}(n - 1)][X(n) - \mathbf{X}^T(n - 1)\mathbf{a}]\} \qquad (11.57a)$$

$$= R_X(0) - \mathbf{R}_X^T(1, N)\mathbf{a} - \mathbf{a}^T \mathbf{R}_X(-1, -N) + \mathbf{a}^T \mathbf{R}_X \mathbf{a} \qquad (11.57b)$$

where $\mathbf{R}^T$ is the transpose of $\mathbf{R}$. Substituting the right-hand side of the optimal weight vector solution of Equation (11.56b) into (11.57b), we have

$$R_d(0) = R_X(0) - \mathbf{R}_X^T(1, N)\mathbf{a}^* - \mathbf{a}^T \mathbf{R}_X(-1, -N) + \mathbf{a}^{*T} \mathbf{R}_X(1, N) \qquad (11.58a)$$

$$= R_X(0) - \mathbf{R}_X^T(-1, -N)\mathbf{a}^* \qquad (11.58b)$$

We can now bring the right-hand side of Equation (11.56) over to the left-hand side, and use Equation (11.58b) to augment the top row of the matrix to obtain the *whitening form* of the optimal predictor. In this form, the only nonzero output of the matrix product occurs at time zero, which is akin to an output impulse.

$$
\begin{bmatrix}
R_X(0) & R_X(-1) & R_X(-2) & R_X(-3) & \cdots & R_X(-N) \\
R_X(1) & R_X(0) & R_X(-1) & R_X(-2) & \cdots & R_X(1-N) \\
R_X(2) & R_X(1) & R_X(0) & R_X(-1) & \cdots & R_X(2-N) \\
R_X(3) & R_X(2) & R_X(1) & R_X(0) & \cdots & R_X(3-N) \\
\vdots & & & & & \\
R_X(N) & R_X(N-1) & R_X(N-2) & R_X(N-3) & \cdots & R_X(0)
\end{bmatrix}
\begin{bmatrix}
1 \\ -a_1 \\ -a_2 \\ -a_3 \\ \vdots \\ -a_N
\end{bmatrix}^{*}
$$

$$
= \begin{bmatrix} R_d(0) \\ 0 \\ 0 \\ 0 \\ \vdots \\ 0 \end{bmatrix} \qquad\qquad (11.59)
$$

The top row of Equation (11.59) states that the power in the prediction error is of the form

$$
R_d(0) = R_X(0)[1 - a_1 C_X(1) - a_2 C_X(2) - \cdots - a_N C_X(N)] \qquad (11.60)
$$

Compare this form to Equation (11.50b). An interesting property of the optimal N-tap predictor filter is this: The coefficient set that obtains the minimum mean-square prediction error also predicts, with zero error, the next $N - 1$ correlation samples from the previous $N - 1$ correlation samples.

For fixed filter coefficients, the DPCM coder can achieve a prediction gain, relative to linear quantizing, of 6 to 8 dB [9]. This prediction gain is essentially independent of filter length once the length exceeds three or four taps. Additional gain is available if the coder has slow adaptive capabilities. Adaptive coders are introduced in Section 11.3.3 and discussed in some detail in Section 11.3.4.

11.3.3 Delta Modulation

Delta modulation, often called *delta mod*, Δ-mod, or DM, is a particularly simple form of one-tap DPCM coding. Equation (11.50c) demonstrates that the prediction gain for a one-tap predictor can be large if the normalized correlation coefficient, $C_X(1)$, is close to unity. Working toward the goal of high sample-to-sample correlation, the predictive filter is generally operated at a rate that far exceeds the Nyquist rate. For example, the sample rate might be chosen at four times Nyquist. Then for a 3.3-kHz bandwidth with a nominal sample rate of 8 kHz, the high correlation prediction filter would operate at a 32-kHz sample rate. The justification for the high correlation, hence the high sample rate, is that the quantizer operating in the error loop can be very simple. The simplest form of the quantizer is a one-bit quantizer, which is in fact, only a comparator which detects and reports the sign of the difference signal. As a result, the prediction error signal is a 1-bit word which has the interesting advantage of not requiring word framing in a communication system. The original attraction of this technique was simplicity of hardware and algorithm structure, neither of which is a major driver in today's integrated circuit marketplace. Delta modulation is still an important option because of its simplicity, ease of analysis, relative ease of embedding adaptive options, and its ability to perform robustly in the presence of channel errors [10, 11].

The block diagram of a delta modulator version of the one-tap linear predictor (presented in Section 11.3.1) is shown in Figure 11.21. Note that the predictor-corrector loop at the receiver is followed by an analog postreconstruction filter. This filter removes the out-of-band quantizing noise and overload distortion which is generated by the two-level coding and which extends beyond the information bandwidth of this coding process. The coder is completely characterized by the sampling frequency, the quantizing step size, Δs, and the postreconstruction filter.

The equations for prediction and for the residual error of the delta modulator are of the form

$$X(n|n-1) = aX(n-1|n-1) \tag{11.61a}$$

$$d(n) = X(n) - X(n|n-1) \tag{11.61b}$$

where n is a sample index. The weighting term a in Equation (11.61a) is nominally set to the correlation coefficient $C_X(1)$, which in light of the high sample rate is very close to unity. Some authors distinguish between delta modulation and 1-bit DPCM on the basis of this parameter, a. When $a = 1$, the system is called delta modulation, and when a is not limited to 1, it is called 1-bit DPCM. The predictor-corrector loop is then characterized as an integrator. Robust recovery from channel-induced errors at the receiver requires that the weighting term not be set to unity, but rather to a smaller value, so that the errors have only a finite

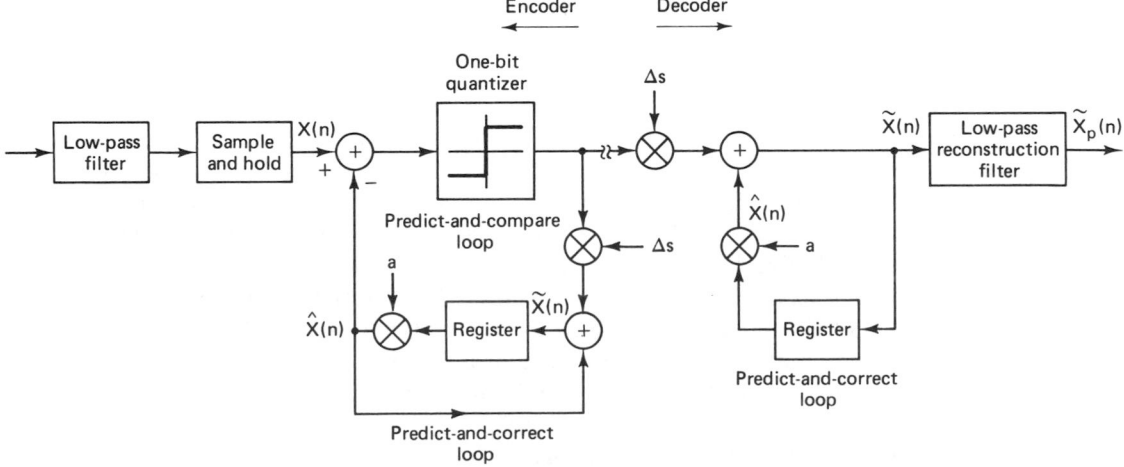

Figure 11.21 One-tap predictor delta modulation process.

persistence. This one-tap predictor is characterized as a leaky integrator; the optimum value of the coefficient to minimize the effects of channel-induced errors while simultaneously controlling prediction errors is [8]

$$a^* = \frac{1 - \sqrt{1 - C_X^2(1)}}{C_X(1)} \tag{11.62}$$

The 1-bit quantized error term is of the form

$$e(n) = \text{sgn } d(n) \tag{11.63a}$$

The correction equation derived from the 1-bit quantized error term is

$$X(n|n) = X(n|n - 1) + \Delta s(n) \text{ sgn } d(n) \tag{11.63b}$$

The step size, $\Delta s(n)$, in Equation (11.63b) is set to a constant for ordinary delta modulation, sometimes called *linear delta mod* (LDM), and is changed by an adaption algorithm in response to the error sequence in the adaptive forms of delta modulation.

An example of the performance of a delta modulator one-tap linear predictor-corrector to an input signal is shown in Figure 11.22. Shown are the input signal, the prediction signal, $\hat{X}(n)$, overlaid, the low-pass filtered version of the prediction signal, and the delta modulation sequence. In this example, the sampling fre-

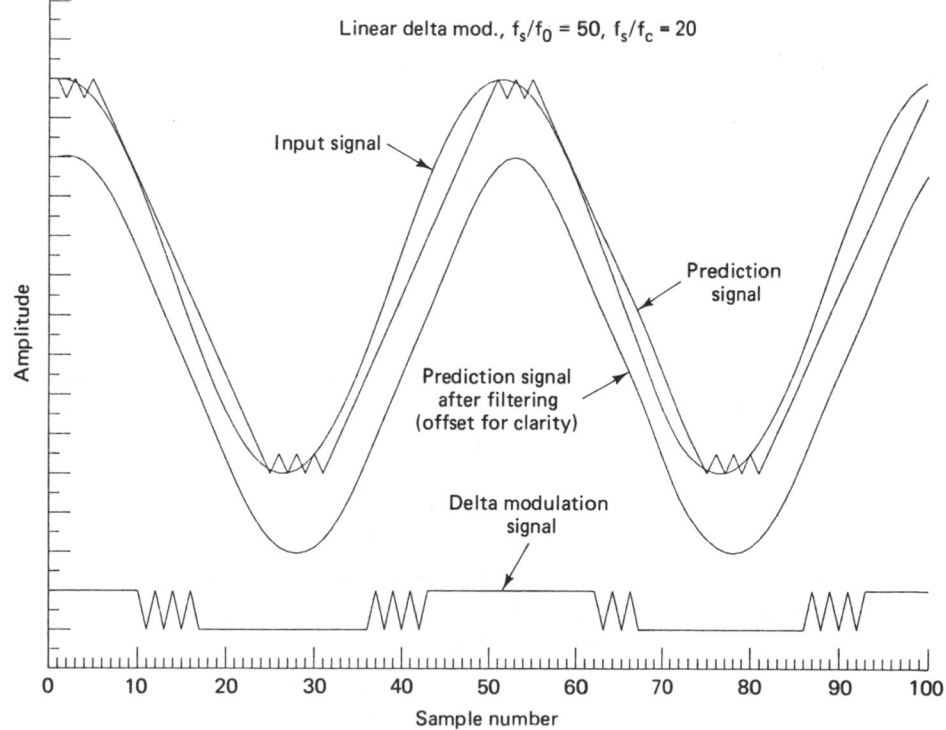

Figure 11.22 Example of waveshapes from a linear delta modulation process.

quency, f_s, is 50 times the frequency of the input sinusoid (i.e., $f_s/f_0 = 50$), and f_s is 20 times the cut-off frequency, f_c, of the low-pass reconstruction filter (i.e., $f_s/f_c = 20$). Note that the difference between the input and the corrected prediction can be characterized by two essentially different types of tracking errors: slope overload and granular noise. *Slope overload* occurs when the slope of the input signal exceeds the delta mod's maximum slope of $\Delta s(1/T_s)$, where T_s is the sampling time. *Granular noise* occurs when the input signal slope falls within the slope capabilities of the delta modulator and the prediction loop attempts to match the local slope with a sequence of positive and negative Δs steps.

We note that the granular noise can be reduced by *decreasing* the size of the increment Δs, and that the slope overload noise can be reduced by *increasing* the size of the increment Δs. This, of course, is the justification for an adaptive step size. For a fixed step size, Δs must be selected to minimize the two sources of error. We first examine the performance of the delta modulator operating without slope overload and then examine a more general case.

Let us assume that the input bandwidth is defined as zero to f_m. This specifies the analog postfilter and the Nyquist sampling rate of $2f_m$. Let the actual sample rate of the LDM exceed the Nyquist rate by the factor F (i.e., $f_s = 2Ff_m$). Further, the step size of the LDM is Δs. The input to the LDM is a sinusoid of amplitude A and of frequency $2\pi f_0$, such as

$$x(t) = A \sin 2\pi f_0 t \tag{11.64}$$

The derivative of the input is

$$\dot{x}(t) = 2\pi f_0 A \cos 2\pi f_0 t \tag{11.65a}$$

and the maximum input slope is

$$\dot{x}_{\max} = 2\pi f_0 A \tag{11.65b}$$

The maximum slope capability of the delta mod is

$$\text{slope}_{\max} = \Delta s\left(\frac{1}{T_s}\right) = \Delta s f_s \tag{11.66}$$

Equating the maximum slope of the predictor to the maximum slope of the input, we have the maximum amplitude of the input signal as a function of step size, sample frequency, and input frequency.

$$A = \frac{\Delta s f_s}{2\pi f_0} \tag{11.67}$$

As expected, the amplitude of the input sinusoid which avoids slope overload is proportional to the step size Δs and the ratio of the sample rate to the input frequency. The signal power, σ_s^2 in the sinusoid is

$$\sigma_s^2 = \frac{A^2}{2} = \frac{1}{2}\left(\frac{\Delta s f_s}{2\pi f_0}\right)^2 \tag{11.68}$$

The only source of reconstruction error is the granular quantizing noise. We can reasonably assume that this noise is uniformly distributed between $+\Delta s$ and $-\Delta s$. As shown in Equation (11.21), the variance, σ_q^2, of this noise is

$$\sigma_q^2 = \tfrac{1}{3}(\Delta s)^2 \tag{11.69}$$

Thus, using Equations (11.68) and (11.69), the prefiltering signal-to-noise ratio is

$$\text{SNR}_{\text{pre}} = \frac{\sigma_s^2}{\sigma_q^2} \frac{\tfrac{1}{2}(\Delta s f_s/2\pi f_0)^2}{\tfrac{1}{3}(\Delta s)^2} = \frac{3}{8\pi^2}\left(\frac{f_s}{f_0}\right)^2 \tag{11.70}$$

We now estimate the filtering gain (improvement in SNR due to the postreconstruction filtering). Since the reconstruction error waveform is a pattern of random bipolar binary pulses, the autocorrelation function, $R_q(\tau)$, of the error pattern is a triangle wave of peak value $\tfrac{1}{3}(\Delta s)^2$, and the error power spectral density, $G_q(f)$, is

$$G_q(f) = \frac{T_s}{3}(\Delta s)^2 \left(\frac{\sin \pi f T_s}{\pi f T_s}\right)^2 \tag{11.71}$$

where T_s is the sampling time. This power spectral density has a main lobe width defined by zeros located at $-1/T_s$ and $+1/T_s$, as shown in Figure 11.23. The mainlobe amplitude can be approximated by a constant over a narrow bandwidth between $-f_m$ and $+f_m$, the bandwidth of the postprocessing filter. If we assume that the postprocessing low-pass filter following the reconstruction is an ideal filter with unity gain between $-f_m$ and $+f_m$, with $f_m \ll 1/T_s$, the integrated output noise power, $(\sigma_q^2)_{\text{out}}$, between these limits is approximately

$$(\sigma_q^2)_{\text{out}} = 2f_m \frac{T_s}{3}(\Delta s)^2 = \frac{2f_m}{f_s}\frac{1}{3}(\Delta s)^2 \tag{11.72a}$$

where $f_s = 1/T_s$. Equation (11.72a) is obtained by assuming that the input quantizing noise power, $(\sigma_q^2)_{\text{in}}$, from Equation (11.69) is

$$(\sigma_q^2)_{\text{in}} = \tfrac{1}{3}(\Delta s)^2 \tag{11.72b}$$

The ratio of the input to output noise powers is the *postfiltering gain*, which is seen from Equation (11.72) to be $f_s/2f_m$. Thus the postfiltering signal-to-noise ratio, SNR_{pst}, is obtained by multiplying Equation (11.70) by this filtering gain, yielding

$$\text{SNR}_{\text{pst}} = \frac{3}{16\pi^2}\frac{f_s^3}{f_0^2 f_m} \tag{11.73}$$

Note that the SNR for the LDM is proportional to the cube of the bit rate (same as sample rate for LDM).

Example 11.8 Sampled Sinusoid

Determine the SNR for a 1-kHz sinusoid, sampled at 32 kHz, without slope overload, and followed by a 4-kHz postreconstruction filter.

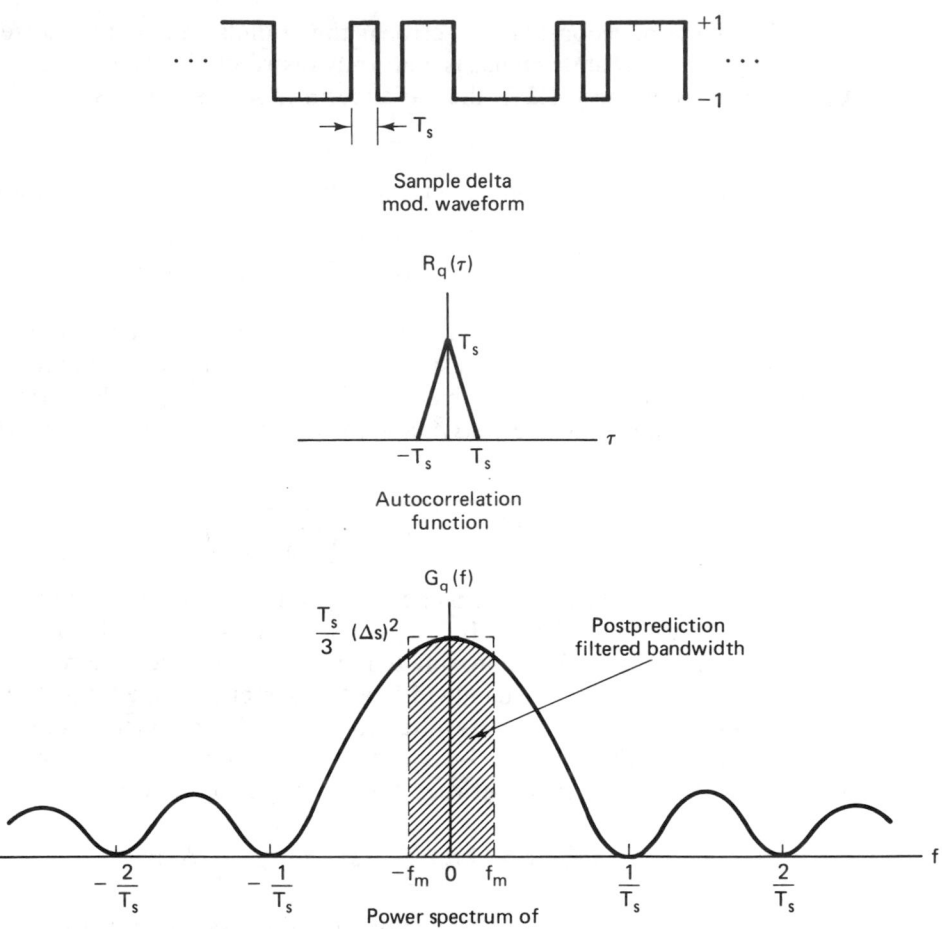

Figure 11.23 Sample time-series, autocorrelation function, and power spectrum for a delta modulation process.

Solution

Substituting in Equation (11.73), we find

$$\text{SNR}_{\text{pst}} = \frac{0.0190(32)^3}{1^2 \times 4} = 155.6$$

$$= 21.9 \text{ dB}$$

We note that the data rate for Example 11.8 is 32 kbits/s, which is the same bit rate obtained by sampling at 8 kHz (Nyquist) with 4 bits/sample. Applying the equations developed in Example 11.4 to the sampling of a full-scale sinusoid, a linear 4-bit PCM quantizer has an average SNR of

$$\text{SNR} = (1.5)2^{2b} = 1.7 + 6b \text{ (dB)} \qquad (11.74\text{a})$$

$$= 25.7 \text{ (dB)} \qquad (11.74\text{b})$$

For all the simplicity of the LDM, it does not perform as well as even a 4-bit linear PCM coder. This is true! The reason the linear PCM does so well is that its SNR increases exponentially with the bit rate while the LDM's SNR increases only as the cube of the bit rate. The gains to be realized with the delta modulator will be treated after incorporating an adaptive step size.

11.3.4 Adaptive Prediction

The prediction gain to be had in classical predictive coders are proportional to the ratio of the *signal variance* to *prediction error variance*. This is because for a fixed quantizing noise level, fewer bits are required to describe a signal with smaller energy. The utility of the predictive coder is limited by possible mismatches between the source signal and the predictor filter. The sources of mismatch are related to the time-varying behavior (i.e., nonstationarity) of the amplitude distribution and of the spectral or correlation properties of the signal. Adaptive encoders incorporate (slow time) auxiliary loops to estimate the parameters required to obtain locally optimal performance. These auxiliary loops periodically schedule modifications to the prediction loop parameters and thus avoid predictor mismatch. The International Telegraph and Telephone Consultative Committee (CCITT) has selected a 32-kbits/s adaptive differential pulse code modulation (ADPCM) coder as a standard for toll-quality speech. This achieves a 2:1 savings in bit rate relative to 64-kbits/s logarithmic compressed PCM.

11.3.4.1 Forward Adaption

In forward adaption algorithms, the input data to be encoded are buffered and processed in order to estimate the local statistics, such as the first N samples of the autocorrelation function. The zero-delay correlation sample, $R_X(0)$, is a short-term estimate of the local variance (zero mean). This estimate is used to adjust the automatic gain control (AGC) in order to obtain an optimal match of the scaled input signal to that of the quantizer dynamic range. This is denoted AQF, for adaptive quantization forward control. The remaining $N - 1$ correlation estimates are used to form new filter coefficients for the prediction filter. This adaption is called *adaptive prediction forward (APF) control*. Figure 11.24 shows this form of the adaptive algorithm. This is an extension of the structure presented in Figure 11.20. Here the predictor coefficients are derived from the input data, now called *side information*, and must be transmitted along with the prediction errors from the encoder to the decoder. The update rate of these adaptive coefficients is related to the length of time the input signal can be considered locally stationary. For example, speech caused by mechanical displacement of the speech articulators (tongue, lips, teeth, etc.) cannot change characteristics more rapidly than 10 or 20 times per second. This suggests an update interval of 50 to 100 ms. Using arithmetically simple, but suboptimal estimating algorithms to compute the local filter parameters makes a higher update rate necessary. An update every 20 ms to compute the parameters of a 10- to 12-tap filter has become a common rate. Prediction gains of 10 to 16 dB can be had with 10-tap filters when feedforward adaption is used with predictive coders [12].

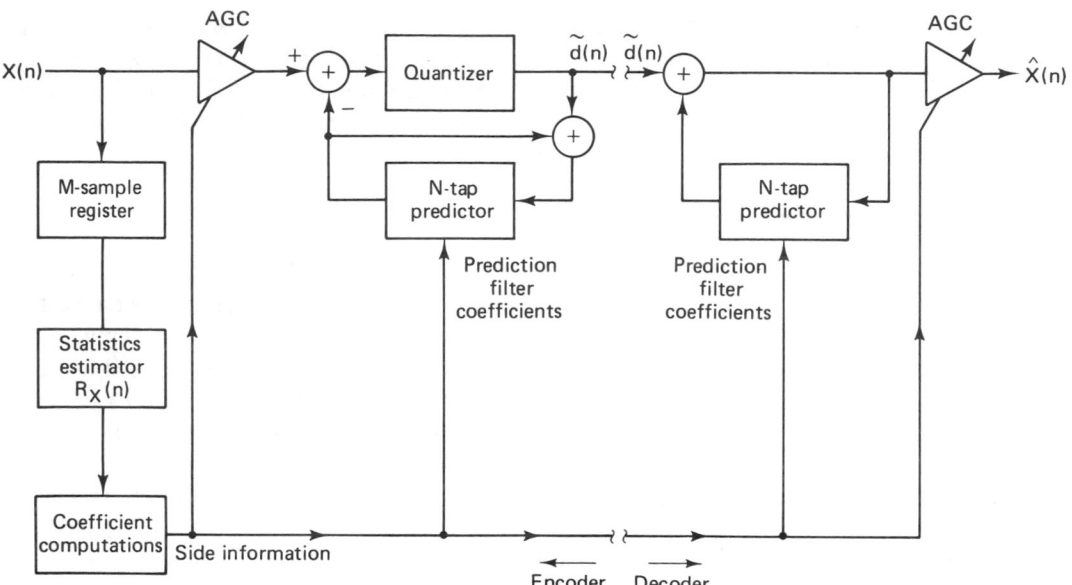

Figure 11.24 Forward adaptive prediction and quantization coding.

11.3.4.2 Backward Adaption

A primary disadvantage of forward adaption algorithms is the need to transmit the side information formed during each adaption cycle. An alternative to the forward adaption algorithms are those that form *backward adaption*. In these algorithms, processing is performed on the output data of the process, as is shown in Figure 11.25, as opposed to being performed on the input data. These same output data are transmitted to the receiver through the communication channel. Thus, applying the same algorithm as the transmitter, the receiver can reconstruct the adaption parameters. Backward adaption of quantizing size is denoted AQB, and backward adaption of predictor coefficients is called APB.

A particularly simple form of the backward adaption is the modification of linear delta modulation (LDM) to form adaptive delta modulation. As described earlier, there are conflicting requirements for large and small LDM step sizes to avoid slope overload and granular noise, respectively. In ADM, a decision algorithm changes the step size based on short run lengths of the successive prediction error polarities. The structure of all ADM algorithms is based on the following argument. When successive errors are of opposing polarity, the delta modulator is successfully tracking and the resultant errors are granular. In this mode, there may be an advantage to reducing the step size $\Delta s(n)$. On the other hand, if a run of successive errors are of the same polarity, the delta modulator may be operating in slope overload. The proper response to the possible slope overload is to increase the step size.

The change of step sizes can be performed by fixed additive scalars or by fixed multiplicative scalars, in which case the algorithm is called *constant factor*

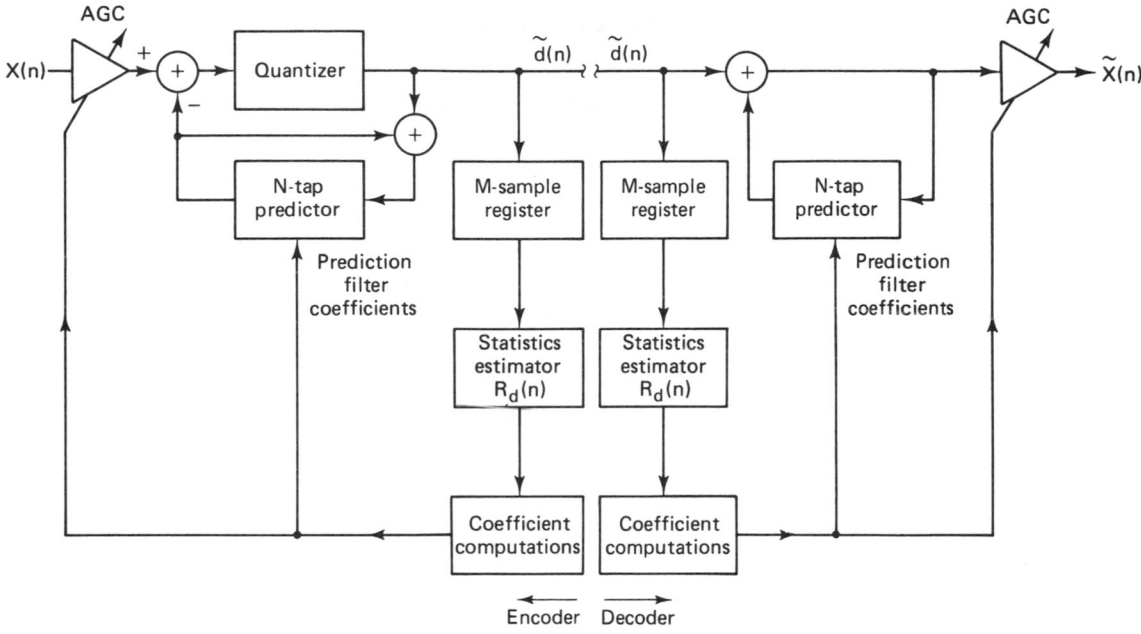

Figure 11.25 Backward adaptive prediction and quantization coding.

delta modulation (CFDM), or may be by a combination of exponential decay and a fixed additive scalar, in which case the algorithm is called *continuously variable slope delta* (CVSD) *modulation.*

Figure 11.26 presents an adaptive delta modulator based on increasing or decreasing the step size by a factor of 50% at each adaption iteration [13]. This is of the form

$$\Delta s(n) = \begin{cases} |\Delta s(n-1)| [d(n) + 0.5d(n-1)] & \text{if } |\Delta s(n-1)| > \Delta s_{\min} \\ \Delta s_{\min} d(n) & \text{if } |\Delta s(n-1)| < \Delta s_{\min} \end{cases}$$

(11.75)

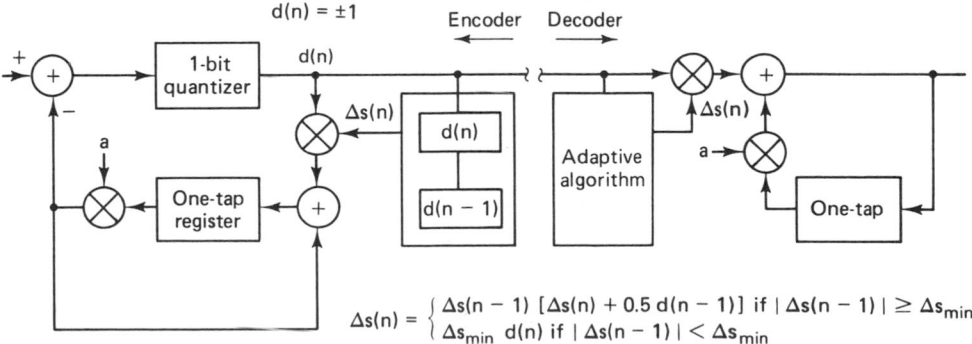

$$\Delta s(n) = \begin{cases} \Delta s(n-1) [\Delta s(n) + 0.5 d(n-1)] & \text{if } |\Delta s(n-1)| \geq \Delta s_{\min} \\ \Delta s_{\min} d(n) & \text{if } |\Delta s(n-1)| < \Delta s_{\min} \end{cases}$$

Figure 11.26 Adaptive delta modulation.

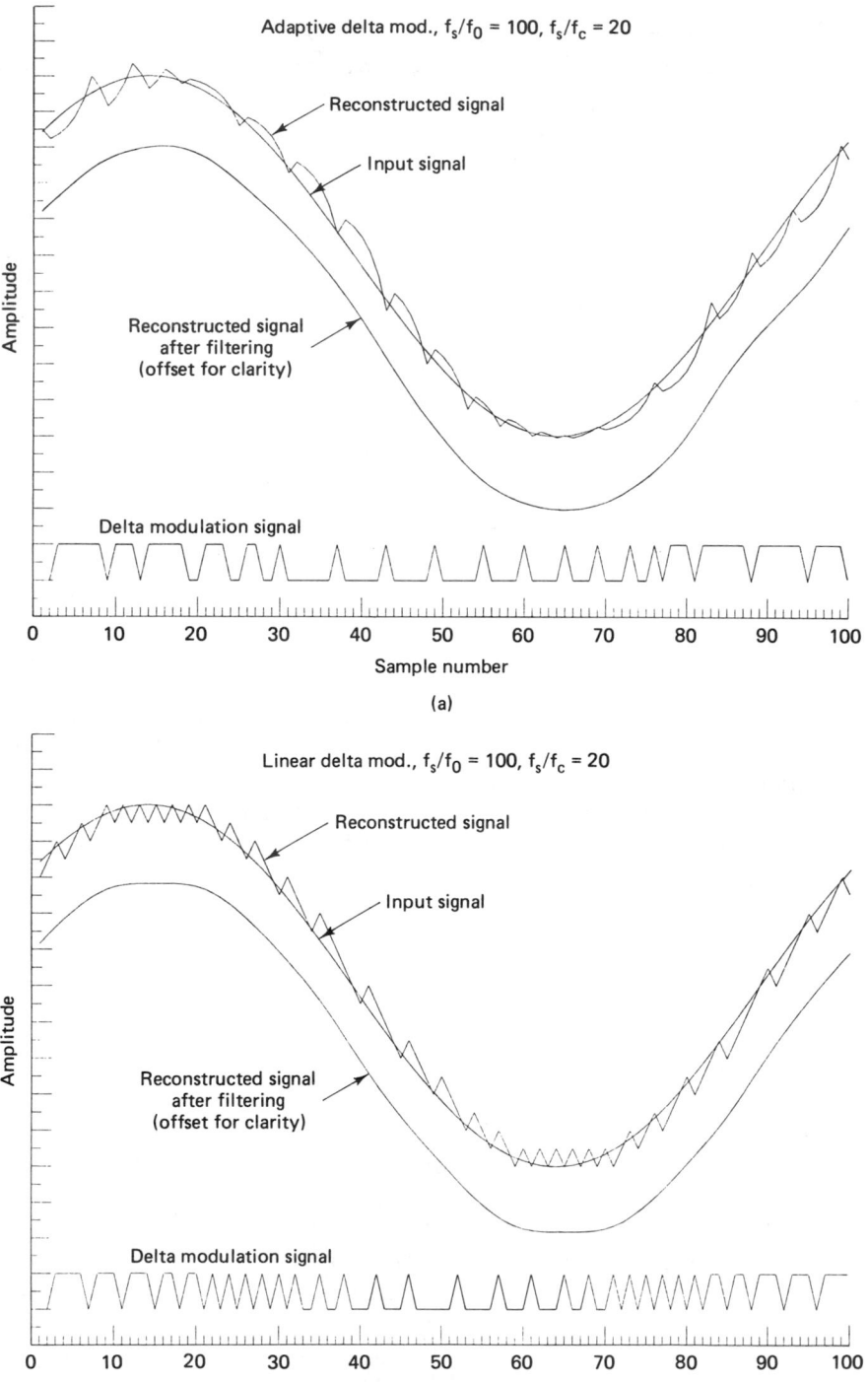

Figure 11.27 Reconstruction performance of an adaptive and linear delta modulation process. (a) Adaptive. (b) Linear.

Figure 11.27a demonstrates the reconstruction performance of this algorithm, and for comparison, Figure 11.27b presents the same signal using a linear delta modulator. The $\Delta s_{\min}$ for this example is $\frac{1}{8}$, the Δs for the LDM is 1, and the amplitude of the input sinusoid in both cases is 10. The ratio of sample frequency to sinusoid frequency is 100 and the cutoff frequency of the analog post filter is $\frac{1}{20}$ of the sample frequency. We see, in Figure 11.27a, that while tracking, the ADM adapts its step sizes, hence its slope, in successive clock intervals. We note that the reduced step size of the ADM results in smaller errors near the extremes of the input signal. We also note that the errors in the two systems are comparable in regions of moderate slope, and while not demonstrated, we can see that the increased slope of the ADM would avoid slope overload if the input signal were to exhibit high slope.

11.4 BLOCK CODING

The quantizers we have examined up to now have been *scalar quantizers*. Scalar quantizers form a *single output sample* based on the present input sample and (possibly) the N previous output samples. Block coders, on the other hand, form a *vector of output samples* based on the present and the N previous input samples. The *coding gain* of a waveform coder is the ratio of the input SNR to the output SNR. When the noise variances of the input and output are equal, this gain is simply the ratio of input-to-output signal variances. The ratio converts directly to 6 dB per bit for the difference between the number of input bits per sample and the average number of output bits per sample. Block coders can achieve impressive coding gains. On the average they can represent sequences quantized to 8 bits with only 1 or 2 bits per sample [8]. Block coding techniques are varied, but a common thread that runs through block coding techniques is the mapping of an input sequence to an alternative coordinate system. This mapping may be to a subspace of a complete space, so that the mapping may not be reversible [8]. Alternatively, a data-dependent editing scheme may be used to identify the subspace of the mapping from which the quantized data are extracted. Block coding techniques are often classified by their mapping techniques, which include, for example, vector quantizers, various orthogonal transform coders, and channelized coders such as the subband coder. Block coders are further described by their algorithmic structures, such as codebook coders, tree coders, trellis coders, discrete Fourier transform, discrete cosine transform, discrete Walsh–Hadamard transform, discrete Karhunen–Loeve transform, and quadrature mirror filter bank. We now examine examples of the various block coding schemes.

11.4.1 Vector Quantizing

Vector quantizers represent an extension of conventional scalar quantization. In scalar quantization a scalar value is selected from a finite list of possible values to represent an input sample. The value is selected to be close (in some sense) to the sample it is representing. The fidelity measures are various weighted mean-square measures which preserve our intuitive concept of distance in terms of ordinary vector lengths. By extension, in vector quantization, a vector is selected from a finite list of possible vectors to represent an input vector of samples. The selected vector is chosen to be close (in some sense) to the vector it is representing.

Each input vector can be visualized as a point in an N-dimensional space. The quantizer is defined by a partition of this space into a set of nonoverlapping volumes [14]. These volumes are called intervals, polygons, and polytopes, respectively, for one-, two-, and N-dimensional vector spaces. The task of the vector quantizer is to determine the volume in which an input vector is located. The output of the optimal quantizer is the vector identifying the centroid of that volume. As in the one-dimensional quantizer, the mean-square error is a function of the boundary locations for the partition and the multidimensional pdf of the input vector.

The description of a vector quantizer can be cast as two distinct tasks. The first is the code design task, which deals with the problem of performing the multidimensional volume quantization (or partition) and selecting the allowable output sequences. The second task is that of using the code and deals with searching for the particular volume in this partition which corresponds (according to some fidelity criterion) to the best description of the source. The two tasks, the partition and the search, may be coupled by the form of the algorithm selected to control the complexity of encoding and decoding. The standard vector coding methods are codebook, tree, and trellis coding algorithms [15].

11.4.1.1 Codebook, Tree, and Trellis Coders

The codebook coders are essentially table look-up algorithms. A list of candidate patterns (codewords) is stored in the codebook memory. Each pattern is identified by an address or pointer index. The coding routine searches through the list of patterns for the one that is closest to the input pattern and transmits to the receiver the address where that pattern can be found in its codebook.

The tree and trellis coders are sequential coders. As such, the allowable codewords of the code cannot be selected independently but must exhibit a node steering structure. This is similar to the structure of the sequential error detection and correction algorithms, which traverse the branches of a graph while forming the branch weight approximation to the input sequence (see Section 6.5.1). A tree graph suffers from exponential memory growth as the dimension or depth of the tree increases. The trellis graph reduces the dimensionality problem by tracking simultaneous contender paths, with an associated path weight metric called *intensity*, through a finite state trellis (see Section 6.3.3).

11.4.1.2 Code Population

The code vectors stored in the codebook, tree, or trellis are the likely or typical vectors. The first task, that of code design, in which the likely code vectors are identified, is called *populating* the code. The methods of determining the code population are classically *deterministic*, *stochastic*, and *iterative*. The deterministic population is a list of preassigned possible outputs based on a simple suboptimal or user perception fidelity criterion or based on a simple decoding algorithm. An example of the former is the coding of the samples in 3-space of the red, green, and blue (RGB) components of a color TV signal. The eye does not have the same resolution to each color and it would appear that the coding can be applied independently to each color to reflect this different sensitivity. The resulting quantizing volumes would be rectangular parallelepipeds. The problem with independent quantizing is that we do not see images in this coordinate system; rather, we see images in the coordinates of luminance, hue, and saturation. A black-and-white photo, for example, uses only the luminance coordinate. Thus quantizing RGB coordinates independently does not result in the smallest amount of user-perceived distortion for a given number of bits. To obtain improved distortion performance, the RGB quantizer should partition its space into regions that reflect the partitions in the alternate space. Alternatively, the quantization could be performed independently in the alternative space by the use of transform coding, treated in Section 11.4.2. Deterministic coding is the easiest to implement but leads to the smallest coding gain (smallest reduction in bit rate for a given SNR).

The stochastic population would be chosen based on an assumed underlying pdf of the input samples. Iterative solutions to the optimal partitions exist and can be determined for any assumed pdf. The overall samples are modeled by the assumed pdf. In the absence of an underlying pdf, iterative techniques based on a large population of training sequences can be used to form the partition and the output population. Training sequences may involve tens of thousands of representative input samples.

11.4.1.3 Searching

Given an input vector and a populated codebook, tree, or trellis, the coder algorithm must conduct a search to determine the best matching contender vector. An exhaustive search over all possible contenders will assure the best match. Coder performance improves for larger-dimensional spaces, but so does complexity. An exhaustive search over a large dimension may be prohibitively time consuming. An alternative is to conduct a nonexhaustive, suboptimal search scheme with acceptably small degradations from the optimal path. Memory requirements and computational complexity are often a driving consideration in the selection of search algorithms. Examples of search algorithms include single-path (best leaving branch) algorithms, multiple-path algorithms, and binary (successive approximation) codebook algorithms. Most of the search algorithms attempt to identify and discard unlikely patterns without having to test the entire pattern.

11.4.2 Transform Coding

In Section 11.4.1 we examined vector quantizers in terms of a set of likely patterns and techniques to determine the one pattern in the set closest to the input pattern. One measure of goodness of approximation is the weighted mean-square error of the form

$$d(\mathbf{X}, \hat{\mathbf{X}}) = (\mathbf{X} - \hat{\mathbf{X}})\mathbf{B}(\mathbf{X})(\mathbf{X} - \hat{\mathbf{X}})^T \tag{11.76}$$

where $B(\mathbf{X})$ is a weight matrix and $\mathbf{X}^T$ is the transpose of $\mathbf{X}$. The minimization may be computationally simpler if the weighting matrix is a diagonal matrix. A diagonal weighting matrix implies a decoupled (or uncorrelated) coordinate set so that the error minimization due to quantization can be performed independently over each coordinate.

Thus transform coding entails the following set of operations, which are shown in Figure 11.28.

1. An invertible transform is applied to the input vector.
2. The coefficients of the transform are quantized.
3. The quantized coefficients are transmitted and received.
4. The transform is inverted with quantized coefficients.

Note that the transform does not perform any source encoding; it merely allows for a more convenient description of the signal vector to permit ease of source encoding. The task of the transform is to map a correlated input sequence into a different coordinate system in which the coordinates have reduced correlation. Recall that this is precisely the task performed by predictive coders. The source encoding occurs with the bit assignment to the various coefficients of the transform. As part of this assignment, the coefficients may be partitioned into subsets which are quantized with different number of bits but not with different quantizing step sizes. This assignment reflects the dynamic range (variance) of each coefficient and may be weighted by a measure that reflects the importance, relative to the human perception [16], of the basis element carried by each coefficient. A subset of the coefficients, for instance, may be set to zero amplitude, or may be quantized with 1 or 2 bits.

The transformation can be chosen to be independent of the data vector. Examples of such transforms are the discrete Fourier transform (DFT), discrete Walsh–Hadamard transform (DWHT), discrete cosine transform (DCT), and the

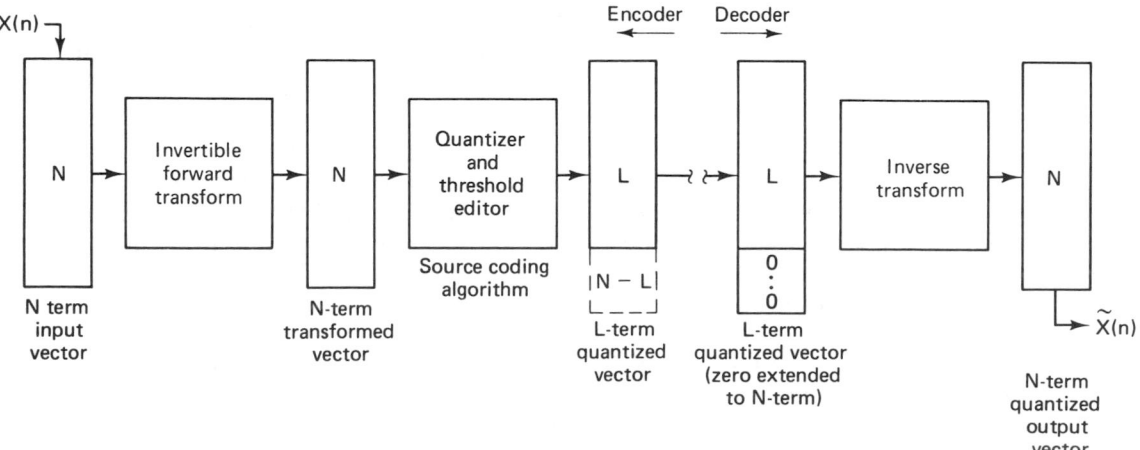

Figure 11.28 Transform coding.

discrete slant transform (DST). The transformation can also be derived from the data vector, as in the discrete Karhunen–Loeve transform (DKLT), sometimes called the principal component transform (PCT) [17].

The data-independent transforms are easiest to implement but do not perform as well as the data-dependent transforms. Often, the attraction of computational simplicity is sufficient justification for using the data-independent transformations. The coding gain penalty for using a good suboptimal transformation is small, typically less than 2 dB, and the degradation is usually cited when demonstrating performance characteristics.

11.4.3 Quantization for Transform Coding

Transform coders are called spectral encoders because the signal is described in terms of a spectral decomposition (in a selected basis set). The spectral terms are computed for nonoverlapped successive blocks of input data. Thus the output of a transform coder can be viewed as a set of time series, one series for each spectral term. The variance of each series can be determined and each can be quantized with a different number of bits. By permitting independent quantization of each transform coefficient we have the option to allocate a fixed number of bits among the transform coefficients to obtain a minimum quantizing error.

11.4.4 Subband Coding

The transform coders of Section 11.4.3 were described as a partition of an input signal into a collection of slowly varying time series, each of which is associated with a particular basis vector of the transform. The spectral terms, the inner product of the data with the basis vectors, are computed by a set of inner products. The set of inner products can be computed by a set of *finite impulse response* (FIR) filters [18]. With this perspective, the transform coder can be considered to be performing a channelization of the input data. By extension, a *subband coder*, which performs a spectral channelization by a bank of contiguous narrowband filters, can be considered a special case of a transform coder. A typical subband coder is shown in Figure 11.29.

Casting the spectral decomposition of the data as a filtering task affords us the option to form a class of custom basis sets (i.e., spectral filters): in particular, basis sets that reflect our user perception preferences and our source models. For example, the quantizing noise generated in a band with large variance will be confined to that band, not spilling into a nearby band with low variance and hence susceptible to low-level noise masking. We also have the option to form filters with equal or with unequal bandwidths (as seen in Figure 11.29). Thus we can independently assign to each subband the sample rate appropriate to its bandwidth and a number of quantizing bits appropriate to its variance. By comparison, in conventional transform coding, each basis vector amplitude is sampled at the same rate.

The subband coder can be designed as a conventional transmultiplexer. Here the input signal is considered to be composed of a number of independent narrow-bandwidth frequency-division-multiplexed subchannels. The encoder dechannel-izes the input frequency-division-multiplexed (FDM) signal into a set of low-data-rate time-division-multiplexed (TDM) channels. After quantization and transmission, the decoder reverses the filtering process, converting the TDM channels back to the original FDM signal. In the classic approach to this process the input signal is demodulated by a bank of single-sideband narrowband filters. This is shown in Figure 11.30. These filters perform a complex (i.e., cosine and sine, or I and Q) heterodyne, thus basebanding the selected center frequency by the $e^{j\phi_k n}$ multiplication, where k is a frequency index and n is a time index. This is a filtering operation that reduces the input bandwidth to the selected channel bandwidth and resamples the signal to the lowest rate that avoids aliasing of the reduced bandwidth channelized data. This down-sampling is often called *deci-mation*, for reasons that escape logic, since nothing happens in tens and nothing is being destroyed. At the receiver, the reverse process is performed. The chan-nelized signals are passed through interpolating filters to increase their sample rate to the desired output sample rate, are heterodyned back to their proper spec-tral position, and are combined to form the original composite signal.

For speech encoding, or more generally, for signals that are related to me-chanical resonances, filter banks with nonequal center frequencies and nonequal bandwidths are desirable. Such filters are called constant-Q, or proportional, filter banks. These filters have logarithmically spaced center frequencies with band-widths proportional to the center frequencies. This proportional spacing appears as uniform spacing and bandwidth when viewed on a log scale and reflects the spectral properties of many physical acoustic sources.

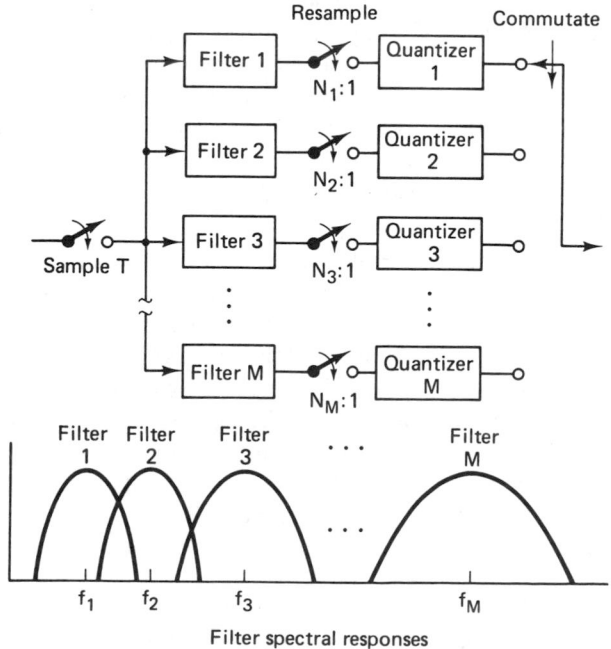

Figure 11.29 Subband coding performed by a channelized spectral decomposition.

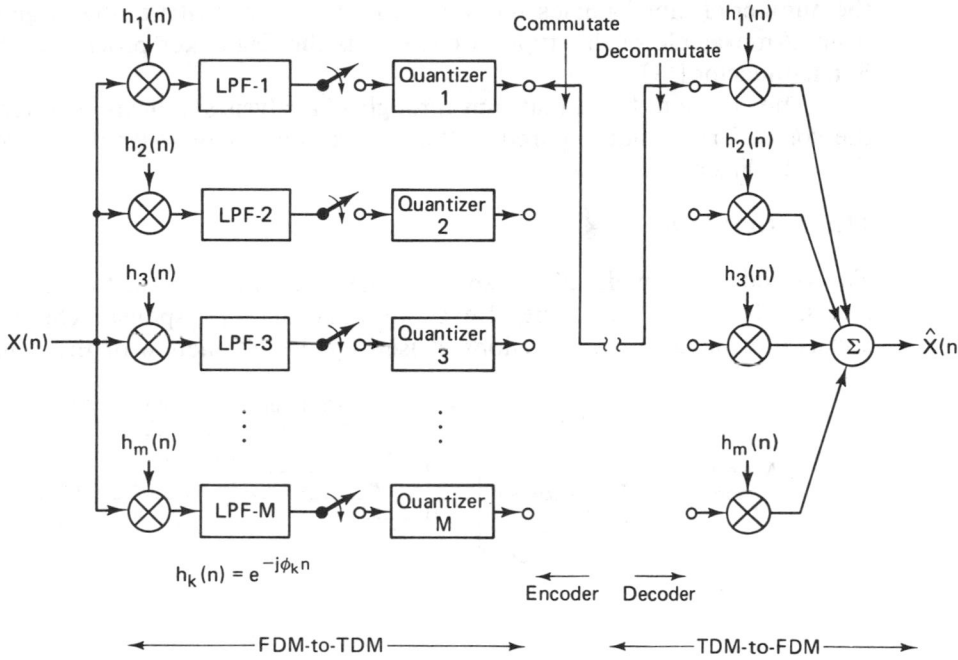

$$h_k(n) = e^{-j\phi_k n}$$

Figure 11.30   Transmultiplexer model of channel coding.

11.5 SYNTHESIS/ANALYSIS CODING

The encoding schemes we have examined until now can be classed as waveform encoders. They construct approximations to the input signals, which minimizes some distance measure between the signal and the approximation. These techniques are very general and can be applied to any signal source. Synthesis/analysis coders, on the other hand, are very signal specific; in particular, they are designed primarily for voice signals. These encoders take advantage of the fact that while the hearing mechanism responds to the amplitude content of a signal's short-term spectrum, it is fairly insensitive to its phase structure. Thus this class of encoder forms a reconstructed signal which approximates the magnitudes and time-varying characteristic of a sequence of the signal's short-term spectra, but makes no attempt to preserve its relative phase.

The spectral characteristics of voice appears to be stationary over periods between 20 and 50 ms. A number of techniques have evolved which analyze the spectral characteristics of voice every 20 ms and use the results of that analysis to synthesize a waveform which exhibits the same short-term power spectrum. Some techniques employ a model of the speech generation mechanism for which model parameters have to be estimated at the update rate. Examples of this type of encoder is the vocoder (voice coder), in its various forms, and the linear predictive coder (LPC). Other techniques manipulate the signal by combinations of spectral modifications and time partitions, which, with side information, reduces the number of time samples required to faithfully reconstruct the original spectrum. An example of this type of encoder is the Digitalker process of National Semiconductor [19].

The common thread that runs through all analysis/synthesis encoders is that the voice signal is not required to "look" like the original signal, but rather, to "sound" like it.

11.5.1 Vocoders

The vocoder is a model of the speech generation mechanism. Speech is a two-step mechanical process, one of excitation and one of response. The excitation consists of the acoustic emissions caused by the interaction of the diaphragm-

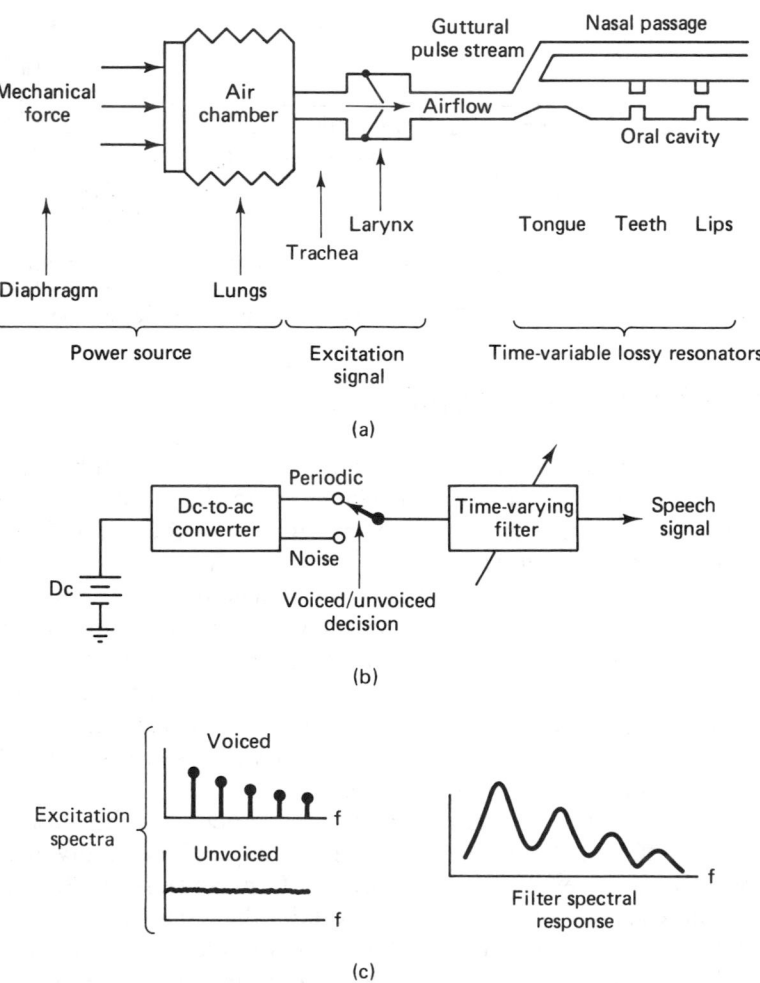

Figure 11.31 Mechanical and electrical modeling of the voice generation process. (a) Mechanical model. (b) Electrical model. (c) Excitation power spectrum and filter response.

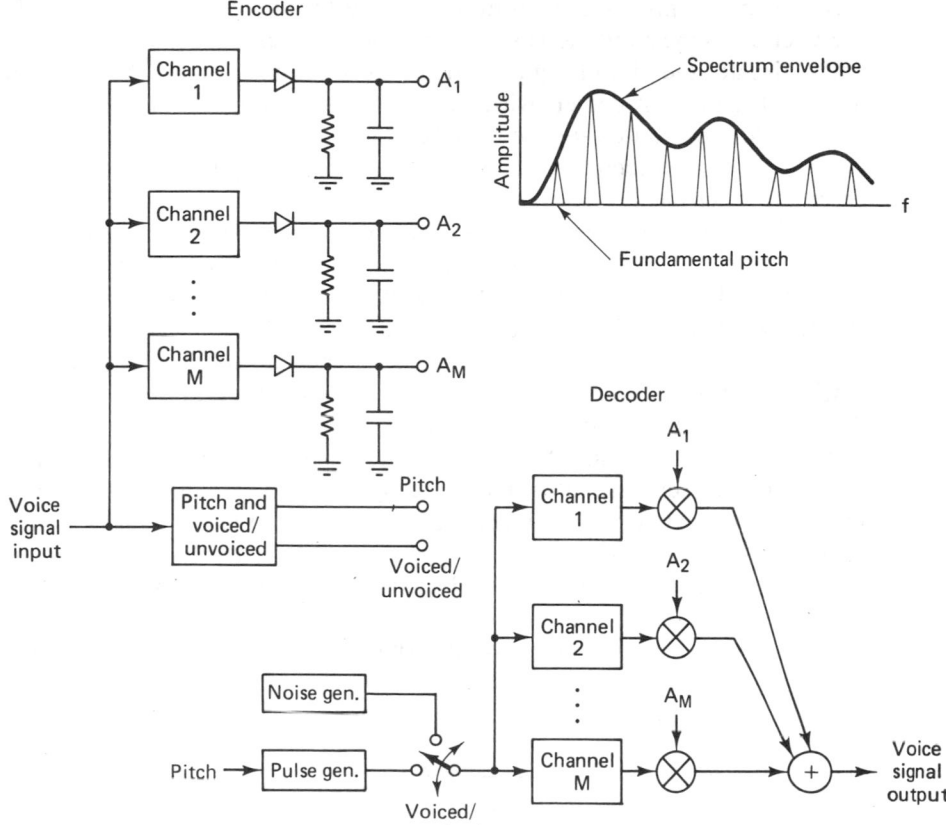

Figure 11.32 Channel vocoder.

supplied air pressure and the valvelike operation of the larynx (or vocal chords). Mechanical and electrical models of the process are shown in Figure 11.31a and b. By control over the vocal chords, the acoustic emissions can be made periodic or turbulent. The periodic emissions are called *voiced sounds*; the turbulent emissions are called *unvoiced*. The frequency of the voiced sound is called pitch. The oral and nasal cavities can be modeled as slowly time-varying lossy resonators which affect the spectral content of the excitation. The changes in the resonators are caused by mechanical displacement of the oral articulators: the lips, teeth, tongue, and so on. A typical excitation power spectrum and filter response are shown in Figure 11.31c. The vocoder models the speech mechanism in a two-step process: the excitation model and the time-varying lossy resonator or filter model.

The first vocoder was described by H. Dudley of Bell Labs and was demonstrated at the 1939 New York World's Fair [20]. The form of his vocoder, now called the *channel vocoder*, is shown in Figure 11.32. The channel vocoder analyzes the signal in what is essentially a channelized spectrum analyzer. This consists of a bank of narrowband filters, detectors, and averagers, which form a local estimate of the power distribution over the frequency band. Also embedded in the vocoder is the excitation estimator, which makes the voiced/unvoiced decision and estimates the pitch frequency if the excitation is voiced. These estimates are transmitted to the receiver, which forms the speech generation model. The spectral envelope estimates are used to alter the channel gains of a bank of narrowband filters, and the voiced/unvoiced decision selects a pulse or noise generator to drive the filter bank.

The bandwidths necessary to describe the parameters of the vocoder can be estimated as follows. Assume that a channel vocoder has 15 filters which span the bandwidth. Each filter is followed by a detector and a low-pass filter with a 20-Hz bandwidth. Thus the channelized bandwidth is 15 times 20 or 300 Hz. Additional bandwidth must be allocated for the voiced/unvoiced and pitch information. The most difficult part of a vocoder implementation is the voiced/unvoiced decision. To preserve speaker naturalness, the lower band of frequencies (200 to 600 Hz), which contains the pitch signal, is simply transmitted to the receiver. The receiver then generates harmonics of this lower band (with a nonlinearity) to excite the channel filters in the remaining spectral region. This implementation is called a *voice-excited vocoder*. Thus, the total vocoder bandwidth is 300 Hz for the channelized data and 400 Hz for the excitation, a total of 700 Hz.

There are many variants of the vocoder. One example is the *formant vocoder*, shown in Figure 11.33, which transmits to the receiver the amplitude and

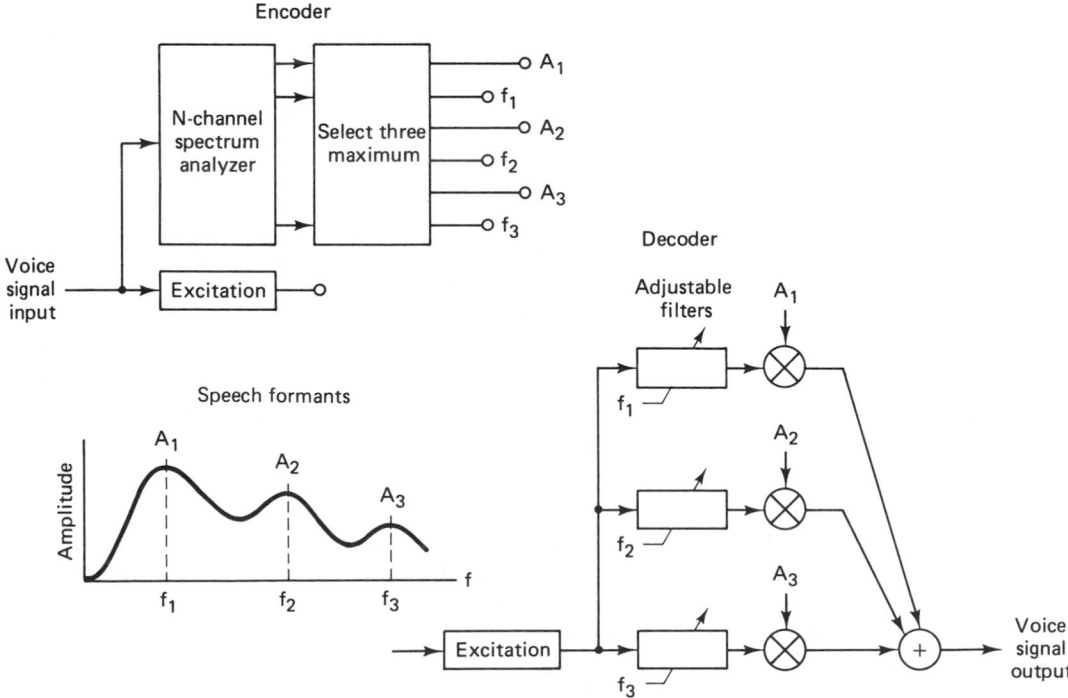

Figure 11.33 Formant vocoder.

spectral position of only the three dominant peaks of the spectral envelope. These dominant peaks, labeled A_1, A_2, and A_3, called formants, represent the dominant resonances of the oral cavity.

11.5.2 Linear Predictive Coding

The adaptive predictors, described in Section 11.3.4, were designed to predict or form good estimates of an input speech signal. In the adaptive form, the prediction coefficients are recomputed as side information from periodic examination of the input data. Then the difference between the input and the prediction is transmitted to the receiver to resolve the prediction error. *Linear predictive coders* (LPCs) are the natural extension of N-tap predictive coders. When the filter coefficients are periodically computed with an optimal algorithm, the prediction is so good that there is (essentially) no prediction error information worth transmitting to the receiver. Rather than transmit these low-level prediction errors, the LPC system transmits the filter coefficients and the voiced/unvoiced excitation decision for the model. Thus the only data sent in LPC is the high-quality side information of the classic adaptive algorithm. An LPC model for voice synthesis is shown in Figure 11.34. The Texas Instruments Speak and Spell learning games use a 12-tap LPC speech synthesizer implemented by a single microchip.

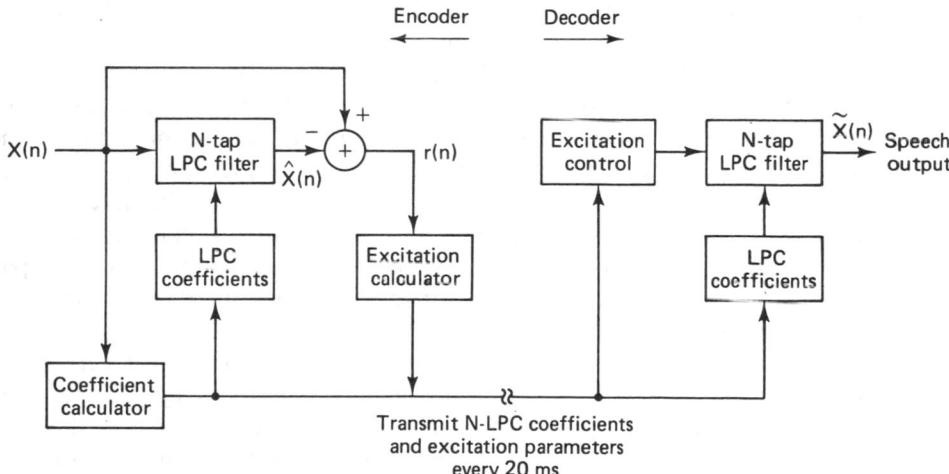

Figure 11.34 Linear predictive coefficient (LPC) speech modeling.

11.6 REDUNDANCY-REDUCING CODING

Coding to reduce the redundancy of a data source entails the selection of an efficient (usually) binary representation of that source. Often this requires the substitution of one binary representation of the source symbols with an alternative representation. The substitution is usually temporary and is performed to achieve an economy of storage or transmission of the discrete source symbols. The binary code assigned to each source symbol must satisfy certain constraints to permit reversal of the substitution. In addition, the code may be further constrained by system considerations such as memory limits or implementation ease.

We are so used to assigning binary codes to represent source symbols that we may lose sight of the arbitrariness of this assignment. The most common example of this is the binary assignments to the cardinal numbers (let's not even consider the negative numbers). We can count in straight binary, binary-coded octal, binary-coded decimal, binary-coded hexadecimal, two-out-of-five decimal, excess-three decimal, and on and on. In this example, ease of computation, error detection, ease of display, or convenience of coding are the considerations for selecting the assignment. For the specific task of data compression, *reduced number of bits* is the primary consideration.

Finite discrete sources are characterized by a set of distinct symbols, $X(n)$ ($n = 1, 2, 3, \ldots, N$), called the source alphabet, where n is a data index. A complete characterization requires the probability of each symbol and the joint probabilities of the symbols taken two at a time, three at a time, and so on. The symbols may represent a two-level (binary) source, such as the black-and-white levels of a facsimile scan, or a many-symbol source, such as the 49 common characters of Sanskrit. Another common many-symbol alphabet is the keyboard of a computer terminal. These nonbinary symbols are mapped, via a dictionary called a character code (see Figure 2.2 for the ASCII code and Figure 2.3 for the EBCDIC code), to a binary alphabet description.

The standard character codes are of fixed length, such as 5, 6, or 7 bits. The length is usually chosen so that there are enough binary characters to assign a unique sequence to each input alphabet character. These may include the upper- and lowercase letters of the alphabet (A, B, C, . . . , Z, a, b, c, . . . , z), numerals (0, 1, 2, . . . , 9), punctuation (!, ?, :, ", ', . . .), special characters (@, #, $, %, &, *, +, /, . . .), and control characters, such as backspace, return, and so on. Fixed-length codes have the property that character boundaries are separated by a fixed bit count. This allows the conversion of a serial data stream to a parallel data stream by a simple bit counter.

Two code standards may define the same symbol in different ways. For example, the ASCII (7-bit) code has enough bits to assign different binary sequences to the upper- and lowercase versions of each letter. On the other hand, the *Baudot* (5-bit) code with only 32 binary sequences cannot do the same. To account for the full character set, the Baudot code defines two control characters, called *letter shift* (LS) and *figure shift* (FS), to be used as prefixes. When used, these control characters reassign the binary-to-symbol mapping. This works very much like the *shift key* on a typewriter; the shift key reassigns a completely new character set to the keyboard. In a similar fashion, the keyboards on some calculators have two prefix character keys, so that each key stroke can have three possible meanings. Also, some word processor instruction codes use double- and triple-stroke command functions. In a very real sense, these two- and three-word instructions represent a variable-length code assignment. These longer code words are assigned to characters (or instructions) that do not occur as often as those assigned single codewords. What we receive in exchange for using the occasional longer words is more efficient storage (smaller keyboard) or transmission of the source.

Data compression codes are often variable-length codes. Intuitively, we would expect the length of a binary sequence assigned to each alphabet symbol to be inversely related to the probability of that symbol. After all, if a symbol occurs with high probability, it contains little information and should not be assigned much of the system resources. In a similar manner, it would not seem unreasonable to find that when all symbols are equally likely, the code should be of fixed length. Perhaps the best known variable-length code is the Morse code. Samuel Morse counted the quantity of letters in a printer's font drawer to determine the relative frequency of letters in normal text. The variable-length code assignment reflects this relative frequency.

A significant amount of *data compression* can be realized when there is a wide difference in the probabilities of the symbols. To achieve this compression there must also be a sufficiently large number of symbols. Sometimes, in order to have a large enough set of symbols, we form a new set of symbols derived from the original set called an *extension code*. We have already seen this trick in Example 11.3 and will examine the general technique in the next section.

11.6.1 Properties of Codes

Earlier we alluded to properties that a code must satisfy for it to be useful. Some of these properties are obvious, some are not. It is worth listing and demonstrating the *desired properties*. We will consider a three-symbol alphabet with the given probability assignment, shown below. Listed with the input alphabet are six binary code assignments. Scan these for a moment and try to determine which codes are practical.

X_i	$P(X_i)$
a	0.73
b	0.25
c	0.02

Symbol	Code 1	Code 2	Code 3	Code 4	Code 5	Code 6
a	00	00	0	1	1	1
b	00	01	1	10	00	01
c	11	10	11	100	01	11

Uniquely Decodeable Property. Uniquely decodeable codes are those that allow us to invert the mapping to the original symbol alphabet. Obviously, code 1 above is not uniquely decodeable because the symbol a and b are assigned the same binary sequence. Thus the first requirement of a useful code is that each symbol be assigned a unique binary sequence. By this condition, all the other codes appear satisfactory until we examine codes 3 and 6 carefully. These codes indeed have unique binary sequences assigned to each symbol. The problem occurs when these code sequences are strung together. For instance, try to decode the binary pattern 1 0 1 1 1 in code 3; is it b, a, b, b, b or b, a, b, c or b, a, c, b? Trying to decode the same sequence in code 6 gives similar difficulties. These codes are not uniquely decodeable even though the individual characters have unique code assignments.

Prefix-Free Property. A sufficient (but not necessary) condition to assure that a code is uniquely decodeable is that no codeword be the prefix of any other code word. Codes that satisfy this condition are called prefix-free codes. Note that code 4 is not prefix-free but is uniquely decodeable. Prefix-free codes also have the property that they are instantaneously decodeable. Code 4 has a property that may be undesirable; it is not instantaneously decodeable. An instantaneously decodeable code is one for which the boundary of the present codeword can be identified by the end of the present codeword rather than by the beginning of the next codeword. For instance, in transmitting the symbol b with the binary sequence 1 0 in code 4, the reciver cannot determine if this is the whole codeword for symbol b or the partial codeword for symbol c.

11.6.1.1 Code Length and Source Entropy

At the beginning of the chapter we described the formal concept of information content and source entropy. We identified the self-information, in bits, about the symbol X_n, denoted $I(X_n)$, as $\log_2 [1/P(X_n)]$. From the perspective that information resolves uncertainty we recognize that the information content of a symbol goes to zero as the probability of that symbol goes to unity. We also defined the *entropy* of a finite discrete source as the average information of that source. From the perspective that information resolves uncertainty, the entropy is the average amount of uncertainty resolved per use of the alphabet. It also represents the average number of bits per symbol required to describe the source. In this sense it is also the lower bound of what can be achieved with some variable-length data compression codes. A number of considerations prevent an actual code from achieving the entropy bound of the input alphabet. These include uncertainty in probability assignment and buffering constraints. The average bit length achieved by a given code is denoted by $\bar{n}$. This average length is computed as the sum of the binary code lengths n_i weighted by the probability of that code symbol $P(X_i)$.

$$\bar{n} = \sum_i n_i P(X_i)$$

A great deal is implied about the performance of a variable-length code when we say *average number of bits*. In a variable-length code assignment some symbols will have code lengths which exceed the average length, while some will have code lengths which are smaller than the average. It may occur that a long pattern of symbols with long codewords are delivered to the coder. The short-term bit rate required to transmit these symbols will exceed the average bit rate of the code. If a channel is expecting data at the average rate, the local excess rate must be buffered in a memory. By the same token, a long pattern of symbols with short codewords may be delivered to the coder. The short-term bit rate required to transmit these symbols will fall short of the average rate of the code. Here the channel will find itself waiting for bits that are not to be had. For this reason, *data buffering* is required to smooth the local statistical variations associated with the input alphabet.

The last caveat is that variable-length codes are designed to operate with a specified list of symbols and proabilities. If the data presented to the coder have a significantly different list of probabilities, the coder buffers may not be able to support the mismatch and underflow or overflow will occur.

11.6.2 Huffman Code

The Huffman code [21] is a prefix-free variable-length code which can achieve the shortest average code length $\bar{n}$ for a given input alphabet. The shortest average code length for a particular alphabet may be significantly greater than the entropy of the source alphabet. This inability to exploit the promised data compression is related to the alphabet, not to the coding technique. Often the alphabet can be modified to form an extension code, and the same coding technique is then reapplied to achieve better compression performance. Compression performance is measured by the *compression* ratio. This measure is equal to the ratio of the average number of bits per sample before compression to the average number of bits per sample after compression.

The Huffman coding procedure can be applied for transforming between any two alphabets. We will demonstrate the application of the procedure between an arbitrary input alphabet and a binary output alphabet. The Huffman code is generated as part of a tree-forming process. The process starts by listing the input alphabet symbols, along with their probabilities (or relative frequencies), in descending order of occurrence. These tabular entries correspond to the branch ends of a tree, as shown in Figure 11.35. Each branch is assigned a branch weight equal to the probability of that branch. The process now forms the tree which supports these branches. The two entries with the lowest relative frequency are merged (at a branch node) to form a new branch with their composite probability. After every merging, the new branch and the remaining branches are reordered (if necessary) to assure that the reduced table preserves the descending probability of occurrence. We call this reordering *bubbling* [22]. During the rearrangement after each merging, the new branch rises through the table until it can rise no further. Thus if we form a branch with a weight of 0.2 and during the bubbling process find two other branches already with the 0.2 weight, the new branch is bubbled to the top of the 0.2 group, as opposed to simply joining it. The bubbling to the top of the group results in a code with reduced code length variance but otherwise a code with the same average length as that obtained by simply joining the group. This reduced code length variance lowers the chance of buffer overflow.

As an example of this part of the code process, we will apply the Huffman procedure to the input alphabet shown in Figure 11.35. The tabulated alphabet and the associated probabilities are shown on the figure. After forming the tree,

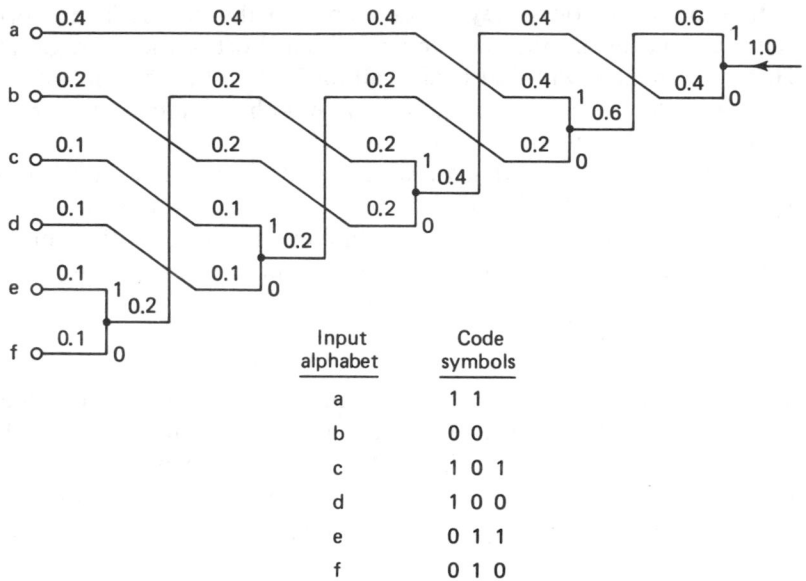

Input alphabet	Code symbols
a	1 1
b	0 0
c	1 0 1
d	1 0 0
e	0 1 1
f	0 1 0

Figure 11.35 Huffman coding tree for a six-character set.

each branch node is labeled with a binary 1/0 decision to distinguish the two branches. The labeling is arbitrary, but for consistency, at each node we will label the branch going up with a "1" and the branch going down with a "0." After labeling the branch nodes we trace the tree path from the base of the tree (far right) to each output branch (far left). The path contains the binary sequence to reach that branch. In the table below, we have listed at each end branch the path sequence corresponding to each path.

X_i	$P(X_i)$	Code	n_i	$n_i P(X_i)$
a	0.4	1 1	2	0.8
b	0.2	0 0	2	0.4
c	0.1	1 0 1	3	0.3
d	0.1	1 0 0	3	0.3
e	0.1	0 1 1	3	0.3
f	0.1	0 1 0	3	0.3

$$\bar{n} = 2.4$$

where $i = 1, \ldots, 6$. We find that the average code length, $\bar{n}$, for this alphabet is 2.4 bits per character. It does not mean that we have to find a way to transmit a noninteger number of bits. Rather, it means that on the average, 240 bits will have to be moved through the communication channel when transmitting 100 input symbols. For comparison, a fixed-length code required to span the six-character input alphabet would be of length 3 bits, and the entropy of the input alphabet, using Equation (11.2) is 2.32 bits. Thus this code offers a compression ratio of 1.25 (3.0/2.4) and achieves 96.7% (2.32/2.40) of the possible compression ratio.

As another example, one for which we can demonstrate the use of code extension, let us examine the three-character alphabet presented in Section 11.6.1.

X_i	$P(X_i)$
a	0.73
b	0.25
c	0.02

The Huffman code tree for this alphabet is shown in Figure 11.36, and the details are tabulated below.

X_i	$P(X_i)$	Code	n_i	$n_i P(X_i)$
a	0.73	1	1	0.73
b	0.27	01	2	0.54
c	0.02	00	2	0.04
			$\bar{n}$ =	1.31

where $i = 1, \ldots, 3$. The average code length for this Huffman code is 1.31 bits; it would be 2 bits for a fixed-length code. The compression ratio for this code is 1.53. Again, using Equation (11.2), the entropy for the alphabet is 0.9443 bit, so that the efficiency (0.944/1.31 = 72%) of the code is significantly smaller than for the preceding example.

To improve coding efficiency or to achieve greater compression gain, we have to redefine the source alphabet. A larger source alphabet holds the promise of increased variability, one requirement to realize a reduction in average code length, and an increased number of tree branches for assigning the variable-length code. We do this by selecting characters two at a time from the source alphabet to be new characters in the extension alphabet. If we assume that the symbols are independent, the probability of each new element is the product of the indi-

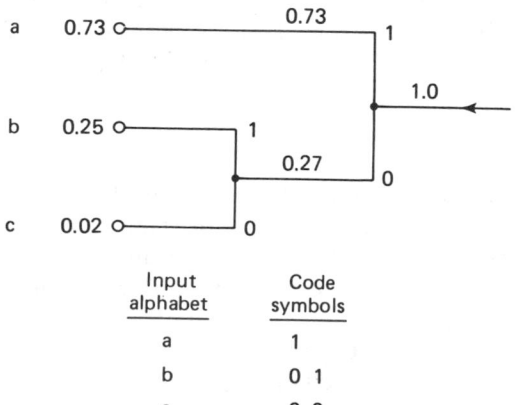

Figure 11.36 Huffman coding tree for a three-character set.

vidual probabilities. The extension alphabet is as follows:

X_i	$P(X_i)$	Code	n_i	$n_i P(X_i)$
aa	0.5329	1	1	0.5329
ab	0.1825	00	2	0.3650
ba	0.1825	011	3	0.5475
bb	0.0625	0101	4	0.2500
ac	0.0146	01000	5	0.0730
ca	0.0146	010011	6	0.0876
bc	0.0050	0100100	7	0.0350
cb	0.0050	01001011	8	0.0400
cc	0.0002	01001010	8	0.0016

$$\bar{n} = 1.9326 \text{ bits/two symbols}$$

$$= 0.9663 \text{ bit/symbol}$$

where $i = 1, \ldots, 9$, and the code sequence for each X_i has been found by the use of the Huffman procedure described above. The compression ratio of this extension code is 2.07 and the coding efficiency is 97.7%.

Extension codes offer a very powerful technique to include the effects of nonindependent symbol sets. For example, in English text, adjacent letters are highly correlated; very common pairs include

th	re	in
sh	he	e_
de	ed	s_
ng	at	r_
te	es	d_

where the dash represents a space. Similarly, common English three tuples include

the	and	for
ing	ion	ess

Thus rather than perform Huffman coding on the individual letters, it is more efficient to extend the alphabet to include all 1-tuples plus common 2-tuples and 3-tuples, and then perform the coding on the extension code.

11.6.3 Run-Length Codes

In many applications, a sequence of symbols to be transmitted or stored is characterized by lengthy runs of specific symbols. Rather than code each symbol of a lengthy run, it makes sense to describe the run with an efficient substitution code. As an example, runs of spaces (the most common symbol in text) are encoded in many communication protocols by a control character followed by the character count. The IBM 3780 BISYNC protocol has an option to replace runs of spaces with an "IGS" character (if EBCDIC; or "GS" if ASCII) followed by a count of 2 to 63. Longer runs are partitioned into successive runs of 63 characters.

The run-length substitution coding can be applied to the original symbol alphabet or the binary representation of that alphabet. Run-length coding is particularly attractive for binary alphabets derived from specific sources. The most important commercial example is facsimile coding used for transmitting documents by instant electronic mail (i.e., U.S. Post Office, *U.S.A. Today*) [23].

11.6.3.1 Huffman Coding for Facsimile Transmission

Facsimile transmission is the process of transmitting a two-dimensional image as a sequence of successive line scans. The most common images are, in fact, documents containing text and figures. The position of the scan lines and the position along a scan line are quantized into spatial locations that define a two-dimensional grid of picture elements called *pixels*. The standard CCITT document is defined to be of width 8.27 in. (20.7 cm) and of length 11.7 in. (29.2 cm), almost 8.5 in. by 11.0 in. The spatial quantization for normal resolution is 1188 pixels/line and 1728 lines/document. The standard also defines a high-resolution quantization of 2376 pixels/line with the same 1728 lines/document. The total number of individual pixels for a normal-resolution facsimile transmission is 2,052,864 and is doubled for high resolution. For comparison, the number of pixels in NTSC (National Television Standard Committee) standard commercial television is 480 × 640 or 307,200. Thus facsimile has 6.7 or 13.4 times the resolution of a standard TV image.

The relative brightness or darkness of the scanned image at each position in the scan is quantized into two levels: B for black and W for white. Thus the signal observed during a scan line is a two-level pattern representing the B and W image intensity under the scan. It is easy to see that a horizontal scan line across this sheet of paper will exhibit a pattern consisting of long runs of B and W levels. The standard CCITT run-length coding scheme to compress the run of B and W levels is based on a modified variable-length Huffman code; it is listed in Table 11.1. Two types of patterns are identified, runs of W and runs of B. Each run length is described by a *partitioned codeword*. The first partition, called the *makeup codeword* or most significant bits (MSB), identifies runs with lengths that are multiples of 64. The second partition, called the *terminating codeword* or least significant bits (LSB), identifies the length of the remaining run. Each run of B (or W) of length from 0 through 63 is assigned a unique Huffman code word, as is each run of length $64 \times K$, $K = 1, 2, \ldots, 27$. A unique END OF LINE (EOL) is also defined in the code, which indicates that no black pixels follow, hence the next line should be started; this is akin to a carriage return on a typewriter.

Example 11.0 Run-Length Code

Use the modified Huffman code to compress the line consisting of 1188 pixel elements

200 W, 10 B, 10 W, 84 B, 884W

TABLE 11.1 Modified Huffman Code for CCITT Facsimile Standard

Run length	White	Black	Run length	White	Black
		Makeup codewords			
64	11011	0000001111	960	011010100	0000001110011
128	10010	000011001000	1024	011010101	0000001110100
192	010111	000011001001	1088	011010110	0000001110101
256	0110111	000001011011	1152	011010111	0000001110110
320	00110110	000000110011	1216	011011000	0000001110111
384	00110111	000000110100	1280	011011001	0000001010010
448	01100100	000000110101	1344	011011010	0000001010011
512	01100101	0000001101100	1408	011011011	0000001010100
576	01101000	0000001101101	1472	010011000	0000001010101
640	01100111	0000001001010	1536	010011001	0000001011010
704	011001100	0000001001011	1600	010011010	0000001011011
768	011001101	0000001001100	1664	011000	0000001100100
832	011010010	0000001001101	1728	010011011	0000001100101
896	011010011	0000001110010	EOL	000000000001	000000000001

Run length	White	Black	Run length	White	Black
		Terminating codewords			
0	00110101	000110111	32	00011011	000001101010
1	000111	010	33	00010010	000001101011
2	0111	11	34	00010011	000011010010
3	1000	10	35	00010100	000011010011
4	1011	011	36	00010101	000011010100
5	1100	0011	37	00010110	000011010101
6	1110	0010	38	00010111	000011010110
7	1111	00011	39	00101000	000011010111
8	10011	000101	40	00101001	000001101100
9	10100	000100	41	00101010	000001101101
10	00111	0000100	42	00101011	000011011010
11	01000	0000101	43	00101100	000011011011
12	001000	0000111	44	00101101	000001010100
13	000011	00000100	45	00000100	000001010101
14	110100	00000111	46	00000101	000001010110
15	110101	000011000	47	00001010	000001010111
16	101010	0000010111	48	00001011	000001100100
17	101011	0000011000	49	01010010	000001100101
18	0100111	0000001000	50	01010011	000001010010
19	0001100	00001100111	51	01010100	000001010011
20	0001000	00001101000	52	01010101	000001000100
21	0010111	00001101100	53	00100100	000000110111
22	0000011	00000110111	54	00100101	000000111000
23	0000100	00000101000	55	01011000	000000100111
24	0101000	00000010111	56	01011001	000000101000
25	0101011	00000011000	57	01011010	000001011000
26	0010011	000011001010	58	01011011	000001011001
27	0100100	000011001011	59	01001010	000000101011
28	0011000	000011001100	60	01001011	000000101100
29	00000010	000011001101	61	00110010	000001011010
30	00000011	000001101000	62	00110011	000001100110
31	00011010	000001101001	63	00110100	000001100111

Solution

Using Table 11.1, we determine the coding for this pattern to be (the spaces are for our reading benefit)

010111	10011	0000100	00111	0000001111	00001101000	000000000001
192 W	8 W	10 B	10 W	64 B	20 B	EOL

Only 56 bits are required to send this line containing a sequence of 1188 bits.

11.7 CONCLUSION

In this chapter we have presented some of the highlights of source coding. We saw that source coding can be applied to digital data and to waveform signals. Digital data can be reconstructed exactly from a reduced data description of a source if the source exhibits correlation between alphabet elements or if the elements are not equally likely. Waveform signals, in general, experience distortion when represented by a digital description. This distortion can be made arbitrarily small by an appropriate increase in bit rate required to describe the source. Source coding can also be applied to waveform sources to obtain reduced-data-rate descriptions if the source exhibits a long correlation interval or if the possible amplitudes are not equally probable.

The system advantage of source coding is the reduced need for the system resources of bandwidth and/or energy per bit required to deliver a description of the source. This advantage is available in exchange for a third system resource—computation and memory. With the cost of these latter resources continuing to fall as they have over the past decade, source coding promises to fill an ever-increasing role in future communication and storage systems. The interested reader is encouraged to examine References [8, 17, 24–26] dealing with source coding.

REFERENCES

1. Papoulis, A., *Probability, Random Variables, and Stochastic Processes,* McGraw-Hill Book Company, New York, 1965.
2. Harris, F. J., "Windows, Harmonic Analysis, and the Discrete Fourier Transform," *Proc. IEEE,* vol. 67, no., Jan. 1979.
3. Martin, G., "Gyroscopes May Cease Spinning," *IEEE Spectrum,* vol. 23, no. 2, Feb. 1986, pp. 48–53.
4. Vanderkooy, J., and Lipshitz, S. T., "Resolution beyond the Least Significant Bit with Dither," *J. Audio Eng. Soc.,* no. 3, Mar. 1984, pp. 106–112.
5. Blesser, B. A., "Digitalization of Audio: A Comprehensive Examination of Theory, Implementation, and Current Practice," *J. Audio Eng. Soc.,* vol. 26, no. 10, Oct. 1978, pp. 739–771.
6. Sluyter, R. J., "Digitalization of Speech," *Philips Tech. Rev.,* vol. 41, no. 7–8, 1983–84, pp. 201–221.

7. Bell Telephone Laboratories Staff, *Transmission Systems for Communications,* Western Electric Co. Technical Publications, Winston-Salem, N.C., 1971.

8. Jayant, N. S., and Noll, P., *Digital Coding of Waveforms,* Prentice-Hall, Inc., Englewood Cliffs, N.J., 1984.

9. Markel, J. D., and Gray, A. H., Jr., *Linear Prediction of Speech,* Springer-Verlag, New York, 1976.

10. Abate, J., "Linear and Adaptive Delta Modulation," *Proc. IEEE,* vol. 55, Mar. 1967, pp. 298–308.

11. Steele, R., *Delta Modulation Systems,* John Wiley & Sons, Inc., New York, 1975.

12. Cummisky, P., Jayant, N., and Flanagan, J., "Adaptive Quantization in Differential PCM Coding of Speech," *Bell Syst. Tech. J.,* vol. 52, 1973, pp. 115–118.

13. Song, C., Garodnick, J., and Schilling, D., "A Variable Step Size Robust Delta Modulator," *IEEE Trans. Commun.,* vol. 19, no. 6, 1971, pp. 1033–1044.

14. Gersho, A., "Asymptotically Optimal Block Quantization," *IEEE Trans. Inf. Theory,* vol. IT25, no. 4, July 1979, pp. 373–380.

15. Gersho, A., "On the Structure of Vector Quantizers," *IEEE Trans. Inf. Theory,* vol. IT28, no. 2, Mar. 1982, pp. 157–166.

16. Jefffress, L., "Masking," in J. Tobias, ed., *Foundations of Modern Auditory Theory,* Academic Press, Inc., New York, 1970.

17. Lynch, T. J., *Data Compression Techniques and Applications,* Lifetime Learning Publications, New York, 1985.

18. Schafer, R. W., and Rabiner, L. R., "Design of Digital Filter Banks for Speech Analysis," *Bell Syst. Tech. J.,* vol. 50, no. 10, Dec. 1971, pp. 3097–3115.

19. DIGITALKER, *Application Note,* National Semiconductor Linear Applications Manual, 1983.

20. Dudley, H., "The Vocoder," *Bell Lab. Rec.,* no. 18, 1919/40, pp. 122–126.

21. Huffman, D. A., "A Method for the Construction of Minimum Redundancy Codes," *Proc. IRE.,* vol. 40, Sept. 1952, pp. 1098–1101.

22. Hamming, R. W., *Coding and Information Theory,* Prentice-Hall, Inc., Englewood Cliffs, N.J., 1980.

23. Usubuchi, T., Omachi, T., and Iinuma, K., "Adaptive Predictive Coding for Newspaper Facsimile," *Proc. IEEE,* vol. 68, July 1980, pp. 807–812.

24. Viterbi, A. J., and J. K. Omura, *Principles of Digital Communication and Coding,* McGraw-Hill Book Company, New York, 1979.

25. Flanagan, J. L., Schroeder, M. R., Atal, B. S., Crochiere, R. E., Jayant, N. S., and Tribolet, J. M., "Speech Coding," *IEEE Trans. Commun.,* COM27, 1979, pp. 710–737.

26. Held, G., *Data Compression, Techniques and Applications, Hardware and Software Considerations,* John Wiley & Sons, Inc., New York, 1983.

PROBLEMS

11.1 A discrete source generates three independent symbols A, B, and C with probabilities 0.9, 0.08, and 0.02, respectively. Determine the entropy of the source.

11.2. A discrete source generates two dependent symbols A and B with conditional probabilities a count of 2 to 63. Longer runs are partitioned into successive runs of 63 characters.

The run-length substitution coding can be applied to the original symbol alphabet or the binary representation of that alphabet. Run-length coding is particularly attractive for binary alphabets derived from specific sources. The most important commercial example is facsimile coding used for transmitting documents by instant electronic mail (i.e., U.S. Post Office, *U.S.A. Today*) [23].

11.6.3.1 Huffman Coding for Facsimile Transmission

Facsimile transmission is the process of transmitting a two-dimensional image as a sequence of successive line scans. The most common images are, in fact, documents containing text and figures. The position of the scan lines and the position along a scan line are quantized into spatial locations that define a two-dimensional grid of picture elements called *pixels*. The standard CCITT document is defined to be of width 8.27 in. (20.7 cm) and of length 11.7 in. (29.2 cm), almost 8.5 in. by 11.0 in. The spatial quantization for normal resolution is 1188 pixels/line and 1728 lines/document. The standard also defines a high-resolution quantization of 2376 pixels/line with the same 1728 lines/document. The total number of individual pixels for a normal-resolution facsimile transmission is 2,052,864 and is doubled for high resolution. For comparison, the number of pixels in NTSC (National Television Standard Committee) standard commercial television is 480 × 640 or 307,200. Thus facsimile has 6.7 or 13.4 times the resolution of a standard TV image.

The relative brightness or darkness of the scanned image at each position in the scan is quantized into two levels: B for black and W for white. Thus the signal observed during a scan line is a two-level pattern representing the B and W image intensity under the scan. It is easy to see that a horizontal scan line across this sheet of paper will exhibit a pattern consisting of long runs of B and W levels. The standard CCITT run-length coding scheme to compress the run of B and W levels is based on a modified variable-length Huffman code; it is listed in Table 11.1. Two types of patterns are identified, runs of W and runs of B. Each run length is described by a *partitioned codeword*. The first partition, called the *makeup codeword* or most significant bits (MSB), identifies runs with lengths that are multiples of 64. The second partition, called the *terminating codeword* or least significant bits (LSB), identifies the length of the remaining run. Each run of B (or W) of length from 0 through 63 is assigned a unique Huffman code word, as is each run of length $64 \times K$, $K = 1, 2, \ldots, 27$. A unique END OF LINE (EOL) is also defined in the code, which indicates that no black pixels follow, hence the next line should be started; this is akin to a carriage return on a typewriter.

Example 11.0 Run-Length Code

Use the modified Huffman code to compress the line consisting of 1188 pixel elements

200 W, 10 B, 10 W, 84 B, 884W

TABLE 11.1 Modified Huffman Code for CCITT Facsimile Standard

Run length	White	Black	Run length	White	Black
		Makeup codewords			
64	11011	0000001111	960	011010100	0000001110011
128	10010	000011001000	1024	011010101	0000001110100
192	010111	000011001001	1088	011010110	0000001110101
256	0110111	000001011011	1152	011010111	0000001110110
320	00110110	000000110011	1216	011011000	0000001110111
384	00110111	000000110100	1280	011011001	0000001010010
448	01100100	000000110101	1344	011011010	0000001010011
512	01100101	0000001101100	1408	011011011	0000001010100
576	01101000	0000001101101	1472	010011000	0000001010101
640	01100111	0000001001010	1536	010011001	0000001011010
704	011001100	0000001001011	1600	010011010	0000001011011
768	011001101	0000001001100	1664	011000	0000001100100
832	011010010	0000001001101	1728	010011011	0000001100101
896	011010011	0000001110010	EOL	000000000001	000000000001

Run length	White	Black	Run length	White	Black
		Terminating codewords			
0	00110101	000110111	32	00011011	000001101010
1	000111	010	33	00010010	000001101011
2	0111	11	34	00010011	000011010010
3	1000	10	35	00010100	000011010011
4	1011	011	36	00010101	000011010100
5	1100	0011	37	00010110	000011010101
6	1110	0010	38	00010111	000011010110
7	1111	00011	39	00101000	000011010111
8	10011	000101	40	00101001	000001101100
9	10100	000100	41	00101010	000001101101
10	00111	0000100	42	00101011	000011011010
11	01000	0000101	43	00101100	000011011011
12	001000	0000111	44	00101101	000001010100
13	000011	00000100	45	00000100	000001010101
14	110100	00000111	46	00000101	000001010110
15	110101	000011000	47	00001010	000001010111
16	101010	0000010111	48	00001011	000001100100
17	101011	0000011000	49	01010010	000001100101
18	0100111	0000001000	50	01010011	000001010010
19	0001100	00001100111	51	01010100	000001010011
20	0001000	00001101000	52	01010101	000000100100
21	0010111	00001101100	53	00100100	000000110111
22	0000011	00000110111	54	00100101	000000111000
23	0000100	00000101000	55	01011000	000000100111
24	0101000	00000010111	56	01011001	000000101000
25	0101011	00000011000	57	01011010	000001011000
26	0010011	000011001010	58	01011011	000001011001
27	0100100	000011001011	59	01001010	000000101011
28	0011000	000011001100	60	01001011	000000101100
29	00000010	000011001101	61	00110010	000001011010
30	00000011	000001101000	62	00110011	000001100110
31	00011010	000001101001	63	00110100	000001100111

$$P(A|A) = 0.8 \quad P(B|A) = 0.2$$
$$P(A|B) = 0.6 \quad P(B|B) = 0.4$$

(a) Determine the probabilities of symbols A and B.

(b) Determine the entropy of the source.

(c) Determine the entropy of the source if the symbols were independent with the same probabilities.

11.3. A 16-bit linear analog-to-digital converter operates over an input range of ± 5.0 V.

(a) Determine the size of a quantile.

(b) Determine the rms quantizing noise voltage.

(c) Determine the average SNR (due to quantizing) for a full-scale sinusoidal input signal.

(d) Consider that the distance traveled on a 100-mile automobile trip is measured to the same accuracy as that of the 16-bit converter. What is the rms error in feet?

11.4. Use the following Microsoft® FORTRAN program to demonstrate the effect of dithering as a linearization technique. The program prompts for an input value to which it adds independent, zero-mean, unit variance random noise prior to truncation and averaging. The subroutine "RAN" is a random number generator. The output is the result of the averaging. See Example 11.5.

```
C   PROGRAM TO DEMONSTRATE THE EFFECT OF DITHERING
        WRITE(*,*) ' '
        WRITE(*,'(A\)') ' ENTER DECIMAL NUMBER TO BE QUANTIZED    -> '
        READ(*,*) ANUMBR
        WRITE(*,'(A\)') ' ENTER NUMBER OF SAMPLES TO BE AVERAGED -> '
        READ(*,*) ICNT
        JSEED=2234
        AVGX=0.0
        VARX=0.0
        ADD=0.5
        IF(ANUMBR.LT.0.0)ADD=-ADD
        WRITE(*,4)
    4   FORMAT(/,30X,'QUANTIZE',/,' INDEX',4X,'INPUT',4X,
      C 'DITHERED',3X,'DITHERED',4X,'RUNNING',4X,'RUNNING',/,9X,
      C 'SAMPLE',5X,'SAMPLE',5X,'SAMPLE',7X,'AVG',8X,'VAR',/)
        DO 15 I=1,ICNT
        XDITHR=ANUMBR+RAN(JSEED)-RAN(JSEED)+RAN(JSEED)-RAN(JSEED)
        IXQUANT=IFIX(XDITHR+ADD)
        X=FLOAT(IXQUANT)
        AVGX=AVGX+(X-AVGX)/FLOAT(I)
        VARX=VARX+((AVGX-ANUMBR)*(AVGX-ANUMBR)-VARX)/FLOAT(I)
        WRITE(*,10)I,ANUMBR,XDITHR,IXQUANT,AVGX,SQRT(VARX)
   10   FORMAT(I4,4X,F7.3,4X,F7.3,5X,I4,6X,F7.3,5X,F6.4)
   15   CONTINUE
        END
```

11.5. A 10-bit A-to-D converter (ADC) is designed to operate over a full-scale range of ± 5.0 V.

(a) Determine the size of a single quantile step.

(b) For a 5.0-V (full-scale) sinusoid, determine the output signal-to-quantizing noise ratio.

(c) For a 0.050-V ($\frac{1}{100}$ of full-scale) sinusoid, determine the output signal-to-quantizing noise ratio.

(d) For an input signal with a Gaussian-distributed amplitude, the probability of saturation is controlled by adjusting the input attenuator so that the saturation level corresponds to four standard deviations. Determine the output signal-to-quantizing noise ratio for this case.

(e) Determine the probability of signal saturation for the signal described in part (d).

11.6. Determine the optimal compression characteristic for the input density function shown in Figure P11.1 (an approximation to a continuous density function).

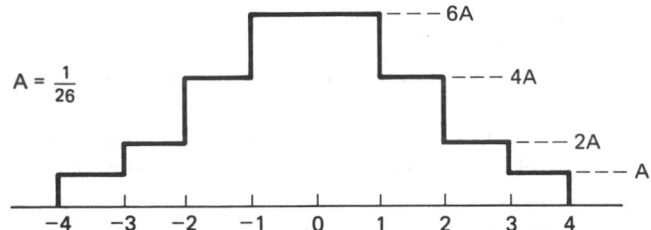

$A = \dfrac{1}{26}$

Figure P11.1

11.7. A 10-bit μ-law converter is designed to operate over a full-scale range of ±5.0 V.
  (a) If $\mu = 100$, determine the output signal-to-quantizing noise ratio for a 5.0-V (full-scale) sinusoid.
  (b) If $\mu = 100$, determine the output signal-to-quantizing noise ratio for a 0.050-V ($\frac{1}{100}$ of full scale) sinusoid.
  (c) Repeat parts (a) and (b) for $\mu = 250$.

11.8. A compact disc (CD) recording system samples each of two stereo signals with a 16-bit A-to-D converter (ADC) at 44.1 kilosamples/s.
  (a) Determine the output signal-to-noise ratio for a full-scale sinusoid.
  (b) If the recorded music is designed to have a crest factor (peak-to-rms ratio) of 20, determine the average output signal-to-quantizing noise ratio.
  (c) The bit stream of digitized data is augmented by the addition of error-correcting bits, substitution bits to aid the clock extraction by a phase-locked loop (PLL), and display and control bit fields. These additional bits represent 100% overhead; that is, 2 bits are stored for each bit generated by the ADC. Determine the output bit rate of the CD recorder system.
  (d) The CD can record an hour's worth of music. Determine the number of bits recorded on a CD.
  (e) For a comparison, a good collegiate dictionary may contain 1500 pages, 2 columns/page, 100 lines/column, 7 words/line, 6 letters/word, and 6 bits/letter. Determine the number of bits required to describe the dictionary and estimate the number of comparable books that can be stored on a CD.

11.9. A 1-bit quantizer is being designed to sample an input sinusoid of amplitude A with uniformly distributed phase. Determine the amplitude X_0, the output level of the 1-bit quantizer, which minimizes the mean-square quantization error.

11.10. A one-step linear predictive filter is to be used to sample a constant-amplitude sinusoid. The ratio of sample frequency to sinusoid frequency is 10.0. Determine the prediction coefficient of the filter. Determine the ratio of output power to input power for the one-tap predictor.

11.11. A two-tap linear predictor filter is being designed to operate in a DPCM system.

Source Coding   Chap. 11

The predictor is of the form

$$\hat{X}(n) = a_1 X(n - 1) + a_2 X(n - 2)$$

(a) Determine the values a_1^* and a_2^* which minimize the mean-square prediction error.

(b) Determine the expression for the mean-square prediction error.

(c) Determine the prediction error power if the correlation coefficient of the input signal is of the form

$$C(n) = \begin{cases} 1 - \dfrac{|n|}{4} & n = 0, 1, 2, 3, 4 \\ 0 & \text{otherwise} \end{cases}$$

(d) Determine the prediction error power if the correlation coefficient of the input signal is of the form

$$C(n) = \cos \theta_0 n$$

11.12. A linear delta modulator is designed to operate at six times the Nyquist rate for a signal with a 3-kHz bandwidth. The modulation step size is 250 mV.

(a) Determine the maximum amplitude of an 800-Hz input signal for which the delta modulator is not in the slope-overload condition.

(b) Determine the prefiltered output signal-to-noise ratio for the signal described in part (a).

(c) Determine the postfiltered output signal-to-noise ratio for the signal described in part (a).

11.13. Design a binary Huffman code for a discrete source of three independent symbols A, B, and C with probabilities 0.9, 0.08, and 0.02, respectively. Determine the average code length for the code.

11.14. Design a binary first-order extension code (two symbols at a time) for the discrete source described in Problem 11.13. Determine the average code length per symbol for this code.

11.15. An input alphabet (a keyboard on a word processor) consists of 100 characters.

(a) If the keystrokes are encoded by a fixed-length code, determine the required number of bits for the encoding.

(b) We make the simplifying assumption that 10 of the keystrokes are equally likely and that each occurs with probability 0.05. We also assume that the remaining 90 keystrokes are equally likely. Determine the average number of bits required to encode this alphabet using a variable-length Huffman code.

11.16. Use the CCITT-modified Huffman facsimile code to encode the following single-line sequence of 2047 black-and-white pixels. Determine the ratio of coded bits to input bits.

1W   1B   2W   2B   4W   4B   8W   8B   16W   16B   32W   32B

64W   64B   128W   128B   256W   256B   512W   512B   1W

CHAPTER 12

Encryption
and
Decryption

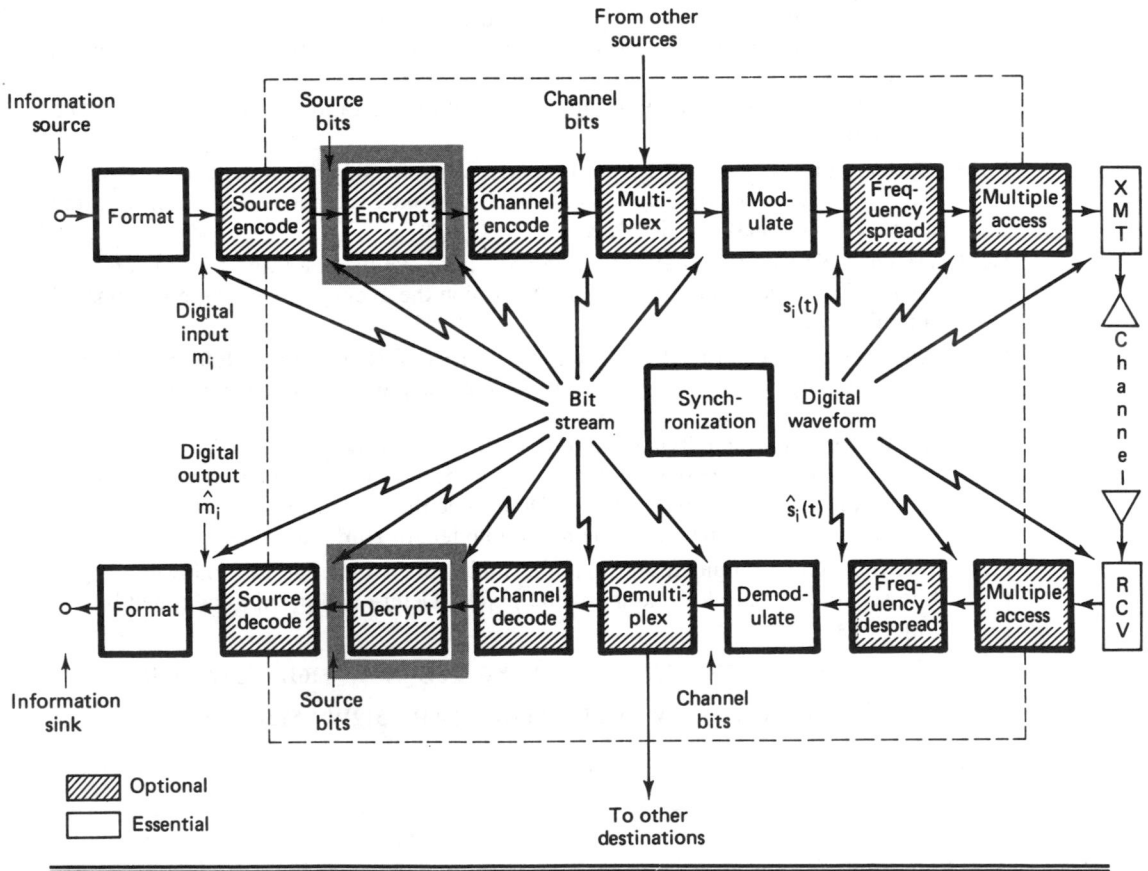

12.1 MODELS, GOALS, AND EARLY CIPHER SYSTEMS

12.1.1 A Model of the Encryption and Decryption Process

The desire to communicate privately is a human trait that dates back to earliest times. Hence the history of secret communications is rich with unique inventions and colorful anecdotes [1]. The study of ways to disguise messages so as to avert unauthorized interception is called *cryptography*. The terms *encipher* and *encrypt* refer to the message transformation performed at the transmitter, and the terms *decipher* and *decrypt* refer to the inverse transformation performed at the receiver. The two primary reasons for using cryptosystems in communications are (1) *privacy*, to prevent unauthorized persons from extracting information from the channel (eavesdropping); and (2) *authentication*, to prevent unauthorized persons from injecting information into the channel (spoofing). Sometimes, as in the case of electronic funds transfer or contract negotiations, it is important to provide the electronic equivalent of a *written signature* in order to avoid or settle any dispute between the sender and receiver as to what message, if any, was sent.

Figure 12.1 illustrates a model of a cryptographic channel. A message, or plaintext, M, is encrypted by the use of an invertible transformation, E_K, that produces a ciphertext, $C = E_K(M)$. The ciphertext is transmitted over an insecure or *public channel*. When an authorized receiver obtains C, he decrypts it with the inverse transformation, $D_K = E_K^{-1}$, to obtain the original plaintext message, as follows:

$$D_K(C) = E_K^{-1}[E_K(M)] = M \qquad (12.1)$$

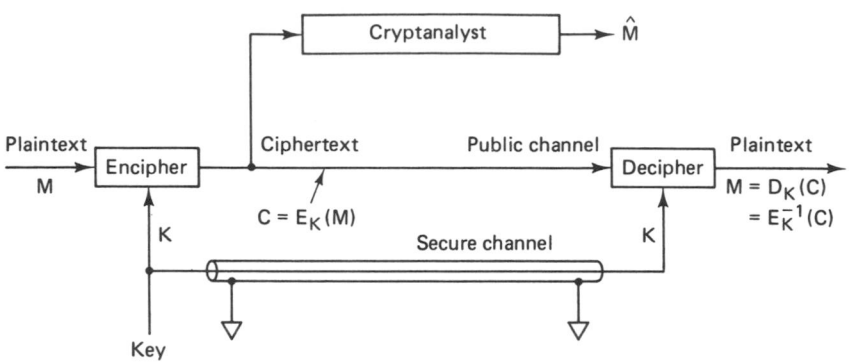

Figure 12.1 Model of a cryptographic channel.

The parameter K refers to a set of symbols or characters called a *key*, which dictates a specific encryption transformation, E_K, from a family of cryptographic transformations. Originally, the security of cryptosystems depended on the secrecy of the entire encryption process, but eventually systems were developed for which the general nature of the encryption transformation or algorithm could be publicly revealed, since the security of the system depended on the specific key. The key is supplied along with the plaintext message for encryption, and along with the ciphertext message for decryption. There is a close analogy here with a general-purpose computer and a computer program. The computer, like the cryptosystem, is capable of a large variety of transformations, from which the computer program, like the specific key, selects one. In most cryptosystems, anyone with access to the key can both encrypt and decrypt messages. The key is transmitted to the community of authorized users over a secure channel (as an example, a courier may be used to hand-carry the sensitive key information); the key usually remains unchanged for a considerable number of transmissions. The goal of the *cryptanalyst* (eavesdropper or adversary) is to produce an estimate of the plaintext, $\hat{M}$, by analyzing the ciphertext obtained from the public channel, without benefit of the key.

Encryption schemes fall into two generic categories: *block encryption*, and *data-stream* or simply *stream encryption*. With block encryption, the plaintext is segmented into blocks of fixed size; each block is encrypted independently from the others. For a given key, a particular plaintext block will therefore be carried into the same ciphertext block each time it appears (similar to block encoding). With data-stream encryption, similar to convolutional coding, there is no fixed block size. Each plaintext bit, m_i, is encrypted with the ith element, k_i, of a sequence of symbols (key stream) generated with the key. The encryption is *periodic* if the key stream repeats itself after p characters for some fixed p; otherwise, it is nonperiodic.

In general, the properties desired in an encryption scheme are quite different from those desired in a channel coding scheme. For example, with encryption, plaintext data should never appear directly in the ciphertext, but with channel coding, codes are often in *systematic form* comprised of unaltered message bits plus parity bits (see Section 5.4.5). Consider another example of the differences between encryption and channel coding. With block encryption, a single bit error at the input of the decryptor might change the value of many of the output bits in the block. This effect, known as *error propagation*, is often a desirable cryptographic property since it makes it difficult for unauthorized users to succeed in spoofing a system. However, in the case of channel coding, we would like the system to correct as many errors as possible, so that the output is relatively unaffected by input errors.

12.1.2 System Goals

The major requirements for a cryptosystem can be stated as follows:

1. To provide an *easy* and *inexpensive* means of encryption and decryption to all authorized users in possession of the appropriate key
2. To ensure that the cryptanalyst's task of producing an estimate of the plaintext without benefit of the key is made *difficult* and *expensive*

Successful cryptosystems are classified as being either *unconditionally secure* or *computationally secure*. A system is said to be *unconditionally secure* when the amount of information available to the cryptanalyst is insufficient to determine the encryption and decryption transformations, no matter how much computing power the cryptanalyst has available. One such system, called a *one-time pad*, involves encrypting a message with a random key that is used one time only. The key is never reused; hence the cryptanalyst is denied information that might be useful against subsequent transmissions with the same key. Although such a system is unconditionally secure (see Section 12.2.1), it has limited use in a conventional communication system, since a new key would have to be distributed for each new message—a great logistical burden. The distribution of keys to the authorized users is a major problem in the operation of any cryptosystem, even when a key is used for an extended period of time. Although some systems can be proven to be unconditionally secure, currently there is no known way to demonstrate security for an arbitrary cryptosystem. Hence the specifications for most cryptosystems rely on the less formal designation of *computational security* for x number of years, which means that under circumstances favorable to the cryptanalyst (i.e., using state-of-the-art computers) the system security could be broken in a period of x years, but could not be broken in less than x years.

12.1.3 Classic Threats

The weakest classification of cryptanalytic threat on a system is called a *ciphertext-only attack*. In this attack the cryptanalyst might have *some* knowledge of the general system and the language used in the message, but the only significant data available to him is the encrypted transmission intercepted from the public channel.

A more serious threat to a system is called a *known plaintext attack*; it involves knowledge of the plaintext *and* knowledge of its ciphertext counterpart. The rigid structure of most business forms and programming languages often provides an opponent with much a priori knowledge of the details of the plaintext message. Armed with such knowledge and with a ciphertext message, the cryptanalyst can mount a known plaintext attack. In the diplomatic arena, if an encrypted message directs a foreign minister to make a particular public statement, and if he does so without paraphrasing the message, the cryptanalyst may be privy to both the ciphertext *and* its exact plaintext translation. While a known plaintext attack is not always possible, its occurrence is frequent enough that a system is not considered secure unless it is designed to be secure against the plaintext attack [2].

When the cryptanalyst is in the position of *selecting* the plaintext, the threat is termed a *chosen plaintext attack*. Such an attack was used by the United States to learn more about the Japanese cryptosystem during World War II. On May 20, 1942, Admiral Yamamoto, Commander-in-Chief of the Imperial Japanese Navy, issued an order spelling out the detailed tactics to be used in the assault of Midway island. This order was intercepted by the Allied listening posts. By this time, the Americans had learned enough of the Japanese code to decrypt most of the message. Still in doubt, however, were some important parts, such as the *place* of the assault. They suspected that the characters "AF" meant Midway island, but to be sure, Joseph Rochefort, head of the Combat Intelligence Unit, decided to use a chosen plaintext attack to trick the Japanese into providing concrete proof. He had the Midway garrison broadcast a distinctive plaintext message in which Midway reported that its fresh-water distillation plant had broken down. The American cryptanalysts needed to wait only two days before they intercepted a Japanese ciphertext message stating that AF was short of fresh water [1].

12.1.4 Classic Ciphers

One of the earliest examples of a monoalphabetic cipher was the *Caesar Cipher*, used by Julius Caesar during the Gallic wars. Each plaintext letter is replaced with a new letter obtained by an *alphabetic shift*. Figure 12.2a illustrates such an encryption transformation, consisting of three end-around shifts of the alphabet. When using this Caesar's alphabet, the message, "now is the time" is encrypted as follows:

Plaintext: N O W I S T H E T I M E

Ciphertext: Q R Z L V W K H W L P H

The decryption key is simply the number of alphabetic shifts; the code is changed by choosing a new key. Another classic cipher system, illustrated in Figure 12.2b, is called the *Polybius square*. Letters I and J are first combined and treated as a single character since the final choice can easily be decided from the context of the message. The resulting 25 character alphabet is arranged in a 5 × 5 array.

Plaintext: A B C D E F G H I J K L M N O P Q R S T U V W X Y Z

Chiphertext: D E F G H I J K L M N O P Q R S T U V W X Y Z A B C

(a)

	1	2	3	4	5
1	A	B	C	D	E
2	F	G	H	IJ	K
3	L	M	N	O	P
4	Q	R	S	T	U
5	V	W	X	Y	Z

(b)

Figure 12.2 (a) Caesar's alphabet with a shift of 3. (b) Polybius square.

Encryption of any character is accomplished by choosing the appropriate row–column (or column-row) number pair. An example of encryption with the use of the Polybius square follows:

Plaintext: N O W I S T H E T I M E

Ciphertext: 33 43 25 42 34 44 32 51 44 42 23 51

The code is changed by a rearrangement of the letters in the 5 × 5 array.

The *Trithemius progressive key*, shown in Figure 12.3, is an example of a *polyalphabetic cipher*. The row labeled shift 0 is identical to the usual arrangement of the alphabet. The letters in the next row are shifted one character to the left with an end-around shift for the leftmost position. Each successive row follows the same pattern of shifting the alphabet one character to the left as compared to the prior row. This continues until the alphabet has been depicted in all possible arrangements of end-around shifts. One method of using such an alphabet is to select the first cipher character from the shift 1 row, the second cipher character from the shift 2 row, and so on. An example of such encryption is

Plaintext: N O W I S T H E T I M E

Ciphertext: O Q Z M X Z O M C S X Q

There are several interesting ways that the Trithemius progressive key can be used. One way, called the *Vigenere key method*, employs a keyword. The key dictates the row choices for encryption and decryption of each successive character in the message. For example, suppose that the word "TYPE" is selected as the key; then an example of the Vigenere encryption method is

Key: T Y P E T Y P E T Y P E

Plaintext: N O W I S T H E T I M E

Ciphertext: G M L M L R W I M G B I

	a	b	c	d	e	f	g	h	i	j	k	l	m	n	o	p	q	r	s	t	u	v	w	x	y	z
Shift:																										
0	A	B	C	D	E	F	G	H	I	J	K	L	M	N	O	P	Q	R	S	T	U	V	W	X	Y	Z
1	B	C	D	E	F	G	H	I	J	K	L	M	N	O	P	Q	R	S	T	U	V	W	X	Y	Z	A
2	C	D	E	F	G	H	I	J	K	L	M	N	O	P	Q	R	S	T	U	V	W	X	Y	Z	A	B
3	D	E	F	G	H	I	J	K	L	M	N	O	P	Q	R	S	T	U	V	W	X	Y	Z	A	B	C
4	E	F	G	H	I	J	K	L	M	N	O	P	Q	R	S	T	U	V	W	X	Y	Z	A	B	C	D
5	F	G	H	I	J	K	L	M	N	O	P	Q	R	S	T	U	V	W	X	Y	Z	A	B	C	D	E
6	G	H	I	J	K	L	M	N	O	P	Q	R	S	T	U	V	W	X	Y	Z	A	B	C	D	E	F
7	H	I	J	K	L	M	N	O	P	Q	R	S	T	U	V	W	X	Y	Z	A	B	C	D	E	F	G
8	I	J	K	L	M	N	O	P	Q	R	S	T	U	V	W	X	Y	Z	A	B	C	D	E	F	G	H
9	J	K	L	M	N	O	P	Q	R	S	T	U	V	W	X	Y	Z	A	B	C	D	E	F	G	H	I
10	K	L	M	N	O	P	Q	R	S	T	U	V	W	X	Y	Z	A	B	C	D	E	F	G	H	I	J
11	L	M	N	O	P	Q	R	S	T	U	V	W	X	Y	Z	A	B	C.	D	E	F	G	H	I	J	K
12	M	N	O	P	Q	R	S	T	U	V	W	X	Y	Z	A	B	C	D	E	F	G	H	I	J	K	L
13	N	O	P	Q	R	S	T	U	V	W	X	Y	Z	A	B	C	D	E	F	G	H	I	J	K	L	M
14	O	P	Q	R	S	T	U	V	W	X	Y	Z	A	B	C	D	E	F	G	H	I	J	K	L	M	N
15	P	Q	R	S	T	U	V	W	X	Y	Z	A	B	C	D	E	F	G	H	I	J	K	L	M	N	O
16	Q	R	S	T	U	V	W	X	Y	Z	A	B	C	D	E	F	G	H	I	J	K	L	M	N	O	P
17	R	S	T	U	V	W	X	Y	Z	A	B	C	D	E	F	G	H	I	J	K	L	M	N	O	P	Q
18	S	T	U	V	W	X	Y	Z	A	B	C	D	E	F	G	H	I	J	K	L	M	N	O	P	Q	R
19	T	U	V	W	X	Y	Z	A	B	C	D	E	F	G	H	I	J	K	L	M	N	O	P	Q	R	S
20	U	V	W	X	Y	Z	A	B	C	D	E	F	G	H	I	J	K	L	M	N	O	P	Q	R	S	T
21	V	W	X	Y	Z	A	B	C	D	E	F	G	H	I	J	K	L	M	N	O	P	Q	R	S	T	U
22	W	X	Y	Z	A	B	C	D	E	F	G	H	I	J	K	L	M	N	O	P	Q	R	S	T	U	V
23	X	Y	Z	A	B	C	D	E	F	G	H	I	J	K	L	M	N	O	P	Q	R	S	T	U	V	W
24	Y	Z	A	B	C	D	E	F	G	H	I	J	K	L	M	N	O	P	Q	R	S	T	U	V	W	X
25	Z	A	B	C	D	E	F	G	H	I	J	K	L	M	N	O	P	Q	R	S	T	U	V	W	X	Y

Figure 12.3 Trithemius progressive key.

where the first letter, T, of the key indicates that the row choice for encrypting the first plaintext character is the row starting with T (shift 19). The next row choice starts with Y (shift 24), and so on. A variation of this key method, called the *Vigenere auto (plain) key method*, starts with a single letter or word used as a *priming key*. The priming key dictates the starting row or rows for encrypting the first or first few plaintext characters, as in the preceding example. Next, the *plaintext characters* themseleves are used as the key for choosing the rows for encryption. An example using the letter ''F'' as the priming key follows:

Key:	F N O W I S T H E T I M
Plaintext:	N O W I S T H E T I M E
Ciphertext:	S B K E A L A L X B U Q

With the auto key method, it should be clear that feedback has been introduced to the encryption process. With this feedback, the choice of the ciphertext is dictated by the contents of the message.

A final variation of the Vigenere method, called the *Vigenere auto (cipher) key method*, is similar to the plain key method in that a priming key and feedback are used. The difference is that after encryption with the priming key, each successive key character in the sequence is obtained from the prior *ciphertext character* instead of from the plaintext character. An example should make this clear. As before, the letter "F" is used as the priming key.

Key:	F S G C K C V C G Z H T
Plaintext:	N O W I S T H E T I M E
Ciphertext:	S G C K C V C G Z H T X

Although each key character can be found from its preceding ciphertext character, it is functionally dependent on *all* the preceding characters in the message plus the priming key. This has the effect of diffusing the statistical properties of the plaintext across the ciphertext, making statistical analysis very difficult for a cryptanalyst. One weakness of the cipher key example depicted here is that the ciphertext contains key characters which will be exposed on the public channel "for all to see." Variations of this method can be employed to prevent such overt exposure [3]. By today's standards Vigenere's encryption schemes are not very secure; his basic contribution was the discovery that nonrepeating key sequences could be generated by using the messages themselves or functions of the messages.

12.2 THE SECRECY OF A CIPHER SYSTEM

12.2.1 Perfect Secrecy

Consider a cipher system with a finite message space $\{M\} = M_0, M_1, \ldots, M_{N-1}$ and a finite ciphertext space $\{C\} = C_0, C_1, \ldots, C_{U-1}$. For any M_i, the a priori probability that M_i is transmitted is $P(M_i)$. Given that C_j is received, the a posteriori probability that M_i was transmitted is $P(M_i|C_j)$. A cipher system is said to have *perfect secrecy* if for every message M_i and every ciphertext C_j, the a posteriori probability is equal to the a priori probability:

$$P(M_i|C_j) = P(M_i) \tag{12.2}$$

Thus for a system with perfect secrecy, a cryptanalyst who intercepts C_j obtains no further information to enable him or her to determine which message was transmitted. A necessary and sufficient condition for perfect secrecy is that for every M_i and C_j,

$$P(C_j|M_i) = P(C_j) \tag{12.3}$$

The schematic in Figure 12.4 illustrates an example of perfect secrecy. In this example, $\{M\} = M_0, M_1, M_2, M_3, \{C\} = C_0, C_1, C_2, C_3, \{K\} = K_0, K_1, K_2, K_3, N = U = 4$, and $P(M_i) = P(C_j) = \frac{1}{4}$. The transformation from message

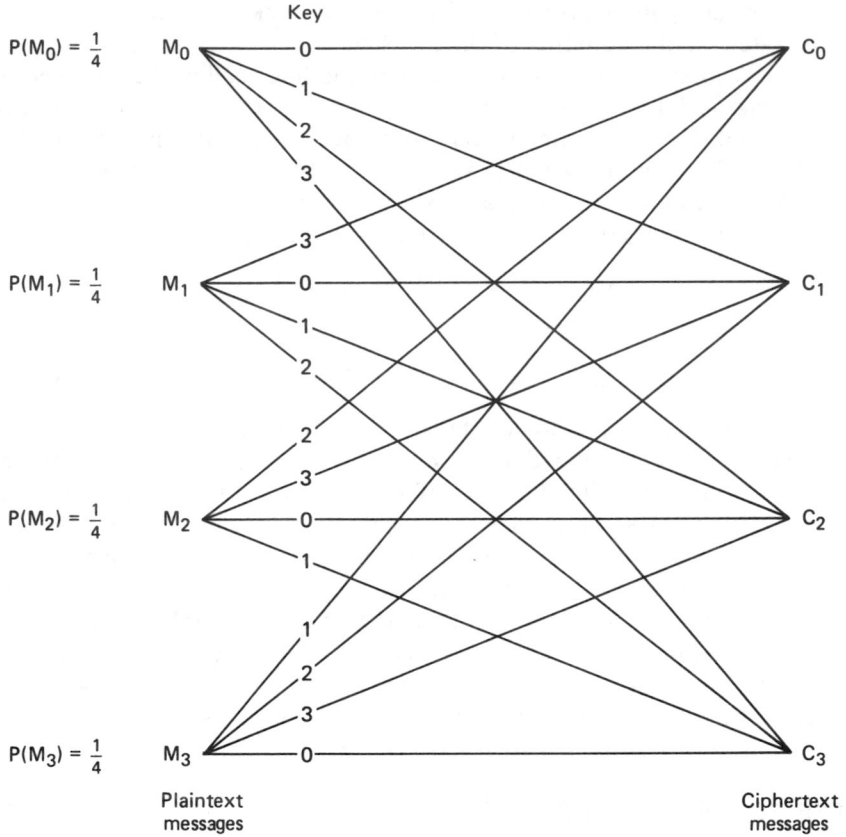

$P(M_0) = \frac{1}{4}$ M_0

$P(M_1) = \frac{1}{4}$ M_1

$P(M_2) = \frac{1}{4}$ M_2

$P(M_3) = \frac{1}{4}$ M_3

Key

C_0

C_1

C_2

C_3

Plaintext
messages

Ciphertext
messages

Figure 12.4   Example of perfect secrecy.

to ciphertext is obtained by

$$C_s = T_{K_j}(M_i) \tag{12.4}$$

$$s = (i + j) \text{ modulo-}N$$

where T_{K_j} indicates a transformation under the key, K_j, and x modulo-y is defined as the remainder of dividing x by y. Thus $s = 0, 1, 2, 3$. A cryptanalyst intercepting one of the ciphertext messages $C_s = C_0, C_1, C_2,$ or C_3 would have no way of determining which of the four keys was used, and therefore whether the correct message is $M_0, M_1, M_2,$ or M_3. A cipher system in which the number of messages, the number of keys, and the number of ciphertext transformations are all equal is said to have perfect secrecy if and only if the following two conditions are met:

1. There is only one key transforming each message to each ciphertext.
2. All keys are equally likely.

If these conditions are not met, there would be some message M_i such that for a given C_j, there is no key that can decipher C_j into M_i, implying that $P(M_i|C_j)$ = 0 for some i and j. The cryptanalyst could then eliminate certain plaintext messages from consideration, thereby simplifying the task. Perfect secrecy is a very desirable objective since it means that the cipher system is unconditionally secure. It should be apparent, however, that for systems which transmit a large number of messages, the amount of key that must be distributed for perfect secrecy can result in formidable management problems, making such systems impractical. Since in a system with perfect secrecy, the number of different keys is at least as great as the number of possible messages, if we allow messages of unlimited length, perfect secrecy requires an infinite amount of key.

Example 12.1 Breaking a Cipher System When the Key Space Is Smaller Than the Message Space

Consider that the 29-character ciphertext

G R O B O K B O D R O R O B Y O C Y P I O C D O B I O K B

was produced by a Caesar cipher (see Section 12.1.4) such that each letter has been shifted by K positions, where $1 \le K \le 25$. Show how a cryptanalyst can break this code.

Solution

Because the number of possible keys (there are 25) is smaller than the number of possible 29-character meaningful messages (there are a myriad), perfect secrecy cannot be achieved. In the original polyalphabetic cipher of Figure 12.3, a plaintext character is replaced by a letter of increasingly higher rank as the row number (K) increases. Hence, in analyzing the ciphertext, we reverse the process by creating rows such that each ciphertext letter is replaced by letters of decreasing rank. The cipher is easily broken by trying all the keys, from 1 to 25, as shown in Figure 12.5, yielding only one key ($K = 10$) that produces the meaningful message: WHERE ARE THE HEROES OF YESTERYEAR (The spaces have been added.)

Example 12.2 Perfect Secrecy

We can modify the key space of Example 12.1 to create a cipher having perfect secrecy. In this new cipher system each character in the message is encrypted using a *randomly selected* key value. The key, K, is now given by the sequence k_1, k_2, ..., k_{29}, where each k_i is a random integer in the range (1, 25) dictating the shift used for the ith character; thus there are a total of $(25)^{29}$ different key sequences. Then the 29-character ciphertext in Example 12.1 could correspond to *any* meaningful 29-character message. For example, the ciphertext could correspond to the plaintext (the spaces have been added)

ENGLISH AND FRENCH ARE SPOKEN HERE

derived by the key: 2, 4, 8, 16, 6, 18, 20, ... Most of the 29-character possibilities can be ruled out because they are not meaningful messages (this much is known without the ciphertext). Perfect secrecy is achieved because interception of the ciphertext in this system reveals no additional information about the plaintext message.

Key											Text																
0	G	R	O	B	O	K	B	O	D	R	O	R	O	B	Y	O	C	Y	P	I	O	C	D	O	B	I	O	K B
1	F	Q	N	A	N	J	A	N	C	Q	N	Q	N	A	X	N	B	X	O	H	N	B	C	N	A	H	N	J A
2	E	P	M	Z	M	I	Z	M	B	P	M	P	M	Z	W	M	A	W	N	G	M	A	B	M	Z	G	M	I Z
3	D	O	L	Y	L	H	Y	L	A	O	L	O	L	Y	V	L	Z	V	M	F	L	Z	A	L	Y	F	L	H Y
4	C	N	K	X	K	G	X	K	Z	N	K	N	K	X	U	K	Y	U	L	E	K	Y	Z	K	X	E	K	G X
5	B	M	J	W	J	F	W	J	Y	M	J	M	J	W	T	J	X	T	K	D	J	X	Y	J	W	D	J	F W
6	A	L	I	V	I	E	V	I	X	L	I	L	I	V	S	I	W	S	J	C	I	W	X	I	V	C	I	E V
7	Z	K	H	U	H	D	U	H	W	K	H	K	H	U	R	H	V	R	I	B	H	V	W	H	U	B	H	D U
8	Y	J	G	T	G	C	T	G	V	J	G	J	G	T	Q	G	U	Q	H	A	G	U	V	G	T	A	G	C T
9	X	I	F	S	F	B	S	F	U	I	F	I	F	S	P	F	T	P	G	Z	F	T	U	F	S	Z	F	B S
10	W	H	E	R	E	A	R	E	T	H	E	H	E	R	O	E	S	O	F	Y	E	S	T	E	R	Y	E	A R
11	V	G	D	Q	D	Z	Q	D	S	G	D	G	D	Q	N	D	R	N	E	X	D	R	S	D	Q	X	D	Z Q
12	U	F	C	P	C	Y	P	C	R	F	C	F	C	P	M	C	Q	M	D	W	C	Q	R	C	P	W	C	Y P
13	T	E	B	O	B	X	O	B	Q	E	B	E	B	O	L	B	P	L	C	V	B	P	Q	B	O	V	B	X O
14	S	D	A	N	A	W	N	A	P	D	A	D	A	N	K	A	O	K	B	U	A	O	P	A	N	U	A	W N
15	R	C	Z	M	Z	V	M	Z	O	C	Z	C	Z	M	J	Z	N	J	A	T	Z	N	O	Z	M	T	Z	V M
16	Q	B	Y	L	Y	U	L	Y	N	B	Y	B	Y	L	I	Y	M	I	Z	S	Y	M	N	Y	L	S	Y	U L
17	P	A	X	K	X	T	K	X	M	A	X	A	X	K	H	X	L	H	Y	R	X	L	M	X	K	R	X	T K
18	O	Z	W	J	W	S	J	W	L	Z	W	Z	W	J	G	W	K	G	X	Q	W	K	L	W	J	Q	W	S J
19	N	Y	V	I	V	R	I	V	K	Y	V	Y	V	I	F	V	J	F	W	P	V	J	K	V	I	P	V	R I
20	M	X	U	H	U	Q	H	U	J	X	U	X	U	H	E	U	I	E	V	O	U	I	J	U	H	O	U	Q H
21	L	W	T	G	T	P	G	T	I	W	T	W	T	G	D	T	H	D	U	N	T	H	I	T	G	N	T	P G
22	K	V	S	F	S	O	F	S	H	V	S	V	S	F	C	S	G	C	T	M	S	G	H	S	F	M	S	O F
23	J	U	R	E	R	N	E	R	G	U	R	U	R	E	B	R	F	B	S	L	R	F	G	R	E	L	R	N E
24	I	T	Q	D	Q	M	D	Q	F	T	Q	T	Q	D	A	Q	E	A	R	K	Q	E	F	Q	D	K	Q	M D
25	H	S	P	C	P	L	C	P	E	S	P	S	P	C	Z	P	D	Z	Q	J	P	D	E	P	C	J	P	L C

Figure 12.5 Example of breaking a cipher system when the key space is smaller than the message space.

12.2.2 Entropy and Equivocation

As discussed in Chapter 7, the amount of information in a message is related to the probability of occurrence of the message. Messages with probability of either 0 or 1 contain no information, since we can be very confident concerning our prediction of their occurrence. The more uncertainty there is in predicting the occurrence of a message, the greater is the information content. Hence when each of the messages in a set is equally likely, we can have *no* confidence in our ability to predict the occurrence of a particular message, and the uncertainty or information content of the message is maximum.

Entropy, $H(X)$, is defined as the average amount of information per message. It can be considered a measure of how much *choice* is involved in the selection of a message, X. It is expressed by the following summation over all possible messages:

$$H(X) = - \sum_X P(X) \log_2 P(X) = \sum_X P(X) \log_2 \frac{1}{P(X)} \qquad (12.5)$$

When the logarithm is taken to the base 2, as shown, $H(X)$ is the *expected number of bits* in an *optimally encoded* message, X. This is not quite the measure that a cryptanalyst desires. He will have intercepted some ciphertext and will want to know how confidently he can predict a message (or key) given that this particular ciphertext was sent. *Equivocation*, $H(X|Y)$, defined as the conditional entropy of X given Y, is a more useful measure for the cryptanalyst in attempting to break the cipher.

$$H(X|Y) = - \sum_{X,Y} P(X, Y) \log_2 P(X|Y)$$

$$= \sum_Y P(Y) \sum_X P(X|Y) \log_2 \frac{1}{P(X|Y)} \qquad (12.6)$$

Equivocation can be thought of as the uncertainty that message X was sent, having received Y. The cryptanalyst would like $H(X|Y)$ to approach zero as the amount of intercepted ciphertext, Y, increases.

Example 12.3   Entropy and Equivocation

Consider a sample message set consisting of eight equally likely messages $\{X\} = X_1, X_2, \ldots, X_8$.

(a) Find the entropy associated with a message from the set $\{X\}$.
(b) Given another equally likely message set $\{Y\} = Y_1, Y_2$. Consider that the occurrence of each message Y narrows the possible choices of X in the following way:

If Y_1 is present: only X_1, X_2, X_3, or X_4 is possible

If Y_2 is present: only X_5, X_6, X_7, or X_8 is possible

Find the equivocation of message X conditioned on message Y.

Solution

(a) $P(X) = \frac{1}{8}$
$H(X) = 8[(\frac{1}{8}) \log_2 8] = 3$ bits/message

(b) $P(Y) = \frac{1}{2}$. For each Y, $P(X|Y) = \frac{1}{4}$ for four of the X's and $P(X|Y) = 0$ for the remaining four X's. Using Equation (12.6), we obtain

$$H(X|Y) = 2[(\frac{1}{2})4(\frac{1}{4} \log_2 4)] = 2 \text{ bits/message}$$

We see that knowledge of Y has reduced the uncertainty of X from 3 bits/message to 2 bits/message.

12.2.3 Rate of a Language and Redundancy

The *true rate* of a language, r, is defined as the average number of *information bits* contained in each character and is expressed for messages of length N by

$$r = \frac{H(X)}{N} \tag{12.7}$$

where $H(X)$ is the message entropy, or the number of bits in the *optimally encoded* message. For large N, estimates of r for written English range between 1.0 and 1.5 bits/character [4]. The *absolute rate* or maximum entropy, r', of a language is defined as the maximum number of information bits contained in each character assuming that all possible sequences of characters are equally likely. The absolute rate is given by

$$r' = \log_2 L \tag{12.8}$$

where L is the number of characters in the language. For the English alphabet r' = $\log_2 26$ = 4.7 bits/character. The true rate of English is, or course, much less than its absolute rate since, like most languages, English is highly redundant and structured.

The *redundancy*, D, of a language is defined in terms of its true rate and absolute rate as follows:

$$D = r' - r \tag{12.9}$$

For the English language with $r' = 4.7$ bits/character and $r = 1.5$ bits/character, $D = 3.2$, and the ratio $D/r' = 0.68$ is a measure of the redundancy in the language.

12.2.4 Unicity Distance and Ideal Secrecy

We stated earlier that perfect secrecy requires an infinite amount of key if we allow messages of unlimited length. With a finite key size, the equivocation of the key $H(K|C)$ generally approaches zero, implying that the key can be uniquely determined and the cipher system can be broken. The *unicity distance* is defined as the smallest amount of ciphertext, N, such that the key equivocation $H(K|C)$ is close to zero. Therefore, the unicity distance is the amount of ciphertext needed to uniquely determine the key and thus break the cipher system. Shannon [5] described an *ideal secrecy* system as one in which $H(K|C)$ does not approach zero as the amount of ciphertext approaches infinity; that is, no matter how much ciphertext is intercepted, the key cannot be determined. The term "ideal secrecy" describes a system that does not achieve perfect secrecy but is nonetheless unbreakable (unconditionally secure) because it does not reveal enough information to determine the key.

Most cipher systems are too complex to determine the probabilities required to derive the unicity distance. However, it is sometimes possible to approximate unicity distance, as shown by Shannon [5] and Hellman [6]. Following Hellman, assume that each plaintext and ciphertext message comes from a finite alphabet of L symbols. Thus there are $2^{r'N}$ possible messages of length N, where r' is the absolute rate of the language. We can consider the total message space partitioned into two classes, meaningful messages, M_1, and meaningless messages M_2:

$$\text{number of meaningful messages} \ = 2^{rN} \tag{12.10}$$

$$\text{number of meaningless messages} = 2^{r'N} - 2^{rN} \tag{12.11}$$

where r is the true rate of the language, and where the a priori probabilities of the message classes are

$$P(M_1) = \frac{1}{2^{rN}} = 2^{-rN} \quad M_1 \text{ meaningful} \tag{12.12}$$

$$P(M_2) = 0 \quad M_2 \text{ meaningless} \tag{12.13}$$

Let us assume that there are $2^{H(K)}$ possible keys (size of the key alphabet), where $H(K)$ is the entropy of the key (number of bits in the key). Assume that all keys are equally likely, that is,

$$P(K) = \frac{1}{2^{H(K)}} = 2^{-H(K)} \tag{12.14}$$

The derivation of the unicity distance is based on a *random cipher* model, which states that for each key K and ciphertext C, the decryption operation $D_K(C)$ yields an independent random variable distributed over all the possible $2^{r'N}$ messages (both meaningful and meaningless). Therefore, for a given K and C, the $D_K(C)$ operation can produce any one of the plaintext messages with equal probability.

Given an encryption described by $C_i = E_{K_i}(M_i)$, a *false solution*, F, arises whenever encryption under another key K_j could also produce C_i either from the message M_i or from some other message M_j; that is,

$$C_i = E_{K_i}(M_i) = E_{K_j}(M_i) = E_{K_j}(M_j) \tag{12.15}$$

A cryptanalyst intercepting C_i would not be able to pick the correct key and hence could not break the cipher system. We are not concerned with the decryption operations that produce *meaningless* messages because these are easily rejected.

For every correct solution to a particular ciphertext there are $2^{H(K)} - 1$ incorrect keys, each of which has the same probability $P(F)$ of yielding a false solution. Because each meaningful plaintext message is assumed equally likely, the probability of a false solution, $P(F)$, is the same as the probability of getting a meaningful message.

$$P(F) = \frac{2^{rN}}{2^{r'N}} = 2^{(r-r')N} = 2^{-DN} \tag{12.16}$$

where $D = r' - r$ is the redundancy of the language. The expected number of false solutions $\overline{F}$ is then

$$\overline{F} = [2^{H(K)} - 1]P(F) = [2^{H(K)} - 1]2^{-DN} \tag{12.17}$$

$$\simeq 2^{H(K) - DN}$$

Because of the rapid decrease of $\overline{F}$ with increasing N,

$$\log_2 \overline{F} = H(K) - DN = 0 \tag{12.18}$$

is defined as the point where the number of false solutions is sufficiently small so that the cipher can be broken. The resulting unicity distance is therefore

$$N = \frac{H(K)}{D} \tag{12.19}$$

We can see from Equation (12.17) that if $H(K)$ is much larger than DN, there will be a large number of meaningful decryptions, and thus a small likelihood of a cryptanalyst distinguishing which meaningful message is the correct message. In a loose sense, DN represents the number of equations available for solving for the key, and $H(K)$ the number of unknowns. When the number of equations is smaller than the number of unknown key bits, a unique solution is not possible and the system is said to be unbreakable. When the number of equations is larger than the number of unknowns, a unique solution is possible and the system can no longer be characterized as unbreakable (although it may still be computationally secure).

It is the predominance of meaningless decryptions that enables cryptograms to be broken. Equation (12.19) indicates the value of using *data compression* techniques prior to encryption. Data compression removes redundancy, thereby increasing the unicity distance. Perfect data compression would result in $D = 0$ and $N = \infty$ for any key size.

Example 12.4 Unicity Distance

Calculate the unicity distance for a written English encryption system, where the key is given by the sequence $k_1, k_2, \ldots, k_{29}$, where each k_i is a random integer in the range $(1, 25)$ dictating the shift number (Figure 12.3) for the ith character. Assume that each of the possible key sequences is equally likely.

Solution

There are $(25)^{29}$ possible key sequences, each of which is equally likely. Therefore, using Equations (12.5), (12.8), and (12.19) we have:

Key entropy: $H(K) = \log_2 (25)^{29} \approx 135$ bits

Absolute rate for English: $r' = \log_2 26 = 4.7$ bits/character

Assumed true rate for English: $r = 1.5$ bits/character

Redundancy: $D = r' - r = 3.2$ bits/character

$$N = \frac{H(K)}{D} = \frac{135}{3.2} \cong 43 \text{ characters}$$

In Example 12.2, perfect secrecy was illustrated using the same type of key sequence described here, with a 29-character message. In this example we see that if the available ciphertext is 43 characters long (which implies that some portion of the key sequence must be used twice), a unique solution may be possible. However, there is no indication as to the computational difficulty in finding the solution. Even though we have estimated the theoretical amount of ciphertext required to break the cipher, it might be computationally infeasible to accomplish this.

12.3 PRACTICAL SECURITY

For ciphertext sequences greater than the unicity distance any system can be solved, in principle, merely by trying each possible key until the unique solution is obtained. This is completely impractical, however, except when the key is extremely small. For example, for a key configured as a permutation of the alphabet, there are $26! \simeq 4 \times 10^{26}$ possibilities (considered small in the cryptographic context). In an exhaustive search, one might expect to reach the right key at about halfway through the search. If we assume that each trial requires a computation time of 1 μs, the total search time exceeds 10^{12} years. Hence techniques other than a brute-force search (e.g., statistical analysis) must be employed if a cryptanalyst is to have any hope of success.

12.3.1 Confusion and Diffusion

A statistical analysis using the frequency of occurrence of individual characters and character combinations can be used to solve many cipher systems. Shannon [5] suggested two encryption concepts for frustrating the statistical endeavors of the cryptanalyst. He termed these encryption transformations confusion and diffusion. *Confusion* involves substitutions that render the final relationship between the key and ciphertext as complex as possible. This makes it difficult to utilize a statistical analysis to narrow the search to a particular subset of the key variable space. Confusion ensures that the majority of the key is needed to decrypt even very short sequences of ciphertext. *Diffusion* involves transformations that smooth out the statistical differences between characters and between character combinations. An example of diffusion with a 26-letter alphabet is to transform a message sequence $M = M_0, M_1, \ldots$ into a new message sequence $Y = Y_0, Y_1, \ldots$ as follows:

$$Y_n = \sum_{i=0}^{s-1} M_{n+i} \qquad \text{modulo-26} \qquad (12.20)$$

where each character in the sequence is regarded as an integer modulo-26, s is some chosen integer, and $n = 1, 2, \ldots$. The new message, Y, will have the same redundancy as the original message, M, but the letter frequencies of Y will be more uniform than in M. The effect is that the cryptanalyst needs to intercept a longer sequence of ciphertext before any statistical analysis can be useful.

12.3.2 Substitution

Substitution encryption techniques, such as the Caesar cipher and the Trithemius progressive key cipher, are widely used in puzzles. Such simple substitution ciphers offer little encryption protection. For a substitution technique to fulfill Shannon's concept of *confusion*, a more complex relationship is required. Figure 12.6 shows one example of providing greater substitution complexity through the use of a nonlinear transformation. In general, *n* input bits are first represented as one of 2^n different characters (binary-to-octal transformation in the example of Figure 12.6). The set of 2^n characters are then permuted so that each character is transposed to one of the others in the set. The character is then converted back to an *n*-bit output.

It can be easily shown that there are $(2^n)!$ different substitution or connection patterns possible. The cryptanalyst's task becomes computationally unfeasible as *n* gets large, say $n = 128$; then $2^n = 10^{38}$, and $(2^n)!$ is an astronomical number. We recognize that for $n = 128$, this substitution box (S-box) transformation is complex (confusion). However, although we can identify the S-box with $n = 128$ as ideal, its implementation is not feasible because it would require a unit with $2^n = 10^{38}$ wiring connections.

To verify that the S-box example in Figure 12.6 performs a *nonlinear transformation*, we need only use the superposition theorem stated below as a test. Let

$$C = Ta + Tb$$
$$C' = T(a + b)$$

(12.21)

where *a* and *b* are input terms, *C* and *C'* are output terms, and *T* is the transformation.

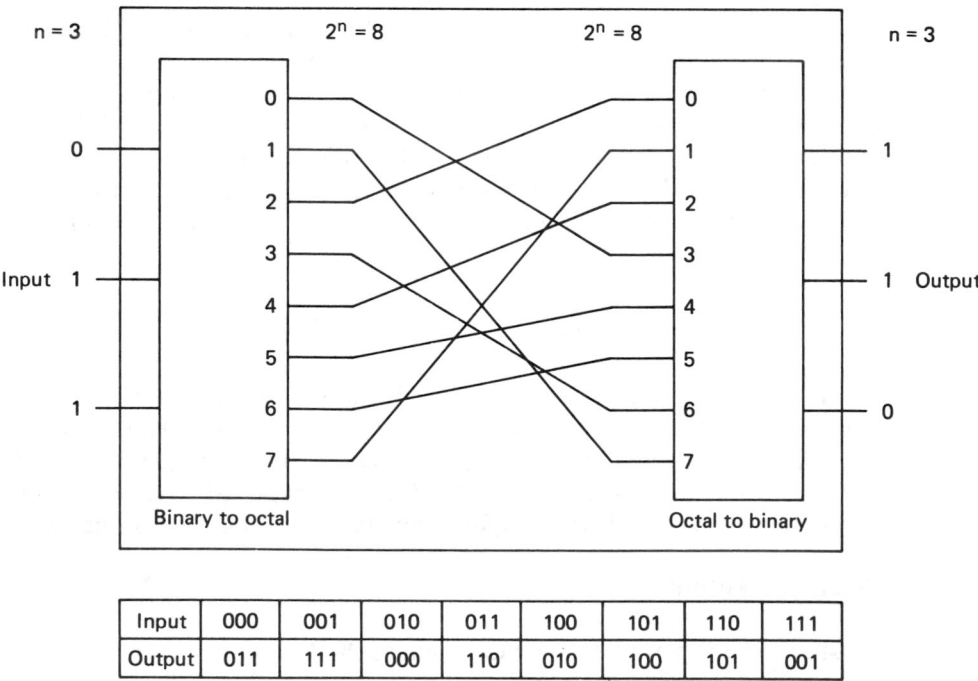

Input	000	001	010	011	100	101	110	111
Output	011	111	000	110	010	100	101	001

Figure 12.6 Substitution box.

$$\text{If } T \text{ is linear: } \quad C = C' \text{ for all inputs}$$

$$\text{If } T \text{ is nonlinear: } \quad C \neq C'$$

Suppose that $a = 001$ and $b = 010$; then using T as described in Figure 12.6,

$$C = T(001) \oplus T(010) = 111 \oplus 000 = 111$$

$$C' = T(001 \oplus 010) = T(011) = 110$$

where the symbol $\oplus$ represents modulo-2 addition. Since $C \neq C'$, the S-box is nonlinear.

12.3.3 Permutation

In permutation (transposition), the positions of the plaintext letters in the message are simply rearranged, rather than being substituted with other letters of the alphabet as in the classic ciphers. For example, the word THINK might appear, after permutation, as the ciphertext HKTNI. Figure 12.7 represents an example of binary data permutation (a linear operation). Here we see that the input data are simply rearranged or permuted (P-box). The technique has one major disadvantage when used alone; it is vulnerable to trick messages. A trick message is illustrated in Figure 12.7. A single 1 at the input and all the rest 0 quickly reveals one of the internal connections. If the cryptanalyst can subject the system to a

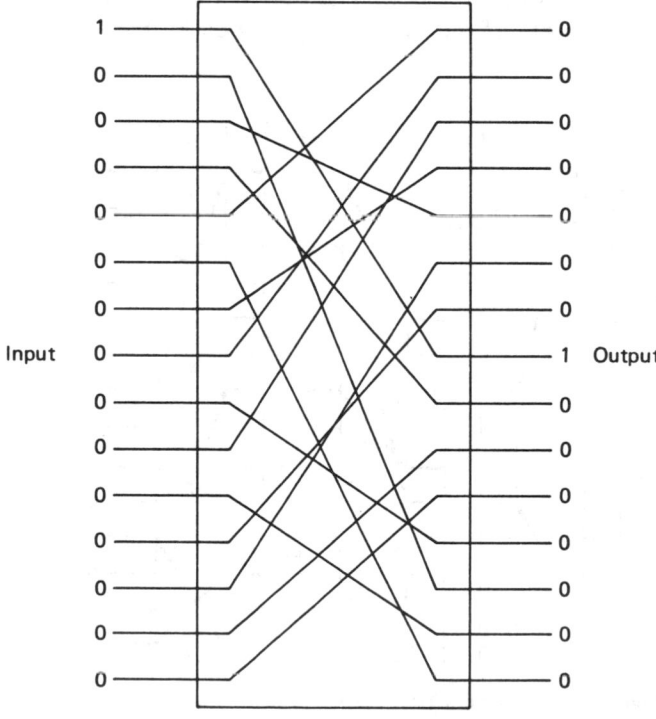

Figure 12.7 Permutation box.

plaintext attack, he will transmit a sequence of such trick messages, moving the single 1 one position for each transmission. In this way, each of the connections from input to output is revealed. This is an example of why a system's security should not depend on its architecture.

12.3.4 Product Cipher System

For transformations involving reasonable numbers of n-message symbols, both of the foregoing cipher systems (the S-box and the P-box) are by themselves wanting. Shannon [5] suggested using a *product cipher* or a combination of S-box and P-box transformations, which together could yield a cipher system more powerful than either one alone. This approach of alternately applying substitution and permutation transformations has been used by IBM in the LUCIFER system [7, 8], and has become the basis for the national Data Encryption Standard (DES) [9]. Figure 12.8 illustrates such a combination of P-boxes and S-boxes. Decryption is accomplished by running the data backward, using the inverse of each S-box. The system as pictured in Figure 12.8 is difficult to implement since each S-box is different, a randomly generated key is not usable, and the system does not lend itself to repeated use of the same circuitry. To avoid these difficulties, the LU-CIFER system [8] used two different types of S-boxes, S_1 and S_0, which could be publicly revealed. Figure 12.9 illustrates such a system. The input data are transformed by the sequence of S-boxes and P-boxes under the dictates of a key. The 25-bit key in this example designates, with a binary one or zero, the choice (S_1 or S_0) of each of the 25 S-boxes in the block. The details of the encryption devices can be revealed since security of the system is provided by the key.

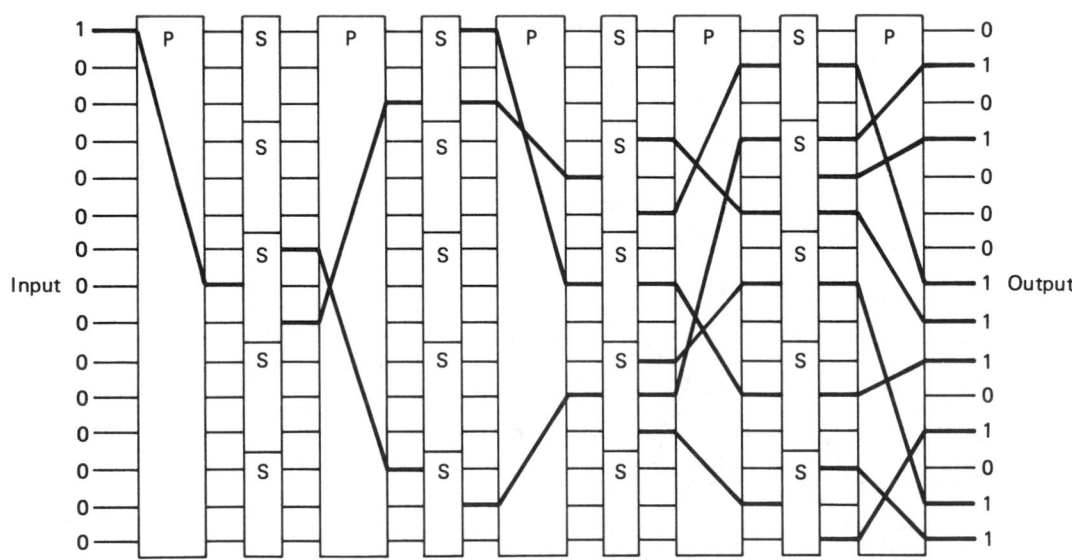

Figure 12.8 Product cipher system.

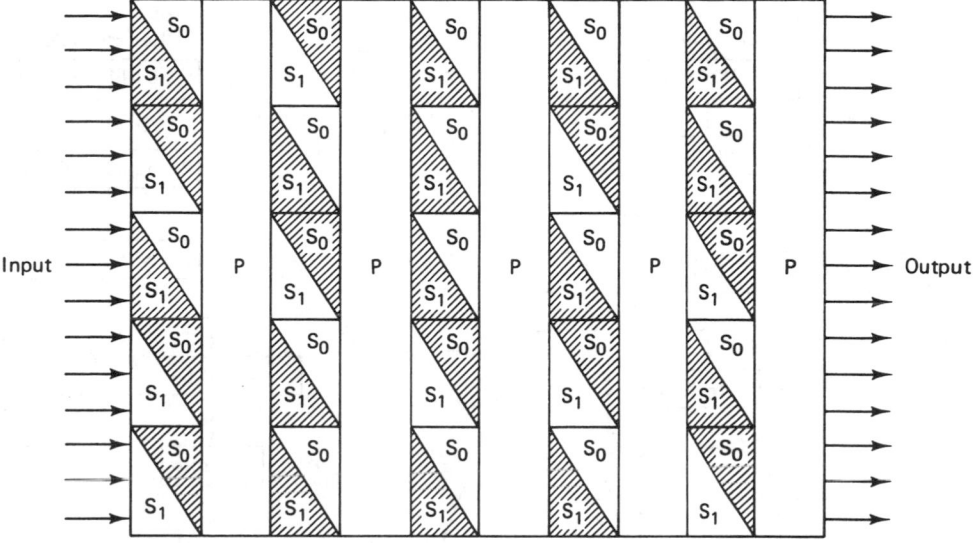

Shaded boxes correspond to the symbols of the binary key below.

Example of binary key

1 0 1 0 0 0 1 0 1 1 1 1 1 0 1 1 0 1 0 1 1 1 0 1 0

Figure 12.9 Individual keying capability.

The iterated structure of the product cipher system in Figure 12.9 is typical of most present-day block ciphers. The messages are partitioned into successive blocks of n bits, each of which is encrypted with the same key. The n-bit block represents one of 2^n different characters, allowing for $(2^n)!$ different substitution patterns. Consequently, for a reasonable implementation, the substitution part of the encryption scheme is performed in parallel on small segments of the block. An example of this is seen in the next section.

12.3.5 The Data Encryption Standard

In 1977, the National Bureau of Standards adopted a modified Lucifer system as the national Data Encryption Standard (DES) [9]. From a system input–output point of view, DES can be regarded as a block encryption system with an alphabet size of 2^{64} symbols, as shown in Figure 12.10. An input block of 64 bits, regarded

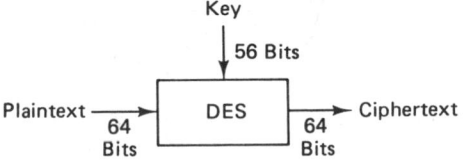

Figure 12.10 Data encryption standard (DES) viewed as a block encryption system.

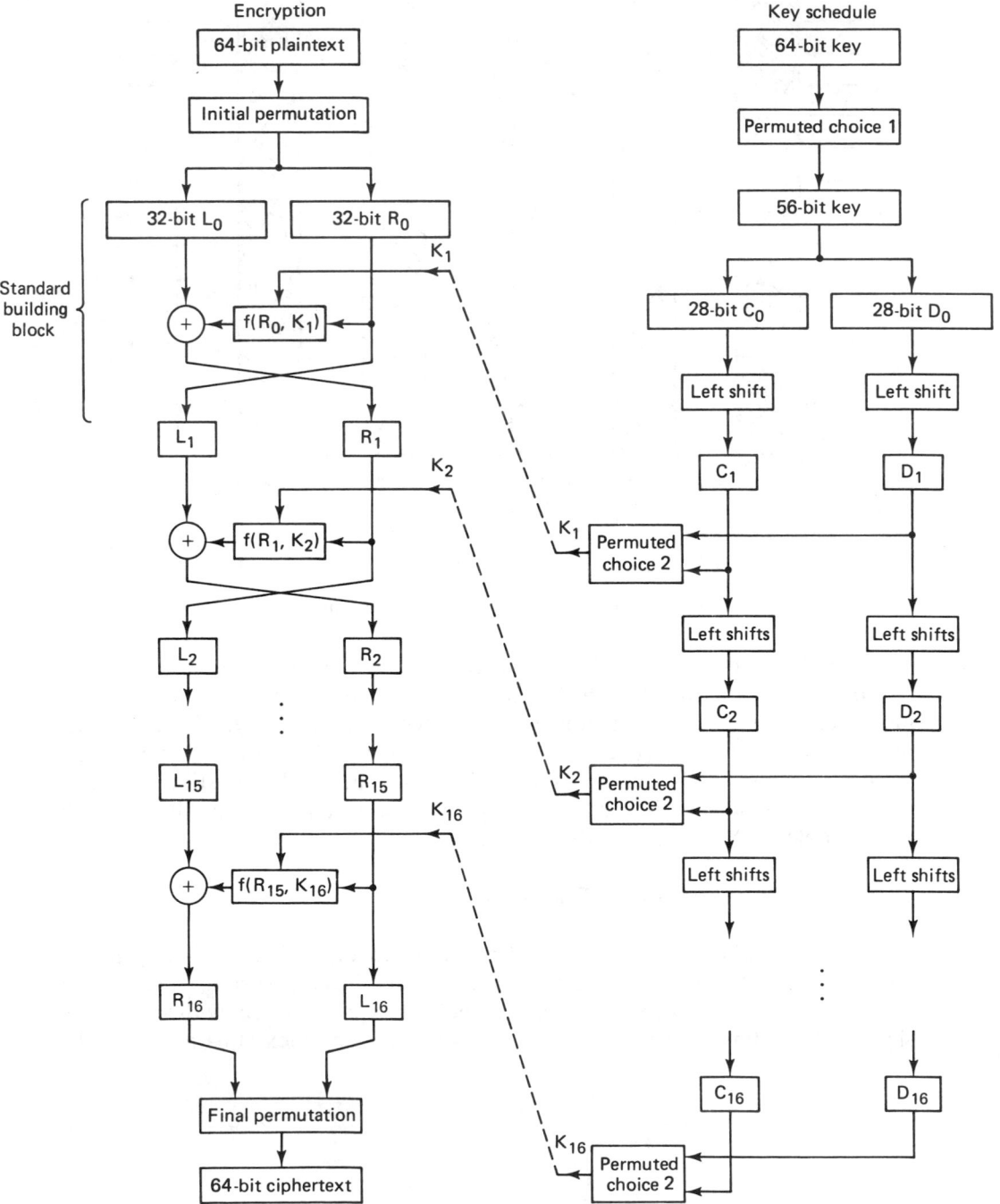

Figure 12.11 Data encryption standard.

as a plaintext symbol in this alphabet, is replaced with a new ciphertext symbol. Figure 12.11 illustrates the system functions in block diagram form. The encryption algorithm starts with an initial permutation (IP) of the 64 plaintext bits, described in the IP-table (Table 12.1). The IP-table is read from left to right and from top to bottom, so that bits $x_1, x_2, \ldots, x_{64}$ are permuted to $x_{58}, x_{50}, \ldots,$ x_7. After this initial permutation, the heart of the encryption algorithm consists of 16 iterations using the standard building block (SBB) shown in Figure 12.12. The standard building block uses 48 bits of key to transform the 64 input data bits into 64 output data bits, designated as 32 left-half bits and 32 right-half bits. The output of each building block becomes the input to the next building block. The input right-half 32 bits (R_{i-1}) are copied unchanged to become the output left-half 32 bits (L_i). The R_{i-1} bits are also *extended* and transformed into 48 bits with the E-table (Table 12.2), and then modulo-2 summed with the 48 bits of the key. As in the case of the IP-table, the E-table is read from left to right and from top to bottom. The table expands bits

$$R_{i-1} = x_1, x_2, \ldots, x_{32}$$

into

$$(R_{i-1})_E = x_{32}, x_1, x_2, \ldots, x_{32}, x_1 \qquad (12.22)$$

Notice that the bits listed in the first and last columns of the E-table are those bit positions that are used twice to provide the 32 bit-to-48 bit expansion.

Next, $(R_{i-1})_E$ is modulo-2 summed with the ith key selection, explained later, and the result is segmented into eight 6-bit blocks

$$B_1, B_2, \ldots, B_8$$

that is,

$$(R_{i-1})_E \oplus K_i = B_1, B_2, \ldots, B_8 \qquad (12.23)$$

Each of the eight 6-bit blocks, B_j, is then used as an input to an S-box function which returns a 4-bit block, $S_j(B_j)$. Thus the input 48 bits are transformed by the S-box to 32 bits. The S-box mapping function, S_j, is defined in Table 12.3. The transformation of $B_j = b_1, b_2, b_3, b_4, b_5, b_6$ is accomplished as follows. The integer corresponding to bits b_1 b_6 selects a row in the table, and the integer corresponding to bits b_2 b_3 b_4 b_5 selects a column in the table. For example, if

TABLE 12.1   Initial Permutation (IP)

58	50	42	34	26	18	10	2
60	52	44	36	28	20	12	4
62	54	46	38	30	22	14	6
64	56	48	40	32	24	16	8
57	49	41	33	25	17	9	1
59	51	43	35	27	19	11	3
61	53	45	37	29	21	13	5
63	55	47	39	31	23	15	7

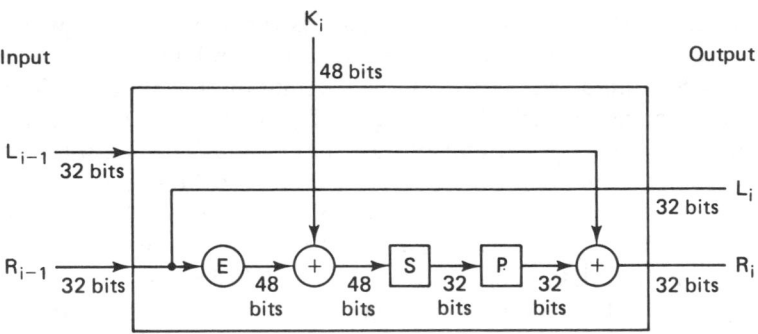

K_i

Input   Output

48 bits

L_{i-1}
32 bits

L_i
32 bits

R_{i-1} E   48   +   48   S   32   P.   32   +   32 bits R_i
32 bits   bits   bits   bits   bits

Figure 12.12 Standard building block (SBB).

$b_1 = 110001$, then S_1 returns the value in row 3, column 8, which is the integer 5 and is represented by the bit sequence 0101. The resulting 32-bit block out of the S-box is then permuted using the P-table (Table 12.4). As in the case of the other tables, the P-table is read from left to right and from top to bottom, so that bits $x_1, x_2, \ldots, x_{32}$ are permuted to $x_{16}, x_7, \ldots, x_{25}$. The 32-bit output of the P-table is modulo-2 summed with the input left-half 32 bits (L_{i-1}), forming the output right-half 32 bits (R_i).

The algorithm of the standard building block can be represented by

$$L_i = R_{i-1} \tag{12.24}$$

$$R_i = L_{i-1} \oplus f(R_{i-1}, K_i) \tag{12.25}$$

where $f(R_{i-1}, K_i)$ denotes the functional relationship comprised of the E-table, S-box, and P-table described above. After 16 iterations of the SBB, the data are transposed according to the final inverse permutation (IP^{-1}) described in the IP^{-1}-table (Table 12.5), where the output bits are read from left to right and from top to bottom, as before.

To decrypt, the same algorithm is used but the key sequence that is used in the standard building block is taken in the reverse order. Note that the value of $f(R_{i-1}, K_i)$ which can also be expressed in terms of the output of the ith block as $f(L_i, K_i)$, makes the decryption process possible.

TABLE 12.2 E-Table Bit Selection

32	1	2	3	4	5
4	5	6	7	8	9
8	9	10	11	12	13
12	13	14	15	16	17
16	17	18	19	20	21
20	21	22	23	24	25
24	25	26	27	28	29
28	29	30	31	32	1

TABLE 12.3 S-Box Selection Functions

Row	0	1	2	3	4	5	6	7	8	9	10	11	12	13	14	15	
									Column								
0	14	4	13	1	2	15	11	8	3	10	6	12	5	9	0	7	
1	0	15	7	4	14	2	13	1	10	6	12	11	9	5	3	8	S_1
2	4	1	14	8	13	6	2	11	15	12	9	7	3	10	5	0	
3	15	12	8	2	4	9	1	7	5	11	3	14	10	0	6	13	
0	15	1	8	14	6	11	3	4	9	7	2	13	12	0	5	10	
1	3	13	4	7	15	2	8	14	12	0	1	10	6	9	11	5	S_2
2	0	14	7	11	10	4	13	1	5	8	12	6	9	3	2	15	
3	13	8	10	1	3	15	4	2	11	6	7	12	0	5	14	9	
0	10	0	9	14	6	3	15	5	1	13	12	17	11	4	2	8	
1	13	7	0	9	3	4	6	10	2	8	5	14	12	11	15	1	S_3
2	13	6	4	9	8	15	3	0	11	1	2	12	5	10	14	7	
3	1	10	13	0	6	9	8	7	4	15	14	3	11	5	2	12	
0	7	13	14	3	0	6	9	10	1	2	8	5	11	12	4	15	
1	13	8	11	5	6	15	0	3	4	7	2	12	1	10	14	9	S_4
2	10	6	9	0	12	11	7	13	15	1	3	14	5	2	8	4	
3	3	15	0	6	10	1	13	8	9	4	5	11	12	7	2	14	
0	2	12	4	1	7	10	11	6	8	5	3	15	13	0	14	9	
1	14	11	2	12	4	7	13	1	5	0	15	10	3	9	8	6	S_5
2	4	2	1	11	10	13	7	8	15	9	12	5	6	3	0	14	
3	11	8	12	7	1	14	2	13	6	15	0	9	10	4	5	3	
0	12	1	10	15	9	2	6	8	0	13	3	4	14	7	5	11	
1	10	15	4	2	7	12	9	5	6	1	13	14	0	11	3	8	S_6
2	9	14	15	5	2	8	12	3	7	0	4	10	1	13	11	6	
3	4	3	2	12	9	5	15	0	11	14	1	7	6	0	8	13	
0	4	11	2	14	15	0	8	13	3	12	9	7	5	10	6	1	
1	13	0	11	7	4	9	1	10	14	3	5	12	2	15	8	6	S_7
2	1	4	11	13	12	3	7	14	10	15	6	8	0	5	9	2	
3	6	11	13	8	1	4	10	7	9	5	0	15	14	2	3	12	
0	13	2	8	4	6	15	11	1	10	9	3	14	5	0	12	7	
1	1	15	13	8	10	3	7	4	12	5	6	11	0	14	9	2	S_8
2	7	11	4	1	9	12	14	2	0	6	10	13	15	3	5	8	
3	2	1	14	7	4	10	8	13	15	12	9	0	3	5	6	11	

TABLE 12.4 P-Table Permutation

16	7	20	21
29	12	28	17
1	15	23	26
5	18	31	10
2	8	24	14
32	27	3	9
19	13	30	6
22	11	4	25

12.3.5.1 Key Selection

Key selection also proceeds in 16 iterations, as seen in the key schedule portion of Figure 12.11. The input key consists of a 64-bit block with 8 parity bits in positions 8, 16, . . . , 64. The permuted choice 1 (PC-1) discards the parity bits and permutes the remaining 56 bits as shown in Table 12.6. The output of PC-1 is split into two halves, C and D, of 28 bits each. Key selection proceeds in 16 iterations in order to provide a different set of 48 key bits to each SBB encryption iteration. The C and D blocks are successively shifted as follows:

$$C_i = LS_i(C_{i-1}) \quad \text{and} \quad D_i = LS_i(D_{i-1}) \quad (12.26)$$

where LS_i is a left circular shift by the number of positions shown in Table 12.7. The sequence C_i, D_i is then transposed according to the permuted choice 2 (PC-2) shown in Table 12.8. The result is the key sequence, K_i, which is used in the ith iteration of the encryption algorithm.

The DES can be implemented as a block encryption system (see Figure 12.11), which is sometimes referred to as a *codebook* method. A major disadvantage of this method is that a given block of input plaintext will always result in the same output ciphertext (under the same key). Another encryption mode, called the *cipher feedback* mode, encrypts single bits rather than characters, resulting in a stream encryption system [3]. With the cipher feedback scheme (described later), the encryption of a segment of plaintext not only depends on the key and the current data, but also on some of the earlier data.

Since the late 1970s, two points of contention have been widely publicized about the DES [10]. The first concerns the key variable length. Some researchers

TABLE 12.5 Final Permutation (IP^{-1})

40	8	48	16	56	24	64	32
39	7	47	15	55	23	63	31
38	6	46	14	54	22	62	30
37	5	45	13	53	21	61	29
36	4	44	12	52	20	60	28
35	3	43	11	51	19	59	27
34	2	42	10	50	18	58	26
33	1	41	9	49	17	57	25

TABLE 12.6 Key Permutation PC-1

57	49	41	33	25	17	9
1	58	50	42	34	26	18
10	2	59	51	43	35	27
19	11	3	60	52	44	36
63	55	47	39	31	23	15
7	62	54	46	38	30	22
14	6	61	53	45	37	29
21	13	5	28	20	12	4

TABLE 12.7 Key Schedule of Left Shifts

Iteration, i	Number of left shifts
1	1
2	1
3	2
4	2
5	2
6	2
7	2
8	2
9	1
10	2
11	2
12	2
13	2
14	2
15	2
16	1

TABLE 12.8 Key Permutation PC-2

14	17	11	24	1	5
3	28	15	6	21	10
23	19	12	4	26	8
16	7	27	20	13	2
41	52	31	37	47	55
30	40	51	45	33	48
44	49	39	56	34	53
46	42	50	36	29	32

felt that 56 bits are not adequate to preclude an exhaustive search. The second concerns the details of the internal structure of the S-boxes, which were never released by IBM. The National Security Agency (NSA), which had been involved in the testing of the DES algorithm, had requested that the information not be publicly discussed, since it was sensitive. The critics feared that NSA had been involved in design selections that would allow NSA to "tap into" any DES-encrypted messages [10].

12.4 STREAM ENCRYPTION

Earlier, we defined a *one-time pad* as an encryption system with a random key, used one time only, that exhibits unconditional security. One can conceptualize a stream encryption implementation of a one-time pad using a truly random key stream (the key sequence never repeats). Thus perfect secrecy can be achieved for an infinite number of messages, since each message would be encrypted with a different portion of the random key stream. The development of stream encryption schemes represents an attempt to emulate the one-time pad. Great emphasis was placed on generating key streams that appeared to be random, yet could easily be implemented for decryption, because they could be generated by algorithms. Such stream encryption techniques use pseudorandom (PN) sequences, which derive their name from the fact that they appear random to the casual observer; binary pseudorandom sequences have statistical properties similar to the random flipping of a fair coin. However, the sequences, of course, are deterministic (see Section 10.2). These techniques are popular because the encryption and decryption algorithms are readily implemented with feedback shift registers. At first glance it may appear that a PN key stream can provide the same security as the one-time pad, since the period of the sequence generated by a maximum-length linear shift register is $2^n - 1$ bits, where n is the number of stages in the register. If the PN sequence were implemented with a 50-stage register and a 1-MHz clock rate, the sequence would repeat every $2^{50} - 1$ microseconds, or every 35 years. In this era of large-scale integrated (LSI) circuits, it is just as easy to provide an implementation with 100 stages, in which case the sequence would repeat every 4×10^{16} years. Therefore, one might suppose that since the PN sequence does not repeat itself for such a long time, it would appear truly random and yield perfect secrecy. There is one important difference between the PN sequence and a truly random sequence used by a one-time pad. The PN sequence is generated by an algorithm; thus, knowing the algorithm, one knows the entire sequence. In Section 12.4.2 we will see that an encryption scheme that uses a linear feedback shift register in this way is very vulnerable to a *known plaintext attack*.

12.4.1 Example of Key Generation Using a Linear Feedback Shift Register

Stream encryption techniques generally employ shift registers for generating their PN key sequences. A shift register can be converted into a pseudorandom sequence generator by including a feedback loop that computes a new term for the

Encryption and Decryption   Chap. 12

first stage based on the previous n terms. The register is said to be linear if the numerical operation in the feedback path is linear. The PN generator example from Section 10.2 is repeated in Figure 12.13. For this example, it is convenient to number the stages as shown in Figure 12.13, where $n = 4$ and the outputs from stages 1 and 2 are modulo-2 added (linear operation) and fed back to stage 4. If the initial state of stages (x_4, x_3, x_2, x_1) is 1 0 0 0, the succession of states triggered by clock pulses would be 1 0 0 0, 0 1 0 0, 0 0 1 0, 1 0 0 1, 1 1 0 0, and so on. The output sequence is made up of the bits shifted out from the rightmost stage of the register, that is, 1 1 1 1 0 1 0 1 1 0 0 1 0 0 0, where the rightmost bit in this sequence is the earliest output and the leftmost bit is the most recent output. Given any linear feedback shift register of degree n, the output sequence is ultimately periodic.

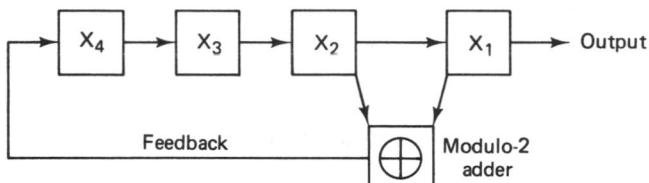

Figure 12.13 Linear feedback shift register example.

12.4.2 Vulnerabilities of Linear Feedback Shift Registers

An encryption scheme that uses a linear feedback shift register (LFSR) to generate the key stream is very vulnerable to attack. A cryptanalyst needs only $2n$ bits of plaintext and its corresponding ciphertext to determine the feedback taps, the initial state of the register, and the entire sequence of the code. In general, $2n$ is very small compared to the period $2^n - 1$. Let us illustrate this vulnerability with the LFSR example illustrated in Figure 12.13. Imagine that a cryptanalyst, who knows nothing about the internal connections of the LFSR, manages to obtain $2n = 8$ bits of ciphertext and its plaintext equivalent. These are shown below, where the rightmost bit is the earliest received and the leftmost bit is the most recent that was received.

Plaintext: 0 1 0 1 0 1 0 1

Ciphertext: 0 0 0 0 1 1 0 0

The cryptanalyst adds the two sequences together, modulo-2, to obtain the segment of the key stream, 0 1 0 1 1 0 0 1, illustrated in Figure 12.14. The key stream sequence shows the contents of the LFSR stages at various times. The rightmost border surrounding four of the key bits shows the contents of the shift register at time t_1. As we successively slide the "moving" border one digit to the left, we see the shift register contents at times $t_2, t_3, t_4, \ldots$. From the linear structure of the four-stage shift register, we can write

$$g_4 x_4 + g_3 x_3 + g_2 x_2 + g_1 x_1 = x_5 \qquad (12.27)$$

where x_5 is the digit fed back to the input and $g_i (= 1 \text{ or } 0)$ defines the ith feedback

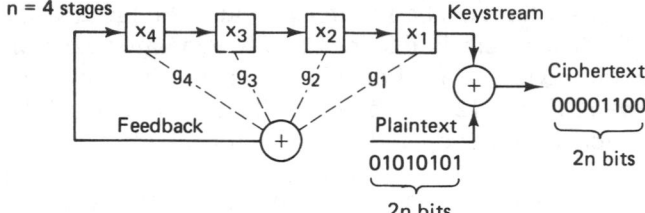

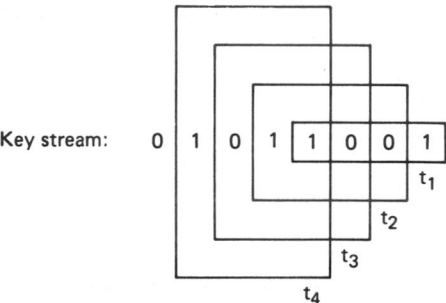

Figure 12.14 Example of vulnerability of a linear feedback shift register.

connection. For this example we can thus write the following four equations with four unknowns, by examining the contents of the shift register at the four times shown in Figure 12.14.

$$g_4(1) + g_3(0) + g_2(0) + g_1(1) = 1$$
$$g_4(1) + g_3(1) + g_2(0) + g_1(0) = 0$$
$$g_4(0) + g_3(1) + g_2(1) + g_1(0) = 1 \tag{12.28}$$
$$g_4(1) + g_3(0) + g_2(1) + g_1(1) = 0$$

whose solution is $g_1 = 1$, $g_2 = 1$, $g_3 = 0$, $g_4 = 0$, corresponding to the LFSR shown in Figure 12.13. The cryptanalyst has thus learned the connections of the LFSR, together with the starting state of the register at time t_1. He can therefore know the sequence for all time [3]. To generalize this example for any n-stage LFSR, we rewrite Equation (12.27) as follows:

$$x_{n+1} = \sum_{i=1}^{n} g_i x_i \tag{12.29}$$

We can write Equation (12.29) as the matrix equation

$$\mathbf{x} = \mathbf{X}\mathbf{g} \tag{12.30}$$

where

$$\mathbf{x} = \begin{bmatrix} x_{n+1} \\ x_{n+2} \\ \vdots \\ x_{2n} \end{bmatrix} \qquad \mathbf{g} = \begin{bmatrix} g_1 \\ g_2 \\ \vdots \\ g_n \end{bmatrix}$$

and

$$\mathbf{X} = \begin{bmatrix} x_1 & x_2 & \cdots & x_n \\ x_2 & x_3 & \cdots & x_{n+1} \\ \vdots & \vdots & & \vdots \\ x_n & x_{n+1} & \cdots & x_{2n-1} \end{bmatrix}$$

It can be shown [3] that the columns of $\mathbf{X}$ are linearly independent; thus $\mathbf{X}$ is nonsingular (its determinant is nonzero) and has an inverse. Hence,

$$\mathbf{g} = \mathbf{X}^{-1}\mathbf{x} \tag{12.31}$$

The matrix inversion requires at most on the order of n^3 operations and is thus easily accomplished by computer for any reasonable value of n. For example, if $n = 100$, $n^3 = 10^6$, and a computer with a 1-μs operation cycle would require 1 s for the inversion. The weakness of a LFSR is caused by the linearity of Equation (12.31). The use of *nonlinear feedback* in the shift register makes the cryptanalyst's task much more difficult, if not computationally intractable.

12.4.3 Synchronous and Self-Synchronous Stream Encryption Systems

We can categorize stream encryption systems as either *synchronous* or *self-synchronous*. In the former, the key stream is generated independently of the message, so that a lost character during transmission necessitates a resynchronization of the transmission and receiver key generators. A synchronous stream cipher is shown in Figure 12.15. The starting state of the key generator is initialized with a known input, I_0. The ciphertext is obtained by the modulo addition of the ith key character, k_i, with the ith message character, m_i. Such synchronous ciphers are generally designed to utilize *confusion* (see Section 12.3.1) but not *diffusion*. That is, the encryption of a character is not diffused over some block length of message. For this reason, synchronous stream ciphers do not exhibit *error propagation*.

In a *self-synchronous* stream cipher, each key character is derived from a fixed number, n, of the preceding ciphertext characters, giving rise to the name *cipher feedback*. In such a system, if a ciphertext character is lost during trans-

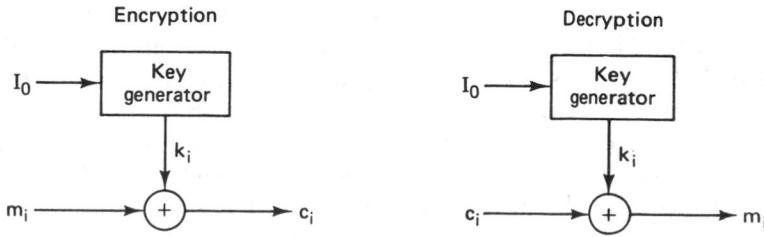

Figure 12.15 Synchronous stream cipher.

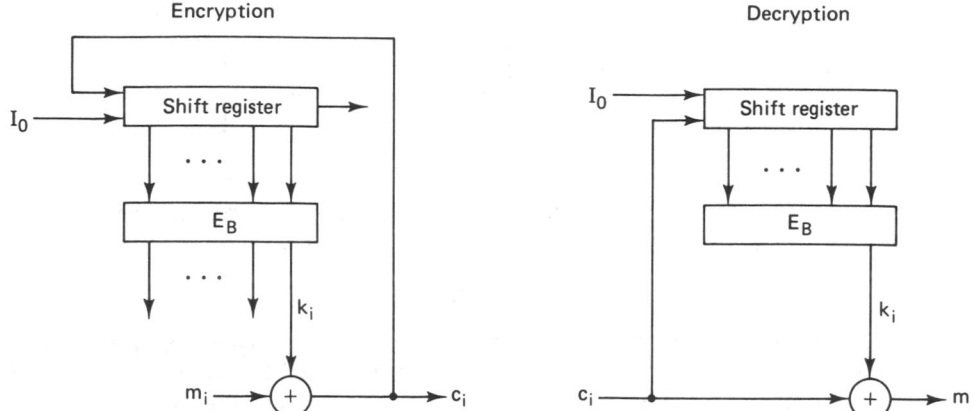

Figure 12.16 Cipher feedback mode.

mission, the error propagates forward for n characters, but the system resynchronizes itself after n correct ciphertext characters are received.

In Section 12.1.4 we looked at an example of cipher feedback in the Vigenere auto key cipher. We saw that the advantages of such a system are that (1) a nonrepeating key is generated, and (2) the statistics of the plaintext message are diffused throughout the ciphertext. However, the fact that the key was exposed in the ciphertext was a basic weakness. This problem can be eliminated by passing the ciphertext characters through a nonlinear block cipher to obtain the key characters. Figure 12.16 illustrates a shift register key generator operating in the cipher feedback mode. Each output ciphertext character, c_i (formed by the modulo addition of the message character, m_i, and the key character, k_i), is fed back to the input of the shift register. As before, initialization is provided by a known input, I_0. At each iteration, the output of the shift register is used as input to a (nonlinear) block encryption algorithm, E_B. The low-order output character from E_B becomes the next key character, k_{i+1}, to be used with the next message character, m_{i+1}. Since, after the first few iterations, the input to the algorithm depends only on the ciphertext, the system is self-synchronizing.

12.5 PUBLIC KEY CRYPTOSYSTEMS

The concept of public key cryptosystems was introduced in 1976 by Diffie and Hellman [11]. In conventional cryptosystems the encryption algorithm can be revealed since the security of the system depends on a safeguarded key. The same key is used for both encryption and decryption. Public key cryptosystems utilize *two different* keys, one for encryption and the other for decryption. In public key cryptosystems, not only the encryption algorithm but also the encryption key can be publicly revealed without compromising the security of the system. In fact, a public directory, much like a telephone directory, is envisioned, which contains the encryption keys of all the subscribers. Only the decryption keys are kept secret. Figure 12.17 illustrates such a system. The important features of a public key cryptosystem are as follows:

1. The encryption algorithm, E_K, and the decryption algorithm, D_K, are invertible transformations on the plaintext, M, or the ciphertext, C, defined by the key K. That is, for each K and M, if $C = E_K(M)$, then $M = D_K(C) = D_K[E_K(M)]$.
2. For each K, E_K and D_K are easy to compute.
3. For each K, the computation of D_K from E_K is computationally intractable.

Such a system would enable secure communication between subscribers who have never met or communicated before. For example, as seen in Figure 12.17, subscriber A can send a message, M, to subscriber B by looking up B's encryption key in the directory and applying the encryption algorithm, E_B, to obtain the ciphertext $C = E_B(M)$, which he transmits on the public channel. Subscriber B is the only party who can decrypt C by applying his decryption algorithm, D_B, to obtain $M = D_B(C)$.

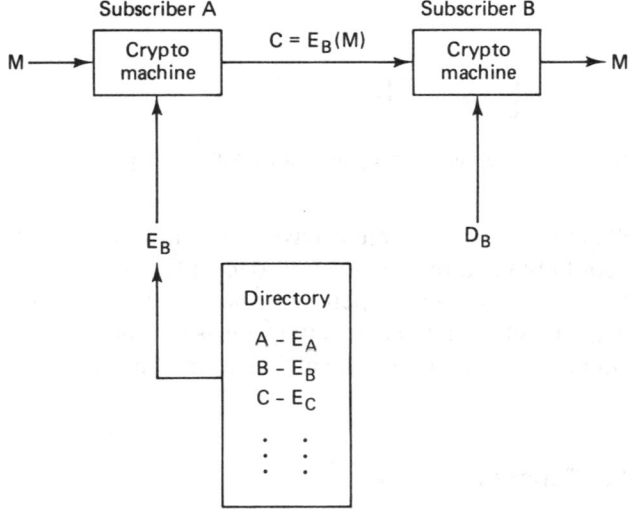

Figure 12.17 Public key cryptosystem.

12.5.1 Signature Authentication Using a Public Key Cryptosystem

Figure 12.18 illustrates the use of a public key cryptosystem for signature authentication. Subscriber A "signs" his message by first applying his decryption algorithm, D_A, to the message, yielding $S = D_A(M) = E_A^{-1}(M)$. Next, he uses the encryption algorithm, E_B, of subscriber B to encrypt S, yielding $C = E_B(S) = E_B[E_A^{-1}(M)]$, which he transmits on a public channel. When subscriber B receives C, he first decrypts it using his private decryption algorithm, D_B, yielding $D_B(C) = E_A^{-1}(M)$. Then he applies the encryption algorithm of subscriber A to produce $E_A[E_A^{-1}(M)] = M$.

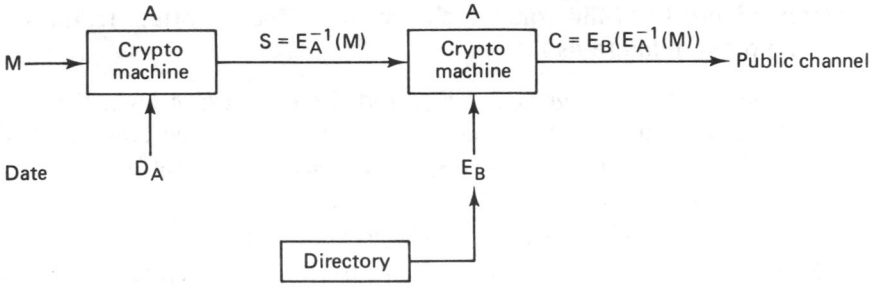

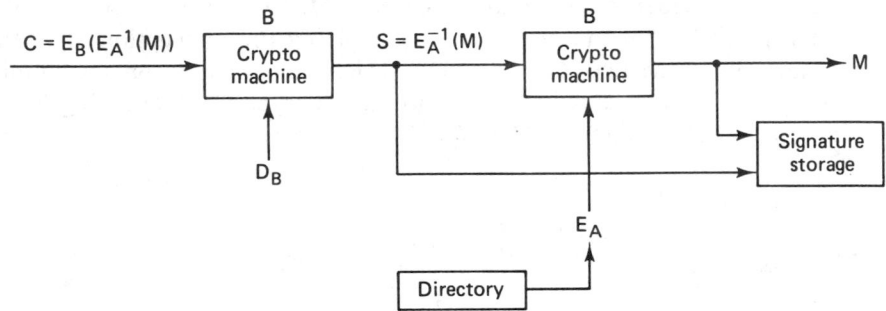

Figure 12.18 Signature authentication using a public key cryptosystem.

If the result is an intelligible message, it must have been initiated by subscriber A, since no one else could have known A's secret decryption key to form $S = D_A(M)$. Notice that S is both message dependent and signer dependent, which means that while B can be sure that the received message indeed came from A, at the same time A can be sure that no one can attribute any false messages to him.

12.5.2 A Trapdoor One-Way Function

Public key cryptosystems are based on the concept of trapdoor one-way functions. Let us first define a *one-way function* as an easily computed function whose inverse is computationally infeasible to find. For example, consider the function $y = x^5 + 12x^3 + 107x + 123$. It should be apparent that given x, y is easy to compute, but given y, x is relatively difficult to compute. A *trapdoor one-way function* is a one-way function whose inverse is easily computed if certain features, used to design the function, are known. Like a trapdoor, such functions are easy to go through in one direction. Without special information the reverse process takes an impossibly long time. We will apply the concept of a trapdoor in Section 12.5.5, when we discuss the Merkle–Hellman scheme.

12.5.3 The Rivest–Shamir–Adelman Scheme

In the Rivest–Shamir–Adelman (RSA) scheme messages are first represented as integers in the range $(0, n - 1)$. Each user chooses his own value of n and another pair of positive integers, e and d, in a manner to be described below. The user places his encryption key, the number pair (n, e), in the public directory. The decryption key consists of the number pair (n, d), of which d is kept secret. Encryption of a message, M, and decryption of a ciphertext, C, are defined as follows:

$$\text{Encryption:} \quad C = E(M) = (M)^e \text{ modulo-}n \qquad (12.32)$$
$$\text{Decryption:} \quad M = D(C) = (C)^d \text{ modulo-}n$$

They are each easy to compute and the results of each operation are integers in the range $(0, n - 1)$. In the RSA scheme, n is obtained by selecting *two large prime numbers*, p and q, and multiplying them together:

$$n = pq \qquad (12.33)$$

Although n is made public, p and q are kept hidden, due to the great difficulty in factoring n. Then

$$\phi(n) = (p - 1)(q - 1) \qquad (12.34)$$

called *Euler's totient function*, is formed. The parameter $\phi(n)$ has the interesting property [12] that for any integer X in the range $(0, n - 1)$ and any integer k,

$$X = X^{k\phi(n)+1} \text{ modulo-}n \qquad (12.35)$$

Therefore, while all other arithmetic is done modulo-n, arithmetic in the exponent is done modulo-$\phi(n)$. A large integer, d, is randomly chosen so that it is relatively prime to $\phi(n)$, which means that $\phi(n)$ and d must have no common divisors other than 1, expressed as

$$\gcd[\phi(n), d] = 1 \qquad (12.36)$$

where gcd means "greatest common divisor." Any prime number greater than the larger of (p, q) will suffice. Then the integer e, where $0 < e < \phi(n)$, is found from the following relationship:

$$ed \text{ modulo-}\phi(n) = 1 \qquad (12.37)$$

which, from Equation (12.35), is tantamount to choosing e and d to satisfy

$$X = X^{ed} \text{ modulo-}n \qquad (12.38)$$

Therefore,

$$E[D(X)] = D[E(X)] = X \qquad (12.39)$$

and decryption works correctly. Given an encryption key (n, e), one way that a cryptanalyst might attempt to break the cipher is to factor n into p and q, compute

$\phi(n) = (p - 1)(q - 1)$, and compute d from Equation (12.37). This is all straightforward except for the factoring of n.

The RSA scheme is based on the fact that it is easy to generate two large prime numbers, p and q, and multiply them together, but it is very much more difficult to factor the result. The product can therefore be made public as part of the encryption key, without compromising the factors that would reveal the decryption key corresponding to the encryption key. By making each of the factors roughly 100 digits long, the multiplication can be done in a fraction of a second, but the exhaustive factoring of the result should take billions of years [2].

12.5.3.1 Use of the RSA Scheme

Using the example in Reference [12], let $p = 47$, $q = 59$. Therefore, $n = pq = 2773$ and $\phi(n) = (p - 1)(q - 1) = 2668$. The parameter d is chosen to be relatively prime to $\phi(n)$. For example, choose $d = 157$. Next, the value of e is computed as follows (the details are shown in the next section):

$$ed \text{ modulo } \phi(n) = 1$$

$$157e \text{ modulo } 2668 = 1$$

Therefore, $e = 17$. Consider the plaintext example

$$\text{ITS ALL GREEK TO ME}$$

By replacing each letter with a two-digit number in the range (01, 26) corresponding to its position in the alphabet, and encoding a blank as 00, the plaintext message can be written as

$$0920\ 1900\ 0112\ 1200\ 0718\ 0505\ 1100\ 2015\ 0013\ 0500$$

Each message needs to be expressed as an integer in the range $(0, n - 1)$; therefore, for this example, encryption can be performed on blocks of four digits at a time since this is the maximum number of digits that will always yield a number less than $n - 1 = 2772$. The first four digits (0920) of the plaintext are encrypted as follows:

$$C = (M)^e \text{ modulo-}n = (920)^{17} \text{ modulo-}2773 = 948$$

Continuing this process for the remaining plaintext digits, we get

$$C = 0948\ 2342\ 1084\ 1444\ 2663\ 2390\ 0778\ 0774\ 0219\ 1655$$

The plaintext is returned by applying the decryption key, as follows:

$$M = (C)^{157} \text{ modulo-}2773$$

12.5.3.2 How to Compute e

A variation of Euclid's algorithm [13] for computing the gcd of $\phi(n)$ and d is used to compute e. First, compute a series $x_0, x_1, x_2, \ldots,$ where $x_0 = \phi(n)$, $x_1 = d$, and $x_{i+1} = x_{i-1} \text{ modulo-}x_i$, until an $x_k = 0$ is found. Then the gcd $(x_0, x_1) = x_{k-1}$. For each x_i compute numbers a_i and b_i such that $x_i = a_i x_0 + b_i x_1$.

If $x_{k-1} = 1$, then b_{k-1} is the multiplicative inverse of x_1 modulo-x_0. If b_{k-1} is a negative number, the solution is $b_{k-1} + \phi(n)$.

Example 12.5 Computation of e from d and $\phi(n)$

For the previous example, with $p = 47$, $q = 59$, $n = 2773$, $\phi(n) = 2688$, and d chosen to be 157, use the Euclid algorithm to verify that $e = 17$.

Solution

i	x_i	a_i	b_i	y_i
0	2668	1	0	
1	157	0	1	16
2	156	1	-16	1
3	1	-1	17	

where

$$y_i = \left\lfloor \frac{x_{i-1}}{x_i} \right\rfloor$$

$$x_{i+1} = x_{i-1} - y_i x_i$$

$$a_{i+1} = a_{i-1} - y_i a_i$$

$$b_{i+1} = b_{i-1} - y_i b_i$$

Hence

$$e = b_3 = 17$$

12.5.4 The Knapsack Problem

The classic knapsack problem is illustrated in Figure 12.19. The knapsack is filled with a subset of the items shown with weights indicated in grams. Given the weight of the filled knapsack (the scale is calibrated to deduct the weight of the empty knapsack), determine which items are contained in the knapsack. For this simple example, the solution can easily be found by trial and error. However, if there are 100 possible items in the set, instead of 10, the problem may become computationally infeasible.

Let us express the knapsack problem in terms of a knapsack vector and a data vector. The knapsack vector is an n-tuple of distinct integers (analogous to the set of possible knapsack items)

$$\mathbf{a} = a_1, a_2, \ldots, a_n$$

The data vector is an n-tuple of binary symbols

$$\mathbf{x} = x_1, x_2, \ldots, x_n$$

The knapsack, S, is the sum of a subset of the components of the knapsack vector

$$S = \sum_{i=1}^{n} a_i x_i \qquad \text{where } x_i = 0, 1$$

$$= \mathbf{ax}$$

(12.40)

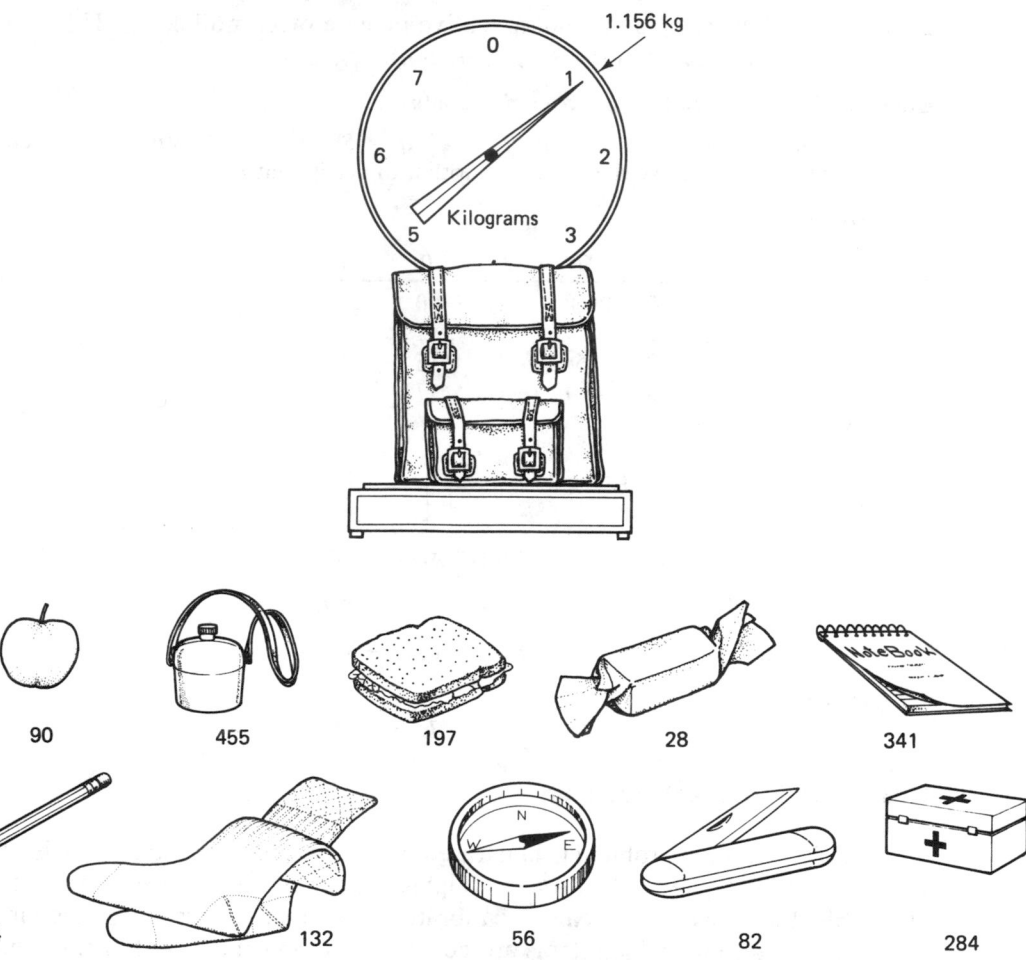

Figure 12.19 Knapsack problem.

The knapsack problem can be stated as follows: Given S and knowing $\mathbf{a}$, determine $\mathbf{x}$.

Example 12.6   Knapsack Example

Given $\mathbf{a} = 1, 2, 4, 8, 16, 32$ and $S = \mathbf{ax} = 26$, find $\mathbf{x}$.

Solution

In this example $\mathbf{x}$ is seen to be the *binary* representation of S. The decimal-to-binary conversion should appear more familiar with $\mathbf{a}$ expressed as 2^0, 2^1, 2^2, 2^3, 2^4, 2^5. The data vector $\mathbf{x}$ is easily found since $\mathbf{a}$ in this example is *super-increasing*, which means that each component of the n-tuple $\mathbf{a}$ is larger than the sum of the preceding components. That is,

$$a_i > \sum_{j=1}^{i-1} a_j \qquad i = 2, 3, \ldots, n \qquad (12.41)$$

When **a** is super-increasing, the solution of **x** is found by starting with $x_n = 1$ if $S \geq a_n$ (otherwise $x_n = 0$), and continuing, as follows:

$$x_i = \begin{cases} 1 & \text{if } S - \sum_{j=i+1}^{n} x_j a_j \geq a_i \\ \\ 0 & \text{otherwise} \end{cases} \tag{12.42}$$

where $i = n - 1, n - 2, \ldots, 1$. From Equation (12.42) it is easy to compute $\mathbf{x} = $ 0 1 0 1 1 0.

Example 12.7   Knapsack Example

Given **a** = 171, 197, 459, 1191, 2410, 4517 and **S** = **ax** = 3798, find **x**.

Solution

As in Example 12.6, **a** is super-increasing; therefore, we can compute **x** using Equation (12.42), which again yields

$$\mathbf{x} = 0 \ 1 \ 0 \ 1 \ 1 \ 0$$

12.5.5 A Public Key Cryptosystem Based on a Trapdoor Knapsack

This scheme, also known as the Merkle–Hellman scheme [14], is based on the formation of a knapsack vector that is not super-increasing and is therefore not easy to solve. However, an essential part of this knapsack is a *trapdoor* that enables the authorized user to solve it.

First, we form a super-increasing n-tuple, **a'**. Then we select a prime number M such that

$$M > \sum_{i=1}^{n} a_i' \tag{12.43}$$

We also select a random number, W, where $1 < W < M$, and we form W^{-1} to satisfy the following relationship:

$$WW^{-1} \text{ modulo-}M = 1 \tag{12.44}$$

The vector **a'** and the numbers M, W, and W^{-1} are all kept hidden. Next, we form **a** with the elements from **a'**, as follows:

$$a_i = Wa_i' \text{ modulo-}M \tag{12.45}$$

The formation of **a** using Equation (12.45) constitutes forming a knapsack vector with a *trapdoor*. When a data vector **x** is to be transmitted, we multiply **x** by **a**, yielding the number S, which is sent on the public channel. Using Equation (12.45), S can be written as follows:

$$S = \mathbf{ax} = \sum_{i=1}^{n} a_i x_i = \sum_{i=i}^{n} (Wa_i' \text{ modulo-}M) x_i \tag{12.46}$$

The authorized user receives S and, using Equation (12.44), converts it to S':

$$S' = W^{-1}S \text{ modulo-}M = W^{-1} \sum_{i=1}^{n} (Wa_i' \text{ modulo-}M)x_i \text{ modulo-}M$$

$$= \sum_{i=1}^{n} (W^{-1}Wa_i' \text{ modulo-}M)x_i \text{ modulo-}M \tag{12.47}$$

$$= \sum_{i=1}^{n} a_i'x_i \text{ modulo-}M$$

$$= \sum_{i=1}^{n} a_i'x_i$$

Since the authorized user knows the secretly held super-increasing vector $\mathbf{a}'$, he or she can use S' to find $\mathbf{x}$.

12.5.5.1 Use of the Merkle–Hellman Scheme

Suppose that user A wants to construct public and private encryption functions. He first considers the super-increasing vector $\mathbf{a}' = (171, 197, 459, 1191, 2410, 4517)$

$$\sum_{i=1}^{6} a_i' = 8945$$

He then chooses a prime number M larger than 8945, a random number W, where $1 \le W < M$, and calculates W^{-1} to satisfy $WW^{-1} = 1$ modulo-M.

$$\left.\begin{array}{l} \text{Choose } M = 9109 \\ \text{choose } W = 2251 \\ \text{then } W^{-1} = 1388 \end{array}\right\} \text{ kept hidden}$$

He then forms the trapdoor knapsack vector as follows:

$$a_i = a_i' \ 2251 \text{ modulo-}9109$$

$$\mathbf{a} = 2343, 6215, 3892, 2895, 5055, 2123$$

User A makes public the vector $\mathbf{a}$, which is clearly not super-increasing. Suppose that user B wants to send a message to user A.

If $\mathbf{x} = 0\ 1\ 0\ 1\ 1\ 0$ is the message to be transmitted, user B forms

$$S = \mathbf{ax} = 14{,}165 \text{ and transmits it to user } A$$

User A, who receives S, converts it to S':

$$S' = \mathbf{a}'\mathbf{x} = W^{-1}S \text{ modulo-}M$$

$$= 1388 \cdot 14{,}165 \text{ modulo-}9109$$

$$= 3798$$

Using $S' = 3798$ and the super-increasing vector $\mathbf{a}'$, user A easily solves for $\mathbf{x}$.
Public key schemes are generally too slow for data encryption. With large

Encryption and Decryption   Chap. 12

encrypted data networks, the biggest problem is how to distribute and manage the keys; the public key systems appear to be very promising for use in key management.

12.6 CONCLUSION

In this chapter we have presented the basic models and goals of the cryptographic process. We looked at some early cipher systems and reviewed the mathematical theory of secret communications established by Shannon. We defined a system that can exhibit perfect secrecy and established that such systems can be implemented but that they are not practical for use where high-volume communications are required. We also considered practical security systems that employ Shannon's techniques known as confusion and diffusion to frustrate the statistical endeavors of a cryptanalyst.

The outgrowth of Shannon's work was utilized by IBM in the LUCIFER system, which later grew into the National Bureau of Standards' Data Encryption Standard (DES). We outlined the DES algorithm in detail. We also considered the use of linear feedback shift registers (LFSR) for stream encryption systems, and demonstrated the intrinsic vulnerability of an LFSR used as a key generator.

Finally, we looked at the novel area of public key cryptosystems and examined two schemes, the Rivest–Shamir–Adelman (RSA) scheme, based on the product of two large prime numbers, and the Merkle–Hellman scheme, based on the classical knapsack problem. The Merkle–Hellman scheme is now considered broken [15], so that today the RSA scheme seems to be the primary means of implementing public key cryptosystems.

REFERENCES

1. Kahn, D., *The Codebreakers*, Macmillan Publishing Company, New York, 1967.
2. Diffie, W., and Hellman, M. E., "Privacy and Authentication: An Introduction to Cryptography," *Proc. IEEE*, vol. 67, no. 3, Mar. 1979, pp. 397–427.
3. Beker, H., and Piper, F., *Cipher Systems*, John Wiley & Sons, Inc., New York, 1982.
4. Denning, D. E. R., *Cryptography and Data Security*, Addison-Wesley Publishing Company, Reading, Mass., 1982.
5. Shannon, C. E., "Communication Theory of Secrecy Systems," *Bell Syst. Tech. J.*, vol. 28, Oct. 1949, pp. 656–715.
6. Hellman, M. E., "An Extension of the Shannon Theory Approach to Cryptography," *IEEE Trans. Inf. Theory*, vol. IT23, May 1978, pp. 289–294.
7. Smith, J. L., "The Design of Lucifer, a Cryptographic Device for Data Communications," *IBM Research Rep. RC-3326*, 1971.
8. Feistel, H. "Cryptography and Computer Privacy," *Sci. Am.*, vol. 228, no. 5, May 1973, pp. 15–23.

9. National Bureau of Standards, "Data Encryption Standard," *Federal Information Processing Standard (FIPS)*, Publication no. 46, Jan. 1977.

10. United States Senate Select Committee on Intelligence, "Unclassified Summary: Involvement of NSA in the Development of the Data Encryption Standard," *IEEE Commun. Soc. Mag.*, vol. 16, no. 6, Nov. 1978, pp. 53–55.

11. Diffie, W., and Hellman, M. E., "New Directions in Cryptography," *IEEE Trans. Inf. Theory*, vol. IT22, Nov. 1976, pp. 644–654.

12. Rivest, R. L., Shamir, A., and Adelman, L., "On Digital Signatures and Public Key Cryptosystems," *Commun. ACM*, vol. 21, Feb. 1978, pp. 120–126.

13. Knuth, D. E., *The Art of Computer Programming*, Vol. 2, *Seminumerical Algorithms*, 2nd ed., Addison-Wesley Publishing Company, Reading, Mass., 1981.

14. Merkle, R. C., and Hellman, M. E., "Hiding Information and Signatures in Trap-Door Knapsacks," *IEEE Trans. Inf. Theory*, vol. IT24, Sept. 1978, pp. 525–530.

15. Shamir, A., "A Polynomial Time Algorithm for Breaking the Basic Merkle-Hellman Cryptosystem," *IEEE 23rd Ann. Symp. Found. Comput. Sci.*, 1982, pp. 145–153.

PROBLEMS

12.1. Let X be an integer variable represented with 64 bits. The probability is $\frac{1}{2}$ that X is in the range $(0, 2^{16} - 1)$, the probability is $\frac{1}{4}$ that X is in the range $(2^{16}, 2^{32} - 1)$, and the probability is $\frac{1}{4}$ that X is in the range $(2^{32}, 2^{64} - 1)$. Within each range the values are equally likely. Compute the entropy of X.

12.2. A set of equally likely weather messages are: sunny (S), cloudy (C), light rain (L), and heavy rain (H). Given the added information concerning the time of day (morning or afternoon), the probabilities change as follows:

$$\text{Morning:} \quad P(S) = \tfrac{1}{8}, P(C) = \tfrac{1}{8}, P(L) = \tfrac{3}{8}, P(H) = \tfrac{3}{8}$$

$$\text{Afternoon:} \quad P(S) = \tfrac{3}{8}, P(C) = \tfrac{3}{8}, P(L) = \tfrac{1}{8}, P(H) = \tfrac{1}{8}$$

(a) Find the entropy of the weather message.
(b) Find the entropy of the message conditioned on the time of day.

12.3. The Hawaiian alphabet has only 12 letters—the vowels, a, e, i, o, u, and the consonants, h, k, l, m, n, p, w. Assume that each vowel occurs with probability 0.116, and that each consonant occurs with probability 0.06. Also assume that the average number of *information bits* per letter is the same as that for the English language. Calculate the unicity distance for an encrypted Hawaiian message if the key sequence consists of a random permutation of the 12-letter alphabet.

12.4. Estimate the unicity distance for an English language encryption system that uses a key sequence made up of 10 random alphabetic characters:
(a) Where each key character can be any one of the 26 letters of the alphabet (duplicates are allowed).
(b) Where the key characters may not have any duplicates.

12.5. Repeat Problem 12.4 for the case where the key sequence is made up of ten integers randomly chosen from the set of numbers 0 to 999.

12.6. (a) Find the unicity distance for a DES system which encrypts 64-bit blocks (eight alphabetic characters) using a 56-bit key.

(b) What is the effect on the unicity distance in part (a) if the key is increased to 128 bits?

12.7. In Figures 12.8 and 12.9, P-boxes and S-boxes alternate. Is this arrangement any more secure than if all the P-boxes were first grouped together, followed by all the S-boxes similarly grouped together? Justify your answer.

12.8. What is the output of the first iteration of the DES algorithm when the plaintext and the key are each made up of zero sequences?

12.9. Consider the 10-bit plaintext sequence 0 1 0 1 1 0 1 0 0 1 and its corresponding ciphertext sequence 0 1 1 1 0 1 1 0 1 0, where the rightmost bit is the earliest bit. Describe the five-stage linear feedback shift register (LFSR) that produced the key sequence and show the initial state of the register. Is the output sequence of maximal length?

12.10. Following the RSA algorithm and parameters in Example 12.5, compute the encryption key, e, when the decryption key is chosen to be 151.

12.11. Given e and d that satisfy ed modulo-$\phi(n) = 1$, and a message that is encoded as an integer number, M, in the range $(0, n - 1)$ such that the gcd $(M, n) = 1$. Prove that $(M^e$ modulo-$n)^d$ modulo-$n = M$.

12.12. Use the RSA scheme to encrypt the message $M = 3$. Use the prime numbers $p = 5$ and $q = 7$. Choose the decryption key, d, to be 11, and calculate the value of the encryption key, e.

12.13. Consider the following for the RSA scheme.
  (a) If the prime numbers are $p = 7$ and $q = 11$, list five allowable values for the decryption key, d.
  (b) If the prime numbers are $p = 13$, $q = 31$, and the decryption key is $d = 37$, find the encryption key, e, and describe how you would use it to encrypt the word "DIGITAL."

12.14. Use the Merkle–Hellman public key scheme with the super-increasing vector, $\mathbf{a'}$ = 1, 3, 5, 10, 20. Use the following additional parameters: a large prime number $M = 51$ and a random number $W = 37$.
  (a) Find the nonsuper-increasing vector, $\mathbf{a}$, to be made public, and encrypt the data vector 1 1 0 1 1.
  (b) Show the steps by which an authorized receiver decrypts the ciphertext.

APPENDIX A

A Review
of Fourier Techniques

A.1 SIGNALS, SPECTRA, AND LINEAR SYSTEMS

Electrical communication signals consist of time-varying voltage or current wave-forms, typically described in the time domain. It is also convenient to describe such signals in the frequency domain. A signal's frequency-domain description is called its *spectrum*. Spectral concepts are important in communication analysis and design; they can describe a signal by its average power or energy content at various frequencies, and they illustrate how much of the electromagnetic spectrum (bandwidth) the signal occupies. Broadcast stations are required by the Federal Communications Commission (FCC) to operate at their assigned frequency with very tight tolerances on the occupied bandwidth; for example, amplitude-modulated (AM) radio channels are spaced 10 kHz apart, and television channels are spaced 6 MHz apart. Our interest in spectra and Fourier techniques has to do with the real-world constraints of ensuring that our communication signals are confined to specified spectral boundaries.

Frequency spectral characteristics can be ascribed to both signal waveforms and to circuits. When we say that a particular spectrum describes a *signal*, we mean that one way of characterizing the signal waveform is to specify its amplitude and phase as a function of frequency. However, when we talk about the spectral attributes of a *circuit* we are referring to the output versus input frequency-domain transfer function of the circuit; in other words, we are characterizing the circuit by how much of a specific input signal spectrum is allowed to pass through it.

710

A.2 FOURIER TECHNIQUES FOR LINEAR SYSTEM ANALYSIS

Fourier techniques are often used for analyzing linear circuits or systems in the following ways: (1) by predicting the system response, (2) by determining the system dynamic specification (transfer function), and (3) by evaluating or interpreting test results. Item 1, predicting system response, is illustrated schematically in Figure A.1. Let the input be an arbitrary periodic waveform with period equal to T_0 seconds. Fourier techniques allow us to describe such an input as a sum of sinusoidal waveforms, as shown in the figure. The lowest-frequency sinusoid, or the *fundamental* frequency of the input periodic, has frequency $1/T_0$ hertz; the balance of the sinusoids have frequencies that are integral *harmonics* ($2/T_0$, $3/T_0$, . . .) of this fundamental frequency. An important attribute of a linear system is that *superposition* applies, which means that the response to the sum of excitations is the sum of the responses to the individually applied excitations. In fact, this is used as a definition of linearity. Specifically, if

$$y_1(t) = \text{system response to } x_1(t)$$

$$y_2(t) = \text{system response to } x_2(t)$$

and

$$ay_1(t) + by_2(t) = \text{system response to } ax_1(t) + bx_2(t)$$

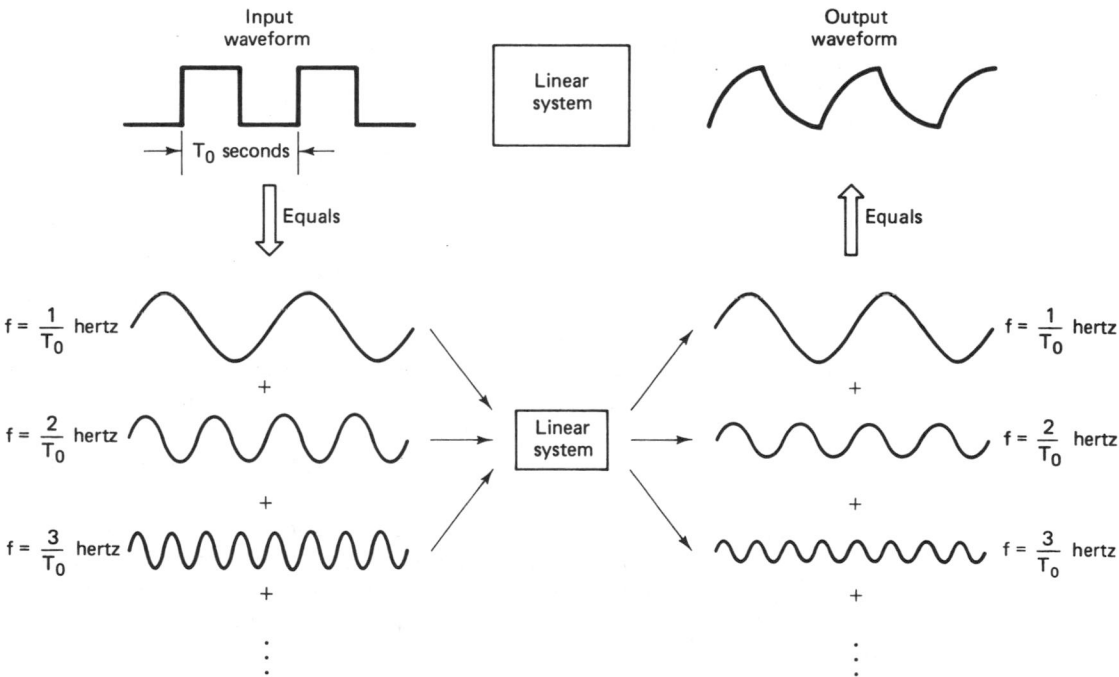

Figure A.1 Predicting system response.

for all a, b, $x_1(t)$, and $x_2(t)$, then the system is linear. A consequence of this definition is that the output response of a *linear* system with sinusoidal input waveforms must be comprised of sinusoidal waveforms having the *same frequencies* as the input waveforms; such a system is typically specified by an output versus input *frequency transfer function* (magnitude and phase versus frequency) as shown in Figure A.2. Figure A.2a illustrates a typical example of signal magnitude versus frequency; similarly, Figure A.2b illustrates a typical example of signal phase versus frequency.

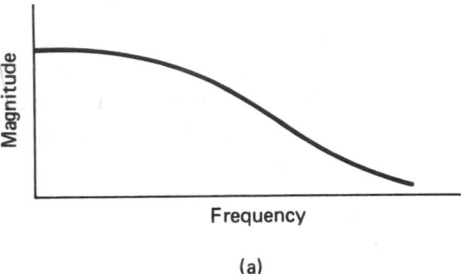

(a)

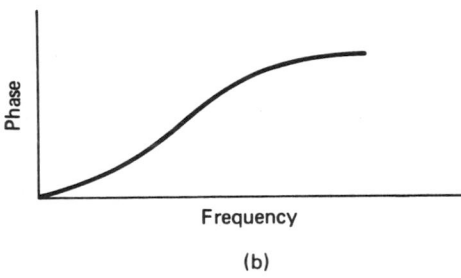

Frequency

(b)

Figure A.2 System transfer function. (a) Magnitude response. (b) Phase response.

The system transfer function serves as a performance specification; it describes the system response to each of the component sinusoids. Therefore, with the system transfer function in hand, one can predict each of the resulting output components. Using the principle of superposition, the final step of the analysis is to sum the individual output responses, thus forming the resulting overall response to the input periodic (see Figure A.1). In a similar manner, one can determine a system's transfer function, or evaluate a system's test results from knowledge of the input and output waveforms.

The development of Fourier methods had a major impact on the analysis of linear systems; it provided the translation between transient phenomena and sinusoidal techniques, and it simplified the analysis of linear systems under the excitation of any arbitrary input waveform. Just as logarithms allow the operation of multiplication to be treated as addition, so Fourier techniques allow the replacement of complex waveforms with sinusoidal components and sinusoidal methods.

A.2.1 Fourier Series Transform

Signals that are periodic with finite energy within each period can be represented by the *Fourier series*. Equation (A.1) describes such an arbitrary periodic waveform, $x(\lambda)$, in terms of an infinite number of increasing harmonic sine and cosine components.

$$x(\lambda) = \tfrac{1}{2}a_0 + a_1 \cos \lambda + a_2 \cos 2\lambda + a_3 \cos 3\lambda$$

$$+ \cdots + b_1 \sin \lambda + b_2 \sin 2\lambda + b_3 \sin 3\lambda + \cdots \quad (A.1)$$

The terms $\cos \lambda$ and $\sin \lambda$ are called the *fundamental terms*; the terms $\cos n\lambda$ and $\sin n\lambda$, for $n > 1$, are called *harmonic terms*, where n is an integer. The terms a_n and b_n represent the coefficients of the fundamental and harmonics, and $\tfrac{1}{2}a_0$ is the constant or dc term.

The function $x(\lambda)$ must have a period of 2π, or a submultiple thereof, and it must be single valued. The Fourier series can be thought of as a "recipe" for synthesizing any *arbitrary periodic* waveform using sinusoidal components. To be useful, the series must converge; that is, the sum of the series, as more and more of the higher harmonics are added, must approach a limit.

The process of synthesizing an arbitrary periodic waveform, from the coefficient values describing the mix of harmonics, is termed *synthesis*. The inverse process of calculating the coefficient values is termed *anaylsis*. Calculation of the coefficients is facilitated by the fact that the average of the sine and cosine cross-products is zero, as well as the average of any sinusoid. Equations (A.2) to (A.4), listed below, illustrate the basic averaging properties of the sine, cosine, their products and cross-products.

$$\left. \begin{aligned} \int_{-\pi}^{\pi} \sin m\lambda \, d\lambda &= 0 \\[6pt] \int_{-\pi}^{\pi} \cos m\lambda \, d\lambda &= 0 \\[6pt] \int_{-\pi}^{\pi} \sin m\lambda \cos n\lambda \, d\lambda &= 0 \end{aligned} \right\} \quad \text{where } m \text{ and } n \text{ are any integers} \quad (A.2)$$

$$\left. \begin{aligned} \int_{-\pi}^{\pi} \sin m\lambda \sin n\lambda \, d\lambda &= 0 \\[6pt] \int_{-\pi}^{\pi} \cos m\lambda \cos n\lambda \, d\lambda &= 0 \end{aligned} \right\} \quad \text{for } m \neq n \quad (A.3)$$

$$\left. \begin{aligned} \int_{-\pi}^{\pi} (\sin m\lambda)^2 \, d\lambda &= \pi \\[6pt] \int_{-\pi}^{\pi} (\cos m\lambda)^2 \, d\lambda &= \pi \end{aligned} \right\} \quad \text{for } m = n \quad (A.4)$$

Consider how one could go about finding the value of the coefficient, a_n or

b_n, in Equation (A.1). To find the coefficient a_3, for example, we can multiply both sides of Equation (A.1) by $\cos 3\lambda \ d\lambda$ and integrate, as follows:

$$\int_{-\pi}^{\pi} x(\lambda) \cos 3\lambda \ d\lambda = \int_{-\pi}^{\pi} \tfrac{1}{2} a_0 \cos 3\lambda \ d\lambda + \int_{-\pi}^{\pi} a_1 \cos \lambda \cos 3\lambda \ d\lambda$$

$$+ \int_{-\pi}^{\pi} a_2 \cos 2\lambda \cos 3\lambda \ d\lambda + \int_{-\pi}^{\pi} a_3 (\cos 3\lambda)^2 \ d\lambda + \cdots$$

$$+ \int_{-\pi}^{\pi} b_1 \sin \lambda \cos 3\lambda \ d\lambda + \int_{-\pi}^{\pi} b_2 \sin 2\lambda \cos 3\lambda \ d\lambda$$

$$+ \int_{-\pi}^{\pi} b_3 \sin 3\lambda \cos 3\lambda \ d\lambda + \cdots$$

$$\int_{-\pi}^{\pi} x(\lambda) \cos 3\lambda \ d\lambda = \int_{-\pi}^{\pi} a_3 (\cos 3\lambda)^2 \ d\lambda = a_3 \pi$$

$$a_3 = \frac{1}{\pi} \int_{-\pi}^{\pi} x(\lambda) \cos 3\lambda \ d\lambda$$

We can generalize the analysis above, to get

$$a_n = \frac{1}{\pi} \int_{-\pi}^{\pi} x(\lambda) \cos n\lambda \ d\lambda \tag{A.5}$$

$$b_n = \frac{1}{\pi} \int_{-\pi}^{\pi} x(\lambda) \sin n\lambda \ d\lambda \tag{A.6}$$

a_0 is found by solving Equation (A.5) with $n = 0$. This results in

$$\tfrac{1}{2} a_0 = \tfrac{1}{2}\pi \int_{-\pi}^{\pi} x(\lambda) \ d\lambda \tag{A.7}$$

which represents the zero-frequency term, or the average value of the periodic waveform. The synthesis process of Equation (A.1) can be expressed in more compact form as follows:

$$x(\lambda) = \tfrac{1}{2} a_0 + \sum_{n=1}^{\infty} (a_n \cos n\lambda + b_n \sin n\lambda) \tag{A.8}$$

There are several ways to express the *transform pair* (analysis and synthesis) of the Fourier series. The most common form makes use of the following identities to express the sine and cosine in exponential form:

$$\cos \lambda = \frac{e^{j\lambda} + e^{-j\lambda}}{2} \tag{A.9}$$

$$\sin \lambda = \frac{e^{j\lambda} - e^{-j\lambda}}{2j} \tag{A.10}$$

A periodic function with period T_0 seconds has frequency components of

A Review of Fourier Techniques   App. A

$f_0, 2f_0, 3f_0, \ldots$, where $f_0 = 1/T_0$ is called the *fundamental frequency.* We also refer to the frequency components as $\omega_0, 2\omega_0, 3\omega_0, \ldots$, where $\omega_0 = 2\pi/T_0$ is called the fundamental *radian* frequency. The terms f and ω are each used to denote frequency. When f is used, frequency in hertz is intended; when ω is used, frequency in radians/second is intended. Let us replace the $n\lambda$ terms of Equations (A.5) to (A.8) with $2\pi n f_0 t = 2\pi n t/T_0$ as the general argument of the sinusoidal components, where n is an integer. For $n = 1$, nf_0 represents the fundamental frequency; for $n > 1$, nf_0 represents harmonics of the fundamental frequency. Using Equations (A.8) to (A.10), we can express $x(t)$ in exponential form as follows:

$$x(t) = \frac{a_0}{2} + \frac{1}{2} \sum_{n=1}^{\infty} [(a_n - jb_n)e^{j2\pi n f_0 t} + (a_n + jb_n)e^{-j2\pi n f_0 t}] \quad (A.11)$$

Let c_n denote the complex coefficients, or spectral components of $x(t)$, related to a_n and b_n by

$$c_n = \begin{cases} \frac{1}{2}(a_n - jb_n) & \text{for } n > 0 \\ \dfrac{a_0}{2} & \text{for } n = 0 \\ \frac{1}{2}(a_n + jb_n) & \text{for } n < 0 \end{cases} \quad (A.12)$$

Then we can simplify Equation (A.11) as follows:

$$x(t) = \sum_{n=-\infty}^{\infty} c_n e^{j2\pi n f_0 t} \quad (A.13)$$

where the coefficients of the exponential harmonics, c_n, are

$$c_n = \frac{1}{T_0} \int_{-T_0/2}^{T_0/2} x(t)e^{-j2\pi n f_0 t} \, dt \quad (A.14)$$

To verify Equation (A.14) we multiply both sides of Equation (A.13) by $e^{-j2\pi m f_0 t} \, dt/T_0$, integrate over the interval $(-T_0/2, T_0/2)$, and use the following relationship:

$$\frac{1}{T_0} \int_{-T_0/2}^{T_0/2} e^{j(n-m)2\pi f_0 t} \, dt = \delta_{nm} = \begin{cases} 1 & \text{for } n = m \\ 0 & \text{for } n \neq m \end{cases} \quad (A.15)$$

where δ_{nm} is known as the *Kronecker delta.* By multiplying and integrating in this way we obtain, for all integers m,

$$\frac{1}{T_0} \int_{-T_0/2}^{T_0/2} x(t)e^{-j2\pi m f_0 t} \, dt = \sum_{n=-\infty}^{\infty} c_n \delta_{nm} = c_m \quad (A.16)$$

In general, the coefficient c_n is a complex number; it can be expressed in the form

$$c_n = |c_n| e^{j\theta_n} \quad (A.17)$$

$$c_{-n} = |c_n| e^{-j\theta_n} \quad (A.18)$$

where

$$|c_n| = \tfrac{1}{2}\sqrt{a_n^2 + b_n^2} \tag{A.19}$$

$$\theta_n = \tan^{-1} -\frac{b_n}{a_n} \tag{A.20}$$

$$b_0 = 0 \quad \text{and} \quad c_0 = \frac{a_0}{2}$$

The value of $|c_n|$ defines the magnitude of the nth harmonic component of the periodic waveform, so that a plot of $|c_n|$ versus frequency, called the *magnitude spectrum*, yields the magnitude of each of the n discrete harmonics in the signal. Similarly, a plot of θ_n versus frequency, called the *phase spectrum*, yields the phase of each harmonic component in the signal.

The Fourier coefficients of a real-valued periodic time function exhibit the following relationship:

$$c_{-n} = c_n^* \tag{A.21}$$

where c_n^* is the complex conjugate of c_n. We therefore have

$$|c_{-n}| = |c_n| \tag{A.22}$$

and the magnitude spectrum is an even function of frequency. Similarly, the phase spectrum θ_n is an odd function of frequency, because from Equation (A.20),

$$\theta_{-n} = -\theta_n \tag{A.23}$$

The Fourier series is particularly useful in characterizing arbitrary periodic waveforms, with finite energy in each period, as presented above. The Fourier series can also be used to characterize nonperiodic signals having finite energy over a finite interval. However, a more convenient frequency-domain representation for such signals uses the Fourier integral transform (presented in Section A.2.3).

A.2.2 Spectrum of a Pulse Train

A signal of great interest in digital communications is an ideal periodic sequence of rectangular pulses, called a *pulse train*, illustrated in Figure A.3. For the pulse train, $x_p(t)$, with pulse amplitude A, pulse width T, and period T_0, the reader can verify, using Equations (A.14) and (A.10), the following expression for the Fourier series coefficients:

$$c_n = \frac{AT}{T_0} \frac{\sin (\pi n T/T_0)}{\pi n T/T_0} = \frac{AT}{T_0} \operatorname{sinc} \frac{nT}{T_0} \tag{A.24}$$

where

$$\operatorname{sinc} y = \frac{\sin \pi y}{\pi y}$$

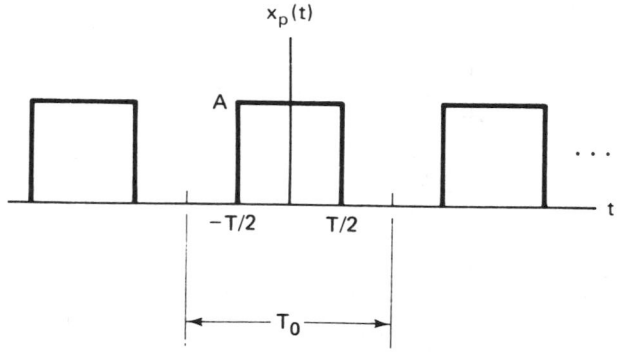

Figure A.3 Pulse train.

The sinc function, as shown in Figure A.4, has a maximum value of unity at $y = 0$ and approaches zero as y approaches infinity, oscillating through positive and negative values. It goes through zero at $y = \pm 1, \pm 2, \ldots$ The pulse train magnitude spectrum, $|c_n|$ as a function of n/T_0, is plotted in Figure A.5a, and the phase spectrum, θ_n, is plotted in Figure A.5b. The positive and negative frequencies of the two-sided spectrum represent a useful way of expressing the spectrum mathematically; of course, only the positive frequencies can be reproduced in a laboratory.

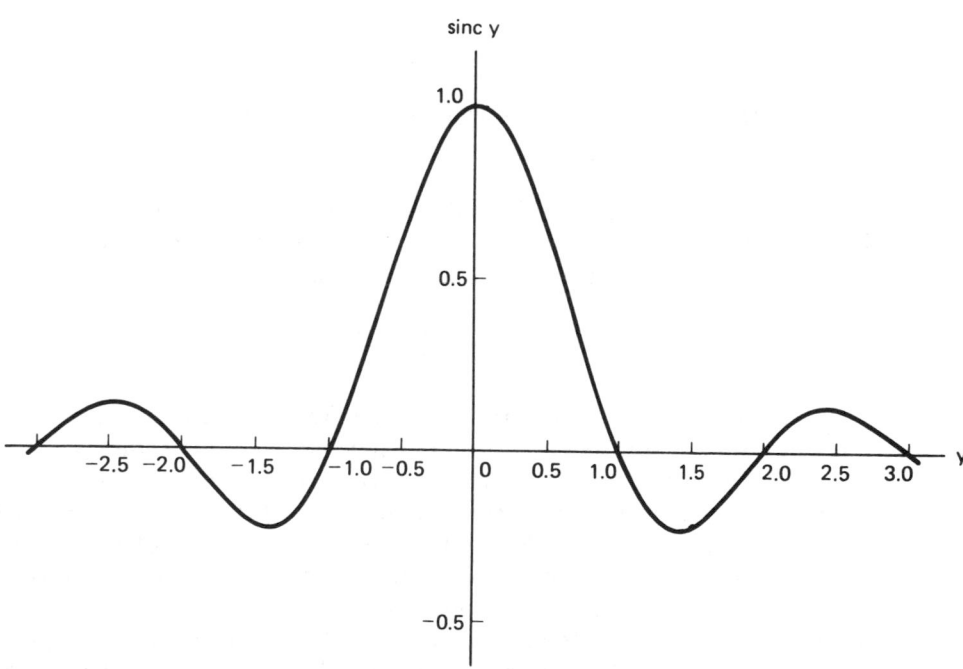

Figure A.4 Sinc function.

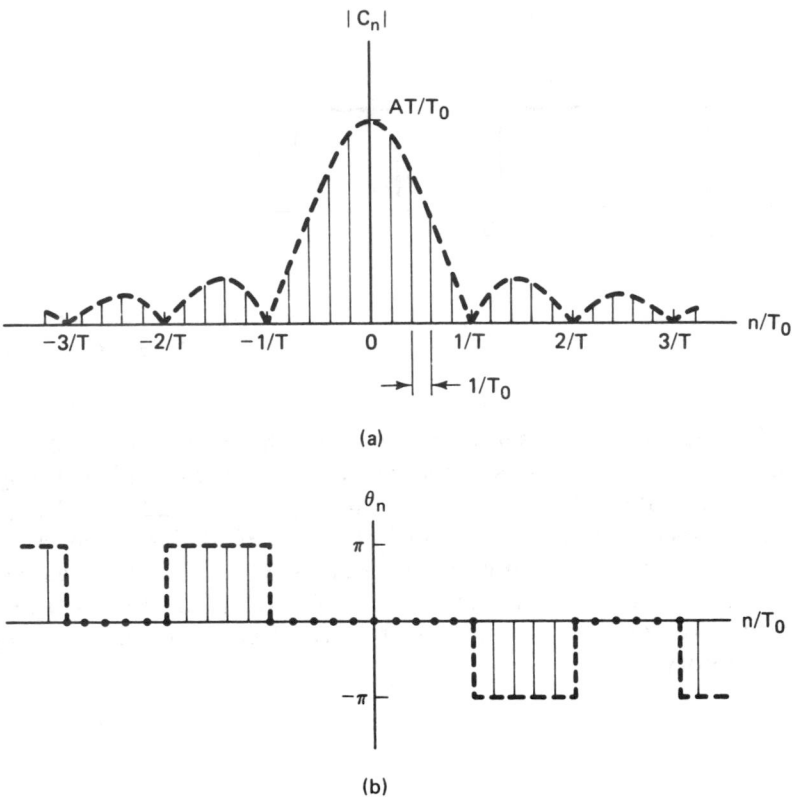

Figure A.5 Spectrum of a pulse train. (a) Magnitude spectrum. (b) Phase spectrum.

Synthesis is performed by substituting the coefficients of Equation (A.24) into Equation (A.13). The resulting series below yields the original ideal pulse train, $x_p(t)$, synthesized from its component parts.

$$x_p(t) = \frac{AT}{T_0} \sum_{n=-\infty}^{\infty} \text{sinc}\, \frac{nT}{T_0}\, e^{j2\pi nf_0 t} \qquad (A.25)$$

The ideal periodic pulse train contains frequency components at all integer multiples of the fundamental. In communication systems the significant portion of a baseband signal's power or energy is often assumed to be contained within the frequencies from zero to the first null of the magnitude spectrum (see Figure A.5a). Therefore, $1/T$ is often used as a measure of signal *bandwidth*, in hertz, for a pulse train with pulse width T. Note that bandwidth is inversely proportional to pulse width; the narrower are the pulses, the wider is the bandwidth associated with these pulses. Also, notice that the spacing between spectral lines $\Delta f = 1/T_0$ is inversely proportional to the pulse period; as the period increases, the lines move closer together.

A Review of Fourier Techniques   App. A

A.2.3 Fourier Integral Transform

In communication systems we often encounter nonperiodic signals having finite energy in a finite interval, and having zero energy outside this interval. Such signals can be conveniently characterized using the Fourier integral transform, or simply the *Fourier transform*. We can describe the nonperiodic signal as a periodic one, in the limiting sense. For example, consider the pulse train shown in Figure A.3. As $T_0 \to \infty$ and the pulse train approaches a single pulse, $x(t)$, the number of spectral lines approaches infinity and the spectral plot approaches a smooth frequency spectrum $X(f)$. For this limiting case, we can define a Fourier integral transform pair. This pair, expressed below, can be used to describe the time–frequency relationship for nonperiodic signals.

$$X(f) = \int_{-\infty}^{\infty} x(t)e^{-j2\pi ft} \, dt \tag{A.26}$$

$$x(t) = \int_{-\infty}^{\infty} X(f)e^{j2\pi ft} \, df \tag{A.27}$$

where f is frequency measured in hertz. Henceforth, the Fourier integral transform operation will be designated by the notation $\mathcal{F}\{\cdot\}$, and the inverse Fourier integral transform will be designated by $\mathcal{F}^{-1}\{\cdot\}$. The relationship between the time and frequency domains will be indicated by using the double arrow as follows:

$$x(t) \leftrightarrow X(f)$$

This notation indicates that $X(f)$ is the Fourier transform of $x(t)$ and that $x(t)$ is the inverse Fourier transform of $X(f)$. In the typical communications context, $x(t)$ is a real-valued function and $X(f)$ is a complex function, having real and imaginary components; in polar form, shown below, the spectrum, $X(f)$, can be specified by a magnitude characteristic and a phase characteristic.

$$X(f) = |X(f)| \, e^{j\theta(f)} \tag{A.28}$$

The properties of $X(f)$, the spectrum of a nonperiodic waveform, are similar to those of the spectrum for a periodic waveform, presented in Equations (A.17) to (A.23); that is, when $x(t)$ is real valued

$$X(-f) = X^*(f) \tag{A.29}$$

$$= |X(f)| \, e^{-j\theta(f)} \tag{A.30}$$

where X^* is the complex conjugate of X. The magnitude spectrum $|X(f)|$ is an even function of f and the phase spectrum is an odd function of f. In many cases $X(f)$ is either purely real or purely imaginary, and only one plot suffices to describe it.

A.3 FOURIER TRANSFORM PROPERTIES

There are many excellent references dealing with the details of Fourier transforms and their properties [1–4]. In this appendix we will emphasize the properties that are fundamental to communication systems. Some of the key features affecting signal transmission in communication systems are time delay, phase shift, multiplication by other signals, frequency translation, waveform convolution, and spectral convolution. We shall focus on the Fourier properties (shifting and convolution) needed to describe these key communication features.

A.3.1 Time Shifting Property

If $x(t) \leftrightarrow X(f)$,

$$\mathcal{F}\{x(t - t_0)\} = \int_{-\infty}^{\infty} x(t - t_0)e^{-j2\pi ft}\, dt$$

Let $\mu = t - t_0$; then

$$\mathcal{F}\{x(t - t_0)\} = \int_{-\infty}^{\infty} x(\mu)e^{-j2\pi f(\mu + t_0)}\, d\mu$$
$$= X(f)e^{-j2\pi f t_0} \tag{A.31}$$

As a signal is delayed in time, the magnitude of its frequency spectrum remains unchanged, but its phase spectrum experiences a phase shift. A time shift of t_0 in the time domain is equivalent to multiplication by $e^{-j2\pi f t_0}$ (a phase shift of $-2\pi f t_0$) in the frequency domain.

A.3.2 Frequency Shifting Property

If $x(t) \leftrightarrow X(f)$,

$$\mathcal{F}\{x(t)e^{j2\pi f_0 t}\} = \int_{-\infty}^{\infty} x(t)e^{j2\pi f_0 t}e^{-j2\pi ft}\, dt$$

$$= \int_{-\infty}^{\infty} x(t)e^{-j2\pi(f - f_0)t}\, dt \tag{A.32}$$

$$= X(f - f_0)$$

This is the basic *frequency translating* property that describes the shifted spectrum resulting from multiplying a signal by $e^{j2\pi f_0 t}$. Equation (A.32) can be used in conjunction with Equation (A.9) to yield the Fourier transform of a waveform multiplied by a cosine wave, as follows:

$$x(t) \cos 2\pi f_0 t = \tfrac{1}{2}[x(t)e^{j2\pi f_0 t} + x(t)e^{-j2\pi f_0 t}] \tag{A.33}$$
$$x(t) \cos 2\pi f_0 t \leftrightarrow \tfrac{1}{2}[X(f - f_0) + X(f + f_0)]$$

This property is also called the *mixing* or *modulation* theorem. Multiplication of an arbitrary signal by a sinusoid of frequency f_0 translates the original signal spectrum by f_0, and also by $-f_0$.

A.4 USEFUL FUNCTIONS

A.4.1 Unit Impulse Function

A useful function in communication theory is the unit impulse or *Dirac delta* function, $\delta(t)$. The impulse function can be developed from any of several fundamental functions (e.g., a rectangular pulse or a triangular pulse). In each development, the impulse function is defined in the limiting sense (the pulse amplitude approaches infinity, the pulse width approaches zero, but the area under the pulse is constrained to be unity) [5]. The unit impulse function has the following important properties:

$$\int_{-\infty}^{\infty} \delta(t)\, dt = 1 \tag{A.34}$$

$$\delta(t) = 0 \qquad \text{for } t \neq 0 \tag{A.35}$$

$$\delta(t) \text{ is unbounded at } t = 0 \tag{A.36}$$

$$\mathcal{F}\{\delta(t)\} = \mathcal{F}^{-1}\{\delta(f)\} = 1 \tag{A.37}$$

$$\int_{-\infty}^{\infty} x(t)\delta(t - t_0)\, dt = x(t_0) \tag{A.38}$$

Equation (A.38) is known as the *sifting* or *sampling property*; the unit impulse multiplier selects a sample of the function $x(t)$ evaluated at $t = t_0$.

In some problems it is useful to use the following equivalent integrals for an impulse function, defined in the time domain or the frequency domain [3]:

$$\delta(t) = \int_{-\infty}^{\infty} e^{j2\pi ft}\, df \tag{A.39}$$

$$\delta(f) = \int_{-\infty}^{\infty} e^{-j2\pi ft}\, dt \tag{A.40}$$

A.4.2 Spectrum of a Sinusoid

For the purpose of representing a sinusoidal waveform by a Fourier transform, the waveform may be assumed to exist only in the interval $(-T_0/2 < t < T_0/2)$. Under these conditions the function has a Fourier transform as long as T_0 is finite. In the limit, T_0 is made very large, but finite. The spectrum of the waveform $x(t) = A \cos 2\pi f_0 t$ can be found by using Equations (A.9) and (A.26):

$$X(f) = \int_{-\infty}^{\infty} \frac{A}{2} (e^{j2\pi f_0 t} + e^{-j2\pi f_0 t}) e^{-j2\pi ft}\, dt$$

$$= \frac{A}{2} \int_{-\infty}^{\infty} e^{-j2\pi(f - f_0)t} + e^{-j2\pi(f + f_0)t}\, dt$$

As described in Equation (A.40), the integral expression above can be equated

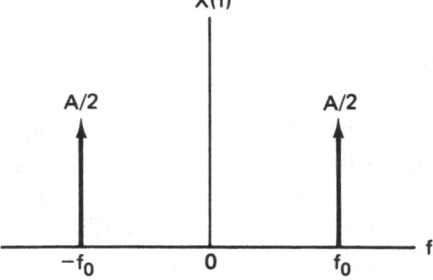

Figure A.6 Spectrum for $x(t) = A \cos 2\pi f_0 t$.

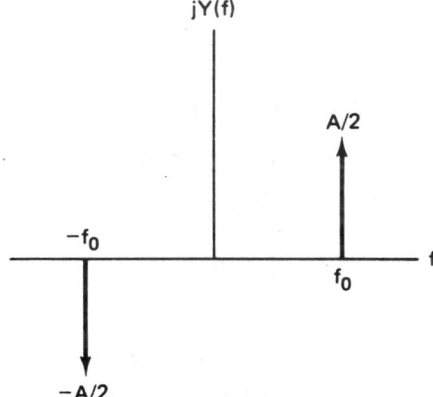

Figure A.7 Spectrum for $y(t) = A \sin 2\pi f_0 t$.

to unit impulse functions located at frequencies $\pm f_0$ as follows:

$$X(f) = \frac{A}{2}[\delta(f - f_0) + \delta(f + f_0)] \qquad (A.41)$$

Similarly, the spectrum of a sine waveform $y(t) = A \sin 2\pi f_0 t$ can be shown to be equal to

$$Y(f) = \frac{A}{2j}[\delta(f - f_0) - \delta(f + f_0)] \qquad (A.42)$$

The cosine waveform spectrum is shown in Figure A.6, and the sine waveform spectrum is shown in Figure A.7. Each of the impulse functions shown on these spectral plots is depicted as a spike with a weight of $A/2$ or $-A/2$.

A.5 CONVOLUTION

Convolution was used by Oliver Heaviside in the late nineteenth century to calculate electrical circuit output current when the input voltage waveform was more complicated than a simple battery source. The use of the methods of Heaviside

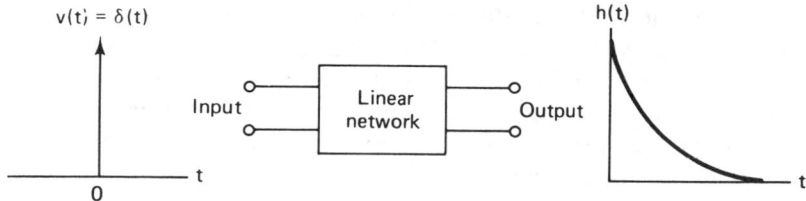

Figure A.8 Impulse response of a linear system.

predates the use of the analytical methods developed by Fourier and Laplace (even though publications by Fourier and Laplace came earlier).

The response of a circuit to an impulse voltage $v(t) = \delta(t)$ is called the *impulse response* and is denoted by $h(t)$, as shown in Figure A.8; it is simply the output voltage that would result if the input were a delta function. Heaviside approximated an arbitrary voltage waveform, like the one shown in Figure A.9a, by a set of equally spaced pulses. Such pulses of finite height and duration are shown in Figure A.9b. In the limit as the pulse width $\Delta\tau$ approaches zero, each pulse approaches an impulse function with weight equal to the area under that

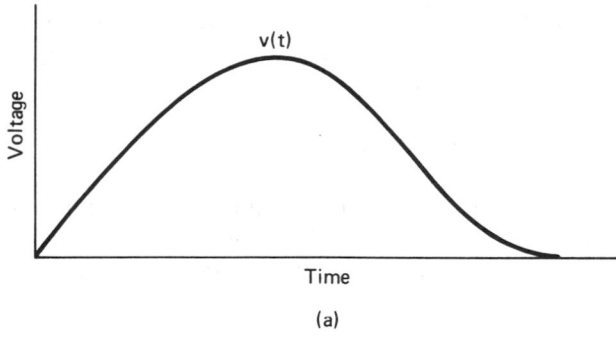

(a)

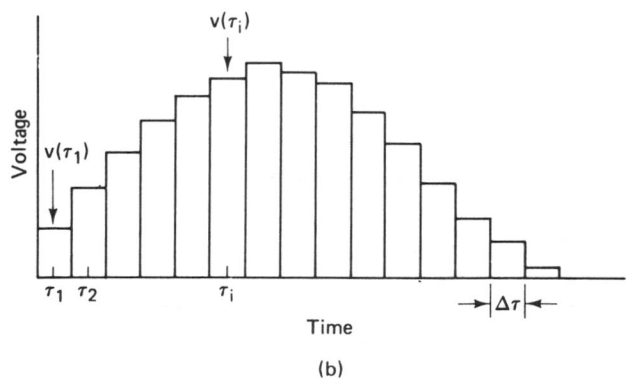

(b)

Figure A.9 (a) Input voltage waveform. (b) Approximate input voltage waveform.

Sec. A.5   Convolution

pulse. In the following discussion we shall refer to these equally spaced pulses as *impulses* even though they are impulses *only in the limit*.

Care needs to be taken with the notation of time, since we are interested in the times at which impulses are applied, and also the times at which their output responses are observed. We need to identify these two different time sequences; we shall use the following notation:

1. Time of the input application will be termed τ, so that the input voltage impulses are designated $v(\tau_1)$, $v(\tau_2)$, . . . , $v(\tau_N)$.
2. Time of the output response will be termed t, so that the output currents are designated $i(t_1)$, $i(t_2)$, . . . , $i(t_N)$.

Heaviside found the response or current produced by each input impulse independently; then he added the individual responses to get the total current. The weight of the impulse produced by the rectangular voltage at time τ_1 is the product $v(\tau_1)\,\Delta\tau$. The series of impulses can approximate the arbitrary input voltage as closely as desired by allowing $\Delta\tau$ to approach zero. Note again that the instant at which an impulse is applied is called τ_i, and the instance at which the system response is determined is called t_i, where τ is the input time variable, t is the output time variable, and $i = 1, \ldots, N$.

Figure A.10 illustrates the output response $i(t) = A_1 h(t - \tau_1)$ to an impulse with height $v(\tau_1)$. Since the input impulse at τ_1 is *not* a *unit* impulse, we weight it with its strength or area, $A_1 = v(\tau_1)\,\Delta\tau$. At some time t_1, where $t_1 > \tau_1$, the output response to the impulse $v(\tau_1)$ is expressed as

$$i(t_1) = A_1 h(t_1 - \tau_1) \qquad \text{for } t_1 > \tau_1$$

as shown in Figure A.10. When there are several input impulses, the total output response for a linear system is simply the sum of the individual responses. Figure A.11 illustrates the response of the network to two input impulses. For N impulses,

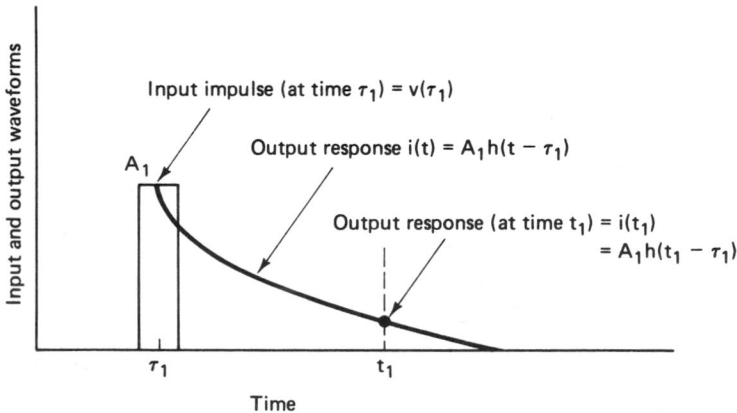

Figure A.10 Output response to an impulse at time τ_1.

A Review of Fourier Techniques   App. A

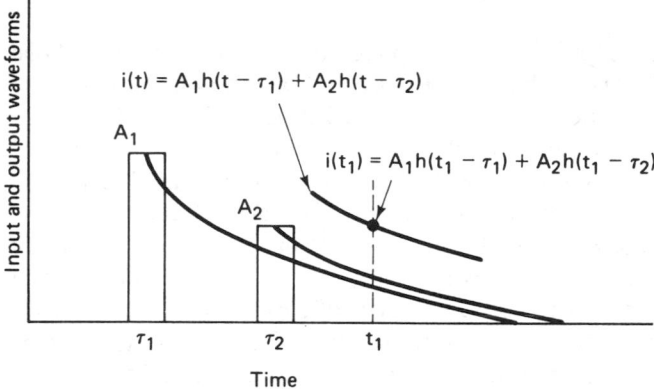

$i(t) = A_1 h(t - \tau_1) + A_2 h(t - \tau_2)$

$i(t_1) = A_1 h(t_1 - \tau_1) + A_2 h(t_1 - \tau_2)$

Input and output waveforms

A_1

A_2

τ_1 τ_2 t_1

Time

Figure A.11 Output response to two impulses.

the output current measured at time t_1 can be expressed as

$$i(t_1) = A_1 h(t_1 - \tau_1) + A_2 h(t_1 - \tau_2) + \cdots + A_N(t_1 - \tau_N)$$

where the impulses are applied at $\tau_1, \tau_2, \ldots, \tau_N$, and where $t_1 > \tau_N$.

Any impulses applied at times greater than t_1 are disregarded, for they contribute nothing to $i(t_1)$. This corresponds to the *causality* requirement for physically realizable systems, which states that the system response must be zero prior to the application of the excitation. By generalizing, we get the output current at any time t,

$$i(t) = A_1 h(t - \tau_1) + A_2 h(t - \tau_2) + \cdots + A_N h(t - \tau_N)$$

or

$$i(t) = \sum_{i=1}^{N} v(\tau_i) \, \Delta\tau h(t - \tau_i) \tag{A.43}$$

since the height of the impulse at τ_i is equal to $v(\tau_i)$. As $\Delta\tau$ approaches zero, the sum of the input impulses approaches the actual applied voltage $v(\tau)$; we can replace $\Delta\tau$ with $d\tau$, and the summation becomes the *convolution integral*:

$$i(t) = \int_{-\infty}^{\infty} v(\tau) h(t - \tau) \, d\tau \tag{A.44}$$

In shorthand notation this is expressed as

$$i(t) = v(t) * h(t) \tag{A.45}$$

In summary, $i(t)$ is the sum of the individual impulse responses as a function of output time t. Each impulse response is due to an impulse applied at some input time τ and is weighted by the strength of that impulse.

Sec. A.5   Convolution

A.5.1 Graphical Illustration of Convolution

Consider that an input square pulse $v(t)$ is applied to a linear network whose impulse response is labeled $h(t)$ as shown in Figure A.12a. The output response is characterized by the convolution integral expressed in Equation (A.44).

The independent variable in the convolution integral is τ. The functions $v(\tau)$ and $h(-\tau)$ are shown in Figure A.12b. Note that $h(-\tau)$ is obtained by folding $h(\tau)$ about $\tau = 0$. The term $h(t - \tau)$ represents the function $h(-\tau)$ shifted by t seconds along the positive τ axis. Figure A.12c shows the function $h(t_1 - \tau)$. The value of the convolution integral at $t = t_1$ is given by Equation (A.44) evaluated at $t = t_1$. This is simply the area under the product curve of $v(\tau)$ and $h(t_1 - \tau)$, shown shaded in Figure A.12d. Similarly, the convolution integral evaluated at $t = t_2$ is equal to the shaded area in Figure A.12e. Figure A.12f is a plot of the output response as a result of the square pulse input to the circuit with impulse response shown in Figure A.12a. Each evaluation of the convolution integral, at some time t_i, yields one point, $i(t_i)$, on the plot of Figure A.12f.

A.5.2 Time Convolution Property

If $x_1(t) \leftrightarrow X_1(f)$ and $x_2(t) \leftrightarrow X_2(f)$,

$$x_1(t) * x_2(t) = \int_{-\infty}^{\infty} x_1(\tau)x_2(t - \tau) \, d\tau$$

$$\mathcal{F}\{x_1(t) * x_2(t)\} = \int_{-\infty}^{\infty} \int_{-\infty}^{\infty} x_1(\tau)x_2(t - \tau) \, d\tau \, e^{-j2\pi ft} \, dt$$

For linear systems, we may exchange the order of integration as follows:

$$\mathcal{F}\{x_1(t) * x_2(t)\} = \int_{-\infty}^{\infty} x_1(\tau) \, d\tau \int_{-\infty}^{\infty} x_2(t - \tau)e^{-j2\pi ft} \, dt \qquad \text{(A.46)}$$

By the Fourier *time shifting property*, the second integral expression of the right-hand side is equal to $X_2(f)e^{-j2\pi f\tau}$:

$$\mathcal{F}\{x_1(t) * x_2(t)\} = X_2(f) \int_{-\infty}^{\infty} x_1(\tau)e^{-j2\pi f\tau} \, d\tau$$

$$= X_1(f)X_2(f) \qquad \text{(A.47)}$$

Therefore, the operation of *convolution* in the time domain can be replaced by *multiplication* in the frequency domain.

A.5.3 Frequency Convolution Property

Because of the symmetry of the Fourier transform pair in Equations (A.26) and (A.27), it can be shown that multiplication in the time domain transforms to convolution in the frequency domain

$$x_1(t)x_2(t) \leftrightarrow X_1(f) * X_2(f) \qquad \text{(A.48)}$$

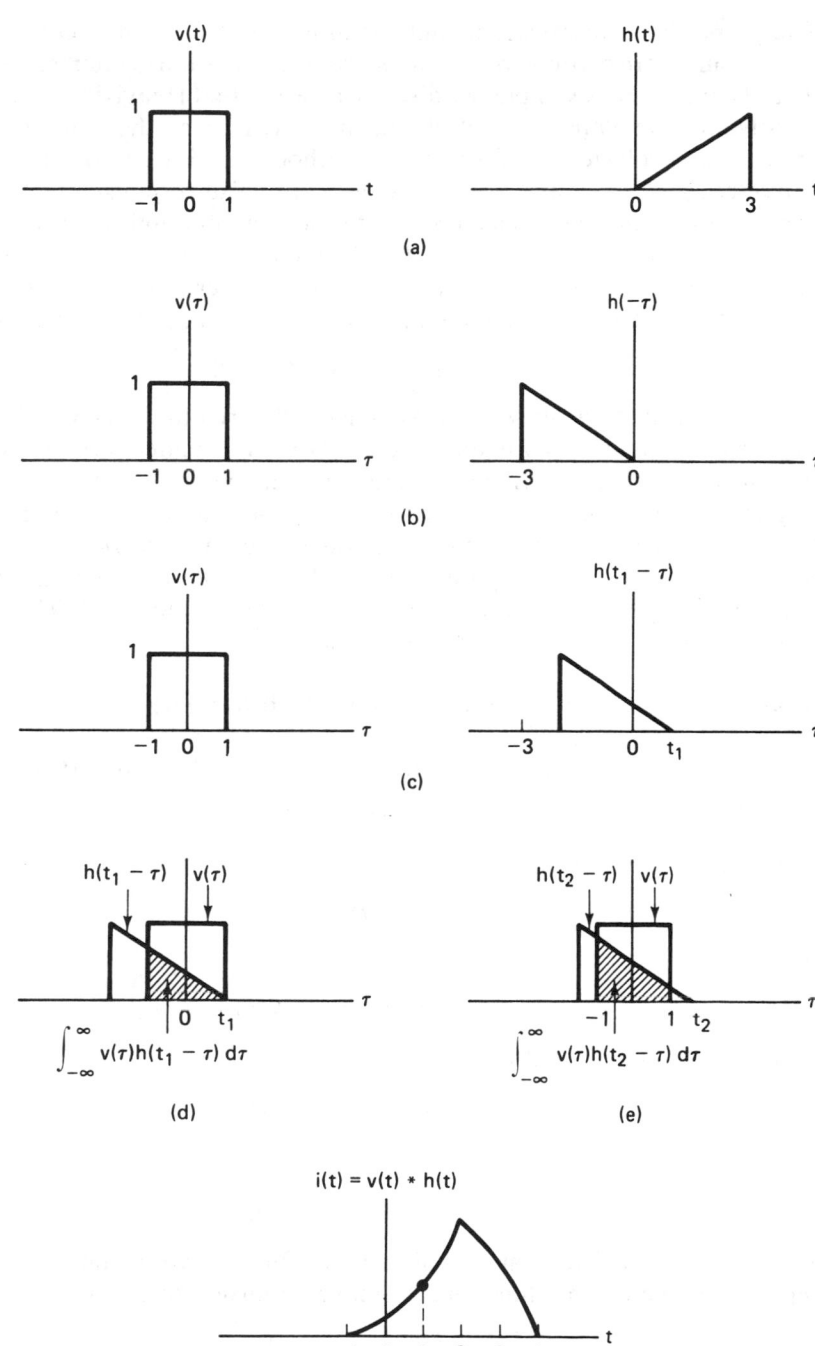

Figure A.12 Graphical example of convolution.

The properties that transform multiplication in one domain to convolution in the other domain are particularly useful, since one operation is often easier to perform than the other. For example, we discussed earlier that Heaviside used convolution to solve for the output current of a linear system when the input was excited by an arbitrary voltage waveform. Such methods involve the (sometimes tedious) convolution of an input waveform with the impulse response of a system. Since convolution in the time domain is transformed into multiplication in the frequency domain, as shown in Equation (A.47), for a linear system we can simply multiply the input waveform spectrum by the system transfer function. The output waveform is then found by taking the inverse Fourier transform of the product.

$$i(t) = \mathcal{F}^{-1}\{V(f)H(f)\} \tag{A.49}$$

Solutions of the form shown in Equation (A.49) are often much easier to perform than those described by Equation (A.45). However, under certain circumstances, the operation of convolution is so simple that it can be performed graphically, by inspection. For example, suppose that we wished to multiply an arbitrary waveform by some fixed frequency cosine wave, such as a carrier wave, in the case of modulation. By applying Equation (A.48), we can convolve the spectrum of the arbitrary waveform with the spectrum of the cosine wave. This is easily accomplished, as is shown in the next section.

A.5.4 Convolution of a Function with a Unit Impulse

By the property shown in Equation (A.47), it should be clear that if

$$x(t) \leftrightarrow X(f)$$

and since

$$\delta(t) \leftrightarrow 1$$

then

$$x(t) * \delta(t) \leftrightarrow X(f) \tag{A.50}$$

It should also be evident that

$$x(t) * \delta(t) = x(t) \tag{A.51}$$

and

$$X(f) * \delta(f) = X(f) \tag{A.52}$$

We therefore conclude that convolution of a function with a unit impulse function reproduces the original function. A simple extension of Equation (A.52) yields

$$X(f) * \delta(f - f_0) = X(f - f_0) \tag{A.53}$$

Figure A.13 illustrates the ease of convolving the spectrum of an arbitrary waveform with the spectrum of a cosine wave. Figure A.13a shows an arbitrary baseband spectrum $X(f)$. Figure A.13b shows a spectrum, $Y(f) = \delta(f - f_0) +$

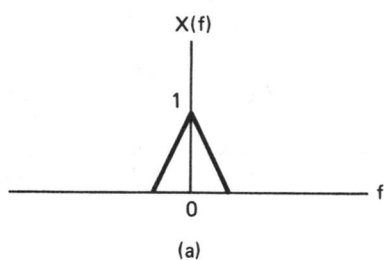

$X(f)$

1

0

f

(a)

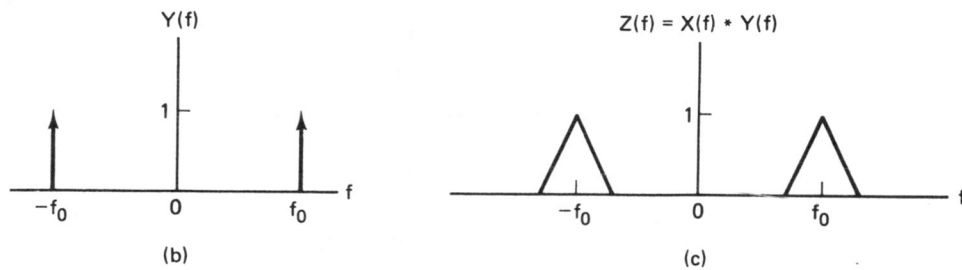

$Y(f)$

1

$-f_0$ 0 f_0

f

(b)

$Z(f) = X(f) * Y(f)$

1

$-f_0$ 0 f_0

f

(c)

Figure A.13   Convolving a signal spectrum with a cosine-wave spectrum.

$\delta(f + f_0) = \mathcal{F}\{2 \cos 2\pi f_0 t\}$. The output, $Z(f) = X(f) * Y(f)$, in Figure A.13c is obtained by convolving the waveform spectrum with the impulse functions of $Y(f)$ according to Equation (A.53), where the impulses act as sampling functions. Hence, for this simple example, convolution can be performed graphically by sweeping the sampling impulses past the waveform spectrum. Multiplication by the impulse functions at each step in the sweep yields replications of the waveform spectrum. The result, shown in Figure A.13c, is a shifted version of the original spectrum $X(f)$ to the locations of the impulse functions in Figure A.13b.

A.5.5 Demodulation Application of Convolution

In Section A.5.4 we examined a waveform multiplied by $2 \cos 2\pi f_0 t$. We illustrated the frequency-domain view of convolving the waveform spectrum with a cosine-wave spectrum. In this section we look at the reverse process. A waveform that has been multiplied by $2 \cos 2\pi f_0 t$ is to be demodulated (the waveform is to be restored to its baseband frequency range).

Figure A.14a represents the spectrum, $Z(f)$, of the waveform that has been upshifted in frequency. We can demodulate this upshifted waveform and recover the baseband waveform, by multiplying it by $2 \cos 2\pi f_0 t$. Instead, we shall illustrate the detection process in the frequency domain by convolving $Z(f)$ with the spectrum of the carrier, $Y(f) = \delta(f - f_0) + \delta(f + f_0)$, shown in Figure A.14b.

A simple extension of Equations (A.52) and (A.53) yields

$$X(f - f_0) * \delta(f - f_1) = X(f - f_0 - f_1) \qquad \text{(A.54)}$$

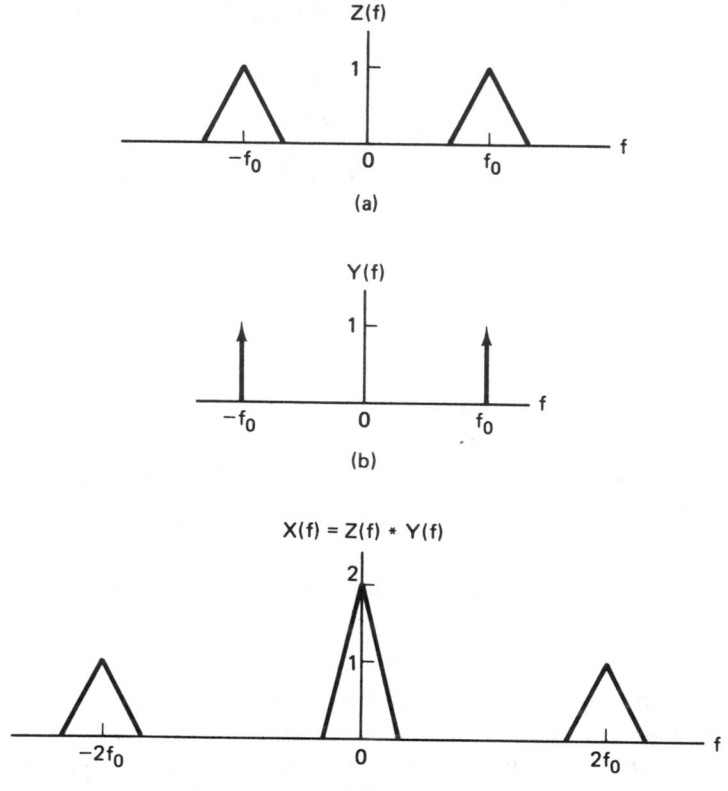

Figure A.14 Demodulation application.

Therefore, the result of demodulation, $X(f) = Z(f) * Y(f)$ is obtained by applying Equation (A.54). The resulting signal spectrum appears at baseband (detected) and also at frequencies $\pm 2f_0$, as shown in Figure A.14c. As in the previous section, the convolution can be performed graphically. The resulting Figure A.14c contains the following terms:

$$[Z(f - f_0) + Z(f + f_0)] * [\delta(f - f_0) + \delta(f + f_0)]$$

$$= Z(f - f_0) * \delta(f - f_0) + Z(f - f_0) * \delta(f + f_0)$$
$$+ Z(f + f_0) * \delta(f - f_0) + Z(f + f_0) * \delta(f + f_0) \qquad \text{(A.55)}$$

$$= 2Z(f) + Z(f - 2f_0) + Z(f + 2f_0)$$

Notice that the resulting terms consist of the baseband spectrum plus terms associated with higher-frequency components. The result is typical of the detection process; the higher-frequency terms are filtered and discarded, leaving the demodulated baseband spectrum.

A.6 TABLES OF FOURIER TRANSFORMS AND OPERATIONS

Commonly used Fourier transforms and operations are tabulated below. The transforms are listed in Table A.1 and the operations in Table A.2.

TABLE A.1 Fourier Transforms

$x(t)$	$X(f)$
1. $\delta(t)$	1
2. 1	$\delta(f)$
3. $\cos 2\pi f_0 t$	$\dfrac{1}{2}[\delta(f - f_0) + \delta(f + f_0)]$
4. $\sin 2\pi f_0 t$	$\dfrac{1}{2j}[\delta(f + f_0) - \delta(f - f_0)]$
5. $\delta(t - t_0)$	$\exp(-j2\pi f t_0)$
6. $\exp(j2\pi f_0 t)$	$\delta(f - f_0)$
7. $\exp(-a\|t\|), \quad a > 0$	$\dfrac{2a}{a^2 + (2\pi f)^2}$
8. $\exp\left[-\pi\left(\dfrac{t}{T}\right)^2\right]$	$T \exp[-\pi(fT)^2]$
9. $u(t) = \begin{cases} 1 & \text{for } t > 0 \\ 0 & \text{for } t < 0 \end{cases}$	$\dfrac{1}{2}\delta(f) + \dfrac{1}{j2\pi f}$
10. $\exp(-at)\,u(t), \quad a > 0$	$\dfrac{1}{a + j2\pi f}$
11. $t\exp(-at)\,u(t), \quad a > 0$	$\dfrac{1}{(a + j2\pi f)^2}$
12. $\text{rect}\left(\dfrac{t}{T}\right)$	$T \,\text{sinc}\, fT$
13. $\cos 2\pi f_0 t\left[\text{rect}\left(\dfrac{t}{T}\right)\right]$	$\dfrac{T}{2}[\text{sinc}\,(f - f_0)T + \text{sinc}\,(f + f_0)T]$
14. $W \,\text{sinc}\, Wt$	$\text{rect}\left(\dfrac{f}{W}\right)$
15. $\begin{cases} 1 - \dfrac{\|t\|}{T} & \text{for } \|t\| \le T \\ 0 & \text{for } \|t\| > T \end{cases}$	$T \,\text{sinc}^2\, fT$
16. $\displaystyle\sum_{m=-\infty}^{\infty} \delta(t - mT_0)$	$\dfrac{1}{T_0}\displaystyle\sum_{n=-\infty}^{\infty} \delta\left(f - \dfrac{n}{T_0}\right)$

Note: $\text{rect}\,(f/2W) = 1$ for $-W < f < W$, 0 for $\|f\| > W$, and $\text{sinc}\, x = (\sin \pi x)/\pi x$.

TABLE A.2 Fourier Operations

Operation	$x(t)$	$X(f)$
1. Scaling	$x(at)$	$\dfrac{1}{|a|} X\left(\dfrac{f}{a}\right)$
2. Time shifting	$x(t - t_0)$	$X(f) \exp(-j2\pi f t_0)$
3. Frequency shifting	$x(t) \exp(j2\pi f_0 t)$	$X(f - f_0)$
4. Time differentiation	$\dfrac{d^n x}{dt^n}$	$(j2\pi f)^n X(f)$
5. Frequency differentiation	$(-jt)^n x(t)$	$\dfrac{d^n X}{df^n}$
6. Time integration	$\displaystyle\int_{-\infty}^{t} x(\tau)d\tau$	$\dfrac{1}{j2\pi f} X(f) + \dfrac{1}{2}X(0)\delta(f)$
7. Time convolution	$x_1(t) * x_2(t)$	$X_1(f)X_2(f)$
8. Frequency convolution	$x_1(t)x_2(t)$	$X_1(f) * X_2(f)$

REFERENCES

1. Papoulis, A., *Signal Analysis*, McGraw-Hill Book Company, New York, 1977.

2. Panter, P. F., *Modulation, Noise, and Spectral Analysis*, McGraw-Hill Book Company, New York, 1965.

3. Bracewell, R., *The Fourier Transform and Its Applications*, McGraw-Hill Book Company, New York, 1978.

4. Haykin, S., *Communication Systems*, John Wiley & Sons, Inc., New York, 1983.

5. Schwartz, M., *Information, Transmission, Modulation, and Noise*, McGraw-Hill Book Company, New York, 1980.

Fundamentals of Statistical Decision Theory

The basic elements of a statistical decision problem are (1) a set of hypotheses that characterize the possible true states of nature, (2) a test in which data are obtained from which we wish to infer the truth, (3) a decision rule that operates on the data to decide in an optimal fashion which hypothesis best describes the true state of nature, and (4) a criterion of optimality. These fundamental steps are treated in the material that follows. The *optimality criterion* we will choose for the decision rule is to minimize the probability of making an erroneous decision, although other criteria are possible [1].

The subject of statistical decision theory and hypothesis testing builds on the mathematical discipline of probability theory and random variables. It is assumed that the reader has a familiarity with these subjects; if not, Reference [2] is a suggested resource.

B.1 BAYES' THEOREM

The mathematical foundations of hypothesis testing rest on Bayes' theorem, which is derived from the definition of the relationship between the conditional and joint probability of the random variables A and B:

$$P(A|B)P(B) = P(B|A)P(A) = P(A, B) \tag{B.1}$$

A statement of the theorem is

$$P(A|B) = \frac{P(B|A)P(A)}{P(B)} \tag{B.2}$$

Bayes' theorem allows us to infer the conditional probability, $P(A|B)$, from the conditional probability $P(B|A)$.

B.1.1 Discrete Form of Bayes' Theorem

Bayes' theorem can be expressed in discrete form, as follows:

$$P(s_i|z_j) = \frac{P(z_j|s_i)P(s_i)}{P(z_j)} \qquad \begin{matrix} i = 1, \ldots, M \\ j = 1, \ldots \end{matrix} \qquad \text{(B.3)}$$

where

$$P(z_j) = \sum_{i=1}^{M} P(z_j|s_i)P(s_i)$$

In a communications application, s_i is the ith signal class, from a set of M classes, and z_j is the jth sample of a received signal. Equation (B.3) can be thought of as the description of an experiment involving a received sample and some statistical knowledge of the signal classes to which the received sample may belong. The probability of occurrence of the ith signal class, $P(s_i)$, before the experiment, is called the *a priori probability*. As a result of examining a particular received sample, z_j, we can find a statistical measure of the *likelihood* that z_j belongs to class s_i from the conditional probability density function (pdf) $P(z_j|s_i)$. *After* the experiment, we can compute the *a posteriori probability*, $P(s_i|z_j)$, which can be thought of as a "refinement" of our prior knowledge. Thus we enter into the experiment with some a priori knowledge concerning the probability of the state of nature, and after examining a sample signal, we are provided with an "after-the-fact" a posteriori probability. The parameter $P(z_j)$ is the probability of the received sample, z_j, over the entire space of signal classes. This term, $P(z_j)$, can be thought of as a scaling factor, since its value is the same for *each* signal class.

Example B.1   Use of Bayes' Theorem (Discrete Form)

Given two boxes of parts. Box 1 contains 1000 parts, of which 10% are defective, and box 2 contains 2000 parts, of which 5% are defective. If a box is randomly chosen and then a part is randomly chosen from it, tested, and found to be good, what is the probability that the part came from box 1?

Solution

$$P(\text{box 1}|\text{GP}) = \frac{P(\text{GP}|\text{box 1})P(\text{box 1})}{P(\text{GP})}$$

where GP means "good part."

$$P(\text{GP}) = P(\text{GP}|\text{box 1})P(\text{box 1}) + P(\text{GP}|\text{box 2})P(\text{box 2})$$

$$= (0.90)(0.5) + (0.95)(0.5)$$

$$= 0.450 + 0.475 = 0.925$$

$$P(\text{box 1}|\text{GP}) = \frac{0.450}{0.925} = 0.486$$

Before the experiment, the a priori probability of having chosen either box 1 or box

2 was equally likely. After obtaining a good part, the Bayesian computation can be regarded as a way of "fine tuning" our thinking that $P(\text{box } 1) = 0.5$ to yield the a posteriori probability of 0.486. The Bayes' theorem is simply a formalization of common sense. Having selected a good part from one of the two boxes, isn't it intuitively reasonable that there is a higher probability that the part came from the box with the larger concentration of good parts, and a lower probability that it came from the box with the smaller concentration of good parts? The Bayes' theorem has refined the a priori statistic into an a posteriori statistic for the probability of box selection.

Example B.2   Decision Theory Applied to a Betting Game

A box has three coins: a fair coin, a two-headed coin, and a two-tailed coin. You are asked to pick one coin at random, look at one side only, and guess head or tail for the other side. What is the optimum decision strategy for this game?

Solution

We can view this problem as a signal detection problem. A signal is transmitted, but because of the channel noise, the received signal is somewhat obscured. Not being able to look at the other side of the coin is tantamount to receiving a noise perturbed signal. Let H_i represent the hypotheses ($i = F, H, T$), where F, H, and T, stand for fair, head, and tail, respectively:

$$H_F: \quad H, T \text{ (fair coin)}$$

$$H_H: \quad H, H \text{ (two-headed coin)}$$

$$H_T: \quad T, T \text{ (two-tailed coin)}$$

Let z_j represent the received sample ($j = H, T$), where z_H is a head and z_T is a tail. Let the a priori probabilities of the hypotheses be equally likely, so that $P(H_F) = P(H_H) = P(H_T) = \frac{1}{3}$. Using Bayes' theorem,

$$P(H_i|z_j) = \frac{P(z_j|H_i)P(H_i)}{\sum\limits_i P(z_j|H_i)P(H_i)}$$

we need to compute the probability for each hypothesis, given each signal class. Thus we need to examine the results of *six* computations before we can establish an optimum decision strategy. In each case the value of $P(z_j|H_i)$ can be obtained from the conditional probabilities drawn in Figure B.1. Consider that we choose a coin and view a head (z_H), we compute the following three a posteriori probabilities:

$$P(H_F|z_H) = \frac{(\frac{1}{2})(\frac{1}{3})}{(\frac{1}{2})(\frac{1}{3}) + (1)(\frac{1}{3}) + 0} = \frac{1}{3}$$

$$P(H_H|z_H) = \frac{(1)(\frac{1}{3})}{(\frac{1}{2})(\frac{1}{3}) + (1)(\frac{1}{3}) + 0} = \frac{2}{3}$$

$$P(H_T|z_H) = 0$$

If the received sample is a tail (z_T), we similarly compute

$$P(H_F|z_T) = \tfrac{1}{3}$$

$$P(H_H|z_T) = 0$$

$$P(H_T|z_T) = \tfrac{2}{3}$$

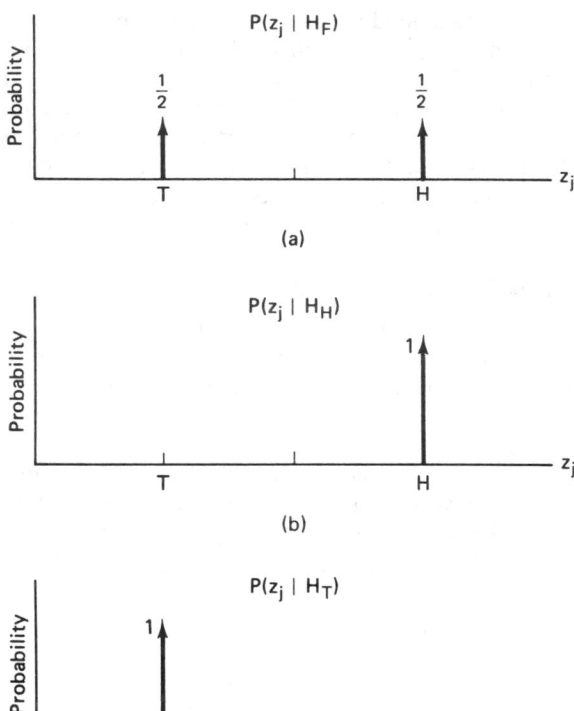

Figure B.1 Conditional probability $P(z_j \mid H_i)$. (a) Conditioned on the fair-coin hypothesis. (b) Conditioned on the two-headed-coin hypothesis. (c) Conditioned on the two-tailed-coin hypothesis.

The optimum decision strategy then is as follows: If a head, z_H, is received, choose hypothesis H_H (that the other side is also a head). If a tail, z_T, is received, choose hypothesis H_T (that the other side is also a tail).

B.1.2 Mixed Form of Bayes' Theorem

For most communication engineering applications of interest, the possible values of the received samples are *continuous* in range, because of the additive Gaussian noise in the channel. Therefore, the most useful form of Bayes' theorem contains a continuous- instead of a discrete-valued pdf. We shall rewrite Equation (B.3) to emphasize this change:

$$P(s_i|z) = \frac{p(z|s_i)P(s_i)}{p(z)} \qquad i = 1, \ldots, M$$

$$p(z) = \sum_{i=1}^{M} p(z|s_i)P(s_i)$$

(B.4)

where $p(z|s_i)$ is the conditional pdf of the received continuous-valued sample, z, conditioned on the signal class, s_i.

Example B.3   A Pictorial View of Bayes' Theorem

Consider two signal classes, s_1 and s_2, characterized by the triangular-shaped conditional pdfs, $p(z|s_1)$ and $p(z|s_2)$, illustrated in Figure B.2. A signal is received; it might have any value on the z-axis. If the pdfs did not overlap, we could classify the signal with certainty. For the example shown in Figure B.2, we need a rule to help us classify received signals, since some signals will fall in the region where the two pdfs overlap. Consider a received signal, z_a. Assume that the two signal classes, s_1 and s_2, are equally likely, and calculate the two alternative a posteriori probabilities. Suggest a decision rule that the receiver should use for deciding to which signal class z_a belongs. Repeat this for signal z_b.

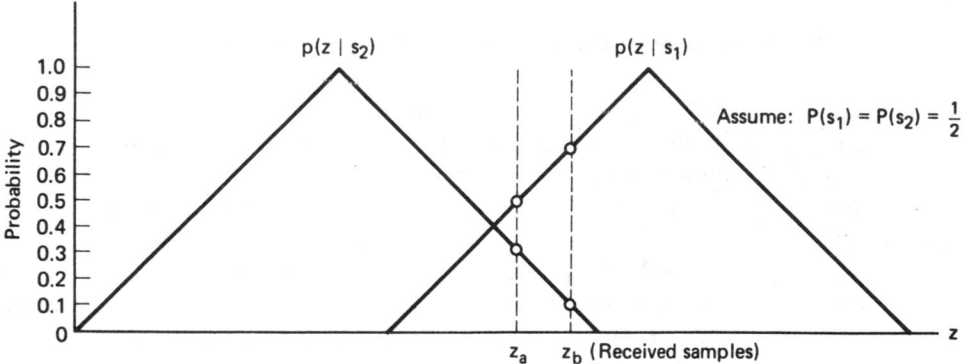

Figure B.2   Pictorial view of Bayes' theorem.

Solution

From Figure B.2 we can see that $p(z_a|s_1) = 0.5$ and $p(z_a|s_2) = 0.3$. Thus

$$P(s_1|z_a) = \frac{p(z_a|s_1)P(s_1)}{p(z_a|s_1)P(s_1) + p(z_a|s_2)P(s_2)}$$

$$= \frac{(0.5)(0.5)}{(0.5)(0.5) + (0.3)(0.5)} = \frac{5}{8}$$

and

$$P(s_2|z_a) = \frac{(0.3)(0.5)}{(0.5)(0.5) + (0.3)(0.5)} = \frac{3}{8}$$

One rule is to decide that the received signal belongs to the class with the maximum a posteriori probability (class s_1). An equivalent rule, for the case of equal a priori probabilities, is to examine the value of the pdf conditioned on each signal class (referred to as the likelihood of the signal class) and choose the class with the maximum. Examine Figure B.2 and notice that this *maximum likelihood rule* parallels our intuition. The likelihood that signal z_a belongs to each class corresponds to an encircled point on each pdf. The maximum likelihood rule is to choose the signal class that yields the largest conditional probability of all the alternatives. We repeat the computations for the received signal z_b, as follows:

$$P(s_1|z_b) = \frac{(0.7)(0.5)}{(0.7)(0.5) + (0.1)(0.5)} = \frac{7}{8}$$

$$P(s_2|z_b) = \frac{(0.1)(0.5)}{(0.7)(0.5) + (0.1)(0.5)} = \frac{1}{8}$$

As before, the maximum likelihood rule dictates that we choose signal class s_1. Notice that in the case of received sample z_b, we can have greater confidence in the correctness of our choice compared to the case of signal z_a. This is because the ratio of $p(z_b|s_1)$ to $p(z_b|s_2)$ is considerably larger than the ratio of $p(z_a|s_1)$ to $p(z_a|s_2)$.

B.2 DECISION THEORY

B.2.1 Components of the Decision Theory Problem

Having reviewed hypothesis testing based on Bayesian statistics, let us examine more carefully the components of the decision theory problem in the context of a communication system, as shown in Figure B.3. The signal source at the transmitter consists of a set $\{s_i(t)\}$, $i = 1, \ldots, M$, of waveforms (or hypotheses). A signal waveform $r(t) = s_i(t) + n(t)$ is received, where $n(t)$ is an additive white Gaussian noise (AWGN) process introduced in the channel. At the receiver, the waveform is reduced to a single number, $z(t = T)$, that may appear anywhere on the z-axis. Because the noise is a Gaussian process and the receiver is assumed linear, the output, $z(t)$, is also a Gaussian process [1], and the number, $z(T)$, is a *continuous-valued random variable*.

$$z(T) = a_i(T) + n_0(T) \tag{B.5}$$

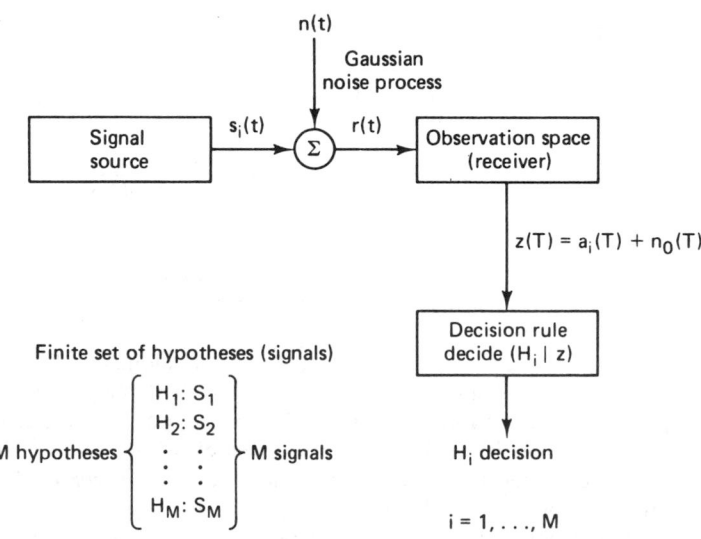

Figure B.3 Components of the decision theory problem in the context of a communication system.

Fundamentals of Statistical Decision Theory   App. B

The sample $z(T)$ is made up of a signal component, $a_i(T)$, and a noise component, $n_0(T)$. The time T is the symbol duration. At each kT, where k is an integer, the receiver uses a decision rule for deciding which signal class has been received. For ease of notation, Equation (B.5) is sometimes written simply as $z = a_i + n_0$, where the functional dependence on T is implicit.

B.2.2 The Likelihood Ratio Test and the Maximum A Posteriori Criterion

A reasonable starting point for establishing the receiver decision rule is shown below for the case of *two* signal classes.

$$P(s_1|z) \underset{H_2}{\overset{H_1}{\gtreqless}} P(s_2|z) \tag{B.6}$$

Equation (B.6) states that we should choose hypothesis H_1 if the a posteriori probability $P(s_1|z)$ is greater than the a posteriori probability $P(s_2|z)$. Otherwise, we should choose hypothesis H_2.

We can replace the a posteriori probabilities of Equation (B.6) with their equivalent expressions from Bayes' theorem [Equation (B.4)], yielding

$$p(z|s_1)\, P(s_1) \underset{H_2}{\overset{H_1}{\gtreqless}} P(z|s_2)\, P(s_2) \tag{B.7}$$

We now have a decision rule in terms of pdfs (likelihoods). If we rearrange Equation (B.7) as shown below

$$\frac{p(z|s_1)}{p(z|s_2)} \underset{H_2}{\overset{H_1}{\gtreqless}} \frac{P(s_2)}{P(s_1)} \tag{B.8}$$

the left-hand ratio is known as the *likelihood ratio*, and the entire equation is often referred to as the *likelihood ratio test*. Equation (B.8) corresponds to making a decision based on a comparison of a measurement of a received signal to a threshold. Since the test is based on choosing the signal class with maximum a posteriori probability, the decision criterion is called the *maximum a posteriori* (MAP) criterion. It is also called the *minimum error criterion*, since on the average, this criterion yields the minimum number of incorrect decisions. It should be emphasized that this criterion is optimum only when each of the error types are equally harmful or costly. When some of the error types are more costly than others, a criterion that incorporates relative cost of the errors should best be employed [1].

B.2.3 The Maximum Likelihood Criterion

Very often there is no knowledge available about the a priori probabilities of the hypotheses or signal classes. Even when such information is available, its accuracy is sometimes mistrusted. In those instances, decisions are usually made by assuming the most conservative a priori probabilities possible; that is, the values of the a priori probabilities are selected so that the classes are *equally likely*. When this is done, the MAP criterion is known as the *maximum likelihood criterion*, and Equation (B.8) can be written as follows:

$$\frac{p(z|s_1)}{p(z|s_2)} \overset{H_1}{\underset{H_2}{\gtrless}} 1 \qquad (B.9)$$

Notice that the maximum likelihood criterion of Equation (B.9) is the same as the maximum likelihood rule that was described in Example B.3.

B.3 SIGNAL DETECTION EXAMPLE

B.3.1 The Maximum Likelihood Binary Decision

The pictorial view of the decision process in Example B.3 dealt with triangular-shaped probability density functions as a convenient example. Figure B.4 illustrates the conditional pdfs for the binary noise-perturbed output signals, $z(T) = a_1 + n_0$ and $z(T) = a_2 + n_0$ from a typical receiver. The signals, a_1 and a_2, are mutually independent and are equally likely. The noise, n_0, is assumed to be an independent Gaussian random variable with zero mean, variance σ_0^2, and pdf $p(n_0)$ given by

$$p(n_0) = \frac{1}{\sigma_0 \sqrt{2\pi}} \exp\left[-\frac{1}{2}\left(\frac{n_0^2}{\sigma_0^2}\right) \right] \qquad (B.10)$$

We can therefore write the likelihood ratio, $L(z)$, described in Equation (B.8) as follows:

$$L(z) = \frac{p(z|s_1)}{p(z|s_2)}$$

$$= \frac{\dfrac{1}{\sigma_0 \sqrt{2\pi}} \exp\left[-\dfrac{1}{2}\left(\dfrac{z - a_1}{\sigma_0}\right)^2 \right]}{\dfrac{1}{\sigma_0 \sqrt{2\pi}} \exp\left[-\dfrac{1}{2}\left(\dfrac{z - a_2}{\sigma_0}\right)^2 \right]} \overset{H_1}{\underset{H_2}{\gtrless}} \frac{P(s_2)}{P(s_1)}$$

$$= \frac{\exp\left(-\dfrac{z^2}{2\sigma_0^2}\right) \exp\left(-\dfrac{a_1^2}{2\sigma_0^2}\right) \exp\left(\dfrac{2za_1}{2\sigma_0^2}\right)}{\exp\left(-\dfrac{z^2}{2\sigma_0^2}\right) \exp\left(-\dfrac{a_2^2}{2\sigma_0^2}\right) \exp\left(\dfrac{2za_2}{2\sigma_0^2}\right)} \overset{H_1}{\underset{H_2}{\gtrless}} \frac{P(s_2)}{P(s_1)} \qquad (B.11)$$

$$= \exp\left[\frac{z(a_1 - a_2)}{\sigma_0^2} - \frac{a_1^2 - a_2^2}{2\sigma_0^2} \right] \overset{H_1}{\underset{H_2}{\gtrless}} \frac{P(s_2)}{P(s_1)}$$

where a_1 is the receiver output signal component when $s_1(t)$ is sent, and a_2 is the output signal component when $s_2(t)$ is sent. The inequality relationship described by Equation (B.11) is preserved for any *monotonically* increasing (or decreasing)

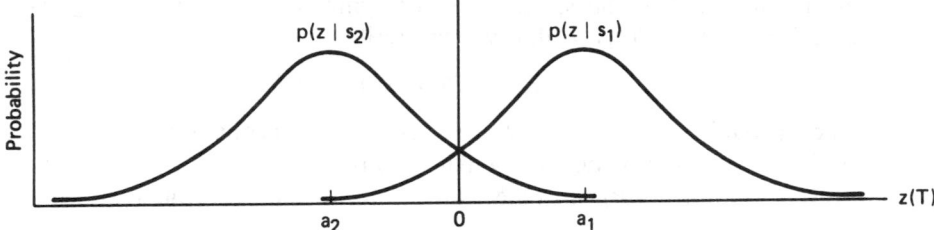

Figure B.4 Conditional pdfs for a typical binary receiver.

transformation. Therefore, to simplify Equation (B.11), we take the natural logarithm of both sides, resulting in the log-likelihood ratio, $l(z)$.

$$l(z) = \frac{z(a_1 - a_2)}{\sigma_0^2} - \frac{a_1^2 - a_2^2}{2\sigma_0^2} \underset{H_2}{\overset{H_1}{\gtrless}} \ln \frac{P(s_2)}{P(s_1)} \qquad \text{(B.12)}$$

When the classes are equally likely,

$$\ln \frac{P(s_2)}{P(s_1)} = 0$$

so that

$$z \underset{H_2}{\overset{H_1}{\gtrless}} \frac{a_1^2 - a_2^2}{2(a_1 - a_2)}$$

$$z \underset{H_2}{\overset{H_1}{\gtrless}} \frac{a_1 + a_2}{2} = \gamma_0 \qquad \text{(B.13)}$$

For *antipodal signals*, $s_1(t) = -s_2(t)$ and $a_1 = -a_2$; thus we can write

$$z \underset{H_2}{\overset{H_1}{\gtrless}} 0 \qquad \text{(B.14)}$$

Therefore, the maximum likelihood rule for the case of equally likely antipodal signals compares the received sample to a zero threshold, which is tantamount to deciding $s_1(t)$ if the sample is positive, and $s_2(t)$ if the signal is negative.

B.3.2 Probability of Bit Error

For the binary example in Section B.3.1, we want to compute the bit error probability, P_B, using the decision rule in Equation (B.13). The probability of an error is calculated by summing the probabilities of the various ways that an error can be made:

$$P_B = P(H_2|s_1)P(s_1) + P(H_1|s_2)P(s_2) \qquad \text{(B.15)}$$

That is, given that class $s_1(t)$ was transmitted, an error results if hypothesis H_2 is chosen; or, given that class $s_2(t)$ was transmitted, an error results if hypothesis

H_1 is chosen. For the special case of symmetric probability density functions, and for $P(s_1) = P(s_2) = 0.5$, we can write

$$P_B = P(H_2|s_1) = P(H_1|s_2) \qquad (B.16)$$

The probability of an error, P_B, is equal to the probability that an incorrect hypothesis, H_1, will be decided when $s_2(t)$ is sent, or that H_2 will be decided when $s_1(t)$ is sent. Thus P_B is numerically equal to the area under the "tail" of either

TABLE B.1 Complementary Error Function $Q(x) = \int_x^\infty (1/\sqrt{2\pi}) \exp(-u^2/2)\, du$

					$Q(x)$					
x	0.00	0.01	0.02	0.03	0.04	0.05	0.06	0.07	0.08	0.09
0.0	0.5000	0.4960	0.4920	0.4880	0.4840	0.4801	0.4761	0.4721	0.4681	0.4641
0.1	0.4602	0.4562	0.4522	0.4483	0.4443	0.4404	0.4364	0.4325	.0.4286	0.4247
0.2	0.4207	0.4168	0.4129	0.4090	0.4052	0.4013	0.3974	0.3936	0.3897	0.3859
0.3	0.3821	0.3783	0.3745	0.3707	0.3669	0.3632	0.3594	0.3557	0.3520	0.3483
0.4	0.3446	0.3409	0.3372	0.3336	0.3300	0.3264	0.3228	0.3192	0.3156	0.3121
0.5	0.3085	0.3050	0.3015	0.2981	0.2946	0.2912	0.2877	0.2843	0.2810	0.2776
0.6	0.2743	0.2709	0.2676	0.2643	0.2611	0.2578	0.2546	0.2514	0.2483	0.2451
0.7	0.2420	0.2389	0.2358	0.2327	0.2296	0.2266	0.2236	0.2206	0.2168	0.2148
0.8	0.2169	0.2090	0.2061	0.2033	0.2005	0.1977	0.1949	0.1922	0.1894	0.1867
0.9	0.1841	0.1814	0.1788	0.1762	0.1736	0.1711	0.1685	0.1660	0.1635	0.1611
1.0	0.1587	0.1562	0.1539	0.1515	0.1492	0.1469	0.1446	0.1423	0.1401	0.1379
1.1	0.1357	0.1335	0.1314	0.1292	0.1271	0.1251	0.1230	0.1210	0.1190	0.1170
1.2	0.1151	0.1131	0.1112	0.1093	0.1075	0.1056	0.1038	0.1020	0.1003	0.0985
1.3	0.0968	0.0951	0.0934	0.0918	0.0901	0.0885	0.0869	0.0853	0.0838	0.0823
1.4	0.0808	0.0793	0.0778	0.0764	0.0749	0.0735	0.0721	0.0708	0.0694	0.0681
1.5	0.0668	0.0655	0.0643	0.0630	0.0618	0.0606	0.0594	0.0582	0.0571	0.0559
1.6	0.0548	0.0537	0.0526	0.0516	0.0505	0.0495	0.0485	0.0475	0.0465	0.0455
1.7	0.0446	0.0436	0.0427	0.0418	0.0409	0.0401	0.0392	0.0384	0.0375	0.0367
1.8	0.0359	0.0351	0.0344	0.0336	0.0329	0.0322	0.0314	0.0307	0.0301	0.0294
1.9	0.0287	0.0281	0.0274	0.0268	0.0262	0.0256	0.0250	0.0244	0.0239	0.0233
2.0	0.0228	0.0222	0.0217	0.0212	0.0207	0.0202	0.0197	0.0192	0.0188	0.0183
2.1	0.0179	0.0174	0.0170	0.0166	0.0162	0.0158	0.0154	0.0150	0.0146	0.0143
2.2	0.0139	0.0136	0.0132	0.0129	0.0125	0.0122	0.0119	0.0116	0.0113	0.0110
2.3	0.0107	0.0104	0.0102	0.0099	0.0096	0.0094	0.0091	0.0089	0.0087	0.0084
2.4	0.0082	0.0080	0.0078	0.0075	0.0073	0.0071	0.0069	0.0068	0.0066	0.0064
2.5	0.0062	0.0060	0.0059	0.0057	0.0055	0.0054	0.0052	0.0051	0.0049	0.0048
2.6	0.0047	0.0045	0.0044	0.0043	0.0041	0.0040	0.0039	0.0038	0.0037	0.0036
2.7	0.0035	0.0034	0.0033	0.0032	0.0031	0.0030	0.0029	0.0028	0.0027	0.0026
2.8	0.0026	0.0025	0.0024	0.0023	0.0023	0.0022	0.0021	0.0021	0.0020	0.0019
2.9	0.0019	0.0018	0.0018	0.0017	0.0016	0.0016	0.0015	0.0015	0.0014	0.0014
3.0	0.0013	0.0013	0.0013	0.0012	0.0012	0.0011	0.0011	0.0011	0.0010	0.0010
3.1	0.0010	0.0009	0.0009	0.0009	0.0008	0.0008	0.0008	0.0008	0.0007	0.0007
3.2	0.0007	0.0007	0.0006	0.0006	0.0006	0.0006	0.0006	0.0005	0.0005	0.0005
3.3	0.0005	0.0005	0.0005	0.0004	0.0004	0.0004	0.0004	0.0004	0.0004	0.0003
3.4	0.0003	0.0003	0.0003	0.0003	0.0003	0.0003	0.0003	0.0003	0.0003	0.0002

pdf, $p(z|s_1)$ or $p(z|s_2)$, falling on the *incorrect side* of the threshold. We can therefore compute P_B by integrating $p(z|s_1)$ between the limits $-\infty$ and γ_0, or as shown below, by integrating $p(z|s_2)$ between the limits γ_0 and ∞.

$$P_B = \int_{\gamma_0 = (a_1 + a_2)/2}^{\infty} p(z|s_2) \, dz$$

$$= \int_{(a_1 + a_2)/2}^{\infty} \frac{1}{\sigma_0 \sqrt{2\pi}} \exp\left[-\frac{1}{2}\left(\frac{z - a_2}{\sigma_0}\right)^2 \right] dz \tag{B.17}$$

Let

$$u = \frac{z - a_2}{\sigma_0}$$

Then $\sigma_0 \, du = dz$, and

$$P_B = \int_{u = (a_1 - a_2)/2\sigma_0}^{u = \infty} \frac{1}{\sqrt{2\pi}} \exp\left(-\frac{u^2}{2} \right) du = Q\left(\frac{a_1 - a_2}{2\sigma_0} \right) \tag{B.18}$$

where $Q(x)$, called the *complementary error function* or *co-error function*,* is tabulated in Table B.1.

Another form of the co-error function which is frequently used is

$$\text{erfc } (x) = \frac{2}{\sqrt{\pi}} \int_x^{\infty} \exp\left(-u^2\right) du \tag{B.19}$$

The two co-error functions, $Q(x)$ and erfc (x), are related as follows:

$$\text{erfc } (x) = 2Q(x\sqrt{2}) \tag{B.20}$$

$$Q(x) = \frac{1}{2} \text{erfc}\left(\frac{x}{\sqrt{2}}\right) \tag{B.21}$$

REFERENCES

1. Van Trees, H. L., *Detection, Estimation, and Modulation Theory*, Part 1, John Wiley & Sons, Inc., New York, 1968.
2. Papoulis, A., *Probability, Random Variables, and Stochastic Processes*, McGraw-Hill Book Company, New York, 1965.

* Note that the co-error function is defined in several ways; however, all definitions are essentially equivalent.

APPENDIX C

Response of Correlators to White Noise

The inputs to a bank of N correlators represent a white Gaussian noise process, $n(t)$, with zero mean and two-sided power spectral density, $N_0/2$. The output of each correlator, n_j, at time $t = T$, is a *Gaussian random variable* defined by

$$n_j = \int_0^T n(t)\psi_j(t) \, dt \qquad j = 1, \ldots, N \tag{C.1}$$

where $\{\psi_j(t)\}$ forms an orthonormal set. Since n_j is Gaussian, it is characterized completely by its mean and variance. The mean, $\bar{n}_j$, is equal to

$$\bar{n}_j = \mathbf{E}\{n_j\} = \mathbf{E}\left\{\int_0^T n(t)\psi_j(t) \, dt\right\} \tag{C.2}$$

where $\mathbf{E}\{\cdot\}$ is the expected value operator. The variance, σ_j^2, of n_j is equal to

$$\sigma_j^2 = \mathbf{E}\{n_j^2\} - \bar{n}_j^2 \tag{C.3}$$

$$= \mathbf{E}\left\{\int_0^T n(t)\psi_j(t) \, dt \int_0^T n(s)\psi_j(s) \, ds\right\} - \bar{n}_j^2 \tag{C.4}$$

$$= \int_0^T \int_0^T \mathbf{E}\{n(t)n(s)\psi_j(t)\psi_j(s)\} \, dt \, ds - \bar{n}_j^2 \tag{C.5}$$

Since $n(t)$ is a zero-mean process, then

$$\mathbf{E}\{n(t)\} = 0 \tag{C.6}$$

which implies that

$$\bar{n}_j = E\{n_j\} = 0 \tag{C.7}$$

The autocorrelation function $R_n(t, s)$ of the process $n(t)$ is equal to

$$R_n(t, s) = E\{n(t)n(s)\} \tag{C.8}$$

If the noise, $n(t)$, is assumed stationary, then $R_n(t, s)$ is only a function of the time difference, $\tau = t - s$. From Equation (C.5) we have

$$\sigma_j^2 = \text{var }\{n_j\} = \int_0^T \int_0^T R_n(\tau)\psi_j(t)\psi_j(s)\, dt\, ds \tag{C.9}$$

For a stationary random process, the power spectral density, $G_n(f)$, and the autocorrelation function, $R_n(\tau)$, form a Fourier transform pair. Thus we can write

$$R_n(\tau) = \int_{-\infty}^{\infty} G_n(f)e^{j2\pi f\tau}\, df \tag{C.10}$$

Since $n(t)$ is white noise, its power spectral density, $G_n(f)$, is $N_0/2$ for all f, and we can write Equation (C.10) as

$$R_n(\tau) = \int_{-\infty}^{\infty} \frac{N_0}{2} e^{j2\pi f\tau}\, df = \frac{N_0}{2} \delta(\tau) \tag{C.11}$$

where $\delta(\tau)$ is the unit impulse function defined in Section A.4.1. Substituting Equation (C.11) into Equation (C.9), we get

$$\sigma_j^2 = \frac{N_0}{2} \int_0^T \int_0^T \delta(t - s)\psi_j(t)\psi_j(s)\, dt\, ds \tag{C.12}$$

$$= \frac{N_0}{2} \int_0^T \psi_j^2(t)\, dt = \frac{N_0}{2} \qquad j = 1, \ldots, N \tag{C.13}$$

where we have utilized the *sifting property* of the unit impulse function (see Section A.4.1) and the fact that $\{\psi_j(t)\}$, $j = 1, \ldots, N$, constitutes an orthonormal set. Thus for white Gaussian noise with two-sided power spectral density $N_0/2$ watts/hertz, the output noise power from each of the N correlators is equal to $N_0/2$ watts.

APPENDIX D

Often Used Identities

$$\cos x \cos y = \tfrac{1}{2} \cos (x + y) + \tfrac{1}{2} \cos (x - y) \qquad \text{(D.1)}$$

$$\sin x \sin y = -\tfrac{1}{2} \cos (x + y) + \tfrac{1}{2} \cos (x - y) \qquad \text{(D.2)}$$

$$\sin x \cos y = \tfrac{1}{2} \sin (x + y) + \tfrac{1}{2} \sin (x - y) \qquad \text{(D.3)}$$

$$\cos x \sin y = \tfrac{1}{2} \sin (x + y) - \tfrac{1}{2} \sin (x - y) \qquad \text{(D.4)}$$

$$\sin (x \pm y) = \sin x \cos y \pm \cos x \sin y \qquad \text{(D.5)}$$

$$\cos (x \pm y) = \cos x \cos y \mp \sin x \sin y \qquad \text{(D.6)}$$

$$\cos^2 x = \tfrac{1}{2}(1 + \cos 2x) \qquad \text{(D.7)}$$

$$\sin^2 x = \tfrac{1}{2}(1 - \cos 2x) \qquad \text{(D.8)}$$

$$\sin x \cos x = \tfrac{1}{2} \sin 2x \qquad \text{(D.9)}$$

$$\sin x + \sin y = 2 \sin \tfrac{1}{2}(x + y) \cos \tfrac{1}{2}(x - y) \qquad \text{(D.10)}$$

$$\sin x - \sin y = 2 \cos \tfrac{1}{2}(x + y) \sin \tfrac{1}{2}(x - y) \qquad \text{(D.11)}$$

$$\cos x + \cos y = 2 \cos \tfrac{1}{2}(x + y) \cos \tfrac{1}{2}(x - y) \qquad \text{(D.12)}$$

$$\cos x - \cos y = -2 \sin \tfrac{1}{2}(x + y) \sin \tfrac{1}{2}(x - y) \qquad \text{(D.13)}$$

$$\sin x = \frac{e^{jx} - e^{-jx}}{2j} \qquad \text{(D.14)}$$

$$\cos x = \frac{e^{jx} + e^{-jx}}{2} \qquad \text{(D.15)}$$

$$P_B = \frac{1}{n} \sum_{j=2}^{n} j \binom{n}{j} p^j (1 - p)^{n-j} = p - p(1 - p)^{n-1} \tag{D.16}$$

Proof:

$$j \binom{n}{j} = j \frac{n!}{j!\,(n-j)!} = \frac{n!}{(j-1)!\,(n-j)!} = n \frac{(n-1)!}{(j-1)!\,[(n-1)-(j-1)]!}$$

$$= n \binom{n-1}{j-1}$$

$$P_B = \sum_{j=2}^{n} \binom{n-1}{j-1} p^j (1 - p)^{n-j} = p \sum_{j=2}^{n} \binom{n-1}{j-1} p^{j-1} (1 - p)^{(n-1)-(j-1)}$$

Change of parameter: $i = (j - 1)$

Therefore, $(j = 2)$ becomes $(i = 1)$, and $(j = n)$ becomes $(i = n - 1)$.

$$P_B = p \sum_{i=1}^{n-1} \binom{n-1}{i} p^i (1 - p)^{(n-1)-i}$$

$$= p \sum_{i=0}^{n-1} \left[\binom{n-1}{i} p^i (1 - p)^{(n-1)-i} - \binom{n-1}{0} p^0 (1 - p)^{(n-1)-0} \right]$$

$$= p[1 - (1 - p)^{n-1}]$$

$$= p - p(1 - p)^{n-1}$$

APPENDIX E

A Convolutional Encoder/Decoder Computer Program

```
      PROGRAM VITALG
C REVISION H, JULY 6, 1989
C
C                    CONVOLUTIONAL ENCODER/DECODER PROGRAM
C                    WRITTEN IN LAHEY FORTRAN FOR IBM PC
C                    BY A. H. YAMADA AND M. A. ROLENZ
C                       THE AEROSPACE CORPORATION
C                       2350 EAST EL SEGUNDO BLVD.
C                         EL SEGUNDO, CA. 90245
C
      COMMON/ENCDE/NCODE(3,128),IPREV(128),NSEQ,IREG,NCRATE,
     +NSQHLF,KL,ITAP(3)
C COMMON BLOCK ENCODE DEFINES CODE STRUCTURES
C        NCODE(I,J)    CODE BIT I FOR PATH J
C        IPREV(I)      STARTING STATE IN TRELLIS FOR PATH I
C        NSEQ          NUMBER OF STATES IN TRELLIS (2**(KL-1))
C        IREG          ENCODER SHIFT REGISTER CONTENTS
C        NCRATE        NUMBER OF CODE GENERATORS (2 OR 3)
C        NSQHLF        HALF OF NSEQ  (2**(KL-2))
C        KL            CONSTRAINT LENGTH
C        ITAP(I)       DECIMAL EQUIVALENT OF CODE GENERATOR TAPS
      COMMON/DECDE/METRIC(64),MLSEQ(64,70),MEMLEN
C COMMON BLOCK DECODE IS FOR BUFFERING VITERBI DECODER TRELLIS HISTOR'
C        METRIC(I)     CURRENT PATH METRIC FOR STATE I
C        MLSEQ(I,J)    TRACE OF SURVIVED LIKELIHOOD SEQUENCE I
C        MEMLEN        LENGTH OF MLSEQ(I,J), PATH MEMORY LENGTH
      COMMON/DECOD2/ILEVEL(3)
C COMMON BLOCK DECODE2 STORES CURRENT INPUTS TO VITERBI DECODER
C        ILEVEL(I)     CURRENT (ITIME) BITS TO BE DECODED
      COMMON/TIME/ITIME,ITIME1,INTIME,IBEST
C COMMON BLOCK HOLDS POINTERS FOR SUBROUTINES DECODE AND DECIDE
C        ITIME         TIMING POINTER FOR CURRENT TIME
C        ITIME1        ITIME-1
C        INTIME        TIMING POINTER FOR THE LAST DECODED BIT
C        IBEST         POINTER FOR THE BEST SURVIVING PATH
```

```
      COMMON/VITEST/IDCBIT(646),ICODED(1938),IBFFER(646),NDEC,LENIBF
C COMMON BLOCK VITEST HOLDS MESSAGE BUFFERS
C       IDCBIT          DECODED ASCII BITS
C       ICODED          ENCODED BIT STREAM
C       IBFFER          INPUT ASCII BITS
C       NDEC            COUNTER FOR DECODED BITS
C       LENIBF          NUMBER OF ASCII BITS AND KL-1 FLUSH BITS
      COMMON/FLUSH/NRECVD,NREP,NTAIL
C COMMON BLOCK FLUSH IS USED TO FLUSH DECODER AT THE END OF MESSAGE
C       NRECVD          COUNTER FOR NUMBER OF BITS INTO THE DECODER
C       NREP            NUMBER OF CODED BITS (LENIBF*NCRATE)
C       NTAIL           DESIGNATES THE BEGINNING OF THE FLUSH BITS
      COMMON/FLAG/IFLAG
      CHARACTER*80 BUFFER,BUFFR2
      CHARACTER*1 ANSWER
      CHARACTER*2 DRIVE
      CHARACTER*6 FILNAM
      CHARACTER*12 FILOUT
C RANDOMIZE THE INITIAL SEED
      SEED=RRAND()
      WRITE(6,1080)
1080  FORMAT(' This program encodes a binary or ASCII message into a',
     +' convolutionally encoded'/' bit stream.  The code parameters',
     +' are user inputs.  The program uses Viterbi'/' decoding (hard',
     +' or soft decision) and compares the transmitted sequence',
     +' against'/' the decoded sequence.'/)
C    INPUT FILE NAME AND OPEN FILE
      PRINT,' ENTER OUTPUT FILE NAME (6 CHARACTERS MAXIMUM) '
      READ(5,*) FILNAM
      PRINT,' ENTER DISK DRIVE FOR OUTPUT (e.g. A:,B:,C: etc.) '
      READ(5,*) DRIVE
      PRINT,' '
      FILOUT=DRIVE//CHARNB(FILNAM)//'.OUT'
      PRINT,' '
      PRINT,'    OUTPUT FILE WILL BE      -        ',FILOUT
      PRINT,' '
      OPEN(UNIT=2,FILE=FILOUT,STATUS='NEW')
C INITIAL DEFAULT CODE PARAMETERS
      NCRATE = 2
      MEMLEN = 15
      KL     = 3
      WRITE(6,1082)
 1082 FORMAT(' INPUT CODE PARAMETERS:'
     +/' THE INITIAL DEFAULT IS THE RATE 1/2, K=3 ODENWALDER CODE')
C    BEGINNING OF MAIN LOOP
      1 CONTINUE
C    INPUT OF CODING PARAMETERS
      CALL PARMIN(NCRATE,MEMLEN,KL,ITAP)
C    INITIALIZE CODE STRUCTURE
      CALL INTIAL
C    INPUT MESSAGE
      CALL MSGIN(BUFFER,LENBUF)
C    CONVERT MESSAGE TO ASCII BITS
      CALL ASCII1(BUFFER,LENBUF,IBFFER,LENIBF,KL)
      NREP = NCRATE*LENIBF
      NTAIL=NREP+(2-KL)*NCRATE
C    ENCODE BITS
      CALL ENCODE
C    INTRODUCE ERRORS INTO BIT STREAM
      PRINT, ' DO YOU WANT RANDOM ERRORS, ASSUMING BPSK MODULATION'
      PRINT, ' AND AN AWGN CHANNEL (Y/N) ? '
      INPUT, ANSWER
      IF((ANSWER.EQ.'Y').OR.(ANSWER.EQ.'y'))THEN
        CALL RANDOM(BUFFER,ICODED,NREP,NCRATE,KL,SNROBS)
      ELSE
        CALL ERRORS(BUFFER,ICODED,NREP,NCRATE,KL)
      ENDIF
```

```
C   DECODER TRELLIS TRACE OUTPUT OPTION
      PRINT,' '
      PRINT,' DO YOU WANT TO SEE THE TRELLIS TRACE (Y/N) ? '
      INPUT, ANSWER
      IF((ANSWER.EQ.'Y').OR.(ANSWER.EQ.'y'))THEN
        WRITE(6,1085)
        WRITE(2,1085)
 1085   FORMAT(/' TRELLIS TRACE'
     +/' The decimal numbers in the array, when converted to binary,'
     +/' represent the trellis states.')
        ITRS=1
      ELSE
        ITRS=0
      ENDIF
      DO 10 I=1,NSEQ
        DO 11 J=1,MEMLEN
          MLSEQ(I,J)=1000
   11   CONTINUE
        METRIC(I)=0
        MLSEQ(I,1)=I
   10 CONTINUE
      IREG=0
      METRIC(1)=1000
      ITIME1=1
      ITIME=1
      INTIME=1
      NRECVD=0
      NDEC=0
      PRINT,' '
      PRINT,' Simulating reception and decoding'
  100 CONTINUE
C   READ NCRATE RECEIVED BITS INTO THE DECODER
      DO 110 I=1,NCRATE
        NRECVD=NRECVD+1
        ILEVEL(I)=ICODED(NRECVD)
  110 CONTINUE
      CALL DECODE(ITRS)
      IF (NRECVD.LT.NREP) GO TO 100
C   CONVERT RECEIVED BITS BACK TO ASCII
      CALL ASCII2(BUFFR2,LENBUF,IDCBIT,LENIBF)
C   COUNT THE NUMBER OF ERRORS
      NBTERR = 0
      DO 200 INDEX = 1 , LENIBF
        IF( IBFFER(INDEX) .NE. IDCBIT(INDEX) ) NBTERR=NBTERR+1
  200 CONTINUE
      MSIZE=LENIBF-KL+1
      BEROBS=FLOAT(NBTERR)/FLOAT(MSIZE)
      WRITE(6,1000) MSIZE,NBTERR,BEROBS
      WRITE(2,1000) MSIZE,NBTERR,BEROBS
      NCHERR = 0
      DO 300 INDEX = 1 , LENBUF
        IF(BUFFER(INDEX:INDEX).NE.BUFFR2(INDEX:INDEX)) NCHERR=NCHERR+1
  300 CONTINUE
C   OUTPUT
      IF (IFLAG.EQ.1) GO TO 402
      WRITE(6,1001) LENBUF,NCHERR
      WRITE(2,1001) LENBUF,NCHERR
  402 WRITE(6,1025)
      WRITE(2,1025)
      NLINES = LENIBF/72 + 1
      DO 400 INDEX = 1 , NLINES
        NBEGIN = (INDEX -1)*72 + 1
        NEND   = NBEGIN + 71
        IF(NEND.GT.MSIZE) NEND=MSIZE
        WRITE(6,1030) (IBFFER(I),I=NBEGIN,NEND)
        WRITE(6,1031) (IDCBIT(I),I=NBEGIN,NEND)
        WRITE(2,1030) (IBFFER(I),I=NBEGIN,NEND)
        WRITE(2,1031) (IDCBIT(I),I=NBEGIN,NEND)
```

```
  400 CONTINUE
      IF (IFLAG.EQ.1) GO TO 401
      WRITE(6,1015)
      WRITE(2,1015)
      WRITE(6,1020)  (BUFFER(I:I),I=1,LENBUF)
      WRITE(6,1020)  (BUFFR2(I:I),I=1,LENBUF)
      WRITE(2,1020)  (BUFFER(I:I),I=1,LENBUF)
      WRITE(2,1020)  (BUFFR2(I:I),I=1,LENBUF)
  401 PRINT,'  '
      PRINT,' START AGAIN (Y/N) ? '
      READ(5,1010) ANSWER
      IF((ANSWER.EQ.'Y').OR.(ANSWER.EQ.'y'))GO TO 1
 1000 FORMAT(//,' ',I4,' decoded bits ',/,
     +' ',I4,' bits in error ',/
     +' ',E10.4,' decoded bit-error rate '/)
 1001 FORMAT(//,' ',I4,' decoded ASCII characters',/,
     +' ',I4,' characters in error',/)
 1010 FORMAT(A1)
 1015 FORMAT(/,' Original vs. decoded ASCII message',/)
 1020 FORMAT(' ',80A1)
 1025 FORMAT(/,' Original vs. decoded bit stream',/)
 1030 FORMAT(' DATA:  ',72I1)
 1031 FORMAT(' DEC''D: ',72I1)
      END
C
      SUBROUTINE INTIAL
      COMMON/ENCDE/NCODE(3,128),IPREV(128),NSEQ,IREG,NCRATE,
     +NSQHLF,KL,ITAP(3)
      COMMON/DECDE/METRIC(64),MLSEQ(64,70),MEMLEN
      NPATH=2**KL
      NSEQ=NPATH/2
      NSQHLF=NSEQ/2
C    LOOP TO GENERATE OUTPUTS FOR EACH PATH
      DO 100 IPATH=1,NPATH
C    IBIT = BIT PATTERN IN THE REGISTER FOR EACH PATH
          IBIT=IPATH-1
          IPREV(IPATH)=MOD(IBIT,NSEQ)+1
C    LOOP TO GENERATE EACH OUTPUT BIT
          DO 10 IOUTN=1,NCRATE
C    IREG=BIT PATTERN AFTER MASKED BY A G VECTOR
              IREGTAP=ITAP(IOUTN)
              IREGBIT=IBIT
C    ICNT COUNTS WEIGHTS
              ICNT=0
              IAMONT=NPATH
C    LOOP TO COUNT WEIGHTS STARTS HERE
              DO 30 IBTCNT=1,KL
                  IAMONT=IAMONT/2
                  IF(IREGBIT.LT.IAMONT) GO TO 31
                  IREGBIT=IREGBIT-IAMONT
                  IF(IREGTAP.LT.IAMONT) GO TO 30
                  ICNT=ICNT+1
   31             CONTINUE
                  IF(IREGTAP.LT.IAMONT) GO TO 30
                  IREGTAP=IREGTAP-IAMONT
   30         CONTINUE
C    CHECK IF WEIGHT IS ZERO
              IF (ICNT.EQ.0) GO TO 43
C    CHECK IF WEIGHT IS  EVEN (TO 43)
              IF (ICNT/2.NE.(ICNT-1)/2) GO TO 43
              NCODE(IOUTN, IPATH)=1
              GO TO 10
   43         NCODE(IOUTN, IPATH)=-1
   10     CONTINUE
  100 CONTINUE
      RETURN
      END
```

```
      SUBROUTINE ENCODE
      COMMON/VITEST/IDCBIT(646),ICODED(1938),IBFFER(646),NDEC,LENIBF
      COMMON/ENCDE/NCODE(3,128),IPREV(128),NSEQ,IREG,NCRATE,
     +NSQHLF,KL,ITAP(3)
      DO 200 INDEX = 1 , LENIBF
        INFO=IBFFER(INDEX)
        IREG=IREG/2+INFO*NSEQ
        IRGPTR=IREG+1
        DO 210 IOUTN=1,NCRATE
          ICODED(IOUTN+(INDEX-1)*NCRATE)=NCODE(IOUTN,IRGPTR)
 210    CONTINUE
 200  CONTINUE
      RETURN
      END
C
      SUBROUTINE DECODE(ITRS)
      DIMENSION IDATED(64),MLST(64)
      COMMON/ENCDE/NCODE(3,128),IPREV(128),NSEQ,IREG,NCRATE,
     +NSQHLF,KL,ITAP(3)
      COMMON/DECDE/METRIC(64),MLSEQ(64,70),MEMLEN
      COMMON/DECOD2/ILEVEL(3)
      COMMON/TIME/ITIME,ITIME1,INTIME,IBEST
      COMMON/FLUSH/NRECVD,NREP,NTAIL
C   METRIC CALCULATION
      DO 200 ISTATE=1,NSEQ
C   FOR EACH STATE, THERE ARE TWO PATHS (IPATH1 AND IPATH2).
        IPATH2=2*ISTATE
        IPATH1=IPATH2-1
        ITMP1=METRIC(IPREV(IPATH1))
        ITMP2=METRIC(IPREV(IPATH2))
        DO 210 IOUTN=1,NCRATE
          ITMP1=ITMP1+NCODE(IOUTN,IPATH1)*ILEVEL(IOUTN)
          ITMP2=ITMP2+NCODE(IOUTN,IPATH2)*ILEVEL(IOUTN)
 210    CONTINUE
C   METRICS:  THE LARGER THE BETTER
        IF(ITMP2-ITMP1) 225,225,223
 223    IDATED(ISTATE)=ITMP2
        MLST(ISTATE)=IPREV(IPATH2)
        GO TO 200
 225    IDATED(ISTATE)=ITMP1
        MLST(ISTATE)=IPREV(IPATH1)
C   MLST HOLDS THE STARTING NODE LOCATION FOR BRANCH TO ISTATE
 200  CONTINUE
C   MAX. LIKELIHOOD BRANCH IS SELECTED FOR EACH NODE AT THIS POINT
C   CHECK FOR BRANCHING NODES
      DO 240 ITPHLF=1,NSQHLF
        IBMHLF=NSQHLF+ITPHLF
        IBRNCH=MLST(ITPHLF)
        IF(IBRNCH.NE.MLST(IBMHLF))GOTO250
        IMISS=IBRNCH-1
        IF(MOD(IBRNCH,2).EQ.1) IMISS=IMISS+2
C   SEARCH FOR BRANCHING AND DROPPED SEQUENCES
        DO 241 ISEQ=1,NSEQ
          IF (MLSEQ(ISEQ,ITIME1).EQ.IMISS) GO TO 242
 241    CONTINUE
        STOP
 242    IMISSQ=ISEQ
        DO 243 ISEQ=1,NSEQ
          IF (MLSEQ(ISEQ,ITIME1).EQ.IBRNCH) GO TO 244
 243    CONTINUE
        STOP
 244    IBRNSQ=ISEQ
C   REPLACE A DROPPED SEQUENCE WITH A BRANCHING SEQUENCE UP TO ITIME1
        JTIME=INTIME
 245    MLSEQ(IMISSQ,JTIME)=MLSEQ(IBRNSQ,JTIME)
        IF(JTIME.EQ.ITIME1) GO TO 249
        JTIME=JTIME+1
```

```
               IF(JTIME.GT.MEMLEN) JTIME=1
               GO TO 245
C     NOW EXTEND THE ML SEQUENCES WITH CURRENT STATES
      250    CONTINUE
               DO 251 ISEQ=1,NSEQ
                  IF (MLSEQ(ISEQ,ITIME1).EQ.MLST(ITPHLF)) GO TO 252
      251    CONTINUE
               STOP
      252    IMISSQ=ISEQ
               DO 256 ISEQ=1,NSEQ
                  IF (MLSEQ(ISEQ,ITIME1).EQ.MLST(IBMHLF)) GO TO 258
      256    CONTINUE
               STOP
      258    IBRNSQ=ISEQ
      249    MLSEQ(IMISSQ,ITIME)=ITPHLF
               MLSEQ(IBRNSQ,ITIME)=IBMHLF
      240 CONTINUE
C    UPDATING METRICS ASSOCIATED WITH EACH STATE
          DO 391 ISTATE=1,NSEQ
      391 METRIC(ISTATE)=IDATED(ISTATE)
C    FORCE METRICS FOR TAIL BIT FLUSHING AT END OF MESSAGE
          IF(NRECVD.LT.NTAIL) GO TO 401
          MNSTATE=2**((NREP-NRECVD)/2)+1
          DO 392 ISTATE=MNSTATE,NSEQ
              METRIC(ISTATE)=-1000
      392 CONTINUE
C    SEARCH OF INPUT BIT STARTS HERE
      401 CONTINUE
C    CHECK IF ANY SEQUENCE MATCHES
          DO 402 ISEQ=1,NSEQ
              IF (MLSEQ(ISEQ,INTIME).NE.MLSEQ(1,INTIME)) GO TO 500
      402 CONTINUE
C    ALL PATHS ARE MERGING
          IBEST=1
          CALL DECIDE(1)
          GO TO 401
      500 CONTINUE
          ITIME1=ITIME
          ITIME=ITIME+1
          IF(ITIME.GT.MEMLEN) ITIME=1
          IF(ITRS.EQ.1) THEN
C THE FOLLOWING 5 COMMANDS ARE USED FOR OBSERVING THE TRELLIS TRACE
              WRITE(6,1110) INTIME,ITIME1
              WRITE(2,1110) INTIME,ITIME1
              DO 299 I=1,NSEQ
              WRITE(6,1111) (MLSEQ(I,J)-1,J=1,MEMLEN)
              WRITE(2,1111) (MLSEQ(I,J)-1,J=1,MEMLEN)
     1110    FORMAT(/' PATHS DIVERGE AT...',I2,8X,'CURRENT INPUT AT...',I2)
     1111    FORMAT(1X,32I2)
      299    CONTINUE
          ENDIF
          IF (ITIME.NE.INTIME) GO TO 599
C    OVERFLOW OF PATH MEMORY DETECTED
          IBEST=1
          DO 520 ISEQ=2,NSEQ
              IF(METRIC(MLSEQ(IBEST,ITIME1)).LT.
     +       METRIC(MLSEQ(ISEQ,ITIME1))) IBEST=ISEQ
      520 CONTINUE
          CALL DECIDE(2)
      599 CONTINUE
          IF(NRECVD.LT.NREP) GO TO 600
C    END OF MESSAGE DETECTED.  FLUSHING THE DECODER
          IBEST=1
          DO 590 ISEQ=2,NSEQ
          IF(METRIC(MLSEQ(IBEST,ITIME1)).LT.
     +METRIC(MLSEQ(ISEQ,ITIME1))) IBEST=ISEQ
      590 CONTINUE
```

```
      CALL DECIDE(3)
  600 CONTINUE
      RETURN
      END
C
      SUBROUTINE DECIDE(MODE)
      COMMON/VITEST/IDCBIT(646),ICODED(1938),IBFFER(646),NDEC,LENIBF
      COMMON/ENCDE/NCODE(3,128),IPREV(128),NSEQ,IREG,NCRATE,
     +NSQHLF,KL,ITAP(3)
      COMMON/DECDE/METRIC(64),MLSEQ(64,70),MEMLEN
      COMMON/TIME/ITIME,ITIME1,INTIME,IBEST
    1 NDEC=NDEC+1
      IDCBIT(NDEC)=1
      IF(MLSEQ(IBEST,INTIME).LE.NSQHLF) IDCBIT(NDEC)=0
      INTIME=INTIME+1
      IF(INTIME.GT.MEMLEN) INTIME=1
      IF(MODE.NE.3) GO TO 20
      IF(NDEC.LT.LENIBF) GO TO 1
   20 CONTINUE
      RETURN
      END
C
      SUBROUTINE MSGIN(BUFFER,LENBUF)
      COMMON/FLAG/IFLAG
C   THIS SUBROUTINE ASKS THE USER FOR AN INPUT MESSAGE
C   IN EITHER BINARY OR CHARACTER FORM
      CHARACTER*80 BUFFER
      CHARACTER*1 ANSWER,ONE
      CHARACTER*8 BITBUF,BLANK
      DATA ONE/'1'/
      DATA BLANK/'        '/
      IFLAG = 0
      PRINT,' BINARY MESSAGE (Y/N) ? '
      READ(5,1000) ANSWER
      IF((ANSWER.EQ.'Y').OR.(ANSWER.EQ.'y'))THEN
        IFLAG = 1
        PRINT,' '
        PRINT,' ENTER 8 BITS (0 OR 1) PER LINE.  FOR AN ALL-ZEROS'
        PRINT,' MESSAGE, HIT ENTER FOR EACH 8-BIT SEQUENCE OF ZEROS.'
        PRINT,' TYPE "END" WHEN DONE'
        PRINT,' '
         KOUNT = 0
  100   CONTINUE
        BITBUF = BLANK
        PRINT,'>'
        READ(5,1020) BITBUF
        IF((BITBUF.EQ.'END     ').OR.(BITBUF.EQ.'end     '))THEN
            GO TO 200
          ELSE
            NCHAR = 0
            DO 10 INDEX = 1 , 8
              IF( BITBUF(INDEX:INDEX) .EQ. ONE ) THEN
                NCHAR = NCHAR + 2**(INDEX-1)
              ENDIF
   10       CONTINUE
            KOUNT = KOUNT + 1
            BUFFER(KOUNT:KOUNT) = CHAR( NCHAR )
          ENDIF
        GO TO 100
  200   CONTINUE
        LENBUF = KOUNT
      ELSE
        DO 1 INDEX = 1 , 80
          BUFFER(INDEX:INDEX) = ' '
    1   CONTINUE
        PRINT,' ENTER ASCII MESSAGE'
        PRINT,' '
```

```
             READ(5,1010) BUFFER
             DO 50 INDEX = 80 , 1 , -1
                 IF( BUFFER(INDEX:INDEX) .NE. ' ') THEN
                     LENBUF = INDEX
                     GO TO 55
                 ENDIF
    50     CONTINUE
    55     CONTINUE
         ENDIF
         RETURN
 1000 FORMAT(A1)
 1010 FORMAT(A80)
 1020 FORMAT(A8)
         END
C
         SUBROUTINE PARMIN(NCRATE,MEMLEN,KL,ITAP)
C    THIS SUBROUTINE ASKS THE USER FOR HIS/HER CHOICE OF CODING
C    PARAMETERS
         DIMENSION ITAP(3),ITAPB(3,2:3,3:7)
C TAP LOCATIONS FOR CODE GENERATORS ARE FROM TABLE 6.4 P.349
         DATA ITAPB/111,101,0,111,111,101,1111,1011,0,1111,1011,1101,
       +10111,11001,0,11111,11011,10101,101111,110101,0,101111,110101,
       +111001,1001111,1101101,0,1001111,1010111,1101101/
 2014 PRINT,' '
         WRITE(6,1000) KL,NCRATE,MEMLEN,(ITAPB(I,NCRATE,KL),I=1,NCRATE)
 1000 FORMAT(' THE DEFAULT CODES ARE THE ODENWALDER CODES LISTED IN',
       +' TABLE 6.4, PAGE 349'//
       +'   1. CONSTRAINT LENGTH (7 MAX.)              ',I1/
       +'   2. NUMBER OF CODE GENERATORS (3 MAX.)      ',I1/
       +'   3. PATH MEMORY LENGTH (70 MAX.)            ',I2//
       +'   4. FIRST GENERATOR                 ',I7/
       +'   5. SECOND GENERATOR                ',I7/
       +'   6. THIRD GENERATOR                 ',I7)
         WRITE(6,1001)
 1001 FORMAT(/' Note that the code generators are shown as a binary',
       +' sequence, where'/' each 1 represents a connected tap, and',
       +' each 0 represents an unconnected tap.'/)
         PRINT,' ENTER PARAMETER NUMBER TO CHANGE (ENTER 0 IF DONE) > '
         READ(5,*) NUMCNG
         IF ( NUMCNG .EQ. 0 ) THEN
             GO TO 2015
         ELSE IF ( NUMCNG .EQ. 1 ) THEN
             PRINT,' ENTER NEW CONSTRAINT LENGTH        > '
             READ(5,*) KL
             GO TO 2014
         ELSE IF ( NUMCNG .EQ. 2 ) THEN
             PRINT,' ENTER NUMBER OF CODE GENERATORS    > '
             READ(5,*) NCRATE
             GO TO 2014
         ELSE IF ( NUMCNG .EQ. 3 ) THEN
             PRINT,' ENTER NEW PATH MEMORY LENGTH       > '
             READ(5,*) MEMLEN
             GO TO 2014
         ELSE IF ( NUMCNG .EQ. 4 ) THEN
             PRINT,' ENTER FIRST GENERATOR              > '
             READ(5,*) ITAPB(1,NCRATE,KL)
             GO TO 2014
         ELSE IF ( NUMCNG .EQ. 5 ) THEN
             PRINT,' ENTER SECOND GENERATOR             > '
             READ(5,*) ITAPB(2,NCRATE,KL)
             GO TO 2014
         ELSE IF ( NUMCNG .EQ. 6 ) THEN
             PRINT,' ENTER THIRD GENERATOR              > '
             READ(5,*) ITAPB(3,NCRATE,KL)
             GO TO 2014
         ELSE
             PRINT,' NUMBER OF CHANGE OUT OF RANGE.  RE-ENTER'
```

```
          GO TO 2014
      ENDIF
2015 CONTINUE
C    CALCULATE ITAP FROM ITAPB
      DO 100 ICRATE=1,NCRATE
         IACCUM=0
         IDIGIT=10
         IBNARY=1
         IWORD=ITAPB(ICRATE,NCRATE,KL)
         DO 110 IBIT=1,KL
           ITEST=MOD(IWORD,IDIGIT)
           IF(ITEST.NE.0) THEN
              IACCUM=IACCUM+IBNARY
              IWORD=IWORD-ITEST
           ENDIF
           IDIGIT=IDIGIT*10
           IBNARY=IBNARY*2
  110    CONTINUE
         ITAP(ICRATE)=IACCUM
  100 CONTINUE
      WRITE(2,1000) KL,NCRATE,MEMLEN,(ITAPB(I,NCRATE,KL),I=1,NCRATE)
  999 FORMAT(A1)
      RETURN
      END
C
      SUBROUTINE ASCII1(BUFFER,LENBUF,IBFFER,LENIBF,KL)
C   THIS SUBROUTINE CONVERTS THE CHARACTER MESSAGE TO A BIT STREAM
      CHARACTER*80 BUFFER
      DIMENSION IBFFER(646)
      KOUNT = 0
      DO 100 INDEX1 = 1 , LENBUF
         NCHAR = ICHAR(BUFFER(INDEX1:INDEX1))
         DO 200 INDEX2 = 1 , 8
           KOUNT = KOUNT + 1
           IBIT = MOD(NCHAR,2**INDEX2)
           IF (IBIT.NE.0) THEN
              IBFFER(KOUNT) = 1
            ELSE
              IBFFER(KOUNT) = 0
           ENDIF
           NCHAR=NCHAR-IBIT
  200    CONTINUE
  100 CONTINUE
      DO 300 INDEX1 = 1 , KL-1
         KOUNT = KOUNT + 1
         IBFFER(KOUNT) = 0
  300 CONTINUE
      LENIBF = KOUNT
      RETURN
      END
C
      SUBROUTINE ASCII2(BUFFER,LENBUF,IBFFER,LENIBF)
C   THIS SUBROUTINE CONVERTS A BIT STREAM TO CHARACTERS
      CHARACTER*80 BUFFER
      DIMENSION IBFFER(646)
      KOUNT = 0
      LENBUF = LENIBF/8
      DO 100 INDEX1 = 1 , LENBUF
         NCHAR = 0
         DO 200 INDEX2 = 1 , 8
           KOUNT = KOUNT + 1
           NCHAR = NCHAR + 2**(INDEX2-1) * (IBFFER(KOUNT))
  200    CONTINUE
         BUFFER(INDEX1:INDEX1) = CHAR(NCHAR)
  100 CONTINUE
      RETURN
      END
```

```
            SUBROUTINE ERRORS (BUFFER,ICODED,NREP,NCRATE,KL)
C    THIS SUBROUTINE QUERIES THE USER WHERE HE/SHE WOULD LIKE ERRORS
C    TO OCCUR IN THE TRANSMITTED BIT STREAM
            CHARACTER*80 BUFFER
            DIMENSION ICODED(NREP)
            CHARACTER*24 CURSOR
            LINESZ = 8 * NCRATE
            NERROR = 0
            NLINES = NREP/LINESZ
            DO 100 INDEX = 1 , NLINES
              DO 110 INDEX2 = 1 , 24
                CURSOR(INDEX2:INDEX2) = ' '
    110     CONTINUE
            PRINT,' '
            PRINT,' CREATE THE DESIRED ERROR PATTERN BY MOVING THE CURSOR'
            PRINT,' TO A POSITION DIRECTLY BELOW THE BIT WHERE AN ERROR'
            PRINT,' IS TO OCCUR, AND TYPING ANY CHARACTER'
            PRINT,' '
            PRINT,' '
            NBEGIN = MIN0( LINESZ * (INDEX-1) + 1 , NREP)
            NEND   = MIN0( NBEGIN + (LINESZ - 1) , NREP)
            WRITE(6,1005) BUFFER(INDEX:INDEX)
            WRITE(2,1005) BUFFER(INDEX:INDEX)
            PRINT,' ENCODED CHARACTER '
            WRITE(6,1010) ((ICODED(I)+1)/2,I=NBEGIN,NEND)
            WRITE(2,1011) ((ICODED(I)+1)/2,I=NBEGIN,NEND)
            WRITE(6,1020)
            READ(5,1000) CURSOR
            DO 120 INDEX2 = 1 , LINESZ
              IF( CURSOR(INDEX2:INDEX2) .NE. ' ' ) THEN
                INDEX3 = NBEGIN + (INDEX2-1)
                ICODED( INDEX3 ) = -ICODED( INDEX3 )
                NERROR=NERROR+1
              ENDIF
    120     CONTINUE
            PRINT,' RECEIVED CHARACTER BIT STREAM'
            WRITE(6,1012) ((ICODED(I)+1)/2,I=NBEGIN,NEND)
            WRITE(2,1013) ((ICODED(I)+1)/2,I=NBEGIN,NEND)
            WRITE(2,1014) CURSOR
            WRITE(2,1020)
    100 CONTINUE
            WRITE(2,1020)
            WRITE(6,1020)
            WRITE(2,1030) NREP-NCRATE*(KL-1),NERROR
            WRITE(6,1030) NREP-NCRATE*(KL-1),NERROR
            RETURN
1000 FORMAT(A24)
1005 FORMAT(' Transmitted character...',A1)
1010 FORMAT(1X,24I1)
1012 FORMAT(1X,24I1)
1011 FORMAT(' Transmitted coded bits  ',24I1)
1013 FORMAT('     Received coded bits  ',24I1)
1014 FORMAT(' Coded bit error pattern ',A24)
1020 FORMAT(' ')
1030 FORMAT(I5,' coded bits sent',I7,' coded bits received in error')
            END
C
            SUBROUTINE RANDOM(BUFFER,ICODED,NREP,NCRATE,KL,SNROBS)
C    THIS SUBROUTINE GENERATES A DEMODULATED BPSK SIGNAL, CORRUPTED
C    BY AWGN, AND EXPRESSED AS A QUANTIZED CODED-SYMBOL STREAM
            CHARACTER*80 BUFFER
            DIMENSION ICODED(NREP),IHIST(32)
            LINESZ = 8 * NCRATE
            NERROR = 0
            VAR=0.
            NLINES = NREP/LINESZ
            PRINT,' ENTER PREDETECTION EB/N0 IN DECIBELS        > '
```

```
          INPUT,SNR
          SNR=10.**(SNR/10.)
          SIGMA=1./SQRT(2.*SNR/NCRATE)
          PRINT,' ENTER NUMBER OF SOFT-DECISION BITS (6 MAX.)  > '
          INPUT,NQ
          NBIN=2.**(NQ-1)
          NLEVEL=2.**NQ-1
          DO 10 K=1,NBIN
            IHIST(K)=0
   10    CONTINUE
          DO 100 INDEX = 1 , NLINES
            NBEGIN = MIN0( LINESZ * (INDEX-1) + 1 , NREP)
            NEND = MIN0( NBEGIN + (LINESZ - 1) , NREP)
            WRITE(6,1005) BUFFER(INDEX:INDEX)
            WRITE(2,1005) BUFFER(INDEX:INDEX)
            WRITE(6,1011) (ICODED(I),I=NBEGIN,NEND)
            WRITE(2,1011) (ICODED(I),I=NBEGIN,NEND)
            DO 120 INDEX2 = 1 , LINESZ
              INDEX3 = NBEGIN + (INDEX2-1)
              CALL NOISE(AWGN)
              VAR=VAR+AWGN*AWGN
              CODED=SIGMA*AWGN + ICODED(INDEX3)
              ERROR=CODED*ICODED(INDEX3)
              IF(ERROR.LE.0.) NERROR=NERROR+1
              Q=CODED*NLEVEL
              IF(Q.GT.NLEVEL) Q=NLEVEL
              IF(Q.LT.-NLEVEL) Q=-NLEVEL
              IQ=INT(Q)
              IQMOD=MOD(IQ,2)
              IF(IQMOD.EQ.0) IQ=IQ+SIGN(1.,Q)
              ICODED(INDEX3)=IQ
              IQP=IABS(IQ)/2+1
              IHIST(IQP)=IHIST(IQP)+1
  120      CONTINUE
            WRITE(6,1013) (ICODED(I),I=NBEGIN,NEND)
            WRITE(2,1013) (ICODED(I),I=NBEGIN,NEND)
  100    CONTINUE
          WRITE(2,1030) NREP-NCRATE*(KL-1),NERROR
          WRITE(6,1030) NREP-NCRATE*(KL-1),NERROR
          VAR=VAR/(NREP-1)
          SNROBS=10.*ALOG10(FLOAT(NCRATE)/2./SIGMA/SIGMA/VAR)
          WRITE(2,1040) SNROBS
          WRITE(6,1040) SNROBS
          WRITE(2,1041) (IHIST(I),I=1,NBIN)
          WRITE(6,1041) (IHIST(I),I=1,NBIN)
          RETURN
 1005 FORMAT(/' Xmitted character...',A1)
 1011 FORMAT(' Xmtd:  ',24I3)
 1013 FORMAT(' Rcvd:  ',24I3)
 1030 FORMAT(/,I5,' coded bits sent',I7,' coded bits received',
      +' in error')
 1040.FORMAT(/,' Observed SNR in dB ',F5.2,/)
 1041 FORMAT(' Quantized likelihood function histogram, where the'
      +/' rightmost entry represents the most assured decisions'//32
      +(1X,I4))
          END
C
          SUBROUTINE NOISE(AWGN)
C THIS SUBROUTINE GENERATES A ZERO-MEAN UNIT-VARIANCE GAUSSIAN RANDOM
C VARIABLE SAMPLE
          AMP=SQRT(-ALOG(RND())*2.)
          PHASE=RND()*6.283185308
          AWGN=AMP*COS(PHASE)
          RETURN
          END
```

List of Symbols

a_{ij}	Coefficient of jth basis function
a_j	Signal component output of jth correlator
A	Peak amplitude of a waveform
A_e	Effective area of an antenna
B_L	Single-sided loop bandwidth
c	Speed of light $\simeq 3 \times 10^8$ m/s
C	Channel capacity
C	Electrical capacitance
$C/\kappa T^\circ$	Ratio of average carrier power to noise power spectral density
d	Distance
d_f	Free distance
$d_{\min}$	Minimum distance
D	Delay time of message
D	Redundancy of a language
D	Decryption transformation
e	The natural number 2.7183
$\mathbf{e}$	Error pattern vector
$e(t)$	Error signal
$\mathbf{e}(X)$	Error pattern polynomial
E	Encryption transformation
E_x	Energy of waveform $x(t)$
$\mathrm{E}\{X\}$	Expected value of the random variable X
EIRP	Effective radiated power with reference to an isotropic source
E_b/J_0	Ratio of bit energy to jammer power spectral density

E_b/N_0	Ratio of bit energy to noise power spectral density
E_c/N_0	Ratio of channel symbol energy to noise power spectral density
f	Frequency (hertz)
f_c	Carrier-wave frequency
f_m	Maximum frequency
f_s	Sampling frequency
f_ℓ	Lower cutoff filter frequency
f_u	Upper cutoff filter frequency
F	Noise figure
$\mathcal{F}\{x\}$	Fourier transform of the function $x(t)$
$\mathcal{F}^{-1}\{X\}$	Inverse Fourier transform of the function $X(f)$
$g(t)$	Pseudorandom code function
$\mathbf{g}(X)$	Generator polynomial for a cyclic code
G	Antenna gain
G	Coding again
$\mathbf{G}$	Generator matrix for a linear block code
G	Normalized total message traffic
G_p	Processing gain
$G_x(f)$	Power spectral density of waveform $x(t)$
$h(t)$	Impulse response of a network
$\mathbf{H}$	Parity-check matrix for a code
H_i	The ith hypothesis
$\mathbf{H}_k$	Hadamard matrix
$H(f)$	Frequency transfer function of a network
$H_0(f)$	Optimum frequency transfer function
$H(X)$	Entropy of information source X
$H(X|Y)$	Conditional entropy (entropy of X, given Y)
$i(t)$	Electrical current waveform
I	Electrical current
$I_0(x)$	Zero-order modified Bessel function of the first kind
$I(X)$	Self-information of information source X
J	Received average jammer power
J_0	Jammer power spectral density
J/S	Ratio of received average jammer power to average signal power
k	Number of bits per M-ary signal set
k/n	Code rate (ratio of number of data bits to total bits in codeword)
K	Constraint length of a convolutional encoder
K	Key, dictating a specific encryption or decryption transformation
ℓ	Number of quantization bits
L	Look-ahead length for convolutional feedback decoding
L	Number of bits in sequence
L	Number of quantization levels
L_s	Space loss
L_o	Other losses
$\mathbf{m}$	Message vector
$\mathbf{m}(X)$	Message polynomial

m_i	Data bit
M	Margin
M	Waveform or signal set size
(n, k)	Code designation by number of total bits (n) and data bits (k) in codeword
$\bar{n}$	Average number of bits per character
n_0	Noise random variable output of correlator at symbol time $t = T$
$n(t)$	Gaussian noise process
N	Noise power
N	Unicity distance
N_0	Level of single-sided power spectral density of white noise
NSR	Ratio of average noise power to average signal power
p	Probability of channel symbol error
p_i	Parity bit
$p(t)$	Instantaneous power
$p(x)$	Probability density function of a continuous random variable
$p(x\|y)$	Probability density function of x conditioned on y
$\mathbf{P}$	Parity array
P_B	Probability of bit error
P_E	Probability of symbol error
P_{FA}	Probability of false alarm
P_m	Probability of miss
P_M	Probability of message or block error
P_{nd}	Probability of undetected error
P_r/N_0	Ratio of received average signal power to noise power spectral density
$P(X)$	Probability of a discrete random variable
P_x	Average power in waveform $x(t)$
q	Quantization step size (quantile interval)
$\mathbf{q}(X)$	Quotient polynomial
$Q(x)$	Complementary error function (integral of the tail beyond x of the Gaussian density function)
r	Filter roll-off factor
r	True rate of a language
r'	Absolute rate of a language
$r(t)$	Received signal waveform
$r(X)$	Remainder polynomial
R	Data rate (bits/second)
R_c	Coded data rate (coded bits/second)
R_p	Code chip rate (chips/second)
R_s	Symbol rate (symbols/second)
$R_x(\tau)$	Autocorrelation function of waveform $x(t)$
$\mathcal{R}$	Electrical resistance
$s(t)$	Signal waveform
$\hat{s}(t)$	Estimate of signal waveform
$\mathbf{s}$	Signal vector
sgn x	Sign function of x

S	Signal power
$\mathbf{S}$	Syndrome vector
SJR	Ratio of average signal power to average jammer power
SNR	Ratio of average signal power to average noise power
S/N	Ratio of signal power to noise power
$S(f)$	Fourier transform of the waveform $s(t)$
$\mathbf{S}(X)$	Syndrome polynomial
t	Number of errors correctable in an error-correcting code
t	Independent time variable
t_0	Time delay
t_{ij}	Amount of message traffic from i to j
T	Pulse width
T	Symbol interval
$T(D)$	Transfer function or generating function of convolutional code
T_{hop}	Duration of a hop
T_s	Sampling interval
T°	Temperature
T_A°	Antenna temperature
T_L°	Effective line temperature
T_R°	Effective receiver temperature
T_S°	System temperature
T_{acq}	Time to acquire
u_i	Code symbol
$u(t)$	Unit step function
$\mathbf{U}$	Codeword vector
$\mathbf{U}(X)$	Codeword polynomial
v	Relative velocity
$v(t)$	Electrical voltage waveform
var (X)	Variance of random variable X
V	Electrical voltage
$w(t)$	Jammer waveform
W	Bandwidth
W_f	Filter bandwidth
W_{DSB}	Double-sideband bandwidth
W_N	Noise equivalent bandwidth
W_{ss}	Spread-spectrum bandwidth
$x(t)$	Normalized VCO output
$y(t)$	VCO input voltage function
$z(t)$	Output of matched filter or correlator
γ	Threshold level
γ_0	Optimum threshold level
δ	Fractional frequency drift per day
δ_{mn}	Kronecker delta function
$\delta(t)$	Impulse (Dirac delta) function
ϵ	Error

ζ	Loop damping characteristic (second-order loop)
η	Antenna efficiency
$\theta(t)$	Time-varying phase
$\Theta(\omega)$	Fourier transform of $\theta(t)$
λ	Wavelength
λ	Packet arrival rate
π	Pi, 3.14159
ρ	Fraction of the frequency band being jammed
ρ	Fraction of the time the jammer is "on"
ρ	Normalized loop signal-to-noise ratio
ρ	Normalized message throughput
ρ	Number of erasures correctable in an error-correcting code
ρ	Time-correlation coefficient
ρ_0	Value of ρ that maximizes bit error probability (worst-case jamming)
σ_X	Standard deviation of random variable X
σ_X^2	Variance of random variable X
τ	Pulse width
τ	Time shift (independent variable of the autocorrelation function)
$\phi(t)$	Time-varying phase
$\psi_j(t)$	Basis function
$\Psi_x(f)$	Energy spectral density of waveform $x(t)$
ω	Radian frequency (radians per second)
κ	Boltzmann's constant, 1.38×10^{-23} J/K

Index

THE COMPLETE HOW TO FIGURE IT

Also by Darrell Huff

How to Lie with Statistics
How to Take a Chance
Cycles in Your Life
Score: The Strategy of Taking Tests
The Complete Book of Home Improvement
How to Work with Concrete and Masonry
Twenty Careers of Tomorrow (with Francis Huff)
The Dog That Came True

CONTENTS

CONTENTS

THE COMPLETE HOW TO FIGURE IT

Chapter A
LIFETIME MONEY STRATEGY

EASY AS FALLING OFF
A LOGARITHM!

What's Future Money Worth Now?

WHICH is worth more, a $5,000 cash gift today or a guaranteed $15,000 twenty years from now?

Even putting aside the unpredictable personal aspects, is it really possible to give a sensible answer to this—when you have no way to predict the inflation rate to come?

Surprisingly, it is.

To turn your answer from a wild guess into a good working approximation, you need only accept one reasonable probability. This is an assumption that the net (meaning after-taxes) interest rate you can expect to earn will exceed inflation by 6 percent.

If the long-range outlook at the moment of decision seems to favor 5, or 4, or 7 percent, go with that instead. (Wherever you're placing your investment, there should be someone about who can give you current figures.)

Then you won't have to think about the actual rates of either inflation or interest. You're reducing your gamble by noting the historic tendency for those two things to move up and down in harmony.

That said, let's try the computation for $5000 to be tucked away for 20 years at 6 percent interest compounded twice a year.

Begin by expressing the chosen rate of interest as a decimal, by dividing that 6 percent by 100 to get 0.06. That's the rate for a year, so it must be divided by 2 to get the rate for each half year, our chosen compounding period.

To this result, add 1.

That sum, 1.03, must now be raised to the power of the number of compounding periods, which is 40 in 20 years. The 40th power of 1.03 is 3.262 (found in 20 seconds on a simple pocket calculator, by entering 1.03, pressing the times key once and the equals key 39 times. Or times once, equals four times, then times once, equals seven times).

Dividing this result into 1 (finding the reciprocal, to use mathspeak) and multiplying by the specified $15,000 produces the answer you're looking for. If our 6 percent premise proves psychic, it would take only $4598.35 invested now to become $15,000 in 20 years.

If that's all it takes to grow a $15,000 crop in 20 years, I'll take that offer of a full $5000 now—and produce a bigger crop than the offered $15,000.

How big? You can wait a couple of decades and then look at your statement from the money fund where you're invested. Or, considerably more quickly and almost as easily, you can apply the tricks just described to this slightly different recipe.

This time all you have to do is find the 40th power of the periodic interest rate and then multiply that by the number of dollars you'd be starting with.

Once again you arrive at 3.262 for the 40th power of 1.03. Multiplying that by the $5000 gift you started with, you find that 20 years from now you could expect to have $16,310.19 if you haven't had any truck with the tax folks in the meantime.

Whither the Dollar?

For as long as we've had inflation—and that's practically forever and everywhere and most of the time—people have made disparaging remarks about our "fifty-cent dollar." Those of us who have been around quite a while are likely to make that "ten-cent."

Actually, though, how long does it take for inflation to cut the value of a currency in half?

The formula is a short one. It says divide the logarithm of 2 by the log of the inflation rate (as a percentage) plus 1.

Given the 6 percent rate of inflation that has been fairly typical of some recent periods, it goes like this. From a table of logarithms, the log of 2 is .30103 and the log of 1 plus .06 (1.06) is 0.025306. Dividing the first of these by the second produces 11.9, which says that the dollar is losing half its value in just under 12 years.

To borrow a term from the physical sciences, its half life is 11.9 years.

It's a happy fact that this kind of problem can be worked backwards. That way you don't have to bother with logarithms at all.

Approaching the previous problem from a slightly different angle, suppose you want to know by what factor the value of your buck is going to be reduced in, say, 12 years.

Just add 1 to the annual rate of depreciation (in decimal form, so that 6 percent is 0.06).

Raise this to the 12th power. Remember that on most simple pocket calculators, this can be done by striking the times key once and the equals key 11 times.

If you've never used this trick, you might test it first by seeing if hitting 2, then times, then equals produces 4. If not, hit times twice instead of once before equals.

It turns out that 1.06 to the 12th power is 2.012, confirming the previous conclusion that in 12 years a dollar declining 6 percent a year will have slipped by a factor of about 2—thus keeping only half its present value.

Now here's how to use this method as an alternative to consulting a log table. We're sometimes told, by apologists for the economic mess we happen to be in at the time, that what we really must worry about is double-digit inflation, meaning 10 percent or more. If 10 percent inflation becomes the norm, how many years will it take to hack a dollar down to half its value?

In terms of your pocket calculator, 1 plus 10 percent of 1 (1.1) must be raised to what power to equal 2?

Enter 1.1 in your calculator. Touch the times key once for the first power (which is the 1.1 you just entered, so the display doesn't change), then the equals for the second power, the equals again for the third, and so on. Keep this up till the displayed amount is about 2. You'll find that on the seventh keystroke (the sixth time you touched equals), the display reads 1.949. That's much closer to 2 than you'll produce with an eighth touch, so you realize that a trifle more than seven years of inflation will cut that buck to half its present worth.

A sobering thought.

And if, by any chance, you doubt the earlier remark about inflation's having been with us just about always, consider the case of Chaucer. Literary historians have sometimes found it amusing that when Chaucer was captured in battle in the mid-1300s, he was ransomed for 16 pounds sterling, which we think of today as maybe twenty-five or thirty dollars. Finally a biographer, John Gardner, has calculated that what this actually amounts to in modern terms is $3840. There's 600 years of off-and-on inflation for you.

Quick Look at Monthly Money

Suppose you had a wad of money under your mattress. Each year, you spent 4 percent, or one-twenty-fifth of it. It would last 25 years, right?

But suppose that same wad were put in a savings account or money fund where it would bear interest. Then how long would it take to use it all up?

You can figure it out from the table for any withdrawal rate from $4 to $10 a year for each $100 invested. Just look opposite the amount chosen and read the figure under the percentage rate of interest that the account bears.

Dollars per $100 invested withdrawn each year	Rate of interest your money is drawing						
	3%	4%	5%	6%	7%	8%	9%
$4	46						
$5	30	41		This is how many years			
$6	23	28	36	you can keep on doing it.			
$7	18	21	25	33			
$8	15	17	20	23	30		
$9	13	14	16	18	22	28	
$10	12	13	14	15	17	20	26

So if you put $15,000 into a 6 percent account and draw out $8 a year for each $100 of that initial amount ($1200 a year or $100 a month) the table says you'll be able to keep it up for just 23 years. And then your money will be all gone.

You probably can see why this table is in a triangular shape instead of the usual rectangle. The missing half represents the situation in which the rate of withdrawal is slower than the rate of increase. If you were to fill it in, it would have to be with infinity signs.

In short, if you were to arrange your affairs that way, you could draw out the chosen amount of money forever and then some. You'd be living on the interest while the principal kept on growing.

How a Retirement Plan Grows

SINCE you don't have to figure in the effects of taxes year by year, it's rather simple to find out how much a retirement fund of the IRA or Keogh type—untaxed till withdrawal—will grow to over the years.

If you have even one of the simplest of business or scientific calculators, its roots-and-powers key will make this computation a cinch.

And it's only moderately more tedious with the ordinary calculator, which will give you the 20th power of the number in its register if you press times once and equals 19 times.

It's more fun—or less agony—to put money aside if you know what it's doing.

To get a handle on this, begin with the constant, or estimated average, rate of return. Express this as a decimal (9 percent becomes 0.09) and divide by the number of compounding periods per year (4 if quarterly).

Add 1.

Raise the result to the power of the total number of compounding periods over the years. Multiply by the number of dollars in the fund.

Using this in our example for 20 years of growth means raising 1.0225 to the 80th power (for the quarterly periods in 20 years). Result: $5.93 for each initial dollar. Encouraging, isn't it?

Figuring Inflation into Retirement

IT'S ONE thing to arrive at an estimate of how much money, or income, you'd need to retire comfortably this year. But if your time to retire is far away, how in the world can you know how much you'll need then—after several more decades of the inflation that has been with us off and on for centuries?

You can educate your guess with even the simplest of pocket calculators, quickly arriving at the inflation factors for 6 percent or any other level you care to postulate. Just enter the percentage, divide by 100, add 1, and multiply the result by itself once for each year after the first.

If you're making a reasonable estimate (it could also be called a wild guess) that there will be 4 percent inflation in the average year from now until your retirement in, let's say, 20 years, this means discovering the 20th power of 1.04.

Found on a typical pocket calculator by entering 1.04 (or entering 4, dividing by 100, and adding 1), then pressing the times key once and equals 19 times, the inflation factor proves to be 2.19.

This tells you that if your guess was good, it will take $2.19 twenty years from now to do the work a dollar does today.

How Much Life Insurance?

T HERE'S no brief formula to tell you how much life insurance you should carry. But people who deal with family economic planning think that a family should figure its insurance plan by adding together these elements:

1. Quick cash. This should be enough to cover funeral expenses, current bills and income-tax obligations, possibly estate or inheritance taxes, costs run up during a last illness.

2. Interim. This should be an amount sufficient to support the surviving family for half a year—probably close to half your annual after-tax earnings in recent years.

3. Family support. This should be sufficient to cover the family's needs indefinitely, after deducting earnings by the survivors. For a young widow or widower with children to care for, this can be quite large, even for an employed survivor who can continue employment with good earning prospects.

4. Education. Remember that college costs have been inflating steadily in recent years, at two or three times the rate of living costs generally.

5. Emergency fund. If a couple of thousand dollars is already held in savings (preferably at the maximum interest rate that permits ready access), no additional amount need be provided through insurance.

Forecasting Your Income

Some of us live on irregular incomes from fees or commissions or from sales of our work. This can be hard on the old nervous system. It doesn't love incomes that are irregular and seem unpredictable. Indeed, many a promising career has been abandoned for no better reason than an illusion of financial insecurity.

In fact, most of the terrors of even a highly irregular income vanish if it can be found to be both sufficient and predictable.

Some years ago, as a free-lance writer with four children to shoe and feed, I finally evolved out of desperation a device for predicting my income for each coming month.

First, I listed all sources from which money might possibly come. Then I estimated the most likely amount of each bit of income and multiplied it by the probability that I would indeed receive that payment and receive it within the period in question.

Thus a check for $75 that had been promised by a source known to be both reliable and prompt would go on my list at full face value or very nearly.

A magazine article written on assignment might go down for $200 if a likely fee for it was $400 and I felt there was an even chance it would fetch its fee within the month.

A purely speculative contribution to an editor I'd never dealt with before might impress me as no better than a 10-percent gamble. So even if the magazine was listed as paying in the thousand-dollar range, I'd put it into the hopper for a mere hundred. Furthermore, if I figured that even if the manuscript did sell with those ten-to-one odds against it, there was an even chance I wouldn't see the check within the month, I'd whack even that modest hundred down to fifty.

In my busy years, when I often had a couple of dozen irons in each month's fire, my forecasts usually proved anywhere from amazingly to uncannily reliable. Many a month they came within a few dollars, for a precision of 95 to 99 percent—even though the individual items on the forecast were never correct.

There's something else this kind of forecast has going for it. Since time is of the essence, as formal agreements sometimes say, my occasional poor forecasts were usually balanced very quickly by comparable differences in the other direction.

I don't think this method can be very reliable unless it is based on enough projects to give the averages a chance to work. In recent years, when I tend to do larger but far fewer projects, I mostly use annual forecasts. These still function well, and they have the additional virtue of persuading me to plan my work more systematically—by the year instead of by the month—now that I can afford that luxury.

Besides, these are just the thing any self-employed laborer needs if he is to make a fair stab at filling out that work sheet each April 15 for pre-payment of estimated federal income tax.

Are You Throwing Away a House?

Having a need at one time to produce a hard-headed view of the relative merits of renting and home ownership, I used some case histories I knew about to make an instructive story. The moral of it, as you will see, is that some folks should rent and some should buy—and the secret of success is knowing which you are and when.

I called my people Bob and Bea Harte and their Uncle Jim. I hope you will find them useful in figuring out moves of your own.

Several decades ago, Uncle Jim and his wife, newly married, concluded that renters have it best. Jim has continued in the same job. They have moved only twice in all those years (one landlord decided to sell; the other was too incredibly stingy with paint). Jim and his wife are living today in essentially the same house as when they started their married life.

After an evening of talk with Uncle Jim, followed by some paper-and-pencil figuring, Bob Harte arrived at a startling conclusion. Before stating that conclusion, let's see how Bob reached it.

When Uncle Jim and his wife moved into their house 23 years earlier, they could have bought it, or something like it, for $36,000. Using his veteran's privileges, Jim would have had no down payment to make, and closing costs would have been minor.

Even at 6 percent for 20 years, Jim's monthly payments would have been $258.12. Taxes and insurance would have added about $70 a month in the early years, possibly $110 by the end (as both rates and the value of the house increased) for an average of $90 a month over the 20 years of the mortgage.

By straining his memory of what his landlord must have laid out, Jim had arrived at a reasonable estimate of maintenance costs for such a house for 20 years. At $500 a year—for small repairs plus an occasional major one, such as exterior painting or a new roof—these came to an average of about $42 a month.

Adding these figures of Jim's together, his nephew Bob arrived at a monthly cost-to-own of something less than $370 for the early years, rising to $410 or so by the twentieth year. Average: $390. But Jim had chosen to rent instead. His landlord, hewing fairly closely to the old rule of thumb that gross monthly rentals should approximate 1 percent of value, had charged him $350 a month when he moved in. As values (and taxes and maintenance) increased over the years, so did the rent Jim paid for his initial rented house and its successors. Jim has no reason to complain about the $550 he currently pays; actually, it represents well under 1 percent of the market value of the house he's in, although this house is similar to one he might have bought years ago for $36,000.

Since Jim's rent increases have been gradual but consistent, following the pattern in most localities, his average rent has been about $450.

And that's how Bob Harte arrived at a startling conclusion: Uncle Jim could have owned the house he's now living in and it wouldn't have cost him an additional cent. There would be no more rent to pay in the future. But now Jim had, as the saying goes, nothing to show for all those years of rent checks, except a bundle of receipts. The difference between Jim's actual rental outlay ($450 a month) and his estimated payments if he had bought ($390) comes to a dazzling $14,400.

On top of that, Uncle Jim could have—depending upon his income and his other deductible expenses—gained by lowering his state and federal income taxes.

Bob Harte had only to multiply Jim's $258 a month by the 240 months the mortgage ran to discover that the total outlay for principal and interest had been $61,920. Taking away the $36,000 that was repayment of principal, $25,920 must have gone for interest. Much or all of this, and property taxes as well, might have been income-tax deductible.

"Gosh," concluded Bob Harte. "Uncle Jim has thrown away a house."

His wife Bea saw it the same way, and the young couple proceeded to buy an attractive new house of their own.

Within the next twelve months, three things happened. Costs of houses over the United States as a whole went up 8 percent. A large government project having to do with space flight

was discontinued along the seashore 10 miles from where the Hartes now lived. And Bob and Bea decided to move to Wyoming, where Bob had found an irresistible teaching job he could combine with the outdoor work he loved, as a park ranger each summer.

The Hartes promptly put their house up for sale, only to discover that because of the local drop in employment, three other families within the small subdivision had just done the same thing. And they had done so at prices reflecting eagerness to sell.

Only by persistence plus some time and money for repairs, and at considerable personal inconvenience, did Bob and Bea sell their house at what was seen locally as a fair market price. The general rise in building costs almost canceled out the effects of the weakness in the local market.

The price their house fetched equaled the full cost of building it a year earlier—what they had paid for it, less the 6 percent the seller had tacked on to cover the real-estate commission. When that selling cost is skimmed off the top of a sale price, it is elementary economics that buyer or seller, or the two jointly, must absorb a loss of that amount. It is a little like the difference between retail and wholesale prices.

The Hartes, having bought in a seller's market and sold in one favoring buyers, were in effect springing for two commissions, for a

12 percent loss. Add normal costs of moving in and closing (twice) and you see an out-of-pocket of about $6000.

"You could call it the Case of the Vanished Down Payment," said Bea, when everything was added up.

"I'd rather think of it as a year's rent," Bob replied. "Is $500 a month so bad?"

Well, yes it is, at the income and expenditure level we're talking about here. Add insurance and the usual amount of first-year landscaping and it all comes to twice what rent would have.

There we have Bea-and-Bob and Uncle Jim, two ways to go wrong, damned if you do and damned if you don't.

Yet we all know that one's own home is a fine investment, at the right time and place. It's necessary only to apply figures like these, as Bob and Jim did not, to boost the odds in your favor.

Living on a Savings Account

IF YOU HAVE a wad of money, one way of living on it is by buying an annuity. You get a guaranteed income, usually for life.

Another way calls for putting the money in a savings account and withdrawing some fixed sum each month. If the interest you can get is 5 percent a year, compounded quarterly, then each $10,000 in your account will permit you to withdraw:

$41.32 per month forever; or

$75 per month for 16 years and 1 month (during which time you'll have taken out not only your initial $10,000 but also $4501.50 interest that it earned); or...

$125 per month for 8 years (including $2115.25 interest).

Although these results are based on a single and apparently arbitrary rate of interest, you may find them quite generally applicable as bases for multiplying or dividing them to fit your own numbers.

What can make a 5 percent figure widely useful is that over a period of time that figure tends to be pretty close to the true rate of interest. True rate, in this connection, means the rate you're told you are getting minus the going level of inflation. Which difference means, in turn, how much the dollars you're getting are really worth when you go to buy something with them.

Appraise a Bank Bargain

In the name of patriotism, a savings-and-loan company in Nassau County, New York, offered a bicentennial special. Just deposit $1,000, wait only eight and one-half years, and you'd receive a historically appropriate $1776.

But some spoilsport with a pocket calculator wondered how much better this was than the offer made by competitors, and even by this same institution, before the bicentennial promotion came along. Even though the actual numbers fluctuate from year to year, anyone inclined to accept such an offer unquestioningly would do well to perform a self-inoculation by learning how this astute critic put the offer under a mathematical microscope.

The standard offer in those days was 7.75 percent interest on any $1000 deposit left for six to ten years; and, as the ad always explained, this was compounded daily—so on an annual basis, it really came to 8.06 percent.

The spoilsport quickly noted that such a deposit must grow to 1 plus 0.0806 during the first year, becoming $1.0806. Multiplied by itself eight times (that is, raised to the eighth power) 1.0806 becomes 1.8592. In eight years, one dollar becomes 1.8592 dollars. And $1000 grows to $1,859 and a few cents.

So with the patriotic plan, you'd wait half a year longer and collect $83 less.

At this point, the savings institution confessed that the patriotic special offer actually paid only 6.65 percent interest. It was their advertising department that thought it all up, they said hastily.

Strategy for Checking

With banks simultaneously fighting for your business and seeking ways to make the maximum profit from it, a little protective arithmetic may be in order.

Does it, for example, pay you to maintain a $500 minimum balance in order to avoid a $5 monthly service charge? If to do so you

have to tie up an extra $500 that could be earning 6 percent somewhere instead of your bank's 1 or 2 or nothing, you're losing the difference on that $500. Divided by 12 that says you're giving up $2.81 to save $5, and that's good.

It may be even better to seek out another bank or other savings institution, possibly a credit union, that omits service charges. Then you can keep the $2.81 too.

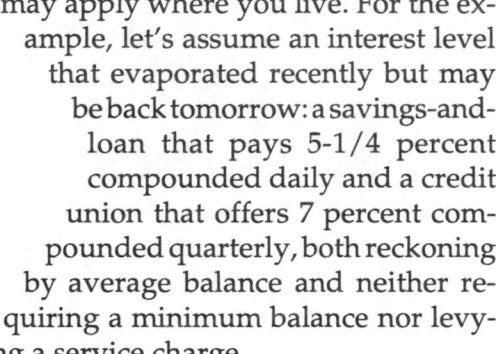

Here's another, somewhat more sophisticated, calculation that may apply where you live. For the example, let's assume an interest level that evaporated recently but may be back tomorrow: a savings-and-loan that pays 5-1/4 percent compounded daily and a credit union that offers 7 percent compounded quarterly, both reckoning by average balance and neither requiring a minimum balance nor levying a service charge.

Let's say that experience and a look through your monthly statements suggest that although your balance sometimes approaches zero it averages about $1000. What will that earn you in a year?

Express the S&L's rate as a decimal (0.0525), divide by the number of times compounded each year (360), add 1, then raise to the 360th power.

Multiply by the number of dollars—$1000 here—and you'll find it grows to $1053.90 in a year.

If the S&L is kind enough to use a 360-day year for dividing and then shift to 365 days for compounding, you'll pick up another 64 cents.

You'll find that an equivalent calculation for the credit union, using 4 for dividing and also for the power, comes up with $1071.86. Even when its higher interest is offset somewhat by the less frequent compounding, it offers you a bonus of $18 a year.

Factoring Inflation into Income

CALCULATIONS having to do with savings and income over a substantial period have come to mean very little unless they take account of inflation.

This is especially true of any scheme to give yourself a fixed monthly income by withdrawal from a single deposit. Withdrawals that today promise to be ample might be hopelessly inadequate in 20 years.

For realistic planning, indexing is the answer. Applied to interest income, that means subtracting the rate of inflation from the rate of interest you expect to receive and then using that difference in your calculations instead of using the stated interest rate.

You may have noticed that when the inflation rate remained near zero over a long period, conservative investments were likely to pay about 4 percent and home mortgages were only a little higher than that.

During some periods back in the 1980s, when inflation was averaging 10 to 12 percent, there were often quite conservative investments around that paid 15 percent or so. And this was more or less the prevailing rate for home mortgages.

For realistic planning for a constant level of effective (meaning constant-dollar) income, it is probably not prudent to assume that, over the long haul, the actual return on an investment will be more than 4 percent. A withdrawal plan based on a 4 percent return will let you adjust regularly the dollar amount taken, so that the purchasing power of your income remains unaffected.

After a period of years in which inflation is substantial, your monthly draw may become many times what it was initially. But its buying power is what counts. If you agree to make 4 percent your estimate of true interest rates, your course is simple. Spend one-twelfth of 4 percent of your principal each month, recalculating at least once a year.

A rather more precise approach would be to adjust your draw each year at the time announcements appear in the newspapers of what the inflation rate has been for the year. An easily available and conservative choice would be the cost-of-living adjustment announced each year for Social Security benefits.

In case of doubt, it should be noted that no one has so far accused the federal government of overstating an inflation rate. Most consumers and quite a few economists believe that by such moves as eliminating car purchases and housing from the index a considerable understatement of the rate has been introduced.

How Much Can You Take Out?

You MAY be planning to live, at some time in your life, at least partly on money withdrawn regularly from accumulated savings. Perhaps in an investment, such as a mutual fund.

Then applying our principles for factoring inflation into calculations about future income take on added importance.

Unless you do some figuring first, you may eventually run into a problem. That's what happened to the worried fellow who wrote to a syndicated investment advisor: "...our $50 monthly withdrawal from a mutual fund has forced the sale of 90 of our 590 shares!"

Value of his shares, he added, was about $7,000.

To pinpoint our man's problem and establish a basis for making a withdrawal plan, let's begin with a little arithmetic. That $50 a month is $600 a year, which divided by $7,000 is 0.0857, or 8.57

percent. As in most times, he couldn't get interest that high from any relatively safe investment.

A more realistic approach would begin with a look at going rates of return. If they're running around 6 percent, you should be able to set up a withdrawal plan at that level with a reasonable hope of seeing your investment hold its dollar value. Let that alarmed investor multiply his $7,000 by 6 percent and scale down his withdrawals to $35 a month, $420 a year.

If the value of the shares he holds, or the dividends the fund's stocks are paying, should rise sharply and stay that way for a while, then he might consider upping his withdrawal.

Any of us in a situation like this should also keep a weather eye peeled for the effects of inflation.

Taking out enough income to keep investment constant may not be prudent. It actually involves eating into capital, since the purchasing power of a fixed number of dollars continues to shrink.

One calculation to counter this is to use a realistic interest rate in specifying withdrawals. Don't use the nominal rate of interest you're told you're getting. Subtract the rate of inflation from the rate of interest, and use that. The dollar figure of your capital will grow, thus keeping its purchasing capacity constant.

Casing Inflation

W ANT TO know in a hurry what inflation does to your dollar? A simple calculation can tell you in seconds.

Subtract the annual inflation rate from 100 and express the result as a decimal—so that, for example, 6 percent inflation gives 0.94. And the 14 percent we were beating our heads against not too long ago becomes 0.86.

Raise that 0.94 (or 0.86) to the power of the number of years you are interested in knowing about.

In the absence of a scientific calculator that finds powers—even fractional ones!—in a flash (using the key labeled x to the power of y, or something like that), remember that this is easily done with most hand calculators by pressing the times key once and the equals key one time less than the number of years. Thus, for five years of inflation, you'd press times once and then equals four times.

With this procedure, you find out what the given rate of inflation does to one dollar in the chosen number of years. So multiply the result by the number of dollars you're dealing with.

If you find that in a mere five years, 14 percent inflation (and we did have a taste of that) will knock a dollar down to 47 cents and reduce the purchasing power of $5000 to a measly $2352.14, you've got it right.

And note that even though we've all had some chastening experiences with inflation, it's still shocking—and a useful warning—to see that a decade of it at 14 percent would shrink the value of $5000 almost down to one G— $1106.51 to be precise

Sailor, take warning!

That Old Bugaboo, Average Rate of Increase

To COMPARE how slowly your salary is rising to what has happened to the price of bread, you need to figure rate of increase. This is a tricky factor to deal with because the rate for any period must be based on the new level at the beginning of that period. It's much like what goes on with compound interest, something that most of us find rather a sticky wicket.

If you were earning $2 an hour thirty years ago and you're making $12 now you would be inclined to call that a 500 percent increase. Which it is, since 500 percent of 2 is 10. And then if you weren't thinking too hard you might go on to divide 500 by 30 and say your earning power in dollars has been going up 17 percent a year on the average. Which it has not.

It's even worse if this state of affairs has gone on with annual inflation running 8 or 10 percent. You've been taking a beating.

A widely applicable way of figuring an average rate of increase starts with dividing the new figure by the old. (That's dividing $12 by $2, in our example.) Having done that, you take the root that is equal to the number of years you're talking about. (In our 30-year example, that means the 30th root.) Now subtract 1 from this and multiply the result by 100 (by moving the decimal point two places to the right) to make it a percentage. You have found the true average rate of increase.

That is equivalent to what happens when a salary (or anything else) starts at 2 and is increased by 6.15 percent each year for 30 years.

Since your gain in earnings turned out to be not 17 percent but just over 6.15, it's no wonder you were falling behind in the face of inflation of 8 percent or more.

Finding a root such as the 30th, by the way, can be done by any of several methods covered in this book.

One is logarithms, if you want to acquire the needed tables and renew your grasp of this rather fascinating concept.

Or you can skip the tables and bring up any log you want instantly with a scientific calculator. If you have such a device, however, you may wish to skip the whole logarithm bit and manage roots the quick and easy way—with a stroke of the key labeled x to the y power.

And finally there's the method that doesn't send you running for either a table of logs or a device that costs more than a couple of dollars—iteration with a simple calculator. That is, you guess at the answer, test your guess with the calculator, then narrow it down with further guesses till it's correct to whatever number of decimal place you fancy.

There's more about this in the calculator chapter, especially in the section on cube roots and more.

The Mortgage that Pays YOU

O NE GREAT idea for some people, for which the time seems—at least as I write this—to have come at last is the RAM, or reverse annuity mortgage.

It's true that through the late 1970s and the 1980s, laws to permit, encourage, and control these novel financial instruments were passed in many of the United States. But only in the nineties have many mortgage lenders offered them, except in the form of pallid imitations.

In case an acceptable RAM is to be had in your neighborhood by the time you read this, you might like to calculate how one would work for you in time of need.

That time most often comes after retirement. And, as is true with so many other calculations and decisions taken in this period, life expectancy should be an important part of the scenario.

Suppose that you, like someone I know, live in a house that is worth $100,000 and on which the mortgage has been paid down to $30,000. You could sell your house and move into something cheaper, but you love the place. At the same time, you resent sitting on this $70,000 equity while scraping by.

What could an RAM do for you? It could permit you to use up to perhaps 80 percent of your equity by taking it in the form of monthly income for a stated number of years. I would regard this as prudent only when that number of years is considered in light of the life expectancy of the owner or owners.

How much money? To find out you'll have to decide how many years you want the RAM to run and either the amount you're willing to commit (up to the possibly 80 percent of equity in the house that a lender might permit) or the amount of monthly income to go for.

Assume you opt for the maximum amount and for a ten-year period. This is how to calculate your monthly income from it.

1. Express as a decimal the annual interest rate available (usually this will be close to that prevailing for ordinary home mortgages) and divide by the number of payments—normally 12— you wish to receive per year. Add 1. Raise this total to the power equal to the total number of payments you are to receive. (If monthly, this will be 12 times the number of years you've chosen.) Now subtract 1. Multiply by the number of payments you are to receive each year.

2. Multiply by the amount of equity available to take from your house—probably 80 percent of the difference between what the house is worth and anything owed on it—by the annual interest rate expressed as a decimal. Dividing this by the result of the preceding paragraph gives the amount of the monthly income you could receive.

Let's use as an example the house mentioned above.

With interest rates hovering around 12.75 percent (as they have done and, heaven forfend, may do again) and income to be taken monthly for 10 years, the computation looks like this. Dividing

0.1275 by 12 and adding 1 gives 1.010625. Raised to the 120th power this is 3.5547. Subtracting 1 and multiplying by 12 gives 30.6564.

When the product of 0.1275 and 80 percent of $70,000 is divided by that 30.6564, you find the monthly income: $232.90.

Similarly, you can make the calculation for some previously chosen amount of income.

If you wish to supplement your present income by $200 a month for 10 years, you make the first part of the calculation just as in Paragraph 1 above. Then divide by the interest rate in decimal form. Multiply by the monthly amount—the $200 you've chosen—to find how much equity you must have available to do this—and how much you'll owe after 10 years.

Lazy Person's Withdrawal Chart

IF YOU deposit a chunk of money and draw out so much each year, how long can you keep it up?

The lazy (or shall we say quick?) way to tell is with a chart like the one below. Reading it, you'll find that if you deposit $10,000 in a 5 percent savings account, you can withdraw $800 a year for 20 years.

Annual Withdrawal	Number of years assets will last at Annual Growth Rate of							
	5%	6%	7%	8%	9%	10%	11%	12%
5%								
6	36							
7	25	33						
8	20	23	30					
9	16	18	22	28				
10	14	15	17	20	26			
11	12	13	14	16	19	25		
12	11	11	12	13	15	18	23	
15	8	8	9	9	10	11	12	14

The $800 a year you are asking to take out is 8 percent of your investment, so you look down the left-hand column till you find 8. Then you read the figure opposite 8 under the heading 5 percent, since this is the return your deposit is collecting. It's 20, so that is how many years you can keep it up. You'll note that if the account paid 6 percent you could continue for 23 years; if 7 percent, 30 years.

You can use this table in reverse. Knowing the rate of return that is being offered, choose a period of years from the column under it. Then read in the left-hand column the annual withdrawal percentage this combination will permit. Multiply this by the amount of your investment to find the number of dollars you can take out each year.

(Say you can get 9 percent interest and want to spread the income over 15 years. Finding 15 in the 9 percent column, you read 12 in the column to the far left. This tells you that you can withdraw 12 percent of your original investment each year.)

What about the blank areas in the table? Any periods of years greater than the ones shown would, if combined with the withdrawal rates opposite them, let you withdraw money forever. Your capital would be growing faster than you were taking it out.

Inflation could knock your picture in the head. What if the amount of income that looks just right today proves hopelessly inadequate in an economy where everything costs five or ten times what it does today?

You can protect yourself against this self-deception by how you choose your interest-rate column in the table.

Instead of using the one that represents your expected rate of return, choose a more realistic one that is equal to the stated rate of return less the predicted annual rate of inflation (and less your income-tax rate as well).

If you invest to receive 12 percent, and can expect income tax to take one-fourth of that, you'll have only a 9 percent net return to work with. If you then assume 5 percent inflation over the years ahead and subtract this 5 from the 9, you'll see that the true return may be nearer 4. The 4 percent column may be your wisest choice for long-range planning.

Money Now or Money Later?

IT CAN BE useful to know what a sum of money to be received in the future is worth today. This is a rather quick process for figuring that out. Since it has a variety of applications in almost everyone's financial life, you'll meet it several times in this book.

One way you can use it is to discover how much you would have to deposit in savings today to have some specified amount at a given time in the future.

Another way is to apply the procedure repeatedly to different numbers of years and rates of interest, thus generating a table for, say, $100 that simple multiplication will convert for other amounts.

Start the process by stating the interest rate as a decimal (15 percent is 0.15) and dividing by the number of compounding periods per year. Add 1. Then raise to the power of the total number of compounding periods. (That's the number per year multiplied by the number of years.) Now divide the number of dollars you're concerned with by this result.

As an example, suppose the interest rate offered is 15 percent, to be compounded quarterly, and you want your account to reach $2000 in five years.

Divide 0.15 by 4 (since quarterly compounding means four periods per year), add 1, raise to the 20th power (four quarters times five years) and divide $2000 by this result of 2.088. You must deposit about $958 (more precisely, $957.78).

By the way, with some calculators you'll find it easier to use the negative of the power (-20th power in this case). Then you can multiply by the wanted dollar amount, instead of going through a division as the final step.

You'll find your handy-dandy little calculator is more friendly when approached in this way.

Prepay That Mortgage?

WHEN INTEREST rates on mortgages soar into the 10 percent range, bankers develop a feeling of nausea when contemplating loans they hold that were written in the 6 percent days. Often they offer homeowners discounts of 5 percent to 15 percent to pay them off at once.

When this happens the bank can then use the money to earn 10 percent on new mortgages or even more than that for consumer-credit deals that bring in 14 to 18 percent.

If you are paying on one of these older mortgages, how can you figure out whether you'd gain by accepting the discount offer? This assumes, of course, that you have the cash or other resources to make the payoff. Your question is, will you gain more by paying the bank or by using the money for some other safe investment. Suppose your mortgage is now paid down to $17,000 and has six years to run at 5-3/4 percent. Your monthly payments are $279.75 for principal and interest.

At this point, your friendly banker offers you a 10 percent discount if you'll take this low-interest investment off his hands by paying it now in full.

It happens that you have the cash available to accept the offer. Should you?

To figure that out, you'll have to know first at what rate of interest you can invest, or already have invested, the $15,300 ($17,000 less 10 percent, most rapidly calculated by multiplying $17,000 by 0.9) if you don't use it to pay off the mortgage.

Using that rate, you treat the payoff amount as if it were a mortgage debt, calculating the monthly payments that it would require over the given period of years.

You can do this by either of the two methods given in this book for payment calculations. Or you can look it up in a table. Let's assume that your cash is now drawing 8 percent a year.

By any method, you'll find that the monthly payoff for $15,300 at that interest rate is $268.28.

So that's how much you could draw each month for the next six years, to use to make the mortgage payments. Since it is not as much as the payments, you'll be richer by $11.47 each month if you accept the offer.

It's only if you can depend on getting more than 9-1/2 percent for your money that you gain by keeping the mortgage alive. At 10 percent, you'll be ahead by $3.69 a month.

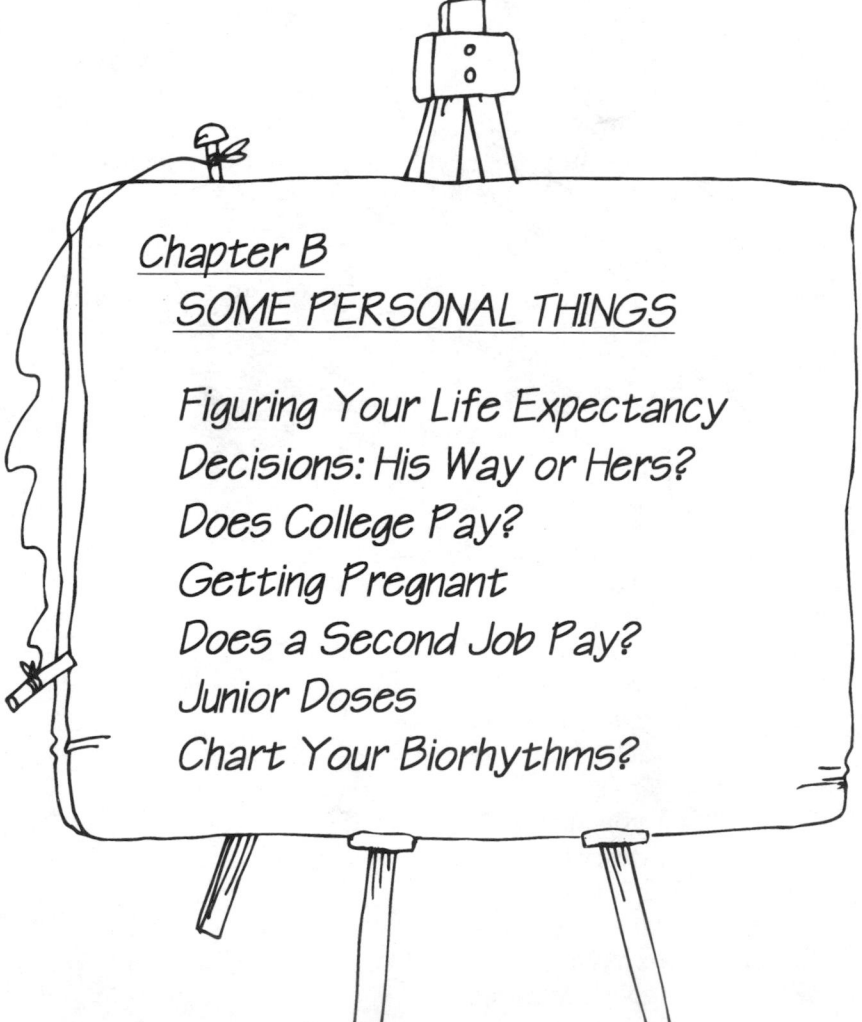

Chapter B
SOME PERSONAL THINGS

Figuring Your Life Expectancy
Decisions: His Way or Hers?
Does College Pay?
Getting Pregnant
Does a Second Job Pay?
Junior Doses
Chart Your Biorhythms?

Figuring Your Life Expectancy

Pᴇᴏᴘʟᴇ at a life insurance company have rated an assortment of factors and arrived at this formula for calculating your life expectancy. They thought you might like to give it a whirl.

But maybe not. Digging into a story on a similar subject in my days as a very young editor (anybody here remember *LOOK* magazine?) I discovered a fact that I tactlessly imparted to my office-mate, a gloomy Britisher. At his age, the years he had coming exactly equaled the number already used up. He didn't speak to me the rest of the day.

Expectancies are so much rosier today that you may find reality less upsetting than my Brit did. So try this.

Start with 72 and subtract 3 if you're male, add 4 if you're female.

(Now you know why a man should marry a woman 7 years older than he is if they aim to finish in a—forgive the expression—dead heat.)

If you live in a city with 2 million or more population, subtract 2. But if you live in the country or in a town of ten thousand or less add 2.

If your work is at a desk, subtract 3; but add 3 if it includes regular hard physical labor.

If you get in at least half an hour of vigorous exercise five times a week or oftener, add 4; for only half that much exercise, add only 2.

Add 5 for living with another person; take off 1 for each 10 years you've lived alone since you were 25.

(It is only sensible to refine some of these items by interpolating. You might reasonably add 1 if you live in a city of 25,000, say, and subtract 1 for residence in a city of half a million. You might also consider how much of your lifetime you expect to have spent in these environments. Use similar logic on many of the items.)

Subtract 4 if you sleep more than 10 hours a night.

Would you describe yourself as an intense, aggressive, angry person? Take off 3. If you're easygoing and relaxed, add 3.

If you consider yourself happy, add 1; unhappy, take away 2.

Subtract 1 for having had a speeding ticket within the last year—because this suggests you might be somewhat prone to accidents, recently the fourth-ranking cause of death.

Subtract 2 if your income exceeds $50,000. (I'll bet that's a surprise and gives you something to think about.)

Add 1 for having finished college, 2 more for holding a professional or graduate degree.

If you are over 65 and still working, add 3.

Add 2 if any grandparent lived past 84. Add 6 if all four of your grandparents reached 80. Subtract 4 if either of your parents died of heart attack or stroke before 50. Subtract 3 if any parent or brother or sister has had cancer or heart trouble before the age of 50 or has had diabetes since childhood.

For smoking more than two packs a day, subtract 8; one to two packs, 6; one half to one, subtract 3.

Take off 1 if you drink the equivalent of a quarter-bottle of liquor a day.

Take off 1 for each 10 pounds you're overweight.

Add 2 if you have annual medical checkups (if male), annual visits to gynecologist (if you're female).

If you're now 30 to 40, add 2; 40 to 50, add 3; 50 to 70, add 4; and add 5 if you're more than 70.

If you come out with a total of 70 to 71, you figure to outlive the average man. To beat out the average woman, you'll have to come up with around 78 or 79.

Decisions: His Way or Hers?

RATING diverse things "on a scale of zero to ten" has become something between a cliché and a parlor game. But psychologists and counselors advocating no-fault marriage have turned this trick to useful ends.

Given a husband who is looking forward to watching football on TV and a wife who thinks a Saturday afternoon picnic and excursion would be a better way to use some sunny hours, how's the matter to

THE COMPLETE HOW TO FIGURE IT

DARRELL HUFF

ILLUSTRATED BY CAROLYN R. KINSEY · DESIGNED BY KRISTY MARIA HUFF

W. W. NORTON & COMPANY
New York   London

Copyright © 1996 by Darrell & Frances Huff, Inc.

First Edition

For information about permission to reproduce selections of this book, write to
Permissions, W. W. Norton & Company, Inc., 500 Fifth Avenue, New York, NY 10110.

Library of Congress Cataloging-in-Publication Data

Huff, Darrell.
The complete how to figure it / Darrell Huff with Kristy Maria Huff :
illustrated by Carolyn R. Kinsey.
p.   cm.
Includes index.
ISBN 0-393-03600-6
1. Mathematics—Popular works.   2. Finance, Personal.
I. Huff, Kristy Maria.   II. Title.
QA93.H785   1996
640'.151—dc20   95–46480
                                                          CIP

W. W. Norton & Company, Inc., 500 Fifth Avenue, New York, NY 10110
http://web.wwnorton.com
W. W. Norton & Company Ltd., 10 Coptic Street, London WC1A 1PU

1 2 3 4 5 6 7 8 9 0

Just a word...

Do you believe with Socrates (who seems to have said it first) and me that the unexamined life is not worth living?

I think you could say the same of the unquantified life. HOW MUCH can be as vital as WHAT.

That, I guess, was the point of a TV show, "Math... Who Needs It?!." With people like Bill Cosby and celebrated teacher Jaime Escolante performing, you knew the answer was going to be "just about everybody."

There lies my justification for offering what may look suspiciously like Everything You Always Wanted to Know About Math But Had the Native Good Sense Not to Ask.

Let me say that I don't think what you hold in your hands is a math book, anyway. I'd rather call it an enabler, and I hope you came upon this copy on the self-help shelves rather than in an obscure corner marked math.

Any mathematical manipulations you learn, or relearn, as you go can be a substantial bonus. But with that miraculous instant-math machine the pocket calculator, a great little enabler itself, you can make use of just about anything in this volume with only cursory recollection of the stuff Miss Fiditch tried to pound into your head in high school.

She didn't have much luck pounding me either. But when I reached a time of hoarding or borrowing or exchanging dollars and choosing timbers and figuring floor loads, the math side of things suddenly came alive for me.

Since so much of what's in this book comes out of my daily life as a family person, it may be relevant that so does the production of the tome itself.

Over the years of collection and creation, my wife Frances made fair copy from foul (when she might have preferred to turn her fingers to her own writing) and thus she performed a very special service in those pre-computer days of typing and retyping. Manually typed copy soon gave way to what emerged from an imposing Selectric—until it in its turn surrendered to the marvels of a 26-pound portable computer called Kaypro. In the end, everything made its way through computer programs variously called Perfect

Writer, Word Star, Word Perfect, and finally to just plain Word for Macintosh. The tool used was Pagemaker. The typewritten part was scanned electronically from paper to computer, acquiring weird errors along the way. These had to be detected and quashed by our granddaughter Lisa Schulman.

Her mother, Kathryn Allan, who is often sharper than I am in matters of math, sedulously read not only proof but raw copy, saving both you and me from many a sad ambiguity or worse. My old college room-mate Jake Coniglio and his wife Faith pitched in here, too.

Kristy Maria Huff, though immersed in computer-oriented gradu-ate study in architecture at the U of Oregon, somehow wrestled a couple of hundred drawings into place within twice that many sub-chapters to design and create a coherent book out of a mishmash—complete with formatting, subheads, legible type, understandable charts, well-placed illustrations, and other good stuff. The body type and heads are Palatino. Chapter summaries are Architect, knocked out (says the designer Hank Gillette) to simulate lettering on a blueprint.

A friend, David Shellabarger, and our grandsons Will Allan (who did many of the charts and tables) and Jesse Huff-Christensen and grandaughter Maia Huff-Owen provided computer savvy and flying keyboard fingers, while keeping some home fires burning with awesome patience.

The major contribution of another daughter, Carolyn Kinsey, speaks for itself all through the book—more than 200 light-hearted drawings.

These, of course, are also the folks responsible for any and all deficiencies you may find on these pages, so it's quite acceptable to bring errors to my attention.

After saying a good word for every math teacher everywhere, I need only add the name of an inordinately patient editor, Hilary Hinzmann, to complete the list of those to whom this volume is hereby dedicated.

I hope you'll put to use the blank pages you find here and there. Recycle them with supplementary notes or computations of your own so the book will know it is being used.

And don't feel mercenary if you make frequent reference to the financial calculations. We have Woody Allen's word that "Money is better than poverty, if only for financial reasons."

be resolved? Why should it go his way rather than hers—or hers rather than his?

Of course, if he doesn't really care what she wants or she doesn't care whether he enjoys himself, then there is no solution. There's no marriage either, when it comes right down to it.

So let's assume both these people want the resolution that will bring the most mutual satisfaction.

How does he rate staying home on this particular afternoon? Undoubtedly there are some things he'd rather do. Play golf, go hunting, visit the French Riviera, possibly; but, for one reason or another, these events that he might rate 9 or 10 are not feasible on this particular afternoon. Though he's not an avid football fan, this promises to be a spectacular game; and he's tired enough after the week's work to welcome sitting. He rates it 7.

As for the picnic, well it certainly beats going to the dentist (zero), visiting his in-laws (1), or even painting the back porch (2). But it doesn't beat them by much, since it involves week-end traffic. He rates it 3.

Looking more closely at the picnic-excursion, the wife finds she can rate it no higher than 7. Although getting away from the house is high on her list at this point, and so is a bit of togetherness after the working week, she doesn't like the highway part either.

And staying home isn't the worst thing she can think of. There are a few chores it would be good to get off her mind. There's a book she's been trying to find time to read. And the football game won't last all afternoon. A semi-idle day around the house falls into a neutral 5.

That 5 added to his stay-home 7 adds up to a combined-satisfaction score of 12 for staying home. Her picnic rating of 7 added to his of 3 comes to only 10. The verdict is stay home today, by a margin of 2...and with a minimum of hard feelings.

This simple technique has some fringe benefits.

In the process of rating the intensity of their affection for their choices, both parties may find that only some of the aspects are significant. Perhaps our man has no objection to a picnic; he just doesn't want to drive on the highway or miss the football. His wife doesn't object to the football-watching or insist on a long excursion. The couple may find that a brief picnic at a nearby park with return home by game time will please them both at levels of 6 or 7 for a score of 13 or so, a clear mutual gain.

Even if that doesn't happen, another self-leveling ingredient can be expected. Having won this decision without rancor or martyrdom to mar the day, our man may be expected to approach the next such decision in the best possible spirit. He is likely to up his own pleasure rating for his wife's preference because he is indeed pleased at the prospect of her pleasure in it. And so the act of losing today enhances the loser's probability of winning tomorrow.

My editor expresses doubt that an arguing couple would actually pause to quantify like this. At your next opportunity, prove him wrong.

Does College Pay?

A NEWS magazine once cited the compiler of this compendium in a favorable sort of way. The following appeared in the "Essay" feature in *Time* :

"In an irreverent study of the numbers game called How to Lie With Statistics, Author Darrell Huff coined the word 'statisticulation'—the art of lying with statistics while seeming objective."

So no doubt it is ungracious to mention that it was in *Time*'s sister publication, *Money*, that I found a fine double-barreled example of statisticulation. Quoted therein was a student who had used World Almanac figures to "prove" that college doesn't pay, even though those figures show a graduate's median earnings as $2,104 a year higher. The estimated $22,000 the ungrad saves by skipping college will compound (at 7 percent) to a sum far greater than the difference in lifetime earnings.

(If these numbers seem low, note that they are for tuition and fees alone and in an unspecified kind of college—and of course our example would still have to pay some kind of bed and board, in school or not.)

The figures are impressive. While the nongrad is sacrificing $42,072 in current earnings (he works four years longer but makes $2,104 less each year) he has that $22,000 compounding away.

Result: He has nearly half a million when he retires. Added to the similar sum he's earned over the years, he might claim lifetime income of more than a million. The college graduate, however, has only a little over half a million to point to.

Convincing this may seem to be, but it is mathematical non-sense—adding oranges to apples. Or worse, totaling watermelons and peanuts.

For a more illuminating comparison let's say grad and nongrad live at the same level until retirement. To keep up, the nongrad will have to draw on his tidily invested $22,000 at the rate of $2,104 a year. My pocket calculator tells me he can do this for just under 20 years before the money runs out.

How can you use a simple calculator to estimate the length of time a chunk of cash, invested and compounded, will go on tossing out a given amount each year? Here's what the basic formula tells you to do.

Add 1 to the interest rate expressed as a decimal. (For 7 percent interest, that's 1 plus 0.07 equals 1.07.)

Raise this to the power of the number of years—20th power for 20 years, and so on.

Divide 1 by this result.

Subtract this from 1.

Divide this by the rate of interest as a decimal.

Multiply by number of dollars to be taken out each year.

But there's one catch. To evolve a procedure that will work with an ordinary calculator, we've required you to know in advance how many years to use in the second step, although that's precisely what you're trying to find out.

What to do? Apply a technique that will often serve you well when a direct method is out of your reach: guess and test.

Using our figures, you'd quickly narrow it down to a guess of 20 years when you found that this period requires an initial investment of $22,290, close enough to the specified $22,000.

It seems, then, that it pays to go to college?

Now that we've acquired a handy formula and used a math trick of wide applicability, let's fire the other barrel. This is the one that exposes the statisticulation behind this and many another correlation.

Even if the figures make it seem that college people do make more money, we have turned up no evidence that it's *because* they went to college.

Bright kids and rich kids are the kinds most likely to go to college. Right? Surely it is also likely that—with their brains and their connections—they are the ones likely to make the most money, with higher education or without.

Getting Pregnant

Knowing the probability of a couple's achieving pregnancy in a month and there's a known average for that—one chance in five—is there a way to tell how likely they are to get a baby started within, let's say, a year?

Clearly you can't just multiply the 30-day number by 12. That could produce an absurd result of more than 100 percent.

This kind of problem is best figured backwards. If there's a 20 percent chance of becoming pregnant in one month—the figure found by research in both Belgium and the U.S. for perfectly fertile couples in their twenties— that's an 80 percent probability of failure.

Express this 80 percent as a decimal: 0.80.

Raise this amount to the power of the number of months—12th power for a year.

(Multiplying 0.80 by itself 11 times isn't too tedious with paper and pencil if you round it off as you go. Of course, with a simple calculator it means merely entering 0.8 and pressing the times key once and the equals key 11 times.)

Now that you've found the probability of *not* becoming pregnant in 12 months (0.07), subtract it from 1 to find that the probability of pregnancy in the course of a year is 0.93, or 93 percent.

For a couple in their late thirties, it has been found that there is about an 8 percent chance per month. So it's 0.92 that you'd raise to the 12th power and subtract from 1, to find a 63 percent chance of pregnancy within a year for normally fertile couples pushing forty.

Five-year figures come in the same way, using the 60th power instead of the 12th. You may find the results a shock, but you'll see why the world is heavily populated.

Does a Second Job Pay?

As I RECALL a novel of the 1930s depression called If I Have Four Apples, it was about the vain efforts of a social worker to convince some employed wives that they were going in the hole every hour they worked.

The whole process of persuasion fell apart on the grounds of simple arithmetic of the kind that begins, "If I have four apples and take away..."

Now, half a century later, with sexism quite rightly suspected around every corner, let me begin with a disclaimer. I'm not about to apply arithmetic to wives' jobs only. Having determined, on whatever grounds, which of the couple shall hold a job, the tests below will help answer the question of whether the partner should do likewise or would be better employed in keeping the home fires blazing.

You can apply the routine as well to monthly or annual figures, but for convenience let's assume weekly pay.

Begin with the additional income the job will produce. Actual salary, not take-home.

Add the value of the fringe benefits, as nearly as you can calculate them. If the first job in the family does not provide adequate medical insurance, what comes with the second job may be worth quite a bit. Value it according to what it would cost you to buy the equivalent independently.

Face a grim fact. A man or woman who is already regularly employed will probably be in a better career position to support self and any children if the other dies or becomes disabled or if the relationship ends.

If a second job permits the prime earner to carry less insurance while the degree of security is kept constant, the reduction of premiums is an immediate saving. But consider only premiums needed for protection in the form of term insurance, remembering that straight life insurance includes an accumulation of savings as well as current protection. Those are the main money gains. Now consider the losses.

Very often, the biggest of these will be increased taxes, not only federal income tax but also social security and very often state and local income and unemployment and workman's-compensation taxes. The income-tax rates on this second income are going to start where the percentage levied on the previous total family income left off...and may jump rapidly from there. Social Security tax is a significant factor when the incomes in question are not very large. It can be bigger than income tax.

For many people, the biggest additional expense created by a new job is transportation. Bus or train fares are usually easy to estimate with precision. For added miles to be driven in a car owned anyway, a reasonable estimate might be around 20 cents a mile; but if an extra car must be owned because of the new job, twice this much may be reasonable. Or an item-by-item addition may produce a sharper estimate.

Lunches, unless included free with the job, can be an important new expense.

This applies also to additional clothing and its maintenance. Taking on, or adding to, outside employment often means paying handymen and plumbers and window-washers and household workers to do what formerly were do-it-yourself chores. Some men or women or couples do so many cost-saving chores around the house that they almost amount to a second job and pay accordingly. Value added to a house by substantial improvement or remodeling—landscaping, a new room, a garage—can be equivalent to earning thousands of tax-free dollars a year. If taking on a new outside job is at the expense of such activities, their value must be included in the reckoning.

Child care, in or out of the home, can be a major new or inflated expense. Deduct from it, however, any tax savings generated.

Taking on more employment may, in the absence of new household help, lead to buying more convenience foods—quick-cooking but costly. There will be less time to shop for household bargains. More frequent dining out is likely. Costs of these effects of taking a job should be included in estimating the new budget. Although it is not necessarily a clear fiscal plus or minus, here's another factor to consider. There's more stress when either a husband or a wife begins to work longer hours to bring in more money, and this may lead to a conviction that a reward is deserved. This can take the form of more costly vacations or luxury clothing in no way related to the job.

These rewards may be precisely what you're taking the job for, so it may not seem to make sense to deduct their cost in estimating your gain.

But however pleasurable these rewards, they are going to reduce the amount you can stick into a well-chosen mutual fund and gloat over when the year is done. For that purpose, be firm and subtract their estimated cost.

How does this work out in an actual situation?

Let's try it. A printing company has some $14,000 jobs open. Your cousin Lulu, who lives near the plant, is considering taking one of them. So is the wife of an industrious young office manager in a nearby town. How well will it pay each of them to go back to work?

Lulu's husband Fred, 65 but retired more by reasons of health than age, draws nearly $8,000 a year in tax-free Social Security benefits. Investments bring in another $4,000. Since their income is too small to be taxed and they own their car and small house clear, Lulu and Fred hope they can live on this, though it doesn't look to be easy. More income would be welcome, at least until Lulu reaches an age where she can collect a spouse's Social Security benefit.

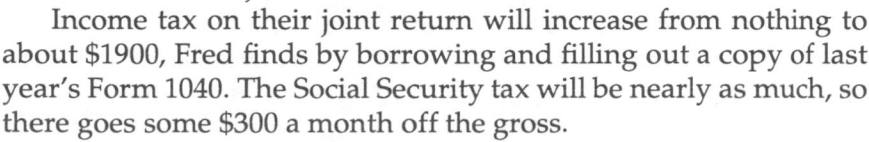

The only useful fringe benefit that comes with the job, Lulu notes, is medical insurance roughly equal to the Blue Cross policy she has been paying $170 a quarter for. Gross value of the job, then, is $1223 a month.

Income tax on their joint return will increase from nothing to about $1900, Fred finds by borrowing and filling out a copy of last year's Form 1040. The Social Security tax will be nearly as much, so there goes some $300 a month off the gross.

Bus fares will run about $20 a month. Since Lulu plans to brown-bag much the same lunch she'd eat at home, there's no added cost there, but it seems prudent to allow $60 a month for additional clothing. A similar amount, she thinks, will take care of any small added food costs produced by more hurried shopping or preparation. There is no question of household help or child care. Fred can easily take up the slack so far as chores around the house are concerned.

A little arithmetic now establishes that the job will pay Lulu about $800 month net, adding two thirds to the family net income. It's well worth considering.

Now what about that young Mrs. Popple, down the street? Since her husband's job already provides comparable medical coverage, there's no fringe benefit there for her. With one good salary already in the family, the $14,000 will be almost fully taxable and at a level averaging 30 percent. Adding the same Social Security levy that Lulu has calculated she would pay, there goes some $5000 off the top.

Fourteen miles each way daily in the new car that Mrs. Popple has decided she will need can be estimated to cost $200 a month—though it will actually come to more than that if the small station wagon she has her eye on doesn't help pay its way with some other uses.

Mrs. Popple next notes what a hole a conservative monthly $40 for lunches (if she holds to her intended starvation diet), $90 for additional clothing, and $100 for hurried shopping for convenience foodstuffs will make in the family budget.

Because of its income-tax deductibility, a monthly $400 outlay for child care will actually put the Popples only $240 more out of pocket if there are no drastic changes in the law.

But a modest amount of a cleaning woman's time (merely four hours twice a week or eight hours once a week at an optimistically estimated $6 an hour and transportation) comes to another $200 a month.

With all these costs added together, Mrs. Popple runs into a bafflement. When she attempts to deduct them from the salary offered, she finds they amount to more than the salary.

Even if she can find a car pool instead of buying a car, and even if she adds to her own working day by eliminating the cleaning woman, Mrs. Popple is going to find herself working for a couple of dollars an hour. That's not much, considering the various stresses and often costly emergencies that pop up when both parents of young children work full time.

There may, of course, be good and even overriding reasons for Mrs. Popple to take a $14,000 job—interesting work, future, she's going nuts taking care of the house and kids all day. But, financially speaking, she'll have to regard it as a hobby.

Junior Doses

I<small>T'S</small> <small>NOT</small> always a safe practice to calculate a medical dosage for a child by using a formula to modify the adult dose. However, with drugs that are in common use and not considered especially dangerous, the methods given here have been used by doctors, in the absence of other information.

The simplest of them is called Clark's Rule. To use it, divide the weight of the child in pounds by 150, then multiply by the adult dose. A 50-pound child will get one-third the dose that would be prescribed for a 150-pound adult.

Young's Rule is a trifle more sophisticated. It instructs you to divide the child's age by his age plus 12 and multiply this by the adult dose. A child of eight would get a dose that is 40 percent of the one recommended for an adult.

Under what's known as Cowling's Rule you would divide the age of the child at its next birthday by 24 and multiply this by the adult dose. With this method, an eight-year-old child (going on nine) would receive 38 percent of the adult allowance. This works out to just about the same as under Young's Rule, a helpful confirmation.

An ultra-cautious approach would seem to be to try each rule for size and then use the one that provides the smallest dose.

Chart Your Biorhythms?

T<small>HERE</small> are many remarkable rhythms in nature and human life. Many are quite well established as facts, even though much of their nature and origin remains mysterious. The most widely discussed and applied of these are called biorhythms.

Whole books about them appear from time to time. A pocket

calculator designed to chart them is on the market, at least in Europe. Computer programs to the same end are widely available.

But not at all necessary. You can quickly figure these rhythms with that handy computing tool, the pencil.

In this theory, each human life is influenced by three cycles that begin at birth. Differing in length by several days, cycles peak at their own times, putting each of us at a different phase of a characteristic on any given day.

These are rhythms of extreme precision. Thus they offer a tool of which the least that might reasonably be expected is a striking reduction in human strife, inefficiency, and accident rates of all kinds. The biorhythm theory is wonderful if true, and interesting to play with in any case.

This theory has emerged from the work of a Dr. Wilhelm Fliess, who though not universally admired today by Freudian scholars was a friend of Sigmund Freud's and practiced in Berlin at the beginning of the century. From a twenty-year study of 80,000 case histories, he found what he believed to be confirmation of something suggested by Hippocrates in the Fifth Century B.C., persistent rhythms in human life, beginning at birth.

One of these rhythms is best thought of as PHYSICAL. It is an ebb and flow of strength and endurance. Its length is 23 days.

A second is EMOTIONAL. Connected with moodiness, creative ability, intuition, sensitivity, ease of expression, and ability to influence others, it runs 28 days from peak to peak.

A third cycle, of 33 days, is INTELLECTUAL. It governs intelligence, concentration, mental alertness, and memory.

Waxing and waning of each of these kinds of power produces critical days, especially in connection with the first two of the cycles. On some days, the proponents say, a person may be seven to eight times more accident-prone than on others.

If you're an airline pilot, for example, information by which you could be grounded on your critical days could obviously be of enormous value; and a similar argument might be made in the case of a surgeon—and patient. Or of a judge or an athlete or a business executive or anyone facing a critical decision.

What makes calculating these cycles most important—assuming the theory is valid—is that each cycle is held to begin on the day of birth and continue in precise length for the life of the person. That has to mean that the exact position of anyone in respect to each cycle may be found at any time in life. It is necessary only to add multiples of any cycle length to the day of birth.

The biorhythm theory has gained enough respectable backing to get it serious consideration in some major enterprises. An enviable safety record for the high-speed transportation system of the Swiss city of Zurich is claimed for application of the Fliess theory to the scheduling of personnel.

In a preface to one of the earlier books on biorhythms, a Dr. F. Wehrli reports with considerable excitement, but few very useful facts, on experiences with the Fliess theory at Kurhaus Villa Montana at Locarno, Switzerland. He says that for fifteen years, it has been the practice at this clinic to "take into consideration our patients' biorhythmic situations." Dr. Wehrli concludes that the "facts and figures" of this experience "offer proof positive of the important relationship of man's resistance to the ebb and flow of his biorhythms, and that in the 10,000 and more cases of our experience where biorhythms were employed, they generally proved to be a valuable aid."

Perhaps, on the basis of all this and anything you may have read on the subject before, you're a believer. Perhaps you're merely curious and would like to observe how the idea proves out in your own life.

Anyway....

To calculate your own biorhythms, you must first determine where you stand today in relation to each of the three cycles.

Find precisely how old you are by reckoning the number of days from your birth to the end of your first year, adding 365 times the number of complete years since, then adding the number of intervening leap years, and finally adding on the number of days that have gone by so far this year.

(Remember that leap day comes in every year evenly divisible by 4 but not by 400.)

Divide this total number of days by 23. Discard the whole-number answer, keeping only the remainder.

Suppose you find that today (April 15, let's say) you are 22,815 days old. Divided by 23, that comes to 991 cycles, plus a remainder of 22 days. Forget the 991.

Since 22 days of your 23-day (physical) cycle have passed since the preceding cycle ended, you can find the beginning of your cycle by subtracting 22 from today's date—or by adding 1 to it. Having marked those two dates on a chart, draw a smooth curve that begins to rise at the initial date and then falls, crossing the baseline at the midpoint and rising to it again at the end. There a new 23-day cycle begins. And so on for life.

For your emotional (some call it sensitivity) cycle, go through the same process, this time dividing by 28.

For your intellectual cycle, divide by 33.

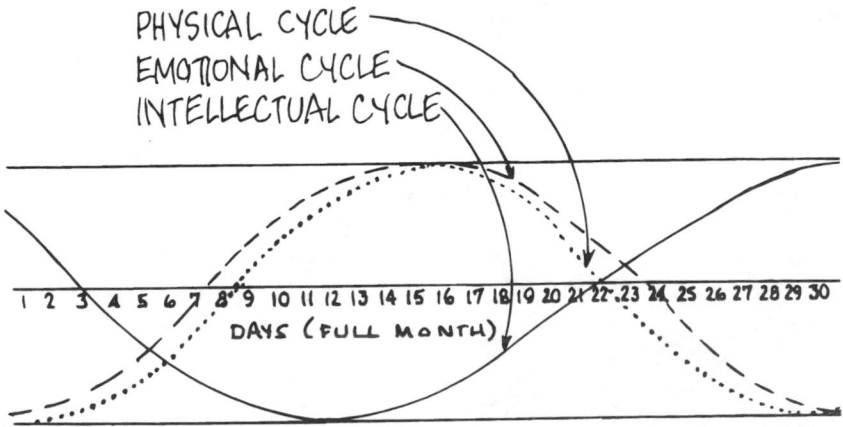

The remainders are 23 and 12 respectively for the person whose physical cycle was just charted.

Some of the writers on biorhythms suggest that you can combine the three cycles for any day by adding the magnitudes of the curves above the baseline and subtracting the amount below. You could even do this just by eye. However, you'll have to decide for yourself just what three unlike characteristics could be said to add up to. A sort of diffuse performance capability, perhaps.

Just to show how it looks, I've charted it. The chart shows the rhythms for our example, who did his calculations on that annual day of reckoning, January 1.

In figuring your cycles you'll very likely want to use your pocket calculator, if you have one around. The only difficulty is that these devices are too fussy. They insist on operating to as many decimals as they have room for, instead of quitting with a quotient plus remainder.

So you must turn the decimal back into a remainder. For the 22,815-day-old fellow just described, the routine goes like this for the physical cycle:

Enter 22,815. Divide by 23. Subtract the whole-number quotient 991. Multiply by 23. Answer is 22. That's the number you're after.

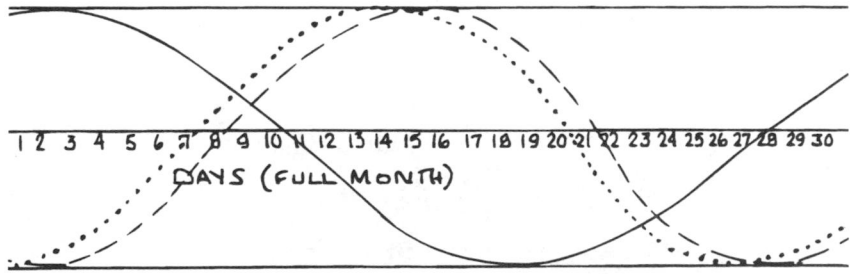

Chapter C
  INTEREST AND SAVING

Yield and the Rule of 72

T HE GOOD thing about the old banking Rule of 72 is its simplicity. It lets you calculate instantly how many years your money will require in order to double itself at any specified rate of interest. And it lets you do this notable deed without recourse to a chart or an electronic device or even a pencil—by merely dividing 72 by the promised interest rate.

Not so good is the fact that it loses some of its usefulness when applied to the abnormal interest rates that turn up from time to time. And as daily or hourly or even continuous compounding has become more usual, this quick trick has become less accurate for estimating how many years it takes for money to double.

But it's still useful, especially if you apply fine tuning. What follows are some rather precise approximating tricks you can apply with a minimum of math.

(Given a scientific calculator, or a computer, you can be even more precise about it, as the next part of this chapter will show.)

For starters, use annual yield if you know how much it is, instead of the nominal interest rate. Savings institutions often give both figures in their advertising. If you must use the interest-rate figure, you may find you're getting closer approximations when you use 70 instead of 72 in applying the rule.

Another thing: The rule we've been talking about is helpful only for questions of doubling. What if you want to know the interest rate or time period required to triple your money or to increase it tenfold? Use these new "Rules of..."

For increasing your money by half, Rule of 40. For fivefold increase, Rule of 160. For sevenfold, Rule of 195.

Use these numbers in the same way as the original Rule. That is, to find how long it will take for a $100 investment to grow sevenfold to $700, with daily compounding, use the Rule of 195. That is, divide

195 by the interest rate and discover that for 15 percent (you should be so lucky—or, more likely, you're the borrower), it's about 13 years.

Care to evolve a few Rules of your own? Just factor the number of years you're after and add together the corresponding Rule numbers given above. (For 6 years, note that 6 is the product of 2, 2, and 1-1/2, so add 70, 70, and 40 to derive a brand new Rule of 180 for a sixfold increase.

Fooling Around with Old 72

IT'S QUICK—the rule that says dividing 72 by the rate of interest will tell you in how many years of compounding you can double your money. But for a precise answer, you'll have to drop the rule of thumb and go to a formula.

Same goes if you're after precision on some increase other than doubling—tripling your money, say. The penalty is that this routine calls for a table of logs or a scientific calculator.

Anyway, here's how to find out how long it will take to multiply your money by any factor you care to name.

Step 1. Find the log of the number by which you want to see your investment multiplied. That is, find log 3 if tripling is the aim, log 4 if quadrupling. Either common or natural log will serve.

Step 2. Express the interest rate as a decimal (11 percent becomes 0.11) and divide by the number of times your invested money is to be compounded in a year. (For daily compounding, the number is 365, or 360 if your bank says so.)

Step 3. Add 1. Now find the log of this number (same type of log as you chose previously) and multiply it by the number of compounding periods in a year—the same number you used at the end of the preceding step.

Divide the result of Step 1 by this result.

For finding how long it will take to triple an investment compounded daily at 15 percent annual interest, the calculation goes like this.

Common log of 3 is 0.477. Save it for later.

Dividing 0.15 (for 15 percent interest) by 365 and adding 1 gives 1.00041, of which the log is 0.000178.

This, multiplied by 365, is 0.06513. Dividing this into 0.477, obtained in Step 1, reveals that it will take 7.3 years to triple your money.

What's That Rate of Return?

YOU HAVE money on deposit in an interest-bearing fund. Your periodic statement comes, telling you that your earnings for 28 days were $18.80 on an initial balance of $2812.59. What rate of return is that?

You need only divide the number of dollars earned by the amount of the beginning balance to find the yield is 0.0067.

Multiplying this by 100 (moving the decimal point two places to the right) converts this to percentage form—as 0.67 percent or about two-thirds of 1 percent.

Call this the periodic yield, the rate of return for a 28-day period.

However, you'll have to convert that periodic yield into an annual one to permit meaningful comparisons with other investments, on which you'll ordinarily be quoted the annual return.

It might appear that you could get this figure by multiplying that periodic rate by the number of periods in a year. But doing that would neglect the fact that in each period you are entitled to interest on not only your initial principal but also on every cent of that delightfully accumulating interest.

What you would get by that method is a fairly good approximation but one that wound underestimate the true figure every time. Multiplying 0.67 percent by 365 and dividing by 28 gives an estimated annual yield of 8.73. Round this upward a bit and call it "around 9 percent."

That's not terribly far from the true figure of 9.1178 percent, most handily obtained with a scientific calculator or a computer.

You get this exact figure by adding 1 to the short-time yield of 0.0067 and then raising that total to the power of the number of periods. In this instance, the number of periods is 365 divided by 28.

The result you have at this point is what each dollar of initial investment will have become at the end of a year, just over $1.09. Subtract the 1 that you added earlier and you convert this to the annual yield.

When the calculation starts, as this one did, with a rather odd number of days, it is most easily done with a scientific or financial calculator, or with a computer.

In the more usual instance where the period is a simple fraction of a year, any calculator will serve. Because you're raising to the twelfth power (for a month) or the fourth (for a quarter) instead of the mix of whole number and decimal you produced when forced to divide 365 by 28.

The Day You Save

Because the ultimate figures are truly impressive, financial institutions love to tell you how much you'll have after so many years if you save so much a month or quarter or year.

Although they don't say so, they usually give you figures that assume you've made the deposit with them at the beginning of the period in question.

But what if you put up the money at the end of the quarter or year instead—as many of us will because that is the last possible moment to get certain tax savings?

Either way, you'll have put in the same amount. But depositing at the beginning of each year instead of the end will mean quite an increase in the amount you'll have for retirement.

A common arrangement at a good rate might be 7-3/4 percent interest compounded daily. After 25 years, this turns $1500 saved each year into more than $100,000. (You'll find the way to calculate this given elsewhere in this book.) The amount will be some 8 percent more if you deposit at the start of each year rather than the end.

If you deposit savings quarterly, the increase you get from doing so at the beginning rather than the end is about 2 percent.

This possibly puzzling difference is one of the effects of compounding interest and doing it more than once a year.

Figuring Simple Interest

I GUESS they call it simple interest because it IS simple—if what you're asking is how much you'll owe if you borrow so-many dollars for so-many years at a known rate of interest.

But it can be utterly confusing when your unknown is one of the other three of the four factors:

> amount of *Principal*,
> amount of *Interest*,
> interest *Rate*,
> and *Time*.

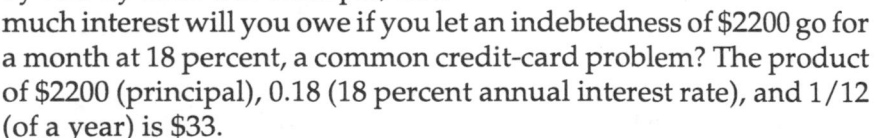

If you know any three of these factors, you can find the fourth by using an arrangement of a basic formula, I=PRT.

To find the amount of *interest*, the formula says, multiply *principal* by *rate* by *time*. For example, how much interest will you owe if you let an indebtedness of $2200 go for a month at 18 percent, a common credit-card problem? The product of $2200 (principal), 0.18 (18 percent annual interest rate), and 1/12 (of a year) is $33.

(Note how easily this turns into a quick mental calculation, if you remember that the common 18 percent carrying charge is 1-1/2 percent a month. Just divide $2200 by 100 and increase by 50 percent.)

To find *principal* amount, divide *interest* amount by *rate* multiplied by *time*. How much need you invest at 10 percent simple interest to earn $100 in six months? Divide the interest of $100 by the rate of 0.10 and by the time of 1/2 year to get $2000.

To find *rate*, divide *interest* by *principal* multiplied by *time*. If $3000 is to produce $22.50 interest in 1 month, what must the rate of interest be? Divide the interest amount of $22.50 by $3000 and by 1/12 year. Result: 0.09, or 9 percent. (Note that the easy way to divide by 1/12 is to multiply by 12.)

To find *time*, divide *interest* by *principal* and by *rate*. How long must you leave $300 at 8 percent for the interest to come to $4? Divide $4 by $300 and by 0.08. Result is 0.1667 years, which decimal multiplied by the number of days in a year gives you 61 days (60 if you use the banker's rounding of a year to 360 days).

Figuring Compound Interest

Compound-interest problems ordinarily give you five factors to deal with. You know four of them and want to find the fifth.

1. There's the initial amount—how many dollars you start with.

2. There's the annual interest percentage. If you're asked for the rate as a decimal fraction, divide the percentage by 100. (Just push the decimal point two places to the left, so 15 percent becomes 0.15.) That's how we'll do it in what follows.

3. There's the number of compounding periods in a year—4 for quarterly compounding, 12 for monthly, and so on. This is the number by which you divide the annual interest rate to find the periodic rate.

4. There's the number of years the compounding goes on.

5. And there's the number of dollars that initial amount will reach by the end of the period.

If you start off knowing four of these things, you can discover the fifth.

How much will it grow to? To find how much compounding will make an investment grow, add 1 to the rate, raise this to the power of the number of compounding periods, and multiply by the initial amount.

(Example: For an investment of $2000 at 15 percent compounded monthly for 20 years, raise 1.0125 to the 240th power and multiply by $2000 to find that it would reach a magnificent $39,430.99 at the end of the period.)

What size investment? To find how much you'd need to start with to accumulate a chosen amount, add 1 to the interest rate (in decimal form), raise to the power of the number of compounding periods, then divide the sum you're going for by this result.

(You should find that $39,430.99 divided by the 240th power of 1.0125 is $2000. Not surprising, since we've chosen our figures to make this a mirror image of the previous exercise.)

What interest rate? How do you choose an interest rate—one that will turn a given sum into a pre-chosen amount in a specified time? Divide the end amount by the beginning amount. Find the root of the figure just obtained that is equal to the number of compounding periods—the fourth root, for example, if there have been four periods, as in quarterly for a year. Then subtract 1.

Turn this into an annual rate by multiplying by 12, and into a percentage by multiplying by 100.

(Example: For the rate needed to turn $2000 into $39,430.99 in 20 years of monthly compounding, divide the larger number by the smaller, find the 240th root, and subtract 1. Multiplied by 12 and by 100, this becomes the annual 15 percent that would be required to do the job.)

O Lord, how long? To find how long it will take to turn one sum into another through compounding at a known rate, divide the amount you aim to end up with by the amount of the initial investment. Find the logarithm of this result by table or calculator. Add 1 to the periodic interest rate (expressed in decimal form) and then find the log of this sum. Divide the first log by the second to find the number of compounding periods.

(In our example, you'd divide the log of 19.715,495 (which is 2.981,405) by the log of 1.0125 (which is 0.0124,225) to get the answer of 240 monthly compounding periods, or 20 years.)

A couple of final notes. In this last caper, you can use either natural or common logarithms, as long as you stick to one kind. The quickest way to obtain them is with a scientific calculator (with power and reciprocal keys to speed up everything). But a log table plus a simple calculator, or even a pencil, will do the job.

Even without a log table, you can crunch your way through by trial-and-error with a calculator of any kind. Same goes for finding a root, such as that 240th. Several methods are discussed in some detail elsewhere in this chapter.

Interest Compounded Continuously

THERE's something pretty fetching about the notion that your hard-earned savings are earning interest for you every second, day and night.

That's why some savings institutions have come to advertising continuous compounding on the money you've deposited with them.

Is this a gimmick or something worth while?

That question is best answered with another question: Compared with what? (As a disillusioned lady is said to have replied when asked how her husband was.)

So, specifically, what will $500 become in 15 years at 8 percent interest?

The standard compounding method tells us to add 1 to the rate (after dividing the percentage form by 100 to turn it into a decimal), raise this to the power of the number of compounding periods, and multiply by the initial amount. Applying it to these figures, we found that if the compounding is quarterly, the $500 will grow to $1640.45.

How can we apply the same calculating procedure to continuous compounding that produces infinitely small compounding periods—and therefore an infinitely large number of them?

What will happen if, for that infinitely large number, we arbitrarily substitute one that is only comparatively large? And then recognize that what we have obtained is an approximation that is somewhat on the low side, and round it up to the next whole dollar?

Since we're taking unauthorized liberties, we might as well choose a number that's easy to work with: 100. The result we get is $1659.26, which we round to $1660. Continuous compounding is significantly better than quarterly, but not dramatically better—or so our approximation method says. A sort of checking run, using 1000 in place of 100, produces $1659.98.

The real way—no approximations—to figure continuous compounding is actually simpler. But I don't know any really expeditious way to do it without a scientific calculator. If you can lay your hands on one of these (and at a sale price that I've seen fall as low as ten dollars, why not?), all you have to do is multiply the interest-rate-as-a-decimal by the number of years, raise the mathematical growth constant e to that power, and then multiply by the number of dollars you started with.

This mathematically impeccable route brings us to $1660.06. Our quick-and-dirty method is proved quite good enough for everyday use.

Effective Rate of Interest

As INTEREST-RATE calculations go, this is a short and simple one. All the same, it's a bit of figuring that has often led to confusion, mostly because it offers a temptingly fertile field for deception by installment-sales operators.

It can hide just about anywhere. If someone tells you of a car loan on which he's paying half the interest rate you know others are paying, you can suspect he's actually been quoted (though not in writing in recent years) the interest charge as a percent of the full initial amount of the indebtedness.

There is a quick route to the truth.

To find the effective rate of interest you are paying, or getting, you need only divide the total amount you are putting out for interest by the number of years and by the average amount that you will owe during the course of the loan.

Typically, you'd work it out this way. Multiply the number of payments to be made by the amount of each payment and subtract from this the original amount of the loan. This difference will be the total amount of interest to be paid.

Dividing this result by the amount borrowed and by the number of years produces the effective rate. Multiplying by 100, a matter of pushing the decimal point two places to the right, turns the rate into a percentage.

Which brings us to a wicked law.

Beware the Iron Law

CLASSICAL economists began talking about the Iron Law of Wages long ago. But it must have been a disgruntled borrower who more recently formulated the tongue-in-cheek Iron Law of Interest for the edification of lenders. This pseudo-law says: Whatever rate you're actually charging, always express it so that it doesn't sound like more than 6 percent.

Soaring interest rates of the seventies and eighties made that a hopeless goal for quite a time but the unprincipled principle survived.

If you consult a car salesman about credit for your purchase, you may be told that an 8 percent loan can be arranged. Or perhaps "8 percent per year." Or, if he's a little more choosy about his language, "8 dollars per hundred per year."

If you accept the offer you may be surprised not to find any figure even faintly resembling 8 percent on the Truth in Lending statement that the law now says you must receive before the deal is concluded. If you read it, you will find that you are paying an APR, or annual percentage rate, of 14.5 percent. Nearly twice what you thought you heard at the oral stage of the deal.

What you have is an add-on loan. The amount you are to pay has been figured by adding the quoted cost (multiplied by the number of years if the loan is to run more than one) to the face amount.

For a one year loan, the $100 becomes $108. This divided by 12 is your monthly payment.

The gimmick lies in the fact that you are repaying by the month. If you were to keep the whole $100 until the end of the year and then repay it, plus $8, that would indeed be 8 percent simple interest.

But you're not borrowing $100 for a year. You are borrowing $100 for one month, then paying $8 of it back with $1 interest. During the second month, you have only $91 of the lender's money. By the last few months, you have only a small amount of it to use.

So, on the average through the year, you had the use of only about half the face amount. Having paid $8 interest for the use of an average of $50 or so for a year, your interest rate is bound to be up towards 16 percent.

It could be worse. There is a close relative of the add-on loan that is called the discounted loan. The difference is that, with this one, when you borrow $100 for a year at $8 per $100, the lender collects his interest off the top—handing you $92. With this kind of loan the true annual interest rate turns out to be even closer to double what the salesman made it sound like. More precisely, it is 15.7 percent.

The reason for the even higher rate is, of course, that this time you're paying for the use of about half of $92 instead of half of $100. And the fee is still $8.

Figuring the exact rate of interest on either an add-on or a discounted loan is not particularly easy.

Books about such things ordinarily offer you one of several methods of approximating the rate. Since the methods are not precise, the answers you get vary quite a bit. It's easier, and just as accurate, to say simply that the true rate of a discounted loan is almost twice what it sounds like when stated in dollars per hundred per year. For an add-on loan, it's only a little less than that.

I haven't been able to construct, or find, a formula for calculating this rate precisely, so I've had to resort to a sort of backward, or checking, method for arriving at the figures in the preceding paragraphs—the 14.5 percent one for the add-on and the 15.7 for the discounted.

The method calls for making an estimate and then checking it. This is handily done by the use of a pocket calculator, but would be so tedious without it that if you don't have one of these things around you will probably have to be satisfied with the estimate. For a discounted loan, a good estimate is simply twice the stated dollars per hundred. For an add-on it's about one-tenth less than double. Then follow these steps if you want to refine your estimate.

1. Express the estimated rate of interest as a decimal and divide it by the number of payments per year.

2. Add 1.

3. Raise this to the power of the total number of payments.

4. Find the reciprocal—that is, divide 1 by the number just obtained.

5. Subtract 1. Drop the minus sign or ignore it from here on.

6. Divide by the interest rate, expressed as a decimal.

7. Multiply by the number of payments per year.

8. Find the reciprocal.

9. Multiply by the amount actually received.

10. Multiply by the number of payments to be made.

For an add-on loan, the result should be very close to the total of your initial indebtedness—that is, the face amount plus the add-on interest dollars. For a discounted loan, the result should be the face amount.

Now run one or more slightly larger or smaller estimates through those same steps until the result is as close as you can make it.

How Much Goes for Interest?

BY READING this graph directly, you can discover how much of your money goes to pay interest during the lifetime of a home mortgage—if the mortgage is for precisely $25,000.

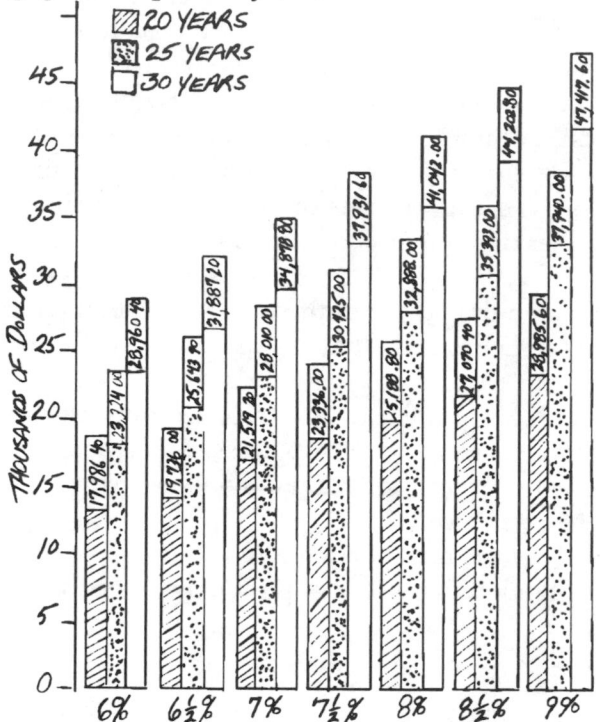

For each of seven interest rates you see a trio of bars. The first of each set of three is for a 20-year mortgage. The second is for 25 years. The third is for a 30-year mortgage (or deed of trust, which is used in some areas to serve the same purpose as a mortgage does in others).

The figures are for interest only. Payments on the $25,000 principal are on top of these.

If the rate of interest or term of years on your mortgage is different from any shown, you can use interpolation or extrapolation to find the interest figure with reasonable accuracy.

Try extrapolating to find the figure for 10 percent interest on a $25,000 mortgage for 30 years. Since the interest at 9 percent is $6,375.60 more than it is at 8 percent, it should be about that much more at 10 percent than at 9 percent. Adding that figure to $47,417.60 gives $53,793.20. Since extrapolating in this situation does not produce extremely precise results, it would be sensible to round off the result and call it "about $54,000."

(If you have a calculator handy you can get a more precise answer by using one of the methods given in this book for calculating mortgage payments, multiplying by number of payments, then subtracting the $25,000 principal. Going this route, you'll arrive at an answer in the neighborhood of $53,984, indicating that you were wise to round off the extrapolated answer.)

By interpolation I can find from this graph how much interest I am putting into a 20-year $30,000 mortgage I've been paying on, since the last previous low-interest days, at 6-1/4 percent.

I find the midpoint between $17,986 (forget the pennies) and $19,736 by adding up the two and dividing by 2. But the result is for $25,000, so for my $30,000 debt I increase it by one-fifth. This is most easily done by multiplying by 1.2. Result of $22,633 rounds off to $22,600.

(Of course, with an actual existing mortgage like this, where I already know the payments, the other route is simpler and may be more precise. Twenty years times 12 months times $219.28 less $30,000 gives me the quite exact figure of $22,627. But the amount I actually mail to the bank is $219, rounded off this way for convenience with the difference to be made up at the time of the final payment. Using $219, I find an interest figure of $22,560, not as close to the true figure as the one I found from the chart here.)

When Interest Rates Change

How CAN you make a shrewd long-term financial arrangement in the face of changing interest rates?

You can get some help from this routine for contending with variable interest rates. It's almost as simple as the formula for finding the future value of money when dealing with a constant rate of return.

Suppose you'd like to have $1000 in a certain account five years from now—perhaps for a gift to help someone start college. You know or estimate that yield for the first year will be 15 percent, dropping to 14 the second year, then to 12, 10, and 8. How much must you deposit now?

For each year express the interest rate as a decimal and add 1 to it. Multiply all these sums together (1.15 times 1.14 times 1.12 times 1.10 times 1.08 equals 1.744,364).

Divide the future value—the amount your plan calls for at the end of the five years—by that result. You'll find that to have $1000 then you'll have to make a deposit today of $573.27.

If you're doing such a calculation by paper and pencil, you may prefer to round off that first product to three decimal places. You can do that without distorting the result by more than a few nickels, 12 cents in this case.

The Interesting e

Sᴇᴄᴏɴᴅ only to pi in usefulness is the mathematical constant called e. At 2.718 it's a bit smaller than pi. It describes the rate at which many things grow. Money, for example, or a tree.

If your savings are in an account that is compounded continuously, you need only raise e to the power of the interest rate expressed as a decimal to discover how much each dollar will turn into in a year. If you were so lucky as to receive 100 percent interest (rated pretty skimpy in the loan-shark trade), you'd raise e to the 1st power. Since the 1st power of a number is just itself, you find that your account would grow in a year to $2.72.

To calculate something more realistic, say $1000 at 7 percent, raise e to the 0.07th power to discover that $1000 will expand to $1072.51 in a year of continuous compounding. (If your bank uses daily compounding instead, the difference will be negligible.)

But how can you find the value of e, especially if you want it to more decimal places? Many calculators now have keys for it. Or you can look it up if you have the right book.

(Just so this volume will be a right book, I hasten to say that the value is approximately 2.718281828. And, by the way, don't be fooled: this is one of those instances where what at first blush looks like a repeating decimal turns out not to be. Carry it out any further and it quits repeating.)

To calculate it for yourself, recall first that a factorial of a number is just that number multiplied by all the smaller ones. And that a reciprocal of a number is just 1 divided by that number.

Reckon e by adding up the reciprocals of the factorials of 0 (which equals 1), 1, 2, 3, and so on. When the total stops changing you have e to as many places as your calculator can go.

Comparing Retirement Plans

A READER of a newspaper column presented a problem in financial strategy to its author to solve. Suppose we let his reply outline an answer to many other strategy problems.

The company that employs the advice seeker has offered him the opportunity to augment his retirement plan by $1000 a year at 65. He is now 38. Accepting the proposal will cost him $3220 now.

What our man wants to know, of course, is whether this is the best investment he can make with his $3220.

One alternative investment he might make is common stocks or their equivalent in a mutual fund. History says that over a period of years this kind of investment will pay an average of about 9 percent.

Another possibility is purchase of industrial bonds. Very safe ones, rated A to AAA, could be had at the moment in question that would pay 8 percent or more. A 10-year government bond would do about the same. This is probably even safer, although we can't be sure what rate of return will be available when the money must be reinvested after 10 years.

The first question to be answered is how much will $3220 grow to in 27 years at 8 or 9 percent compounded annually?

General ways to deal with compounding are covered elsewhere in some depth under figuring compound interest. There's no need to repeat the general method here, since we can use the special, and especially simple, routine applicable to annual compounding.

It goes like this.

Express the rate of interest as a decimal (8 percent becomes 0.08). Add 1 to this. Raise the result to the power equal to the number of years of compounding. Multiply by the amount being invested.

In our example, we raise 1.08 to the 27th power and multiply this by $3220.

It turns out that by the magic of compounding $3220 grows in 27 years to $25,721 at 8 percent interest. At 9 percent, it becomes $32,989.

Taking the more conservative of these figures and rounding it downward to $25,000, we find our man would have a sum that would give him $2,000 a year income if he invested it at 8 percent.

This is double the additional pension he'd draw if he took the alternative route. And the capital is his forever, in the end becoming part of his estate.

Interest by Calculator

MAYBE you haven't tucked into interest problems since you left school and some admirable people invented pocket calculators. Since these relate to mortgages and other time payments that most of us are plagued with from time to time, it may be useful to squint at them specifically as seen by your calculator.

Simple interest: This tells you what you'll owe a month or a year from now if you borrow money today. To find it, multiply the interest rate by the time, add 1, then multiply by the principal amount. For $250 borrowed for 6 months at 12.5 percent simple interest, first turn that rate percentage into a decimal by moving the point two places to the left. Then divide that 0.125 by 2, add 1, and then multiply by the $250 to find that you'll owe $265.63 when the loan comes due.

When interest is compounded, it becomes more exciting—and more worthy of your electronic helper's talents. The formula for compounding tells you to add 1 to the interest rate (divided by 100 to express it as a decimal), then raise this total to a power equal to the number of periods involved, then multiply this factor by the amount of money. Just remember that "number of periods" means how many times compounding occurs, and the interest rate you want is the one for the period used for compounding...not the annual rate.

How much will $1200 be worth if you leave it in savings to be compounded semi-annually for 5 years at 6-1/2 percent?

Thinking of 6-1/2 as the decimal 6.5, divide it by 100 (because it's a percentage) and then by 2, because compounding is to be twice a year.

Now add 1, and the display will read 1.0325. Multiply that by itself 9 times to raise it to the 10th power, because there are 10 half-yearly interest periods in 5 years. The resulting factor, multiplied by the amount of money, should tell you that your $1200 will swell to $1652.27.

How Regular Savings Grow

Lᴇᴛ'ssᴀʏ there's a new baby at your house and you want to start right now to save for his, or her, education. How much will a few dollars a month come to in 18 years?

Or suppose your household has resolved to cut its grocery bill by shrewd shopping—and save the difference for a family trip to Europe in 10 years—or a retirement fund 40 years from now. How can you find out how many dollars the savings will amount to?

Just multiplying the monthly amount by the number of months won't give you even an approximation. Your money will be drawing interest, and interest on interest, during the whole period, swelling to a much larger sum than you might expect.

To take the extreme example, suppose you can save $50 a month for the next 40 years. That's $24,000 actually deposited.

By the way, that's no impossibility. While helping my wife write a book on supermarket strategy, back when a dollar was worth twice what it is now, I learned the surprising fact that a great many families were spending at least $50 a month more than they needed to, just by bad shopping methods.

But when I wanted to tell readers how much this would come to by retirement after 40 years of family spending, I discovered that nobody at my bank or savings and loan had any idea how to figure it. They merely supposed that the computer at headquarters far away could do it. Well, once I got hold of my first mini calculator, I found it wasn't all that difficult to compute for myself.

Let's first take the method, then apply it to our example.

1. Express the annual interest rate as a decimal. Divide it by the number of times compounded per year. Add 1.

2. Raise this result to a power equal to the total number of times compounded.

3. Subtract 1. Multiply by the number of deposits per year. Then divide by the annual interest rate (in decimal form).

4. Multiply by the amount you're saving.

Applied to a monthly saving of $50 for 40 years in a bank that pays 6 percent interest compounded monthly, the procedure with your calculator is:

1. Enter 0.06, divide by 12, add 1 to get 1.005.

2. Raise this to 480th power (40 years times 12 months per year) by method explained below. Result: 10.9575.

3. Subtract 1 and multiply by 12, to get 119.4894.

4. Multiply by $50, then divide by 0.06, to discover that your savings have swelled to nearly $100,000. My mini calculator gives precisely $99,574.54. Because of early rounding off, yours may disagree by a few dollars.

(Now, about that raising to a power. It's an instant operation with a scientific or financial calculator, but quick enough with any model. Just turn that 480 into its factors—any handy set of numbers that multiply to 480. Say you use 8, 6, and 10. That means you hit the times key, then the equals 7 times; then the times, then the equals 5 times; then the times once more and the equals 9 times.)

Saving: How Much per Month?

How much must you put aside each month to accumulate the price of a Mercedes in ten years or a beach condominium when you're 65?

To find out, begin by deciding first the size of your goal and how long you're prepared to take to reach it. Find out, or estimate, the rate of interest you can count on. From this you can figure out precisely how much you must save each month.

If the procedure that follows seems familiar, there's a reason. Up to the last two steps it is the same as for figuring how much a known sum saved each month will come to over the years.

1. Divide the interest rate, expressed as a decimal, by the number of times the institution where you are placing your savings compounds per year.

2. Add 1.

3. Raise this total to the power equal to the number of times compounding will occur. (This is the number of times compounded per year multiplied by the number of years of your program.)

4. Subtract 1.

5. Multiply by the number of deposits you're to make per year.

6. Divide by the annual rate of interest.

7. Divide by the amount of your dollar goal.

8. Find the reciprocal of the result. (This means divide 1 by the number you've found. On an electronic calculator, you can do this by pressing the divide key once and then the equals key twice. Advanced calculators let you do this in a single step with a key marked 1/x.)

Let's run through an example. You want a $100,000 stake. You've got 24 years to get it together. You are prepared to bet that bonds, stocks, or real estate investment will bring you a 9 percent return with quarterly compounding, but since taxes in your income bracket will take one-third of the interest, you treat this as a net of 6 percent.

Expressing that percentage rate as its decimal form, 0.06 and dividing by the 4 times a year that quarterly compounding will occur, you have 0.015. Add 1 to this and raise the sum to the 96th power (compounded 4 times a year for 24 years).

From this 4.1758, subtract 1. Then multiply by 12 for the number of times a year you plan to add to your investment. Dividing by the annual interest rate of 0.06 gives 635.1607; dividing this by your goal of $100,000 or bust, produces 0.00635, of which the reciprocal turns out to be 157.44.

So you'll have to sock away $157.44 a month to reach an accumulation of $100,000 in two dozen years at the realistic rate of return we've postulated.

In serious planning of this nature, involving a long span of years, it's important to take into account what inflation may do. In short, ask yourself what purchasing power the $100,000 in our example may have after 24 years.

A way to get a handle on this is to subtract your estimate of the coming rate of inflation from the nominal interest rate at the same time that you subtract the income tax rate. If you see inflation as likely to be around 2 percent, this reduces your true rate of return in the example from 6 percent to 4.

In practice, you may want to use current figures for nominal return, tax bracket, and inflation—instead of trying to guess for the long period. Then recalculate at intervals as these variables change, and adjust the amount of your monthly deposit.

Figuring a Set-Aside

WHAT PROPORTION of your income must you set aside each year in an interest-bearing account to accumulate a fund equal to a year's income?

That's what this table is about.

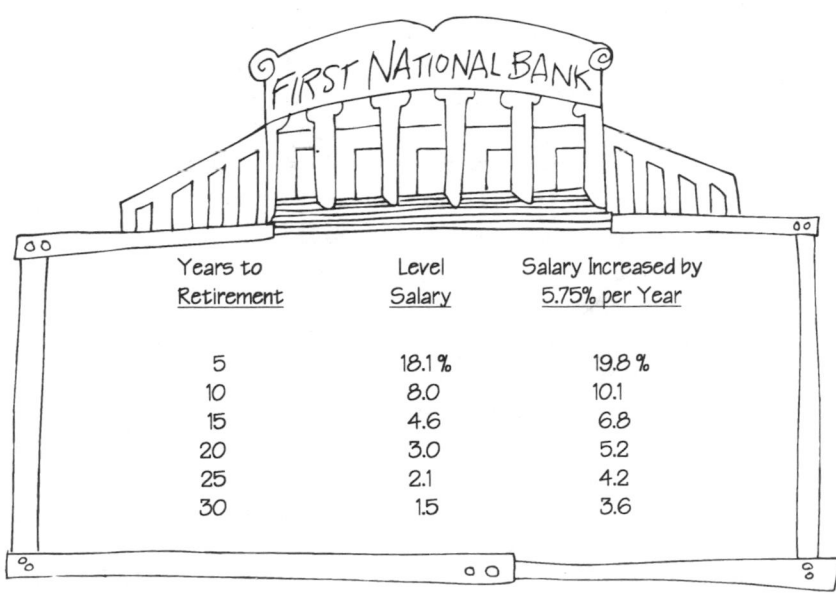

Years to Retirement	Level Salary	Salary Increased by 5.75% per Year
5	18.1%	19.8%
10	8.0	10.1
15	4.6	6.8
20	3.0	5.2
25	2.1	4.2
30	1.5	3.6

You might want to use it, as my family did when we gave ourselves a sabbatical year in Europe...traveling, studying, researching, writing stuff that couldn't be expected to produce immediate money.

If you expect your income to remain level through the period, use the "level income" column. (If you take "level" to mean rising at the same rate as inflation, you won't have to allow for inflation in your reckoning.)

The other column is the one to use if you can expect a fairly normal 5.75 percent increase each year on the average.

Here's an example of what the table tells you. If you want to have a fund at retirement 20 years from now that is equal to your final year's salary—your earnings having risen 5.75 percent a year— you'll have to set aside 5.2 percent of each year's income.

This assumes your money is earning 5 percent. You may be able to do much better than that, but remember that the 5 percent means after state and federal income taxes have taken their bite.

How Much Interest Are You Getting?

ALMOST everybody knows how to figure a certain percentage of something, but reversing the calculation is somehow ultimately confusing.

You have $41,106.75 invested in a fund that pays interest monthly. Payoff is at whatever rate the fund is able to earn currently. You receive an interest check amounting to $216.67.

You may want to know what rate of interest that represents so you can decide whether to switch to some other form of investment. Basically, to find an interest rate you just divide the amount of return by the amount invested, and move the decimal point two places to the right to change to percent.

But there is usually one complication. For purposes of comparison, you want to know the annual interest rate, while the return is more likely to come monthly or quarterly.

So, to work out the example above, you must divide the monthly income by the amount of the principal and multiply by 12, the number of months in a year. Move the decimal two places and your result should be just over 6.3 percent.

Sometimes you need to be even more precise than this. If your investment pays, as some do, from the first (or perhaps the second) business day of one month to the next, the number of days for which you are getting interest may vary markedly from one month to the next.

Then this is what you might go through.

Let's say you've just received your check for February. The accompanying slip tells you that you own 4,108.071 shares and their present price is $10.02. Date of the slip is March 3, and date of the previous slip was February 3. What's the interest rate?

You can find it most directly by multiplying the number of days in a year by the number of dollars in the check, then dividing by the product of three things: The number of shares, the price of each, and the number of days in the month in question.

Steps with an electronic calculator are: 365 times 228.24 divided by 4108.07 divided by 10.02 divided by 28 equals 0.0722804. So you find you're getting just under 7-1/4 percent interest.

INCREDIBLY INTEREST-ING!

Chapter D
INVESTING

Finding Earned-Growth Rate

IT'S A VALUABLE measure of true growth that can be used for any kind of business, from food processing to steel. It's called the EGR, or earned-growth rate.

To determine it, subtract the dividends per share from the earnings per share and divide by the book value (also known as stockholders' equity) at the beginning of the year.

The result will be the percentage by which the value of each share of common stock rose during the year. The price of the stock may not reflect this rise immediately, but it will tend to do so in the long run.

Tailoring an Investment

ONE MAN'S winning investment can be another man's loser. Or woman's, of course.

An important aspect of tailoring an investment to your own needs is considering the impact of your personal income-tax bracket. This can be pretty tricky when you must consider the effects of a tax that comes at the end of a long period rather than year by year. Or take into account an alternate investment that is not taxed at all—but bears a lower nominal rate of return.

For purposes of illustration assume that your tax bracket (federal, state, and local combined) comes to about 40 percent and, so far as you can predict, will remain about the same.

You might find yourself considering three options.

One option is a single-premium deferred annuity for 20 years at 9 percent.

Another is a 20-year corporate bond at 10 percent.

The third is a tax-exempt bond paying a mere 7 percent.

What you might want to know, for each option, is how many after-tax dollars you'll have at the end of 20 years for each dollar you invest today.

For the single-premium deferred annuity, begin by expressing the percentage as a decimal (so that 9 percent is 0.09) and adding that to 1. Raise this sum to the power of the number of years and then subtract 1. What this will tell you is the amount that original dollar will have earned before you pay the tax on those earnings.

In our example you've raised 1.09 to its 20th power and then taken away 1, to get 4.6044. But if your financial environment is that 40-percent bracket you'll get to keep only 60 percent of the earnings. Multiplying the previous figure by 0.60 and adding 1 (for the dollar you put in to begin with) gives 3.7626. That $3.76 is the after-tax amount to which each of your annuity dollars will have grown in 20 years.

Next question. Can you beat that 9-percent single-premium deferred annuity with a 10-percent corporate bond?

To find out, express the percentage rate the corporate bond pays as a decimal and multiply by the fraction of your income you retain after taxes (0.10 times 0.60 in our example), add 1, and raise this sum to the power of the number of years—20 in this case. Your buck will have grown to 3.2071 dollars that you get to keep. The annuity did better than that.

But if you are dealing with a 12 percent bond instead, you'll discover through a similar calculation that the figure becomes a more salubrious 4.0169.

For a tax-exempt bond the calculation is even quicker. Just express the interest rate as a decimal, add it to 1, and raise that sum to the power of the number of years.

Suppose the bond offers 7 percent. Again express the percentage rate of return as a decimal and add 1. Raise this 1.07 to the 20th power. Since this comes to 3.8697, you've found how many dollars one dollar will grow to in 20 years at 7 percent interest compounded (or reinvested) annually, if there are no taxes to consider. Just about $3.87.

By using these three formulas you can compare the three different kinds of investments, whatever the interest rate and whatever your own tax bracket.

By modifying any of the formulas slightly you can deal with a bond or other investment that pays oftener than once a year. To apply the formulas above, just use the number of compounding periods instead of the number of years; and divide the interest rate by the number of periods in a year.

That probably sounds more complicated than it is. It just means that for the bond that offers semiannual payments at 10 percent a year for 20 years, you'll use 40 as the number of periods and 0.05 as the rate per period. You'll find that with semiannual payments your dollar grows to 3.2620 in 20 years. For a $10,000 investment, that's an improvement of $549.

Local Taxes and Investing

IN THE FIRST place, whether to put money into U.S. Government E bonds should depend in part on where you live.

If you have to pay local income tax on bank interest but not on earnings from these bonds, you'll want to start by figuring the net return after local tax (state, county, city).

To do this, subtract the tax bracket this local agency has you in from 1. Then divide this into the rate of return the investment gives. Use decimals to do this.

For instance, suppose your state income tax bracket is a rather high 7-1/2 percent. As a decimal that's .075, which subtracted from 1 equals .925. When you state the E bond rate of, say, 6 percent, as the decimal .06 and divide that by .925 you get .065 approximately, or 6-1/2 percent equivalent.

You can see that if at the time bank and savings & loan accounts yield between 5 and 6 percent, that's pretty good. On the other hand, if you are dealing in sufficient sums and long enough periods of time to buy certificates of deposit from these same institutions at yields of 7 to 8 percent or a trifle more, those bonds are not all that appealing as an investment.

Keogh? Ira?

WHAT DO YOU gain by putting money for retirement into an IRA or Keogh Plan fund instead of just tucking it away in an ordinary bank account?

One way to tell is by looking at this little table an insurance company worked out.

Another way is by calculating it yourself for your own precise situation. Begin with the amount you'd like to put away each month. Multiply that by the tax bracket you're in (federal plus state, if your state allows deductions similar to the federal ones). The result is how much you reduce your tax bill for the year.

But keep in mind, with this and all other such calculations, that the money you take out after retiring will be taxable income. The idea is that by then your earnings will be far smaller and your exemption larger, so your tax rate, if any, will be much lower. This is in addition to what the table tells you about how much you profit by the delay in taking taxes out of your earnings.

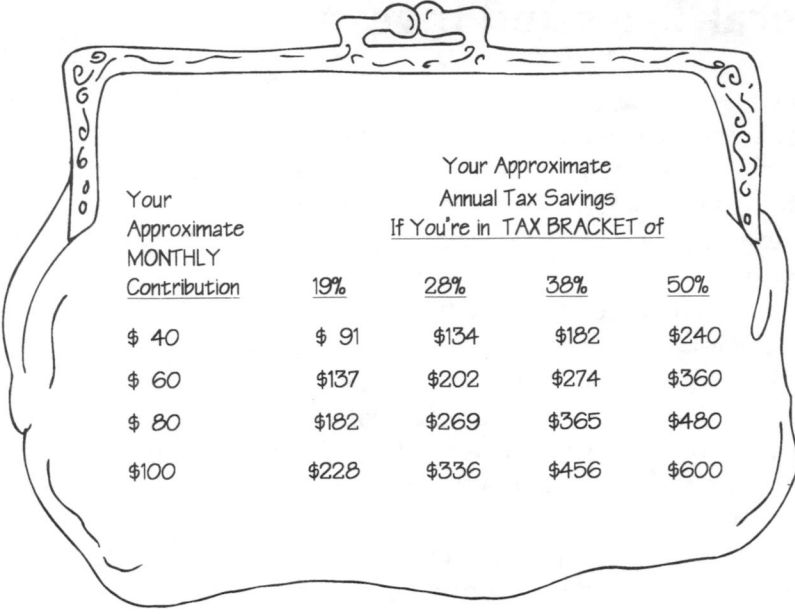

Your Approximate MONTHLY Contribution	Your Approximate Annual Tax Savings If You're in TAX BRACKET of			
	19%	28%	38%	50%
$ 40	$ 91	$134	$182	$240
$ 60	$137	$202	$274	$360
$ 80	$182	$269	$365	$480
$100	$228	$336	$456	$600

A different way of looking at the matter is by figuring how much more you'll have in the end for the same out-of-pocket cost if you do your saving under a Keogh or IRA plan.

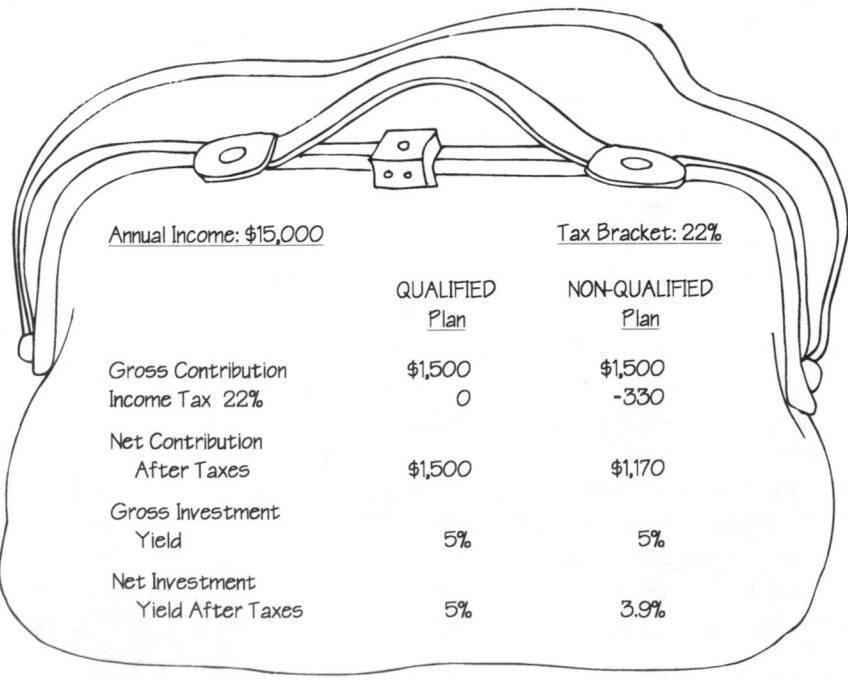

Annual Income: $15,000		Tax Bracket: 22%
	QUALIFIED Plan	NON-QUALIFIED Plan
Gross Contribution	$1,500	$1,500
Income Tax 22%	0	-330
Net Contribution After Taxes	$1,500	$1,170
Gross Investment Yield	5%	5%
Net Investment Yield After Taxes	5%	3.9%

This table applies to a person with an income around $15,000, or to a family with several thousand dollars more, which would produce the same amount subject to taxes.

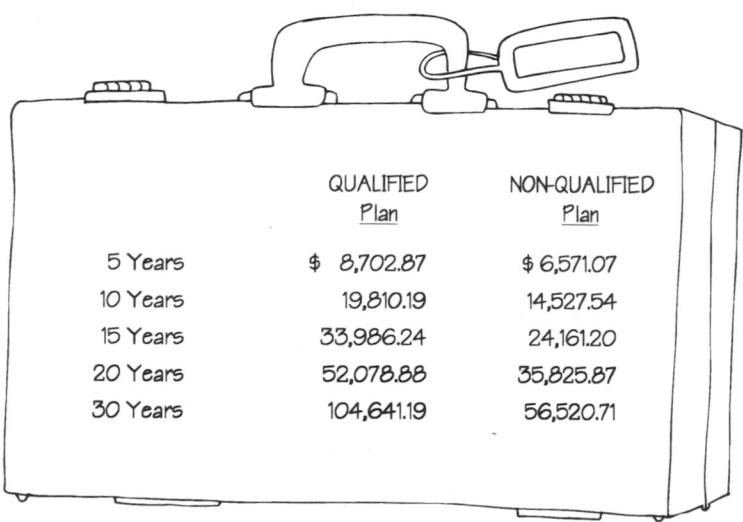

	QUALIFIED Plan	NON-QUALIFIED Plan
5 Years	$ 8,702.87	$ 6,571.07
10 Years	19,810.19	14,527.54
15 Years	33,986.24	24,161.20
20 Years	52,078.88	35,825.87
30 Years	104,641.19	56,520.71

Most convincing of all is the third little table. It shows how much you'll have after various periods if you put aside $1,500 each year and are in a 22 percent (federal plus whatever) tax bracket.

Using a tax-deductible plan instead of an ordinary savings account bearing the same interest rate, means the difference between having $104,641 after 30 years and having little more than half that amount.

Compounding Your Money

T HIS TABLE offers a quick way—meaning hardly any figuring—to find out how much money you'll have someday if you put aside a fixed amount each year.

If the sum you plan to save is $100 a year, you can read the result for any of a number of periods and interest rates directly from the table. You can see that $100 invested each year and compounded annually will grow in half a century to $128,030.

For any other amount, divide the figure in the table by 100 and multiply by the number of dollars to be put aside each year.

The table works backwards too.

Suppose your goal is to accumulate $20,000 over 30 years and you can assume an 8 percent net return annually. Where the 30-year and 8 percent lines intersect you'll find $12,234.

Divide the wanted amount of $20,000 by that and you'll get a quotient of 1.63. Multiplying that by the table's assumed $100, you'll find your annual investment must be $163.

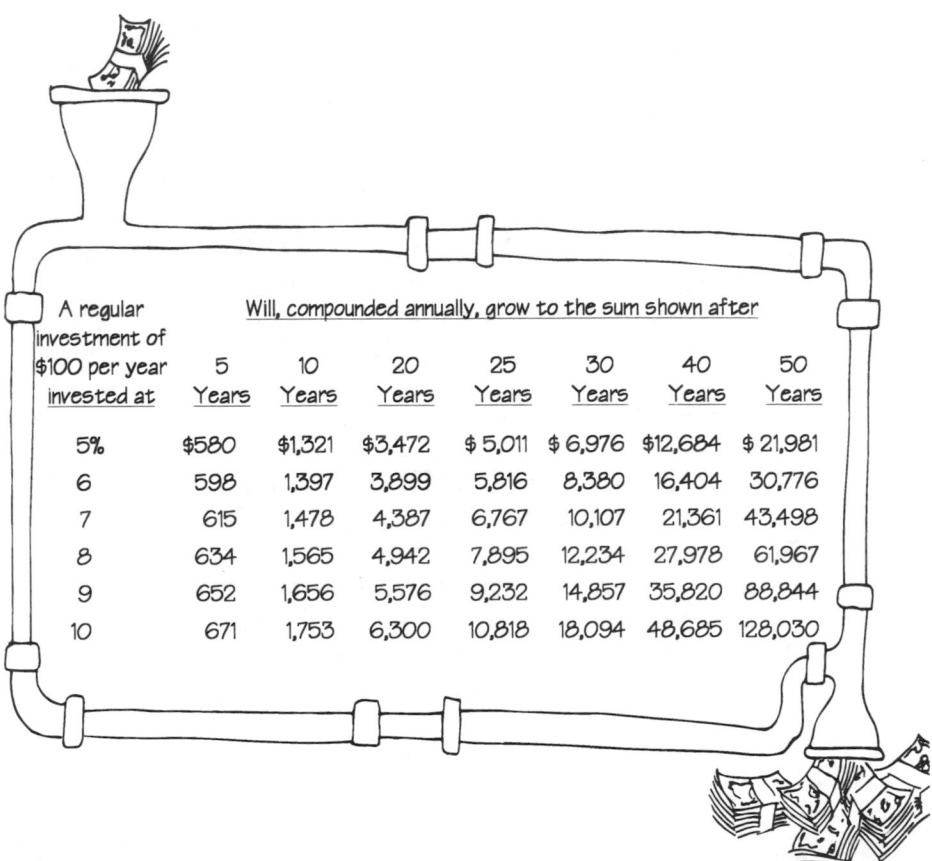

A regular investment of $100 per year invested at	Will, compounded annually, grow to the sum shown after						
	5 Years	10 Years	20 Years	25 Years	30 Years	40 Years	50 Years
5%	$580	$1,321	$3,472	$ 5,011	$ 6,976	$12,684	$ 21,981
6	598	1,397	3,899	5,816	8,380	16,404	30,776
7	615	1,478	4,387	6,767	10,107	21,361	43,498
8	634	1,565	4,942	7,895	12,234	27,978	61,967
9	652	1,656	5,576	9,232	14,857	35,820	88,844
10	671	1,753	6,300	10,818	18,094	48,685	128,030

Retirement-Fund Accumulation

THIS IS one for regular savers.

For each dollar you put into a tax-sheltered IRA or Keogh retirement fund each year for the next 25, how many will you have available for use when the time comes?

If it's a fund that pays 12 percent interest—a ridiculously optimistic level in some eras but normal in fairly heavy inflation— compounded twice a year, you'll have $145.16.

At $1000 a year, it'll come to a nice $145,160.

That's the kind of thing this table figures for you.

That's for semiannual compounding. For other compounding periods, interest rates, and periods of time, you can figure it yourself with a hand calculator by methods given elsewhere in this compendium.

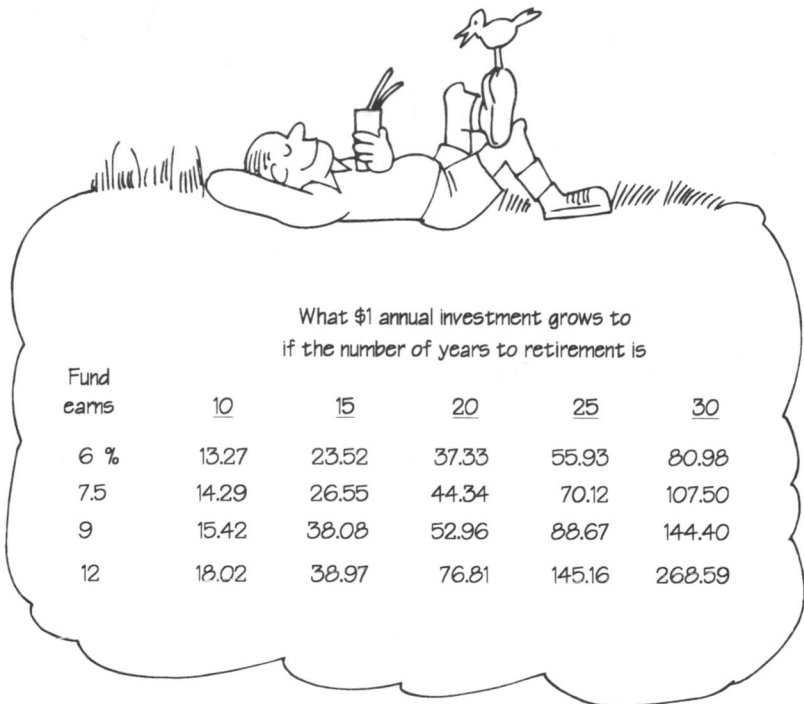

Fund earns	What $1 annual investment grows to if the number of years to retirement is				
	10	15	20	25	30
6 %	13.27	23.52	37.33	55.93	80.98
7.5	14.29	26.55	44.34	70.12	107.50
9	15.42	38.08	52.96	88.67	144.40
12	18.02	38.97	76.81	145.16	268.59

What Does an Annuity Cost?

Have you run through the procedure for calculating how big an annuity a given sum will buy? Reversing the routine will tell you how much an annuity of any given amount will cost you.

Having decided what monthly income you'll need, you can then learn what it will cost you to buy it. You then multiply by a fixed sum that depends upon your age.

For a straight-life annuity bought when you turn 65, the multiplier has recently been 115.

So if at 65 you want an annuity that will bring you $1,000 a month for life—and die when you do—you'll have to plunk down $115,000.

For an annuity bought at the same age but guaranteed to keep paying not only for your lifetime but for a minimum of ten years no matter what happens to you, the multiplier is 123.8.

How Much Annuity Can You Buy?

A STRAIGHT-LIFE annuity is a monthly income you buy with cash in advance. It is guaranteed to go on as long as you live, and it dies when you do, leaving nothing for heirs.

Because of this limitation it produces a larger income than any ordinary safe investment, especially for someone getting along in years. Such an annuity has little appeal for anyone under retirement age.

You can get a pretty close approximation of what a straight-life annuity will produce in the way of monthly income if bought at the

age of 65 by dividing its cost by 115. This will tell you that if, for example, you had $45,000 to invest at 65, it would bring you $391 a month as long as you lived.

By comparison, the same $45,000, if put into an 8 percent savings account or bonds, would produce only $300 a month. But, of course, the principal would still be there for your lucky heirs to collect. If you heartily resent the idea that almost every cent you put into an annuity would be lost if you should die shortly thereafter, you might prefer some other type. One variety, called "ten years certain," not only pays as long as you live but goes on for ten years even if you die the next day. To figure the typical return on this type, if you should buy one at the age of 65, divide by 123.8 instead of 115. You'll find the monthly return on your $45,000 would then be $363.49.

What's an Annuity Worth?

LET'S SAY you want to arrange for yourself a supplementary income of $5,000 a year for the next twenty years...total of $100,000.

Tucking a thousand $100 bills into a mattress and pulling out fifty of them every January would be one way to do it.

It would be a wasteful way, of course, since the part of the money you hadn't used yet would be sitting there drawing no interest. Clearly, it should be possible to "buy" the income we're talking about for less than the whole $100,000 if interest can help.

The question, then, is this: What amount must you deposit today in order to have it last just twenty years if withdrawn at $5,000 each year?

Assume you choose to put the money into a bond fund that will pay 7 percent interest.

Then the calculation proceeds this way.

Express the interest rate as a decimal. Add 1. Raise this sum to the power of the number of years you've chosen. Find the reciprocal (that is, divide 1 by this result). Subtract this from 1. Divide by the interest rate expressed as a decimal. Multiply by the number of dollars you wish to withdraw each year. For the example, the computation runs:

Multiplying 1.07 by itself 20 times gives 3.8696833, which divided into 1 is 0.258419. Subtracted from 1 and then divided by 0.07, this becomes 10.594014.

Multiply this by $5,000 and you'll find the cost of this self-created annuity to be $52,970.

So at 7 percent interest, you can buy now for little more than half the amount that will be returned to you.

Make the same calculation for 8 percent net bond return and you'll find the cost to be less than half.

Decoding the Phony "Average Return"

RATES of return on savings are often quoted in the form of a pair of figures. One figure may be described as the *interest* rate and the other, which is a bit higher, as the annual rate of *return*.

Nothing wrong with that. When interest is compounded at short intervals, and especially when compounded daily or hourly or continuously, you gain an extra edge above the mere annual rate.

Thus $100 held for a year at 8.21 percent simple interest will become $108.21. But if compounded continuously at that rate it will turn into $108.56. You are getting what can legitimately be called an annual rate of return of 8.56 percent. Note that the difference between those two numbers (8.21 and 8.56) is only a fraction of a percentage point.

Suppose you pick up the morning paper and see, as I did one day at a time when interest rates were comparatively high, that a savings institution is offering a certificate paying an annual rate of 8.21

percent, for which an "average annual return" of 9.31 percent is claimed for a 3-year certificate.

How can that be? How can the return be more than a whole percentage point above the rate? The trick lies in the deceptive use of that misunderstood word average. A bank or S&L can offer you literally any average annual return it cares to name simply by choosing a sufficiently large number of years over which to average it.

If the S&L in question had wished to offer you the even more appealing (but perhaps suspiciously tasty) bait of 12.73 percent instead, it could easily have done so by mentioning 10 years instead of 3.

(For 100 years, if you're curious, the average annual return for an 8.21 percent account, compounded continuously, is 3,677 percent. Even for a mere 20 years it's a bit more than 20 percent.)

Calculating these returns for yourself is not difficult if you don't mind dipping into the table of logarithms to do it. You won't need that fat table, of course, if you have at hand a scientific or financial calculator or a personal computer.

In that case, this is what you do.

Enter the value of that widely useful constant called e, which your computer program or advanced calculator will have ready. Or simply type in 2.728. Then express the annual interest rate as a decimal and multiply this by the number of years to find the power by which you will then raise e. Subtract 1. Divide by the number of years. Multiply by 100.

For the 10-year example above, this means multiplying 0.0821 by 10 and raising 2.728 to that power (0.821), obtaining the result 2.2728. Subtracting 1, dividing by 10, and multiplying by 100 brings you to the 12.73 percent average cited.

This is an interesting calculation and sometimes a useful one. What's more to the point, though, is remembering that average annual returns are just so many deceptive and essentially meaningless numbers...unless they are used solely to compare two results for precisely the same number of years.

How Does Dollar Averaging Work?

Having chosen a stock in which you intend to invest at regular intervals, should you plan to:

...buy a fixed number of shares each time; or

...put up the same number of dollars at each purchase?

There's a trick for investing in stocks over a period of years that follows the rules of arithmetic but somehow seems more like magic. It's a device that guarantees that you can, without any exercise of judgment, buy your stocks for less than the average price they are selling at over the period.

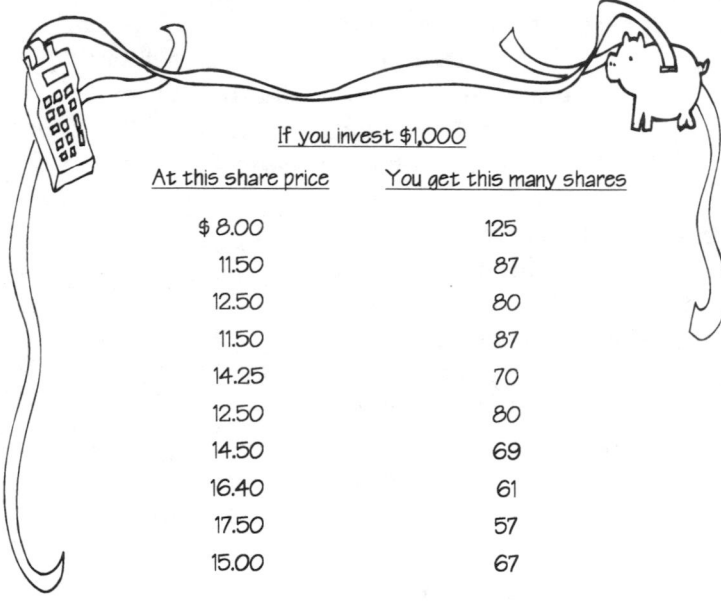

If you invest $1,000	
At this share price	You get this many shares
$ 8.00	125
11.50	87
12.50	80
11.50	87
14.25	70
12.50	80
14.50	69
16.40	61
17.50	57
15.00	67

All you have to do is invest periodically, keeping the amount you invest the same each time without regard to the price at which the stock is selling.

Called dollar averaging, it's actually easier and simpler to do than buying a preset number of shares at intervals—because it takes the same amount out of each paycheck.

You can apply this routine just as well to buying mutual-fund shares as to purchasing a single stock.

This table sums up what occurs—assuming you have a thousand dollars to invest every month, or every second month, or whatever.

Add up those share prices and divide by the number of purchasing occasions and you'll find that, on the average, the stock was selling at $13.57.

Yet your 783 shares will have cost you $10,000 divided by 783, or $12.77 a share.

Ain't it wonderful?

How can this be? Well, the thing is, the lower the price of the stock on your purchase date, the more shares you get for a given investment.

And that's why putting up a constant sum of money each time works out better than buying some stipulated number of shares at regular intervals.

Super Dollar-Cost Averaging

I F YOU like the idea of dollar-cost-averaging your investing, you may want to carry it even further—into super-averaging. By this rather cute method, you invest fewer dollars when the price is high, more dollars when it's lower.

To do this, you must first calculate the average cost of the shares you own. Divide this by the price that is currently quoted, then multiply by the amount you've determined to invest monthly.

Say you find that your stock has cost you an average of $9 and the market price right now is $8. Dividing 9 by 8 gives 1.125. If your plan is to invest $150 a month on the average, multiply this by that previous result (1.125) to find that your investment this month should be $168.75.

If, however, the market price had been $10 you'd have divided 9 by 10 and multiplied by $150 to discover that your strategy called for a $135 investment this month.

Value-Averaging Your Investment

VALUE-AVERAGING is calculated to make you an even shrewder trader—and automatically at that—than the dollar and super-dollar averagers. It is best suited to purchasers of no-load mutual funds, since brokers' commissions can make small purchases of stocks costly.

Choose a fund that permits a low initial investment.

Decide the number of dollars by which you wish to increase the value of your holdings each month. (Let's say you choose $200.)

At the end of each month, put in enough to bring your fund's total value to this amount ($200) more than it stood at when the month started. When your holding in the fund has grown by $120 in a month, it is time to put in $80. If it has diminished by $10, you'll put in $210. And if it has leaped by $250, you should—to be consistent— take out $50. Better, take this as a signal that it is time to lift your goal by leaving that extra $50 in place—and perhaps to increase the monthly $200.

What's a Tax-Exempt Bond Worth?

WHEN YOU read about people who have great big incomes and pay little bitty income taxes you wonder how they do it. One way is with tax-exempt bonds.

Will it work for you?

The simplest rule of thumb is: probably not unless your combination of federal, state, and local taxes comes to more than 30 percent.

This is because, historically, market value of securities of approximately equal safety has usually worked out so that the return from a tax-exempt is about 30 percent lower than from an ordinary bond. If an A-rated corporate bond is producing an 8 percent return, a comparable tax-exempt will sell at a price to yield about 30 percent less, or somewhere around 5.5 percent.

And don't forget that you find your bracket for this purpose by adding together all applicable federal, state, and local income levies.

At a time when this tax total works out to around 30 percent or so for you, you may find a formula useful. Here is one for calculating how much an ordinary bond must yield to be worth as much to you, after you've paid taxes on the income, as a given tax-exempt one.

First find how much Uncle Sam leaves you by subtracting your top total tax bracket from 100 percent.

Divide the percentage yield of the tax-free bond by that figure. Result is the level of yield from an ordinary security you'll need for an equal spendable return.

For a 36 percent tax bracket and a 4 percent tax-exempt bond, it goes like this: After taking assorted deductions, you get to keep 64 percent of your income (100 percent minus the 36 percent tax liability). That is, you retain your income multiplied by 0.64. The word multiply makes this sound jollier than it is.

Divide the 4 of the bond by that 0.64 to get 6.25. You have determined that an ordinary bond yielding 6-1/4 percent will do as much for you as a tax-exempt 4.

If you have the opposite question, turn the formula around so it says that tax-free yield is equal to taxable yield multiplied by after-tax percentage.

If you're doing the figuring on an electronic calculator you may be surprised to run into a small hitch. Although all you're doing is dividing one figure by the difference between the two others, this is an operation the ordinary calculator can't perform without some help.

Here's how to give it that help.

Instead of trying to divide one figure by the other, multiply that first figure by the reciprocal of the other.

These are the steps:

Enter 1 and subtract from it your tax bracket expressed as a decimal, using the equals key.

Find the reciprocal—by pressing the divide key once and the equals key twice.

Multiply this by the coupon rate of the tax-free bond.

The answer is the rate that an industrial or other bond that is not tax-free must yield to give you as much after-tax income.

Income from Capital

I's NOT hard to figure out how much monthly income a specified amount of money will produce forever. Just multiply the amount by the interest percent and divide by 100 and by 12 to find the monthly income. For a weekly income, make that 100 and 52.

With the table, it's even easier, for the half-dozen percentage rates it covers.

Amount of capital	Monthly income at rate of					
	5%	6%	7%	8%	9%	10%
$ 10,000	$ 42	$ 50	$ 58	$ 67	$ 75	$ 83
20,000	83	100	117	133	150	167
30,000	125	150	175	200	225	250
40,000	167	200	233	267	300	333
50,000	208	250	292	333	375	417
75,000	312	375	438	500	562	625
100,000	417	500	583	667	750	833

But both the table and your own calculations have a weakness to beware of. They don't mean much until you have subtracted the amount of any tax bite for which you'll be liable on the income. And then there's inflation. Your true investment return is only the amount by which the yield exceeds the current inflation rate. So if you have an 8 percent yield and a 25 percent tax, there goes one-fourth of 8, or 2 percent. In a time of 6 percent annual inflation, the remaining interest is wiped out. You can't take out a dime without, as my grandmother always put it, dipping into capital.

How'd That Investment Do?

THERE's a simple and standardized way to compare one investment with another.

Applied to something like a mutual fund, it works like this.

Begin with what's called its NAV (net asset value) at the beginning of the year. Let's say that was $10.

Now add to the NAV at the end of the year ($12, let's say) all that the fund has made in dividends and capital gains during the year—whether paid to you or reinvested. Say the dividends totaled 30 cents and the capital gain 20 cents. Add these to the $12 to get $12.50.

Subtract the NAV at the beginning from this total. Gain comes to $2.50. To find what percentage this is of what you started with, divide the gain by the initial NAV and multiply by 100.

You've had an increase of 25 percent.

To compare this with some other kind of investment, you should also allow for the tax aspects. These, in turn, depend upon what bracket you're in. Remember, too, that—depending upon what has happened in Washington lately—part of dividend income may be excluded from taxation and capital gains may be taxed at a lower rate than ordinary income.

The Moving Average

ONE KEY indicator that many stock-market professionals watch closely, and often act upon with real money, is the moving average. It's also called the current mean.

These pros believe they have found that market moves commonly exhaust themselves after going up or down by about 25 percent at most from the 40-week moving average of a group of stocks such as the Dow-Jones industrials.

To calculate this 40-week moving average, you add up the closing figures on the index for each of the past 40 weeks and then divide by 40.

After that it gets easier. Each week you add the new figure to the old total and subtract the earliest one. Then divide by 40.

The further the new figure is from the original average, the more likely it is that the stock market is shifting direction. Or so say those who call the turns for a living.

Of course, anyone using this indicator will want to consider all other available information as well. Whatever any investment type you talk to might give you to believe, there's no magic in it.

How Much Does a Treasury Bill Yield?

It is often possible to get a better rate of interest by investing in one or more U.S. Treasury Bills than by putting money into a savings account. Of course, that's if you have a chunk of money available for investment at one time.

You can make the purchase at no charge from any Federal Reserve Bank, or you can pay your bank or a stockbroker a small fee to do it for you.

The way it works, you pay the Treasury the face amount of the Bill—say, $10,000—and get back, in a few days, a discount check that constitutes your interest.

To find the yield, divide the amount of the discount check by the amount the Bill has actually cost you. You can convert this to an annual rate by multiplying it by the number of times a year your money could be turned over.

Consider a $10,000 Bill that brought you a discount check of $150.

Your cost is then $9,850. Dividing 150 by 9,850 gives .0152. If this was a three-month Bill (most are for three, six, or twelve months) multiply by 4, for one-fourth of a year. The result is .0608, which means 6.08 percent annual interest.

If you pay a fee to bank or broker, you really should subtract that from the discount. If it were $10 in this example, you'd divide 140 by 9,860 before multiplying by 4, to find the true annual yield rate to be only about 5.8 percent.

Here's something to consider, if you want to be truly canny about your investing—especially if you are going to buy many Bills at once or over a period. If you are paying with a certified or bank check on which you are sacrificing interest from the date it is drawn to the date the Fed must have it, add this into your cost.

Compare this loss with the cost of having your bank make the purchase for you.

Is Tax-Free for You?

THIS IS a quick-and-easy routine for finding out if a tax-free investment, such as a municipal bond, is a good choice for someone in your income-tax bracket.

From 100 subtract the highest tax percentage bracket that applies to you. (If you're considering a bond that is issued in your state and is exempt from state tax also, add your combined state and federal brackets.)

Multiply the yield of the tax-exempt security by 100 and divide by the figure from the previous paragraph.

This gives you the yield you'd have to get from a taxable investment in order to come out as well.

So now let's suppose your top dollar of income is taxed at 30 percent and you're considering a tax-free bond yielding 8 percent.

Subtracting 30 from 100 and dividing into 800 gives 11.4. So 11.4 percent is the yield you'll need from a taxable bond to equal an 8 percent return that you receive free of taxes.

What Percent Does a Stock Yield?

You know what percentage you can get on your money if you put it into a bank or some such place. The bank tells you: three or four, maybe, or seven or eight percent on a time deposit.

To compare the yield from a common stock, you have to figure it in the same terms.

To find out what a stock is returning to you for what you paid for it, divide the total of dividends for one year by the cost of the stock and multiply by 100. Thus a stock for which you paid $40 a share, including commission, that brings 25 cents a share in dividends four times a year, is yielding 2.5 percent on your investment.

The stock may be going up in price and making you potential profits that way, as well, but this is not what we're talking about at the moment.

A better indicator of whether a stock is worth hanging onto is yield relative to present value.

If that $40 stock has dropped in value to $25 or so, its annual dividend of $1 represents a 4 percent return on the amount of money you could get if you sold it. The fact that this is just 2.5 percent on what you paid for the stock is a matter merely of historical interest now.

What does a p/e tell?

There are a couple of figures by which the value of a stock is often judged.

One of these, yield, is the relationship between the price of the stock and its dividend for the year. A stock that is selling at 50 and producing four $1 dividend checks a year is yielding 8 percent.

If you are investing primarily for income, yield is pretty important to you.

But if you're primarily hoping for growth in the value of the stock, you'll probably learn more from a calculation of the stock's price/earnings ratio. This is determined from the company's earnings per share (the number of dollars the company earned last year divided by the number of shares of stock outstanding). Current price of the stock is then divided by this figure.

For any stock regularly traded, a broker will have the p/e on tap.

You need to know more about a stock than its p/e to make an intelligent decision about it, of course. But in general a stock with a p/e of 10 or less is considered to be underpriced and a good buy.

P/e and yield, for any given stock, could correspond closely, since the more a company earns the more it could distribute to its owners, the stockholders. But what actually happens is that a fast-growing company is likely to hang onto most of its earnings to use in expansion and in research.

So a stock with a low yield is likely to be one with good growth prospects. A high-yielding stock usually indicates a company that is making money but not necessarily going anywhere. (Maybe it's already there.) It's useful in interpreting a price-earnings ratio, once you've calculated it, to know something about typical levels in the past.

For example, for that representative group of sound, stodgy, reliable industrial stocks used to compute the Dow-Jones averages, the p/e hit a high point of 17.3 back in 1971, then began a three-year slide that brought it to about one-third of that figure before it started up again.

What are sometimes called the glamour stocks are quite another story. Since they have glamour—that is, promise of rising sharply in value and earnings—they normally sell for far more than they could possibly be regarded as worth on a p/e basis. P/e levels of 20 to 50 are quite common for them. Polaroid has had a p/e as high as 114,

IBM as high as 161. Eventually, those ratios slid precipitately, and so did the selling prices of the stocks.

For what it may be worth in making decisions based in part on price-earnings ratios, here's a rule of thumb that is safely conservative and certainly simple to apply. It is one reached by some analysts in the mid-part of the 1970–80 decade. It declares that, no matter how glamorous a company's prospects, its stock should never sell at a p/e in excess of 25.

Figuring Percentage Gain

If you've made an investment, in stock for example, you might want to know how well it has done for you in comparison to putting your money in the bank. There are several mathematically equivalent ways to find this out.

The easy formula for figuring it begins with what it's worth now. Just the stock, if you want to calculate the capital gain. Or the stock plus dividends you've received.

Divide the present value by what it cost you (or what it was worth) at the beginning of the period.

From this, subtract 1.

To get a percentage answer, multiply by 100 by moving the decimal point two places to the right.

If the period in question is a year, the answer will be the annual percentage return.

It goes like this. If you own some stock that cost you $32.25 a share a year ago, but is now worth $35.75, you divide the latter figure by the former to get 1.1085, subtract 1 to get 0.1085, and multiply by 100 to find the capital gain of 10.85 percent.

(If the answer is negative, you'll have discovered your percentage of loss. Sorry.)

If you want to know your total gain, and that stock has produced a total of $1 in dividends in a year, you'll add that dollar to the present value and begin with $36.75 instead, finding the total gain to have been 13.95 percent.

(If great precision is important, you should add on not only the dividends but also any interest earned by those dividends during the year.)

Total Return and the Deceptive Average

A LARGE mutual fund has announced that its total return to investors for its first 10 years was 413 percent.

That figure is not deceptive in itself and it does convey some information.

But for comparison with a return produced by a competing fund, this number is useful only if the figure from the second fund is also for 10 years (and preferably the same 10).

For making comparisons generally, you might find it more useful to know the annual rate...on the average. In a careless moment, however, this could lead you to use ordinary averaging methods that don't apply to measuring rates of increase. That is, you might—but you shouldn't—say that if the total return for 10 years was 413 percent, then the average yearly return would be one tenth of that, or 41.3 percent.

A large, glossy, and widely circulated consumer-finance magazine blithely went at it that way once and produced a brew of figures that were, at the same time, highly impressive and totally meaningless.

Being more experienced than the editors with numbers like these, the mutual fund has calculated that theirs was an average annual return of 17.8 percent. The fund's method produced a far less exciting conclusion and it's harder to understand at a glance.

However, it does have the virtue of being correct.

What's the trick for translating the total return for a period of years into an annual average?

First turn the percentage into a decimal by dividing by 100. Add 1.

Derive the root corresponding to the number of years (for example, tenth root if you're dealing with ten years).

Subtract 1.

Return this result to percentage form by multiplying by 100.

Doing this to the 413 percent 10 year return, you'll be taking the 10th root of 5.13, which proves to be 1.178. With 1 subtracted and the decimal point pushed 2 places to the right, you'll find you've confirmed the fund's figure of 17.8 percent average annual return.

Don't be alarmed by having to derive something like a tenth root. You can do it almost effortlessly with (take your choice) a modest scientific calculator, a computer, or a log table.

If you lack the first two and have a not unnatural aversion to the third, there's also a great do-it-yourself approach that can make you feel pretty clever. Call it iteration. Or just call it cut-and-fit.

To find any root, including a tenth, make a guess. Then make another guess or two till you have it bracketed. Then close in with guesses that fall between.

Seeking the tenth root of 5.13, you might guess safely low at 1.1. Testing that guess, by hitting the times button once and the equals button nine times, for a total of 10 hits, will give you 2.59. Not big enough.

Try 1.2. This series of ten punches produces 6.19, too big by a trifle, but when you then try 1.19, you have it nicely bracketed.

So you're not surprised when 1.18 hits it so nearly on the nose that you may not bother to refine it further to 1.178.

Understanding Bond Language

IF YOU are interested in buying U.S. Treasury notes and bonds, and bonds of government agencies, you'll have to learn bondish—the language of the business.

Bonds are described by their maturity dates and rate of interest—based on the par value. So if the bonds you're interested in are described as "4s of June 1999," you know they mature—become payable—on that date and bear interest at 4 percent.

Now about price, which is not so simple. Bonds are quoted in points, which means 1 percent of par value—$10 per point in a $1,000 bond. But what looks like a decimal isn't. It's a 32nd. So when a bond is quoted at an asking price of $99.16, that means it is 99 and 16/32, or 99-1/2. For a $1,000-par-value bond, that means a price of $995.

As you can see, this is a pretty crazy system, but it's what the bond world uses. So if the price is given as 99.8, that means it is cheaper than one quoted at 99.16, since 99.8 is 99 plus 8/32, or 99-1/4. Sorry.

Chapter E
OTHER PEOPLE'S MONEY

The Everlasting Payoff

THIS ONE comes pretty close to being the prince of personal calculations. It figures out the amount of the monthly payment for any size mortgage at any rate of interest for any period of years.

So it replaces an entire bookful of closely printed tables comprising roughly half a million digits. That's what people in the real estate and mortgage businesses carry around to tell you how much something is going to cost you each time the first of the month rolls around.

Actually, this calculation method is far more versatile than your average hefty book of tables, which will cover only a limited range of interest rates—say 6 to 12 percent if it is designed for the mortgage business.

When you do your own calculating you can handle lower or higher rates, such as may be needed for automobile or other installment loans. You can figure payments for odd rates of interest (7-1/8 percent for example) not given in most tables. And you can get exact results for paying off amounts that are not round numbers—all this with no additional arithmetic.

Although the method is reasonably simple, the actual computation is horrendously tedious if tackled by straight paper-and-pencil arithmetic.

A table of logarithms will quickly cut the job down to modest size, as will a slide rule—if I may offer what amounts to a historic note. But most people these days will use pocket calculators.

Whatever your method, these are the steps:

1. Express the stated rate of interest as a decimal and divide it by the number of payments per year.

2. Add 1.

3. Raise this sum to the power of the total number of payments.

4. Find the reciprocal—that is, divide 1 by the number just obtained.

5. Subtract 1. Drop the minus sign or ignore it from here on.

6. Divide by the interest rate, expressed as a decimal.

7. Multiply by the number of payments per year.

8. Find the reciprocal.

9. Multiply by the amount of the mortgage or debt.

All this may sound pretty fierce, but these steps will become easier to follow after you've seen them applied to an actual debt payment.

I once encountered a loan company second mortgage in the odd amount of $4,791.36. It was to be paid off in three years at what the legally required disclosure form admits is an annual percentage rate of 18.25. Finding the nine steps above reveals the size of the monthly payments.

1. There are 12 payments per year and 0.1825 (18.25 percent divided by 100 to change it into a decimal) divided by12 is 0.0152.

2. Adding 1 gives 1.0152.

3. The previous sum when raised to the 36th power (because there are 36 months in 3 years) becomes 1.7218.

4. When 1 is divided by that number, the answer is 0.5808.

5. Subtracting 1 gives 0.4192 when you ignore or change the minus sign.

6. Dividing by 0.1825 gives 2.2971.

8. Dividing this into 1 produces 0.4353.

9. You now have the amount of each payment per dollar of debt. Multiply it by the size of the mortgage ($4,791.36) to find that the monthly payments will be $173.82.

A couple of these steps require techniques not covered by the little booklets that come with ordinary calculators.

Even with this mere 3-year loan, Step 3 calls for raising something to a fairly high power, the 36th. That's not too hard to do by giving the times key 36 punches (or the times key one and the equals key 35) and, I find, takes under half a minute. But the 360 key pokes for a 30-year loan will take several minutes or longer, since you may have to repeat the process several times before you're confident that you've counted right.

Better you should factor. Small numbers that multiplied together give 360 are, for example, 6, 6, and 10. To find the 360th power the quick way, touch the times key once and the equals key 5 times for that first 6, to get the 6th power. Repeat for that second 6. And then touch times once and equals 9 times for the 10th power.

For a 20-year loan, convenient factors of 240 are 3, 4, 4, and 5. And so on.

A couple of the steps call for finding a reciprocal. Some calculators have a special button for this, often marked 1/x. With most others, you just punch divide and then equals. Test this on your own machine before relying on it. These machines often have ways all their own.

If you'd like another example to use in checking your technique, one of the fat payment books the real estate people use says that a $42,000 mortgage at 7 percent interest will pay off in 20 years at $325.63 a month. (My calculator agrees.)

Credit: How Much Is Too Much?

Accorpning to one "family resources management specialist," there is a simple formula to warn you when you are getting too far into debt. I've found this kind of warning worth respecting, since piling up unwieldy debt is a pretty sure way to surrender control of your life.

In the absence of the guts to tough it out, this has been known to entrap whole countries, even the richest in the world.

To keep control of your credit, says the formula, first add up all your monthly payments, excluding housing. Multiply by 100 (by tacking on a pair of zeros) and divide this figure by your monthly income (not gross—take-home, or after taxes, pay).

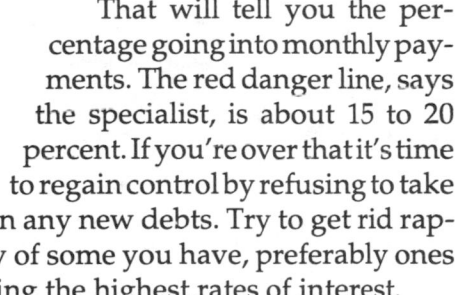

That will tell you the percentage going into monthly payments. The red danger line, says the specialist, is about 15 to 20 percent. If you're over that it's time to regain control by refusing to take on any new debts. Try to get rid rapidly of some you have, preferably ones carrying the highest rates of interest.

Of course, you should keep in mind that our specialist is offering only a very rough rule of thumb for families in normal situations. If yours is out of the ordinary, you'll have to cut the generalization to fit. A family paying rent on a rather expensive residence should be pretty wary about letting its payments reach even 15 percent of income. Another family, owning a home clear or buying one in the late years of a low-payment mortgage, might find it tolerable to put more than 20 percent of income into payments for really worthwhile (and durable) things.

How's Your Credit Rating?

HOW CREDITWORTHY—as they say in that line of business—are you?
Each loan or time-payment officer has his own way of figuring
that, but many rely heavily on some kind of prefabricated scoring

Years on job	Below .5/ unemployed	.5-1.49	1.5-2.49	2.5-6.49	6.5-10.49	10.5-20.49	20.5-up/ retired
	5	15	18	20	27	39	53

Home phone	Yes	No	Close by
	35	0	30

Years of address	Below .5 yrs.	.5-2.49	2.5-5.49	5.5-2.49	12.5-over
	2	8	19	25	30

Dept. store / major CC	None given	Dept. store	Major/both
	0	11	27

Age of auto	No auto	1-2 yrs.	3 yrs.	4-5/ under 1	over 5 yrs.
	0	22	10	9	9

Finance Co. ref.	Yes	No
	5	11

Source	New customer	Present customer	Former customer
	5	5	18

STEP 2 - CREDIT BUREAU INFORMATION

Excel. & sat.	Derog only	No record	One	Two	Three & up
	-15	-4	0	8	18

sheet. The specimen you see here is a typical example of ones that may have been used many times to rate you.

If you wonder why credit managers use such seemingly crude instruments instead of their own good sense, here's why. Sears Roebuck uses one of these systems and finds it can keep its losses down to 1 percent. But where it has experimented with allowing credit managers to override the rulings of the system, 95 percent of the accounts to which the system had said "no" proved difficult or impossible to collect.

You may wonder at the absence of a salary or income item on the schedule. That's because lenders have discovered that size of income claimed is no help in predicting whether you'll pay your bill.

Suppose you've worked at the same job for 7 years, have a home phone, have lived at the same address 2 years, possess an American Express as well as a department store credit card, don't own a car, have not given a finance company reference, are a new customer— and the local credit bureau has never heard of you.

Your scores will be 20, 35, 8, 27, 0, 11, and 5 for a subtotal of 106, from which you must subtract 4 for lack of a credit bureau record. That's probably passing, since the folks who offer this scoring system say that only by rejecting people who score less can a lender keep his delinquency rate below 4 percent.

Real Cost of a Discounted Loan

I<small>T HAS BEEN</small> a historical, and pernicious, practice (and not just among car salesmen) to quote the cost of a discounted loan in "dollars per hundred per year"—and then to refer to this as the interest rate.

Here's the way to find the true rate, which you might need to know in order to compare it with alternatives.

Multiply the number of installment payments to be made per year by 2 and also by the cost of the loan. (Cost of the loan is the so-called discount, the difference between the amount you actually receive and the sum you agree to repay.)

Now divide by the amount you actually receive. Also divide by 1 more than the total number of payments you are to make.

For a loan of $1000 at $6 per hundred discounted to be paid monthly over 2 years, the calculation works out like this.

At the quoted $6 per hundred per year, the discount amount is $60 per thousand per year, or $120 for 2 years. Subtracted from $1000, this leaves $880 as the amount you receive. At one payment each month, there will be 24 payments in 2 years.

So you must multiply 2 by 12 by $120. And divide by $880 and by 24 plus 1, or 25. Which comes out to 0.1309, or almost 13.1 percent interest. That sounds quite a bit different from "6 percent per hundred." And is.

Calculating the Cost of Credit

THERE ARE three methods in common use for figuring the cost of consumer credit. Even when the stated interest rate is the same, each produces a different effect on your checkbook.

Under the federal truth-in-lending law, you must be told which method is used—as well as the annual percentage rate, which over the years has often been around 18 percent, or 1-1/2 percent per month.

Suppose that by your billing date of August 1, you have paid enough on your bill at an auto-parts house to bring it down to $155. On August 5, you pay $50. On August 10, you charge a pair of tires for $90, and on the 25th a cassette player for $180. It's clear you're going to be billed September 1 for $375 plus interest.

But how much interest?

Read the fine print. If it says adjusted balance method, interest will be simply 1-1/2 percent of the old balance after your payment was subtracted, or 0.015 times $105. This is $1.58.

But suppose you charge your car stuff at a store, or on a credit card, that uses the average daily balance method, excluding new purchases. Your average balance was $155 for 5 days, then $105 for the remaining 26 days of the month. Adding the product of the first two figures to the product of the second two and dividing by 31 days gives your average daily balance as $113.06. Multiplied by 0.015 this becomes $1.70.

If you use a credit card specifying average daily balance plus new purchases, it's quite a different ball game. Now you have 5 days times $155, 5 times $105, 15 days at $195, and 6 days at $180. Average daily balance, when these products have been added together and divided by 31, is $171.13, of which 1-1/2 percent is $2.57.

On large purchases—or even on small ones over the years—it could make quite a difference. Compared to the first method, the second has cost you about 8 percent more interest and the third has elevated the cost of credit by 63 percent.

Pretty slick, that.

On the whole, however, people who run unpaid credit card balances month after month apparently do not concern themselves about where their money goes. So why are we bothering?

Figuring an Early Payoff

Let's assume you're making an early payoff of some kind of loan. It's one in which the total amount of interest is stated in dollars and for which the agreement has specified that the Rule of 78s applies.

This widely used rule specifies how much rebate you're entitled to for paying off the loan before it is due. To discover this amount, you multiply the total interest charge by a fraction.

You can find the upper term of this fraction by subtracting from the total number of payments called for the number you've already made—and then multiplying this by itself plus 1.

(That is, if yours is a 24-month loan on which you have made 19 payments, you'll multiply 24 less 19 by 1 more than 24 less 19—or 5 times 6.)

To find the lower term of the fraction, multiply the total number of payments by itself plus 1. For this 24-month loan, that means multiplying 24 by 25.

So your fraction is 30 over 600, or 1/20.

If your loan was $1,500 at 12.75 percent, the 24 payments would each have been $71.14, for a total of $1707.36, of which $207.36 is interest. Using the fraction in the preceding paragraph, take 1/20 of that total interest amount to find that your rebate is $10.37.

Payoff Rule of 78s

THE BUSINESS world has an odd way of figuring how much of the interest on a loan you've paid off at any given time. You can make use of it for any kind of loan or installment payment where the interest has been figured as a dollar amount—and the Rule of 78s is specified.

While the rule most often comes into play when you're making an early payoff, it is also useful for computing (for income-tax or other purposes) how much interest you've paid during a calendar year.

Here's how it goes.

Starting with the last payment that your agreement calls for, and labeling it number one, number all the payments—two for the next to last and so on.

Add up those label numbers for all the payments made during the tax year. Use this total as a numerator (that's the top number of a fraction—remember? I never can.)

Now add up the numbers for all the payments, past and future. Make this the denominator of your fraction.

Multiply the total amount of interest by this fraction. The result is the amount of interest you paid during the year.

Example. If you're being charged $1,000 interest on a 12-month loan and made the first 9 payments during the tax year, you'll multiply that $1,000 by 72/78 to find that you paid $923.08 interest during the year.

When You Make Partial Payment

YOU'VE borrowed $3000 for 6 months. After a month has gone by and you've become acutely aware that the 18 per-cent interest you've agreed on is mounting up at a fierce rate, you find you have some money to spare. You seize an opportunity to pay back $1200 of the debt.

How should this be credited?

What is known as the United States Rule has been in common use since the U.S. Supreme Court embodied it in a ruling in 1839. It says that whatever you pay goes first to interest, with anything remaining being applied against the principal amount.

It works like this.

After a month you owed only $45 interest ($3000 times 1/12 year times 0.18). So $45 of your payment takes care of interest to date, leaving $1155 to be credited to principal.

Subtracting $1155 from $3000 leaves $1845—now the amount of your debt.

If you make no further payments, you'll owe this amount, plus 5 months interest on it, at the end of the 6 months. This is $1845 plus the product of $1845 times 5/12 of a year times 0.18. Which comes to $1983.38.

Adding this to your earlier payment gives a total of $3183.38—or $183.38 interest compared to the $270 the loan would have cost you in interest if you hadn't paid anything on it until you had to.

How Big Is That Carrying Charge?

THE SIGNIFICANCE of 18 percent annual interest, or 1-1/2 percent per month, is that it's a fair approximation of what most of us pay for a good part of the credit we use.

Department stores commonly assess this amount on time accounts, as do most credit-card organizations. Many installment purchases and loans take this rate in most states.

Take a look at your credit cards or their disclosures. You may be bluntly informed that the cost in your state is 18 percent, although it may be half that in some others. You might ask yourself, or your legislators, why some states protect their citizens against shocking rates of interest while yours doesn't. But anyway, to know quickly what such interest is going to cost you in pennies-out-of-pocket on a specific transaction, just keep this handy figure in mind: 18 percent a year (1-1/2 percent a month) is precisely a nickel a day on each $100.

It was a banker that helpfully pointed this out to me, in connection with the fact that 18 percent happens to be what many banks charge for automatic loans to cover overdrafts. Though insidious and potentially costly, this kind of plan seems to me well worth the price—if you repay that automatic loan in a hurry.

Perhaps that's just how you will use it, if you keep those nickels-a-day in mind. Ponying up five cents a day for a week to avoid a rush trip to the bank when you're a hundred in hock is a bargain. It's even more so compared to being hit with a ten- or fifteen-dollar penalty for a single careless overdraft just because you haven't signed up for this kind of plan.

One warning! Don't waste a useful amount, as I did once to the tune of $11, by forgetting that the interest will run on and on until you speak to the bank about it, even if meanwhile you put a million dollars in your checking account.

One of my kids, by the way, tells me that her bank in Maryland charges only 10 percent, which comes to almost exactly 2-3/4 cents per day per $100. That's 83 cents a month.

To make your calculation for one of these Executive Check or Balance Plus overdraft accounts, you need to know one other thing. In what increments are the transfers made? Some banks will move $100 the minute you need a cent, which can make a tiny overdraft disproportionately expensive. I can forgive my own bank for its high rate when I note that it lends me the money in $5 bits so I'm not charged for more money than I'm using.

By the way, the statement permitted under the federal truth-in-lending law, that 1-1/2 percent per month is 18 percent per year is actually true only when you're talking about a single month. If you let your account run, the interest you'll actually pay is 18 percent per year compounded monthly and this comes to 19.56 percent simple interest. If you owe $100 and don't pay it for a year, you'll find that you now owe not $118 but $119.56.

(This is figured by using the same formula that would apply to a savings account at 18 percent compounded monthly. Because 1-1/2 percent is 0.015 and there are 12 months in a year, the calculation is 1 added to .015, that sum then being raised to the 12th power. Run that off on your mini-calculator and you'll get 1.1956. Multiply by $100 and, lo, the result is $119.56.

What, Borrow on My Life Insurance?

I'M RIGHT sick of coming upon the same silly statement in every discussion of whether it's smart to borrow against the cash value of your life insurance.

In one way or another, you're cautioned against doing so on these grounds (to quote a recent magazine article): "If you die while the loan is outstanding, the amount going to your beneficiaries will be reduced by the amount of the loan and interest unpaid."

Sure it will. But how much you leave behind is going to be reduced by the amount of any other type of loan you get instead, so what's the difference? Either way, your heirs aren't going to get it because you already got it.

With this cliché out of the way, you're free to consider a life insurance loan with the knowledge that the only thing against it is that there is no pressure on you to pay it back. If that's a disadvantage.

Going for it is the fact that it is probably easier, and most certainly cheaper, to borrow against your insurance than any other way. Well, it ought to be cheaper—it's your money.

I've been happily paying interest on an insurance loan for upwards of ten years. It's costing me less than 4 percent and I don't see any reason to pay it back with money that has meanwhile been invested to fetch me two or three times that for an easy profit of several dollars a year on each hundred.

Chapter F
GETTING A HOME LOAN

EASY AS FALLING OFF A LOGARITHM!

Hiring the Money

THE MOST costly purchase you'll ever make is probably not your new home—it's the mortgage on it. A $100,000 loan at 8 percent for 30 years will cost you a total of $164,152 in interest alone. Those payments of $733.77 you'll be making for 360 months come to $264,152. Even with a 20-year loan at 8 percent, the total you'll repay will be more than twice the sum you borrowed.

You can figure this out for the mortgage you have, or are contemplating, in a couple of steps—for any interest rate and period of time.

Multiply the monthly payment by 12 times the number of years. From this, which is the total you've promised to put out, subtract the amount of the loan to find out how much is interest.

Here's another way to look at it. On a 30-year loan at 7-1/2 percent interest, you will pay $1.50 interest—as well as $1 in repayment of principal—for each $1 you borrow. And that $1.50 becomes $2.00 as interest rates pass 9-1/4 percent.

Following are the factors to look for and weigh in shopping for the best mortgage deal. The essentials are the same, whether it's called a mortgage or, as in some areas, a deed of trust.

What kind of loan is it? Home loans come in two main types: government sponsored and conventional. In most areas, the latter are more usual today, except where groups of houses are involved.

Best of all are the subsidized kind, if your state has such a program (in California, there's Calvet for veterans). If you're eligible, it can save you a lot of money. VA (sometimes called GI) or Veterans Administration loans are also a bargain for those who qualify through military service. The government partially guarantees them and they're often made with little or nothing down, if the appraiser agrees that the house is worth the full selling price.

Next best may be an FHA loan, insured, but not subsidized, by a government agency. This insurance will cost an additional 1/2 percent, which may be partly refundable. Be sure to include this insurance cost in your comparisons. Banks and lending institutions make the FHA loans, not the government.

The majority of home loans, those without subsidies, are the kind called conventional. However, interest rates, amount of loan offered, and subsidiary clauses on these loans vary a great deal from one bank or other lending institution to another.

There are also private mortgage-insurance agencies. The biggest are Guarantee Insurance Corporation (known as MGIC), American Mortgage Insurance Corporation, and Continental Mortgage Insurance, Inc.

Here is how this type of insurance might help you get your home. You have found the house you want, at $80,000, but can manage only 10 percent down. After failing to get either a conventional or FHA loan for the needed $72,000, you find a savings-and-loan that will lend you the money if an insurance company will cover most of the risk.

Private insurance is obtained. The cost (paid by you, of course) is a small appraisal-review fee, plus a first-year cost of perhaps 1/2 of 1 percent of the amount of the loan. This buys a guarantee to the lender that either you or the guarantor will pay the top 20 percent of your loan. Since the lender still has the house as security for the rest of his money, he is quite well protected—and happy to make the loan that he otherwise might rule out as too risky.

Your loan-insurance premium will decrease each year. You may even be able to persuade the lender to let you drop the insurance after a few years as the loan becomes smaller and you build up a record as a reliable payment-maker.

How Much Can You Get?

Banks sometimes lend a severely limited percentage of the appraised value of a property, perhaps as little as 65 to 75 percent of the value unless the loan is insured. Savings-and-loans and mortgage companies may go higher.

Appraisal policies of banks tend to be conservative. Rules for estimating your financial capacity for a given loan may also vary among lending institutions. A private lender may be most flexible of all, though likely to require a high rate of interest, unless he is the seller of the house.

FHA has rigid rules that you must meet before you can get an insured loan. Since many lenders have similar qualifying rules, it is helpful to know what some of them are.

First, you must be able to show a dependable income. A second income will count only if it comes from what appears to be a permanent job. Casual moonlighting, overtime pay, plans to rent out a room in the new house—all these may not count at all, even when they should.

Your present indebtedness, credit rating, and standard of living should be within your income. Your total financial obligations must leave room for the new payments. And beware: Your mere owner-ship of several high-limit credit cards, even if unused, may be construed as such an obligation.

If your children are approaching college age, you may be viewed as in a more stressful financial period than if they were very young or through college.

Your total housing expenses will be considered, including pay-ments, taxes, insurance, repairs, utilities—and even transportation if the house you are buying is in a rural area or far from your work.

Considered are all such factors and these as well: your age (30-40 is excellent, 25-29 good, 41-55 fair, over 55 or under 25 poor); number of dependents (one or two children best); substantial life insurance (about twice the amount of the proposed mortgage).

That Important Percent

Interest rates vary, just as prices of other commodities do. Shop around. The buyer of a former home of ours saved substantially on a loan by shopping carefully so that he was ready to make a decisive move. His own bank was asking 7-1/4 percent, but another bank he dealt with was still lending at the previous week's 7 percent. He closed his loan deal just hours before the second bank boosted its own rates.

What you do for a living can also be a factor when you come to borrow a large sum. Some years ago I found four rates of interest offered within a radius of a few miles. Savings-and-loans wanted 6-1/4 percent; a branch bank serving wealthy depositors (and over-loaded with cash) was asking only 5-3/4 percent; an insurance company, 5-1/2 percent. I qualified for the 5-3/4 percent and saved $20 to $40 a month for a long time. I was denied the even more favorable 5-1/2 percent loan because the insurance company wouldn't risk its money on a free-lance writer. They didn't seem to take my point that a writer's income may be somewhat unpredictable but he's never unemployed.

How important to you is a jump from 8 percent, say, to 9 percent? For a typical mortgage each such jump will add $6 to $7 to every monthly payment on each $10,000 of the debt.

Here's another way to look at it. If you find yourself paying 9-1/2 percent when, with more shopping around, you might have been paying 8-1/2 percent, it will take you 25 years to pay off a loan that otherwise would have taken only 20 years. The higher rate will eventually add $18,400 to the payoff cost of a $40,000 loan—the price of a trip for two around the world (if you want to look at it that way—and I do!) or a new car or two.

Variable or Fixed?

SHOULD you go for a variable-rate mortgage?

The term can apply to any home loan on which the interest rate (and your payments) will go up or down with changes in the level of interest rates generally. The measuring standard may be any of several widely published rates, such as banks' cost of funds, the prime lending rate for business customers, and so on.

This choice between fixed and variable can be a nasty one to make. It can involve a lot of money—yours. It's a forced gamble with a family's economic security that no one should have to make. For most of us it's a once- or twice-in-a-lifetime undertaking, so we face it with little or no previous experience. It's justified only by the happy circumstance that the values of home ownership generally make it worthwhile.

And, happily, some rules-of-thumb come down on your side.

Rule 1. Begin with the basic, and obvious, principle that if you believe interest rates are going to fall, you should go variable. If it appears that you are borrowing at a period of relatively low interest, you will prefer a fixed rate.

Rule 2. Temper that truism if you expect to keep the mortgage for no more than five or six years, which is not far from the average life of home loans. Since most variables start out, for tactical reasons, with a temporary low rate, the cost for the first few years usually averages out quite attractively. This may be your most important consideration.

Rule 3. If the combined force of the first two rules is not decisive, consider this. It is something lenders may neglect to emphasize; they have a natural bias in the direction of a variable, since it guarantees that the interest they collect will never drop below a profitable rate.

If a time comes when mortgages are offered at a rate well below your fixed one, you can probably refinance yours at that lower figure. You can go to a new source of money, unless your lender meets the competition.

With a variable, however little you may have come to love it, there is usually no way out.

Rule 4. Look for special-incentive variables that might fit your own situation. Ask about plans that combine features of both fixed-rate and variable. Example: A variable providing that if at any time within the first three years the going rate should drop, you could ask to have your rate adjusted to that level—but only once.

Rule 5. Start doing the arithmetic as far as possible before you must make your decision final. Here is how a variable 8 percent 30-year mortgage for $25,000 might go. Get a preview of your position by multiplying each of the following figures by the number of times $25,000 goes into the sum you are about to borrow.

Monthly payments on $25,000 would be about $183. If after five years the going rate for similar mortgages had risen to 8-1/2 percent, your payments would rise to $191 a month. Another 5 years and the rate goes to, let's say, 9 percent. Your payments would then be $198.

Those Tricky Points

A PROSPECTIVE mortgage lender may charge you one or more points, in additional to other fees. This happens when he's not willing to make the loan at a stated rate but finds it undiplomatic or illegal to raise the interest directly.

"Point" is short for percentage charge. You pay it only once. A 2-point charge, for example, on a $122,000 loan means you'll pay (in advance) a one-time fee of 2 percent of $122,000, or $2440, in addition to all other costs.

FHA specifically forbids charging points to the buyer. The seller, knowing he is stuck with the points, must manage to work the charge into the selling price of the house some way or other. Of course, it's usually the buyer who really pays after all.

Points are essentially just another way of raising the interest rate on a mortgage and collecting the whole bonus right now. If you happen to sell the house in a year or two, 2 points will have cost you an additional 2 percent. In other words your 9 percent mortgage will have been an 11 percent mortgage. But if you spread the cost of the points over 30 years of payments, those points will have cost you only a fraction of 1 percent a year.

A useful quick-calc rule to go by: assuming you keep the mortgage 20 years, each point equals 1/8 of 1 percent added to the stated rate of interest.

So when you hold a 20-year mortgage at 8-1/2 percent plus 4 points all the way to maturity, its cost to you is approximately the same as for a 9 percent loan with no points.

Penalties, Options, Balloons

A<small>FTER</small> a lender has gone to the trouble of making a loan, he doesn't want it paid off before he has had time to collect quite a bit of interest. He doesn't want you to pay him off by borrowing elsewhere for less if interest rates decline. So he may stipulate a penalty if you pay off early.

Keep any prepayment-penalty clause out of your mortgage if possible. Then if you want to make bigger payments, refinance, or sell the house, you won't be stuck with a big bill. FHA cut off prepayment penalties way back in 1972.

An open-end clause is a very desirable provision in a mortgage. It lets you re-borrow the money you've paid in and do so at the same interest rate the original mortgage carries.

Let's say you've been paying on a $100,000, 30-year mortgage for 12 years. Balance owed is $79,068. Now you need some extra cash to finance home improvements or tuition fees for your college-bound children. An open-end mortgage will permit you to borrow the difference between $100,000 and $79,068—$20,932. You'll repay this new loan—with interest—by increased monthly payments or by an extension in the length of time the mortgage is to run.

An open-end offers an unfortunate temptation to use up a considerable chunk of money you've safely invested in your home. It is, however, such a convenience that it has spawned a monster with an even greater potential for evil. This is that all-too-familiar home-equity loan, which can offer useful savings but which lacks the advantage of a previously set interest rate that might be a good bit lower.

Caution! Balloon!

A mortgage that is scheduled to come to a sudden end with a great big final payment, appropriately called a balloon, may be the only way to keep the monthly costs low. But unless you know where the cash is coming from, you must be prepared to refinance at what may prove to be a time that is both awkward and expensive. Remember the distress of the widow in the old melodramas as the villainous banker approaches with foreclosure papers.

Closing Costs

ALL THOSE petty annoyances and minor disasters totted up as closing costs can come to a substantial figure. Here are a few that you may have to pay before you move in: credit report, survey, title fee, title insurance, attorney fees, document preparation, recording fee, transfer taxes, escrow fees.

Depending mostly upon where you live, the total amount on even a modest house can come to several thousand dollars that you'll have to put up in addition to the down payment.

You can learn what these costs are through your bank, real-estate agent, or loan source. However, the hardest part may be to pry this information out of them as far ahead of the fatal day as you may feel you need it.

Often, the biggest single cost of closing is for title insurance, a protection for the lender against loss arising from a defect in your ownership title. Similar protection for you will usually come from a title policy purchased by the seller—though in some areas the buyer also pays for this, or divides cost with the seller. You naturally should try to get this arranged to your advantage, unless local custom is too fixed to buck.

At closing time, you can expect yet another financial shock. Any taxes or insurance costs the seller has paid in advance for months yet to come must be refunded to him.

By you, of course.

Assume or Buy Subject To?

IS THE PROPERTY under an existing mortgage? If so, find out if you can assume it. The interest rate may be much lower than you could get on a new loan, say 6 or 7 percent when the going rate on new loans is 8 to 10.

Such an assumption at a favorable rate will put you well ahead in the long run, if you can tolerate the short-run drawbacks. An assumable mortgage very often is not as big as you'd like and has to be paid off more quickly than you might wish.

Consider this situation. A man and wife came upon a 10-year old house they liked very much. Price arrived at after some haggling was $98,400. The existing mortgage, initially $72,000 for 20 years, had been paid down to $52,800. At a modest 7 percent, compared to the going rate of 9, it sounded good. The hitch showed up when they realized they'd have to put down $45,600 instead of the $19,680 (20 percent) they'd hoped to get by with. Happily, the pair had enough savings to handle the difference, and the bargain interest rate convinced them this was the best investment around.

If they hadn't had this much cash, or wanted to keep it intact, they could have tried to interest the seller in taking back a second mortgage for the amount lacking. Even if this second were written at a rate somewhat higher than the 9 percent going rate for firsts, the combination would still have been a bargain if they could handle the comparatively high total monthly outlay. Their reward would be an early mortgage burning, traditionally one of the more joyous bonfires.

Instead of assuming, or taking over, a mortgage, you might buy a house "subject to" an existing loan. This is a perfectly good alternative from the point of view of lender and buyer alike.

However, since the seller remains responsible for any loss if you default (instead of fully getting out from under) he may not be altogether enthusiastic unless you can make it a condition of purchase.

How About a Second?

THE VERY words second mortgage have taken on an unpleasant ring from association with high interest rates and buyers who are in over their heads. Caution is called for, as with any enterprise requiring either new clothes (Thoreau) or borrowing (Franklin, probably).

When is taking on a second justified? Generally, a second mortgage is an acceptable tool when it permits going ahead with a transaction that would be impossible without it and that in all other respects is prudent. Of course, there must be assurance that the double monthly obligation can be met.

The more usual second is the kind called purchase money. The lender is the seller of the house. In order to make the sale he is accepting the down-payment partly in cash and partly in a second mortgage. Because of the same incentive, he may go along with an interest rate on the second that is no higher than the bank or other prime lender has offered on the first.

The Land Contract

Mortgage? Deed of trust? There is another kind of financing most buyers or sellers of building sites or houses never think of. It is called a contract of sale or, in some parts of the country, a land contract. It doesn't involve an actual mortgage or trust deed, but it takes the place of one.

There's really nothing weird about it. It just means buying a home the way you've normally bought a car or a refrigerator: on time payments. The difference is that with real estate there are more safeguards. In some states, partly because of past abuses, legislation and court decisions have made land contracts more protective of the buyer than the alternatives have proved to be. In short, safer than an ordinary mortgage.

Before buying or selling on a contract of sale, consult a lawyer who works with real estate. He can see that the agreement is properly drawn up and can tell you whether there are any hidden dangers in this arrangement in your state. As a buyer, one thing that you are interested in is this: if you should prove unable to make one or more payments at some time, you will not be treated more harshly than you could be under an ordinary mortgage. Make sure that the terms of the contract meet this requirement.

One advantage of a land contract is that, being an installment sale, it can permit the seller to postpone, or spread out, payment of much of the income tax on any profit he makes in selling the house. This concerns you because it may offer an incentive to him to accept a lower price. I've used it as a regular thing, to avoid formalities and the nuisance and cost of a lot of paper work in transactions that didn't call for those money-borrowing instruments called mortgages and deeds of trust.

Not everybody—especially lawyers—may see this my way. I went into contracts of sale in writing a book on housing some years ago and darned if the publishers didn't come back to me with alarm after their lawyers had given the manuscript a routine going-over. Said their legal talent had never heard of such a thing. I never heard of anyone complaining, however, after the book came out.

Trying a Mortgage for Size

THIS IS a special kind of mortgage calculation. It works in reverse. Instead of starting with a mortgage of known size, it uses interest rate, payment amount, and number of years to discover the face amount of the mortgage.

You might want to use it to estimate how big a mortgage you can handle, and from that the price range of houses to look at.

It will serve equally to reveal the original size of an existing loan for which you have the other information.

Suppose you are looking at a 9 percent 20-year mortgage with payments of $965.55 a month. What is the face amount?

1. Divide the interest percentage by 12 to get a monthly figure and by 100 to make it decimal. Store this for later.

2. Add 1 to the previous result, then raise this total to the power equal to the number of months (not years!) the loan is to run. Store this result, too.

3. Subtract 1.

4. Divide by the result saved in Step 1.

5. Divide by the result saved in Step 2.

6. Multiply by the amount of the monthly payment.

Here's how it runs for that 9 percent, 20-year mortgage with its payments of $965.55 a month.

Dividing 9 by 12 and by 100 gives 0.0075. Adding 1 to that result and raising to the 360th power gives 14.73. Subtracting 1 and dividing by 0.0075 and 14.73 produces 124.29. Multiplying by $965.55 shows the face amount of the loan, what it came to in the beginning, to be approximately $120,000.

How Much per Year?

HERE IS a different kind of table from the usual one dealing with home mortgages.

Using this table, you can instantly find what percentage of the face amount of the mortgage you will have to dig up in payments each year.

You can then quickly calculate how much that will be in dollars each year for any size mortgage. Then you multiply the figure in the table by the size of the mortgage.

START WITH THE PERCENT

	6%	6.5%	7%	7.5%	8%	8.5%	9%	9.5%	10%	11%	12%	15%
5	23.20	23.48	23.76	24.05	24.33	24.62	24.91	25.20	25.50	26.09	26.69	28.55
6	19.89	20.17	20.46	20.75	21.04	21.33	21.63	21.93	22.23	22.84	23.46	25.37
7	17.53	17.82	18.11	18.41	18.70	19.00	19.31	19.61	19.92	20.55	21.18	23.16
8	15.77	16.06	16.36	16.66	16.96	17.27	17.58	17.89	18.21	18.85	19.50	21.53
9	14.41	14.71	15.01	15.31	15.62	15.94	16.25	16.57	16.89	17.55	18.22	20.31
10	13.32	13.63	13.93	14.24	14.56	14.88	15.20	15.53	15.86	16.53	17.22	19.36
11	12.44	12.75	13.06	13.38	13.70	14.02	14.35	14.69	15.02	15.71	16.41	18.61
12	11.71	12.02	12.34	12.66	12.99	13.32	13.66	14.00	14.34	15.04	15.76	18.01
13	11.10	11.41	11.74	12.06	12.40	12.73	13.08	13.42	13.77	14.49	15.22	17.52
14	10.57	10.90	11.22	11.56	11.90	12.24	12.59	12.94	13.30	14.03	14.78	17.12
15	10.13	10.45	10.79	11.12	11.47	11.82	12.17	12.53	12.90	13.64	14.40	16.80
16	9.74	10.07	10.41	10.75	11.10	11.45	11.81	12.18	12.55	13.31	14.08	16.52
17	9.40	9.73	10.08	10.42	10.78	11.14	11.51	11.88	12.25	13.02	13.81	16.29
18	9.10	9.44	9.79	10.14	10.50	10.87	11.24	11.61	12.00	12.78	13.58	16.10
19	8.83	9.18	9.53	9.89	10.25	10.63	11.00	11.39	11.78	12.57	13.38	15.94
20	8.60	8.95	9.30	9.67	10.04	10.41	10.80	11.19	11.58	12.39	13.21	15.80
25	7.73	8.10	8.48	8.87	9.26	9.66	10.07	10.48	10.90	11.76	12.64	15.37
30	7.20	7.58	7.98	8.39	8.81	9.23	9.66	10.09	10.53	11.43	12.34	15.17
35	6.84	7.25	7.67	8.09	8.52	8.96	9.41	9.86	10.32	11.24	12.19	15.08
40	6.60	7.03	7.46	7.90	8.34	8.80	9.26	9.72	10.19	11.14	12.10	15.04

Going further, you can closely approximate what the monthly payments will be—just by dividing that last figure by 12.

Say yours is a $20,000 mortgage at 9 percent for 40 years. Opposite 40 and under 9 percent you will find the figure 9.26. That's the percentage of $20,000 which you are obligated to pay each year—$1,852.

Divided by 12, that becomes your monthly payment of $154.33.

Figuring a FLIP

A FLIP IS A special form of graduated-payment home mortgage, offering Flexible Interest Payments. It takes less out of your pocket in the early years of a loan and more in the later years.

Since paying later is always more costly than paying now, using a FLIP increases the total paid. This may be worth while if it permits owning a house that, although affordable in the long run, would otherwise be temporarily out of reach. That's why this kind of loan may be especially useful for a young couple who can count on a rising income.

The next table compares three possible sets of figures with those for a traditional loan, for a $45,000, 30-year mortgage at 9 percent. Even in a period when mortgages are running far from 9 percent, these figures can be useful for making comparisons. Likewise, the use of $45,000, as for a $50,000 house after a 10-percent down payment, makes it easy to calculate the result for a house selling for two, three, or more times $50,000.

For the most extreme version shown, you can see that the initial monthly payments would be nearly $100 less than for a conventional loan, increasing until the sixth year, and remaining level after that. Even for a house where a family might be required to show a $22,656 income for a conventional loan, the same family might be reckoned able to handle a FLIP on an income of $18,132. The same relationship would hold for purchase of a more typical house by a family with a larger income.

The feasibility of a FLIP is based on two factors that tend to raise a young family's dollar income as the years go by—increased real earnings and inflation. Add these expectations together. If they come to as much as the percentage by which the payments will increase each year, meeting the increase should not be a special problem.

(How can you calculate the percentage of increase? Subtract this

Year	Traditional	FLIP 4%	FLIP 5%	FLIP 6%
1	$362.09	$292.64	$280.01	$267.75
2	362.09	310.14	300.33	290.76
3	362.09	328.24	321.65	315.12
4	362.09	347.27	344.05	340.91
5	362.09	366.95	367.56	368.20
6-30	362.09	387.43	392.25	397.08
First-year income needed	$22,656	$19,296	$18,720	$18,132

year's monthly payment from next year's and divide by this year's. Multiply by 100 to make this decimal a percentage.)

Cost of a Mortgage Bargain

WHEN A house is offered with a conventional mortgage carrying an interest rate below the going rate, you naturally wonder how the lender can be so generous.

The most likely answer is that the seller is paying the bank or other lender a fee in return for the bargain rate that is being dangled before you. And since he has to get this additional money somewhere the seller adds it to the price he charges you for the house.

If you were to pay cash or get your own financing you should expect to buy this or a comparable house at a lower price.

The question is—how much lower?

You can get a pretty close idea by using a rule of thumb. It says that for every 1-point loan fee (1 percent of the amount of the mortgage) he receives, the lender can afford to drop his interest rate by one-eighth percent.

So if you are offered a $100,000 mortgage at 10 percent when 11 is normal, that's nearly $1000 in your pocket the first year, shrinking only slowly thereafter. But it means that the seller is paying the lender a fee of about eight points which he had to get by inflating the house price by something like $8000.

It's another case of there isn't any free lunch.

Those being the facts, what should your strategy be?

Unless you have reason to assume otherwise, it's reasonable to regard it as a tossup whether you'll stay with the mortgage to the end, thus taking full advantage of the lower rate you've paid for.

If, as is the more usual case, you should refinance or sell in a few years you'll probably be better off accepting higher interest in return for a lower price.

Hidden Side of a Home Loan

IN DECIDING whether you can afford the home you covet, you'll naturally begin by ascertaining the monthly payment. That's easily done by using the tables or the calculating methods given in this volume. Either will give you the exact monthly figure for each of the usual interest rates and time-payment periods, for loans of various amount. They apply equally to conventional and guaranteed mortgages and to homes purchased on contract of sale.

To determine whether you can afford a payment of a given size you must, of course, add other shelter costs: property taxes, insurance, maintenance. You may pay these directly, or they may be added to your basic payments.

Maintenance is the great variable. For people short of manual skills and interests who have the misfortune to acquire a house in poor condition, upkeep can be a financial disaster. A house designed, built, and landscaped for easy maintenance, on the other hand, may require only the attention that a willing owner can easily give, at a cost that may average but a few dollars a month.

A typical maintenance chore is replacement of the thermocouple in a gas water heater or furnace, which tends to wear out within 10 years. A new part commonly costs less than $5, and a householder of average savvy can install it even if he has never done this particular task before. But put in by a repairman the little thermocouple inevitably is accompanied by a charge of $30 to $50 or so—because the trip alone may have taken up an hour of the repairman's time.

To estimate maintenance costs on your prospective home, begin with a look at history. How much have you (and your landlord, if you've been renting) been spending? How much, if any, more should have been spent for fully adequate maintenance? Judging by the amount of mechanical equipment and its condition, exterior and interior finishes, and status of landscaping, how much more or less will the new house require? By this route, you can arrive at a fairly precise and realistic expectation—and thus avoid a major money trap.

Consider also what is an even bigger figure in many cases: automobile or other daily travel costs to the extent that they will be greater in your new location than in your old.

Many a budget has foundered because that idyllic new home site added 40 miles a day (at 20 to 40 cents a mile) to the daily home-to-job round trip. Other budgets have suffered because an isolated location necessitated an extra car for someone. Car cost, for many families, belongs under "shelter" in the budget.

How Much Will Your Payments Be?

I HAD THOUGHT of including in this volume a set of tables of mortgage payments. They would have told you how much you would pay each month for each size of mortgage at each rate of interest for each number of years.

Let's see now. I could have included at least 25 mortgage sizes, 25 interest rates, and upwards of 30 terms of years. Fifteen or 20 thousand figures! Fifty to a hundred pages of this book, produced the easy way!

Tempting.

But even if I'd done that, you'd still have had to do some arithmetic to fit the table to the exact dollar size of the mortgage you're interested in.

So it seems to me to make a lot more sense to give one or two ways to calculate an installment payment yourself, precisely for your own situation, whether payments on a car or furniture purchase or on a personal loan or a home mortgage.

Here's a real do-it-yourself alternative to digging figures out of a great big table and then interpolating. At a glance, you might deem it hopelessly tedious to tackle without some kind of shortcut device, such as a scientific or financial electronic calculator or a slide rule or logarithms.

Not so. The only threatening element is in Step 2, and it's a devil only when payments will extend for many years.

To find a 360th power (for a 30-year mortgage) you can strike the times key once and then the equals key 359 times for the required 360. But if you note that 6 times 6 times 10 equals 360, you can leap to the same result in a few seconds by...

...punching the times key once, then equals five times;

...doing the same sequence again;

...then hitting times once and equals nine times.

Here are the steps, for payments to be made monthly or at any other interval.

1. Express the interest rate as a decimal (9 percent becomes 0.09) and divide by the number of payments per year.

2. Add 1, then raise this sum to the power equal to the total number of payments you are to make. (For a 6-year loan paid monthly, this is the 72nd power.)

3. Divide the result of Step 2 by itself less 1.

4. Multiply by the amount of the original debt.

5. Multiply by the figure found in Step 1 (or multiply by interest rate as a decimal and divide by number of payments a year).

Here's how it goes for monthly payments over six years on a $5,000 loan.

1. Dividing .09 by 12 gives .0075. Add 1 to this.

2. Raised to the 72nd power, 1.0075 becomes 1.71255.

3. Dividing 1.71255 by 0.71255 is easy on any calculator, since you have the digits before you until you begin to enter the divisor. Result: 2.4034102.

4. Multiplied by $5,000, this becomes 12,017.05.

5. Final multiplication by 0.0075 gives a result that rounds off to $90.13 monthly payments.

(For a final step, multiply that last figure by the number of payments (72) to learn that repaying this $5000 loan in this way costs you $6,489.19. Subtracting the loan amount of $5,000 reveals that almost $1,500 is for interest.)

Setting Up a Payment Schedule

IF YOU'RE buying or selling something on installments—house, car, whatever—it can be very convenient to have a payment schedule all laid out on a piece of paper or in a little book.

To create one, you'll need to know the amount of the debt, the payment period (usually one month), and the rate of interest. With these you can calculate the payment by using any of the methods given in this chapter or the one on computers.

You can then fill in the figures for each month of each year under column headings of INTEREST, PRINCIPAL, and BALANCE.

To find the first of these, multiply the balance owed by the interest rate, divide by the number of payments per year, and move the decimal point two places to the left. This is the part of your first payment that will go for interest.

Subtract this interest from the amount of the payment and you'll have the sum to be applied to the principal.

Subtract this principal from the amount owed. This is the balance.

Here it is, as I set up the schedule when my wife and I sold our house to our oldest daughter and her husband after building a new one a mile away. I commend, to anyone who can use it, this shrewd form of family planning. It assures that as we grow too old and cranky for anyone else to come near, there will be someone delightful who is stuck with us. Just a little twist on the cynical definition of home as where, when you have to go there, they have to take you in.

Here's the routine for a 30-year, 9 percent, $120,000 mortgage, calling for monthly payments of $965.55. Multiplying $120,000 by 9, dividing by 12, and moving the decimal point two hops gives $900 paid to interest the first month and $65.55 to principal, with a balance owing of $119,934.45. This operation is repeated for the second month's figures, using that $119,934.45 instead of $120,000. And so on.

As you can see, this is quite feasible as a paper-and-pencil operation, for problems covering a short span of years. But if you start that way you'll surely spring for a pocket calculator very soon.

With a programmable calculator, it's a cinch. Putting in pauses for the needed figures lets you make a payment table as fast as you can note the figures down. You're saved even that chore if yours is a calculator that prints its results on paper, or if you are using a computer hooked to a printer.

Everything so far assumes equal payments made every week or month. If the buyer has the option of paying more when he likes, and exercises that option, you can't use a schedule made up in advance. You'll have to do each monthly line as payment is made. The other effect this has is that you'll want to provide an additional column, marked PAYMENT, to the left of the others, to show the sum actually paid. Naturally, it is from this amount that you will subtract the sum attributable to interest as you find how much to credit to principal.

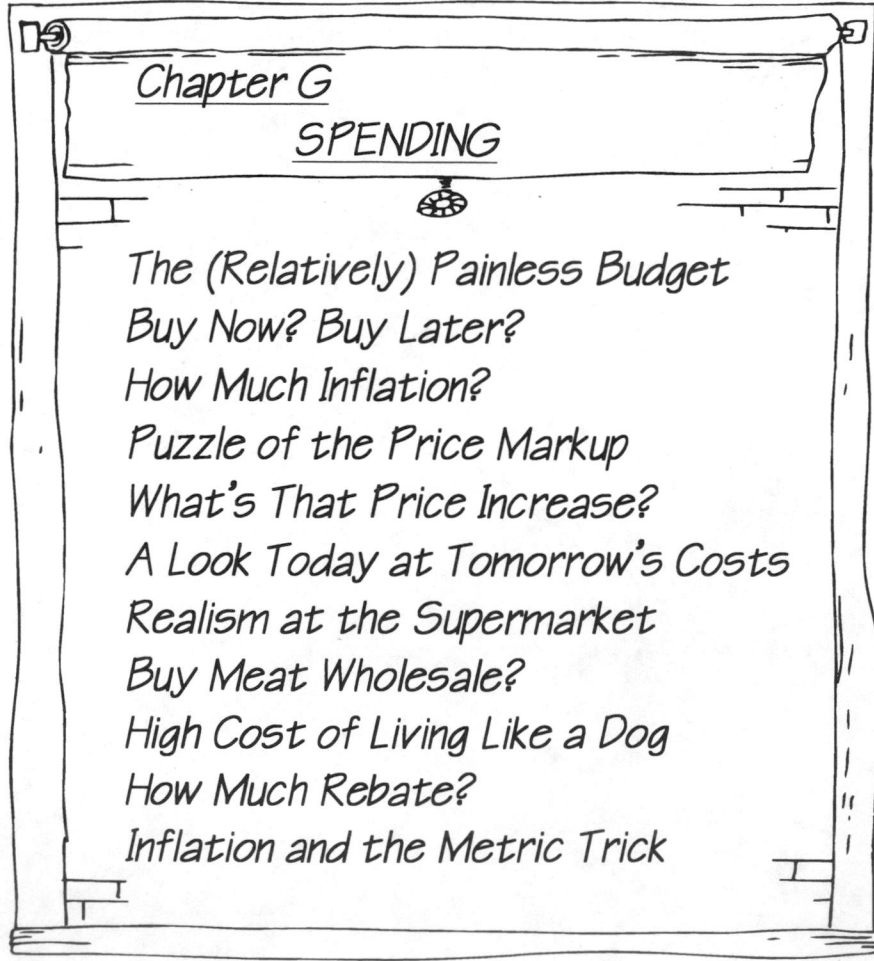

Chapter G
SPENDING

The (Relatively) Painless Budget
Buy Now? Buy Later?
How Much Inflation?
Puzzle of the Price Markup
What's That Price Increase?
A Look Today at Tomorrow's Costs
Realism at the Supermarket
Buy Meat Wholesale?
High Cost of Living Like a Dog
How Much Rebate?
Inflation and the Metric Trick

The (Relatively) Painless Budget

THERE ARE whole books full of forms and formulas for setting up a household budget. They can be quite useful, especially if you want to make a real production out of it.

If you're given to impatience, however, you may be glad to know it's feasible to set up a fine, workable system by just following the steps outlined in this limited space. And you'll have that budget in full flow in less time than you might spend reading how to do it.

1. List all income you can reasonably count on for the next 12 months—wages, interest, dividends, gifts, bonuses, rent from property you own. Subtract any expenses connected with getting the income and also subtract federal, state, and local income taxes, whether paid directly or withheld. This will give you your net annual after-tax income.

2. List your fixed expenses for the next year—all those things that are quite predictable and in amounts that aren't up to you. These may include rent, property taxes, insurance premiums, tuition, installment payments for which you are already committed, alimony or support, even such other things as utilities if you can estimate them closely enough. And one final item you probably hadn't expected to list under fixed expenses—savings.

Maybe this calls for some explanation. In the first place, there must be savings in your program—if only in preparation for the inevitable emergencies. In the second place, it's essential to get the habit going even if the amount is pretty small. Equally important, once you get the savings fund up beyond what you must keep for emergencies, you can use it in wondrous ways.

For one thing, it can ward off the never-never, as the English are wont to label their hire-purchase (our installment) plan.

Say it's time to buy a $400 washing machine, which on the never-never, with its accompanying 18 percent piggyback, could actually run you $500 by the time it's paid for. Pull the $400 in cash out of your savings when you buy, then take the amount of the installments out of the household-operation budget till the $500 is repaid, giving your savings a fine boost with it. Why let someone else grab those high-rate interest dollars when your own account can just as well earn them?

Eventually, you should be able to buy even such things as new cars and other durables free of interest costs. As you use up your car, put the amount by which you estimate it has depreciated that month into the savings fund to earn interest till it's time to buy your new car—for cash.

So how much shall savings be? If you have no established habits, try to begin with at least 5 percent of your net income. I don't think you need me to tell you that you'll be happier in the long run if you can start with 10 percent or more.

3. Now list your other expenses by major categories such as: food; household operation; furniture; clothing; transportation; medical care; education and recreation; gifts and contributions; personal (including allowances); miscellaneous and investment. Put down the amount you intend to spend.

4. From your net income, first subtract the total of your fixed expenses you've just listed.

5. If there's money left, check the list again to make sure you've been realistic in each category. Use any remaining money to expand the last category with an investment program in mind. Either now or after your budget has stood the test of a year, you may find yourself in the happy position of being able to expand the savings item substantially. Perhaps recreation, too.

If however, as budgeters more often discover, the total of variable spending is larger than the money available, something's got to give. Cut as much as you can from each item. But don't kid yourself. There's no use taking a new suit out of the paper budget if, in fact, you're still going to buy that suit.

If the budget won't balance now, and it isn't feasible to increase your income, you will have to recognize that your fixed expenses are out of line. Look for long-range changes that will reduce them.

6. Unfortunately, there is a second paper chore connected with successful budgeting. You must keep track of your expenditures. Do this with special care for the first couple of months. This doesn't mean writing down every penny. Whatever may have been the case in B. Franklin's time, records to the nearest dollar are close enough in our day.

And you can take Robert Frost's word for it, endorsed by suffering members of my own family when I become over-zealous, that "Nobody was ever meant/To remember or invent/What he did with every cent."

The ideal time to start keeping records is a month or several before starting your budget. The sad truth, however, is that most people who postpone budgeting while they compile a history never get around to budgeting at all. Better to strike while the iron of impulse is hot.

If you tough it out in the beginning, your rough working budget will refine itself into a flexible and precise tool for helping you get the most out of your money. It may even teach you and your family so much about the arcane art of handling money that you won't need to maintain a formal budget unless you feel like it.

Buy Now? Buy Later?

THAT PANICKY impulse to buy now to beat inflation in a period of rising prices needs a sober second look.

But so does the urge toward thrift that says don't buy it as long as you can get along without it.

In short, some calculation is called for.

Will the purchase have some immediate and continuing use? That's one vote in favor of acting now. If it will not wear out rapidly or deteriorate from the passage of time, it deserves another pro vote.

With those aspects considered you're in better shape to make a decision on the sound basis of alternate use of your money. Do you have a safe alternate lodging place for your money, one that pays a true return? How much?

This brings you to questions of inflation and taxes.

Begin by supposing you're in a time when inflation is running at an annual rate of 10 percent. Something—a painting, an antique, a building lot—that you can buy today for $50,000 will stand to cost you $55,000 a year from now.

You know that if you do not make the purchase, you can invest your money safely for a year at a yield of 10 percent. But keep in mind that this is the gross return. In your combined state and federal tax bracket of 40 percent, you'll retain only 60 percent of that return, or 6 percent net. That comes to $3000.

So at the end of the year your $50,000 will have grown to $53,000, but it will take $55,000 to make the buy. If you can trust those suppositions, buy now.

It could also cost you something, in property taxes or storage and insurance outlay, to hold the purchase for a year. Consider both the amounts of these and when you'll incur the added costs.

And don't forget to factor in the value of use you can make of earlier ownership. Which brings us back to where we came in.

How Much Inflation?

WHEN INFLATION leads a manufacturer to boost a list price he sometimes takes the opportunity to slip a hidden boost behind the apparent one. What follows may seem a trivial instance, but it has its point.

On the occasion when Luden's Menthol Cough Drops went up from 15 to 20 cents a box, the maker no doubt hoped you'd see this—if you noticed it—as just another nickel in a 20-cent purchase, a price jump of one-fourth or 25 percent. You, however, might reasonably complain that this was a jump of 33-1/3 percent—a nickel being one-third of the 15 cents you'd been used to paying.

But there's more to this tiny inflation story. At the same time that the price went up, the contents of the box went down from 16 cough drops to 15. So how much has the true price risen?

One way to get at this without finding yourself mired down in confusion is through a unit-pricing calculation. A cough drop used to cost 15/16 of a cent. After the increase, it cost 20/15. So we ask, as we did in the simpler calculation above, what percentage of the old price is the increase?

Again, we find out by first dividing the increase by the old price, to find the ratio of increase, and multiplying that by 100 to make it a percentage of increase.

One way to perform the operation comes under the head of Fun With Fractions, complete with that grade-school enemy the Lowest (or did we call it Least?) Common Denominator. I guess it is allowable to remind you how to do it since I've just gone to considerable trouble to recall it myself. Starting with the new price of 20/15ths and subtracting 15/16ths from it (after turning them into 225/240ths and 320/240ths) produces 95/240ths. Dividing that by 320/240ths and multiplying by 100 gets you to 9500/240ths. Dividing by 15/16ths is now guaranteed to bring you to 42.2.

Doing it like this, with fractions, is the old-fashioned way, for sure, but it tends to make the principle reasonably clear.

Most calculators abhor fractions, love decimals. So with one of these little fellows, you just divide 20 by 15, from this subtract (15 divided by 16) and then divide what you now have by that same quantity (15 divided by 16). Multiplying by 100 turns this into a percentage of increase—42.2.

So whichever way you look at it, the price of not coughing has gone up a rather formidable 42.2 percent.

When the magazine *Consumer Reports* noted this occurrence, the editors calculated the increase as 41.5 percent. The discrepancy is minor, but it suggests that we might well check our result by using an alternate method to see if we're wrong.

Consider that 240 drops would now be 16 boxes at 60 cents each, for an expenditure of $3.20. In the good old days, you could get 240 of the things by buying only 15 boxes at 15 cents each, for $2.25. The difference is 95 cents which, times 100 and divided by the old cost of $2.25, again comes to 42.2 percent. How the magazine got its slightly different figure is not readily apparent.

But the virtue of rounding is.

If the editors had eschewed unnecessary precision by reporting the rise as "about 42 percent," they'd have been right, and quite precise enough as well.

Puzzle of the Price Markup

Bᴜsɪɴᴇss price markups would be a pretty simple matter if it weren't that they come in two dissimilar forms.

Wholesalers are in the habit of figuring markups as a percentage of what the item cost them. Retailers base their markup figures on what they charge the customer.

This creates no misunderstandings as long as markups are expressed as so many dollars and cents. When an item goes from $3 to $4 it has been marked up $1.

But suppose a wholesaler who would tell you he operates on a markup of 50 percent has passed along a $10 (his cost) item to the retailer for $15. What will the retailer put on the price tag if he also adheres to a 50 percent markup? Will he charge $22.50?

Not if he follows the more usual retail practice of expressing the markup as a percentage not of his cost but of the selling price.

But how can the retailer go about adding 50 percent of the retail price to his wholesale cost when he hasn't yet figured out what that retail price is? He goes about it somewhat backhandedly, by dividing his cost price by the markup percentage—expressed as a decimal.

As a decimal, his 50 percent markup becomes 0.50, since percent means one hundredth and moving the decimal point two places to the left is dividing by 100.

Dividing his cost of $15 by 0.50 doubles it, giving a retail price of $30. The amount of the markup is $15, or 50 percent of the selling price.

Which also is, of course, 100 percent of the amount he paid for it.

It's confusing, but that's the way it's done. For one thing, it sounds better this way.

What's That Price Increase?

Oɴᴇ ᴅᴀʏ some time ago, along with other news good and bad, the day's mail brought word that my modest Blue Cross plan had gone to billing every two months instead of quarterly. Reason: "Because so many members have asked for more convenient billing."

I'll admit that my immediate response was to recall reading that on one social occasion the Duke of Wellington, imposing and impressively dress-uniformed, was accosted by a lady. "Mr. Smith, I believe," she said. "Madam," replied the Iron Duke sternly, "if you believe that you'll believe anything."

And if writing six checks a year instead of four is to my convenience, why not make it twelve?

Convenient or not, the new billing was accompanied by a price increase, as the covering literature mentioned incidentally.

How much of a jump is neatly concealed by the billing change? All that is evident at first glance is that the next check I mail will be gratifyingly smaller than previous ones.

Instead of $107.46 quarterly I'll be paying $85.76 six times a year. A little multiplication tells me that my annual cost has just risen from $429.84 to $514.56.

What's that increase as a percentage? This jump of $84.72, divided by the old cost and multiplied by 100, proves to be almost 20 percent at one clip, which could be called whopping. No wonder Blue Cross deemed it worthy of some obfuscation of presentation.

Too bad, though, that frankness wasn't the order of the day. Considering where medical costs went from then on, we all could have used a sharp warning.

Since the same figures applied to my wife's policy, we were being hit by a 40 percent combined increase, weren't we? If there were three more of us, it would come to 100 percent, wouldn't it? You can best put your finger on this bit of confusion by asking 100 percent of what. In a typical month when you buy a thousand things and read that costs went up 1 percent while you were doing it, your total expenditure was inflated by 1 percent—not 1000 percent.

(Don't laugh. Silly as this unwarranted adding up of percentages may be, no less than our nation's most respected Sunday book review once came up with something equally silly. It found that "plant and book manufacturing expenses alone have risen as much as 10 to 12 percent over the last decade, materials are up 6 to 9 percent, selling and advertising expenses have climbed upwards of 10 percent. Combined boosts add up to a minimum of 33 percent....")

A Look Today at Tomorrow's Costs

THIS IS a quick-reference table showing how much you can expect to pay in the future for each dollar the same purchase would cost today.

It tells you that an apartment that rents for $540 today can be expected to cost 6.85 times that much ($3699 a month!) 25 years from now if inflation meanwhile has amounted to 8 percent a year.

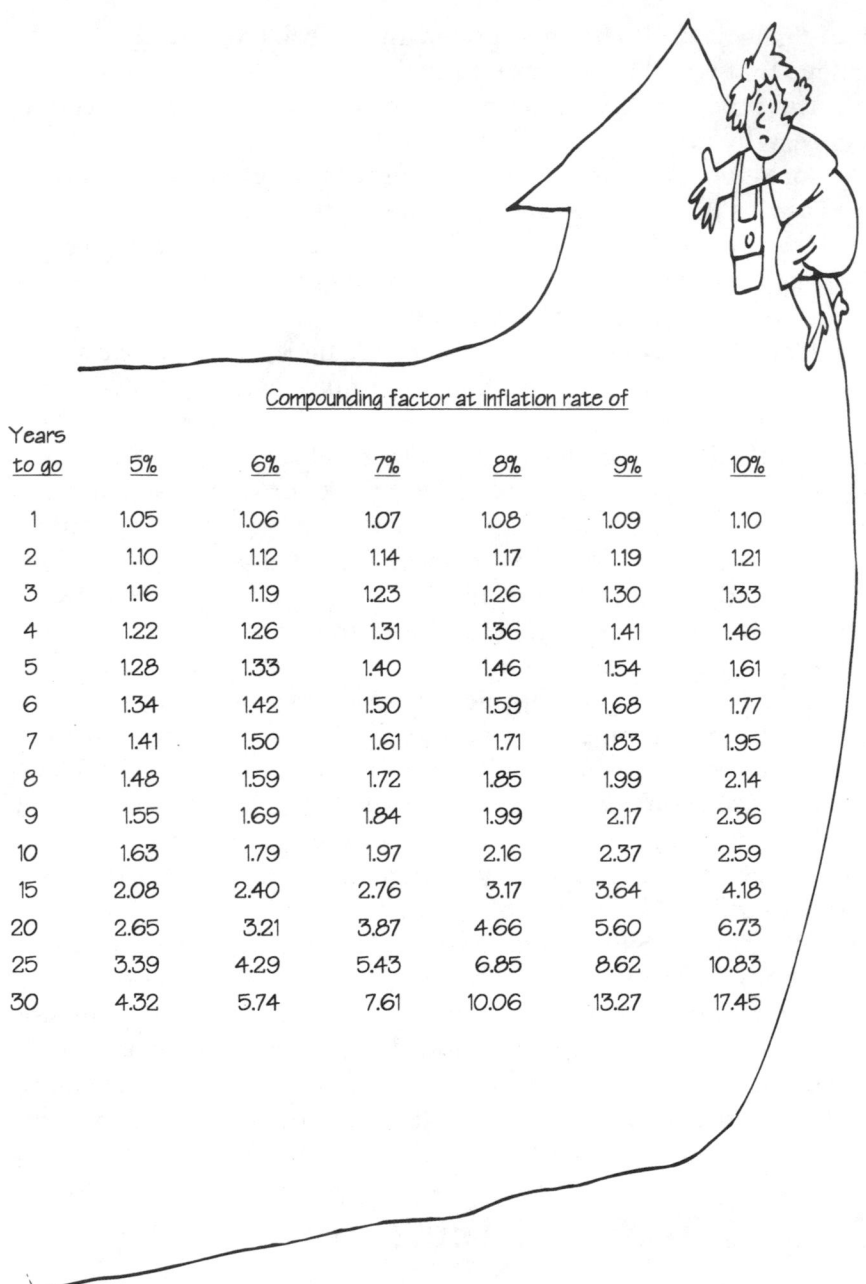

Years to go	Compounding factor at inflation rate of					
	5%	6%	7%	8%	9%	10%
1	1.05	1.06	1.07	1.08	1.09	1.10
2	1.10	1.12	1.14	1.17	1.19	1.21
3	1.16	1.19	1.23	1.26	1.30	1.33
4	1.22	1.26	1.31	1.36	1.41	1.46
5	1.28	1.33	1.40	1.46	1.54	1.61
6	1.34	1.42	1.50	1.59	1.68	1.77
7	1.41	1.50	1.61	1.71	1.83	1.95
8	1.48	1.59	1.72	1.85	1.99	2.14
9	1.55	1.69	1.84	1.99	2.17	2.36
10	1.63	1.79	1.97	2.16	2.37	2.59
15	2.08	2.40	2.76	3.17	3.64	4.18
20	2.65	3.21	3.87	4.66	5.60	6.73
25	3.39	4.29	5.43	6.85	8.62	10.83
30	4.32	5.74	7.61	10.06	13.27	17.45

Regarded in another way, it tells the future value of a present asset.

It's no crystal ball, of course. To make it work you must correctly guess the future course of inflation. In the case of future purchases, the table assumes that the object will not be an exception to the general level of inflation. With non-money assets, there is an assumption that the thing will not deteriorate or otherwise depreciate with the passing of time. The assumption with money is that it will bear interest at about the rate of inflation.

You might like to know the multiplier for some rate or period not shown in the table, such as 30 years at 18 percent (a rate not unknown in the U.S. economy and vastly exceeded at times in many others). It's a neat 143.37. Today's $8000 car would then cost a good bit over a million dollars.

There's a familiar formula by which you can figure this and other rates and periods not in the table.

Just express the predicted annual rate of inflation or rate of interest rate as a decimal and add 1 (so that 18 percent, for example, becomes 1.18), then raise this to the power of the number of years. The 30th power of 1.18 is the 143.37 of the preceding paragraph.

Realism at the Supermarket

IN TODAY's supermarket, you rarely have to ask the price of an item. It's marked on the package, tacked to the edge of the shelf, or posted nearby.

What takes figuring is far more important—the unit price. That's the amount you pay for each ounce, pound, pint, quart, or whatever—and somewhere in the receding future, gram, kilogram, or liter, when we join the rest of the world by going on the metric system. In recent years, supermarkets have come to the aid of the shopper by posting unit prices, usually by attaching a sticker to the shelf where the item is displayed. Unit prices are printed right on the label in the case of some items such as meat and cheese. You are given the total price of the package of hamburger, as well as the weight and price per pound. It's a cinch then to compare the price per pound of this ground meat with that of any other type of meat or grade of hamburger.

Likewise, you can quickly compare prices of two brands of canned applesauce, or two sizes of corn flakes. If the ounce price on a small can or package is 2-1/2 cents and 2 cents on the giant family size, it's easy to figure that the smaller one is one-fourth, or 25 percent, more per ounce than the large one.

You're then in a position to decide whether the saving in price is more important to you than any inconvenience in having a large can or box that takes up more shelf space at home and may not get eaten as quickly. There's no saving if the super-size deteriorates because it's used slowly. A part of the item might even have to be thrown out.

It is unfortunate that so few shoppers seem to make use of these posted unit prices. Yet they offer the simplest, most convenient way of cutting food costs. If your grocer doesn't post unit prices (or if they're printed too dimly to read or are too close to the floor to please your knees) you can figure the arithmetic mentally—or easier yet, take along one of those inexpensive pocket calculators to do the job for you.

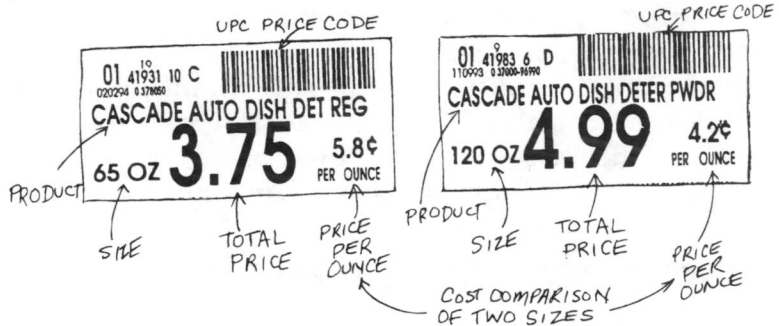

All you need do to calculate a unit price is divide the total cost of the item by the number of units—pounds, ounces, pints—it contains. Be careful to use the same unit in both cases. The unit of one item may have to be changed—usually by multiplying the number of ounces by 16 to find the number of pounds or pints. Simplify the task by rounding off. If 12-1/2 ounces of an item cost 62 cents, forget precision, and call it 13 ounces for 65 cents, or 5 cents an ounce.

Unit-cost calculations may lead you to discover that a small loaf of compact bread costs little or no more per ounce than an air-filled and apparently more economical sandwich loaf.

Consider an incident my wife reported after a supermarket trip.

Two shoppers arrive at the checkout counter. One woman has a small loaf of a rather dense, widely distributed bread called, I think, Wheatberry . The price is $1.38.

"Golly," remarks the checker, "that's a lot of money for such a small loaf of bread."

Next in line is a shopper carrying a foot-long loaf of family sandwich bread, price 39 cents.

"With the size of my family, I have to buy the biggest and the cheapest," she remarks.

It's a slack time at the checkout counter, so all three take a minute to look over the labels of the two loaves. The small wheat loaf weighing a pound and a half, contains high-protein cracked-wheat, freshly milled wheat flour, high-score creamery butter, honey, and not a single preservative.

That mighty long sandwich loaf weighs but a pound and its contents are primarily white flour, softeners, emulsifiers, mold-retardant and a chemical freshener.

A quick calculation shows the difference in price per ounce to be a fraction of a cent; the woman with the big family was buying neither a nutritional bargain nor an economic one.

Let's say you decide to buy canned sausage and find three comparable brands on the shelf. They vary in price and weight. You can easily run a unit-cost comparison test of the 4-ounce size. A little more than one-fourth is saved by taking a 50-cent can instead of a 68-cent can of the same size—so forget the 68-center.

But what of another brand that contains 5 ounces for 70 cents?

In this case, a direct comparison is easy. Since 5 ounces is one-fourth more than 4 ounces, it should cost one-fourth more than 50 cents—or 62-1/2 cents. At 68 cents, forget it too. It's overpriced by 5-1/2 cents.

Often, informal price comparison is easy. If frozen orange-juice concentrate in the small size is less than half the large-can price you can choose small almost without thinking. You can, that is, if you look at the labels and find that the big can has exactly twice the content of the small one.

Buy Meat Wholesale?

Buying meat in quantity for freezer storage can save you 15 percent or more on your meat bill, home economists find, IF....

IF you're a good judge of the quality of meat as well as of the reputation of the farmer or dealer who offers you a good buy. Watch out for fast-pitch operators who may take a whole page in your local paper and promise an attractively low price per pound without telling you that this price is before cutting and before removal of waste. Extra charges for wrapping may also be added.

We early discovered the importance of taking into account the eating habits of the family. All cuts from steak to stew are best if used in a year or less. Hamburger should be consumed within a few months after it is frozen. If you happen to buy a tough side of beef— and this occurs all too often—you may have to grind the entire side into hamburger to make it edible.

Do you own a freezer? Will you need to rent a locker? Will you have a long drive to pick up meat from a locker? Facing such questions will reveal costs that add to the price per pound.

What do you get from a side of beef? A typical half-ton steer will dress out to about 600 pounds, but only 430 of these pounds are edible meat.

Half of the 430 will become stew meat, hamburger, liver, pot roasts, kidneys. Tender steaks and oven roasts, preferred by most families, add up to about 180 pounds. If kidneys, liver, and soup bones are thrown out or become dog food in your family, Rover is getting some expensive meals.

High Cost of Living Like a Dog

You CAN FIGURE—roughly—that a fully grown dog will require an ounce of food for each two pounds of his weight. That means a terrier weighing 30 pounds or a little more is going to eat a pound a day. You can find yourself going broke buying dog food at the rate of as much as five pounds a day if you acquire a St. Bernard, or any other large breed.

How Much Rebate?

If you have an installment loan of the add-on type and some cash you can spare, there may be a bonus lurking for you in the banker's Rule of 78s (which also figures a couple of other places in this volume).

Your loan is an add-on if the lender has simply added the amount of the finance charge to the face amount of your loan when making it, then divided by the number of months to calculate your monthly payment.

Here's one way to find how much rebate you have coming for an early payoff.

Divide the number of payments already made by the number you were supposed to make in the course of the loan. Subtract that figure from 1.

Now divide the number of payments you have made by 1 more than the number you were supposed to make. Subtract this from 1.

Multiply together those two results and the amount of the finance charge. The result is how much rebate you can get by paying up now.

Here's a sample.

If you've made seven out of 24 payments on a loan that includes $70 interest, your rebate for paying off the whole thing now is figured like this. Divide 7 by 24 and subtract from 1. Then divide 7 by 25 and subtract from 1. Multiply these two results together and by $70. Pay now and save $35.70.

Inflation and the Metric Trick

A READER of the Los Angeles Times wrote to alert us all to what he is pleased to call metricized inflation:

I went to get a gallon of my favorite cream sherry the other day at my favorite wine store. The smiling clerk informed me that, due to government directives, they had to go to metric size bottles. They now would furnish three-liter bottles instead of the gallon size.

"For this they would drop the price 20 cents from the gallon price. I demurred, finally capitulated and bought the three-liter size for $5.75, plus tax.

"On my way home, I mentally juggled the relationship between the liter and the gallon. It slowly dawned on me that the price of my favorite wine had jumped over 30 percent."

Although the gentleman's hasty mental arithmetic is—and not surprisingly—imperfect, his point is valid. Changeovers to metric have offered a lot of temptations . It behooves us all to watch out for concealed price increases.

Here's the calculation for this one.

Divide the new price by the number of liters (3) to get the new cost per liter: $1.92.

Divide the old price by the number of liters in a gallon (3.7853) to find the old cost per liter: $1.57.

Subtract one from the other to find the increase: $.35.

Divide the increase by the old liter price to discover the percentage of increase to be 22 percent. Not 30 percent, but bad enough.

Same routine works, of course, for any packaging change from ounces or pounds or pints or quarts to grams or kilograms or liters, with or without an alteration in the dollar price.

Chapter H
MEASURING THINGS

Human Yardstick
How's the Wind Up There?
Power of a Wave
Weighing with Coins
Comparing TV Picture Sizes
Measuring the Unmeasurable
How Tall Is It?
Height by Trig
How Big Is a Tree?
Speed of Sound in Water
How Much Does It Slope?
Light Through a Prism
Estimating Water Flow
How Far Away Is It?
How Fast the Fall?
How Deep the Well?
The Metric Nickel
How High Is Up?
Measuring Height by Angle
How Big a Pinhole?
How Fast Will It Cool?
How Cold Will It Get?
Recipes for Odd Numbers

Human Yardstick

IF YOU memorize a few of your body dimensions, you'll always have with you a means of making approximate measurements.

For example, your span, or wingspread—fingertip to fingertip with your arms outstretched to the sides—will be 6 feet, give or take a few inches. In most situations, this is easier to use than your height and rather more dignified.

Elbow to tip of index finger will be about 18 inches. Half a yard should be easy to remember. (But check to see if using a finger other than the index won't give you even more precisely a foot and a half.)

If you're a man of average stature your shod foot is probably about a foot long— naturally enough. But if you're not a man or not of average dimension, that may not hold. Measure a shoe now for a more exact figure.

Your stride is most useful of all, as you've probably noted many times. It will be more useful in the future if you measure it now or practice stretching it to a consistent three feet or restricting it to two.

Another good move is measuring the length and breadth of the tip joint of your thumb so you'll know these dimensions when you need them.

How's the Wind Up There?

FOR SUCH purposes as planning a windmill (and—who knows?— you might; there's a lot of cheap energy there to be harvested) it's useful to know wind velocity at various altitudes.

A rule of thumb says wind speeds increase as the one-seventh power of the distance from the surface of the earth.

Suppose you have measured wind speed at the highest point you can reach, say 14 feet above the ground, and want to know what it will be at 60 feet.

Dividing 60 by 14 tells you the greater height is 4.2857 times the lesser.

Then a scientific calculator (or resort to the log table) will find the one-seventh power (also called seventh root) of this to be 1.231. Multiplying this by the measured speed for the 14-foot altitude will give you a close approximation of what you would find at 60 feet if you had some way to get your anemometer up there.

This is a rather arcane bit of calculating—a fractional root, for goodness' sake. So I guess it won't hurt to mention here, and not for the only time in this book, that you can do the trick almost as quickly with an ordinary hand calculator as with the flossiest of scientific models.

Or in a matter of minutes with pencil and paper.

By guessing and testing, in either case.

You know your root must be greater than 1. So you might start by guessing 1.1. When you find that to be far too small, you might try again with 1.2 and quickly discover you're well on your way. Punching in 1.2 followed by times once and equals six times brings you so close that in a couple more tries you'll probably hit on 1.23, which is as precise an answer as you'll ever need.

Power of a Wave

Expressed in horsepower, how much potential energy is there in an ocean wave?

In general, a wave twice as high as another packs four times the energy. But one twice as long as another offers only twice the energy.

To find the potential horsepower in a wave, express its height and breadth in feet and also express in feet the length of the wave between successive crests. Square the height of the wave. (Save this square to use again later.) Divide this square by the square of the length of the wave between crests.

Multiply by 4.935.

Subtract that product from 1.

Multiply by length of the wave between successive crests.

Find the square root of that result.

Multiply by the square of the wave height. (That's the figure you saved during the first step.)

Now multiply by 0.0329.

And, finally, multiply by the breadth of the wave to discover its energy potential in horsepower.

Run through this for a wave that is 1 foot high and 25 feet long and you'll find that the next to last step brings you to a horsepower figure of 0.1638. So for, say, 10 feet of wave breadth you'd have 1.638 horsepower.

Weighing with Coins

As LONG as you have a good mix of coins in pocket or purse you can do an emergency weighing job with surprising precision.

Just rig up any kind of simple balance—such as a pair of paper or plastic bags hung from the ends of a stick balanced over a sharp edge.

With the object to be weighed in one bag, an easily recalled clue to the heft of the coins you place in the other is the 4-to-1 relationship between U.S. coins and weight in grains.

Since a silver dollar weighs 400 grains, a half-dollar is 200, a quarter-dollar 100, and a dime 40 grains.

Nine U.S. cents weigh 1 ounce. An ounce is 437.5 grains.

Comparing TV Picture Sizes

A MAGAZINE given to testing consumer products expressed surprise at the discovery that two inches can make quite a difference in the apparent size of a TV picture.

The editors might have noted that this applies equally to other cathode-ray tubes, including computer monitors.

The magazine was comparing 19- and 21-inch (diagonal measurement) sets.

Well, the difference in these diagonal measurements is, of course, only about 10 percent. But what you're buying is area, or square measurement. So you really should compare the squares of those dimensions. Do it by subtracting the square of the smaller diagonal from the square of the greater, then dividing that quantity by the square of the smaller.

That will tell you you're getting an increase of slightly more than 18 percent when you jump to the 21-incher.

Measuring the Unmeasurable

IT'S EASY enough to discover the weight of an irregular or recalcitrant object. If it's something you can't put on scales or can't keep there long enough—a cat, for instance, or perhaps a baby—the classic trick is to hold it while standing on the scales yourself, then weigh yourself separately and subtract. This is handy also with packed soft-side luggage too amorphous to sit on scales unaided.

But how can you measure the volume of an irregular object? If water won't hurt it, do what Archimedes did. Dunk it under water and measure the amount by which the water rises. Cooks use this trick with butter since it also avoids having a greasy measuring cup to wash. After using this method, you are entitled to flash down the street naked, shouting "Eureka!"

How Tall Is It?

AN AMATEUR explorer returned from sunny Baja California to report finding what he was convinced was the tallest cactus of its type ever discovered—the variety sometimes called the boojum tree, I think it was. He was most regretful that he hadn't had some kind of instrument along to measure its height with.

But of course he did have all the instruments he needed, right on his feet and in his head.

He might have started by measuring his shadow, preferably mid-morning or mid-afternoon when it would be fairly long. This he could have done with the length of his boot sole as a module.

Then he could have measured the length of the shadow of the cactus in the same way.

Having found his own shadow to be, say 8-1/2 boot soles long, and having lain down in the sand long enough to prove his shod height to be an even 6 "soles," he'd have had a fairly simple calculation to make after measuring the cactus shadow length and finding it to be 72 "soles."

Since the length of each object would bear the same relationship to the length of its shadow, he'd have had a little proportion like this to scratch in the sand with a stick: 6 over 8-1/2 = ? over 72.

He could look at an equation like this in several ways, but the simplest thing to do with it is cross-multiply. Starting with 6 times 72 is 432, he needed merely to divide that result by 8-1/2 to find that he had a 51-sole boojum tree. (Quickest way to do division like this is by doubling both numbers, since 864 divided by 17 is easy enough.)

Another way to look at the same math is that the fellow's height (6 boot soles) bears the same relationship to the length of his shadow (8-1/2 soles) as the boojum's unknown height bears to the known 72-sole length of its shadow. Since 8-1/2 divided by 6 can be treated as 17 divided by 12, or 1-5/12, this is the number by which 72 must be divided to produce that same result, about 51.

Our explorer could assume for the moment that his boot sole was about a foot long, making his cactus roughly a 50-footer. On return to civilization, he could check the precise length of his boot and correct for any discrepancy from 12 inches.

Height by Trig

You're placing a guy wire or a prop to steady a pole. It is to go at an angle of 65 degrees to the ground, and be secured 14 feet from the pole. Without climbing or otherwise measuring how high on the pole the prop is to be fastened, you'd like a means of calculating the length of that prop.

Drawing a little picture of the situation, you'll see that you have a right triangle, of which you know a leg and an angle. From that information, you want to find out the length of the longest of the three sides, the hypotenuse.

The rule to apply is that the hypotenuse of a triangle can be found by dividing the length of a side by the cosine of the angle between it and the hypotenuse. In this case, you're dividing 14 by 0.4226, to find an answer of 33.13 feet. No climbing.

Also no use, of course, unless you have either a scientific calculator or a few pages of trig tables at hand.

How Big Is a Tree?

Trees have quite specific size designations. The unit of measurement is dbh, which means diameter breast high.

A sapling has a dbh of 4 inches or less.

From 4 inches to 10 inches, the tree is called a pole.

A standard tree has a dbh of 1 to 2 feet.

When the diameter exceeds 2 feet, the tree is a veteran.

Speed of Sound in Water

Thanks to Leroy, you can readily calculate the velocity in feet per second with which sound will move through water.

Combine the six quantities below. But omit the two parts of the fifth step unless you are dealing with salt water—or do not know if it is fresh or salt or what the degree of salinity is.

This procedure uses something called Leroy's equation, which is easy enough to apply, wonderfully mysterious though it is (like the man himself, about whom I cannot tell you more).

1. Start with 4755.2.

2. Add the temperature of the water in degrees Celsius multiplied by 15.067.

3. Subtract 0.1765 times the square of the temperature.

4. Add 0.00085 times the cube of the temperature, still using Celsius, of course.

5. With fresh water, that finishes the calculation. With salt water add 3.9 times the result of subtracting 35 from the salinity in parts per thousand; and also add the result of dividing the depth in feet by 61.

You might want to check yourself on the use of this equation. Try it for water at 20 degrees C., salinity 50 parts per thousand, depth 250 feet. The result you get should be 5055 feet per second. Note, however, that if the water is fresh, the indicated speed would be 4993 feet per second.

How Much Does It Slope?

Measuring the slope of land can be important for many projects.

Building steps, for example. Or excavating for a swimming pool. Or in applying for a building permit if your town, like the sometimes over-fussy one I live in, demands that you go through an additional process for an application for a use permit—even before you can apply for a building permit—if you're on a site having a slope in excess of 30 percent.

You can measure a slope most precisely with a surveyor's level costing many hundreds of dollars. Or a simpler dumpy or transit level at a few hundred. You'll need a tripod with any of these.

For similar results, with less precision but more speed, you can use a sort of pocket-size monocular called a hand level. Metal types cost several times as much, but you should be able to find a wooden version for just a few dollars. You could make one yourself, for that matter. This kind is often used by farmers doing contour plowing to help them plow furrows that may wander all over yet remain near dead level so rains won't wash soil away.

With any of these, you can sight on a yardstick or a measuring tape stretched along a stick held vertically—or on the convenient but more costly equivalent thereof called a sliding-section leveling builder's rod.

Or you can dispense with the stick, and with a helper as well, by sighting directly into the slope. If the point at which the ground sights out to eye level is 20 feet away and your eye level is 5 feet, you know the land is sloping 5 feet in 20 feet. Divide 5 by 20 and you get 0.25. Your slope is 25 percent.

(Slope measuring is a little confusing. A slope of 20 feet in 20 feet is a 100 percent slope. And a straight-down cliff is an infinite slope, not 100 percent as you might think.)

But you don't need even a sighting level if you have time to rig up a more primitive and bulky device that has served me well.

What you can do is fasten an ordinary carpenter's level, preferably one at least a couple of feet long, onto a photographer's tripod or other type of stand or table.

Since the standard tripod screw is 1/4", 20 thread, you can fix the level tightly in place by drilling a hole through the rim of the level and putting on a 1/4" nut.

You don't have cross-hairs to use in sighting with this kind of home-made device, but you'll find you can sight quite well along the top surface of the level. If you want to refine the procedure a little, you may be able to find a hardware store that carries little clamp-on devices for converting an ordinary carpenter's level into a sighting level.

Even without such fancy stuff, however, some friends and I once applied this method to excavate for a 50-foot swimming pool that took us weeks of pick-and-shovel work. With the help of the level-on-a-photographer's-tripod, we were able to pile up what we dug out to form a precisely horizontal bank, even though we were working on a rather steep side hill.

Light Through a Prism

As a ray of light passes through a triangular prism, the path of the ray is bent. How much will the deviation be? Several steps are required to find out.

The procedure calls for using trigonometric functions, a prospect that may seem grim but need not be. The job can be done with a table of sines and either an ordinary calculator or merely paper and pencil. It's far easier, though, if you use a scientific calculator with its built-in trig functions.

The data you will need are the angle of incidence of the light, the prism angle, and the refractive index of the material of which the prism is made.

Step 1. Divide the sine of the angle of incidence by the refractive index, and then find the arc sine of this result.

Step 2. Subtract this from the prism angle, find the sine of the result, and multiply by the refractive index.

Step 3. Find the arc sine of the previous result, then add the angle of incidence and subtract the prism angle. Result is the deviation you were looking for.

To check the procedure with an example, use 30 degrees as the angle of incidence, 45 degrees as the prism angle, and 1.8 as the refractive index. The deviation should come out 45.3598441 degrees.

Estimating Water Flow

You can get a pretty good estimate pretty quickly of how much water is flowing in a stream you plan to dam or use for production of power. To do this, you need two figures.

One is the cross-section area of the stream.

The other is how fast the water is moving.

To get the first one, begin by placing a board from bank to bank just above the surface of the water. At, say, ten to twenty regular intervals along the board, beginning at one bank and ending at the other, measure the depth of the water.

Take an average by adding these depth measurements together and dividing by the number of them.

Multiply by the width of the stream to find the cross-section area. To change this figure to square feet, divide by 12 if one of your measurements has been in inches rather than feet, by 144 if both have been.

Next, find out how fast the water is moving. Drop in some fairly heavy object, such as a corked bottle or plastic jug with some water in it, and measure the distance it goes in one minute.

Multiply this by the area figure and then by 0.83, or 5/6. This is to allow for the fact that water flows faster at the surface than below.

Result is the flow of the stream in cubic feet per minute.

To be sure this is all clear, let's run through an instance.

Your stream is 13 feet wide. Measured every foot, the depths of the stream, in inches, are: 0, 4, 9, 11, 8, 8, 12, 13, 15, 12, 10, 11, 10, 2.

Their total of 125 must then be multiplied by the width of 13 and divided by the number of measurements, 14, and by 12 (to convert to feet) to produce a cross-section area of almost 10.5 square feet.

If you've found the float to move 88 feet in one minute, you multiply 10.5 by 88 and by the correction factor of 0.83. The result is about 767.

Since conditions vary with the season and the year, and since none of your measurements will have been quite precise, it would be safer to think of the result in round terms. In a season like the present, your stream flows at seven or eight hundred cubic feet per minute.

How Far Away Is It?

T O MEASURE the distance from here to there—when "there" is 6 to 100 feet away—a tape and someone to hold the end of it constitute the normal way to go.

If you don't have that someone along, you'll need a device for fixing the end of the tape into place. A spike you can push into the ground works pretty well. Even a twig can serve.

If you don't have a tape either—or only a short one and you're in a hurry—pacing can be surprisingly accurate. But only if the ground is fairly even and you know the length of your stride. Best thing is to practice occasionally on a marked distance so you know just how to hold your stride to a fairly exact yard. You can check the precision of your one-yard pace right on the job if you happen to be carrying even the shortest pocket tape. Or if you know the length of your shoe (which, not surprisingly, may turn out to be just about 1 foot) you can use that.

Measure reasonably short distances quite accurately by taking heel-to-toe little steps. If you know your sole length and it's, say, 11 inches, you'll find it isn't hard to estimate for the inch gap between toe and heel.

For quicker results over somewhat longer distances, try using the range finder in a good camera.

How Fast the Fall?

A SAN FRANCISCAN made headlines nationally by surviving a 29-story drop down an elevator shaft in the Transamerica Building, "hitting the cement floor at the bottom at an estimated speed of more than 100 mph."

How did the reporter know how fast the man was moving?

To find out, you could start with a couple of formulas.

One says the speed, in feet per second, of a freely falling object is found by multiplying the elapsed seconds by 32, although in practice it varies a trifle from that with changes in altitude and geology. The centimeter equivalent is 975.

The other formula tells how far an object will have fallen after a given number of seconds. To get this answer in feet, square the number of seconds and multiply by 16. (Multiply by 490 instead for centimeters.)

Assuming a reasonable 12 feet of height for each story gives a fall of 348 feet.

Reversing the first formula, divide 348 by 16 and then find the square root, 4.66. That's how many seconds the fall must have taken. Multiply that by 32 to find the man's velocity at the end of his fall. It's 149 feet per second, which multiplied twice by 60 and divided by 5280 gives his mph as just about 102.

That checks so neatly with the newspaper story that we may conclude that the news man, or perhaps a police source, also used the assumption of 12 feet to the floor.

If you want to penetrate this subject a bit more deeply, you'll find more stuff about it in the later chapter on the outdoors.

How Deep the Well?

YOU CAN pretty accurately measure the height of a cliff you're standing on, or the depth of a well, by dropping a stone. All you need is the information that acceleration from gravity is about 32 feet or 975 centimeters per second per second ("about" because it varies a little from place to place). Those numbers tell you how fast a freely falling object will be going at the end of the first second. At the end of the second second it will be going twice 32 feet, and at the end of the third second three times 32 feet per second.

I realize this "second second" stuff is a mite confusing but you'll have to blame that on the English language. Even my computer program took an instant dislike to what it regarded as the double-word fault. *Zweite Sekunde* or *deuxieme second* does sound more sensible.

Well, anyway, the speed of our falling object during each second will be the average of starting and ending velocities. So it will cover 16 feet (the average of 0 and 32) during the first second, 48 feet (the average of 32 and 64) during the next, then 80, and so on.

From this it follows that to find out how far an object has fallen you square the number of seconds elapsed and multiply by 16.

So a stone that reaches the bottom of the cliff or well after 5 seconds must have traveled some 400 (25 times 16) feet. This ignores the effect of air resistance, which isn't great for a small distance and an object as dense as a stone.

Also left out here is the time it has taken for the sound to reach your ear. Sound traveling at some 1100 feet per second would need about one third of a second to go that 400 feet to reach your ear. We really should subtract that one third of a second from the 5, calculating 16 times the square of 4-2/3 instead. Then the height of the cliff or the depth of the well comes out very close to 350 feet.

The Metric Nickel

As LONG AS you have one coin—a U.S. five-cent piece—in your pocket, you have a key to the metric system.

A nickel weighs 5 grams. So two weigh a decagram. And a collection of nickels can give you any multiple of 5 grams, including the useful 100 grams.

The diameter of a nickel is 2 centimeters. String out five and you have a decimeter. Multiply the length of the five nickels by 10 and that's a meter.

How High Is Up?

WITH THE help of a few figures you can use an ordinary barometer to do the work of an altimeter.

The result you get will be quite close, even though the relationship between pressure and altitude actually involves quite a number of factors.

Since pressure at sea level is 760 millimeters of mercury, you start by dividing that figure by the barometer reading—also in millimeters of mercury, of course. Then find the natural log of the result, from a table or by pushing a key (probably marked LN) on a scientific calculator. Multiply by 7620. The result is a close approximation of your altitude in meters.

For example, suppose your barometer reads 238 mm. Dividing 760 by that gives 3.1933, of which the natural log turns out to be 1.1610. Multiplying by 7620 gives just over 8847 meters.

If, by the way, you ever produce just those figures in an actual situation you'll know not only how high you are but also just where in the world you probably are. The summit of Mount Everest is at 8848 meters.

Measuring Height by Angle

You can determine the height of an object at a known distance away even though it is inaccessible. It might, for instance, be a building on the other side of a river.

You can do this by using any device that will measure the angle between two lines of sight.

Place yourself directly opposite one end, so that the length you wish to calculate is at a right angle to your line of sight. Sight on that end and on the other end, finding the angle between the two sightings. If you've read the angle in degrees, minutes, and seconds, convert to decimal form in degrees.

Multiply the tangent of the angle by the known distance from you to the near point of the object. The result is its length.

Say the object is half a mile away and you find the angle to be 3 degrees, 15 minutes, 7 seconds. Dividing the minutes by 60 and the seconds by 3600 and adding both results to the number of degrees will give you the decimal form of the angle: 3.2519.

By a trig table or scientific calculator, you find the tangent of that angle to be 0.0568. Multiplying by the number of feet in half a mile (5280 divided by 2) you find that the distant object has a length of just about 150 feet.

How Big a Pinhole?

Anyone with the faintest interest in the technical side of photography must sooner or later build a version of the camera that started it all.

It's a simple light-tight box that holds a piece of film, or even photographic paper, at one end and has a pinhole at the other. Because a pinhole—although in its own way it does the work of a lens—has no focal length, a pinhole camera can produce kinds of pictures that an ordinary camera can't. Everything's in focus at the same time.

Before you prick, with a needle, that bit of black paper or foil or whatever you choose to produce the pinhole "lens" in, you'll want to know what its diameter should be.

Here's how to figure it. Real scientific-like.

Start with the wave length of the light you want to work with. If you have no preference, you might use the figure for light in the green region, 0.00055 mm.

Multiply that by the length of your camera, from the plane of the film to the pinhole, again in millimeters.

Take the square root.

Multiply by 1.9 and you'll have the optimum diameter for your pinhole photography.

How Fast Will It Cool?

Howw long will it take an object to cool to some predetermined temperature?

For example, in a 75-degree room, how long will it take a glass of tea to drop from boiling to 132 degrees Fahrenheit if you've found it drops to 172 in 5 minutes?

It's a nice exercise in the use of logarithms, if nothing else. And they are worth knowing about, even if back in school when the whole subject baffled you, you could hardly have cared less.

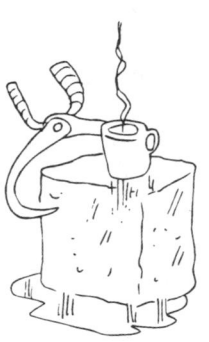

Begin with some simple subtractions.

Subtract room temperature from final temperature desired; from original temperature; and from temperature after first interval, calling these three results A, B, and C. Call the measured time interval D.

You now have three ways to go.

You can get the result by using a pocket calculator that does logarithms. That's the easiest way.

Or you can use a log table. That's almost as easy and it makes a fine introduction to the use of logarithms—or a fine brush-up course. If you're interested.

Or you can use a trial-and-error technique, if you have at hand a calculator with a powers key (marked x to the power of y) on it. No logs needed.

For either log approach (the first two above), subtract log B from log A and multiply by D. Divide this result by log C minus log B.

You can check your use of the log table with the example. You should find yourself multiplying 5 by the difference between 1.7559 and 2.1367, then dividing by the difference between 1.9868 and 2.1367. Result: 12.7 minutes to cool that tea.

The guess-and-check method is both quick and interesting. First you divide A by B. Then divide C by B. Then try raising the latter quotient to various powers until you find the one that most closely approximates the first quotient. Multiply that by D to find the number of minutes the tea will have to cool.

In the example, the two quotients (57 and 97 each divided by 137) are 0.416 and 0.708. You can see that if you square something that is close to 0.7 the result will be about 0.49, which is too big. If you cube it, you'll get around 0.35, which is too little. Now, using the calculator, you try raising 0.708 to a power in between these, perhaps 2.5. This proves to be close, so you now try 2.6 and find you have the answer bracketed. Another try or two brings you to 2.54, the power of .708 that turns out to be equal to 0.416. Multiplied, this 2.54 by 5 (as before) gives 12.7 minutes.

By the way, you might want to take note that this and the companion formula are accurate only where the object in question is not in closely confined space. In cooling, it must not appreciably raise the surrounding temperature.

How Cold Will It Get?

THIS CALCULATION will tell you how cold your coffee will be after sitting a certain number of minutes. Of course, the method applies to all sorts of situations where cooling takes place.

First you'll need a test figure. This is obtained by measuring the temperature of the object at the beginning and again after some convenient interval. From this, you can figure what the temperature will be after any other length of time.

Of course you must also know the temperature of the room.

Then proceed like this. Subtract the room temperature from that of object after the interval.

Subtract the room temperature from the original temperature.

Divide the first of these results by the second.

Raise this to the power obtained by dividing total time by time of measured interval.

Multiply this by the difference between original temperature and room temperature.

Add room temperature.

For an example, take that cup of coffee served to you at full boiling, as it always seems to be when you're in a hurry. Your thermometer tells you that after 5 minutes in a 65-degree room the coffee has cooled to 190. What will its temperature be after another 10 minutes?

You subtract 65 from 190, divide that by the difference between 212 and 65, to get 0.85. You raise this to the power of 15 divided by 5; that is, you cube it.

Multiplying by 147 and adding 65 tells you that after 15 minutes, your coffee will have cooled to about 155 degrees. Still pretty hot.

(Note that there is one precaution you should take unless you have a pocket calculator of the type that has a powers key on it, or are prepared to resort to logarithms. Be sure the total time is a simple multiple of the measured time. Since we wanted to know the temperature after 15 minutes, we measured it after 5. We could have used 3 minutes or 7-1/2 minutes without creating difficulties. But if we had begun with 4 minutes the power called for would have been 15/4, which is not especially amenable to paper-and-pencil arithmetic, although with a scientific calculator it is easy enough to handle.)

Recipes for Odd Numbers

MOST RECIPES are published for four or six eaters. If you're feeding one or two or eight, you may have some figuring to do. Often the easiest way to do it is through using equivalents such as those offered here.

A dash or a few drops or a few grains is taken to be less than 1/8 of a teaspoon.

A tablespoon is 3 teaspoons or 1/2 fluid ounce.

One-fourth cup is 4 tablespoons or 2 fluid ounces.

It follows that 1/2 cup is 8 tablespoons or 4 fluid ounces, 3/4 cup is 12 tablespoons or 6 fluid ounces, and 1 cup is 16 tablespoons or 8 fluid ounces.

Four cups is 1 quart or 32 fluid ounces.

A gallon is 4 quarts.

These equivalents should simplify dividing in two a recipe meant for 4 or adding one half to it to serve 6. For the odd problem, such as serving 7, it's usually better to stay with the figure for 6 or double everything as if for 8. It's safer to make a little less than you should, or a little more, than to get into hasty arithmetic that could lead to a serious error.

Incidentally, if your refigured recipe demands half an egg there's a simple answer. Beat yolk and white together with a fork and then put half into the recipe and half aside to enrich your next omelet. Or just figure that 2 tablespoons is half a large egg.

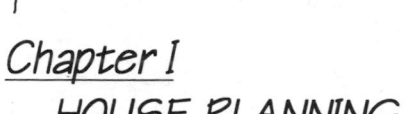

Chapter I
HOUSE PLANNING

Where To Live
How Big a House?
How Much Wind
   Can This Building Take?
Figuring Window Location
Costly Window Placement
How Much Glass?
How Much Solar Overhang?
Choosing a Solar Angle
Fighting Noise Pollution
Figuring Plumbing Plans

Where To Live

BEFORE you buy a house—or even rent one—you might want to investigate how far it is from various facilities.

The diagram you see here is based on one used by real-estate developers. They have found that most people have pretty clearly established limits of how far they find it acceptable to live from school, church, shopping, and the like. So it should help you figure things about your prospective location while it is still prospective.

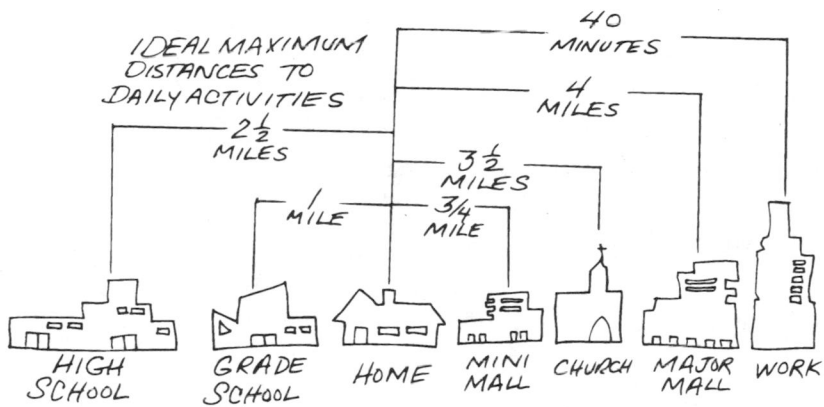

You'll note that the diagram is in terms of distances—except for job location, where time is the critical factor. Time and convenience of transportation are involved in the other distances too, of course, so some of them can safely be stretched if those factors are unusually favorable.

How Big a House?

WITHOUT getting into actual bedroom count or square footage, there are some generalizations about housing worth keeping in mind. The idea is to avoid the fundamental mistakes families often commit in their housing decisions.

A study made at the Pratt Institute School of Architecture tells us how families change as years go by.

During the first five years of a marriage, a one-bedroom house or studio apartment is tolerable for most people.

During the second five years, as income grows and children are added, the need changes to a two- or three-bedroom home, with more storage and living space as well.

During the third five years, still more living space is needed, with more bedrooms and an additional bathroom or two.

In the fourth and fifth five-year periods, needs are stabilized, but a change in the family income or job status often occurs.

In the sixth, seventh, and eighth five-year periods, retirement looms, needs dwindle, and a large home may become a liability. Yet it is often just at this period, the study warns, that so many families buy too-large, too-expensive houses.

How Much Wind
Can This Building Take?

IF YOU'RE building something in a windy spot, you may need to figure out how much wind the structure will take.

A common requirement in building codes is that a one- or two-story structure must be braced to withstand a pressure of 20 pounds to the square foot.

Beginning with this figure, or one specified for your own area, you can calculate how strong a wind can be (in miles per hour) before it produces excessive stresses. Or you can find how much pressure (in pounds per square foot) will be exerted on the building by a wind of any given force.

For the first calculation you divide the pressure in pounds

per square foot (20, for example) by 0.002558 and then find the square root. To see how this works, try it for 20 pounds. Your answer should be that to produce this pressure on the walls of the building will require a wind of 88.42 miles per hour.

Now try going the other way. How much pressure will a 60-mile wind produce? Multiply 60 by itself (that is, square it) and then multiply it by 0.002558. It seems that a 60-mile wind will produce a pressure of 9.2 pounds per square foot.

How much pressure is generated by a hundred-mile wind? Note that you can do this calculation by merely moving a decimal point four places to the right, thus multiplying the constant, 0.002558, by 100 twice (same as once by 100 squared) and getting the answer: 25.58 pounds to the square foot.

Figuring Window Location

NEW ENERGY-SAVING requirements rather drastically limit window areas in houses built today. This makes arrangement and location all the more important.

A window typi-cally influences all four of these im-portant aspects of a room simulta-neously: light, view, heat, ventila-tion. Windows high in a wall are most helpful in lighting a room because they throw light to the far wall. But

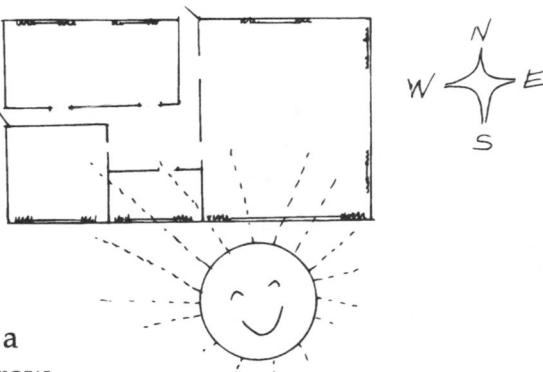

there's a limiting factor: a window too close under an overhang may not receive much light. Effectiveness of a window in respect to daylight can be judged visually, even in an existing house on a dull day, since the amount of light admitted will be roughly proportionate to the area of sky that can be seen through the window.

To admit light without creating excessive glare, avoid small windows widely spaced. A single window of equivalent area will be easier on the eyes, since glare comes largely from contrast between the windows and the darker wall areas around and between them.

A window as a source of a view is most effective if barriers to sight are avoided at eye level. That's obvious enough; but where designers often go wrong is in overlooking the fact that eye level changes from room to room.

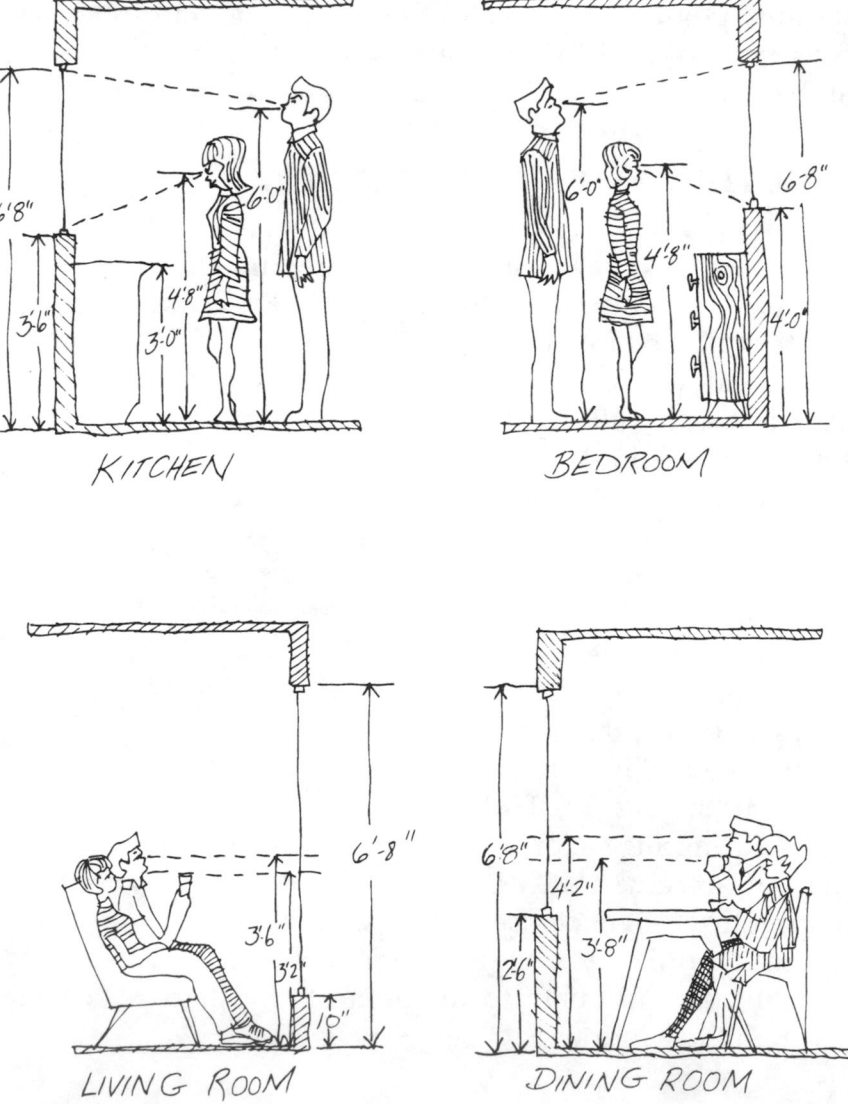

KITCHEN   BEDROOM

LIVING ROOM   DINING ROOM

The four drawings with human figures show what levels are important for various rooms. Between these levels, try to avoid horizontal divisions more than 4 inches thick.

The other four drawings have to do with through ventilation. Each of these structures has the same pair of openings in both its walls, but comfort will vary greatly from one to another as a result of window location.

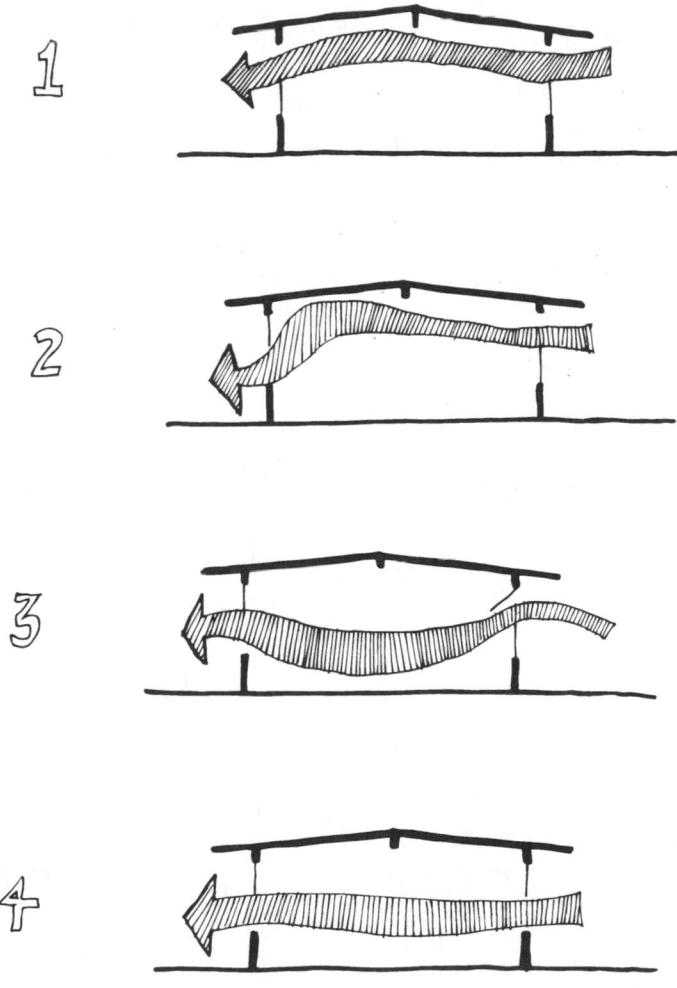

Suppose you supply a pair of high openings, as in the first drawing. Such windows will function fairly well, by carrying off warm air stored at ceiling level.

For a better result you might try placing the openings low in the wall on the side from which breezes come and high in the other. As Sketch 3 shows, deflecting the breezes to body level increases comfort.

Simplest solution of all is to keep openings fairly low in the walls, as in the fourth drawing. When views or light requirements dictate otherwise, a possible solution is to separate the functions—as by placing the windows where you want them and building louvered and screened ventilating openings below.

Costly Window Placement

WHEN ADDING a room or when building a house or cabin, how much more will it cost you to put windows in some arbitrary

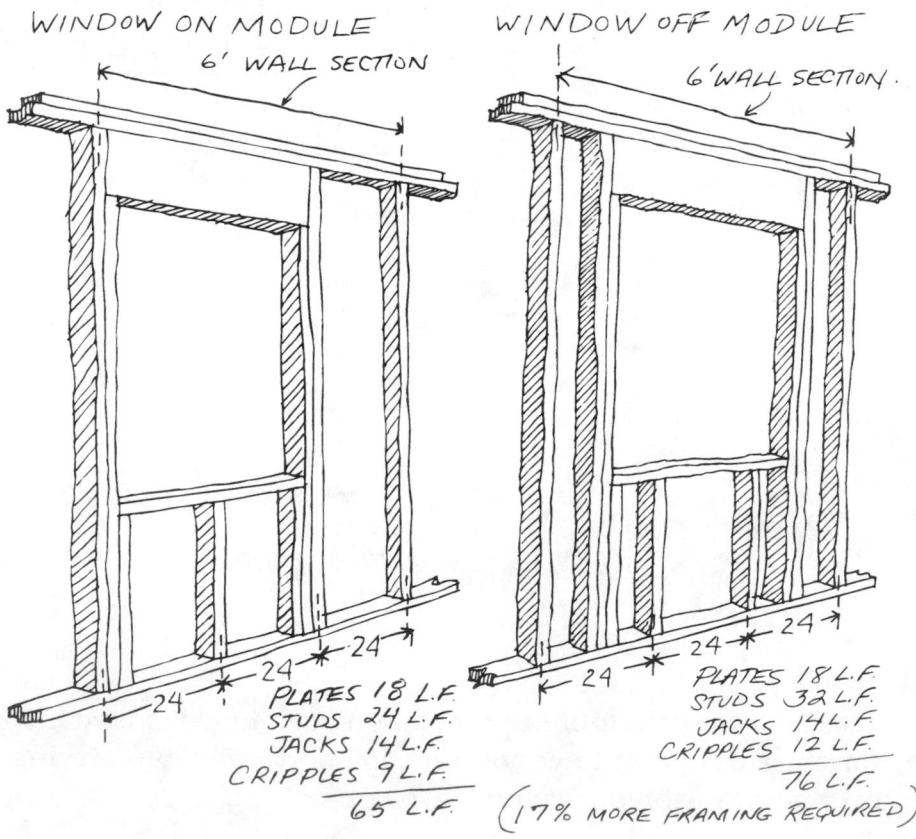

WINDOW ON MODULE

6' WALL SECTION

WINDOW OFF MODULE

6' WALL SECTION.

PLATES 18 L.F.
STUDS 24 L.F.
JACKS 14 L.F.
CRIPPLES 9 L.F.
65 L.F.

PLATES 18 L.F.
STUDS 32 L.F.
JACKS 14 L.F.
CRIPPLES 12 L.F.
76 L.F.

(17% MORE FRAMING REQUIRED)

location? Compared, that is, with putting them on a module—meaning right where a wall stud must go anyway.

The drawings show how it works out. Putting a 4-foot window in an arbitrary place instead of on module adds 17 percent to the quantity of framing lumber needed for a 6-foot section of wall.

So good planning saves you the cost of that additional 17 percent of framing lumber, and the labor of putting it up. Might even help save a tree too.

How Much Glass?

IN BUILDING or remodeling a house in the solar age, window area becomes an increasingly critical factor.

For north walls, and generally for east and west as well, the rule is the less glass you can get by with the better.

South is where to concentrate your attention—and your glass.

A southern window will pay its way in free heat and do far better than that if you can provide means to reduce heat loss when the sun is not shining through it. However, window area beyond what will bring the house to 70 on a partially overcast midwinter day is an embarrassment in excessive heat gains at times, paid for by excessive heat losses at others.

From these evident but regularly neglected principles, some useful calculations emerge. To figure out the optimum window area for a south-facing wall, first find the surface area of the house. This is wall area plus twice the floor area. Divide this figure by 35 and you'll have a good basic number for a very well insulated house in an average climate.

Fine-tune that figure to fit your house.

Double it if the house is poorly caulked, poorly sealed, and poorly insulated. Multiply by 1.5 if it's up to average standards. With unusually tight construction, you can use the figure as is.

Adjust again, for sunshine factor. Multiply by 0.7 if your locality's percentage of sunshine is very high—60 percent or more. For a very low percentage, 30 to 40, multiply by 1.3. For average sunshine, leave the number alone.

You should be able to get that sunshine figure from the local weatherman or from a public utility, a likely source also for the next one you'll want: annual number of degree days.

A degree day is just the number of degrees by which the average (mean) temperature of a day is below the comfort standard of 68 degrees Fahrenheit. Add all those deviations together and you have the figure for the heating season.

If the number of degree days where you live is about 5000, regard it as average—and your calculation is concluded without it. Otherwise, multiply the figure you have by the number of degree days and divide by 5000.

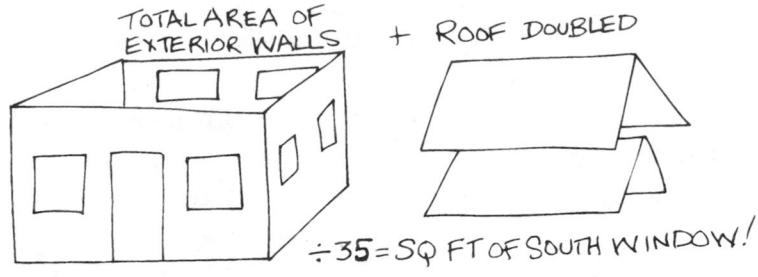

What with all the complications and chances of error—yours or mine or both—we need an example. Let's use a two-story rectangular house 20 by 40 feet with walls 16 feet high.

Adding total wall area (16 times 120) to twice the floor area of 800 gives 3520. Divided by 35, that calls for about 100 square feet of south window.

Now you need to consider the thermal quality of construction—essentially whether this will be an easy or a demanding house to heat. Assuming weather-stripping and insulation are not up to modern standards, you might choose a construction-level factor of 1.9. In doing so, you've decided that your house will demand 90 percent more heat than a well-built and insulated similar one.

For the sunshine factor, if yours is an area getting around 45 percent of the possible sun, you'd assign 1.1.

For degree-days, let us say that the records show 3500.

So that's the rest of the calculation, with 100 to multiply by 1.9 and by 1.1 and by 3500 before dividing by 5000.

It seems that 146 square feet of glass will be about right for that southern exposure. It is well, however, to think of the result as simply 140 to 150, if only as a reminder that this is just a good working approximation based on a series of estimates and is not to be taken as gospel.

How Much Solar Overhang?

WHEN YOU build a new house or modify an old one, proper respect for sunshine is in order.

It is not enough merely to put a lot of glass on the south. You should also protect that glass, and the people living behind it, from too much solar heat in hot months.

What you want is an overhang that will cast a shadow over a window from 8 in the morning till 4 in the afternoon from May through August. You can't do this perfectly, of course. But you can do it for 90 percent of the time merely by making the width of your overhang equal to two thirds of the distance from the bottom of the window to the overhang.

That's if you live in New York City, Pittsburgh, Denver, Salt Lake City, or anywhere else near 40 degrees north latitude. For 45 degrees north (Montreal, Minneapolis, Yellowstone, Portland, Ore.) increase the overhang by about one eighth. Shorten it by one eighth for the southernmost United States, which is at about 32 degrees.

Interpolate for in-between latitudes. Suppose you live in Tulsa, about halfway between latitudes 32 and 40, and you have an 8-foot-high window on the south. Two thirds of 8 is 5.3. Shortening that by one sixteenth gives you just 5 feet for the width of the ideal overhang.

Choosing a Solar Angle

IN PLACING a solar collector to help heat your house—or hot water for it—how should you aim it?

In respect to principal direction, it's south, of course. A variation of as much as 15 degrees from true south won't hurt the efficiency seriously, but if you have a choice in the matter, a little east of south is usually better than west of south.

And keep in mind that you're concerned with true, not magnetic, north.

An easy way to discover true north-south is by observing the shadow of a vertical post. When it reaches its shortest—around noon, standard time—the shadow line is running north and south.

After direction, your next concern is elevation. What angle of elevation will gain you the most heat? You can get it from tables or through intricate calculations, but there is an easy approximation that's good enough for practical purposes. Just add 12 to the latitude of your locality (see any atlas). This works fine for most of the northern hemisphere. For Denver or Philadelphia, at about 40 degrees north latitude, your solar collector should be at about 52 degrees from horizontal.

Here's some consolation, in case the calculation or the actual construction gives you a problem. Variation of 10 degrees either way is not especially important, so the easy angle of 45 degrees—a foot of rise for each foot of run—will be quite acceptable for most of the United States.

Fighting Noise Pollution

Quiet is a rare boon in the modern world. So it's worth while to know a little about how to protect your ears from sounds you don't want to hear.

From that point of view there are two kinds of sounds—noise originating in the room you're in and noise coming from outside it. Protection methods against the two types are almost totally different.

To muffle sounds within a room you need walls and floor and ceiling that are as soft and rough and porous as you can get. This means heavy carpets and draperies and acoustic ceiling tiles.

However, none of these will do very much to stop sound from going right on through into another room, or to stop it coming into your room. They can most effectively deaden what is being produced right there. In fact, if you soft-surface a room very thoroughly, it will become so dead acoustically that listening to music will become a new problem.

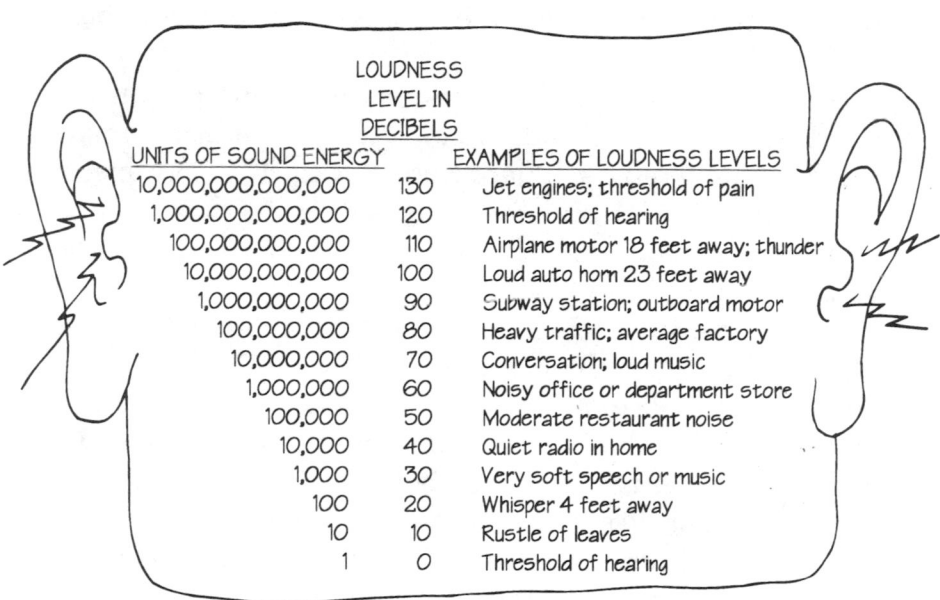

UNITS OF SOUND ENERGY	LOUDNESS LEVEL IN DECIBELS	EXAMPLES OF LOUDNESS LEVELS
10,000,000,000,000	130	Jet engines; threshold of pain
1,000,000,000,000	120	Threshold of hearing
100,000,000,000	110	Airplane motor 18 feet away; thunder
10,000,000,000	100	Loud auto horn 23 feet away
1,000,000,000	90	Subway station; outboard motor
100,000,000	80	Heavy traffic; average factory
10,000,000	70	Conversation; loud music
1,000,000	60	Noisy office or department store
100,000	50	Moderate restaurant noise
10,000	40	Quiet radio in home
1,000	30	Very soft speech or music
100	20	Whisper 4 feet away
10	10	Rustle of leaves
1	0	Threshold of hearing

You can estimate the value of a wall in stopping noise penetration by its weight. Sheer mass is the most important factor in keeping sound out. A poured concrete or stone wall is excellent at doing this, and even hollow masonry blocks are very good.

With ordinary frame walls, where there's not much mass, just putting in ordinary insulation will do disappointingly little. You'll accomplish more by separating the two wall surfaces so there are no mechanical connections between them except the unavoidable ones at floor, ceiling, and wall ends.

Standard way to do this is with staggered studs. Floor and ceiling plates are typically two-by-sixes instead of the more usual two-by-fours, so that every second two-by-four stud can be offset so it doesn't transmit sound across the stud space. An insulating blanket woven between the studs gives some additional muffling.

There is another arrangement that, although a little less effective, permits use of standard two-by-four materials. Using studs turned flatwise, alternate ones flush with each side, permits an insulating blanket, yet does not increase the space occupied by the wall. You might use it for the common situation between rooms where the wall is not required to bear a load.

To figure a floor that won't let much sound get through, you'd again be best off with masonry, such as poured concrete. But fortunately most standard floors are fairly heavy and will do an acceptable job on airborne noises.

The stickler comes with impact sounds—the dread Chinese water torture of having someone pace a bare floor over your head, wearing stiletto heels, while you're trying to fall asleep.

Most effective nonviolent remedy is reducing the impact that produces the sound. Stealing the spike heels will help, but a more feasible solution is installation of a heavy carpet. Or a heavy pad under the carpet. Or both.

A layer of fairly soft insulation board, fastened to the floor before pad and carpet are laid, adds further to the effect. And, of course, each of these does help a little in reducing transmission of airborne sounds from both above and below.

To get further into these important problems and deal with materials and partition systems that have calculated sound-transmission losses, you'll need to know a little about the methods and terms used.

The common unit of sound, the decibel, is most simply described as the least amount of sound your ear can detect.

When the energy producing a sound is multiplied by 10 the sound will seem only slightly louder to you.

To keep the measuring unit, the decibel, proportionate to what your ear tells you, it is on a logarithmic scale. All that means to you is that each perceptible increase in sound level is called 10 decibels.

Here's a typical instance of how decibel measurements work out.

SOUND REDUCTION

30 DECIBELS OR LESS
NORMAL SPEECH HEARD/UNDERSTOOD THRU WALL

30-35 DECIBELS
LOUD SPEECH HEARD FAIRLY WELL IF OTHERWISE QUIET

35 TO 40 DECIBELS
LOUD SPEECH AUDIBLE BUT NOT INTELLIGIBLE

40-45 DECIBELS
NORMAL SPEECH NOT AUDIBLE - LOUD SPEECH FAINT, UNINTELLIGIBLE

While you're trying to enjoy music in one room, somebody's talking loudly in the next room, at a 75-decibel level. By the time his voice reaches the wall between you and him the loudness will have dropped to perhaps 65 db. Distance helps.

Now on your side of the partition, you will benefit by the masking effect of the music, to the tune of about 15 db. You can tolerate 10 db sound, no louder than the whisper of leaves (as the chart indicates). You'll hear it but you won't really notice it.

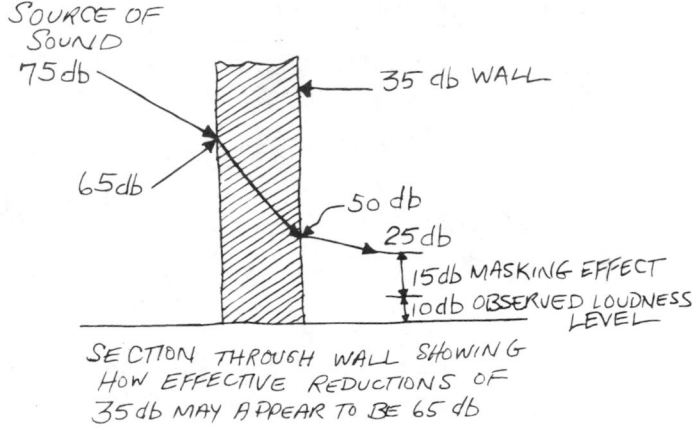

SECTION THROUGH WALL SHOWING
HOW EFFECTIVE REDUCTIONS OF
35 db MAY APPEAR TO BE 65 db

Add those two figures to the 5-db drop you can count on as the sound moves from the wall to your ear, and you get a total of 30 db.

The difference between the 65-db sound that will reach the partition and the 30 that you can accept is 35. So you need a wall with a 35-db rating.

For an idea of what kind of wall that means, note that ordinary two-by-four stud construction with half-inch gypsum board on both sides won't do it. But the same wall with a mineral-wool blanket in the stud space will. The blanket makes the difference between a 31- and a 39-db rating.

Figuring Plumbing Plans

MOST OF WHAT you might need to know about plumbing as an amateur house-builder or remodeler, you can pick up by talking to the people who sell you your supplies.

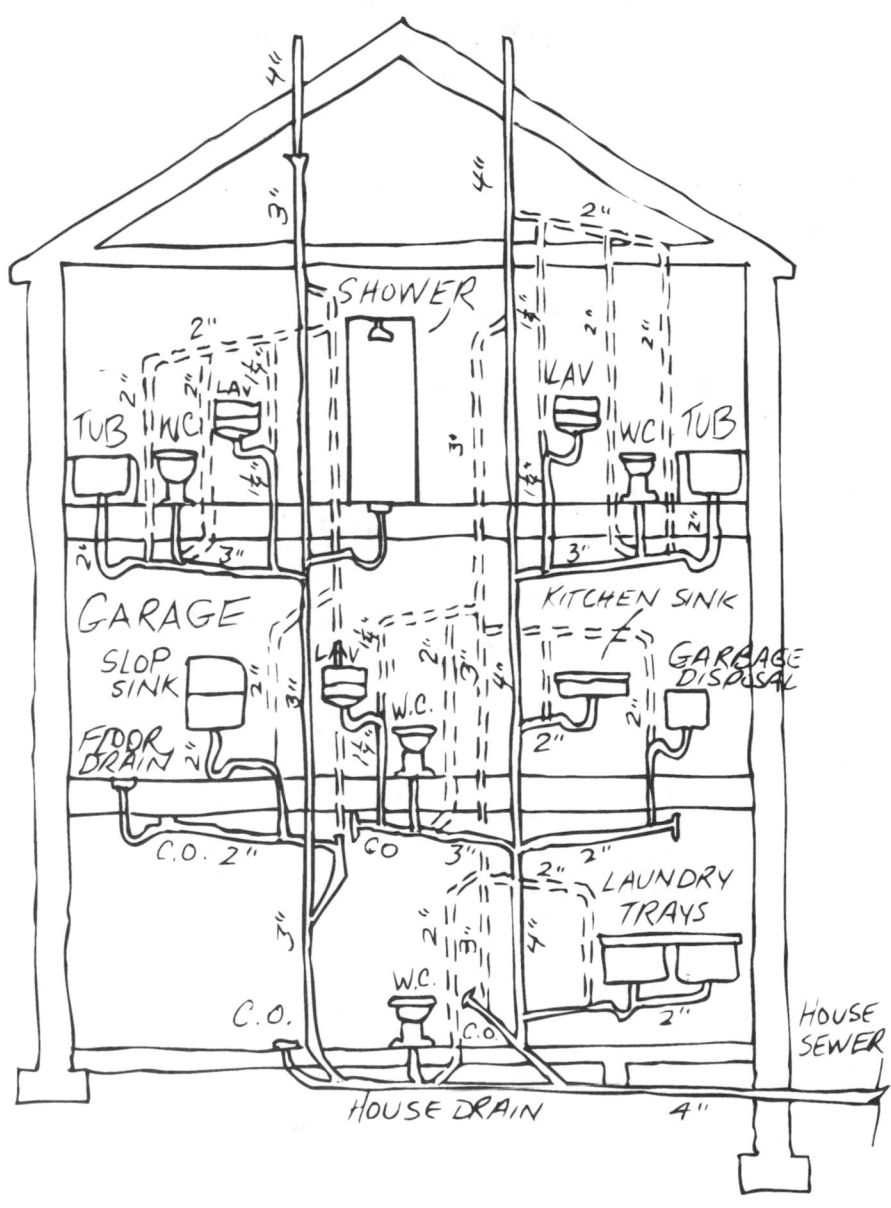

But the hard part—to me, anyway—is plumbing the mysteries of drains and vents. That's why I was delighted to come upon this inelegant little drawing (now neatened up a little for you) of an old-fashioned house that shows just this part.

As you will see, solid lines indicate drains. Dashed lines are vents. Clean-outs are shown, too. A few minutes with this drawing should give you a good start in designing plumbing for your own job, which I trust will be a simpler collection of pipes than this conglomeration.

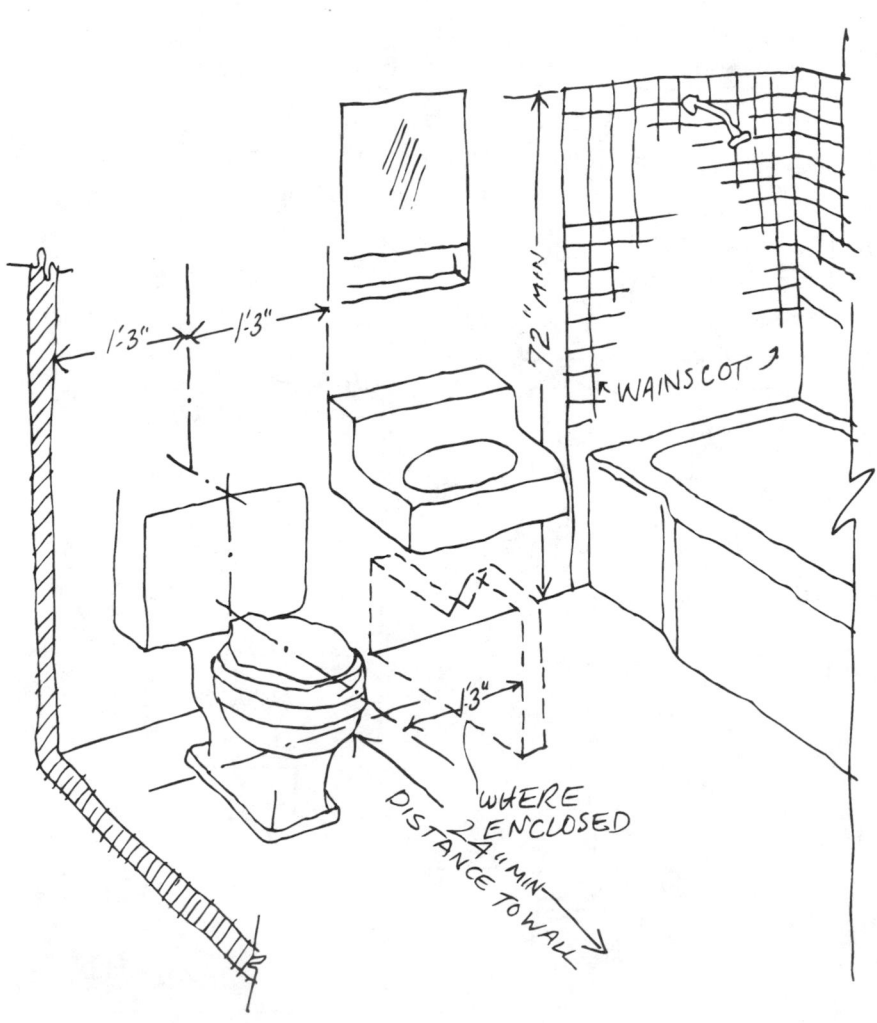

Water lines aren't shown. That makes this section drawing easier to follow. And water lines are the obvious part anyway.

I'm throwing in the other drawing because it, too, has to do with figuring plumbing. Those dimensions aren't just suggestions. They're code minimums in many parts of the country, though sometimes I wonder whose business it is how much space I provide around my john.

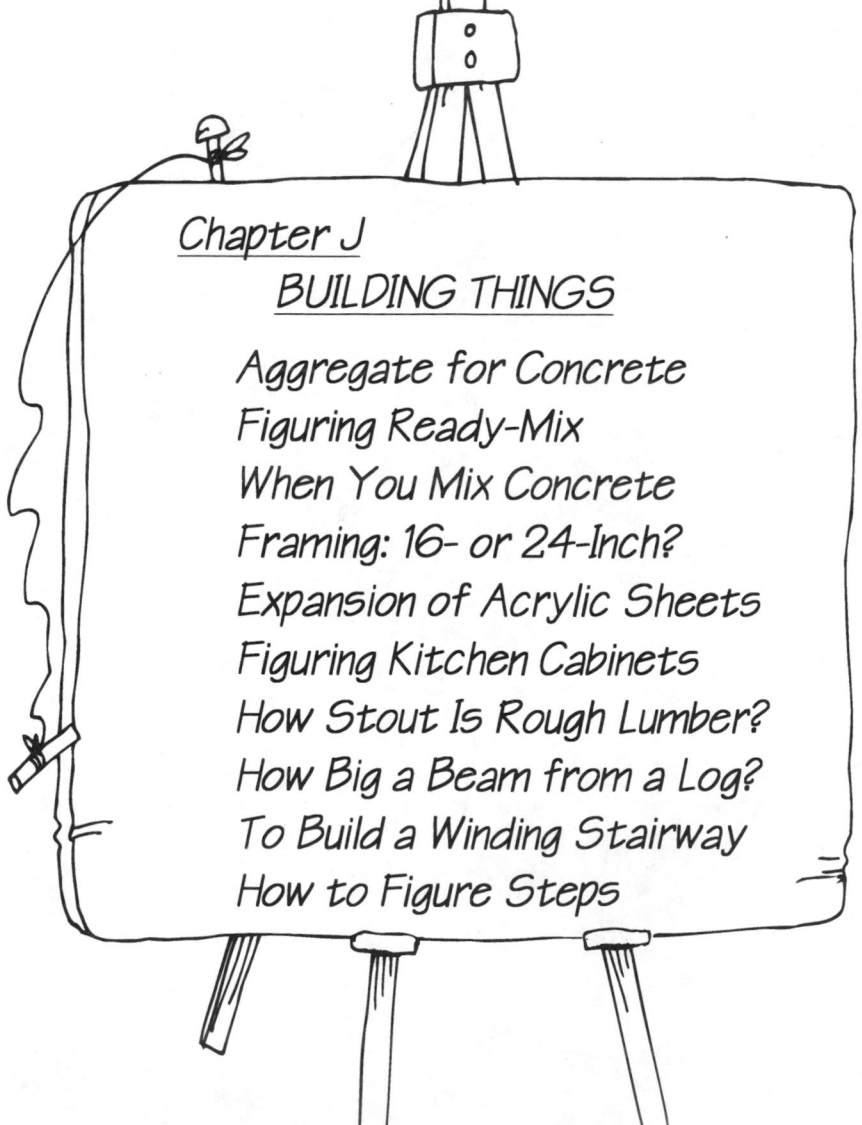

Chapter J

BUILDING THINGS

Aggregate for Concrete

How big should the biggest stones or hunks of crushed gravel be in concrete? One good rule to figure from is that when you're pouring a slab, the biggest ones should be not more than one-third the thickness of the slab.

Within this limit, however, the bigger the better.

With larger aggregate, you can get concrete of the strength you need while using less cement. Which means fewer dollars.

For the usual 4-inch slab (probably 3-1/2 inches actually, that being the width of the one-by or two-by-fours used for so many forms) you would order aggregate to maximum size of 1 inch. Going to 1-1/2 inches will affect the strength and make the stuff balkier to work with at this thickness.

For driveways, walks, and patios, you'll want concrete giving a compressive strength of 3500 pounds to the square inch when tested after 28 days.

To accomplish this with aggregate of 3/8-inch maximum size, your concrete must contain 610 pounds of cement to the cubic yard.

With 1/2-inch aggregate, this requirement drops to 590 pounds.

For 3/4-inch, it's 540.

For l-inch, it's 520.

Let the aggregate go as high as 1-1/2 inches and the mix may be a little more difficult to place but it will require only 470 pounds of cement to the yard of concrete to meet the strength test.

Figuring Ready-Mix

You won't find a bigger boon than transit-mixed concrete for the man, woman, or child engaged—however reluctantly—in doing a really substantial project.

When it comes to concrete, half of the fun is not getting it there in the form of sand, gravel, water, and cement. It's having somebody bring concrete to you already mixed and dumping it at least close to where it will probably remain into the next century.

What's no fun is to miscalculate your order and run just short— or be stuck with a considerable part of a ton of gooey stuff and no place to put it.

So calculate with care before you place your order. You'll have to figure out how much concrete you need and—with advice from the supplier—what kind.

To find the quantity, first multiply the length of the area by its breadth, both in feet. Then divide this by the number of thicknesses it would take to make a foot. Finally, divide your product by 27, the number of cubic feet in a cubic yard.

Suppose you're about to pour a 4-inch-thick sidewalk that is 4 feet wide and 26-1/2 feet long. Since 4 inches is one third of a foot, you multiply 4 by 26-1/2 to get 106, and then divide that by 3.

Dividing this result by 27 gives just over 1.3 cubic yards.

To include the usual 5 to 10 percent allowance for errors in measuring and for waste in pouring, you'll probably order a yard and a half. Then avoid waste by having a few small wooden forms on hand to use in casting some useful stepping stones with any spare concrete.

If you see any possibility of running short, have a small pile of rocks on hand. You can push them into the slab toward the end of the pour if you need to make bulk.

When You Mix Concrete

Concrete is the original handy-dandy plastic, infinitely useful. When you learn to work with it efficiently, you're halfway toward completion of many a major project.

Making concrete that is easy to work, strong, and yet economical calls for figuring rather precisely the amount of each ingredient.

For smallish home jobs, about the handiest measuring device is an ordinary galvanized water pail, although a wooden box is often used.

For each pail of cement, use 2-1/2 pails of sand (preferably wet) and 1/2 pail of water.

The amount of coarse aggregate (stones or crushed gravel) to use will depend upon the maximum size of the stones, which you will probably have specified in ordering the material. This size, by the way, should be not more than 1/3 the thickness of the slab you are pouring—1 inch for a 3-inch to 4-inch sidewalk, for example.

For 3/8-inch aggregate, use 1-1/2 pails. For 1/2-inch, use 2 pails. For 3/4-inch, 2-1/2 pails. For 1-inch, 2-3/4 pails. And for aggregate that ranges to 1-1/2-inch, you should use 3 pails.

As you can see, the bigger the aggregate the more gravel you can use for a given amount of cement, the costliest part of the mix. That's why it is good economy to get gravel as large as the rule about thickness of slab allows.

Loading the mixer is best done by putting in the aggregate and half the water, then starting the mixer and then adding the sand, the cement, and finally the rest of the water.

Framing: 16- or 24-Inch?

Hᴏᴡ ᴍᴜᴄʜ can you save by framing a building with the studs and joists on 24-inch centers instead of the more traditional 16 inches?

This is worth knowing because it is now generally accepted that the wider spacing is perfectly satisfactory for a one-story structure or for the upper story of a taller one.

For the wall-framing part, you'll reduce the labor cost (or hours spent) some 17 percent. Material cost will go down 8 percent. The two combined will drop nearly 15 percent.

BUT if you're covering that framing with plywood and the wider spacing forces you to go to thicker plywood, you'll almost wipe out the saving on materials. Over-all saving will slip to 7 percent.

Works out much the same with floors. Putting the joists on 24-inch centers saves 17 percent but if you must then go to thicker subflooring, figure the saving drops to 4 percent for the whole floor-framing and subfloor job.

The 24-inch framing will save on labor costs for drywall installation and hole drilling for wiring, too.

With these things included, you should wind up with a saving of 6 to 7 percent. The amount in one instance was $219 saved on a $3,175 job of framing walls and floor of a small house.

If there had been no need to shift to heavier wall and floor plywood, the saving would have been half again as much—some $307 or about 10 percent.

Either way, it's enough to make up for quite a bit of inflation while helping to save a forest.

Expansion of Acrylic Sheets

W HEN GLAZING with acrylics—Lucite, Plexiglas—you should calculate the possible expansion and allow for it. This problem comes up most often when replacing a glass pane in a storm door or window because it has been broken—or as a safety precaution.

For each 1 foot that the opening measures, subtract 1/32 of an inch to allow for the thermal expansion.

Figuring Kitchen Cabinets

I F LIKE SO MANY of us you're remodeling your house or itching to, you very likely want to start with the kitchen. It helps a lot if you can draw a picture of what you have in mind. And there's no other detail as complex, and as tricky to draw, as the working walls of a kitchen.

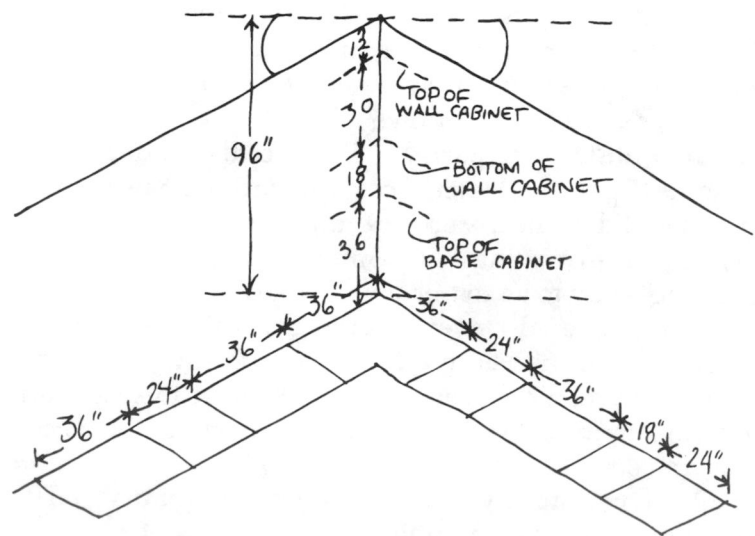

Here's the idea. You begin with the vertical corner line of the room, working to any convenient scale. Mark off location of wall and base cabinets. Those right and left walls take off from the vertical corner line at 60-degree angles (or 30 degrees from the horizontal dotted lines).

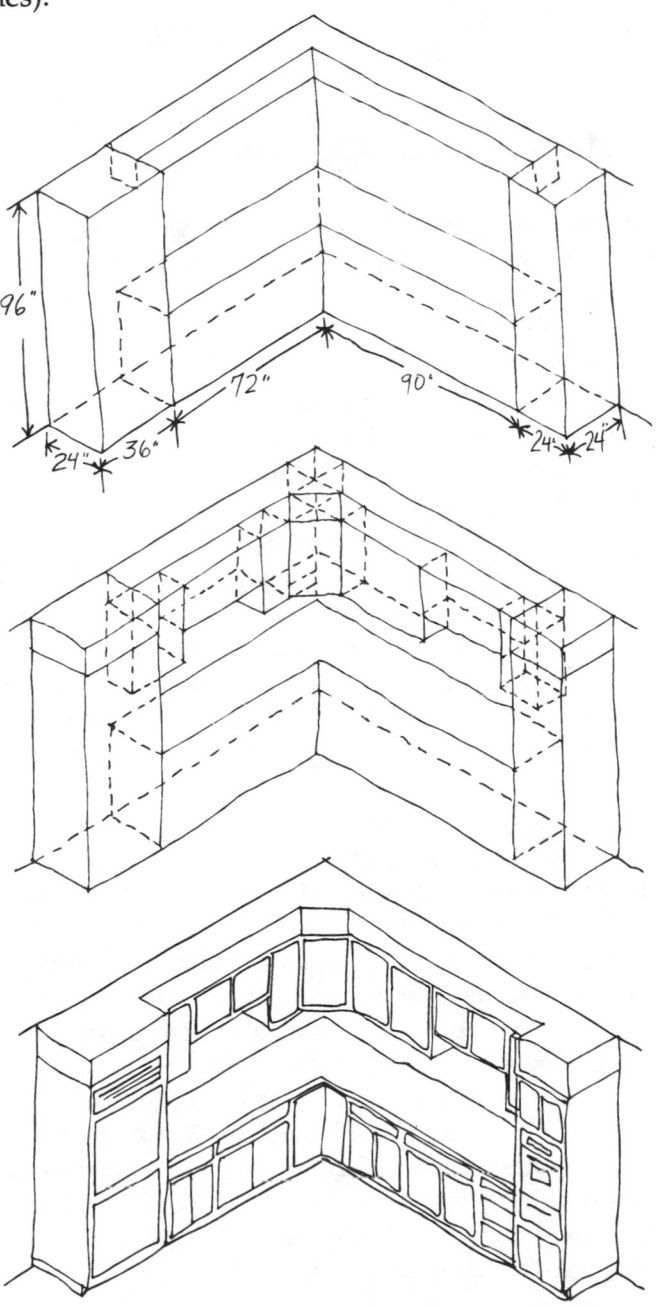

In the second drawing, you see how to block in outlines of base cabinets, oven, fridge, and the soffit. Lines are parallel because your drawing is an isometric, easier for this purpose and more useful, too, than a perspective.

Step 3 adds more details in the upper cabinets. These three steps are most easily done on a single piece of paper, but for the final step you may prefer to tape on a sheet of tracing paper and work on that.

How Stout Is Rough Lumber?

YOU MAY have a choice between using rough and surfaced lumber in some building job. The rough stuff is especially appealing if you happen to have timber of your own or can buy from a small local mill where the rough is cheaper than the smooth.

One really nice thing about rough lumber is that its width and thickness average about what its nominal dimensions say they are. A rough four-by-four is actually 4 inches thick, not the 3-1/2, more or less, of a surfaced piece. This extra wood means more strength. The question is—how much more?

If you're talking about strength in tension and compression, the rough is about one-third stronger. Exact gain depends upon the size of the board or timber you're dealing with, but many decisions can be made on the basis of that approximation.

More likely, you're concerned with bending strength: how a rafter will behave when the load goes on, for example.

With 1- and 2-inch lumber the gain will be at least one-third and may be as much as one-half.

Here's how to figure it. First calculate the relative strength of the surfaced lumber by squaring the dimension that is in the direction of stress. Then multiply by the other dimension. Use the actual dimensions, not the bigger nominal ones.

For the rough lumber, do the same, but use the nominal dimensions, or the actual measured ones if they differ and are available.

Thus if you're using two-by-four rafters for a small building and they have been surfaced to 1-1/2 by 3-1/2 inches, you square 3-1/2 to get 12-1/4 and multiply by 1-1/2 to get 18-3/8, or 18.375.

Squaring 4 and multiplying by 2 gives 32 for the rough lumber. Those are your relative strengths. To discover the percentage of gain from using rough instead of surfaced two-by-fours, subtract to get the difference (13.625) and divide this by 18.375 to discover a gain of 74 percent or almost three-fourths.

(If you want to check yourself on the method, take the nominal dimensions of a two-by-four to be 1-5/8 and 3-5/8 inches this time. See if the improvement from going to full 2-inch by 4-inch lumber comes out 50 percent.)

How Big a Beam from a Log?

Y OU HAVE a log. Or a pole. Or any chunk of wood that is circular in cross section.

You want to trim off the four sides to make it into a beam or girder to hold up something, such as a part of a building.

How do you make your saw cuts so as to get the stoutest possible beam from it?

It's no good just taking equal amounts off the four sides. That would give you a square beam, such as a four-by-four, which would not be the strongest you could get.

This is true because what counts most in making a strong beam is its depth—the measurement from top to bottom.

Its width counts less. A beam that's twice as wide as another beam will also be twice as strong—as you might expect.

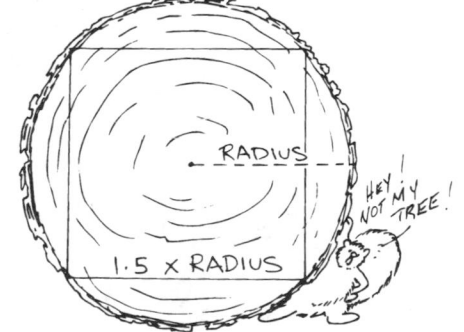

But one that's twice as deep (high) as another will be four times as stout, since strength varies as the square of the depth. (A lot more use of this relationship and the formula expressing it is made elsewhere in this book, where beams are calculated.)

Unless you have more fondness for a page of calculus than I have you'll be glad just to accept the result of someone else's work.

The strongest beam you can cut is the one whose width is 1.15 times the radius (or 0.575 times the diameter) of the log.

Give it a try for a log or post just 6 inches in diameter. You should find that the strongest beam of rectangular cross section it will yield is one 3.45 inches wide and 4.9 inches deep.

To Build a Winding Stairway

THIS BOOK'S no place to try to tell the world all about how to build a spiral staircase. But this drawing impressed me as something that tells its story all by itself and is so useful to anyone with such a problem in building that I wanted to preserve it.

Anyway, the drawing doesn't take up a lot of space and it should get a first-time stair builder off to a good start.

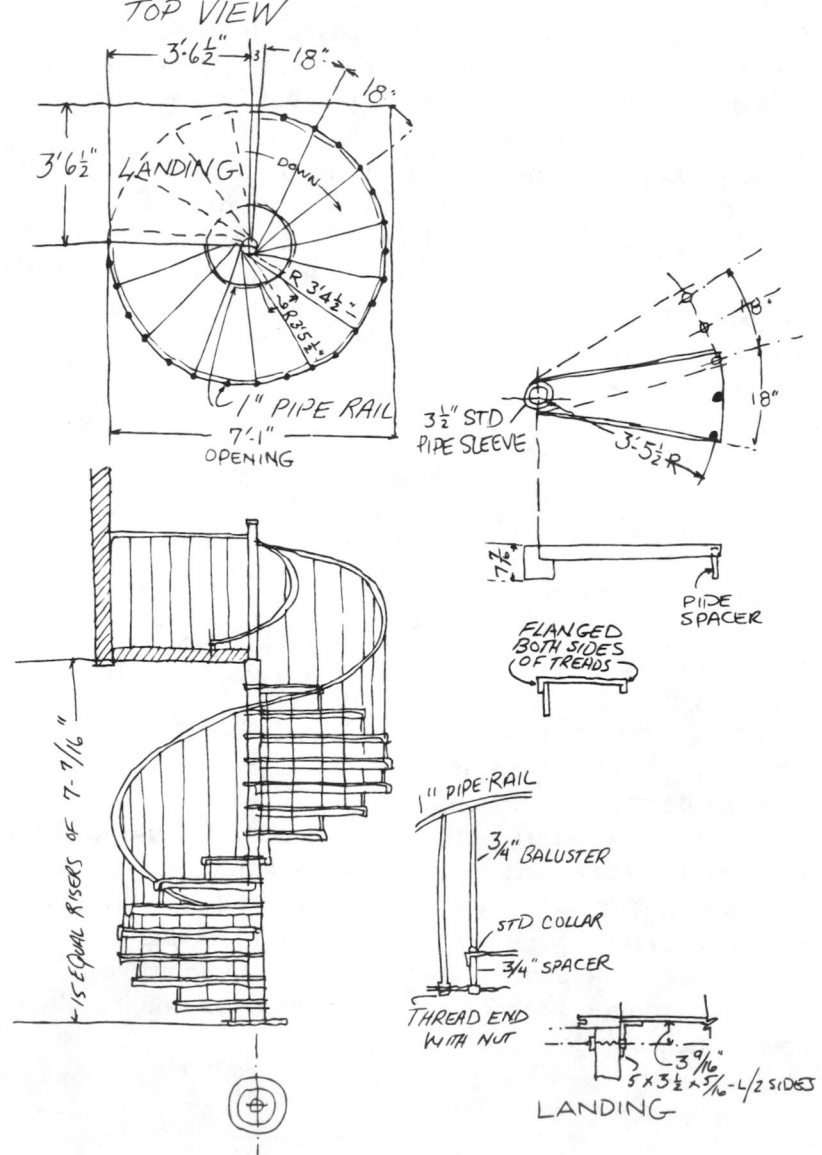

How to Figure Steps

IF YOU'RE going to build a stairway, or even a few steps to a porch, you can find all the dimensions and angles involved in a book of special tables. Or you can figure 'em for yourself very rapidly. Like this, working with four basic dimensions: total height; run; riser; tread.

Total height is the vertical distance you're working with, such as how much above the first floor of a house the second floor is. Since this is generally fixed by circumstances beyond your control, it is likely to be the figure you start your calculations from.

Run means the horizontal length of the whole stairway.

Riser is how much higher the surface of one step is than the one below it. It should be not more than 8-1/4 inches.

Tread, the horizontal measurement from one riser to the next, should be at least 9 inches. It is usually somewhat less than the actual width of the board you use for the step, since the step commonly sticks out an inch or so beyond the riser. This protrusion is called the nosing and you don't count it in your calculation.

Decide first how far down you prefer to go with each step you take. Usual minimum is around 6 inches, which makes for a very easy climb; 8 inches is pretty steep.

Divide the total height by that figure (riser height) to find the number of risers. Adjust it slightly to make the number of risers come out a whole number. (You can't, or anyway shouldn't, try to build a stairway with a half step somewhere.)

Now subtract riser height from 17-1/2 inches to get tread. (That 17-1/2 inches is a figure that long experience has established as making for a stairway most people find a comfortable compromise between ease and speed of use.)

Multiply tread by number of steps to find run. Check to make sure the run is short enough for the space available. From the figures you have at this point, you can work out a couple of others you may want.

One of these is the settings to use on a framing square to cut a carriage. The other is the angle at which this carriage—the slanting piece of framing that holds up the steps—will be placed.

The first is simple. Use the square so the tongue is at the height you've chosen for each riser (between 6 and 8 inches) and the body is at tread width.

The second is not so simple to figure out, but probably you won't need it. It is the angle at which the carriage will go—its incline. You could find it convenient to know this when setting a radial-arm or table saw for cutting the carriage. I've found it a good thing to know when building a special type of stairway having no risers—just treads set into grooves cut at an angle across a 2-by-8 to make a carriage piece without notches.

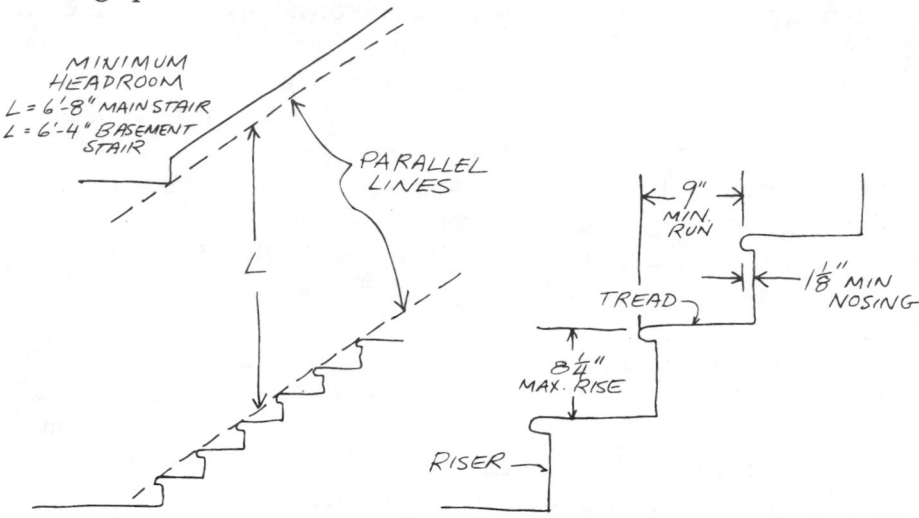

To discover this angle, first divide the height of the riser by the width of the tread. Then find the angle of which this is the tangent with a pocket calculator of the scientific type that delivers trig functions at the push of a button. Or...

If you have a computer to work with, you can use the program called BASIC (it comes with IBM-compatibles) for trig stuff. Or...

In a math reference book, find the nearest number in the table of tangents (within the section on natural trigonometric functions) to the one obtained by your division. The angle shown for this tangent is the one you want.

I think all this stuff will become clearer when you see it applied to a stair I've built in my house.

Total height, measured from the finished floor surface of my second floor to ditto on the first floor, is precisely 8 feet 9-1/8 inches. Turning that into inches and dividing by various possible number of risers, I find that 13 or less would give me risers greater than 8 inches, making for a pretty steep climb. With 18 or more, I'd get an excessively gentle climb and too many steps.

Wanting a comparatively quick climb, I choose 14 and divide to find that will produce 7-1/2-inch risers. There's one eighth of an inch left over in my calculations, but it has been found by experience that making one riser up to one-fourth inch more or less than the others won't be noticeable to eye or foot. That's about the maximum discrepancy allowable, however.

Subtracting 7-1/2 from the arbitrary, or God-given, 17-1/2 inches, says treads should be 10 inches.

Those two figures tell the marks to use on a framing square. They also tell how to figure the angle by trigonometry. When 7-1/2 is divided by 10, the closest figure to the resulting 0.75 that you'll find in a standard tangent table is at 36 degrees, 52 minutes. So that's my saw setting if I need one.

A stair with 14 risers will have 13 treads, so I find the run by multiplying 10 inches by 13. If this 130 inches, or 10 feet 10 inches, is within the space available, I'm all set to go. If not, I'll have to recalculate, using higher risers, so that the treads will be shorter and fewer.

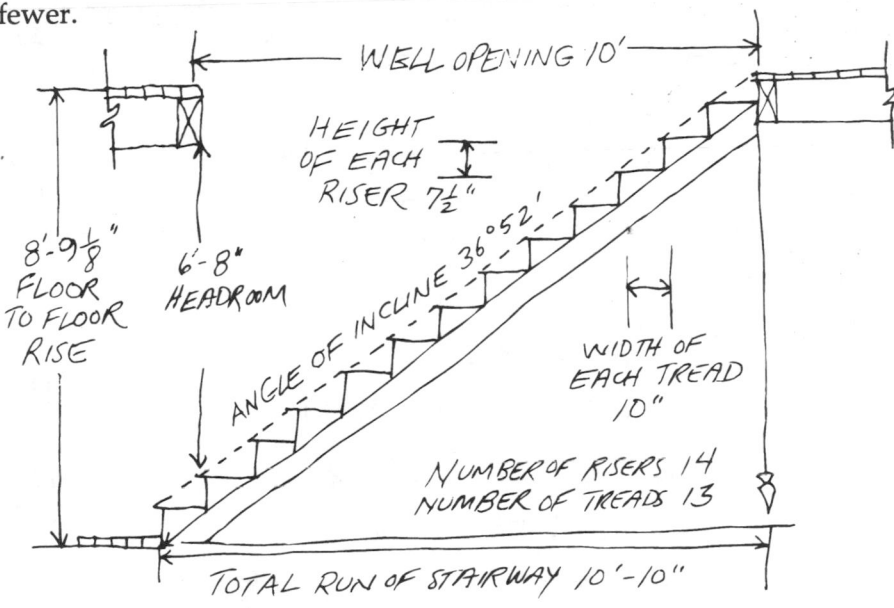

With rather steep 8-1/2 inch risers, for instance, I'd need only 13 of them, hence only 12 treads, of 9-3/8 inches each. That would reduce the run by almost a foot and a half.

If you're figuring things like this on a newly acquired pocket calculator that doesn't favor fractions, it may not be immediately evident to you how to multiply something like 9-3/8 by 12. The trick is to start with the fraction. Enter 3, divide by 8, add 9, then multiply by 12.

The one question that remains is amount of nosing. This overhang of each step surface for better footing is usually a little more than an inch, so the 10-inch tread we've figured on is fine if we're using standard lumber. A surfaced 1-by-12 or 2-by-12 will be about 11-1/4 inches wide. Just enough for nosing without any waste.

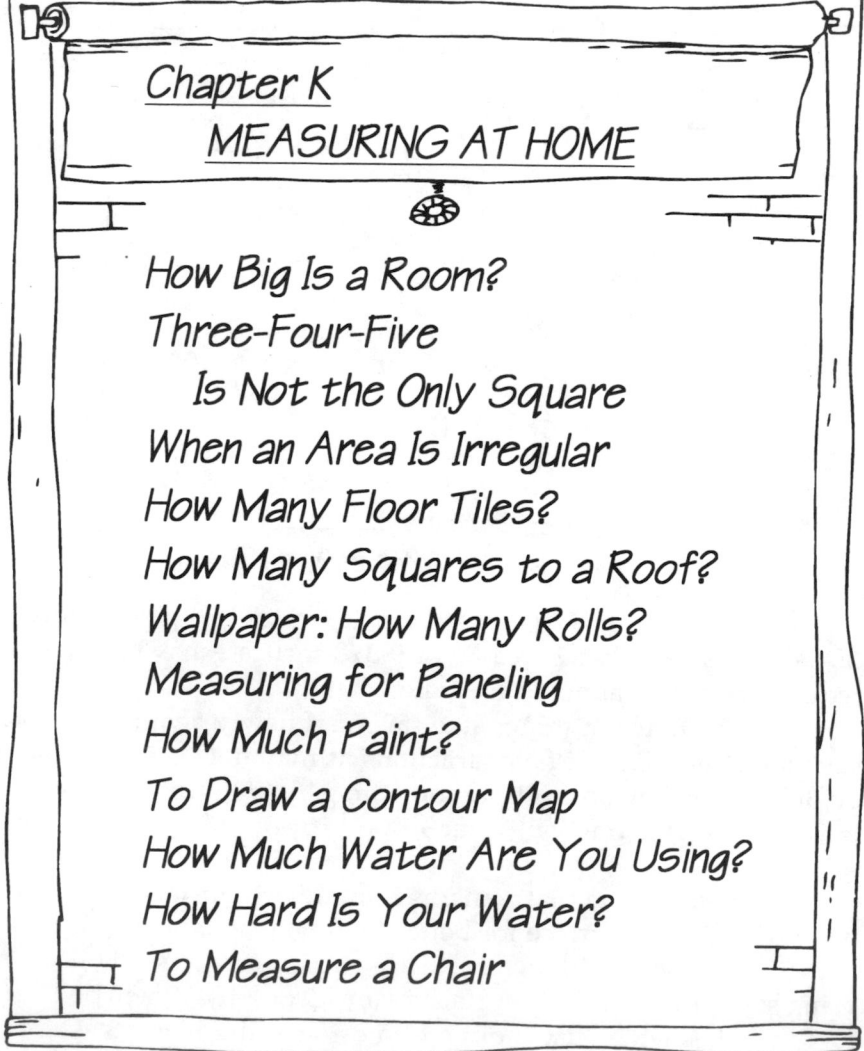

Chapter K
MEASURING AT HOME

How Big Is a Room?
Three-Four-Five
   Is Not the Only Square
When an Area Is Irregular
How Many Floor Tiles?
How Many Squares to a Roof?
Wallpaper: How Many Rolls?
Measuring for Paneling
How Much Paint?
To Draw a Contour Map
How Much Water Are You Using?
How Hard Is Your Water?
To Measure a Chair

How Big Is a Room?

Sıze of a room is usually pretty obvious to figure. You multiply the length by the breadth to get the area. If there is a closet or other jog within it that you don't want to include, you figure it the same way and subtract the one from t'other.

If two corners are square but the opposite wall is at an angle, then you multiply the length of the wall between the square corners by the average length of the two adjoining walls.

But if it's a room you're building as part of a house construction or remodeling and the ceiling is too low, you may have another problem. Under most building codes your room is not the size it looks. Some of it may not count at all, legally speaking.

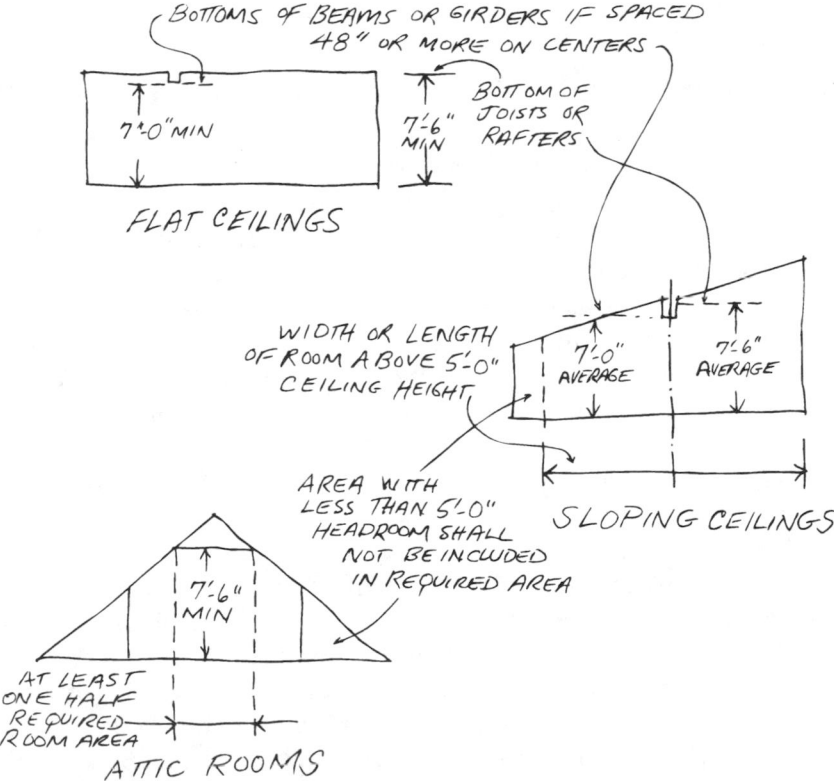

That can be important when the code specifies certain room dimensions, as do most of the various building codes. In general, every dwelling must have at least one room of 120 square feet or more; other habitable rooms except kitchens must have at least 70 square feet; and no habitable room other than a kitchen may be less than 7 feet in any dimension.

Assuming it's any of the code's business what size your rooms are (and I'm not so sure it is) these dimensions seem reasonable. But there's a gimmick. According to code, part of the room doesn't exist—if the ceiling is too low.

The UBC begins by telling you that halls and bathrooms must have ceilings at least 7 feet high measured to the lowest projection from the ceiling. Habitable rooms and those used for storage or laundry must have at least 7-foot-6 ceilings.

But what about a ceiling that slopes, as in so much modern post-and-beam construction? Then, as the drawings help explain, the prescribed ceiling height is required in only half the area. But "no portion of the room measuring less than 5 feet from the finished floor to the finished ceiling shall be included in any computation of the minimum area thereof."

Three-Four-Five Is Not the Only Square

IF YOU'RE laying out a building and would rather like the corners to be square, you may mutter, "Three, four, five."

Because they're handy in size and easy to remember, these three dimensions are in common use by builders and lots of other people. Those dimensions describe the two shorter sides and the hypotenuse, of a right triangle. But they're not the only set of numbers that produces a right triangle. In fact, there's no limit to the number of sets of digits you can use for this purpose...and not just multiples of 3, 4, and 5 either.

You can choose any even number you wish for one side of the triangle.

To find the other side, select any two numbers whose product is half the length of the long side. Square each number and subtract the smaller square from the larger. That's your second side.

Add those same two squares together to find the hypotenuse.

(For example, say you choose 28. Half of that is 14, which is 7 times 2. Square of 7 less square of 2 is 45. Square of 7 plus square of 2 is 53. You have a 28-45-53 right triangle. To check it, add squares of the first two numbers and note that their sum equals the square of the third, the hypotenuse.)

Now, for a working example, let's suppose you've driven four stakes where you'd like the corners of a 20-by-30-foot rectangular building to be. A most convenient triangle for checking squareness would be one having 30 feet for its long side, letting you use the side of the building as the long side of your square.

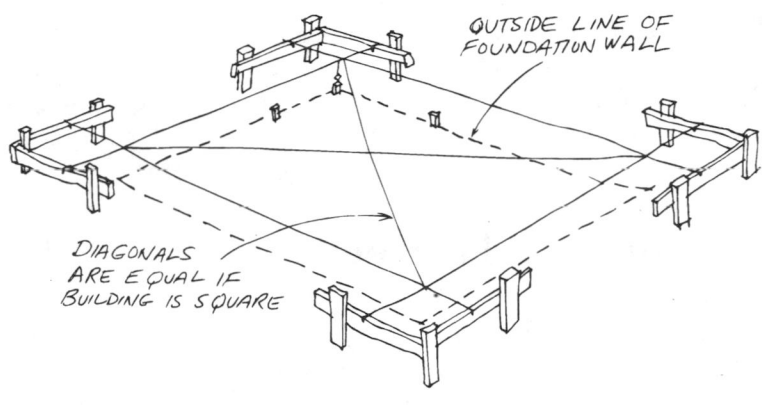

OUTSIDE LINE OF
FOUNDATION WALL

DIAGONALS
ARE EQUAL IF
BUILDING IS SQUARE

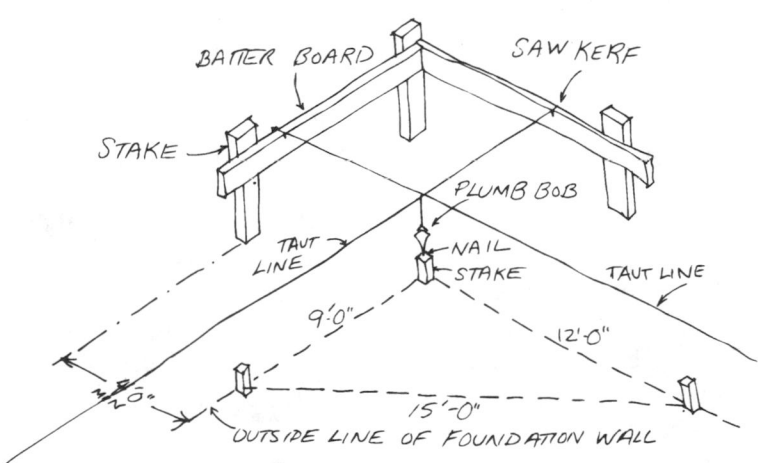

BATTER BOARD   SAW KERF

STAKE

PLUMB BOB

TAUT
LINE   NAIL
STAKE   TAUT LINE

9'0"   12'-0"

2'0"

15'-0"

OUTSIDE LINE OF FOUNDATION WALL

You can do this with 3, 4, 5 proportions by a little arithmetic. Since the long side is 7-1/2 times 4, multiply 3 by this quantity to find the short side and do the same to 5 for the hypotenuse. But this produces a short side greater than the width of your building layout....

So back to our scheme. Long side is to be 30. Half of that is 15, which can be factored into 5 and 3. Squaring both, then subtracting, you get 16 for the short side. Adding gives 34 for the hypotenuse.

Our numbers are keeping us within the dimensions of the building, so measuring shouldn't be difficult.

You measure 16 feet along the short side and stretch a line to find out whether the distance from there to the far corner is precisely 34 feet. If it is, your building is on its way to being properly rectangular.

The drawings show how these notions are put into practice. In fact, they do a fine job of showing just how to lay out a simple building. This consists of first staking corners, then building batter boards a couple of feet away from which to stretch lines after the stakes have come out during foundation excavating.

When an Area Is Irregular

WHETHER it's to order flooring, buy paint, or pour concrete, you often need to calculate an area. For a rectangle or other regular shape, this is easy enough. But in real life, not all shapes are regular.

They can, however, be broken up into a series of quite regular—and easily calculable—parts with whatever degree of precision the situation demands.

The basic shape for calculation is the rectangle. You'll recall, almost without thinking, that its area is just the length times the width.

The area of a triangle is half the length of the base times the height, the height being the vertical distance from base to opposite vertex, or corner.

(When it's a right triangle, as it usually is, the area is half what you find by multiplying together the two sides next to the square corner.)

What about a trapezoid, a shape you might find in a short driveway? Take an average of the two parallel sides and multiply by the distance between them. If "take an average" seems vague, it just means you add those lengths and divide by 2. Or think of it as the distance halfway between the two .

The area you are measuring, or some portion of that area, may be a circle or part of a circle. For a full circle, remember that you first find the radius, which is the distance from the center to the rim, multiply that number by itself, and finally multiply by the value of pi.

(For most purposes, the value of pi can be taken as 3-1/7 or 22/7, whichever seems simpler. For greater precision, or to eschew fractions in favor of decimals, use 3.1416. With a calculator handy and a wish to indulge yourself with even greater precision, without appreciable extra effort, help yourself to 3.1415927.)

Naturally, a half-circle has half the area of a full circle and a quarter-circle one-fourth.

If these round and three- and four-sided figures don't cover all your needs, here are a couple of others:

A five-sided figure, or pentagon, has an area 1.721 times the square of the length of any side.

Since the area of a regular hexagon, or six-sided figure, is 2.598 (as near to 2.6 as usually makes no difference) times the square of a side, you can proceed by first multiplying the length of a side by itself and then by 2.6. A hexagon with 20-foot sides, like the living area of a house I once built, will have area of 1040 square feet.

Here is another good method of approximating an area that can be combined with this sum-of-parts trick. It also may be used alone, especially when you're stuck with a highly irregular area that may have been laid out long ago b'guess and b'gosh. When the geometric methods prove tedious, map the area with pencil and ruler, and possibly compass, on a sheet of graph paper. Then count squares, half-squares, quarter-squares and so on—and add 'em up.

How Many Floor Tiles?

T HIS IS a quick on-the-job way to estimate how many 9-inch-square floor tiles you'll need for a job.

Multiply the length in feet by the width in feet to find the square footage. Double this. Then take off 10 percent.

A floor 12 by 18 feet has 216 square feet. Doubled, it's 432. Taking away 43 leaves 389. That's how many tiles to figure on.

Take the time to make the calculation by a more precise method and you'll find the figure is actually 386. But then you'd probably want to add a few tiles for waste, anyway, or to keep on hand in case you had to replace a damaged tile some day. So the quick estimate turns out to be more useful than an exact figure would be.

In short, it's better than perfect.

How Many Squares to a Roof?

T HE SUBJECT isn't square meals but the measurement commonly used by contractors to figure areas for roofing and siding jobs.

A square is merely 100 square feet, or the equivalent of a wall or roof section 10 feet by 10 feet.

To estimate the cost of a siding job, multiply the height of the walls by their combined length and from this subtract the total area of windows, doors, and such things as fireplace chimneys or other masonry features. Divide by 100 to get the number of squares.

For a roof, multiply the actual area of surface and proceed in the same way. You can't use the square footage of the house, because that would overlook the effect of roof slope, which increases the area, and of overhang, which can be substantial.

Wallpaper: How Many Rolls?

IF YOU'RE using American-made wall covering, divide the square footage you've found by measuring the walls (and ceiling if you're papering that) by 30. This will tell you how many single rolls to order. It includes an allowance of 6 square feet per single roll for waste.

For European-made paper you might use 22 as a guess to divide by if you don't know the number of square feet in a roll. But since size varies, better to find out the correct number if you can and then subtract 6 to get the number by which to divide the square footage.

Even after all this it's safest to add another roll to the total to allow for minor miscalculations or excessive waste. Allow more yet if the pattern repeat is unusually large or if your room contains few spaces above doors or windows where scraps can be used.

A reason for being ultraconservative is that if you have to go back to the store for a final roll or two you may wind up with wallpaper from a different batch in which ink colors are not quite the same.

Measuring for Paneling

To figure how many panels of plywood or hardboard or other big-sheet wall materials to buy when doing the walls of a room, first find the perimeter.

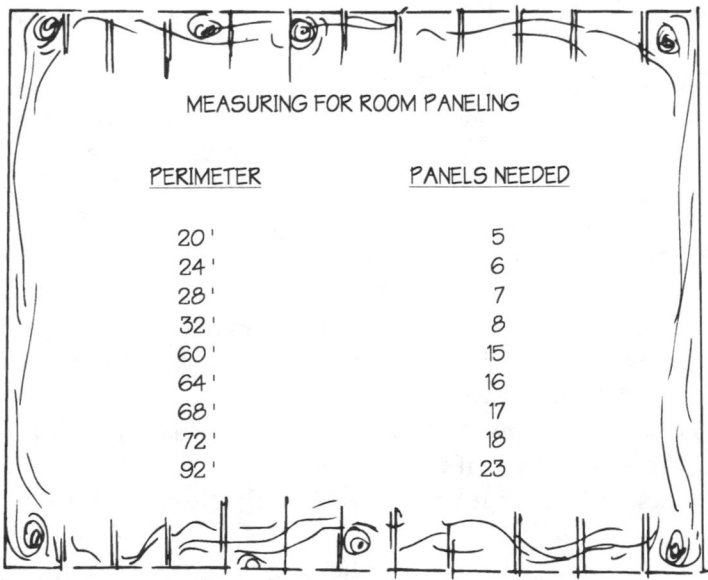

MEASURING FOR ROOM PANELING

PERIMETER	PANELS NEEDED
20'	5
24'	6
28'	7
32'	8
60'	15
64'	16
68'	17
72'	18
92'	23

Find number of panels in the table. Use the next higher number if yours falls between figures on the table.

Deduct for doors, windows, fireplaces.

Typical deductions are half a panel for a window or fireplace, two-thirds for a door.

Subtract the total of deductions from the number of panels originally determined, rounding off to the next higher number if necessary.

How Much Paint?

Figuring out how much paint you'll need for a job is essentially an area calculation, even though several surfaces may be involved. Add up all the areas you're going to cover and divide by the coverage estimate given on the paint can.

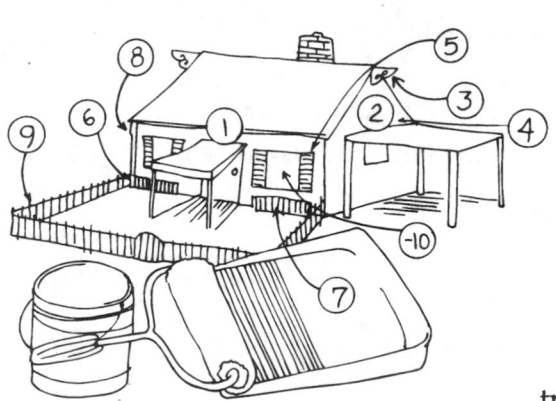

Here is a procedure, as worked out by a paint manufacturer, for quite accurately estimating the total area.

1. Measure the rectangular areas and multiply length by width or height.

2. Do the same for triangular areas, such as gables, but then divide by 2.

3. For plain cornices, multiply length by width and double the result. For ornamental cornices, triple the result.

4. For plain eaves, multiply width of overhang by length and add 50 percent. Double that result if the rafters are exposed.

5. For each face of slatted shutters and blinds, multiply length by width and double the result.

6. For columns, measure the circumference and multiply by the height. Add 50 percent if they're fluted.

7. Rails, lattices, and balustrades: multiply the length by 4 times the height.

8. Gutters and down spouts: having found the circumference in inches, divide by 12 to get feet, then multiply by length. For gutters to be painted both inside and out, double this.

9. For picket fences to be painted on both sides, multiply the length by 4 times the height.

10. Add everything together and subtract any areas you've included that are not to be painted—doors, windows, and such.

To Draw a Contour Map

IN THE BEGINNING, I learned how to draw a contour map with inspiration from one of the most effective incentives in this wicked world—greed.

I had to turn in such a map to get city permission to build my house on a steeply sloping lot. To follow standard practice, I'd have had this prepared by a surveyor who would then have deprived me of three or four hundred dollars.

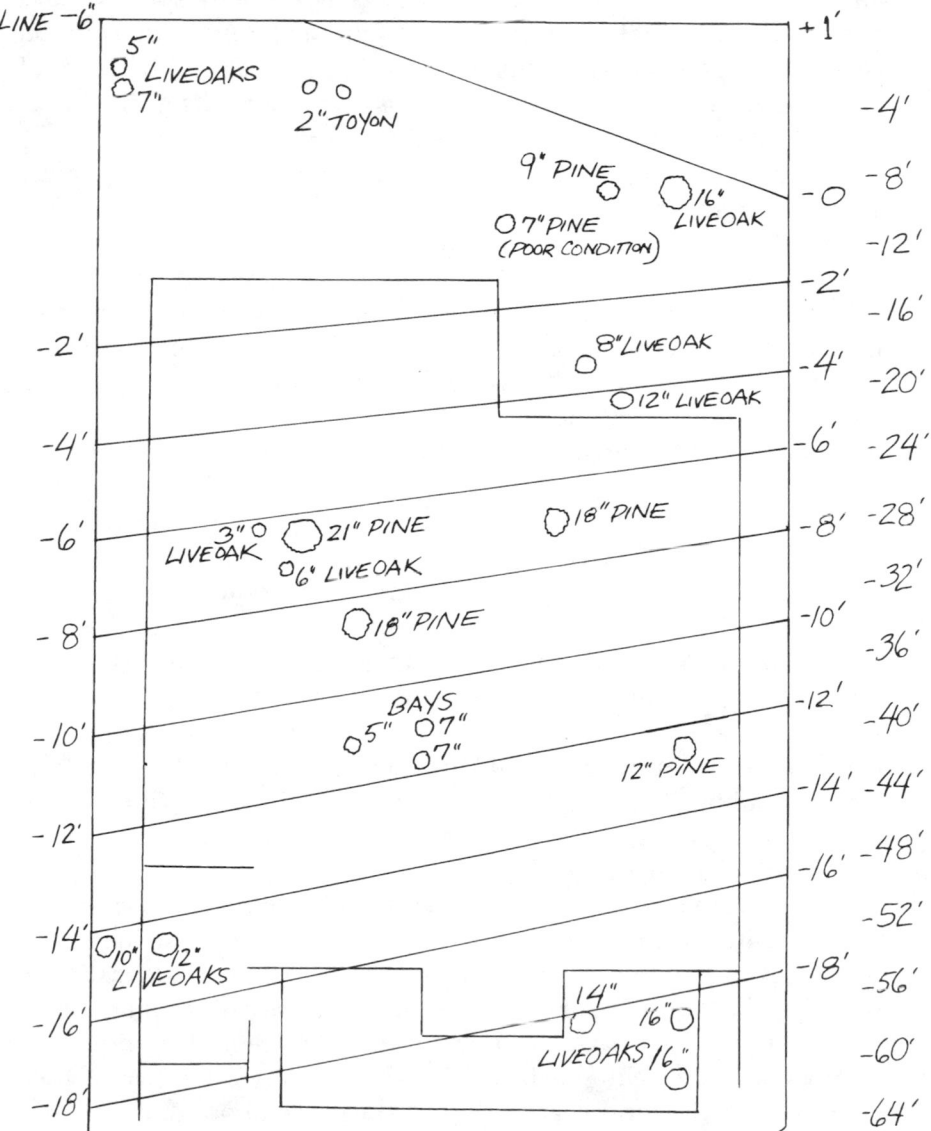

Since I was about to build the house myself—design, plans, concrete work, carpentry, plumbing, all the sweaty stuff—it went against the grain even to contemplate farming out one of the cleanest parts of the task, which I was convinced needn't take more than a few hours. My confidence was enhanced from having gleaned a few tips over the years from a relative by marriage, Gerry Kinsey, who knows a great deal about surveying but unfortunately lived too many miles away to be of personal help in my crisis.

I didn't produce the same kind of contour map a surveyor would have drawn. In the first place, my requirements were special—all I needed was information with which to show a planning commission what my slope was like and how my house would fit onto it. And I didn't have surveyors' instruments. Just a hand-held sighting level (borrowed—but purchasable for as little as $8).

I decided to let my map show 2-foot changes along the critical slope, so I held my sighting level on a 2-foot length of dowel to keep it consistently this distance from the ground.

I started as near the bottom of the slope as a steep gully permitted, along one side of the property, and worked my way up. Each time, with help of a young assistant who placed and held his toe where told, I could find how far it was from my sighting point to a point on a level with my instrument.

That point became the new sighting point on which I placed my dowel.

I marked off the distances along one side of a map of the property that I'd drawn on a piece of ruled (graph) paper.

Then I went back to my original point and sighted across to find the point on the other sideline that was at the same level as my original point. And worked my way up that side in the same fashion.

And then did the same thing up the middle of the lot.

From one side to the other of my 40-foot lot, the slope was uniform. If it had not been, or if I had needed a more detailed map—for landscaping?—I might have drawn several lines within the lot, parallel to sidelines, and sighted them, too. But one line was enough to show the nature of my slopes.

To make my map as useful as possible I then chose a zero point more significant for construction purposes than my starting point. Since easy driveway access was my most important consideration in planning the level of my house, I took as my base point a place in the driveway-to-be that was an even number of feet (18) above my original starting point. I called this 0. Each of my other points then became -2, -3 feet, and so on, down to the original point, which was now -18.

Connecting the points of equal level then gave me the contour map you see. There are better ways to make contour maps (sighting on "break" points—where contours make sudden changes) but this one is the simplest I know, and it serves such purposes as mine well enough. This one not only satisfied a rather stuffy planning commission (after the city engineer told them it would serve its purpose) but also proved to be just what I needed when I drew my house plans and when I did the actual building.

The second drawing you see here shows how I was able to use the contour figures in a drawing of a section through the house, very helpful to me in completing the design.

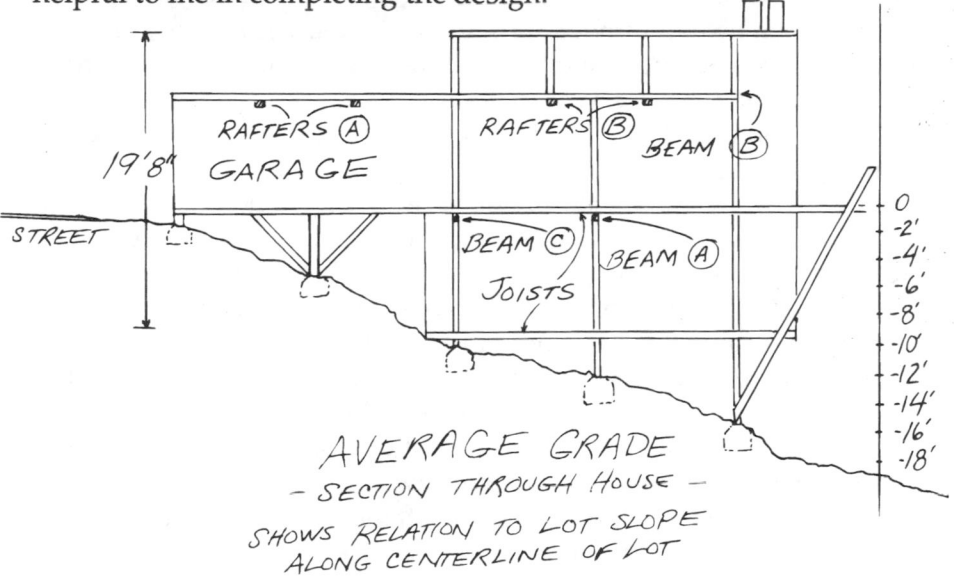

– SECTION THROUGH HOUSE –
SHOWS RELATION TO LOT SLOPE ALONG CENTERLINE OF LOT

How Much Water Are You Using?

IN MANY places, it's so much a duty these days to conserve water that we've even got the water companies believing in it. It may surprise you, as it did me, to discover that your water meter is a sufficiently delicate instrument to help you measure water wastage.

First, you have to know how to read it. That's easy if yours is digital, whether reading in gallons or cubic feet. You'll notice that with the help of its little dial you can read right down to a fraction of a gallon or one-or two-tenths of a cubic foot.

Same goes for the old-fashioned meters with their clusters of dials, but they're trickier to read—as you've likely discovered. You must tackle them a dial at a time.

If your meter reads in cubic feet and you'd like to have the figure in gallons, multiply by 7.5 to find the result in gallons. Of course, you can change in the other direction by dividing by 7.5. If you want to be fussy, use the more precise 7.48, or even 7.4805194—so says my calculator.

One of these days, you won't have to bother about such calculations. You'll read your meters in liters, which has a nice lilt to it. There are 3.7853 of them in a gallon, by the way.

To use your meter in a way you may not have thought of, watch it during periods of specific water consumption. That is, keep an eye on it for one minute of lawn sprinkling. Multiply what you read by 60 and then by the number of hours you usually leave the sprinkler running. Do you find that lots of gallons go on your lawn and garden (or run down the street) each week?

You could do the same thing for a typical shower and for a tub filling. Also for a cycle of your washing machine.

You can use the meter for a leak test when nobody else is home and you aren't using any water and no automatic water-using device, such as a humidifier, the chickens' drinking fountain, or an ice maker is in operation. Observe or mark the meter reading and then do the same thing an hour later. If a change in reading indicates leakage or drip somewhere and you can't find it, suspect a toilet.

The good old w.c., as flush toilets are known almost worldwide (it's one of the few truly international terms, like O.K. and kaput) most often wastes water at the overflow. If this is set too high, gently bend the float arm so the valve shuts off the water half an inch below the top of the overflow pipe.

If that's not the trouble, you may have a plunger-ball leak, which is not so easy to detect. Try putting a little food coloring in the tank water. If it soon shows up in the bowl, you likely have a ball that needs replacing or putting back into alignment.

How Hard Is Your Water?

IF YOU HAVE a public water supply, the easiest way to find out how hard your water is may be just to phone the water company and ask.

You may discover, however, that this factor varies from time to time, or from one part of town to another.

If the water company can't give you a fairly precise figure, you'll have to get a water-testing kit or have someone test your water for you.

When it comes to judging hardness for purposes of choosing a detergent or determining the softening capacity your household needs, you may have to convert from one measuring standard to the other. The two standards are gpg and ppm. The first is grains of hardness per gallon and the second is parts per million. What is referred to is carbonate hardness—the usual problem with water being calcium carbonate or magnesium carbonate—or equivalent.

To convert grains per gallon to parts per million, multiply by 17.1. To convert the other way, you naturally divide the number of parts per million by 17.1.

So if you find, or are told, that your water is 8.6 grains hard, you can calculate that this means it contains about 147 parts of hardness per million. In each one million grams (1000 kilograms or liters) there are 147 grams of carbonates or equivalent.

Anything under 1 grain of carbonate per gallon is regarded as soft water. Anything over 3.5 grains is hard, though it takes several times that carbonate content to produce the kind of very hard water you sometimes encounter. This is the stuff that so totally refuses to make lather that you have the feeling it is throwing the cake of soap back at you.

To Measure a Chair

"Rule of 3" is what they call it in the furniture trade. It has a lot to do with seating comfort.

It seems the manufacturers start with the standard that a woman is 5 feet 4 inches tall and will be most comfortable on a chair with a seat level 16-1/2 inches from the floor.

However, for every 3 inches of height above 5 feet 4, the sitter—female or male—will welcome another half-inch of height in the chair seat.

Same rule goes for chair depth, the distance between the front edge of the chair and the back upholstery. The seat should be 21 inches deep for that 5-foot-4 woman. And it should increase half an inch for each additional 3 inches of human height.

Since so many chairs are selected, usually unconsciously, by women to conform to these standards, it is no wonder that the man of the house often finds them uncomfortable. He may fit better into a sofa, which probably was made and selected to inflict minimum discomfort on a maximum range of sitters. Thus a common height is 17-1/2 inches.

It probably also follows that if you expect to occupy a chair while barefoot or in slippers, you shouldn't shop for it while wearing heavy walking shoes or high heels.

Chapter L
RAFTERS AND BEAMS

To Cut a Rafter
The Tangent and the Rafter
Beams: How Strong, How Stiff?
Strength of a Beam
When Beams Get B-I-G
Case of the Unequally Loaded Beam
Figuring Deflection
Anyone for Cantilevers?
Horizontal Shear

EASY AS FALLING OFF A LOGARITHM!

To Cut a Rafter

WE BEGIN this excursion into trigonometry gently, in the company of those most familiar of old acquaintances named Math. That's right—the Pythagorean theorem and the square root.

You're building something—a cabin, let's say. It's to have a shed roof, meaning a roof that slopes just one way. The structure is 10 feet, 8 inches high on one side, 7 feet high on the other. Those two walls are 12 feet apart.

How long must you cut the rafters, not including any overhang?

Look at each rafter as hypotenuse of a right triangle, of which the sides are 12 feet and 3 feet, 8 inches. Only difficult part of such a problem is that feet-and-inches stuff. Probably the easiest out, until the metric system rescues us all, is turning everything into inches. This produces larger numbers—but with a calculator, who cares?

So the 12-foot horizontal distance becomes 144 inches, which squared is 20,736. That, added to the square of 44 inches (1936), makes 22,672 inches. Since that's the square of the hypotenuse, the next step is to derive its square root—the rafter length we're going for.

You no doubt recall, probably with loathing, that you were required in school to master—momentarily—a routine for finding a square root with paper and pencil. You can reasonably be assumed to having totally forgotten that algorithm, or recipe. So, in the absence of a pocket calculator equipped with a square-root button (or a key for roots and powers generally), it's time for a cut-and-fit procedure described in our calculator chapter.

Since the horizontal distance is 144 inches, you know the rafter will have to be just a few inches longer than that. As it happens, if you make your initial guess the logical 150 (smallest round number greater than 144) and then try 150.6 and 150.57, that third trial will produce an answer so near perfect there'll be no need for further guesses; the square of 150.57 is 22,671. Cut your rafter to 12 feet, 6 inches plus 0.57 of an inch.

That 0.57 looks like about 9/16, verifiable by dividing 9 by 16. So the final figure, outside to outside is 12 feet, 6-9/16 inches. Be sure to add the two overhangs when you make your cuts.

If you now want to figure out how to set your saw to make that cut so that it will be parallel to floor or wall, you'll need a trig table.

As they told us all in high school, the tangent of either of the other angles in a right triangle is what you get when you divide the length of the far side by the length of the near side—in this case 44 divided by 144. Look up the resulting 0.3055555 in the tan column of a trig table. You'll find it comes as near to 17 degrees as makes no difference. So that's where to set your radial saw—there or at 90 minus 17, depending on which rafter cut you're making.

The Tangent and the Rafter

THERE I WAS, about to build a house for my family to live in. We're a sun-worshipping lot, right down to Ms Blue, our Siamese.

To pick up the sun from the south without getting a view of my all-too-near neighbor's house along with it, mine needed a bank of clerestory windows. These had to be in a wall jutting above the flat roof of a bedroom.

This gives the living room a shed roof, one that slopes in just one direction, rising 4 feet in a horizontal distance of 18 feet.

One little problem I had in building it was cutting off the rafters for this sloping part so that the rafter ends would be plumb. Which is to say vertical.

The answer to this kind of problem, and dozens of others in building, is in some simple trigonometry.

The level distance of this part of my house and the vertical one that holds the windows form a right triangle of the proportions 4 and 18. When you divide the far side of a triangle like this by the near side, you get a proportion, or ratio, called the tangent of the angle formed by the near side and the hypotenuse (or, in this case, the roof).

Everybody learns about tangents in school, and then nearly everybody forgets.

It follows that the tangent of this angle is 4 divided by 18, or 0.222. If you were to look through a table of tangents, you would find that the quantity in the table that came the closest to 0.222 is something like 0.221999, which the table declares to be the tangent of 12 degrees, 31 minutes, also known as 12-31/60 degrees. For the purpose of setting your saw you would think of that as 12-1/2 degrees.

(But sometimes in setting a tool, you're working not from zero but from 90 degrees. In that case, subtract the 12-1/2 from 90 and set the saw at 77-1/2 degrees.)

What do you do if you don't have a table of trig functions lying around? Well, you could wish you owned one of the pocket calculators that have trig buttons on them and sell for only a few dollars more than the simpler machines.

Or you could do a little arithmetic. I don't mean that there is an easy way to calculate trig functions. But there does exist an approximation method that will give you results that are far closer to exact than you'll ever need for cutting a rafter.

In fact there are several such methods, of which one is suited to finding the angle for which you know the number and it is less than 1.

(If the tangent of the angle you're working with is more than 1, shift to the other acute angle of the triangle. It doesn't make any difference to you which one you calculate, since you can then quickly find the other by subtracting from 90.)

For the following procedure I'm assuming the use of a simple electronic calculator, although naturally the steps will work if you follow them by paper-and-pencil. The parenthetical suggestions are to aid someone using a calculator.

Multiply the tangent by itself and then by 0.32383. Subtract from 1 (by changing the sign of that decimal to negative and adding 1). Multiply by the square of the tangent and then by 0.52678. Again, subtract from 1.

Multiply by the square of the tangent and then divide by 3. Once more subtract from 1.

Multiply by the tangent and by 57.29578.

That's it.

Just to be sure, though, let's run through it for that same triangle with its sides of 4 and 18 feet, giving tangent of 0.2222222 on the usual eight-digit calculator.

Squared, it's 0.0493827. Multiplied by 0.32383 and subtracted from 1, it's 0.9840085.

Multiplied by the tangent twice, this becomes 0.0485929. Times 0.52678, it becomes 0.1170622. Subtracting that from 1 gives 0.8829378.

When you multiply that by the tangent twice and then divide by 3, you get 0.0145339, which subtracted from 1 is 0.9854661.

Multiplying once more by tan and then by 57.29578, you get the answer 12.54734, or a trifle more than 12-1/2 degrees. Check.

This trig method is the quick and easy way to find this number—but not if it involves you in a search for a trig table. In some case, a carpenter's alternative may be a better idea. Put one rafter in place and hold a level against it while you pencil in your cutoff line. Set your saw to this angle by eye.

Beams: How Strong, How Stiff?

Strength of a beam refers to how heavy a load it will take before it gives way. For a given species and grade of lumber, this is affected by both width and thickness.

Thickness means the direction of the stress on the beam—the direction in which the load will try to bend it. Bearing capacity of a beam is in proportion to the square of the thickness.

At the same time, it is in direct proportion to the width.

That is, two two-by-fours on edge will have twice the strength of one of them. Naturally enough.

But one two-by-eight on edge will not be merely twice as strong as a two-by-four, even though it contains about twice as much wood. It will be four times as strong. And a two-by-twelve will bear about nine times the load that a two-by-four can carry.

Stiffness is a somewhat different characteristic from strength, even though it sounds about the same. Stiffness refers to how much a beam will bend when loaded, not with how much weight it will handle before giving way. As with strength, stiffness varies proportionately to the width, so two beams of a given size, placed side by side, are twice as strong as one.

But in the direction of the stress, the proportion is not as the square but as the cube. A two-by-eight on edge, merely twice as deep as a two-by-four, will have not twice nor four times, but eight times the stiffness.

Strength of a Beam

$\mathbf{W}$ILL TWO-BY-SIX rafters support that carport roof?

For floor joists in a cabin or a room addition—will two-by-eights serve, or must you go to bigger and more expensive lumber?

If there's a building code to be met, you can get the answer from a book of tables. But that's no help when the book is not around, or if there is no book, or if you are departing in any way from conventional construction—as in a post-and-beam design, for example.

In the course of building several houses to live in over the years, I've saved a lot of money in engineers' fees and a lot of nuisance as well, by being able to figure a beam for myself. I can consider many alternatives quickly in trying to arrive at the choice of a girder or arrangement of joists that will cost me the least in dollars and sweat.

Best of all, I can handle unconventional construction methods as well as the more usual ones—post-and-beam, cantilevers, and all that.

To figure the strength of a joist, beam, girder, or rafter, you must first assemble four figures; the actual width and depth of the beam, the distance the beam is to span, and the fiber stress in bending allowable for the species and grade of lumber you have in mind.

This fiber-stress factor is often referred to as "f." The table gives it for many kinds of lumber. Note that there are two columns of figures. The first is for a beam used by itself. "Repetitive" is for a series of three or more members, as would be the case with joists or rafters used in parallel and connected by floor or roof boards nailed to them.

The "f" value is expressed in pounds per square inch. So width and depth must be inches, too, and so must the length of span.

And keep in mind that depth means the direction in which the load will produce the stress—the longer of the two dimensions for lumber placed on edge, as it usually is. This is important, because strength increases as the square of the dimension that is in the direction of the stress.

Multiply the "f" value by the width and by the square of the depth. Divide by the span length in inches. This will give you a basic strength value that is sometimes called Z. It's a value that can be modified to fit all sorts of building situations. This makes it useful in calculating beams that are unequally loaded and also beams that are cantilevered.

For now, though, we'll stick to the condition that applies most of the time: a simple beam, freely supported, with evenly distributed loading. This describes the majority of house girders, beams, joists, and rafters.

SPECIES OR GROUP	GRADE	EXTREME FIBER STRESS IN BENDING	
		SINGLE	REPETITIVE
DOUGLAS FIR - LARCH	Select structural	1800	2050
	No. 1 / Appearance	1500	1750
	No. 2	1250	1450
	No. 3	725	850
DOUGLAS FIR SOUTH	Select structural	1700	1950
	No. 1 / Appearance	1450	1650
	No. 2	1200	1350
	No. 3	700	800
HEM - FIR	Select structural	1400	1650
	No. 1 / Appearance	1200	1400
	No. 2	1000	1150
	No. 3	575	675
MOUNTAIN HEMLOCK	Select structural	1500	1700
	No. 1 / Appearance	1250	1450
	No. 2	1050	1200
	No. 3	625	700
SUBALPINE FIR	Select structural	1050	1200
(White Woods)	No. 1 / Appearance	900	1050
(Western Woods)	No. 2	750	850
	No. 3	425	500
ENGELMANN SPRUCE	Select structural	1150	1350
(Engelmann Spruce -	No. 1 / Appearance	975	1150
Lodgepole Pine)	No. 2	800	925
	No. 3	475	550
LODGEPOLE PINE	Select structural	1300	1500
	No. 1 / Appearance	1100	1300
	No. 2	925	1050
	No. 3	525	625
PONDEROSA PINE -	Select structural	1200	1400
SUGAR PINE	No. 1 / Appearance	1050	1200
(Ponderosa Pine -	No. 2	850	975
Lodgepole Pine)	No. 3	500	575
WESTERN CEDARS	Select structural	1250	1450
	No. 1 / Appearance	1050	1200
	No. 2	875	1000
	No. 3	525	600
WESTERN HEMLOCK	Select structural	1550	1800
	No. 1 / Appearance	1350	1550
	No. 2	1100	1250
	No. 3	650	750

And the modification for such simple beams is: multiply by 4 and divide by 3. (Or multiply by 4/3 or by 1-1/3 if either equivalent method seems quicker.) Call this result W, for the weight the beam can safely carry.

Let's try a rather typical example. You're building a room with a wood floor. The joists must be designed to carry 40 pounds to the square foot. (If this is the "live load" requirement only, you will have to add 5 to 10 pounds to it for "dead load," which means the weight of the beams or joists themselves and the floor boards and coverings they hold up—or the roofing materials in the case of a roof calculation.) That's the usual code requirement and a good safe standard to meet for a solid structure even if you're not governed by a code. You wish to place the joists 24 inches apart, center to center. The distance to be spanned is 10 feet. So each joist will have to carry the weight of an area 2 feet wide and 10 feet long. Twenty square feet times 40 pounds gives a load of 800 pounds for each joist.

Question: Will No. 2 Douglas fir two-by-sixes be strong enough to handle the weight—800 pounds?

The table says that under repetitive conditions, which is what you have with a series of similar joists to which a floor is fastened, the "f" value of No. 2 Doug fir is 1450. Four times this, multiplied by the width (1.5 inches) and the square of the depth (5.5 times 5.5 inches) and then divided by 3 times the span, 120 inches, is 731 pounds.

That's almost strong enough, but not quite. If you were constructing a simple cabin somewhere outside code jurisdiction, you might decide to use the two-by-sixes anyway, especially if you didn't plan heavy furniture or big parties (are any of your friends big parties?) or had noted that 30 pounds has been called an adequate allowance for sleeping rooms in some codes and FHA specs. You would, however, select your lumber with more than usual care to avoid stock with major defects near span centers. More likely, though, you'd stick to your standards or be required to do so. You'd then recalculate either for a higher grade of lumber or a fatter dimension. Or move your joists closer together, so each would have less weight to carry.

Since No. 1 lumber shows "f" of 1750 instead of 1450, it can be considered 300/1450 stronger. The earlier figure of 731 (when multiplied by 300, divided by 1450 and added to 731) becomes 882— same result as you'll get if you calculate it directly.

So upgrading lumber quality will take care of the problem with more than 10 percent to spare.

Going to two-by-eight lumber, while sticking to the original No. 2 grade, should work too. You can recalculate from scratch—or make the change by multiplying 731 by the square of 7.25 (actual dimension of what is called 8-inch lumber) and dividing twice by 5.5. Either way you get 1270.

This is far more strength than you need. But if cost remains comparable you may well prefer to go this route. Recalling that stiffness increases as the cube of the dimension stressed, you calculate that the shift to 8-inch joists—grade remaining the same—considerably more than doubles the stiffness. In a floor, that much added stiffness is a worthwhile bonus.

(How do you make that stiffness calculation? The obvious way is to multiply 7.25 by itself twice to find its cube, then divide by the cube of 5.5. But the easier way, by pencil or calculator, is to divide 7.25 by 5.5 and then cube that result.)

When Beams Get B-I-G

Wʜᴇɴ ʙᴇᴀᴍs get big, you can call them girders.

Most of what you need to know to figure what size hunk of wood will hold up how much weight is covered in the adjacent item on strength of a beam, but the big ones call for a few special common-sense considerations.

Let's say you're building something that needs a long central support—a beam or girder to run right along the spine of the structure and hold up the joists that hold up the floor...that holds up you and the furniture, and so on. This beam is to sit on rough six-by-six posts spaced 6 feet apart, so you decide you'd prefer that central girder to be 6 inches across, for easy and secure fastening to the posts.

How deep must it be to hold its share of the weight of the floor and the contents thereof?

First of all, you have to find out what that share of weight is. If joists run from girder to foundation wall, then half their weight must be on the girder. If the structure is 24 feet wide, that means 12 feet for the girder to hold up in each direction. Each 6-foot section of girder—the distance between posts—is responsible for an area 6 feet by 24 feet, or 144 square feet.

(Would you use pencil or calculator to do this multiplication rapidly? Not if you're practicing your mental arithmetic. Simplest trick here is to halve one number and double the other, so you need only say 12 times 12 is 144.)

If the floor is for a house, you'll probably want to figure on a loading of 40 pounds to the square foot. That's what building codes usually say. For a roof, it may be much less—perhaps 20 pounds unless there's a heavy snow load to think about. For a garage floor, it may be 100 pounds.

But let's say it's for a room of a house. We multiply the square footage of 144 by the specified 40 pounds to find that each 6-foot span of girder has 5,760 pounds to carry. Not a loading to fool around with.

Will a Western cedar six-by-six, No.3 grade, rough, do it? The strength-of-a-beam table says the fiber strength (the "f" value) of a single beam of the kind of cedar we've specified is 525. That, multiplied by the width of the beam and multiplied twice by its depth (6 inches in both cases since we're dealing with rough lumber where real dimensions approximate the nominal ones), divided by the span, and finally multiplied by 4/3, gives 2,100 pounds.

That's not even half what we need.

What about a six-by-eight on edge? Multiplying the previous result by 8 times 8 and dividing it by 6 times 6 produces 3,711. (If you don't trust that approach, do the new calculation from the beginning.) Nope. Still not enough strength.

If we're determined to stick to No.3 rough cedar, we can only move on to something bigger—say a six-by-ten. That brings us to 5,833 pounds. Just does it.

Why not consider something like a three-by-fourteen instead? Using just over two-thirds as much lumber, it would still give almost as much strength. In general, it's a good idea to make supporting members comparatively narrow and deep, for economy. But for a girder like this, stability is important, too, and the squat shape gives that while still letting it sit nicely upright. And it gives a good surface for attaching the joists.

Case of the Unequally Loaded Beam

Under "Strength of a Beam"—which may very well be the most useful single building item in this book—I promised to tell how to modify the calculation to fit an unusual case.

This is the case of the unequally loaded beam. It refers to the timber or girder or joist or rafter that must carry a lot more weight at some known point or points than at others.

There are several different ways the weight may be arranged and still fit some standard methods of calculating. With any of these, you begin by finding what is called Z in the more general strength article. In that article, you were told to multiply Z by 1-1/3 to find how much evenly distributed weight could be safely supported by a simple beam.

For other arrangements, you multiply Z by a different quantity to find W, the weight the beam can carry.

If the ends of the simple beam are firmly held—not merely nailed or bolted in place—and the load is evenly distributed, multiply by 2 instead of 1-1/3. For a freely supported simple beam with one concentrated loading at the center of the span, multiply Z by 2 and divide by 3 to find W.

If the ends of the beam are firmly held in place and the loading is concentrated at the center of the span, use 2/3 as the multiplier; that is, multiply by 2 and divide by 3.

When loading on a freely supported simple beam is at its greatest near the ends of the span and decreases to zero at the middle, just multiply Z by 2 to find W.

With the opposite condition—heaviest load in the middle, dropping to nothing at the ends—don't multiply at all. Z and W are equal.

Figuring Deflection

I_T's NOT always enough to know that beams and joists and rafters are strong enough to do their job safely. Sometimes you want to limit deflection—bending—to some minimum. The purpose may be to keep a plaster ceiling from cracking. Or you may be concerned with meeting the terms of a building code that limits deflection of the wood member in question to 1/360 of the span.

Well, anyway, you figure deflection by first multiplying the width of the rafter, beam, or joist by the cube of its depth. Then you divide by the square of its length. Be sure to use the same unit of measurement (inches, for example) for the length that you use for the width and depth.

(Keep in mind that width means from side to side of the beam as it will be placed, a nominal 2 inches in the case of a two-by-four used on edge, as it most often will be. And depth, then, is the nominal 4-inch dimension that is the direction of the stress in carrying the load.)

Now you need only multiply the result by a certain quantity to find out how many pounds of loading the timber will bear before it will deflect as much as 1/360 of the length of the span.

For the most usual situation—a simple beam, freely supported, with evenly distributed loading—the quantity is 31,290. For a simple two-by-eight joist spanning 10 feet, you would first multiply the width (1-1/2 inches in actuality) by the cube of the 7-1/4-inch depth, to get 572. Dividing by the square of 10 feet (120 inches squared is

14,400) gives 0.0397. Multiply this by the factor of 31,290 to find that the joist will hold 1243 pounds without excess bending.

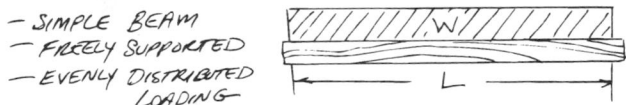

If the joist is to hold up a floor that can carry 40 pounds to the square foot (a usual code requirement for living rooms in houses) and the joists are to be placed 2 feet center to center, the design load is 2 times 10 times 40, or 800 pounds. Since the joist will handle 1234 pounds stiffly enough, there's no deflection problem. You have a margin of rather more than 50 percent.

If your simple beam with evenly distributed loading is firmly fixed at both ends, the 31,290 multiplier does not apply. You can use the much larger 156,440, but don't without qualified engineering assurance that this is indeed the case; "firmly fixed" means something much stronger, and rarer, than mere bolting or nailing. When in doubt, use the smaller multiplier.

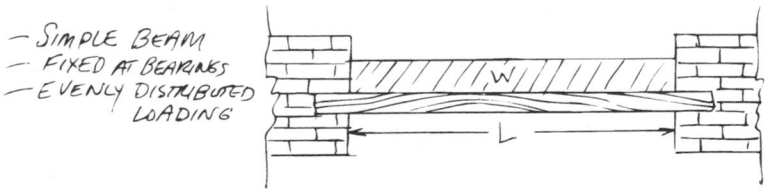

Here are multipliers for other situations:

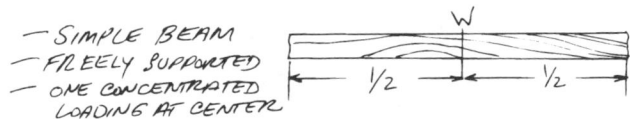

For a simple beam, freely supported, with one concentrated loading at center of span: 19,555. (But if it is firmly fixed at bearing points, use 78,222.)

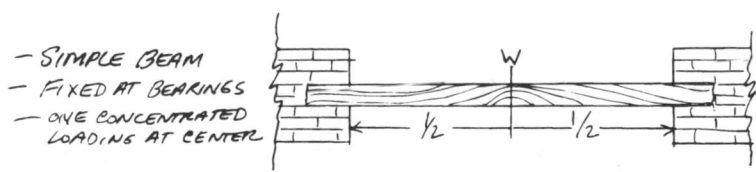

Simple beam freely supported, with loading that decreases uniformly to zero at center of span: 43,460. (But if load decreases uniformly from maximum at center to zero at both bearings, use 24,440.)

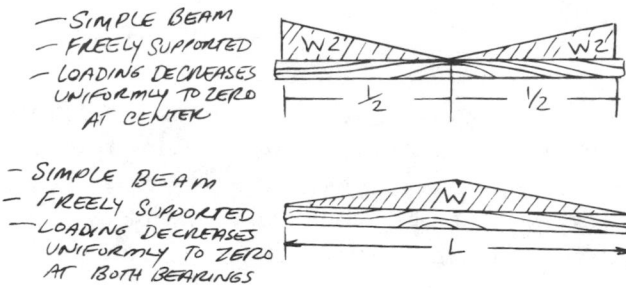

Simple beam freely supported, with loading that decreases uniformly from a maximum at one end of the span to zero at the other: 31,100.

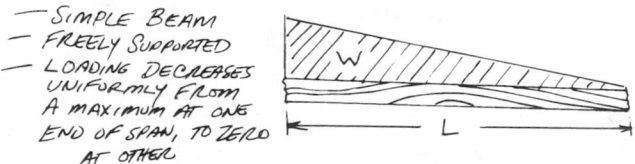

Cantilevers deflect much more than ordinary beams and rafters that are supported at both ends. So the factors by which you multiply to calculate loads for deflection are smaller.

For a cantilever beam with evenly distributed loading, multiply by 3258.

For a cantilever with concentrated loading near the free end, multiply by 1222. If loading is at a maximum at the bearing point and decreases uniformly to zero at the free end, use 3258.

Anyone for Cantilevers?

A CANTILEVER is elegant and useful. An upstairs deck that extends out from a house, with no posts underneath it, is an example.

A cantilever takes a little engineering, however, as I found out many years ago when I wanted to build a house with my bedroom jutting out over a driveway.

"How," asked the building inspector who had to pass on the plans, "do you know those joists that stick out there are strong enough?"

It was a good question. I didn't really know. So I spent a lot of time looking into the matter, finding out how the calculation could be made, and finally doing it to the satisfaction of the inspector. I am glad to say that the room is still there, jutting out as firmly as ever, and with no objectionable bounce when some heavyweight jumps up and down on the floor.

There are three important procedures for calculating the load-bearing capacity of a cantilevered beam. The one to choose depends upon the way the load on the cantilever will be distributed.

All begin with the basic load-bearing figure that is called Z in the article on strength of a beam. When you've found Z, as described there, you must divide by 3, 2, or 6 to find out how much weight your cantilever can be trusted to support.

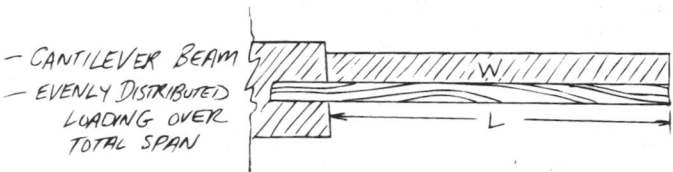

If the load on the cantilever will be evenly distributed along the whole of the span, you must divide Z by 3 to find the allowable bending load.

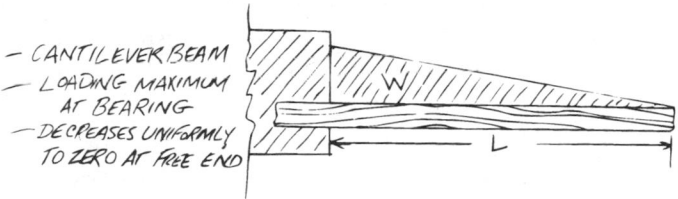

If the load is at its maximum at the point of support of the cantilever, and decreases uniformly until it is zero at the free end, the stress will naturally be somewhat less. You divide by 2.

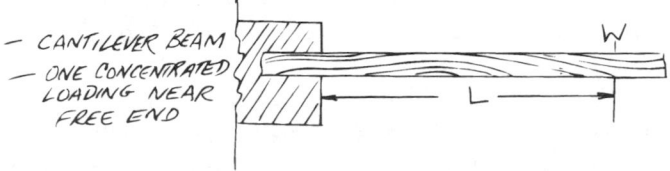

But when the whole load is concentrated near the free end, the downward push created by this load becomes much greater. You must divide that Z figure by 6.

Horizontal Shear

I<small>N SOME</small> situations, the loading on a wood beam used in construction will cause it to come apart in the middle—parallel to the grain.

This can happen even when the beam is strong enough to carry its load without breaking in two. But it is likely only when the depth of the beam is one-tenth or more of the span for which it is used, and the beam is heavily loaded.

If you have a situation that seems to meet this description, you should use a shear calculation to check on the loading. You could do this after you've used the method described earlier in this section on figuring the strength (W value) of a beam.

What you want to discover now is whether a shear will be produced that is more than 120 pounds to the square inch. This 120 pounds is the amount of sheer stress considered safe when you're using the common grades of most kinds of lumber.

For any simple beam or joist, having found the W value, multiply by 3 and divide by 4 and then by the width and by the depth of the beam.

To put it another way, the intensity of the shearing stress is 3 times W, divided by the product of the width of the lumber, times its depth, times 4.

If the result you find is more than 120, you should change to a beam of different dimension or otherwise modify your design.

The calculation as described applies to all simple beams with symmetrical loads. It goes for beams with evenly distributed loading (the usual thing in house construction) whether the beam is freely supported, merely nailed in place, or firmly fastened at its ends. It works for a beam with its loading concentrated at its center or one with most of the load at the ends.

About the only beams it doesn't apply to are those that have off-center loading or are cantilevered.

Where the load decreases from a maximum at one end to nothing at the other, shear value is found by dividing W by the product of the width and depth of the beam.

Where there is a single load concentrated neither at an end nor at the center of the span, the formula becomes just a little more complicated. Here you first multiply W by 3 times the distance from the point of loading to the more distant point of support. Then divide this by the product of the depth of the beam times its width times twice its length. And remember to keep all dimensions in inches.

For all cantilevers, whether the load is concentrated at the free end or the fixed end or is evenly distributed, the calculation is the same. You multiply W by 3 and then divide by twice the product of the width and the depth of the cantilevered beam.

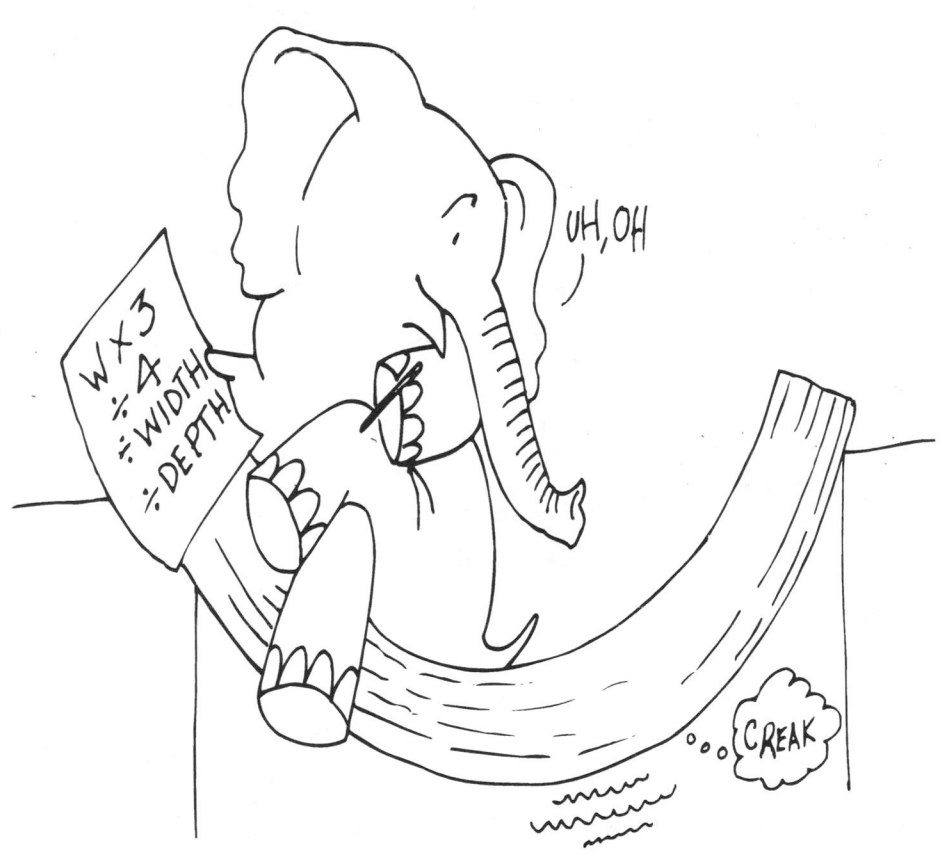

Chapter M
AROUND THE HOUSE

How Many Degree Days?
Sizing a Wood Stove
Cord of Wood, Ton of Coal
Wood or Coal?
Comparing Furnace Efficiencies
What Does Heat Cost?
How Much Heat Loss?
Costing a Heat Pump
Heat Storage, Heat Loss
Trying a Cooler for Size
Meeting Insulation Requirements
Figuring Areas for Insulation
How Much Insulation?
Shading a House
To Ventilate a House
What's a Square Foot of Ventilation?
How Much Can the Sun Save You?
Capacity of a Solar Water Heater
What Will a Circuit Take?
What's a Switch Worth?
Working Out an Island Sink
Loading a Waterbed
Aiming a Satellite Dish

How Many Degree Days?

Dᴇɢʀᴇᴇ-ᴅᴀʏ figures, so useful when estimating heating bills or insulation needs, are usually to be had from fuel dealers or government or utility-company sources. They were devised in the nineteen-thirties by heating-industry people who required a number as nearly proportional to fuel use as possible for comparing areas and periods.

Cumulatively, at any time during the season, they represent the load for the period to date.

If you need degree-day totals for periods or localities for which they're not available, you can calculate your own. Record maximum and minimum temperature each day and subtract the average of the two numbers from 65.

Like this. In a 24-hour period when day temperature reached 73 degrees and night dropped as low as 45, the average found by adding the temperatures together and dividing by 2 is 59. Subtracted from 65, that's 6 degree days.

All it takes is patience.

Sizing a Wood Stove

A ᴡᴏᴏᴅ sᴛᴏᴠᴇ that's too large will be difficult to manage. It will produce too much heat or will have to be operated at such a low temperature that a potentially dangerous amount of creosote will be deposited in the chimney.

A stove that is too small will have to be fed too frequently or used at an excessively high temperature if it is to provide an acceptable level of heat.

Those are among the reasons you may find a use for this rough formula for determining ideal stove size.

Begin with an estimate of heat load (or heat loss) for the house.

If you don't have a source from which you can get a more precise figure, use the following to get a good working approximation.

Multiply the number of square feet of floor area by 30. This assumes a house with an average amount of insulation and window area. For a well-insulated building reduce the factor to 25, and for a poorly insulated one increase it to 35.

Your answer will be in Btus (British thermal units) per hour. Multiplying it by 0.172 will tell you the approximate firebox volume, in cubic inches, that your wood stove should offer.

Cord of Wood, Ton of Coal

THOUGH naturally this figure varies considerably with species and condition, burning a cord of wood produces about 26 million Btus of energy.

So does burning about one and one-third tons of coal.

A lot of this energy is wasted, of course. A typical airtight stove might have an efficiency of around 50 percent.

So in calculating the net heat produced, the figure you'd use would be more like 13 million Btus.

Wood or Coal?

HARDWOOD or soft coal—which is the better buy for burning?

To calculate this you must first know the heating values of the two fuels and the cost of each.

A pound of bituminous coal has a heating value of about 11,000 British thermal units, so a ton has 2000 times this much, or 22 million Btus.

A typical hardwood is good for 27 million British thermal units to the cord.

So dividing the price of a ton of coal by 22 will give you your cost per million Btus.

And dividing the price of a cord of hardwood by 27 will give its cost for each million Btus.

If you pay $170 for a cord of oak you'll get a million Btus for each $6.30.

Coal at $90 a ton will cost you only $4.09 per million Btus.

Of course the true difference in cost will also have to reflect the relative efficiency with which your stove or furnace will burn the competing fuels.

If your modern wood stove can burn wood with 60 percent efficiency, you'll be paying $10.50 for each million Btus you actually benefit by ($6.50 divided by 0.60).

And if your ancient coal furnace has an efficiency of only 45 percent (ask a furnace dealer for his best guess on this), you must divide that figure of $4.09 by 0.45 to discover an effective cost of $9.09 for each million Btus obtained from coal and actually used.

Comparing Furnace Efficiencies

IF YOUR present furnace has an efficiency rating (known or estimated, perhaps with help from a dealer or your utility company) of 55 percent and you contemplate buying one with a rating of 80 percent, divide the old figure by the new one.

Since 55 divided by 80 is 0.69 , multiply your present heating bill by this decimal. The result is how much your bill should be with the more efficient new furnace. If you've been spending $500 a season, you should be as comfortable from here on out with a cozier $344.

What Does Heat Cost?

THESE rough formulas can give you an estimate of the relative costs of different fuels for home heating. The answers are in British thermal units per dollar. (Your best sources for the various base figures you'll need usually are public utilities and dealers in heating devices.)

For a *gas furnace,* multiply the efficiency in percent by 100,000 and divide by the average cost in cents per therm.

An example: If gas costs you 60 cents per therm (equal to 100,000 of those ubiquitous Btus, remember?) and your furnace offers the fairly typical efficiency of 65 percent, you'll multiply 65 by 100,000 and divide by 60 to find you can expect 108,333 Btus of heat for each dollar spent.

For *electric heat,* multiply 3413 by the seasonal efficiency of the heater and divide by the cost in cents per kilowatt hour.

Efficiency of most electrical heating devices is almost a thrifty 100 percent. However, a radiant-heating device that warms you directly, rather than indirectly by warming the room, may give you more for your money. A heat pump may even offer 150 percent. If you are working with a heat pump of this efficiency where electricity averages 10 cents, you'll multiply 3413 by 150 and divide by 10 to get 204,780 Btus to the buck.

If a *heat pump* with more than 100 percent efficiency impresses you as being about as likely as perpetual motion, take note that this device does not merely turn electrical energy into heat. It extracts heat from well water or the surrounding air to make you comfy.

For *wood,* multiply the Btus per cord for the type of wood you're using by the percentage of efficiency of the stove or other heating device, and divide by the price you pay for a cord. Unless your wood is practically bone dry, deduct an additional 15 percent or more for the heat lost in evaporating the water content.

The number of million Btus you can expect from a cord of fully dried wood varies from 13 for pine, 17 for elm, 20 for ash or maple, 22 for oak or birch to 25 for hickory.

A stove with a thermostat should give around 40 percent efficiency, a reasonable approximation to use if you don't have a specific figure for the stove in question. A wood furnace should do even better—say 50 to 70 percent. Therefore, if you burn air-dried (not completely dry) hickory at $125 a cord in a good stove, multiply 25 million by 40 and by 0.85 (as a way of deducting 15 percent moisture content.) And then divide by the cord price in cents.

If those figures—including what in areas like yours may be a wildly optimistic guess for the price of wood if you don't cut it yourself—fit your situation, you're getting about 68,000 Btus for your wood dollar.

How Much Heat Loss?

YOU CAN calculate the total heat loss through a wall, floor, or ceiling of a house for given temperature conditions and heating season. Follow up by feeding in the cost of the energy (from a fuel dealer or utility) and you can find what the heat loss is costing you in dollars per year.

This kind of calculation is particularly useful in estimating how much you can save each year by adding insulation. The formula works equally well for comparing the money saved by one kind or amount of insulation with another.

Here is a list of some common building materials, with the thermal conductivity of each in millions of British thermal units per second per foot per degree Fahrenheit. (You don't really need to absorb that definition, however. You can simply use the number where called for in the calculation.)

A second number you'll need is the difference between the temperature to be maintained inside the wall or ceiling and the temperature outside it. If, for example, you assume 75 degrees inside and 5 below outside, the figure to use will be 80.

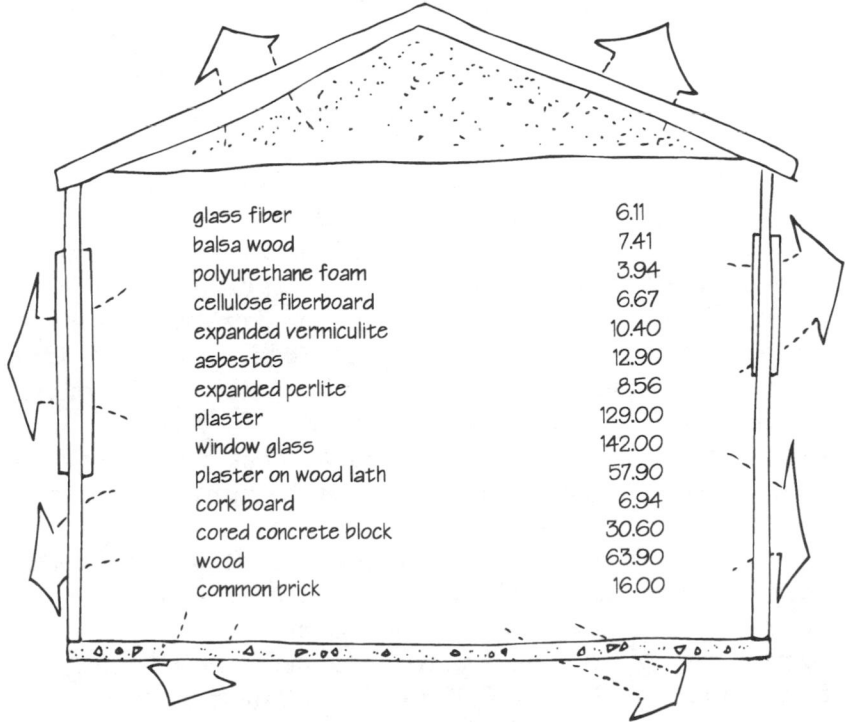

glass fiber	6.11
balsa wood	7.41
polyurethane foam	3.94
cellulose fiberboard	6.67
expanded vermiculite	10.40
asbestos	12.90
expanded perlite	8.56
plaster	129.00
window glass	142.00
plaster on wood lath	57.90
cork board	6.94
cored concrete block	30.60
wood	63.90
common brick	16.00

Having obtained the cost figure for fuel from dealer or utility, divide it by the estimated efficiency percentage of your furnace or stove. If you've been given the cost in therms, multiply it by 10 to turn it into dollars per million British thermal units.

To apply this, assume your cost for natural gas is 30 cents per therm (or $3 for a million Btus) and the efficiency of your furnace is 75 percent. Dividing $3 by 0.75 gives the figure of $4 for each million Btus actually benefiting you.

With this information, you can make a sample calculation that will demonstrate how to apply the procedure to any situation you may have. Assume a 5-month heating season with an 80-degree difference between the inside temperature you plan to maintain at ceiling level and the outside one. The ceiling is 40 feet by 50 feet and consists of 3/4-inch plaster on wood lath. It is not insulated.

1. Express thickness of wall, ceiling, or floor in feet. (Divide 3/4 inch by 12 to find a thickness of 0.0625 feet.)

2. Use the table to find the conductivity of the material, and divide this by its thickness. (Dividing 57.90 by 0.0625 gives 926.)

3. Multiply this by area in thousands of square feet and by temperature difference (926 times 2 times 80 is 148,160).

4. Find the number of seconds in 3 months and divide by 1,000,000 (60 times 60 times 24 times 90 divided by one million is 7.776). Multiply the previous result by this and divide by 1000, getting as a rounded-off result 1150. That's the annual energy consumption in millions of Btus.

5. Multiply by your cost per million ($4) to find that about $4600 goes down the drain—or out the wall or ceiling in question.

To compare this heat loss and expense with the result if 4 inches of glass-fiber insulation were added, do the calculation again with a different Step 2. This time divide the *thickness* of each material by its *conductivity*, add the two results together, and find the reciprocal. (If you've forgotten, you find the reciprocal of a figure by dividing it into 1, or by pressing the 1/x key found on many calculators.)

As before, the thickness and conductivity figures for the lath-and-plaster are 0.0625 and 57.9. For 4 inches, or 0.33 feet, of glass fiber, they are 0.33 and 6.11. Dividing 1 by the sum of these quotients gives approximately 18.

From here you could go on with the calculation as before. But if you've already done it for the first material, you need only divide the previous Step 2 result by the new one to find how many times more the plain ceiling is costing than the insulated one. (Dividing 926 by 18 gives about 50. That $4600 figure found earlier, and divided by 50 is $92, which should be the cost of the winter's heat loss through the new ceiling. The difference of more than $4500 is the remarkable annual saving in gas bills produced by the insulation.)

Costing a Heat Pump

Here's a calculation that will mean something to you only if you should get into comparing the cost of heating with a heat pump with that of a gas furnace.

A useful rule of thumb says that if the price for one therm of gas is less than that for 8 kilowatts of electricity, then gas used in a heating device offering 70 percent efficiency is cheaper than power for a heat pump having a COP of 2.5. That COP is an acronym used in the field of thermodynamics for coefficient of performance.

Heat Storage, Heat Loss

This calculation has to do with how much heat can be stored and released by bulk materials.

This is an important element in comfort and energy saving in any house. It is especially significant for a house that has been properly designed to take in a lot of free heat on sunny days. In fact, the first tax credit offered to encourage solar design made it mandatory.

How much heat can a brick or concrete floor tuck away on a sunny winter day to help keep the whole house warm overnight?

To find this you need to know first of all the heat capacity per cubic foot for the material.

In British thermal units per degree of temperature change, it is 28 for concrete or brick, the materials for which you may be most interested in having a figure.

It is 62.5 for water, a most efficient heat-storage medium though not always an easy one to work with since, unlike masonry, it doesn't store itself.

Let's suppose you have a room sitting on an 8-inch concrete slab 20 by 40 feet. You find that it reaches an average temperature of 75 degrees from sun shining on it through a couple of hundred square feet of windows.

Find the cubic footage of the slab (20 times 40 times 8 divided by 12 equals 533) and multiply by the figure of 28 for concrete. The result of about 15,000 is the number of British thermal units of heat that the slab stores with each rise of 1 degree in its temperature.

And of course that's also how many it will give back later in the process of dropping 1 degree Fahrenheit.

If the temperature of the slab goes down to 67 overnight, a drop of 8 degrees, it will supply 8 times the previous per-degree figure, or 120,000 Btus during the night. Is this enough to make up for what is lost to the outdoors during those sunless hours?

To figure the heat loss, multiply the number of hours by the loss per hour by the difference between the indoor and outdoor temperatures during the night.

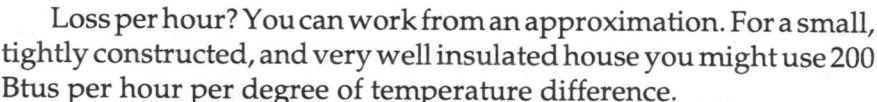

You might take 15 as the number of hours between one solar heating day and the next.

If it's 65 degrees indoors, 25 outdoors, difference is 40.

Loss per hour? You can work from an approximation. For a small, tightly constructed, and very well insulated house you might use 200 Btus per hour per degree of temperature difference.

(For comparison, I recently estimated heat loss for a generously windowed 2,200-square-foot house of average construction and—by standards of these energy-conscious times—minimum insulation as just under 800. My justification for going ahead and building to this standard is that I'm in a mild California climate where Minnesota-level insulation isn't warranted. Also, much of what otherwise could be deemed excessive heat loss will be balanced by the large solar intake from the very windows producing much of that loss.)

So there are three numbers to multiply together—15 hours, 40-degree temperature difference, and loss per hour of 200—and their product is 120,000.

This being the same figure produced by the previous bit of arithmetic, we have described circumstances in which heat gain and heat loss are neatly balanced.

Trying a Cooler for Size

AN AIR CONDITIONER that is too small won't do its job adequately when it's needed most. One that's too big will cost more than necessary and will complete its cooling job too rapidly to dehumidify the air properly.

The proper thing to do when choosing such a device, you're told, is to have it sized scientifically by means of detailed calculations. But what with various imponderables and personal choices not allowed for by the formulas, you may come out as well with a quick estimate.

Make it this way.

First figure the volume of the room in the standard way—length times width times height, all these dimensions in feet.

If the windows are few and the attic overhead is insulated (or if there's a room above) multiply that cubic-foot figure by 10. With many windows and not much insulation, use 20 instead.

For an in-between situation, choose an in-between number.

Now multiply the result you have by 16 if the longest wall faces north, by 17 if it faces east, by 18 if it faces south, or by 19 if it faces west.

Divide by 60 to find the number of Btus (British thermal units) that will be needed to cool the space.

Meeting Insulation Requirements

O<small>N</small> ADEQUATE insulation, people's ideas and the building codes have both been changing lately—in the direction of more.

Code minimums shown in the drawing are adequate for mild climates, but thickness should be increased if possible where climate is extreme or heating or cooling especially costly.

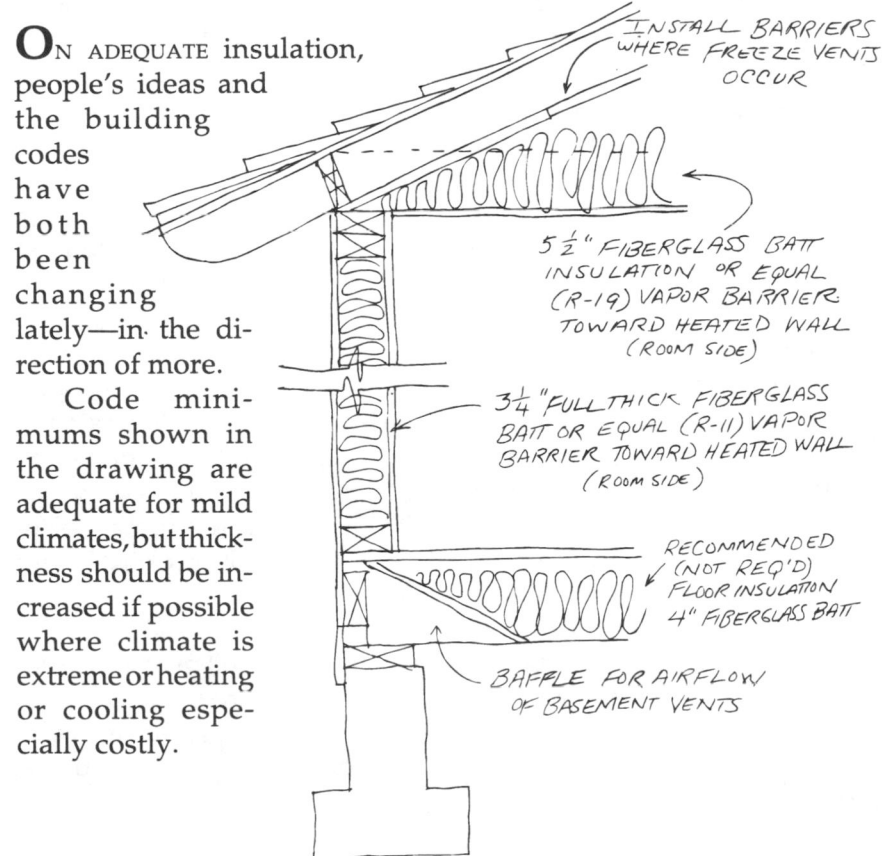

INSTALL BARRIERS WHERE FREEZE VENTS OCCUR

5½" FIBERGLASS BATT INSULATION OR EQUAL (R-19) VAPOR BARRIER TOWARD HEATED WALL (ROOM SIDE)

3¼" FULL THICK FIBERGLASS BATT OR EQUAL (R-11) VAPOR BARRIER TOWARD HEATED WALL (ROOM SIDE)

RECOMMENDED (NOT REQ'D) FLOOR INSULATION 4" FIBERGLASS BATT

BAFFLE FOR AIRFLOW OF BASEMENT VENTS

Figuring Areas for Insulation

A SHORTCUT calculation when buying insulation is to multiply the gross area by 90 percent, if framing is on 16-inch centers. Use 94 percent if framing members are on 24-inch centers.

To put it another way, knock off 10 percent for the closer spacing, 6 percent for the 24-inch.

This formula applies to any ceiling, wall, or floor area.

So: with 1,000 square feet of ceiling to insulate, joists spaced 16 inches, you'd buy 900 square feet of insulation.

How Much Insulation?

IN GENERAL terms, it's not hard to find out how much insulation your house should have—a lot! Magazines dealing with home ownership tell you this regularly, and a phone call to your utility company in our energy-conscious era will also evoke a figure.

Instead of the old standard of R9, you're now likely to be told to try for R19. As heating and cooling costs escalate further, you're going to hear more about that as a minimum, with something up towards R-25 to R-30 for ceilings. Not that such a mass of insulation pays for itself in most climates with most fuels—just that it soon will do so. And we might as well start getting the benefits now. After all, insulation is not going to wear out.

What does all this R-stuff mean? R-value is the figure you get when you add up all the components of a section of wall or ceiling or floor or roof.

To figure the heat resistance of any part of your structure you need only find the R-value of each layer—and add 'em all together.

This directly proportional cumulative effect simplifies calculations—but it can be deceptive, too. Although when you double the thickness of insulation you double the insulating value, you by no means double the savings. Look at it this way. If the first layer of insulation cuts heat loss 60 percent, there's no way the second similar layer can shave off more than 40 percent of the original loss, since that's all there is left to work on.

What it will do, of course, is reduce the loss by 60 percent of that remaining 40 percent, for an actual figure of 24 percent. Each increment of insulation is bound to save you fuel money, but it is also bound to save you less than the preceding one. The worse the wall, the greater the result!

Here are some typical R-values for the structural parts of an ordinary frame wall:

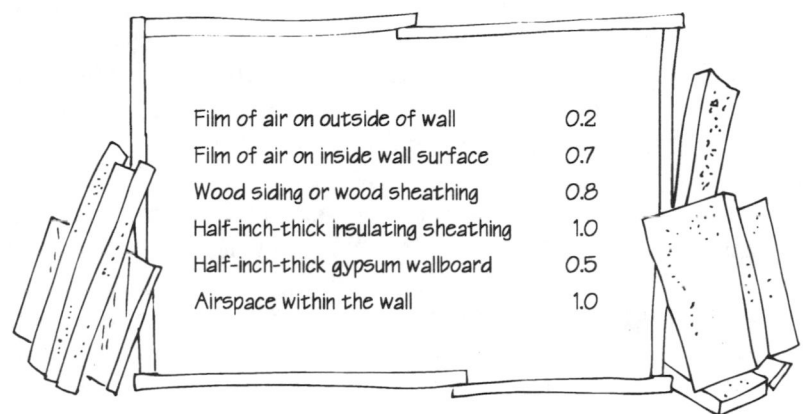

Film of air on outside of wall	0.2
Film of air on inside wall surface	0.7
Wood siding or wood sheathing	0.8
Half-inch-thick insulating sheathing	1.0
Half-inch-thick gypsum wallboard	0.5
Airspace within the wall	1.0

A typical combination of these, forming an uninsulated frame wall, would use one layer of each. Add those six values together and you get a total of R4.2, which is not very good. So consider the following table.

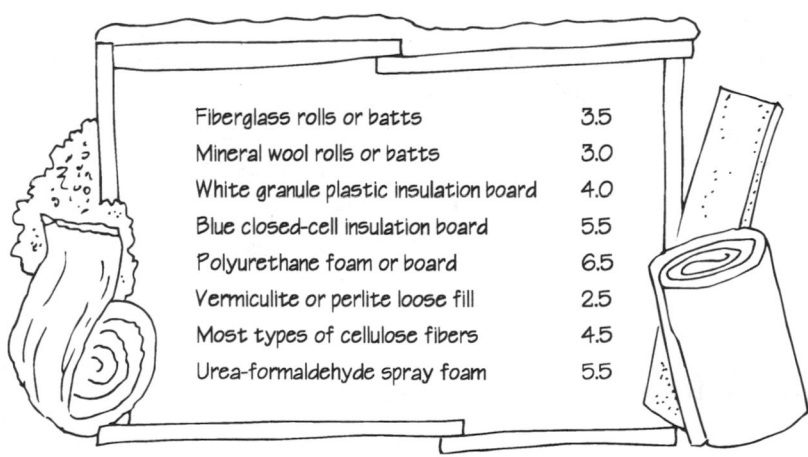

Fiberglass rolls or batts	3.5
Mineral wool rolls or batts	3.0
White granule plastic insulation board	4.0
Blue closed-cell insulation board	5.5
Polyurethane foam or board	6.5
Vermiculite or perlite loose fill	2.5
Most types of cellulose fibers	4.5
Urea-formaldehyde spray foam	5.5

A typical two-by-four stud wall has 3-1/2 inches of air space.

As you can see, substituting 3-1/2 times one of those figures (since they are for each inch of thickness) for the mere 1 produced by an empty airspace will give your wall a far more respectable resistance rating.

To make the actual calculation, first multiply the R value (3.5 for fiberglass, for example) by the thickness in inches (presumably 3-1/2 again, since that's the thickness of the cavity you are filling) and add the figures for the other components. The total you'll get is 15.45. (If you come up with 16.45 instead you probably forgot to subtract or omit the value of the airspace, which you've now lost by stuffing it full of fiberglass.)

If 15.45 isn't good enough, and in many climates it truly is not, you may consider going to a more efficient insulation—comparing costs and not overlooking flammability. In new construction, you might go to a wall framed with two-by-six instead of two-by-four lumber. By simply making room for these two additional inches of fiberglass, you can add a potent 7 to the R value at relatively small cost.

Another excellent possibility in new construction, and sometimes in a major remodeling, is adding a layer of additional insulation on the outside of the wall. Since this layer of insulation board covers not only the wall cavities but the framing lumber (which can transmit significant heat) as well, it is especially worth while.

Shading a House

AT MIDDAY in midsummer in mid-America (40 degrees north latitude) the sun hurls an average of 230 Btus of heat against each square foot of a south-facing wall or window.

In the instance of the window you can estimate that 80 percent of this comes through the glass and into the house to increase discomfort or add to the air-conditioning load.

Figure that fiberglass solar screening (which is not the same thing as plain old-fashioned insect screen) installed one inch outside the window will reduce this intake by about 80 percent, to 185 Btus.

To Ventilate a House

FOR COMFORTABLE summer cooling by natural ventilation, you might choose as your goal one change of air every three minutes.

This will call for a rather generous collection of openings— windows, doors, louvers, vents.

How generous? There is a formidable formula for finding out, but happily it can be simplified to the following routine.

First, determine the wind-direction factor. For breezes that customarily blow directly into the opening, this is 0.5.

For diagonal winds, it is 0.3.

From these numbers, you can estimate factors for other angles—0.4 if the wind on a south wall is from south-southwest, for example, or 0.1 if it comes along almost parallel to the wall containing the opening.

Multiply this factor by the wind speed in feet per minute. Lacking an official figure for wind speed, estimate it in the more familiar form of miles per hour, as you hear it given in weather bulletins. Multiply by 88 to convert it to feet per minute.

Now multiply by the number of minutes you're allowing for each air change, perhaps 3.

Find the cubic footage of your house (length times width times 8 for each story, unless the height is unusual). Divide by the previous figure. This is the total square feet of openings needed.

Here's a typical calculation, for a one-story building 25 by 40 feet with 8-foot ceilings, for a volume of 8000 cubic feet. Ventilation is through south windows, by prevailing breezes from south-southeast averaging 2 miles per hour.

Multiplying this 2 by 88, by the direction factor of 0.4, and by 3 gives 211. Dividing this into 8000 gives almost 38. So that is how many square feet of windows and other openings you'll need in that wall to ventilate by breezes alone—helped, naturally, by outlets in other walls.

If those outlets are high and the intake openings are low, you can get another physical principle working for you—again free of charge. Temperature difference between hot air near the ceiling and cooler stuff in what should be a shaded spot just outside the ventilating openings will move air even with little breeze.

To figure what this will produce, begin with the height difference in feet between centers of your low air inlets and high outlets. Multiply by the temperature difference between the air just outside and the hot air under the ceiling. Find the square root of this product, and multiply that by 9.4 and by the number of minutes in which you wish to make a complete change of air.

If this is 3 again and there's a height difference of 7 feet and a temperature difference of 10 degrees, you should get a figure of about 236. Dividing a house cubage of 8000 by this gives 34 square feet that the inlet openings should add up to.

Perhaps yours is a house that has some of both things going for it. Then you should get by nicely with openings less than either calculation alone would call for.

If 38 square feet will do the job by breeze alone and 34 by temperature differential, 25 should suffice when both are working for you.

What's a Square Foot of Ventilation?

Building codes and manufacturers' instructions often warn you to allow a specified area of opening to ventilate a crawl space under a house, or an attic, or to provide combustion air for a furnace.

A typical prescription for crawl space is 1 square foot of opening for each 25 linear feet of wall.

So a house that is 32 by 40 feet, with 144 feet of perimeter, will need openings totaling nearly 6 square feet. But there's a catch to this you'd better take account of: such openings usually are screened or louvered to keep creatures out.

A rough rule is that louvered openings count only half. So double the area of ventilating openings. For our example that means 12 square feet.

If an opening is both louvered and screened (with 16-mesh screen) multiply by 3 instead of 2.

Better—since screen may be clogged by paint, cobwebs, and dirt—is to use 1/2-inch mesh instead. Then you don't have to allow anything extra.

How Much Can the Sun Save You?

The sun's heat is free—but it costs money to use it. That's the dilemma if you're contemplating—as you probably should be—solar heating for the hot water for your house.

Before putting in a solar device, you naturally want to know if it will pay its way, and how many years it will take to do so.

One installer of such things has worked out this calculation:

A typical family in Hawaii has an electric bill of about $200 every two months. Let's say this family uses 2349 kilowatt hours per billing period, or 39 kwh per day, of which 50 percent are estimated to be for hot water: 19.5 kwh times 365 days times 7.76 cents per kwh equals an estimated savings of $552.30 per year, $46.02 per month. A system for this typical family costs $4600, installed. The payback period is between eight and one-third and nine years—or so the installer we're quoting concludes.

But is it?

(Note first that since a couple of his base figures are necessarily approximations, he has couched his conclusion similarly. That's very much to his credit. Especially in making sales pitches, savings estimates are often presented as though they could possibly be known to the last cent.)

But that's not where the trouble lies. As the contest puzzles in magazines used to say, what's wrong with this picture?

Nothing's wrong with the method up to the final four words.

That's why I've quoted it as a useful example for anyone giving sober consideration to solar water heating or to other devices having a long but not unlimited life.

But—the payoff period here is a faulty figure. True, if you divide $2300 by $276.15, you will indeed get 8-1/3 for the number of years. But will the maker of the equipment install it on your roof and let you pay for it over that period—without interest? Not likely.

So, to find a realistic answer, you need a more elaborate calculation, ready-made or made by you.

For the ready-made answer, consult any time-payment table for mortgages and installment loans, to be found in the top drawer of any banker or money-lender's desk.

Assume any reasonable rate of interest. Say 8 percent. That's about as much as you've been able to get for your money in a guaranteed deposit in recent times and is somewhat less than you could borrow at if you wished to finance a home improvement.

Turn $276.15 a year into monthly payments of $23.01.

Look up $2000 and $300 in the book and you'll find $19.83 and $2.98, respectively, as the monthly payments for a 14-year loan. So (doing a little adding) $22.81 a month will pay off $2,300 in 14 years. That's close enough to $23.01 for any reasonable purpose.

So it seems the true payback period is 14 years—close to double what the man said.

If Hawaii seems remote, here's a somewhat different calculation, from a solar-hot-water man in New York:

"A family of four using 80 gallons of hot water per day will require about 20 kwh to heat the water electrically. At Lilco (Long Island Lighting Company) rates (five cents per kwh then) it costs that family $1 per day for hot water—$365 a year, not including fuel-cost adjustments, and assuming a constant utility rate."

This time, the system would cost $1600 and provide 80 percent of the family's hot water. The payback period is given as between five and six years.

Again, this is a method you may find useful. But, also again, don't accept the figure given in the example for payback period and don't estimate your own by dividing the cost of the equipment by the annual savings, as obviously was done. Instead of five to six years it would really take more than seven.

Capacity of a Solar Water Heater

THE FIGURE to start from, in sizing tanks and collectors for solar water heating, is daily consumption.

Per person this can be as high as 50 gallons, or so the utility companies used to like to tell us. In grimmer times it perhaps had better be more like 5 or 10 gallons except in households making prudent use of the free heat from the sun.

With solar assistance a good figure to work from is 20 gallons of hot water for each person each day. Unless there will be an auxiliary tank in the system, double that figure. Multiply by 1.1 to allow for heat loss. If you will have a heat exchanger, again multiply by 1.1 to allow for its losses.

This will tell you the approximate size of tank to obtain.

This result will also help you estimate the area of solar collectors you will need.

To find this in square feet, divide the tank size (which you have found in gallons) by 2 if you live in the south or southwest. This assumes you have just one layer of glass or plastic over the heat-absorbing element in your collector.

If you live in Canada or the northern U.S., multiply the tank size by 1.5 and also use two transparent layers.

If you live somewhere between, use either one or two transparent layers and make the collector size in square feet equal to the tank capacity in gallons.

What Will a Circuit Take?

To FIND OUT what you can load onto a single household lighting or appliance circuit, first examine the fuse box or breaker panel. You'll probably find the circuit is fused for either 15 or 20 amperes, this capacity being marked on the fuse or on the breaker handle.

Multiply this by the voltage, usually 110 to 120, that your power people deliver. If you don't know this figure, you can call the electric company and ask for it. If they're not sure what it is at your house, they'll come out and measure it for you. Or you can just assume 115 and be close enough for most purposes.

At 115 volts, a 20-amp circuit will give you 2,300 watts to work with.

Name plates on appliances usually tell you the wattages they require. If you have a coffee maker rated at 600 watts and a small electric heater rated at 1,000, already in use on such a circuit as just described, there goes 1,600 watts. With only 700 watts of capacity remaining, you know what will happen if you plug in a 1000-watt toaster as well. The fuse will blow or the breaker will trip.

Heavy appliances, such as permanent heating devices, ranges, ovens, or large air conditioners, commonly require more voltage. One of these will have a 220- to 240-volt circuit to itself. The calculation is made in the same manner, so a 20-amp heater wired for 230 volts will consume 4600 amperes.

What's a Switch Worth?

I WAS SHAVING under a couple of hundred watts of lighting that illuminated a whole bathroom. It occurred to me that a lot of electrical energy is wasted in such arrangements. If directed at my face, a fraction of the light I was using would have served as well.

But in wiring such a bathroom would it have paid to install two switches instead of one? Would the power saving have been more than balanced by the cost of the extra switch and wire?

With my next house in mind, I ran through these figures.

Extra switch box, 75 cents; extra switch, ditto; additional 12 feet of cable, $1.50. Total outlay, $3.00. (If I weren't doing the work myself I should double this total, to allow for the several extra minutes of the electrician's time.)

Continuing to ruminate while shaving, I guessed that the bathroom light is in use two hours a day, including the time it's burning because someone neglected to shut it off. Even if both bulbs would still be used half the time, the additional switch would save a 100-watt bulb one hour each day, or one-tenth of a kilowatt hour each day.

That comes to 36 kilowatt hours a year.

At a power rate of about 11 cents per kwh, that's $4 a year. The extra switch would pay for itself in a year or two in most localities. And saving electrical power is a good deed in itself.

Working Out an Island Sink

KNOWING that sinks and other plumbing contrivances require vents that go up through the roof, I've occasionally wondered how the problem is solved when the sink is in a counter in the middle of a room.

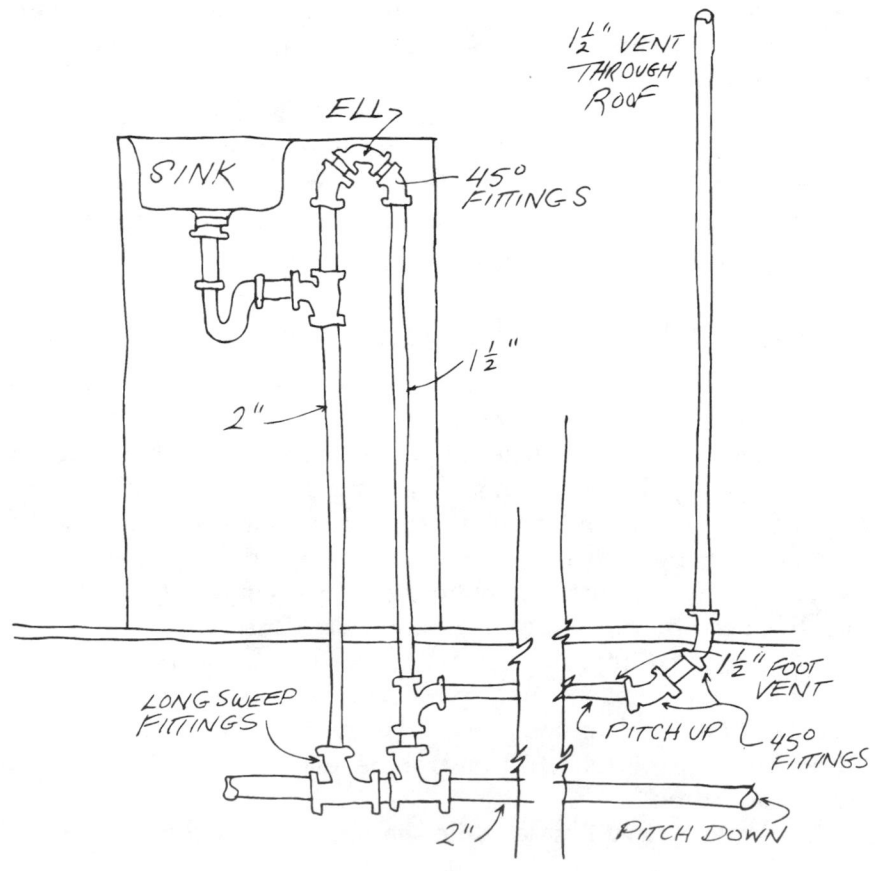

Well, now I know. And you do too. The trick includes a long sweep fitting below the floor with a pipe connection running horizontally to the nearest partition, where 45-degree fittings you see here make for easy cleaning of any stoppage that might develop.

The return vent you see beside the sink should be as high under the drain board as possible. The loop is made as gradual as possible with a pair of 45-degree fittings and one long-radius elbow.

Loading a Waterbed

A WATERBED can weigh a ton or more, enough to break through some floors. If you're putting in one or renting to a tenant who might, a little arithmetic may be in order. So says an expert on landlording.

First you should figure the area of the bed. For queen size, which is most usual, it's 60 by 80 inches. You can figure the area directly in feet or multiply those two dimensions together and divide by 144. Either way you do it, you get 33-1/3.

Multiply the water capacity in gallons by 8.33. Then, to find the total weight you're dealing with, add about 4 percent to this for the weight of the bed parts if it's an all-water bed. Make that about 20 percent if it's a hybrid that uses urethane foam or something of that sort as well as water.

For an all-water queen, 9-inch fill, capacity may be given as 196 gallons. Multiplied by 8.33, and by 1.04 to add that 4 percent, gives a total weight of 1698 pounds. With a couple of occupants, that's just about a ton. King size will be even heavier and some of the big round beds heavier yet.

Divide your weight figure by the area of the base, calculated in feet as above.

If the weight of the bed plus sleepers is a ton, you'll find the loading on the bedroom floor to be about 50 pounds to the square foot.

A result like this might be taken as a caution warning, since residential floors are most often designed to accept a loading of 40 pounds to the square foot, in some areas 30.

Consultation with the building department in your town might be the next step.

Aiming a Satellite Dish

THE ANGLES you need to know when aiming a communications-satellite dish can be found in four routines. They're all based on this trio of basic figures:

A. *Longitude of the site* of your dish. If it's west longitude, put a minus sign in front of the figure.

B. *Longitude of the satellite.* Again, if it's west longitude make the figure negative.

C. *Latitude of the site* of the dish. In the Southern Hemisphere, make this negative.

(If these figures are in degrees, fine. But if any include minutes as well, change those minutes to decimal degrees by dividing them by 60. That way 40 degrees, 15 minutes becomes 40.25 degrees.)

From these three location points, you can find the great-circle angle between your site and the point on the Equator directly below the satellite. You can also find the site azimuth from due north.

Then you can use that great-circle angle to find the range from earth-station site to satellite. Use the range, in turn, to find the elevation-from-horizon of the satellite.

To find great-circle angle:

1. From A, subtract B. If the result is not between minus and plus 180 degrees, add or subtract 360 to make it so.

2. Find the cosine.

3. Multiply by the cosine of C.

4. Take the arc-cosine. (If this is more than 81.3 degrees, you're in trouble; the satellite will not be visible from that location.)

To find azimuth:

1. This step is the same as Step 1 above.

2. Find the tangent.

3. Divide by the sine of C.

4. Find the arc-tangent.

5. If the site is in the Northern Hemisphere, add 180.

6. If the result is negative, add 360.

To find the range in kilometers:

1. Multiply cosine of great-circle angle by 536,950,000.

2. Subtract this result from 1,818,590,000.

3. Find the square root.

To find the degree of elevation from the horizon:

1. Square the range and subtract 1,737,517,200.

2. Divide by the range and also by 12,734.

3. Take the arc-cosine.

4. Subtract 90.

Applying these procedures to aiming an earth-station dish at Orlando, Florida (longitude 81 degrees 20 minutes west, latitude 28 degrees 32 minutes north) for Satcom 1, with west longitude of 135 degrees, first change minutes to decimal degrees. Your working values then are: A = -81.33, B = -135, C = 28.53.

Great-circle angle: A minus B is 53.67, of which the cosine is 0.59. Cosine of C (28.53) is 0.88. Multiplying the two cosines together gives 0.52, of which the arc-cosine is 58.67 degrees. (If you do all this with a calculator or computer, not rounding off anything, you'll get the more precise 58.6346 degrees.)

Azimuth: The tangent of 53.67 is 1.36, which divided by the sine of C is 2185. Arc-tangent of 2185 is 70.65. Adding 180 degrees for the Northern Hemisphere location makes it 250.65 degrees, or more precisely 250.64.

Range: The cosine of the great-circle angle 58.63 degrees is 0.52. Multiplying by 536,950,000 and subtracting from 1,818,590,000 gives 1,539,097,476, of which the square root is 39,231 (kilometers).

Elevation: Beginning with the square of the range and subtracting 1,737,517,200 and then dividing by the range (39,231) and by 12,734, taking the arc-cosine and then subtracting 90 gives 23.40 degrees elevation.

Since this is all a little involved, you might like to check the method against the results calculated for a second location. It's a spot in the Carmel Valley in central California. A friend of mine has been installing and aiming an earth-station dish there. It's at 121.8 degrees west longitude, 36.5 north latitude. The calculations are for the same Satcom 1, at 135 degrees west.

My four results are 36.50 degrees, 210.75 degrees, 37,394 kilometers, and 45.42 degrees.

The basic routines for these calculations came to me from Stephen Reed of Maitland, Florida, by way of David Nelsen of Lethbridge, Alberta, who mentioned another formula that could come in handy in conjunction with these others. It's one you can use to figure how far a dish must be set back from a structure or tree to avoid interference.

Having obtained the degree of elevation by the last of the four calculations above, subtract an arbitrary 5 degrees as a margin for avoiding noise. Find the tangent of the result. Divide the height of the object by that tangent.

For instance, using the 23.4-degree elevation found for Orlando, you'd subtract the 5-degree margin to get 18.4. Dividing 50 by the tangent of 18.4 (which is 0.33) warns you to keep your dish at least 150 feet back from that 50-foot tree.

Chapter N
OPERATING YOUR SHELTER

To Build a House
Managing a Mortgage
Shrink a Mortgage?
Your Depreciating House
How to Sell a House
Property-Tax Strategy
How Much Home Insurance?
Cost-Effective Energy Saver?
Fueling the Water Heater
How Much Heat in Firewood?
What Does Fuel Cost?
Mobile Home—Does It Figure?
Free Color TV?
Cork in the Eye

To Build a House

As MANY a family has discovered to its sorrow, becoming the owner of the kind of house you dream about can be a shockingly expensive proposition. But a custom-built house is still what most of us want, so it's worth taking a hard look at how to figure the cost—and how to reduce it.

Making decisions about such major matters as this is, after all, one of the things that figuring is for.

Cal and Midge Mason are about to build a new home for themselves and young Mark, who is seven. They know what they want—pretty much—and they have already searched their town in vain for anything like it.

"No wonder!" Midge sighs. "If I had a house like the one I'm dreaming about, I wouldn't part with it either."

Having consulted one architect, several real-estate salesmen, and a couple of custom builders, the Masons have learned that a house of the size and quality they want will cost about $110,000 in their small city.

At that, they're lucky. In applying this and the cost estimates to follow in many larger towns, or any city in most areas, the dollar figures would have to be doubled or worse. On the thesis that it is easier to multiply than divide, a near-minimum example may be the most useful.

The arithmetic behind that $110,000 price is simple. The least-expensive suitable lots the Masons have seen are listed at $22,000. They want a house of about 1600 square feet, and the going rate for custom construction is a minimum $50 a square foot. That comes to $80,000.

Few architects will design a small house and supervise its construction for a fee of less than 10 percent, $8000 in this case. And to get a house that fits both site and occupants, there has got to be a designer in the picture.

Before coming to any decision, Cal and Midge list the parts that add up to $110,000, using round numbers:

Land: $22,000.

Design: $4800.

Supervision: Architect, $3200; time, overhead, and profit for contractor and sub-contractors, $20,000. Total: $23,200.

On-site labor (carpenters, plumbers, etc.):$30,000.

Materials: $30,000.

For that amount of money, $110,000, they can have their special house. They need only choose their architect well and make certain she understands their needs and desires. She will help them choose a lot, draw a detailed sheaf of plans and specifications, first submitting rough sketches. They can be reasonably sure of getting exactly what they want and can afford, provided they don't add too many extras along the way.

They can also see that, although the down payment is within possibility, the monthly mortgage payments will be excessive.

The first place to save is on the land.

"If we can buy a lot directly from the owner, we can avoid a real-estate commission, and that's a flat 10 percent on vacant land," Midge points out.

It's a good idea, though it means some work. It means touring the area they're interested in, looking for vacant lots that fit their specifications, noting location so they can identify them on tax maps at city hall or county court house. And then obtaining names and addresses of owners from tax records, and phoning or writing to find out if the owners are prepared to sell.

For the Masons, this procedure produced two level lots and a third that sloped sharply down from the street.

"Looks like we can buy either level lot for about $19,000 or the steep one for a couple thousand less," Cal says.

"From what I've been reading, the saving on the steep lot is about what it will cost us extra to build on it."

Either way, the Masons' do-it-yourself land-buying approach should save them about $3000, mostly from bypassing commission at the cost of some shoe leather.

Now how can they save on the next item, design? They can skip the architect and go instead to a student or a recent graduate who is unemployed or moonlighting or find a designer who is less than a full-fledged architect.

Midge and Cal plump for an alternate economy solution: ready-drawn plans. Home magazines offer such plans for $50 or $100 or so. Many have been drawn in the first place for clients who paid a lot of money for them.

A ready-made plan may be excellent, but this solution is not without its risks. For one thing, it takes good judgment to choose a plan, especially if it must be adapted to an unusual building site with special engineering requirements or code problems.

Whether Midge and Cal can safely save more of the planning and supervising costs by acting as their own general contractor will depend upon their experience and how much they are willing to learn in a hurry. They will have to hire people to do concrete work, framing, finish carpentry, plumbing, wiring—instead of paying a general contractor to do this or to subcontract the work.

If the Masons go this far—finding their lot, bargaining for it, drawing their own plan or buying a stock plan, then supervising construction—they stand to save as much as $30,000. Their dream, hopelessly expensive at $110,000, is within reach if they can hold to this plan.

You, or anyone capable and willing to go this far, might consider going even further. You just might tackle some or most—or conceivably all—of the construction work itself, as I have done more than once and with highly satisfying consequences.

What will you save? Well, for a house like the one just described for the Masons, labor was estimated at $30,000. Providing it yourself can knock anything up to that off the $80,000. For a very rough estimate, the house you build yourself should cost half what it would sell for.

I can back up that generalization. Thirty-some years ago, when my family was in somewhat the position of the Masons, we slowly built a house for ourselves—big one, five bedrooms, 3,000 square feet. When finished, it was valued at just about three times the amount we had in it. With a modest boost from inflation, it sold a dozen years later for four times our investment. Loss of interest over that period on our invested money was more than balanced by the rent we weren't paying.

Meanwhile, as our kids began going off to college, we built something better but smaller. On a rather costly lot, it ran us some $26,000 and was worth about $60,000. We lived in it for 10 years, rented it out for 3, then sold it to our oldest daughter and her husband. Of course, the appraisal of $120,000 had 13 years of inflation in it.

Some people don't know when to stop, so by this time we were living in another homemade home. This time we used a good bit of hired handyman help, but contracted out nothing except the tar-and-gravel roof. Materials came to $15,000, help to $10,300. Appraisal by the tax man, two years after completion, was $51,000, not counting land.

Here's a final interesting figure to toy with.

You own a lot that is acceptable as a down payment. You want to build a $104,000 house on it...a $52,000 house if you build it yourself.

You can afford payments of $800 a month for principal and interest. The going rate is 8-1/2 percent. Consulting a table of mortgage payments, you find that your $800 a month will pay off $104,000 in just 30 years. But it will pay off one of half that size in a remarkably brief seven years plus a couple of months. By going the sweat-equity route and slicing your cost of building in half, you've cut your total outlay over the years to one-fourth.

Managing a Mortgage

IF YOU'RE now carrying a mortgage that was tailored for you two or ten years ago, it is like a suit of clothes of the same age. It may be uncomfortably tight in its payments or too short in its amortization period.

The mortgage may need alterations because your house does. Rates that have dropped appreciably may translate into lower monthly payments or an earlier payoff.

Two other situations might help you.

You may have had to accept a high rate of interest because your equity was small. By now, that equity has grown through your payments to principal, improvements you've made, development of the neighborhood, wider acceptance of an unconventional design.

Small changes in interest rates are more significant than they may sound. Over the life of a 30-year $25,000 mortgage, reducing a 9-1/2 percent rate to 8-1/2 means a healthy $5,397 in your pocket. It cuts monthly payments about $18.

In refinancing, you need to offset potential savings by the cost of the change. These possible expenses include appraisal and loan fees, title-insurance costs, and any penalty provided for in your present mortgage. With the help of your loan source, you can quickly discover whether a change is worthwhile now or may be in the future.

If so, you'll have some pleasant, though possibly difficult, choices. You can reduce the monthly payments. You can shorten the payoff time. You can turn some of your equity into cash.

Example 1. When interest on your mortgage runs 10 percent and the going rate has dropped to 9, is it time to refinance? If associated costs, which may include new title insurance, survey, appraisal, legal fees, and taxes, eat up the advantage, you have little to lose by waiting in hope of a larger drop.

Example 2. When your mortgage can be refinanced at a lower rate, reducing payments even after closing costs are added, are there additional hidden advantages or drawbacks? Yes, if you need a further break, you may be able at the same time to reduce monthly payments by extending the life of the loan. And this time you may be able to include an open-end clause, which is handy if you want another loan in the future for home improvements.

Example 3. If your present mortgage is high-interest but it wouldn't pay you to change just now, are there any steps it would be prudent to take? Prepay it as fast as you can afford to. Making extra payments is really a way of investing money safely at a favorable rate.

Example 4. What are the rules about prepayments? Read the fine print and consult the lender. Sometimes you're allowed to pay extra only on the principal, and you may be required to give written notice. The lender may be willing to agree that the pre-payments can be used later on (in an emergency) to count as payments on the principal—especially if you're unemployed or ill for a period.

Example 5. Having come into or accumulated some extra cash, should you prepay part or all of a mortgage—or invest it elsewhere? Paying off a loan is the safest kind of investment, but by no means the most liquid. If you can safely park the cash elsewhere at a rate higher than the loan bears, you might as well let the mortgage stand and reap the difference. Moreover, if the mortgage is large and assumable, it might be an asset if you should sell.

Example 6. For some years, you've been paying on first and second mortgages, as well as making improvements on your home. The second runs out in a few months, when you must make a final payment of $2,000. Where can you find the money? You've probably built up a hefty equity. Refinance with a present or new lender if you can get a cheaper first that covers the balloon and closing.

Example 7. You suspect your loan payments, manageable now, may become a burden when the kids reach college age. What to do? Prepayment on your mortgage is one possible move, provided the interest rate is as high as you could get by stashing the same amounts in a savings account. Or maybe you can work out a prepayment schedule with the lender that will make it feasible to retire the mortgage before those college years arrive.

If you hope to retire before the mortgage is paid off, look for a prepayment schedule that will permit you and the mortgage to retire at the same time. For more about this, take a look at the last item in this book.

Example 8. You've lost your job and you're falling behind on mortgage payments. How can you keep your house?

Get together with the lender. He doesn't want a foreclosure either. Together you can probably work out a plan to catch up when you go back to work. If yours is an FHA loan, the lender may wait a year before foreclosing, provided the house is being kept up. VA also has helpful rules.

Example 9. You've kept up your payments so far, but now you see stormy times ahead. Your lender is not prepared to carry you if you fall behind.

He might be amenable to an entirely new loan, extending the time while reducing the payments. You may have better luck with a new source, although you might encounter heavier closing costs. Either way, your good record to date and your prudent anticipation of trouble should work for you.

Shrink a Mortgage?

ONE OF THE SILLIER of the notions that keep cropping up in books on budgeting, and in newspaper and magazine columns of financial advice has to do with the wisdom of keeping your home mortgaged because the interest is a big tax deduction.

See the fallacy?

This option arises only if you have spare money for making a bigger down payment than required. Or reducing or paying off an existing mortgage. Your alternatives are to put the money into the house or into some interest-paying investment.

Say there's $20,000 remaining on your 8 percent mortgage. Under federal tax law the interest on this reduces the income on which you must pay tax at a rate of $1,600 a year. Pay it off and you lose that benefit (and perhaps one under your state's tax laws as well). But if instead of paying off the mortgage, you invest the $20,000 in bonds or something at 8 percent, you immediately increase your income subject to taxes by an offsetting $1600 a year, gaining nothing.

So in figuring out whether to put your extra money into your own mortgage, you can forget the advice about hanging onto the tax deduction. You can decide purely on other factors.

It's pretty clear that if you can safely invest the money at a rate of return higher than you're paying on your mortgage, it makes sense to let the mortgage ride and gain the difference. It's equally apparent that if you can't find a safe investment at an equal rate, you'll be ahead in dollars if you invest in your own mortgage instead, prepaying to the extent that is permitted free of penalty.

Don't forget that a comparatively large existing mortgage can be a help in selling your house if the buyer is allowed to take it over. However, a mortgage that is small relative to the value of the house is usually no help at all. Most buyers won't have enough cash to make up the difference.

Your Depreciating House

IF YOU OWN a house and take care of it, it probably gets better every year. The landscaping matures, you update its energy efficiency and its mechanical parts as they wear out.

All the same, from the income-tax point of view it's depreciating. Which means that if it—or any part of it—is producing income, it is also generating an income-tax saving for you. This applies if some or all of it is rented, full- or part-time, or used as an office or shop or garage or photographic darkroom that produces taxable income.

Since you are allowed to depreciate an income-producing room or dwelling over precisely 27-1/2 years, the first step in figuring how this works is division by that cumbersome number. What you divide by it is the cost to you of that structure (land doesn't count, mostly because it doesn't ordinarily wear out).

Dividing something by 27-1/2 is equivalent to multiplying it by 0.03636, or taking 3.636 percent of it. That operation gives you the depreciation for each year after the first...until 27-1/2 years have passed.

Which brings us back to that first year in which you are entitled to take depreciation, the year that produces an interesting calculation.

Depreciation begins in the month when the property is put into service. But no matter which day of the month this is, tax law assumes it was halfway through the month. Not surprisingly, this assumption is dubbed the mid-month convention. It means that regardless of whether you are figuring from the first of January or the 31st, you still are credited with the same 11-1/2 months. And if you start any time in December, you get half a month.

Divide that mystic 0.03636 by 12 and multiply by the number of months in which the use occurred—a partial month counting as one half. Multiply by the cost of the building. That's the first-year depreciation.

Let's say you've spent $100,000 (not including land cost) to build a house. You rent it out, beginning sometime between March 2 and 30th. Multiplying $100,000 by 0.03636 will give you the depreciation you may take each year after the first: $3636.

In the first year, however, you must multiply that result by 9-1/2 divided by 12, coming to $2878.79.

If, by the way, you're renting out only part of the property, keep in mind that only that fraction counts. If you're occupying one-third of the property yourself, the $100,000 in our example becomes $66,667 instead. You would probably be allowed to choose the fraction in the way most helpful to you—by area, or by number of rooms, for example.

For something other than real estate, the time period, and hence the numbers, will be different. And while even the details of the real-estate computation may change from year to year, understanding the basics gone into here should still help.

How to Sell a House

WHAT KIND of figuring will persuade a prospect that he can afford the payments involved in buying the house you want to sell? In his syndicated real-estate column, Robert J. Bruss answers this very nicely.

Dear Bob: I am a new real estate saleswoman. Although I've been working for four months now, I've only closed one sale. I could have sold many more, but the buyers balked at the high monthly payments. Yet I see other agents selling homes with high payments. Do you have any suggestions for overcoming the sales objection of high monthly payments? Bettina W., Washington, D.C.

Dear Bettina: You've got one of the best jobs in the world, real-estate sales. You need to show your clients why they can't afford not to buy from you. Once you find a home that your client likes, the key to getting a sale is to show the buyer's true net housing cost. For example, suppose a client is considering buying a $60,000 home and he has $15,000 for the down payment. You can arrange a $45,000 mortgage for 25 years at 9 percent interest. The monthly mortgage payment will be $377.65, plus $100 per month for property taxes and $25 for insurance. But the client doesn't like the $502.65 monthly total, although he can afford it.

At this point, ask the client what tax bracket he is in. Suppose he says 40 percent fed-state-local, including his wife's earnings. Then show him his first-year monthly tax deductions: $373.50 interest, $100 taxes, for a $473.50 total. He saves income tax on 40 percent, for $189.40 tax savings. That reduces the monthly housing cost to $313.25, probably less than the cost of a comparable rental. In addition, the mortgage principal is increasing while the value of the house is probably appreciating. Show your clients why they can't afford *not* to buy a home from you.

(In figuring something like this for your own purposes, you should include one factor overlooked here. From now on, the buyer will lose the interest he could be getting on the $15,000 he put down. Call it 8 percent. So the loss is $1200 a year, $100 a month, leaving a net after taxes of $60. Realistically, that must be added to the $313.25. And in dealing with a $240,000 house, that $60 becomes a more impressive $240. Bettina will still have a winning argument in most cases, though.)

Property-Tax Strategy

I GO ALONG, and no doubt you do too, with Oliver Wendell Holmes, Jr.: Taxes are what we pay for civilized society.

But that doesn't imply any eagerness to pay more than our fair share, even if it takes a certain amount of figuring to insure that this doesn't happen.

Given neighbors, with houses comparable to yours, who are willing to discuss their tax bills, you can quickly estimate whether the valuation placed on yours is out of line. Evaluating property is far from an exact science, and an error or simple misjudgment is not at all unlikely. Essentially, it is up to the homeowner to spot an instance of this and get it rectified.

If you seem to have a case worth pursuing, consult city-hall or courthouse records, open to the public. You get a precise comparison on as large a scale as you choose.

Property taxes are usually in proportion to estimated market value, with land and improvements often calculated separately.

Full value, or some arbitrary percentage of it, becomes the "assessed valuation." Your mathematical approach begins with dividing this assessed valuation by that percentage to see exactly how much the appraiser has said your place is worth.

If you don't agree, it's up to you to say so. If you can show, by actual instances (and a lot of exact numbers down on paper—always effective) that comparable homes aren't selling for that much today, you have a much more telling argument to present.

In the end, you may have to appeal formally to get a reduction, but an informal appeal is less of a nuisance for everyone and often gets more hospitable treatment.

Over the years, I've followed the latter course four times and four times had my taxes reduced. Since my complaints were legitimate and typical of those of many householders, you may find some suggestive value in the details.

Forty-five years ago, in a rural area, I built my first house, brashly using simpler methods than some in general use at the time. My framing of two-by-fours spaced 24 inches and my masonry of hollow blocks not plastered or stuccoed on either side, both commonplace today, were radical (and economical) short-cuts at the time. Having produced a house at far less than normal costs of materials, I was flattered—but not particularly pleased—to find it valued by the tax people at the same level as houses of similar footage but far more elaborate construction.

I addressed a mild note of protest to the office of the county assessor. One of his assistants turned up a few days later, examined my case and my house, and ruled that my structure should be reclassified—not at all flatteringly this time—as "partially substandard." This cut my tax bill by about 10 percent.

The house, by the way, is still standing sturdy and would be regarded, if anyone questioned it today, as conventional in construction.

My second brush with tax people came after a move to where taxes are assessed in three parts: land, improvement, personal property. There I built a rather elaborate large house. I received a proportionately huge tax bill, but found unwarranted only the very heavy levy on the personal-property portion. The $900 valuation, implying nearly $5,000 market value, didn't square with our scanty array of home-made and makeshift furniture that was all we needed in a house loaded with built-ins. It wouldn't have fetched $500 at auction.

I asked how come, and the county sent a man to explain. "To save time, we assess personal property in proportion to house value," he explained. "That way we don't have to poke through your closets and everything. A house as big as this, we figure you'd have maybe a grand piano and other expensive furnishings. Or you're going to buy them."

"Well, I don't," I said. "And I'm not. And so I shouldn't pay tax on something I don't own."

The man eventually agreed that there was nothing in the law to justify taxing expectations—even his. When I received my final tax bill, the $900 was crossed out and $300 written in—still ample for our modest possessions but quite a tax saving for us.

A little later when we built another house, a similar informal protest brought agreement that full valuation was premature on an unfinished structure, even though occupied. (Occupancy before completion was permitted in our town, which was small, still friendly, and not overfond of red tape.)

My companion argument, that our lot should be valued at less than its neighbors because an extreme slope added to foundation cost, fared less well.

"Right you are," said the assessor. "So we didn't charge you for the cost of the foundation in valuing the structure."

He had me there.

A few years back, the assessment and taxes nearly doubled in a single year on five rocky, waterless, unimproved acres we own in northern California. To buttress our opinion that the new valuation was more than the land was worth, we listed it for sale at that figure with a reputable real-estate office. It failed even to get an offer, and we used this fact in our letter of complaint. The only reply was a form acknowledgment, but the next tax bill showed a drop in "full value" from $15,000 to a reasonable $10,000.

To me, these experiences say that where there is an avenue for an informal request, one well-based and carefully worded letter may produce a substantial benefit that goes on year after year. If you are not confident of your grounds, avoid formal protest if possible. There's always a risk, however slight, of what happened to our dentist. He told me about it over the whir of the drill.

"I demanded a hearing on my tax bill," he said. "The county went over all the figures and jumped my taxes nearly $100 a year. And there's nothing I can do about it."

How Much Home Insurance?

Balancing costs against benefits to decide how much insurance to carry on your home starts with estimating its value.

The first thing to keep in mind is that you are valuing the structure—the part that can burn or otherwise be destroyed—not the house-and-lot combination that you would buy or sell. And for insurance purposes, you will ordinarily count foundation and pipes that are below the surface of the ground as part of the land, not as part of the house.

Here is how I went about putting a value on my house.

I began by measuring the number of square feet of floor space in the house. I did not count open porches or patios. I did count the garage, but only half its area since it is generally assumed that garage space costs about half as much to build as living space. If there were any complicated, roofed porches, I'd have counted half their area as well.

I found my house to have 1576 square feet of living space, plus an attached single garage of the usual 12-by-20-foot size. Adding 1576 to half of 240 gave me roughly 1700 square feet.

A phone call to my insurance broker told me that their current figures suggested replacement costs in the range of $65 to $75 a square foot. An architect I was playing tennis with a couple of weeks earlier told me he figured local building costs at $70 on up. (Never pass up a chance to get free professional advice—except possibly from M.D.s, who are more inclined to resent your maneuvers than to be flattered by them.)

Multiplying 1700 by $70 gave me a valuation of $119,000.

Since my house had been reassessed this year for tax purposes, I consulted my county tax bill to see what the assessor's person thinks my house would fetch if put up for sale. I found the assessment for "improvements" corresponded so remarkably to my own estimate that I couldn't help suspecting it may have been reached in precisely the same way.

Although I know that assessments tend to run just a trifle on the low side (perhaps to minimize ratepayer squawks) and also lag a year behind, I decided to go for the $70 figure. I'd have made it more, except for three things:

1. Mathematically speaking, insurance is equivalent to gambling, so the same rules apply. Insurance is a relatively poor buy for those with less than average chance of needing it. These are people known in the jargon as good risks and are beloved of insurance companies. I counted myself as one of these, since my house is solidly constructed, lived in by two fearful people, and is very close to our fire department.

2. Since I built the house myself I could rebuild it more economically than a contractor could. Hence, my apparent loss would be somewhat greater than my real loss.

3. I needn't concern myself with the effects of inflation on the value of my house during the period of the policy, since it contains an "inflation guard endorsement". This ups my coverage every three months.

Putting all this together, I can figure my house as safely insured at that $70 figure.

But now I have another question to face. Do I want to pay a premium big enough to keep my house fully insured? The chance that it will be totally destroyed is pretty tiny. I'll be paying a lot to cover a very unlikely eventuality—smart only if the loss would be crippling. I review my financial state and conclude that I can afford to cover a part of this minor risk myself.

The question is: how much of it? Perhaps I'll decide to insure my house for just half its value.

Clever of me? No. It's good economy from my point of view, but not from that of the insurance people. They've figured out a way to forestall this move. It's the 80 percent rule.

This stipulation says that if you insure your property for less than 80 percent of value, you can wind up collecting less than you expect. You can be limited to the same proportion of replacement cost as the insurance you're carrying is of 80 percent of the value.

In other words, insure at 50 percent of value and you may collect only 50/80 of your loss. Have yourself a little $1000 fire and 5/8 of $1000 is what you get; the other $375 of the repair bills is on you.

For most people, this makes it sensible to keep coverage up to 80 percent of real value.

On that basis, I multiply the full value of $119,000, arrived at a few paragraphs back, by 80 percent, to get $95,200. I round that off and insure for $95,000. Or maybe an even rounder $100,000.

Cost-Effective Energy Saver?

INNUMERABLE formulas have been devised to estimate the cost-effectiveness of an energy-saving home improvement. But here's one that is both simpler and more realistic than most.

It applies to any home you plan to live in for quite a while, and assumes that if you sell your house the buyer will recognize the value of the improvement—even if, as with such vital but prosaic contributions as insulation, it may not be readily seen.

Begin by estimating how many years the new item will last. For most energy-savers, such as double window glass or attic insulation, this may be the life of the house, perhaps 35 or 50 years. For a mechanical device, it may be nearer 10 or 15. To find the annual percentage of depreciation, divide 100 by that number of years.

Now figure the interest on your investment. It's better not to use an interest rate as incomplete and as temporary as the rate you'll pay if you borrow or the return you're sacrificing by spending the money instead of keeping it invested. True long-haul return may be better estimated by subtracting from the going rate both the tax you'd have to pay on the interest and that part of the return that is going down the drain because of inflation.

Add that cost-of-money figure to the depreciation percentage calculated according to the second paragraph above. Multiply by the cash price of the project. Subtract the annual cost of upkeep, if any. This will tell you how much the improvement will really cost you each year. To find how cost-effective it will be, compare it with the dollar saving that will be produced by the improvement.

Let's try the procedure, for a $2500 insulation job in a house you think should be standing for 50 years to come. Dividing 100 by 50 gives you 2 percent annual depreciation.

If you can get 12 percent on your money, with a quarter of that being taxed away, you're sacrificing a net of 9 percent. If inflation looks to average 4 percent, the difference is 5 percent. And, in fact, experience in the '70s, '80s, and early '90s suggests that 5 percent is as good a guess as any at the amount by which return on invested money can hope to exceed the erosion by inflation.

To that 5 percent, you should add the 2 percent from depreciation. Since 7 percent of $2500 is $175 (and there's no maintenance cost to add in) you now know the annual cost of the improvement.

Compare that amount to what you can expect to save in fuel and you'll know whether your energy-saving measure is likely to prove cost-effective.

If the price of gas, oil, or electricity goes up faster than inflation in the future, your picture will become increasingly favorable. And, of course, many improvements yield an intangible bonus in comfort or convenience as long as you're in the house.

Fueling the Water Heater

THERE IS an easy formula for comparing the cost of gas and electricity for use in heating water:

Electricity is cheaper than gas only when the cost per therm (100,000 British thermal units, or Btus) of gas is more than 22 times the price of a kilowatt hour of electricity.

If you cannot tell from your bill how much you are paying for a therm of gas, a phone call to the gas company will get the information.

Similarly, the power company can tell you about your electricity cost.

These costs may differ according to quantity consumed, so keep in mind that the cost you need for this purpose is the highest rate you pay. The lower-cost units of power aren't involved in this, so you look only at the top level of cost.

Example: If the cost of the last increment of power on your bill is

10 cents per kilowatt hour, electricity is cheaper only if gas is costing you $2.20 or more per therm.

The same figures apply to other comparisons of gas and electricity, but you may have to allow for how the fuel is used. Electric room heaters, for instance, may not be as costly as the formula says because it may be easier to leave some rooms unheated part of the time than it is with a gas furnace.

How Much Heat in Firewood?

CONTRARY to what many people think, you can figure on just about as much heat from softwood as hardwood, pound for pound. The thing is, oak and hickory weigh twice as much to the cord as pine does. So there's twice the heat to the cord.

Kind of wood (air dried)	Weight in pounds per cord	Equivalent in gallons of fuel oil
hickory	3900	191
white oak	3600	175
beech	3500	165
hard maple	3500	160
yellow birch	3500	160
ash	3300	154
elm	3000	135
soft maple	2800	132
tamarack	2500	132
cherry	2550	129
spruce	2100	99
hemlock	2100	98
aspen	1900	97
basswood	1900	93
white pine	1800	94

I don't know how it is where you live, but in my central California coastal country, when oak runs $200 a cord (green, unstacked) to $225 (dry and put where you want it) pine costs at least $110. You can see that it's only the unknowing among us, deluded by the numbers and the impressive bulk, that buy the pine. Anytime softwood costs more than half what hardwood does, it's a bad buy so far as heat is concerned. It's not very good for your chimney either; use much pithy wood and a chimney that's not frequently cleaned can become a fire hazard.

Having figured the comparative values of kinds of firewood, you may wonder about cost of wood compared with other fuel. The table offers a reasonable approximation of how many gallons of fuel oil it takes to give you the heat to be found in a cord of firewood. Multiply 191, for example, by the gallon price of oil, and you'll learn how much you can pay for a cord of hickory and come out even—until you take efficiency into account.

What Does Fuel Cost?

THIS IS an alternative way to compare true costs of using such fuels as coal, wood, electricity, natural gas, liquid-petroleum gas, and oil.

To use it, you need to know three things: heating value of the fuel, cost per unit of the fuel, and the level of efficiency at which the heating device operates.

Coal: Heating values are 8000 to 11,000 British thermal units per pound, which is 16 to 22 million per ton.

Wood averages 7000 to the pound, or 14 (for some softwoods) to 31 million (for the densest hardwoods) to the cord.

Electricity: One kilowatt hour produces 3413 Btus.

Natural gas: A cubic foot has a heating value of 850 to 1000. Your public utility company can no doubt provide a more precise Btu figure for your area. However, even for a given location, it's a so-called constant that in fact will vary from time to time.

L-P Gas: Figure on 95,500 Btus to the gallon.

Oil yields 140,000 to the gallon.

Efficiency of electricity used to heat by resistance is virtually 100 percent.

A good airtight stove will be 40 to 60 percent efficient. For a circulator fireplace, the figure is 20 to 30 percent, but count on only 5 to 15 percent for an ordinary fireplace. Furnaces usually burn their fuel with efficiency of 50 to 85 percent.

If yours is one of the high-efficiency kind, you probably know it and have the actual test figure to use.

With these figures, or more precise ones if you can obtain them for your own fuel and equipment, you can predict relative costs.

Multiply the number of Btus per unit by the efficiency, expressed as a decimal.

(Remember that this means that 60 percent becomes 0.60.)

Divide by the dollar cost per unit. This tells you the number of Btus you can expect for each dollar spent.

Thus the best firewood, used in a good stove and costing $150 a cord gives 31 million times 0.60 divided by 150, or 124,000 Btus for each dollar of cost.

And electricity at 10 cents a kilowatt hour gives 34,130 of those cozy British units for a dollar (3413 times 1 divided by 0.10). At last, here's a calculation for which you don't need even paper and pencil since dividing by 0.10, or 1/10, is the same as multiplying by 10.

Mobile Home—Does It Figure?

THERE ARE a lot of personal considerations in buying a mobile home rather than a conventional one plunked down permanently on a lot by itself. Nobody can help you much with those decisions.

But the financial aspects are something else again. They can be clarified by a little paper-and-pencil work.

To begin with the plus side, the typical mobile home costs perhaps one-fourth as much as the average conventional house. That is partly because it is smaller. But even when you compare homes of the same size, the mobile comes in at half the price.

There is more to the story, beginning with three financial drawbacks.

• Mobile homes depreciate fast. Just as they are commonly licensed much like motor vehicles, they depreciate almost as fast as cars do. Even in our worst inflationary times, mobiles were losing half their market value in the first 6-1/2 years. After that, the value drops more slowly until, by the time about 15 years have gone by, they are worth one-fifth to one-third their original cost. Regular houses, as you know, have tended in our time to appreciate in value instead of losing; and, despite short-term setbacks, this trend has great staying power in regions of population growth.

• Most mobiles are financed at automobile interest rates—10 to 12 percent when home mortgages are going at 7 or 8.

• When you buy a house you are also investing in land, which can be expected to increase in value year after year. As an owner in a mobile park, you rent the use of land and cannot expect to share in the appreciation of its value. (But if you buy a lot for your mobile home, on your own or in a mobile-home subdivision, you eliminate this disadvantage.)

Since no general rule will tell you whether a mobile is financially advantageous to your own special situation, you'll have to add up some figures and alternatives to find out. You can do this by following the procedure worked out for one western family. Let's call them the Wrights.

Anyone using these figures will have to adapt them to their own tax bracket, current interest rates in the area, and housing level. For instance, although the Wrights are comparing a $20,000 mobile to a $48,000 conventional house, a larger family in a more expensive area might find a need to double or triple these numbers.

The Wrights began by looking at the newest mobile models, averaging 12 feet wide by 60 feet long—pretty big, but much smaller than the houses formed by putting two or even three basic units side by side. They chose a "doublewide" costing about $20,000. (Half the units sold in California were "doublewides," they learned.)

This unit could be purchased for 20 percent down and payments spread over 7 years at 10 percent interest. Their payments would be $265.62 a month, according to a standard loan table. Since there are 84 monthly payments in 7 years, and $16,000 divided by 84 is $190.48, they could find their average monthly interest payment by subtracting $190.48 from $265.62. Rent for space in the trailer park they favored would be $120 a month for the unit in question.

From residents in the park, they learned they could expect to pay a modest $400 a year vehicle and personal-property tax. There would be no ordinary county or city property tax, of course, except to the extent that one of the things their site rental was calculated to cover was a portion of the property tax paid by the owner of the land on which the trailer park sat.

They were also able to make some reasonably good estimates of maintenance, heating and other utility bills and insurance costs by talking to owners of units similar to the one they were considering.

Interest and property taxes are deductible for income-tax purposes. But this is helpful only to those who can't advantageously use the standard deduction. This being true for them, the Wrights added these together ($75.14 plus $33.34 is $108.48). They estimated 25 percent of this amount as their average monthly tax saving from purchase and ownership of the mobile home. (Actual saving would be more than this in early years, when most of each monthly payment would be for interest, and less in the later years of the loan.)

Then they made a similar list of costs for owning a house comparable in living qualities to the mobile home. They did this for the 7-year period in which the mobile would be paid off. Although it was only slightly larger than the mobile, the selling price of this house was more than twice as much, because it included land and involved a construction cost per square foot about double that of the factory-built mobile. Its price came to $48,000, of which they estimated $8,000 represented value of the building site.

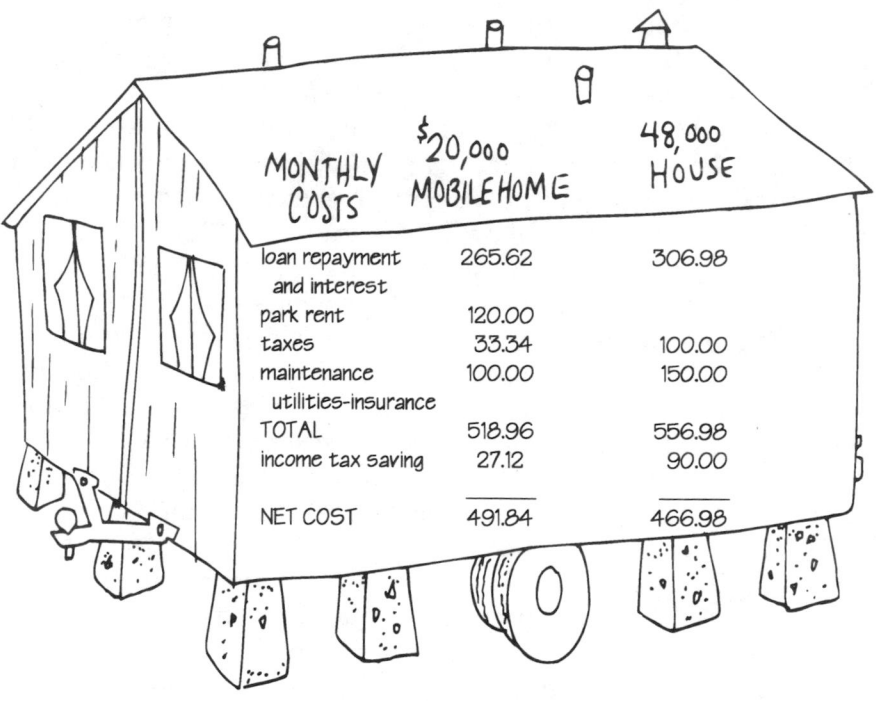

MONTHLY COSTS	$20,000 MOBILE HOME	48,000 HOUSE
loan repayment and interest	265.62	306.98
park rent	120.00	
taxes	33.34	100.00
maintenance utilities-insurance	100.00	150.00
TOTAL	518.96	556.98
income tax saving	27.12	90.00
NET COST	491.84	466.98

Beginning with the same amount of cash, $4,000—to keep costs as comparable as possible—they would require a mortgage of nearly $45,000 for a total of $49,000. The extra $1000 was for some closing costs not involved in a mobile-home purchase, and furnishings they would need in a new house but not in a largely prefurnished mobile.

At 7-1/4 percent, a 30-year $45,000 mortgage requires monthly payments of $306.98. Since very little goes toward principal in the early years, they estimated that about $260 of this would be for interest, on the average, during those initial 7 years.

Where as owners of a mobile home they would have monthly site rent, as house buyers they would have none. But in a typical town or city their taxes as homeowners would be much higher; in this instance about $1200, as they learned from owners of a house similar to the one they were contemplating.

For maintenance-utilities-insurance, they estimated an average monthly cost of $150 compared with $100 for the mobile home.

They added their estimated monthly property-tax cost ($100) to the figure for mortgage interest ($260) to find they'd have $360 a month to deduct from income for tax purposes—a net saving of about $90 a month for people in their bracket.

With the same initial expense, the comparison for monthly costs worked out as shown in the table.

The first thing these figures show is that for conventional housing of comparable size, monthly costs would, in this instance, be pretty much the same. Although the conventional house came out slightly ahead in the tabulation, this was true only because of the greater income tax saving. In any year in which it was no more costly to take the standard deduction in filing their federal return, this family would run somewhat ahead with a mobile home.

Like all taxpayers, the Wrights could either list and deduct from taxable income such things as mortgage interest and property taxes paid and charitable contributions and possibly some medical costs— or they could deduct a flat amount based on a percentage of their income up to a maximum. Deductible expenses produce a saving only to the extent that all of them together add up to more than the standard deduction would be.

They also noted that by dropping to a smaller mobile home, they could reduce their net cost for most items about 60 percent—the natural consequence of buying a $12,000 unit instead of a $20,000 one. This could cut their outlay per month to as little as $340. But with single-family houses, it might be hard to find anything that much smaller and cheaper and in an acceptable neighborhood.

Being forward-looking people, the Wright family considered the position in which they would find themselves after 7 years. If they bought a mobile home it would then be fully paid for, and their expenses would drop by $246.84 a month ($265.62 loan payments ended, but $18.78 tax saving lost). If they had bought a house instead, their payments might go on for 23 more years.

Since that is actually the more realistic way to look at it—the average American family moves at least that often—our people finally zeroed in on the most helpful comparison of all.

Where, they asked, will we stand 7 years from now? With mobile homes depreciating 50 percent in the first 6-1/2 years, it is clear that they would be owners of a unit worth about $10,000. With the conventional house, the Wrights would have paid $3800 on the principal in 7 years. How much that equity would be worth would depend upon the real-estate market in their area. In recent years, their equity would have grown by more than $10,000.

But that's clearly a gamble based on continuing inflation. The economic aspect remains something of a toss-up. The Wrights could make their decision on the basis of what they really preferred to do, without predictable financial indiscretion. Clearly, that is the most comfortable position to be in.

Free Color TV?

SOME BANKS have been making an exciting offer. They'll give you a color television set if you'll deposit a mere $1000 and leave it with them for 55 months. Virtually no interest is paid, of course; at the end of the four years and seven months your account will be $1003.76.

Is it a good offer?

A simple way to look at the question is by asking how much you would be willing to pay today to have this amount at that future date. Use either a table or this formula that is often useful in a "today's worth" calculation. For rate of interest, use the highest rate being offered for any equally safe account in any financial institution.

Briefly, the formula says: Express the rate of interest as a decimal and divide by the number of compounding periods in a year. Add 1. Raise this to the power of the total number of dollars you will then have in the bank.

When you subtract the result from the $1000 you deposit today, you'll learn what you are truly paying for the television set.

The example given here is an offer made by a bank in Oakland, California. Using their usual rate of interest for similar accounts at the time, 7.25 percent, it turns out that what you are paying for a 19-inch Hitachi is just under $280. If comparison shopping tells you that's a bargain, so is the offer (although you'd also want to look into any other stipulations attached, such as what happens if you need to withdraw any part of the money before those 55 months are up).

By the way, since California has a sales tax, there is in this instance, another—and simpler—way to check what you're really paying. The Oakland bank says it must collect $18.20 sales tax. Knowing that the sales tax rate is 6-1/2 percent, you cannily proceed to divide $18.20 by 0.065 to get—surprise!—$280.

Evidently, your banker knows how to do interest-bearing arithmetic—and so must you.

Cork in the Eye

Having myself known two celebrants who suffered from champagne even before drinking it, I was pleased to come upon a valuable calculation offered by Dr. Richard Gordon of "Doctor in the House" fame.

"In swinging 1965, when Britons drank 5,181,185 bottles of champagne, eight Londoners between 24 and 72 years of age were taken to Moorfields Eye Hospital with injuries from the corks. Seven were serious enough for admission, three developed cataracts. Four of the eight patients were waiters. The relevant physics should be in the mind of anyone broaching a bottle of Cliquot.

"Chilled champagne (47 degrees F) has an intra-bottle pressure of 90 pounds per square inch, rising at room temperature (65 degrees F) and with shaking about. This can shoot the cork 40 feet (a bottle of claret needs 300 pounds of force on the corkscrew to open it)."

Dr. Gordon goes on to a calculation that the cork from an ill-aimed bottle "strikes the eye with a velocity of 45 feet per second. It arrives in half the time of a blink, with a force of 100 atmospheres. This is comparable with blast injuries to the eye in mines and quarries.

"To avoid ruining the party, champagne should be opened with a linen napkin over its mouth from the first breaking of the golden foil, the cork eased out pointing at someone else."

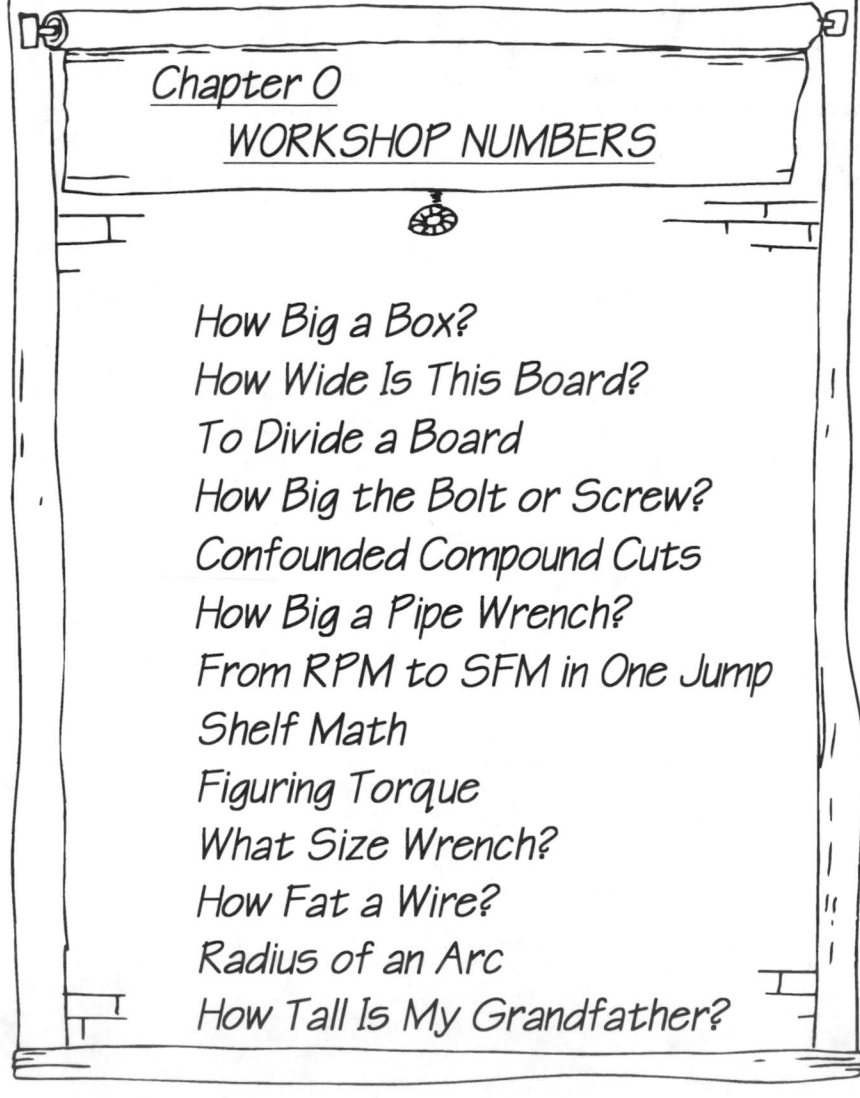

Chapter 0
   WORKSHOP NUMBERS

How Big a Box?
How Wide Is This Board?
To Divide a Board
How Big the Bolt or Screw?
Confounded Compound Cuts
How Big a Pipe Wrench?
From RPM to SFM in One Jump
Shelf Math
Figuring Torque
What Size Wrench?
How Fat a Wire?
Radius of an Arc
How Tall Is My Grandfather?

How Big a Box?

You HAVE a sheet of tin. You want to cut the corners, removing the four little squares or folding them over, so you can fold the sheet into a box open on the top. You want to construct a box with the largest volume you can get from a square sheet.

How much should you cut out?

It's not an unusual problem, actually. You might be making little drawers or open boxes to store bolts or other metal parts. The solution, however, calls for elaborate graphing or the use of calculus—a page full of equations.

You may be, as I was when the question first arose, happier just to accept the solution ready made. Which is that the little squares you cut out should have sides just one sixth the length of the side of the original square.

Maybe that's not clear. Well, given a sheet of tin a foot each way, the biggest box (the one with the most cubic inches of capacity) that you can fold from it is one that is 2 inches high and has a base 8 inches square.

That works out so neatly you might have hit it by guess. But note that with a sheet of length not evenly divisible by 6, things become less obvious. The height that will give you the maximum volume with a 10-inch square is one-sixth of 10, or approximately 1.67 inches.

How Wide Is This Board?

CALLING a hunk of wood an inch and a half thick and three and a half inches wide a two-by-four seems faintly fraudulent, but that's the way it is.

The idea, of course, is that this is what's left when you start with something two inches thick and four inches wide and smooth it up.

It's true that when you buy rough lumber, you often get the dimensions you ask for, though not precisely.

Anyway, you frequently need to know true dimensions when building or when making calculations about strength of beams.

So it is useful to memorize the simple rule that lumber called 2 to 6 inches thick or wide dresses down to half an inch less than that.

From nominal 8 inches on, the amount you pay for but don't get is three-quarters of an inch. These figures go for seasoned—dried—lumber.

If you buy your lumber unseasoned—green—it can be expected to shrink to these dimensions. But while it is green, it will measure a bit more. The 2-, 3-, and 4-inch dimensions will be shorted by 7/16 of an inch. A nominal 6-inch dimension will be off by 3/8 of an inch, making it really 5-5/8 inches. Anything beyond that will be off by half an inch.

To Divide a Board

Given a plank that you wish to cut into two (or three or any other number) equal lengths, how do you measure it if you're stuck somewhere without any measuring device?

A piece of string as long as the plank (or the garden plot or whatever you're dealing with) will do fine—if you've got a piece of string.

If you have no string and if the plank seems consistent in weight and is light enough to lift, just balance it on a finger and cut at the point of balance.

If you're dealing with a heavy timber or a section of garden or anything else you can't lift, you'll have to find another way. A useful way is one called successive approximations and the beauty of it is it will work for quite a number of situations. It will serve to divide a length or distance into three or more parts just about as easily as into two. It will even solve practical mathematical problems, as you will see elsewhere in this volume in material dealing with finding square roots.

Use the principle this way.

To mark a long garden into three equal plots, break off a stick to a length that you can see is at least one third the length of the garden. Lay it down three times and then pick it up and break off and throw away what looks to be not quite one third of the part that stuck out beyond the length of the garden. Repeat the process with your shortened stick. Go on repeating as many times as seems necessary.

After a few trials, you'll have that garden marked off into precise thirds, right down to a gnat's eyelash.

How Big the Bolt or Screw?

KNOWING the number size of a bolt or a wood screw you can easily determine its diameter without chasing down a chart or handbook.

This is useful when you have to drill a hole for the screw or bolt, using a bit for which the diameter is given as a decimal or fraction of an inch.

The routine called for is a trifle weird but not complicated. Multiply the size number of the screw or bolt by 0.013 and add 0.060. Don't ask me why.

Anyway, you'll find that a No. 10 wood screw has a diameter of 0.190 inches (10 times 0.013 plus 0.060 is 0.19).

Confounded Compound Cuts

HAVE YOU ever tried to make one of those popular planter boxes with many tapered sides?

If you have, you've encountered the intricacies of compound-miter cuts and perhaps discovered how hard they are to work out by trial and error.

Actually, setting a circular saw to cut tricky compound miters for such home-workshop projects can be a cinch.

So says R. F. Gyger, Houma, La., who burrowed through a fairish bit of trigonometry to supply this straightforward routine, which he has sent to me to share with the world. Or with such part of it as hankers to cut compound miters but was afraid to ask how.

Decide first how many sides your box is to have and how many degrees from vertical you'd like the sides to slope. You might, for example, choose to build a hexagonal planter with a side slope of 10 degrees.

The geometric calculations that follow can be handled nicely with paper and pencil—and even more nicely with a simple hand-calculator used in the presence of a table of trig functions.

But there's no denying that one of the basic scientific calculators makes it all faster. And, to be honest about it, a good bit more fun.

So:

Divide 180 by the number of sides. Find the tangent of this, then multiply by the sine of the angle of slope. The arc tangent of this is the setting for your saw arm or miter. For the hex planter with 10-degree slope just mentioned, this means multiplying tan 30 by sin 10 and then getting the arc tan, which turns out to be 5.7 degrees.

After again dividing 180 by the number of sides, this time multiply the sine of this result by the cosine of the slope.

The arc sine of this result gives the tilt angle, which proves to be 29.5 degrees.

And there you are with the two saw settings you need for your power saw.

How Big a Pipe Wrench?

Since any size pipe wrench will work on any pipe within reason, you might think size doesn't make any difference.

But a wrench that's too small won't do the work, at least not without excessive effort. And—this is important—one that's too big will multiply your muscle power so effectively you may over-tighten a connection. Besides making it excessively difficult to undo later on, this overpowering may split a fitting. If you discover this you'll have the job to do over again, possibly after a trip to the plumbing-supply shop. If you don't, you could end up with a concealed leak—of water, which is bad, or gas, which is worse.

Here are recommended sizes:

10-inch wrench for pipe up to 1-inch diameter.

18-inch wrench for pipe from 1-inch to 2-inch diameter.

24-inch wrench for pipe 3-inch diameter or over.

Or use a chain wrench for pipe over 3-inch or of irregular shape.

(Note: a midget wrench 6 to 8 inches long is a good auxiliary for working with pipe under 1-inch when space is short. Note also that for taking apart old pipe that is frozen with corrosion, there's no law against using a bigger wrench than the table indicates. If that's not adequate, or you don't have a big wrench, try slipping a length of pipe or tubing over the handle to extend it and increase your leverage. Your strength will be as the strength of ten, even if your heart isn't pure.)

From RPM to SFM in One Jump

It's quite common under workshop conditions to know the speed of a tool in revolutions per minute—and need to find what the surface speed is.

The surface speed is frequently expressed in surface feet per minute. Multiply the known rpm by the diameter of the pulley in inches and divide by 4 to get a quick approximation of the sfm. What you are doing actually is multiplying the diameter of the tool or work by pi to find the circumference and dividing by 12 to change inches to feet.

And that is indeed the routine to follow if you need a precise answer. Which will prove to be not quite 5 percent more than your quick approximation.

Shelf Math

ABOUT THE ONLY sensible way to determine the depth (distance from front to back) of a shelf or set of shelves is by first measuring whatever is to be stored. You may be surprised to discover that shelves only six inches deep will do nicely for most books. You can save a lot of lumber and some space by making them only that deep—if you also provide one or more deeper shelves for outsize books.

The only other rule applying specifically to book shelves is that they should be far enough apart to allow 1-1/2 inches above a typical book for finger room.

More important—and harder to determine by logic or guess—is how long a span to allow either for book or general storage. The problem is to avoid sagging.

Research at the University of Illinois has produced some useful rules of thumb for this:

When using 3/8-inch plywood, limit the spans to 26 inches for normal loads such as toys, clothing, light dishes. For books and records and other heavy loads, keep spans under 16 inches to avoid sagging.

But 3/4-inch plywood will usually handle normal loads on 46-inch spans, heavy loads on 29-inch.

If you're using solid wood 3/4 of an inch thick (nominal 1-inch lumber, that is) you can stretch these spans slightly. Figure on spans to 48 inches for normal loads, 30 inches for heavy ones.

Thicker lumber usually makes much better looking shelves as well as stouter ones. When it's 1-1/8 inches thick, you can increase spans for shelves to bear normal loads to 66 inches; 42 inches for heavy loads.

For a major shelving project, or for shelves that are to bear unusual loads, you may wish to be more scientific. Use the table for acceptable loading for shelves of any width made of the usual thicknesses of lumber (nominal 1-inch and 2-inch) or of plywood (3/8-inch, 1/2-inch, 5/8-inch, 3/4-inch).

The figures in pounds are for the loads, in pounds per square foot, that the shelf would support without sagging more than 1/240 of the span, which works out to 1/20 inch for each foot of length.

FIGURING SHELVES

Distance Between Supports	Pine and Cedar Lumber		Douglas Fir and Larch Lumber		Douglas Fir Plywood			
	$3/4$"	$1\,1/2$"	$3/4$"	$1\,1/2$"	$3/8$"	$1/2$"	$5/8$"	$3/4$"
12"	670 lb.	2,685 lb.	1,134 lb.	4,537 lb.	163 lb.	330 lb.	593 lb.	817 lb.
16"	377 lb.	1,510 lb.	638 lb.	2,552 lb.	68 lb.	138 lb.	249 lb.	374 lb.
20"	222 lb.	967 lb.	356 lb.	1,633 lb.	35 lb.	71 lb.	127 lb.	191 lb.
24"	129 lb.	667 lb.	206 lb.	1,126 lb.	20 lb.	41 lb.	75 lb.	111 lb.
32"	54 lb.	378 lb.	87 lb.	638 lb.	9 lb.	17 lb.	30 lb.	46 lb.
36"	38 lb.	306 lb.	61 lb.	489 lb.	6 lb.	12 lb.	22 lb.	33 lb.
40"	28 lb.	223 lb.	44 lb.	356 lb.	4 lb.	9 lb.	16 lb.	24 lb.
48"	16 lb.	130 lb.	26 lb.	206 lb.	2 lb.	5 lb.	10 lb.	14 lb.
60"	8 lb.	66 lb.	13 lb.	106 lb.	–	–	–	7 lb.
72"	5 lb.	38 lb.	8 lb.	61 lb.	–	–	–	–

The table tells you, for example, what to expect if you use cedar boards 3/4 of an inch thick, supported at 32-inch intervals. Without objectionable sag, they'll hold 54 pounds per square foot. So if each shelf is to have three spans, for a total length of 8 feet, and to be 2 feet wide, the calculation is 8 times 2 times 54. Evenly distributed, 864 pounds should be an acceptable load for the whole shelf.

Figuring Torque

WHEN YOU get into many mechanical things, even ordinary car repairs, you are likely to be told how energetically to do such things as tighten nuts.

An extreme example is a flywheel gland nut on some Volkswagens. Torque it to 220 foot pounds, they tell you.

"They" think you have a torque wrench that measures how hard you twist. But what if you don't?

Estimate by understanding what foot pounds of torque really is all about. This is just the number of pounds of pressure you're putting on, multiplied by the leverage.

Suppose the point at which your hand is exerting pressure on a wrench (or on the pipe sleeve you've slipped over the wrench to extend its handle) is 18 inches from the center of the nut. You're putting all your 150 pounds behind the push. Multiply the leverage distance in feet (18 inches equals 1-1/2 feet) by your 150 pounds of pull and you get 225. Which is almost exactly what that VW nut needs.

Though by no means as accurate as a torque wrench, this calculation at least brings you to a well-educated estimate.

If you'd rather be metric about it, the term is meter kilograms. It takes about 7.2 foot pounds to make one of them, so that gland nut has to be torqued to 30 mkg.

What Size Wrench?

IF YOU MUST choose a wrench to fit a nut or bolt that's not in the same place as your wrenches, you may be glad to know that there's a fairly reliable and consistent pattern to sizes.

And happily it's a relationship that's not hard to remember.

Multiply the diameter of the bolt—its thread diameter, that is—by one and one-half. For a half-inch bolt you'll want a three-quarter-inch wrench.

How Fat a Wire?

IF THAT MYSTIFYING unidentified copper electrical wire that you come upon in the course of household wiring is the thickness of a penny, it is No. 14. This is the size often used in wiring a household circuit for lighting only, and its capacity is limited to 15 amps.

Thickness equal to that of a nickel will identify No. 12 wire for you. This is the size required for appliance circuits so that they can safely be fused at 20 amps; and it is often used for 15-amp circuits as well, just to avoid having to fool with an extra size during wiring.

If the wire is as thick as two dimes, it's No. 10. Since this size can be fused to carry 30 amps, it is useful for single-purpose, fairly heavy-duty circuits on either 110-120 or 220-240 volts.

Radius of an Arc

AN EXPERIENCED woodworker says he understands there are various ways to calculate the radius of an arc. But this is the one he finds simplest.

First, draw a line connecting the ends of the arc.

(In doing this, if you place your yardstick or whatever you are measuring with so that this distance is an even number, your arithmetic will be simplified.)

Measure to the center of this line. Call the distance A.

At a right angle to the line at this point, measure the distance to the arc. Call that distance B.

Square A and square B and add those results together.

Divide by two times B.

You can now reproduce the circle whose arc you began with, because you've found its radius.

How Tall Is My Grandfather?

GRANDFATHER clock, that is.

When building a clock of the pendulum variety, one thing you'll need to know is precisely how long that pendulum must be.

You can find out from a rather simple formula. Square the period (time per swing) of the pendulum. Multiply by the acceleration due to gravity. Divide by 4 times the square of pi.

That's not so hard. You may recall pi as 3.1416 or, have it on your pocket calculator. For gravity, use 32 feet (or for a metric answer, 975 centimeters) per second per second.

If you want greater precision you'll have to find someone to tell you the figure for your own little portion of the earth. It does vary just a trifle from place to place.

So for the length of a pendulum with a period of 2 seconds, square 2, multiply by 32, divide by 4 and by the square of pi.

If you don't have a pi-bearing calculator at hand, you can save time by using a figure I've just pulled out of mine. To find the length of the pendulum in feet, multiply the square of the time period by 0.81057 (or, for most purposes, just by 0.81). For the length in centimeters, use 24.7 instead.

The formula behind this works in reverse, naturally. So you can calculate the period (swing time) of the pendulum from its length.

Just divide the length by that acceleration constant, take the square root, and multiply by 2 times pi.

Quick example: The square root of 8/32 is 1/2, which multiplied by twice pi has to be just plain pi. So an 8-foot pendulum will have a period of 3.1416 seconds.

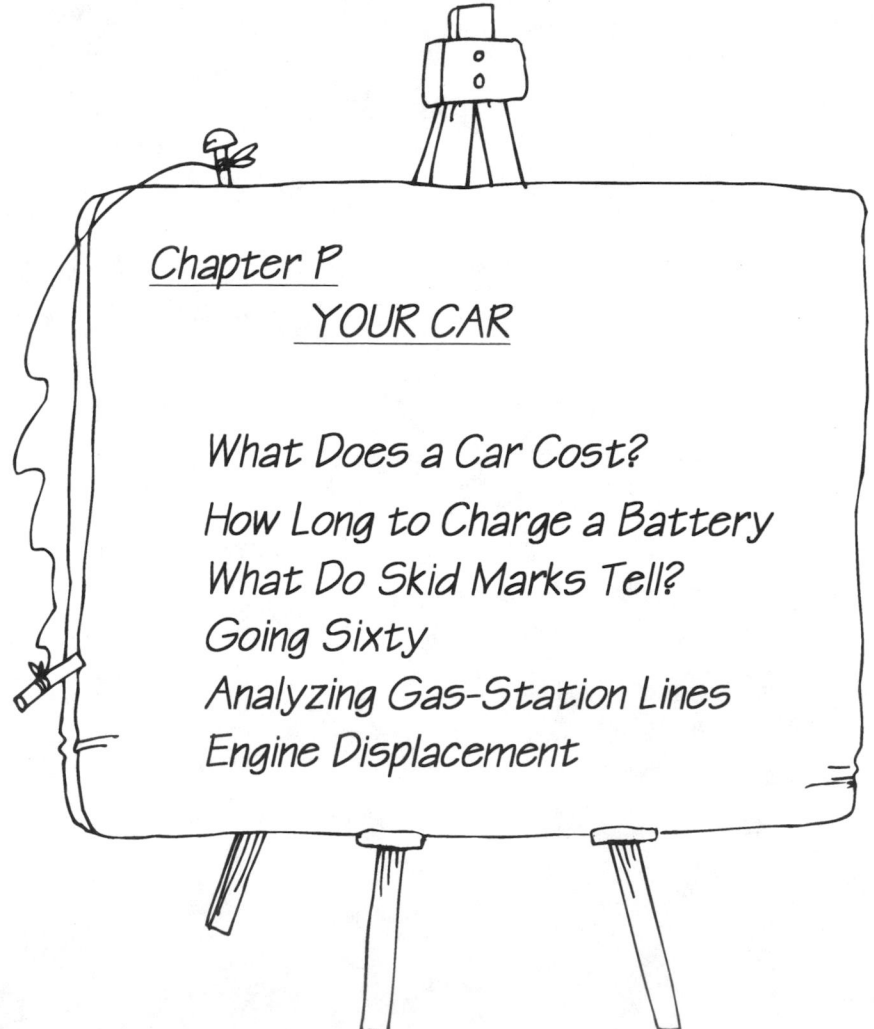

Chapter P
   YOUR CAR

What Does a Car Cost?

How Long to Charge a Battery

What Do Skid Marks Tell?

Going Sixty

Analyzing Gas-Station Lines

Engine Displacement

What Does a Car Cost?

Wʜᴀᴛ's ᴛʜᴇ ʙɪɢɢᴇsᴛ item of expense in automobile ownership? That's right—for most of us it's depreciation.

What's second? Gas? Insurance? Repairs?

Actually, for most owners it's none of the above. It's the cost of money. This refers not only to the interest you and I pay but also the hidden interest we lose on the hard-earned stuff that's tied up in our cars.

Surprisingly, this robust cost figure is missing not only from most published material—but even from that offered by statistical organizations and by the federal Department of Transportation. It's a neglect that harks back to those happy days when interest costs were half what they've become and fully deductible to boot.

All of which makes interest a good place to start in adding up what it costs you to operate your car for a year.

Interest. Financially speaking, your car is divided into two parts. There's the part that the bank or other loan agency owns; and there's your equity, the part you own.

To find the cost of the first, add together the sums owing on the car at the beginning of the year and the end of the year and divide by two. This is the average amount you pay interest on, so multiply it by the APR or average percentage rate in your contract.

Next add the market value of your car (what you could reasonably expect it to fetch if put up for sale) at the beginning and at the end of the year in question and divide by two. From this, subtract the average amount of the bank's equity, found in the previous paragraph, to determine your average equity. Multiply this by the rate of return your money could be earning if not tied up in a car.

To find your total interest cost, add this interest you lose to the amount you pay.

Depreciation. This is both a biggie and a toughie. It varies enormously from one make of car to another. So the loose generalizations tossed off for estimating this are to the real world as one-size-fits-all shirts are to actual people.

However, if yours is a car of average price and popularity and you've paid full sticker for it, these approximations may prove close enough to be of some use: 35 percent of new price the first year, then

17, 13, 11, and 8. After the fifth year how well you've maintained the car will count far more than age.

Another fairly quick way to arrive at a meaningful figure for depreciation is to browse classified ads and compare cars like yours with similar ones a year earlier and a year later.

However, there's another tactic I've found much quicker and less tedious than perusing used-car ads. By consulting an automobile blue book, you can arrive at a quite reliable estimate. Bank loan officers and many libraries have these compact and detailed guides. By consulting two issues dated a year apart, you can find out how much value a specific model has lost in a year.

Since it's not always easy to get hold of two issues a year apart, I've worked out another way to produce an excellent estimate of depreciation for the year gone by or the year to come.

All it requires is one up-to-date blue book.

At one time, a current blue-book figure told me that my daughter's aging Civic station wagon should retail at $5445. A similar car newer by a year, also a 4-door with average mileage and standard equipment, was valued at $6465. So a quick subtraction suggested depreciation of $1020 for the last 12 months.

Similarly, when I looked up a Honda of my own that was three years older, I found its retail value given as $4255, which was $510 less than the figure for an equivalent model one year newer. This confirmed my suspicion that an average 6-year-old car tends to run up depreciation only about half as fast as a 3-year-old.

To estimate depreciation for the year to come, I could subtract the book value for the previous year's model from the current figure for my car.

The blue book will also tell you how to refine your figures by adding or subtracting for miles racked up, major accessories, and so on. But if you can keep these elements consistent within the group of cars, you can safely ignore them.

Blue books give two values for each car—low and high. Either will serve for making comparisons, but an average of the low and high—wholesale and retail—will be especially useful. It will be close to what the car might go for in a private deal. The main thing is to consider what you're going to use the result for—and then be consistent.

Gas and oil. This one's easier. Even without records you may know about how often you put in a fairly consistent number of gallons and how much you pay for each on the average. You may do better by making use of the odometer. Estimate yearly gallons from what that instrument says divided by the miles-per-gallon figure you've probably been keeping track of—or occasionally sampling— all along.

Insurance. Ay, there's the rub—for some of us anyway. Only when I began totting up figures did I fully realize what a major chunk of my driving expense insurance, especially liability, had become in recent years.

The cost of keeping a car conservatively insured can be right up there with interest and depreciation. For a young urban bachelor, it can top all other costs, especially if he drives something old (so that depreciation is negligible), high-powered (scares insurers to death), and sporty (which they don't especially admire either).

At least it's easy to predict the premiums. Just look at your bills or check stubs. And it's not hard to get quotations on a car you're planning to add.

Remember, though, that if you've got more than one car on a policy, taking one off may affect the rate on any other.

You could find, as I did, that the insurance premium is Cost No. 1 on an old, low-depreciation car driven sparingly. This may well be the primary motive for getting rid of a second or third car that is not really needed.

Maintenance and fees. Since these costs, like those for insurance, tend to stay high however little you drive, this is as good a place as any to list and add in the registration fees and the charges for any required periodic safety checks.

Your maintenance budget, including tires and cosmetic as well as mechanical care, might start at a couple of hundred dollars or less the first year and rise to three or four times that by the third or fourth.

Very low mileage drivers will spend less, of course. If you're one of those, you'll make a better guess by suspecting somewhere between five and fifteen cents a mile, depending mostly upon size and age of the vehicle, who made it—and how you drive it.

Your own habits, experience, and any records you've kept will tell you the most about where your costs fit into the all-purpose estimate.

Garage, parking, tolls. A nickel or dime a mile is not an unusual figure for this trio.

However, for a low-mileage driver who incurs two or all three of these costs regularly, the total may be several times that. And for the many drivers who seldom encounter any of them, it may be too small to reckon.

Adding up parking and tolls is not hard, but drivers who do not keep track often seriously underestimate this category. Don't overlook fines.

As for garage, if your car is kept in your own space on which you pay rent or taxes whether there's a car in it or not, you may have no outlay here to charge against your car expenses. But if in the absence of a car the garage could and would become a rentable storage space, there's a cost worth counting. And that's equally true if in the absence of a car the garage space could be turned into a workshop for which you would willingly pay $50 to $100 or so a month.

That bottom line. Realistic analysis of your car costs, both visible and hidden, can lead to profitable decisions.

You may find that you can easily part with a car that your figures prove is costing you more than you thought—and more than it's worth.

You may happily discover that trading up to something far more comfortable or more fun than what you're driving will cost you little more in the long run. This can most easily happen when you move to a model notable for its low maintenance or low depreciation.

If you've been considering any of the touted leasing plans you'll now have sound comparative data. You can save quite a bit by not erroneously believing you're in one of the comparatively rare situations where leasing is thrifty. (More about this toward the end of this chapter.)

If you're putting business miles on a car, you'll now know whether to use actual costs for income-tax purposes. The alternative, of course, is to go with the per-mile figure the government allows.

As car prices and operating costs creep upward and some insurance premiums double in a single year—you have the best of times and the best of reasons for calculating precisely where you stand when you sit behind the wheel.

How Long to Charge a Battery

THERE's a simple formula for estimating how long it will take to charge a storage battery in your car and other equipment.

First you read the battery's capacity in ampere hours—on the battery. For most batteries it's between 30 and 130.

Then you divide that by the rating of your charger—3 to 10 amperes with most inexpensive home-type chargers.

The answer is how long it should take to charge the battery after it is run down. As you can see, this will take anywhere from about three hours to a couple of days.

Another use for this calculation is in choosing a charger. If you're likely to be charging a battery of 100 ampere hours or so, and you'd like to be able to do it overnight, you're going to prefer a charger rated at 8 to 10 amperes.

(Incidental intelligence: Since a typical battery will require about 1.2 kilowatt hours of power to charge it, the cost per charging operation will be around a nickel or a dime at most utility rates.)

What Do Skid Marks Tell?

NEXT TO the shriek of brakes, there's no more disturbing reflection of highway disaster than those long black stripes of rubber you see on the pavement. And like a sentence printed in an unfamiliar language, when translated they have a story to tell.

They do a pretty good job of revealing how fast a car was going when its driver hit his brakes.

To interpret those cryptic skid marks, you use a physical law that says that braking distance varies as the square of the velocity. It takes a potentially disastrous 9 times as long a distance to stop a car going 90 miles an hour as one going 30.

In cases of major importance, law officers may use elaborate methods for converting skid-mark measurements into estimated speeds. These may include measuring the actual coefficient of friction of the road surface. They may take into account special circumstances when the friction was not the same for all wheels, as when one side of the car skidded along a muddy shoulder.

But generally the calculation will depend on coefficients established by measurements on typical surfaces. Here are some figures in miles per hour for skid marks 10 feet long.

Ice, 6
Packed snow, 8
Wet oiled gravel or wet brick or concrete, 12
Wet asphalt or wet or dry gravel, 13
Dry brick, concrete, asphalt, or oiled gravel, 15.

To apply the table, divide the measured length of the skid mark by 10 and find the square root. Multiply this by the figure listed for the pavement and its condition.

For an 88-foot mark on wet asphalt, divide 88 by 10 and then find the square root of 8.8, which is 2.97. Multiplying by 13 gives 38.56. In recognition that all the assumptions and measurements you're working with are approximations, it would be sensible to round this off to 39.

If you'd taken the quite reasonable course of rounding 88 feet to 90 in the first place, 39 mph is the result you'd have obtained anyway, in a matter of seconds.

Going Sixty

Remembering that 60 miles an hour is 88 feet a second makes various mental calculations about driving easy.

At 55 miles per hour, the figure becomes an even simpler 80 feet a second (80-2/3 to be precise).

It may be salutary sometimes to contemplate at 55 how much closer than 80 feet you are to a car or anything else ahead—and how brief a second is for responding to an emergency.

Analyzing Gas-Station Lines

WHY IS IT that at certain periods, as in 1974, motorists found it necessary to wait in lines a block or more long to buy gasoline?

Was it because of a shortage of gas, as most of us assumed at the time?

No. In fact, during much of the apparent crisis there was more gas than usual in customers' tanks. For fear of not being able to buy the stuff tomorrow or the next day, motorists were keeping their tanks nearly full at all times, to the tune of half a billion gallons in what might be called mobile storage.

The only real shortage was in service-station time.

Since it takes nearly as long to dispense a couple of gallons as a tankful, hysterical customers strained station time before they began to exhaust fuel supplies.

We·may be able to avert future crises, now that the manner in which all this works has been figured out by students of queuing theory. That's a branch of the mathematics of statistics and probability.

What follows is an example that can serve as a pattern for forecasting many comparable occurrences. Here is how it goes for a typical situation.

A small gas station able to serve two cars at a time is open 60 hours a week. It takes 5 minutes to serve a car, so the station can handle as many as 24 customers an hour.

In a typical region, it might be found that such a station serves 300 customers who drive, on the average, 200 miles a week in cars that go 300 miles on a tankful.

Normally such a customer would stop for gas once a week. To meet this demand, the station would have to serve only 5 customers an hour, which is almost ridiculously far below its capacity of 24. Allowing for occasional chance stack-ups and for the likelihood of certain times of day being busier than others, the probabilities still remain more than 9 out of 10 that when you drive into the station you could be served instantly.

But suppose those same customers respond to rumors of a gas famine by topping up every time their gauges show them down a couple of gallons. They stop 5 times a week instead of once and the station is now being asked to handle 25 customers an hour. There will be very long lines. In fact, this condition calculates to a 30-minute average wait with 1 chance in 6 that your time in line would be more than an hour.

And this without any gasoline shortage at all!

Now suppose occasional very slight shortages of fuel do develop. To ensure—in all fairness—that every customer gets at least something, a station operator may begin limiting each sale. Result: even customers who were not coming in every time they burned a couple of gallons will have to turn up more frequently than had been their custom, compounding the problem.

Of course, if people will respond more rationally, curb their gimmes, fill tanks only when necessary, and meet any shortfall by curbing use proportionately...then, as these queuing-theory calculations show, we'll all do just fine.

Engine Displacement

SINCE THERE are almost exactly 61 cubic inches in 1 liter, you can use that factor to make the leap between the two measuring systems.

If the displacement is given in liters, multiply by 61 to find it in cubic inches.

If the displacement is given in cubic centimeters, divide by 1000 first—that is, slide the decimal point three places to the left, converting a 2000cc figure, for example, to 2 liters—and then do the multiplication by 61.

To go the other way, from liters (or cubic centimeters divided by 1000) to cubic inches, divide by 61 instead of multiplying.

If all you want is a reasonable equivalent in a hurry, you can estimate it from this:

3000cc is about 183 cubic inches; 5000cc is about 305 cubic inches.

To put it another way, multiply the cc figure by 6 and divide by 100 (set the decimal point two places to the left).

In the other direction (turning cubic inches into cubic centimeters) you just divide by 6 and add on two zeros.

I don't own a car with a rotary engine. Very likely, you don't either. But some folks do, so bear with me.

To figure the displacement of a Wankel rotary engine, multiply 3 times the square root of 3 by all of these factors: rotor eccentricity, rotor radius, rotor width, number of rotors.

(For occasions when you're doing a Wankel calculation without a square-root-finding calculator or a computer, that square root is 1.732 and 3 times that is 5.196.)

For the Mazda RX-2, later RX-3, and early RX-7 automobiles, these factors are 1.5, 10.5, and 7 centimeters and there are 2 rotors. So it seems that all these models have 1146-cc engines.

For RX-4 and American Cosmo models, the 7 becomes 8. For the Japanese Cosmo Sport, the R-100, and the early RX-3, the 7 becomes 6. For the NSU Ro80, the factors are 1.4, 10, 6.7, with two rotors. Same for NSU Wankel Spider, but only one rotor.

If you want the displacement in cubic inches, divide the cubic-centimeter result by 16.387.

Chapter Q
TRAVEL

EASY AS FALLING OFF A LOGARITHM!

Jet Lag

T HE MAIN thing most of us need to know about jet lag is how to minimize the effect of an air trip that drags your reluctant body across several time zones in a few hours. Some doctors at the National Institute of Mental Health have suggested a kind of fool-the-eye approach. It comes down to visually simulating the light conditions at your destination from the time you get on the plane till some days after you get off.

During the hours when it's daylight where you're going, keep yourself exposed to as much light of any kind as you can where you are. Similarly, during the hours when your destination is in darkness, simulate that condition by wearing an eye shade or dark glasses when you can.

From my own experiences over the years, I'd advocate trying to start this some days before getting on that plane.

But it isn't easy.

You may be delighted to find that science and informed observation have sprung to our aid in respect to another facet of the same problem. How many days rest do you need after a long air trip before you're fit for further serious travel or business?

Something called the International Civil Aviation Organization has worked out a formula for estimating this. These folks advise you to apply it to avoid making bad business or diplomatic or personal decisions while your body is out of tune with your new surroundings.

First, divide the number of hours of travel time by 2.

Second, subtract 3 from the number of time zones crossed.

Third, find your departure-time coefficient. If you leave between 8 a.m. and noon, it's 0; noon to 6 p.m., it's 1; for 6 to 10 p.m., it's 3; for 10 p.m. to 1 a.m., it's 4; and 1 to 8 a.m., it's 5.

Fourth, find your arrival-time coefficient. If you arrive between 8 a.m. and noon, the coefficient is 4; noon to 6 p.m. it's 2; for 6 to 10 p.m., it's 0; for 10 p.m. to 1 a.m., it's 1; and for 1 to 8 a.m., it's 3.

(Some of these seem a little too arbitrary. I'd suggest that if your timing is borderline, you split the difference; arriving at 1 a.m., for instance, use 2 as the coefficient.)

Now add your four figures together. Divide by 10. The result is the recommended number of recovery days.

To see how it works, consider a flight leaving New York at 7 in the evening and reaching Rome at 9 the next morning, crossing seven time zones in seven hours. The figures you get will be 3.5, 4, 3, and 4. Their sum divided by 10 is 1.35.

Your body will need about a day and one-third to get over this flight.

And that's not even considering the possible effects of airline cuisine and the sleep you lost if you haven't bought first-class space or learned to curl up like a kitten.

Tactics for Travel Funds

On a vacation or business trip you'll probably use travelers checks to pay-as-you-go or charge cards to go-now-pay-later. Cash is too risky and the traditional bank letter-of-credit is outdated as too much trouble for most circumstances.

Between checks and cards, which is the better economy?

If you carry checks with you, you'll have paid for them in advance. When you use cards you postpone the outlay until the chits drift in days or weeks later and then get billed to you—when you still have something like a month of free interest coming before you have to pay. You can count on gaining at least a couple of months interest, which could be worth $100 to you on a $5000 trip.

Using cards may also save you the fee that is sometimes still charged—though frequently waived—for travelers checks. Against this you should reckon the annual fee for the card unless you can get it free or would be paying rent on it anyway.

There's another offset for some travelers. Since not so many of the less-expensive establishments accept the cards, wanting to use yours may persuade you into a costlier place than you'd had in mind. (Now you know why pricey hotels and restaurants are willing to split their take with card issuers. On the other hand, a few of the tonier establishments, notably fine restaurants, can get all the customers they can handle without offering plastic bait, so they don't.)

For foreign travel, another factor enters—rate of exchange. It was my clear impression during the 1970s that the charge-card people generally used less favorable rates than I could have obtained by exchanging cash or travelers checks as I went. But a shift seems to have occurred by the middle 1980s and my careful records show that exchanges in more recent years have been consistently in my favor.

The last I knew, both MasterCard and Visa were using a standard international rate, but where Master was adding a 1 percent charge to it when figuring exchange, Visa was taking only 1/4 of 1 percent. If you want to save a bit by bargain hunting even before your foreign trek begins, ask your card suppliers about their policies in this respect.

The difference in rates between card and check will be even more likely to favor use of the card if the alternative is cashing a check in a hotel or restaurant. When the merchant sets the rate he is likely to round it off pretty roughly and not in your favor, either.

(How much this can hurt I found out one rainy night in a Shell station on the German Autobahn. With my gas running low I was forced to accept the offered rate for my dollars even though it was some 10 percent worse than I'd been getting at banks. I wouldn't have minded that percentage levied on the cost of the gas, but it applied to the whole $100 travelers check that unfortunately was the only kind I had left.)

When exchange rates are fluctuating, the card user won't know how much he's spending till the bill comes months later. The exchange rate is not set at the time of the purchase but days or even weeks later when the ticket turns up at the bank. You could win. You could lose.

To discover, when you receive your bill, how much you've lost or gained by using the charge card, you might try this sequence I've used to check a bill of mine.

At the time and at my bank, anyway, MasterCard (unlike Visa) bills did not show the rate of exchange being used. Fortunately, I'd retained the original charge slip, which came to 860 Portuguese escudos. I had only to divide this figure into the $20.48 that MasterCard was billing me for to discover the exchange rate used: 0.0238.

My trip diary showed that, on approximately the day of the charge, I was getting a rate of 0.0225 at Portuguese banks—a difference of 0.0013. Dividing this difference by the 0.0225 rate told me that I was paying a penalty of 0.0578—or 5.78 percent. In return for paying by card I'd had use of the money for a couple of months. That's one-sixth of a year, so I multiplied the penalty by 6 to discover I was being assessed interest at an annual rate of about 35 percent. Too much.

This experience, you'll note, was contrary to most recent ones. This time, I'd lost instead of gained by using the card.

If you snobbishly distinguish between tourists and travelers, you'd have to say that all this is becoming less important, at least for the former. The big cities and other highly developed places, where tourists throng, commonly offer access to those remarkable machines that read your bank card and thrust cash into your hand. And, amazingly, do it at lower cost to you.

Unless you have evidence to the contrary, you probably should give the nod to the iron teller—if you can find one where and when you need it. Even dedicated travelers, those who disdain the beaten paths, are now coming upon them even in many unlikely places.

What Does Devaluation Cost?

You could be traveling, or importing, or investing. How can you figure what happens to you when there's a change in the value of one or the other of two currencies you're dealing with?

Let's say you were in Germany when the Deutsche mark was revalued upward 10 percent. That's simple enough. Everything you bought in Germany would still take the same number of marks—but you'd have to pay precisely 10 percent more dollars for each mark you bought.

Now suppose you were in New Zealand (as I was) on the day in 1973 when the U.S. dollar was devalued by a flat 10 percent. You might suppose that your tour down under had just become 10 percent more expensive, and indeed that's what most newspapers and bankers told people.

But they were wrong. It was worse than that. From that moment on, the New Zealand dollar began to cost Americans 11.1 percent more. And so it followed that costs for people traveling with American dollars instantly increased by that 11.1 percent.

How come? Well, from that point on, the tourist was using a 90-cent dollar. Add 10 percent of 90 to that and you're up to 99. It takes 11.1-plus percent more than 90 cents to make 100.

If that's a little hard to swallow—and it is—try an exaggerated case.

When your currency drops to half its former value you will need twice as much of it to do the same job. So a devaluation of 50 percent (half) means a cost increase of 100 percent (double). For calculations, you'll need to figure out how much more the old dollar was worth than the new one—instead of the other way around. When devaluation is 10 percent the calculation you want is this: What percentage of 90 is 10?

That calls for the usual formula—multiply 10 by 100 and divide by 9.

Which gives 111.1 percent, or 11.1 percent more than before.

Global Distances

$\mathbf{Y}$OU CAN FIND the distance between any two points on the earth's surface by using either of two methods—bearing-and-range or rectangular-coordinates. Either way, the procedure is unavoidably pretty complex.

You won't need either of those methods, and you might prefer to avoid them, when the two points happen to lie on the same meridian. Here's one for that situation. It involves no tools more difficult or unfamiliar than the value of pi.

If pi is not built into your calculator, you'll probably recall that it is commonly expressed as 3.1416. You can easily type it in.

If both of your points lie in the same direction, subtract one from the other to find the number of degrees they are apart. If they're in different directions—one north of the equator and one south, for instance—add them together. Change minutes and seconds to decimal degrees.

Multiply this distance in degrees by 7917.6 and by pi and divide by 360. For example, if you've started with points 6 degrees and 30 minutes north and 18 degrees south of the equator on the same meridian, you'll find they're just over 1692 miles apart.

If, by the way, you don't recognize that odd number, 7917.6, it's the diameter of the earth in miles.

The Tricky Average Speed

IF YOU DRIVE from here to anywhere at a steady 20 miles an hour and return at an equally steady 30, what's your average speed?

Sounds simple, but the problem's difficult enough to have found its way into college entrance examinations. As you might guess from this, the answer is not 25. That's the average of the speeds, true enough, but it's not the average speed.

Since you have spent more of the trip driving 20 than 30, you must find a way to give the slower speed more weight in computing the average.

How it works is easier to understand if you assign some arbitrary distance to the trip.

It's neatest if you make it a distance that works easily with the figures you have.

So call the one-way distance 60 miles.

Then you must have driven at 20 mph for 3 hours and at 30 mph for 2 hours. It took you 5 hours to go 120 miles. That's 24 mph.

Since there are many work problems ("If John can build a fence in 5 hours and Bill in 3, how long will it take the two of them together?") and pump problems ("If one pump, or pipe, can fill a tank in 40 minutes and another...") that follow the same principle, it may be worth while to look at a quicker, though less instantly clear, way of solving such problems.

To begin with, remember that the reciprocal of a number is what you get when you divide 1 by that number. Many hand calculators will find this value for you at a single punch of the key marked $1/x$.

The formula is brief: Divide the number of trips by the sum of the reciprocals of the speeds.

In our example, divide 1 by 20 and 1 by 30 and add the two fractions together. Divide that result into the number of trips—two—and you'll again have 24.

Your Clothing Size

NOT ALL Continental and British clothing sizes can be transmuted into their American counterparts by simple formulas, but many can. As follows:

Men's suits, overcoats, sweaters, pajamas: English and American sizes are the same. Add 10 to get the Continental size. (Example:

English 40 is U.S. 40 is Continental 50.) For U.S. small, medium, large, extra large, try 38, 40, 42, 44 in Britain.

Men's hats: Add the fraction one-eighth to the English size to find the American. (That is, if you wear a size-7 hat in the U.S., buy a 6-7/8 hat when in London. But consider it risky to buy for anyone not there to try it on.) Continental sizes are 8 times U.S., our 7 being a European 56.

A shirt in London and a shirt in New York have the same collar size. To find your Continental size, just multiply by 2-1/2, and round off upwards if necessary. At home, when I buy a decently marked shirt—not one of the small-medium-large abominations—I choose something with a 15 collar. In Europe, I multiply by 2-1/2 the easy way, getting 37-1/2. That rounds up to 38, which has proved to be just right.

(The easy way? Multiply by 5 and divide by 2.)

Shoes: Subtract 1 from the U.S. size to find the British equivalent. To discover the Continental size, try adding 30 to 33 to the U.S. size. (But this won't work with some of the smaller sizes and is not entirely consistent from one country to another.)

Women's blouses, dresses, sweaters: Try adding 2 to U.S. sizes in Britain, 8 on the Continent. Where that produces ridiculous results, make it 22 and 28 instead.

More Dangerous Way To Go?

"IF YOU ARE worried about your chances of being killed on a coast-to-coast trip, you won't get much relevant information by asking whether trains, planes, or cars killed the greatest number of people last year. Get the rate, by inquiring into the number of fatalities for each million passenger miles. That will come closest to telling you where your greatest risk lies."

This advice, published in a book of mine about how to lie with statistics, fetched a letter from an engineering student in Atlanta: "I strongly disagree on rating fatalities in passenger miles. It seems to me they should be plotted with respect to time exposed to the danger...."

This is by no means hairsplitting. You'll come to quite a different conclusion (you should forgive the expression) if you use one method than you will if you use the other. It is evident that airline people are going to prefer to figure by the mile and train or bus or car folks by the hour.

I think this is a case where you can make sense only by applying real-world, or human standards.

Perhaps your inner dialogue runs like this. "I'm going to fly or drive for the next eight hours. Which is more dangerous?" If that's what you want to know, seek figures using passenger hours.

I think it far more likely, however, that you will be deciding between air and surface for a trip in which the constant is the distance—here to Chicago, say, or across the country, as in my example.

Then it's passenger miles that count, not hours.

How Warm Is a Sleeping Bag?

WARMTH of a sleeping bag is pretty tricky to measure objectively. It depends on bag design and size and stuffing material—and of course on who is sleeping in it.

I can testify to this after hundreds of nights in down- and Dacron-stuffed bags of various shapes and designs, in climates from those of Fiji, New Zealand, and Baja California to Romania and Finland.

The common method of comparison for down bags—by total weight of stuffing used—doesn't really tell the whole story. Four pounds of down in an ineptly designed rectangular bag may not keep you as warm as two pounds in a mummy bag of overlapping-tube design.

In comparing bags of similar construction, figure goose down as the warmest insulation, duck down next, and the more costly of the synthetic fibers as an excellent third.

To compare bags of varied construction (mummy, rectangular, and semi-rectangular shapes; different methods of arranging the down) the best criteria I've found are quality of down—and loft.

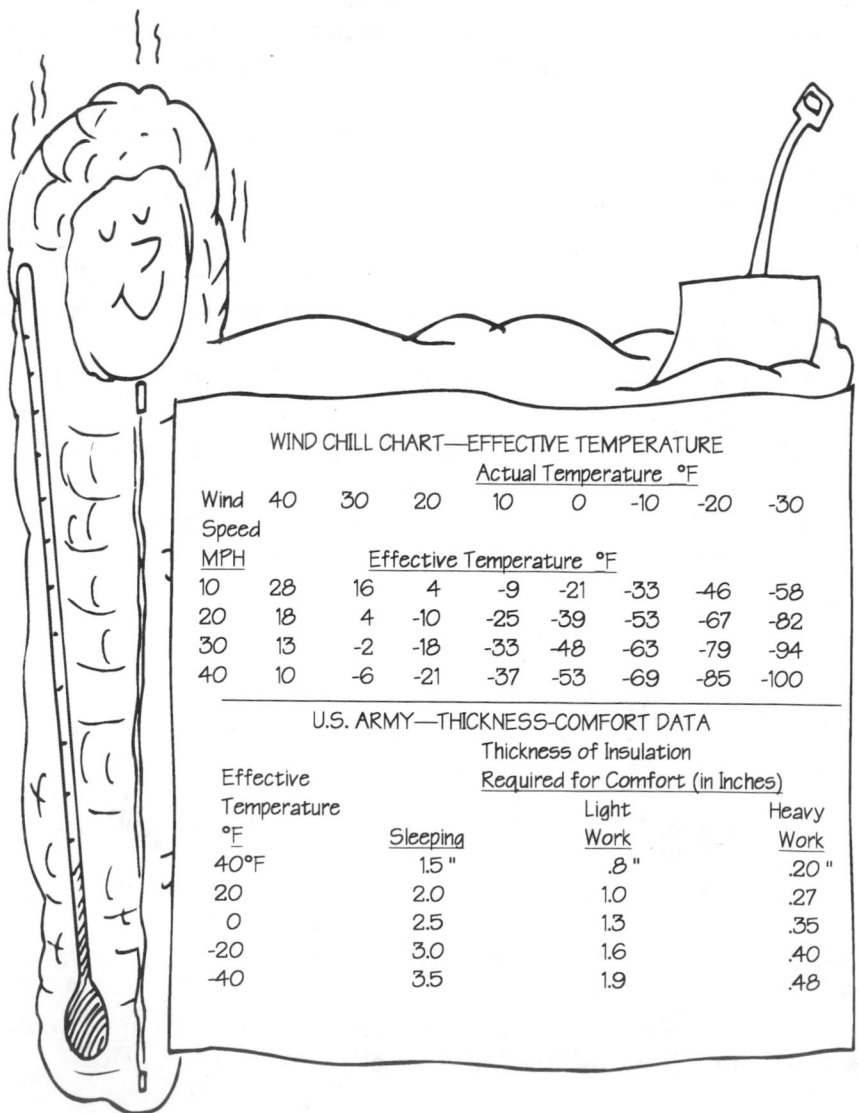

WIND CHILL CHART—EFFECTIVE TEMPERATURE

Wind Speed MPH	Actual Temperature °F							
	40	30	20	10	0	-10	-20	-30
	Effective Temperature °F							
10	28	16	4	-9	-21	-33	-46	-58
20	18	4	-10	-25	-39	-53	-67	-82
30	13	-2	-18	-33	-48	-63	-79	-94
40	10	-6	-21	-37	-53	-69	-85	-100

U.S. ARMY—THICKNESS-COMFORT DATA

Effective Temperature °F	Thickness of Insulation Required for Comfort (in Inches)		
	Sleeping	Light Work	Heavy Work
40°F	1.5 "	.8 "	.20 "
20	2.0	1.0	.27
0	2.5	1.3	.35
-20	3.0	1.6	.40
-40	3.5	1.9	.48

Loft is how thick the wall of the bag is when you've fluffed it up—as you should when getting into it on a cold night.

A bag that lofts to 3 inches (for each wall, not the total height of the bag) should keep you comfortable down to 20 degrees.

Add enough down to increase the loft by half an inch and you should be able to take a temperature drop to zero.

These calculations should hold for average wind conditions. If you're well protected or there's no wind at all, you'll be comfortable at even lower temperatures.

To be more scientific about it while allowing for extreme conditions, the U.S. Army has come up with a thickness-comfort chart, to be used along with a simplified wind-chill calculator. It takes into account the familiar fact that when the wind blows, it can turn chilly weather into freezing.

(As a kid in the American middle west I used to enjoy February tennis occasionally—after brooming snow off the court—on still days well below freezing. But more recently a gale wind helped frostbite my ears painfully in the mere minute or so it took me to dash from my camping bus into a Turkish border station. At 9 in the morning, the guards were—wisely—still sacked out.)

To use this pair of charts, read across from wind speed to the column under the actual temperature in degrees Fahrenheit.

This figure for effective temperature will tell you how thick the army thinks your insulation should be. You will note that the chart is for clothing as well as sleeping equipment.

Example: Opposite wind speed of 20 mph and under the 0 degrees temperature column, you will find that the effective temperature for this combination is 39 degrees below zero. The nearest temperature to this in the thickness-comfort chart is minus 40, which calls for 3-1/2 inches of insulation for sleeping.

Liters, Kilometers, Gallons

Americans touring abroad find gasoline quantities reasonably easy to understand. Distances in kilometers, too. But fuel consumption is something else again, since it depends on what you might call a double or simultaneous translation—between liters and gallons and between kilometers and miles.

Figures of the kind you see here in the third column are often listed in car comparisons in motor magazines. That's why a traveler put together this little table.

I've emphasized one handy set of figures that would be easy to memorize and use to compare one car's appetite with the hunger of another.

KM per liter	KM per gallon	Liters per 100 KM
3	7.1	33.3
4	9.4	25.0
5	11.8	20.0
6	14.1	16.7
7	16.5	14.3
8	18.8	12.5
9	21.2	11.2
10	23.5	10.0
11	25.9	9.1
12	28.2	8.3
13	30.6	7.7
14	33.0	7.2
15	35.3	6.7
16	37.7	6.3
17	40.0	5.9
18	42.4	5.5

What's That in Real Money?

T HE TRAVELER, real or armchair, has frequent need to figure the value of foreign currencies in American dollars. Doing this with a paper and pencil is not difficult. With a pocket calculator, even a $2 model, it's downright easy...even when extreme precision is wanted.

So let's look at the various procedures that may be called for when there's time to make a calculation. Spur-of-the-moment estimates that must be made during actual shopping in a foreign country are a bit trickier, but we'll get around to those in a moment.

Dollar exchange for the Danish crown may be listed as 0.1759. This gives you the decimal fraction of a dollar that a crown is worth: in other words, about 17-1/2 U.S. cents. So that Kr 10,000 fur coat in the window will set you back by a mere $1759. You have found this figure by multiplying the price in Danish currency by the value of the dollar today.

Sometimes you have to go the other way. Your Danish cousin has unblushingly asked you how much the Volkswagen you're driving cost you back home. (Danes, even more than most Europeans, are delightfully unencumbered by those feelings of delicacy about prices that afflict Yanks.) About $9,500 you tell him. "Okay, okay," he says, "but how much is that in real money?"

To calculate in this direction, you divide the price by the rate: 9,500 divided by 0.1759 is 54,008 *kroner* and never mind the few odd Danish cents.

The wistful glint that comes into his eye reminds you that the car would run him far, far more than that, with Danish taxes.

Sometimes you'll find currency quotations in a different form: the number of units of the foreign currency required to make a dollar. That crown will be listed not as about 17-1/2 cents but as about 5.7 to the dollar. Then the route from U.S. price to Danish is by multiplication.

Quite often you'll find a quotation in one form when you'd rather have it in the other. To make the swap, just divide 1 by the number. To get technical about it, the two rates are reciprocals of one another.

With a pocket calculator having a constant function built in (most do) you just punch in the number and then hit the division key three times. More elaborate calculators have a reciprocal key to save you half a second; it's the key marked 1/x.

A budget-conscious tourist abroad will usually need speed more than precision. What he really wants to know is whether that pizza or *croque monsieur* on the menu is in the two-dollar neighborhood or more like five or ten. Never mind exactness.

So his first fiscal move upon reaching a new country will be to learn the rate in whichever form is more manipulable by mental arithmetic. His second will be to reduce it to a simple approximation formula.

Sometimes it's easy to do this. If you find the French franc at about 20 cents (where it is as I edit these lines) you need only divide by 5. (Or multiply by 2 and divide by 10. Take your choice.)

When German marks, Swiss francs and Dutch guilders were hovering close to 40 cents, they were pretty easy to figure. Multiply by 4 and divide by 10. That 60-D-Mark tripod you were considering became $24 worth almost instantly in your mind. As you multiply by 4, your brain shoves the decimal over by a point without your consciously thinking about it.

But what do you do with a franc that's quoted at 22 or 23 cents or some other unfriendly figure? At 22 cents or thereabouts, I'd say divide by 5 for a rough approximation, getting more precision (if called for) by then increasing that result by 10 percent or one-tenth.

At an exchange rate of anywhere around 0.225 dollars, it's more accurate, and almost as easy, to divide by 4 and then reduce the result by one-tenth. When you're dealing with a rate around 24 to 26 cents, you'll come close if you divide by 4 and—on major amounts—then reduce or increase by half of one-tenth.

Finding a Mach Number

SINCE THE Mach number for the speed of an aircraft is simply that speed divided by the speed of sound, the idea is simple enough. Mach 2 merely means twice as fast as sound. This is convenient, since aviation buffs love Mach numbers.

The apparent simplicity, however, can mislead. It overlooks the fact that the velocity of sound changes with altitude, making the actual computation more complex than it would at first appear to be. But tackled a step at a time it is quite straightforward and, with the aid of a pocket calculator having the powers key found on even the most modest scientific models, not difficult.

Given instrument readings of pressure altitude (PALT) of 25,500 feet and calibrated airspeed (CAS) of 350 knots, here's how it goes.

1. Divide the CAS by 661.5 and square the result (getting, in our example, 0.2799).

2. Divide by 5, add 1, raise to the 3.5 power, and subtract 1 (getting 0.2101). Save this result for later use.

3. Multiply the PALT by 6.875 and divide by 1 million by moving the decimal point 6 places to the left (getting 0.1753).

4. Subtract this from 1 and then raise the result to the power of -5.2656. (In our example this should give you 2.7592.)

5. Multiply by the number stored in Step 2. Add 1. Raise to the 0.286 power. Subtract 1. Multiply by 5 (getting 0.6984).

6. The square root of this (0.8357 in our example) is the Mach number. The plane is flying at a little more than eight-tenths of the speed of sound.

Thinking About Time Travel

WHAT WOULD HAPPEN if you were to mount a time machine, as in the famous H.G. Wells story and ride a few years into your past? Paradoxes would fly like confetti.

You might meet your younger former self and start a disturbing conversation. You would understand the situation but he would not. About the healthiest conclusion he could come to, everything considered, is that although you know an amazing lot about him, you are probably crazy.

A way to look at such a situation from the outside was devised by Richard Feynman, a Nobel Prize winning physicist. To follow the life of a time traveler, just place a ruler or even a piece of paper horizontally on the drawing.

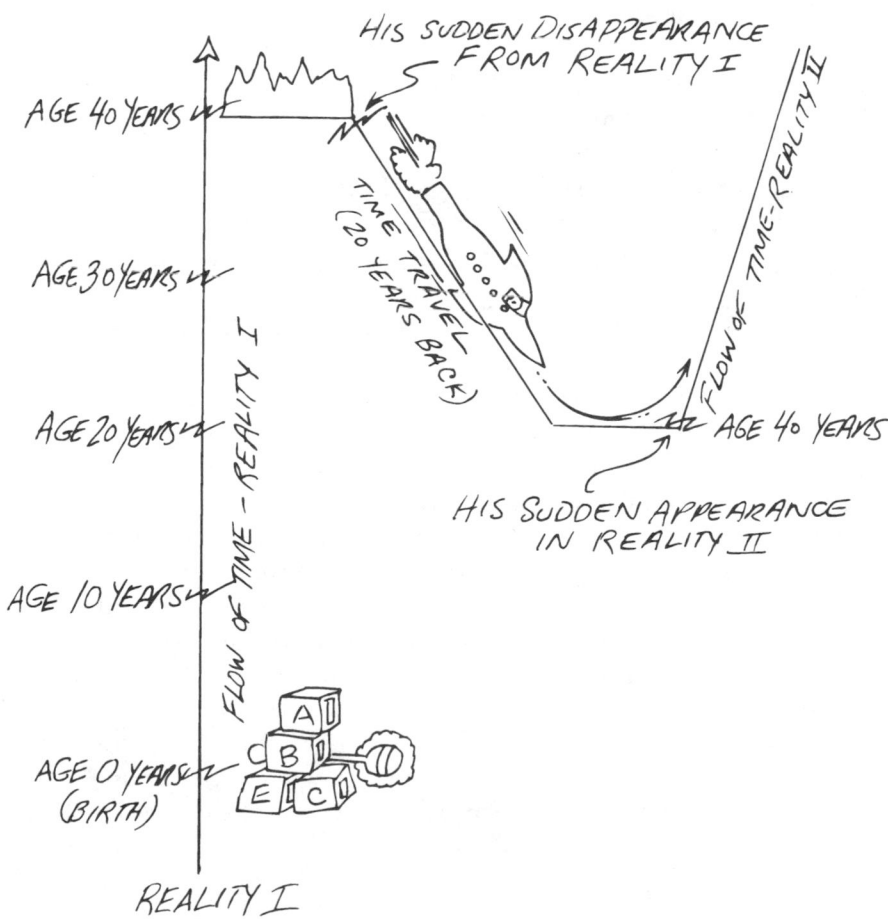

Move it slowly upwards. You begin with the man's birth, following his life until he's about twenty. Suddenly, then, he appears simultaneously on another track at the age of forty. Now as the ruler slides upward you are following both his lives until he reaches forty on the left-hand track—and vanishes. By this time, he's sixty on the right-hand track, and he goes on from there.

Our man's time travel is represented by just a line on the diagram, since on this he's traveling backward in time and so is invisible.

Interesting.

Einstein: Mass at Speed

ACCORDING TO relativity, the mass of an object changes with the speed of its motion. This is described numerically by what I believe is called the Lorenz-Fitzgerald contraction.

The phenomenon is not especially important on the highway, or even at supersonic air speeds, but it is for spaceships as they approach the speed of light (186,000 miles per second) so perhaps it behooves us to have this formula ready. You never know when you're going to need something.

Step 1. Square the speed of light. From this, subtract the square of the velocity of the object. Find the square root.

Step 2. Multiply the speed of light by the mass of the object (at rest).

Step 3. Divide that result by the result of Step 1. This answer tells you the perceived mass relative to the given velocity. In the calculations, you naturally must be consistent in the measures used. If you use the speed of light in miles per second, then you must express the velocity of the object in miles per second as well. If you feed in mass in pounds, the result will be in pounds.

Check the procedure against these results: A 5-kilogram object, traveling at 93,150 miles per second (just half the speed of light) will have a perceived mass of 5.7735 kilograms, which is rather a significant weight gain. It suggests what would happen as a spaceship and its passengers came closer and closer to the speed of light.

If you find any of this confusing or doubtful, recall the fine old anonymous limerick (of which the version here is probably as authentic as any) from the days when the writer Gertrude and the sculptor Jacob were almost as much in the public eye as Albert was:

There's a wonderful family called Stein.
There's Gert and there's Ep and there's Ein;
Gert's poems are bunk,
Ep's statues are junk,
And no one can understand Ein.

Chapter R OUTDOORS

Weigh a Mountain, Measure a Pond

To ESTIMATE the weight of a great mountain—or a mere pile of dirt—think of it as a series of cylinders piled one on the other like a stack of pancakes of varying sizes. Then apply to each cylinder the standard formula of height times pi times the square of the radius to get an approximation of the volume.

To turn your volume figure into weight, you need to have some idea of what the stuff it's made of weighs per cubic foot or meter. In pounds per cubic foot, that's chalk 143, hard coal 94, concrete 137, earth 75 if loose and 100 if rammed, granite 168, gravel 109, gypsum 150, ice 56, iron slag or marble 168, limestone or quartz, 162, sand 100 dry and 125 wet, sandstone 143, slate 175.

Sometimes you can get a sample of the material and weigh a cubic foot or other measure of it.

If your mass is more or less in the shape of a cone, you'll save time and possibly come closer by treating it as such. Handiest version of the formula for the volume of a cone: Multiply the height by the square of the diameter and by pi (3.1416 or 22/7 will do for this) and divide by 3.

Same approach works well, naturally, for finding the area of any irregular surface, such as a body of water. Draw it on square-ruled paper. To the full squares, add estimates of each of the partial ones.

In anything like this, it's important to round off your final figure. A precise conclusion founded on estimated data is an illusion and a snare.

Water Over the Dam

FOR MANY purposes, including construction of a water-powered electrical generating system, it is necessary to measure the flow of a stream with some accuracy.

(If you don't need precision, it's easier to use the estimating method described in the next item.)

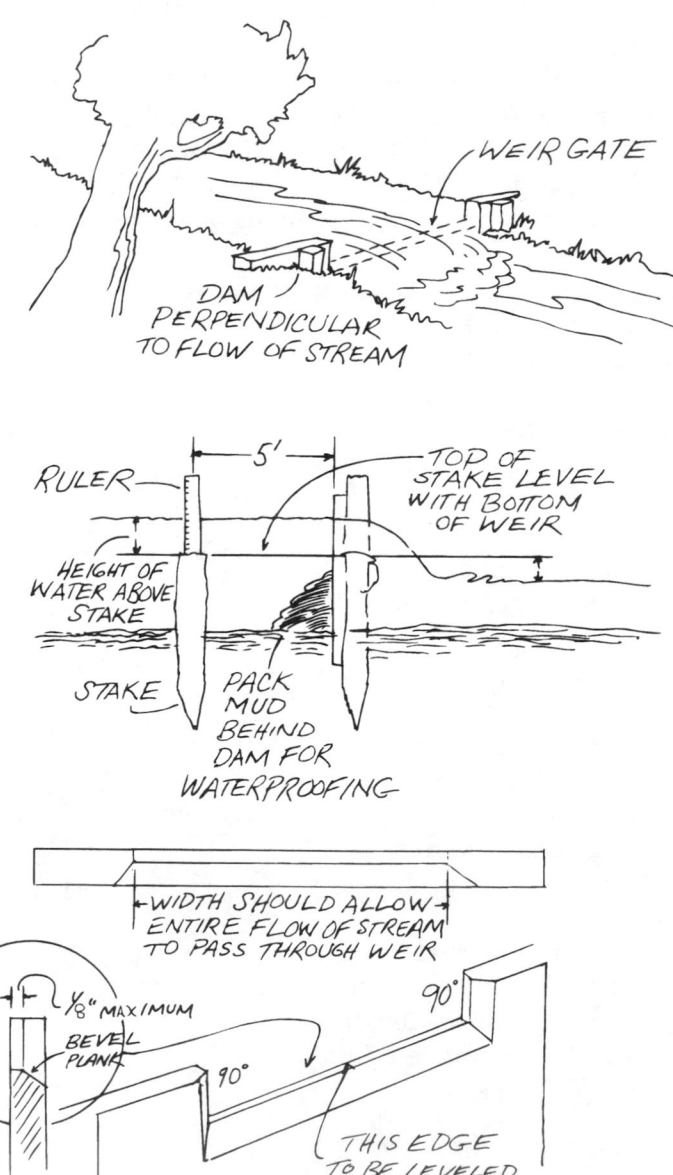

The method giving the more accurate result begins with construction of a kind of simple dam with what is called a weir gate.

The weir is a rectangular opening in the top of the dam, at its center. This opening should be at least several times as wide as it is high. It should be level and its bottom must be beveled to 45 degrees so that the edge the water passes over is no more than 1/8-inch wide. Width and depth must be such that the whole flow of the stream can pass through the opening.

Of course your dam must be directly across the stream—that is, perpendicular to the flow of the water.

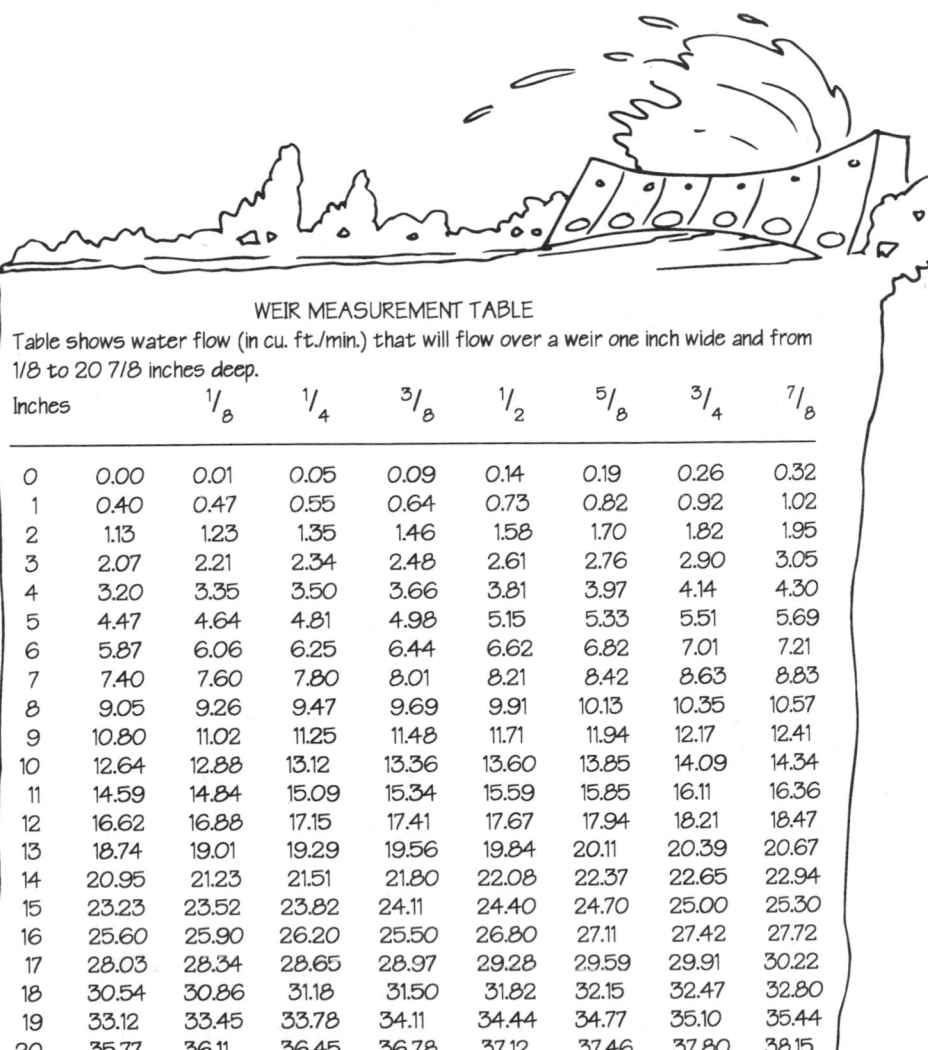

WEIR MEASUREMENT TABLE

Table shows water flow (in cu. ft./min.) that will flow over a weir one inch wide and from 1/8 to 20 7/8 inches deep.

Inches	$\frac{1}{8}$	$\frac{1}{4}$	$\frac{3}{8}$	$\frac{1}{2}$	$\frac{5}{8}$	$\frac{3}{4}$	$\frac{7}{8}$
0	0.00	0.01	0.05	0.09	0.14	0.19	0.26	0.32
1	0.40	0.47	0.55	0.64	0.73	0.82	0.92	1.02
2	1.13	1.23	1.35	1.46	1.58	1.70	1.82	1.95
3	2.07	2.21	2.34	2.48	2.61	2.76	2.90	3.05
4	3.20	3.35	3.50	3.66	3.81	3.97	4.14	4.30
5	4.47	4.64	4.81	4.98	5.15	5.33	5.51	5.69
6	5.87	6.06	6.25	6.44	6.62	6.82	7.01	7.21
7	7.40	7.60	7.80	8.01	8.21	8.42	8.63	8.83
8	9.05	9.26	9.47	9.69	9.91	10.13	10.35	10.57
9	10.80	11.02	11.25	11.48	11.71	11.94	12.17	12.41
10	12.64	12.88	13.12	13.36	13.60	13.85	14.09	14.34
11	14.59	14.84	15.09	15.34	15.59	15.85	16.11	16.36
12	16.62	16.88	17.15	17.41	17.67	17.94	18.21	18.47
13	18.74	19.01	19.29	19.56	19.84	20.11	20.39	20.67
14	20.95	21.23	21.51	21.80	22.08	22.37	22.65	22.94
15	23.23	23.52	23.82	24.11	24.40	24.70	25.00	25.30
16	25.60	25.90	26.20	25.50	26.80	27.11	27.42	27.72
17	28.03	28.34	28.65	28.97	29.28	29.59	29.91	30.22
18	30.54	30.86	31.18	31.50	31.82	32.15	32.47	32.80
19	33.12	33.45	33.78	34.11	34.44	34.77	35.10	35.44
20	35.77	36.11	36.45	36.78	37.12	37.46	37.80	38.15

With this done, drive a stake into the stream bed at least 5 feet upstream from the weir. Make its top just level with the bottom edge of the weir. With the steam flowing freely, use a ruler to measure the depth of water over the top of the stake.

The table figures water flow over a weir one inch wide with depths up to 20-7/8 inches.

Choose the number from the table that represents this depth of water. You'll notice that the first column is in exact inches, while each other column is for that number of inches plus the fraction given at the top of the table.

Multiply the number you've found by the width of the weir, the opening, in inches. The result will be the flow of your stream in cubic feet per minute.

Here's an example. You've built a weir 4 feet wide. You find that the water flow establishes itself at a depth 5-1/2 inches above the top of the stake. Opposite 5 inches in the column headed by 1/2 you find the number 5.15. Multiplying this by the weir width of 48 inches gives you 247. Your stream is flowing 247 cubic feet per minute.

Water Power Calculated Quickly

W ITH A DAM and a water turbine, how much power can you develop from a small stream?

How can you do the measuring and calculating quickly—in short, without first building one of those temporary mini-dams called weirs?

Take these steps:

1. Measure the width of the stream in feet.

2. Measure the depth of the stream in feet.

3. Find the average velocity.

(One way is by stretching strings across the stream, one 20 feet downstream from t'other and counting the seconds a cork takes to float from the first string to the second. Divide 20 by this number. Multiply by 0.7 to reflect the fact that the bottom and edges of a stream move more slowly.)

4. Measure the depth in feet of the fall at the dam.

5. Multiply these four figures together and divide by 23. (This assumes your turbine will have the usual efficiency of about 50 percent. If you know otherwise, divide this answer by 50 and multiply by the actual percent figure to compensate.) Result is how many kilowatt hours of power you can expect to develop.

For the answer in horsepower, divide the kilowatt result by 0.75. For a monthly total, multiply either figure by 720.

Try it for a small stream about 3 feet wide and 6 inches deep with a flow of 2.5 feet per second (that's 1.7 miles per hour) and a 3-foot drop. Multiplying together 3 and 0.5 and 2.5 and 3 and dividing by 23 gives the figure of 0.65 horsepower. In kilowatt hours per month it comes to about 360.

With a very tiny stream, by the way, it may be easier to measure the flow with a bucket. Divide the bucket capacity in gallons by the number of seconds required to fill it. Multiply by 0.14 to convert from gallons to cubic feet. Use this result instead of the three figures from the first three steps above.

Horsepower in a Stream

HAVING DETERMINED the flow of a stream that you're planning to dam in order to generate electric power, you'll very likely want to calculate how much usable power is there.

In addition to the flow (in cubic feet per minute) you'll need to know the head—the vertical distance in feet that the water falls.

Multiply these together.

Now multiply that product by the weight in pounds of a cubic foot of water (62.4) and divide by the number of foot pounds per minute that equals one horsepower (33,000).

(Instead of those last two operations, you can get almost the same result more quickly if you just multiply by 190 and move the decimal point in your result five places to the left.)

Either way, your answer is in available horsepower.

But what does that mean in terms of the amount of horsepower you can actually generate? Since a reasonable assumption is that your water wheel will be at least 75 percent efficient and your generator 80 percent, multiply the available horsepower by the product of these, 0.6, to find what might be called real, or switchboard, power.

Water and Kilowatts

THERE IS more than one way to estimate the power to be had from running water, and this may be just the formula you're looking for.

How much electric power you can develop from a stream depends upon two things. One is how fast the water is flowing. The other is how far the water drops, which is called the head.

Having measured the rate of flow in cubic feet per minute, express in feet the distance the water drops.

To find the power available, in kilowatts, multiply those two figures together and divide the result by 709.

So if your stream flows at 20 cubic feet each minute and has a head of 50 feet, it could generate just over 1.4 kilowatts (20 times 50 divided by 709).

To this information you might want to apply a rule of thumb that says a combination of flow rate and head that produces no more than half a kilowatt (500 watts) probably isn't worth developing.

But you'll decide that for yourself. It is, after all, enough power to keep a pretty bright light going indefinitely.

How Far Can You See at Sea?

THE RULE for figuring out how many miles you can see at sea (which naturally depends upon how high up you are) goes like this:

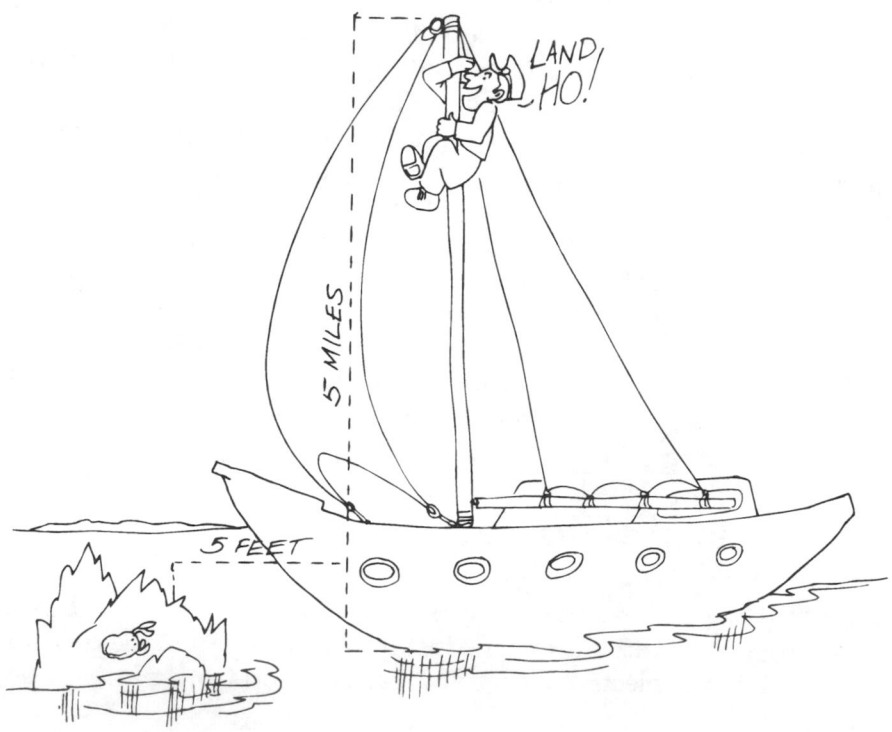

Multiply the height of your eye above sea level in feet by 3, divide by 2, then take the square root.

Say you're on a big ship or up in the rigging, or whatever, of a lesser one, so your eye is 96 feet above the water line. Divide 96 by 2, multiply by 3, and you get 144—of which the square root is an easy 12. From 96 feet in the air, you can see about 12 miles. Of course, you won't often come out with a perfect square like 144, but your trial-and-error estimate of the square root will be close enough for most purposes.

The original formula created for a slightly different purpose, the one from which I derived the above, is equally useful. It says you can find how high you must climb, in feet, by squaring the number of miles you wish to see and then multiplying by two-thirds.

The Accurate Cricket

THE PALE green snowy-tree cricket, common in the Americas, is often called the thermometer or temperature cricket. He (or, for all I know, she) is so sensitive to temperature that you need only count chirps for 15 seconds and add 39 to determine the temperature in degrees Fahrenheit with remarkable precision.

If you think better in Celsius, you might want to count some chirps and work out a formula of your own. Or simply add 7 instead of 39 to that 15-second count, then multiply by 5 and divide by 9.

Or take up the matter with your local cricket or entomologist.

Camera's Hyperfocal Distance

IN THE GREATER part of the photos most of us take, we want the distant background sharp and as much as possible of the foreground as well. Even an autofocus camera won't always do that for us.

A way to insure that result when we want it is to take over command from the camera's workings and set the lens at its hyperfocal distance.

This refers to the nearest distance for which you can set your camera and still keep far objects sharp. With this setting, everything from infinity down to half the hyperfocal distance will be sharp.

This fact makes hyperfocal distance a sort of multi-purpose setting, a good one on which to leave your camera between uses. If you then must take a picture in a hurry—or if you forget to check the focus—there's a good chance all will still turn out well.

Hyperfocal distance depends on two things: focal length of the lens and the f-stop you're using. To find it, choose any f-stop and set your distance at infinity. Read the near limit of the depth of field. This, the closest point that is sharp when the camera is set for far distance, is the hyperfocal distance.

Now set your camera so it is focused at the hyperfocal distance. Everything from half that distance to infinity will be in focus.

Here are a couple of examples.

With the distance scale on my old twin-lens reflex set to the farthest point possible, the depth scale on the side tells me that at f/22 everything will be sharp from 15 feet on. So I reset the focus to 15 feet, knowing that everything should be sharp from half of 15 feet to infinity. The depth-of-field scale confirms this.

Applying this to a 35-millimeter camera with a 50mm lens, I find that at f/8 with focus at infinity, the nearest point of sharpness is 10 meters. If I set the camera at 10 meters, I can be confident of sharpness from about 5 meters to infinity.

With the same lens set on f/16 to get greater depth, I find that the hyperfocal distance is only 5 meters. Everything should then be sharp—from 2.5 meters to the moon.

Sometimes it's more convenient to find hyperfocal distance by calculation.

The focal length of a lens is engraved on its barrel. Multiply this quantity by itself and then divide by the f-stop you're planning to use.

Divide also by what you consider to be an acceptable circle of confusion, or blur. A commonly used figure is 0.03 millimeters.

So for a typical small-camera 50-mm lens, to be used at f/8, the calculation proceeds like this:

The square of 50 divided by 8 and by 0.03 is 10,416. Changed to meters, that's 10.4, which confirms the 10 meters found by the direct experiment. And you know that if you use a 10-meter setting while working at f/8, everything from half that distance to infinity will be in focus.

Need a Ped Xing?

THE SUBJECT of pedestrian crossings, those marked rights of way for foot traffic the British call zebras (they like to stripe them) reminds me of something I heard about a foreign visitor to the U.S. The only American phrase he encountered repeatedly and couldn't find in his phrase book, he said later, was ped xing.

If you were to come into contention with traffic authorities about whether a pedestrian crossing is needed somewhere that you go, you might have use for a formula. It could also serve you if you should find yourself on a town planning committee of some kind. Or in a P.T.A. where safety of children was involved.

A formula has been arrived at by the Department of Environment, in England. It is called PV2.

It says that to justify establishing a marked pedestrian crossing, you must first count the number of pedestrians crossing the road in one hour. Then count the number of vehicles, square that number, and multiply it by the first number. If the product exceeds one hundred million, the ped xing is justified.

If your count shows 100 people crossing in an hour at a point used by one vehicle every four seconds...well, that isn't quite enough. (Fifteen cars a minute is 900 in an hour; 900 squared times 100 is only 81 million.). But make that a car or truck every third second and PV2 is more than satisfied.

By the way, you're not permitted to count organized kids in this. That is, children on the way to school are presumed to be protected by special traffic wardens with those little round signs on sticks. (I am delighted to discover that the British call these guards lollipop men.)

So much for zebra crossings.

As for what are known in England as pelican crossings—traffic lights used on straight stretches of road to stop traffic when a pedestrian pushes a button—there PV2 has to reach 500 million.

Acceleration from Gravity

IGNORING air resistance (which, of course, you can't with relatively light objects unless they are in a vacuum) you can figure that a freely falling object increases its velocity by about 32 feet per second during each second of its fall.

Using this figure you can find out just about how fast an object will be falling after a given number of seconds. Or you can find out how far it has fallen if you know long the fall has taken.

Such a useful figure may be worth knowing more precisely. It is generally given as 32.173 feet per second per second, for those who want it more exactly than 32. However, because the earth is not a perfect sphere, even this may not be precise enough to please you, since acceleration from gravity varies according to latitude by as much as one half of one percent. The figure of 32.173 is close only for latitudes of about 45 degrees.

For 0 (the Equator) to 5 degrees, add 0.09 to the working approximation of 32. For 10 or 15 degrees, make that 0.1; for 20 degrees, 0.11; for 25, 0.12; for 30, 0.13; and for 35, 0.15.

For 40 degrees, use 0.16; for 45, 0.17, as mentioned; for 50, 0.19; 55, 0.2; 60, 0.21; 65, 0.23; 70 to 75, 0.24; 80 to 85, 0.25. For 90 degrees, use 0.26. You are at one of the poles.

With a table of trig functions, or even more quickly with a scientific calculator or a computer, you can dig out yet more precise acceleration figures. All you have to do is subtract the cosine of twice the latitude from 32.173.

How To Bank a Road

Hᴏᴡ ᴍᴜᴄʜ slope should be built into a curve in a road to counteract the tendency (that we call centrifugal force) for the car to run off the road?

Though you may not build many roads yourself, the calculation might interest you the next time this happens to your car. Sloppy road engineering—or was your speed different from what the road builder had in mind?

The gradient in degrees can be found by multiplying the square of the velocity in miles per hour by 0.067 and dividing by the radius of the curve in feet.

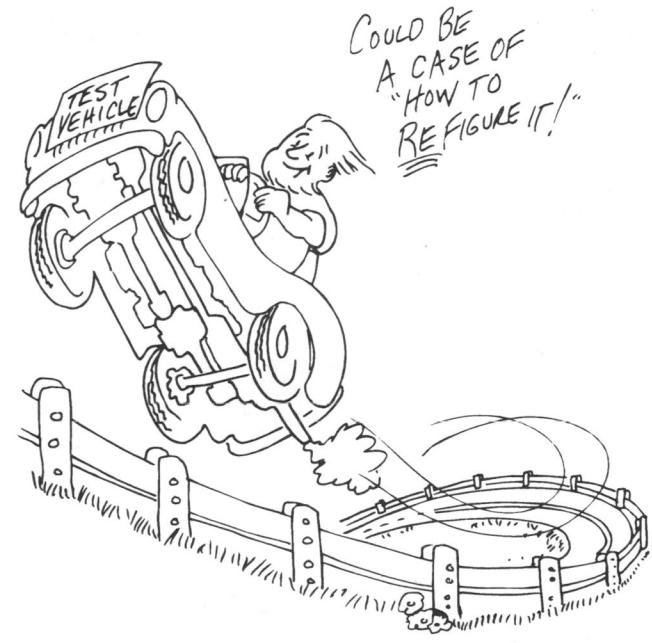

Thus if you are planning a curve of radius 2500 feet to handle traffic at 60 miles an hour maximum, you would multiply 60 by 60 and by 0.067, and divide by 2500.

Result is the information that you'll need a lateral gradient, as it's called, of 0.09648. Which times 100 is 9.648 percent.

If the road is to be used by heavy vehicles as well, the builder may have to compromise. With slope more than 6 percent, a big, slow-moving truck might slide off the road when it's icy.

With a road that is already designed or built you can work backwards to find the ideal, or balancing, speed for a vehicle taking a curve of known radius and gradient.

Say the radius is 2400 feet and the gradient 6 percent. The balancing velocity will be 3.86 times the square root of the product of the gradient and the radius of the curve.

Multiplying 2400 by 0.06 gives 144, of which the square root is 12. And 12 times 3.86 is 46 miles per hour.

Underpass for a Bicycle

Wɪᴛʜᴏᴜᴛ going into just how many of us are likely ever to be called upon to design an underpass, it is worth noting that one of the things the world needs is more and better bike paths. They're major ammunition in our continuing, and I like to think, accelerating battle to wrest the world back from the automobile. Well, anyway, here's how Kenneth Claxton—architect of some of the best cycleways anywhere—at the English new town of Stevenage—sums up the formula:

"Two up, one down."

That means elevate the roadway for cars two yards and drop the bike path one yard.

Make that meters and it should still be a close enough approximation of the ideal, in our metric future.

I hope quite a few people will have occasion to use the formula soon. *Vive la bicyclette!*

When the Tide?

For boating or other salubrious purposes it is often useful to know at just what minute the tide will reach some specified height.

There is a procedure for finding this out, starting from the information offered in widely available tide tables. The calculation is straightforward arithmetic, except for one step that calls for the use of a trigonometric ratio called the arc cosine or reverse cosine.

A cosine, as most of us can recall from high school, is simply the number you get when you divide the base of a right triangle by the hypotenuse (longest of the three sides). The arc cosine of a number is the reverse. It's simply the angle of which your number is the cosine.

The cosine of 30 degrees is 0.866; so the arc cosine of 0.866 is 30 degrees. A scientific calculator will get it for you at the touch of a key or two. So will a computer. Or you can read it from a table. You don't actually need to know all that stuff in the previous paragraph to use it. They do say, though, that the unobserved life is not worth living.

For navigational or other reasons, you might like to know at what time on a morning shortly before Christmas the tide will be at 4 feet at Sandy Hook, N.J. It is a day for which a tide table gives high-water mark as 4.8 feet at 6:21 and low water of 0.3 feet at 12:40.

If those heights were given in feet and inches, you'd have to change them to decimal feet or express them as feet plus twelfths of a foot (5 feet 4 inches becoming 5 plus 4/12 feet). If times are not given on a 24-hour clock, it would be necessary to add 12 to any p.m. hour. Any minutes given should be treated as fractions of an hour, for example by expressing the time of 6:21 as 6 plus 21/60 and 12:40 as 12 plus 40/60.

With that out of the way, here's the procedure.

From the chosen height—the one for which you want to know the time—subtract the low-water height. (In our example that means taking a minus 0.3 from 4, giving the result 4.3.) Next, divide that by the result of subtracting from the high-water height the low-water height. (Divide 4.3 by 5.1, getting 0.843.)

Now multiply by 2 and subtract the result from 1 (1 minus 1.686 is -0.686).

That's the number for which you must find the arc cosine by calculator or table. (Arc cos -0.686 is 133.336.)

Divide by 180 (getting, in our example, 0.7408).

Multiply by the difference between the time of high water and the time of low water (which here means taking 12 plus 40/60 from 6 plus 21/60 to get -6.317, and multiplying by the previous result to come up with -4.679).

Add the time of low water to get the answer in decimal form. You can change the decimal part to clock time in hour and minutes by dropping any decimals after the first two and multiplying those remaining by 60. (In this example, adding together 12 plus 40/60 and -4.679 gives, when rounded to two decimal places, 7.99, or almost 8 o'clock. Multiplying the decimal part of that by 60 gives 59 minutes. So the 4-foot tide we've been running through all these steps to predict can be expected at 7:59 in the morning.)

How High the Tide?

WHEN YOU'RE about to sail under a bridge, or go fishing, you might want to predict the height of a tide at some specific hour.

To do so you could begin by taking, from a tide table, the times and heights of low and high water. You would convert any of these times that were expressed in hours and minutes into hours and fractions (common or decimals as you prefer) of hours on a 24-hour clock. You would do the same thing with the time for which you wish the information. (So 8:15 p.m. would become 20 and 15/60 hours. Or 20.25.)

Heights should be in feet and fractions or decimals, not in feet and inches. Of course if you're working with meters there won't be any problem. Good for you.

So here we go to find the height of the tide in Boston harbor at 8 o'clock on a December evening.

From the published table, we know that high tide will be 10.3 feet at 9:13 p.m., and low water will be -2.1 feet at 5:02 p.m. (Don't overlook the minus sign.) The table also tells us the time of high water, 23 hours 13 minutes, and the time of low water, 17 hours 2 minutes.

Now proceed.

From the time for which you need to know the height, subtract the time of low water (20 minus 17 and 2/60 equals 2.97).

Divide that result by the difference between high-water time and low-water time (23 and 13/60 hours minus 17 and 2/60 is 6.19), getting the result 0.48.

Multiply by 180 (getting 86.40).

Using a computer, a scientific calculator, or a trig table (even a slide rule if you really want to) find the cosine of that result. (You'll find the cosine of 86.40 to be 0.06.)

Next multiply by the difference between the high-water height and the low-water height (10.3 feet minus a minus 2.1 feet comes to 12.4, so now you have 0.78).

Divide by 2 (bringing your result so far to 0.39).

Subtract this from the average height of the tide, found by adding high-water height to low-water height and then dividing by 2.

(In our example, you're adding 10.3 to a negative 2.1, and dividing by 2 to get 4.1. Subtracting the previously obtained 0.39 from this gives you about 3.7 feet. And that's the height you may expect the tide to reach at the hour you've chosen to fish or to sail your boat under a bridge.)

Ship's Propulsion Power

How much power does a ship need?

To figure that out, you naturally have to know some things about the ship in question. You must know its displacement tons and the required normal speed in knots—as well as a constant, called the Admiralty Coefficient, for ships of its particular class.

Then you can find the horsepower needed by squaring the displacement tons and taking the cube root of that result, then multiplying by the cube of the speed in knots and dividing by the Admiralty Coefficient.

It's not as difficult as it sounds, if you have a table of cube roots and an ordinary calculator. You don't even need a table if you have an ordinary calculator and employ the method given in this book for calculating cube roots.

With a scientific calculator, this formidable little operation is actually a cinch; you do the first two steps in one by asking for the 2/3 power.

You might like to give the procedure a trial run. See how it goes for the U.S.S. Constitution, displacing 30,090 tons, having a normal speed of 22.5 knots and an Admiralty Coefficient of 370.

Square of 30,090 is 905,408,100, for which the cube root is found to be 967.419. Cube root of 22.5 is 2.823.

Multiplying these two results together, and dividing by the coefficient of 370, tells you that this 1951 ship requires 29,782 horsepower.

How Far the Projectile?

You fire a shot.... The gun crew of a cruiser shoots a cannon....

How far will these or any other projectiles travel when launched at a given speed and a known angle with the surface of the water or flat terrain? At what angle will the projectile travel the farthest?

To answer either question, begin by squaring the velocity, in feet per second, and dividing by 32.

With the velocity expressed in meters per second, square as above but divide by 9.8 instead of 32.

Either way, call this Answer 1.

(Those divisors, by the way, are the acceleration per second, per second, with which a body falls—in a vacuum—at a point near the earth's surface.) The figures are approximations, but don't worry about that. They are as good as any for most calculating purposes, since ignoring air resistance has already introduced a modest error.

(If you'd like a more precise average value for the earth's surface generally, use 32.174 feet instead of 32; divide by 9.80665 instead of 9.8 meters. If you know you're dealing with a point near the poles, use 32.2575 feet or 9.8321 meters. At the equator, use 32.086 feet or 9.7799 meters.)

Next step:

Multiply the launching angle by 2 and find the sine of this result (in a trig table or, more easily, by pressing that handy SIN button on a scientific calculator). Multiply this sine by that Answer 1 you obtained previously, and you will know how far the projectile will travel.

If you want to find out what angle to choose for the longest possible flight, you can use a trial-and-error technique. It is not at all tedious with a calculator.

Double each of several possible launching angles, multiplying the sine of each by Answer 1. You'll soon find the maximum result comes when you use 90 degrees—not surprising if you've noted that the sine of 90 is 1 and that they don't get any bigger than that. So the elevation for maximum flight is 45 degrees.

Which you may not find too amazing, either.

Quick Height Measurement

T<small>HIS IS ONE WEIRD</small> way to determine the height of a distant object, such as a tree. But it works, so it must be anatomically correct.

Hold a foot ruler vertically at arm's length as you approach the tree or other object to be measured. Stop when one end of the ruler lines up with the top of the tree and the other with its base. Now pace off the distance to the tree, taking normal strides. The number of paces will be the approximate height of the tree in feet.

This trick obviously can be adapted to various circumstances. Clearly your measuring stick can be longer or shorter as long as you multiply its length in feet by the number of strides.

You can refine the method for greater precision by tuning your stride until you get consistent precise results for an object of known height.

What if you can't conveniently get far enough away from the tree to sight it in this manner? Well, there's a trick for this situation too.

Stand a yardstick or other object of known height against the trunk. Back off just far enough to sight on this stick with your ruler. Raise the ruler up the tree in jumps. The number of increments it takes to bring you to the top will tell you how many times as tall the tree is as the length of the stick.

How High the Shot?

A PLANE is going by up there and you are on the deck of a cruiser with a cannon. You determine that the plane is at an angle of elevation of 70 degrees. Given a muzzle velocity of 700 feet per second, can a shell from your cannon reach the plane if it is flying at 4000 feet? How about 6000 feet? Or 8000?

To find the maximum height of a projectile, you must square the velocity, multiply by the square of the sine of the angle of elevation, then divide by twice the acceleration of gravity.

That acceleration constant is 32 if you're working in feet, 9.8 if you're working in meters.

Whether you find the sine from a trig table or by the touch of a key on a scientific calculator, you can use either the square of the sine or the sine of the square of the angle. They come to the same thing.

So in this example you multiply 490,000 by -0.8830 and divide by 64 to get 6761 feet as the maximum height your shell can reach under the conditions given.

How Long to Fall?

IF YOU DROP something down a well, or fall off a mountain, how many seconds will it take the object, or you, to hit bottom?

Answer: one-fourth the square root of the distance in feet.

Drop a rock off a 400-foot cliff and it should arrive in 5 seconds, the square root of 400 being 20, one-fourth of which is 5. A fall from a mile-high airplane should take one-fourth of the square root of 5280, or just over 18 seconds.

Actually, it will take longer, since our formula neglects the resistance of the air.

A large, dense object will fall almost as fast as the formula says it will, especially until it reaches quite high speeds and the resistance increases. Even when both are made of the same stony stuff, a small pebble will be slowed more than a huge boulder will. A pillow will not fall so fast as a heavy suitcase of similar size and shape.

J.B.S. Haldane describes the difference vividly in a classic essay, On Being the Right Size.

"You can drop a mouse down a thousand-yard mine shaft; and, on arriving at the bottom, it gets a slight shock and walks away. A rat would probably be killed, though it can fall safely from the eleventh story of a building; a man is killed, a horse splashes."

The formula we're using here is based on an acceleration of 32 feet per second per second, which is a slightly imperfect approximation. The true figure, in fact, varies from one place on the face of the earth to another.

Would you rather do your figuring in metric? Start with the square root of the distance in meters. Divide by 2.214.

Where on Earth Is It?

THERE ARE TWO common ways of describing where one point on the earth, whether on land or sea, is in respect to another. It is often necessary to translate one of these descriptions into the other, especially in navigation.

One of the methods is called bearing and range, which means simply in what direction and how far away. An example is "14.14 miles northeast of Podunk Post Office."

The other familiar method is by rectangular coordinates: "10 miles east of the post office and 10 miles north."

Both describe the same point. But how do you get from one to the other? Here's an example, courtesy of the U.S. Navy.

"A contact is at bearing 075 degrees, range 20,000 yards. What are the rectangular coordinates?"

The north (also called the x-axis) coordinate is found by multiplying the cosine of the angle (75 degrees in this example) by the range (20,000 yards here) to find that the object is 5176 yards to the north. Repeating, but with sine instead of cosine, tells you it is 19,319 yards east—on the y-axis, that is.

If, on the other hand, you know the coordinates and wish to find the range and bearing, you naturally reverse the procedure. But perhaps the simplest approach is to start by squaring each of the coordinates, adding these squares together, then taking the square root. That's the range. The bearing, in degrees from north, is found by dividing the east coordinate by the north coordinate and then deriving the arc tangent.

Some of this may sound formidable, but it can be done with any calculator, or a home computer, or even paper and pencil. The easiest way is on a scientific calculator having rectangular-polar capability built in, making for a fast five-step solution.

Your calculator may be one (very likely from Hewlett-Packard) that uses what is called reverse-Polish logic, and is quickly identifiable by the absence of an equals sign.

To translate from bearing and range with this breed of calculator, start by entering the angle. Putting in the range and pressing the rectangular-conversion key produces the first coordinate. Pressing the x-y exchange key gives the other.

To go the opposite direction, enter one coordinate, then the other, press the polar key for range and the exchange key for the bearing.

If you're using the more common type of scientific calculator, one with what's called algebraic logic, enter the range and then press the x-exchange-t key and enter the bearing. Press the polar-rectangular conversion key for one coordinate and that same exchange key to find the other. For the reverse routine, successively enter x-coordinate, exchange, y-coordinate, rectangular-polar, exchange.

Speed and Power

W<small>HEN YOU</small> know the propulsion power a ship requires at one speed, how can you determine how much that power must be beefed up to permit some higher speed?

Or, knowing the normal power and speed, you wonder how much that speed will change when available power is increased or decreased.

The rule for this is one that says the power is proportional to the velocities cubed.

So what will happen to the normal speed of the vessel when there's a change in horsepower from, say, 29,782 to 35,000? To find out, divide the proposed horsepower of 35,000 by the present horsepower of 29,782, find the cube root, and multiply by the present speed of 22.5 knots.

Turns out the new velocity is 23.7 knots.

Not much improvement.

What if the horsepower were doubled from the present 29,782? Instead of applying that cube rule to the speed itself, it's quicker and easier to apply the rule to that simple ratio of 2:1. First, find the cube root of 2 and multiply that by the present speed of 22.5 knots. Available speed goes up respectably to 28.348 knots.

Suppose you know the speed available at present (22.5 knots, say) and the speed you need to power for (let's make that 25 knots). If you're the U.S.S. Constitution with a present horsepower of 29,782, to what must that be increased? To find out, divide the present velocity of 22.5 by the 25 knots you want, cube it, and divide the present horsepower by that result.

To get this modest increase in speed, up your horsepower must go—all the way to 40,853.

Figuring Air Pressure

IF YOU KNOW air pressure in pounds per square inch and need it in kilograms per square centimeter, what do you do?

You can consult a log table, if you can find one.

You can convert each of the factors separately and then divide.

Or you can make use of a remarkably simple formula:

Multiply by 7 and then move the decimal point two places to the left.

Thus 16 pounds per square inch becomes 1.12 kilograms per square centimeter.

Siting a Windmill

IN CHOOSING a site for a windmill, you'll presumably wish to avoid the turbulence that would interfere with its smooth operation. Proximity to buildings or other obstructions is a frequent cause of this condition.

Here are some rather crude general formulas that have been found to deal with this problem.

If the windmill is upwind, raise it above the ground by twice the height of the obstruction. You should have little interference when placing a windmill downwind if its distance is at least ten times the height of the obstruction. If you must place the windmill very close, go up to three times the building height.

How Much Wind Is Enough?

TO MAKE a windmill a cost-effective power source, it has been found by experience that you'll need an annual average wind speed of at least 10 miles an hour.

If this information leads you to a difficult decision, however, it would be prudent to check it against the latest research in novel windmill designs.

You will probably find that it could take you a long time to gather data for your own particular location.

However, if there is a weather station anywhere near you in surroundings at all like yours, data from it that shows 12 miles an hour or more is a pretty good guarantee that you have suitable conditions.

If it's a little less—10 to 12, say—you will probably want to make some tests.

To avoid making your tests all year round, you can run several over a short period and use them to establish the relationship between your wind speeds and those at the nearest airport or other weather station.

If you divide the average figure measured at your site by the one reported at the same time by the station, you'll find a factor by which you can multiply the official records from other times of year—to estimate seasonal conditions at your site.

Thus you might make three measurements over a short period in the spring and find 11, 13, and 14 miles an hour, for an average of 12.67. (The kind of average we're after here is the simple one, called an arithmetic mean, that you get by adding together all the quantities and dividing by the number of them, in this case 3.)

If observations made at the same time at your nearest airport are 12, 13.5, and 15, your average divided by theirs turns out to be 0.94.

Multiplying by this factor will tell you that if the airport averages 15-mile winds in December, yours should be about 14.

That Wind-Chill Factor

THE WEATHER sometimes feels a lot colder than the thermometer says it is. This is especially true when the wind is blowing, for the same reason that food solidifies faster in your freezer or locker plant if a fan is blowing on it.

You will recognize this as the phenomenon familiarly known as the wind-chill factor.

To get a good idea of the effective temperature in Fahrenheit, multiply the wind speed in miles per hour by one and one-half and subtract that result from what the mercury is trying to make you believe.

That's why 10 degrees above zero with a 50-mile gale can freeze you as fast as a quiet 65 below.

That's why it may be prudent to believe what our ears, as long as they are unmuffed, are trying to tell us.

How Many Stars?

SOMEDAY, you might—for all I know—want a trifle of information to support an argument favoring the existence of extra-terrestrial life.

Most useful for this, perhaps, would be knowing how many stars there are. Here are some handy—if immense—figures to work from.

Sir Arthur Eddington's old rule still holds up well enough in light of modern findings: The number of stars in a galaxy is about 10 to the 11th power, and—coincidentally, I suppose—so is the number of galaxies in the universe.

Since this easily written and easily remembered number, the 11th power of 10, is shorthand for 1 followed by 11 zeros, it signifies that there are quite a few of whatever it is that you apply it to.

Taking the whole thing a step further, we can multiply the 11th power of 10 by itself to find that the number of stars in the universe is on the order of 10 to the 22nd power.

This, if you've forgotten what the teacher told you more than once, follows from the rule for multiplying together the powers of any quantity. Just add those superscripts that indicate powers. Thus 10 to the 11th times 10 to the 11th equals 10 to the 22nd.

Which comes to 1 followed by 22 zeros or, in plain American English, ten thousand billion billion.

In plain British English, where billion is U.S. trillion, I guess it comes to ten thousand million billion or perhaps ten thousand milliard milliards. In any case I am clearly getting out of my depth, and I doubt that a billion or two will affect the argument anyway.

To Rate a Quake

W<small>HEN</small> T<small>HE</small> B<small>IG</small> O<small>NE</small> comes and half of California drops into the sea, some of us living along the Coast may want to rate the magnitude of the earthquake.

Here is a recipe for converting comparative intensities into magnitudes on the widely used Richter scale. By knowing how many times more powerful one quake was than another and also having the Richter rating of one, you are in a position to determine the Richter figure for the other.

Divide the known or estimated intensity of the second quake by that of the first. Find the common logarithm of this figure, by using a scientific calculator (that's the easiest way) or a table. Subtract this from the Richter rating of the second quake.

It goes like this. The famous San Francisco event that people there prefer to think of as the 1906 fire (but that nevertheless has been rated 8.25 Richter) was 105 times stronger than the Nicaraguan earthquake of 1972. That statement in effect puts the relative intensities at 1:105, so it is 105 of which the common log (not the natural log, please note) is needed. That log is 2.02. Subtracting from 8.25 puts the Nicaraguan quake at 6.23 on the Richter scale.

Figuring Binoculars

S<small>OME</small> <small>CALCULATIONS</small> are useful in selecting binoculars.

To begin with, binoculars are usually identified by two performance figures: 7x35. This tells you first the magnification (objects seen through it will appear to be only one seventh as far away as they actually are) and, second, the diameter of the objective lens—the one farthest from your eye—in millimeters.

You can also use these two figures to find a third one, called exit pupil. This is the diameter of the image or light-disc as it leaves the eyepiece. Since the pupil of the human eye ordinarily has a diameter between 4 and 5 millimeters, binoculars with exit pupils under 4 millimeters don't take full advantage of the seeing power of the eye. On the other hand, binoculars with exit pupils of more than 5 millimeters are providing more light than the eye may be able to make use of. Note, though, that after you've been in the dark for a while your pupils get larger and can make use of a somewhat greater exit pupil.

Since the exit pupil is found by dividing the second of our figures by the first, it comes to 5 for the common 7x35 binoculars. This corresponds well to the capability of the typical eye.

Something called "relative brightness" is often used to compare binoculars. It is found by squaring the figure for exit pupil. Do this for 7x35 binoculars and you are first dividing 35 by 7 to get 5 and then squaring that to get 25. For 7x50 glasses, it works out to 51. This tells you the 7x50 gives twice the brightness of the 7x35.

That's all very well, but the human eye doesn't see doubled brightness as doubled. It will perceive the one as no more than one-third brighter than the other. Apparently it is for this reason that some optical manufacturers, especially in Germany, prefer to use what is called the "twilight factor."

To find this factor you multiply the power by the objective diameter, then take the square root.

For the 7x35 it works out to 7 times 35 is 245, of which the square root is 15.7. But 7 times 50 is 350, with a square root of 18.7. This relationship comes closer to what your eye might tell you, so perhaps is more useful in making a decision.

(The 7x50 instrument is likely to be a good bit costlier than the 7x35, as well as bulkier and heavier. In making a decision it may be helpful to know that in exchange for the added bulk, weight, and cost you are getting an improvement that your eye will see as being in the ratio of 15.7 to 18.7—roughly one-fifth brighter.)

Another thing you may have need to calculate in order to compare binoculars is field. You may find this printed on binoculars, along with magnification and diameter of the objective lens. It is usually expressed in a rather eccentric fashion: as the width in feet of the field you can see at a distance of 1,000 yards. Some makers, however, prefer to express it in degrees, so in comparing instruments you may need a conversion factor.

To convert degrees to field in feet at 1,000 yards, multiply by 52.5. Thus the fairly common 7-degree field becomes 367 feet. Binoculars with fields of more than about 9 degrees fall within the loose designation of wide angle.

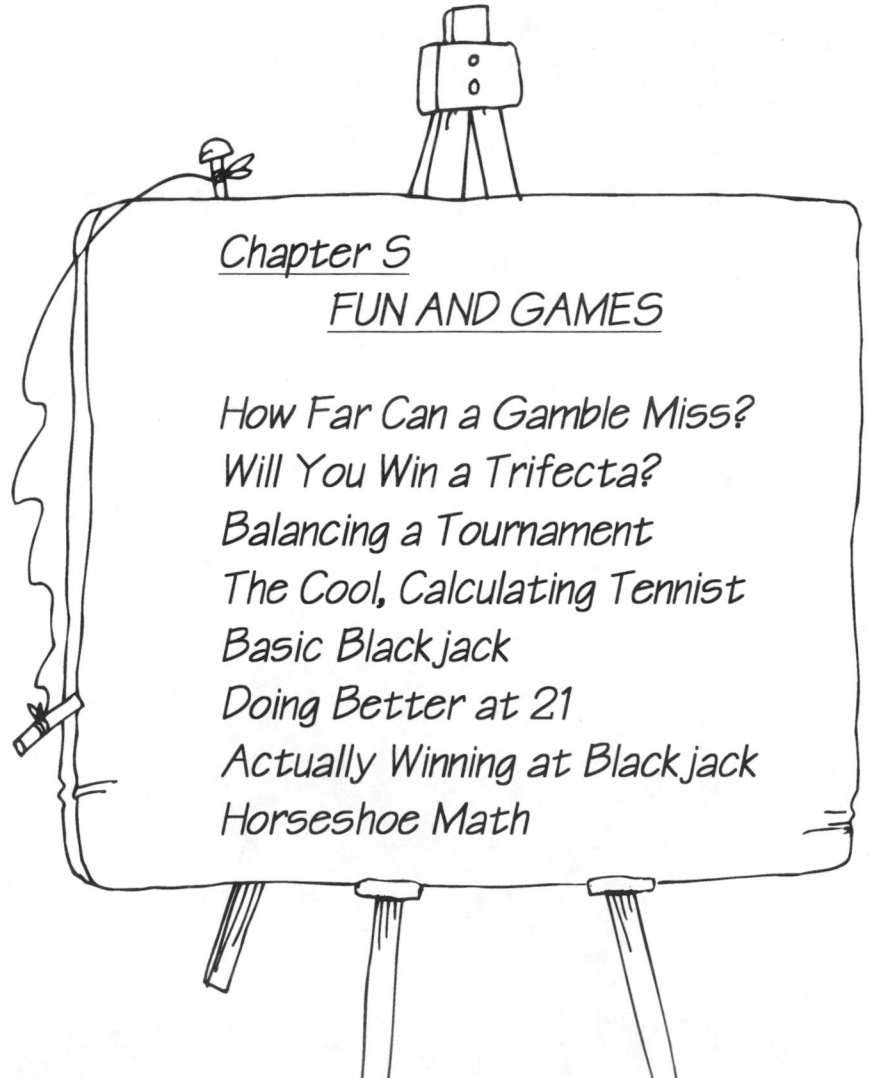

Chapter 5
FUN AND GAMES

How Far Can a Gamble Miss?
Will You Win a Trifecta?
Balancing a Tournament
The Cool, Calculating Tennist
Basic Blackjack
Doing Better at 21
Actually Winning at Blackjack
Horseshoe Math

How Far Can a Gamble Miss?

AₗₜₕₒᵤGₕ millions are eager to try, no one can predict what a roulette wheel or pack of cards or pair of dice or any other instrument of chance will do today or tomorrow.

Good thing, too, else where's the fun?

There are, however, ways to predict with considerable success what each of those instruments will do in that well-known period, the long run. (But remember: we live in the short run.)

The wheel, for instance, will produce, within a very small range of deviation, an equal quantity of reds and blacks, of odds and evens, and of each of the various numbers.

As for the short run, well, there is a kind of predictability there, too, that has been reduced to numbers. It is considered reasonable and proper to say that when a result turns up that is further off than would be expected to occur 99 percent of the time, that is unreasonable, and we should strongly suspect that something other than chance is at work.

By this test, what would you say of a roulette wheel that has been found to have produced red 6578 and black 6442 times over a period of observation? That's fairly close to the expected fifty-fifty, but is it close enough?

The deviation from even-steven can easily be calculated to be half the difference between the two figures, or 68. (Put in another way, 6578 minus 6442 is 136, and half of that is 68.) Does this mean anything? Can it be called significant?

Try this simple test.

Multiply the total number of trials by the product of the chance of winning and the chance of losing and then take the square root. Any deviation from fifty-fifty that is larger than the result given by this calculation can be expected to occur no more than 1 percent of the time.

Let's apply that general statement to the problem at hand.

Multiplying the number of trials (6578 plus 6442) by the product of the chance of winning and the chance of losing (one-half times one-half, or one-fourth, here) and then taking the square root gives an answer of 57.

But the deviation of 68 found earlier is somewhat greater than 57, so we have to call it highly unusual and perhaps even suspicious. Such a thing would not occur even once in a hundred times over the long run.

Will You Win a Trifecta?

A TRIFECTA is something that happens at race tracks. You and I don't hang around such places. But they do offer instructive mathematical models that can apply to many things. Anyway....

To win a trifecta, I'm told, you must name the nags that will come in first, second, and third—and name them in the right order, to boot. In an eight-horse race what would you say would be your chances of doing so?

Assuming you know nothing about these horses or the opinions of the odds makers on them, you'd have to give yourself 1 chance in 8 of predicting the winner. Then you'd have 1 chance in 7 of choosing the horse that will finish second among the remaining 7 horses. And 1 chance in 6 of predicting the third finisher.

That gives you 1 chance in 8 times 7 times 6—or 1 chance in 336, of hitting all three.

Expressed another way, the odds against you are 335 to 1.

When you combine things this way in a prescribed order you are dealing with what are called permutations. In this case it is the permutation of 8 things taken 3 at a time. The number of permutations is found by dividing the factorial of the number of things by the factorial of the difference between the number of things and the number of them to be taken at a time.

(A factorial of a number—for the benefit of those who may have forgotten or nodded off just long enough to have escaped learning—is simply the quantity you get when you multiply the number in question by all the smaller whole numbers. Factorial 5, which is usually written 5!, is 5 times 4 times 3 times 2 times 1, or 120.)

Thus the number of ways our 8-horse trifecta could come out is equal to factorial 8 divided by factorial 5. This is perhaps not easier to understand than the method we used first, but it is often a lot faster.

In real-life wagering you would probably have some information, or opinion at least, about the relative merits of the eight horses. You might estimate that the chance of the best of them winning was no worse than 1 in 2. You might believe the next horse of your choosing to have 1 chance in 3 of placing precisely second to your first-choice horse. And you might be able to choose another horse with what impressed you as 1 chance in 5 of coming in right after the other two. This would lead you to calculate that you have 1 chance in 2 times 3 times 5, or 30, of winning the trifecta.

You might also be wrong about the quality of your information or even your judgment. But that's the risk you assume.

If you don't like your gambles too wild, you might prefer to bet on what I am told is called a trifecta box. For this wager you choose the three horses to finish in the first three places but you do not have to say in what order.

What interests us about this, since we have agreed we're not railbirds, is this. As a mathematician would put it, we have changed a permutation problem into one dealing with combinations. The formula for finding the number of combinations calls for you to divide the factorial of the number of things by the factorial of the difference between the number of things and the number of them taken at a time (so far, just as in figuring permutations) and then also dividing by the factorial of the number taken at a time.

For an eight-horse race, that means factorial 8 (which comes to 40,320) divided by factorial 5 and factorial 3. It seems you have 1 chance in 56 of winning the trifecta box—just on chance alone—which is at least a lot better than 1 in 336.

For a shortcut in doing this kind of computation, you might note that factorial 8 is factorial 5 times 6 times 7 times 8. Canceling this factorial 5 with the one you're dividing by gets you the 336 of the previous paragraph by a simple multiplication of 6, 7, and 8. Brings the problem down to a level easily handled by mental arithmetic.

Balancing a Tournament

A SOCIABLE sort of tennis tournament with players of varied ability will be a lot more fun if you handicap it to even things up a bit. The difficulty with any ordinary form of handicapping is that you have to know your players' abilities in advance, although presumably this is what a tournament is supposed to find out.

Also you can get people sore at you unless your handicapping is done with care, and maybe even if it is.

Consider automatic handicapping. It works this way.

Suppose player A (or doubles team A) wins the first game. A then must start the second game down 0-15.

If he still wins, he starts the next game down 0-30.

If he wins again, he starts the next game down 0-40. His handicap remains 0-40 (or three points) as long as he continues to win.

Whenever he loses a game, the handicap decreases by one point. That is, when a player starts a game down 0-30 and loses it, he starts the next one down 0-15. If he loses that, the next game is played from scratch. And so on.

This system—which, by the way, can be applied to some other games than tennis—may remind you a little of vulnerability in bridge. The difference, however, is that being vulnerable at bridge increases the potential rewards as well as the penalties.

The Cool, Calculating Tennist

Are you the conservative player who gets his first serve in practically every time? You're wasting that second serve you're allowed.

On the other hand, are you the belter who comes through with an average of a double-fault per game? You're beating yourself.

Your pocket calculator can help you work out a combination of first and second serves that will give you both a maximum first-ball attack and an acceptable minimum of double faults. A little later, we'll see how to apply the same approach to other games and gambling problems.

By keeping track for a while, you may find that you miss your first serve 33 times out of 71 and your second 10 times out of 33. Dividing the number of faults by the number of tries and multiplying by 100 gives you your failure percentage: 46 on the first serve, 30 on the second.

Multiply those percentages together and you'll find you're serving doubles nearly 14 percent of the time. No wonder you lose a lot of service games!

How can you win more often? Ask your calculator what will happen if you steady down those first serves until three out of four go in, second serves until eleven out of twelve do so. Result is equivalent to about one double fault in two sets—acceptable on any neighborhood court or even on the pro circuit.

The principle involved in this calculation, by the way, is one with many other applications. It says that the easy way to figure the chances of succeeding in at least one of two things is by multiplying together the chances of failing at each. Having done so, subtract that result from one (or from 100 percent, if you're working with percentages) to find the probability of success.

Basic Blackjack

For anyone interested in the strategies that follow but not too familiar with the game, here are the principal rules of blackjack as played in most casinos. Some places vary them a little.

Bets are placed first. Then each player gets two cards face down, while dealer gets one down and one up.

Each card counts its face value. Face cards count as 10s. Aces count 1 or 11, whichever gives a better total.

Winning hand is the one closest to 21—but not over 21.

A "natural" or "blackjack" is a 21 total reached with two cards. When the player receives it, he wins 1-1/2 times his bet. When the dealer gets it, the player loses his stake. When both get it, no money changes hands. (As you can see, this 1-1/2 rule gives the players an advantage over the house. They'll need it.)

The player now decides whether to accept more cards, one at a time. If he "busts" (goes over 21) he loses at once.

After the players, the dealer similarly "stands" on what he has or draws more cards. Unlike the player, he has no discretion but must keep drawing until his total is more than 16. When dealer has an ace and counting it as 11 brings his total to 17 to 21, he must stand (in many casinos it's 18 to 21).

If dealer busts he pays each player who did not bust an amount equal to his bet. If neither goes over 21, higher total wins. With ties, no money changes hands.

The house advantage, more than balancing the player's edge mentioned above, comes from one aspect of this rule: If the player busts, he pays even if the dealer later busts also. The customer has the substantial disadvantage of having to choose first.

I hope you'll forgive the digression, but I am irresistibly reminded of the story of a lone traveler who attempted to cross China during the revolution. Captured by soldiers and asked which side he favored, he answered that he was loyal to the government. The soldiers, being rebels, beat him severely before turning him loose. When captured again a little later, he hastily avowed support of the rebellion. The soldiers, this time government troops, beat him with even greater vigor. A few days later, having recovered sufficiently to resume his journey, he was taken prisoner a third time and again questioned about his loyalties.

This time his reply was, "You say first."

In addition to basic play, the player has some options he may take if he wishes.

One of these is called doubling down. After seeing his first two cards, the player may double his bet if he wishes. However, he must then turn over his two cards, accept one more card, and then stand.

The other option is splitting pairs.

If the player's first two cards are of equal value, he may turn them face up and receive an additional card on each, at the same time putting up a stake the size of his initial bet on the second hand.

He is now playing two hands. The usual rules apply to each, with one exception: If he has split aces (a good thing to do) he must then stand. A count of 21 reached after splitting is not scored as a "natural."

Another option, called "insurance," is usually unprofitable and better ignored. When dealer's up card is an ace, a player may make a side bet of anything up to half his original bet. Dealer then checks his own hole card to see if he has a natural, often called a blackjack. If yes, the player wins 1-1/2 times the amount of his side bet at once; if not, he loses it. Play of the hand then continues.

Doing Better at 21

Even before someone I love took a turn as a blackjack, or 21, dealer, I'd found the game the French call vingt-et-un (remarkably corrupted by the English to Van-John) especially fascinating. Something I said about blackjack in a little book I wrote years ago, on some applications of probability theory, even led me to become a footnote to an historic article in a journal of mathematics. That article launched its author, Edward O. Thorp, on a most unlikely avocation for a university professor. He began hanging around gambling casinos in Reno, Las Vegas, and elsewhere. He was bent on showing that a system had at last been evolved that would beat the house.

Unlike any past systems player who stuck around long enough to give his theory a real test, Ed Thorp didn't go broke. He made money so fast that eventually no casino management would let him anywhere near a 21 table.

How does Ed Thorp—and anyone else who is prepared to learn his methods—beat the house? Two ways: optimum strategy, and counting.

The first is what this article is about. A separate piece deals with the second.

Most of the profit the house makes at blackjack is contributed by customers who play badly, violating simple rules that any book on the subject could warn them about. Even the 21 dealers will often point out to players their obvious mistakes. They do this out of simple decency, but the casino managements do not mind; they know that the player who gambles well will enjoy himself more, stay longer—and lose as much or more in the end.

But when a player uses the very best basic strategy, something different happens.

For the first time in the history of commercial gambling, there comes a time when the customer has the edge. True, his edge is so small as to be negligible over any reasonable period of play. The point is not that the customer has gained a tiny advantage, but that the house no longer has the rather substantial advantage it holds over the ordinary player.

The result is that it is extremely unlikely that the player will, over a period of time, either lose or win much; the casino is giving him free entertainment at the table. This is quite remarkable when you consider how much it is costing the casino in overhead and wages to run a 21 table.

To realize what a triumph this new, computer-generated, strategy is, you have to compare it with previous tactics.

The popular "never bust" tactic of simply taking no more cards whenever your total has reached 12 or more (17 if your hand includes an ace) gives the house an edge of about 6 percent.

"Mimic the dealer," which means keep drawing cards till your total reaches 17 or more, shaves the house advantage to 5.7 percent.

By using a little more judgment—and studying the older books on gambling strategies—you can do even better. You can cut the house percentage to a mere 3.2 percent.

Now to Professor Thorp's "best basic strategy" for completely wiping out the house advantage. It consists of doing the following:

1. If your hand is "hard" (does not contain an ace) use the following method. When the dealer shows a seven or bigger card or an ace, keep drawing until your total reaches 17. When dealer shows

4, 5, or 6, stand when you reach 12 or more. When dealer shows 2 or 3, stand at 13.

2. If your hand contains an ace: Keep drawing until your total reaches at least 18. Exception is that if the dealer is showing 9 or 10, you draw until your total is 19. (If you go over 21 by counting your ace as 11, you automatically count it as 1—and follow the rule above for a hard hand.)

Merely following these rules cuts the house advantage down to about 2 percent. You can eliminate even this small edge by properly doubling down and splitting. (In some locations, casino rules will permit doubling down only on hands totaling 10 or 11, whittling further the player's tiny edge with the best strategy.)

3. First, the strategy for doubling down when your hand is hard: Never double down when holding 12 or more. Always double down with 11. Double down with 10 unless dealer is showing 10 or ace. Do so with 9 if dealer is showing a 6 or smaller card. If your total is 8, double down unless it consists of 6 and 2.

4. If your hand is soft: If dealer shows 2, double down only if the card with your ace is a 6. If dealer shows 3, double down if you have a 6 or 7. Always double down when dealer shows 4, unless you have a pair of aces. Always double down when dealer shows 5 or 6; never when he has 7 or more. (But if you hold a pair of aces, split them rather than doubling down, unless house rules forbid.)

5. Now about splitting pairs. Split aces or 8s always, 10s, 5s, or 4s never. Split other pairs when dealer shows 2 through 7.

This is an abbreviated version of the perfect splitting strategy, but is almost as good as the detailed form which is considerably more difficult to memorize. Thorp gives you a choice, along with a vast amount of other information in his Beat the Dealer, published in paperback by Vintage Books.

He has even calculated for you what you can expect will happen if you apply this basic strategy over a period of time. Suppose you were to play 1,000 hands at $1 a hand. This would take you from three hours up. You could lose a lot of money or win a lot, but either eventuality is very unlikely.

There is only one chance in a thousand that you'd lose or win as much as $50; and the odds are about two to one that you'd wind up within $15 of where you started.

But that's the nature of any gamble where, as when you play blackjack with the strategy outlined here, you're getting an even break. In the long run, you'll come out pretty close to where you started, which in a casino is quite an accomplishment.

Actually Winning at Blackjack

As EXPLAINED earlier in this section, you need only adopt an easy strategy to convert a casino's blackjack, or 21, table into a nonprofit institution. You can cut the house's winning percentage right down to zero.

To swing the casino's edge clear around to an advantage for you—in short, to beat the house at blackjack—you have to go a step further in your tactics. You have to "count."

A deck from which most of the high cards have been played on the first few hands builds up the casino's advantage. One that is rich in high cards favors the player. If you can remember which cards have been used (of those you had a chance to see) you can make heavy bets when the situation is favorable. When it's not, you can stick to minimum bets to keep your place at the table.

That is the essence of counting—but when it was first discovered and publicized, it was enough to throw the world of casino gambling into an uproar.

It's hardly possible to remember every card that has gone by and to remember just which strategy is then called for. There are several systems to get around the problem. Possibly the simplest of these is the point-count worked out by Edward O. Thorp. To use this system, you regard a 6 or less as a low card and 10s, face cards, and aces as high cards.

As play progresses, you count each high card you see as +1 and each low card as -1. You ignore 7s, 8s, and 9s. Thus after observing a 10, a jack, two 5s, a 9, and an ace, you'd have a running total of +1.

At all times when that running total is zero or minus, you make the smallest permissible bet, just to hold your place at the table. The rest of the time, you multiply the minimum bet by the total to determine your bet. At a $1 table when the total in your head reaches +5 you bet $5.

Elementary as that is, it's all there is to the simple point-count system.

You can increase the odds in your favor by various small modifications. One way is to take account of the fact that the fewer cards that remain in the deck, the more significant is your knowledge of their high-low proportion.

The mathematical way to take advantage of this fact is by dividing your mental point-count by how many decks are left. At a table where only half a deck was left when you reached the point-count of +5, you'd divide 5 by 1/2 to get 10. And you'd make your next bet 10 units, or 10 times the minimum.

If the casino is one where two decks are used at once and one is about half gone, you'd divide by 1-1/2 and wager only about $3 at a $1 table. (To divide 5 by 1-1/2 readily you can double both and say one-third of 10 is 3 plus.)

Horseshoe Math

YOU ARE, let's say, a horseshoe pitcher of competition class. You know, therefore, just what your season's percentage of ringers is. We'll say it's 42.75. However, you've just had a lucky string. You suddenly discover that you've tied the local record and will set a new one if you can make at least one ringer in your remaining four tosses. What's the probability that you can do it?

Applying the logic also used for tennis in this section, you can figure that your chance of a ringer is 42.75 percent on each toss—or (subtracting from 100 percent) a 57.25 percent probability of missing. Converting from percentage to decimal, you make this computation: 0.5725 times 0.5725 times 0.5725 times 0.5725 equals 0.1074. Which says there's a little more than a 10 percent chance you'll miss all four times. So there's better than 90 percent probability that your name will go down in the local record books.

Of course, if your calculator has a constant, as it most likely does, you won't have to enter .5725 four times. You'll merely punch it once, then hit the times key four times or the times key once and the equals three times. The precise procedure with your calculator is between you and your instruction booklet. If the booklet isn't helpful, try experimenting.

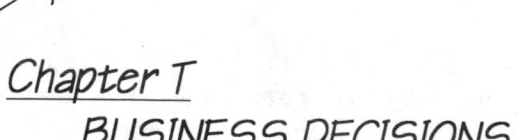

Chapter T
BUSINESS DECISIONS

Real-Estate Cash Flow
How Cash Flow Works
What's the Profit Rate?
Buying Earned-Growth Rate
Figuring Fiscal Ratios
How Durable the Building?
Counting Absenteeism

Real-Estate Cash Flow

MORE OF US are real-estate investors than realize it. Merely retaining ownership of a house from which for one reason or another you have departed, can make you one. This is an analysis that's useful in both choosing to invest and deciding whether to hang on or get out.

To analyze the cash flow from income property, run through four connected steps. These steps will successively calculate the potential gross income, the effective gross income, the net operating income, and the cash throw-off to equity.

Here they are, with results for a typical example, a 30-unit apartment building with rentals of $500 a month, a 5 percent vacancy rate, and an annual operation cost of $76,855. The property has just been financed with an 11 percent mortgage for $700,000 to be amortized in level monthly payments over 20 years.

First calculate the *potential gross income* by adding together all the monthly rental amounts and multiplying by 12. (Here we have $500 times 30 units times 12 months, $180,000.)

From that figure, subtract the vacancy loss and add any other income produced. (Reducing the potential income by the 5 percent vacancy loss of $9000 gives *effective gross income* of $171,000.)

Now subtract operating expenses ($76,855) to find that the *net operating income* (in this example) is $94,145.

Now subtracting the annual debt payments will tell you the cash throw-off. With these payments, taken from a table or calculated by one of the procedures given in this book, coming to $86,705, the *cash throw-off to equity* comes to $7440.

How Cash Flow Works

FINDING the amount of cash flow is a basic calculation in analyzing an income-producing property.

It is useful as the first step in determining such vital measures as rates of return. And it tells you whether an investment, however promising in the long run, is manageable in the short run.

The sample analysis that follows is founded on real figures for a 20-unit apartment building but of course the percentages used might be quite different for another time or place or structure.

The idea here is that you're considering buying an unfurnished apartment building in a suburb. Cost is $550,000 and you find that you can finance $450,000 of it with a 25-year mortgage at a fixed rate of 8.5 percent.

What percentage return can you expect to get during the first year on the $100,000 you must put up?

Gross income is the first thing to figure. With rents of $450 a month for each of eight studios, $550 for each of eight one-bedroom apartments, and $750 for each of four two-bedroom units, you find the maximum yearly rental income to be $132,000. Parking and laundry user fees promise to add $4,140 for a total gross of $136,140.

Vacancies and defaulting tenants will cost you something. Using local experience figures, you estimate this at 2.6 percent of gross. Multiplying the gross by 0.026 and subtracting expense of $3540 leaves net revenue of $132,600.

(When figuring such a discount, save time by subtracting the percentage from 100, then multiplying by the 0.974.)

Operating expenses will include maintenance (estimated at 12.2 percent of gross), utilities (12 percent), professional administration (8 percent of gross, a national average), and taxes and insurance (25.8 percent, based on a regional average).

These percentages of $136,140 ($16,610; $16,336; $10,892; $35,124) come to $78,962. With this deducted from net revenue of $132,600, what's left is net operating income of $53,638.

Mortgage payments, figured by one of the methods in this volume, come to $3623.50 a month, or $43,482 a year. Take that away from the net operating income.

Pre-tax cash flow is the result: $10,156.

And that's what you wanted to know.

Since you've gone this far, however, you might as well note also the return on investment. You can see at a glance that on the $100,000 you're putting up you'll be getting a little more than a 10 percent return. The precise figure, 10.156 percent, is found by dividing the cash flow by the investment, and then multiplying by 100 to produce a percentage.

Since you were after the immediate cash flow, you haven't added in the build-up of equity produced by that part of each mortgage payment that went toward the principal. It's of no use to you now, is tiny in the early years anyway and probably dwarfed by the appreciation—or depreciation—in value of the $550,000 worth of land and building.

What's the Profit Rate?

PR—PROFIT RATE—is an investor's or broker's term for a handy measurement of the ability of a company to make money with your money.

It can reasonably be described as measuring the profitability of a corporation. It shows the rate of return on the shareholder's equity capital at corporate book value.

To find the PR of one company for comparison with that of another, do the following:

Divide the profit per share by the book value per share as given at the start of the year.

Then shift the decimal point two places to the right to convert the result into a percentage.

Buying Earned-Growth Rate

THE WORLD of money is rife with terms that sound formidable on first encounter, but can prove to be quite simple and useful concepts.

One flourishing example is EGR, or earned-growth rate. It's defined as the annual percentage rate at which a company's equity capital behind each share of common stock is increased by net earnings after payment of dividends.

It can be important in your buying decisions. It is one of the measures used in comparing the worth of the stock of one company with that of another.

To find an EGR, first subtract the dividends paid per share from the earnings per share.

Then divide by the book value per share at the start of the year. (This can be found in all annual reports and most financial summaries.)

Multiply by 100 to convert this result to a percentage.

For example, if a stock earned $3.53 per share and paid $1.05 in dividends, while starting the year with a book value of $19.83, you'd begin with the first of these figures. Then you would subtract the second and divide by the third, and move the decimal point two places to the right.

This procedure should inform you that the value of a share grew just over 12.5 percent during the year.

Figuring Fiscal Ratios

AUTHORITIES warn us that we can deceive ourselves by judging the health of a business on its balance sheet or income statement alone. A business hovering on the verge of bankruptcy might produce a salubrious statement.

Easily calculated ratios can help us dig more deeply into the truth of things. Some of the most helpful of these have been isolated by Dennis P. Curtin, Jeffrey R. Alves, and Anne K. Briggs in their book, Controlling Financial Performance for Higher Profits. It's a good source of further information about the meaning and usefulness of the ratios that follow.

Ten ratios are used to answer seven questions.

The first question deals with liquidity: Can the company pay its bills?

Current ratio, which can be described as a general test of solvency, gives one answer to this question. To find the current ratio, divide the current assets of the company by its current liabilities. A result of 2 or more is commonly felt to indicate solid financial condition.

Quick ratio, sometimes called the acid-test ratio, is regarded as a conservative test of liquidity. Like the current ratio it can be calculated monthly from the balance sheet. There are two ways to figure it.

One is to add together cash, marketable securities, and accounts receivable and divide by current liabilities.

The other is to subtract inventory from current assets and divide by current liabilities.

An acceptable quick ratio for manufacturers is 1. Other kinds of businesses—wholesale, retail, and service companies—will usually show a ratio of less than 1.

Average collection period is a measure of how long a company waits—on the average—from the time it makes a sale until it has the cash from that sale in hand. You might wish to calculate it monthly or yearly.

Let's assume it's done monthly. Waiting till the end of a year is pretty late to find out how a business is behaving.

Begin by dividing the month's credit sales by the number of days in the month. Call this result daily credit sales and divide the amount of the receivables by it. Call that the average collection period, or ACP. A rule of thumb is that an ACP should not be greater than 1.3

times the collection period that has been established by the business. If that established period is the rather usual one of 30 days, then an ACP greater than 39 (1.3 times 30) suggests that the business is suffering from slow collections and may be extending credit unwisely.

Inventory turnover. This is a ratio that tells a business how fast its merchandise is moving. It is an indication of how efficiently a company's assets are being used and is sometimes called stock turn or stock-to-sales. Because of seasonal variations it is well to calculate it monthly.

Look at the income statement to find cost of goods sold. This is the expense item that includes the purchase price of stock—in a retail business. Or it is total cost of goods for sale—raw materials plus labor plus overhead— in the case of a manufacturing firm. Divide by inventory, as found on the balance sheet, to find the ratio called inventory turnover.

Generally, the higher the turnover the better. However, an unusually high rate can be a warning that the business may frequently encounter an out-of-stock problem that could lead to customer dissatisfaction. It can also be a warning that a condition of low-raw-material inventory may sometimes exist and lead to disruption of production.

During a period of substantial growth a year-end figure may be misleading. It may be better to use the average inventory for the period, found by adding together the beginning and ending inventories and dividing by two. The cost of goods sold is then divided by this to produce a ratio of inventory turnover.

Safety (leverage) ratios indicate a company's ability to pay its debts. A safety ratio should be figured annually and quarterly and, for a highly seasonal business, monthly as well. The basic such ratio is debt-to-worth and is found by dividing total liabilities by total owner equity. It is often considered desirable that this ratio be no greater than 1 to 1. A stable and predictable business, however, may operate quite soundly with a larger ratio. A ratio as large as 3 to 1, or even 4 to 1, is not necessarily out of line for a utility.

A misleading safety ratio may be produced in a company's books when some assets are carried at a depreciated value for accounting or tax reasons. A hidden-assets adjustment will then produce a more revealing debt-to-worth ratio. Get this by adding the value of the hidden assets to owner equity and using this total as the figure by which to divide total liabilities.

Profitability ratios are most usefully calculated relative to sales and relative to invested capital or assets.

One of these ratios is the profit margin, which it is well to calculate monthly. To find it, divide income before taxes (excluding extraordinary or nonrecurring items) by net sales. The decimal result becomes a percentage when multiplied by 100.

Thus an income of $40,000 divided by net sales of $250,000 gives a result of 0.16, read as a profit margin of 16 percent. Which says that after all expenses are paid, except taxes, there's 16 cents left over out of each dollar of sales.

Return on assets is a ratio that measures how much profit a business is generating with each dollar of assets. Calculated monthly it tells a good deal about both profitability and asset management. Large corporations often use this ratio as a yardstick of performance.

To calculate return on assets, first find income before taxes, on the income statement. Divide that figure by total assets as found on the balance sheet. I've been told that, in a rough way, 20 percent has generally been considered good. But this should not be applied blindly to all industries.

How Durable the Building?

Most buildings could last almost forever, but they don't. The time comes when it pays someone to tear them down, often more's the pity. The truth is, far more buildings are torn down than fall down.

Anyway, the United States Treasury Department long ago decided that the figures herewith are reasonable ones to use in estimating the lives of various kinds of buildings.

BUILDING TYPE	Good Construction	Average Construction	Cheap Construction
Apartment	40 years	40 years	33 years
Bank	50	50	40
Dwelling	50	40	33
Factory	44	40	33
Farm	50	50	40
Garage	50	40	33
Grain elevator	67	50	40
Hotel	40	40	33
Loft	50	50	33
Machine shop	50	40	33
Office	50	40	33
Store	50	50	40
Theater	40	33	29
Warehouse	67	50	40

Counting Absenteeism

THE STANDARD method of looking at the absentee rate in a business is by percentage.

To find it, multiply the number of lost days by 100 and divide by the number of employees and also by the number of working days in the period.

(Of course the same method works for figuring the rate by hours as long as you use hours for both time lost and working time available.)

Suppose 8 days of absence occur during a month having 21 working days, in an office with 15 people employed. Dividing 800 by 21 and then by 15 (or dividing 800 by the product of 21 and 15 if you prefer) shows the absenteeism rate to have been just a little more than 2.5 percent.

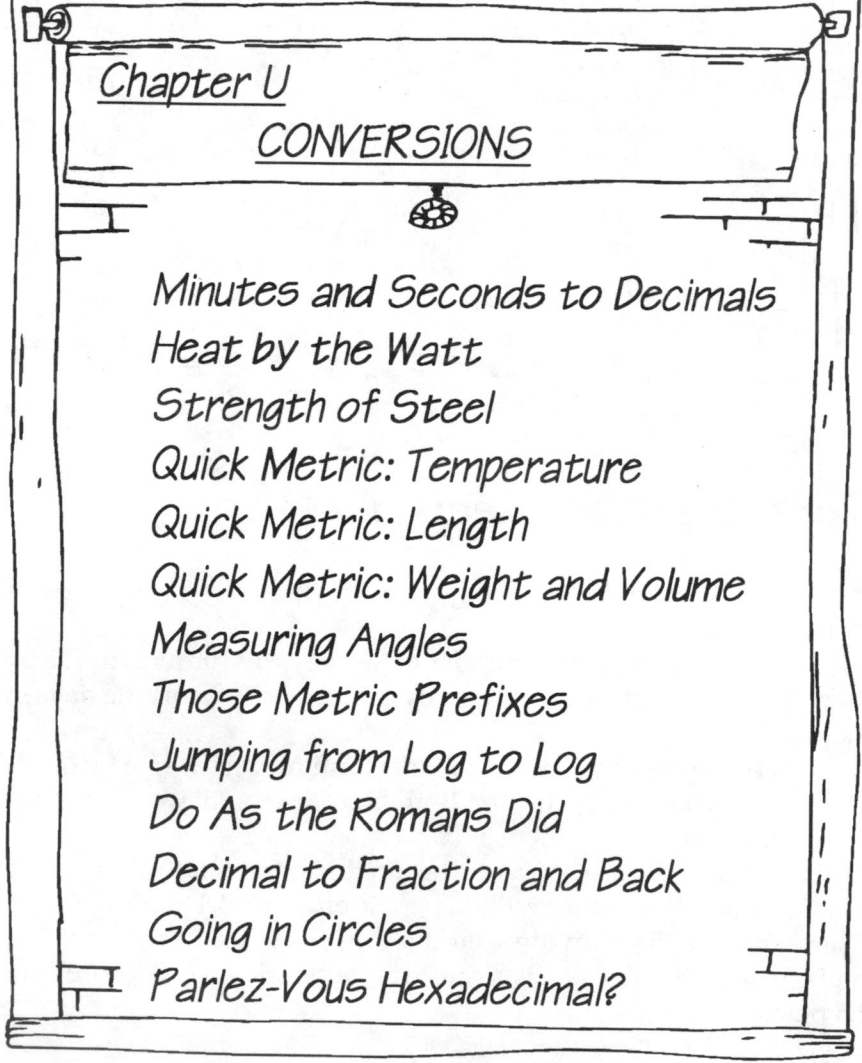

Chapter U
CONVERSIONS

Minutes and Seconds to Decimals
Heat by the Watt
Strength of Steel
Quick Metric: Temperature
Quick Metric: Length
Quick Metric: Weight and Volume
Measuring Angles
Those Metric Prefixes
Jumping from Log to Log
Do As the Romans Did
Decimal to Fraction and Back
Going in Circles
Parlez-Vous Hexadecimal?

Minutes and Seconds to Decimals

You MAY find yourself doing calculations that involve hours (or degrees—it comes to the same thing), minutes, and seconds. Could even happen while you're navigating a boat.

It's often simplest to convert minutes and seconds into decimals and add them to the hours (or degrees). Then for each measurement you have only one quantity to deal with.

To accomplish this, divide the minutes by 60 and the seconds by 3600 and add these sums to the number of hours.

If you start with 40 degrees (or hours), 30 minutes, 32 seconds, you'll add 30 divided by 60 (which is 0.5) and 32 divided by 3600 (which comes to 0.00889) to 40 to get the result 40.50889 hours or degrees.

Heat by the Watt

How MUCH heat, in British thermal units per hour, does an electric heater produce? The factor you need, derived from something called Joule's Law, is almost exactly 3.412. Merely multiply the wattage of the heater by it.

You'll find, for example, that a heater rated at 1500 watts will turn out 5118 Btus per hour.

Strength of Steel

If YOU KNOW the Brinell hardness number of the type of steel you're dealing with, you can get a very close approximation of its tensile strength by a simple calculation.

Just divide the hardness number by 2. Then move the decimal point three places to the right—or multiply by 1000, which comes to the same thing, of course.

Thus a steel with a hardness of 186 BHN will have a tensile strength of very close to 93,000 pounds to the square inch (psi).

From the Brinell hardness number you can also derive the Equivalent Hardness of a material. Divide the BHN by 10 to find the Equivalent Hardness on the Rockwell "C" hardness scale.

So 50 Rockwell "C" is about 500 Brinell; and 350 Brinell is about 35 Rockwell "C," found by dividing by 10.

Quick Metric: Temperature

As LONG AS we're still afflicted with remnants of the old English system, while trying to wean our minds off the familiar and get them more or less onto metric, there'll be occasional need for quick-and-dirty conversions.

Here are a few fast approximations.

Take half the Fahrenheit temperature and subtract 15. Be warned: This quick approximation is good enough for thinking about the weather—it's on the nose at 50 degrees F—but it's too far off in the region of 100 F. to use in deciding whether you have a serious fever.

To go the other direction, add 15 to the Celsius temperature, then double it.

Long before metrication loomed up for all of us, Americans living in Europe had these problems, and temperatures had bothered them much more than distances. For the more usual direction of conversion, C. to F., the rigmarole Americans abroad have found easiest to use is: multiply Celsius by 2, deduct 10 percent, add 32.

These suggestions, of course, are just rough ways to make the conversions in your head.

The real formulas are not very complex, either. Celsius times 9, divide by 5, add 32 yields Fahrenheit. And F. minus 32, multiplied by 5, and divided by 9 gives C.

Quick Metric: Length

DOUBLE the number of inches, then add half again, to get centimeters. (This is an approximation of the exact method, multiplying by 2.54.)

To go the other way, multiply the centimeters by 4 and divide by 10.

To convert miles to kilometers, add 60 percent. Or multiply by 16 and divide by 10. (To be more precise, you can multiply by 1.609.)

For kilometers to miles, multiply by 6, divide by 10—which means moving the decimal point one place to the left. (What you're doing here is approximating multiplication by 6.214.)

Quick Metric: Weight and Volume

To GET FROM pounds to kilograms, take half and then knock off a bit more. You'll be quite close if the amount you knock off is about 10 percent. Thus 10 pounds will become 4.5 kilos. (More exact is 4.536.)

Kilos to pounds: Double, then add 10 percent. Ten kilos will become 22 pounds, which is pretty darn close to the more precise 22.05.

I ONLY WEIGH 124 (... KILOGRAMS) AND I JUST LOST 12 (..POUNDS)

For ounces to grams, multiply by 30 and deduct 5 percent. (More exact: multiply by 28.35.)

Quart and liter are nearly equal but, like the critters in George Orwell's Animal Farm, some are more equal than others. Take 5 percent off the number of quarts to get the number of liters. Add 5 percent to go the other way. (Or be quite precise by multiplying by 0.9464 or 1.0567.)

Measuring Angles

THE USUAL way of measuring an angle is in degrees, of which there are 360 in a circle. Another rather common measure is the radian, which is 180 degrees divided by pi. That comes to about 57.2958 degrees, giving a handy figure to use if you're working with paper and pencil or a simple calculator. Given an angle expressed in radians, just multiply by that figure and you have it in degrees.

Of course, if you're using a calculator and it has a key for pi, you can more simply—and precisely—do that same operation of turning radians into degrees by multiplying by 180 and dividing by pi.

To go the other direction, multiply the number of degrees by pi and divide by 180.

The one-step paper-and-pencil routine for turning degrees into radians is to multiply by 0.0174532. (This works because that ugly decimal is pi divided by 180.)

Those Metric Prefixes

W<small>HEN YOU'RE</small> working with metric measurements—as you very slowly but pretty surely will be doing more and more—you have to know your prefixes.

To keep them straight, use this silly little memory helper: Kitty, How Does My Dog Catch Mice?

And, thanks to those initial letters, K, H, D, M, D, C, M, you'll know the order of the metric prefixes as the days of fuller metrication come along.

Each denotes one-tenth of its predecessor: Kilo, Hecto, Deka, (just plain Meter fits in here) Deci, Centi, Milli.

Jumping from Log to Log

T<small>WO KINDS</small> of logarithms are often used. What is called the common logarithm takes 10 for its base. Natural logarithms use as base the number designated by the letter e, the value of which is approximately 2.718.

A problem arises if you have a table of one kind of log and need the other kind. Fortunately, there are quick fixes.

To find the natural log of any number, divide its common log by 0.4343. When it's more convenient, you can get the same result by multiplying by the reciprocal of that number, which is 2.303.

The reverse of this naturally follows. Divide a natural log by 2.303 and you have the common log. Or multiply by 0.4343.

Those key numbers are the natural log of 10 and its reciprocal. So if you are working with a calculator that has natural logs only, rather than with a table of logs, you can enter the multiplier more efficiently by asking the calculator for the natural log of 10.

Do As the Romans Did

I<small>F</small> <small>YOU'VE</small> gotten a little rusty on the finer points of reading Roman numerals, you can be fooled quite easily—especially when someone would be happy if you didn't read a date. Titles of movies, especially ancient ones being shown on TV, are commonly accompanied by the obligatory copyright notice—with the year briefly flashed in Roman numerals. This practically guarantees that most viewers won't discover how old the picture is.

It reminds us of the importance of reading Roman.

So did an ad I recall for a rather costly publication that offered as a premium to subscribers a "Consumer Encyclopedia." You might assume you were to receive a new and original volume unless you grasped the significance of a tiny footnote (in 4 or 5 point, I'd guess, or half the height of most newspaper or book type faces). It told you that this premium was also published as "CONSUMERS ALL" DEPT. OF AGRIC. MCMLXVIII.

Aha! It happens that each year the U.S.D.A. puts out a book of useful advice and information, topics varying from year to year, that you can purchase or, by merely writing to your Congressman, get free.

What's being offered here as a subscription incentive is evidently a reprint of a free government book—but from what year?

As we all learned in school, Roman Numeral I is 1, V is 5, X is 10, L is 50, C is 100, M is 1000. To read 'em you add 'em together...except that when a lesser number precedes a greater, you subtract it.

So the first date here is 1000 plus (100 less than 1,000) plus 50, plus twice 10, plus 5.

And the other is made up of MCM, or 1900 and LXVIII, or 68. Decoded, the message reads that the folks offering this premium back in '75 had reprinted a 1968 Federal publication, the use of which is free to all, and were not all-out eager for you to notice that detail or the vintage of the volume. Did they get the idea from those brief flashes on the TV screens?

(It doesn't apply here, but there is one more rule for reading and writing Roman that should be inserted for completeness and because it isn't widely known. When a line is drawn above a numeral, its value is multiplied by 1000.)

Maybe you've wondered how in the name of Jove ancient peoples ever found a way to add and subtract with those unlikely Roman symbols. I don't think they did, or could have. I like to believe they computed mechanically, with scratches in sand or with the beads of an abacus, and only after reaching a result expressed it in those funny alphabetic numerals. That makes me feel better. How about you?

Decimal to Fraction and Back

HAVE YOU ever tried to figure building jobs in feet and inches—and fractions of inches—using a hand calculator?

Most calculators being decimal-oriented, they can cope with the monstrosities of our traditional system only when you help them with routines like those that follow.

Decimals to fractions. Temporarily ignoring the number of whole feet, multiply the decimal part by 12 to turn it into inches. Now, temporarily ignoring the whole inches, multiply the new decimal part by the denominator you want (by 8, for example, to find the answer in eighths). String the three results together.

Say you need a plank 12.85 feet long, which you must measure off to the nearest 1/16 of an inch using an ordinary feet-and-inches tape. Multiplying the decimal part (0.85) by 12 gives 10.20. Multiplying the new decimal part (0.20) by 16, since you want the answer in six-teenths, gives 3.20. Your answer is the original 12 feet plus the whole inches (10) plus the 3/16, or 12 feet, 10-3/16 inches.

Fractions to decimals. Naturally this is the reverse of the other operation. Temporarily ignoring the feet, divide the number of whole inches by 12. Divide the upper number of the fractional inches by the lower and also by 12. Add these results together and append them to the number of feet.

Starting with 12 feet 5-7/16 inches, you'd divide 5 by 12 to get 0.4167, then 7 by 16 and by 12 to get 0.0365. Adding these two results gives 0.4531, for an answer of 12.45 feet.

Going in Circles

IF YOU work with round things you will probably have no trouble recalling that to find the area of a circle you must multiply its radius by itself, then by something called pi.

I don't know about your school, but in mine it was rated fairly hilarious to greet the formula with a well-worn witticism along the lines of, "Like heck pie are square—pie are round!"

Since the value of pi is one of those confounded decimals that go on forever and so are deservedly called irrational, it has to be rounded off somewhere for practical use. Usually it is called 3.1416. Slightly less precise, but handier if you're working with fractions, is calling it 22/7 or 3-1/7.

If what you know is not the radius but the diameter, you can divide that diameter by 2 to find the radius. Or you can use a slightly different formula, one that says the area of a circle is equal to one-fourth the product of pi and the diameter squared. You can save a step by starting with the information that one-fourth of pi is 0.7854 (which is exactly as precise as using 3.1416 for pi) or approximately 11/14.

Either of the values of pi is also useful in working with circumferences of circles. The circumference, you may recall, is equal to the diameter (or twice the radius) multiplied by pi.

Thus if your city, like some, requires a permit for removal from a building site of any tree with a diameter of, say, 6 inches at a point four feet from the ground, you can check with a flexible tape. Diameter of 6 inches means circumference of 6 times pi, which comes to about 18-7/8 inches.

Parlez-Vous Hexadecimal?

IF WE WERE listening at all, I think most of us were pretty shocked the day an arithmetic teacher first told us there was nothing sacred about our 10-based number system. If it weren't for our having ten fingers, the base might have been almost anything. When I was told of a civilization that used 20 as its base number, I guessed that it might have existed in a climate so warm that people didn't wear shoes much.

Computers work in a primitive number system called binary (bi-as in bicycle or biped), which means it is based on the number 2. Since the decimal system doesn't translate readily into binary, people who work directly with computers, in machine language, must adopt either binary or some system reasonably compatible with it. Binary is slow and cumbersome, so most computer systems have come to hexadecimal, which is 16-based. A few use octal, based on 8, or 32.

You can change a number from one system to another quite readily, because they all use the same rules.

The last, or right-hand, number has the same value in all systems; that is, just what you'd expect it to have. The value of the next to last number, however, must be multiplied by the base of the system you're using. The digit before it is multiplied by the base number squared. And so on.

That's all there is to binary or octal or decimal systems. Since binary requires two characters, you need only 0 and 1. Octal employs the eight characters 0 through 7. Decimal, as we all know, uses the 10 digits 0 through 9, combining them to create any larger number wanted.

But hexadecimal calls for 16 distinct digits. Since it is helpful for these characters to be familiar and to be available on a standard typewriter keyboard, the letters A through F have been pressed into service to follow in sequence after 7, 8, 9. So the number of eggs in a dozen becomes C.

To see how all this works, let's try putting the number of eggs in two dozen into each of these number systems.

Decimal is pretty easy for us, since we've all grown up with it. When we add 4 in the right-hand column to 2 multiplied by the decimal base, 10, we get the required 24.

As we move from right to left in binary, we must multiply the value of each digit by 2, then 2 times 2 or 4, then twice that, and so on. We write 11000, which read from right to left means no 1s, no 2s, no 4s, one 8, and one 16. Adding 8 to 16 produces the two dozen we're looking for.

The next-to-last digit in octal takes the value of the base, 8. So the number by which it must be multiplied to give 24 is 3. Two dozen eggs in octal is written 30, for three 8s plus no 1s.

Now for the important hexadecimal. Placing 1 in the next to last column will produce 16 (following the rule that anything in that column must be multiplied by the number that is the base of the system). One 16 leaves us eight short of the needed 24, so 8 goes in the right-hand column. In hex, two dozen is written 18.

For a final touch let's examine the largest quantity that can be written in binary with no more than eight digits. In decimal terms, it's 255. In binary, it's eight 1s in a row, representing the sum of 1, 2, 4, 8, 16, 32, 64, and 128. Since hexadecimal values from right to left are 1, 16, 16 times 16, and so on, we write FF, representing 15 with the F on the right and 16 times 15 with the other one. Note that the next larger quantity, 256, is written simply 100.

Chapter V
MATH IN A HURRY

Breaking Down Numbers

ANY TIME you can work with 10 or 100 instead of a more stubborn number, you've turned a tough calculation into an easy one. Some examples:

Add 39 and 54. Takes a little thought, does it? Try adding 1 to the first number and taking it away from the second. Now all you must do is add 40 to 53, which is no great mental stretch.

Add 98 to 49. Same trick works, except this time increase the 98 by 2 and decrease the 49 by the same amount, so you can say 147, hardly doing any addition at all.

In mental multiplication, make changeovers by multiplying and dividing.

Like this: How much is 32 times 15?

Numbers ending in zero are the easiest to work with. So double that 15 to convert it into a handy 30, balancing this (fair's fair!) by halving the first term to 16. The process becomes simply 3 times 16, tack on a zero.

When one of your numbers is close to 10, 20, and so on, or to 100, turn it into that number plus (or minus) something. Regard 9 as 10 minus 1. Treat 21 as 20 plus 1. And so on.

Then how much is 28 times 9? That's not easy at first glance, so change it to 28 times (10 minus 1) and say to yourself, 280 minus 28 is 252. (You may find yourself simplifying even this little bit of subtraction at the end, by taking 30 from 280, then adding 2.)

Any number evenly divisible into 100 (or, for that matter, 1000) can be converted into simpler form. Treat 25, for instance, as 100 divided by 4.

So to multiply 48 by 25, all you do is multiply the 48 by 100 (that is, add two zeros to make 48 into 4800) and divide by 4.

To multiply by 75, do the same thing—but include a further multiplication by 3.

In division problems, you can often simplify things by turning a single division process into several. That may not sound very probable, but take a look at this.

You are asked to divide 336 by 8. They're both even numbers so divide both by 2.

The result, 168 to be divided by 4, is still even, so divide by 2 again. And then again (84 divided by 2 is 42 and that's your answer).

Great Chief Sohcahtoa

W_HAT FOLLOWS_ is not exactly a calculation. You might call it a bit of nonsense. But it's also just about the handiest bit of mathematical help I've ever had handed to me from half a dozen directions at once.

While conducting a mathematical column in a magazine, I confessed in print to a distressing inability to recall which of the common trigonometric ratios was which. That, I said, was why I'd taped the information to the back of my favorite calculator. Several dozen readers quickly responded with word about the great Chief Sohcahtoa, the letters of whose weird but rememberable name to this day keep me reminded that in any right triangle:

The Sine is equal to the Opposite side over the Hypotenuse. The Cosine is equal to the Adjacent side over the Hypotenuse. And the Tangent is equal to the Opposite over the Adjacent.

Of course you may prefer an alternative that one reader suggested: Simple Old Herman Carries A Huge Tub Of Apples.

The Squaring Trick

T_HERE'S_ often need to multiply a number by itself. Comes up so frequently that even medium-priced pocket calculators often have a special button just for squaring.

In a way, so does your brain.

Begin with a rather specialized trick. You can use it only with numbers that end in 5 or 0.5 (or one-half, which comes to the same thing).

Among photographers, however, those who eschew automation can learn to love it. Many lens stops end in .5, and squaring a stop value is a common need, since the amount of light admitted by a lens varies as the square of the stop.

To square a whole number ending in 5, first drop the 5. Multiply what remains by itself plus one. Then append 25. (Thus squaring 95 means multiplying 9 by 10 to get 90 and tacking on 25 to get 9025.)

Similarly, to square a mixed number ending in one-half, just multiply the whole number by 1 more than itself and add on one-fourth (or .25). The square of 10.5 (or 10-1/2) is 110.25 (or 110-1/4).

Another breed of number easy to square mentally is anything ending in 1. To the number in question add the next smaller number and the square of that smaller number. That's easy to do, since that next lower number always ends in zero.

So to square 31 you just add 31 to 30 to 900.

You can apply to any number the general rule from which these special rules have sprung. Maybe you remember this squaring rule from algebra: square of the first times twice the first by the second, plus square of the second.

To use it you must first turn the number you're working with into the sum of two numbers, making the first of these end in zero. Thus 86 becomes 80 plus 6. And 122 becomes 120 plus 2.

Applying the rule to the latter gives the square of 120 (14,400, as you know from the twelves times table) plus twice the product of 120 and 2 (480) plus the square of 2. If you write it out it looks like this: 14,400 plus 480 plus 4 equals 14,884.

For numbers like 38 or 129 with a last digit closer to 10 than to zero, you may prefer another version of the rule. Express your original number as a difference instead of a sum. See 38 as 40 minus 2. The rule then asks for the square of the first MINUS twice the first by the second plus the square of the second.

Halving a Mixed Fraction

Mᴇɴᴛᴀʟʟʏ dividing in half a number plus a fraction can be confusing if the number is odd. It's a problem that comes up frequently, as in woodworking where a nominal 4-inch board may have an actual measured width of 3-1/2 or 3-5/8 or 3-9/16 inches.

There's a shortcut that will save time and make an error less likely. It's a good bit faster and easier to do than it is to explain.

Subtract 1 from the whole number and divide by 2 to get the new whole number.

Add the parts of the fraction together to get the new numerator.

Double the old denominator to get the new one.

So this is how you find half of 9-13/16:

Nine minus 1 is 8, of which half is 4.

Adding 13 and 16 gives 29.

Twice 16 is 32.

And the answer is 4-29/32.

With a little practice, you'll find that you can perform this minor operation almost instantly.

(Adding mixed numbers has its perils, too. Following the standard routine of reducing everything to a common denominator is tedious. It often leads to errors when the list is long. A more efficient approach starts by seeking out combinations that add up to 1 and crossing those out.)

Pint's a Pound

JUST REMEMBER "a pint's a pound the world around" and you'll never have to wonder how many fluid ounces there are in a pint (16) or a quart (32) or a gallon (128).

Actually, if you want to be fussy about it, a pint of water weighs just a trifle more than a pound, and such fluids as oil, gasoline, turpentine, and kerosene weigh a bit less.

Illusion of the Shifting Base

LET'S SUPPOSE you had income one year of $19,444 and the next year of $23,111. Subtracting the lesser amount from the greater, then dividing by the earlier figure, you find the increase to be at the rate of 18.86 percent.

Percentage increases being a little confusing to calculate, you might want to check that result in a way that is easy to understand.

You could do that check by multiplying last year's figure by 1.1886.

Or you could multiply $19,444 by 0.1886 and then add on the original $19,444. Whichever makes more sense to you.

What goes around comes around, they say—perhaps the only observation on which dedicated optimists and pessimists agree.

So perhaps at the end of the next year you will find that your income has declined by 18.86 percent or thereabouts. You'll be right back right where you started, won't you?

I'm afraid not. If you subtract 18.86 percent of $23,111 from $23,111 you'll find yourself way down to $18,607.

You're up against the illusion of the shifting base. It is the reason that a 10 percent pay cut followed by a 10 percent raise constitutes a net loss of 1 percent.

And, just to rub it all in, if the 10 percent raise comes first and the 10 percent cut afterwards, you're still the loser by 1 percent.

(Here, by the way, is another kind of calculation of percentage of increase that may prove useful in other contexts. Suppose that knowing only that this year's income of $23,111 is an increase of 18.86 percent over last year's, how would you calculate how much you earned last year? Expressing the percentage in decimal form, adding 1 (0.1886 plus 1 equals 1 1886) and then dividing the $23,111 by this gives you $19,444.)

The Catch-22 Deduction

YOU MAY think you're in Wonderland with Alice the first time you encounter a strange rule that turns up now and then in such likely places as U.S. income-tax regulations. It has been found in, for example, the limits on Keogh Plan pension contributions by self-employed people and in deductions for contributions to 501(b) and 401(k) tax-deferred savings plans and in the limitations on charitable contributions by certain small businesses classified as "C" corporations.

The rule may tell you that you are allowed to deduct no more than 5 percent of net income. And then proceed to define net income as what you have after subtracting not only your other expenses but the contribution as well. Ay, there's the rub: how can you calculate 5 percent of a figure you haven't figured out yet?

As you will see, it's tough even when you're working with round numbers. Suppose net income before contributions is $10,000 and you're allowed a 5 percent deduction. Take 5 percent off $10,000 and there's $9500 left.

So you're allowed 5 percent of $9500, or $475.

But when you take that from $10,000, you discover that.... Seems there's no end to it.

But, fortunately, there is a formula to guide you through this thicket.

Just state the allowed percentage as a number (5 if you're permitted 5 percent) and divide by 1 plus the percentage stated as a decimal (1 plus 0.05, or 1.05). This tells you what percentage of the net income you may deduct. In our instance this comes out 4.7619 percent. This times the $10,000 you started with comes to $476.19, so that's how much you may deduct.

Totaling Fibonaccis

THE NUMBER series named for Leonardo of Pisa, known as Fibonacci, has inspired not only whole books but clubs and magazines as well. Just one of its fascinating aspects is that it precisely describes the course of rabbit breeding if the frequent, if harsh, custom is followed of eating the parent bunnies after two litters have been produced.

Start with a pair that has a litter after one month and a second after another month. The third month there are two litters, one from the old pair, one from the young. The fourth month there are three litters, the fifth month five, and so it goes.

The number series then is one in which each term is the sum of the two preceding it and that's the Fibonacci: 0 1 2 3 5 8 13 21 34 55 89....

(If you happen to have a programmable calculator, it will probably allow you to generate this series for as many terms as you like. With Texas Instruments' TI-59, try the program Sum, 0, Esc, 0, Pause, RST. Enter 1, press RST and R/S, and watch the Fibonacci numbers fly. For several other calculators, including some from Hewlett-Packard, you can find programs for this series in accompanying handbooks.)

There may be no news in this so far for people who habitually fool around with numbers. Not so widely known, though, is a quick trick for finding the total of any ten consecutive numbers in the series. Just multiply the seventh number in any Fibonacci series by 11. Try it with the part of the series given above and you'll indeed find that 11 times 21 is equal to the sum of the ten Fibonacci numbers that end with 89.

Multiply Like a Peasant?

Do you find it as surprising as I do to learn that you can multiply together large numbers by a process of halving, doubling, and adding? That means getting by with a multiplication table ending with the twos. None of that 7 times 9 stuff that made us so much trouble in grade school.

And that, I suppose, is precisely why this method was used by European peasants three centuries ago. Presumably both illiterate and virtually innumerate, they'd have found it pretty hard to use the routine familiar to us.

As you will see, the peasant way is slow—but wonderfully undemanding.

To use it, first place the numbers to be multiplied at what will become the tops of two columns.

Divide the first number by 2 and that result by 2 and so on until you reach 1, putting the results in a column under the first number. Ignore fractions; as far as you're concerned half of 9 is 4.

Under the second number that you're working with, put the results of successively doubling it, stopping when you have placed a number opposite the 1 in the first column. Select from the second column every number that is opposite an odd number in the first column. Adding these numbers will give you your answer.

No kidding.
Here's how all this looks when you multiply 36 by 47.

36 x	47	
18	94	
9	188	188
4	376	
2	752	
1	1504	1504

1692 is the answer

When you calculate in this manner, you're actually using binary, or 2-based, arithmetic. That's the kind computers use today, so don't elevate your nose at it.

Would we have computers at the dawn of the Twentieth Century if the great mathematician Gottfried von Leibnitz had not learned this trick from watching those dumb peasants?

Easy Iteration

FANATICAL puzzle-hounds have been known to muddle away at problems for weeks, or even years, because the solution boiled down to one of those miserable third-order or fourth-order equations. Meaning they contain x to the third or fourth power.

Such equations are almost hopelessly tedious to solve by ordinary and proper mathematical methods. Indeed, a peculiarly formidable math handbook that I use mostly to intimidate myself gives only limited treatment to the solution of fourth-order (quartic) equations because the formula for the general solution is so complicated it is "hardly used."

Yet by the power of the repetitive process called iteration, which is really just trial and error, applied by way of an ordinary hand calculator, you can tame one of these beasts in minutes.

I can offer as an example a puzzle I once published that tormented readers by leading them into a nasty fourth-order equation.

Expressed as it would sound if read aloud, here's the (correct) equation most came up with:

Four times x to the fourth power, plus 12 times x cubed, plus 9 times x squared, minus 486 times x, minus 729 equals zero.

You can see how this problem can be solved by trial and error. Guess a value of x that makes sense in this problem. Substitute this value in the equation. Then try another number. As you find the range of the guesses that bring the value of the left side close to zero, you'll close in on an exact solution.

I don't think you'd want to try it without a calculator. Maybe not even with one.

Fortunately, there is a marvelous variation on this iteration method. Tolerable without a hand calculator, with one it's a piece of cake. Simply go through the equation from left to right, totally ignoring the exponents (the business about fourth power, cubed, squared) as you replace x with your best guess of what it might be.

Even if you have to make your first guess pretty much at random, further trials will rapidly bring you closer and closer until you have a solution that is correct to as many decimal places as you choose to go for.

When this procedure produces a left side of the equation that is equal to zero, you've solved the equation.

Clues in this particular problem indicated that the answer should fall in the region of 4 or 5. If you were to substitute for x those two numbers in turn, you'd see that the answer must fall close to midway between them. And, in fact, it does turn out to be just 4.5 when you make that your guess.

On the chance that I've muddied this manipulation too badly, let me run over it step by step: guess times 4; add 12; times guess; add 9; times guess; subtract 486; times guess; subtract 729. (If any power of x is absent from the equation you're working with, supply it with zero as coefficient. Thus if the equation had no term between x cubed and x, you'd fill the gap with 0 times x squared.)

Smidgens to Oodles

Though books of arithmetic note it not, there is a system of measurement that may well survive even after metrics have knocked out feet and pounds altogether.

Its units are handfuls and splashes, bits and pieces, jots and tittles, bunches and oodles.

Its informal coinages include the glug, passed along to me by one of my machetunim.

(Forgive the interruption, but this valuable word demands instant adoption into everyday English to fill a gap. It's Yiddish for those numerous relatives by marriage for which not every tongue provides a precise name.)

He got "glug" from a servant in the Southern household in which he grew up who, in orally passing along her best recipes (receipts, she called them) used the term to denote the quantity of molasses poured between each pair of glugging sounds from the jug.

There's little use offering a scale for such measurements as long as no two of us has bunches, hunks, passels, or tittles of the same size. All the same, a laborer in the field, Joe Ecclesine, has done his best for us.

PAPRIKA

PAPRIKA MEASURE

2 dashes = 1 smidgen
2 smidgens = l pinch
3 pinches = l soupçon
2 soupçons = too much paprika

ALCOHOLIC BEVERAGE MEASURE

2 fingers = 1 tot
2 tots = 1 shot
2 shots = 1 slug
4 slugs = 1 snootful
2 snootfuls = 1 night in jail

A sex-linked differential has been noted in at least one popular measure. The dollop, which is how much sour cream you add to the soup, has been found to turn into the double-dollop when it's a man who is doing the kitchen duty.

"Too good to miss," said the editor, of the contribution of a reader of *Consumer Reports* who sent in a Cheez Doodles package and "with dcft analytical thrusts," pointed out that the cheese appears after the salt in the list of ingredients, which, under the law, means there's less cheese than salt. The package also indicates that the product contains about 0.002 ounces of salt. Ergo, Cheez Doodles contains less than 0.002 ounces of cheese.

"Since 'oodles' is plural, the product must contain at least two oodles. So one oodle of cheese would equal less than 0.001 ounces." With inexorable logic, the reader concluded: "For those interested in obscure measurements, an oodle would be about two-tenths of a scruple, so what we seem to have is a case of almost no scruples."

What, Me Solve Quadratics?

You MAY think solving quadratic equations (you know, the kind that start with x squared) is something done only in algebra classes.

But these equations turn up frequently in practical work, to say nothing of mathematical recreation. I found that out the other day, when I tackled one of those fiendish little mathematical puzzles that look so simple.

I found out that the formulas you meet for solving quadratics look pretty awful, especially the sequential version offered for use with an electronic calculator.

But actually here's all you have to do.

Arrange the equation in the normal form—something times the square of x, plus something times x, plus something. Give these somethings convenient names: first number, second number, third number.

Multiply 4 times the first number by the third number. Subtract that from the square of the second number.

Take the square root of the result.

Add the square root to the second number.

Divide by twice the first number.

Change the sign in front of the result.

Now from that same square root subtract the second number and then divide by twice the first number.

Each of these last two operations will give you one answer to the problem, since all quadratic equations have two roots. Both answers will be mathematically correct, although you may have to discard one on common-sense grounds. When a problem deals with a real distance, for instance, you'll naturally toss aside any solution that is a negative number.

Add-On Taxes and Take-Off Discounts

CALCULATING the amount of a sales tax and then adding it on, or of a discount and then subtracting it, is the slow way—if all you want to know is the final price.

In either case, it's simpler when done in a single step.

With a sales tax such as 6 percent, merely multiply the price by 1.06.

With a discount such as 10 percent, just multiply by 0.9.

Depreciation by Sum of Years' Digits

THIS ROUTINE may be more or less a historical note for most of us. It appears that the income-tax people no longer encourage its use for new purchases. But it is so wonderfully weird that it seems worth preserving for the benefit of those who appreciate an odd way to depreciate—if that makes sense. Other than that, I can only say that anyone who does need it will need it badly.

One long-accepted method of writing off the cost of a car or piece of business equipment uses a method called the sum of the years' digits. It may seem a strange way to figure depreciation, but in practice it often works out rather well. To use it, begin by determining the number of years over which the depreciation is to be taken. Add to that number all the smaller whole numbers.

Use that total as the lower part of a set of fractions. For the first year the upper part is the number of years over which you've chosen to take the depreciation. For the second year it is the next smaller number. And so on. Multiply the original value by each of these fractions to find the amount of write-off for each year.

If you're depreciating a $9000 computer over 5 years, the digits will be 5, 4, 3, 2, 1, for a total of 15. First-year depreciation is 5/15 of $9000, or $3000; second is 4/15 of $9000, or $2400. It will be fully depreciated after 5 years.

Addition Helper

OF ALL arithmetic shortcuts, this one is perhaps the biggest time saver, because you'll use it the most often.

Adding 10 is easy, so watch for 10-pairs when adding a column of figures. These are the number combinations that add up to 10, such as 3 and 7 or 5 and 5. You'll soon learn to spot and use them even when there's another digit or two in between.

You may eventually come to handling trios that total 10, like 2, 3, and 5, in the same way.

Done efficiently, mental addition is often faster than keying a list into a calculator—even assuming you have one of those lovable little machines right at hand.

Add Without Carrying

Do you get confused by carrying when adding long columns of multi-digit numbers? Maybe you've lost the knack, what with using a calculator whenever there's one around.

By totaling columns separately, you can avoid carrying.

As you can see, you could equally well start with the left-hand column. Advantage of doing it that way is that when an approximate answer will serve, you can stop as soon as you reach one that you can see will meet your need.

```
   12,344
   54,323
   98,765
   59,789
      21
      20
      20
      23
      20
  225,221
```

Subtract Without Subtracting

If you'd like an alternative to the usual way of subtracting several-digit numbers, try this.

Write the number from which you are to subtract.

Under it, instead of writing the number you are taking away, write what is called the nines complement of each digit.

(The nines complement is the number which, added to the number you have, will make 9. For instance, the nines complement of 6 is 3.)

Then add.

Now separate the left-hand digit of that answer and add it to the rest of the answer.

Suppose that from 2372 you wish to subtract 1894. The nines complements of the four digits 1, 8, 9, and 4 are 8, 1, 0, and 5.

So to 2372 you add 8105, getting 10,477. When the first digit, 1, is separated from the rest of the result and then added to it, you get your final answer, 478.

This may not be easier than the good old way, except maybe for those folks who had such a dreadful time when they first met subtraction in grammar school. But at least it could say, as Jimmy Durante said of himself, "Me, I'm a novelty!"

As a point of further interest, I'm told that this is how a computer would handle the problem, since computers are pretty simple-minded and not at all good at subtracting. A consoling thought for all of us.

Mix It the Midpoint Way

THE TRADITIONAL mixing problem is a nice exercise in elementary algebra. It's not especially difficult to handle, once you've formed the simultaneous equations and recalled how to solve them.

But not all of us have retained our algebraic skills in rust-free condition.

So a kindly soul has worked out something that might be called the midpoint method.

It's a method for solving mixture problems without recourse to algebra, and who can hate that?

Here it is, applied to a typical question.

With peanuts worth 65 cents a pound and cashews 90 cents, how much of each do you need if you wish to produce a 40-pound sack of mixed nuts worth 75 cents a pound?

1. Find the midpoint or average of the two values (adding 65 to 90 and dividing by 2 gives 77.5).

2. From this result, subtract the wanted value (77.5 minus 75 is 2.5).

3. From the midpoint result, subtract the smaller value (77.5 minus 65 is 12.5).

4. Multiply half the wanted amount by the result of step 2. Divide by the result of step 3. (Forty divided by 2, multiplied by 2.5, and divided by 12.5 is 4.)

5. To this, add one-half the quantity wanted (4 plus half of 40 is 24).

This is the amount to use of the lower-valued ingredient—24 pounds of peanuts.

6. Subtracting this from the total amount gives the figure for the higher-valued ingredient.

Forty less 24 says you should use 16 pounds of cashews.

Better Way to Mix It?

AFTER TRYING both the traditional algebraic and the novel "midpoint" approaches to mixture problems, I wondered if there might not be a simpler approach than either. Maybe you'll think this one that I've worked out fills the bill.

As an example I'm using the same nutty problem: With peanuts at 65 cents a pound and cashews at 90, how much of each kind will you require to produce a 40-pound sack of mixed nuts costing 75 cents a pound?

1. From wanted value, subtract smaller value (75 minus 65 is 10).

2. Multiply this by quantity wanted (10 times 40 is 400).

3. Now divide by the difference between the two values (400 divided by the difference between 90 and 65 is 16).

So there you have the quantity needed of the higher-valued ingredient. The remaining 24 of the required 40 pounds must be the other kind of nut.

This seems so simple it might be worth trying on another practical problem. Given an 18 percent solution and a 24 percent solution, how much should you use of each to make up 90 ounces of a 22 percent solution?

From 22, subtract 18. Multiply by 90. Divide by 6, which is the difference between 24 and 18. Result: Use 60 ounces of the higher valued (24 percent) solution, making up the rest of the required 90 ounces with 30 ounces of the 18 percent stuff.

Triangle a Problem

A SURPRISING number of measurements and calculations can be made and problems solved by converting what doesn't sound at all like a triangle into just that familiar geometric figure.

One example is estimating the area of an odd-shaped piece of land. Subdivide a map of it into rectangles and right triangles. The areas of these are easily calculated.

Here's another example. I'm putting it in partly because it illustrates a basic method...but mostly because it produces an answer I find quite surprising.

The Golden Gate bridge that takes you northward out of San Francisco is about a mile long. On a warm day (they don't hardly have hot ones there) it expands about two feet. If there were no expansion strips to compensate, how high a bulge would this make? Assume the whole bridge will bulge upward in a uniform curve.

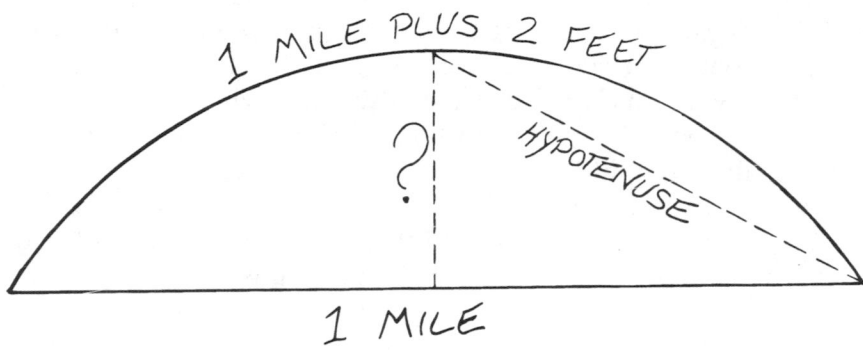

Have you made your guess?

An easy way to calculate this is by taking the height of the bulge as one side of a right triangle that has half a mile as another side and 1 foot plus half a mile as the hypotenuse.

Going back to Pythagoras, as you usually do with right triangles, you remember that the square of the hypotenuse is equal to the sum of the squares of the other two sides. To find any side, then, you square the hypotenuse and the known side and take the square root of the difference between them.

This is quickly done with a calculator (or slide rule, for that matter, if you happen to be the last of the die-hard engineers). To spare you the agony of pencil arithmetic if you have neither of those tools at hand, I'll tell you that 2,641 squared is 6,974,881 and 2,640 squared is 6,969,600. The difference is 5,281, of which the square root is 72 plus.

So the extra two feet of length would produce a bulge more than 72 feet high.

I find that remarkable. It must be a problem to engineers.

How Big Is a Triangle?

W HETHER doing a little land measuring or cutting a sail, you may have reason to calculate the area of a triangle. If it's a right triangle— the kind that contains a right angle—it's easy enough. Just multiply together the lengths of the legs and divide by 2.

There are formulas that work for triangles of any shape. Here is one that doesn't require a trig table, though it does demand finding a square root. Your calculator probably has a key for this function. If it doesn't, you can always chase it down through a table of square roots, the good old paper-and-pencil method from school days, or through the trial-and-error method that the simplest hand calculators have turned into a cinch.

Add together the lengths of the three sides of the triangle to find its perimeter. Divide the result by 2. Since this is one half of the perimeter, let's call it half-p.

From half-p, subtract the length of one side of the triangle. Also from half-p, subtract the second side. And from half-p, subtract the third side.

Multiply those three results together. Then multiply that product by half-p.

Find the square root. That's the area of the triangle.

Give it a try. If you're sewing a sail having sides 8, 10, and 12 feet, the perimeter will be 30 and half-p will be 15. Subtracting each side from 15 will give you 7, 5, and 3 to multiply together and by 15 to produce 1575, of which the square root is 39.7 square feet or (when divided by 9) about 4.4 square yards.

Approximate a Time Payment

L OOK, MA! No algebra. No calculating device. No fat book of tables to refer to.

No logarithms, either.

Without any of those helps, you can figure out—with surprising accuracy that's more than good enough for most decision-making— how much your monthly payments will be for a home mortgage or a car installment loan.

A rather weird approximation formula lets you immediately, with reasonable use of pencil and paper, get this useful result for any preliminary purpose. Other methods given in this book can be used at leisure to find payment amounts to the penny.

All you need to supply in order to use this approximation method is the number of monthly payments to be made, the annual interest rate, and the amount of the loan. Juggling these will produce three quantities to add together.

1. First divide 1000 by the number of payments.

2. Now multiply the interest rate by 0.4286

3. Multiply the interest rate by itself and by half the number of payments. Move the decimal point four places to the left.

Add these amounts together. The result is the estimated monthly payment for a $1000 loan.

Multiply by the number of thousands of dollars of installment loan. The answer will be a close approximation of the monthly payment.

Does it work?

Try it for a $100,000 mortgage at 9 percent to run for 20 years.

Divide 1000 by the 240 payments to get 4.17.

Multiply 0.4286 by 9 to get 3.86.

Squaring 9, multiplying by half of 240, and dividing by 10,000 gives 0.97.

Adding up these three quantities and multiplying by the number of thousands of dollars of the loan (100) says the payments will approximate $900 a month. You're within a few dimes of what an actual calculation or the fat book in your banker's desk says it should be—$899.72.

Tested for a 25-year, 6 percent $100,000 loan, this method gives $644. Not bad, since the precise answer is $644.31.

It seems to work equally well for a short-term personal or auto loan, too. The book says $1000 at 12 percent for 4 years will cost you $26.34 a month. Our method estimates $26.31. How about that?

You may wonder where such unfathomable approximations as this come from. Beats me. Someone must have doodled a long time for this one. In somewhat different form, it was passed along to me by James Booher who was my first-year college-math instructor lo! these many years ago.

What Percent of...?

WHAT PERCENT of 16 is 4?
   What percent is 4 of 16?
   What percent of 4 is 16?
   It's easy to be confused by these "percent of" questions.
   Maybe it will help if you simply remember to start with the number after "is" and multiply it by 100, then divide by the other number.
   Approached that way, the first two questions prove to be identical at 25 percent. And the third one comes to 400 percent.
   You may not need the rule for those familiar fractions—but without it might you not fumble a bit before replying to "What percent is 73 of 37?"

Percents, Decimals, Fractions

CONVERTING percentages and decimals to fractions and the other way around is child's play. That's because children, the older ones anyway, have probably just been studying that kind of stuff in school. For the rest of us it may not be so easy.
   It helps to keep in mind that all these things are fractions. A percent is just a fraction in which the lower number, or denominator, is 100. A decimal is a fraction in which that denominator is 10 or some power of 10. And all fractions can be regarded as merely uncompleted divisions.
   If you want to turn 3/40 into a decimal, just do the division indicated by that slash mark. Divide 3 by 40 to get 0.075, which is 75/1000.
   If you want the result as a percent, turn that fraction into hundredths by dividing each part by 10. The 75/1000 that started as 3/40 now can be seen as 7.5 percent.
   Making the trip the other way, from a decimal to a fraction, requires a decision. What do you want the lower number to be? You might choose 64, since 1/64 of an inch is a useful and quite precise measurement for many things.
   To convert a decimal such as 0.35 to 64ths, just multiply the two numbers together. Comes out about 22/64.
   Similarly, you can turn one fraction into another. What is 3/40 in 64ths? Divide 64 by 40 and multiply by 3. The closest whole number to your 4.8 result is 5. Your answer rounds to 5/64.

Quick Fractions

W HAT YOU'RE most usually called upon to do with a pair of fractions is to divide, multiply, add, or subtract them. You can do any of these things in a hurry and almost without thinking, even if you're out of practice.

To multiply, just write the product of the numbers above the line and then, below the line, write the product of the numbers below the line.

The same goes for dividing, except that first you must lay hold of the fraction you are to divide by—and turn it upside down.

Quick way to add a pair of fractions begins with cross-multiplying and adding the products to get your new numerator. Then, below it put the new denominator, which you obtain by multiplying the old denominators together. (Save time by canceling first wherever you can.)

Subtracting one fraction from another proceeds much like adding. But this time you subtract one product from the other instead of adding the products together.

That Tricky Percentage Increase

Y OUR INCOME from some source in one year was $19,444. For the next year it was $23,111. This being during a period of notably high inflation, you wonder what the trend of your real return was. Checking back, you learn that the published rate of inflation for the year in question was 13.4 percent.

How did you fare by comparison? Try this for size.

Subtract the earlier year's income from the later year's larger one. Divide that difference by the earlier year's amount. Push the decimal point two notches to the right, thus multiplying by 100 and converting the fraction into a percentage. Well, it seems you had an 18.86 percent increase in the number of dollars coming in.

Not bad, although inflation gobbled up more than two thirds of it. To be more precise, it ate some 71 percent (13.4 divided by 18.86 and multiplied by 100).

To look at this event another way, subtract the inflation percentage from your rate of return. The effective rate of return, then, was 5.46. A rule of thumb used by some investment seers says this is pretty much what you might expect over the years.

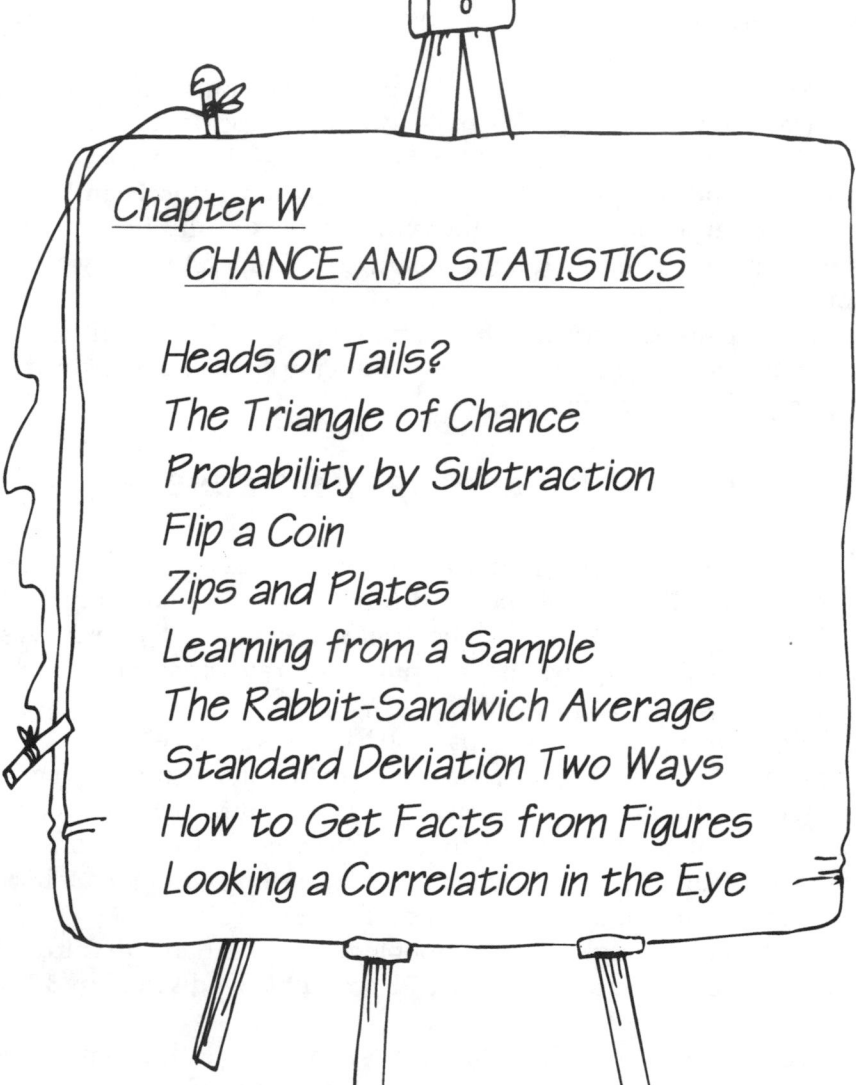

Chapter W
CHANCE AND STATISTICS

Heads or Tails?
The Triangle of Chance
Probability by Subtraction
Flip a Coin
Zips and Plates
Learning from a Sample
The Rabbit-Sandwich Average
Standard Deviation Two Ways
How to Get Facts from Figures
Looking a Correlation in the Eye

Heads or Tails?

THIS LITTLE item gets into a how-to-figure-it book under the heading of how not to figure it.

For background, let's begin by establishing that you're too sophisticated to fall for what is called the maturity of the chances, or the gambler's fallacy.

That error is simply a belief that if a coin has come up heads most of the time in a series of tosses, it must thereafter favor tails until things even out. Or some similar nonsense to the effect that a run of red at a roulette table indicates that black is "overdue."

The error is based on a misunderstanding of what the law of averages really means when it says that in the long run heads and tails will tend to appear with equal frequency. So they will—from here on at any given moment.

What they will not do is tend to show up unequally in order to rectify whatever has already happened.

Unless they are defective, coins and dice and roulette wheels and other instruments of chance have no memories. (No consciences either, in respect to such matters as whether your baby needs new shoes.)

What has set me off on this is reading an article in a national magazine. Its author, having explained all this, takes off from it to a conclusion that could cost you money if you went with him. Since you may have encountered this thesis there or elsewhere, or even come to the conclusion yourself, it is worth exploring in the interests of protecting your pocketbook.

Says the author: "Psychologists and mathematicians have discovered that seven out of ten times, a person predicting the flip of a coin will call "heads."

I think that's very likely true. But now let's follow the author to his misguided conclusion.

"Therefore, make sure that you are the tosser, and your opponent the caller. Scientifically, you see, the chances are even for heads or tails to come up. But if he calls heads 70 percent of the time, you have a 3 to 2 chance of coming out ahead."

Somehow it sounds convincing, doesn't it? But the fact is that if the pigeon calls heads and tails with equal frequency he'll tend to be right half the time. If he calls heads on every flip he'll be right half the time, in the long run. Just remember that each flip is independent of the past and of the future—and for that matter of the call.

The Triangle of Chance

You MIGHT read that somewhere a woman has given birth to her eighth child, and for the eighth time it's a boy. How remarkable is this? In mathematical terms, what are the odds against it?

"When you toss a coin..." is a phrase that turns up in all sorts of scientific writing, on subjects as diverse as card playing and genetics. The toss of an "honest" coin is a perfect model for calculating probabilities.

Flip a coin once and the possible outcomes are two, with an equal chance for each.

Toss two coins (or one twice) and you may get a head and a tail or two heads or two tails. But not with equal probability, since there are two paths to head-and-tail: head then tail, tail then head. In somewhat more complex instances, this distinction caused a lot of confusion among learned mathematical people at one time, as it still does among amateur and professional gamblers—often to their financial woe.

If you flip a quarter 10 times, how probable is it that you'll get precisely four heads and six tails? Assuming boys and girls appear with equal frequency, which is close enough to true for purposes like ours, how often will a brood of eight children be equally divided?

You can list all the possible paths and add up the number that meet the requirement. But as the numbers dealt with become larger, you'll find it an increasingly tedious chore.

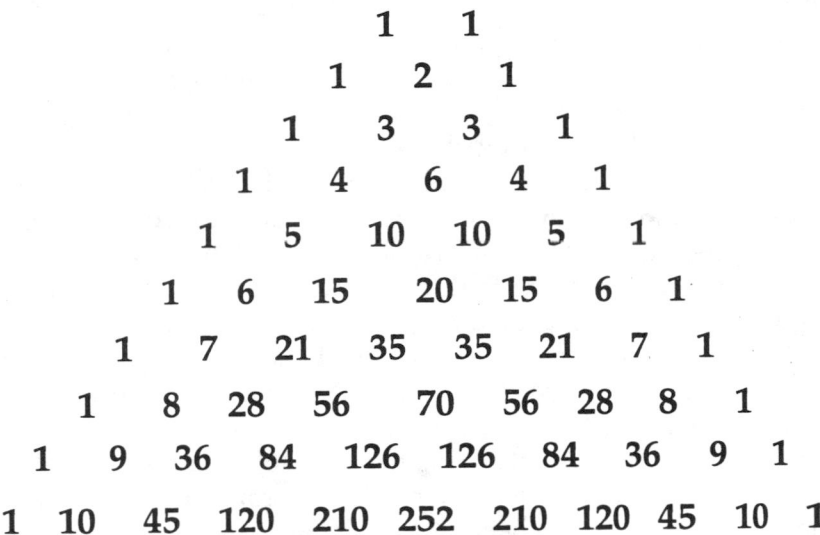

That's what makes a triangle devised by Blaise Pascal, more than 300 years ago, both fun and useful. As you will discover, the basic form displayed here can be expanded indefinitely to calculate the distributions of even longer runs of heads or tails—or girl or boy babies.

You'll quickly detect the pattern that produces the triangle. Each added number is the sum of the two numbers most nearly above it. You can go on adding lines indefinitely, if you wish to consider cases of more than ten coin tosses or children.

Now what does the triangle say about that family of eight kids— or any other number of youngsters?

After the zero line, which says in effect that all groups of no children are the same, line 1 says that with 1 child there are 2 possibilities (1 plus 1), which we hardly needed to be told.

Line 2 (for 2 children or 2 coin tosses) totals 4 when added across. Of these 4 possibilities, the first and last numbers represent the single possibility that both will be boys or both will be girls. The 2 in the middle is for the two chances in 4 that there will be one child of each sex.

Line 3, for 3 kids, totals 8 for the 8 possibilities. There is 1 chance in 8 of all boys, 3 of 2 boys and a girl, 3 of 1 boy and 2 girls, and 1 of 3 girls.

Jumping down to line 8 we find the answer to the question about a brood of 8. The total across is 256. So there is only 1 chance in 256 that all 8 will be boys and a similar probability—represented by the 1 at the right-hand end of the line—that all will be girls. Similarly, a 1-boy 7-girl mix will occur 8 times in 256 for a 1-in-32 chance, same being true for 7 boys and 1 girl.

The next number in the line says there'll be 2 boys and 6 girls 28 times out of 256. The 56 that follows is how many chances there are in 256 of getting precisely 3 boys and 5 girls, and the 70 says there are that many chances in 256 that the 4-and-4 mix that many parents might prefer can be expected that many times in 256.

Note that, as bridge players already know, while the equal distribution is the single most likely one it will occur far less often than the other possibilities combined.

The tenth line is, naturally, the one to examine to find the likelihood of heads coming up precisely 4 times in 10 tosses of a quarter. Adding across gives 1024. The numbers in line 10 say there is the usual 1 chance of all heads, 10 chances of 2, 45 of 3, and 120 of 4 heads. A quick exercise in division says that this probability of 120 in 1024 means about 1 chance in 8 or 9.

You don't need much math to use this wonderful triangle. But those who remember—probably not favorably—encountering the binomial theorem in elementary algebra will find that here it is again.

Think of the simple binomial a + b as one a plus one b, that is 1a + 1b. In the triangle, those 1s, the coefficients of a and b, appear as Line 1.

Now consider the second power, or square of a + b. Multiplying a + b by itself gives a result of which the coefficients are 1, 2, and 1. Sure enough, Line 2 of that remarkable triangle.

The coefficients of the third power of a + b prove to be 1, 3, 3, 1, and so it goes.

But you won't need to continue this expansion to find out how likely 8 boys in a row would be. The eighth line of the triangle says this is to be expected once in 256 runs of eight births. And, since a similar run of girls is equally likely, the chances that a brood of 8 will be same sex are 2 in 256 or 1 in 128.

Probability by Subtraction

W<small>HETHER</small> we're buying, selling, investing—or merely crossing a busy street—life has us constantly calculating probabilities. Unfortunately, this is a often a tricky thing to do.

How maddeningly true this can be was nicely demonstrated by a newspaper columnist. Marilyn Vos Savant, in the weekly Parade magazine section, handed her readers this killer one Sunday.

You're on a game show. The host offers you the choice of three doors. One hides a car, the others hide goats. After you've chosen the left door—let's say—the host asks, "Aren't you glad you didn't choose the middle one?" (or it could be the right-hand one). And opens it to reveal a goat, as he knew it would.

"Now," he says, "would you like to change your choice?"
Should you?

When the "Ask Marilyn" columnist gave the answer Yes, 92 percent of the readers who contributed to the bushels of mail that followed disagreed, This included almost two-thirds of the replies that came from university campuses.

Perhaps despairing of proving her conclusion by mere explanation, the columnist eventually convinced many of the skeptics by challenging them to model the game with three paper cups and a coin. More sophisticated readers came around after writing computer simulations—by no means overkill when the problem is one dealing with probability, the branch of math that the great mathematician Charles Sanders Peirce has described as "the only one, I believe, in which good writers frequently get results entirely erroneous."

By both kinds of test—cup and computer—changing was proved to be the successful strategy almost exactly 67 percent of the time.

At the end of the controversy, all that was lacking was a direct, simple verbal explanation of why changing the guess was the way to improve the 33 percent chance—not merely to 50 percent, as a great many contended, but all the way to two thirds.

As I see it, at this point the plot demands a final scene in which none other than everybody's favorite master of logic intervenes.

Buried in the heaps of mail from readers is found an odd-shaped envelope, weathered with age, its postmark indecipherable.

As I have had occasion to remark to the sometimes obtuse Watson [it began, in a regrettably patronizing way] when you have eliminated the impossible, whatever remains, however improbable, must be the truth.

Could the signature be? It was! Sherlock Holmes!

In a trice, the great man had made it all clear. Given a 100 percent chance that a car crouched behind one of the doors and a 33 percent chance that this was Door 1, the remaining 67 percent chance belonged to the second and third doors jointly. One of these having been eliminated by the sight of a pair of hairy ears, giving it a probability of zero, the contestant merely need opt for the other to win 67 percent of the time.

Knowing that the chances are 2 in 3 that his first choice is a goat, our contestant should stand pat only if he prefers a Toggenberg to a Toyota.

Flip a Coin

How good are the chances of getting heads exactly half the time in a series of, let's say, 10 flips of a coin? Not so good as you might think. Not even 1 in 4.

Of the ways of figuring that out, one is especially interesting because both the method as a whole and its initial step are adaptable to so many other calculations.

You can begin by finding how many ways 10 objects (or coin flips) can be combined into sets of 5. This is the same as discovering how many different sets of 5 volumes each can be made from a 10-volume encyclopedia.

You figure this by dividing the factorial of the number of objects by the factorial of the number in the set you're assembling and by the factorial of the difference between these two numbers. Since the factorial of a number is simply the product of that number and all the smaller positive whole numbers, you'll find yourself dividing factorial 10 (3,628,800) by factorial 5 (120) twice, to get 252. (By the way, if you've forgotten, factorial 10, or 10 factorial, is commonly written 10!)

For your second step, take the probability of getting heads on any single flip and raise it to the power of the number of heads the problem calls for. Since heads is always 1 chance in 2, this means raising 1/2 (or 0.5) to the fifth power, which is 0.03125.

Third step: Raise the probability of NOT getting heads (in this example it is also 1/2) to the power of the number of times this is to happen (again 5) to get the same 0.03125.

Multiplying together the results of Steps 1, 2, and 3 gives the probability that the event in question will occur. It is 0.246, or not quite a 25 percent chance.

You can extend this routine to finding answers to questions of other probabilities.

If you use it to find the chance of precisely 6 heads in 10 flips and then of precisely 4 (which will be the same, naturally) and then add both these results to the previous one for 5 heads, you'll discover that the probability of getting between 4 and 6 heads in 10 flips of an honest coin is very close to 2 out of 3.

Zips and Plates

THE INVENTION by the U.S. Postal Service of the nine-digit zip code may just possibly approach the ultimate human limit. Anyway, the ability to recall—briefly—a nine-digit number is found at the "superior adult" level of a widely used intelligence test.

Designing such things as zip codes and license-plate numbers calls for a calculation of permutations. All this has to do with is the number of ways a set of things can be arranged and rearranged.

How this works is especially easy to follow when the things are digits. In our decimal system there are 10 of these, 0 through 9. You can use each followed by the same or another digit to form a two-digit code, producing all the numbers from 0 through 99, giving 10 times 10 possibilities. You can call that 10 squared and call the possibilities with three digits 10 cubed.

The five-digit postal zip code thus permits 10 to the fifth power, or 100,000, different zips. With the introduction of a 9-digit zip, this became a billion, which is several times as many zips as there are Americans. Even if the population triples, each citizen can still have his own personal postal delivery code.

With license plates that use letters as well as numbers, the possibilities multiply. If each letter and each number can occupy any position, a single character may have 34 variations (10 plus 26 minus 2—to exclude the letters O and I which might be confused with digits).

Such a 5-character zip or plate could come in the number of variations equal to 34 to the fifth power. This is upwards of 45 million and should serve any likely purpose. Such mixtures might be easier to remember than all-digit codes.

When letters and digits are not made fully interchangeable, as with California's license plates and the zip codes used in many countries, the calculation changes slightly. A rule that has been used for the 7-character California plates is 4 digits followed by 3 letters. So the number of variations available is 10 times 10 times 10 times 10 times 26 times 26 times 26, more simply expressed as 10 to the fourth power times 26 cubed. Either way, you'll find it comes out to more than 175 million.

Learning from a Sample

A GREAT PART of the figures you encounter in your reading about incomes, inflation, living costs, unemployment, health, and innumerable other matters are based on samples.

Although conclusions drawn from samples can be highly misleading and should be taken with several grains of salt, there are many things that cannot be found out except from a sample. As a maker of electric irons, you might want to test acceptance of a new and radically different model before investing large sums in producing it. You can't have it tried by everybody in the country; that would cost too much. So you test it on a sample of perhaps a thousand.

Or suppose you make light bulbs. You want to know how many hours they will last before burning out. Again you must test a sample, because if you tested them all to destruction you wouldn't have anything left to sell.

But you must be careful to choose a sample that is typical of the whole. Otherwise you may do something equivalent to what the Kinsey people did in their first famous study of sex: They issued figures based largely on college graduates and prison inmates, the two groups they found most accessible and willing to answer questions. How well these represent the nation as a whole is open to question—and statisticians have indeed been questioning it ever since.

A Gallup poll on adoption of the metric system produced the conclusion that 33 percent of people who had been to college had never heard of it. A large majority of those with less education were also in the dark.

So what would you make of the findings of a California newspaper that 98 percent of its readers knew about the metric system? The paper then boasted "how much more knowledgeable" its readers were than Americans generally.

But the discrepancy is much more easily explained: a badly biased sample. While Gallup had taken a cross-section of the public, the newspaper had relied on coupons clipped out, filled in, and mailed by readers. It is fair to suppose that most folks who had never heard of the metric system had simply selected themselves out of the poll by not bothering to participate.

Even such popular conclusions as the one that criminals are subnormal in intelligence are dangerous to believe, based as they are on biased samples. All studies "proving" this have been made on prison populations. What they prove, of course, is that criminals who get caught are not terribly bright. But we must accept the fact that we remain in ignorance about the IQs of those who contrive to stay at large.

The undersized sample must share guilt along with the biased one. Most of us use it informally, coming to conclusions about the intelligence of other races or the combativeness of Irishmen or the dumbness of blondes on the basis of the very small sample of each we have personally known—and possibly observed with something less than scientific objectivity.

This is foolish, true—but scientific teams have done the same thing many, many times. This is attested to by the horde of medical miracles, including cold cures, that have come and gone over the years. As a great 19th century physician once wryly advised, "Make haste to use a new remedy before it is too late."

How big must a sample be? There is no exact answer. A handful of cases may be indicative in some matters, while hundreds tell little in others. In an early test of polio vaccine, 450 children were vaccinated, while 680 were left unvaccinated as controls. Nothing was learned, because no cases developed in either group. Even at its worst, polio was a disease of such low incidence that groups of 7,500 to 25,000 would have been needed to produce significant conclusions.

In that word significant lies the answer for us when we want to judge a conclusion from a report of a medical experiment or other

sampling study. A good report will give us what is called the
significance of the result. If we find such words as "statistically
significant at the five percent level," we can conclude that the odds
are 95 to 5, or 19 to 1, that the result found is meaningful. That is, there
are 95 chances in 100 that the treatment really did the work. And 5 in
100 that the result was pure chance and could not be relied upon to
occur again.

The Rabbit-Sandwich Average

THE IMPORTANT thing when calculating an average is to avoid coming
up with a rabbit sandwich. As we shall see.

You're not a snob and I'm not in the real estate business. But let's
pretend for the moment that you are and I am. And that I want to sell
you some property along a rural road I happen to be familiar with.

You'd like to know the average income of your prospective
neighbors. I run my fingers over my calculator, entering some fairly
accurate educated guesses. Not less than $100,000, I tell you, what
with the presence of several doctors and lawyers and substantial
business people.

Having bought on the strength of this snob-appeal, you are
amazed a few months later when I ask you to join a campaign against
a school-bond issue. The basis of our fight, I tell you, will be that
people of this area just can't afford it. After all, their incomes average
only about $25,000 a year.

Which figure is the phony?

Actually, I've told you the truth both times. But not the whole
truth.

My newer figure is the kind of average called a median. It tells
you that half the people in the neighborhood receive less than $25,000
a year, and half of them receive more. Since ours is an area of about
ninety percent wage earners and retired people, this figure does
represent their economic level rather accurately. And it is generally
true that the median is the type of average that gives the most useful
information about income distribution that can be crowded into a
single figure.

As for my first figure, it too is an average correctly arrived at. It
is the more familiar type, the arithmetic average or mean, reached by
adding everybody's income together and dividing by the number of
incomes. The weakness of this kind of figure for dealing with
incomes is that a few extreme cases influence it to a highly dispropor-

tionate degree. This kind of average works out to that impressive $100,000 because at one end of our little country road are clustered the week-end retreats of three millionaires. City folk, you might say.

The way a few large figures can distort a mean might remind you of an old joke about the operator of a diner who offered rabbit sandwiches at a remarkably low price. When questioned about it, he admitted that he used some horse meat to keep his costs down. "But I mix 'em fifty-fifty," he avowed. "One horse to one rabbit."

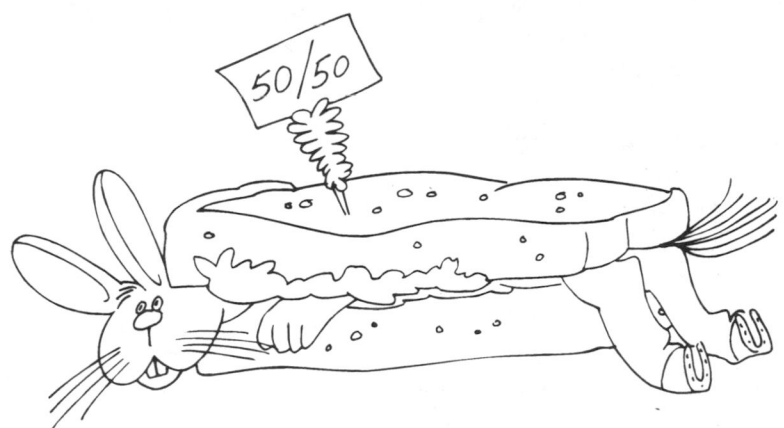

It should be noted in passing that a third kind of average also crops up occasionally. It is called the mode, and it means the most common case. It is this kind of average we have in mind when we say, or did at one time, that the "average advertising man" wears a gray flannel suit. (It's not a subject with which I have been keeping up.) Or when we pass along the truism that the "average high school graduate" can't spell.

Standard Deviation Two Ways

IF THE MOST common of all statistical calculations is the simple average (or mean) the second most often wanted is the standard deviation. Here is a choice of two routines for finding it.

Along the way, each method produces another result that is often wanted—the variance.

In the first method, you begin by finding out how much each of the scores or values you're working with differs from the mean (which no doubt you've already calculated by adding the values together and dividing by the number of them).

Now square each difference. Add up these squares. Divide by the number of scores or values. (This result is the variance.)

Now find the square root. That's the standard deviation.

To use the second method, begin by adding up the values. Square this total. Divide by the number of values. Subtract this result from a quantity obtained by squaring each value and adding all these squares together. Divide by the number of values (noting that this step gives you the variance). Once again, you can now obtain the standard deviation, or sigma, by deriving the square root.

If you'd like to test either routine, try these seven figures: 2, 2, 3, 5, 7, 9, 14. You should find a variance of 16.57 and a standard deviation of 4.07.

How to Get Facts from Figures

ALL OF US are reaching conclusions and taking action from reports of a statistical nature—newspaper and magazine articles, business presentations, advertising... That's why it's important to develop critical reading methods that spot such symptoms of unreliability as inconsistent reporting, unwarranted precision, and figures not quite attached to what they purport to talk about.

Inconsistent reporting methods lie behind many things we read. Newspaper stories once brought out the surprising "fact" that in Connecticut 2 percent of traffic accidents are fatal, while the figure for Georgia was 17 percent. This seems to suggest something important about the driving laws or habits of these states. But all it really meant, when examined more carefully, was that reporting methods differ. Connecticut had listed practically all accidents, however minor, and the fatal ones were a tiny number compared with mere fender bendings. In Georgia, only major crashes got on the list. Naturally a larger proportion of these involved fatalities.

Better reporting is one of the reasons for the alarming apparent increase in lung cancer in recent decades, when the true incidence has been shocking enough. More of the cases get into the published figures as diagnostic and reporting methods improve.

At one time, annual deaths recorded in New York State from coronary artery disease jumped from 8,119 to 36,390 in just six years. When the raw figures were published, a great wave of alarm quite naturally followed. But what had really occurred was a sudden change in the methods of reporting causes of death. Only after the earlier of the years was coronary artery disease regularly listed as such on death certificates.

The overly precise figure is another warning that what you are encountering is less than meets the eye.

A news report once said that the average Yale graduate of the class of '24 was now earning $25,111 a year—a princely income at the time. If you wondered how anyone could have uncovered such private information right down to the dollar, you might have read on and learned that the figure came from replies to a mail questionnaire. Since it is a fair guess that the greater part of those sons of Eli who had wound up in poverty were unreachable or unenthusiastic about telling their incomes, it is unlikely that the average was close to truth. And, anyway, few of us can calculate our own incomes down to the dollar.

Yet note how effectively those last three digits manage to imply that exact information is being imparted.

In the same category comes the sociological study that long ago found that a working woman in New York needed $40.13 a week to get by. The amount has changed as the years have passed—but using over-precision to lend respectability to a wild generalization has not.

Some figures are foolishly precise without at first seeming to be. Note the National Board of Fire Underwriters' announcement that fire losses in one recent year totaled $1,107,824,000. Impressively precise but not suspiciously so—isn't it?

In fact, it is being given down to less than one part in a million—

an implied accuracy of one ten-thousandth of one percent. Yet the figure was reached by adding to reported claims (which in themselves may be presumed to contain many exaggerations of actual losses) the Board's estimate of unreported losses. Surely such a total must be thousands of times as imprecise as it has been made to seem.

All overprecise figures, and especially this last one, belong in the same bag as a fine old joke. A farmer shows a tourist some bones found on his land and remarks that they are one million and three years old.

How can he know so precisely?

"Because," the farmer says, "an archaeologist told me they were a million years old—and that was three years ago in May."

The semi-attached figure is a handy device for those who would deceive. It's a source of a great many innocent, but costly, misunderstandings as well.

Thus you might have read about an overwhelmingly great increase in the populations of metropolitan areas of the United States in a mere ten years. Official census figures were cited by a news writer to back up the conclusions.

But what had really happened was that the Bureau of the Census had changed its definition of a metropolitan area. The new use of the term took in much more territory. What was probably a real but modest increase became all too striking. It should be noted that many a businessman makes potentially costly decisions on the basis of such population figures.

A San Francisco political figure, a Naval Reserve admiral, polled his constituents about attitudes toward the war in Vietnam. He merrily concluded that the people were predominantly hawks, because 42 percent voted for more use of military power, 27 for the current stand, and only 28 percent for immediate withdrawal.

The gimmick, as you can see, is that the politician's choice of questions forced the answers to an unknown degree. There was no resting place for the people who favored a position less aggressive than the administration's but short of simple withdrawal.

Moral: In reading the results of any poll, it is as important to study the questions as to study the answers.

Both unions and employers have been known to use semi-attached figures against each other. Thus an employer "makes a survey" by asking each workman if he has a complaint against the union. He probably has. By adding up all the assorted irritations, minor or otherwise, the employer is able to announce that "87 percent of the employees in this plant are opposed to the union."

Semi-attached figures have been with us for a long time. The death rate recorded in the Navy during the Spanish-American War was 9 per thousand. For civilians in New York City during the same period it was 16. Navy recruiters used these figures to prove that it was safer to be in the Navy than out of it.

These figures, of course, were not comparable. Those for civilians included old people and infants and the ailing, all of whom have relatively high death rates. Those for the Navy were based on a population from which all such people had been excluded.

Looking a Correlation in the Eye

I CAN'T THINK of anything to beat *post hoc* at leading people up the well-known garden path.

When two things are found to occur together regularly, they are, in statisticians' terms, correlated. The danger comes in figuring that this proves one has caused the other.

Post hoc, ergo propter hoc was the ancient logicians' way of reminding us that "after this, therefore because of it" is a piece of dangerously faulty reasoning.

It was once demonstrated that college students who smoke make somewhat lower grades on the average than those who don't. Aha! Smoking dulls the mind! For all I know, or anybody does, that may be the truth. But the correlation in question doesn't prove it. There are too many equally plausible explanations. Perhaps being smart has some tendency to keep you from smoking. Perhaps the sociable sort of fellow who takes his books less seriously than he might is also inclined to smoke. Who knows?

These three possible explanations of a relationship between smoking and low grades illustrate a useful rule. When two things, A and B, are associated, there are three possibilities: A causes B; B causes A; or both are caused by some third factor.

A leading medical journal reported that, according to a careful large-scale study, radiologists died 5.2 years earlier than did other physicians. This was immediately attributed to the effects of radiation, and it naturally produced a heavy dose of terror in radiology offices. Only sometime later did the voice of reason point out that radiology was such a new specialty that any of its practitioners who had died by that time were bound to have died young. There were no old radiologists.

What can you make of the conclusion of the Fair Campaign

Practices Committee that dirty politics doesn't pay off? On the basis of a national election the committee's chairman declared: "In 68 percent of the cases the candidate involved in unfair campaign tactics lost the election."

What the committee overlooked is the likelihood that the dirty tactics were often desperation moves on the part of a candidate who was already trailing badly. It is not unlikely that here B has caused A.

As J. Edgar Hoover (with growth of his FBI and his personal power at stake) and others since have pointed out many times over several decades, our national crime rate is growing. Something must have gone wrong in our schools or our law enforcement system or somewhere to produce this distressing change? Not necessarily. There are more young people in our population every year, and young people have always had a higher crime rate than adults.

The rising crime rate, then, must be seen at least in part as another symptom of the population explosion. As one expert has pointed out, the only sure way we know of fighting crime is birth control.

You can keep your guard up against unwarranted assumptions of cause and effect by remembering some of the amusing but spurious correlations that have been discovered.

One is a close association between the cost of rum in Havana and the salary level of ministers in Massachusetts. No one is going seriously to suggest that either of these things is a cause of the other. They are, of course, simply two products of a third factor, the historic increase over the years in the price of practically everything.

And then there is the correlation between the number of stork nests on Dutch or Danish houses and the number of children in the household. You can accept this as scientific verification of an ancient myth. Or you can note the presence of a third factor. Big families choose big houses; big houses have more of those toasty chimney pots in which storks love to nest.

How trickily confusing cause-and-effect relationships can be is nicely illustrated by what I hope you will forgive me for calling a lousy example.

The natives of the New Hebrides long believed that body lice bring good health—so no one should be without them. The basis was centuries of observation that had taught them that people in good health usually had lice, while sick people often did not. It took medical investigators a while to get this one straightened out. Almost everyone in the islands, it was discovered, had lice as a matter of course. When a man took a fever and became too warm for comfortable habitation, his lice (which may have brought him his feverish ailment in the first place) left him.

So here was a case in which cause and effect were not only the reverse of what observation indicated but also most confusingly intermingled.

"It ain't so much the things we don't know that get us in trouble," remarked the American humorist Artemus Ward long ago. "It's the things we know that ain't so."

Chapter X
  REALITY AND ILLUSION

How to Draw a Fact
Hornswoggled by a Bar Chart?
The Illusory Line Graph
Figures Don't Lie?
How to Find a Prominent Number
The Prisoner's Dilemma
Conquering a Map Chart
Taking a Problem Apart
How Lethal Is My Warhead?
Those Awful Work Problems

How to Draw a Fact

Pictographs constitute an especially useful device for transmitting complicated information simply. But when used carelessly or dishonestly, can they ever distort and misinform!

All too common is an illegitimate version of the pictorial graph in which your eye is forced to compare dissimilar shapes. A large utility company once presented a comparison of its annual tax bill with its expenditures for salaries and wages, indicating their relative size by pictures of moneybags. By drawing one fat bag and one sadly slack one, the artist managed to make a two-to-one ratio look like far, far more than that. If these were actual bags, the one at the left would hold eight or ten times as much as the slack one.

The artist should not be blamed too much. He was up against a problem knotty enough to have produced a great many misleading pictographs.

How would you solve it? How, that is, can you draw two moneybags so they represent 98 and 49 million dollars respectively? Actually, you can't. The ordinary beholder has difficulty in comparing bulks precisely. If you draw a true representation of a pair of bags of such relative size that one would indeed contain twice as much as the other, they will appear so nearly the same size that they will understate your case.

Draw them so that one occupies precisely twice the area on the page that the other does? Again no—because some eyes will see them as the solid objects they represent, and the bag covering twice the area would have much more than twice the volume.

Make one bag twice the height of the other? This is a common practice, and has been done with everything from moneybags to steel mills. But the answer is decidedly no. A bag that is twice as high will be twice as wide and twice as thick. The volume of one such bag will be to that of the other as the cubes of any dimension. One would actually contain not twice as many coins as the other, but eight times as many.

There is just one way out. Keep the bags uniform in size and vary their numbers. Draw two bags for the larger sum, one for the smaller. Or let each bag represent, say, 10 million dollars, using for the first quantity ten bags and for the second, five. For precision, you can omit a small part of the last of the bags in each set.

Whether your story involves populations or tools or telephones, you can tell it by using an appropriate number of little symbols.

Hornswoggled by a Bar Chart?

BAR CHARTS can be deceptive in their simplicity.

A very tall and narrow one will tend to exaggerate, a short and squatty one to minimize.

A bar chart from which the zero line—and consequently some part of the area of the chart itself—is missing will exaggerate even more. Hence, with a chopped-off bar chart, as with a truncated line graph, you are entitled to a visual clue that something is missing. When a bar chart must be amputated, a bread-knife-shaped wavy line at the bottom is at least better than no warning that this unfortunate operation has been performed.

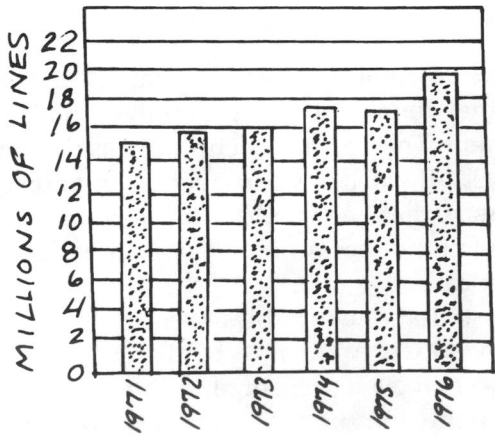

How chopping exaggerates can be seen by comparing a proper bar chart with its truncated brother. (These figures have to do with lines of advertising carried by a magazine.) As a careful look at the numbers shows, both of these represent the same data, but you'd never guess it at first glance. One of them makes the trend seem much more marked than it is. And that, of course, is the one that appeared in an ad.

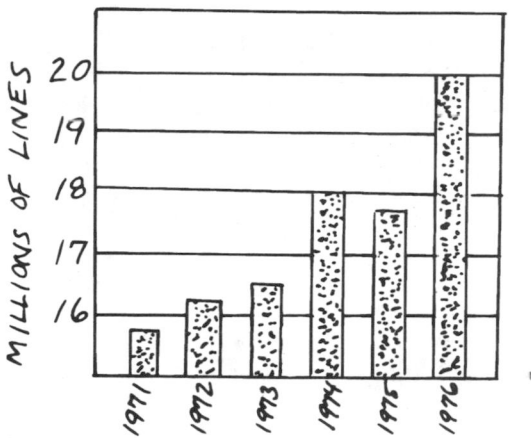

The trend or comparison offered by a bar chart can also be minimized by an unfair trick of geometry. The bars of a proper chart vary in only one dimension. The eye can easily compare these bars. But charts are sometimes drawn with all bars being of the same shape and proportion. The eye is forced to compare areas, which it cannot do well.

The Illusory Line Graph

LINE GRAPHS are widely used for pictorial presentation of everything from production and sales trends to medical and scientific data. Proper ones include numbers or times along both axes. True information can always be gleaned from them by careful attention to these numbers.

But the impression these graphs give is subject to distortion through use of a bagful of tricks, both intended and otherwise. How true this is can be understood by a glance at the set of graphs here. Since all four of them are based on precisely the same data, they should in all honesty create the same impression.

But, as you can see, they do not.

The first of the four is all that a line graph should be. It represents a substantial but not overwhelming trend, and that is just the impression it produces.

The second one employs horizontal expansion to make exactly the same trend look like something less than it is.

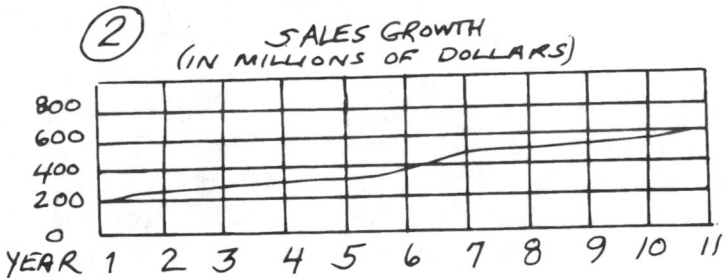

The third has been expanded vertically to do precisely the opposite. It makes the trend appear much more marked than it is. The shape of these last two charts should be taken as a warning signal. Whenever you meet a graph that is very far from square, it is likely to produce an impression different from what you would have obtained from the data themselves. So examine those figures and take your impression from them, not from the misshapen chart.

Even more deceptive is Trick 4, in which the same information is cast once more. Chopping a little off the top of the graph gives an impression of a trend that is all but bursting the top. More important, slicing off all the bottom below the curve itself leads the eye of the observer to believe that this trend goes from nothing to maximum.

It is not unreasonable to call the biggy on the opposite page a gee-whiz chart.

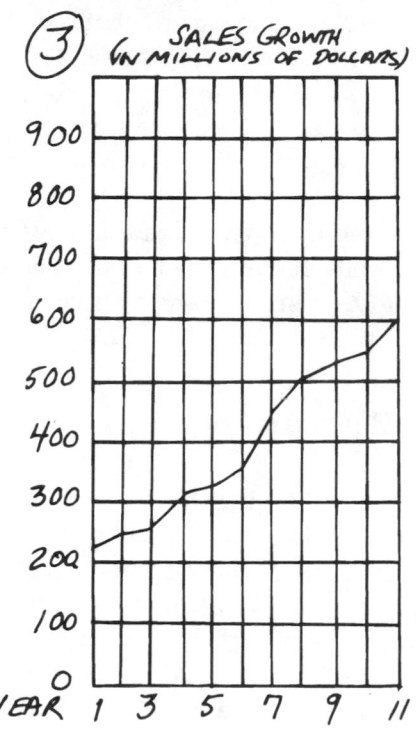

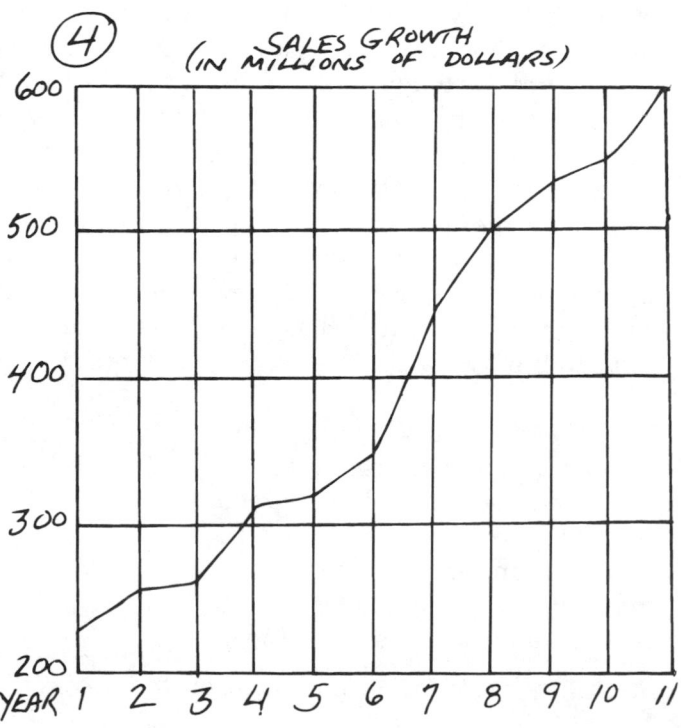

To guard against its misleading effect, it is well to remember that almost any graph in which the vertical scale does not begin with zero is deceptive.

If it should become necessary for reasons of space to violate this rule, the producer of the graph should feel obligated to offer a visual clue.

The most effective clue he can give is a break in the chart—such as a couple of wavy lines that might be easily drawn with the help of a bread knife—to give instant warning that some of the chart is missing.

Figures Don't Lie?

Whᴀᴛ ꜰᴏʟʟᴏᴡs the impact of statistics, or indeed any mass of figures, upon the unprepared human mind is likely to be a big buzzing blur.

That's why graphic methods were born. It's also why they offer a useful adjunct to any consideration of how-to-figure-anything.

A graph or chart or pictograph is a splendid device for converting a complex trend into a picture the eye can grasp almost instantly. Unfortunately, it is also a device by which figures can be distorted, trends exaggerated, differences minimized, and the eye fooled as effectively as by a practitioner of sleight of hand. When it comes to matters as serious as those about which we receive so much of our information from graphs today, it is not fun to be fooled. A little sophistication in how to draw an honest graph or, when reading, to distinguish one from its disreputable brothers is well worth gaining.

The distorted curve makes the simplest example. Let us suppose you are engaged in the manufacture of widgets of a prescribed weight. To discover how closely they hew to this weight, you ask that a number of widgets be snatched at random from production and individually weighed. A graph of these weights will probably fall into what is called the normal curve. It should look pretty much like this (A).

It will look like that instead of like either of the other two drawings because that is the proportion in which statisticians have agreed to agree that the normal curve should be drawn—the top of the curve being about three-fourths as high as the base line is long. How well this accepted shape is hewed to in

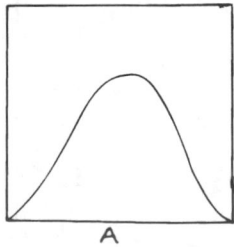

A

the graphing process is determined by how distances are selected on the horizontal and vertical scales.

Let us now compress the horizontal scale or expand the vertical one. We'll produce a curve that is tall and narrow, like this (B). The new curve seems to say that our widget production is very well controlled, with minimum variation from unit to unit. If you are Big Brass, this is the kind of chart a slightly shifty production manager might bring to impress you.

By contrary treatment, expansion of the horizontal scale, a curve with a flatter look is produced. What this one seems to say is that our widgets vary a great deal (C).

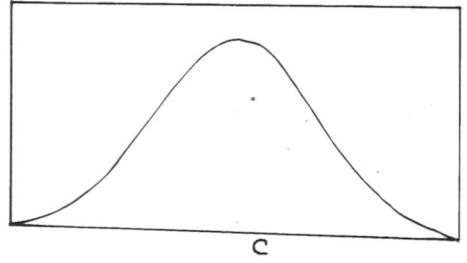

The data behind these three curves are the same. The information they transmit should be identical, and it is—if you take care to examine the numbers along the side and bottom that ordinarily accompany such curves. But where one of these graphs speaks to you in normal tones, the second whispers—and the third shouts.

Can you imagine a curve that reveals a quirk in human nature? Here is one that does this by catching your eye with a significant anomaly. It's a normal curve (D) with an odd little bump. Since the curve was produced by graphing the scores of a large number of students on a test, its normal shape was to be expected.

But why that bump?

Well, it occurs so often that it has been given a name, peak of charity. It is a product of the tendency on the part of most test graders to give the benefit

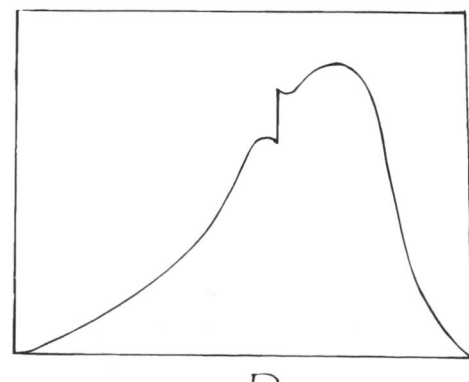

of any doubt to a student who is almost but not quite passing. This naturally pushes a significant number of grades from the area just below passing to that just above.

So the graphing process has revealed a phenomenon that might otherwise slip by unnoticed.

How to Find a Prominent Number

IT's NOT likely to happen to you. I don't suppose it has ever happened to anyone, except as part of an experiment done by a psychologist. But suppose you were one of a small group of people promised a large reward if, being shown some numbers such as 7, 13, 99, 100, 261, 555, they could all independently hit upon the same one.

You would have to say to yourself: Which of these numbers has a special prominence that will catch the eye of the other persons in the group? Assuming they're all much like yourself in background and intelligence, it may well be the same one that catches your fancy—if you make your choice by a thoroughly logical process the others are likely to duplicate.

You can put up an argument for almost any of these numbers but if you think about it long enough you'll probably agree that only one of them really is unique. Two of them are of superstitious significance, two are symmetrical, one is basic in our decimal system. But only one is negatively prominent; 261 is meaningless to almost everyone.

Perhaps the only thing you can do with this idea is to turn it into a parlor game, but it is not without importance. It illustrates the idea of interdependence in thinking—the process involved when you lose your husband or your wife in a department store and try to handle the problem by using your head.

Let's say you're the husband. You begin by looking in the places your wife might logically go. Then you go where you should logically go, on the assumption that it will occur to her to look for you there. Then you ask yourself where she would expect you to expect her to go. And so on.

As Thomas Schelling, the experimenter responsible for the list of numbers, puts it: "What would I do if I were she wondering what she would do if she were I wondering what I would do if I were she...?"

It helps if the couple share much the same sense of humor, which could send them both promptly to the lost-and-found office.

Now suppose, to make it more fun, you are a secret agent

supposed to meet a colleague on a certain day in a certain city, but neither of you has been told place or hour. What do you have to pin your hopes on beyond sheer luck or coincidence?

Begin with place. Some cities might be an exception, but in most the city hall and the police station and the post office would all come quickly to mind. The first two having some drawbacks from the point of view of a spy, surely a spot in front of the main entrance to the central post office would be a good bet.

Time is possibly easier. Noon and midnight are the most distinctive hours. Unless at least one of you has reason to avoid being seen in daylight, the first of these is probably the choice which you and your accomplice would make.

See you in front of the post office at high noon, 007.

The Prisoner's Dilemma

Y OU DON'T have to be behind bars to face the prisoner's dilemma. It is the very model of many marriages and business affairs and relationships between governments.

It usefully illustrates why it is difficult for a couple of sensible people, or two great nations, to come to agreements even when they obviously offer a lot of benefit to both parties. A sad thing.

As formulated by Professor Albert W. Tucker, a Princeton mathematician, it goes roughly like this.

Two men charged with jointly committing an armed robbery are brought to the district attorney's office. He confirms what they already know. For lack of evidence, there's no way of convicting either of them of robbery unless at least one of them confesses.

"But," says the D.A., "I can get you both sent up for six months for illegal possession of firearms. And that's just what I'm going to do if I don't get a confession.

"If one of you confesses, he goes scot-free as a state's witness, and I'll see to it that the other gets the maximum of 20 years. If both of you confess, however, I won't need a witness. I won't need an elaborate trial either. So I'm prepared to offer you both the minimum sentence of two years for armed robbery."

The prisoners are returned to their cells.

If you were one of them what would you decide to do?

(We're going on the assumption that you have no opportunity to talk the thing over with your accomplice and have made no prior agreements covering an eventuality like this. But this doesn't really make very much difference, since then it would come down to whether you could trust him to keep his word—in short, to whether you are dealing with an honest crook.)

The simple, sensible answer is for both of you to keep mum and get off with six-month sentences. You see that right away and surely so does your partner.

But if your partner concludes that you will remain silent, will he take advantage of this to confess and get off free? If so, you'd better confess too. After all, two years is far better than twenty.

Also, will your partner—having realized that you will have followed this course of reasoning—dare remain silent even if his first impulse was to accept the six-months alternative?

So it goes, around and around in your head, with no real solution. What you may in the end decide to do will depend not only upon how much you trust your accomplice but also upon how much you think he trusts you. To say nothing of how much you think he thinks you think he trusts you...and so ad infinitum. It's like the picture on the oatmeal box that portrays an oatmeal box with a picture of an oatmeal box on it containing a picture....

By no means frivolous, or limited to criminal or to unreal conditions, this model problem takes on added significance every time disarmament talks resume. For decades, while the United States was spending itself into potentially crippling public debt on military preparedness, it was facing a U.S.S.R. that was producing weapons at the expense of tolerable living conditions for its people. Why in the world couldn't the leaders of these countries agree at least to slack off a bit for mutual benefit? Some of the reasons that the only sane solution, mutual disarmament, long proved impossible are explainable by the prisoner's dilemma.

In figuring out the best answer, if any, to a problem involving trust, whether it is between man and wife or between business partners or even between two world powers desperately seeking a safe way to reduce armaments, a handy device is a four-cell matrix.

Since this approach is all part of what is called the theory of games, it is usual to refer to the parties as players. Thus Player A represents you (or your country). You have your choice of strategies a1 and a2. The other fellow chooses between strategies b1 and b2.

In graphic, or matrix, form it looks like the accompanying illustration.

You might enjoy using such a matrix to outline other dilemmas and to play with this one and others of its kind with a real opponent.

See if the results differ, depending upon whether the opponent is a friend (or spouse) or not.

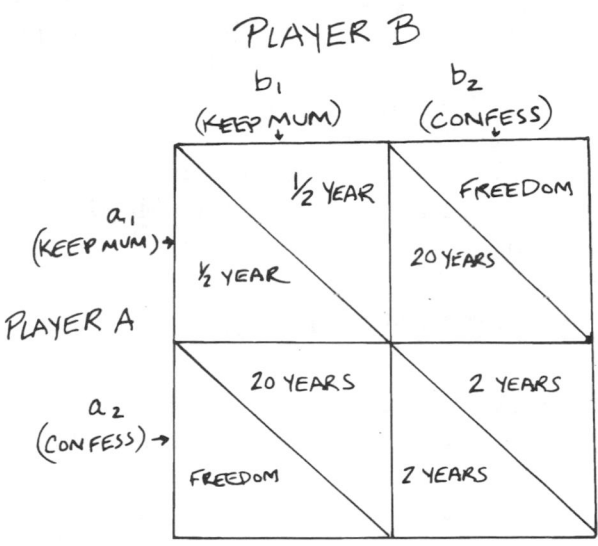

Conquering a Map Chart

W<small>ANT TO</small> show some fact clearly by illustrating it with a map chart?

Or are you just trying to read one of the things and understand what it truly means?

Either way, some discretion is required.

Take the classic example of The Darkening Shadow.

It was distributed by a large bank and widely reproduced by taxpayers' groups and newspapers and magazines. It was intended to impress us all with what proportion of our national income was being taken and spent by the federal government. By shading almost

the entire areas of the states west of the Mississippi, it told us that federal spending had become equal to the combined incomes of the people of those states. And so it had.

But the trick is as unfair as it is old.

By choosing states of large

area and generally sparse population, the map maker has been able to cast his shadow over more than half the nation.

Had he chosen to start at the East Coast instead, his shadow would have covered such a tiny area that no one would have been impressed.

Yet this is a story that is worth telling, and by no means unimpressive even when presented objectively. All a map maker need do to cast an honest shadow is choose a block of states having total income in the same ratio to the national that their combined areas bear to that of the whole country.

Taking a Problem Apart

Sᴛᴀʀᴛɪɴɢ with a situation like the one in this drawing of a belt passing around a couple of pulleys, how do you figure out the length of the belt you'd need?

A good approach is to break up the complex problem into two or three simple ones you probably already know how to solve.

The length of the belt you see here is approximately equal to a couple of half-circles plus the hypotenuses of a pair of right triangles.

Suppose these pulleys have radii of 3 inches and 5 inches and their centers are 10 inches apart.

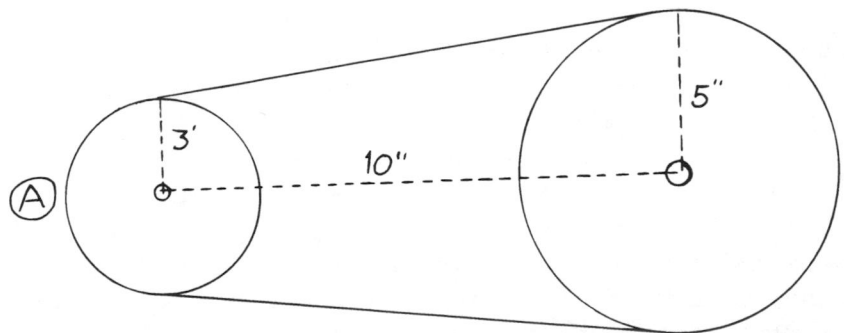

Since the circumference of a circle is equal to its radius times pi times two, half its circumference (what you're after here) is found by multiplying its radius by pi. So you can find the length of the two curved segments combined by adding 3 times pi to 5 times pi—that is, 8 times pi.

The straight segments are the hypotenuses of right triangles. Each triangle has a long side that is 10 inches and a short side that is

2 inches. Squaring each of these sides and finding the square root of the sum (10 squared plus 2 squared is 104) gives the length of each straight segment. Add 'em all up for the answer.

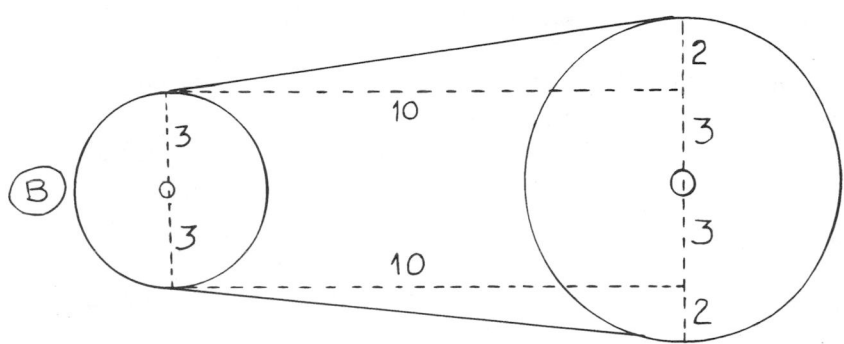

How Lethal Is My Warhead?

IN AN ERA in which we all must live with decisions about nuclear armaments and overkill, the calculation that follows is not without its grim interest—Cold War or no Cold War. It comes from Robert C. Aldridge, an aerospace engineer who helped design weapons for the Trident submarine of the '80s.

It seems that you can find the comparative deadliness, or lethality of warhead, of a submarine armed with ballistic missiles by this modest operation: Square the warhead yield in megatons and take the cube root; twice divide this result by the accuracy, expressed in nautical miles, and multiply by the number of missiles.

(If your accuracy figure is given in feet, divide it by 6080.2 to change it to nautical miles.)

It works like this for the older Poseidons, the subs of the '60s and '70s. Figure each as having ten 40-kiloton MIRVs (Multiple Individually Targeted Re-Entry Vehicles) with accuracy of 0.3 nautical miles. Since it takes a thousand kilotons to make a megaton, each Poseidon weapon rates 0.04 megatons.

Squaring that and taking the cube root (that is, raising 0.04 to the 2/3 power, which is the quick approach when working with computer or scientific calculator) and twice dividing by 0.3 gives a figure of 1.3. For ten of these MIRVs, the total lethality (and isn't that a great word to contemplate?) then becomes 13.

If you'd find it educational to compare various seagoing weapons, present and future, in this way, here are some figures to toy with.

Russian SS-N-8 (an earlier weapon): one 1-megaton warhead; accuracy, 0.8 nautical miles.

Soviet SS-NX-18 (a later entry): three 500-kiloton MIRVs; 0.25 nautical miles.

United States Trident 1: eight 100-kiloton MIRVs; 0.25 nautical miles.

Second-generation U.S. Trident 2 may be armed with fourteen 150-kiloton or seven 300-kiloton MIRVs (accuracy figure, 300 feet) or with seventeen 75-kiloton MARVs (Maneuvering Re-entry Vehicles), accuracy figure 90 feet.

If working with these last numbers gives you a calculated total lethality of more than 13,000, you've got the hang of it. It's sensible to round off your results since you're working with data that are only estimates of what the future could hold.

Those Awful Work Problems

IT CAN BE a bit on the nasty side. It's called a work problem, and most of us have encountered its kind, in school if nowhere else. It can go something like this.

You're an old hand with a lathe, on which you can make a little wooden bowl in 20 minutes. Your shop partner needs 40 minutes to do the same job. How many man-minutes should you charge against each bowl in your system of cost accounting?

You might guess it is the average—the mean—of the two figures, or 30, but it is not. It's a less familiar average, the harmonic mean. You find it by adding together the reciprocals of the time periods, then dividing into the number of these periods. In this instance, you divide 1 by 20 and 1 by 40 and add together those two results, which you then divide into 2, to get the result of 26.67. Which is quite a bit different from 30. It is so much different, in fact, that if you're serious about this lathe thing, it could tell you why the business is going broke.

What you don't know can't hurt you? Well, it can dangerously mislead you when the subject is averages.

With numbers as simple as these, you could have an exercise in mental arithmetic. The sum of the reciprocals is seen to be 3/40. To divide that into 2, invert and multiply—as we learned to do in school—so the final step is dividing 80 by 3.

Or, for a slight variation, try dividing the sum of the reciprocals by 2 instead of t'other way around. Then find the reciprocal (that is, complete the division of 3 by 80).

And that's still a harmonic mean.

Chapter Y
STRETCHING A CALCULATOR

Calculator Counting
Dividing into a Number
Percentages by Calculator
Square Roots and Diagonals
Cube Root and Beyond
Divide by Multiplying
Finding the Remarkable e
The Hyperbolic Rope
Squares and More
Calculating Reciprocals
Two Reciprocal Tricks
Finding Sums of Products
Coping with Big, Bad Numbers
Bigger, Badder Numbers
Memory at the Market
To Make a Payment Book
Easy Calculator Subtraction
Testing a Scientific Calculator
Calculator Superstretcher

EASY AS FALLING OFF A LOGARITHM!

Calculator Counting

You won't get any help on this from the booklet that comes with your calculator. More's the pity, since this is a task each machine does in its own way.

It's a useful little trick for counting supermarket items, or traffic, or the box office. Or keeping track of Scouts you're trying to get onto the bus without leaving one or two behind.

With some calculators, generally those with an automatic constant that works with all functions, you can count by just hitting the one and then the plus repeatedly.

If your calculator is an old timer with a constant (K) switch, turn that on first. If it has a K key, the procedure will probably be to hit successively the 1, equals, K, and plus keys.

Often you punch 1, plus, and then equals repeatedly.

To count by 2s, just use the 2 key instead of the 1. To count backwards or correct a mistake, touch the minus.

If these methods fail, it may be that the constant on your machine works only with multiplication and division. This situation demands a crafty trick. Load up the entry with zeros and ones in such a way that each time you multiply, the product increases by slightly more than 1—but so slightly more that the difference is at the tail end of so many decimals that the machine will drop it overboard.

You do this, simply enough, by multiplying 1.0000001 by 10000000. (You needn't count those zeros; just put in enough to fill the machine.) Now each time you touch the equals key, the digit at far right will increase by 1.

That's how you do it for a model designed so the first number you enter becomes the constant. For machines with which it's the second number, you do the same multiplication but in reverse order. If instead, you divide 10,000,000 by 1.0000001 you will be able to perform the opposite operation—count down from 9,999,999.

Dividing into a Number

A common bugaboo of calculator users is finding they need to divide a number already on the display of the machine into another number.

Suppose, for example, you've added up a long series of numbers and now want to divide the total into 365. This will stump the ordinary pocket calculator.

How can you do it without copying down the total and then re-entering it after feeding in 365 and the division sign?

Answer: Divide the total by itself twice, accomplished on most calculators by pressing divide and then equals. Then multiply by the 365. When you multiply by the reciprocal (that's what you got by the double division of a number) it's the same as dividing by the number.

Percentages by Calculator

Many personal calculators—those just above the bottom price range—boast percentage keys.

For regular business use, you may find that additional key well worth its extra cost. But for occasional use, you can get along just fine without it. In effect, you teach the simple electronic helper you already own how to handle percentages without recourse to a special key.

Basic rule: Instead of hitting the number and then the percent key, just hit the decimal point and then the number, so that you've moved the point two places to the left.

Thus 25 percent becomes 0.25; 12-1/2 percent becomes 0.125; and 4 percent becomes 0.04.

To complete the conversion of your mini calculator into an efficient percentager, keep in mind that a discounted price can be found directly by first subtracting the discount percentage from 1. For example:

You're thinking of ordering a $127.50 bicycle on which you've been offered a 22-1/2 percent discount. Sounds complicated, but....
Just enter 1 into your machine, subtract 0.225, multiply by 127.50—and read off that you'll be writing a check for $98.81. (And if that price is subject to 6 percent sales tax, multiply by 1.06 to find the grand total the easiest way.)

Square Roots and Diagonals

You're laying out a cabin and want its corners square. Well, how do you find the length of the diagonal of a rectangle? You find it, as you probably recall, by squaring the lengths of two adjacent sides, adding these squares together, and finding the square root.

That's great unless there's no square root button on your calculator and you're not one of those precious few who can recall from school days how to paper-and-pencil it.

Happily, there's a neat way to approximate a square root. It's said to go back to Sir Isaac Newton, who would have given an arm and a leg for your little electronic marvel.

You begin by guessing what the square root might be. Any reasonable number will do, but a careful guess will speed things up a little.

Then, on your calculator, you divide the original number by the guessed root, add the amount of the guess, then divide by 2. What is the square root of 139, for example?

Recalling that 12 times 12 is 144, you might take 12 for your guess. You divide 139 by 12, add 12, divide by 2, to get 11.791666. Check by squaring that on your machine to get 139.04338, surely close enough for most purposes.

If you're really fussy, repeat, using the new root instead of your guess. Divide 139 by 11.791666, add 11.791666, divide by 2, and you'll get a square root that squares to 138.9999. Closer than that you can't get, as you'll prove if you try taking it another step.

Since square root is an important addition to the partnership between your mini and you, try another example: square root of 150 this time. Divide 150 by the same guess, 12, add 12, divide by 2, to get 12.25. Squaring that to check it yields 150.0625. If that's not close enough to suit you, divide 150 by 12.25 this time, add 12.25, divide by 2, and you'll get an approximate square root that squares to almost precisely 150, as you will see.

Cube Root and Beyond

There is a rather complicated formula you can use for finding cube roots, fourth roots, and so on. But it's often easier to go the route of plain guess-and-try if you have a calculator handy. That way you don't have to remember a formula.

It goes like this.

To cut-and-fit your way to the cube root of 1,234, for instance, you might reasonably begin by guessing 10.5 since you can see that 10 is the cube root of 1,000 and so is a bit small.

When you cube 10.5, you find that's too small too, so you could try 10.7, which cubes to 1225 plus. You're close, so you might try 10.72. When that proves too small but 10.73 proves too big, you know the answer's between them. Within a couple more tries, you'll find that the cube of 10.726 comes to 1233.9948, giving you a result that's as on-the-nose as an answer can be unless you want to get into a ridiculous number of decimal places. You've gone as far as your calculator can go, and quickly too.

For a really extreme example of how well this somewhat random walk works, I decided to go for the 10th root of 98,765 (a number with no significance, just some digits strung in descending order). In nineteen minutes I had the answer to the seventh decimal.

A reference to this in a magazine column I was conducting brought me a stack of mail, mostly from readers who were pleased to note that they had beat me all hollow.

Divide by Multiplying

Faced by the chore of dividing quite a few different quantities by the same many-digit number, you'll probably reach for your pocket calculator.

In the absence of such a device, however, you may find the following trick worth knowing, and interesting as well.

It is based on a couple of facts. One is that it is easier to multiply than to divide. The other is that you get the same result when you divide by a number as when you multiply by its reciprocal. You can see why this is when you recall that a reciprocal is just 1 divided by the number.

Say you have many numbers to be divided by 1.20671. Find its reciprocal by dividing 1 by 1.20671. The answer, rounded to 0.8287 (assuming that's far enough to carry it for your purpose) becomes the basis for a table you construct by multiplying it by each number from 2 to 9.

Now suppose the first number you need to divide by 1.20671 is 92. Set it up as the product of 92 and the reciprocal of 1.20671, like this:

$$
\begin{array}{r}
.8287 \\
\times\ 92 \\
\hline
1\ .6574 \\
74\ .583 \\
\hline
76\ .2404
\end{array}
$$

Since those two partial products came right out of your table, all you've had to do is add them together.

Finding the Remarkable e

THE REMARKABLE constant called e that is used in so many scientific, financial, and engineering calculations can be defined in many ways. It is, for instance, the number of dollars you'll have at the end of a year if you deposit one dollar at 100 percent interest compounded continuously.

You can approximate its value in quite a short operation. Just increase 1 by an extremely small amount and then raise this total to the power of the reciprocal of that small amount. Easy IF you have a calculator.

If the "small amount" you use is 0.0001, your result will be correct to three decimal places, the form in which the value of e is usually given: 2.718. Having added 0.0001 to 1 you can complete the calculation (raising that total to the power of the reciprocal of 0.0001) in an instant with a scientific calculator.

Using a simple calculator, you must first find that reciprocal by dividing 1 by 0.0001 to get 10,000. To find that 10,000th power, you first find the tenth power (with most calculators, by pressing times once and equals nine times) then repeat three more times. If the result your calculator gives is off in the third decimal place, an error from internal rounding, you may want to try 0.001 as the "small amount" to see if the result of this shorter procedure isn't just as good.

The Hyperbolic Rope

IF YOU should ever wish to graph the shape taken by a heavy rope or chain, you'll find a need for something called hyperbolic functions. Only a few of the more elaborate scientific calculators have these built in, but you can calculate them quite readily on your own. Even a simple calculator will do, although the process is a little faster with a scientific model.

If your calculator has that useful constant e on it, begin by raising e to the power of the angle you're working with and also to the minus power of that angle. Subtracting the second of these from the first and dividing by 2 gives you the hyperbolic sine (commonly abbreviated sinh). Adding the two together instead of subtracting produces the hyperbolic cosine (cosh) after the division by 2.

To find tanh, divide the difference between the two quantities by their sum. Dividing the sum by the difference produces the hyperbolic cotangent, or coth.

The gimmick in all this is that you may be working with a calculator that lacks an e key or a means of finding negative powers (to say nothing of the powers that are not whole numbers, called for if the angle you're working with is not a whole number of degrees).

Handle the absence of an e key by entering its value, which is 2.71828. To get a negative power, take the reciprocal of the positive power, either with the special 1/x key or by dividing 1 by the positive power. To work with angles that are not whole numbers of degrees, you'll need the help of logarithms.

Squares and More

MOST CALCULATORS can be taught to square and cube and all that, without your having to enter the number more than once. That's important if it's a number like 3.1415927, as it not infrequently is. When you've taught your mini the procedure, you'll find it makes life almost as simple as if you had a special squaring button.

You'll need squaring, and the higher powers, too, for interest calculations—both the kind you pay when you buy a car or a house and the nicer kind that the bank or savings and loan pays you.

You'll need squaring for geometrical figuring that you do in laying out shop and building projects and for almost anything having to do with light and optics. For instance the amount of exposure your camera lens gives is calculated by comparing the squares of various "f" stops. At f/1.2 you're getting 14 times as much exposure at a given shutter speed as at f/4.5, and squaring the two stop numbers and dividing the bigger one by the littler is a way to find that out.

With the automatic constant found on most calculators, all you need do is hit the number, the times, and then the equals (repeatedly for higher powers). If the constant is governed by a switch, turn that on first.

Calculating Reciprocals

A RECIPROCAL may not sound like anything you'd ever want to calculate. But, as using this book will demonstrate, a reciprocal has its uses. The 1/x button isn't on scientific calculators for nothing— but the truth is you don't really need it. As we shall see.

You can employ reciprocals to figure electrical resistance easily and quickly, since the joint conductance of several conductors is found by adding together the reciprocals of the resistances.

And there are those wonderfully pesky tank problems too: If one hose could fill the tank in two hours and the second hose.... And work problems: Jim could do the job alone in 8 days; John could do it alone in 5. How long will it take them working together? (Please, no jokes.)

Since the reciprocal of a number is the fraction or decimal you get when you divide 1 by that number, there's a special trick for getting it. You could, of course, just ring up 1 and then divide by the number. But often the number of which you want the reciprocal is the one already on your machine and you don't want to have to copy it off on a piece of paper (humiliating, anyway) and then put it back into the machine after punching the 1.

The quick trick is this: Just divide the number by itself twice. The first step automatically produces the 1 you need, and the second step gives you the reciprocal.

Carry this operation one step further. Punch the key you're using one more time. You'll get the reciprocal of the square of the number you started with. And so on.

Two Reciprocal Tricks

T HERE ARE two sly ways to tackle one whole family of problems that cannot be solved by direct and obvious use of the ordinary small calculator. Problems in this family are represented by the example that follows.

You've dug out the information that in a certain small county, total personal income for last year was $98,765,432. You know the population consisted of two cities having populations respectively of 12,345 and 4321 and of the remaining rural areas with 12,321 persons.

What's the average (mean) income of the inhabitants of this county? Obviously it is the dollar amount divided by the sum of the three population figures.

If your calculator has a memory, you can use it to solve this type of problem. Or you can resort to pencil and paper.

But how can you solve it quickly, without paper and pencil, without memory feature, using a typical mini calculator that has a constant and 8-digit capacity , as most do?

Two ways. Take your choice.

One is to add up the population, find the reciprocal of this total, then multiply that by the dollar figure. (Dividing by a number and multiplying by its reciprocal come to the same thing.) Remember that you get the reciprocal of a number by dividing it by itself, which gives 1, then by itself again. The constant lets you do this, usually by hitting the divide key three times or the divide key once and equals key twice.

The other method also makes use of reciprocals.

All you do is turn the problem upside down to work it. Then you do the same with the answer to restore it to propriety.

In our example you add the populations together and divide by the dollars (instead of the reverse) and then restore things by getting the reciprocal. As in the other method, you do this by hitting the divide key three times (or, depending upon how your tiny miracle machine operates, the divide key once and the equals twice).

You may be dismayed to discover that these methods will give you results differing very slightly from each other and from the precise result obtainable by other approaches. This is the result of some digits having dropped off by the wayside as the calculation proceeded. They are probably no loss.

If you need extreme precision, test the method first on the kind of figures you're working with. On approximate data of the sort we're talking about—population counts and income estimates—the final result cannot be exact anyway and should be rounded off. Whether the method selected gives $3397, $3407, or $3408 (and some odd cents, in each case) rounding off to $3400 is advisable. Otherwise you're guilty of offering, or using, figures laden with false precision.

Finding Sums of Products

ONE TIME you are likely to encounter sums-of-products is as assistant shopper in charge of numbers at the supermarket. You might want to know the total cost of several items for which you have unit prices. Two pounds of eggplant at so much a pound, plus 3 eggs at some price per dozen...and so on. The complete eggplant dish is going to cost you a sum of products.

You can solve this kind of problem by rearranging it into a straight-line task.

Suppose you have these three products to add up: 5888 times 4; 6981 times 166; 59.4487 times 5. How do you go about it?

With any calculator, but especially with the common types having a constant that works on addition, there's a terrific time saver. It works best when at least one number in all but one of the pairs is small.

In the example given, begin by multiplying together the pair in the middle, because they're both big. Then just add 5888 four times and 59.4487 five times.

There's another approach. It may be a little more trouble, at least the first time, but it will work with large numbers as well as small and the simplest electronic calculator can handle it.

The idea is that a problem like 2 times 3 plus 4 times 5, though it may lead to utter confusion, is at least well arranged for working with paper and pencil or with an advanced calculator. But a simple calculator finds it devilish until it is rearranged by dividing both parts by the last number, 5, and then multiplying the whole thing by 5. As in the first picture. Now you can proceed straightforwardly to multiply, divide, add, multiply, an easy routine for any calculator.

$$\left[\frac{2 \times 3}{5} + 4 \right] \times 5$$

When you handle that earlier problem by this system, the second drawing shows the form it takes. The answer you should get is 1,182,695. This method can produce a very slight inaccuracy from early rounding.

$$\left[\left(\frac{5888 \times 4}{166} + 6981 \right) \times \frac{166}{5} + 59.4487 \right] \times 5 =$$

Use a similar rearrangement to make a sum of quotients feed in smoothly. To do 32 divided by 8 plus 39 divided by 3 plus 21 divided by 7, rewrite the problem as you see it in the third picture.

$$\left[\left(\frac{32 \times 3}{8} + 39 \right) \times \frac{7}{3} + 21 \right] - 7$$

That may look formidable until you've become used to it, and possibly not worth while for this simple computation. But for big numbers, it's a trick that can ease life considerably.

Coping with Big, *Bad* Numbers

WHAT CAN you do when there's need to multiply numbers too big for your calculator?

Suppose you were to ask almost any calculator to multiply 456,789 by 987,654. It would give you an answer with zeros shown in place of the last digits.

Despair not. There is a trick by which you can hogtie numbers of any size, as long as the smaller of them has no more than seven digits.

Using this method, you multiply just as you were taught in grade school: multiplying by one digit at a time, then adding up the partial products. But you do this with a calculator, thus saving a lot of effort and avoiding mistakes.

(You can do the adding step by machine, too, if you like. But you probably can do this part with pencil and paper faster than by re-entering into the machine. Try it both ways.)

Ordinarily, if your shorter number has seven digits, you should multiply it by the other number one digit at a time. If, however, it has six digits, you can speed up things greatly by multiplying by pairs of digits.

Multiplying by one digit at a time, you'll get the result you see in the first calculation.

```
        456789
    x   987654
      1827156
     2283945
    2740734
   3197523
  3654312
 4111101
451149483006
```

That's simple enough. It's the routine you've known since grade school, except that a calculator is doing most of the work. Let's look at the special case mentioned, where one of the numbers has exactly six digits. Speed the work by doubling up, taking the digits in pairs, as in the calculation below.

456789	x	54 :	24666606
	x	76 :	34715964
	x	98 :	44765322

$$451149483006$$

The calculator can handle paired digits as easily as single ones, cutting your work in half.

Be sure to take advantage of the constant on your calculator, or the memory feature, to save finger work. In doing the problem you see here, you'll have to enter 456789 only once.

IT'S A CINCH!

Bigger, Badder Numbers

A POCKET CALCULATOR is great on big numbers but not on big, big ones. Past eight (on some machines, six) digits, something funny happens. Work stops and a warning sign pops on. With some calculators, there's no warning and digits begin falling off at the right like lemmings off a cliff.

Expensive machines solve this problem with something called scientific notation, and so can you. What the scientific machine does is show the first eight digits—all it has room for—and tell you how many zeros you must tack on to make your answer the right size.

To get the same result, you can reduce the problem to manageable size and keep track of the decimal places.

For a problem involving the number of days in a year, you might want to raise 365 to the 6th power, a number with far too many digits for ordinary calculators. Move the decimal point two places to the left and operate with 3.65 instead. The answer you get will be 2364.5972.

But you know you've swiped two places off the original number and then used the number six times. Add those 12 places you owe and you have the result: 2,364,597,200,000,000. (In scientific notation, which is often the easiest way to record and use such a number, you could call it 2.3645972 times 10 to the 15th power.)

Sometimes you can bypass the problem altogether. The formula for figuring out how many different poker hands can be dealt calls for dividing the product of all the numbers from 52 to 48 by the product of all the numbers from 5 to 1. If you write that down, you'll discover you can reduce the size of the numerator and get rid entirely of the denominator, by some simple canceling. This process brings the chore down to the capacity of the calculator and tells you there are 2,598,960 things that could happen to you on the first deal in your next draw session.

Would you like to try both canceling and mental scientific notation on a single problem? Calculate the number of possible hands at bridge—all the possible groups of 13 cards that can be dealt from a deck of 52.

Following the rule for combinations, the denominator will consist of the digits from 1 through 13 multiplied together. And above the line you put the numerator, consisting of the 13 digits from 52 down through 40.

The number you come up with will tell you that the chance of getting 13 spades in your next bridge hand is 1 in 635,013,559,600.

Memory at the Market

IF YOUR pocket calculator has a memory built in, you can make it act like two machines at once. And two is how many you really need for some chores, such as grocery shopping cannily done.

Way to go is this. Having found, with the help of your calculator, that the little can of sausage is running one third more per ounce than the big one, and welcoming a chance to save, you flip the bigger can into your basket and its price into the memory bank of your calculator.

Now you can clear the machine for price-checking the next item, while keeping a running total in the memory. It will be all ready to compare with what you're told you owe at the cash register. Which is known as checking the checker.

To Make a Payment Book

WITH A POCKET calculator to do the hard work, it's a simple chore to set up a payment book. You can do one for any kind of installment loan or purchase, including a home mortgage. The routine described here includes alternate approaches to discovering the amount of each payment, to use if you don't have that figure to begin with.

Including this item led to warm discussions with some of the people who helped edit these pages. They found enough repetition of earlier installment-payment ideas to suggest the word redundant. I won on the basis that this word for a worker who must be fired because he isn't needed also applies to a fail-safe machine in which if one part breaks down a spare takes over at once.

Step 1. To find the payment amount, express the interest rate as a decimal and divide by the number of payments a year. (Save this result separately for later use.) Add 1. Raise this sum to the power of the total number of payments to be made over the period of the loan.

Step 2. Divide this number by itself less 1. Multiply by the amount of the loan and by the amount you saved in Step 1. (If you'd like to check yourself, try this computation for a mortgage loan of $45,500 at 12.5 percent for 30 years. The monthly payments should come out $485.60.)

For an alternate route to the same result, follow Step 1 above. But instead of Step 2, find the reciprocal (that is, divide 1 by the number you have). Then subtract 1 and change the sign. (If your calculator lacks a change-sign key, probably marked CHS or +/-, just ignore the minus sign from here on.) Next, divide by the interest rate in decimal form, which would make 12.5 percent become 0.125. Multiply by the number of payments per year. Find the reciprocal. Multiply by the amount of the debt.

Use these headings: PAYMENT, INTEREST, PRINCIPAL, BAL-ANCE.

The amount calculated by either of the methods just given will go in the first column.

For the second column, multiply the amount of the debt by the annual interest rate expressed as a decimal and divide by the number of payments per year.

Subtract the second figure from the first and enter it in the third column.

Column 3 subtracted from the previous balance gives you the new balance for Column 4.

Although the simplest of pocket calculators will serve, a change-sign key is useful and a memory even more so. A programmable calculator is a great help. If you've never used one and want to get started, this is an excellent introductory exercise. That's also true of the BASIC program in a personal computer.

Easy Calculator Subtraction

You've added up a set of figures and you want to subtract their total from 365. How can you do this without copying down the total on a scrap of paper and then re-entering it after entering the 365?

Use the change-sign key, if your calculator has one, then add to 365.

On some mini calculators, the minus key can be used to make this sign change.

But if your calculator has no special features at all, just work backwards: subtract the 365 from the total you have. Then mentally change the sign of the result.

Testing a Scientific Calculator

Scientific or "slide rule" pocket calculators are the ones that can come up with a logarithm or a cosine as blithely as the ordinary machine can multiply your income by your tax rate.

They used to cost an arm and a leg, meaning a three-digit sum. Then they came down, down, down until anyone with the price of an ordinary calculator could own one by looking around a bit for a bargain offer.

Choosing among the models offered is a little tougher.

You can begin by running through the same routines you'd use to compare ordinary electronic calculators. Test for convenience in key operation, visibility of results, presence of the keys you want most, type of memory, and so on.

Less obvious are these three specific tests you might want to try.

1. To test for speed, punch in sine 45 degrees. Some machines will dredge this up in about half a second, while others can take as long as three seconds. The reason speed is important is that with one of the slower calculators you may enter the next step in a problem too soon and be misled by a wrong answer as a result.

2. To test for capability in one field where some calculators are

lacking, try this. Ask for the sines of 45, 135, 225, and 315 degrees. You should get the same answer each time, except that two will have minus signs. If, instead, the machine you're testing gives zero answers in two cases, that means it's not capable of dealing with negative functions of angles.

3. It may surprise you to find how inaccurate a scientific calculator can be. Enter some easily remembered number that fills the display—eight digits with most calculators. Then hit successively either the lnX and e to the power of X keys or those for log X and 10 to the power of X.

Theoretically, each pair of demands will give you back the number you started with. In practice, there'll be some error, and naturally you'll prefer as little of that as possible.

For example a slightly antiquated "slide rule" calculator that I own changed 88,888,888 into 88,889,050 with one of these routines and into 88,888,980 when I used the other.

Only one of six calculators I tried returned the last four digits of my telephone number perfectly. Four converted 8052 into 8051.98 and the sixth came up with a remarkably unprecise 7578.

Calculator Superstretcher

SOONER or later, every calculating device runs out of working space. Most, even the expensive scientific types, are limited to eight or ten digits.

As one of the books on using electronic calculators says, "if you wanted to multiply something like 55,612.31 by 58.412 and wanted to know the answer to more than one decimal place, you would be out of luck, because with that calculation the machine would run out of display space after 3,248,426.2. Back to the pencil-and-paper method you go."

Well, yes—and no.

The author is right, in a way. You can't do the problem directly on your eight-digit calculator. But you can do it with the help of your calculator and only limited recourse to paper and pencil. The method will save you some tedium and a good deal of time, while reducing the chances of making a mistake.

Think of the problem cited as being (55,612 plus 0.31) times (58 plus 0.412).

Now multiply each of the first pair of numbers by each of the second pair, doing the chore on your calculator and writing down each product, like this:

$$
\begin{array}{llr}
55,612 \text{ times } 58 & = & 3,225,496. \\
55,612 \text{ times } 0.412 & = & 22,912.144 \\
0.31 \text{ times } 58 & = & 17.98 \\
\underline{0.31 \text{ times } 0.412} & = & \underline{0.12772} \\
& & 3,248,426.25172
\end{array}
$$

The calculator has done the hard work. All you've had to do with paper, pencil, and brain is the comparatively simple addition.

You may have noticed that the last two multiplication steps in this problem could have been done in one. I've taken the longer way here to demonstrate the routine for even bigger numbers than these.

Something else may occur to you. Given numbers with as many digits as these but no decimal points, the principle still works. You could divide a number like 5,561,231 by 100 as a preliminary step. Just don't forget that the final result will then have to be multiplied by 100.

For even longer numbers, use 1,000 instead. And so on.

I guess the important principle here is that what your calculator can't do alone, it and your brain can do together.

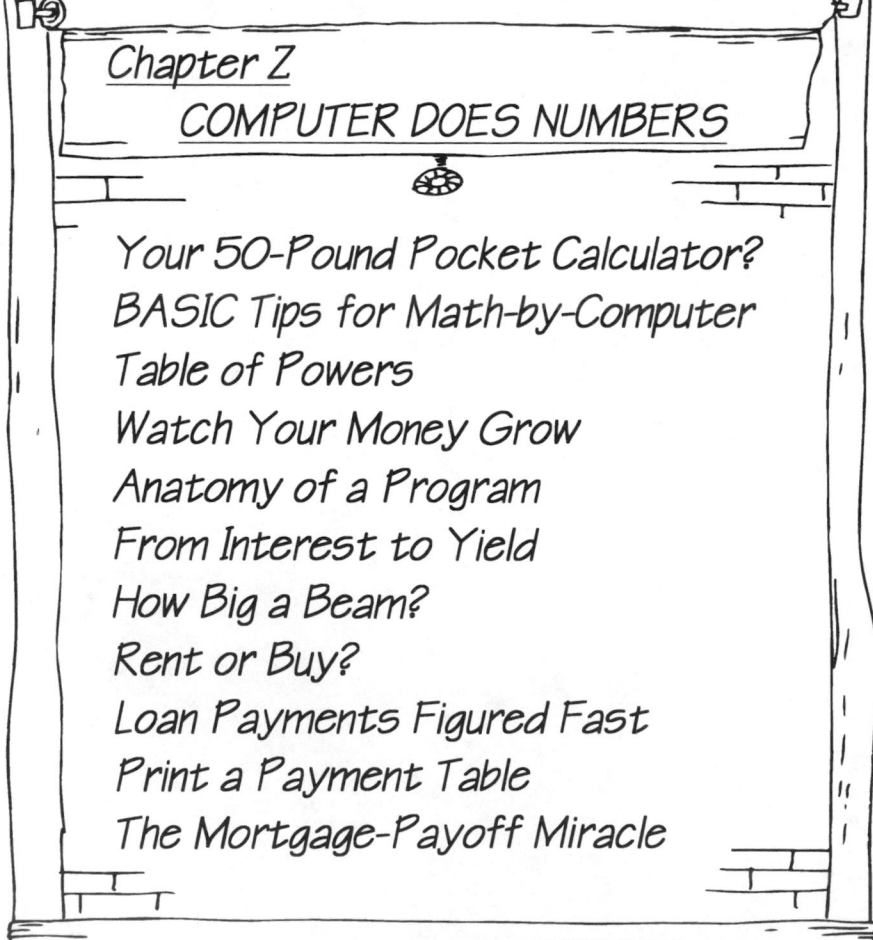

Chapter Z

COMPUTER DOES NUMBERS

Your 50-Pound Pocket Calculator?
BASIC Tips for Math-by-Computer
Table of Powers
Watch Your Money Grow
Anatomy of a Program
From Interest to Yield
How Big a Beam?
Rent or Buy?
Loan Payments Figured Fast
Print a Payment Table
The Mortgage-Payoff Miracle

Your 50-Pound Pocket Calculator?

THE TIMES they are achangin' for sure.

When I started to amass this compendium, the only tools I dared assume a reader would have besides the book itself were a pencil and a piece of paper.

Those are still all you'll need for most of the stuff covered herein.

But in a few cases—including some pretty useful calculations—this assumes that you're willing to embrace such sticky buddies as tables of logarithms and trigonometric functions, requiring you to relearn many things since schooldays happily forgot.

I believe that most people in the closing decade of our high-tech century will prefer to find an electronic route.

For most of the figuring with which this book deals, you can leap from data to conclusions with a simple *pocket calculator* at two or three dollars (a miracle). For the rest of the stuff, a *scientific calculator* at anything from ten dollars up (an even greater miracle) will facilitate things immensely.

Still better, for those willing to learn to use it, is a calculator that is not only loaded with scientific functions but is programmable as well.

Programmability is especially desirable—almost essential—for otherwise repetitious chores, such as comparing results for various angles, dimensions, sums of money, or interest rates. But a versatile *programmable calculator* can cost more than many people are eager to pay for a device likely to be used only now and then.

And the outlay will be doubled or more if a page-size calculator printer is added so that detailed tables can be created.

That's where a *personal computer* comes in. Not, perhaps, as something to buy for our limited purpose but as something to turn to an additional important use if you already own it.

Turn your hand to it and you will quickly discover that any computer will do plain arithmetic in a trice, even when faced with pretty staggering numbers. It will also do complex scientific and financial figuring of a kind that, if approached by the calculator route, would demand an expensive and specialized variety.

Combine computer with the printer that most owners already have and it can hand you your results spread out on paper in whatever format you fancy. This is especially nice if you're creating a bulky table of financial calculations—a schedule of loan payments, for instance.

One particularly nice aspect of doing your calculating by computer is that you already own the darn thing and won't have to lay out one single additional dime to add it to your calculating arsenal.

In making use of this fabulous capacity you're just using a little of the potential for which the idea of a computer was dreamed up in a previous era and brought to practicability in this century—number crunching. It's a talent too easily overlooked in the maze of spread-sheeting, data-basing, and word-processing, not to mention game-playing, that were all most of us were thinking of when we acquired our computers.

I'm mostly talking about the IBMs and their innumerable clones that predominate. Tucked into their operating systems, alongside DOS, is something called BASIC. It can be used for all sorts of programming chores, including the simple and straightforward ones we're concerned with. It sits there in DOS computers, waiting patiently, and hardly anybody pays any attention to it.

As I write this, Macintosh computers come with an application called Calculator that's fine for ordinary arithmetic. For more demanding applications, a Mac user group or a dealer can steer you to low-cost or free software that sits on your screen and looks and works just like a sophisticated scientific calculator.

But although I work on a Mac myself, I regret to say I have yet to uncover a simple and affordable way to use it for math programming. I find this, like other aspects of our national allergy to quantitative thinking, deplorable. But this Mac deficiency is belatedly beginning to heal itself with the arrival of an operating system compatible with DOS software. Or so I'm told.

There's one further way out.

If you use a spread sheet in any breed of computer, my friends assure me that there lies your best answer. You may own such a program, stand-alone or as part of an integrated "works" program. Some cost little, as shareware or freeware.

Once called The Universal Machine (I understand that was long before our forebears actually got around to inventing it) the computer has a few of its own ground rules for doing arithmetic. Some are in the established language of math. A couple are pure computerese.

• If a statement contains parentheses, you're supposed to deal first with what's inside them. If there's more than one set of parentheses or brackets, work from the inner ones out.

• If there's anything to be raised to a power, that operation comes ahead of ordinary arithmetic. Then the school kid's memory aid, My Dear Aunt Sally, takes over to remind you that—working from the left—you Multiply and Divide, then Add and Subtract, using the familiar plus and minus signs on the keyboard.

• But note that confusion with the letter X is avoided by using the asterisk or star (*) instead of the usual sign for multiplying.

• The right slash (/) says divide.

• The tiny tent (^) we know as a caret (or occasionally the upward-pointing arrow or a pair of asterisks) says raise-to-the-power-of and is used to get squares, cubes, cube roots, and all that kind of stuff. Thus the cube, or 3rd power, of 10 is written 10^3 and the cube root is 10^1/3.

If you're new at the game, you'll probably want to start by practicing some typical maneuvers right from the keyboard—without programming.

How much is 3 times 75? Type PRINT 3*75 and a return. You'll be rewarded by the answer.

Try using a question mark instead of PRINT. Works the same in most BASICs. If you use this time-saver in programs and play them back with the command LIST, note that the question mark has undergone a sea change.

As you will have noticed, the PRINT command means print on the screen. With your printer on-line and a send-to-printer command, such as LPRINT, substituted for just plain PRINT, the answer will be typed out for you.

Sometimes you'll want to record the problem as well as the answer. Use quotation marks to do this. To demonstrate that what follows quotes doesn't function but merely prints, type PRINT (or LPRINT) followed by "3 X 75 =" 3*75. With most programs, your computer will print out the stuff in quotes and then perform the quoteless instruction, giving you 3 x 75 = 225.

You've probably noted that until you type a return, you can correct errors—in either direct mode or a program. But when you've finished a line, there's a new problem. You can't get back to change it. What to do in that case is just type in the whole line all over again, including the original line number, wherever you happen to be.

The erroneous line will still show on the screen but it will be as dead as a dodo and the corrected line will function in its place.

If you then view the program by typing LIST and a return you'll find things nicely in order.

BASIC Tips for Math-by-Computer

Pencil and paper are the only tools you need to perform most math chores you meet in the course of spending money, running a business, designing or building a house. But each further step—adding a simple calculator, a scientific calculator, then a programmable one—speeds up the figuring. A personal computer, desk top or lap top, running the program called BASIC, wraps up the virtues of all these approaches into one. Some form of it comes with most computers.

For most calculations, you can use BASIC to turn a set of instructions, or a formula, into a program that will run through a calculation for you. As you progress, you'll eventually find use for such commands as these:

GOTO (to jump the program—which is just a series of steps—directly to whatever line you specify);

GOSUB (to send the program through a set of commands ending with RETURN);

FOR NEXT, a valuable tool when you encounter certain equations you wouldn't care to tangle with barehanded.

IF THEN, a command taking a form like IF A=1 THEN 90, that jumps execution provisionally to a new place, Line 90 in this instance.

There are also a couple of housekeeping commands you may find especially handy:

NEW clears memory and gets you off to a fresh start. Your clue that you should have used it is often the appearance of an asterisk after a line number, indicating that the number is already in use.

To clean up a messy screen so you can start fresh, type CLR.

Oddly missing from indexes of any books on this subject I've looked into is the word roots. I guess we should all be flattered by the assumption that we remember, from an algebra class or somewhere, that to find the cube root of a number, we need merely raise it to the 1/3 power (fourth root, 1/4 power, and so on).

It's not easy to extract information about such fundamentals as these from manuals. Most of the time, their authors are all too eager to rush you on into the complex stuff at which they shine.

With this modest boost, you should be able to employ many others of the math functions built into BASIC.

Some examples:

Square root is wanted so often it rates a command of its own, as in SQR(99) for finding the square root of 99.

Used similarly, LOG finds natural logarithms, from which common logs can be obtained by multiplying the natural log by 0.434295. (To go the other way, from common logs to natural, divide by that same decimal fraction. Or multiply by its reciprocal.)

SIN, COS, TAN, and ATN (that last is arc tangent) are built into at least some forms of BASIC in radians, for those who want to know the angle at which to saw a rafter—or to use trigonometry in any other way. The rest of the trig functions can be derived from these.

Take it from there.

Table of Powers

ONE OF THE THINGS a personal computer and its BASIC program can do that you may not have thought of is write mathematical tables.

Try this one.

First bring up BASIC.

When your computer responds (most likely you'll be greeted by an "OK" as a prompt) type the following. End each line with the usual return.

```
10 FOR N=1 TO 15
20 PRINT N,N^2,N^3,N^4,N^5
30 NEXT N
RUN
```

This should give you a loverly screenful of table. For the numbers 1 through 15 you have all but instantaneously compiled a chart of squares, cubes, and fourth and fifth powers. Think what would have happened if, instead of 15 you'd typed 150! (And stop a moment for respectful contemplation in honor of the fellow who first ground out any halfway comparable table with pencil and paper or—for all I know—sharp stick and clay tablet.)

Now, if you have a printer hooked up to your computer, write that tiny program all over again. But this time instead of PRINT in the second line type LPRINT.

Oh, by the way, don't be put off by what may seem the eccentricity of numbering program lines not consecutively but by tens. That's merely so you'll have unused numbers galore if you want to insert more lines later. As for LPRINT, that's just BASIC'S odd way of telling the computer not to put stuff on the screen but to feed it to the printer instead.

Watch Your Money Grow

YOU MAY BE relieved to find that this whole dread subject of interest compounding comes down to a single short, if nasty, expression.

The sum of money you start with is called PV, for present value. How much you'll end up with is called FV (naturally) for future value.

That nasty expression (Line 130 in the program) is the quantity by which the known PV must be multiplied to tell you the unknown FV.

Interest can be compounded at any interval and any rate that you, or your lender or your borrower, happens to choose—and the market will bear and the law will allow. Over any period of time.

In seeing how this works, let's make our example a $1749.45 savings deposit that is to be compounded daily at 8.75 percent. What will that amount to after 27.5 years?

And, what is 8.75 percent interest when expressed as annual yield? This is valuable to know for comparison with any other opportunities described to you in terms of annual yield.

To begin with, examine that nasty-but-short multiplier mentioned a few paragraphs back. Its formula tells you to express the interest rate per period as a decimal, add 1, and then raise this sum to the power of the total number of periods.

Sounds like a chore? Have no fear. Your computer is here.

If you're interested in making this kind of calculation only once you can do it directly and forget programming.

But you're likely to want to do this kind of calculation many times, so you'll save time in the long run if you'll type in a neat program you can save and use repeatedly. By tucking in new figures as you run it, you can get practically instant answers whenever you need them.

Type the following.

(If you're working with MBASIC and you first type AUTO and a return, you won't have to remember to supply the line numbers; the little devils will appear automatically, thus filling you with delight the first time or two it happens. To cancel AUTO when you're done with it, when Line 160 shows, hit control-C and a return.)

```
10 REM (FILE NAME: DY360INT)
20 PRINT "RATE=?"
30 INPUT R
40 PRINT R
50 PRINT "AMOUNT=?"
60 INPUT A
70 PRINT A
80 PRINT "YEARS=?"
90 INPUT Y
100 PRINT Y
110 LET N=360*Y
120 LET I=R/36000
130 LET B=A*(I+1)^N
140 PRINT "BALANCE=$";B
```

After the return following the last line, type RUN and then another return. Answer the queries that appear, using the rate, time, and amount suggested earlier.

(By the way, don't use commas to set off trios of digits when you plug in amounts such as the $1749.45. Commas having a special meaning in BASIC, your computer will be confused and will be inclined to speak sharply to you.)

If you've done everything right, you should find your for-example nest egg swollen to a nice $19,389.40. And you'll know you have a working program you can apply to your own fiscal life.

Anatomy of a Program

Sᴏᴍᴇᴛʜɪɴɢ for which you may well wish to use your daily-interest program "Watch Your Money Grow" is to compare alternative rates or periods, when making a choice among investments or loans.

For this purpose, a printed record is helpful. In order to make one, LIST the BASIC program (that is, type LIST and a return) and then change PRINT to LPRINT wherever it appears.

Take the time to examine the daily-interest program, called

DY360INT, line by line. You'll see how each one of those lines does its work. The 360 in its name, by the way, means that it follows the convenient, if slightly weird, banking convention of calling 360 days a year.

The first line has been used for a remark. Whenever a line starts with REM (or a single quotation mark, which your computer will have turned into REM by the next time you see it) the program will retain that line and LIST it as a sort of memo but won't RUN it. Use such lines as reminders to yourself—or any other user—of what the program is up to.

The next three lines ask you for the interest rate; stow it away in a specified place you are calling R, and display it.

The two sets of three lines each that follow do the same thing for the amount of money involved and for the number of years.

Line 110 multiplies the number of years by 360, to change it to the number of days, stowing this result in a place called N.

In a single step, Line 120 does two things. It divides the interest rate you put into R by 100 to change it from percentage (the form in which you entered it) to the decimal form called for by the formula. At the same time it divides by 360, to change the rate from yearly to daily.

Now all the data are in and converted to the form the standard compounding formula asks for.

That formula, as you can see from Line 130, says that you start by adding 1 to the daily interest rate in decimal form.

Then you raise this sum to the power of the number of compounding periods—which for daily interest is the total number of days the money will be working.

Next you multiply by the number of dollars you started with, designated as A for amount. Line 140 instructs the computer to print that quantity, called B, as your new balance.

From Interest to Yield

A COMPUTER program for calculating compound interest has another use. It can convert an interest figure into an annual yield. You'll recall yield as that puzzling second figure that crops up alongside interest in such places as bank ads touting certificates of deposit.

It's a useful number for making different investments directly comparable. It has the virtue of combining the effects of rate of interest and frequency of compounding into a single figure.

To find it, using a compound-interest formula, enter the interest rate as usual when asked. But for amount and for number of years, simply enter 1.

The "balance" the program will print will be the amount your single dollar will have grown to in a year of daily compounding at the interest rate you have named.

Now drop the 1 from the left, and move the decimal point two places to the right. For 6 percent interest, you will discover the annual yield, with daily compounding, to be 6.1824 percent.

(You might find it interesting to try a series of guesses to determine, by trial and error, the level of interest rate at which yield becomes a whole percentage point higher than rate.)

Would you like your compound-interest program ("Watch Your Money Grow") to show yield directly? Add these two lines:

```
150 LET L=(B-1)*100
160 PRINT "YIELD=;L;"PERCENT"
```

When you're using your daily-interest program to compare alternative rates or periods (perhaps for making a choice among investments or loans) a printed record is helpful. In order to make one, LIST the BASIC program. That is, type LIST and a return and then change PRINT to LPRINT wherever it appears.

How Big a Beam?

HERE, in the form of an easy-to-use computer program, is a fundamental building formula that has saved me time and money.

I've used it many times over the years, whenever the urge overtook me to build a new house for my family or to put up a cabin or other smaller structure. It's a calculation that has done even more for me since I enlisted the aid of a computer and turned it into a program that does the job in a trice or two.

Yes, it is a chore to type in the rather bulky set of instructions—unless you enjoy doing it. But you could think of it as an investment; it has to be done only once.

This figure-a-beam program comes from formulas that calculate how much weight a beam can carry without danger of collapsing and without bending more than 1/360 of the length of the span.

(Bending, or deflection, doesn't have much of anything to do with safety. It deals with the less urgent matter of avoiding bounciness that might be unpleasant in a floor or unattractive in a roof, or cause cracking in the plaster of a ceiling.)

Notice that this program uses LPRINT commands. In the strange language of BASIC, that means your results will be printed out for you on paper—assuming that your computer is hooked up to a printer. If not, or if you'd rather read the figures off the screen, just change each LPRINT command to PRINT as you copy the program.

Before getting into the program for finding how much a beam can hold, you may want to calculate how much weight you're going to ask it to hold.

To do this, you should first figure out how many square feet of floor or roof the beam—or girder or rafter or joist—is to hold up.

First, you need to know the span—the distance in feet between the points of support. (This is the same span you will use in the program, but note that there you have to multiply the feet by 12 since the program calls for the span in inches.)

And you need to know the width of the area to be supported. Normally this is the same as the center-to-center distance, less the width of a beam, since the beam or joist is supporting half of each of two spans.

However, at the edge of an area, you'll have a different deal. There's only half of one interval.

(If the floor or roof to be supported extends to form a cantilever, you have to count in that whole distance, since it has no other support. An edge joist or the last beam in a series may also have another chore you must allow for. It may be supporting a wall, and possibly a second floor and a roof, and you must add in all the weights involved.)

Having found the area (the span in feet multiplied by the interval in feet) you multiply by the bearing capacity required. Building codes commonly call for 40 pounds to the square foot for house floors. Less may be acceptable for second floors or roofs.

This result is the total load that you are going to ask each span of the beam to carry.

There's still another factor that comes up occasionally. It becomes significant only when the depth of a beam is about one tenth or more of the span. It's called horizontal shear and refers to the loss of load-bearing capacity as the beam tends to split apart internally.

Shear is rarely anything to be concerned about. Generally a safe limit for this is 120 pounds per square foot. To see if your simple beam (not a cantilever) meets this test, first divide the calculated load by 4 times the product of the breadth of the beam and its height; and then multiply by 3. This applies to any simple beam, whether fixed at the ends or freely supported and whether evenly loaded or with the load concentrated at the center of the span.

For the cantilevers covered by the program, the numbers are slightly different. Whether the cantilever has its load evenly distributed over the span or concentrated near the free end, divide the weight to be carried by 2 (instead of 4) times the product of the breadth and depth of the beam. Then multiply by 3.

If the result in either case is no more than 120, there should be no problem with horizontal shear.

After you've worked with the program that follows you probably could write one for shear to use with it, if you need it. Base it on the information in the preceding three paragraphs. If that gives you any problems, you'll find elsewhere in this section a dual-capacity program for calculating beam load and horizontal shear.

```
10'PRINTS STRENGTH & DEFLECTION OF WOOD BEAM
   (BEAMCALC)
20 INPUT "WIDTH IN INCHES";B
30 INPUT "DEPTH IN INCHES";H
40 INPUT "f=(USE 1100 FOR COAST DF #2)";F
50 INPUT "SPAN IN INCHES";L
60 Z=B*H*H*F/L:LPRINT CHR$(26)
70 LPRINT   "For   SIMPLE   BEAM   or   CANTILEVER
   BEAM"B"by"H"inches,
80 LPRINT "f= "F", span "L inches, allowable bending loading
90 LPRINT "in pounds and (given in parentheses) loading that
100 LPRINT "would cause deflection of 1/360 of the span are:
110 LPRINT:LPRINT "Evenly loaded, freely supported:
120 LPRINT INT(Z*4/3) "( DEF "INT(31290*B*H*H*H/L/L) ")
130 LPRINT:LPRINT "Evenly loaded, fixed bearings:
140 LPRINT INT(Z*2) "( DEF "INT(156440!*B*H*H*H/L/L) ");LPRINT
150 LPRINT "Load concentrated at span center, freely supported:
160 LPRINT INT(Z*2/3) "(DEF "INT(19555*B*H*H*H/L/L) ");LPRINT
170 LPRINT "Load concentrated at center of span, fixed bearings
180 LPRINT INT(Z*4/3) "(DEF "INT(78222!*B*H*H*H/L/L) ")
190 LPRINT:LPRINT "CANTILEVER, evenly loaded:
200 LPRINT INT(Z/3) "(DEF "INT(3258*B*H*H*H/L/L) ");LPRINT
210 LPRINT "CANTILEVER, load concentrated at free end:
220 LPRINT INT(Z/6) "(DEF "INT(1222*B*H*H*H/L/L) ")
230 END
```

When you start this useful program running, it will ask you four questions. If yours is, for example, a 6-inch-by-8-inch beam placed on edge, your first two answers will be 6 and 8. For a surfaced beam, enter actual dimensions in decimal form, 5.5 and 7.25.

Lumber grades for which you would enter 1100 for f include Coast region Douglas fir and No.2 southern pine joists and planks. Select and structural grades may run as high as 1500 to 1950.

Span is the actual unsupported distance in inches. If you were placing joists across a series of beams 6 inches wide that were on 8-foot centers, the span would be entered as 90 (inches) since the actual distance between the beams would be 8 feet minus 6 inches.

When you run this program for a beam of actual width 6 inches and depth 8 inches, stress grade 1400, for a 90-inch span, you should get these useful results:

Such a beam, when it is evenly loaded with its ends not fastened in place, will safely support 7,964 pounds. Fixed in place, it will handle 11,946 pounds.

It will support a 3,982-pound load concentrated at the center of the span if freely supported, 7,964 pounds if ends are held.

In each instance, you'll also be given a DEF figure that is larger than the one for load, showing deflection is not a problem.

If the beam is a cantilever, you'll learn that it can handle 1,991 evenly loaded pounds—but only 995 pounds if that load is concentrated at the free end.

Rent or Buy?

Most of the time, when it comes to a choice between remaining a renter and becoming a homeowner, the bulk of the long-run considerations favor buying.

But we live in the short run, too. For a real-world decision, it is often necessary to compare the payments about to be undertaken with the amount now going for rent. Or with the maximum you could lay out each month if you had to. What follows is a comparative calculation. It estimates the monthly mortgage payment that will put no greater strain on your budget than the rent you now pay.

```
10 'RENTBUY
20 INPUT "RENT= ";R
30 INPUT "ANNUAL INCOME= ";M
40 INPUT "INCOME TAXES= ";T
50 V=T/M
60 X=R*12*(1+V)
70 Y=X/10000*.043
80 N=V-Y
90 A=(M-X)*N
100 P=R+(T-A)/12
110 PRINT "PAYMENT=$ ";P
```

In running this program, if you enter instead of your actual rent the heftiest rent you could handle, the calculation takes on a slightly different usefulness. Now it gives you an educated estimate of the largest mortgage payment you dare sign up for.

Naturally there have to be some assumptions embedded in any crystal-ball calculation like this. One is that your income is not about to change drastically. Another is that income-tax rates and the allowable deductions for mortgage interest and other itemized expenses are not drastically changing.

A further assumption built into this program is that you will benefit substantially from the mortgage-interest deduction. If your income, interest payments, and deductible expenses are low, this is not so. Taking the interest deduction at the cost of the standard deduction may gain you little or nothing.

At higher levels of income and mortgage payments, however, this luxury-housing subsidy in the law functions very well, creating enormous tax breaks.

Considering all these imponderables, it would be well to confirm estimates from this program by using totally independent calculations before taking any irrevocable action.

The program asks for three bits of information. In addition to the rent you're handling now (or could manage if put to it) it wants to know your annual income and income taxes.

Note that the .043 figure in Line 70 is for a couple. To run the program for a single person, enter .052 instead.

Replying to the program's questions by successively entering present rent of $900, income $50,000, taxes $9573, a couple should find that mortgage payments of about $1277 would entail no added budget stress. For a single person, it's $1362.

Loan Payments Figured Fast

ALL THIS useful schedule of time payments—for car loans, home mortgages, or whatever—asks you to do is fill in three variables.

Since you're dealing with monthly payments, the formula then expresses (in Steps 120 and 130) the period of years and the rate of interest in monthly terms.

It applies a standard formula, one that has in its day generated whole books of bankers' tables.

Written in BASIC language as I present it here, with a separate line for each tiny step in the interest of clarity, it's less compact than it could be.

But I think you'll find it easy to follow and to type in.

```
10 REM AMORTIZATION
20 REM (FILE NAME 'PAYMENTS')
30 PRINT "RATE=?"
40 INPUT A
50 PRINT A
60 PRINT "AMOUNT=?"
70 INPUT B
80 PRINT B
90 PRINT "YEARS=?"
100 INPUT C
110 PRINT C
120 LET D=C*12
130 LET I=A/1200
140 LET R=B*I/(1-(1+I)^(-D))
150 PRINT "PAYMENTS=$";R
160 PRINT "TOTAL=$";R*D
```

Line 140, the formula that does the work, calls for—take a deep breath now—raising 1 plus the monthly interest rate (in decimal form) to the negative power of the number of payments, subtracting this from 1, then using this mishmash as a quantity by which to divide a figure you get by multiplying that monthly interest rate by the amount of the loan.

Don't be alarmed even if you're not used to math by computer. You don't have to do anything with the program or that formula. Just type it in. Ah, the beauty of computer calculating!

Print a Payment Table

```
10 LPRINT "AMORTIZATION TABLE"
15 LPRINT
20 INPUT "INTEREST PERCENT(TYPE 0 TO QUIT)";A
30 IF A=0 THEN STOP
40 INPUT "PRINCIPAL AMOUNT";B
50 INPUT "NUMBER OF YEARS";C
60 INPUT "PAYMENTS PER YEAR";D
65 INPUT "PAYMENT AMOUNT";R
70 LPRINT A"% " "$"B  C"YRS "D"PER  YR " "PAYMENTS
     $"R
80 I=0
85 LPRINT
90 LPRINT "PMT", "INTEREST", "APPL TO PRIN", "BALANCE"
100 A1=A/100
105 A2=A1/D:B1=B*A2:B2=R-B1:B=B-B2
110 I=I+1
120 LPRINT I,B1,B2,B
130 IF B>.1 THEN 100
140 LPRINT
150 LPRINT
160 GOTO 20
```

What this table shows is amount of payment, how much of it goes for interest, how much for principal, and balance owing.

You'll probably find most of this program easy enough to follow. However, you might wonder why Line 130 doesn't use B>0 as its cue to stop running. Reason is that if the payments, because rounded to the nearest cent, wind up a penny or two short, the table will call for a tiny additional payment. This being silly, it seems sensible to alter the command to say that when the balance drops to less than a dime, let's just forget it.

After saving this table-printing program and using it, you may become critical of one aspect. You may find it a nuisance to have to dig out a separate program to calculate the payment amount you'll need for Step 65. So why not combine the programs?

All you need do is LIST the program you already have and then type below it the supplementary steps that follow.

Now when you LIST or SAVE the program, the new steps will have inserted themselves between Steps 60 and 70.

Since one of these new steps uses its line number, Step 65 finds itself automatically canceled—an easy way to get rid of it now that it is no longer needed.

```
62 LET I=A/(100*D)
64 LET R=B*I/(1-(1+I)^(-D*C))
65 LPRINT "TOTAL PRINCIPAL AND INTEREST=$";R*D*C
67 LPRINT
68 LPRINT  "TOTAL INTEREST=$ ";R*D*C-B
69 LPRINT
```

The Mortgage-Payoff Miracle

THERE IS no free lunch. But for millions of home buyers, there is a plan that can produce results that verge on the miraculous. You may view it as a computer-assisted bit of self-hypnosis, but all the same it could save you $80,000 on a $100,000 mortgage. And that ain't hay.

Here's the idea: By making your loan payments every two weeks, in an amount less than half your present or contemplated monthly ones, you can pay off a 30-year mortgage in less than 21 years.

For anyone who gets paid every week or two weeks, this can enormously simplify budgeting. Your paycheck has become smaller but on a consistent basis that's not hard to live with—since you're compensated by never having to think about your home-loan payments. If your bank frowns on payments other than monthly, there's still a way to get this big saving. Make the payments into an interest-earning savings account and pay the mortgage holder from that. Make an additional payment on the principal whenever enough accumulates. Work out with the bank how this is to be done.

In setting up the loan, try to find a bank that will give you a reduced rate of interest because it's a 19-year loan rather than a

30-year one. (Some will, some won't.) Then you'll have an even earlier payoff and an even greater saving.

You can see why many people, not quite sure how they can arrange a payment program like this for themselves, have paid promoters substantial sums to do it for them.

You can test it for yourself by using one of the mortgage-payment (amortization) routines in this book. The easiest way, given the aid of a home computer, is to type in and use the following program, written in BASIC, that will tell all. Unlike most programs you might come across—and most tables of payment schedules—this one can process payments made at any interval.

```
10 REM FILENAME AMORTANY
20 INPUT "INTEREST PERCENT";A
30 INPUT "PRINCIPAL AMOUNT";B
40 INPUT "NUMBER OF YEARS";C
50 INPUT "NUMBER OF PAYMENTS PER YEAR";D
60 LET I=A/(100*D)
70 LET R=B*I/(1-(1+I)^(-D*C))
80 PRINT "TOTAL=$ ";R*D*C
90 PRINT "TOTAL INTEREST=$";R*D*C-B
100 PRINT A"% " "$"B C"YRS "D"PER YEAR" "PAYMENTS $"R
```

Try it first for 30 years with 12 payments a year, and then for 21 years with payments every two weeks—26 payments a year.

Try other combinations till you find the one that fits you best.

Think of the great trips you can make to Paris and Tahiti during those years of no mortgage payments.

INDEX